Magnifies or reduces a portion of the curve being viewed and can "square" the graph to reduce distortion.

Controls the values that are used when creating a table.

Determines the portion of the curve(s) shown and the scale of the graph.

Used to enter the equation(s) that is to be graphed.

Controls whether graphs are drawn sequentially or simultaneously and if the window is split.

Activates the secondary functions printed above many keys.

Used to delete previously entered characters.

Used to write the variable, x.

These keys are similar to those found on a scientific calculator (see inside back cover).

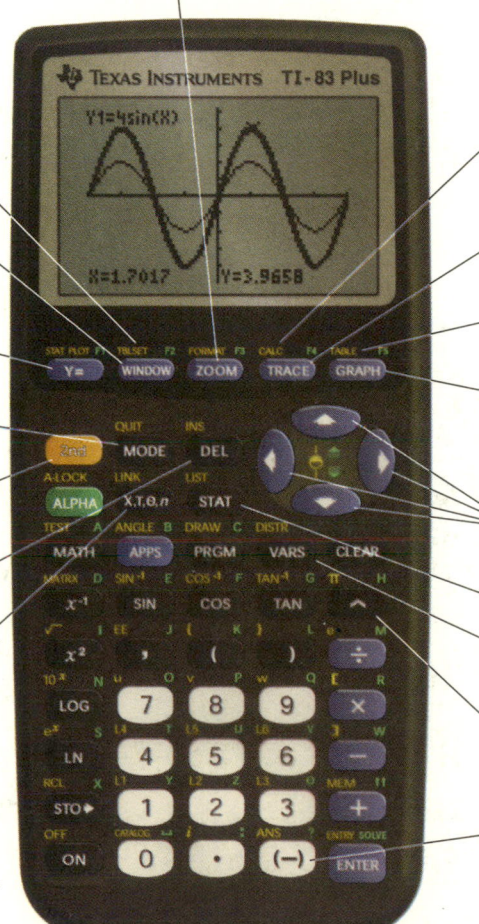

Used to determine certain important values associated with a graph.

Used to display the coordinates of points on a curve.

Used to display x- and y-values in a table.

Used to graph equations that were entered using the Y= key.

Used to move the cursor and adjust contrast.

Used to fit curves to data.

Used to access a previously named function or equation.

Used to raise a base to a power.

Used as a negative sign.

The use of a graphing calculator is optional in this text.

Elementary and Intermediate Algebra

CONCEPTS AND APPLICATIONS
A COMBINED APPROACH

THIRD EDITION

MARVIN L. BITTINGER
*Indiana University—
Purdue University at Indianapolis*

DAVID J. ELLENBOGEN
Community College of Vermont

BARBARA L. JOHNSON
*Indiana University—
Purdue University at Indianapolis*

Boston • San Francisco • New York • London • Toronto • Sydney • Tokyo • Singapore
Madrid • Mexico City • Munich • Paris • Cape Town • Hong Kong • Montreal

Publisher	Jason A. Jordan
Acquisitions Editor	Jennifer Crum
Project Manager	Kari Heen
Assistant Editor	Greg Erb
Managing Editor	Ron Hampton
Production Supervisor	Kathleen A. Manley
Text Design	Geri Davis/The Davis Group, Inc.
Editorial and Production Services	Martha Morong/Quadrata, Inc.
Art Editor	Geri Davis/The Davis Group, Inc.
Marketing Manager	Dona Kenly
Illustrators	Network Graphics and Jim Bryant
Compositor	The Beacon Group
Cover Design	Dennis Schaefer
Cover Photo	Stuart Westmorland/Index Stock Imagery (background)
	Ron Sanford/Index Stock Imagery (foreground)
Prepress Supervisor	Caroline Fell
Print Buyer	Evelyn Beaton

Photo Credits appear at the back of the book.

The product name "TI-83 Plus" and the likeness of the same product are used by permission of Texas Instruments.

Library of Congress Cataloging-in-Publication Data
Bittinger, Marvin L.
   Elementary and intermediate algebra: concepts and applications: a combined approach / Marvin L. Bittinger, David J. Ellenbogen, Barbara L. Johnson.—3rd ed.
     p. cm.
   Includes index.
   ISBN 0-201-71966-5 (alk. paper)
   1. Algebra. I. Ellenbogen, David. II. Johnson, Barbara L. (Barbara Loreen), 1962-III. Title.
QA152.2 .B5797   2001
512.9—dc21

                                         2001041375

3 4 5 6 7 8 9 10—DOW—05 04 03 02 01

To Ruth and Loren
For communicating to others their lifelong love of learning

Contents

Preface

We are pleased to present the third edition of *Elementary and Intermediate Algebra: Concepts and Applications: A Combined Approach*. Each time we work on a new edition, it's a balancing act. On the one hand, we want to preserve the features, applications, and explanations that faculty have come to rely on and expect. On the other hand, we want to blend our own ideas for improvement with the many insights that we receive from faculty and students throughout North America. The result is a living document in which new features and applications are developed while successful features and popular applications from previous editions are updated and refined. Our goal, as always, is to present content that is easy to understand and has the depth required for success in this and future courses.

Appropriate for a course combining the study of elementary and intermediate algebra, this text covers both elementary and intermediate algebra topics without the repetition of instruction necessary in two separate texts. It is the third of three texts in an algebra series that also includes *Elementary Algebra: Concepts and Applications*, Sixth Edition, and *Intermediate Algebra: Concepts and Applications*, Sixth Edition, by Bittinger/Ellenbogen.

Approach

Our goal, quite simply, is to help today's students both learn and retain mathematical concepts. To achieve this goal, we feel that we must prepare developmental-mathematics students for the transition from "skills-oriented" elementary and intermediate algebra courses to more "concept-oriented" college-level mathematics courses. This requires that we teach these same students critical thinking skills: to reason mathematically, to communicate mathematically, and to identify and solve mathematical problems. Following are some aspects of our approach that we use in this revision to help meet the challenges we all face teaching developmental mathematics.

Problem Solving

One distinguishing feature of our approach is our treatment of and emphasis on problem solving. We use problem solving and applications to motivate the material wherever possible, and we include real-life applications and problem-solving techniques throughout the text. Problem solving not only encourages students to think about how mathematics can be used, it helps to prepare them for more advanced material in future courses.

- In Chapter 2, we introduce the five-step process for solving problems: (1) Familiarize, (2) Translate, (3) Carry out, (4) Check, and (5) State the answer. These steps are then used consistently throughout the text whenever we encounter a problem-solving situation. Repeated use of this problem-solving strategy gives students a sense that they have a starting point for any type of problem they encounter, and frees them to focus on the mathematics necessary to successfully translate the problem situation. We often use estimation and carefully checked guesses to help with the Familiarize and Check steps (see pp. 103, 118, and 477). In this edition, we also use dimensional analysis as a quick check of certain problems (see pp. 162 and 164).

Applications

Interesting applications of mathematics help motivate both students and instructors. Solving applied problems gives students the opportunity to see their conceptual understanding put to use in a real way. In the third edition of *Elementary and Intermediate Algebra: Concepts and Applications: A Combined Approach*, not only have we increased the total number of applications and real-data problems overall, nearly 20 percent of our applications are new, and we have increased the number of source lines to better highlight the real-world data. As in the past, art is integrated into the applications and exercises to aid the student in visualizing the mathematics. (See pp. 140, 190, 330, and 392.)

Pedagogy

New! **Connecting the Concepts.**   To help students understand the "big picture," Connecting the Concepts subsections within each chapter (and highlighted in the table of contents) relate the concept at hand to previously learned and upcoming concepts. Because students may occasionally "lose sight of the forest because of the trees," we feel confident that this feature will help them keep better track of their bearings as they encounter new material. (See pp. 177, 317, 377, and 578.)

New! **Study Tips.**   Most plentiful in the first three chapters when students are still establishing their study habits, Study Tips are found in the margins and interspersed throughout the first nine chapters. Our Study Tips range from how to approach assignments, to reminders of the various study aids that are available, to strategies for preparing for a final exam. (See pp. 75, 100, and 134.)

New! *Aha!* Exercises. Designated by *Aha!*, these exercises can be solved quickly if the student has the proper insight. The *Aha!* designation is used the first time a new insight can be used on a particular type of exercise and indicates to the student that there is a simpler way to complete the exercise that requires less lengthy computation. It's then up to the student to find the simpler approach and, in subsequent exercises, to determine if and when that particular insight can be used again. Occasionally the *Aha!* exercise is easily answered by looking at the preceding odd exercise. Our hope is that the *Aha!* exercises will discourage rote learning and reward students who "look before they leap" into a problem. (See pp. 90, 190, 251, and 310.)

Technology Connections. Throughout each chapter, optional Technology Connection boxes help students use graphing calculator technology to better visualize a concept that they have just learned. To connect this feature to the exercise sets, certain exercises are marked with a graphing calculator icon and reinforce the use of this optional technology. (See pp. 80, 85, 112, 217, 225, 234, and 281.)

Skill Maintenance Exercises. Retaining mathematical skills is critical to a student's success in future courses. To this end, nearly every exercise set includes six to eight Skill Maintenance exercises that review skills and concepts from preceding chapters of the text. In this edition, not only have the Skill Maintenance exercises been increased by 50 percent, but they are now designed to provide extra practice with the specific skills needed for the very next section of the text. We also now list answers to both odd- and even-numbered skill maintenance exercises, along with their section references, in the answer section. (See pp. 182, 228, 299, and 376.)

Synthesis Exercises. Following the Skill Maintenance section, each exercise set ends with a group of Synthesis exercises designated by their own heading. These exercises offer opportunities for students to synthesize skills and concepts from earlier sections with the present material, and often provide students with deeper insights into the current topic. Synthesis exercises are generally more challenging than those in the main body of the exercise set and occasionally include *Aha!* exercises. (See pp. 76, 211, 245, and 376.)

Writing Exercises. In this edition, every set of exercises includes at least four writing exercises. Two of these are more basic and appear just before the Skill Maintenance exercises. The other writing exercises are more challenging and appear as Synthesis exercises. All writing exercises are marked with ▱ and require answers that are one or more complete sentences. This type of problem has been found to aid in student comprehension, critical thinking, and conceptualization. Because some instructors may collect answers to writing exercises, and because more than one answer may be correct, answers to writing exercises are not listed at the back of the text. (See pp. 76, 140, 235, and 376.)

Collaborative Corners. In today's professional world, teamwork is essential. We continue to provide optional Collaborative Corner features throughout the text that require students to work in groups to explore and solve mathematical problems. There are one to three Collaborative Corners per chapter, each one appearing after the appropriate exercise set. (See pp. 10, 124, 169, and 254.)

Cumulative Review. After Chapters 3, 6, 9, 12, and 14, we have included a Cumulative Review, which reviews skills and concepts from all preceding chapters of the text. (See pp. 201, 406, 593, 803, and 912.)

Review Chapter. This text includes a valuable review appendix for those students who require a brief overview of elementary algebra topics before moving on to the intermediate algebra portion of the text.

What's New in the Third Edition?

We have rewritten many key topics in response to user and reviewer feedback and have made significant improvements in design, art, pedagogy, and an expanded supplements package. Detailed information about the content changes is available in the form of a conversion guide. Please ask your local Addison-Wesley sales consultant for more information. Following is a list of the major changes in this edition.

New Design

You will see that the page dimension for this edition is larger, which allows for an open look and a typeface that is easier to read. In addition, we continue to pay close attention to the pedagogical use of color to make sure that it is used to present concepts in the clearest possible manner.

Content Changes

A variety of content changes have been made throughout the text. Some of the more significant changes are listed below.

- Chapter 3 now includes a gentle introduction to interpolation and extrapolation as well as a new section that stresses the connection between rate of change and graphing. Slope–intercept and point–slope form also now appear in this chapter.
- Chapter 6 has expanded material on rates and units. Section 6.8 now also provides practice in recognizing and solving all the various types of equations previously discussed.

- Chapter 7 includes an expanded discussion of functions, including a section on domain and range. The types of expressions already studied in Chapters 1–6 are linked to function notation in Section 7.3.
- The topic of variation has been moved from Section 11.6 to Section 7.5. As a result, Chapter 11 has been shortened to 9 sections.
- Chapter 10 has been rewritten so that Section 10.3 is strictly multiplying radical expressions. Section 10.4 is now strictly division of radical expressions. Section 10.5 is now devoted to expressions with two or more radical terms.

Supplements for the Instructor

New! Annotated Instructor's Edition
(ISBN 0-201-65876-3)

The *Annotated Instructor's Edition* includes all the answers to the exercise sets, usually right on the page where the exercises appear, and Teaching Tips in the margins that give insights and classroom discussion suggestions that will be especially useful for new instructors. These handy answers and ready Teaching Tips will help both new and experienced instructors save classroom preparation time.

New! MyMathLab

MyMathLab.com is a complete, on-line course for Addison-Wesley mathematics textbooks that integrates interactive, multimedia instruction correlated to the textbook content. MyMathLab can be easily customized to suit the needs of students and instructors and provides a comprehensive and efficient on-line course-management system that allows for diagnosis, assessment, and tracking of students' progress.

MyMathLab features the following:

- Fully interactive multimedia textbooks are built in CourseCompass, a version of Blackboard™ designed specifically for Addison-Wesley.
- Chapter and section folders from the textbook contain a wide range of instructional content: videos, software tools, audio clips, animations, and electronic supplements.
- Hyperlinks take you directly to on-line testing, diagnosis, tutorials, and gradebooks in MathXL—Addison-Wesley's tutorial and testing system for mathematics and statistics.
- Instructors can create, copy, edit, assign, and track all tests for their course as well as track student tutorial and testing performance.
- With push-button ease, instructors can remove, hide, or annotate Addison-Wesley preloaded content, add their own course documents, or change the order in which material is presented.
- Using the communication tools found in MyMathLab, instructors can hold on-line office hours, host a discussion board, create communication groups within their class, send e-mails, and maintain a course calendar.
- Print supplements are available on-line, side-by-side with their textbooks.

For more information, visit our Web site at www.mymathlab.com or contact your Addison-Wesley sales representative for a live demonstration.

Printed Test Bank/ Instructor's Resource Guide (ISBN 0-201-73405-2)

The Instructor's Resource Guide portion of this supplement contains the following:

- Extra practice problems
- Black-line masters of grids and number lines for transparency masters or test preparation
- A videotape index and section cross-references to our tutorial software packages
- A syllabus conversion guide from the Fifth Edition to the Sixth Edition

The Printed Test Bank portion of this supplement contains the following:

- Six new alternate free-response test forms for each chapter, organized with the same topic order as the chapter tests in the main text. Each form includes synthesis questions, as appropriate, at the end of each test.
- Two new multiple-choice versions of each chapter test
- Eight new alternate test forms for the final examination: Alternate Test Forms A, B, and C of the final examinations are organized by chapter and D, E, and F are organized by problem type.
- Answers to all tests

Instructor's Solutions Manual
(ISBN 0-201-73406-0)

The *Instructor's Solutions Manual* contains fully worked-out solutions to the odd-numbered exercises and brief solutions to the even-numbered exercises in the exercise sets.

TestGen-EQ/QuizMaster-EQ
(ISBN 0-201-74639-5)

Available on a dual-platform Windows/Macintosh CD-ROM, this fully networkable software enables instructors to build, edit, print, and administer tests using a computerized test bank of questions organized according to the contents of each chapter. Tests can be printed or saved for on-line testing via a network on the Web, and the software can generate a variety of grading reports for tests and quizzes.

InterAct Math Plus
(ISBN 0-201-72140-6)

Available to Windows users of *Elementary and Intermediate Algebra Concepts and Applications: A Combined Approach*, Third Edition, this networkable software provides course management and on-line administration for Addison-Wesley's InterAct Math® tutorial software (see "Supplements for the Student"). InterAct Math Plus enables instructors to create and administer on-line tests, summarize students' results, and monitor students' progress in the tutorial software, providing an invaluable teaching and tracking resource.

InterAct MathXL: www.mathxl.com

(12-month registration ISBN 0-201-71111-7, stand-alone)

The MathXL Web site provides diagnostic testing and tutorial help, all on-line using InterAct Math® tutorial software and TestGen-EQ testing software. Students can take chapter tests correlated to the text, receive individualized study plans based on those test results, work practice problems and receive tutorial instruction for areas in which they need improvement, and take further tests to gauge their progress. Instructors can create and customize tests and track all student test results, study plans, and practice work.

Supplements for the Student

New! Web Site: www.MyMathLab.com

Ideal for lecture-based, lab-based, and on-line courses, this state-of-the-art Web site provides students with a centralized point of access to the wide variety of on-line resources available with this text. The pages of the actual book are loaded into MyMathLab.com, and as students work through a section of the on-line text, they can link directly from the pages to supplementary resources (such as tutorial software, interactive animations, and audio and video clips) that provide instruction, exploration, and practice beyond what is offered in the printed book. MyMathLab.com generates personalized study plans for students and allows instructors to track all student work on tutorials, quizzes, and tests. Complete course-management capabilities, including a host of communication tools for course participants, are provided to create a user-friendly and interactive on-line learning environment.

Student's Solutions Manual

(ISBN 0-201-64211-5)

The *Student's Solutions Manual* by Judith A. Penna contains completely worked-out solutions with step-by-step annotations for all the odd-numbered exercises in the text, with the exception of the writing exercises. This manual also lists, without complete solutions, the answers for even-numbered text exercises.

InterAct Math® Tutorial CD-ROM

(ISBN 0-201-74625-5)

This interactive tutorial software provides Windows users with algorithmically generated practice exercises that correlate at the objective level to the odd-numbered exercises in the text. Each practice exercise is accompanied by both an example and a guided solution designed to involve students in the solution process. Selected problems also include a video clip that helps students visualize concepts. The software recognizes common student errors and provides appropriate feedback. Instructors can use InterAct Math Plus course-management software to create, administer, and track on-line tests and monitor student performance during practice sessions.

InterAct MathXL www.mathxl.com
(12-month registration ISBN 0-201-71630-5, stand-alone)
The MathXL Web site provides diagnostic testing and tutorial help, all on-line, using InterAct Math® tutorial software and TestGen-EQ testing software. Students can take chapter tests correlated to the text, receive individualized study plans based on those test results, work practice problems and receive tutorial instruction for areas in which they need improvement, and take further tests to gauge their progress.

Videotapes (ISBN 0-201-74207-1)
Developed and produced especially for this text, the videotapes feature an engaging team of instructors, including the authors. These instructors present material and concepts by using examples and exercises from every section of the text in a format that stresses student interaction.

Digital Video Tutor
(ISBN 0-201-74643-3, stand-alone)
The videotapes for this text are now available on CD-ROM, making it easy and convenient for students to watch video segments from a computer at home or on campus. The complete digitized video set, now affordable and portable for students, is ideal for distance learning or supplemental instruction.

AW Math Tutor Center
(ISBN 0-201-72170-8, stand-alone)
The AW Math Tutor Center is staffed by qualified mathematics instructors who provide students with tutoring on examples and odd-numbered exercises from the textbook. Tutoring is available via toll-free telephone, fax, or e-mail.

Acknowledgments

No book can be produced without a team of professionals who take pride in their work and are willing to put in long hours. Laurie A. Hurley deserves special thanks for her well-thought-out suggestions and uncanny eye for detail. Judy Penna's outstanding work in organizing and preparing the printed supplements and the indexes amounts to an inspection of the text that goes far beyond the call of duty and for which we will always be extremely grateful. Thanks to Tom Schicker for authoring the *Printed Test Bank.* Dawn Mulheron, Jeremy Pletcher, and Daphne Bell provided enormous help, often in the face of great time pressure, as accuracy checkers. We are also indebted to Chris Burditt and Jann MacInnes for their many fine ideas that appear in our Collaborative Corners and Vince McGarry and Janet Wyatt for their recommendations for Teaching Tips featured in the *Annotated Instructor's Edition.*

Martha Morong, of Quadrata, Inc., provided editorial and production services of the highest quality imaginable—she is simply a joy to work with. Geri Davis, of the Davis Group, Inc., performed superb work as designer, art editor, and photo researcher, and always with a disposition that can brighten an otherwise gray day. Network Graphics generated the graphs, charts, and many of the illustrations. Not only are the people at Network reliable, but they clearly take pride in their work. The many hand-drawn illustrations appear thanks to Jim Bryant, a gifted artist with true mathematical sensibilities. Tom and Pam Hansen, of Copy Ship Fax Plus, consistently went the extra yard in providing the best in copying services.

Our team at Addison-Wesley deserves special thanks. Assistant Editor Greg Erb coordinated all the reviews, tracked down countless pieces of information, and managed many of the day-to-day details—always in a pleasant and reliable manner. Executive Project Manager Kari Heen expertly provided a steadying influence along with gentle prodding at just the right moments. Senior Acquisitions Editor Jenny Crum provided many fine suggestions along with unflagging support. Senior Production Supervisor Kathy Manley exhibited patience when others would have shown frustration. Designer Dennis Schaefer's willingness to listen and then creatively respond resulted in a book that is beautiful to look at. Marketing Manager Dona Kenly skillfully kept us in touch with the needs of faculty; Executive Technology Producer Lorie Reilly provided us with the technological guidance so necessary for our many supplements; and Media Producer Tricia Mescall remains the steady hand responsible for our fine video series. Our publisher, Jason Jordan, deserves credit for assembling this fine team and remaining accessible to us on both a professional and personal level. To all of these people we owe a real debt of gratitude.

A special thanks to the students at the Community College of Vermont and Professor Tony Julianelle of the University of Vermont for their thoughtful comments and suggestions. We also thank the following professors for their thoughtful reviews and insightful comments.

Prerevision Diary Reviewers (Second Edition)
Judy Godwin, *Collin County Community College*
Meredith Anne Higgs, *Middle Tennessee State University*
Sybil MacBeth, *Tidewater Community College*
Mindy Schonberg, *CUNY College of Staten Island*
Janet Wyatt, *Longview Community College*

Manuscript Reviewers
Mark Bates, *Oxnard College*
Shari Bennett, *Jefferson Community and Technical Colleges*
Jon Blakely, *College of the Sequoias*
Paul Blankenship, *Lexington Community College*
Debra Bryant, *Tennessee Technological University*
Tony Bower, *St. Philip's College*
Barbara Burke, *Hawaii Pacific University*
Margret Hathaway, *Kansas City Community College*
Alan Hayashi, *Oxnard College*
Harvey Johnson, *St. Philip's College*
Mike Kirby, *Tidewater Community College*

Linda Lohman, *Jefferson Community College*
Ben Mayo, *Yakima Valley Community College*
Robert McCoy, *University of Alaska Anchorage*
Valerie Morgan-Krick, *Tacoma Community College*
Kim Nunn, *Northeast State Technical Community College*
Annette Smith, *South Plains College*
Dwight Smith, *Prestonsburg Community College*
Mark Tom, *College of the Sequoias*

Finally, a special thank you to all those who so generously agreed to discuss their professional use of mathematics in our chapter openers. These dedicated people, none of whom we knew prior to writing this text, all share a desire to make math more meaningful to students. We cannot imagine a finer set of role models.

M.L.B.
D.J.E.
B.L.J.

Feature Walkthrough

6
Rational Expressions and Equations

CHAPTER OPENERS
Each chapter opens with a list of the sections covered and a real-life application that includes a testimonial from a person in that field to show how integral mathematics is in problem solving. Real data are often used in these applications, as well as in many other exercises, and in "on the job" examples (like those students might find in the workplace) to increase student interest.

AN APPLICATION

In South Africa, the design of every woven handbag, or *gipatsi*, is created by repeating two or more patterns around the bag. If a weaver uses a four-strand, a six-strand, and an eight-strand pattern, what is the smallest number of strands needed in order for all three patterns to repeat a whole number of times?

This problem appears as Exercise 80 in Section 6.3.

*W*eavers use math when designing patterns, from calculating the size of the pattern and number of times it repeats to figuring the number of threads per inch and the weight of the thread. I also use math to convert between English and metric units when dyeing warps and woofs.

ESTELLE CARLSON
Handweaver and Designer
Los Angeles, California

Example 4

Teaching Tip

Students should suspect that they might have to solve by factoring if the equation has a term of degree ≥ 2.

Solve: $x^2 + 5x + 6 = 0$.

Solution This equation differs from those sol... like terms to combine, and there is a squared te... mial. Then we use the principle of zero products

$$x^2 + 5x + 6 = 0$$
$$(x + 2)(x + 3) = 0$$ Fac...
$$x + 2 = 0 \quad or \quad x + 3 = 0$$ Usi...
$$x = -2 \quad or \quad x = -3.$$

Check: For -2:
$$\frac{x^2 + 5x + 6 = 0}{(-2)^2 + 5(-2) + 6 \overset{?}{=} 0}$$
$$4 - 10 + 6$$
$$-6 + 6$$
$$0 \mid 0 \text{ TRUE}$$

For -3:
$$\frac{x^2 + 5x + 6 = 0}{(-3)^2 + 5(-3) + 6 \overset{?}{=} 0}$$
$$9 - 15 + 6$$
$$-6 + 6$$
$$0 \mid 0 \text{ TRUE}$$

The solutions are -2 and -3.

The principle of zero products is used even if the factoring consists of only removing a common factor.

Study Tip

Immediately after each quiz or test, write out a step-by-step solution to any questions you missed. Visit your professor during office hours or consult with a tutor for help with problems that are still giving you trouble. Misconceptions tend to resurface if they are not corrected as soon as possible.

for every horizontal distance of 100 ft, the ... of grade also occurs in skiing or snowboard... very tame, but a 40% grade is considered s...

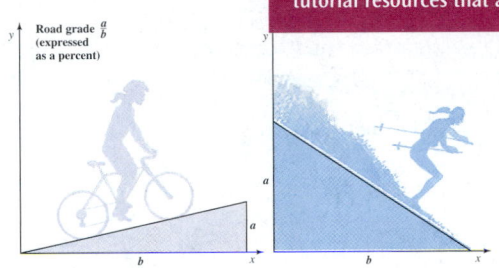

Road grade $\frac{a}{b}$ (expressed as a percent)

Example 5

Skiing. Among the steepest skiable terrain in North America, the Headwall on Mount Washington, in New Hampshire, drops 720 ft over a horizontal distance of 900 ft. Find the grade of the Headwall.

Example 1

technology connection

Sometimes we may wish to re-call and modify a calculation. For example, suppose that after calculating $30 \cdot 1800$ we wish to find $30 \cdot 1870$. Pressing **2nd** **ENTRY** gives the following

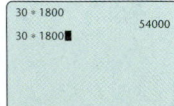

```
30 * 1800
                    54000
30 * 1800▪
```

Moving the cursor left, we change 1800 to 1870 and press **ENTER** .

```
30 * 1800
                    54000
30 * 1870
                    56100
```

1. Verify the work above and then use **2nd** **ENTRY** to find $39 \cdot 1870$.

Furnace output. Contractors in the Northeast use the formula $B = 30a$ to determine the minimum furnace output B, in British thermal units (Btu's), for a well-insulated house with a square feet of flooring (*Source*: U.S. Department of Energy). Determine the minimum furnace output for an 1800-ft^2 house that is well insulated.

Solution We substitute 1800 for a and calculate B:

$$B = 30a = 30(1800) = 54,000.$$

The furnace should be able to provide at least 54,000 Btu's of h...

Solving for a Letter

Suppose that a contractor has an extra furnace and wants to de... of the largest (well-insulated) house in which it can be used. Th... substitute the amount of the furnace's output in Btu's—say, 63,... then solve for a:

$$63,000 = 30a \quad \text{Replacing } B \text{ with } 63,000$$
$$2100 = a. \quad \text{Dividing both sides by 30:}$$

$$\begin{array}{r} 2\,100 \\ 30\overline{)63,000} \\ \underline{60} \\ 3\,0 \\ \underline{3\,0} \\ 000 \end{array}$$

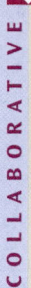

CORNER

Determining Depreciation Rates

Focus: Modeling, graphing, and rates
Time: 30 minutes
Group size: 3
Materials: Graph paper and straightedges

From the minute a new car is driven out of the dealership, it *depreciates*, or drops in value with the passing of time. The N.A.D.A.® Official Used Car Guide (often called the "blue book" despite its orange cover) is a monthly listing of the trade-in values of used cars. The data below are taken from the N.A.D.A. guides for June and December of 2000.

Car	Trade-in Value in June 2000	Trade-in Value in December 2000
1998 Chevrolet Camaro convertible V6	$16,200	$14,900
1998 Ford Mustang convertible, 2 door, GT	$16,425	$15,325
1998 VW Passat GLS Turbo, 4 cylinder	$15,975	$14,850

ACTIVITY

1. Each group member should select a different one of the cars listed in the table above as his or her own. Assuming that the values are dropping linearly, each student should draw a line representing the trade-in value of his or her car. Draw all three lines on the same graph. Let the horizontal axis represent the time, in months, since June 2000, and let the vertical axis represent the trade-in value of each car. Decide as a group how many months or dollars each square should represent. Make the drawings as neat as possible.
2. At what *rate* is each car depreciating and how are the different rates illustrated in the graph of part (1)?
3. If one of the three cars had to be sold in June 2001, which one would your group sell and why? Compare answers with other groups.

COLLABORATIVE CORNER FEATURE
A popular feature from the previous edition, optional Collaborative Corners are inserted throughout the text. Collaborative Corners give students the opportunity to work as a group to solve problems or to perform specially designed activities. There are one to three Collaborative Corners per chapter, each one appearing after the appropriate exercise set.

Visualizing Rates

Graphs allow us to visualize a rate of change. As a rule, the quantity listed in the numerator appears on the vertical axis and the quantity listed in the denominator appears on the horizontal axis.

Example 3

Communication. In 1998, there were approximately 69 million cellular telephone subscribers in the United States, and the figure was growing at a rate of about 14 million per year (*Source*: *Statistical Abstract of the United States*, 1999). Draw a linear graph to represent this information.

Solution To decide which labels to use on the axes, we note that the rate is given in millions of customers per year. Thus we list *Number of customers, in millions*, on the vertical axis and *Year* on the horizontal axis.

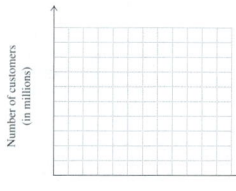

14 million per year is 14 million/yr; millions of customers is the vertical axis; year is the horizontal axis.

Next, we must decide on a scale for each axis that will allow us to plot the given information. If we count by 10's of millions on the vertical axis, we can easily reach 69 million without needing a terribly large graph. On the horizontal axis, we list several years, making certain that 1998 is included (see the figure on the left below).

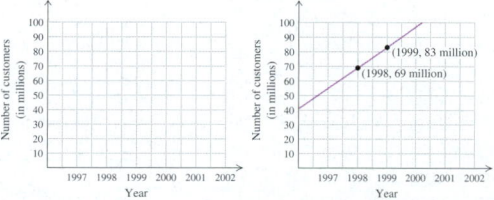

Finally, we display the given information. To do so, we plot the point that corresponds to (1998, 69 million). Then, to display the rate of growth, we move from that point to a second point that represents 14 million more users one year later. The coordinates of this point are (1998 + 1, 69 + 14 million), or (1999, 83 million). We draw a line passing through the two points, as shown in the figure on the right above.

REAL-DATA APPLICATIONS
Applications have always been a strength of this text, and now the authors bring you even more of a good thing. This edition includes 20% new application and real-data problems, along with an increase in the total number of applications overall.

EXERCISES

NEW!

Sample page (5.2 FACTORING TRINOMIALS OF THE TYPE $x^2 + bx + c$ 289)

47. $27 + 12y + y^2$

48. $50 + 15x + x^2$

49. $t^2 - 0.3t - 0.10$

50. $y^2 - 0.2y - 0.08$

51. $p^2 + 3pq - 10q^2$

52. $a^2 - 2ab - 3b^2$

53. $m^2 + 5mn + 5n^2$

54. $x^2 - 11xy + 24y^2$

55. $s^2 - 2st - 15t^2$

56. $b^2 + 8bc - 20c^2$

57. $6a^{10} - 30a^9 - 84a^8$

58. $7x^9 - 28x^8 - 35x^7$

59. Marge factors $x^3 - 8x^2 + 15x$ as $(x^2 - 5x)(x - 3)$. Is she wrong? Why or why not? What advice would you offer?

60. Without multiplying $(x - 17)(x - 18)$, explain why it cannot possibly be a factorization of $x^2 + 35x + 306$.

SKILL MAINTENANCE

Solve.

61. $3x - 8 = 0$

62. $2x + 7 = 0$

Multiply.

63. $(x + 6)(3x + 4)$

64. $(7w + 6)^2$

65. In a recent year, 29,090 people were arrested for counterfeiting. This figure was down 1.2% from the year before. How many people were arrested the year before?

66. The first angle of a triangle is four times as large as the second. The measure of the third angle is 30° greater than that of the second. How large are the angles?

SYNTHESIS

67. When searching for a factorization, why do we list pairs of numbers with the correct *product* instead of pairs of numbers with the correct *sum*?

68. What is the advantage of writing out the prime factorization of c when factoring $x^2 + bx + c$ with a large value of c?

69. Find all integers b for which $a^2 + ba - 50$ can be factored.

70. Find all integers m for which $y^2 + my + 50$ can be factored.

Factor each of the following by first factoring out -1.

71. $30 + 7x - x^2$

72. $45 + 4x - x^2$

73. $24 - 10a - a^2$

74. $36 - 9a - a^2$

75. $84 - 8t - t^2$

76. $72 - 6t - t^2$

Factor completely.

77. $x^2 + \frac{1}{4}x - \frac{1}{8}$

78. $x^2 + \frac{1}{2}x - \frac{3}{16}$

79. $\frac{1}{3}a^3 - \frac{1}{3}a^2 - 2a$

80. $a^7 - \frac{25}{7}a^5 - \frac{30}{7}a^6$

81. $x^{2m} + 11x^m + 28$

82. $t^{2n} - 7t^n + 10$

Aha! **83.** $(a + 1)x^2 + (a + 1)3x + (a + 1)2$

84. $ax^2 - 5x^2 + 8ax - 40x - (a - 5)9$
(*Hint:* See Exercise 83.)

Find a polynomial in factored form for the shaded area in each figure. (Leave answers in terms of π.)

85.

86.

87. Find the volume of a cube if its surface area is $6x^2 + 36x + 54$ square meters.

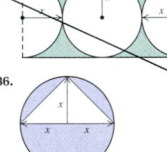

88. A census taker asks a woman, "How many children do you have?"

"Three," she answers.

"What are their ages?"

She responds, "The product of their ages is 36. The sum of their ages is the house number next door."

The math-savvy census taker walks next door, reads the house number, appears puzzled, and returns to the woman, asking, "Is there something you forgot to tell me?"

"Oh yes," says the woman. "I'm sorry. The oldest child is at the park."

The census taker records the three ages, thanks the woman for her time, and leaves.

How old is each child? Explain how you reached this conclusion. (*Hint:* Consider factorizations.) (*Source:* Adapted from Harnadek, Anita, *Classroom Quickies.* Pacific Grove, CA: Critical Thinking Press and Software)

Elementary and Intermediate Algebra

CONCEPTS AND APPLICATIONS
A COMBINED APPROACH

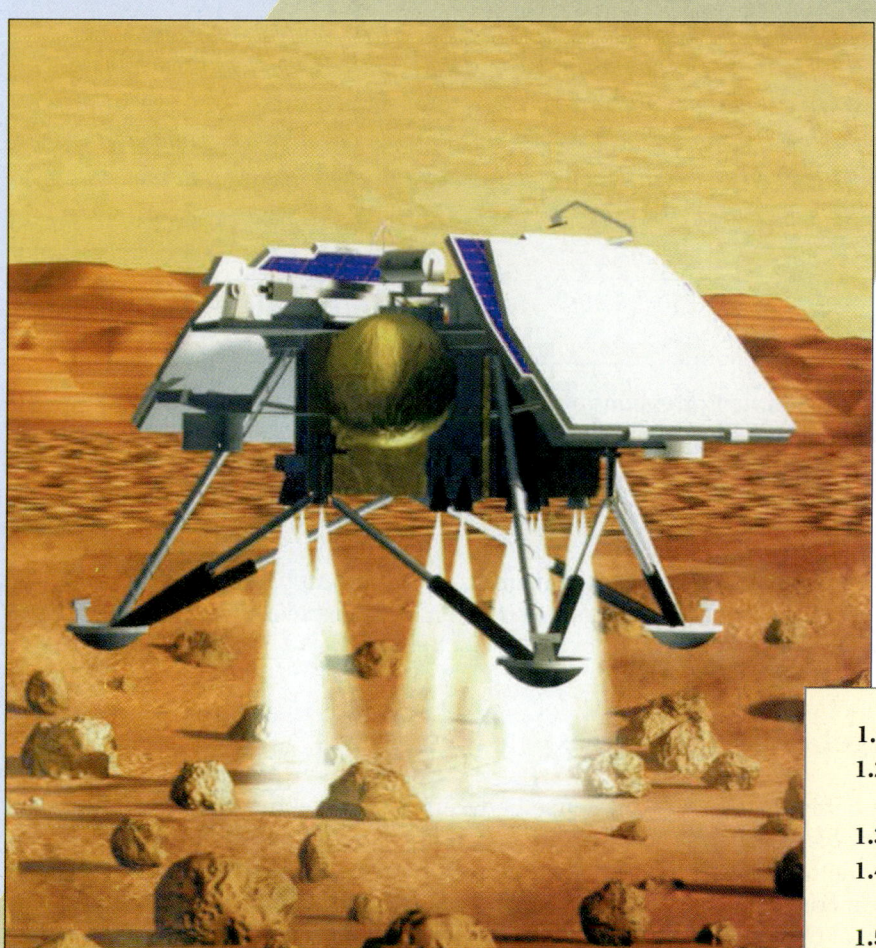

1

Introduction to Algebraic Expressions

AN APPLICATION

The Viking 2 Lander spacecraft has determined that temperatures on Mars range from −125° Celsius (C) to 25°C (*Source*: The Lunar and Planetary Institute). Find the temperature range on Mars.

This problem appears as Example 11 in Section 1.6.

*T*he math we use every day is mostly algebra and geometry. We determine the constraints, like weights and dimensions, and then design robotic equipment that works within those constraints. Using the math correctly makes all the difference in whether the equipment does its job or not.

SHANNON CROWELL
Development Engineer
Richland, Washington

P*roblem solving is the focus of this text. Chapter 1 presents some preliminaries that are needed for the problem-solving approach that is developed in Chapter 2 and used throughout the rest of the book. These preliminaries include a review of arithmetic, a discussion of real numbers and their properties, and an examination of how real numbers are added, subtracted, multiplied, divided, and raised to powers.*

Introduction to Algebra

1.1

Algebraic Expressions • Translating to Algebraic Expressions • Translating to Equations

This section introduces some of the basic concepts and expressions used in algebra. Solving real-world problems is an important part of algebra, so we will concentrate on the wordings and mathematical expressions that often arise in applications.

Algebraic Expressions

Probably the greatest difference between arithmetic and algebra is the use of *variables* in algebra. When a letter can represent a variety of different numbers, that letter is a **variable**. For example, if n represents the number of tickets sold for a Los Lobos concert, then n will vary, depending on factors like price and day of the week. Thus the number n is a variable. If every ticket costs \$25, then a total of $25 \cdot n$ dollars will be paid for tickets. Note that $25 \cdot n$ means 25 *times* n. The number 25 is called a **constant** because it does not change.

Cost per Ticket (in dollars)	Number of Tickets Sold	Total Collected (in dollars)
25	n	$25 \cdot n$

The expression $25 \cdot n$ is a **variable expression** because its value varies with the choice of n. In this case, the total amount collected, $25 \cdot n$, will change with the number of tickets sold. In the following chart, we replace n with a variety of values and compute the total amount collected. In doing so, we are **evaluating the expression** $25 \cdot n$.

Cost per Ticket (in dollars), 25	Number of Tickets Sold, n	Total Collected (in dollars), $25 \cdot n$
25	400	\$10,000
25	500	12,500
25	600	15,000

Variable expressions are examples of *algebraic expressions*. An **algebraic expression** consists of variables and/or numerals, often with operation signs and grouping symbols. Examples are

$$t + 97, \quad 5 \cdot x, \quad 3a - b, \quad 18 \div y, \quad \frac{9}{7}, \quad \text{and} \quad 4r(s + t).$$

Recall that a fraction bar is a division symbol: $\frac{9}{7}$, or 9/7, means $9 \div 7$. Similarly, multiplication can be written in several ways. For example, "5 times x" can be written as $5 \cdot x$, $5 \times x$, $5(x)$, or simply $5x$.

To evaluate an algebraic expression, we **substitute** a number for each variable in the expression. This replaces each variable with a number.

E x a m p l e   1

Evaluate each expression for the given values.

a) $x + y$ for $x = 37$ and $y = 28$
b) $5ab$ for $a = 2$ and $b = 3$

Solution

a) We substitute 37 for x and 28 for y and carry out the addition:

$$x + y = 37 + 28 = 65.$$

The number 65 is called the **value** of the expression.
b) We substitute 2 for a and 3 for b and multiply:

$$5ab = 5 \cdot 2 \cdot 3 = 10 \cdot 3 = 30.$$

E x a m p l e   2

The area A of a rectangle of length l and width w is given by the formula $A = lw$. Find the area when l is 17 in. and w is 10 in.

Solution   We evaluate, using 17 in. for l and 10 in. for w, and carry out the multiplication:

$$
\begin{aligned}
A = lw &= (17 \text{ in.})(10 \text{ in.}) \\
&= (17)(10)(\text{in.})(\text{in.}) \\
&= 170 \text{ in}^2, \text{ or } 170 \text{ square inches.}
\end{aligned}
$$

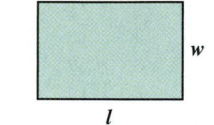

Note that we use square units for area and $(\text{in.})(\text{in.}) = \text{in}^2$. Exponents like the 2 in the expression in^2 are discussed in detail in Section 1.8.

E x a m p l e   3

The area of a triangle with a base of length b and a height of length h is given by the formula $A = \frac{1}{2}bh$. Find the area when b is 8 m and h is 6.4 m.

Solution   We substitute 8 m for b and 6.4 m for h and then multiply:

$$
\begin{aligned}
A = \tfrac{1}{2}bh &= \tfrac{1}{2}(8 \text{ m})(6.4 \text{ m}) \\
&= \tfrac{1}{2}(8)(6.4)(\text{m})(\text{m}) \\
&= 4(6.4) \text{ m}^2 \\
&= 25.6 \text{ m}^2, \text{ or } 25.6 \text{ square meters.}
\end{aligned}
$$

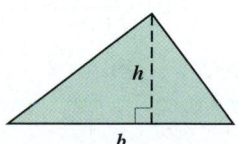

Translating to Algebraic Expressions

Before attempting to translate problems to equations, we need to be able to translate certain phrases to algebraic expressions.

Key Words

Addition (+)	Subtraction (−)	Multiplication (·)	Division (÷)
add	subtract	multiply	divide
sum of	difference of	product of	quotient of
plus	minus	times	divided by
more than	less than	twice	ratio of
increased by	decreased by	of	per

Example 4

Translate each phrase to an algebraic expression.

a) Four less than Jean's height, in inches
b) Eighteen more than a number
c) A week's pay, in dollars, divided by three

Solution   To help think through a translation, we sometimes begin with a specific number in place of a variable.

a) If the height were 60, then 4 less than 60 would mean $60 - 4$. If the height were 70, the translation would be $70 - 4$. If we use h to represent "Jean's height, in inches," the translation of "Four less than Jean's height, in inches" is $h - 4$.

b) If we knew the number to be 10, the translation would be $10 + 18$, or $18 + 10$. If we use t to represent "a number," the translation of "Eighteen more than a number" is

$$t + 18, \quad \text{or} \quad 18 + t.$$

c) We let w represent "a week's pay, in dollars." If the pay were \$450, the translation would be $450 \div 3$, or $\frac{450}{3}$. Thus our translation of "a week's pay, in dollars, divided by three" is

$$w \div 3, \quad \text{or} \quad \frac{w}{3}.$$

Caution!   Because the order in which we subtract and divide affects the answer, answering $4 - h$ or $3 \div w$ in Examples 4(a) and 4(c) is incorrect.

E x a m p l e   5   Translate each of the following.

a) Kate's age increased by five
b) Half of some number
c) Three more than twice a number
d) Six less than the product of two numbers
e) Seventy-six percent of the town's population

Solution

Phrase	*Algebraic Expression*
a) Kate's age increased by five	$a + 5$, or $5 + a$
b) Half of some number	$\frac{1}{2}t$, or $\frac{t}{2}$, or $t/2$, or $t \div 2$
c) Three more than twice a number	$2x + 3$
d) Six less than the product of two numbers	$mn - 6$
e) Seventy-six percent of the town's population	76% of p, or $0.76p$

Translating to Equations

The symbol $=$ ("equals") indicates that the expressions on either side of the equals sign represent the same number. An **equation** is a number sentence with the verb $=$. Equations may be true, false, or neither true nor false.

E x a m p l e   6   Determine whether each equation is true, false, or neither.

a) $8 \cdot 4 = 32$ **b)** $7 - 2 = 4$ **c)** $x + 6 = 13$

Solution

a) $8 \cdot 4 = 32$ The equation is *true*.
b) $7 - 2 = 4$ The equation is *false*.
c) $x + 6 = 13$ The equation is *neither* true nor false, because we do not know what number x represents.

> **Solution**
> A replacement or substitution that makes an equation true is called a *solution*. Some equations have more than one solution, and some have no solution. When all solutions have been found, we have *solved* the equation.

To determine whether a number is a solution, we evaluate all expressions in the equation. If the values on both sides of the equation are the same, the number is a solution.

Example 7   Determine whether 7 is a solution of $x + 6 = 13$.

Solution

$$
\begin{array}{r|l}
x + 6 = 13 & \text{Writing the equation} \\
\hline
7 + 6 \; ? \; 13 & \text{Substituting 7 for } x \\
13 \;\mid\; 13 & 13 = 13 \text{ is TRUE.}
\end{array}
$$

Since the left-hand and the right-hand sides are the same, 7 is a solution.

Although we do not study solving equations until Chapter 2, we can translate certain problem situations to equations now. The words "is the same as," "equal," "is," and "are" translate to "=."

Example 8   Translate the following problem to an equation.

What number plus 478 is 1019?

Solution   We let y represent the unknown number. The translation then comes almost directly from the English sentence.

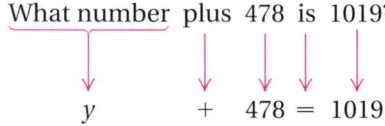

Note that "plus" translates to "+" and "is" translates to "=."

Sometimes it helps to reword a problem before translating.

Example 9   Translate the following problem to an equation.

The Petronas Twin Towers in Kuala Lumpur are the world's tallest buildings. At 1483 ft, they are 33 ft taller than the Sears Tower. (*Source*: *New York Times*) How tall is the Sears Tower?

Solution   We let h represent the height, in feet, of the Sears Tower. A rewording and translation follow:

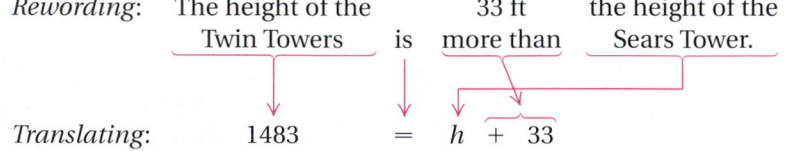

technology connection

Technology Connections are activities that make use of the graphing calculator as a tool for using algebra. These activities use only basic features that are common to most graphing calculators. **(Henceforth in this text we will refer to all graphing utilities as graphers.)** In some cases, students may find the user's manual for their particular grapher helpful for exact keystrokes.

Although all graphers are not the same, most share the following characteristics.

Screen. The large screen can show graphs and tables as well as several operations at once. The screen has a different layout for different functions. Computations are performed in the **home screen**. On many calculators, the home screen is accessed by pressing $\boxed{\text{2nd}}$ $\boxed{\text{QUIT}}$. The **cursor** shows location on the screen, and the **contrast** determines how dark the characters appear.

Keypad. There are often options written above the keys as well as on them. To access those options, we press $\boxed{\text{2nd}}$ or $\boxed{\text{ALPHA}}$ and then the key. Expressions are usually entered as they would appear in print. For example, to evaluate $3xy + x$ for $x = 65$ and $y = 92$, we press 3 $\boxed{\times}$ 65 $\boxed{\times}$ 92 $\boxed{+}$ 65 and then $\boxed{\text{ENTER}}$ or $\boxed{\text{EXE}}$. The value of the expression, 18005, will appear at the right of the screen.

```
3*65*92+65
                              18005
■
```

Evaluate each of the following.

1. $27a - 18b$, for $a = 136$ and $b = 13$
2. $19xy - 9x + 13y$, for $x = 87$ and $y = 29$

Exercise Set 1.1

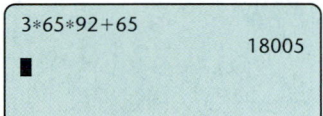

Evaluate.

1. $3a$, for $a = 9$

2. $8x$, for $x = 7$

3. $t + 6$, for $t = 2$

4. $13 - r$, for $r = 9$

5. $\dfrac{x + y}{4}$, for $x = 2$ and $y = 14$

6. $\dfrac{p + q}{7}$, for $p = 15$ and $q = 20$

7. $\dfrac{m - n}{2}$, for $m = 20$ and $n = 6$

8. $\dfrac{x - y}{6}$, for $x = 23$ and $y = 5$

9. $\dfrac{a}{b}$, for $a = 45$ and $b = 9$

10. $\dfrac{m}{n}$, for $m = 54$ and $n = 9$

11. $\dfrac{9m}{q}$, for $m = 6$ and $q = 18$

12. $\dfrac{5z}{y}$, for $z = 9$ and $y = 15$

Substitute to find the value of each expression.

13. *Hockey.*   The area of a rectangle with base b and height h is bh. A regulation hockey goal is 6 ft wide and 4 ft high. Find the area of the opening.

14. *Orbit time.*   A communications satellite orbiting 300 mi above the earth travels about 27,000 mi in one orbit. The time, in hours, for an orbit is

$$\frac{27,000}{v},$$

where v is the velocity, in miles per hour. How long will an orbit take at a velocity of 1125 mph?

15. *Zoology.*   A great white shark has triangular teeth. Each tooth measures about 5 cm across the base and has a height of 6 cm. Find the surface area of the front side of one such tooth. (See Example 3.)

16. *Work time.*   Enrico takes five times as long to do a job as Rosa does. Suppose t represents the time it takes Rosa to do the job. Then $5t$ represents the time it takes Enrico. How long does it take Enrico if Rosa takes **(a)** 30 sec? **(b)** 90 sec? **(c)** 2 min?

17. *Olympic softball.*   A softball player's batting average is h/a, where h is the number of hits and a is the number of "at bats." In the 2000 Summer Olympics, Crystl Bustos had 10 hits in 37 at bats. What was her batting average? Round to the nearest thousandth.

18. *Area of a parallelogram.*   The area of a parallelogram with base b and height h is bh. Find the area of the parallelogram when the height is 4 cm (centimeters) and the base is 6.5 cm.

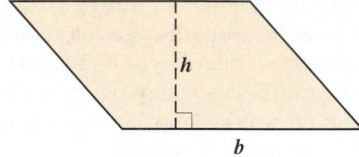

Translate to an algebraic expression.

19. 8 more than Jan's age

20. The product of 4 and a

21. 6 more than b

22. 7 more than Lou's weight

23. 9 less than c **24.** 4 less than d

25. 6 increased by q **26.** 11 increased by z

27. 9 times Phil's speed

28. c more than d

29. x less than y

30. 2 less than Lorrie's age

31. x divided by w

32. The quotient of two numbers

33. m subtracted from n

34. p subtracted from q

35. The sum of the box's length and height

36. The sum of d and f

37. The product of 9 and twice m

38. Paula's speed minus twice the wind speed

39. One quarter of some number

40. One third of the sum of two numbers

41. 64% of the women attending

42. 38% of a number

43. Lita had $50 before paying x dollars for a pizza. How much remains?

44. Dino drove his pickup truck at 65 mph for *t* hours. How far did he go?

Determine whether the given number is a solution of the given equation.

45. 15; $x + 17 = 32$

46. 75; $y + 28 = 93$

47. 93; $a - 28 = 75$

48. 12; $8t = 96$

49. 63; $\dfrac{t}{7} = 9$

50. 52; $\dfrac{x}{8} = 6$

51. 3; $\dfrac{108}{x} = 36$

52. 7; $\dfrac{94}{y} = 12$

Translate each problem to an equation. Do not solve.

53. What number added to 73 is 201?

54. Seven times what number is 2303?

55. When 42 is multiplied by a number, the result is 2352. Find the number.

56. When 345 is added to a number, the result is 987. Find the number.

57. *Chess.*   A chess board has 64 squares. If you control 35 squares and your opponent controls the rest, how many does your opponent control?

58. *Hours worked.*   A carpenter charges $25 an hour. How many hours did she work if she billed a total of $53,400?

59. *Recycling.*   Currently, Americans recycle or compost 27% of all municipal solid waste. This is the same as recycling or composting 56 million tons. What is the total amount of waste generated?

60. *Travel to work.*   In the Northeast, the average commute to work is 24.5 min. The average commuting time in the West is 1.8 min less. How long is the average commute in the West?

To the student and the instructor: Writing exercises, denoted by ▯, *should be answered using one or more English sentences. Because answers to many writing exercises will vary, solutions are not listed in the answers at the back of the book.*

▯ 61. What is the difference between a variable, a variable expression, and an equation?

▯ 62. What does it mean to evaluate an algebraic expression?

SYNTHESIS

To the student and the instructor: Synthesis exercises *are designed to challenge students to extend the concepts or skills studied in each section. Many synthesis exercises will require the assimilation of skills and concepts from several sections.*

▯ 63. If the lengths of the sides of a square are doubled, is the area doubled? Why or why not?

▯ 64. Write a problem that translates to $1998 + t = 2006$.

65. Signs of Distinction charges $90 per square foot for handpainted signs. The town of Belmar commissioned a triangular sign with a base of 3 ft and a height of 2.5 ft. How much will the sign cost?

66. Find the area that is shaded.

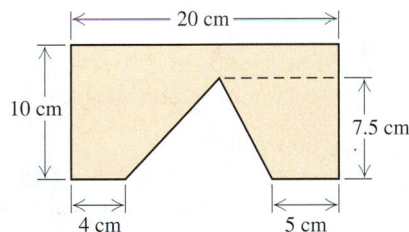

67. Evaluate $\dfrac{x - y}{3}$ when *x* is twice *y* and $x = 12$.

68. Evaluate $\dfrac{x + y}{2}$ when *y* is twice *x* and $x = 6$.

69. Evaluate $\dfrac{a + b}{4}$ when *a* is twice *b* and $a = 16$.

70. Evaluate $\dfrac{a - b}{3}$ when *a* is three times *b* and $a = 18$.

Answer each question with an algebraic expression.

71. If $w + 3$ is a whole number, what is the next whole number after it?

72. If $d + 2$ is an odd number, what is the preceding odd number?

Translate to an algebraic expression.

73. One third of one half of the product of two numbers

74. The perimeter of a rectangle with length l and width w (perimeter means distance around)

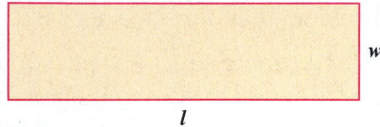

75. The perimeter of a square with side s (perimeter means distance around)

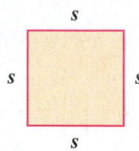

76. Ray's age 7 yr from now if he is 2 yr older than Monique and Monique is a years old

77. If the length of the height of a triangle is doubled, is its area also doubled? Why or why not?

CORNER

Teamwork

COLLABORATIVE

Focus: Group problem solving; working collaboratively

Time: 15 minutes

Group size: 2

Working and studying as a team often enables students to solve problems that are difficult to solve alone.

ACTIVITY

1. The left-hand column below contains the names of 12 colleges. A scrambled list of the names of their sports teams is on the right. As a group, match the names of the colleges to the teams.

1. University of Texas	**a.** Antelopes
2. Western State College of Colorado	**b.** Fighting Banana Slugs
3. University of North Carolina	**c.** Sea Warriors
4. University of Massachusetts	**d.** Gators
5. Hawaii Pacific University	**e.** Mountaineers
6. University of Nebraska	**f.** Sailfish
7. University of California, Santa Cruz	**g.** Longhorns
8. University of Southern Louisiana	**h.** Tarheels
9. Grand Canyon University	**i.** Seawolves
10. Palm Beach Atlantic College	**j.** Ragin' Cajuns
11. University of Alaska, Anchorage	**k.** Cornhuskers
12. University of Florida	**l.** Minutemen

2. After working for 5 min, confer with another group and reach mutual agreement.

3. Does the class agree on all 12 pairs?

4. Do you agree that group collaboration enhances our ability to solve problems?

1.2 The Commutative, Associative, and Distributive Laws

Equivalent Expressions • The Commutative Laws • The Associative Laws • The Distributive Law • The Distributive Law and Factoring

In order to solve equations, we must be able to manipulate algebraic expressions. The commutative, associative, and distributive laws discussed in this section enable us to write *equivalent expressions* that will simplify our work. Indeed, much of this text is devoted to finding equivalent expressions.

Equivalent Expressions

The expressions $4 + 4 + 4$, $3 \cdot 4$, and $4 \cdot 3$ all represent the same number, 12. Expressions that represent the same number are said to be **equivalent**. The equivalent expressions $t + 18$ and $18 + t$ were used on p. 4 when we translated "eighteen more than a number." To check that these expressions are equivalent, we make some choices for t:

$$\text{When } t = 3, \quad t + 18 = 3 + 18 \quad \text{and} \quad 18 + t = 18 + 3$$
$$= 21 \qquad\qquad\qquad\qquad = 21.$$

$$\text{When } t = 40, \quad t + 18 = 40 + 18 \quad \text{and} \quad 18 + t = 18 + 40$$
$$= 58 \qquad\qquad\qquad\qquad = 58.$$

The Commutative Laws

Recall that changing the order in addition or multiplication does not change the result. Equations like $3 + 78 = 78 + 3$ and $5 \cdot 14 = 14 \cdot 5$ illustrate this idea and show that addition and multiplication are **commutative**.

> ### The Commutative Laws
> *For Addition.* For any numbers a and b,
> $$a + b = b + a.$$
> (Changing the order of addition does not affect the answer.)
>
> *For Multiplication.* For any numbers a and b,
> $$ab = ba.$$
> (Changing the order of multiplication does not affect the answer.)

E x a m p l e   1   Use the commutative laws to write an expression equivalent to each of the following: **(a)** $y + 5$; **(b)** $9x$; **(c)** $7 + ab$.

Solution

a) $y + 5$ is equivalent to $5 + y$ by the commutative law of addition.

b) $9x$ is equivalent to $x9$ by the commutative law of multiplication.

c) $7 + ab$ is equivalent to $ab + 7$ by the commutative law of *addition*.

　　$7 + ab$ is also equivalent to $7 + ba$ by the commutative law of *multiplication*.

　　$7 + ab$ is also equivalent to $ba + 7$ by both commutative laws.

The Associative Laws

Parentheses are used to indicate groupings. We normally simplify within the parentheses first. For example,

$$3 + (8 + 4) = 3 + 12 \quad \text{and} \quad (3 + 8) + 4 = 11 + 4$$
$$= 15 \qquad\qquad\qquad\qquad = 15.$$

Similarly,

$$4 \cdot (2 \cdot 3) = 4 \cdot 6 \quad \text{and} \quad (4 \cdot 2) \cdot 3 = 8 \cdot 3$$
$$= 24 \qquad\qquad\qquad\qquad = 24.$$

Note that, so long as only addition or only multiplication appears in an expression, changing the grouping does not change the result. Equations such as $3 + (7 + 5) = (3 + 7) + 5$ and $4(5 \cdot 3) = (4 \cdot 5)3$ illustrate that addition and multiplication are **associative**.

> ### The Associative Laws
>
> *For Addition.* For any numbers a, b, and c,
>
> $$a + (b + c) = (a + b) + c.$$
>
> (Numbers can be grouped in any manner for addition.)
>
> *For Multiplication.* For any numbers a, b, and c,
>
> $$a \cdot (b \cdot c) = (a \cdot b) \cdot c.$$
>
> (Numbers can be grouped in any manner for multiplication.)

E x a m p l e   2   Use an associative law to write an expression equivalent to each of the following: **(a)** $y + (z + 3)$; **(b)** $(8x)y$.

Solution

a) $y + (z + 3)$ is equivalent to $(y + z) + 3$ by the associative law of addition.

b) $(8x)y$ is equivalent to $8(xy)$ by the associative law of multiplication.

When only additions or only multiplications are involved, parentheses do not change the result. For that reason, we sometimes omit them altogether. Thus,

$$x + (y + 7) = x + y + 7, \quad \text{and} \quad l(wh) = lwh.$$

A sum such as $(5 + 1) + (3 + 5) + 9$ can be simplified by pairing numbers that add to 10. The associative and commutative laws allow us to do this:

$$(5 + 1) + (3 + 5) + 9 = 5 + 5 + 9 + 1 + 3$$
$$= 10 + 10 + 3 = 23.$$

E x a m p l e   3

Use the commutative and/or associative laws of addition to write at least two expressions equivalent to $(x + 5) + y$.

Solution

a) $(x + 5) + y = x + (5 + y)$ Using the associative law; $x + (5 + y)$ is one equivalent expression.

$ = x + (y + 5)$ Using the commutative law

b) $(x + 5) + y = y + (x + 5)$ Using the commutative law; $y + (x + 5)$ is one equivalent expression.

$ = y + (5 + x)$ Using the commutative law again

E x a m p l e   4

Use the commutative and/or associative laws of multiplication to rewrite $2(x3)$ as $6x$. Show and give reasons for each step.

Solution

$$2(x3) = 2(3x) \quad \text{Using the commutative law}$$
$$= (2 \cdot 3)x \quad \text{Using the associative law}$$
$$= 6x \quad \text{Simplifying}$$

The Distributive Law

The *distributive law* is probably the single most important law for manipulating algebraic expressions. Unlike the commutative and associative laws, the distributive law uses multiplication together with addition.

You have already used the distributive law although you may not have realized it at the time. To illustrate, try to multiply $3 \cdot 21$ mentally. Many people find the product, 63, by thinking of 21 as $20 + 1$ and then multiplying 20 by 3 and 1 by 3. The sum of the two products, $60 + 3$, is 63. Note that if the 3 does not multiply both 20 and 1, the result will not be correct.

E x a m p l e 5

Compute in two ways: $4(7 + 2)$.

Solution

a) As in the discussion of $3(20 + 1)$ above, to compute $4(7 + 2)$, we can multiply both 7 and 2 by 4 and add the results:

$$4(7 + 2) = 4 \cdot 7 + 4 \cdot 2 \qquad \text{Multiplying both 7 and 2 by 4}$$
$$= 28 + 8 = 36. \qquad \text{Adding}$$

b) By first adding inside the parentheses, we get the same result in a different way:

$$4(7 + 2) = 4(9) \qquad \text{Adding; } 7 + 2 = 9$$
$$= 36. \qquad \text{Multiplying}$$

The Distributive Law

For any numbers a, b, and c,

$$a(b + c) = ab + ac.$$

(The product of a number and a sum can be written as the sum of two products.)

E x a m p l e 6

Multiply: $3(x + 2)$.

Solution   Since $x + 2$ cannot be simplified unless a value for x is given, we use the distributive law:

$$3(x + 2) = 3 \cdot x + 3 \cdot 2 \qquad \text{Using the distributive law}$$
$$= 3x + 6. \qquad \text{Note that } 3 \cdot x \text{ is the same as } 3x.$$

The expression $3x + 6$ has two *terms*, $3x$ and 6. In general, a **term** is a number, a variable, or a product or quotient of numbers and/or variables. Thus, t, 29, $5ab$, and $2x/y$ are terms in $t + 29 + 5ab + 2x/y$. Note that terms are separated by plus signs.

E x a m p l e 7

List the terms in $7s + st + \dfrac{3}{t}$.

Solution   Terms are separated by plus signs, so the terms in $7s + st + 3/t$ are $7s$, st, and $3/t$.

The distributive law can also be used when more than two terms are inside the parentheses.

Example 8

Multiply: $6(s + 2 + 5w)$.

Solution

$$6(s + 2 + 5w) = 6 \cdot s + 6 \cdot 2 + 6 \cdot 5w \qquad \text{Using the distributive law}$$
$$= 6s + 12 + (6 \cdot 5)w \qquad \text{Using the associative law}$$
$$\text{for multiplication}$$
$$= 6s + 12 + 30w$$

Because of the commutative law of multiplication, the distributive law can be used on the "right": $(b + c)a = ba + ca$.

Example 9

Multiply: $(c + 4)5$.

Solution

$$(c + 4)5 = c \cdot 5 + 4 \cdot 5 \qquad \text{Using the distributive law on the right}$$
$$= 5c + 20$$

Caution!   To use the distributive law for removing parentheses, be sure to multiply *each* term inside the parentheses by the multiplier outside:

$$a(b + c) \neq ab + c.$$

The Distributive Law and Factoring

If we use the distributive law in reverse, we have the basis of a process called **factoring**: $ab + ac = a(b + c)$. To **factor** an expression means to write an equivalent expression that is a product. The parts of the product are called **factors**. Note that "factor" can be used as either a verb or a noun.

Example 10

Use the distributive law to factor each of the following.

a) $3x + 3y$ 　　　　　　　　　　　　　　**b)** $7x + 21y + 7$

Solution

a) By the distributive law,

$$3x + 3y = 3(x + y). \qquad \text{The } \textit{common factor} \text{ is 3.}$$

b) $7x + 21y + 7 = 7 \cdot x + 7 \cdot 3y + 7 \cdot 1 \qquad \text{The common factor is 7.}$
$$= 7(x + 3y + 1) \qquad \text{Using the distributive law}$$

Be sure not to omit the 1 or the common factor, 7.

To check our factoring, we multiply to see if the original expression is obtained. For example, to check the **factorization** in Example 10(b), note that

$$7(x + 3y + 1) = 7x + 7 \cdot 3y + 7 \cdot 1$$
$$= 7x + 21y + 7.$$

FOR EXTRA HELP

Exercise Set 1.2

Digital Video Tutor CD 1
Videotape 1

InterAct Math

Math Tutor Center

MathXL

MyMathLab.com

Use the commutative law of addition to write an equivalent expression.

1. $7 + x$ **2.** $a + 2$

3. $ab + c$ **4.** $x + 3y$

5. $9x + 3y$ **6.** $3a + 7b$

7. $5(a + 1)$ **8.** $9(x + 5)$

Use the commutative law of multiplication to write an equivalent expression.

9. $2 \cdot a$ **10.** xy

11. st **12.** $4x$

13. $5 + ab$ **14.** $x + 3y$

15. $5(a + 1)$ **16.** $9(x + 5)$

Use the associative law of addition to write an equivalent expression.

17. $(a + 5) + b$ **18.** $(5 + m) + r$

19. $r + (t + 7)$ **20.** $x + (2 + y)$

21. $(ab + c) + d$ **22.** $(m + np) + r$

Use the associative law of multiplication to write an equivalent expression.

23. $(8x)y$ **24.** $(9a)b$

25. $2(ab)$ **26.** $9(rp)$

27. $3[2(a + b)]$ **28.** $5[x(2 + y)]$

Use the commutative and/or associative laws to write two equivalent expressions. Answers may vary.

29. $r + (t + 6)$ **30.** $5 + (v + w)$

31. $(17a)b$ **32.** $x(3y)$

Use the commutative and/or associative laws to show why the expression on the left is equivalent to the expression on the right. Write a series of steps with labels, as in Example 4.

33. $(5 + x) + 2$ is equivalent to $x + 7$

34. $(2a)4$ is equivalent to $8a$

35. $(m3)7$ is equivalent to $21m$

36. $4 + (9 + x)$ is equivalent to $x + 13$

Multiply.

37. $4(a + 3)$ **38.** $3(x + 5)$

39. $6(1 + x)$ **40.** $6(v + 4)$

41. $3(x + 1)$ **42.** $9(x + 3)$

43. $8(3 + y)$ **44.** $7(s + 5)$

45. $9(2x + 6)$ **46.** $9(6m + 7)$

47. $5(r + 2 + 3t)$ **48.** $4(5x + 8 + 3p)$

49. $(a + b)2$ **50.** $(x + 2)7$

51. $(x + y + 2)5$ **52.** $(2 + a + b)6$

List the terms in each expression.

53. $x + xyz + 19$ **54.** $9 + 17a + abc$

55. $2a + \dfrac{a}{b} + 5b$ **56.** $3xy + 20 + \dfrac{4a}{b}$

Use the distributive law to factor each of the following. Check by multiplying.

57. $2a + 2b$ **58.** $5y + 5z$

59. $7 + 7y$ **60.** $13 + 13x$

61. $18x + 3$ **62.** $20a + 5$

63. $5x + 10 + 15y$ **64.** $3 + 27b + 6c$

65. $12x + 9$

66. $6x + 6$

67. $3a + 9b$

68. $5a + 15b$

69. $44x + 11y + 22z$

70. $14a + 56b + 7$

71. Is subtraction commutative? Why or why not?

72. Is division associative? Why or why not?

SKILL MAINTENANCE

To the student and the instructor: Skill maintenance exercises *review skills studied in earlier sections. Often these exercises are preparation for the next section. Answers at the back of the book indicate the section in which that material originally appeared.*

Translate to an algebraic expression.

73. Twice Kara's salary

74. Half of m

SYNTHESIS

75. Are terms and factors the same thing? Why or why not?

76. Explain how the distributive, commutative, and associative laws can be used to show that $2(3x + 4y)$ is equivalent to $6x + 8y$.

Tell whether the expressions in each pairing are equivalent. Then explain why or why not.

77. $8 + 4(a + b)$ and $4(2 + a + b)$

78. $7 \div 3m$ and $m3 \div 7$

79. $(rt + st)5$ and $5t(r + s)$

80. $yax + ax$ and $xa(1 + y)$

81. $30y + x15$ and $5[2(x + 3y)]$

82. $[c(2 + 3b)]5$ and $10c + 15bc$

83. Evaluate the expressions $3(2 + x)$ and $6 + x$ for $x = 0$. Do your results indicate that $3(2 + x)$ and $6 + x$ are equivalent? Why or why not?

84. Factor $15x + 40$. Then evaluate both $15x + 40$ and the factorization for $x = 4$. Do your results *guarantee* that the factorization is correct? Why or why not? (*Hint*: See Exercise 83.)

COLLABORATIVE CORNER

Mental Addition

Focus: Application of commutative and associative laws

Time: 10 minutes

Group size: 2–3

Legend has it that while still in grade school, the mathematician Carl Friedrich Gauss (1777–1855) was able to add the numbers from 1 to 100 mentally. Gauss did not add them sequentially, but rather paired 1 with 99, 2 with 98, and so on.

ACTIVITY

1. Use a method similar to Gauss's to simplify the following:

$$1 + 2 + 3 + 4 + 5 + 6 + 7 + 8 + 9 + 10.$$

One group member should add from left to right as a check.

2. Use Gauss's method to find the sum of the first 25 counting numbers:

$$1 + 2 + 3 + \cdots + 23 + 24 + 25.$$

Again, one student should add from left to right as a check.

3. How were the associative and commutative laws applied in parts (1) and (2) above?

4. Now use a similar approach involving both addition and division to find the sum of the first 10 counting numbers:

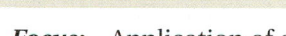

5. Use the approach in step (4) to find the sum of the first 100 counting numbers. Are the associative and commutative laws applied in this method, too? How is the distributive law used in this approach?

1.3

Fraction Notation

Factors and Prime Factorizations • Fraction Notation •
Multiplication, Division, and Simplification • More Simplifying •
Addition and Subtraction

This section covers multiplication, addition, subtraction, and division with fractions. Although much of this may be review, note that fractional expressions that contain variables are also included.

Factors and Prime Factorizations

In order to be able to study addition and subtraction using fraction notation, we first review how *natural numbers* are factored. **Natural numbers** can be thought of as the counting numbers:

$$1, 2, 3, 4, 5, \ldots .*$$

(The dots indicate that the established pattern continues without ending.) To factor a number, we simply express it as a product of two or more numbers.

Example 1   Write several factorizations of 12. Then list all factors of 12.

Solution   The number 12 can be factored in several ways:

$$1 \cdot 12, \quad 2 \cdot 6, \quad 3 \cdot 4, \quad 2 \cdot 2 \cdot 3.$$

The factors of 12 are 1, 2, 3, 4, 6, and 12.

Some numbers have only two factors, the number itself and 1. Such numbers are called **prime**.

Prime Number

A *prime number* is a natural number that has exactly two different factors: the number itself and 1.

Example 2   Which of these numbers are prime?   7, 4, 1

Solution

7 is prime. It has exactly two different factors, 7 and 1.

4 is not prime. It has three different factors, 1, 2, and 4.

1 is not prime. It does not have two *different* factors.

---

*A similar collection of numbers, the **whole numbers**, includes 0: 0, 1, 2, 3, ...

If a natural number, other than 1, is not prime, we call it **composite**. Every composite number can be factored into a product of prime numbers. Such a factorization is called the **prime factorization** of that composite number.

Example 3

Find the prime factorization of 36.

Solution   We first factor 36 in any way that we can. One way is like this:

$$36 = 4 \cdot 9.$$

The factors 4 and 9 are not prime, so we factor them:

$$36 = 4 \cdot 9$$
$$= 2 \cdot 2 \cdot 3 \cdot 3. \qquad \text{2 and 3 are both prime.}$$

The prime factorization of 36 is $2 \cdot 2 \cdot 3 \cdot 3$.

Fraction Notation

An example of **fraction notation** for a number is

$$\frac{2}{3} \cdot \quad \begin{array}{l} \longleftarrow \text{Numerator} \\ \longleftarrow \text{Denominator} \end{array}$$

The top number is called the **numerator**, and the bottom number is called the **denominator**. When the numerator and the denominator are the same nonzero number, we have fraction notation for the number 1.

> ### *Fraction Notation for 1*
> For any number a, except 0,
> $$\frac{a}{a} = 1.$$
> (Any nonzero number divided by itself is 1.)

Multiplication, Division, and Simplification

Recall from arithmetic that fractions are multiplied as follows.

> ### *Multiplication of Fractions*
> For any two fractions a/b and c/d,
> $$\frac{a}{b} \cdot \frac{c}{d} = \frac{ac}{bd}.$$
> (The numerator of the product is the product of the two numerators. The denominator of the product is the product of the two denominators.)

Example 4

Multiply: **(a)** $\dfrac{2}{3} \cdot \dfrac{7}{5}$; **(b)** $\dfrac{4}{x} \cdot \dfrac{8}{y}$.

Solution We multiply numerators as well as denominators.

a) $\dfrac{2}{3} \cdot \dfrac{7}{5} = \dfrac{2 \cdot 7}{3 \cdot 5} = \dfrac{14}{15}$
b) $\dfrac{4}{x} \cdot \dfrac{8}{y} = \dfrac{4 \cdot 8}{x \cdot y} = \dfrac{32}{xy}$

Two numbers whose product is 1 are **reciprocals**, or **multiplicative inverses**, of each other. All numbers, except zero, have reciprocals. For example,

the reciprocal of $\frac{2}{3}$ is $\frac{3}{2}$ because $\frac{2}{3} \cdot \frac{3}{2} = \frac{6}{6} = 1$;

the reciprocal of 9 is $\frac{1}{9}$ because $9 \cdot \frac{1}{9} = \frac{9}{9} = 1$; and

the reciprocal of $\frac{1}{4}$ is 4 because $\frac{1}{4} \cdot 4 = 1$.

Reciprocals are used to rewrite division as multiplication.

> ### Division of Fractions
>
> To divide two fractions, multiply by the reciprocal of the divisor:
>
> $$\frac{a}{b} \div \frac{c}{d} = \frac{a}{b} \cdot \frac{d}{c}.$$

Example 5

Divide: $\dfrac{1}{2} \div \dfrac{3}{5}$.

Solution

$$\dfrac{1}{2} \div \dfrac{3}{5} = \dfrac{1}{2} \cdot \dfrac{5}{3} \qquad \dfrac{5}{3} \text{ is the reciprocal of } \dfrac{3}{5}$$

$$= \dfrac{5}{6}$$

When one of the fractions being multiplied is 1, multiplying yields an equivalent expression because of the *identity property of* 1. A similar property could be stated for division, but there is no need to do so here.

> ### The Identity Property of 1
>
> For any number a,
>
> $$a \cdot 1 = a.$$
>
> (Multiplying a number by 1 gives that same number.)

Example 6

Multiply $\dfrac{4}{5} \cdot \dfrac{6}{6}$ to find an expression equivalent to $\dfrac{4}{5}$.

Solution   We have

$$\frac{4}{5} \cdot \frac{6}{6} = \frac{4 \cdot 6}{5 \cdot 6} = \frac{24}{30}.$$

Since $\frac{6}{6} = 1$, the expression $\frac{4}{5} \cdot \frac{6}{6}$ is equivalent to $\frac{4}{5} \cdot 1$, or simply $\frac{4}{5}$. Thus, $\frac{24}{30}$ is equivalent to $\frac{4}{5}$.

The steps of Example 6 are reversed by "removing a factor equal to 1"—in this case, $\frac{6}{6}$. By removing a factor that equals 1, we can *simplify* an expression like $\frac{24}{30}$ to an equivalent expression like $\frac{4}{5}$.

To simplify, we factor the numerator and the denominator, looking for the largest factor common to both. This is sometimes made easier by writing prime factorizations. After identifying common factors, we can express the fraction as a product of two fractions, one of which is in the form a/a.

Example 7

Simplify: **(a)** $\dfrac{15}{40}$; **(b)** $\dfrac{36}{24}$.

Solution

a) Note that 5 is a factor of both 15 and 40:

$$\frac{15}{40} = \frac{3 \cdot 5}{8 \cdot 5}$$ Factoring the numerator and the denominator, using the common factor, 5

$$= \frac{3}{8} \cdot \frac{5}{5}$$ Rewriting as a product of two fractions; $\dfrac{5}{5} = 1$

$$= \frac{3}{8} \cdot 1 = \frac{3}{8}.$$ Using the identity property of 1 (removing a factor equal to 1)

b) $\dfrac{36}{24} = \dfrac{2 \cdot 2 \cdot 3 \cdot 3}{2 \cdot 2 \cdot 2 \cdot 3}$ Writing the prime factorizations and identifying common factors; 12/12 could also be used.

$$= \frac{3}{2} \cdot \frac{2 \cdot 2 \cdot 3}{2 \cdot 2 \cdot 3}$$ Rewriting as a product of two fractions; $\dfrac{2 \cdot 2 \cdot 3}{2 \cdot 2 \cdot 3} = 1$

$$= \frac{3}{2} \cdot 1 = \frac{3}{2}$$ Using the identity property of 1

It is always wise to check your result to see if any common factors of the numerator and the denominator remain. (This will never happen if prime factorizations are used correctly.) If common factors remain, repeat the process by removing another factor equal to 1 to simplify your result.

More Simplifying

"Canceling" is a shortcut that you may have used for removing a factor equal to 1 when working with fraction notation. With *great* concern, we mention it as a

Study Tip

Take the time to include all the steps when working your homework problems. Doing so will help you organize your thinking and avoid computational errors. It will also give you complete, step-by-step solutions of the exercises that can be used when studying for quizzes and tests.

possible way to speed up your work. Canceling can be used only when removing common factors in numerators and denominators. Canceling *cannot* be used in sums or differences. Our concern is that "canceling" be used with understanding. Example 7(b) might have been done faster as follows:

$$\frac{36}{24} = \frac{2 \cdot 2 \cdot 3 \cdot 3}{2 \cdot 2 \cdot 2 \cdot 3} = \frac{3}{2}, \quad \text{or} \quad \frac{36}{24} = \frac{3 \cdot 12}{2 \cdot 12} = \frac{3}{2}, \quad \text{or} \quad \frac{\overset{3}{\overset{18}{\cancel{36}}}}{\underset{2}{\underset{12}{\cancel{24}}}} = \frac{3}{2}.$$

Caution!   Unfortunately, canceling is often performed incorrectly:

$$\frac{\cancel{2} + 3}{\cancel{2}} = 3, \qquad \frac{\cancel{4} - 1}{\cancel{4} - 2} = \frac{1}{2}, \qquad \frac{1\cancel{5}}{\cancel{5}4} = \frac{1}{4}.$$

Wrong!   Wrong!   Wrong!

$$\frac{2 + 3}{2} = \frac{5}{2} \qquad \frac{4 - 1}{4 - 2} = \frac{3}{2} \qquad \frac{15}{54} = \frac{5 \cdot 3}{18 \cdot 3} = \frac{5}{18}$$

In each of these situations, the expressions canceled are *not* factors. Factors are parts of products. For example, in $2 \cdot 3$, the numbers 2 and 3 are factors, but in $2 + 3$, 2 and 3 are *not* factors. **If you can't factor, you can't cancel! If in doubt, don't cancel!**

Sometimes it is helpful to use 1 as a factor in the numerator or the denominator when simplifying.

Example 8

Simplify: $\dfrac{9}{72}$.

Solution

$$\frac{9}{72} = \frac{1 \cdot 9}{8 \cdot 9} \qquad \text{Factoring and using the identity property of 1 to write 9 as } 1 \cdot 9$$

$$= \frac{1 \cdot \cancel{9}}{8 \cdot \cancel{9}} = \frac{1}{8} \qquad \text{Simplifying by removing a factor equal to 1: } \frac{9}{9} = 1$$

Addition and Subtraction

When denominators are the same, fractions are added or subtracted by adding or subtracting numerators and keeping the same denominator.

Addition and Subtraction of Fractions
For any two fractions a/d and b/d,

$$\frac{a}{d} + \frac{b}{d} = \frac{a + b}{d} \quad \text{and} \quad \frac{a}{d} - \frac{b}{d} = \frac{a - b}{d}.$$

E x a m p l e   9

Add and simplify: $\dfrac{4}{8} + \dfrac{5}{8}$.

Solution   The common denominator is 8. We add the numerators and keep the common denominator:

$$\frac{4}{8} + \frac{5}{8} = \frac{4+5}{8} = \frac{9}{8}.$$

You can think of this as
$$4 \cdot \frac{1}{8} + 5 \cdot \frac{1}{8} = 9 \cdot \frac{1}{8}, \text{ or } \frac{9}{8}.$$

In arithmetic, we often write $1\frac{1}{8}$ rather than the "improper" fraction $\frac{9}{8}$. In algebra, $\frac{9}{8}$ is generally more useful and is quite "proper" for our purposes.

When denominators are different, we use the identity property of 1 and multiply to find a common denominator. Then we add, as in Example 9.

E x a m p l e   1 0

Add or subtract as indicated: **(a)** $\dfrac{7}{8} + \dfrac{5}{12}$; **(b)** $\dfrac{9}{8} - \dfrac{4}{5}$.

Solution

a) The number 24 is divisible by both 8 and 12. We multiply both $\frac{7}{8}$ and $\frac{5}{12}$ by suitable forms of 1 to obtain two fractions with denominators of 24:

$$\frac{7}{8} + \frac{5}{12} = \frac{7}{8} \cdot \frac{3}{3} + \frac{5}{12} \cdot \frac{2}{2}$$

Multiplying by 1.
Since $8 \cdot 3 = 24$, we multiply $\frac{7}{8}$ by $\frac{3}{3}$.
Since $12 \cdot 2 = 24$, we multiply $\frac{5}{12}$ by $\frac{2}{2}$.

$$= \frac{21}{24} + \frac{10}{24}$$

Performing the multiplication

$$= \frac{31}{24}.$$

Adding fractions

b) $\dfrac{9}{8} - \dfrac{4}{5} = \dfrac{9}{8} \cdot \dfrac{5}{5} - \dfrac{4}{5} \cdot \dfrac{8}{8}$

Using 40 as a common denominator

$$= \frac{45}{40} - \frac{32}{40} = \frac{13}{40}$$

Subtracting fractions

After adding, subtracting, multiplying, or dividing, we may still need to simplify the answer.

E x a m p l e   1 1

Perform the indicated operation and, if possible, simplify.

a) $\dfrac{7}{10} - \dfrac{1}{5}$

b) $8 \cdot \dfrac{5}{12}$

c) $\dfrac{\frac{5}{6}}{\frac{25}{9}}$

 **technology connection**

Some graphers can perform operations using fraction notation. Others may be able to convert answers given in decimal notation to fraction notation. Often this conversion is done using a command found in a **menu**, or a list of options that appears when a key is pressed. To select an item from a menu, we highlight its number and press ENTER or simply press the number of the item.

For example, to find fraction notation for $\frac{2}{15} + \frac{7}{12}$, we enter the expression as $2/15 + 7/12$. The answer is given in decimal notation. To convert this to fraction notation, we press MATH and select the Frac option. In this case, the notation Ans▶Frac shows that the grapher will convert .7166666667 to fraction notation.

```
2/15+7/12
              .7166666667
Ans▶Frac
                    43/60
```

We see that $\frac{2}{15} + \frac{7}{12} = \frac{43}{60}$.

Solution

a) $\frac{7}{10} - \frac{1}{5} = \frac{7}{10} - \frac{1}{5} \cdot \frac{2}{2}$ Using 10 as the common denominator

$= \frac{7}{10} - \frac{2}{10}$

$= \frac{5}{10} = \frac{1 \cdot 5}{2 \cdot 5} = \frac{1}{2}$ Removing a factor equal to 1: $\frac{5}{5} = 1$

b) $8 \cdot \frac{5}{12} = \frac{8 \cdot 5}{12}$ Multiplying numerators and denominators. Think of 8 as $\frac{8}{1}$.

$= \frac{2 \cdot 2 \cdot 2 \cdot 5}{2 \cdot 2 \cdot 3}$ Factoring; $\frac{4 \cdot 2 \cdot 5}{4 \cdot 3}$ can also be used.

$= \frac{2 \cdot 2 \cdot 2 \cdot 5}{2 \cdot 2 \cdot 3}$ Removing a factor equal to 1: $\frac{2 \cdot 2}{2 \cdot 2} = 1$

$= \frac{10}{3}$ Simplifying

c) $\dfrac{\frac{5}{6}}{\frac{25}{9}} = \frac{5}{6} \div \frac{25}{9}$ Rewriting horizontally. Remember that a fraction bar indicates division.

$= \frac{5}{6} \cdot \frac{9}{25}$ Multiplying by the reciprocal of $\frac{25}{9}$

$= \frac{5 \cdot 3 \cdot 3}{2 \cdot 3 \cdot 5 \cdot 5}$ Writing as one fraction and factoring

$= \frac{5 \cdot 3 \cdot 3}{2 \cdot 3 \cdot 5 \cdot 5}$ Removing a factor equal to 1: $\frac{5 \cdot 3}{3 \cdot 5} = 1$

$= \frac{3}{10}$ Simplifying

FOR EXTRA HELP

Exercise Set 1.3

 Digital Video Tutor CD 1 Videotape 1

 InterAct Math

 Math Tutor Center

 MathXL

MyMathLab.com

To the student and the instructor: *Beginning in this section, selected exercises are marked with the icon* Aha! . *These "Aha!" exercises can be answered most easily if the student pauses to inspect the exercise rather than proceed mechanically. This is done to discourage rote memorization. Some "Aha!" exercises are left unmarked,* *to encourage students to always pause before working a problem.*

Write at least two factorizations of each number. Then list all the factors of the number.

1. 50   **2.** 70   **3.** 42   **4.** 60

Find the prime factorization of each number. If the number is prime, state this.

5. 26   **6.** 15   **7.** 30

8. 55   **9.** 20   **10.** 50

11. 27   **12.** 98   **13.** 18

14. 54   **15.** 40   **16.** 56

17. 43   **18.** 120   **19.** 210

20. 79   **21.** 115   **22.** 143

Simplify.

23. $\dfrac{10}{14}$ **24.** $\dfrac{14}{21}$ **25.** $\dfrac{16}{56}$

26. $\dfrac{72}{27}$ **27.** $\dfrac{6}{48}$ **28.** $\dfrac{12}{70}$

29. $\dfrac{49}{7}$ **30.** $\dfrac{132}{11}$ **31.** $\dfrac{19}{76}$

32. $\dfrac{17}{51}$ **33.** $\dfrac{150}{25}$ **34.** $\dfrac{170}{34}$

35. $\dfrac{75}{80}$ **36.** $\dfrac{42}{50}$ **37.** $\dfrac{120}{82}$

38. $\dfrac{75}{45}$ **39.** $\dfrac{210}{98}$ **40.** $\dfrac{140}{350}$

Perform the indicated operation and, if possible, simplify.

41. $\dfrac{1}{2} \cdot \dfrac{3}{7}$ **42.** $\dfrac{11}{10} \cdot \dfrac{8}{5}$ **43.** $\dfrac{9}{2} \cdot \dfrac{3}{4}$

Aha! **44.** $\dfrac{11}{12} \cdot \dfrac{12}{11}$ **45.** $\dfrac{1}{8} + \dfrac{3}{8}$ **46.** $\dfrac{1}{2} + \dfrac{1}{8}$

47. $\dfrac{4}{9} + \dfrac{13}{18}$ **48.** $\dfrac{4}{5} + \dfrac{8}{15}$ **49.** $\dfrac{3}{a} \cdot \dfrac{b}{7}$

50. $\dfrac{x}{5} \cdot \dfrac{y}{z}$ **51.** $\dfrac{4}{a} + \dfrac{3}{a}$ **52.** $\dfrac{7}{a} - \dfrac{5}{a}$

53. $\dfrac{3}{10} + \dfrac{8}{15}$ **54.** $\dfrac{7}{8} + \dfrac{5}{12}$ **55.** $\dfrac{9}{7} - \dfrac{2}{7}$

56. $\dfrac{12}{5} - \dfrac{2}{5}$ **57.** $\dfrac{13}{18} - \dfrac{4}{9}$ **58.** $\dfrac{13}{15} - \dfrac{8}{45}$

Aha! **59.** $\dfrac{20}{30} - \dfrac{2}{3}$ **60.** $\dfrac{5}{7} - \dfrac{5}{21}$ **61.** $\dfrac{7}{6} \div \dfrac{3}{5}$

62. $\dfrac{7}{5} \div \dfrac{3}{4}$ **63.** $\dfrac{8}{9} \div \dfrac{4}{15}$ **64.** $\dfrac{9}{4} \div 9$

65. $12 \div \dfrac{3}{7}$ **66.** $\dfrac{1}{10} \div \dfrac{1}{5}$ *Aha!* **67.** $\dfrac{7}{13} \div \dfrac{7}{13}$

68. $\dfrac{17}{8} \div \dfrac{5}{6}$ **69.** $\dfrac{\frac{2}{5}}{\frac{3}{3}}$ **70.** $\dfrac{\frac{3}{8}}{\frac{1}{5}}$

71. $\dfrac{9}{\frac{1}{2}}$ **72.** $\dfrac{\frac{7}{3}}{5}$

73. Under what circumstances would the sum of two fractions be easier to compute than the product of the same two fractions?

74. Under what circumstances would the product of two fractions be easier to compute than the sum of the same two fractions?

SKILL MAINTENANCE

Use a commutative law to write an equivalent expression. There can be more than one correct answer.

75. $5(x + 3)$ **76.** $7 + (a + b)$

SYNTHESIS

77. Bryce insists that $(2 + x)/8$ is equivalent to $(1 + x)/4$. What mistake do you think is being made and how could you demonstrate to Bryce that the two expressions are not equivalent?

78. Use the word *factor* in two sentences—once as a noun and once as a verb.

79. *Packaging.* Tritan Candies uses two sizes of boxes, 6 in. and 8 in. long. These are packed end to end in bigger cartons to be shipped. What is the shortest-length carton that will accommodate boxes of either size without any room left over? (Each carton must contain boxes of only one size; no mixing is allowed.)

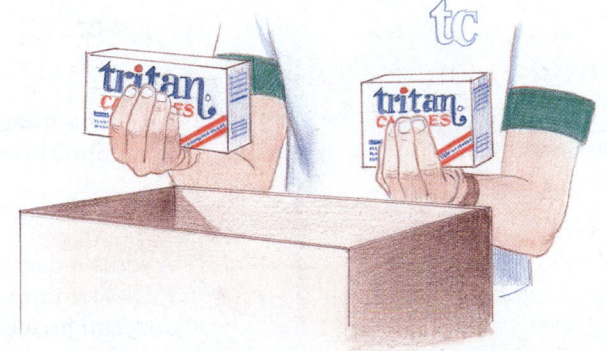

80. In the following table, the top number can be factored in such a way that the sum of the factors is the bottom number. For example, in the first column, 56 is factored as $7 \cdot 8$, since $7 + 8 = 15$, the bottom number. Find the missing numbers in each column.

Product	56	63	36	72	140	96	168
Factor	7						
Factor	8						
Sum	15	16	20	38	24	20	29

Simplify.

81. $\dfrac{16 \cdot 9 \cdot 4}{15 \cdot 8 \cdot 12}$

82. $\dfrac{9 \cdot 8xy}{2xy \cdot 36}$

83. $\dfrac{27pqrs}{9prst}$

84. $\dfrac{512}{192}$

85. $\dfrac{15 \cdot 4xy \cdot 9}{6 \cdot 25x \cdot 15y}$

86. $\dfrac{10x \cdot 12 \cdot 25y}{2 \cdot 30x \cdot 20y}$

87. $\dfrac{\dfrac{27ab}{15mn}}{\dfrac{18bc}{25np}}$

88. $\dfrac{\dfrac{45xyz}{24ab}}{\dfrac{30xz}{32ac}}$

Find the area of each figure.

89.

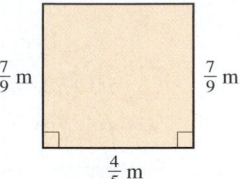

90.

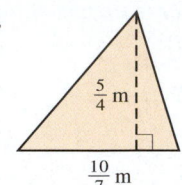

91. Find the perimeter of a square with sides of length $3\frac{5}{9}$ m.

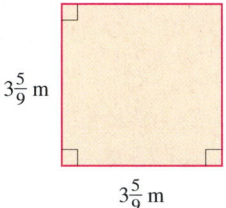

92. Find the perimeter of the rectangle in Exercise 89.

93. Make use of the properties and laws discussed in Sections 1.2 and 1.3 to explain why $x + y$ is equivalent to $(2y + 2x)/2$.

1.4
Positive and Negative Real Numbers

The Integers • The Rational Numbers • Real Numbers and Order • Absolute Value

A **set** is a collection of objects. The set containing 1, 3, and 7 is usually written $\{1, 3, 7\}$. In this section, we examine some important sets of numbers. More on sets can be found in Appendix A.

The Integers

Two sets of numbers were mentioned in Section 1.3. We represent these sets using dots on a number line.

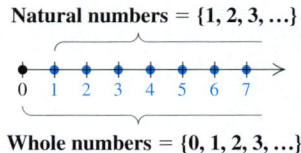

To create the set of *integers*, we include all whole numbers, along with their *opposites*. To find the opposite of a number, we locate the number that is the same distance from 0 but on the other side of the number line. For example,

the opposite of 1 is negative 1, written −1;

and

the opposite of 3 is negative 3, written −3.

The **integers** consist of all whole numbers and their opposites.

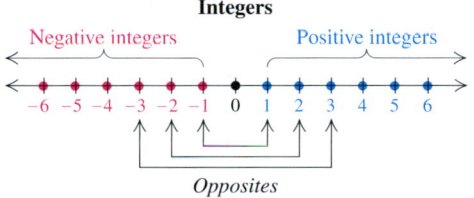

Opposites are discussed in more detail in Section 1.6. Note that, except for 0, opposites occur in pairs. Thus, 5 is the opposite of −5, just as −5 is the opposite of 5. Note that 0 acts as its own opposite.

> ### Set of Integers
> The set of integers = $\{\ldots, -4, -3, -2, -1, 0, 1, 2, 3, 4, \ldots\}$.

Integers are associated with many real-world problems and situations.

E x a m p l e   1   State which integer(s) corresponds to each situation.

a) In 1997, Tiger Woods set a U.S. Masters tournament record by finishing 18 strokes under par.

b) Death Valley is 280 ft below sea level.

c) Jaco's Bistro made $329 on Sunday, but lost $53 on Monday.

In 1997, Tiger Woods set a U.S. Masters tournament record by finishing 18 strokes under par.

Solution

a) The integer -18 corresponds to 18 under par.

b) The integer -280 corresponds to the situation (see the figure below). The elevation is -280 ft.

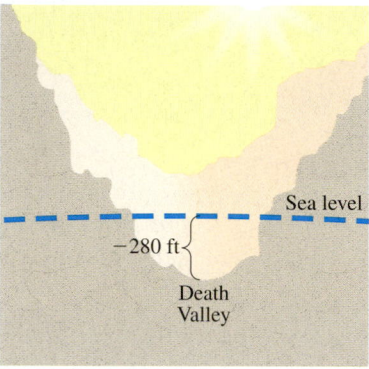

c) The integer 329 corresponds to making \$329 on Sunday and -53 corresponds to losing \$53 on Monday.

The Rational Numbers

Although numbers like $\frac{5}{9}$ are built out of integers, these numbers are not themselves integers. Another set, the **rational numbers**, contains fractions and decimals, as well as the integers. Some examples of rational numbers are

$$\frac{5}{9}, \quad -\frac{4}{7}, \quad 95, \quad -16, \quad 0, \quad \frac{-35}{8}, \quad 2.4, \quad -0.31.$$

In Section 1.7, we show that $-\frac{4}{7}$ can be written as $\frac{-4}{7}$ or $\frac{4}{-7}$. Indeed, every number listed above can be written as an integer over an integer. For example, 95 can be written as $\frac{95}{1}$ and 2.4 can be written as $\frac{24}{10}$. In this manner, any *ratio*nal number can be expressed as the *ratio* of two integers. Rather than attempt to list all rational numbers, we use this idea of ratio to describe the set as follows.

> ### Set of Rational Numbers
>
> The set of rational numbers $= \left\{ \dfrac{a}{b} \,\middle|\, a \text{ and } b \text{ are integers and } b \neq 0 \right\}.$
>
> This is read "the set of all numbers $\dfrac{a}{b}$, where a and b are integers and $b \neq 0$."

In Section 1.7, we explain why b cannot equal 0.

To *graph* a number is to mark its location on a number line.

E x a m p l e   2   Graph each of the following rational numbers.

a) $\frac{5}{2}$ **b)** -3.2 **c)** $\frac{11}{8}$

Solution

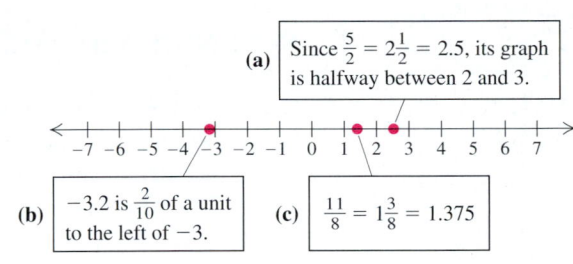

(a) | Since $\frac{5}{2} = 2\frac{1}{2} = 2.5$, its graph is halfway between 2 and 3.

(b) | -3.2 is $\frac{2}{10}$ of a unit to the left of -3.

(c) | $\frac{11}{8} = 1\frac{3}{8} = 1.375$

Every rational number can be written as a fraction or a decimal.

E x a m p l e   3   Convert to decimal notation: $-\frac{5}{8}$.

Solution   We first find decimal notation for $\frac{5}{8}$. Since $\frac{5}{8}$ means $5 \div 8$, we divide.

$$
\begin{array}{r}
0.6\ 2\ 5 \\
8\overline{)5.0\ 0\ 0} \\
\underline{4\ 8\ 0\ 0} \\
2\ 0\ 0 \\
\underline{1\ 6\ 0} \\
4\ 0 \\
\underline{4\ 0} \\
0 \quad \longleftarrow \text{The remainder is 0.}
\end{array}
$$

Thus, $\frac{5}{8} = 0.625$, so $-\frac{5}{8} = -0.625$.

Because the division in Example 3 ends with the remainder 0, we consider -0.625 a **terminating decimal**. If we are "bringing down" zeros and a remainder reappears, we have a **repeating decimal**, as shown in the next example.

E x a m p l e   4   Convert to decimal notation: $\frac{7}{11}$.

Solution   We divide:

$$
\begin{array}{r}
0.6\ 3\ 6\ 3... \\
1\ 1\overline{)7.0\ 0\ 0\ 0} \\
\underline{6\ 6} \\
4\ 0 \\
\underline{3\ 3} \\
7\ 0 \\
\underline{6\ 6} \\
4\ 0
\end{array}
$$

$-$ 4 reappears as a remainder.

We abbreviate repeating decimals by writing a bar over the repeating part—in this case, $0.\overline{63}$. Thus, $\frac{7}{11} = 0.\overline{63}$.

Although we do not prove it here, every rational number can be expressed as either a terminating or repeating decimal, and every terminating or repeating decimal can be expressed as a ratio of two integers.

Real Numbers and Order

Some numbers, when written in decimal form, neither terminate nor repeat. Such numbers are called **irrational numbers**.

What sort of numbers are irrational? One example is π (the Greek letter *pi*, read "pie"), which is used to find the area and circumference of a circle: $A = \pi r^2$ and $C = 2\pi r$.

Another irrational number, $\sqrt{2}$ (read "the square root of 2"), is the length of the diagonal of a square with sides of length 1. It is also the number that, when multiplied by itself, gives 2. No rational number can be multiplied by itself to get 2, although some approximations come close:

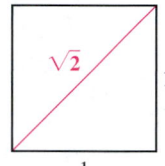

1.4 is an *approximation* of $\sqrt{2}$ because $(1.4)(1.4) = 1.96$;

1.41 is a better approximation because $(1.41)(1.41) = 1.9881$;

1.4142 is an even better approximation because $(1.4142)(1.4142) = 1.99996164$.

To approximate $\sqrt{2}$ on some calculators, simply press 2 and then $\sqrt{}$. With other calculators, press $\sqrt{}$, 2, and ENTER, or consult a manual.

Example 5   Graph the real number $\sqrt{3}$ on a number line.

Solution   We use a calculator and approximate: $\sqrt{3} \approx 1.732$ ("$\approx$" means "approximately equals"). Then we locate this number on a number line.

The rational numbers and the irrational numbers together correspond to all the points on a number line and make up what is called the **real-number system**.

Set of Real Numbers

The set of real numbers = The set of all numbers corresponding to points on the number line.

The following figure shows the relationships among various kinds of numbers.

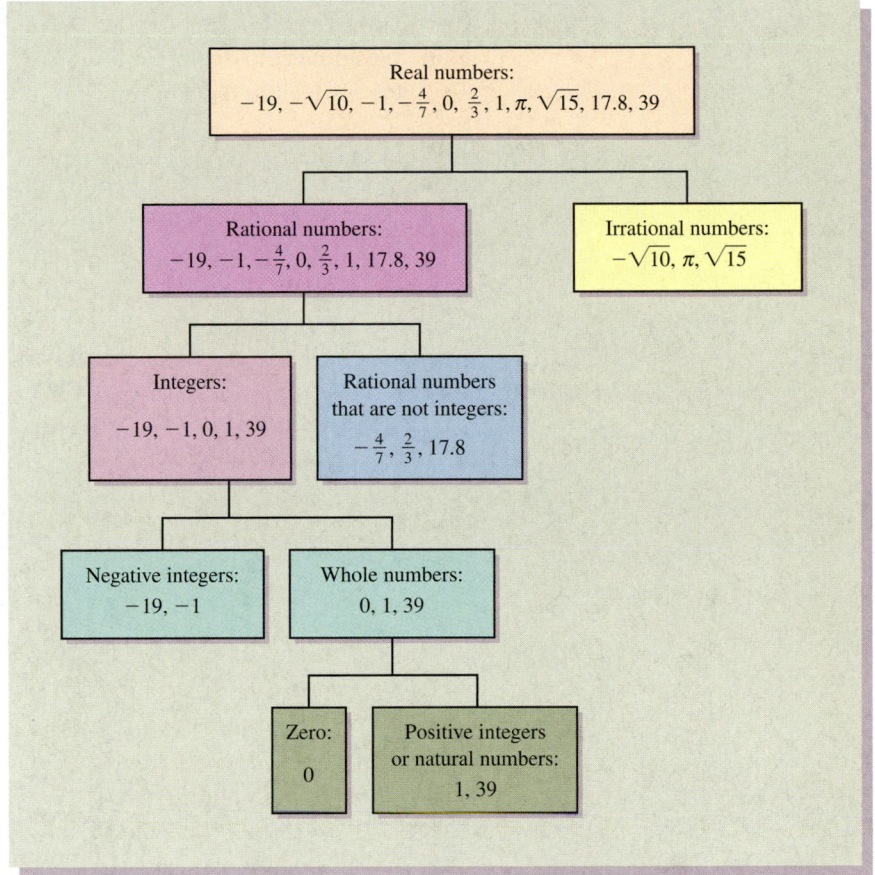

technology connection

To approximate $\sqrt{3}$ on most graphers, we press $\boxed{\sqrt{}}$ and then enter 3 enclosed by parentheses. Some graphers will supply the left parenthesis automatically when $\boxed{\sqrt{}}$ is pressed.

Many graphers use the notation abs (2) to indicate the absolute value of 2 (see Example 8). Thus, abs(2) = |2| = 2.

The $\boxed{(-)}$ and the $\boxed{-}$ keys on a grapher are not interchangeable. We use $\boxed{(-)}$ to indicate a negative number and $\boxed{-}$ for subtraction.

Simplify.

1. $5.9 + |\sqrt{2 + 7} - 7.2|$
2. $|3.5 - \sqrt{3^2 + 4^2}|$

Real numbers are named in order on the number line, with larger numbers further to the right. For any two numbers, the one to the left is less than the one to the right. We use the symbol **<** to mean "**is less than**." The sentence $-8 < 6$ means "**−8 is less than 6**." The symbol **>** means "**is greater than**." The sentence $-3 > -7$ means "−3 is greater than −7."

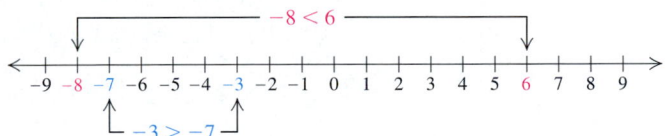

E x a m p l e   6   Use either $<$ or $>$ for ▨ to write a true sentence.

a) 2 ▨ 9   **b)** -3.45 ▨ 1.32 **c)** 6 ▨ -12
d) -18 ▨ -5 **e)** $\frac{7}{11}$ ▨ $\frac{5}{8}$

Solution

a) Since 2 is to the left of 9 on a number line, we know that 2 is less than 9, so $2 < 9$.
b) Since -3.45 is to the left of 1.32, we have $-3.45 < 1.32$.
c) Since 6 is to the right of -12, we have $6 > -12$.

d) Since -18 is to the left of -5, we have $-18 < -5$.

e) We convert to decimal notation: $\frac{7}{11} = 0.\overline{63}$ and $\frac{5}{8} = 0.625$. Thus, $\frac{7}{11} > \frac{5}{8}$.

We also could have used a common denominator: $\frac{7}{11} = \frac{56}{88} > \frac{55}{88} = \frac{5}{8}$.

Sentences like "$a < -5$" and "$-3 > -8$" are **inequalities**. It is useful to remember that every inequality can be written in two ways. For example,

$$-3 > -8 \quad \text{has the same meaning as} \quad -8 < -3.$$

It may be helpful to think of an inequality sign as an "arrow" with the smaller side pointing to the smaller number.

Note that $a > 0$ means that a represents a positive real number and $a < 0$ means that a represents a negative real number.

Statements like $a \leq b$ and $b \geq a$ are also inequalities. We read $a \leq b$ as "a **is less than or equal to** b" and $a \geq b$ as "a **is greater than or equal to** b."

E x a m p l e 7

Classify each inequality as true or false.

a) $-3 \leq 5$ **b)** $-3 \leq -3$ **c)** $-5 \geq 4$

Solution

a) $-3 \leq 5$ is *true* because $-3 < 5$ is true.

b) $-3 \leq -3$ is *true* because $-3 = -3$ is true.

c) $-5 \geq 4$ is *false* since neither $-5 > 4$ nor $-5 = 4$ is true.

Absolute Value

There is a convenient terminology and notation for the distance a number is from 0 on a number line. It is called the **absolute value** of the number.

> ### Absolute Value
>
> We write $|a|$, read "the absolute value of a," to represent the number of units that a is from zero.

E x a m p l e 8

Find each absolute value: **(a)** $|-3|$; **(b)** $|7.2|$; **(c)** $|0|$.

Solution

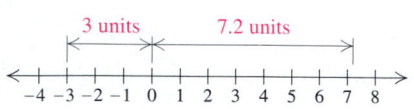

a) $|-3| = 3$ since -3 is 3 units from 0.

b) $|7.2| = 7.2$ since 7.2 is 7.2 units from 0.

c) $|0| = 0$ since 0 is 0 units from itself.

Distance is never negative, so numbers that are opposites have the same absolute value. If a number is nonnegative, its absolute value is the number itself. If a number is negative, its absolute value is its opposite.

Exercise Set **1.4**

FOR EXTRA HELP

Digital Video Tutor CD 1
Videotape 1

InterAct Math

Math Tutor Center

MathXL

MyMathLab.com

Tell which real numbers correspond to each situation.

1. In Burlington, Vermont, the record low temperature for Washington's Birthday is 19° Fahrenheit (F) below zero. The record high for the date is 59°F above zero.

2. Karissa's golf score was 2 under par, while Alex's score was 5 over.

3. Using a NordicTrack exercise machine, Kit burned 150 calories. She then drank an isotonic drink containing 65 calories.

4. A painter earned $1200 one week and spent $800 the next.

5. The Dead Sea is 1286 feet below sea level, whereas Mt. Everest is 29,029 feet above sea level.

6. In bowling, the Jets are 34 pins behind the Strikers after one game. Describe the situation from the viewpoint of each team.

7. Janice deposited $750 in a savings account. Two weeks later, she withdrew $125.

8. In 1998, the world birthrate, per thousand, was 22. The death rate, per thousand, was 9. (*Source: Time Almanac 2000,* 1999)

9. During a video game, Cindy intercepted a missile worth 20 points, lost a starship worth 150 points, and captured a base worth 300 points.

10. Ignition occurs 10 seconds before liftoff. A spent fuel tank is detached 235 seconds after liftoff.

Graph each rational number on a number line.

11. $\frac{10}{3}$ 12. $-\frac{17}{5}$ 13. -4.3

14. 3.87 15. -2 16. 5

Find decimal notation.

17. $\frac{7}{8}$ 18. $-\frac{1}{8}$ 19. $-\frac{3}{4}$

20. $\frac{5}{6}$ 21. $\frac{7}{6}$ 22. $\frac{5}{12}$

23. $\frac{2}{3}$ 24. $\frac{1}{4}$ 25. $-\frac{1}{2}$

26. $-\frac{3}{8}$ *Aha!* 27. $\frac{13}{100}$ 28. $-\frac{7}{20}$

Write a true sentence using either $<$ *or* $>$.

29. -8 ▦ 2 30. 9 ▦ 0

31. 7 ▦ 0 32. 8 ▦ -8

33. -6 ▦ 6 34. 0 ▦ -7

35. -8 ▦ -5 36. -4 ▦ -3

37. -5 ▦ -11 38. -3 ▦ -4

39. -12.5 ▦ -9.4 40. -10.3 ▦ -14.5

41. $\frac{5}{12}$ ▦ $\frac{11}{25}$ 42. $-\frac{14}{17}$ ▦ $-\frac{27}{35}$

For each of the following, write a second inequality with the same meaning.

43. $-7 > x$ 44. $a > 9$

45. $-10 \le y$ 46. $12 \ge t$

Classify each inequality as true or false.

47. $-3 \ge -11$ 48. $5 \le -5$

49. $0 \ge 8$ 50. $-5 \le 7$

51. $-8 \le -8$ 52. $8 \ge 8$

Find each absolute value.

53. $|-23|$ 54. $|-47|$ 55. $|17|$

56. $|3.1|$ 57. $|5.6|$ 58. $\left|-\frac{2}{5}\right|$

59. $|329|$ 60. $|-456|$ 61. $\left|-\frac{9}{7}\right|$

62. $|8.02|$ 63. $|0|$ 64. $|-1.07|$

65. $|x|$, for $x = -8$ 66. $|a|$, for $a = -5$

For Exercises 67–72, consider the following list:

$$-83, -4.7, 0, \tfrac{5}{9}, \pi, \sqrt{17}, 8.31, 62.$$

67. List all rational numbers.

68. List all natural numbers.

69. List all integers.

70. List all irrational numbers.

71. List all real numbers.

72. List all nonnegative integers.

73. Is every integer a rational number? Why or why not?

74. Is every integer a natural number? Why or why not?

SKILL MAINTENANCE

75. Evaluate $3xy$ for $x = 2$ and $y = 7$.

76. Use a commutative law to write an expression equivalent to $ab + 5$.

SYNTHESIS

77. Is the absolute value of a number always positive? Why or why not?

78. How many rational numbers are there between 0 and 1? Justify your answer.

79. Does "nonnegative" mean the same thing as "positive"? Why or why not?

List in order from least to greatest.

80. $13, -12, 5, -17$

81. $-23, 4, 0, -17$

82. $\tfrac{4}{5}, \tfrac{4}{3}, \tfrac{4}{8}, \tfrac{4}{6}, \tfrac{4}{9}, \tfrac{4}{2}, -\tfrac{4}{3}$

83. $-\tfrac{2}{3}, \tfrac{1}{2}, -\tfrac{3}{4}, -\tfrac{5}{6}, \tfrac{3}{8}, \tfrac{1}{6}$

Write a true sentence using either $<$, $>$, or $=$.

84. $|-5| \;\rule{12pt}{8pt}\; |-2|$

85. $|4| \;\rule{12pt}{8pt}\; |-7|$

86. $|-8| \;\rule{12pt}{8pt}\; |8|$

87. $|23| \;\rule{12pt}{8pt}\; |-23|$

88. $|-3| \;\rule{12pt}{8pt}\; |5|$

89. $|-19| \;\rule{12pt}{8pt}\; |-27|$

Solve. Consider only integer replacements.

90. $|x| = 7$

91. $|x| < 3$

92. $2 < |x| < 5$

Given that $0.3\overline{3} = \tfrac{1}{3}$ and $0.6\overline{6} = \tfrac{2}{3}$, express each of the following as a ratio of two integers.

93. $0.1\overline{1}$

94. $0.9\overline{9}$

95. $5.5\overline{5}$

96. $7.7\overline{7}$

To the student and instructor: *The calculator icon, 🔲, is used to indicate those exercises designed to be solved with a calculator.*

97. When Helga's calculator gives a decimal value for $\sqrt{2}$ and that value is promptly squared, the result is 2. Yet when that same decimal approximation is entered by hand and then squared, the result is not exactly 2. Why do you suppose this is?

Addition of Real Numbers

1.5

Adding with a Number Line • Adding without a Number Line • Problem Solving • Combining Like Terms

We now consider addition of real numbers. To gain understanding, we will use a number line first. After observing the principles involved, we will develop rules that allow us to work more quickly without a number line.

Adding with a Number Line

To add $a + b$ on a number line, we start at a and move according to b.

a) If b is positive, we move to the right (the positive direction).
b) If b is negative, we move to the left (the negative direction).
c) If b is 0, we stay at a.

E x a m p l e   1

Add: $-4 + 9$.

Solution   To add on a number line, we locate the first number, -4, and then move 9 units to the right. Note that it requires 4 units to reach 0. The difference between 9 and 4 is where we finish.

$$-4 + 9 = 5$$

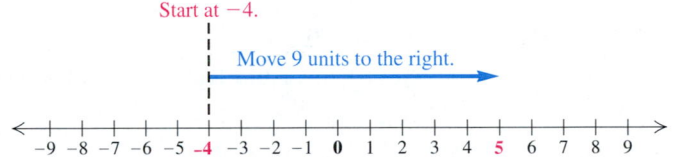

E x a m p l e   2

Add: $3 + (-5)$.

Solution   We locate the first number, 3, and then move 5 units to the left. Note that it requires 3 units to reach 0. The difference between 5 and 3 is 2, so we finish 2 units to the left of 0.

$$3 + (-5) = -2$$

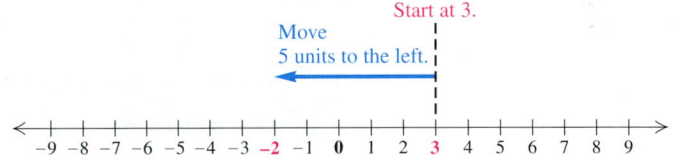

E x a m p l e   3

Add: $-4 + (-3)$.

Solution   After locating -4, we move 3 units to the left. We finish a total of 7 units to the left of 0.

$$-4 + (-3) = -7$$

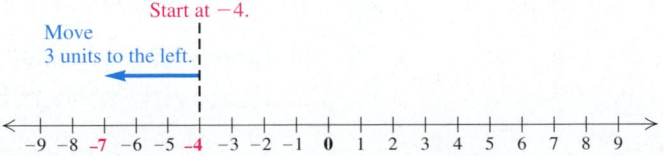

E x a m p l e   4

Add: $-5.2 + 0$.

Solution   We locate -5.2 and move 0 units. Thus we finish where we started, at -5.2.

$$-5.2 + 0 = -5.2$$

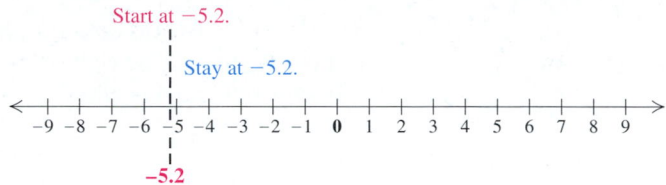

From Examples 1–4, we observe the following rules.

> ### Rules for Addition of Real Numbers
> 1. *Positive numbers*: Add as usual. The answer is positive.
> 2. *Negative numbers*: Add absolute values and make the answer negative (see Example 3).
> 3. *A positive number and a negative number*: Subtract the smaller absolute value from the greater absolute value. Then:
>   a) If the positive number has the greater absolute value, the answer is positive (see Example 1).
>   b) If the negative number has the greater absolute value, the answer is negative (see Example 2).
>   c) If the numbers have the same absolute value, the answer is 0.
> 4. *One number is zero*: The sum is the other number (see Example 4).

Rule 4 is known as the **identity property of 0**. It says that for any real number a, we have $a + 0 = a$.

Adding without a Number Line

The rules listed above can be used without drawing a number line.

E x a m p l e   5

Add without using a number line.

a) $-12 + (-7)$ b) $-1.4 + 8.5$

c) $-36 + 21$ d) $1.5 + (-1.5)$

e) $-\frac{7}{8} + 0$ f) $\frac{2}{3} + \left(-\frac{5}{8}\right)$

Study Tip

Solution

a) $-12 + (-7) = -19$ Two negatives. *Think*: Add the absolute values, 12 and 7, to get 19. Make the answer *negative*, -19.

b) $-1.4 + 8.5 = 7.1$ A negative and a positive. *Think*: The difference of absolute values is $8.5 - 1.4$, or 7.1. The positive number has the larger absolute value, so the answer is *positive*, 7.1.

c) $-36 + 21 = -15$ A negative and a positive. *Think*: The difference of absolute values is $36 - 21$, or 15. The negative number has the larger absolute value, so the answer is *negative*, -15.

d) $1.5 + (-1.5) = 0$ A negative and a positive. *Think*: Since the numbers are opposites, they have the same absolute value and the answer is 0.

e) $-\dfrac{7}{8} + 0 = -\dfrac{7}{8}$ One number is zero. The sum is the other number, $-\frac{7}{8}$.

f) $\dfrac{2}{3} + \left(-\dfrac{5}{8}\right) = \dfrac{16}{24} + \left(-\dfrac{15}{24}\right)$ This is similar to part (b) above.

$$= \dfrac{1}{24}$$

If we are adding several numbers, some positive and some negative, the commutative and associative laws allow us to add all the positives, then add all the negatives, and then add the results. Of course, we can also add from left to right, if we prefer.

Example 6

Add: $15 + (-2) + 7 + 14 + (-5) + (-12)$.

Solution

$$15 + (-2) + 7 + 14 + (-5) + (-12)$$
$$= 15 + 7 + 14 + (-2) + (-5) + (-12) \qquad \text{Using the commutative law of addition}$$

$$= (15 + 7 + 14) + [(-2) + (-5) + (-12)] \qquad \text{Using the associative law of addition}$$

$$= 36 + (-19) \qquad \text{Adding the positives; adding the negatives}$$

$$= 17 \qquad \text{Adding a positive and a negative}$$

Problem Solving

Addition of real numbers occurs in many real-world applications.

Example 7

Lake level. Lake Champlain straddles the border of New York and Vermont. From mid-November 1998 through mid-November 2000, the water level dropped $1\frac{1}{2}$ ft, rose $\frac{1}{4}$ ft, and dropped 1 ft (*Source*: *Burlington Free Press*, Nov. 21, 2000). By how much did the level change over the 2 yr?

Solution   The problem translates to a sum:

Rewording: The 1st   the 2nd   the 3rd   the total
                 change plus change plus change is change.

Translating: $-1\frac{1}{2}$ $+$ $\frac{1}{4}$ $+$ (-1) $=$ Total change.

Adding from left to right, we have

$$-1\tfrac{1}{2} + \tfrac{1}{4} + (-1) = -1\tfrac{2}{4} + \tfrac{1}{4} + (-1) = -1\tfrac{1}{4} + (-1) = -2\tfrac{1}{4}.$$

The lake level dropped $2\frac{1}{4}$ ft between mid-November 1998 and mid-November 2000.

Combining Like Terms

When two terms have variable factors that are exactly the same, like $5ab$ and $7ab$, the terms are called **like**, or **similar**, **terms**.* The distributive law enables us to **combine**, or **collect**, **like terms**. The above rules for addition will again apply.

Example 8

Combine like terms.

a) $-7x + 9x$ **b)** $2a + (-3b) + (-5a) + 9b$
c) $6 + y + (-3.5y) + 2$

Solution

a) $-7x + 9x = (-7 + 9)x$ Using the distributive law
          $= 2x$ Adding -7 and 9
b) $2a + (-3b) + (-5a) + 9b$
    $= 2a + (-5a) + (-3b) + 9b$ Using the commutative law of addition
    $= (2 + (-5))a + (-3 + 9)b$ Using the distributive law
    $= -3a + 6b$ Adding

*Like terms are discussed in greater detail in Section 1.8.

c) $6 + y + (-3.5y) + 2 = y + (-3.5y) + 6 + 2$ Using the commutative
law of addition

$$= (1 + (-3.5))y + 6 + 2$$ Using the distributive law

$$= -2.5y + 8$$ Adding

With practice we can leave out some steps, combining like terms mentally. Note that numbers like 6 and 2 in the expression $6 + y + (-3.5y) + 2$ are constants and are also considered to be like terms.

Exercise Set 1.5

FOR EXTRA HELP

  Digital Video Tutor CD 1 Videotape 2   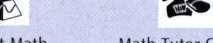 InterAct Math   Math Tutor Center   MathXL   MyMathLab.com

Add using a number line.

1. $4 + (-7)$

2. $2 + (-5)$

3. $-5 + 9$

4. $-3 + 8$

5. $8 + (-8)$

6. $6 + (-6)$

7. $-3 + (-5)$

8. $-4 + (-6)$

Add. Do not use a number line except as a check.

9. $-15 + 0$

10. $-6 + 0$

11. $0 + (-8)$

12. $0 + (-2)$

13. $12 + (-12)$

14. $17 + (-17)$

15. $-24 + (-17)$

16. $-17 + (-25)$

17. $-15 + 15$

18. $-18 + 18$

19. $18 + (-11)$

20. $8 + (-5)$

21. $10 + (-12)$

22. $9 + (-13)$

23. $-3 + 14$

24. $13 + (-6)$

25. $-14 + (-19)$

26. $11 + (-9)$

27. $19 + (-19)$

28. $-20 + (-6)$

29. $23 + (-5)$

30. $-15 + (-7)$

31. $-23 + (-9)$

32. $40 + (-8)$

33. $40 + (-40)$

34. $-25 + 25$

35. $85 + (-65)$

36. $63 + (-18)$

37. $-3.6 + 1.9$

38. $-6.5 + 4.7$

39. $-5.4 + (-3.7)$

40. $-3.8 + (-9.4)$

41. $\frac{-3}{5} + \frac{4}{5}$

42. $\frac{-2}{7} + \frac{3}{7}$

43. $\frac{-4}{7} + \frac{-2}{7}$

44. $\frac{-5}{9} + \frac{-2}{9}$

45. $-\frac{2}{5} + \frac{1}{3}$

46. $-\frac{4}{13} + \frac{1}{2}$

47. $\frac{-4}{9} + \frac{2}{3}$

48. $\frac{-1}{6} + \frac{1}{3}$

49. $35 + (-14) + (-19) + (-5)$

50. $28 + (-44) + 17 + 31 + (-94)$

Aha! **51.** $-4.9 + 8.5 + 4.9 + (-8.5)$

52. $24 + 3.1 + (-44) + (-8.2) + 63$

Solve. Write your answer as a complete sentence.

53. *Class size.* During the first two weeks of the semester, 5 students withdrew from Elisa's algebra class, 8 students were added to the class, and 6 students were dropped as "no-shows." By how many students did the original class size change?

54. *Telephone bills.* Maya's telephone bill for July was $82. She sent a check for $50 and then made $37 worth of calls in August. What was her new balance?

55. *Profits and losses.* The following table shows the profits and losses of Fax City over a 3-yr period.

Find the profit or loss after this period of time.

Year	Profit or loss
1998	$-$26,500
1999	$-$10,200
2000	$+$32,400

56. *Yardage gained.*   In a college football game, the quarterback attempted passes with the following results.

First try	13-yd gain
Second try	12-yd loss
Third try	21-yd gain

Find the total gain (or loss).

57. *Account balance.*   Leah has $350 in a checking account. She writes a check for $530, makes a deposit of $75, and then writes a check for $90. What is the balance in the account?

58. *Credit card bills.*   Lyle's credit card bill indicates that he owes $470. He sends a check to the credit card company for $45, charges another $160 in merchandise, and then pays off another $500 of his bill. What is Lyle's new balance?

59. *Stock growth.*   In the course of one day, the value of a share of America Online rose $\$\frac{3}{16}$, dropped $\$\frac{1}{2}$, and then rose $\$\frac{1}{4}$. How much had the stock's value risen or fallen at the end of the day?

60. *Peak elevation.*   The tallest mountain in the world, as measured from base to peak, is Mauna Kea in Hawaii. From a base 19,684 ft below sea level, it rises 33,480 ft. (*Source*: *The Guinness Book of Records*, 1999) What is the elevation of its peak?

Combine like terms.

61. $7a + 5a$

62. $3x + 8x$

63. $-3x + 12x$

64. $2m + (-7m)$

65. $5t + 8t$

66. $5a + 9a$

67. $7m + (-9m)$

68. $-4x + 4x$

69. $-5a + (-2a)$

70. $10n + (-17n)$

71. $-3 + 8x + 4 + (-10x)$

72. $8a + 5 + (-a) + (-3)$

Find the perimeter of each figure.

73.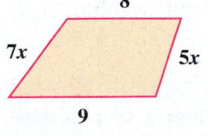

74.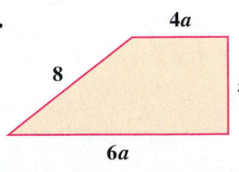

75.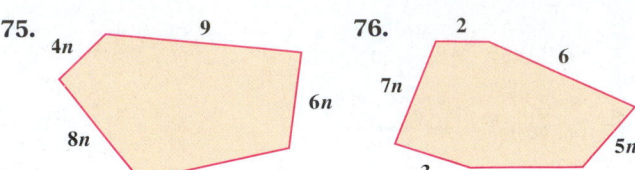

76.

77. Explain in your own words why the sum of two negative numbers is negative.

78. Without performing the actual addition, explain why the sum of all integers from -10 to 10 is 0.

SKILL MAINTENANCE

79. Multiply: $7(3z + y + 2)$.

80. Divide and simplify: $\frac{7}{2} \div \frac{3}{8}$.

SYNTHESIS

81. Under what circumstances will the sum of one positive number and several negative numbers be positive?

82. Is it possible to add real numbers without knowing how to calculate $a - b$ with a and b both nonnegative and $a \geq b$? Why or why not?

83. *Stock prices.*   The value of EKB stock rose $\$2\frac{3}{8}$ and then dropped $\$3\frac{1}{4}$ before finishing at $\$64\frac{3}{8}$. What was the stock's original value?

84. *Sports card values.*   The value of a sports card dropped $12 and then rose $17.50 before settling at $61. What was the original value of the card?

Find the missing term or terms.

85. $4x + \underline{\hspace{1cm}} + (-9x) + (-2y) = -5x - 7y$

86. $-3a + 9b + \underline{\hspace{1cm}} + 5a = 2a - 6b$

87. $3m + 2n + \underline{\hspace{1cm}} + (-2m) = 2n + (-6m)$

88. $\underline{\hspace{1cm}} + 9x + (-4y) + x = 10x - 7y$

Aha! **89.** $7t + 23 + \underline{\hspace{1cm}} + \underline{\hspace{1cm}} = 0$

90. *Geometry.*   The perimeter of a rectangle is $7x + 10$. If the length of the rectangle is 5, express the width in terms of x.

91. *Golfing.*   After five rounds of golf, a golf pro was 3 under par twice, 2 over par once, 2 under par once, and 1 over par once. On average, how far above or below par was the golfer?

Subtraction of Real Numbers

1.6

Opposites and Additive Inverses • Subtraction • Problem Solving

In arithmetic, when a number b is subtracted from another number a, the difference, $a - b$, is the number that when added to b gives a. For example, $45 - 17 = 28$ because $28 + 17 = 45$. We will use this approach to develop an efficient way of finding the value of $a - b$ for any real numbers a and b. Before doing so, however, we must develop some terminology.

Opposites and Additive Inverses

Numbers such as 6 and -6 are *opposites,* or *additive inverses,* of each other. Whenever opposites are added, the result is 0; and whenever two numbers add to 0, those numbers are opposites.

E x a m p l e   1   Find the opposite of each number: **(a)** 34; **(b)** -8.3; **(c)** 0.

Solution

a) The opposite of 34 is -34: $34 + (-34) = 0$.
b) The opposite of -8.3 is 8.3: $-8.3 + 8.3 = 0$.
c) The opposite of 0 is 0: $0 + 0 = 0$.

To write the opposite, we use the symbol $-$, as follows.

> **Opposite**
>
> The *opposite,* or *additive inverse,* of a number a is written $-a$ (read "the opposite of a" or "the additive inverse of a").

Note that if we take a number, say 8, and find its opposite, -8, and then find the opposite of the result, we will have the original number, 8, again. Thus, for any number a,

$$-(-a) = a.$$

E x a m p l e   2   Find $-x$ and $-(-x)$ when $x = 16$.

Solution

If $x = 16$, then $-x = -16$.   The opposite of 16 is -16.

If $x = 16$, then $-(-x) = -(-16) = 16$.   The opposite of the opposite of 16 is 16.

E x a m p l e   3   Find $-x$ and $-(-x)$ when $x = -3$.

Solution

If $x = -3$, then $-x = -(-3) = 3$.   The opposite of -3 is 3.

If $x = -3$, then $-(-x) = -(-(-3)) = -(\ 3\ ) = -3$.

Note in Example 3 that an extra set of parentheses is used to show that we are substituting the negative number -3 for x. The notation $-\ -x$ is not used.

A symbol such as -8 is usually read "negative 8." It could be read "the additive inverse of 8," because the additive inverse of 8 is negative 8. It could also be read "the opposite of 8," because the opposite of 8 is -8.

A symbol like $-x$, which has a variable, should be read "the opposite of x" or "the additive inverse of x" and *not* "negative x," since to do so suggests that $-x$ represents a negative number. As we saw in Example 3, $-x$ can represent a positive number. This notation can be used to restate a result from Section 1.5 as *the law of opposites*:

> **The Law of Opposites**
>
> For any two numbers a and $-a$,
>
> $$a + (-a) = 0.$$
>
> (When opposites are added, their sum is 0.)

A negative number is said to have a "negative *sign*." A positive number is said to have a "positive *sign*." When we replace a number with its opposite, or additive inverse, we can say that we have "changed or reversed its sign."

E x a m p l e   4   Change the sign (find the opposite) of each number: **(a)** -3; **(b)** -10; **(c)** 14.

Solution

a) When we change the sign of -3, we obtain 3.
b) When we change the sign of -10, we obtain 10.
c) When we change the sign of 14, we obtain -14.

Subtraction

Opposites are helpful when subtraction involves negative numbers. To see why, look for a pattern in the following:

Subtracting		*Adding the Opposite*
$9 - 5 = 4$	since $4 + 5 = 9$	$9 + (-5) = 4$
$5 - 8 = -3$	since $-3 + 8 = 5$	$5 + (-8) = -3$
$-6 - 4 = -10$	since $-10 + 4 = -6$	$-6 + (-4) = -10$
$-7 - (-10) = 3$	since $3 + (-10) = -7$	$-7 + 10 = 3$
$-7 - (-2) = -5$	since $-5 + (-2) = -7$	$-7 + 2 = -5$

technology connection

On nearly all graphers, it is essential to distinguish between the key for negation and the key for subtraction. To enter a negative number, we use $\boxed{(-)}$ and to subtract, we use $\boxed{-}$. This said, be careful not to rely on a calculator for computations that you will be expected to do by hand.

The matching results suggest that we can subtract by adding the opposite of the number being subtracted. This can always be done and often provides the easiest way to subtract real numbers.

> ### Subtraction of Real Numbers
>
> For any real numbers a and b,
>
> $$a - b = a + (-b).$$
>
> (To subtract, add the opposite, or additive inverse, of the number being subtracted.)

Example 5

Subtract each of the following and then check with addition.

a) $2 - 6$
b) $4 - (-9)$
c) $-4.2 - (-3.6)$

Solution

a) $2 - 6 = 2 + (-6) = -4$

The opposite of 6 is -6. We change the subtraction to addition and add the opposite. *Check*: $-4 + 6 = 2$.

b) $4 - (-9) = 4 + 9 = 13$

The opposite of -9 is 9. We change the subtraction to addition and add the opposite. *Check*: $13 + (-9) = 4$.

c) $-4.2 - (-3.6) = -4.2 + 3.6$

$= -0.6$

Adding the opposite of -3.6.
Check: $-0.6 + (-3.6) = -4.2$.

The symbol "$-$" is read differently depending on where it appears. For example, $-5 - (-x)$ is read "negative five minus the opposite of x."

Example 6

Read each of the following and then subtract.

a) $3 - 5$
b) $-4.6 - (-9.8)$

Solution

a) $3 - 5$;   Read "three minus five"
$3 - 5 = 3 + (-5) = -2$ Adding the opposite

b) $-4.6 - (-9.8)$;   Read "negative four point six minus negative nine point eight"
$-4.6 - (-9.8) = -4.6 + 9.8 = 5.2$ Adding the opposite

E x a m p l e   7   Subtract $-\frac{3}{5}$ from $\frac{1}{5}$.

Solution   A common denominator exists. We subtract as follows:

$$\frac{1}{5} - \left(-\frac{3}{5}\right) = \frac{1}{5} + \frac{3}{5} \qquad \text{Adding the opposite}$$

$$= \frac{1+3}{5} = \frac{4}{5}.$$

Check: $\frac{4}{5} + \left(-\frac{3}{5}\right) = \frac{4}{5} + \frac{-3}{5} = \frac{4 + (-3)}{5} = \frac{1}{5}.$

E x a m p l e   8   Simplify: $8 - (-4) - 2 - (-5) + 3$.

Solution

$$8 - (-4) - 2 - (-5) + 3 = 8 + 4 + (-2) + 5 + 3 \qquad \text{To subtract, we add the opposite.}$$

$$= 18$$

Recall from Section 1.2 that the terms of an algebraic expression are separated by plus signs. This means that the terms of $5x - 7y - 9$ are $5x$, $-7y$, and -9, since $5x - 7y - 9 = 5x + (-7y) + (-9)$.

E x a m p l e   9   Identify the terms of $4 - 2ab + 7a - 9$.

Solution   We have

$$4 - 2ab + 7a - 9 = 4 + (-2ab) + 7a + (-9), \qquad \text{Rewriting as addition}$$

so the terms are 4, $-2ab$, $7a$, and -9.

E x a m p l e   1 0   Combine like terms.

a) $1 + 3x - 7x$ **b)** $-5a - 7b - 4a + 10b$
c) $4 - 3m - 9 + 7m$

Solution

a) $1 + 3x - 7x = 1 + 3x + (-7x)$ Adding the opposite

$\qquad\qquad\quad = 1 + (3 + (-7))x$ Using the distributive law.
$\qquad\qquad\quad = 1 + (-4)x$ Try to do this mentally.

$\qquad\qquad\quad = 1 - 4x$ Rewriting as subtraction to be more concise

b) $-5a - 7b - 4a + 10b = -5a + (-7b) + (-4a) + 10b$ Adding the opposite

$$= -5a + (-4a) + (-7b) + 10b$$ Using the commutative law of addition

$$= -9a + 3b$$ Combining like terms mentally

c) $4 - 3m - 9 + 7m = 4 + (-3m) + (-9) + 7m$ Rewriting as addition

$$= 4 + (-9) + (-3m) + 7m$$ Using the commutative law of addition

$$= -5 + 4m$$

Problem Solving

Subtraction is used to solve problems involving differences.

E x a m p l e   1 1

Changes in temperature.   The Viking 2 Lander spacecraft has determined that temperatures on Mars range from $-125°$ Celsius (C) to $25°C$ (*Source*: The Lunar and Planetary Institute). Find the temperature range on Mars.

Solution   It is helpful to make a drawing of the situation.

To find the difference between two temperatures, we always subtract the lower temperature from the higher temperature:

$$25 - (-125) = 25 + 125$$

$$= 150.$$

The temperature range on Mars is $150°C$.

Exercise Set 1.6

FOR EXTRA HELP

 Digital Video Tutor CD 1 Videotape 2   InterAct Math   Math Tutor Center   MathXL   MyMathLab.com

Find the opposite, or additive inverse.

1. 39 **2.** −17 **3.** −9

4. $\frac{7}{2}$ **5.** −3.14 **6.** 48.2

Find −x when x is each of the following.

7. 23 **8.** −26 **9.** $-\frac{14}{3}$

10. $\frac{1}{328}$ **11.** 0.101 **12.** 0

Find −(−x) when x is each of the following.

13. 72 **14.** 29

15. $-\frac{2}{5}$ **16.** −9.1

Change the sign. (Find the opposite.)

17. −1 **18.** −7

19. 7 **20.** 10

Write words for each of the following and then perform the subtraction.

21. −3 − 5 **22.** −4 − 7

23. 2 − (−9) **24.** 5 − (−8)

25. 4 − 6 **26.** 9 − 12

27. −5 − (−7) **28.** −2 − (−5)

Subtract.

29. 6 − 8 **30.** 4 − 13

31. 0 − 5 **32.** 0 − 8

33. 3 − 9 **34.** 3 − 13

35. 0 − 10 **36.** 0 − 7

37. −9 − (−3) **38.** −9 − (−5)

Aha! **39.** −8 − (−8) **40.** −10 − (−10)

41. 14 − 19 **42.** 12 − 16

43. 30 − 40 **44.** 20 − 27

45. −7 − (−9) **46.** −8 − (−3)

47. −9 − (−9) **48.** −40 − (−40)

49. 5 − 5 **50.** 7 − 7

51. 4 − (−4) **52.** 6 − (−6)

53. −7 − 4 **54.** −6 − 8

55. 6 − (−10) **56.** 3 − (−12)

57. −14 − 2 **58.** −4 − 15

59. −4 − (−3) **60.** −6 − (−5)

61. 5 − (−6) **62.** 5 − (−12)

63. 0 − 6 **64.** 0 − 5

65. −3 − (−1) **66.** −5 − (−2)

67. −9 − 16 **68.** −7 − 14

69. 0 − (−1) **70.** 0 − (−5)

71. −9 − 0 **72.** −8 − 0

73. 12 − (−5) **74.** 3 − (−7)

75. 18 − 63 **76.** 2 − 25

77. −18 − 63 **78.** −42 − 26

79. −45 − 4 **80.** −51 − 7

81. 1.5 − 9.4 **82.** 3.2 − 8.7

83. 0.825 − 1 **84.** 0.072 − 1

85. $\frac{3}{7} - \frac{5}{7}$ **86.** $\frac{2}{11} - \frac{9}{11}$

87. $\frac{-2}{9} - \frac{5}{9}$ **88.** $\frac{-1}{5} - \frac{3}{5}$

89. $-\frac{2}{13} - \left(-\frac{5}{13}\right)$ **90.** $-\frac{4}{17} - \left(-\frac{9}{17}\right)$

Translate each phrase to mathematical language and simplify. See Example 11.

91. The difference between 3.8 and −5.2

92. The difference between −2.1 and −5.9

93. The difference between 114 and −79

94. The difference between 23 and −17

95. Subtract 37 from −21.

96. Subtract 19 from −7.

97. Subtract −25 from 9.

98. Subtract −31 from −5.

Simplify.

99. 25 − (−12) − 7 − (−2) + 9

100. 22 − (−18) + 7 + (−42) − 27

101. −31 + (−28) − (−14) − 17

102. −43 − (−19) − (−21) + 25

103. −34 − 28 + (−33) − 44

104. $39 + (-88) - 29 - (-83)$

Aha! **105.** $-93 + (-84) - (-93) - (-84)$

106. $84 + (-99) + 44 - (-18) - 43$

Identify the terms in each expression.

107. $-7x - 4y$

108. $7a - 9b$

109. $9 - 5t - 3st$

110. $-4 - 3x + 2xy$

Combine like terms.

111. $4x - 7x$

112. $3a - 14a$

113. $7a - 12a + 4$

114. $-9x - 13x + 7$

115. $-8n - 9 + n$

116. $-7 + 9n - 8$

117. $3x + 5 - 9x$

118. $2 + 3a - 7$

119. $2 - 6t - 9 - 2t$

120. $-5 + 3b - 7 - 5b$

121. $5y + (-3x) - 9x + 1 - 2y + 8$

122. $14 - (-5x) + 2z - (-32) + 4z - 2x$

123. $13x - (-2x) + 45 - (-21) - 7x$

124. $8x - (-2x) - 14 - (-5x) + 53 - 9x$

Solve.

125. *Record temperature drop.*   The greatest recorded temperature change in one day occurred in Browning, Montana, when the temperature fell from 44°F to −56°F (*Source*: *The Guinness Book of Records*, 1999). How much did the temperature drop?

126. *Loan repayment.*   Gisela owed Ramon $290. Ramon decides to "forgive" $125 of the debt. How much does Gisela owe?

127. *Elevation extremes.*   The lowest elevation in Asia, the Dead Sea, is 1312 ft below sea level. The highest elevation in Asia, Mount Everest, is 29,028 ft. (*Source*: *The World Almanac and Book of Facts* 2000) Find the difference in elevation.

128. *Elevation extremes.*   The elevation of Mount Whitney, the highest peak in California, is 14,776 ft more than the elevation of Death Valley, California (*Source*: 1999 *Information Please Almanac*). If Death Valley is 282 ft below sea level, find the elevation of Mount Whitney.

129. *Changes in elevation.*   The lowest point in Africa is Lake Assal, which is 156 m below sea level. The lowest point in South America is the Valdes Peninsula, which is 40 m below sea level. How much lower is Lake Assal than the Valdes Peninsula?

130. *Underwater elevation.*   The deepest point in the Pacific Ocean is the Marianas Trench, with a depth of 10,415 m. The deepest point in the Atlantic Ocean is the Puerto Rico Trench, with a depth of 8648 m. What is the difference in elevation of the two trenches?

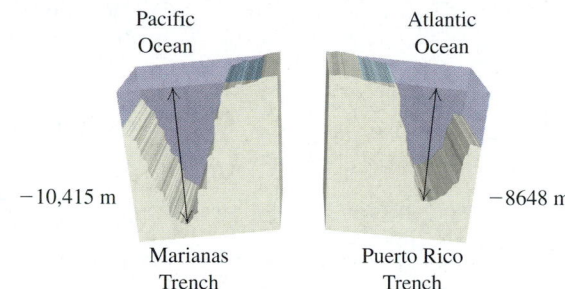

131. Jeremy insists that if you can *add* real numbers, then you can also *subtract* real numbers. Do you agree? Why or why not?

132. Are the expressions $-a + b$ and $a + (-b)$ opposites of each other? Why or why not?

SKILL MAINTENANCE

133. Find the area of a rectangle when the length is 36 ft and the width is 12 ft.

134. Find the prime factorization of 864.

SYNTHESIS

135. Why might it be advantageous to rewrite a long series of additions and subtractions as all additions?

136. If a and b are both negative, under what circumstances will $a - b$ be negative?

Tell whether each statement is true or false for all real numbers m and n. Use various replacements for m and n to support your answer.

137. If $m > n$, then $m - n > 0$.

138. If $m > n$, then $m + n > 0$.

139. If m and n are opposites, then $m - n = 0$.

140. If $m = -n$, then $m + n = 0$.

141. A gambler loses a wager and then loses "double or nothing" (meaning the gambler owes twice as much) twice more. After the three losses, the gambler's assets are −$20. Explain how much the gambler originally bet and how the $20 debt occurred.

142. If n is positive and m is negative, what is the sign of $n + (-m)$? Why?

Multiplication and Division of Real Numbers

1.7

Multiplication • Division

We now develop rules for multiplication and division of real numbers. Because multiplication and division are closely related, the rules are quite similar.

Multiplication

We already know how to multiply two nonnegative numbers. To see how to multiply a positive number and a negative number, consider the following pattern in which multiplication is regarded as repeated addition:

This number → $4(-5) = (-5) + (-5) + (-5) + (-5) = -20$ ← This number
decreases by $3(-5) = \qquad (-5) + (-5) + (-5) = -15$ increases by
1 each time. $2(-5) = \qquad\qquad (-5) + (-5) = -10$ 5 each time.
                    $1(-5) = \qquad\qquad\qquad (-5) = -5$
                    $0(-5) = \qquad\qquad\qquad\qquad 0 = 0$

This pattern illustrates that the product of a negative number and a positive number is negative.

> ### *The Product of a Negative Number and a Positive Number*
> To multiply a positive number and a negative number, multiply their absolute values. The answer is negative.

Example 1

Multiply: **(a)** $8(-5)$; **(b)** $-\frac{1}{3} \cdot \frac{5}{7}$.

Solution

a) $8(-5) = -40$ *Think*: $8 \cdot 5 = 40$; make the answer negative.

b) $-\frac{1}{3} \cdot \frac{5}{7} = -\frac{5}{21}$ *Think*: $\frac{1}{3} \cdot \frac{5}{7} = \frac{5}{21}$; make the answer negative.

The pattern developed above includes not just products of positive and negative numbers, but a product involving zero as well.

> ### *The Multiplicative Property of Zero*
> For any real number a,
> $$0 \cdot a = a \cdot 0 = 0.$$
> (The product of 0 and any real number is 0.)

E x a m p l e   2

Multiply: $173(-452)0$.

Solution   We have

$$173(-452)0 = 173[(-452)0]$$ Using the associative law of multiplication

$$= 173[0]$$ Using the multiplicative property of zero

$$= 0.$$ Using the multiplicative property of zero again

Note that whenever 0 appears as a factor, the product will be 0.

We can extend the above pattern still further to examine the product of two negative numbers.

This number → decreases by 1 each time.

$$
\begin{aligned}
2(-5) &= (-5) + (-5) = -10 \\
1(-5) &= (-5) = -5 \\
0(-5) &= 0 = 0 \\
-1(-5) &= -(-5) = 5 \\
-2(-5) &= -(-5) - (-5) = 10
\end{aligned}
$$

← This number increases by 5 each time.

According to the pattern, the product of two negative numbers is positive.

> **The Product of Two Negative Numbers**
>
> To multiply two negative numbers, multiply their absolute values. The answer is positive.

E x a m p l e   3

Multiply: **(a)** $(-6)(-8)$; **(b)** $(-1.2)(-3)$.

Solution

a) The absolute value of -6 is 6 and the absolute value of -8 is 8. Thus,

$$(-6)(-8) = 6 \cdot 8$$ Multiplying absolute values. The answer is positive.

$$= 48.$$

b) $(-1.2)(-3) = (1.2)(3)$ Multiplying absolute values. The answer is positive.

$$= 3.6$$ Try to go directly to this step.

When three or more numbers are multiplied, we can order and group the numbers as we please, because of the commutative and associative laws.

E x a m p l e   4

Multiply: **(a)** $-3(-2)(-5)$; **(b)** $-4(-6)(-1)(-2)$.

Solution

a) $-3(-2)(-5) = 6(-5)$ Multiplying the first two numbers. The product of two negatives is positive.

$\qquad\qquad\qquad\quad = -30$ The product of a positive and a negative is negative.

b) $-4(-6)(-1)(-2) = 24 \cdot 2$ Multiplying the first two numbers and the last two numbers

$\qquad\qquad\qquad\qquad\quad = 48$

We can see the following pattern in the results of Example 4.

The product of an even number of negative numbers is positive.

The product of an odd number of negative numbers is negative.

Division

Recall that $a \div b$, or $\frac{a}{b}$, is the number, if one exists, that when multiplied by b gives a. For example, to show that $10 \div 2$ is 5, we need only note that $5 \cdot 2 = 10$. Thus division can always be checked with multiplication.

E x a m p l e   5

Divide, if possible, and check your answer.

a) $14 \div (-7)$ **b)** $\dfrac{-32}{-4}$ **c)** $\dfrac{-10}{9}$ **d)** $\dfrac{-17}{0}$

Solution

a) $14 \div (-7) = -2$ We look for a number that when multiplied by -7 gives 14. That number is -2. *Check*: $(-2)(-7) = 14$.

b) $\dfrac{-32}{-4} = 8$ We look for a number that when multiplied by -4 gives -32. That number is 8. *Check*: $8(-4) = -32$.

c) $\dfrac{-10}{9} = -\dfrac{10}{9}$ We look for a number that when multiplied by 9 gives -10. That number is $-\frac{10}{9}$. *Check*: $-\frac{10}{9} \cdot 9 = -10$.

d) $\dfrac{-17}{0}$ is **undefined**.   We look for a number that when multiplied by 0 gives -17. There is no such number because the product of 0 and *any* number is 0, not -17.

The sign rules for division are the same as those for multiplication: The quotient of a positive number and a negative number is negative; the quotient of two negative numbers is positive.

Rules for Multiplication and Division

To multiply or divide two real numbers:

1. Using the absolute values, multiply or divide, as indicated.
2. If the signs are the same, the answer is positive.
3. If the signs are different, the answer is negative.

Had Example 5(a) been written as $-14 \div 7$ or $-\frac{14}{7}$, rather than $14 \div (-7)$, the result would still have been -2. Thus from Examples 5(a)–5(c), we have the following:

$$\frac{-a}{b} = \frac{a}{-b} = -\frac{a}{b} \quad \text{and} \quad \frac{-a}{-b} = \frac{a}{b}.$$

E x a m p l e   6

Rewrite each of the following in two equivalent forms: **(a)** $\frac{5}{-2}$; **(b)** $-\frac{3}{10}$.

Solution   We use one of the properties just listed.

a) $\dfrac{5}{-2} = \dfrac{-5}{2}$ and $\dfrac{5}{-2} = -\dfrac{5}{2}$

b) $-\dfrac{3}{10} = \dfrac{-3}{10}$ and $-\dfrac{3}{10} = \dfrac{3}{-10}$

Since $\dfrac{-a}{b} = \dfrac{a}{-b} = -\dfrac{a}{b}$

When a fraction contains a negative sign, it may be helpful to rewrite (or simply visualize) the fraction in an equivalent form.

E x a m p l e   7

Perform the indicated operation: **(a)** $\left(-\frac{4}{5}\right)\left(\frac{-7}{3}\right)$; **(b)** $-\frac{2}{7} + \frac{9}{-7}$.

Solution

a) $\left(-\dfrac{4}{5}\right)\left(\dfrac{-7}{3}\right) = \left(-\dfrac{4}{5}\right)\left(-\dfrac{7}{3}\right)$ Rewriting $\dfrac{-7}{3}$ as $-\dfrac{7}{3}$

$\qquad\qquad\qquad = \dfrac{28}{15}$ Try to go directly to this step.

b) Given a choice, we generally choose a positive denominator:

$$-\dfrac{2}{7} + \dfrac{9}{-7} = \dfrac{-2}{7} + \dfrac{-9}{7} \qquad \text{Rewriting both fractions with a common denominator of 7}$$

$$= \dfrac{-11}{7}, \text{ or } -\dfrac{11}{7}.$$

E x a m p l e   8

Find the reciprocal: **(a)** -27; **(b)** $\frac{-3}{4}$; **(c)** $-\frac{1}{5}$.

Solution

a) The reciprocal of -27 is $\frac{1}{-27}$. More often, this number is written as $-\frac{1}{27}$.

b) The reciprocal of $\frac{-3}{4}$ is $\frac{4}{-3}$, or, equivalently, $-\frac{4}{3}$.

c) The reciprocal of $-\frac{1}{5}$ is -5.

Recall that the opposite, or additive inverse, of a number is what we add to the number to get 0, whereas a reciprocal is what we multiply the number by to get 1. Compare the following.

Number	Opposite (Change the sign.)	Reciprocal (Invert but do not change the sign.)
$-\dfrac{3}{8}$	$\dfrac{3}{8}$	$-\dfrac{8}{3}$
19	-19	$\dfrac{1}{19}$
0	0	Undefined

$$\left(-\frac{3}{8}\right)\left(-\frac{8}{3}\right) = 1$$

$$-\frac{3}{8} + \frac{3}{8} = 0$$

When dividing with fraction notation, it is usually easier to multiply by a reciprocal. With decimal notation, it is usually easier to carry out division.

Example 9

Divide: **(a)** $-\frac{2}{3} \div \left(-\frac{5}{4}\right)$; **(b)** $-\frac{3}{4} \div \frac{3}{10}$; **(c)** $27.9 \div (-3)$.

Solution

a) $-\dfrac{2}{3} \div \left(-\dfrac{5}{4}\right) = -\dfrac{2}{3} \cdot \left(-\dfrac{4}{5}\right) = \dfrac{8}{15}$ Multiplying by the reciprocal

Be careful not to change the sign when taking a reciprocal!

b) $-\dfrac{3}{4} \div \dfrac{3}{10} = -\dfrac{3}{4} \cdot \left(\dfrac{10}{3}\right) = -\dfrac{30}{12} = -\dfrac{5}{2} \cdot \dfrac{6}{6} = -\dfrac{5}{2}$ Removing a factor equal to 1: $\frac{6}{6} = 1$

c) $27.9 \div (-3) = \dfrac{27.9}{-3} = -9.3$ Dividing: $3\overline{)27.9}$ with quotient 9.3. The answer is negative.

In Example 5(d), we explained why we cannot divide -17 by 0. This also explains why *no* nonzero number b can be divided by 0: Consider $b \div 0$. Is there a number that when multiplied by 0 gives b? No, because the product of 0 and any number is 0, not b. We say that $b \div 0$ is **undefined** for $b \neq 0$. In the special case of $0 \div 0$, we look for a number r such that $0 \div 0 = r$ and $r \cdot 0 = 0$. But, $r \cdot 0 = 0$ for *any* number r. For this reason, we say that $b \div 0$ is undefined for any choice of b.*

Finally, note that $0 \div 7 = 0$ since $0 \cdot 7 = 0$. This can be written $0/7 = 0$.

Example 10

Divide, if possible: **(a)** $\frac{0}{-2}$; **(b)** $\frac{5}{0}$.

Solution

a) $\dfrac{0}{-2} = 0$ *Check*: $0(-2) = 0$.

b) $\dfrac{5}{0}$ is undefined.

---

*Sometimes $0 \div 0$ is said to be *indeterminate*.

> ### Division Involving Zero
>
> For any real number a,
>
> $$\frac{a}{0} \text{ is undefined,}$$
>
> and for $a \neq 0$,
>
> $$\frac{0}{a} = 0.$$

Exercise Set 1.7

Multiply.

1. $-4 \cdot 9$

2. $-3 \cdot 7$

3. $-8 \cdot 7$

4. $-9 \cdot 2$

5. $8 \cdot (-3)$

6. $9 \cdot (-5)$

7. $-9 \cdot 8$

8. $-10 \cdot 3$

9. $-6 \cdot (-7)$

10. $-2 \cdot (-5)$

11. $-5 \cdot (-9)$

12. $-9 \cdot (-2)$

13. $17 \cdot (-10)$

14. $-12 \cdot (-10)$

15. $-12 \cdot 12$

16. $-13 \cdot (-15)$

17. $-25 \cdot (-48)$

18. $39 \cdot (-43)$

19. $-3.5 \cdot (-28)$

20. $97 \cdot (-2.1)$

21. $6 \cdot (-13)$

22. $7 \cdot (-9)$

23. $-7 \cdot (-3.1)$

24. $-4 \cdot (-3.2)$

25. $\frac{2}{3} \cdot \left(-\frac{3}{5}\right)$

26. $\frac{5}{7} \cdot \left(-\frac{2}{3}\right)$

27. $-\frac{3}{8} \cdot \left(-\frac{2}{9}\right)$

28. $-\frac{5}{8} \cdot \left(-\frac{2}{5}\right)$

29. $(-5.3)(2.1)$

30. $(-4.3)(9.5)$

31. $-\frac{5}{9} \cdot \frac{3}{4}$

32. $-\frac{8}{3} \cdot \frac{9}{4}$

33. $3 \cdot (-7) \cdot (-2) \cdot 6$

34. $9 \cdot (-2) \cdot (-6) \cdot 7$

Aha! **35.** $-27 \cdot (-34) \cdot 0$

36. $-43 \cdot (-74) \cdot 0$

37. $-\frac{1}{3} \cdot \frac{1}{4} \cdot \left(-\frac{3}{7}\right)$

38. $-\frac{1}{2} \cdot \frac{3}{5} \cdot \left(-\frac{2}{7}\right)$

39. $-2 \cdot (-5) \cdot (-3) \cdot (-5)$

40. $-3 \cdot (-5) \cdot (-2) \cdot (-1)$

41. $(-14) \cdot (-27) \cdot 0$

42. $7 \cdot (-6) \cdot 5 \cdot (-4) \cdot 3 \cdot (-2) \cdot 1 \cdot 0$

43. $(-8)(-9)(-10)$

44. $(-7)(-8)(-9)(-10)$

45. $(-6)(-7)(-8)(-9)(-10)$

46. $(-5)(-6)(-7)(-8)(-9)(-10)$

Divide, if possible, and check. If a quotient is undefined, state this.

47. $28 \div (-7)$

48. $\dfrac{24}{-3}$

49. $\dfrac{36}{-9}$

50. $26 \div (-13)$

51. $\dfrac{-16}{8}$

52. $-32 \div (-4)$

53. $\dfrac{-48}{-12}$

54. $-63 \div (-9)$

55. $\dfrac{-72}{9}$

56. $\dfrac{-50}{25}$

57. $-100 \div (-50)$

58. $\dfrac{-200}{8}$

59. $-108 \div 9$

60. $\dfrac{-64}{-7}$

61. $\dfrac{400}{-50}$

62. $-300 \div (-13)$

63. $\dfrac{28}{0}$

64. $\dfrac{0}{-5}$

65. $-4.8 \div 1.2$

66. $-3.9 \div 1.3$

67. $\dfrac{0}{-9}$

Aha! **68.** $\dfrac{(-4.9)(7.2)}{0}$

69. $0 \div 7$

70. $0 \div (-47)$

Write each number in two equivalent forms, as in Example 6.

71. $\dfrac{-8}{3}$

72. $\dfrac{-12}{7}$

73. $\dfrac{29}{-35}$

74. $\dfrac{9}{-14}$

75. $-\dfrac{7}{3}$

76. $-\dfrac{4}{15}$

77. $\dfrac{-x}{2}$

78. $\dfrac{9}{-a}$

Find the reciprocal of each number.

79. $\dfrac{4}{-5}$

80. $\dfrac{2}{-9}$

81. $-\dfrac{47}{13}$

82. $-\dfrac{31}{12}$

83. -10

84. 13

85. 4.3

86. -8.5

87. $\dfrac{-9}{4}$

88. $\dfrac{-6}{11}$

89. -1

90. $3/5$

Perform the indicated operation and, if possible, simplify. If a quotient is undefined, state this.

91. $\left(\dfrac{-7}{4}\right)\left(-\dfrac{3}{5}\right)$

92. $\left(-\dfrac{5}{6}\right)\left(\dfrac{-1}{3}\right)$

93. $\left(\dfrac{-6}{5}\right)\left(\dfrac{2}{-11}\right)$

94. $\left(\dfrac{7}{-2}\right)\left(\dfrac{-5}{6}\right)$

95. $\dfrac{-3}{8} + \dfrac{-5}{8}$

96. $\dfrac{-4}{5} + \dfrac{7}{5}$

Aha! **97.** $\left(\dfrac{-9}{5}\right)\left(\dfrac{5}{-9}\right)$

98. $\left(-\dfrac{2}{7}\right)\left(\dfrac{5}{-8}\right)$

99. $\left(-\dfrac{3}{11}\right) + \left(-\dfrac{6}{11}\right)$

100. $\left(-\dfrac{4}{7}\right) + \left(-\dfrac{2}{7}\right)$

101. $\dfrac{7}{8} \div \left(-\dfrac{1}{2}\right)$

102. $\dfrac{3}{4} \div \left(-\dfrac{2}{3}\right)$

103. $\dfrac{9}{5} \cdot \dfrac{-20}{3}$

104. $\dfrac{-5}{12} \cdot \dfrac{7}{15}$

105. $\left(-\dfrac{18}{7}\right) + \left(-\dfrac{3}{7}\right)$

106. $\left(-\dfrac{12}{5}\right) + \left(-\dfrac{3}{5}\right)$

Aha! **107.** $-\dfrac{5}{9} \div \left(-\dfrac{5}{9}\right)$

108. $-\dfrac{5}{4} \div \left(-\dfrac{3}{4}\right)$

109. $-44.1 \div (-6.3)$

110. $-6.6 \div 3.3$

111. $\dfrac{1}{9} - \dfrac{2}{9}$

112. $\dfrac{2}{7} - \dfrac{6}{7}$

113. $\dfrac{-3}{10} + \dfrac{2}{5}$

114. $\dfrac{-5}{9} + \dfrac{2}{3}$

115. $\dfrac{7}{10} \div \left(\dfrac{-3}{5}\right)$

116. $\left(\dfrac{-3}{5}\right) \div \dfrac{6}{15}$

117. $\dfrac{5}{7} - \dfrac{1}{-7}$

118. $\dfrac{4}{9} - \dfrac{1}{-9}$

119. $\dfrac{-4}{15} + \dfrac{2}{-3}$

120. $\dfrac{3}{-10} + \dfrac{-1}{5}$

121. Most calculators have a key, often appearing as $\boxed{1/x}$, for finding reciprocals. To use this key, enter a number and then press $\boxed{1/x}$ to find its reciprocal. What should happen if you enter a number and then press the reciprocal key twice? Why?

122. Multiplication can be regarded as repeated addition. Using this idea and a number line, explain why $3 \cdot (-5) = -15$.

SKILL MAINTENANCE

123. Simplify: $\dfrac{264}{468}$.

124. Combine like terms: $x + 12y + 11x - 14y - 9$.

SYNTHESIS

125. If two nonzero numbers are opposites of each other, are their reciprocals opposites of each other? Why or why not?

126. If two numbers are reciprocals of each other, are their opposites reciprocals of each other? Why or why not?

127. Show that the reciprocal of a sum is *not* the sum of the two reciprocals.

128. Which real numbers are their own reciprocals?

Tell whether each expression represents a positive number or a negative number when m and n are negative.

129. $\dfrac{m}{-n}$

130. $\dfrac{-n}{-m}$

131. $-m \cdot \left(\dfrac{-n}{m}\right)$

132. $-\left(\dfrac{n}{-m}\right)$

133. $(m + n) \cdot \dfrac{m}{n}$

134. $(-n - m)\dfrac{n}{m}$

135. What must be true of m and n if $-mn$ is to be **(a)** positive? **(b)** zero? **(c)** negative?

136. The following is a proof that a positive number times a negative number is negative. Provide a reason for each step. Assume that $a > 0$ and $b > 0$.

$$a(-b) + ab = a[-b + b]$$
$$= a(0)$$
$$= 0$$

Therefore, $a(-b)$ is the opposite of ab.

137. Is it true that for any numbers a and b, if a is larger than b, then the reciprocal of a is smaller than the reciprocal of b? Why or why not?

Exponential Notation and Order of Operations

1.8

Exponential Notation • Order of Operations • Simplifying and the Distributive Law • The Opposite of a Sum

Algebraic expressions often contain *exponential notation*. In this section, we learn how to use exponential notation as well as rules for the *order of operations*, in performing certain algebraic manipulations.

Exponential Notation

A product like $3 \cdot 3 \cdot 3 \cdot 3$, in which the factors are the same, is called a **power**. Powers occur often enough that a simpler notation called **exponential notation** is used. For

$$\underbrace{3 \cdot 3 \cdot 3 \cdot 3}_{4 \text{ factors}}, \quad \text{we write} \quad 3^4.$$

This is read "three to the fourth power," or simply, "three to the fourth." The number 4 is called an **exponent** and the number 3 a **base**.

Expressions like s^2 and s^3 are usually read "s squared" and "s cubed," respectively. This comes from the fact that a square with sides of length s has an area A given by $A = s^2$ and a cube with sides of length s has a volume V given by $V = s^3$.

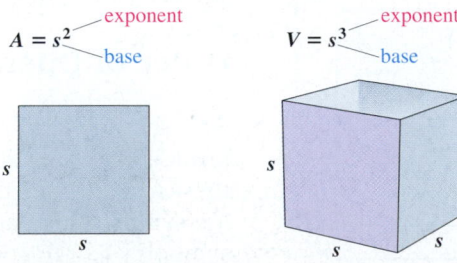

E x a m p l e   1   Write exponential notation for $10 \cdot 10 \cdot 10 \cdot 10 \cdot 10$.

Solution

Exponential notation is 10^5.   5 is the exponent.
                                      10 is the base.

E x a m p l e   2   Evaluate: **(a)** 5^2; **(b)** $(-5)^3$; **(c)** $(2n)^3$.

Solution

a) $5^2 = 5 \cdot 5 = 25$ The second power indicates two factors of 5.

b) $(-5)^3 = (-5)(-5)(-5)$ The third power indicates three factors of -5.

   $= 25(-5)$ Using the associative law of multiplication

   $= -125$

c) $(2n)^3 = (2n)(2n)(2n)$ The third power indicates three factors of $2n$.

   $= 2 \cdot 2 \cdot 2 \cdot n \cdot n \cdot n$ Using the associative and commutative laws of multiplication

   $= 8n^3$

To determine what the exponent 1 will mean, look for a pattern in the following:

$$7 \cdot 7 \cdot 7 \cdot 7 = 7^4$$
$$7 \cdot 7 \cdot 7 = 7^3$$
$$7 \cdot 7 = 7^2$$
$$7 = 7^?$$

We divide by 7 each time.

The exponents decrease by 1 each time. To continue the pattern, we say that

$$7 = 7^1.$$

Exponential Notation

For any natural number n,

$$n \text{ factors}$$
$$b^n \quad \text{means} \quad \overbrace{b \cdot b \cdot b \cdot b \cdots b}.$$

Order of Operations

How should $4 + 2 \times 5$ be computed? If we multiply 2 by 5 and then add 4, the result is 14. If we add 2 and 4 first and then multiply by 5, the result is 30. Since these results differ, the order in which we perform operations matters. If grouping symbols such as parentheses ( ), brackets [ ], braces { }, absolute-value symbols | |, or fraction bars are used, they tell us what to do first. For example,

---

technology connection

On most graphers, grouping symbols, such as a fraction bar, must be replaced with parentheses. For example, to calculate

$$\frac{12(9-7) + 4 \cdot 5}{2^4 + 3^2},$$

we enter $(12(9-7) + 4 \cdot 5) \div$ $(2^4 + 3^2)$. To enter an exponential expression, we enter the base, press $\boxed{\wedge}$ and then enter the exponent. The $\boxed{x^2}$ key can be used to enter an exponent of 2. We can also convert to fraction notation if we wish.

```
(12(9−7)+4∗5)/(2^4+3²)
                    1.76
Ans▶Frac
                   44/25
■
```

$$(4 + 2) \times 5 = 6 \times 5 = 30$$

and

$$4 + (2 \times 5) = 4 + 10 = 14.$$

In addition to grouping symbols, the following conventions exist for determining the order in which operations should be performed.

Rules for Order of Operations

1. Calculate within the innermost grouping symbols.
2. Simplify all exponential expressions.
3. Perform all multiplication and division, working from left to right.
4. Perform all addition and subtraction, working from left to right.

Thus the correct way to compute $4 + 2 \times 5$ is to first multiply 2 by 5 and then add 4. The result is 14.

E x a m p l e   3

Simplify: $15 - 2 \times 5 + 3$.

Solution   When no groupings or exponents appear, we always multiply or divide before adding or subtracting:

$$15 - 2 \times 5 + 3 = 15 - 10 + 3 \qquad \text{Multiplying}$$
$$= 5 + 3 \qquad \text{Subtracting and adding}$$
$$= 8. \qquad \text{from left to right}$$

Always calculate within parentheses first. When there are exponents and no parentheses, simplify powers before multiplying or dividing.

E x a m p l e   4

Simplify: **(a)** $(3 \cdot 4)^2$; **(b)** $3 \cdot 4^2$.

Solution

a) $(3 \cdot 4)^2 = (12)^2 \qquad$ Working within parentheses first
$$= 144$$

b) $3 \cdot 4^2 = 3 \cdot 16 \qquad$ Simplifying the power
$$= 48 \qquad \text{Multiplying}$$

Note that $(3 \cdot 4)^2 \ne 3 \cdot 4^2$.

Caution!   Example 4 illustrates that, in general, $(ab)^2 \ne ab^2$.

E x a m p l e   5

Evaluate for $x = 5$: **(a)** $(-x)^2$; **(b)** $-x^2$.

Solution

a) $(-x)^2 = (-5)^2 = (-5)(-5) = 25$ We square the opposite of 5.

b) $-x^2 = -5^2 = -25$ We square 5 and then find the opposite.

Caution!   Example 5 illustrates that, in general, $(-x)^2 \neq -x^2$.

E x a m p l e   6

Evaluate $-15 \div 3(6 - a)^3$ for $a = 4$.

Solution

$$
\begin{aligned}
-15 \div 3(6 - a)^3 &= -15 \div 3(6 - 4)^3 & \text{Substituting 4 for } a \\
&= -15 \div 3(2)^3 & \text{Working within parentheses first} \\
&= -15 \div 3 \cdot 8 & \text{Simplifying the exponential expression} \\
&= -5 \cdot 8 \\
&= -40 & \left.\begin{array}{c}\\\\\end{array}\right\} \text{Dividing and multiplying from left to right}
\end{aligned}
$$

When combinations of grouping symbols are used, the rules still apply. We begin with the innermost grouping symbols and work to the outside.

E x a m p l e   7

Simplify: $8 \div 4 + 3[9 + 2(3 - 5)^3]$.

Solution

$$
\begin{aligned}
8 \div 4 + 3[9 + 2(3 - 5)^3] &= 8 \div 4 + 3[9 + 2(-2)^3] & \text{Doing the calculations in the innermost parentheses first} \\
&= 8 \div 4 + 3[9 + 2(-8)] & \begin{array}{l}(-2)^3 = (-2)(-2)(-2) \\ \quad = -8\end{array} \\
&= 8 \div 4 + 3[9 + (-16)] \\
&= 8 \div 4 + 3[-7] & \text{Completing the calculations within the brackets} \\
&= 2 + (-21) & \text{Multiplying and dividing from left to right} \\
&= -19
\end{aligned}
$$

Example 8

Calculate: $\dfrac{12(9 - 7) + 4 \cdot 5}{3^4 + 2^3}$.

Solution   An equivalent expression with brackets is

$$[12(9 - 7) + 4 \cdot 5] \div [3^4 + 2^3].$$

In effect, we need to simplify the numerator, simplify the denominator, and then divide the results:

$$\frac{12(9 - 7) + 4 \cdot 5}{3^4 + 2^3} = \frac{12(2) + 4 \cdot 5}{81 + 8}$$

$$= \frac{24 + 20}{89} = \frac{44}{89}.$$

Simplifying and the Distributive Law

Sometimes we cannot simplify within parentheses. When a sum or difference is within the parentheses, the distributive law often allows us to simplify the expression.

Example 9

Simplify: $5x - 9 + 2(4x + 5)$.

Solution

$$5x - 9 + 2(4x + 5) = 5x - 9 + 8x + 10 \qquad \text{Using the distributive law}$$

$$= 13x + 1 \qquad \text{Combining like terms}$$

Now that exponents have been introduced, we can make our definition of *like* or *similar terms* more precise. **Like**, or **similar, terms** are either constant terms or terms containing the same variable(s) raised to the same power(s). Thus, 5 and -7, $19xy$ and $2yx$, and $4a^3b$ and a^3b are all pairs of like terms.

Example 10

Simplify: $7x^2 + 3(x^2 + 2x) - 5x$.

Solution

$$7x^2 + 3(x^2 + 2x) - 5x = 7x^2 + 3x^2 + 6x - 5x \qquad \text{Using the distributive law}$$

$$= 10x^2 + x \qquad \text{Combining like terms}$$

The Opposite of a Sum

When a number is multiplied by -1, the result is the opposite of that number. For example, $-1(7) = -7$ and $-1(-5) = 5$.

> **The Property of -1**
>
> For any real number a,
>
> $$-1 \cdot a = -a.$$
>
> (Negative one times a is the opposite of a.)

When grouping symbols are preceded by a "$-$" symbol, we can multiply the grouping by -1 and use the distributive law. In this manner, we can find the *opposite*, or *additive inverse*, of a sum.

E x a m p l e   1 1   Write an expression equivalent to $-(3x + 2y + 4)$ without using parentheses.

Solution

$$
\begin{aligned}
-(3x + 2y + 4) &= -1(3x + 2y + 4) && \text{Using the property of } -1 \\
&= -1(3x) + (-1)(2y) + (-1)4 && \text{Using the distribu-} \\
& && \text{tive law} \\
&= -3x - 2y - 4 && \text{Using the property of } -1
\end{aligned}
$$

Example 11 illustrates an important property of real numbers.

> **The Opposite of a Sum**
>
> For any real numbers a and b,
>
> $$-(a + b) = -a + (-b).$$
>
> (The opposite of a sum is the sum of the opposites.)

To remove parentheses from an expression like $-(x - 7y + 5)$, we can first rewrite the subtraction as addition:

$$
\begin{aligned}
-(x - 7y + 5) &= -(x + (-7y) + 5) && \text{Rewriting as addition} \\
&= -x + 7y - 5. && \text{Taking the opposite of a sum}
\end{aligned}
$$

This procedure is normally streamlined to one step in which we find the opposite by "removing parentheses and changing the sign of every term":

$$-(x - 7y + 5) = -x + 7y - 5.$$

E x a m p l e   1 2   Simplify: $3x - (4x + 2)$.

Solution

$$
\begin{aligned}
3x - (4x + 2) &= 3x + [-(4x + 2)] && \text{Adding the opposite of } 4x + 2 \\
&= 3x + [-4x - 2] && \text{Taking the opposite of } 4x + 2 \\
&= 3x + (-4x) + (-2) && \\
&= 3x - 4x - 2 && \text{Try to go directly to this step.} \\
&= -x - 2 && \text{Combining like terms}
\end{aligned}
$$

In practice, the first three steps of Example 12 are usually skipped.

E x a m p l e   1 3

Simplify: $5t^2 - 2t - (4t^2 - 9t)$.

Solution

$$5t^2 - 2t - (4t^2 - 9t) = 5t^2 - 2t - 4t^2 + 9t \qquad \text{Removing parentheses and changing the sign of each term inside}$$

$$= t^2 + 7t \qquad \text{Combining like terms}$$

Expressions such as $7 - 3(x + 2)$ can be simplified as follows:

$$7 - 3(x + 2) = 7 + [-3(x + 2)] \qquad \text{Adding the opposite of } 3(x + 2)$$

$$= 7 + [-3x - 6] \qquad \text{Multiplying } x + 2 \text{ by } -3$$

$$= 7 - 3x - 6 \qquad \text{Try to go directly to this step.}$$

$$= 1 - 3x. \qquad \text{Combining like terms}$$

E x a m p l e   1 4

Simplify: **(a)** $3n - 2(4n - 5)$; **(b)** $7x^3 + 2 - [5(x^3 - 1) + 8]$.

Solution

a) $3n - 2(4n - 5) = 3n - 8n + 10 \qquad \text{Multiplying each term inside the parentheses by } -2$

$$= -5n + 10 \qquad \text{Combining like terms}$$

b) $7x^3 + 2 - [5(x^3 - 1) + 8] = 7x^3 + 2 - [5x^3 - 5 + 8] \qquad \text{Removing parentheses}$

$$= 7x^3 + 2 - [5x^3 + 3]$$

$$= 7x^3 + 2 - 5x^3 - 3 \qquad \text{Removing brackets}$$

$$= 2x^3 - 1 \qquad \text{Combining like terms}$$

C O N N E C T I N G   T H E   C O N C E P T S

Algebra is a tool that can be used to solve problems. We have seen in this chapter that certain problems can be translated to algebraic expressions that can in turn be simplified (see Section 1.6). In Chapter 2, we will solve problems that require us to solve an equation.

As we progress through our study of algebra, it is important that we be able to distin-

guish between the two tasks of **simplifying an expression** and **solving an equation.** In Chapter 1, we did not solve equations, but we did simplify expressions. This enabled us to write *equivalent expressions* that were simpler than the given expression. In Chapter 2, we will continue to simplify expressions, but we will also begin to solve equations.

Exercise Set 1.8

Write exponential notation.

1. $4 \cdot 4 \cdot 4$

2. $6 \cdot 6 \cdot 6 \cdot 6$

3. $x \cdot x \cdot x \cdot x \cdot x \cdot x \cdot x$

4. $y \cdot y \cdot y \cdot y \cdot y \cdot y$

5. $3t \cdot 3t \cdot 3t \cdot 3t \cdot 3t$

6. $5m \cdot 5m \cdot 5m \cdot 5m \cdot 5m$

Simplify.

7. 2^4

8. 5^3

9. $(-3)^2$

10. $(-7)^2$

11. -3^2

12. -7^2

13. 4^3

14. 9^1

15. $(-5)^4$

16. 5^4

17. 7^1

18. $(-1)^7$

19. $(3t)^4$

20. $(5t)^2$

21. $(-7x)^3$

22. $(-5x)^4$

23. $5 + 3 \cdot 7$

24. $3 - 4 \cdot 2$

25. $8 \cdot 7 + 6 \cdot 5$

26. $10 \cdot 5 + 1 \cdot 1$

27. $19 - 5 \cdot 3 + 3$

28. $14 - 2 \cdot 6 + 7$

29. $9 \div 3 + 16 \div 8$

30. $32 - 8 \div 4 - 2$

Aha! **31.** $84 \div 28 - 84 \div 28$

32. $18 - 6 \div 3 \cdot 2 + 7$

33. $4 - 8 \div 2 + 3^2$

34. $3(-10)^2 - 8 \div 2^2$

35. $9 - 3^2 \div 9(-1)$

36. $8 - (2 \cdot 3 - 9)$

37. $(8 - 2 \cdot 3) - 9$

38. $(8 - 2)(3 - 9)$

39. $(-24) \div (-3) \cdot \left(-\frac{1}{2}\right)$

40. $32 \div (-2)^2 \cdot 4$

41. $13(-10)^2 + 45 \div (-5)$

42. $5 \cdot 3^2 - 4^2 \cdot 2$

43. $2^4 + 2^3 - 10 \div (-1)^4$

44. $40 - 3^2 - 2^3 \div (-4)$

45. $5 + 3(2 - 9)^2$

46. $9 - (3 - 5)^3 - 4$

47. $[2 \cdot (5 - 8)]^2$

48. $3(5 - 7)^4 \div 4$

49. $\dfrac{7 + 2}{5^2 - 4^2}$

50. $\dfrac{5^2 - 3^2}{2 \cdot 6 - 4}$

51. $8(-7) + |6(-5)|$

52. $|10(-5)| + 1(-1)$

53. $\dfrac{(-2)^3 + 4^2}{3 - 5^2 + 3 \cdot 6}$

54. $\dfrac{7^2 - (-1)^5}{3 - 2 \cdot 3^2 + 5}$

55. $\dfrac{27 - 2 \cdot 3^2}{8 \div 2^2 - (-2)^2}$

56. $\dfrac{(-5)^2 - 4 \cdot 5}{3^2 + 4 \cdot 2(-1)^5}$

Evaluate.

57. $7 - 5x$, for $x = 3$

58. $1 + x^3$, for $x = -2$

59. $24 \div t^3$, for $t = -2$

60. $20 \div a \cdot 4$, for $a = 5$

61. $45 \div 3 \cdot a$, for $a = -1$

62. $50 \div 2 \cdot t$, for $t = -5$

63. $5x \div 15x^2$, for $x = 3$

64. $6a \div 12a^3$, for $a = 2$

Aha! **65.** $(12 \cdot 17) \div (17 \cdot 12)$

66. $-30 \div t(t + 4)^2$, for $t = -6$

67. $-x^2 - 5x$, for $x = -3$

68. $(-x)^2 - 5x$, for $x = -3$

69. $\dfrac{3a - 4a^2}{a^2 - 20}$, for $a = 5$

70. $\dfrac{a^3 - 4a}{a(a - 3)}$, for $a = -2$

Write an equivalent expression without using parentheses.

71. $-(9x + 1)$

72. $-(3x + 5)$

73. $-(7 - 2x)$

74. $-(6x - 7)$

75. $-(4a - 3b + 7c)$

76. $-(5x - 2y - 3z)$

77. $-(3x^2 + 5x - 1)$

78. $-(8x^3 - 6x + 5)$

Simplify.

79. $5x - (2x + 7)$

80. $7y - (2y + 9)$

81. $2a - (5a - 9)$

82. $11n - (3n - 7)$

83. $2x + 7x - (4x + 6)$

84. $3a + 2a - (4a + 7)$

85. $9t - 5r - 2(3r + 6t)$

86. $4m - 9n - 3(2m - n)$

87. $15x - y - 5(3x - 2y + 5z)$

88. $4a - b - 4(5a - 7b + 8c)$

89. $3x^2 + 7 - (2x^2 + 5)$

90. $5x^4 + 3x - (5x^4 + 3x)$

91. $5t^3 + t - 3(t + 2t^3)$

92. $8n^2 + n - 2(n + 3n^2)$

93. $12a^2 - 3ab + 5b^2 - 5(-5a^2 + 4ab - 6b^2)$

94. $-8a^2 + 5ab - 12b^2 - 6(2a^2 - 4ab - 10b^2)$

95. $-7t^3 - t^2 - 3(5t^3 - 3t)$

96. $9t^4 + 7t - 5(9t^3 - 2t)$

97. $5(2x - 7) - [4(2x - 3) + 2]$

98. $3(6x - 5) - [3(1 - 8x) + 5]$

99. Some students use the mnemonic device PEM-DAS to help remember the rules for the order of operations. Explain how this can be done.

100. Jake keys $18/2 \cdot 3$ into his calculator and expects the result to be 3. What mistake is he probably making?

SKILL MAINTENANCE

Translate to an algebraic expression.

101. Nine more than twice a number

102. Half of the sum of two numbers

SYNTHESIS

103. Write the sentence $(-x)^2 \neq -x^2$ in words. Explain why $(-x)^2$ and $-x^2$ are not equivalent.

104. Write the sentence $-|x| \neq -x$ in words. Explain why $-|x|$ and $-x$ are not equivalent.

Simplify.

105. $5t - \{7t - [4r - 3(t - 7)] + 6r\} - 4r$

106. $z - \{2z - [3z - (4z - 5z) - 6z] - 7z\} - 8z$

107. $\{x - [f - (f - x)] + [x - f]\} - 3x$

108. Is it true that for any real numbers a and b,

$$-(ab) = (-a)b = a(-b)?$$

Why or why not?

109. Is it true that for any real numbers a and b,

$$ab = (-a)(-b)?$$

Why or why not?

If $n > 0$, $m > 0$, and $n \neq m$, determine whether each of the following is true or false.

110. $-n + m = m - n$

111. $-n + m = -(n + m)$

112. $m - n = -(n - m)$

113. $n(-n - m) = -n^2 + nm$

114. $-m(n - m) = -(mn + m^2)$

115. $-m(-n + m) = m(n - m)$

116. $-n(-n - m) = n(n + m)$

Evaluate.

Aha! **117.** $[x + 3(2 - 5x) \div 7 + x](x - 3)$, for $x = 3$

Aha! **118.** $[x + 2 \div 3x] \div [x + 2 \div 3x]$, for $x = -7$

119. In Mexico, between 500 B.C. and 600 A.D., the Mayans represented numbers using powers of 20 and certain symbols. For example, the symbols

represent $4 \cdot 20^3 + 17 \cdot 20^2 + 10 \cdot 20^1 + 0 \cdot 20^0$. (*Source*: National Council of Teachers of Mathematics, 1906 Association Drive, Reston, VA 22091) Evaluate this number.

120. Examine the Mayan symbols and the numbers in Exercise 119. What numbers do

  •,   ▬,   and   ⬭

each represent?

COLLABORATIVE

CORNER

Select the Symbols

Focus: Order of operations

Time: 15 minutes

Group size: 2

One way to master the rules for the order of operations is to insert symbols within a display of numbers in order to obtain a predetermined result. For example, the display

$$1 \quad 2 \quad 3 \quad 4 \quad 5$$

can be used to obtain the result 21 as follows:

$$(1 + 2) \div 3 + 4 \cdot 5.$$

Note that without an understanding of the rules for the order of operations, solving a problem of this sort is impossible.

ACTIVITY

1. Each group should prepare an exercise similar to the example shown above. (Exponents are not allowed.) To do so, first select five single-digit numbers for display. Then insert operations and grouping symbols and calculate the result.
2. Pair with another group. Each group should give the other its result along with its five-number display, and challenge the other group to insert symbols that will make the display equal the result given.
3. Share with the entire class the various mathematical statements developed by each group.

Summary and Review 1

Key Terms

Important Properties and Formulas

Area of a rectangle:	$A = lw$
Area of a triangle:	$A = \frac{1}{2}bh$
Area of a parallelogram:	$A = bh$
Commutative laws:	$a + b = b + a; \quad ab = ba$
Associative laws:	$a + (b + c) = (a + b) + c; \quad a(bc) = (ab)c$
Distributive law:	$a(b + c) = ab + ac$
Identity property of 1:	$1 \cdot a = a \cdot 1 = a$
Identity property of 0:	$a + 0 = 0 + a = a$
Law of opposites:	$a + (-a) = 0$
Multiplicative property of 0:	$0 \cdot a = a \cdot 0 = 0$
Property of -1:	$-1 \cdot a = -a$
Opposite of a sum:	$-(a + b) = -a + (-b)$
Division involving 0:	$\dfrac{0}{a} = 0; \quad \dfrac{a}{0}$ is undefined

$$\frac{-a}{b} = \frac{a}{-b} = -\frac{a}{b}, \qquad \frac{-a}{-b} = \frac{a}{b}$$

Rules for Order of Operations

1. Calculate within the innermost grouping symbols.
2. Simplify all exponential expressions.
3. Perform all multiplication and division, working from left to right.
4. Perform all addition and subtraction, working from left to right.

Review Exercises

Evaluate.

1. $5t$, for $t = 3$

2. $\dfrac{x - y}{3}$, for $x = 17$ and $y = 5$

3. $10 - y^2$, for $y = 4$

4. $-10 + a^2 \div (b + 1)$, for $a = 5$ and $b = 4$

Translate to an algebraic expression.

5. 7 less than z

6. The product of x and z

7. One more than the product of two numbers

8. Determine whether 35 is a solution of $x/5 = 8$.

9. Translate to an equation. Do not solve.

> In 1999, $4.6 billion worth of tea was sold wholesale in the United States. This was $2.8 billion more than the amount sold in 1990. How much tea was sold wholesale in 1990?

10. Use the commutative law of multiplication to write an expression equivalent to $2x + y$.

11. Use the associative law of addition to write an expression equivalent to $(2x + y) + z$.

12. Use the commutative and associative laws to write three expressions equivalent to $4(xy)$.

Multiply.

13. $6(3x + 5y)$

14. $8(5x + 3y + 2)$

Factor.

15. $21x + 15y$

16. $35x + 14 + 7y$

17. Find the prime factorization of 52.

Simplify.

18. $\dfrac{20}{48}$

19. $\dfrac{18}{8}$

Perform the indicated operation and, if possible, simplify.

20. $\dfrac{5}{12} + \dfrac{4}{9}$

21. $\dfrac{9}{16} \div 3$

22. $\dfrac{2}{3} - \dfrac{1}{15}$

23. $\dfrac{9}{10} \cdot \dfrac{16}{5}$

24. Tell which integers correspond to this situation: Renir has a debt of $45 and Raoul has $72 in his savings account.

25. Graph on a number line: $\dfrac{-1}{3}$.

26. Write an inequality with the same meaning as $-3 < x$.

27. Classify as true or false: $8 \geq 8$.

28. Classify as true or false: $0 \leq -1$.

29. Find decimal notation: $-\dfrac{7}{8}$.

30. Find the absolute value: $|-1|$.

31. Find $-(-x)$ when x is -7.

Simplify.

32. $4 + (-7)$

33. $-\dfrac{2}{3} + \dfrac{1}{12}$

34. $10 + (-9) + (-8) + 7$

35. $-3.8 + 5.1 + (-12) + (-4.3) + 10$

36. $-2 - (-7)$

37. $-\dfrac{9}{10} - \dfrac{1}{2}$

38. $-3.8 - 4.1$

39. $-9 \cdot (-6)$

40. $-2.7(3.4)$

41. $\dfrac{2}{3} \cdot \left(-\dfrac{3}{7}\right)$

42. $2 \cdot (-7) \cdot (-2) \cdot (-5)$

43. $35 \div (-5)$

44. $-5.1 \div 1.7$

45. $-\dfrac{3}{5} \div \left(-\dfrac{4}{5}\right)$

46. $|-3 \cdot 4 - 12 \cdot 2| - 8(-7)$

47. $|-12(-3) - 2^3 - (-9)(-10)|$

48. $120 - 6^2 \div 4 \cdot 8$

49. $(120 - 6^2) \div 4 \cdot 8$

50. $(120 - 6^2) \div (4 \cdot 8)$

51. $\dfrac{4(18 - 8) + 7 \cdot 9}{9^2 - 8^2}$

Combine like terms.

52. $11a + 2b + (-4a) + (-5b)$

53. $7x - 3y - 9x + 8y$

54. Find the opposite of -7.

55. Find the reciprocal of -7.

56. Write exponential notation for $2x \cdot 2x \cdot 2x \cdot 2x$.

57. Simplify: $(-5x)^3$.

Remove parentheses and simplify.

58. $2a - (5a - 9)$ **59.** $3(b + 7) - 5b$

60. $3[11x - 3(4x - 1)]$ **61.** $2[6(y - 4) + 7]$

62. $[8(x + 4) - 10] - [3(x - 2) + 4]$

63. Explain the difference between a constant and a variable.

64. Explain the difference between a term and a factor.

SYNTHESIS

65. Describe at least three ways in which the distributive law was used in this chapter.

66. Devise a rule for determining the sign of a negative quantity raised to a power.

67. Evaluate $a^{50} - 20a^{25}b^4 + 100b^8$ for $a = 1$ and $b = 2$.

68. If $0.090909\ldots = \frac{1}{11}$ and $0.181818\ldots = \frac{2}{11}$, what rational number is named by each of the following?

   a) $0.272727\ldots$ **b)** $0.909090\ldots$

Simplify.

69. $-\left| \frac{7}{8} - \left(-\frac{1}{2}\right) - \frac{3}{4} \right|$

70. $(|2.7 - 3| + 3^2 - |-3|) \div (-3)$

Match the phrase in the left column with the most appropriate choice from the right column.

_____ **71.** A number is nonnegative.   A. a^2

_____ **72.** The reciprocal of a sum   B. $a + b = b + a$

_____ **73.** A number squared   C. $a < 0$

_____ **74.** The opposite of a sum

                                    D. $a + \dfrac{1}{a}$

_____ **75.** The opposite of an opposite is the original number.   E. $|ab|$

_____ **76.** The order in which numbers are added does not change the result.   F. $(a + b)^2$

    G. $|a| < |b|$

    H. $-(a + b)$

_____ **77.** A number is negative.   I. $a \geq 0$

_____ **78.** The absolute value of a product   J. $\dfrac{1}{a + b}$

_____ **79.** A sum of a number and its reciprocal   K. $-(-a) = a$

_____ **80.** The square of a sum

_____ **81.** The absolute value of one number is less than the absolute value of another number.

Chapter Test 1

1. Evaluate $\dfrac{2x}{y}$ for $x = 10$ and $y = 5$.

2. Write an algebraic expression: Nine less than some number.

3. Find the area of a triangle when the height h is 30 ft and the base b is 16 ft.

4. Use the commutative law of addition to write an expression equivalent to $3p + q$.

5. Use the associative law of multiplication to write an expression equivalent to $x \cdot (4 \cdot y)$.

6. Determine whether 3 is a solution of $96 - a = 93$.

7. Translate to an equation. Do not solve.

On a hot summer day, Green River Electric met a demand of 2518 megawatts. This is only 282 megawatts less than its maximum production capability. What is the maximum capability of production?

Multiply.

8. $5(6 - x)$ **9.** $-5(y - 1)$

Factor.

10. $11 - 44x$ **11.** $7x + 21 + 14y$

12. Find the prime factorization of 300.

13. Simplify: $\frac{10}{35}$.

Write a true sentence using either < or >.

14. $-4 \quad \blacksquare \quad 0$

15. $-3 \quad \blacksquare \quad -8$

Find the absolute value.

16. $\left|\frac{9}{4}\right|$

17. $|-2.7|$

18. Find the opposite of $\frac{2}{3}$.

19. Find the reciprocal of $-\frac{4}{7}$.

20. Find $-x$ when x is -8.

21. Write an inequality with the same meaning as $x \leq -2$.

Compute and simplify.

22. $3.1 - (-4.7)$

23. $-8 + 4 + (-7) + 3$

24. $\frac{2}{5} + \frac{3}{8}$

25. $2 - (-8)$

26. $3.2 - 5.7$

27. $\frac{1}{8} - \left(-\frac{3}{4}\right)$

28. $4 \cdot (-12)$

29. $-\frac{1}{2} \cdot \left(-\frac{3}{8}\right)$

30. $-45 \div 5$

31. $-\frac{3}{5} \div \left(-\frac{4}{5}\right)$

32. $4.864 \div (-0.5)$

33. $-2(16) - |2(-8) - 5^3|$

34. $6 + 7 - 4 - (-3)$

35. $256 \div (-16) \div 4$

36. $2^3 - 10[4 - (-2 + 18)3]$

37. Combine like terms: $18y + 30a - 9a + 4y$.

38. Simplify: $(-2x)^4$.

Remove parentheses and simplify.

39. $5x - (3x - 7)$

40. $4(2a - 3b) + a - 7$

41. $4\{3[5(y - 3) + 9] + 2(y + 8)\}$

SYNTHESIS

42. Evaluate $\dfrac{5y - x}{4}$ when $x = 20$ and y is 4 less than x.

Simplify.

43. $\dfrac{13,800}{42,000}$

44. $|-27 - 3(4)| - |-36| + |-12|$

45. $a - \{3a - [4a - (2a - 4a)]\}$

2

Equations, Inequalities, and Problem Solving

AN APPLICATION

To cater a party, Curtis' Barbeque charges a $50 setup fee plus $15 per person. The cost of Hotel Pharmacy's end-of-season softball party cannot exceed $450. How many people can attend the party?

This problem appears as Example 1 in Section 2.7.

Most people might not associate math with barbeque, but in fact I use math in everything from ordering ingredients for my secret sauce to keeping the books for my business.

CURTIS TUFF
Chef and Owner,
Curtis' Barbeque
Putney, VT

*S*olving equations and inequalities is a recurring theme in much of mathematics. In this chapter, we will study some of the principles used to solve equations and inequalities. We will then use equations and inequalities to solve applied problems.

Solving Equations

2.1

Equations and Solutions • The Addition Principle • The Multiplication Principle

Solving equations is essential for problem solving in algebra. In this section, we study two of the most important principles used for this task.

Equations and Solutions

We have already seen that an equation is a number sentence stating that the expressions on either side of the equals sign represent the same number. Some equations, like $3 + 2 = 5$ or $2x + 6 = 2(x + 3)$, are *always* true and some, like $3 + 2 = 6$ or $x + 2 = x + 3$, are *never* true. In this text, we will concentrate on equations like $x + 6 = 13$ or $7x = 141$ that are *sometimes* true, depending on the replacement value for the variable.

> **Solution of an Equation**
>
> Any replacement for the variable that makes an equation true is called a *solution* of the equation. To *solve* an equation means to find all of its solutions.

To determine whether a number is a solution, we substitute that number for the variable throughout the equation. If the values on both sides of the equals sign are the same, then the number that was substituted is a solution.

Example 1

Determine whether 7 is a solution of $x + 6 = 13$.

Solution We have

$$x + 6 = 13 \qquad \text{Writing the equation}$$
$$7 + 6 \ ? \ 13 \qquad \text{Substituting 7 for } x$$
$$13 \ | \ 13 \quad \text{TRUE} \qquad \text{Note that 7, not 13, is the solution.}$$

Since the left-hand and the right-hand sides are the same, 7 is a solution.

E x a m p l e 2   Determine whether 19 is a solution of $7x = 141$.

Solution   We have

$$
\begin{array}{c|c}
7x = 141 & \text{Writing the equation} \\
\hline
7(19) \ ? \ 141 & \text{Substituting 19 for } x \\
133 \ | \ 141 \ \ \text{\small FALSE} & \text{The statement } 133 = 141 \text{ is false.}
\end{array}
$$

Since the left-hand and the right-hand sides differ, 19 is not a solution.

The Addition Principle

Consider the equation

$x = 7.$

We can easily see that the solution of this equation is 7. Replacing x with 7, we get

$7 = 7,$ which is true.

Now consider the equation

$x + 6 = 13.$

In Example 1, we found that the solution of $x + 6 = 13$ is also 7. Although the solution of $x = 7$ may seem more obvious, the equations $x + 6 = 13$ and $x = 7$ are **equivalent**.

Equivalent Equations

Equations with the same solutions are called *equivalent equations*.

There are principles that enable us to begin with one equation and end up with an equivalent equation, like $x = 7$, for which the solution is obvious. One such principle concerns addition. The equation $a = b$ says that a and b stand for the same number. Suppose this is true, and some number c is added to a. We get the same result if we add c to b, because a and b are the same number.

The Addition Principle

For any real numbers a, b, and c,

$$a = b \ \ \text{is equivalent to} \ \ a + c = b + c.$$

To visualize the addition principle, consider a balance similar to one a jeweler might use. (See the figure on the following page.) When the two sides of the balance hold quantities of equal weight, the balance is level. If weight is added or removed, equally, on both sides, the balance will remain level.

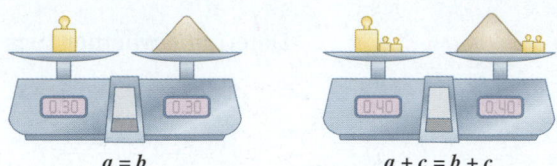

$$a = b \qquad\qquad a + c = b + c$$

When using the addition principle, we often say that we "add the same number to both sides of an equation." We can also "subtract the same number from both sides," since subtraction can be regarded as the addition of an opposite.

Example 3

Solve: $x + 5 = -7$.

Solution   We can add any number we like to both sides. Since -5 is the opposite, or additive inverse, of 5, we add -5 to each side:

$$x + 5 = -7$$
$$x + 5 - 5 = -7 - 5 \qquad \text{Using the addition principle: adding } -5 \text{ to both sides or subtracting 5 from both sides}$$
$$x + 0 = -12 \qquad \text{Simplifying; } x + 5 - 5 = x + 5 + (-5) = x + 0$$
$$x = -12. \qquad \text{Using the identity property of 0}$$

It is obvious that the solution of $x = -12$ is the number -12. To check the answer in the original equation, we substitute.

Check: $$\frac{x + 5 = -7}{-12 + 5 \ ? \ -7}$$
$$\qquad\qquad -7 \ | \ -7 \ \text{TRUE}$$

The solution of the original equation is -12.

In Example 3, note that because we added the *opposite*, or *additive inverse*, of 5, the left side of the equation simplified to x plus the *additive identity*, 0, or simply x. These steps effectively replaced the 5 on the left with a 0. When solving $x + a = b$ for x, we simply add $-a$ to (or subtract a from) both sides.

Example 4

Solve: $-6.5 = y - 8.4$.

Solution   The variable is on the right side this time. We can isolate y by adding 8.4 to each side:

$$-6.5 = y - 8.4 \qquad \text{This can be regarded as } -6.5 = y + (-8.4).$$

$$-6.5 + 8.4 = y - 8.4 + 8.4 \qquad \text{Using the addition principle: Adding 8.4 to both sides "eliminates" } -8.4 \text{ on the right side.}$$

$$1.9 = y. \qquad \begin{aligned} y - 8.4 + 8.4 &= y + (-8.4) + 8.4 \\ &= y + 0 = y \end{aligned}$$

Check:

$$-6.5 = y - 8.4$$

$$\overline{-6.5 \; ? \; 1.9 - 8.4}$$

$$-6.5 \; | \; -6.5 \qquad \text{TRUE}$$

The solution is 1.9.

Note that the equations $a = b$ and $b = a$ have the same meaning. Thus, $-6.5 = y - 8.4$ could have been rewritten as $y - 8.4 = -6.5$.

Example 5

Solve: $-\frac{2}{3} + x = \frac{5}{2}$.

Solution   We have

$$-\frac{2}{3} + x = \frac{5}{2}$$

$$-\frac{2}{3} + x + \frac{2}{3} = \frac{5}{2} + \frac{2}{3} \qquad \text{Adding } \tfrac{2}{3} \text{ to both sides}$$

$$x = \frac{5}{2} + \frac{2}{3}$$

$$= \frac{5}{2} \cdot \frac{3}{3} + \frac{2}{3} \cdot \frac{2}{2} \qquad \text{Multiplying by 1 to obtain a common denominator}$$

$$= \frac{15}{6} + \frac{4}{6}$$

$$= \frac{19}{6}.$$

The check is left to the student. The solution is $\frac{19}{6}$.

The Multiplication Principle

A second principle for solving equations concerns multiplying. Suppose a and b are equal. If a and b are multiplied by some number c, then ac and bc will also be equal.

> ### The Multiplication Principle
> For any real numbers a, b, and c, with $c \neq 0$,
> $$a = b \quad \text{is equivalent to} \quad a \cdot c = b \cdot c.$$

Example 6

Solve: $\frac{5}{4}x = 10$.

Solution   We can multiply both sides by any nonzero number we like. Since $\frac{4}{5}$ is the reciprocal of $\frac{5}{4}$, we multiply each side by $\frac{4}{5}$:

$$\frac{5}{4}x = 10$$

$$\frac{4}{5} \cdot \frac{5}{4}x = \frac{4}{5} \cdot 10 \qquad \text{Using the multiplication principle: Multiplying both sides by } \tfrac{4}{5} \text{ "eliminates" the } \tfrac{5}{4} \text{ on the left.}$$

$$1 \cdot x = 8 \qquad \text{Simplifying}$$

$$x = 8. \qquad \text{Using the identity property of 1}$$

Check: $\frac{5}{4}x = 10$

$$\frac{\frac{5}{4} \cdot 8 \;?\; 10}{\qquad 10 \;\bigm|\; 10 \;\text{TRUE}}$$

The solution is 8.

In Example 6, to get x alone, we multiplied by the *reciprocal*, or *multiplicative inverse* of $\frac{5}{4}$. We then simplified the left-hand side to x times the *multiplicative identity*, 1, or simply x. These steps effectively replaced the $\frac{5}{4}$ on the left with 1.

Because division is the same as multiplying by a reciprocal, the multiplication principle also tells us that we can "divide both sides by the same nonzero number." That is,

$$\text{if } a = b, \text{ then } \quad \frac{1}{c} \cdot a = \frac{1}{c} \cdot b \quad \text{and} \quad \frac{a}{c} = \frac{b}{c} \qquad (\text{provided } c \neq 0).$$

In a product like $3x$, the multiplier 3 is called the **coefficient**. When the coefficient of the variable is an integer or a decimal, it is usually easiest to solve an equation by dividing on both sides. When the coefficient is in fraction notation, it is usually easier to multiply by the reciprocal.

E x a m p l e   7   Solve: **(a)** $-4x = 92$;   **(b)** $12.6 = 3t$;   **(c)** $-x = 9$;   **(d)** $\dfrac{2y}{9} = \dfrac{8}{3}$.

Solution

a) $-4x = 92$

$\dfrac{-4x}{-4} = \dfrac{92}{-4}$ Using the multiplication principle: Dividing both sides by -4 is the same as multiplying by $-\frac{1}{4}$.

$1 \cdot x = -23$ Simplifying

$x = -23$ Using the identity property of 1

Check: $\dfrac{-4x = 92}{-4(-23) \;?\; 92}$
$$\qquad\qquad 92 \;\bigm|\; 92 \;\text{TRUE}$$

The solution is -23.

b) $12.6 = 3t$

$\dfrac{12.6}{3} = \dfrac{3t}{3}$ Dividing both sides by 3 or multiplying both sides by $\frac{1}{3}$

$4.2 = 1t$

$4.2 = t$ Simplifying

Check: $\dfrac{12.6 = 3t}{12.6 \;?\; 3(4.2)}$
$$\qquad\quad 12.6 \;\bigm|\; 12.6 \quad\text{TRUE}$$

The solution is 4.2.

c) To solve an equation like $-x = 9$, remember that when an expression is multiplied or divided by -1, its sign is changed. Here we multiply both sides by -1 to change the sign of $-x$:

$$-x = 9$$

$$(-1)(-x) = (-1)9 \qquad \text{Multiplying both sides by } -1 \text{ (Dividing by } -1 \text{ would also work)}$$

$$x = -9. \qquad \text{Note that } (-1)(-x) \text{ is the same as } (-1)(-1)x.$$

Check:

$$\begin{array}{c|c} -x = 9 \\ \hline -(-9) \; ? \; 9 \\ 9 \; | \; 9 \quad \text{TRUE} \end{array}$$

The solution is -9.

d) To solve an equation like $\frac{2y}{9} = \frac{8}{3}$, we rewrite the left-hand side as $\frac{2}{9} \cdot y$ and then use the multiplication principle:

$$\frac{2y}{9} = \frac{8}{3}$$

$$\frac{2}{9} \cdot y = \frac{8}{3} \qquad \text{Rewriting } \frac{2y}{9} \text{ as } \frac{2}{9} \cdot y$$

$$\frac{9}{2} \cdot \frac{2}{9} \cdot y = \frac{9}{2} \cdot \frac{8}{3} \qquad \text{Multiplying both sides by } \frac{9}{2}$$

$$1y = \frac{3 \cdot 3 \cdot 2 \cdot 4}{2 \cdot 3} \qquad \text{Removing a factor equal to 1: } \frac{3 \cdot 2}{2 \cdot 3} = 1$$

$$y = 12.$$

Check:

$$\begin{array}{c|c} \dfrac{2y}{9} = \dfrac{8}{3} \\ \hline \dfrac{2 \cdot 12}{9} \; ? \; \dfrac{8}{3} \\ \dfrac{24}{9} \\ \dfrac{8}{3} \; | \; \dfrac{8}{3} \quad \text{TRUE} \end{array}$$

The solution is 12.

FOR EXTRA HELP

Exercise Set 2.1

 Digital Video Tutor CD 1 Videotape 3   InterAct Math   Math Tutor Center   MathXL   MyMathLab.com

Solve using the addition principle. Don't forget to check!

1. $x + 8 = 23$

2. $x + 5 = 8$

3. $t + 9 = -4$

4. $y + 9 = 43$

5. $y + 7 = -3$

6. $t + 9 = -12$

7. $-5 = x + 8$

8. $-6 = y + 25$

9. $x - 9 = 6$

10. $x - 8 = 5$

11. $y - 6 = -14$

12. $x - 4 = -19$

13. $9 + t = 3$

14. $3 + t = 21$

15. $12 = -7 + y$

16. $15 = -9 + z$

17. $-5 + t = -9$

18. $-6 + y = -21$

19. $r + \frac{1}{3} = \frac{8}{3}$

20. $t + \frac{3}{8} = \frac{5}{8}$

21. $x + \frac{3}{5} = -\frac{7}{10}$

22. $x + \frac{2}{3} = -\frac{5}{6}$

23. $x - \frac{5}{6} = \frac{7}{8}$

24. $y - \frac{3}{4} = \frac{5}{6}$

25. $-\frac{1}{5} + z = -\frac{1}{4}$

26. $-\frac{1}{8} + y = -\frac{3}{4}$

27. $m + 3.9 = 5.4$

28. $y + 5.3 = 8.7$

29. $-9.7 = -4.7 + y$

30. $-7.8 = 2.8 + x$

Solve using the multiplication principle. Don't forget to check!

31. $5x = 80$

32. $3x = 39$

33. $9t = 36$

34. $6x = 72$

35. $84 = 7x$

36. $56 = 7t$

37. $-x = 23$

38. $100 = -x$

Aha! **39.** $-t = -8$

40. $-68 = -r$

41. $7x = -49$

42. $9x = -36$

43. $-12x = 72$

44. $-15x = 105$

45. $-3.4t = -20.4$

46. $-1.3a = -10.4$

47. $\frac{a}{4} = 13$

48. $\frac{y}{-8} = 11$

49. $\frac{3}{4}x = 27$

50. $\frac{4}{5}x = 16$

51. $\frac{-t}{5} = 9$

52. $\frac{-x}{6} = 9$

53. $\frac{2}{7} = \frac{x}{3}$

54. $\frac{1}{9} = \frac{z}{5}$

Aha! **55.** $-\frac{3}{5}r = -\frac{3}{5}$

56. $-\frac{2}{5}y = -\frac{4}{15}$

57. $\frac{-3r}{2} = -\frac{27}{4}$

58. $\frac{5x}{7} = -\frac{10}{14}$

Solve. The icon ▦ indicates an exercise designed to give practice using a calculator.

59. $4.5 + t = -3.1$

60. $\frac{3}{4}x = 18$

61. $-8.2x = 20.5$

62. $t - 7.4 = -12.9$

63. $12 = y + 29$

64. $96 = -\frac{3}{4}t$

65. $a - \frac{1}{6} = -\frac{2}{3}$

66. $-\frac{x}{7} = \frac{2}{9}$

67. $-24 = \frac{8x}{5}$

68. $\frac{1}{5} + y = -\frac{3}{10}$

69. $-\frac{4}{3}t = -16$

70. $\frac{17}{35} = -x$

▦ **71.** $-483.297 = -794.053 + t$

▦ **72.** $-0.2344x = 2028.732$

🗒 **73.** When solving an equation, how do you determine what number to add, subtract, multiply, or divide by on both sides of that equation?

🗒 **74.** What is the difference between equivalent expressions and equivalent equations?

SKILL MAINTENANCE

Simplify.

75. $9 - 2 \cdot 5^2 + 7$

76. $10 \div 2 \cdot 3^2 - 4$

77. $16 \div (2 - 3 \cdot 2) + 5$

78. $12 - 5 \cdot 2^3 + 4 \cdot 3$

SYNTHESIS

🗒 **79.** To solve $-3.5 = 14t$, Anita adds 3.5 to both sides. Will this form an equivalent equation? Will it help solve the equation? Explain.

🗒 **80.** Explain why it is not necessary to state a subtraction principle: For any real numbers a, b, and c, $a = b$ is equivalent to $a - c = b - c$.

Some equations, like $3 = 7$ or $x + 2 = x + 5$, have no solution and are called **contradictions**. *Other equations, like $7 = 7$ or $2x = 2x$, are true for all numbers and are called* **identities**. *Solve each of the following and if an identity or contradiction is found, state this.*

81. $2x = x + x$

82. $x + 5 + x = 2x$

83. $5x = 0$

84. $4x - x = 2x + x$

85. $x + 8 = 3 + x + 7$

86. $3x = 0$

Aha! **87.** $2|x| = -14$

88. $|3x| = 6$

Solve for x. Assume a, c, $m \neq 0$.

89. $mx = 9.4m$

90. $x - 4 + a = a$

91. $\dfrac{7cx}{2a} = \dfrac{21}{a} \cdot c$

92. $5c + cx = 7c$

93. $5a = ax - 3a$

94. $|x| + 6 = 19$

95. If $x - 4720 = 1634$, find $x + 4720$.

96. Lydia makes a calculation and gets an answer of 22.5. On the last step, she multiplies by 0.3 when she should have divided by 0.3. What should the correct answer be?

97. Are the equations $x = 5$ and $x^2 = 25$ equivalent? Why or why not?

2.2

Using the Principles Together

Applying Both Principles • Combining Like Terms • Clearing Fractions and Decimals

CONNECTING THE CONCEPTS

We have stated that most of algebra involves either simplifying expressions (by writing equivalent expressions) or solving equations (by writing equivalent equations). In Section 2.1, we used the addition and multiplication principles to produce equivalent equations, like $x = 5$, from which the solution—in this case, 5—is obvious. Here in Section 2.2, we will find that more complicated equations can be solved by using both principles together and by using the commutative, associative, and distributive laws to write equivalent expressions.

An important strategy for solving a new problem is to find a way to make the new prob-

lem look like a problem we already know how to solve. This is precisely the approach taken in this section. You will find that the last steps of the examples in this section are nearly identical to the steps used for solving the examples of Section 2.1. What is new in this section appears in the early steps of each example.

Before reading this section, make sure that you thoroughly understand the material in Section 2.1. Without a solid grasp of how and when to use the addition and multiplication principles, the problems in this section will seem much more difficult than they really are.

Applying Both Principles

In the expression $5 + 3x$, the variable x is multiplied by 3 and then 5 is added. To reverse these steps, we first subtract 5 and then divide by 3. Thus, to solve $5 + 3x = 17$, we first subtract 5 from each side and then divide both sides by 3.

E x a m p l e   1   Solve: $5 + 3x = 17$.

Solution   We have

$$5 + 3x = 17$$

$$5 + 3x - 5 = 17 - 5 \qquad \text{Using the addition principle: subtracting 5 from both sides (adding } -5)$$

$$5 + (-5) + 3x = 12 \qquad \text{Using a commutative law. Try to perform this step mentally.}$$

| First isolate the *x*-term. | $3x = 12$ | Simplifying |

$$\frac{3x}{3} = \frac{12}{3} \qquad \text{Using the multiplication principle: dividing both sides by 3 (multiplying by } \tfrac{1}{3})$$

| Then isolate *x*. | $x = 4.$ | Simplifying |

Check:
$$\frac{5 + 3x = 17}{5 + 3 \cdot 4 \ ? \ 17} \qquad \text{We use the rules for order of operations:}$$
$$5 + 12 \qquad\qquad \text{Find the product, } 3 \cdot 4, \text{ and then add.}$$
$$17 \ \mid \ 17 \ \text{TRUE}$$

The solution is 4.

Multiplication by a negative number and subtraction are handled in much the same way.

E x a m p l e   2   Solve: $-5x - 6 = 16$.

Solution   In $-5x - 6$ we multiply first and then subtract. To reverse these steps, we first add 6 and then divide by -5.

$$-5x - 6 = 16$$

$$-5x - 6 + 6 = 16 + 6 \qquad \text{Adding 6 to both sides}$$

$$-5x = 22$$

$$\frac{-5x}{-5} = \frac{22}{-5} \qquad \text{Dividing both sides by } -5$$

$$x = -\frac{22}{5}, \text{ or } -4\frac{2}{5} \qquad \text{Simplifying}$$

Check:
$$\frac{-5x - 6 = 16}{-5\left(-\frac{22}{5}\right) - 6 \ ? \ 16}$$
$$22 - 6 \qquad\qquad$$
$$16 \ \mid \ 16 \ \text{TRUE}$$

The solution is $-\frac{22}{5}$.

Example 3

Solve: $45 - t = 13$.

Solution   We have

$$45 - t = 13$$
$$45 - t - 45 = 13 - 45 \qquad \text{Subtracting 45 from both sides}$$
$$\left.\begin{array}{l}45 + (-t) + (-45) = 13 - 45 \\ 45 + (-45) + (-t) = 13 - 45\end{array}\right\} \qquad \text{Try to do these steps mentally.}$$
$$-t = -32 \qquad \text{Try to go directly to this step.}$$
$$(-1)(-t) = (-1)(-32) \qquad \begin{array}{l}\text{Multiplying both sides by } -1 \\ \text{(Dividing by } -1 \text{ would also work.)}\end{array}$$
$$t = 32.$$

Check:
$$\begin{array}{c}45 - t = 13 \\ \hline 45 - 32 \ ?\ 13 \\ 13 \ \bigm|\ 13 \quad \text{TRUE}\end{array}$$

The solution is 32.

As our skills improve, many of the steps can be streamlined.

Example 4

Solve: $16.3 - 7.2y = -8.18$.

Solution   We have

$$16.3 - 7.2y = -8.18$$
$$-7.2y = -8.18 - 16.3 \qquad \begin{array}{l}\text{Subtracting 16.3 from both sides. We} \\ \text{write the subtraction of 16.3 on the} \\ \text{right side and remove 16.3 from the} \\ \text{left side.}\end{array}$$
$$-7.2y = -24.48$$
$$y = \frac{-24.48}{-7.2} \qquad \begin{array}{l}\text{Dividing both sides by } -7.2. \text{ We write} \\ \text{the division by } -7.2 \text{ on the right side} \\ \text{and remove the } -7.2 \text{ from the left side.}\end{array}$$
$$y = 3.4.$$

Check:
$$\begin{array}{c}16.3 - 7.2y = -8.18 \\ \hline 16.3 - 7.2(3.4) \ ?\ -8.18 \\ 16.3 - 24.48 \ \bigm| \\ -8.18 \ \bigm|\ -8.18 \quad \text{TRUE}\end{array}$$

The solution is 3.4.

Combining Like Terms

If like terms appear on the same side of an equation, we combine them and then solve. Should like terms appear on both sides of an equation, we can use the addition principle to rewrite all like terms on one side.

Example 5

technology
connection

Most graphers have a TABLE feature that enables the calculator to evaluate a variable expression for different choices of x. For example, to evaluate $6x + 5 - 7x$ for $x = 0, 1, 2, \dots$, we first use $\boxed{\text{Y=}}$ to enter $6x + 5 - 7x$ as y_1. We then use $\boxed{\text{2nd}}$ $\boxed{\text{TBLSET}}$ to specify which x-values will be used. Using TblStart = 0, ΔTbl = 1, and selecting Auto twice, we can generate a table in which the value of $6x + 5 - 7x$ is listed for values of x starting at 0 and increasing by ones.

X	Y₁
0	5
1	4
2	3
3	2
4	1
5	0
6	−1

X = 0

1. Create the above table on your grapher. Scroll up and down to extend the table.
2. Enter $10 - 4x + 7$ as y_2. Your table should now have three columns.
3. For what x-value is y_1 the same as y_2? Compare this with the solution of Example 5(c). Is this a reliable way to solve equations? Why or why not?

Solve.

a) $3x + 4x = -14$ **b)** $2x - 4 = -3x + 1$
c) $6x + 5 - 7x = 10 - 4x + 7$ **d)** $2 - 5(x + 5) = 3(x - 2) - 1$

Solution

a)
$$3x + 4x = -14$$
$$7x = -14 \qquad \text{Combining like terms}$$
$$x = \frac{-14}{7} \qquad \text{Dividing both sides by 7}$$
$$x = -2$$

The check is left to the student. The solution is -2.

b) To solve $2x - 4 = -3x + 1$, we must first write only variable terms on one side and only constant terms on the other. This can be done by adding 4 to both sides, to get all constant terms on the right, and $3x$ to both sides, to get all variable terms on the left. We can add 4 first, or $3x$ first, or do both in one step.

> Isolate variable terms on one side and constant terms on the other side.

$$2x - 4 = -3x + 1$$
$$2x - 4 + 4 = -3x + 1 + 4 \qquad \text{Adding 4 to both sides}$$
$$2x = -3x + 5 \qquad \text{Simplifying}$$
$$2x + 3x = -3x + 3x + 5 \qquad \text{Adding } 3x \text{ to both sides}$$
$$5x = 5 \qquad \text{Combining like terms and simplifying}$$
$$x = \frac{5}{5} \qquad \text{Dividing both sides by 5}$$
$$x = 1 \qquad \text{Simplifying}$$

Check:
$$\frac{2x - 4 = -3x + 1}{}$$
$$2 \cdot 1 - 4 \; ? \; -3 \cdot 1 + 1$$
$$2 - 4 \; | \; -3 + 1$$
$$-2 \; | \; -2 \qquad \text{TRUE}$$

The solution is 1.

c)
$$6x + 5 - 7x = 10 - 4x + 7$$
$$-x + 5 = 17 - 4x \qquad \text{Combining like terms on both sides}$$
$$-x + 5 + 4x = 17 - 4x + 4x \qquad \text{Adding } 4x \text{ to both sides}$$
$$5 + 3x = 17 \qquad \text{Simplifying. This is identical to Example 1.}$$
$$3x = 12 \qquad \text{Subtracting 5 from both sides and simplifying}$$
$$x = 4 \qquad \text{Dividing both sides by 3 and simplifying}$$

Check:
$$\frac{6x + 5 - 7x = 10 - 4x + 7}{}$$
$$6 \cdot 4 + 5 - 7 \cdot 4 \; ? \; 10 - 4 \cdot 4 + 7$$
$$24 + 5 - 28 \; | \; 10 - 16 + 7$$
$$1 \; | \; 1 \qquad \text{TRUE}$$

The solution is 4.

d) $2 - 5(x + 5) = 3(x - 2) - 1$

$\quad\quad 2 - 5x - 25 = 3x - 6 - 1$ Using the distributive law. This is now similar to part (c) above.

$\quad\quad\quad -5x - 23 = 3x - 7$ Combining like terms on both sides

$\left.\begin{array}{r} -5x - 23 + 7 = 3x \\ -23 + 7 = 3x + 5x \end{array}\right\}$ Adding 7 and 5x to both sides. This isolates the x-terms on one side and the constant terms on the other.

$\quad\quad\quad\quad\quad\quad -16 = 8x$ Simplifying

$\quad\quad\quad\quad\quad\quad\quad -2 = x$ Dividing both sides by 8

The student can confirm that -2 checks and is the solution.

Clearing Fractions and Decimals

Equations are generally easier to solve when they do not contain fractions or decimals. The multiplication principle can be used to "clear" fractions or decimals, as shown here.

Clearing Fractions	Clearing Decimals
$\frac{1}{2}x + 5 = \frac{3}{4}$	$2.3x + 7 = 5.4$
$4\left(\frac{1}{2}x + 5\right) = 4 \cdot \frac{3}{4}$	$10(2.3x + 7) = 10 \cdot 5.4$
$2x + 20 = 3$	$23x + 70 = 54$

In each case, the resulting equation is equivalent to the original equation, but easier to solve.

The easiest way to clear an equation of fractions is to multiply *both sides* of the equation by the smallest, or *least*, common denominator.

E x a m p l e 6

Solve: **(a)** $\frac{2}{3}x - \frac{1}{6} = 2x$; **(b)** $\frac{2}{5}(3x + 2) = 8$.

Solution

a) The number 6 is the least common denominator, so we multiply both sides by 6.

$$6\left(\frac{2}{3}x - \frac{1}{6}\right) = 6 \cdot 2x$$ Multiplying both sides by 6

$$6 \cdot \frac{2}{3}x - 6 \cdot \frac{1}{6} = 6 \cdot 2x$$

> ***Caution!*** Be sure the distributive law is used to multiply *all* the terms by 6.

$$4x - 1 = 12x$$ Simplifying. Note that the fractions are cleared.

$$-1 = 8x$$ Subtracting 4x from both sides

$$-\frac{1}{8} = x$$ Dividing both sides by 8

The number $-\frac{1}{8}$ checks and is the solution.

b) To solve $\frac{2}{5}(3x + 2) = 8$, we can multiply both sides by $\frac{5}{2}$ (or divide by $\frac{2}{5}$) to "undo" the multiplication by $\frac{2}{5}$ on the left side.

$$\frac{5}{2} \cdot \frac{2}{5}(3x + 2) = \frac{5}{2} \cdot 8 \qquad \text{Multiplying both sides by } \tfrac{5}{2}$$

$$3x + 2 = 20 \qquad \text{Simplifying; } \tfrac{5}{2} \cdot \tfrac{2}{5} = 1 \text{ and } \tfrac{5}{2} \cdot \tfrac{8}{1} = 20$$

$$3x = 18 \qquad \text{Subtracting 2 from both sides}$$

$$x = 6 \qquad \text{Dividing both sides by 3}$$

The student can confirm that 6 checks and is the solution.

To clear an equation of decimals, we count the greatest number of decimal places in any one number. If the greatest number of decimal places is 1, we multiply both sides by 10; if it is 2, we multiply by 100; and so on.

Example 7

Solve: $16.3 - 7.2y = -8.18$.

Solution   The greatest number of decimal places in any one number is *two*. Multiplying by 100 will clear all decimals.

$$100(16.3 - 7.2y) = 100(-8.18) \qquad \text{Multiplying both sides by 100}$$

$$100(16.3) - 100(7.2y) = 100(-8.18) \qquad \text{Using the distributive law}$$

$$1630 - 720y = -818 \qquad \text{Simplifying}$$

$$-720y = -818 - 1630 \qquad \text{Subtracting 1630 from both sides}$$

$$-720y = -2448 \qquad \text{Combining like terms}$$

$$y = \frac{-2448}{-720} \qquad \text{Dividing both sides by } -720$$

$$y = 3.4$$

In Example 4, the same solution was found without clearing decimals. Finding the same answer two ways is a good check. The solution is 3.4.

An Equation-Solving Procedure

1. Use the multiplication principle to clear any fractions or decimals. (This is optional, but can ease computations.)
2. If necessary, use the distributive law to remove parentheses. Then combine like terms on each side.
3. Use the addition principle, as needed, to get all variable terms on one side and all constant terms on the other.
4. Combine like terms again, if necessary.
5. Multiply or divide to solve for the variable, using the multiplication principle.
6. Check all possible solutions in the original equation.

Exercise Set 2.2

Solve and check.

1. $5x + 3 = 38$

2. $3x + 6 = 30$

3. $8x + 4 = 68$

4. $6z + 3 = 57$

5. $7t - 8 = 27$

6. $6x - 3 = 15$

7. $3x - 9 = 33$

8. $5x - 9 = 41$

9. $8z + 2 = -54$

10. $4x + 3 = -21$

11. $-39 = 1 + 8x$

12. $-91 = 9t + 8$

13. $9 - 4x = 37$

14. $12 - 4x = 108$

15. $-7x - 24 = -129$

16. $-6z - 18 = -132$

17. $48 = 5x + 7x$

18. $4x + 5x = 45$

19. $27 - 6x = 99$

20. $32 - 7x = 11$

21. $4x + 3x = 42$

22. $6x + 19x = 100$

23. $-2a + 5a = 24$

24. $-4y - 7y = 33$

25. $-7y - 8y = -15$

26. $-10y - 2y = -48$

27. $10.2y - 7.3y = -58$

28. $3.4t - 1.2t = -44$

29. $x + \frac{1}{3}x = 8$

30. $x + \frac{1}{4}x = 10$

31. $9y - 35 = 4y$

32. $4x - 6 = 6x$

33. $6x - 5 = 7 + 2x$

34. $5y - 2 = 28 - y$

Aha! **35.** $6x + 3 = 2x + 3$

36. $5y + 3 = 2y + 15$

37. $5 - 2x = 3x - 7x + 25$

38. $10 - 3x = 2x - 8x + 40$

39. $7 + 3x - 6 = 3x + 5 - x$

40. $5 + 4x - 7 = 4x - 2 - x$

41. $4y - 4 + y + 24 = 6y + 20 - 4y$

42. $5y - 10 + y = 7y + 18 - 5y$

Clear fractions or decimals, solve, and check.

43. $\frac{5}{4}x + \frac{1}{4}x = 2x + \frac{1}{2} + \frac{3}{4}x$

44. $\frac{7}{8}x - \frac{1}{4} + \frac{3}{4}x = \frac{1}{16} + x$

45. $\frac{2}{3} + \frac{1}{4}t = 6$

46. $-\frac{1}{2} + x = -\frac{5}{6} - \frac{1}{3}$

47. $\frac{2}{3} + 4t = 6t - \frac{2}{15}$

48. $\frac{1}{2} + 4m = 3m - \frac{5}{2}$

49. $\frac{1}{3}x + \frac{2}{5} = \frac{4}{15} + \frac{3}{5}x - \frac{2}{3}$

50. $1 - \frac{2}{3}y = \frac{9}{5} - \frac{1}{5}y + \frac{3}{5}$

51. $2.1x + 45.2 = 3.2 - 8.4x$

52. $0.91 - 0.2z = 1.23 - 0.6z$

53. $0.76 + 0.21t = 0.96t - 0.49$

54. $1.7t + 8 - 1.62t = 0.4t - 0.32 + 8$

55. $\frac{2}{5}x - \frac{3}{2}x = \frac{3}{4}x + 2$

56. $\frac{5}{16}y + \frac{3}{8}y = 2 + \frac{1}{4}y$

Solve and check.

57. $7(2a - 1) = 21$

58. $5(2t - 2) = 35$

59. $35 = 5(3x + 1)$

60. $9 = 3(5x - 2)$

61. $2(3 + 4m) - 6 = 48$

62. $3(5 + 3m) - 8 = 88$

63. $7r - (2r + 8) = 32$

64. $6b - (3b + 8) = 16$

65. $13 - 3(2x - 1) = 4$

66. $5(d + 4) = 7(d - 2)$

67. $3(t - 2) = 9(t + 2)$

68. $8(2t + 1) = 4(7t + 7)$

69. $7(5x - 2) = 6(6x - 1)$

70. $5(t + 3) + 9 = 3(t - 2) + 6$

71. $19 - (2x + 3) = 2(x + 3) + x$

72. $13 - (2c + 2) = 2(c + 2) + 3c$

73. $\frac{1}{4}(3t - 4) = 5$

74. $\frac{1}{3}(2x - 1) = 7$

75. $\frac{4}{3}(5x + 1) = 8$

76. $\frac{3}{4}(3t - 6) = 9$

77. $\frac{3}{2}(2x + 5) = -\frac{15}{2}$

78. $\frac{1}{6}\left(\frac{3}{4}x - 2\right) = -\frac{1}{5}$

79. $\frac{3}{4}\left(3x - \frac{1}{2}\right) - \frac{2}{3} = \frac{1}{3}$

80. $\frac{2}{3}\left(\frac{7}{8} - 4x\right) - \frac{5}{8} = \frac{3}{8}$

81. $0.7(3x + 6) = 1.1 - (x + 2)$

82. $0.9(2x + 8) = 20 - (x + 5)$

83. $a + (a - 3) = (a + 2) - (a + 1)$

84. $0.8 - 4(b - 1) = 0.2 + 3(4 - b)$

85. When an equation contains decimals, is it essential to clear the equation of decimals? Why or why not?

86. Why must the rules for the order of operations be understood before solving the equations in this section?

SKILL MAINTENANCE

Evaluate.

87. $3 - 5a$, for $a = 2$

88. $12 \div 4 \cdot t$, for $t = 5$

89. $7x - 2x$, for $x = -3$

90. $t(8 - 3t)$, for $t = -2$

SYNTHESIS

91. What procedure would you follow to solve an equation like $0.23x + \frac{17}{3} = -0.8 + \frac{3}{4}x$? Could your procedure be streamlined? If so, how?

92. Dave is determined to solve the equation $3x + 4 = -11$ by first using the multiplication principle to "eliminate" the 3. How should he proceed and why?

Solve. If an equation is an identity or a contradiction (see p. 76), state this.

93. $8.43x - 2.5(3.2 - 0.7x) = -3.455x + 9.04$

94. $0.008 + 9.62x - 42.8 = 0.944x + 0.0083 - x$

95. $-2[3(x - 2) + 4] = 4(5 - x) - 2x$

96. $0 = y - (-14) - (-3y)$

97. $3(x + 4) = 3(4 + x)$

98. $5(x - 7) = 3(x - 2) + 2x$

99. $2x(x + 5) - 3(x^2 + 2x - 1) = 9 - 5x - x^2$

100. $x(x - 4) = 3x(x + 1) - 2(x^2 + x - 5)$

101. $9 - 3x = 2(5 - 2x) - (1 - 5x)$

102. $2(7 - x) - 20 = 7x - 3(2 + 3x)$

Aha! **103.** $[7 - 2(8 \div (-2))]x = 0$

104. $\dfrac{x}{14} - \dfrac{5x + 2}{49} = \dfrac{3x - 4}{7}$

105. $\dfrac{5x + 3}{4} + \dfrac{25}{12} = \dfrac{5 + 2x}{3}$

CORNER

Step-by-Step Solutions

Focus: Solving linear equations

Time: 20 minutes

Group size: 3

In general, there is more than one correct sequence of steps for solving an equation. This makes it important that you write your steps clearly and logically so that others can follow your approach.

ACTIVITY

1. Each group member should select a different one of the following equations and, on a fresh sheet of paper, perform the first step of the solution.

$4 - 3(x - 3) = 7x + 6(2 - x)$

$5 - 7[x - 2(x - 6)] = 3x + 4(2x - 7) + 9$

$4x - 7[2 + 3(x - 5) + x] = 4 - 9(-3x - 19)$

2. Pass the papers around so that the second and third steps of each solution are performed by the other two group members. Before writing, make sure that the previous step is correct. If a mistake is discovered, return the problem to the person who made the mistake for repairs. Continue passing the problems around until all equations have been solved.

3. Each group should reach a consensus on what the three solutions are and then compare their answers to those of other groups.

COLLABORATIVE

Formulas

2.3

Evaluating Formulas • Solving for a Letter

Many applications of mathematics involve relationships among two or more quantities. An equation that represents such a relationship will use two or more letters and is known as a **formula**. Although most of the letters in this book represent variables, some—like c in $E = mc^2$ or π in $C = \pi d$—represent constants.

Evaluating Formulas

E x a m p l e   1

Sometimes we may wish to recall and modify a calculation. For example, suppose that after calculating $30 \cdot 1800$ we wish to find $30 \cdot 1870$. Pressing **2nd** **ENTRY** gives the following

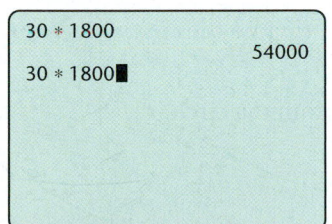

```
30 * 1800
                54000
30 * 1800■
```

Moving the cursor left, we change 1800 to 1870 and press **ENTER**.

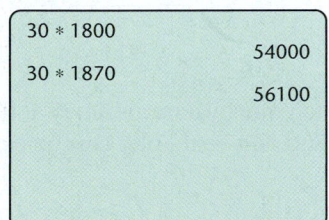

```
30 * 1800
                54000
30 * 1870
                56100
```

1. Verify the work above and then use **2nd** **ENTRY** to find $39 \cdot 1870$.

Furnace output. Contractors in the Northeast use the formula $B = 30a$ to determine the minimum furnace output B, in British thermal units (Btu's), for a well-insulated house with a square feet of flooring (*Source*: U.S. Department of Energy). Determine the minimum furnace output for an 1800-ft^2 house that is well insulated.

Solution   We substitute 1800 for a and calculate B:

$$B = 30a = 30(1800) = 54{,}000.$$

The furnace should be able to provide at least 54,000 Btu's of heat.

Solving for a Letter

Suppose that a contractor has an extra furnace and wants to determine the size of the largest (well-insulated) house in which it can be used. The contractor can substitute the amount of the furnace's output in Btu's—say, 63,000—for B, and then solve for a:

$$63{,}000 = 30a \qquad \text{Replacing } B \text{ with 63,000}$$

$$2100 = a. \qquad \text{Dividing both sides by 30:}$$

$$\begin{array}{r} 2\ 100 \\ 30\overline{)63{,}000} \\ 60 \\ \hline 3\ 0 \\ 3\ 0 \\ \hline 000 \end{array}$$

Were these calculations to be performed for a variety of furnaces, the contractor would find it easier to first solve $B = 30a$ for a, and *then* substitute values for B. This can be done in much the same way that we solved equations in Sections 2.1 and 2.2.

E x a m p l e   2

Solve for a: $B = 30a$.

Solution   We have

$$B = 30a \qquad \text{We want this letter alone.}$$

$$\frac{B}{30} = a. \qquad \text{Dividing both sides by 30}$$

The equation $a = B/30$ gives a quick, easy way to determine the floor area of the largest (well-insulated) house that a furnace with B Btu's could heat.

To see how the addition and multiplication principles apply to formulas, compare the following. In (A), we solve as usual; in (B), we do not simplify; and in (C), we *cannot* simplify since a, b, and c are unknown.

A. $5x + 2 = 12$

$5x = 12 - 2$

$5x = 10$

$x = \dfrac{10}{5} = 2$

B. $5x + 2 = 12$

$5x = 12 - 2$

$x = \dfrac{12 - 2}{5}$

C. $ax + b = c$

$ax = c - b$

$x = \dfrac{c - b}{a}$

E x a m p l e   3

Circumference of a circle.   The formula $C = 2\pi r$ gives the *circumference C* of a circle with radius r. Solve for r.

Solution   The **circumference** is the distance around a circle.

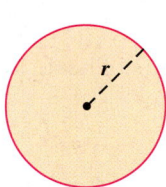

Given a radius r, we can use this equation to find a circle's circumference C. _____

Given a circle's circumference C, we can use this equation to find the radius r. _____

$$C = 2\pi r \qquad \text{We want this letter alone.}$$

$$\frac{C}{2\pi} = \frac{2\pi r}{2\pi} \qquad \text{Dividing both sides by } 2\pi$$

$$\frac{C}{2\pi} = r$$

E x a m p l e   4

Nutrition.   The number of calories K needed each day by a moderately active woman who weighs w pounds, is h inches tall, and is a years old, can be estimated using the formula

$$K = 917 + 6(w + h - a).*$$

Solve for h.

*Based on information from M. Parker (ed.), *She Does Math*! (Washington DC: Mathematical Association of America, 1995), p. 96.

Study Tip

Solution   We reverse the order in which the operations occur on the right side:

We want h alone.

$$K = 917 + 6(w + h - a)$$

$$K - 917 = 6(w + h - a) \qquad \text{Subtracting 917 from both sides}$$

$$\frac{K - 917}{6} = w + h - a \qquad \text{Dividing both sides by 6}$$

$$\frac{K - 917}{6} + a - w = h. \qquad \text{Adding } a \text{ and subtracting } w \text{ on both sides}$$

This formula can be used to estimate a woman's height, if we know her age, weight, and caloric needs.

The above steps are similar to those used in Section 2.2 to solve equations. We use the addition and multiplication principles just as before. The main difference is the need to factor when combining like terms.

To Solve a Formula for a Given Letter

1. If the letter for which you are solving appears in a fraction, use the multiplication principle to clear fractions.
2. Get all terms with the letter for which you are solving on one side of the equation and all other terms on the other side.
3. Combine like terms, if necessary. This may require factoring.
4. Multiply or divide to solve for the letter in question.

E x a m p l e   5

Solve for x: $y = ax + bx - 4$.

Solution   We solve as follows:

$$y = ax + bx - 4 \qquad \text{We want this letter alone.}$$

$$y + 4 = ax + bx \qquad \text{Adding 4 to both sides}$$

$$y + 4 = x(a + b) \qquad \text{Combining like terms by factoring out } x$$

$$\frac{y + 4}{a + b} = x. \qquad \text{Dividing both sides by } a + b, \text{ or multiplying both sides by } 1/(a + b)$$

We can also write this as

$$x = \frac{y + 4}{a + b}.$$

Caution! Had we performed the following steps in Example 5, we would *not* have solved for x:

$$y = ax + bx - 4$$

$$y - ax + 4 = bx \qquad \text{Subtracting } ax \text{ and adding 4 to both sides}$$

Two occurrences of x

$$\frac{y - ax + 4}{b} = x. \qquad \text{Dividing both sides by } b$$

The mathematics of each step is correct, but since x occurs on both sides of the formula, *we have not solved the formula for x.* Remember that the letter being solved for should be alone on one side of the equation, with no occurrence of that letter on the other side!

FOR EXTRA HELP

Exercise Set 2.3

 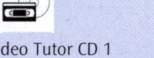 Digital Video Tutor CD 1 Videotape 3

 InterAct Math

  Math Tutor Center

 MathXL

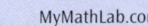 MyMathLab.com

1. *Distance from a storm.* The formula $M = \frac{1}{5}t$ can be used to determine how far M, in miles, you are from lightning when its thunder takes t seconds to reach your ears. If it takes 10 sec for the sound of thunder to reach you after you have seen the lightning, how far away is the storm?

2. *Electrical power.* The power rating P, in watts, of an electrical appliance is determined by

$$P = I \cdot V,$$

where I is the current, in amperes, and V is the voltage, measured in volts. If a kitchen requires 30 amps of current and the voltage in the house is 115 volts, what is the wattage of the kitchen?

3. *College enrollment.* At many colleges, the number of "full-time-equivalent" students f is given by

$$f = \frac{n}{15},$$

where n is the total number of credits for which students have enrolled in a given semester. Determine the number of full-time-equivalent students on a campus in which students registered for a total of 21,345 credits.

4. *Surface area of a cube.* The surface area A of a cube with side s is given by

$$A = 6s^2.$$

Find the surface area of a cube with sides of 3 in.

5. *Calorie density.* The calorie density D, in calories per ounce, of a food that contains c calories and weighs w ounces is given by

$$D = \frac{c}{w}.^*$$

Eight ounces of fat-free milk contains 84 calories. Find the calorie density of fat-free milk.

6. *Wavelength of a musical note.* The wavelength w, in meters per cycle, of a musical note is given by

$$w = \frac{r}{f},$$

where r is the speed of the sound, in meters per second, and f is the frequency, in cycles per second. The speed of sound in air is 344 m/sec. What is the

---

*Source: *Nutrition Action Healthletter*, March 2000, p. 9. Center for Science in the Public Interest, Suite 300; 1875 Connecticut Ave NW, Washington, D.C. 20008.

wavelength of a note whose frequency in air is 24 cycles per second?

7. *Absorption of ibuprofen.*   When 400 mg of the painkiller ibuprofen is swallowed, the number of milligrams n in the bloodstream t hours later (for $0 \le t \le 6$) is estimated by

$$n = 0.5t^4 + 3.45t^3 - 96.65t^2 + 347.7t.$$

How many milligrams of ibuprofen remain in the blood 1 hr after 400 mg has been swallowed?

8. *Size of a league schedule.*   When all n teams in a league play every other team twice, a total of N games are played, where

$$N = n^2 - n.$$

If a soccer league has 7 teams and all teams play each other twice, how many games are played?

Solve each formula for the indicated letter.

9. $A = bh$, for b
(Area of parallelogram with base b and height h)

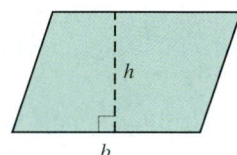

10. $A = bh$, for h

11. $d = rt$, for r
(A distance formula, where d is distance, r is speed, and t is time)

12. $d = rt$, for t

13. $I = Prt$, for P
(Simple-interest formula, where I is interest, P is principal, r is interest rate, and t is time)

14. $I = Prt$, for t

15. $H = 65 - m$, for m
(To determine the number of heating degree days H for a day with m degrees Fahrenheit as the average temperature)

16. $d = h - 64$, for h
(To determine how many inches d above average an h-inch-tall woman is)

17. $P = 2l + 2w$, for l
(Perimeter of a rectangle of length l and width w)

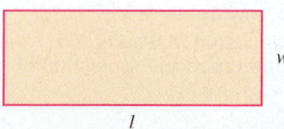

18. $P = 2l + 2w$, for w

19. $A = \pi r^2$, for π
(Area of a circle with radius r)

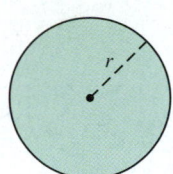

20. $A = \pi r^2$, for r^2

21. $A = \frac{1}{2}bh$, for h
(Area of a triangle with base b and height h)

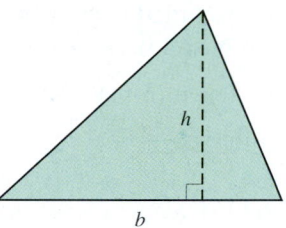

22. $A = \frac{1}{2}bh$, for b

23. $E = mc^2$, for m
(A relativity formula from physics)

24. $E = mc^2$, for c^2

25. $Q = \dfrac{c + d}{2}$, for d

26. $Q = \dfrac{p - q}{2}$, for p

27. $A = \dfrac{a + b + c}{3}$, for b

28. $A = \dfrac{a + b + c}{3}$, for c

29. $M = \dfrac{A}{s}$, for A
(To compute the Mach number M for speed A and speed of sound s)

30. $P = \dfrac{ab}{c}$, for b

31. $A = at + bt$, for t

32. $S = rx + sx$, for x

33. *Area of a trapezoid.*   The formula

$$A = \tfrac{1}{2}ah + \tfrac{1}{2}bh$$

can be used to find the area A of a trapezoid with bases a and b and height h. Solve for h. (*Hint:* First clear fractions.)

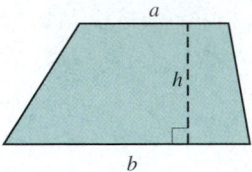

34. *Compounding interest.*   The formula

$$A = P + Prt$$

is used to find the amount A in an account when simple interest is added to an investment of P dollars (see Exercise 13). Solve for P.

35. *Chess rating.*   The formula

$$R = r + \frac{400(W - L)}{N}$$

is used to establish a chess player's rating R after that player has played N games, won W of them, and lost L of them. Here r is the average rating of the opponents (*Source*: The U.S. Chess Federation). Solve for L.

36. *Angle measure.*   The angle measure S, of a sector of a circle, is given by

$$S = \frac{360A}{\pi r^2},$$

where r is the radius, A is the area of the sector, and S is in degrees. Solve for r^2.

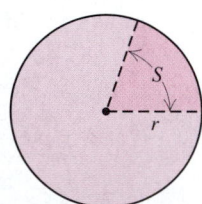

37. Naomi has a formula that allows her to convert Celsius temperatures to Fahrenheit temperatures. She needs a formula for converting Fahrenheit temperatures to Celsius temperatures. What advice can you give her?

38. Under what circumstances would it be useful to solve $d = rt$ for r? (See Exercise 11.)

SKILL MAINTENANCE

Multiply.

Aha! **39.** $0.79(38.4)0$ **40.** $(0.085)(108)$

Simplify.

41. $20 \div (-4) \cdot 2 - 3$ **42.** $5|8 - (2 - 7)|$

SYNTHESIS

43. The equations

$$P = 2l + 2w \quad \text{and} \quad w = \frac{P}{2} - l$$

are equivalent formulas involving the perimeter P, length l, and width w of a rectangle. Devise a problem for which the second of the two formulas would be more useful.

44. Describe a circumstance for which the answer to Exercise 34 would be useful.

45. The number of calories K needed each day by a moderately active man who weighs w kilograms, is h centimeters tall, and is a years old, can be determined by

$$K = 19.18w + 7h - 9.52a + 92.4.*$$

If Janos is moderately active, weighs 82 kg, is 185 cm tall, and needs to consume 2627 calories a day, how old is he?

46. *Altitude and temperature.*   Air temperature drops about 1° Celsius (C) for each 100-m rise above ground level, up to 12 km (*Source: A Sourcebook of School Mathematics*, Mathematical Association of America, 1980). If the ground level temperature is t°C, find a formula for the temperature T at an elevation of h meters.

47. *Dosage size.*   Clark's rule for determining the size of a particular child's medicine dosage c is

$$c = \frac{w}{a} \cdot d,$$

where w is the child's weight, in pounds, and d is the usual adult dosage for an adult weighing a pounds. (*Source*: Olsen, June Looby, et al., *Medical Dosage Calculations*. Redwood City, CA: Addison Wesley, 1995). Solve for a.

---

*Based on information from M. Parker (ed.), *She Does Math!* (Washington DC: Mathematical Association of America, 1995), p. 96.

48. *Weight of a fish.*   An ancient fisherman's formula for estimating the weight of a fish is

$$w = \frac{lg^2}{800},$$

where w is the weight, in pounds, l is the length, in inches, and g is the girth (distance around the midsection), in inches. Estimate the girth of a 700-lb yellow tuna that is 8 ft long.

Solve each formula for the given letter.

49. $\dfrac{y}{z} \div \dfrac{z}{t} = 1$, for y

50. $ac = bc + d$, for c

51. $qt = r(s + t)$, for t

52. $3a = c - a(b + d)$, for a

53. *Furnace output.*   The formula

$$B = 50a$$

is used in New England to estimate the minimum furnace output B, in Btu's, for an old, poorly insulated house with a square feet of flooring. Find an equation for determining the number of Btu's saved by insulating an old house. (*Hint*: See Example 1.)

54. Revise the formula in Example 4 so that a woman's weight in kilograms (2.2046 lb = 1 kg) and her height in centimeters (0.3937 in. = 1 cm) are used.

55. Revise the formula in Exercise 45 so that a man's weight in pounds (2.2046 lb = 1 kg) and his height in inches (0.3937 in. = 1 cm) are used.

Applications with Percent

2.4

Converting Between Percent Notation and Decimal Notation •
Solving Percent Problems

Recently Middlesex Toy and Hobby installed a new cash register and the sales clerks inadvertently set up the machine to print out "totals" on each receipt without separating each transaction into "merchandise" and the five-percent "sales tax." For tax purposes, the shop needs a formula for separating each total into the amount spent on merchandise and the amount spent on tax. Before developing such a formula, we need to review the basics of percent problems.

Converting Between Percent Notation and Decimal Notation

Nutritionists recommend that no more than 30% of the calories in a person's diet come from fat. This means that of every 100 calories consumed, no more than 30 should come from fat. Thus, 30% is a ratio of 30 to 100.

Calories consumed

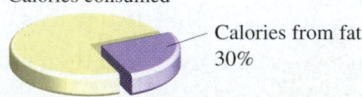

Calories from fat
30%

The percent symbol % means "per hundred." We can regard the percent symbol as part of a name for a number. For example,

$$30\% \quad \text{is defined to mean} \quad \frac{30}{100}, \quad \text{or} \quad 30 \times \frac{1}{100}, \quad \text{or} \quad 30 \times 0.01.$$

Percent Notation

$$n\% \quad \text{means} \quad \frac{n}{100}, \quad \text{or} \quad n \times \frac{1}{100}, \quad \text{or} \quad n \times 0.01.$$

E x a m p l e   1

Convert to decimal notation: **(a)** 78%; **(b)** 1.3%.

Solution

a) $78\% = 78 \times 0.01$ Replacing % with ×0.01
     $= 0.78$

b) $1.3\% = 1.3 \times 0.01$ Replacing % with ×0.01
     $= 0.013$

As shown above, multiplication by 0.01 simply moves the decimal point two places to the left.

To convert from percent notation to decimal notation, move the decimal point two places to the left and drop the percent symbol.

E x a m p l e   2

Convert 43.67% to decimal notation.

Solution

     43.67% $0.43.67$ $43.67\% = 0.4367$

Move the decimal point two places to the left.

The procedure used in Examples 1 and 2 can be reversed:

     $0.38 = 38 \times 0.01$

         $= 38\%.$ Replacing ×0.01 with %

To convert from decimal notation to percent notation, move the decimal point two places to the right and write a percent symbol.

E x a m p l e   3

Convert to percent notation: **(a)** 1.27; **(b)** $\frac{1}{4}$; **(c)** 0.3.

Solution

a) We first move the decimal point
two places to the right: $1.27.$

and then write a % symbol: 127% This is the same as multiply-
                                              ing 1.27 by 100 and writing %

b) Note that $\frac{1}{4} = 0.25$. We move the decimal point two places to the right: 0.25.

and then write a % symbol: 25%   Multiplying by 100 and writing %

c) We first move the decimal point two places to the right (recall that $0.3 = 0.30$): 0.30.

and then write a % symbol: 30%   Multiplying by 100 and writing %

Solving Percent Problems

To solve problems involving percents, we translate to mathematical language and then solve an equation.

E x a m p l e   4   What is 11% of 49?

Solution

Translate: What is 11% of 49?

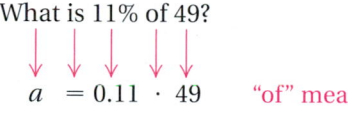

$a = 0.11 \cdot 49$ "of" means multiply; 11% = 0.11

$a = 5.39$

A way of checking answers is by estimating as follows:

$$11\% \times 49 \approx 10\% \times 50$$
$$= 0.10 \times 50 = 5.$$

Since 5 is close to 5.39, our answer is reasonable.

Thus, 5.39 is 11% of 49. The answer is 5.39.

E x a m p l e   5   3 is 16 percent of what?

Solution

Translate: 3 is 16 percent of what?

$3 = 0.16 \cdot y$

$\dfrac{3}{0.16} = y$ Dividing both sides by 0.16

$18.75 = y$

Thus, 3 is 16 percent of 18.75. The answer is 18.75.

Example 6

What percent of $50 is $16?

Solution

Translate: What percent of $50 is $16?

$$n \quad \cdot \quad 50 = 16$$

$$n = \frac{16}{50} \qquad \text{Dividing both sides by 50}$$

$$n = 0.32 = 32\% \qquad \text{Converting to percent notation}$$

Thus, 32% of $50 is $16. The answer is 32%.

Examples 4–6 represent the three basic types of percent problems.

Example 7

Retail sales.   Recently, receipts from Middlesex Toy and Hobby indicated the total amount paid (including tax), but not the price of the merchandise. Given that the sales tax was 5%, find the following.

a) The cost of the merchandise when the total read $31.50
b) A formula for the cost of the merchandise c when the total reads T dollars

Solution

a) When tax is added to the cost of an item, the customer actually pays more than 100% of the item's price. When sales tax is 5%, the total paid is 105% of the price of the merchandise. Thus if $c =$ the cost of the merchandise, we have

$31.50 is 105% of c

$$31.50 = 1.05 \cdot c$$

$$\frac{31.50}{1.05} = c \qquad \text{Dividing both sides by 1.05}$$

$$30 = c. \qquad \text{Simplifying}$$

The merchandise cost $30 before tax.

b) When the total is T dollars, we modify the approach used in part (a):

$$T = 1.05c$$

$$\frac{T}{1.05} = c. \qquad \text{Dividing both sides by 1.05}$$

As a check, note that when T is $31.50, we have $31.50 ÷ 1.05 = $30. Since this matches the result of part (a), our formula is probably correct.

The formula $c = T/1.05$ can be used to find the cost of the merchandise when the total T is known and the sales tax is 5%.

Exercise Set 2.4

Find decimal notation.

1. 82% **2.** 49% **3.** 9% **4.** 91.3%

5. 43.7% **6.** 2% **7.** 0.46% **8.** 4.8%

Find percent notation.

9. 0.29 **10.** 0.78 **11.** 0.998

12. 0.358 **13.** 1.92 **14.** 1.39

15. 2.1 **16.** 9.2 **17.** 0.0068

18. 0.0095 **19.** $\frac{3}{8}$ **20.** $\frac{3}{4}$

21. $\frac{7}{25}$ **22.** $\frac{4}{5}$ **23.** $\frac{2}{3}$ **24.** $\frac{5}{6}$

Solve.

25. What percent of 68 is 17?

26. What percent of 150 is 39?

27. What percent of 125 is 30?

28. What percent of 300 is 57?

29. 14 is 30% of what number?

30. 54 is 24% of what number?

31. 0.3 is 12% of what number?

32. 7 is 175% of what number?

33. What number is 35% of 240?

34. What number is 1% of one million?

35. What percent of 60 is 75?

Aha! **36.** What percent of 70 is 70?

37. What is 2% of 40?

38. What is 40% of 2?

Aha! **39.** 25 is what percent of 50?

40. 8 is 2% of what number?

41. *Student loans.* To finance her community college education, Sarah takes out a Stafford loan for $3500. After a year, Sarah decides to pay off the interest, which is 8% of $3500. How much will she pay?

42. *Student loans.* Paul takes out a subsidized federal Stafford loan for $2400. After a year, Paul decides to pay off the interest, which is 7% of $2400. How much will he pay?

43. *Votes for president.* In 2000, Al Gore received 48.62 million votes. This accounted for 48.36% of all votes cast. How many people voted in the 2000 presidential election?

44. *Lotteries and education.* In 1997, $6.2 billion of state lottery money was used for education. This accounted for 52% of all lottery proceeds for the year (*Source: Statistical Abstract of the United States, 1999*). What was the total amount of lottery proceeds in 1997? (Round to the nearest tenth of a billion.)

45. *Infant health.* In a study of 300 pregnant women with "poor" diets, 8% had babies in good or excellent health. How many women in this group had babies in good or excellent health?

46. *Infant health.* In a study of 300 pregnant women with "good-to-excellent" diets, 95% had babies in good or excellent health. How many women in this group had babies in good or excellent health?

47. *Nut consumption.* Each American consumes, on average, 2.25 lb of tree nuts each year (*Source: USA Today, 2/17/00*). Of this amount, 25% is almonds. How many pounds of almonds does the average American consume each year?

48. *Junk mail.* The U.S. Postal Service reports that we open and read 78% of the junk mail that we receive. A business sends out 9500 advertising brochures. How many of them can the business expect to be opened and read?

49. *Left-handed bowlers.* It has been determined by sociologists that 17% of the population is left-handed. Each week 160 bowlers enter a tournament conducted by the Professional Bowlers Association. How many would you expect to be left-handed? (Round to the nearest one.)

50. *Kissing and colds.* In a medical study, it was determined that if 800 people kiss someone else who has a cold, only 56 will actually catch the cold. What percent is this?

51. On a test of 88 items, a student got 76 correct. What percent were correct?

52. A baseball player had 13 hits in 25 times at bat. What percent were hits?

53. A bill at Officeland totaled $37.80. How much did the merchandise cost if the sales tax is 5%?

54. Doreen's checkbook shows that she wrote a check for $987 for building materials. What was the price of the materials if the sales tax is 5%?

55. *Deducting sales tax.*   A tax-exempt school group received a bill of $157.41 for educational software. The bill incorrectly included sales tax of 6%. How much should the school group pay?

56. *Deducting sales tax.*   A tax-exempt charity received a bill of $145.90 for a sump pump. The bill incorrectly included sales tax of 5%. How much does the charity owe?

57. *Cost of self-employment.*   Because of additional taxes and fewer benefits, it has been estimated that a self-employed person must earn 20% more than a non–self-employed person performing the same task(s). If Roy earns $15 an hour working for Village Copy, how much would he need to earn on his own for a comparable income?

58. Refer to Exercise 57. Clara earns $12 an hour working for Round Edge stairbuilders. How much would Clara need to earn on her own for a comparable income?

59. *Calorie content.*   Pepperidge Farm Light Style 7 Grain Bread® has 140 calories in a 3-slice serving. This is 15% less than the number of calories in a serving of regular bread. How many calories are in a serving of regular bread?

60. *Fat content.*   Peek Freans Shortbread Reduced Fat Cookies® contain 35 calories of fat in each serving. This is 40% less than the fat content in the leading imported shortbread cookie. How many calories of fat are in a serving of the leading shortbread cookie?

61. Campus Bookbuyers pays $30 for a book and sells it for $60. Is this a 100% markup or a 50% markup? Explain.

62. If Julian leaves a $12 tip for a $90 dinner, is he being generous, stingy, or neither? Explain.

Translate to an algebraic expression.

63. 5 more than some number

64. 4 less than Tino's weight

65. The product of 8 and twice *a*

66. 1 more than the product of two numbers

67. Does the following advertisement provide a convincing argument that summertime is when most burglaries occur? Why or why not?

68. Erin is returning a tent that she bought during a 25%-off storewide sale that has ended. She is offered store credit for 125% of what she paid (not to be used on sale items). Is this fair to Erin? Why or why not?

69. The community of Bardville has 1332 left-handed females. If 48% of the community is female and 15% of all females are left-handed, how many people are in the community?

70. Rollie's Music charges $11.99 for a compact disc. Sound Warp charges $13.99 but you have a coupon for $2 off. In both cases, a 7% sales tax is charged on the *regular* price. How much does the disc cost at each store?

71. The new price of a car is 25% higher than the old price. The old price is what percent lower than the new price?

72. Claude pays 26% of his pretax earnings in taxes. What percentage of his *post*-tax earnings is this?

 73. *U.S. birth rate.* There were 3.88 million births in 1997 and 3.94 million births in 1998 (*Source*: *Burlington Free Press,* page 2A, March 29, 2000). By what percentage did the number of births increase?

Aha! **74.** Would it be better to receive a 5% raise and then an 8% raise or the other way around? Why?

75. Herb is in the 30% tax bracket. This means that 30¢ of each dollar earned goes to taxes. Which would cost him the least: contributing $50 that is tax-deductible or contributing $40 that is not tax-deductible? Explain.

COLLABORATIVE

CORNER

Sales and Discounts

Focus: Applications and models using percent

Time: 15 minutes

Group size: 3

Materials: Calculators are optional.

Often a store will reduce the price of an item by a fixed percentage. When the sale ends, the items are returned to their original prices. Suppose a department store reduces all sporting goods 20%, all clothing 25%, and all electronics 10%.

ACTIVITY

1. Each group member should select one of the following items: a $50 basketball, an $80 jacket, or a $200 portable sound system. Fill in the first three columns of the first three rows of the chart below.

2. Apply the appropriate discount and determine the sale price of your item. Fill in the fourth column of the chart.

3. Next, find a multiplier that can be used to convert the sale price back to the original price and fill in the remaining column of the chart. Does this multiplier depend on the price of the item?

4. Working as a group, compare the results of part (3) for all three items. Then develop a formula for a multiplier that will restore a sale price to its original price, p, after a discount r has been applied. Complete the fourth row of the table and check that your formula will duplicate the results of part (3).

5. Use the formula from part (4) to find the multiplier that a store would use to return an item to its original price after a "30% off" sale expires. Fill in the last line on the chart.

6. Inspect the last column of your chart. How can these multipliers be used to determine the percentage by which a sale price is increased when a sale ends?

Original Price, p	Discount, r	$1 - r$	Sale Price	Multiplier to convert back to p
p	r	$1 - r$		
	.30			

Problem Solving

2.5

Five Steps for Problem Solving • Applying the Five Steps

Probably the most important use of algebra is as a tool for problem solving. In this section, we develop a problem-solving approach that will be used throughout the remainder of the text.

Five Steps for Problem Solving

In Section 2.4, we solved a problem in which Middlesex Toy and Hobby needed a formula. To solve the problem, we *familiarized* ourselves with percent notation so that we could then *translate* the problem into an equation. At the end of the section, we *solved* the equation, *checked* the solution, and *stated* the answer.

> ### Five Steps for Problem Solving in Algebra
> 1. *Familiarize* yourself with the problem.
> 2. *Translate* to mathematical language. (This often means writing an equation.)
> 3. *Carry out* some mathematical manipulation. (This often means *solving* an equation.)
> 4. *Check* your possible answer in the original problem.
> 5. *State* the answer clearly.

Of the five steps, the most important is probably the first one: becoming familiar with the problem. Here are some hints for familiarization.

> ### To Become Familiar with a Problem
> 1. Read the problem carefully. Try to visualize the problem.
> 2. Reread the problem, perhaps aloud. Make sure you understand all important words.
> 3. List the information given and the question(s) to be answered. Choose a variable (or variables) to represent the unknown and specify what the variable represents. For example, let L = length in centimeters, d = distance in miles, and so on.
> 4. Look for similarities between the problem and other problems you have already solved.
> 5. Find more information. Look up a formula in a book, at a library, or on-line. Consult a reference librarian or an expert in the field.
>
> *(continued)*

> **6.** Make a table that uses all the information you have available. Look for patterns that may help in the translation.
> **7.** Make a drawing and label it with known and unknown information, using specific units if given.
> **8.** Think of a possible answer and check the guess. Observe the manner in which the guess is checked.

Applying the Five Steps

E x a m p l e   1

Hiking. In 1998, at age 79, Earl Shaffer became the oldest person to hike all 2100 miles of the Appalachian Trail—from Springer Mountain, Georgia, to Mount Katahdin, Maine (*Source*: Appalachian Trail Conference). At one point, Shaffer stood atop Big Walker Mountain, Virginia, which is three times as far from the northern end as from the southern end. How far was Shaffer from each end of the trail?

Solution

1. Familiarize. It may be helpful to make a drawing.

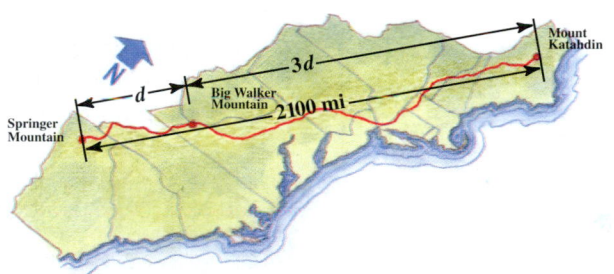

To gain some familiarity, let's suppose that Shaffer stood 600 mi from Springer Mountain. Three times 600 mi is 1800 mi. Since 600 mi + 1800 mi = 2400 mi and 2400 mi > 2100 mi, we see that our guess is too large. Rather than guess again, we let

d = the distance, in miles, to the southern end,

and

$3d$ = the distance, in miles, to the northern end.

(We could also let x = the distance to the northern end and $\frac{1}{3}x$ = the distance to the southern end.)

2. Translate. From the drawing, we see that the lengths of the two parts of the trail must add up to 2100 mi. This leads to our translation.

Rewording: Distance to southern end plus distance to northern end is 2100 mi

Translating: $\quad d \quad + \quad 3d \quad = \quad 2100$

3. **Carry out.** We solve the equation:

$$d + 3d = 2100$$

$$4d = 2100 \qquad \text{Combining like terms}$$

$$d = 525. \qquad \text{Dividing both sides by 4}$$

4. **Check.** As expected, d is less than 600 mi. If $d = 525$ mi, then $3d = 1575$ mi. Since 525 mi + 1575 mi = 2100 mi, we have a check.

5. **State.** Atop Big Walker Mountain, Shaffer stood 525 mi from Springer Mountain and 1575 mi from Mount Katahdin.

E x a m p l e   2

Page numbers. The sum of two consecutive page numbers is 305. Find the page numbers.

Solution

1. **Familiarize.** If the meaning of the word consecutive is unclear, we should consult a dictionary or someone who might know. Consecutive numbers are integers that are one unit apart. Thus, 18 and 19 are consecutive numbers, as are -24 and -23. Let's "guess and check": If the first page number is 40, the next would be 41. Since $40 + 41 = 81$ and $81 < 305$, our guess is much too small. Suppose the first page number is 130. The next page would then be 131. Since $130 + 131 = 261$ and $261 < 305$, our guess is still a bit too small. We could continue guessing, but algebra offers a more direct approach. Let's have

$$x = \text{the first page number}$$

and, since the two numbers must be one unit apart,

$$x + 1 = \text{the next page number.}$$

2. **Translate.** We reword the problem and translate as follows.

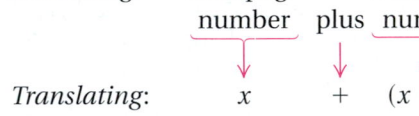

Rewording:	First page number	plus	next page number	is	305
Translating:	x	$+$	$(x + 1)$	$=$	305

3. **Carry out.** We solve the equation:

$$x + (x + 1) = 305$$

$$2x + 1 = 305 \qquad \text{Using an associative law and combining like terms}$$

$$2x = 304 \qquad \text{Subtracting 1 from both sides}$$

$$x = 152. \qquad \text{Dividing both sides by 2}$$

If x is 152, then $x + 1$ is 153.

4. **Check.** Our possible answers are 152 and 153. These are consecutive integers and their sum is 305, so the answers check in the original problem.

5. **State.** The page numbers are 152 and 153.

$\mathsf{Study\ Tip}$

Do not be surprised if your success rate drops some as you work through the exercises in this section. *This is normal.* Your success rate will increase as you gain experience with these types of problems and use some of the study tips already listed.

E x a m p l e   3

Taxi rates.   In Bermuda, a taxi ride costs $4.80 plus $1.68 for each mile traveled. Debbie and Alex have budgeted $18 for a taxi ride (excluding tip). How far can they travel on their $18 budget?

Solution

1. **Familiarize.**   Suppose the taxi takes Debbie and Alex 5 mi. Such a ride would cost $4.80 + 5($1.68), or $4.80 + $8.40 = $13.20. Since 13.20 < 18, we know that the actual answer exceeds 5 mi. Rather than guess again, we let

   s = the distance, in miles, driven by the taxi for $18.

2. **Translate.**   We rephrase the problem and translate.

   Rewording: The initial charge plus the mileage charge is $18.

   Translating: $4.80   + $s($1.68)   = $18

3. **Carry out.**   We solve as follows:

   $$4.80 + 1.68s = 18$$
   $$1.68s = 13.20 \qquad \text{Subtracting 4.80 from both sides}$$
   $$s = \frac{13.20}{1.68} \qquad \text{Dividing both sides by 1.68}$$
   $$s \approx 7.8. \qquad \text{Simplifying}$$

   In the work above, the symbol $\approx$ means _is approximately equal to_. We use it here because we rounded off our result.

4. **Check.**   A 7.8-mi taxi ride would cost $4.80 + 7.8($1.68), or $17.90. Since we rounded down when finding s, the check will not be precise.

5. **State.**   Debbie and Alex can take a 7.8-mi taxi ride and stay within their budget.

   The division in Example 3 would normally be rounded _up_ to 7.9. In the circumstances of the problem, however, this would lead to a taxi fare slightly in excess of $18. When solving problems, be careful whenever you round off to round in the appropriate direction.

E x a m p l e   4

Gardening.   A rectangular community garden is to be enclosed with 92 m of fencing. In order to allow for compost storage, the garden must be 4 m longer than it is wide. Determine the dimensions of the garden.

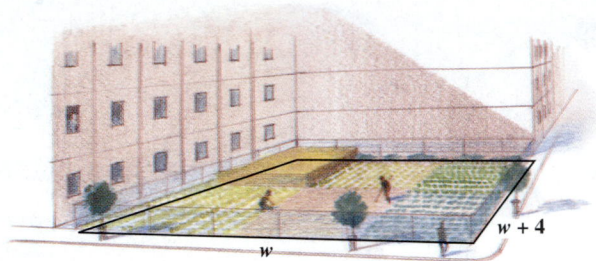

Solution

1. **Familiarize.**   Recall that the perimeter of a rectangle is twice the length plus twice the width. Suppose the garden were 30 m wide. The length would then be 30 + 4 m, or 34 m, and the perimeter would be 2 · 30 m + 2 · 34 m, or 128 m. This shows that for the perimeter to be 92 m, the width must be less than 30 m. Instead of guessing again, we let w = the width of the garden, in meters. Since the garden is "4 m longer than it is wide," we let $w + 4$ = the length of the garden, in meters.

2. **Translate.**   To translate, we use $w + 4$ as the length and 92 as the perimeter.

   Rewording: Twice the   twice the
                   length   plus   width   is 92 m.

   Translating: $2(w + 4)$ + $2w$ =   92   To double the length, $w + 4$, parentheses are essential.

3. **Carry out.**   We solve the equation:

   $$2(w + 4) + 2w = 92$$
   $$2w + 8 + 2w = 92 \qquad \text{Using the distributive law}$$
   $$4w + 8 = 92 \qquad \text{Combining like terms}$$
   $$4w = 84$$
   $$w = 21.$$

   The dimensions appear to be $w = 21$ m and l, or $w + 4$, = 25 m.

4. **Check.**   If the width is 21 m and the length 25 m, then the garden is 4 m longer than it is wide. The perimeter is 2(25 m) + 2(21 m), or 92 m, and since 92 m of fencing is available, we have a check.

5. **State.**   The garden should be 21 m wide and 25 m long.

> *Caution!*   Always be sure to answer the original problem completely.
> For instance, in Example 1 we need to find *two* numbers: the distances
> from *each* end of the trail to the hiker. Similarly, in Example 2 we needed
> to find two page numbers and in Example 4 we needed to find two
> dimensions, not just the width.

E x a m p l e   5

Selling a home.   The McCanns are planning to sell their home. If they want to
be left with $117,500 after paying 6% of the selling price to a realtor as a com-
mission, for how much must they sell the house?

Solution

1. **Familiarize.**   Suppose the McCanns sell the house for $120,000. A 6%
   commission can be determined by finding 6% of $120,000:

   6% of $120,000 = 0.06($120,000) = $7200.

   Subtracting this commission from $120,000 would leave the McCanns with

   $120,000 − $7200 = $112,800.

   This shows that in order for the McCanns to clear $117,500, the house must
   sell for more than $120,000. To determine what the sale price must be, we
   could check more guesses. Instead, we let x = the selling price, in dollars.
   With a 6% commission, the realtor would receive $0.06x$.

2. **Translate.**   We reword the problem and translate as follows.

   Rewording: Selling price less commission is amount remaining.

   Translating: x − $0.06x$ =   117,500

3. **Carry out.** We solve the equation:

$$x - 0.06x = 117{,}500$$

$$1x - 0.06x = 117{,}500$$

$$0.94x = 117{,}500 \qquad \text{Combining like terms. Had we noted that after the commission has been paid, 94\% remains, we could have begun with this equation.}$$

$$x = \frac{117{,}500}{0.94} \qquad \text{Dividing both sides by 0.94}$$

$$x = 125{,}000.$$

4. **Check.** To check, we first find 6% of $125,000:

$$6\% \text{ of } \$125{,}000 = 0.06(\$125{,}000) = \$7500. \qquad \text{This is the commission.}$$

Next, we subtract the commission to find the remaining amount:

$$\$125{,}000 - \$7500 = \$117{,}500.$$

Since, after the commission, the McCanns are left with $117,500, our answer checks. Note that the $125,000 sale price is greater than $120,000, as predicted in the *Familiarize* step.

5. **State.** To be left with $117,500, the McCanns must sell the house for $125,000.

Example 6

Angles in a triangle. The second angle of a triangle is 20° greater than the first. The third angle is twice as large as the first. How large are the angles?

Solution

1. **Familiarize.** We make a drawing. In this case, the measure of the first angle is x, the measure of the second angle is $x + 20$, and the measure of the third angle is $2x$.

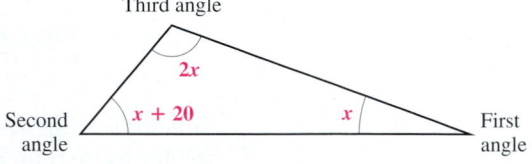

2. **Translate.** To translate, we need to recall that the sum of the measures of the angles in a triangle is 180°.

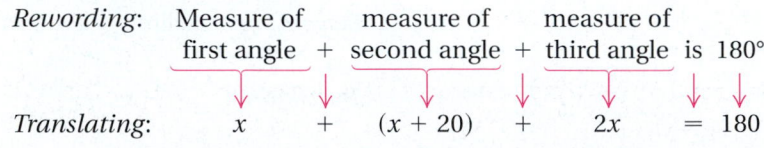

3. **Carry out.**   We solve:

$$x + (x + 20) + 2x = 180$$
$$4x + 20 = 180$$
$$4x = 160$$
$$x = 40.$$

The measures for the angles appear to be:

First angle: $x = 40°,$
Second angle: $x + 20 = 40 + 20 = 60°,$
Third angle: $2x = 2(40) = 80°.$

4. **Check.**   Consider 40°, 60°, and 80°. The second angle is 20° greater than the first, the third is twice the first, and the sum is 180°. These numbers check.

5. **State.**   The measures of the angles are 40°, 60°, and 80°.

We close this section with some tips to aid you in problem solving.

Problem-Solving Tips

1. The more problems you solve, the more your skills will improve.
2. Look for patterns when solving problems. Each time you study an example in a text, you may observe a pattern for problems that you will encounter later in the exercise sets or in other practical situations.
3. When translating in mathematics, consider the dimensions of the variables and constants in the equation. The variables that represent length should all be in the same unit, those that represent money should all be in dollars or all in cents, and so on.
4. Make sure that units appear in the answer whenever appropriate and that you have completely answered the original problem.

FOR EXTRA HELP

Exercise Set **2.5**

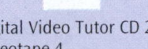
Digital Video Tutor CD 2
Videotape 4

InterAct Math

Math Tutor Center

MathXL

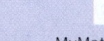
MyMathLab.com

Solve. Even though you might find the answer quickly in some other way, practice using the five-step problem-solving process.

1. Three less than twice a number is 19. What is the number?

2. Two fewer than ten times a number is 78. What is the number?

3. Five times the sum of 3 and some number is 70. What is the number?

4. Twice the sum of 4 and some number is 34. What is the number?

5. *Price of sneakers.*   Amy paid $63.75 for a pair of New Balance 903 running shoes during a 15%-off sale. What was the regular price?

6. *Price of a CD player.*   Doug paid $72 for a shockproof portable CD player during a 20%-off sale. What was the regular price?

7. *Price of a textbook.*   Evelyn paid $89.25, including 5% tax, for her biology textbook. How much did the book itself cost?

8. *Price of a printer.*   Jake paid $100.70, including 6% tax, for a color printer. How much did the printer itself cost?

9. *Running.*   In 1997, Yiannis Kouros of Australia set the record for the greatest distance run in 24 hr by running 188 mi (*Source*: *Guinness World Records 2000 Millennium Edition*). After 8 hr, he was approximately twice as far from the finish line as he was from the start. How far had he run?

10. *Sled-dog racing.*   The Iditarod sled-dog race extends for 1049 mi from Anchorage to Nome. If a musher is twice as far from Anchorage as from Nome, how many miles has the musher traveled?

11. The sum of three consecutive page numbers is 60. Find the numbers.

12. The sum of three consecutive page numbers is 99. Find the numbers.

13. The sum of two consecutive odd numbers is 60. Find the numbers. (*Hint*: Odd numbers, like even numbers, are separated by two units.)

14. The sum of two consecutive odd integers is 108. What are the integers?

15. The sum of two consecutive even integers is 126. What are the integers?

16. The sum of two consecutive even numbers is 50. Find the numbers.

17. *Oldest groom.*   The world's oldest groom was 19 yr older than his bride (*Source*: *Guinness World Records 2000 Millennium Edition*). Together, their ages totaled 187 yr. How old were the bride and the groom?

18. *Oldest divorcees.*   In the world's oldest divorcing couple, the woman was 6 yr younger than the man (*Source*: *Guinness World Records 2000 Millennium Edition*). Together, their ages totaled 188 yr. How old were the man and the woman?

19. *Angles of a triangle.*   The second angle of a triangle is three times as large as the first. The third angle is 30° more than the first. Find the measure of each angle.

20. *Angles of a triangle.*   The second angle of a triangle is four times as large as the first. The third angle is 45° less than the sum of the other two angles. Find the measure of each angle.

21. *Angles of a triangle.*   The second angle of a triangle is three times as large as the first. The third angle is 10° more than the sum of the other two angles. Find the measure of the third angle.

22. *Angles of a triangle.*   The second angle of a triangle is four times as large as the first. The third angle is 5° more than the sum of the other two angles. Find the measure of the second angle.

23. *Page numbers.*   The sum of the page numbers on the facing pages of a book is 385. What are the page numbers?

24. *Page numbers.*   The sum of the page numbers on the facing pages of a book is 281. What are the page numbers?

25. *Perimeter of a triangle.*   The perimeter of a triangle is 195 mm. If the lengths of the sides are consecutive odd integers, find the length of each side.

26. *Hancock Building dimensions.*   The top of the John Hancock Building in Chicago is a rectangle whose length is 60 ft more than the width. The perimeter is 520 ft. Find the width and the length of the rectangle. Find the area of the rectangle.

27. *Dimensions of a state.*   The perimeter of the state of Wyoming is 1280 mi. The width is 90 mi less than the length. Find the width and the length.

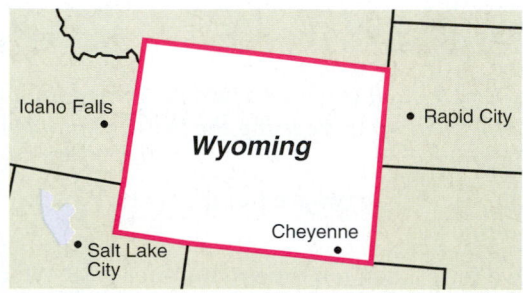

28. *Copier paper.*   The perimeter of standard-size copier paper is 99 cm. The width is 6.3 cm less than the length. Find the length and the width.

29. *Stock prices.*   Sarah's investment in America Online stock grew 28% to $448. How much did she invest?

30. *Savings interest.*   Sharon invested money in a savings account at a rate of 6% simple interest. After 1 yr, she has $6996 in the account. How much did Sharon originally invest?

31. *Credit cards.*   The balance in Will's Mastercard® account grew 2%, to $870, in one month. What was his balance at the beginning of the month?

32. *Loan interest.*   Alvin borrowed money from a cousin at a rate of 10% simple interest. After 1 yr, $7194 paid off the loan. How much did Alvin borrow?

33. *Taxi fares.*   In Beniford, taxis charge $3 plus 75¢ per mile for an airport pickup. How far from the airport can Courtney travel for $12?

34. *Taxi fares.*   In Cranston, taxis charge $4 plus 90¢ per mile for an airport pickup. How far from the airport can Ralph travel for $17.50?

35. *Truck rentals.*   Truck-Rite Rentals rents trucks at a daily rate of $49.95 plus 39¢ per mile. Concert Productions has budgeted $100 for renting a truck to haul equipment to an upcoming concert. How far can they travel in one day and stay within their budget?

36. *Truck rentals.*   Fine Line Trucks rents an 18-ft truck for $42 plus 35¢ per mile. Judy needs a truck for one day to deliver a shipment of plants. How far can she drive and stay within a budget of $70?

37. *Complementary angles.*   The sum of the measures of two *complementary* angles is 90°. If one angle measures 15° more than twice the measure of its complement, find the measure of each angle.

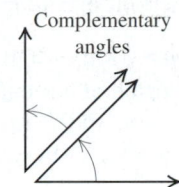

Complementary angles

38. *Supplementary angles.*   The sum of the measures of two *supplementary* angles is 180°. If one angle measures 45° less than twice the measure of its supplement, find the measure of each angle.

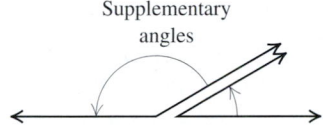

Supplementary angles

39. *Cricket chirps and temperature.*   The equation $T = \frac{1}{4}N + 40$ can be used to determine the temperature T, in degrees Fahrenheit, given the number of times N a cricket chirps per minute. Determine the number of chirps per minute for a temperature of 80°F.

40. *Race time.*   The equation $R = -0.028t + 20.8$ can be used to predict the world record in the 200-m dash, where R is the record in seconds and t is the number of years since 1920. In what year will the record be 18.0 sec?

41. Sean claims he can solve most of the problems in this section by guessing. Is there anything wrong with this approach? Why or why not?

42. When solving Exercise 17, Beth used a to represent the bride's age and Ben used a to represent the groom's age. Is one of these approaches preferable to the other? Why or why not?

SKILL MAINTENANCE

Write a true sentence using either $<$ or $>$.

43. -9 ▮ 5 **44.** 1 ▮ 3

45. -4 ▮ 7 **46.** -9 ▮ -12

SYNTHESIS

47. Write a problem for a classmate to solve. Devise it so that the problem can be translated to the equation $x + (x + 2) + (x + 4) = 375$.

48. Write a problem for a classmate to solve. Devise it so that the solution is "Audrey can drive the rental truck for 50 mi without exceeding her budget."

49. *Discounted dinners.* Kate's "Dining Card" entitles her to $10 off the price of a meal after a 15% tip has been added to the cost of the meal. If, after the discount, the bill is $32.55, how much did the meal originally cost?

50. *Test scores.* Pam scored 78 on a test that had 4 fill-ins worth 7 points each and 24 multiple-choice questions worth 3 points each. She had one fill-in wrong. How many multiple-choice questions did Pam get right?

51. *Gettysburg Address.* Abraham Lincoln's 1863 Gettysburg Address refers to the year 1776 as "four *score* and seven years ago." Determine what a score is.

52. One number is 25% of another. The larger number is 12 more than the smaller. What are the numbers?

53. *Perimeter of a rectangle.* The width of a rectangle is three fourths of the length. The perimeter of the rectangle becomes 50 cm when the length and the width are each increased by 2 cm. Find the length and the width.

54. *Angles in a quadrilateral.* The measures of the angles in a quadrilateral are consecutive odd numbers. Find the measure of each angle.

55. *Angles in a pentagon.* The measures of the angles in a pentagon are consecutive even numbers. Find the measure of each angle.

56. *Sharing fruit.* Apples are collected in a basket for six people. One third, one fourth, one eighth, and one fifth of the apples are given to four people, respectively. The fifth person gets ten apples, and one apple remains for the sixth person. Find the original number of apples in the basket.

57. *Discounts.* In exchange for opening a new credit account, Filene's Department Stores® subtracts 10% from all purchases made the day the account is established. Julio is opening an account and has a coupon for which he receives 10% off the first day's reduced price of a camera. If Julio's final price is $77.75, what was the price of the camera before the two discounts?

58. *Winning percentage.* In a basketball league, the Falcons won 15 of their first 20 games. In order to win 60% of the total number of games, how many more games will they have to play, assuming they win only half of the remaining games?

59. *Music-club purchases.* During a recent sale, BMG Music Service® charged $8.49 for the first CD ordered and $3.99 for all others. For shipping and handling, BMG charged $2.47 for the first CD, $2.28 for the second CD, and $1.99 for all others. The total cost of a shipment (excluding tax) was $65.07. How many CD's were in the shipment?

60. *Test scores.* Ella has an average score of 82 on three tests. Her average score on the first two tests is 85. What was the score on the third test?

61. *Taxi fares.* In New York City, a taxi ride costs $2 plus 30¢ per $\frac{1}{5}$ mile and 20¢ per minute stopped in traffic. Due to traffic, Glenda's taxi took 20 min to complete what is usually a 10-min drive. If she is charged $13 for the ride, how far did Glenda travel?

62. A school purchases a piano and must choose between paying $2000 at the time of purchase or $2150 at the end of one year. Which option should the school select and why?

Aha! **63.** Annette claims the following problem has no solution: "The sum of the page numbers on facing pages is 191. Find the page numbers." Is she correct? Why or why not?

64. The perimeter of a rectangle is 101.74 cm. If the length is 4.25 cm longer than the width, find the dimensions of the rectangle.

65. The second side of a triangle is 3.25 cm longer than the first side. The third side is 4.35 cm longer than the second side. If the perimeter of the triangle is 26.87 cm, find the length of each side.

Solving Inequalities

2.6

Solutions of Inequalities • Graphs of Inequalities •
Solving Inequalities Using the Addition Principle •
Solving Inequalities Using the Multiplication Principle •
Using the Principles Together

Many real-world situations translate to *inequalities*. For example, a student might need to register for *at least* 12 credits; an elevator might be designed to hold *at most* 2000 pounds; a tax credit might be allowable for families with incomes of *less than* $25,000; and so on. Before solving applications of this type, we must adapt our equation-solving principles to the solving of inequalities.

Solutions of Inequalities

Recall from Section 1.4 that an inequality is a number sentence containing $>$ (is greater than), $<$ (is less than), $\geq$ (is greater than or equal to), or $\leq$ (is less than or equal to). Inequalities like

$$-7 > x, \qquad t < 5, \qquad 5x - 2 \geq 9, \quad \text{and} \quad -3y + 8 \leq -7$$

are true for some replacements of the variable and false for others.

E x a m p l e   1   Determine whether the given number is a solution of $x < 2$: **(a)** -3; **(b)** 2.

Solution

a) Since $-3 < 2$ is true, -3 is a solution.
b) Since $2 < 2$ is false, 2 is not a solution.

E x a m p l e   2   Determine whether the given number is a solution of $y \geq 6$: **(a)** 6; **(b)** -4.

Solution

a) Since $6 \geq 6$ is true, 6 is a solution.
b) Since $-4 \geq 6$ is false, -4 is not a solution.

Graphs of Inequalities

Because the solutions of inequalities like $x < 2$ are too numerous to list, it is helpful to make a drawing that represents all the solutions. The **graph** of an inequality is such a drawing. Graphs of inequalities in one variable can be drawn on a number line by shading all points that are solutions. Open dots are used to indicate endpoints that are *not* solutions and closed dots indicate endpoints that *are* solutions.

E x a m p l e   3

Graph each inequality: **(a)** $x < 2$; **(b)** $y \geq -3$; **(c)** $-2 < x \leq 3$.

Solution

a) The solutions of $x < 2$ are those numbers less than 2. They are shown on the graph by shading all points to the left of 2. The open dot at 2 and the shading to its left indicates that 2 is *not* part of the graph, but numbers like 1.2 and 1.99 are.

b) The solutions of $y \geq -3$ are shown on the number line by shading the point for -3 and all points to the right of -3. The closed dot at -3 indicates that -3 *is* part of the graph.

c) The inequality $-2 < x \leq 3$ is read "-2 is less than x *and* x is less than or equal to 3," or "x is greater than -2 *and* less than or equal to 3." To be a solution of $-2 < x \leq 3$, a number must be a solution of both $-2 < x$ *and* $x \leq 3$. The number 1 is a solution, as are -0.5, 1.9, and 3. The open dot indicates that -2 is *not* a solution, whereas the closed dot indicates that 3 *is* a solution. The other solutions are shaded.

Solving Inequalities Using the Addition Principle

Consider a balance similar to one that appears in Section 2.1. When one side of the balance holds more weight than the other, the balance tips in that direction. If equal amounts of weight are then added to or subtracted from both sides of the balance, the balance remains tipped in the same direction.

The balance illustrates the idea that when a number, such as 2, is added to (or subtracted from) both sides of a true inequality, such as $3 < 7$, we get another true inequality:

$$3 + 2 < 7 + 2, \quad \text{or} \quad 5 < 9.$$

Similarly, if we add -4 to both sides of $x + 4 < 10$, we get an *equivalent* inequality:

$$x + 4 + (-4) < 10 + (-4), \quad \text{or} \quad x < 6.$$

We say that $x + 4 < 10$ and $x < 6$ are **equivalent**, which means that both inequalities have the same solution set.

> **The Addition Principle for Inequalities**
>
> For any real numbers a, b, and c:
>
> $$a < b \text{ is equivalent to } a + c < b + c;$$
> $$a \leq b \text{ is equivalent to } a + c \leq b + c;$$
> $$a > b \text{ is equivalent to } a + c > b + c;$$
> $$a \geq b \text{ is equivalent to } a + c \geq b + c.$$

As with equations, our goal is to isolate the variable on one side.

E x a m p l e   4   Solve $x + 2 > 8$ and then graph the solution.

Solution   We use the addition principle, subtracting 2 from both sides:

$$x + 2 - 2 > 8 - 2 \qquad \textcolor{magenta}{\text{Subtracting 2 from, or adding } -2 \text{ to, both sides}}$$
$$x > 6.$$

From the inequality $x > 6$, we can determine the solutions easily. Any number greater than 6 makes $x > 6$ true and is a solution of that inequality as well as the inequality $x + 2 > 8$. The graph is as follows:

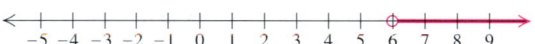

Because most inequalities have an infinite number of solutions, we cannot possibly check them all. A partial check can be made using one of the possible solutions. For this example, we can substitute any number greater than 6—say, 6.1—into the original inequality:

$$\frac{x + 2 > 8}{6.1 + 2 \; ? \; 8}$$
$$8.1 \mid 8 \quad \text{\small TRUE} \qquad \textcolor{magenta}{8.1 > 8 \text{ is a true statement.}}$$

Since $8.1 > 8$ is true, 6.1 is a solution. Any number greater than 6 is a solution.

Although the inequality $x > 6$ is easy to solve (we merely replace x with numbers greater than 6), it is worth noting that $x > 6$ is an *inequality*, not a *solution*. In fact, the solutions of $x > 6$ are numbers. To describe the set of all solutions, we will use **set-builder notation** to write the *solution set* of Example 4 as

$$\{x \mid x > 6\}.$$

This notation is read

"The set of all x such that x is greater than 6."

Thus a number is in $\{x \mid x > 6\}$ if that number is greater than 6. From now on, solutions of inequalities will be written using set-builder notation.

Example 5

technology connection

As a partial check of Example 5, we can let $y_1 = 3x - 1$ and $y_2 = 2x - 5$. By scrolling up or down, you can note that for $x \leq -4$, we have $y_1 \leq y_2$.

X	Y₁	Y₂
−5	−16	−15
−4	−13	−13
−3	−10	−11
−2	−7	−9
−1	−4	−7
0	−1	−5
1	2	−3

X = −5

Solve $3x - 1 \leq 2x - 5$ and then graph the solution.

Solution   We have

$$3x - 1 \leq 2x - 5$$
$$3x - 1 + 1 \leq 2x - 5 + 1 \qquad \text{Adding 1 to both sides}$$
$$3x \leq 2x - 4 \qquad \text{Simplifying}$$
$$3x - 2x \leq 2x - 4 - 2x \qquad \text{Subtracting } 2x \text{ from both sides}$$
$$x \leq -4. \qquad \text{Simplifying}$$

The graph is as follows:

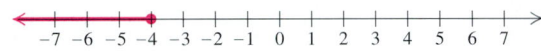

Any number less than or equal to -4 is a solution, so the solution set is $\{x \mid x \leq -4\}$.

Solving Inequalities Using the Multiplication Principle

There is a multiplication principle for inequalities similar to that for equations, but it must be modified when multiplying both sides by a negative number. Consider the true inequality

$$3 < 7.$$

If we multiply both sides by a *positive* number, say 2, we get another true inequality:

$$3 \cdot 2 < 7 \cdot 2, \quad \text{or} \quad 6 < 14. \qquad \text{TRUE}$$

If we multiply both sides by a negative number, say -2, we get a *false* inequality:

$$3 \cdot (-2) < 7 \cdot (-2), \quad \text{or} \quad -6 < -14. \qquad \text{FALSE}$$

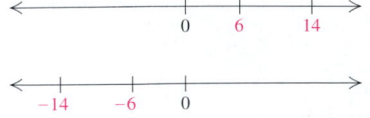

The fact that $6 < 14$ is true, but $-6 < -14$ is false, stems from the fact that the negative numbers, in a sense, mirror the positive numbers. Whereas 14 is to the *right* of 6, the number -14 is to the *left* of -6. Thus if we reverse the inequality symbol in $-6 < -14$, we get a true inequality:

$$-6 > -14. \qquad \text{TRUE}$$

> **The Multiplication Principle for Inequalities**
>
> For any real numbers a and b, and for any *positive* number c:
>
> $$a < b \quad \text{is equivalent to} \quad ac < bc, \quad \text{and}$$
> $$a > b \quad \text{is equivalent to} \quad ac > bc.$$
>
> For any real numbers a and b, and for any *negative* number c:
>
> $$a < b \quad \text{is equivalent to} \quad ac > bc, \quad \text{and}$$
> $$a > b \quad \text{is equivalent to} \quad ac < bc.$$
>
> Similar statements hold for $\leq$ and $\geq$.

E x a m p l e   6   Solve and graph each inequality: **(a)** $\frac{1}{4}x < 7$; **(b)** $-2y < 18$.

Solution

a) $\frac{1}{4}x < 7$

  $4 \cdot \frac{1}{4}x < 4 \cdot 7$ Multiplying both sides by 4, the reciprocal of $\frac{1}{4}$

                  The symbol stays the same, since 4 is positive.

  $x < 28$ Simplifying

The solution set is $\{x \mid x < 28\}$. The graph is as follows:

b) $-2y < 18$

  $\dfrac{-2y}{-2} > \dfrac{18}{-2}$ Multiplying both sides by $-\frac{1}{2}$, or dividing both sides by -2

                  At this step, we reverse the inequality, because $-\frac{1}{2}$ is negative.

  $y > -9$ Simplifying

As a partial check, we substitute a number greater than -9, say -8, into the original inequality:

$$\frac{-2y < 18}{-2(-8) \; ? \; 18}$$
$$16 \mid 18 \;\; \text{TRUE} \qquad 16 < 18 \text{ is a true statement.}$$

The solution set is $\{y \mid y > -9\}$. The graph is as follows:

Using the Principles Together

We use the addition and multiplication principles together to solve inequalities much as we did when solving equations.

E x a m p l e   7   Solve: **(a)** $6 - 5y > 7$; **(b)** $2x - 9 \leq 7x + 1$.

Solution

a) $6 - 5y > 7$

$-6 + 6 - 5y > -6 + 7$ Adding -6 to both sides

$-5y > 1$ Simplifying

$-\frac{1}{5} \cdot (-5y) < -\frac{1}{5} \cdot 1$ Multiplying both sides by $-\frac{1}{5}$, or dividing both sides by -5

⎣———————— Remember to reverse the inequality symbol!

$y < -\frac{1}{5}$ Simplifying

As a check, we substitute a number smaller than $-\frac{1}{5}$, say -1, into the original inequality:

$$\frac{6 - 5y > 7}{6 - 5(-1) \ ? \ 7}$$
$$6 - (-5) \ \Big|$$
$$11 \ \Big| \ 7 \quad \text{TRUE} \qquad 11 > 7 \text{ is a true statement.}$$

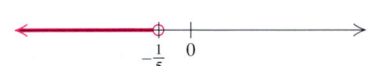

The solution set is $\left\{ y \,\middle|\, y < -\frac{1}{5} \right\}$. We show the graph in the margin for reference.

b) $2x - 9 \leq 7x + 1$

$2x - 9 - 1 \leq 7x + 1 - 1$ Subtracting 1 from both sides

$2x - 10 \leq 7x$ Simplifying

$2x - 10 - 2x \leq 7x - 2x$ Subtracting $2x$ from both sides

$-10 \leq 5x$ Simplifying

$\dfrac{-10}{5} \leq \dfrac{5x}{5}$ Dividing both sides by 5

$-2 \leq x$ Simplifying

The solution set is $\{x \,|\, -2 \leq x\}$, or $\{x \,|\, x \geq -2\}$.

All of the equation-solving techniques used in Sections 2.1 and 2.2 can be used with inequalities provided we remember to reverse the inequality symbol when multiplying or dividing both sides by a negative number.

E x a m p l e   8

Solve: **(a)** $16.3 - 7.2p \leq -8.18$; **(b)** $3(x - 2) - 1 < 2 - 5(x + 6)$.

Solution

a) The greatest number of decimal places in any one number is *two*. Multiplying both sides by 100 will clear decimals. Then we proceed as before.

$$16.3 - 7.2p \leq -8.18$$

$100(16.3 - 7.2p) \leq 100(-8.18)$	Multiplying both sides by 100
$100(16.3) - 100(7.2p) \leq 100(-8.18)$	Using the distributive law
$1630 - 720p \leq -818$	Simplifying
$-720p \leq -818 - 1630$	Subtracting 1630 from both sides
$-720p \leq -2448$	Simplifying
$p \geq \dfrac{-2448}{-720}$	Dividing both sides by -720

Remember to reverse the symbol.

$$p \geq 3.4$$

The solution set is $\{ p \mid p \geq 3.4 \}$.

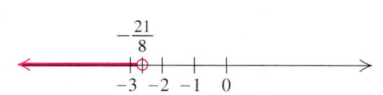

b) $3(x - 2) - 1 < 2 - 5(x + 6)$

$3x - 6 - 1 < 2 - 5x - 30$	Using the distributive law to remove parentheses
$3x - 7 < -5x - 28$	Simplifying
$3x + 5x < -28 + 7$	Adding $5x$ and also 7 to both sides. This isolates the x-terms on one side.
$8x < -21$	Simplifying
$x < -\dfrac{21}{8}$	Dividing both sides by 8

The solution set is $\left\{ x \mid x < -\frac{21}{8} \right\}$.

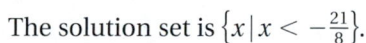

FOR EXTRA HELP

Exercise Set 2.6

 Digital Video Tutor CD 2 Videotape 4   InterAct Math   Math Tutor Center   MathXL   MyMathLab.com

Determine whether each number is a solution of the given inequality.

1. $x > -2$

   a) 5   **b)** 0   **c)** -1.9

   d) -7.3 **e)** 1.6

2. $y < 5$

   a) 0   **b)** 5   **c)** 4.99

   d) -13 **e)** $7\frac{1}{4}$

3. $x \geq 6$

   a) -6 **b)** 0 **c)** 6

   d) 6.01 **e)** $-3\frac{1}{2}$

4. $x \leq 10$

   a) 4 **b)** -10 **c)** 0

   d) 10.2 **e)** -4.7

Graph on a number line.

5. $x \leq 7$ **6.** $y < 2$

7. $t > -2$ **8.** $y > 4$

9. $1 \leq m$ **10.** $0 \leq t$

11. $-3 < x \leq 5$ **12.** $-5 \leq x < 2$

13. $0 < x < 3$ **14.** $-5 \leq x \leq 0$

Describe each graph using set-builder notation.

15.

16.

17.

18.

19.

20.

21.

22.

Solve using the addition principle. Graph and write set-builder notation for the answers.

23. $y + 2 > 9$ **24.** $y + 6 > 9$

25. $x + 8 \leq -10$ **26.** $x + 9 \leq -12$

27. $x - 3 < 7$ **28.** $x - 3 < 14$

29. $5 \leq t + 8$ **30.** $4 \leq t + 9$

31. $y - 7 > -12$ **32.** $y - 10 > -16$

33. $2x + 4 \leq x + 9$ **34.** $2x + 4 \leq x + 1$

Solve using the addition principle. Write the answers in set-builder notation.

35. $5x - 6 \geq 4x - 1$ **36.** $3x - 9 \geq 2x + 11$

37. $y + \frac{1}{3} \leq \frac{5}{6}$ **38.** $x + \frac{1}{4} \leq \frac{1}{2}$

39. $t - \frac{1}{8} > \frac{1}{2}$ **40.** $y - \frac{1}{3} > \frac{1}{4}$

41. $-9x + 17 > 17 - 8x$ **42.** $-8n + 12 > 12 - 7n$

Aha! **43.** $-23 < -t$ **44.** $19 < -x$

Solve using the multiplication principle. Graph and write set-builder notation for the answers.

45. $5x < 35$ **46.** $8x \geq 32$

47. $9y \leq 81$ **48.** $350 > 10t$

49. $-7x < 13$ **50.** $8y < 17$

51. $-24 > 8t$ **52.** $-16x < -64$

Solve using the multiplication principle. Write the answers in set-builder notation.

53. $7y \geq -2$ **54.** $5x > -3$

55. $-2y \leq \frac{1}{5}$ **56.** $-2x \geq \frac{1}{5}$

57. $-\frac{8}{5} > -2x$ **58.** $-\frac{5}{8} < -10y$

Solve using the addition and multiplication principles.

59. $7 + 3x < 34$ **60.** $5 + 4y < 37$

61. $6 + 5y \geq 26$ **62.** $7 + 8x \geq 71$

63. $4t - 5 \leq 23$ **64.** $5y - 9 \leq 21$

65. $13x - 7 < -46$ **66.** $8y - 4 < -52$

67. $16 < 4 - 3y$ **68.** $22 < 6 - 8x$

69. $39 > 3 - 9x$ **70.** $40 > 5 - 7y$

71. $5 - 6y > 25$ **72.** $8 - 2y > 14$

73. $-3 < 8x + 7 - 7x$ **74.** $-5 < 9x + 8 - 8x$

75. $6 - 4y > 4 - 3y$ **76.** $7 - 8y > 5 - 7y$

77. $7 - 9y \leq 4 - 8y$ **78.** $6 - 13y \leq 4 - 12y$

79. $33 - 12x < 4x + 97$

80. $27 - 11x > 14x - 18$

81. $2.1x + 43.2 > 1.2 - 8.4x$

82. $0.96y - 0.79 \leq 0.21y + 0.46$

83. $0.7n - 15 + n \geq 2n - 8 - 0.4n$

84. $1.7t + 8 - 1.62t < 0.4t - 0.32 + 8$

85. $\dfrac{x}{3} - 4 \leq 1$

86. $\dfrac{2}{3} - \dfrac{x}{5} < \dfrac{4}{15}$

87. $3 < 5 - \dfrac{t}{7}$

88. $2 > 9 - \dfrac{x}{5}$

89. $4(2y - 3) < 36$

90. $3(2y - 3) > 21$

91. $3(t - 2) \geq 9(t + 2)$

92. $8(2t + 1) > 4(7t + 7)$

93. $3(r - 6) + 2 < 4(r + 2) - 21$

94. $5(t + 3) + 9 > 3(t - 2) + 6$

95. $\frac{2}{3}(2x - 1) \geq 10$

96. $\frac{4}{5}(3x + 4) \leq 20$

97. $\frac{3}{4}\left(3x - \frac{1}{2}\right) - \frac{2}{3} < \frac{1}{3}$

98. $\frac{2}{3}\left(\frac{7}{8} - 4x\right) - \frac{5}{8} < \frac{3}{8}$

99. Are the inequalities $x > -3$ and $3 > -x$ equivalent? Why or why not?

100. Are the inequalities $t > -7$ and $7 < -t$ equivalent? Why or why not?

SKILL MAINTENANCE

Translate to an algebraic expression.

101. The sum of 3 and some number

102. Twice the sum of two numbers

103. Three less than twice a number

104. Five more than twice a number

SYNTHESIS

105. Explain in your own words why it is necessary to reverse the inequality symbol when multiplying both sides of an inequality by a negative number.

106. Explain how it is possible for the graph of an inequality to consist of just one number. (*Hint*: See Example 3c.)

Solve.

107. $6[4 - 2(6 + 3t)] > 5[3(7 - t) - 4(8 + 2t)] - 20$

108. $27 - 4[2(4x - 3) + 7] \geq 2[4 - 2(3 - x)] - 3$

Solve for x.

109. $-(x + 5) \geq 4a - 5$

110. $\frac{1}{2}(2x + 2b) > \frac{1}{3}(21 + 3b)$

111. $y < ax + b$ (Assume $a > 0$.)

112. $y < ax + b$ (Assume $a < 0$.)

113. Determine whether each number is a solution of the inequality $|x| < 3$.

    a) 3.2 **b)** -2 **c)** -3
    d) -2.9 **e)** 3 **f)** 1.7

114. Graph the solutions of $|x| < 3$ on a number line.

Aha! **115.** Determine the solution set of $|x| > -3$.

116. Determine the solution set of $|x| < 0$.

Solving Applications with Inequalities

2.7

Translating to Inequalities • Solving Problems

The five steps for problem solving can be used for problems involving inequalities.

Translating to Inequalities

Before solving problems that involve inequalities, we list some important phrases to look for. Sample translations are listed as well.

Important Words	Sample Sentence	Translation
is at least	Bill is at least 21 years old.	$b \geq 21$
is at most	At most 5 students dropped the course.	$n \leq 5$
cannot exceed	To qualify, earnings cannot exceed $12,000.	$r \leq 12{,}000$
must exceed	The speed must exceed 15 mph.	$s > 15$
is less than	Tucker's weight is less than 50 lb.	$w < 50$
is more than	Boston is more than 200 miles away.	$d > 200$
is between	The film is between 90 and 100 minutes long.	$90 < t < 100$
no more than	Bing weighs no more than 90 lb.	$w \leq 90$
no less than	Valerie scored no less than 8.3.	$s \geq 8.3$

Solving Problems

Example 1

Catering costs.   To cater a party, Curtis' Barbeque charges a $50 setup fee plus $15 per person. The cost of Hotel Pharmacy's end-of-season softball party cannot exceed $450. How many people can attend the party?

Solution

1. **Familiarize.**   Suppose that 20 people were to attend the party. The cost would then be $50 + $15 · 20, or $350. This shows that more than 20 people could attend without exceeding $450. Instead of making another guess, we let *n* represent the number of people in attendance.

2. **Translate.**   The cost of the party will be $50 for the setup fee plus $15 times the number of people attending. We can reword as follows:

   Rewording: The setup fee   plus   the cost of the meals   cannot exceed $450.

   Translating: 50   +   15 · n   ≤   450

3. **Carry out.**   We solve for *n*:

   $$50 + 15n \leq 450$$
   $$15n \leq 400 \qquad \text{\color{magenta}Subtracting 50 from both sides}$$
   $$n \leq \frac{400}{15} \qquad \text{\color{magenta}Dividing both sides by 15}$$
   $$n \leq 26\frac{2}{3}. \qquad \text{\color{magenta}Simplifying}$$

4. **Check.**   Although the solution set of the inequality is all numbers less than or equal to $26\frac{2}{3}$, since *n* represents the number of people in attendance, we round *down* to 26. If 26 people attend, the cost will be $50 + $15 · 26, or $440, and if 27 attend, the cost will exceed $450.

5. **State.**   At most 26 people can attend the party.

> **Caution!**   Solutions of problems should always be checked using the original wording of the problem. In some cases, answers might need to be whole numbers or integers or rounded off in a particular direction.

Example 2

Nutrition.   The U.S. Department of Health and Human Services and the Department of Agriculture recommend that for a typical 2000-calorie daily diet, no more than 65 g of fat be consumed. In the first three days of a four-day vacation, Phil consumed 70 g, 62 g, and 80 g of fat. Determine (in terms of an inequality) how many grams of fat Phil can consume on the fourth day if he is to average no more than 65 g of fat per day.

Solution

1. **Familiarize.**   Suppose Phil consumed 64 g of fat on the fourth day. His daily average for the vacation would then be

$$\frac{70\text{ g} + 62\text{ g} + 80\text{ g} + 64\text{ g}}{4} = 69\text{ g}.$$

This shows that Phil cannot consume 64 g of fat on the fourth day, if he is to average no more than 65 g of fat per day. Let's have x represent the number of grams of fat that Phil consumes on the fourth day.

2. **Translate.**   We reword the problem and translate as follows:

Rewording: The average consumption of fat   should be   no more than   65 g.

Translating: $\dfrac{70 + 62 + 80 + x}{4}$ $\leq$ 65

3. **Carry out.**   Because of the fraction, it is convenient to use the multiplication principle first:

$$\frac{70 + 62 + 80 + x}{4} \leq 65$$

$$4\left(\frac{70 + 62 + 80 + x}{4}\right) \leq 4 \cdot 65 \qquad \text{Multiplying both sides by 4}$$

$$70 + 62 + 80 + x \leq 260$$

$$212 + x \leq 260 \qquad \text{Simplifying}$$

$$x \leq 48. \qquad \text{Subtracting 212 from both sides}$$

4. **Check.**   As a partial check, we show that Phil can consume 48 g of fat on the fourth day and not exceed a 65-g average for the four days:

$$\frac{70 + 62 + 80 + 48}{4} = \frac{260}{4} = 65.$$

5. **State.**   Phil's average fat intake for the vacation will not exceed 65 g per day if he consumes no more than 48 g of fat on the fourth day.

Exercise Set 2.7

FOR EXTRA HELP

Digital Video Tutor CD 2
Videotape 4

InterAct Math

Math Tutor Center

MathXL

MyMathLab.com

Translate to an inequality.

1. A number is at least 7.

2. A number is greater than or equal to 5.

3. The baby weighs more than 2 kilograms (kg).

4. Between 75 and 100 people attended the concert.

5. The speed of the train was between 90 and 110 mph.

6. At least 400,000 people attended the Million Man March.

7. At most 1,200,000 people attended the Million Man March.

8. The amount of acid is not to exceed 40 liters (L).

9. The cost of gasoline is no less than $1.50 per gallon.

10. The temperature is at most −2°C.

Use an inequality and the five-step process to solve each problem.

11. *Blueprints.*   To make copies of blueprints, Vantage Reprographics charges a $5 setup fee plus $4 per copy. Myra can spend no more than $65 for the copying. What numbers of copies will allow her to stay within budget?

12. *Banquet costs.*   The women's volleyball team can spend at most $450 for its awards banquet at a local restaurant. If the restaurant charges a $40 setup fee plus $16 per person, at most how many can attend?

13. *Truck rentals.*   Ridem rents trucks at a daily rate of $42.95 plus $0.46 per mile. The Letsons want a one-day truck rental, but must stay within a budget of $200. What mileages will allow them to stay within budget? Round to the nearest tenth of a mile.

14. *Phone costs.*   Simon claims that it costs him at least $3.00 every time he calls an overseas customer. If his typical call costs 75¢ plus 45¢ for each minute, how long do his calls typically last?

15. *Parking costs.*   Laura is certain that every time she parks in the municipal garage it costs her at least $2.20. If the garage charges 45¢ plus 25¢ for each half hour, for how long is Laura's car generally parked?

16. *Furnace repairs.*   RJ's Plumbing and Heating charges $25 plus $30 per hour for emergency service. Gary remembers being billed over $100 for an emergency call. How long was RJ's there?

17. *College tuition.*   Angelica's financial aid stipulates that her tuition not exceed $1000. If her local community college charges a $35 registration fee plus $375 per course, what is the greatest number of courses for which Angelica can register?

18. *Van rentals.*   Atlas rents a cargo van at a daily rate of $44.95 plus $0.39 per mile. A business has budgeted $250 for a one-day van rental. What mileages will allow the business to stay within budget? (Round to the nearest tenth of a mile.)

19. *Grade average.*   Nadia is taking a literature course in which four tests are given. To get a B, a student must average at least 80 on the four tests. Nadia scored 82, 76, and 78 on the first three tests. What scores on the last test will earn her at least a B?

20. *Quiz average.*   Rod's quiz grades are 73, 75, 89, and 91. What scores on a fifth quiz will make his average quiz grade at least 85?

21. *Nutrition.*   Following the guidelines of the Food and Drug Administration, Dale tries to eat at least 5 servings of fruits or vegetables each day. For the first six days of one week, she had 4, 6, 7, 4, 6, and 4 servings. How many servings of fruits or vegetables should Dale eat on Saturday, in order to average at least 5 servings per day for the week?

22. *College course load.*   To remain on financial aid, Millie needs to complete an average of at least 7 credits per quarter each year. In the first three quarters of 2001, Millie completed 5, 7, and 8 credits. How many credits of course work must Millie complete in the fourth quarter if she is to remain on financial aid?

23. *Music lessons.*   Band members at Colchester Middle School are expected to average at least 20 min of practice time per day. One week Monroe practiced 15 min, 28 min, 30 min, 0 min, 15 min, and 25 min. How long must he practice on the seventh day if he is to meet expectations?

24. *Electrician visits.*   Dot's Electric made 17 customer calls last week and 22 calls this week. How many calls must be made next week in order to maintain an average of at least 20 for the three-week period?

25. *Perimeter of a rectangle.*   The width of a rectangle is fixed at 8 ft. What lengths will make the perimeter at least 200 ft? at most 200 ft?

26. *Perimeter of a triangle.*   One side of a triangle is 2 cm shorter than the base. The other side is 3 cm longer than the base. What lengths of the base will allow the perimeter to be greater than 19 cm?

27. *Perimeter of a pool.*   The perimeter of a rectangular swimming pool is not to exceed 70 ft. The length

is to be twice the width. What widths will meet these conditions?

28. *Volunteer work.*   George and Joan do volunteer work at a hospital. Joan worked 3 more hr than George, and together they worked more than 27 hr. What possible numbers of hours did each work?

29. *Cost of road service.*   Rick's Automotive charges $50 plus $15 for each (15-min) unit of time when making a road call. Twin City Repair charges $70 plus $10 for each unit of time. Under what circumstances would it be more economical for a motorist to call Rick's?

30. *Cost of clothes.*   Angelo is shopping for a new pair of jeans and two sweaters of the same kind. He is determined to spend no more than $120.00 for the clothes. He buys jeans for $21.95. How much can Angelo spend for each sweater?

31. *Area of a rectangle.*   The width of a rectangle is fixed at 4 cm. For what lengths will the area be less than 86 cm²?

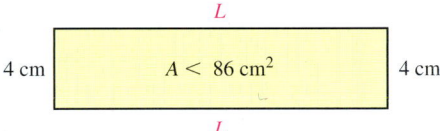

32. *Area of a rectangle.*   The width of a rectangle is fixed at 16 yd. For what lengths will the area be at least 264 yd²?

33. *Insurance-covered repairs.*   Most insurance companies will replace a vehicle if an estimated repair exceeds 80% of the "blue-book" value of the vehicle. Michelle's insurance company paid $8500 for repairs to her Subaru after an accident. What can be concluded about the blue-book value of the car?

34. *Insurance-covered repairs.*   Following an accident, Jeff's Ford pickup was replaced by his insurance company because the damage was so extensive. Before the damage, the blue-book value of the truck was $21,000. How much would it have cost to repair the truck? (See Exercise 33.)

35. *Body temperature.*   A person is considered to be feverish when his or her temperature is higher than 98.6°F. The formula $F = \frac{9}{5}C + 32$ can be used to convert Celsius temperatures C to Fahrenheit temperatures F. For which Celsius temperatures is a person considered feverish?

36. *Melting butter.* Butter stays solid at Fahrenheit temperatures below 88°. Use the formula in Exercise 35 to determine those Celsius temperatures for which butter stays solid.

37. *Fat content in foods.* Reduced Fat Skippy® peanut butter contains 12 g of fat per serving. In order for a food to be labeled "reduced fat," it must have at least 25% less fat than the regular item. What can you conclude about the number of grams of fat in a serving of the regular Skippy peanut butter?

38. *Fat content in foods.* Reduced Fat Chips Ahoy!® cookies contain 5 g of fat per serving. What can you conclude about the number of grams of fat in regular Chips Ahoy! cookies (see Exercise 37)?

39. *Well drilling.* All Seasons Well Drilling offers two plans. Under the "pay-as-you-go" plan, they charge $500 plus $8 a foot for a well of any depth. Under their "guaranteed-water" plan, they charge a flat fee of $4000 for a well that is guaranteed to provide adequate water for a household. For what depths would it save a customer money to use the pay-as-you-go plan?

40. *Running.* In the course of a week, Tony runs 6 mi, 3 mi, 5 mi, 5 mi, 4 mi, and 4 mi. How far should his next run be if he is to average at least 5 mi per day?

41. *Track records.* The formula $R = -0.012t + 20.8$, where R is in seconds, can be used to predict the world record in the 200-m dash t years after 1920. For what years will the world record be less than 19.8 sec?

42. *Track records.* The formula $R = -0.0084t + 3.85$, where R is in minutes, can be used to predict the world record in the 1500-m run t years after 1930. For what years will the world record be less than 3.22 min?

43. *Pond depth.* On July 1, Garrett's Pond was 25 ft deep. Since that date, the water level has dropped $\frac{2}{3}$ ft per week. For what dates will the water level not exceed 21 ft?

44. *Weight gain.* A 9-lb puppy is gaining weight at a rate of $\frac{3}{4}$ lb per week. When will the puppy's weight exceed $22\frac{1}{2}$ lb?

45. *Area of a triangular flag.* As part of an outdoor education course, Wanda needs to make a bright-colored triangular flag with an area of at least 3 ft². What heights can the triangle be if the base is $1\frac{1}{2}$ ft?

46. *Area of a triangular sign.* Zoning laws in Harrington prohibit displaying signs with areas exceeding 12 ft². If Flo's Marina is ordering a triangular sign with an 8-ft base, how tall can the sign be?

47. *Toll charges.* The equation $y = 0.027x + 0.19$ can be used to determine the approximate cost y, in dollars, of driving x miles on the Indiana toll road. For what mileages x will the cost be at most $6?

48. *Price of a movie ticket.* The average price of a movie ticket can be estimated by the equation $P = 0.1522Y - 298.592$, where Y is the year and P is the average price, in dollars. The price is lower than what might be expected due to senior-citizen discounts, children's prices, and special volume discounts. For what years will the average price of a movie ticket be at least $6? (Include the year in which the $6 ticket first occurs.)

49. If f represents Fran's age and t represents Todd's age, write a sentence that would translate to $t + 3 < f$.

50. Explain how the meanings of "Five more than a number" and "Five is more than a number" differ.

SKILL MAINTENANCE

Simplify.

51. $\dfrac{9 - 5}{6 - 4}$

52. $\dfrac{8 - 5}{12 - 6}$

53. $\dfrac{8 - (-2)}{1 - 4}$

54. $\dfrac{7 - 9}{4 - (-6)}$

SYNTHESIS

55. Write a problem for a classmate to solve. Devise the problem so the answer is "The Rothmans can drive 90 mi without exceeding their truck rental budget."

56. Write a problem for a classmate to solve. Devise the problem so the answer is "At most 18 passengers can go on the boat." Design the problem so that at least one number in the solution must be rounded down.

57. *Parking fees.*   Mack's Parking Garage charges $4.00 for the first hour and $2.50 for each additional hour. For how long has a car been parked when the charge exceeds $16.50?

58. *Ski wax.*   Green ski wax works best between 5° and 15° Fahrenheit. Determine those Celsius temperatures for which green ski wax works best. (See Exercise 35.)

Aha! 59. The area of a square can be no more than 64 cm². What lengths of a side will allow this?

Aha! 60. The sum of two consecutive odd integers is less than 100. What is the largest pair of such integers?

61. *Nutritional standards.*   In order for a food to be labeled "lowfat," it must have fewer than 3 g of fat per serving. Reduced fat Tortilla Pops® contain 60% less fat than regular nacho cheese tortilla chips, but still cannot be labeled lowfat. What can you conclude about the fat content of a serving of nacho cheese tortilla chips?

62. *Parking fees.*   When asked how much the parking charge is for a certain car (see Exercise 57), Mack replies "between 14 and 24 dollars." For how long has the car been parked?

63. Alice's Books allows customers to select one free book for every 10 books purchased. The price of that book cannot exceed the average cost of the 10 books. Neoma has bought 9 books that average $12 per book. How much should her tenth book cost if she wants to select a $15 book for free?

64. After 9 quizzes, Blythe's average is 84. Is it possible for Blythe to improve her average by two points with the next quiz? Why or why not?

65. Arnold and Diaz Booksellers offers a preferred-customer card for $25. The card entitles a customer to a 10% discount on all purchases for a period of one year. Under what circumstances would an individual save money by purchasing a card?

CORNER

Calling Plans

Focus: Problem solving and inequalities

Time: 20 minutes

Group size: 4

Materials: Calculators

A recent ad for "Five Line" is represented below.

	Rate per minute	Monthly fee	Distance	Restrictions
MCI 5¢ Everyday	25¢ in the day, 5¢ evening and weekends	$1.95	State-to-state	Minimum monthly bill of $5.00
Five Line	5¢	None	State-to-state AND in-state	Each completed call costs a minimum of 50¢.
AT&T Seven Sense	7¢	$5.95 ($4.95 with internet billing)	State-to-state	
Sprint Nickel Nights	10¢ in the day, 5¢ in the evening	$5.95	State-to-state	

ACTIVITY

The following table lists one month of Kate's calls. There are 34 calls for a total of 173 minutes.

Length of call (in minutes)	Number of calls of given length
1	13
2	6
3	5
4	3
5	2
9	1
16	1
18	1
33	1
35	1

1. Assume all calls are state-to-state calls. Each group member should choose a different one of the four plans and compute the bill (not including taxes and other fees) for each of the following situations.
   a) All 34 calls are evening (off-peak) calls.
   b) All 34 calls are daytime (peak) calls.
2. Is there a best plan for evening phone use? Is there a best plan for daytime phone use?
3. For each plan, describe the type of caller, if one exists, for whom that plan works best.
4. How can inequalities be used to compare these plans?

Summary and Review 2

Key Terms

Important Properties and Formulas

Solving Equations

Addition principle:

For any real numbers a, b, and c,

$$a = b \quad \text{is equivalent to} \quad a + c = b + c.$$

Multiplication principle:

For any real numbers a, b, and c, with $c \neq 0$,

$$a = b \quad \text{is equivalent to} \quad a \cdot c = b \cdot c.$$

Solving Inequalities

Addition principle:

For any real numbers a, b, and c,

$$a < b \quad \text{is equivalent to} \quad a + c < b + c;$$
$$a > b \quad \text{is equivalent to} \quad a + c > b + c.$$

Multiplication principle:

For any real numbers a and b and any *positive* number c,

$$a < b \quad \text{is equivalent to} \quad ac < bc;$$
$$a > b \quad \text{is equivalent to} \quad ac > bc.$$

For any real numbers a and b and any *negative* number c,

$$a < b \quad \text{is equivalent to} \quad ac > bc;$$
$$a > b \quad \text{is equivalent to} \quad ac < bc.$$

Similar statements hold for $\leq$ and $\geq$.

An Equation-Solving Procedure

1. Use the multiplication principle to clear any fractions or decimals. (This is optional, but can ease computations.)
2. If necessary, use the distributive law to remove parentheses. Then combine like terms on each side.
3. Use the addition principle, as needed, to get all variable terms on one side and all constant terms on the other.
4. Combine like terms again, if necessary.
5. Multiply or divide to solve for the variable, using the multiplication principle.
6. Check all possible solutions in the original equation.

To Solve a Formula for a Given Letter

1. If the letter for which you are solving appears in a fraction, use the multiplication principle to clear fractions.
2. Get all terms with the letter for which you are solving on one side of the equation and all other terms on the other side.
3. Combine like terms, if necessary. This may require factoring.
4. Multiply or divide to solve for the letter in question.

Percent Notation

$n\%$ means $\dfrac{n}{100}$, or $n \times \dfrac{1}{100}$, or $n \times 0.01$.

Five Steps for Problem Solving in Algebra

1. *Familiarize* yourself with the problem.
2. *Translate* to mathematical language. (This often means writing an equation.)
3. *Carry out* some mathematical manipulation. (This often means *solving* an equation.)
4. *Check* your possible answer in the original problem.
5. *State* the answer clearly.

Review Exercises

Solve.

1. $x + 9 = -16$

2. $-8x = -56$

3. $-\dfrac{x}{4} = 17$

4. $n - 7 = -6$

5. $15x = -60$

6. $x - 0.1 = 1.01$

7. $-\frac{2}{3} + x = -\frac{1}{6}$

8. $\frac{4}{5}y = -\frac{3}{16}$

9. $5z + 3 = 41$

10. $5 - x = 13$

11. $5t + 9 = 3t - 1$

12. $7x - 6 = 25x$

13. $\frac{1}{4}x - \frac{5}{8} = \frac{3}{8}$

14. $14y = 23y - 17 - 10$

15. $0.22y - 0.6 = 0.12y + 3 - 0.8y$

16. $\frac{1}{4}x - \frac{1}{8}x = 3 - \frac{1}{16}x$

17. $3(x + 5) = 36$

18. $4(5x - 7) = -56$

19. $8(x - 2) = 5(x + 4)$

20. $-5x + 3(x + 8) = 16$

Solve each formula for the given letter.

21. $C = \pi d$, for d

22. $V = \dfrac{1}{3}Bh$, for B

23. $A = \dfrac{a + b}{2}$, for a

24. Find decimal notation: 0.9%.

25. Find percent notation: $\frac{11}{25}$.

26. What percent of 60 is 12?

27. 42 is 30% of what number?

Determine whether the given number is a solution of the inequality $x \le 4$.

28. -3

29. 7

30. 4

Graph on a number line.

31. $5x - 6 < 2x + 3$

32. $-2 < x \le 5$

33. $y > 0$

Solve. Write the answers in set-builder notation.

34. $t + \frac{2}{3} \ge \frac{1}{6}$

35. $9x \ge 63$

36. $2 + 6y > 20$

37. $7 - 3y \ge 27 + 2y$

38. $3x + 5 < 2x - 6$

39. $-4y < 28$

40. $3 - 4x < 27$

41. $4 - 8x < 13 + 3x$

42. $-3y \ge -36$

43. $-4x \le \frac{1}{3}$

Solve.

44. An ink jet printer sold for $139 in July. This was $28 less than the cost in February. Find the cost in February.

45. A can of powdered infant formula makes 120 oz of formula. How many 6-oz bottles of formula will the can make?

46. A 12-ft "two by four" is cut into two pieces. One piece is 2 ft longer than the other. How long are the pieces?

47. About 52% of all charitable contributions are made to religious organizations. In 1997, $75 billion was given to religious organizations (*Source: Statistical Abstract of the United States,* 1999). How much was given to charities in general?

48. The sum of two consecutive odd integers is 116. Find the integers.

49. The perimeter of a rectangle is 56 cm. The width is 6 cm less than the length. Find the width and the length.

50. After a 25% reduction, a picnic table is on sale for $120. What was the regular price?

51. In 1997, the average male with a bachelor's degree or above earned about $66,000. This was 57% more than the average woman with a comparable background. How much did the average woman with a comparable background earn?

52. The measure of the second angle of a triangle is 50° more than that of the first. The measure of the third angle is 10° less than twice the first. Find the measures of the angles.

53. Jason has budgeted an average of $95 a month for entertainment. For the first five months of the year, he has spent $98, $89, $110, $85, and $83. How much can Jason spend in the sixth month without exceeding his average budget?

54. The length of a rectangle is 43 cm. For what widths is the perimeter greater than 120 cm?

55. How does the multiplication principle for equations differ from the multiplication principle for inequalities?

SYNTHESIS

56. Explain how checking the solutions of an equation differs from checking the solutions of an inequality.

57. The combined length of the Nile and Amazon Rivers is 13,108 km (*Source*: *Statistical Abstract of the United States*, 1999). If the Amazon were 234 km longer, it would be as long as the Nile. Find the length of each river.

58. Consumer experts advise us never to pay the sticker price for a car. A rule of thumb is to pay the sticker price minus 20% of the sticker price, plus $200. A car is purchased for $15,080 using the rule. What was the sticker price?

Solve.

59. $2|n| + 4 = 50$

60. $|3n| = 60$

61. $y = 2a - ab + 3$, for a

Chapter Test 2

Solve.

1. $x + 8 = 17$

2. $t - 3 = 12$

3. $3x = -18$

4. $-\frac{4}{7}x = -28$

5. $3t + 7 = 2t - 5$

6. $\frac{1}{2}x - \frac{3}{5} = \frac{2}{5}$

7. $8 - y = 16$

8. $-\frac{2}{5} + x = -\frac{3}{4}$

9. $3(x + 2) = 27$

10. $-3x + 6(x + 4) = 9$

11. $\frac{5}{6}(3x + 1) = 20$

Solve. Write the answers in set-builder notation.

12. $x + 6 > 1$

13. $14x + 9 > 13x - 4$

14. $\frac{1}{3}x < \frac{7}{8}$

15. $-2y \geq 26$

16. $4y \leq -32$

17. $-5x \geq \frac{1}{4}$

18. $4 - 6x > 40$

19. $5 - 9x \geq 19 + 5x$

Solve each formula for the given letter.

20. $A = 2\pi rh$, for r

21. $w = \dfrac{P + l}{2}$, for l

22. Find decimal notation: 230%.

23. Find percent notation: 0.054.

24. What number is 32% of 50?

25. What percent of 75 is 33?

Graph on a number line.

26. $y < 4$

27. $-2 \leq x \leq 2$

Solve.

28. The perimeter of a rectangle is 36 cm. The length is 4 cm greater than the width. Find the width and the length.

29. Kari is taking a 240-mi bicycle trip through Vermont. She has three times as many miles to go as she has already ridden. How many miles has she biked so far?

30. The perimeter of a triangle is 249 mm. If the sides are consecutive odd integers, find the length of each side.

31. Wholesale tea sales in 1999 were 255% of what they were in 1990 (*Source*: *Burlington Free Press*, 2/14/00). In 1990, there was $1.8 billion in wholesale tea sales. How much tea was sold wholesale in 1999?

32. Find all numbers for which six times the number is greater than the number plus 30.

33. The width of a rectangular ballfield is 96 yd. Find all possible lengths so that the perimeter of the ballfield will be at least 540 yd.

SYNTHESIS

Solve.

34. $c = \dfrac{2cd}{a - d}$, for d

35. $3|w| - 8 = 37$

36. A movie theater had a certain number of tickets to give away. Five people got the tickets. The first got one third of the tickets, the second got one fourth of the tickets, and the third got one fifth of the tickets. The fourth person got eight tickets, and there were five tickets left for the fifth person. Find the total number of tickets given away.

3
Introduction to Graphing

AN APPLICATION

As part of an ill-fated expedition to climb Mt. Everest, the world's tallest peak, author Jon Krakauer departed "The Balcony," elevation 27,600 ft, at 7:00 A.M. Krakauer reached the summit, elevation 29,028 ft, at 1:25 P.M. (*Source*: Krakauer, Jon, *Into Thin Air, the Illustrated Edition*. New York: Random House, 1998) Determine Krakauer's average rate of ascent, in feet per minute and in minutes per foot.

This problem appears as Exercise 11 in Section 3.4.

*I*n expeditions, we are always converting between feet and meters. Also, when we are working in exotic markets with parallel markets, we must often calculate the percentage difference between, say, a 65-rupee exchange rate and an 80-rupee exchange rate.

ALAN BURGESS
Mountain Expedition Guide
Salt Lake City, Utah

*W*e now begin our study of graphing. First we will examine graphs as they commonly appear in newspapers or magazines and develop some terminology. Following that, we will graph certain equations and study the connection between rate and slope. We will also learn how graphs can be used as a problem-solving tool for certain applications.

Our work in this chapter centers on solving equations that contain two variables.

Reading Graphs, Plotting Points, and Estimating Values

3.1

Problem Solving with Graphs • Points and Ordered Pairs • Estimations and Predictions

Today's print and electronic media make almost constant use of graphs. This can be attributed to the widespread availability of graphing software and the large quantity of information that a graph can display. In this section, we consider problem solving with bar graphs, line graphs, and circle graphs. Then we examine graphs that use a coordinate system.

Problem Solving with Graphs

A *bar graph* is a convenient way of showing comparisons. In every bar graph, certain categories, such as body weight in the example below, are paired with certain numbers.

E x a m p l e 1

Driving under the influence.   A blood-alcohol level of 0.08% or higher makes driving illegal in the United States. This bar graph shows how many drinks a person of a certain weight would need to consume in 1 hr to achieve a blood-alcohol level of 0.08% (*Source*: Adapted from soberup.com and vsa.vassar.edu/~source/drugs/alcohol.html). Note that a 12-oz beer, a 5-oz glass of wine, or a cocktail containing $1\frac{1}{2}$ oz of distilled liquor all count as one drink.

a) Approximately how many drinks would a 200-lb person have consumed if he or she had a blood-alcohol level of 0.08%?

b) What can be concluded about the weight of someone who can consume 3 drinks in an hour without reaching a blood-alcohol level of 0.08%?

Solution

a) We go to the top of the bar that is above the body weight 200 lb. Then we move horizontally from the top of the bar to the vertical scale listing numbers of drinks. It appears that approximately 4 drinks will give a 200-lb person a blood-alcohol level of 0.08%.

b) By moving up the vertical scale to the number 3, and then moving horizontally, we see that the first bar to reach a height of 3 corresponds to a weight of 140 lb. Thus an individual should weigh over 140 lb if he or she wishes to consume 3 drinks in an hour without exceeding a blood-alcohol level of 0.08%.

Circle graphs, or *pie charts*, are often used to show what percent of the whole each particular item in a group represents.

E x a m p l e   2

Color preference.   The circle graph below shows the favorite colors of Americans and the percentage that prefers each color (*Source*: Vitaly Komar and Alex Melamid: The Most Wanted Paintings on the Web). Because of rounding, the total is slightly less than 100%. There are approximately 272 million Americans, and three quarters of them live in cities. How many urban Americans choose red as their favorite color?

What is Your Favorite Color?

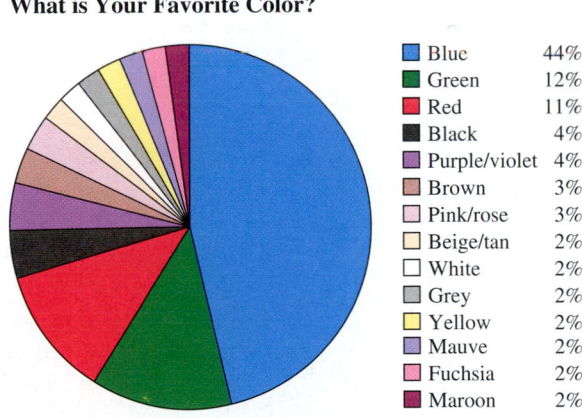

Blue	44%
Green	12%
Red	11%
Black	4%
Purple/violet	4%
Brown	3%
Pink/rose	3%
Beige/tan	2%
White	2%
Grey	2%
Yellow	2%
Mauve	2%
Fuchsia	2%
Maroon	2%

Solution

1. Familiarize.   The problem involves percents, so if we were unsure of how to solve percent problems, we might review Section 2.4. We are told that three quarters of all Americans live in cities. Since there are about 272 million Americans, this amounts to

$$\frac{3}{4} \cdot 272, \quad \text{or 204 million urban Americans.}$$

The chart indicates that 11% of the U.S. population prefers red. We let $r =$ the number of urban Americans who select red as their favorite color.

For the purpose of this problem, we will assume that the color preferences of urban Americans are typical of all Americans.

2. **Translate.** We reword and translate the problem as follows:

    Rewording: What is 11% of 204 million?

    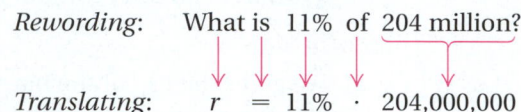

    Translating: $r = 11\% \cdot 204{,}000{,}000$

3. **Carry out.** We solve the equation:

    $r = 0.11 \cdot 204{,}000{,}000 = 22{,}440{,}000.$

4. **Check.** The check is left to the student.

5. **State.** About 22,440,000 urban Americans choose red as their favorite color.

E x a m p l e 3

Exercise and pulse rate. The following line graph shows the relationship between a person's resting pulse rate and months of regular exercise.*

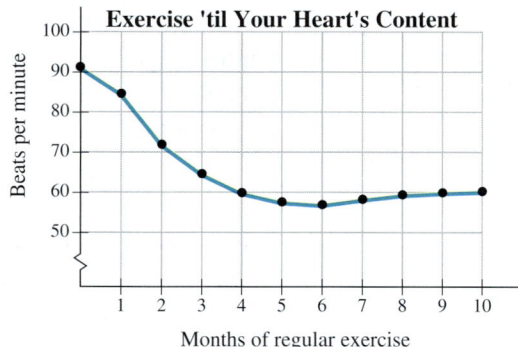

a) How many months of regular exercise are required to lower the pulse rate as much as possible?

b) How many months of regular exercise are needed to achieve a pulse rate of 65 beats per minute?

Solution

a) The lowest point on the graph occurs above the number 6. Thus, after 6 months of regular exercise, the pulse rate is lowered as much as possible.

b) We locate 65 on the vertical scale and then move right until the line is reached. At that point, we move down to the horizontal scale and read the information we are seeking.

*Data from *Body Clock* by Dr. Martin Hughes (New York: Facts on File, Inc.), p. 60.

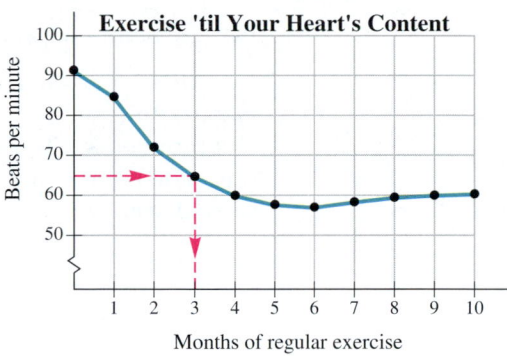

The pulse rate is 65 beats per minute after 3 months of regular exercise.

Points and Ordered Pairs

The line graph in Example 3 contains a collection of points. Each point pairs up a number of months of exercise with a pulse rate. To create such a graph, we **graph**, or **plot**, pairs of numbers on a plane. This is done using two perpendicular number lines called **axes** (pronounced "ak-sēz"; singular, **axis**). The point at which the axes cross is called the **origin**. Arrows on the axes indicate the positive directions.

Consider the pair (3, 4). The numbers in such a pair are called **coordinates**. The **first coordinate** in this case is 3 and the **second coordinate** is 4. To plot (3, 4), we start at the origin, move horizontally to the 3, move up vertically 4 units, and then make a "dot." Thus, (3, 4) is located above 3 on the first axis and to the right of 4 on the second axis.

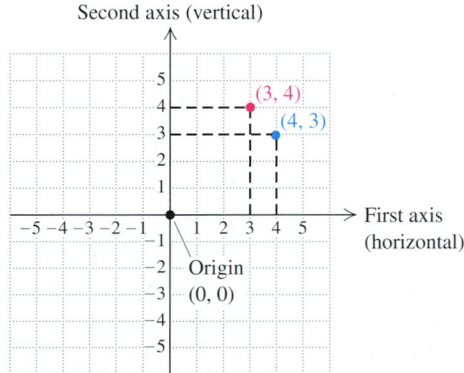

The point (4, 3) is also plotted in the figure above. Note that (3, 4) and (4, 3) are different points. For this reason, coordinate pairs are called **ordered pairs**—the order in which the numbers appear is important.

The portion of the coordinate plane shown by a grapher is called a viewing window. We indicate the **dimensions** of the window by setting a minimum *x*-value, a maximum *x*-value, a minimum *y*-value, and a maximum *y*-value. The **scale** by which we are counting must also be chosen. Window settings are often abbreviated in the form [L, R, B, T], with the letters representing **L**eft, **R**ight, **B**ottom, and **T**op endpoints. The window $[-10, 10, -10, 10]$ is called the **standard viewing window**. On most graphers, a standard viewing window can be set up using an option in the ZOOM menu.

In order to set up a $[-100, 100, -5, 5]$ viewing window, we use the following settings. We choose a scale for the *x*-axis of 10 units. If the scale chosen is too small, the marks on the *x*-axis will not be distinct.

```
WINDOW
  Xmin=-100
  Xmax=100
  Xscl=10
  Ymin=-5
  Ymax=5
  Yscl=1
  Xres=1
```

Set up the following viewing windows, choosing an appropriate scale.

1. $[-10, 10, -10, 10]$
2. $[-5, 5, 0, 100]$
3. $[-1, 1, -0.1, 0.1]$

Example 4   Plot the point $(-3, 4)$.

Solution   The first number, -3, is negative. Starting at the origin, we move 3 units in the negative horizontal direction (3 units to the left). The second number, 4, is positive, so we move 4 units in the positive vertical direction (up). The point $(-3, 4)$ is above -3 on the first axis and to the left of 4 on the second axis.

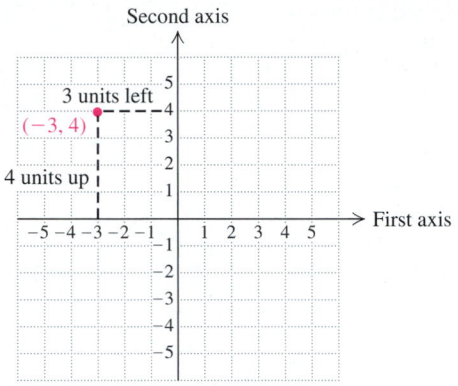

To find the coordinates of a point, we see how far to the right or left of the origin the point is and how far above or below the origin it is. Note that the coordinates of the origin itself are $(0, 0)$.

Example 5   Find the coordinates of points *A, B, C, D, E, F,* and *G.*

Study Tip

If you are finding it difficult to master a particular topic or concept, talk about it with a classmate. Verbalizing your questions about the material might help clarify it for you. If your classmate is also finding the material difficult, it is possible that the majority of the students in your class are confused and you can ask your instructor to explain the concept again.

Solution   Point *A* is 4 units to the right of the origin and 3 units above the origin. Its coordinates are $(4, 3)$. The coordinates of the other points are as follows:

B: $(-3, 5)$; C: $(-4, -3)$; D: $(2, -4)$;

E: $(1, 5)$; F: $(-2, 0)$; G: $(0, 3)$.

The horizontal and vertical axes divide the plane into four regions, or **quadrants**, as indicated by Roman numerals in the following figure. In region I (the *first quadrant*), both coordinates of any point are positive. In region II (the *second quadrant*), the first coordinate is negative and the second is positive. In region III (the *third quadrant*), both coordinates are negative. In region IV (the *fourth quadrant*), the first coordinate is positive and the second is negative.

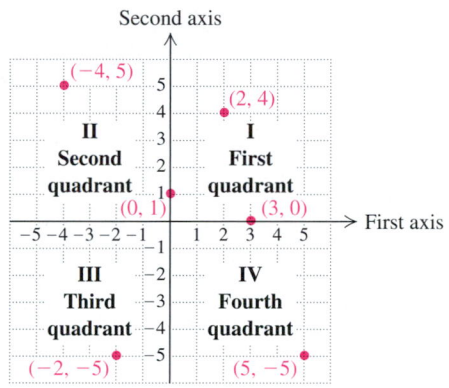

Note that the point $(-4, 5)$ is in the second quadrant and the point $(5, -5)$ is in the fourth quadrant. The points $(3, 0)$ and $(0, 1)$ are on the axes and are not considered to be in any quadrant.

Estimations and Predictions

It is possible to use line graphs to estimate real-life quantities that are not already known. To do so, we approximate the coordinates of an unknown point by using two points with known coordinates. When the unknown point is located *between* the two points, this process is called **interpolation**.* Sometimes a graph passing through the known points is *extended* to predict future values. Making predictions in this manner is called **extrapolation**.*

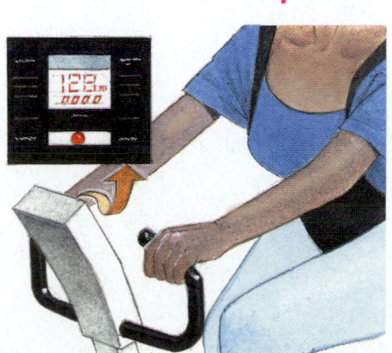

Example 6

Aerobic exercise.   A person's target heart rate is the number of beats per minute that bring the most aerobic benefit to his or her heart. The target heart rate for a 20-year-old is 150 beats per minute and for a 60-year-old, 120 beats per minute.

a) Estimate the target heart rate for a 35-year-old.
b) Predict what the target heart rate would be for a 75-year-old.

*Both interpolation and extrapolation can be performed using more than two known points and using curves other than lines.

Solution

a) We first draw a horizontal axis for "Age" and a vertical axis for "Target heart rate" on a piece of graph paper. Next, we number the axes, using a scale that will permit us to view both the given and the desired data. The given information allows us to then plot (20, 150) and (60, 120).

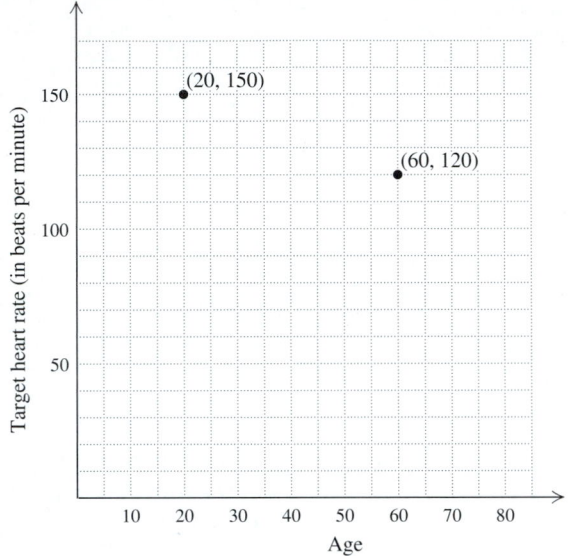

Next, we draw a line segment connecting the points. To estimate the target heart rate for a 35-year-old, we locate 35 on the horizontal axis. From there we move vertically to a point on the line and then left to the other axis. We see then that the target heart rate for a 35-year-old is approximately 140 beats per minute.

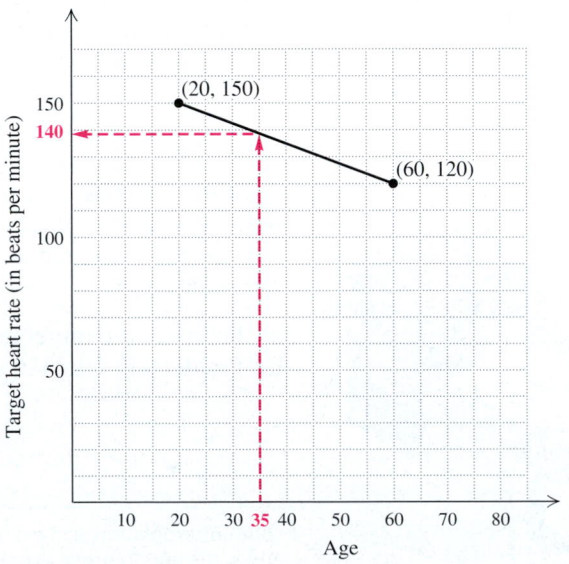

b) To predict the target heart rate for a 75-year-old, we extend the line segment on the graph until it is above 75 on the horizontal axis. Next, we proceed as in part (a) above, moving vertically to the line and then to the vertical axis. We predict that the target heart rate for a 75-year-old is approximately 110 beats per minute.

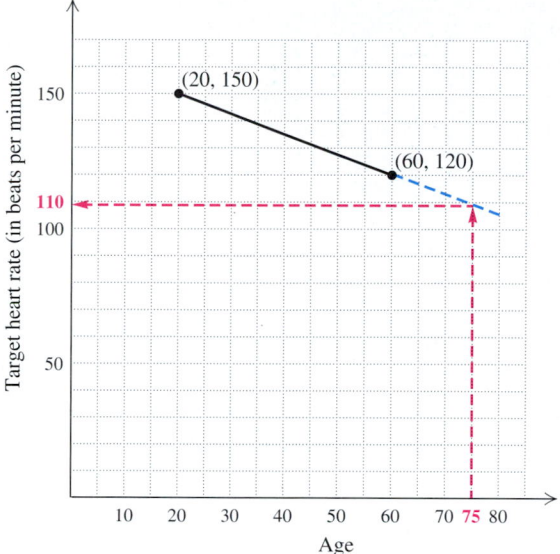

Exercise Set **3.1**

 Digital Video Tutor CD 2 Videotape 5    InterAct Math    Math Tutor Center   MathXL    MyMathLab.com

Blood alcohol level. *Use the bar graph in Example 1 to answer Exercises 1–4.*

1. Approximately how many drinks would a 100-lb person have consumed in 1 hr to reach a blood-alcohol level of 0.08%?

2. Approximately how many drinks would a 160-lb person have consumed in 1 hr to reach a blood-alcohol level of 0.08%?

3. What can you conclude about the weight of someone who has consumed 4 drinks in 1 hr without reaching a blood-alcohol level of 0.08%?

4. What can you conclude about the weight of someone who has consumed 5 drinks in 1 hr without reaching a blood-alcohol level of 0.08%?

Favorite color. *Use the information in Example 2 to answer Exercises 5–8.*

5. About one third of all Americans live in the South. How many Southerners choose brown as their favorite color?

6. About 50% of all Americans are at least 35 years old. How many Americans who are 35 or older choose purple/violet as their favorite color?

7. About one eighth of all Americans are senior citizens. How many senior citizens choose black as their favorite color?

8. About one fourth of all Americans are minors (under 18 years old). How many minors choose yellow as their favorite color?

Sorting solid waste.   Use the following pie chart to answer Exercises 9–12.

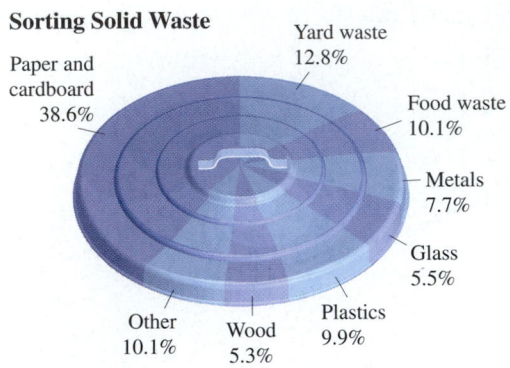

Sorting Solid Waste

Paper and cardboard 38.6%

Yard waste 12.8%

Food waste 10.1%

Metals 7.7%

Glass 5.5%

Plastics 9.9%

Wood 5.3%

Other 10.1%

Source: Statistical Abstract of the United States, 1999

9. In 1998, Americans generated 210 million tons of waste. How much of the waste was plastic?

10. In 2000, the average American generated 4.4 lb of waste per day. How much of that was paper and cardboard?

11. Americans are recycling about 26% of all glass that is in the waste stream. How much glass did Americans recycle in 1998? (See Exercise 9.)

12. Americans are recycling about 5% of all plastic waste. How much plastic does the average American recycle each day? (Use the information in Exercise 10.)

Recorded music.   Use the following line graphs to answer Exercises 13–18. The graphs show the percentage of all recordings sold that were CD or cassette. (Source: Recording Industry Association of America, Washington, D.C.)

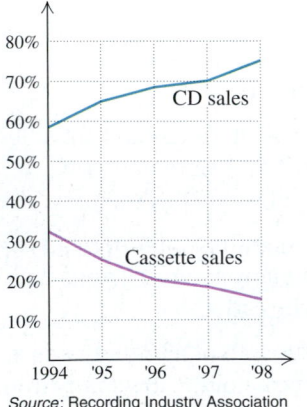

CD sales

Cassette sales

1994   '95   '96   '97   '98

Source: Recording Industry Association of America, Washington, D.C.

13. Approximately what percent of the recordings sold in 1997 were CDs?

14. Approximately what percent of the recordings sold in 1996 were cassettes?

15. In what year were approximately 25% of the recordings sold as cassettes?

16. In what year were approximately 75% of the recordings sold as CDs?

17. In what year did sales of CDs increase the most?

18. In what year did sales of cassettes change the least?

Plot each group of points.

19. $(1, 2), (-2, 3), (4, -1), (-5, -3), (4, 0), (0, -2)$

20. $(-2, -4), (4, -3), (5, 4), (-1, 0), (-4, 4), (0, 5)$

21. $(4, 4), (-2, 4), (5, -3), (-5, -5), (0, 4), (0, -4),$ $(3, 0), (-4, 0)$

22. $(2, 5), (-1, 3), (3, -2), (-2, -4), (0, 4), (0, -5),$ $(5, 0), (-5, 0)$

In Exercises 23–26, find the coordinates of points A, B, C, D, and E.

23.

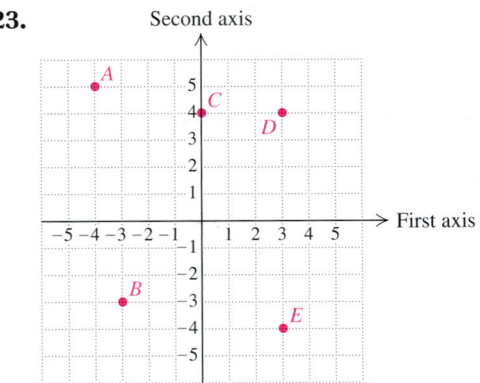

24.

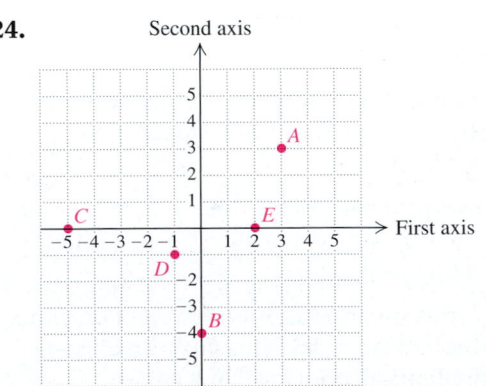

25.

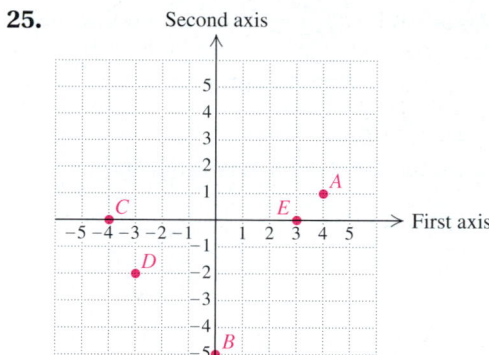

26.

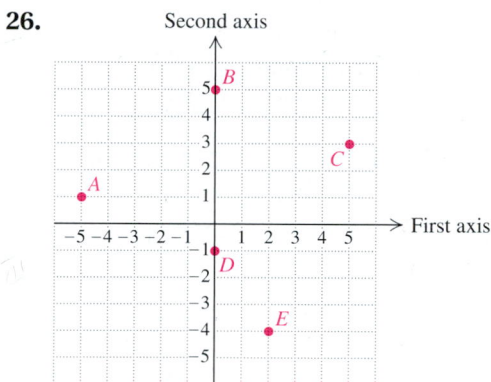

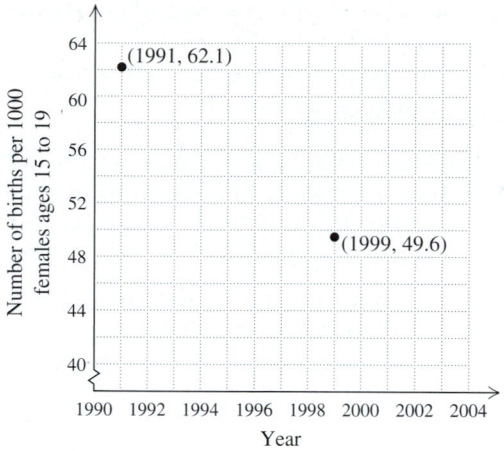

a) Estimate the birth rate among teenagers in 1995.

b) Predict the birth rate among teenagers in 2003.

40. *Food-stamp program participation.*   Due to changing rules and a booming economy, monthly participation in the U.S. food-stamp program dropped from approximately 27.5 million people in 1994 to approximately 20 million in 1999 (*Source*: U.S. Department of Agriculture).

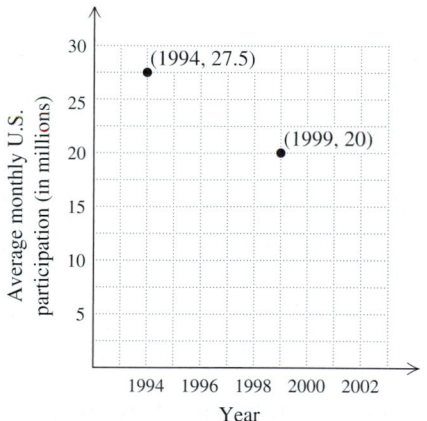

In which quadrant is each point located?

27. $(7, -2)$

28. $(-1, -4)$

29. $(-4, -3)$

30. $(1, -5)$

31. $(2, 1)$

32. $(-4, 6)$

33. $(-4.9, 8.3)$

34. $(7.5, 2.9)$

35. In which quadrants are the first coordinates positive?

36. In which quadrants are the second coordinates negative?

37. In which quadrants do both coordinates have the same sign?

38. In which quadrants do the first and second coordinates have opposite signs?

39. *Birth rate among teenagers.*   The birth rate among teenagers, measured in births per 1000 females age 15–19, fell steadily from 62.1 in 1991 to 49.6 in 1999 (*Source*: National Center for Health Statistics; 1999 figure is preliminary).

a) Estimate the number of participants in 1997.

b) Predict the number of participants in 2003.

In Exercises 41–46, display the given information in a graph similar to those used in Exercises 39 and 40. Use the horizontal axis to represent time. Then answer parts (a) and (b).

41. *Cigarette smoking.*   The percentage of people age 26 to 34 who smoke has dropped from 45.7% in 1985 to 32.5% in 1998 (*Source*: The World Almanac and Book of Facts 2000, 1999).

a) Approximate the percentage of people age 26–34 who smoked in 1990.

b) Predict the percentage of people age 26–34 who will smoke in 2003.

42. *Cigarette smoking.* The percentage of people age 18–25 who smoke has changed from 47.4% in 1985 to 41.8% in 1998 (*Source*: *The World Almanac and Book of Facts 2000*, 1999).

a) Approximate the percentage of people age 18–25 who smoked in 1990.

b) Predict the percentage of people age 18–25 who will smoke in 2003.

43. *College enrollment.* U.S. college enrollment has grown from approximately 60.3 million in 1990 to 68.3 million in 2000 (*Source*: *The World Almanac and Book of Facts 2000*, 1999).

a) Approximate the U.S. college enrollment for 1996.

b) Predict the U.S. college enrollment for 2005.

44. *High school enrollment.* U.S. high school enrollment has changed from approximately 12.5 million in 1990 to 14.9 million in 2000 (*Source*: *The World Almanac and Book of Facts 2000*, 1999).

a) Approximate the U.S. high school enrollment for 1996.

b) Predict the U.S. high school enrollment for 2005.

45. *Aging baby boomers.* The number of U.S. residents over the age of 65 was approximately 31 million in 1990 and 34.4 million in 2000. (*Source*: *Statistical Abstract of the United States*).

Aha! **a)** Estimate the number of U.S. residents over the age of 65 in 1995.

b) Predict the number of U.S. residents over the age of 65 in 2010.

46. *Urban population.* The percentage of the U.S. population that resides in metropolitan areas increased from about 78% in 1980 to about 80% in 1996.

a) Estimate the percentage of the U.S. population residing in metropolitan areas in 1992.

b) Predict the percentage of the U.S. population residing in metropolitan areas in 2008.

47. What do all of the points on the vertical axis of a graph have in common?

48. The following graph was included in a mailing sent by Agway® to their oil customers in 2000. What information is missing from the graph and why is the graph misleading?

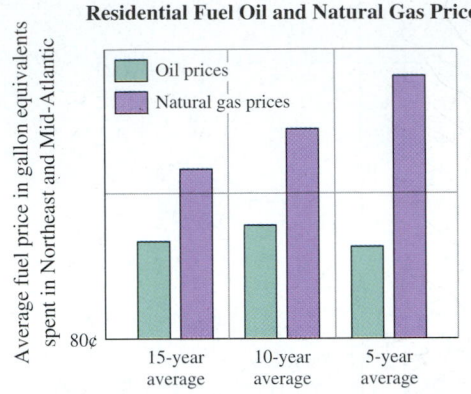

Residential Fuel Oil and Natural Gas Prices

Source: Energy Research Center, Inc. *3/1/99–2/29/00

SKILL MAINTENANCE

Simplify.

49. $4 \cdot 3 - 6 \cdot 5$

50. $5(-2) + 3(-7)$

51. $-\frac{1}{2}(-6) + 3$

52. $-\frac{2}{3}(-12) - 7$

Solve for y.

53. $3x - 2y = 6$

54. $7x - 4y = 14$

SYNTHESIS

55. Describe what the result would be if the first and second coordinates of every point in the following graph of an arrow were interchanged.

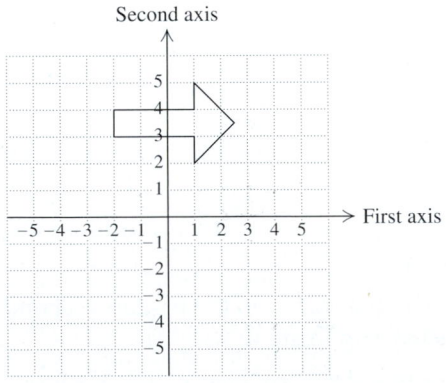

56. What advantage(s) does the use of a line graph have over that of a bar graph?

57. In which quadrant(s) could a point be located if its coordinates are reciprocals of each other?

58. In which quadrant(s) could a point be located if the point's second coordinate is $-\frac{2}{3}$ times the value of the first coordinate?

59. The points $(-1, 1)$, $(4, 1)$, and $(4, -5)$ are three vertices of a rectangle. Find the coordinates of the fourth vertex.

60. The pairs $(-2, -3)$, $(-1, 2)$, and $(4, -3)$ can serve as three (of four) vertices for three different parallelograms. Find the fourth vertex of each parallelogram.

61. Graph eight points such that the sum of the coordinates in each pair is 7.

62. Graph eight points such that the first coordinate minus the second coordinate is 1.

63. Find the perimeter of a rectangle if three of its vertices are $(5, -2)$, $(-3, -2)$, and $(-3, 3)$.

64. Find the area of a triangle whose vertices have coordinates $(0, 9)$, $(0, -4)$, and $(5, -4)$.

Coordinates on the globe. *Coordinates can also be used to describe the location on a sphere: 0° latitude is the equator and 0° longitude is a line from the North Pole to the South Pole through France and Algeria. In the figure*

shown here, hurricane Clara is at a point about 260 mi northwest of Bermuda near latitude 36.0° North, longitude 69.0° West.

65. Approximate the latitude and the longitude of Bermuda.

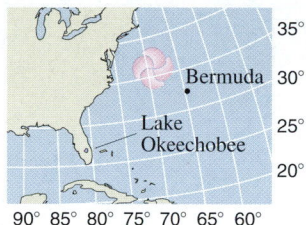

66. Approximate the latitude and the longitude of Lake Okeechobee.

67. In the *Star Trek* science-fiction series, a three-dimensional coordinate system is used to locate objects in space. If the center of a planet is used as the origin, how many "quadrants" will exist? Why? If possible, sketch a three-dimensional coordinate system and label each "quadrant."

68. The graph accompanying Example 3 flattens out. Why do you think this occurs?

COLLABORATIVE

CORNER

You Sank My Battleship!

Focus: Graphing points

Time: 15–25 minutes

Group size: 3–5

Materials: Graph paper

In the game Battleship®, a player places a miniature ship on a grid that only that player can see. An opponent guesses at coordinates that might "hit" the "hidden" ship. The following activity is similar to this game.

ACTIVITY

1. Using only integers from -10 to 10 (inclusive), one group member should secretly record the coordinates of a point on a slip of paper. (This point is the hidden "battleship.")

2. The other group members can then ask up to 10 "yes/no" questions in an effort to determine the coordinates of the secret point. Be sure to phrase each question mathematically (for example, "Is the *x*-coordinate negative?")

3. The group member who selected the point should answer each question. On the basis of the answer given, another group member should cross out the points no longer under consideration. All group members should check that this is done correctly.

4. If the hidden point has not been determined after 10 questions have been answered, the secret coordinates should be revealed to all group members.

5. Repeat parts (1)–(4) until each group member has had the opportunity to select the hidden point and answer questions.

Graphing Linear Equations

3.2

Solutions of Equations • Graphing Linear Equations

We have seen how bar, line, and circle graphs can represent information. Now we begin to learn how graphs can be used to represent solutions of equations.

Solutions of Equations

When an equation contains two variables, solutions must be ordered pairs in which each number in the pair replaces a letter in the equation. Unless stated otherwise, the first number in each pair replaces the variable that occurs first alphabetically.

E x a m p l e   1

Determine whether each of the following pairs is a solution of $4b - 3a = 22$: **(a)** $(2, 7)$; **(b)** $(1, 6)$.

Solution

a) We substitute 2 for a and 7 for b (alphabetical order of variables):

$$\begin{array}{c} 4b - 3a = 22 \\ \hline 4 \cdot 7 - 3 \cdot 2 \ ? \ 22 \\ 28 - 6 \ \Big| \\ 22 \ \Big| \ 22 \quad \text{TRUE} \end{array}$$

Since $22 = 22$ is *true*, the pair $(2, 7)$ *is* a solution.

b) In this case, we replace a with 1 and b with 6:

$$\begin{array}{c} 4b - 3a = 22 \\ \hline 4 \cdot 6 - 3 \cdot 1 \ ? \ 22 \\ 24 - 3 \ \Big| \\ 21 \ \Big| \ 22 \quad \text{FALSE} \end{array}$$

Since $21 = 22$ is *false*, the pair $(1, 6)$ is *not* a solution.

E x a m p l e   2

Show that the pairs $(3, 7)$, $(0, 1)$, and $(-3, -5)$ are solutions of $y = 2x + 1$. Then graph the three points to determine another pair that is a solution.

Solution   To show that a pair is a solution, we substitute, replacing x with the first coordinate and y with the second coordinate of each pair:

$$\begin{array}{c} y = 2x + 1 \\ \hline 7 \ ? \ 2 \cdot 3 + 1 \\ \Big| \ 6 + 1 \\ 7 \ \Big| \ 7 \qquad \text{TRUE} \end{array} \qquad \begin{array}{c} y = 2x + 1 \\ \hline 1 \ ? \ 2 \cdot 0 + 1 \\ \Big| \ 0 + 1 \\ 1 \ \Big| \ 1 \qquad \text{TRUE} \end{array} \qquad \begin{array}{c} y = 2x + 1 \\ \hline -5 \ ? \ 2(-3) + 1 \\ \Big| \ -6 + 1 \\ -5 \ \Big| \ -5 \qquad \text{TRUE} \end{array}$$

In each of the three cases, the substitution results in a true equation. Thus the pairs $(3, 7)$, $(0, 1)$, and $(-3, -5)$ are all solutions. We graph them below, labeling the "first" axis x and the "second" axis y. Note that the three points appear to "line up." Will other points that line up with these points also represent solutions of $y = 2x + 1$? To find out, we use a ruler and sketch a line passing through $(-3, -5)$, $(0, 1)$ and $(3, 7)$.

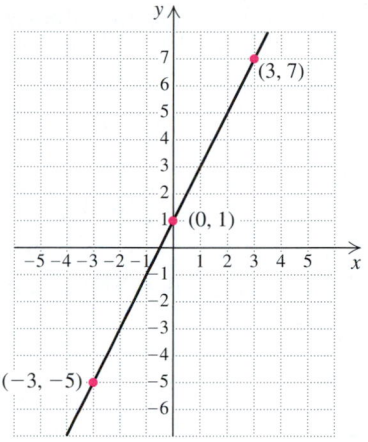

The line appears to pass through $(2, 5)$. Let's check if this pair is a solution of $y = 2x + 1$:

$$\begin{array}{c|l} \multicolumn{2}{l}{y = 2x + 1} \\ \hline 5 & ?\ 2 \cdot 2 + 1 \\ & 4 + 1 \\ 5 & 5 \qquad \text{TRUE} \end{array}$$

We see that $(2, 5)$ *is* a solution. You should perform a similar check for at least one other point that appears to be on the line.

Example 2 leads us to suspect that *any* point on the line passing through $(3, 7)$, $(0, 1)$, and $(-3, -5)$ represents a solution of $y = 2x + 1$. In fact, every solution of $y = 2x + 1$ is represented by a point on this line and every point on this line represents a solution. The line is called the **graph** of the equation.

Graphing Linear Equations

Equations like $y = 2x + 1$ or $4b + 3a = 22$ are said to be **linear** because the graph of each equation is a line. In general, any equation that can be written in the form $y = mx + b$ or $Ax + By = C$ (where m, b, A, B, and C are constants and A and B are not both 0) is linear.

To *graph* an equation is to make a drawing that represents its solutions. Linear equations can be graphed as follows.

> ### *To Graph a Linear Equation*
> 1. Select a value for one coordinate and calculate the corresponding value of the other coordinate. Form an ordered pair. This pair is one solution of the equation.
> 2. Repeat step (1) to find a second ordered pair. A third ordered pair can be used as a check.
> 3. Plot the ordered pairs and draw a straight line passing through the points. The line represents all solutions of the equation.

E x a m p l e 3

Graph: $y = -3x + 1$.

Solution   We select a convenient value for x, compute y, and form an ordered pair. Then we repeat the process for other choices of x.

$$\text{If } x = 2, \quad \text{then } y = -3 \cdot 2 + 1 = -5, \quad \text{and } (2, -5) \text{ is a solution.}$$
$$\text{If } x = 0, \quad \text{then } y = -3 \cdot 0 + 1 = 1, \quad \text{and } (0, 1) \text{ is a solution.}$$
$$\text{If } x = -1, \quad \text{then } y = -3(-1) + 1 = 4, \quad \text{and } (-1, 4) \text{ is a solution.}$$

Results are often listed in a table, as shown below. The points corresponding to each pair are then plotted.

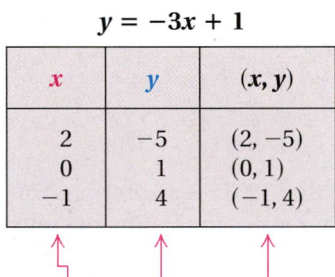

$y = -3x + 1$

x	y	(x, y)
2	-5	$(2, -5)$
0	1	$(0, 1)$
-1	4	$(-1, 4)$

(1) Choose x.
(2) Compute y.
(3) Form the pair (x, y).
(4) Plot the points.

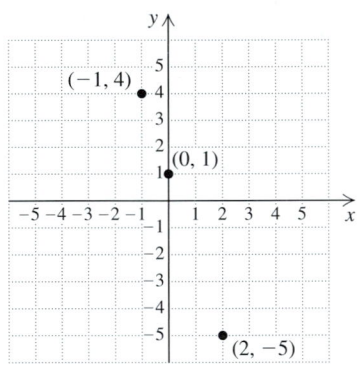

Note that all three points line up. If they didn't, we would know that we had made a mistake, because the equation is linear. When only two points are plotted, an error is more difficult to detect.

Finally, we use a ruler or other straight-edge to draw a line. Every point on the line represents a solution of $y = -3x + 1$.

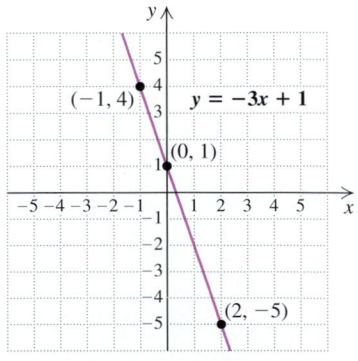

E x a m p l e   4

Graph: $y = 2x - 3$.

Solution   We select some x-values and compute y-values.

If $x = 4$,   then $y = 2 \cdot 4 - 3 = 5$,   and $(4, 5)$ is a solution.

If $x = 1$,   then $y = 2 \cdot 1 - 3 = -1$,   and $(1, -1)$ is a solution.

If $x = 0$,   then $y = 2 \cdot 0 - 3 = -3$,   and $(0, -3)$ is a solution.

$y = 2x - 3$

x	y	(x, y)
4	5	$(4, 5)$
1	-1	$(1, -1)$
0	-3	$(0, -3)$

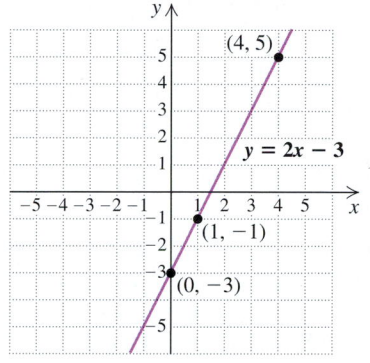

E x a m p l e   5

Graph: $4x + 2y = 12$.

Solution   To form ordered pairs, we can replace either variable with a number and then calculate the other coordinate:

If $y = 0$, we have $4x + 2 \cdot 0 = 12$

$$4x = 12$$

$$x = 3,$$

so $(3, 0)$ is a solution.

If $x = 0$, we have $4 \cdot 0 + 2y = 12$

$$2y = 12$$

$$y = 6,$$

so $(0, 6)$ is a solution.

If $y = 2$, we have $4x + 2 \cdot 2 = 12$

$$4x + 4 = 12$$

$$4x = 8$$

$$x = 2,$$

so $(2, 2)$ is a solution.

$4x + 2y = 12$

x	y	(x, y)
3	0	$(3, 0)$
0	6	$(0, 6)$
2	2	$(2, 2)$

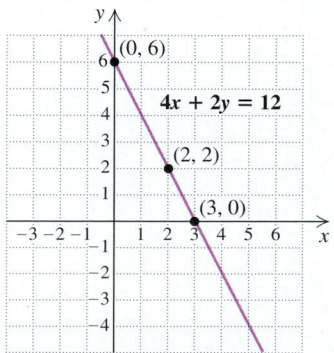

Note that in Examples 3 and 4 the variable y is isolated on one side of the equation. This generally simplifies calculations, so it is important to be able to solve for y before graphing.

E x a m p l e   6

Graph $3y = 2x$ by first solving for y.

Solution   To isolate y, we divide both sides by 3, or multiply both sides by $\frac{1}{3}$:

$$3y = 2x$$

$$\frac{1}{3} \cdot 3y = \frac{1}{3} \cdot 2x \qquad \text{Using the multiplication principle to multiply both sides by } \frac{1}{3}$$

$$\left. \begin{array}{l} 1y = \frac{2}{3} \cdot x \\ y = \frac{2}{3}x. \end{array} \right\} \qquad \text{Simplifying}$$

Because all the equations above are equivalent, we can use $y = \frac{2}{3}x$ to draw the graph of $3y = 2x$.

To graph $y = \frac{2}{3}x$, we can select x-values that are multiples of 3. This will allow us to avoid fractions when the corresponding y-values are computed.

$$\left. \begin{array}{lll} \text{If } x = 3, & \text{then } y = \frac{2}{3} \cdot 3 = 2. \\ \text{If } x = -3, & \text{then } y = \frac{2}{3}(-3) = -2. \\ \text{If } x = 6, & \text{then } y = \frac{2}{3} \cdot 6 = 4. \end{array} \right\} \qquad \begin{array}{l} \text{Note that when multiples of 3 are} \\ \text{substituted for } x, \text{ the } y\text{-coordinates} \\ \text{are not fractions.} \end{array}$$

The following table lists these solutions. Next, we plot the points and see that they form a line. Finally, we draw and label the line.

$3y = 2x, \quad$ **or** $\quad y = \dfrac{2}{3}x$

x	y	(x, y)
3	2	$(3, 2)$
-3	-2	$(-3, -2)$
6	4	$(6, 4)$

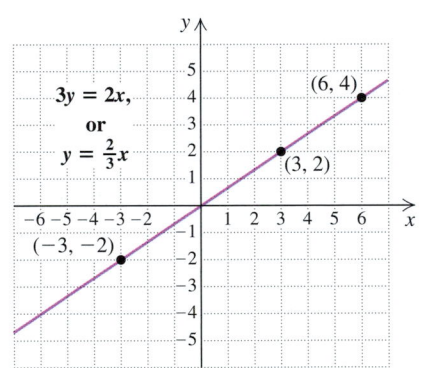

E x a m p l e   7

Graph $x + 5y = -10$ by first solving for y.

Solution   We have

$$x + 5y = -10$$

$$5y = -x - 10 \qquad \text{Adding } -x \text{ to both sides}$$

$$y = \frac{1}{5}(-x - 10) \qquad \text{Multiplying both sides by } \frac{1}{5}$$

$$y = -\frac{1}{5}x - 2. \qquad \text{Using the distributive law}$$

> ***Caution!***   It is very important to multiply *both* $-x$ and -10 by $\frac{1}{5}$.

Thus, $x + 5y = -10$ is equivalent to $y = -\frac{1}{5}x - 2$. If we choose x-values that are multiples of 5, we can avoid fractions when calculating the corresponding y-values.

If $x = 5$,   then $y = -\frac{1}{5} \cdot 5 - 2 = -1 - 2 = -3$.

If $x = 0$,   then $y = -\frac{1}{5} \cdot 0 - 2 = 0 - 2 = -2$.

If $x = -5$,   then $y = -\frac{1}{5}(-5) - 2 = 1 - 2 = -1$.

$x + 5y = -10$, or $y = -\dfrac{1}{5}x - 2$

x	y	(x, y)
5	-3	$(5, -3)$
0	-2	$(0, -2)$
-5	-1	$(-5, -1)$

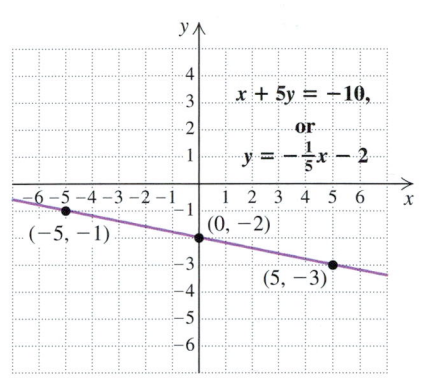

Linear equations appear in many real-life situations.

Example 8

Value of an office machine. The value, in thousands of dollars, of Dupligraphix's color copier t years after purchase is given by $v = -\frac{1}{2}t + 3$. Graph the equation and then use the graph to estimate the value of the copier $2\frac{1}{2}$ yr after the date of purchase.

Solution   We graph $v = -\frac{1}{2}t + 3$ by selecting values for t and then calculating the associated value v. Since time cannot be negative in this case, we select nonnegative values for t.

If $t = 0$,   then $v = -\frac{1}{2} \cdot 0 + 3 = 3$.

If $t = 4$,   then $v = -\frac{1}{2} \cdot 4 + 3 = 1$.

If $t = 8$,   then $v = -\frac{1}{2} \cdot 8 + 3 = -1$.

t	v
0	3
4	1
8	-1

We label the axes and plot the points (see the figure on the left at the top of the next page). The points line up, so our calculations are probably correct. However, including any points below the horizontal axis seems unrealistic since the value of the copier cannot be negative. Thus, when drawing the graph, we end the solid line at the horizontal axis.

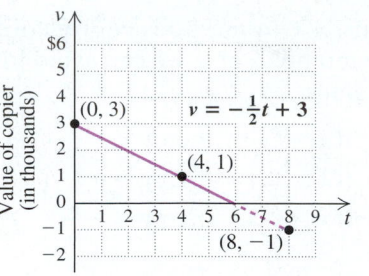
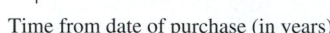
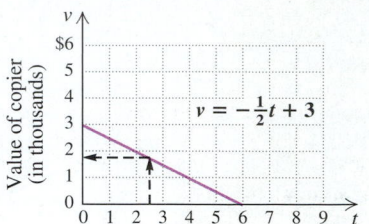

To estimate the value of the copier after $2\frac{1}{2}$ yr, we need to determine what second coordinate is paired with $2\frac{1}{2}$. To do this, we locate the point on the line that is above $2\frac{1}{2}$ and then find the value on the vertical axis that corresponds to that point (see the figure on the right above). It appears that after $2\frac{1}{2}$ yr, the copier is worth about $1800.

Caution!   When the coordinates of a point are read from a graph, as in Example 8, values should not be considered exact.

Many equations in two variables have graphs that are not straight lines. Three such graphs are shown below. As before, each graph represents the solutions of the given equation. Graphing calculators are especially helpful when drawing these *nonlinear* graphs.

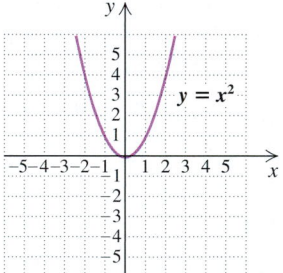
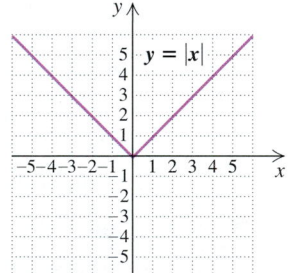
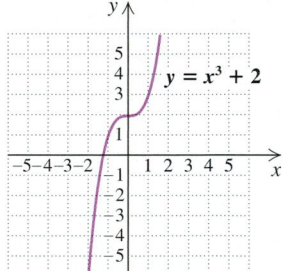

technology
connection

Most graphers require that y be alone on one side before the equation is entered. For example, to graph $5y + 4x = 13$, we would first solve for y. The student can check that solving for y yields the equation $y = -\frac{4}{5}x + \frac{13}{5}$.

We enter $-\frac{4}{5}x + \frac{13}{5}$ as Y1 and press GRAPH . The standard viewing window $[-10, 10, -10, 10]$ results in the graph shown.

Using a grapher, graph each of the following. Select the "standard" $[-10, 10, -10, 10]$ window.

1. $y = -5x + 6.5$ **2.** $y = 3x + 4.5$
3. $7y - 4x = 22$ **4.** $5y + 11x = -20$
5. $2y - x^2 = 0$ **6.** $y + x^2 = 8$

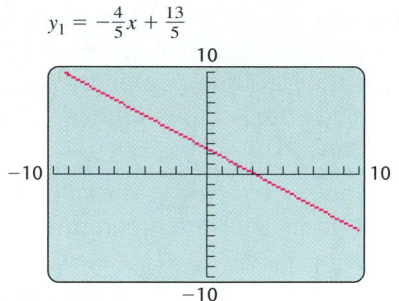

$y_1 = -\frac{4}{5}x + \frac{13}{5}$

Exercise Set 3.2

Determine whether each equation has the given ordered pair as a solution.

1. $y = 7x + 1$; $(0, 2)$

2. $y = 2x + 3$; $(0, 3)$

3. $3y + 2x = 12$; $(4, 2)$

4. $5x - 3y = 15$; $(0, 5)$

5. $4a - 3b = 11$; $(2, -1)$

6. $3q - 2p = -8$; $(1, -2)$

In Exercises 7–14, an equation and two ordered pairs are given. Show that each pair is a solution of the equation. Then graph the two pairs to determine another solution. Answers may vary.

7. $y = x - 2$; $(3, 1), (-2, -4)$

8. $y = x + 3$; $(-1, 2), (4, 7)$

9. $y = \frac{1}{2}x + 3$; $(4, 5), (-2, 2)$

10. $y = \frac{1}{2}x - 1$; $(6, 2), (0, -1)$

11. $y + 3x = 7$; $(2, 1), (4, -5)$

12. $2y + x = 5$; $(-1, 3), (7, -1)$

13. $4x - 2y = 10$; $(0, -5), (4, 3)$

14. $6x - 3y = 3$; $(1, 1), (-1, -3)$

Graph each equation.

15. $y = x - 1$ **16.** $y = x + 1$

17. $y = x$ **18.** $y = -x$

19. $y = \frac{1}{2}x$ **20.** $y = \frac{1}{3}x$

21. $y = x + 2$ **22.** $y = x + 3$

23. $y = 3x - 2$ **24.** $y = 2x + 2$

25. $y = \frac{1}{2}x + 1$ **26.** $y = \frac{1}{3}x - 4$

27. $x + y = -5$ **28.** $x + y = 4$

29. $y = \frac{5}{3}x - 2$ **30.** $y = \frac{5}{2}x + 3$

31. $x + 2y = 8$ **32.** $x + 2y = -6$

33. $y = \frac{3}{2}x + 1$ **34.** $y = -\frac{2}{3}x + 4$

35. $6x - 3y = 9$ **36.** $8x - 4y = 12$

37. $8y + 2x = -4$ **38.** $6y + 2x = 8$

Solve by graphing. Label all axes, and show where each solution is located on the graph.

39. *Bottled water.* The number of gallons of bottled water w consumed by an average American in one year is given by $w = \frac{1}{2}t + 5$, where t is the number of years since 1990 (based on an article in *New York Times Magazine*, 8/30/98). Graph the equation and use the graph to predict the number of gallons consumed per person in 2004.

40. *Value of computer software.* The value v of a shopkeeper's inventory software program, in hundreds of dollars, is given by $v = -\frac{3}{4}t + 6$, where t is the number of years since the shopkeeper first bought the program. Graph the equation and use the graph to estimate what the program is worth 4 yr after it was first purchased.

41. *Increasing life expectancy.* A smoker is 15 times more likely to die from lung cancer than a non-smoker. An ex-smoker who stopped smoking t years ago is w times more likely to die from lung cancer than a nonsmoker, where

$$t + w = 15.*$$

Graph the equation and use the graph to estimate how much more likely it is for Sandy to die from lung cancer than Polly, if Polly never smoked and Sandy quit $2\frac{1}{2}$ years ago.

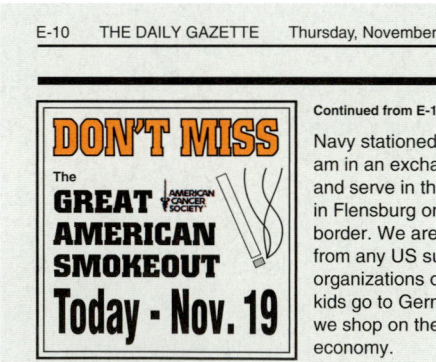

42. *Price of printing.* The price p, in cents, of a photocopied and bound lab manual is given by $p = \frac{7}{2}n + 20$, where n is the number of pages in the manual. Graph the equation and use the graph to estimate the cost of a 25-page manual. (*Hint*: Count by 5's on both axes.)

Source: Data from *Body Clock* by Dr. Martin Hughes, p. 60. New York: Facts on File, Inc.

43. *Cost of college.* The cost T, in hundreds of dollars, of tuition and fees at many community colleges can be approximated by $T = \frac{6}{5}c + 1$, where c is the number of credits for which a student registers (based on information provided by the Community College of Vermont). Graph the equation and use the graph to estimate the cost of tuition and fees when a student registers for 4 three-credit courses.

44. *Cost of college.* The cost C, in thousands of dollars, of a year at a private four-year college (all expenses) can be approximated by $C = \frac{5}{4}t + 21$, where t is the number of years since 1995 (based on information in *Statistical Abstract of the United States*, 1998). Graph the equation and use the graph to estimate the cost of a year at a private four-year college in 2005.

45. *Coffee consumption.* The number of gallons of coffee n consumed each year by the average U.S. consumer can be approximated by $n = \frac{5}{2}d + 20$, where d is the number of years since 1994 (based on information in *Statistical Abstract of the United States*, 1998). Graph the equation and use the graph to estimate what the average coffee consumption was in 2000.

46. *Record temperature drop.* On January 22, 1943, the temperature T, in degrees Fahrenheit, in Spearfish, South Dakota, could be approximated by $T = -2m + 54$, where m is the number of minutes since 9:00 A.M. that morning (*Source*: 2000 *Information Please Almanac*). Graph the equation and use the graph to estimate the temperature at 9:15 A.M.

47. The equations $3x + 4y = 8$ and $y = -\frac{3}{4}x + 2$ are equivalent. Which equation would be easier to graph and why?

48. Suppose that a linear equation is graphed by plotting three points and that the three points line up with each other. Does this *guarantee* that the equation is being correctly graphed? Why or why not?

SKILL MAINTENANCE

Solve and check.

49. $5x + 3 \cdot 0 = 12$ **50.** $2x - 5 \cdot 0 = 9$

51. $7 \cdot 0 - 4y = 10$

Solve.

52. $pq + p = w$, for p

53. $Ax + By = C$, for y

54. $A = \dfrac{T + Q}{2}$, for Q

SYNTHESIS

55. Janice consistently makes the mistake of plotting the x-coordinate of an ordered pair using the y-axis, and the y-coordinate using the x-axis. How will Janice's incorrect graph compare with the appropriate graph?

56. Explain how the graph in Example 8 can be used to determine when the value of the color copier has dropped to $1500.

57. *Bicycling.* Long Beach Island in New Jersey is a long, narrow, flat island. For exercise, Lauren routinely bikes to the northern tip of the island and back. Because of the steady wind, she uses one gear going north and another for her return. Lauren's bike has 14 gears and the sum of the two gears used on her ride is always 18. Write and graph an equation that represents the different pairings of gears that Lauren uses. Note that there are no fractional gears on a bicycle.

In Exercises 58–61, try to find an equation for the graph shown.

58.

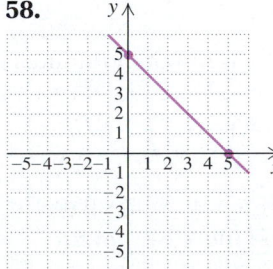

59.

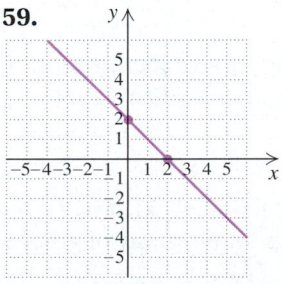

60.

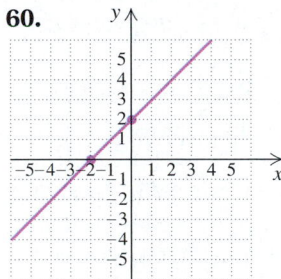

61.

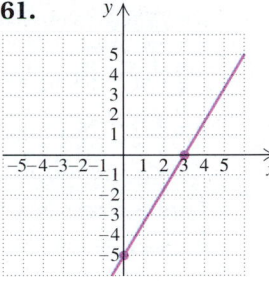

62. Translate to an equation:

　　　d dimes and n nickels total $1.75.

Then graph the equation and use the graph to determine three different combinations of dimes and nickels that total $1.75 (see also Exercise 75).

63. Translate to an equation:

　　　d $25 dinners and l $5 lunches total $225.

Then graph the equation and use the graph to determine three different combinations of lunches and dinners that total $225 (see also Exercise 75).

Use the suggested x-values $-3, -2, -1, 0, 1, 2,$ *and* 3 *to graph each equation.*

64. $y = |x|$

Aha! **65.** $y = -|x|$

Aha! **66.** $y = |x| - 2$

67. $y = -|x| + 2$

68. $y = |x| + 3$

For Exercises 69–74, use a grapher to graph the equation. Use a $[-10, 10, -10, 10]$ *window.*

69. $y = -2.8x + 3.5$

70. $y = 4.5x + 2.1$

71. $y = 2.8x - 3.5$

72. $y = -4.5x - 2.1$

73. $y = x^2 + 4x + 1$

74. $y = -x^2 + 4x - 7$

75. Study the graph of Exercise 62 or 63. Does *every* point on the graph represent a solution of the associated problem? Why or why not?

COLLABORATIVE

CORNER

Follow the Bouncing Ball

Focus: Graphing and problem solving

Time: 25 minutes

Group size: 3

Materials: Each group will need a rubber ball, graph paper, and a tape measure.

Does a rubber ball always rebound a fixed percentage of the height from which it is dropped? The following activity attempts to answer this. Please be sure to read all steps before beginning.

ACTIVITY

1. One group member should hold a rubber ball 5 ft above the floor. A second group member should measure this height for accuracy. The ball should then be dropped and caught at the peak of its bounce. The second group member should measure this rebound height and the third group member should record the measurement.

2. Repeat part (1) two more times from the same height. Then find the average of the three rebound heights and use that number to form the ordered pair (original height, rebound height).

3. Repeat parts (1) and (2) four more times at heights of 4 ft, 3 ft, 2 ft, and 1 ft to find five ordered pairs. Graph the five pairs on graph paper and draw a straight line starting at (0, 0) that comes as close as possible to all six points.

4. Use the graph in part (3) to predict the rebound height for a ball dropped from a height of 7 ft. Then perform the drop and check your prediction.

5. Repeat part (4), dropping the ball from a "new" height of the group's choice.

6. Does it appear that a rubber ball will always rebound a fixed percentage of its original height? Why or why not?

7. Compare results from other class groups. What conclusions can you draw?

Graphing and Intercepts

3.3

Intercepts • Using Intercepts to Graph • Graphing Horizontal or Vertical Lines

Unless a line is horizontal or vertical, it will cross both axes. Knowing where the axes are crossed gives us another way of graphing linear equations.

Intercepts

In Example 5 of Section 3.2, we graphed $4x + 2y = 12$ by plotting the points $(3, 0)$, $(0, 6)$, and $(2, 2)$ and then drawing the line.

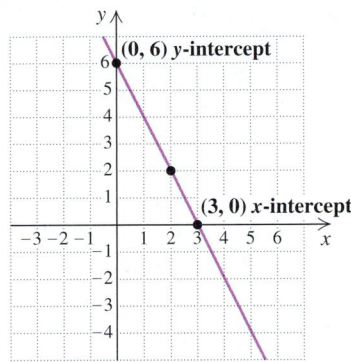

The point at which a graph crosses the y-axis is called the **y-intercept**. In the figure above, the y-intercept is $(0, 6)$. The x-coordinate of a y-intercept is always 0.

The point at which a graph crosses the x-axis is called the **x-intercept**. In the figure above, the x-intercept is $(3, 0)$. The y-coordinate of an x-intercept is always 0.

It is possible for the graph of a curve to have more than one y-intercept or more than one x-intercept.

E x a m p l e   1   For the graph shown below, **(a)** give the coordinates of any x-intercepts and **(b)** give the coordinates of any y-intercepts.

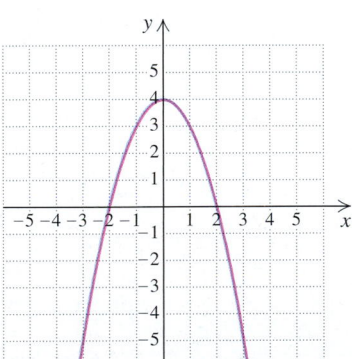

Solution

a) The x-intercepts are the points at which the graph crosses the x-axis. For the graph shown, the x-intercepts are $(-2, 0)$ and $(2, 0)$.

b) The y-intercept is the point at which the graph crosses the y-axis. For the graph shown, the y-intercept is $(0, 4)$.

Using Intercepts to Graph

It is important to know how to locate a graph's intercepts from the equation being graphed.

> ### To Find Intercepts
>
> To find the y-intercept(s) of an equation's graph, replace x with 0 and solve for y.
>
> To find the x-intercept(s) of an equation's graph, replace y with 0 and solve for x.

E x a m p l e   2   Find the y-intercept and the x-intercept of the graph of $2x + 4y = 20$.

Solution   To find the y-intercept, we let $x = 0$ and solve for y:

$$2 \cdot 0 + 4y = 20 \qquad \text{Replacing } x \text{ with 0}$$
$$4y = 20$$
$$y = 5.$$

Thus the y-intercept is $(0, 5)$.

To find the x-intercept, we let $y = 0$ and solve for x:

$$2x + 4 \cdot 0 = 20 \qquad \text{Replacing } y \text{ with 0}$$
$$2x = 20$$
$$x = 10.$$

Thus the x-intercept is $(10, 0)$.

As we saw in Example 5 of Section 3.2, intercepts can be used to graph a linear equation.

E x a m p l e   3   Graph $2x + 4y = 20$ using intercepts.

Solution   In Example 2, we showed that the y-intercept is $(0, 5)$ and the x-intercept is $(10, 0)$. Before drawing a line, we plot a third point as a check. We substitute any convenient value for x and solve for y.

If we let $x = 5$, then

$$2 \cdot 5 + 4y = 20 \qquad \text{Substituting 5 for } x$$
$$10 + 4y = 20$$
$$4y = 10 \qquad \text{Subtracting 10 from both sides}$$
$$y = \tfrac{10}{4}, \text{ or } 2\tfrac{1}{2}. \qquad \text{Solving for } y$$

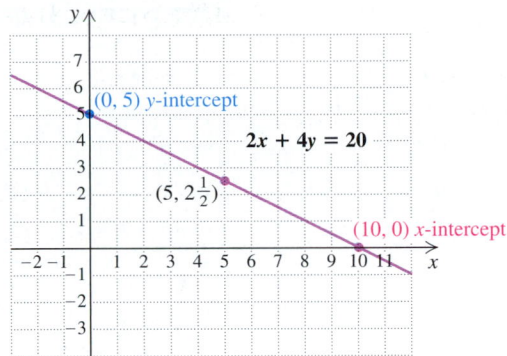

The point $\left(5, 2\frac{1}{2}\right)$ appears to line up with the intercepts, so our work is probably correct. To finish, we draw and label the line.

Note that when we solved for the y-intercept, we simplified $2x + 4y = 20$ to $4y = 20$. Thus, to find the y-intercept, we can momentarily ignore the x-term and solve the remaining equation.

In a similar manner, when we solved for the x-intercept, we simplified $2x + 4y = 20$ to $2x = 20$. Thus, to find the x-intercept, we can momentarily ignore the y-term and then solve this remaining equation.

Example 4

Graph $3x - 2y = 6$ using intercepts.

Solution To find the y-intercept, we let $x = 0$. This amounts to temporarily ignoring the x-term and then solving:

$$-2y = 6 \qquad \text{For } x = 0, \text{ we have } 3 \cdot 0 - 2y, \text{ or simply } -2y.$$
$$y = -3.$$

The y-intercept is $(0, -3)$.

To find the x-intercept, we let $y = 0$. This amounts to temporarily disregarding the y-term and then solving:

$$3x = 6 \qquad \text{For } y = 0, \text{ we have } 3x - 2 \cdot 0, \text{ or simply } 3x.$$
$$x = 2.$$

The x-intercept is $(2, 0)$.

To find a third point, we replace x with 4 and solve for y:

$$3 \cdot 4 - 2y = 6 \qquad \text{Numbers other than 4 can be used for } x.$$
$$12 - 2y = 6$$
$$-2y = -6$$
$$y = 3. \qquad \text{This means that } (4, 3) \text{ is on the graph.}$$

The point $(4, 3)$ appears to line up with the intercepts, so we draw the graph.

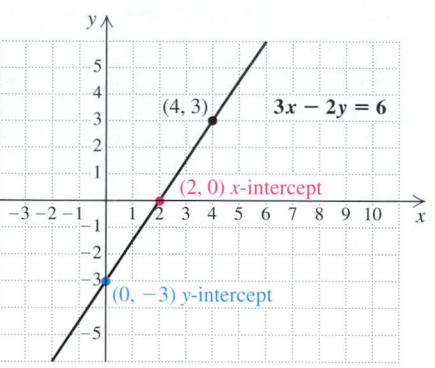

technology connection

When an equation has been entered into a grapher, we may not be able to see both intercepts. For example, if $y = -0.8x + 17$ is graphed in the $[-10, 10, -10, 10]$ window, neither intercept is visible.

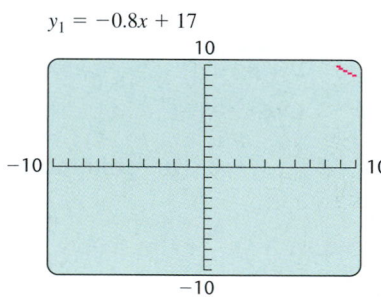

To better view the intercepts, we can change the window dimensions or we can zoom out. The ZOOM feature

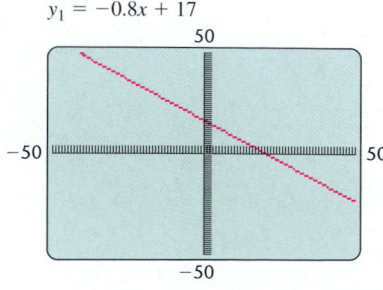

allows us to reduce or magnify a graph or a portion of a graph. Before zooming, the ZOOM *factors* must be set in the memory of the ZOOM key. If we zoom out with factors set at 5, both intercepts are visible but the axes are heavily drawn, as shown in the preceding figure.

This suggests that the *scales* of the axes should be changed. To do this, we use the WINDOW menu and set Xscl to 5 and Yscl to 5. The resulting graph has tick marks 5 units apart and clearly shows both intercepts. Other choices for Xscl and Yscl can also be made.

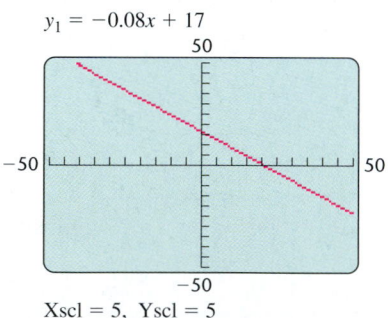

Xscl = 5, Yscl = 5

Graph each equation so that both intercepts can be easily viewed. Zoom or adjust the window settings so that tick marks can be clearly seen on both axes.

1. $y = -0.72x - 15$ **2.** $y - 2.13x = 27$
3. $5x + 6y = 84$ **4.** $2x - 7y = 150$
5. $19x - 17y = 200$ **6.** $6x + 5y = 159$

Graphing Horizontal or Vertical Lines

The equations graphed in Examples 3 and 4 are both in the form $Ax + By = C$. We have already stated that any equation in the form $Ax + By = C$ is linear, provided A and B are not both zero. What if A or B (but not both) is zero? We will find that when A is zero, there is no x-term and the graph is a horizontal line. We will also find that when B is zero, there is no y-term and the graph is a vertical line.

Example 5

Graph: $y = 3$.

Solution   We can regard the equation $y = 3$ as $0 \cdot x + y = 3$. No matter what number we choose for x, we find that y must be 3 if the equation is to be solved. Consider the following table.

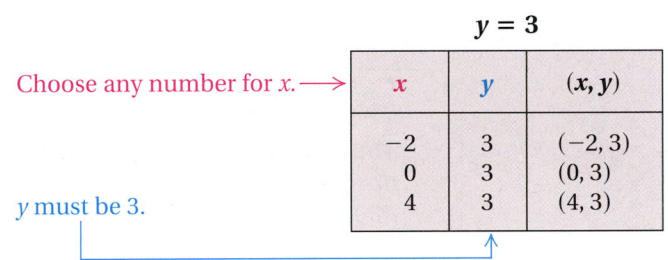

Choose any number for x. →

y must be 3.

$y = 3$

x	y	(x, y)
-2	3	$(-2, 3)$
0	3	$(0, 3)$
4	3	$(4, 3)$

All pairs will have 3 as the y-coordinate.

When we plot the ordered pairs $(-2, 3)$, $(0, 3)$, and $(4, 3)$ and connect the points, we obtain a horizontal line. Any ordered pair $(x, 3)$ is a solution, so the line is parallel to the x-axis with y-intercept $(0, 3)$.

Example 6

Graph: $x = -4$.

Solution   We can regard the equation $x = -4$ as $x + 0 \cdot y = -4$. We make up a table with all -4's in the x-column.

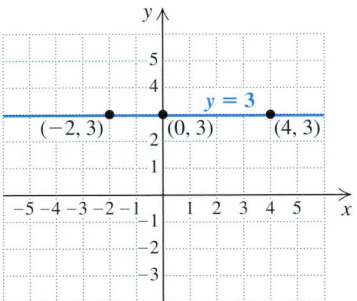

x must be -4. →

Choose any number for y.

$x = -4$

x	y	(x, y)
-4	-5	$(-4, -5)$
-4	1	$(-4, 1)$
-4	3	$(-4, 3)$

All pairs will have -4 as the x-coordinate.

When we plot the ordered pairs $(-4, -5)$, $(-4, 1)$, and $(-4, 3)$ and connect them, we obtain a vertical line. Any ordered pair $(-4, y)$ is a solution. The line is parallel to the y-axis with x-intercept $(-4, 0)$.

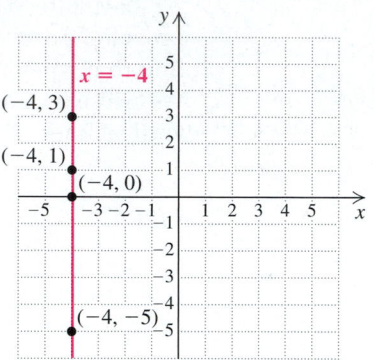

Linear Equations in One Variable

The graph of $y = b$ is a horizontal line, with y-intercept $(0, b)$.

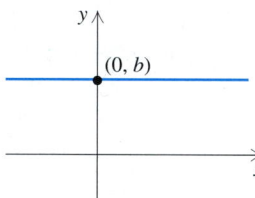

The graph of $x = a$ is a vertical line, with x-intercept $(a, 0)$.

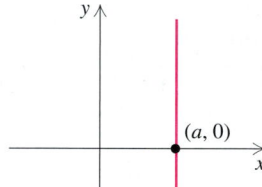

E x a m p l e   7   Write an equation for each graph.

a)

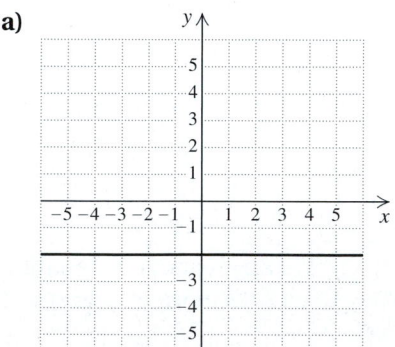

b)

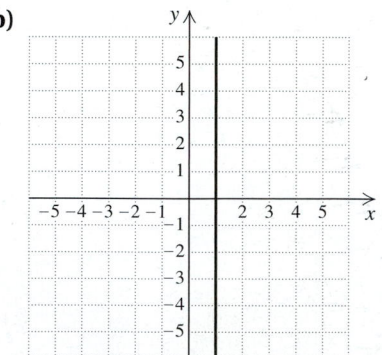

Solution

a) Note that every point on the horizontal line passing through $(0, -2)$ has -2 as the y-coordinate. Thus the equation of the line is $y = -2$.

b) Note that every point on the vertical line passing through $(1, 0)$ has 1 as the x-coordinate. Thus the equation of the line is $x = 1$.

Exercise Set **3.3**

  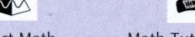

For Exercises 1–8, find **(a)** *the coordinates of the y-intercept and* **(b)** *the coordinates of all x-intercepts.*

1.
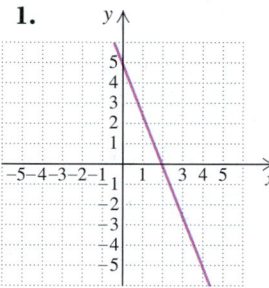

2.
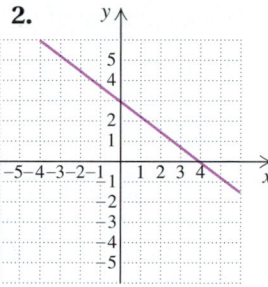

3.
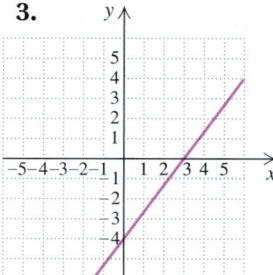

4.
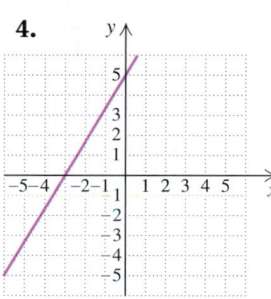

5.
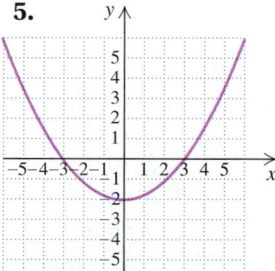

6.
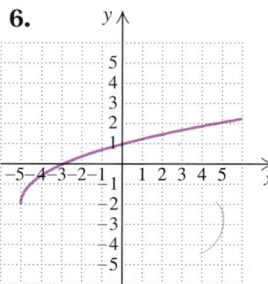

7.
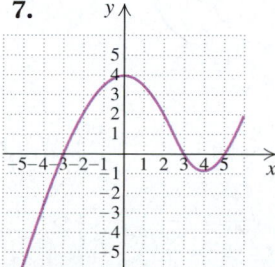

8.
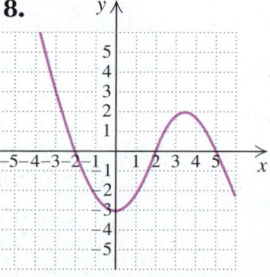

For Exercises 9–16, find **(a)** *the coordinates of any y-intercept and* **(b)** *the coordinates of any x-intercept. Do not graph.*

9. $5x + 3y = 15$

10. $5x + 2y = 20$

11. $7x - 2y = 28$

12. $4x - 3y = 24$

13. $-4x + 3y = 10$

14. $-2x + 3y = 7$

Aha! **15.** $y = 9$

16. $x = 8$

Find the intercepts. Then graph.

17. $x + 2y = 6$

18. $3x + 2y = 12$

19. $6x + 9y = 36$

20. $x + 3y = 6$

21. $-x + 3y = 9$

22. $-x + 2y = 8$

23. $2x - y = 8$

24. $3x + y = 9$

25. $y = -3x + 6$

26. $y = 2x - 6$

27. $5x - 10 = 5y$

28. $3x - 9 = 3y$

29. $2x - 5y = 10$

30. $2x - 3y = 6$

31. $6x + 2y = 12$

32. $4x + 5y = 20$

33. $x - 1 = y$

34. $3x + 2y = 8$

35. $2x - 6y = 18$

36. $2x + 7y = 6$

37. $4x - 3y = 12$

38. $3x - y = 2$

39. $-3x = 6y - 2$

40. $y = -3 - 3x$

41. $3 = 2x - 5y$

42. $-4x = 8y - 5$

43. $x + 2y = 0$

44. $y - 3x = 0$

Graph.

45. $y = 5$

46. $y = 2$

47. $x = 4$

48. $x = 6$

49. $y = -2$

50. $y = -4$

51. $x = -1$

52. $x = -6$

53. $y = 7$

54. $y = 1$

55. $x = 1$

56. $x = -2$

57. $y = 0$

58. $y = \frac{3}{2}$

59. $x = -\frac{5}{2}$

60. $x = 0$

 61. $-5y = 15$

62. $12x = -36$

63. $35 + 7y = 0$

64. $-3x - 24 = 0$

Write an equation for each graph.

65.

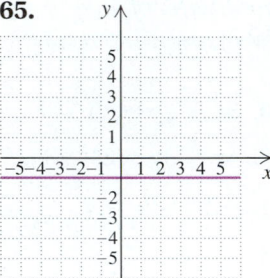

66.

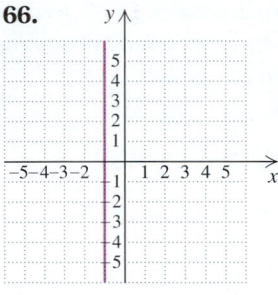

67.

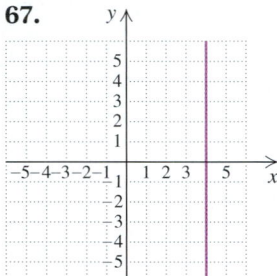

68.

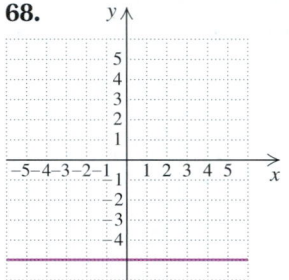

69.

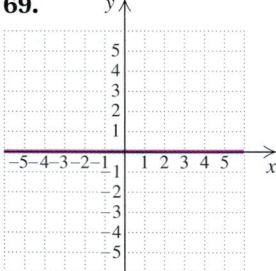

70.

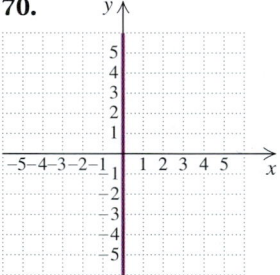

 71. Explain in your own words why the graph of $y = 8$ is a horizontal line.

 72. Explain in your own words why the graph of $x = -4$ is a vertical line.

SKILL MAINTENANCE

Translate to an algebraic expression.

73. 7 less than d

74. 5 more than w

75. The sum of 2 and a number

76. The product of 3 and a number

77. Twice the sum of two numbers

78. Half of the sum of two numbers

SYNTHESIS

 79. Describe what the graph of $x + y = C$ will look like for any choice of C.

 80. If the graph of a linear equation has one point that is both the x- and the y-intercepts, what is that point? Why?

81. Write an equation for the x-axis.

82. Write an equation of the line parallel to the x-axis and passing through $(3, 5)$.

83. Write an equation of the line parallel to the y-axis and passing through $(-2, 7)$.

84. Find the coordinates of the point of intersection of the graphs of $y = x$ and $y = 6$.

85. Find the coordinates of the point of intersection of the graphs of the equations $x = -3$ and $y = x$.

86. Write an equation of the line shown in Exercise 1.

87. Write an equation of the line shown in Exercise 4.

88. Find the value of C such that the graph of $3x + C = 5y$ has an x-intercept of $(-4, 0)$.

89. Find the value of C such that the graph of $4x = C - 3y$ has a y-intercept of $(0, -8)$.

 90. For A and B nonzero, the graphs of $Ax + D = C$ and $By + D = C$ will be parallel to an axis. Explain why.

In Exercises 91–96, find the intercepts of each equation algebraically. Then adjust the window and scale so that the intercepts can be checked graphically with no further window adjustments.

91. $3x + 2y = 50$

92. $2x - 7y = 80$

93. $y = 0.2x - 9$

94. $y = 1.3x - 15$

95. $25x - 20y = 1$

96. $50x + 25y = 1$

3.4

Rates

Rates of Change • Visualizing Rates

Rates of Change

Because graphs make use of two axes, they allow us to visualize how two quantities change with respect to each other. A number accompanied by units is used to represent this type of change and is referred to as a *rate*.

> ### Rate
>
> A *rate* is a ratio that indicates how two quantities change with respect to each other.

Rates occur often in everyday life:

A town that grows by 3400 residents over a period of 2 yr has a *growth rate* of $\frac{3400}{2}$, or 1700, residents per year.

A person running 150 m in 20 sec is moving at a *rate* of $\frac{150}{20}$, or 7.5, m/sec (meters per second).

A class of 25 students pays a total of $93.75 to visit a museum. The *rate* is $\frac{\$93.75}{25}$, or $3.75, per student.

Caution!   To calculate a rate, it is important to keep track of the units being used.

E x a m p l e   1   On January 3, Nell rented a Ford Focus with a full tank of gas and 9312 mi on the odometer. On January 7, she returned the car with 9630 mi on the odometer.* If the rental agency charged Nell $108 for the rental and needed 12 gal of gas to fill up the gas tank, find the following rates:

a) The car's rate of gas consumption, in miles per gallon
b) The average cost of the rental, in dollars per day
c) The car's rate of travel, in miles per day

---

*For all rental problems, assume that the pickup time was later in the day than the return time so that no late fees were applied.

Solution

a) The rate of gas consumption, in miles per gallon, is found by dividing the number of miles traveled by the number of gallons used for that amount of driving:

$$\text{Rate, in miles per gallon} = \frac{9630 \text{ mi} - 9312 \text{ mi}}{12 \text{ gal}} \qquad \text{\color{magenta}The word "per" indicates division.}$$

$$= \frac{318 \text{ mi}}{12 \text{ gal}}$$

$$= 26.5 \text{ mi/gal} \qquad \text{\color{magenta}Dividing}$$

$$= 26.5 \text{ miles per gallon.}$$

b) The average cost of the rental, in dollars per day, is found by dividing the cost of the rental by the number of days:

$$\text{Rate, in dollars per day} = \frac{108 \text{ dollars}}{4 \text{ days}} \qquad \text{\color{magenta}From January 3 to January 7 is } 7 - 3 = 4 \text{ days.}$$

$$= 27 \text{ dollars/day}$$

$$= \$27 \text{ per day.}$$

c) The car's rate of travel, in miles per day, is found by dividing the number of miles traveled by the number of days:

$$\text{Rate, in miles per day} = \frac{318 \text{ mi}}{4 \text{ days}} \qquad \text{\color{magenta}9630 mi} - 9312 \text{ mi} = 318 \text{ mi} \\ \text{From January 3 to January 7 is } 7 - 3 = 4 \text{ days.}$$

$$= 79.5 \text{ mi/day}$$

$$= 79.5 \text{ mi per day.}$$

Many problems involve a rate of travel, or *speed*. The **speed** of an object is found by dividing the distance traveled by the time required to travel that distance.

Example 2

Transportation.   The Atlantic City Express is a bus that makes regular trips between New York City and Atlantic City, New Jersey. At 6:00 P.M., the bus is at mileage marker 70 on the Garden State Parkway, and at 8:00 P.M. it is at marker 200. Find the average speed of the bus.

Solution   Speed is the distance traveled divided by the time spent traveling:

$$\text{Bus speed} = \frac{\text{Distance traveled}}{\text{Time spent traveling}}$$

$$= \frac{\text{Change in mileage}}{\text{Change in time}}$$

$$= \frac{130 \text{ mi}}{2 \text{ hr}} \qquad \text{\color{magenta}200 mi} - 70 \text{ mi} = 130 \text{ mi}; \\ \text{8:00 P.M.} - 6:00 \text{ P.M.} = 2 \text{ hr}$$

$$= 65 \frac{\text{mi}}{\text{hr}}$$

$$= 65 \text{ miles per hour} \qquad \text{\color{magenta}This } \textit{average} \text{ speed does not indicate by how much the bus speed may vary along the route.}$$

Visualizing Rates

Graphs allow us to visualize a rate of change. As a rule, the quantity listed in the numerator appears on the vertical axis and the quantity listed in the denominator appears on the horizontal axis.

E x a m p l e   3

Communication. In 1998, there were approximately 69 million cellular telephone subscribers in the United States, and the figure was growing at a rate of about 14 million per year (*Source: Statistical Abstract of the United States*, 1999). Draw a linear graph to represent this information.

Solution   To decide which labels to use on the axes, we note that the rate is given in millions of customers per year. Thus we list *Number of customers, in millions*, on the vertical axis and *Year* on the horizontal axis.

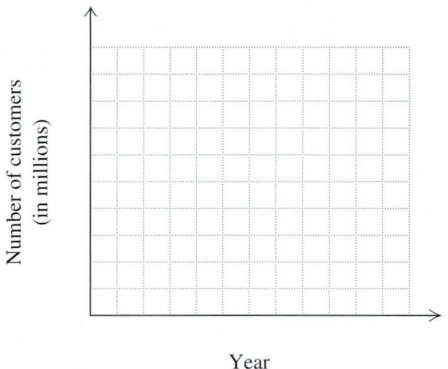

14 million per year is 14 million/yr; millions of customers is the vertical axis; year is the horizontal axis.

Next, we must decide on a scale for each axis that will allow us to plot the given information. If we count by 10's of millions on the vertical axis, we can easily reach 69 million without needing a terribly large graph. On the horizontal axis, we list several years, making certain that 1998 is included (see the figure on the left below).

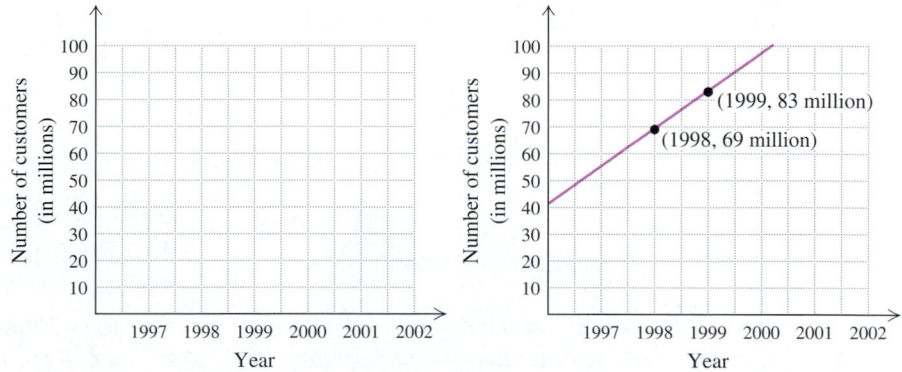

Finally, we display the given information. To do so, we plot the point that corresponds to (1998, 69 million). Then, to display the rate of growth, we move from that point to a second point that represents 14 million more users one year later. The coordinates of this point are (1998 + 1, 69 + 14 million), or (1999, 83 million). We draw a line passing through the two points, as shown in the figure on the right above.

E x a m p l e   4

Haircutting.   Gary's Barber Shop has a graph displaying data from a recent day of work.

a) What rate can be determined from the graph?
b) What is that rate?

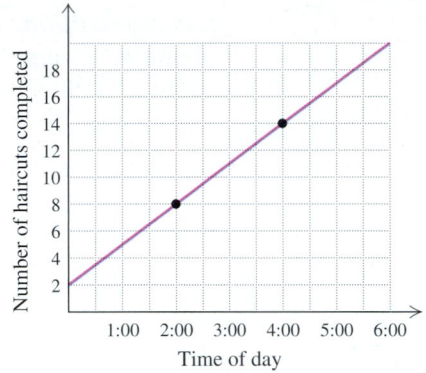

Solution

a) Because the vertical axis shows the number of haircuts completed and the horizontal axis lists the time in hour-long increments, we can find the rate *Number of haircuts per hour.*

b) The points (2:00, 8 haircuts) and (4:00, 14 haircuts) are both on the graph. This tells us that in the 2 hr between 2:00 and 4:00, there were $14 - 8 = 6$ haircuts completed. Thus the rate is

$$\frac{14 \text{ haircuts} - 8 \text{ haircuts}}{4{:}00 - 2{:}00} = \frac{6 \text{ haircuts}}{2 \text{ hours}}$$

$$= 3 \text{ haircuts per hour.}$$

FOR EXTRA HELP

Exercise Set 3.4

Digital Video Tutor CD 2
Videotape 5

InterAct Math

Math Tutor Center

MathXL

MyMathLab.com

Solve. For Exercises 1–8, round answers to the nearest cent. On Exercises 1 and 2, assume that the pickup time was later in the day than the return time so that no late fees were applied.

1. *Van rentals.*   Late on July 1, Frank rented a Dodge Caravan with a full tank of gas and 13,741 mi on the odometer. On July 4, he returned the van with 14,014 mi on the odometer. The rental agency

charged Frank $118 for the rental and needed 13 gal of gas to fill up the tank.

a) Find the van's rate of gas consumption, in miles per gallon.
b) Find the average cost of the rental, in dollars per day.
c) Find the rate of travel, in miles per day.
d) Find the rental rate, in cents per mile.

2. *Car rentals.*   On February 10, Maggie rented a Chevy Blazer with a full tank of gas and 13,091 mi on the odometer. On February 12, she returned the vehicle with 13,322 mi on the odometer. The rental agency charged $92 for the rental and needed 14 gal of gas to fill the tank.

a) Find the Blazer's rate of gas consumption, in miles per gallon.

b) Find the average cost of the rental, in dollars per day.

c) Find the rate of travel, in miles per day.

d) Find the rental rate, in cents per mile.

3. *Bicycle rentals.*   At 2:00, Denise rented a mountain bike from The Slick Rock Cyclery. She returned the bike at 5:00, after cycling 18 mi. Denise paid $10.50 for the rental.

a) Find Denise's average speed, in miles per hour.

b) Find the rental rate, in dollars per hour.

c) Find the rental rate, in dollars per mile.

4. *Bicycle rentals.*   At 9:00, Blair rented a mountain bike from The Bike Rack. He returned the bicycle at 11:00, after cycling 14 mi. Blair paid $12 for the rental.

a) Find Blair's average speed, in miles per hour.

b) Find the rental rate, in dollars per hour.

c) Find the rental rate, in dollars per mile.

5. *Temporary help.*   A typist from Jobsite Services, Inc., reports to AKA International for work at 9:00 A.M. and leaves at 5:00 P.M. after having typed from the end of page 12 to the end of page 48 of a prospectus. AKA International pays $128 for the typist's services.

a) Find the rate of pay, in dollars per hour.

b) Find the average typing rate, in number of pages per hour.

c) Find the rate of pay, in dollars per page.

6. *Temporary help.*   A typist for Kelly Services reports to 3E's Properties for work at 10:00 A.M. and leaves at 6:00 P.M. after having typed from the end of page 8 to the end of page 50 of a proposal. 3E's pays $120 for the typist's services.

a) Find the rate of pay, in dollars per hour.

b) Find the average typing rate, in number of pages per hour.

c) Find the rate of pay, in dollars per page.

7. *Two-year-college tuition.*   The average tuition at a public two-year college was $1239 in 1996 and $1318 in 1998 (*Source*: *Statistical Abstract of the United States,* 1999). Find the rate at which tuition was increasing.

8. *Four-year-college tuition.*   The average tuition at a public four-year college was $2977 in 1995 and $3489 in 1998 (*Source*: *Statistical Abstract of the United States,* 1999, p. 199). Find the rate at which tuition was increasing.

9. *Elevators.*   At 2:38, Serge entered an elevator on the 34th floor of the Regency Hotel. At 2:40, he stepped off at the 5th floor.

a) Find the elevator's average rate of travel, in number of floors per minute.

b) Find the elevator's average rate of travel, in seconds per floor.

10. *Snow removal.*   By 1:00 P.M., Erin had already shoveled 2 driveways, and by 6:00 P.M., the number was up to 7.

a) Find Erin's shoveling rate, in number of driveways per hour.

b) Find Erin's shoveling rate, in hours per driveway.

11. *Mountaineering.*   As part of an ill-fated expedition to climb Mt. Everest in 1996, author Jon Krakauer departed "The Balcony," elevation 27,600 ft, at 7:00 A.M. and reached the summit, elevation 29,028 ft, at 1:25 P.M. (*Source*: Krakauer, Jon, *Into Thin Air, the Illustrated Edition*. New York: Random House, 1998)

a) Find Krakauer's average rate of ascent, in feet per minute.

b) Find Krakauer's average rate of ascent, in minutes per foot.

12. *Mountaineering.*   The fastest ascent of Mt. Everest was accomplished by Kaji Sherpa on October 17, 1998. Kaji Sherpa climbed from base camp, elevation 17,552 ft, to the summit, elevation 29,028 ft, in 20 hr 24 min (*Source*: *Guinness Book of World Records* 2000, Millenium Edition).

a) Find Kaji Sherpa's rate of ascent, in feet per minute.

b) Find Kaji Sherpa's rate of ascent, in minutes per foot.

In Exercises 13–20, draw a linear graph to represent the given information. Be sure to label and number the axes appropriately (see Example 3).

13. *Law enforcement.*   In 1996, there were approximately 37 million crimes reported in the United States, and the figure was dropping at a rate of about 2.5 million per year (*Source*: *Statistical Abstract of the United States*, 1999).

14. *Fire fighting.*   In 1996, there were approximately 2 million fires in the United States, and the figure was dropping at a rate of about 0.2 million per year (*Source*: *Statistical Abstract of the United States*, 1999).

15. *Train travel.*   At 3:00 P.M., the Boston–Washington Metroliner had traveled 230 mi and was cruising at a rate of 90 miles per hour.

16. *Plane travel.*   At 4:00 P.M., the Seattle–Los Angeles shuttle had traveled 400 mi and was cruising at a rate of 300 miles per hour.

17. *Wages.*   By 2:00 P.M., Diane had earned $50. She continued earning money at a rate of $15 per hour.

18. *Wages.*   By 3:00 P.M., Arnie had earned $70. He continued earning money at a rate of $12 per hour.

19. *Telephone bills.*   Roberta's phone bill was already $7.50 when she made a call for which she was charged at a rate of $0.10 per minute.

20. *Telephone bills.*   At 3:00 P.M., Larry's phone bill was $6.50 and increasing at a rate of 7¢ per minute.

In Exercises 21–30, use the graph provided to calculate a rate of change in which the units of the horizontal axis are used in the denominator.

21. *Hairdresser.*   Eve's Custom Cuts has a graph displaying data from a recent day of work. At what rate does Eve work?

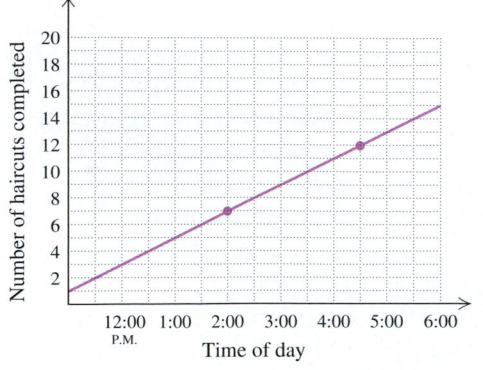

22. *Manicures.*   The following graph shows data from a recent day's work at the O'Hara School of Cosmetology. At what rate do they work?

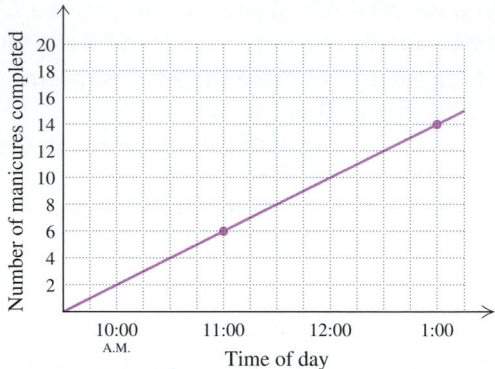

23. *Train travel.*   The following graph shows data from a recent train ride from Chicago to St. Louis. At what rate did the train travel?

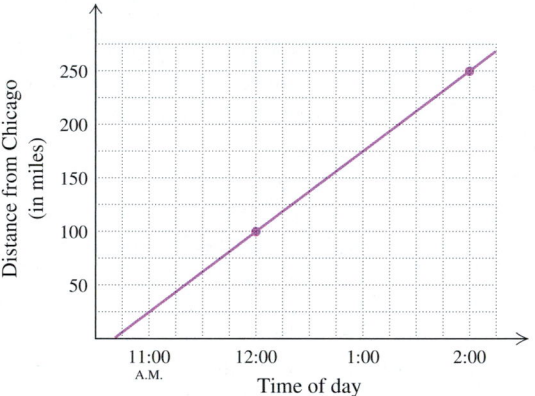

24. *Train travel.*   The following graph shows data from a recent train ride from Denver to Kansas City. At what rate did the train travel?

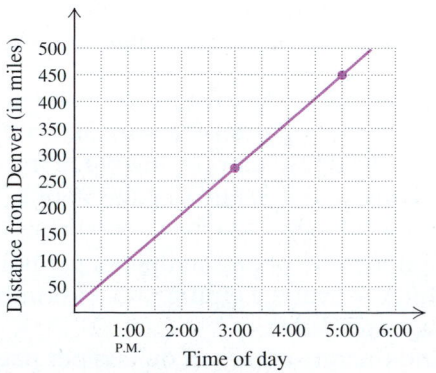

25. *Cost of a telephone call.*   The following graph shows data from a recent AT&T phone call between Burlington, VT, and Austin, TX. At what rate was the customer being billed?

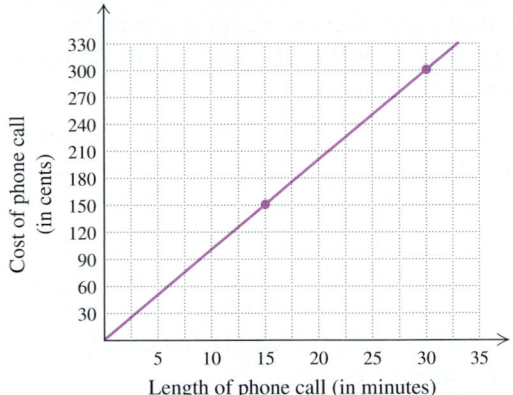

Length of phone call (in minutes)

26. *Cost of a telephone call.*   The following graph shows data from a recent MCI phone call between San Francisco, CA, and Pittsburgh, PA. At what rate was the customer being billed?

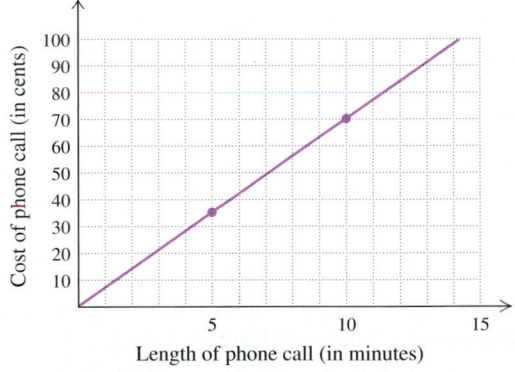

Length of phone call (in minutes)

27. *Depreciation of an office machine.*   Data regarding the value of a particular color copier is represented in the following graph. At what rate is the value changing?

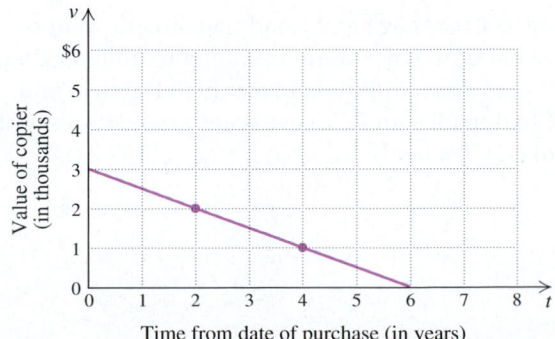

Time from date of purchase (in years)

28. *NASA spending.*   Data regarding the amount spent on the National Aeronautics and Space Administration (NASA) is represented in the following graph (based on information in the *Statistical Abstract of the United States*, 1999). At what rate is the amount spent on NASA changing?

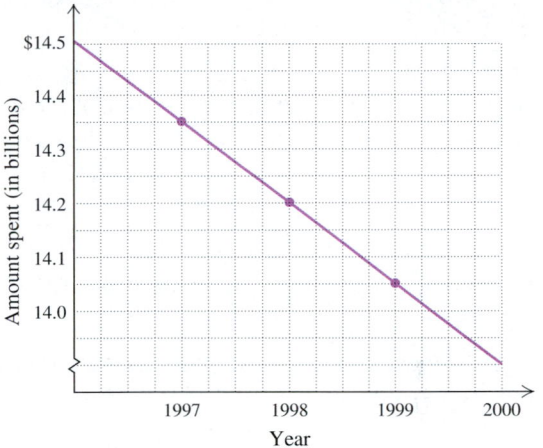

Year

29. *Gas mileage.*   The following graph shows data for a Honda Odyssey driven on interstate highways. At what rate was the vehicle consuming gas?

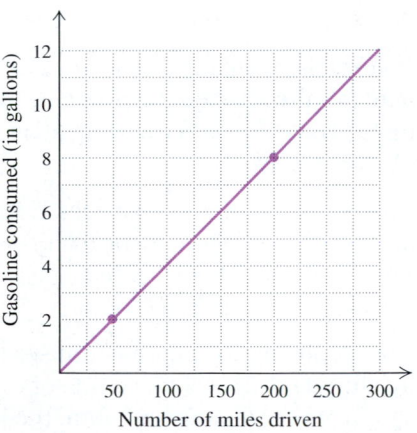

Number of miles driven

30. *Gas mileage.*   The following graph shows data for a Ford Explorer driven on city streets. At what rate was the vehicle consuming gas?

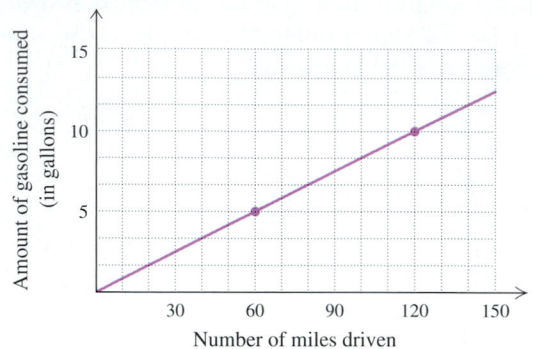

31. What does a negative rate of travel indicate? Explain.

32. Explain how to convert from kilometers per hour to meters per second.

SKILL MAINTENANCE

33. $-2 - (-7)$

34. $-9 - (-3)$

35. $\dfrac{5 - (-4)}{-2 - 7}$

36. $\dfrac{8 - (-4)}{2 - 11}$

37. $\dfrac{-4 - 8}{7 - (-2)}$

38. $\dfrac{-5 - 3}{6 - (-4)}$

SYNTHESIS

39. Write an exercise similar to Exercises 21–30 for a classmate to solve. Design the problem so that the solution is "The plane was traveling at a rate of 300 miles per hour."

40. Write an exercise similar to Exercises 1–12 for a classmate to solve. Design the problem so that the solution is "The motorcycle's rate of gas consumption was 65 miles per gallon."

41. *Aviation.*   A Boeing 737 climbs from sea level to a cruising altitude of 31,500 ft at a rate of 6300 ft/min. After cruising for 3 min, the jet is forced to land, descending at a rate of 3500 ft/min. Represent the flight with a graph in which altitude is measured on the vertical axis and time on the horizontal axis.

42. *Wages with commissions.*   Each salesperson at Mike's Bikes is paid $140 a week plus 13% of all sales up to $2000, and then 20% on any sales in excess of $2000. Draw a graph in which sales are measured on the horizontal axis and wages on the vertical axis. Then use the graph to estimate the wages paid when a salesperson sells $2700 in merchandise in one week.

43. *Gas mileage.*   Suppose that a Honda motorcycle goes twice as far as a Honda Odyssey on the same amount of gas (see Exercise 29). Draw a graph that reflects this information.

44. *Taxi fares.*   The driver of a New York City Yellow Cab recently charged $2 plus 50¢ for each fifth of a mile traveled. Draw a graph that could be used to determine the cost of a fare.

45. *Navigation.*   In 3 sec, Penny walks 24 ft, to the bow (front) of a tugboat. The boat is cruising at a rate of 5 feet per second. What is Penny's rate of travel with respect to land?

46. *Aviation.*   Tim's F-16 jet is moving forward at a deck speed of 95 mph aboard an aircraft carrier that is traveling 39 mph in the same direction. How fast is the jet traveling, in minutes per mile, with respect to the sea?

47. *Running.*   Annette ran from the 4-km mark to the 7-km mark of a 10-km race in 15.5 min. At this rate, how long would it take Annette to run a 5-mi race?

48. *Running.*   Jerod ran from the 2-mi marker to the finish line of a 5-mi race in 25 min. At this rate, how long would it take Jerod to run a 10-km race?

49. Marcy picks apples twice as fast as Ryan. By 4:30, Ryan had already picked 4 bushels of apples. Fifty minutes later, his total reached $5\frac{1}{2}$ bushels. Find Marcy's picking rate. Give your answer in number of bushels per hour.

50. By 3:00, Catanya and Chad had already made 46 candles. Forty minutes later, the total reached 64 candles. Find the rate at which Catanya and Chad made candles. Give your answer as a number of candles per hour.

C O R N E R

Determining Depreciation Rates

Focus: Modeling, graphing, and rates

Time: 30 minutes

Group size: 3

Materials: Graph paper and straightedges

From the minute a new car is driven out of the dealership, it *depreciates,* or drops in value with the passing of time. The N.A.D.A.® Official Used Car Guide (often called the "blue book" despite its orange cover) is a monthly listing of the trade-in values of used cars. The data below are taken from the N.A.D.A. guides for June and December of 2000.

Car	Trade-in Value in June 2000	Trade-in Value in December 2000
1998 Chevrolet Camaro convertible V6	$16,200	$14,900
1998 Ford Mustang convertible, 2 door, GT	$16,425	$15,325
1998 VW Passat GLS Turbo, 4 cylinder	$15,975	$14,850

ACTIVITY

1. Each group member should select a different one of the cars listed in the table above as his or her own. Assuming that the values are dropping linearly, each student should draw a line representing the trade-in value of his or her car. Draw all three lines on the same graph. Let the horizontal axis represent the time, in months, since June 2000, and let the vertical axis represent the trade-in value of each car. Decide as a group how many months or dollars each square should represent. Make the drawings as neat as possible.

2. At what *rate* is each car depreciating and how are the different rates illustrated in the graph of part (1)?

3. If one of the three cars had to be sold in June 2001, which one would your group sell and why? Compare answers with other groups.

3.5

Slope

Rate and Slope • Horizontal and Vertical Lines •
Applications

In Section 3.4, we introduced *rate* as a method of measuring how two quantities change with respect to each other. In this section, we will discuss how rate can be related to the slope of a line.

Rate and Slope

Suppose that a car manufacturer operates two plants: one in Michigan and one in Pennsylvania. Knowing that the Michigan plant produces 3 cars every 2 hours and the Pennsylvania plant produces 5 cars every 4 hours, we can set up tables listing the number of cars produced after various amounts of time.

Michigan Plant	
Hours Elapsed	**Cars Produced**
0	0
2	3
4	6
6	9
8	12

Pennsylvania Plant	
Hours Elapsed	**Cars Produced**
0	0
4	5
8	10
12	15
16	20

By comparing the number of cars produced at each plant over a specified period of time, we can compare the two production rates. For example, the Michigan plant produces 3 cars every 2 hours, so its *rate* is $3 \div 2 = 1\frac{1}{2}$, or $\frac{3}{2}$ cars per hour. Since the Pennsylvania plant produces 5 cars every 4 hours, its rate is $5 \div 4 = 1\frac{1}{4}$, or $\frac{5}{4}$ cars per hour.

Let's now graph the pairs of numbers listed in the tables, using the horizontal axis for time and the vertical axis for the number of cars produced. Note that the rate in the Michigan plant is slightly greater so its graph is slightly steeper.

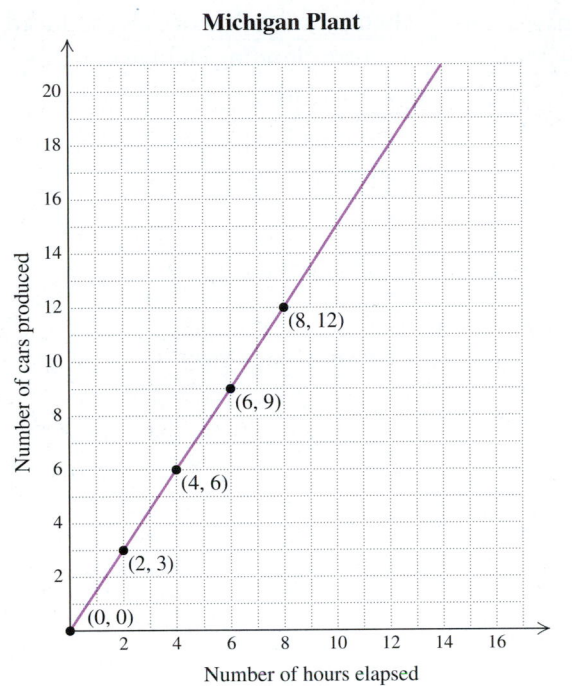

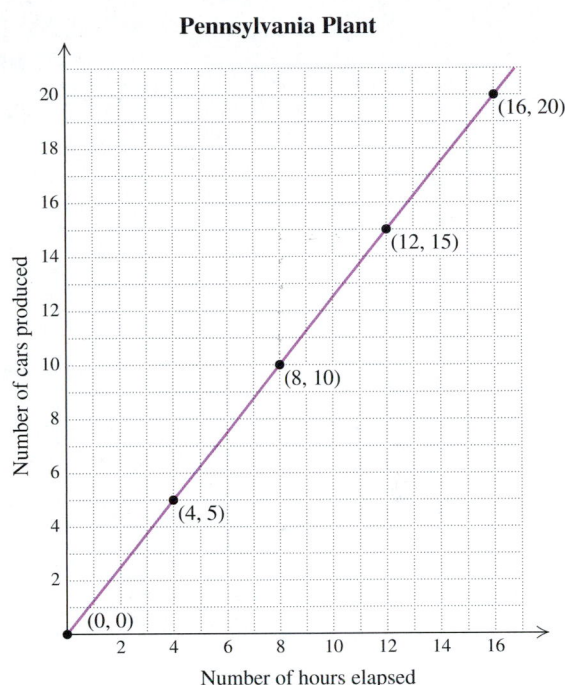

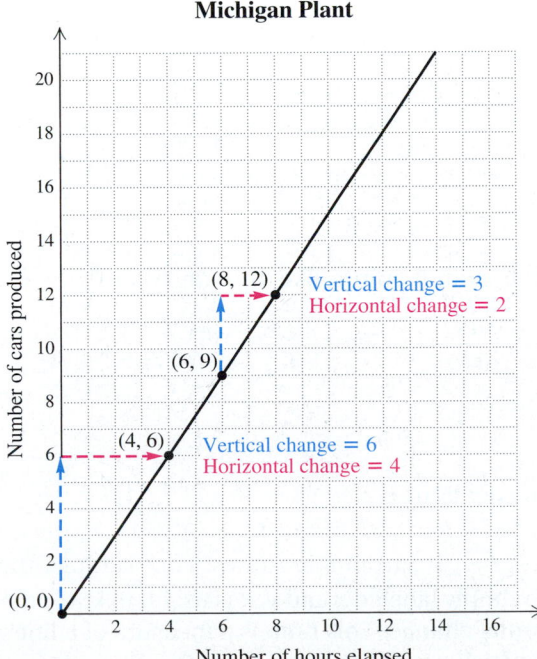

The rates $\frac{3}{2}$ and $\frac{5}{4}$ can also be found using the coordinates of any two points that are on the line. For example, we can use the points $(6, 9)$ and $(8, 12)$ to find the production rate for the Michigan plant. To do so, remember that these coordinates tell us that after 6 hr, 9 cars have been produced, and after 8 hr, 12 cars have been produced. In the 2 hr between the 6-hr and 8-hr points, $12 - 9$, or 3 cars were produced. Thus,

$$\begin{matrix} \text{Michigan} \\ \text{production rate} \end{matrix} = \frac{\text{change in number of cars produced}}{\text{corresponding change in time}}$$

$$= \frac{12 - 9 \text{ cars}}{8 - 6 \text{ hr}}$$

$$= \frac{3 \text{ cars}}{2 \text{ hr}} = \frac{3}{2} \text{ cars per hour.}$$

Because the line is straight, the same rate is found using *any* pair of points on the line. For example, using $(0, 0)$ and $(4, 6)$, we have

$$\begin{matrix} \text{Michigan} \\ \text{production rate} \end{matrix} = \frac{6 - 0 \text{ cars}}{4 - 0 \text{ hr}} = \frac{6 \text{ cars}}{4 \text{ hr}} = \frac{3}{2} \text{ cars per hour.}$$

Note that the rate is always the vertical change divided by the associated horizontal change.

Example 1   Use the graph of car production at the Pennsylvania plant to find the rate of production.

Solution   We can use any two points on the line, such as $(12, 15)$ and $(16, 20)$:

$$\begin{aligned} \text{Pennsylvania} \atop \text{production rate} &= \frac{\text{change in number of cars produced}}{\text{corresponding change in time}} \\[6pt] &= \frac{20 - 15 \text{ cars}}{16 - 12 \text{ hr}} \\[6pt] &= \frac{5 \text{ cars}}{4 \text{ hr}} \\[6pt] &= \frac{5}{4} \text{ cars per hour.} \end{aligned}$$

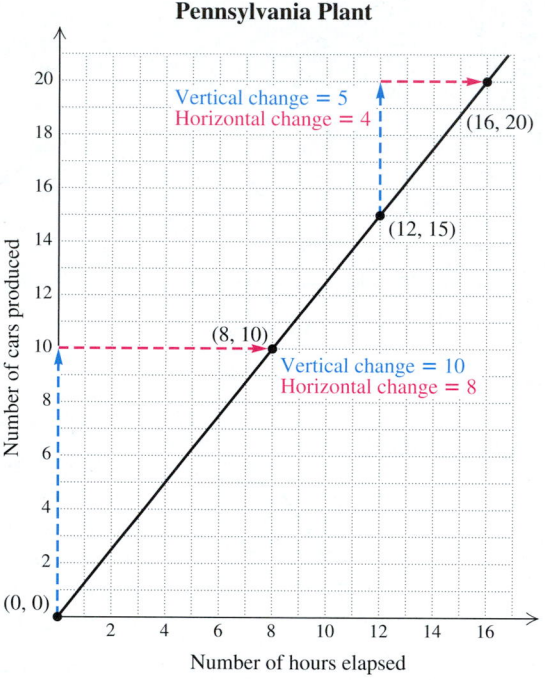

Pennsylvania Plant

As a check, we can use another pair of points, like $(0, 0)$ and $(8, 10)$:

$$\begin{aligned} \text{Pennsylvania} \atop \text{production rate} &= \frac{10 - 0 \text{ cars}}{8 - 0 \text{ hr}} \\[6pt] &= \frac{10 \text{ cars}}{8 \text{ hr}} \\[6pt] &= \frac{5}{4} \text{ cars per hour.} \end{aligned}$$

When the axes of a graph are simply labeled x and y, it is useful to know the ratio of vertical change to horizontal change. This ratio is a measure of a line's slant, or **slope**, and is the rate at which y is changing with respect to x.

Consider a line passing through $(2, 3)$ and $(6, 5)$, as shown below. We find the ratio of vertical change, or *rise*, to horizontal change, or *run*, as follows:

$$\text{Ratio of vertical change to horizontal change} = \frac{\text{change in } y}{\text{change in } x} = \frac{\text{rise}}{\text{run}}$$

$$= \frac{5 - 3}{6 - 2}$$

$$= \frac{2}{4}, \text{ or } \frac{1}{2}.$$

Note that these calculations can be performed without viewing a graph.

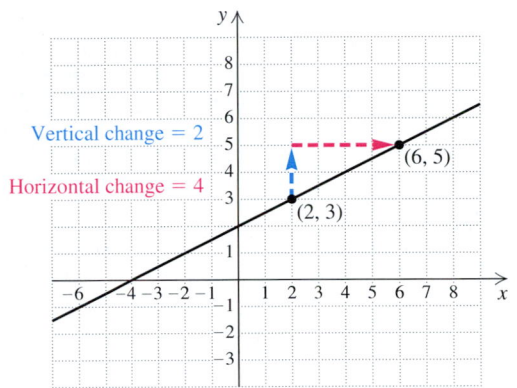

Thus the y-coordinates of points on this line increase at a rate of 2 units for every 4-unit increase in x, 1 unit for every 2-unit increase in x, or $\frac{1}{2}$ unit for every 1-unit increase in x. The slope of the line is $\frac{1}{2}$.

Slope

The *slope* of the line containing points (x_1, y_1) and (x_2, y_2) is given by

$$m = \frac{\text{change in } y}{\text{change in } x} = \frac{\text{rise}}{\text{run}} = \frac{y_2 - y_1}{x_2 - x_1}.$$

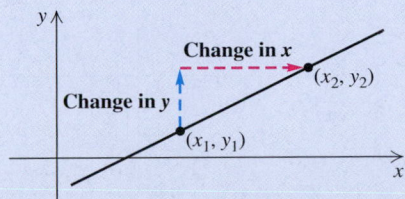

E x a m p l e   2   Graph the line containing the points $(-4, 3)$ and $(2, -6)$ and find the slope.

Solution   The graph is shown below. From $(-4, 3)$ to $(2, -6)$, the change in y, or rise, is $-6 - 3$, or -9. The change in x, or run, is $2 - (-4)$, or 6. Thus,

$$\text{Slope} = \frac{\text{change in } y}{\text{change in } x}$$

$$= \frac{\text{rise}}{\text{run}}$$

$$= \frac{-6 - 3}{2 - (-4)}$$

$$= \frac{-9}{6}$$

$$= -\frac{9}{6}, \text{ or } -\frac{3}{2}.$$

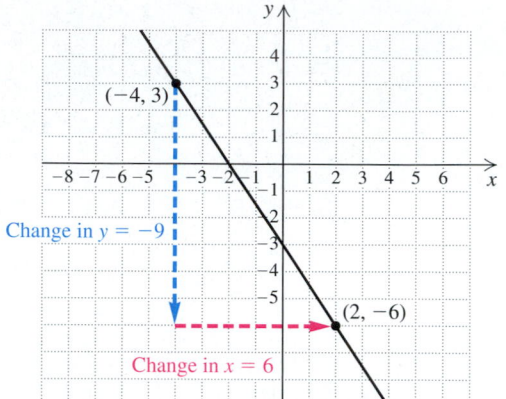

Change in $y = -9$

Change in $x = 6$

Caution!   When we use the formula

$$m = \frac{y_2 - y_1}{x_2 - x_1},$$

it makes no difference which point is considered (x_1, y_1). What matters is that we subtract the y-coordinates in the same order that we subtract the x-coordinates.

To illustrate, we reverse *both* of the subtractions in Example 2. The slope is still $-\frac{3}{2}$:

$$\text{Slope} = \frac{\text{change in } y}{\text{change in } x} = \frac{3 - (-6)}{-4 - 2} = \frac{9}{-6} = -\frac{3}{2}.$$

If a line has a positive slope, it slants up from left to right. The larger the slope, the steeper the slant. A line with negative slope slants down from left to right.

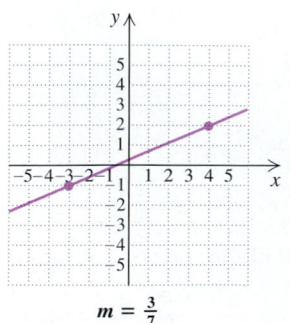

$m = \frac{3}{7}$

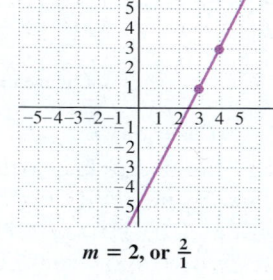

$m = 2, \text{ or } \frac{2}{1}$

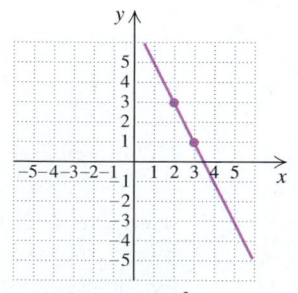

$m = -2, \text{ or } \frac{-2}{1}$

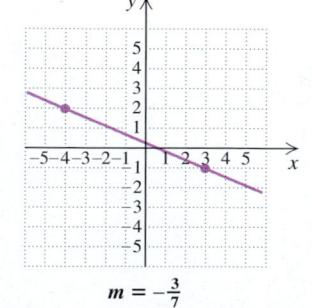

$m = -\frac{3}{7}$

Horizontal and Vertical Lines

What about the slope of a horizontal or a vertical line?

E x a m p l e   3   Find the slope of the line $y = 4$.

Solution   Consider the points $(2, 4)$ and $(-3, 4)$, which are on the line. The change in y, or the rise, is $4 - 4$, or 0. The change in x, or the run, is $-3 - 2$, or -5. Thus,

$$m = \frac{4 - 4}{-3 - 2}$$

$$= \frac{0}{-5}$$

$$= 0$$

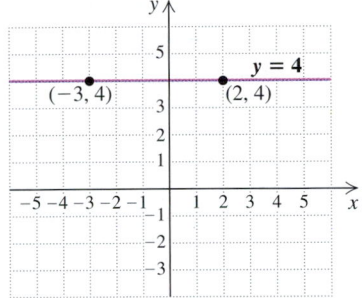

Any two points on a horizontal line have the same y-coordinate. Thus the change in y is 0, so the slope is 0.

A horizontal line has slope 0.

E x a m p l e   4   Find the slope of the line $x = -3$.

Solution   Consider the points $(-3, 4)$ and $(-3, -2)$, which are on the line. The change in y, or the rise, is $-2 - 4$, or -6. The change in x, or the run, is $-3 - (-3)$, or 0. Thus,

$$m = \frac{-2 - 4}{-3 - (-3)}$$

$$= \frac{-6}{0} \quad \text{(undefined)}$$

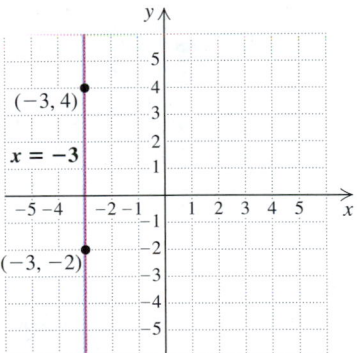

Since division by 0 is not defined, the slope of this line is not defined. The answer to a problem of this type is "The slope of this line is undefined."

The slope of a vertical line is undefined.

Applications

We have seen that slope has many real-world applications, ranging from car speed to production rate. Some applications use slope to measure steepness. For example, numbers like 2%, 3%, and 6% are often used to represent the **grade** of a road, a measure of a road's steepness. That is, a 3% grade means that for every horizontal distance of 100 ft, the road rises or drops 3 ft. The concept of grade also occurs in skiing or snowboarding, where a 4% grade is considered very tame, but a 40% grade is considered steep.

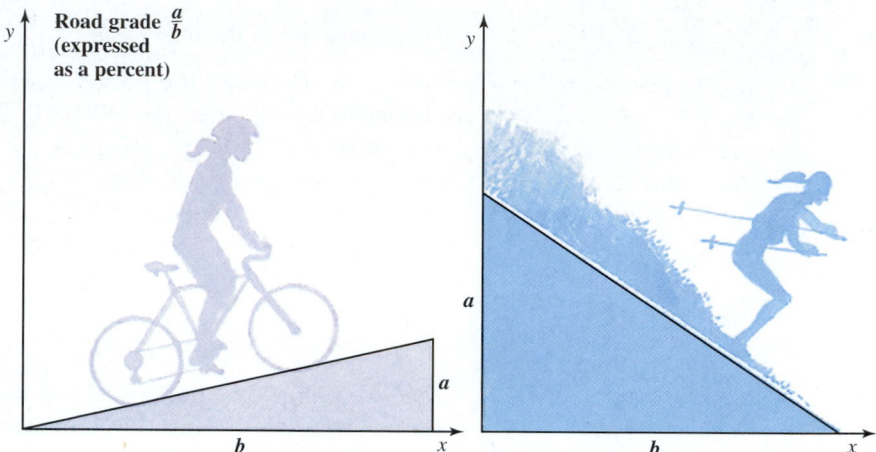

E x a m p l e   5

Skiing.   Among the steepest skiable terrain in North America, the Headwall on Mount Washington, in New Hampshire, drops 720 ft over a horizontal distance of 900 ft. Find the grade of the Headwall.

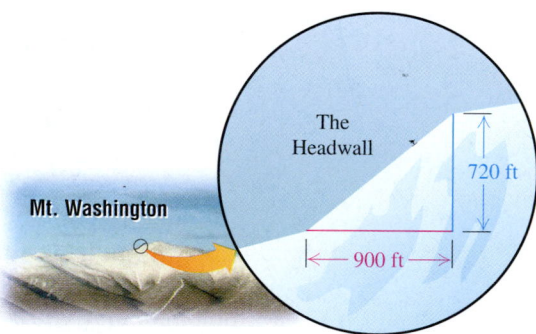

Solution   The grade of the Headwall is its slope, expressed as a percent:

$$m = \frac{720}{900}$$
$$= \frac{8}{10}$$
$$= 80\%.$$

Grade is slope expressed as a percent.

Carpenters use slope when designing stairs, ramps, or roof pitches. Another application occurs in the engineering of a dam—the force or strength of a river depends on how much the river drops over a specified distance.

CONNECTING THE CONCEPTS

In Chapters 1 and 2, we simplified expressions and solved equations. In Sections 3.1–3.3 of this chapter, we learned how a graph can be used to represent the solutions of an equation in two variables, like $y = 4 - x$ or $2x + 5y = 15$.

Graphs are used in many important applications. Sections 3.4 and 3.5 have shown us that the slope of a line can be used to represent the rate at which the quantity measured on the vertical axis changes with respect to the quantity measured on the horizontal axis.

In Section 3.6, we will return to the task of graphing equations. This time, however, our understanding of rates and slope will provide us with the tools necessary to develop shortcuts that will streamline our work.

FOR EXTRA HELP

Exercise Set 3.5

Digital Video Tutor CD 2
Videotape 6

InterAct Math

Math Tutor Center

MathXL

MyMathLab.com

1. Find the rate at which a runner burns calories.

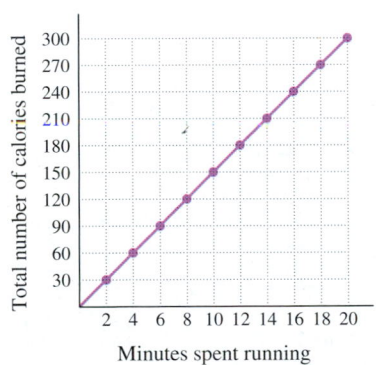

2. Find the rate of change of the U.S. population (based on information in the *Statistical Abstract of the United States*, 1999).

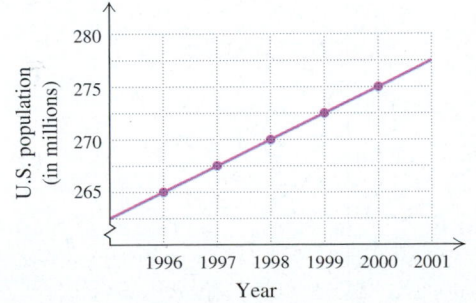

3. Find the rate of change in SAT math scores with respect to family income (based on data from 1999 college-bound seniors in Massachusetts).

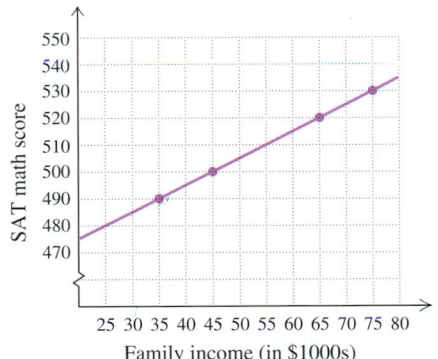

4. Find the rate of change in SAT verbal scores with respect to family income (based on data from 1999 college-bound seniors in Massachusetts).

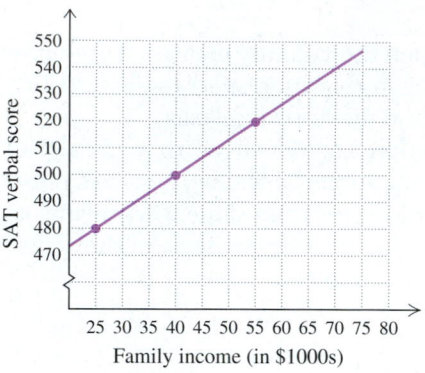

5. Find the rate of change in the percent of the U.S. budget spent on defense (based on data from the U.S. Office of Management and Budget).

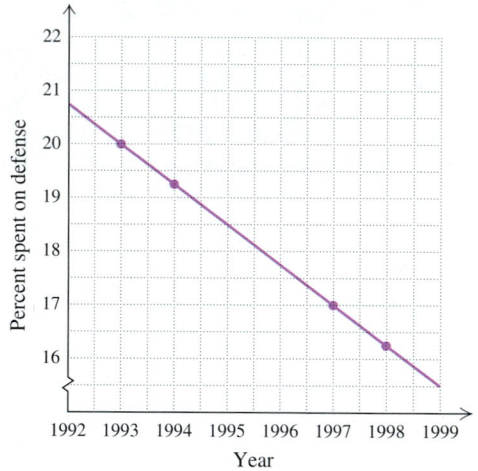

6. Find the rate of change in the percent of U.S. workers that are unemployed (*Source*: U.S. Bureau of Labor Statistics).

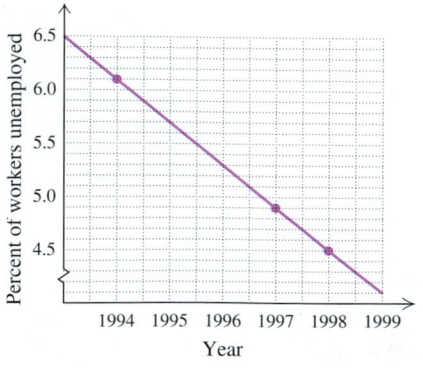

Find the slope, if it is defined, of each line. If the slope is undefined, state this.

7.

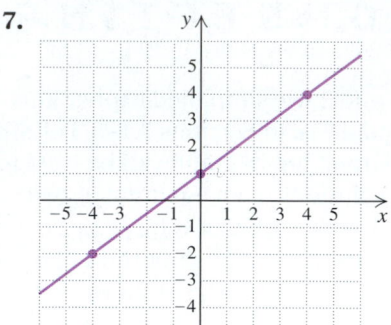

8.

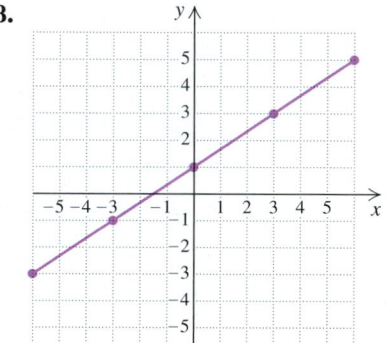

9.

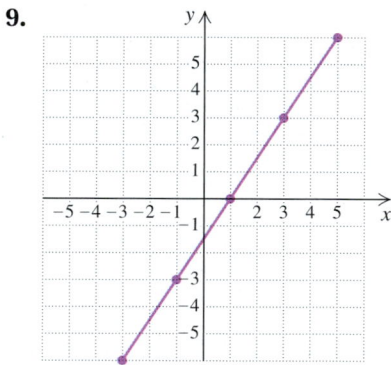

10.

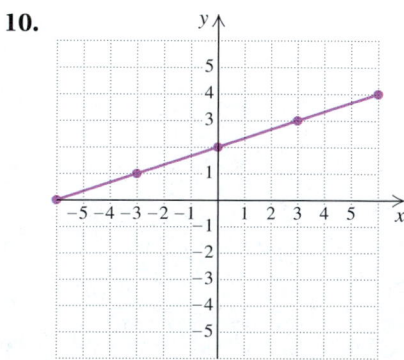

11.

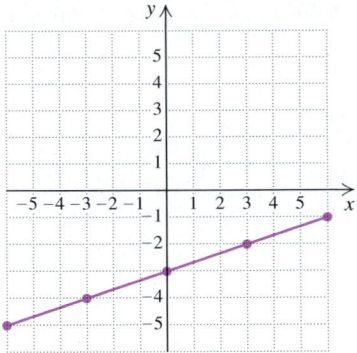

15.

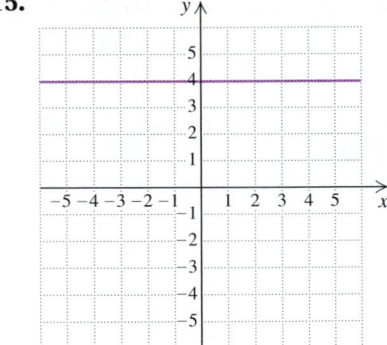

12.

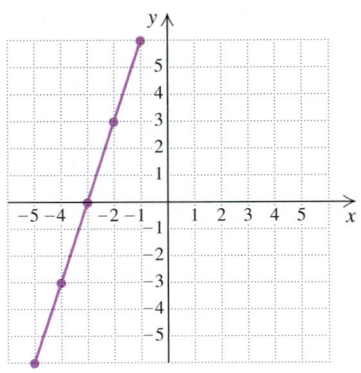

16.

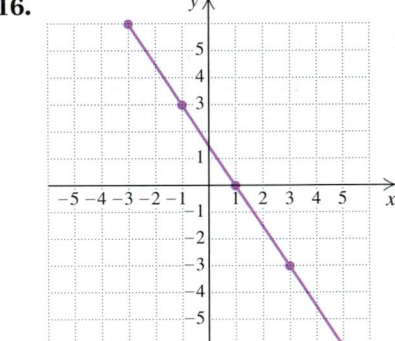

13.

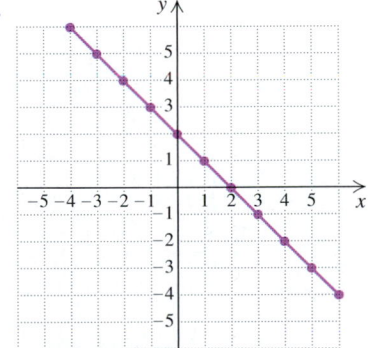

17.

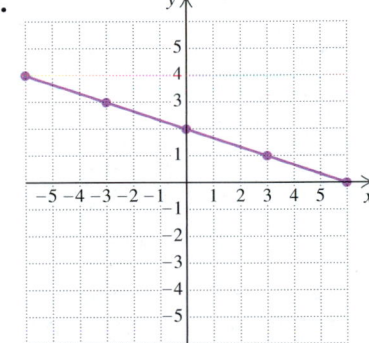

14.

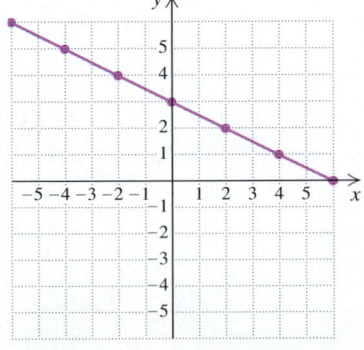

18.

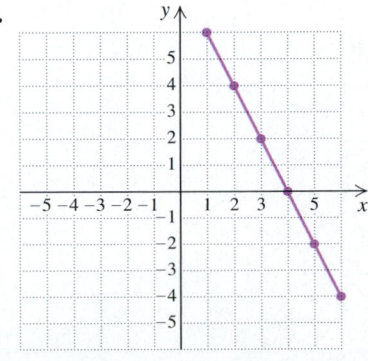

19.

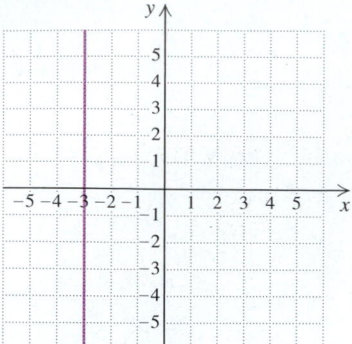

20.

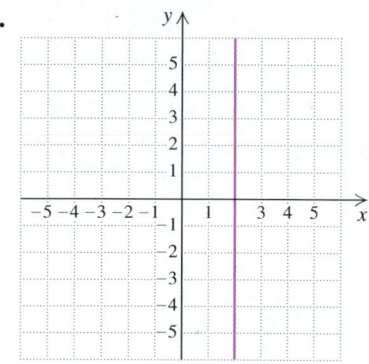

21.

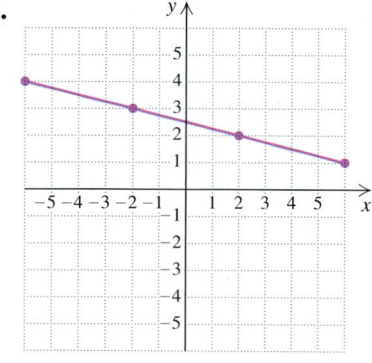

22.

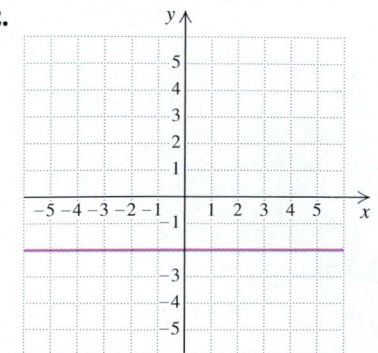

23.

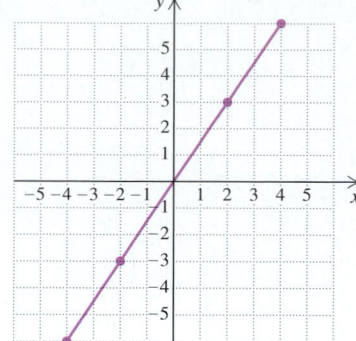

24.

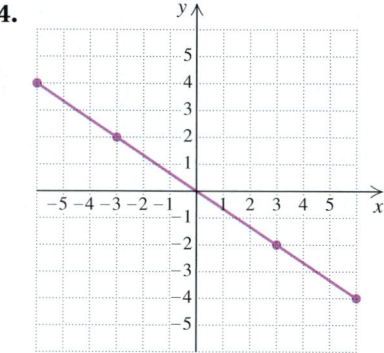

25.

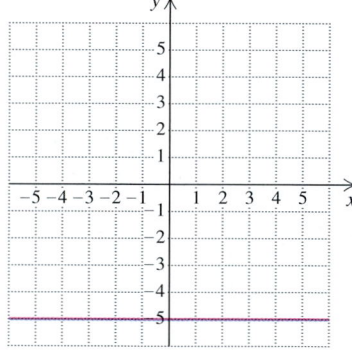

26.

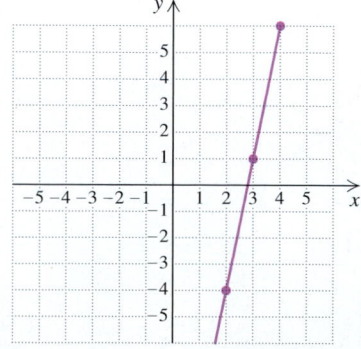

27.

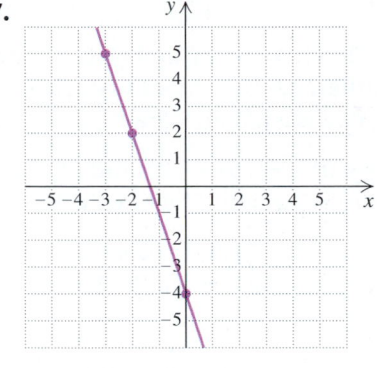

28.

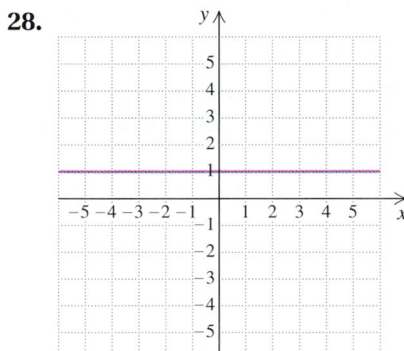

Find the slope of the line containing each given pair of points. If the slope is undefined, state this.

29. $(1, 2)$ and $(5, 8)$

30. $(2, 1)$ and $(6, 9)$

31. $(-2, 4)$ and $(3, 0)$

32. $(-4, 2)$ and $(2, -3)$

33. $(-4, 0)$ and $(5, 7)$

34. $(3, 0)$ and $(6, 2)$

35. $(0, 8)$ and $(-3, 10)$

36. $(0, 9)$ and $(4, 7)$

37. $(-2, 3)$ and $(-6, 5)$

38. $(-2, 4)$ and $(6, -7)$

Aha! **39.** $\left(-2, \frac{1}{2}\right)$ and $\left(-5, \frac{1}{2}\right)$

40. $(-5, -1)$ and $(2, 3)$

41. $(3, 4)$ and $(9, -7)$

42. $(-10, 3)$ and $(-10, 4)$

43. $(6, -4)$ and $(6, 5)$

44. $(5, -2)$ and $(-4, -2)$

Find the slope of each line. If the slope is undefined, state this.

45. $x = -3$ **46.** $x = -4$

47. $y = 4$ **48.** $y = 17$

49. $x = 9$ **50.** $x = 6$

51. $y = -9$ **52.** $y = -4$

53. *Surveying.*   Tucked between two ski areas, Vermont Route 108 rises 106 m over a horizontal distance of 1325 m. What is the grade of the road?

54. *Navigation.*   Capital Rapids drops 54 ft vertically over a horizontal distance of 1080 ft. What is the slope of the rapids?

55. *Architecture.*   To meet federal standards, a wheelchair ramp cannot rise more than 1 ft over a horizontal distance of 12 ft. Express this slope as a grade.

56. *Engineering.*   At one point, Yellowstone's Beartooth Highway rises 315 ft over a horizontal distance of 4500 ft. Find the grade of the road.

57. *Carpentry.*   Find the slope (or pitch) of the roof.

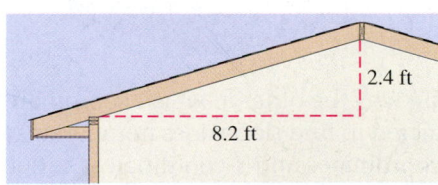

58. *Exercise.*   Find the slope (or grade) of the treadmill.

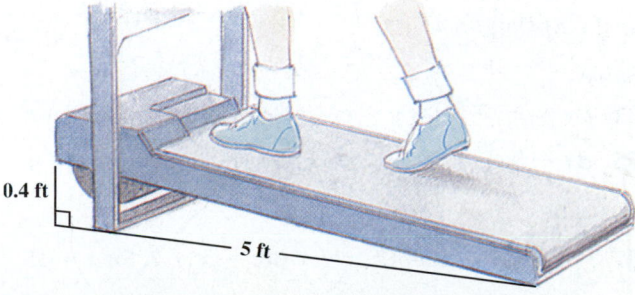

59. *Surveying.* From a base elevation of 9600 ft, Longs Peak, Colorado, rises to a summit elevation of 14,255 ft over a horizontal distance of 15,840 ft. Find the grade of Longs Peak.

60. *Construction.* Public buildings regularly include steps with 7-in. risers and 11-in. treads. Find the grade of such a stairway.

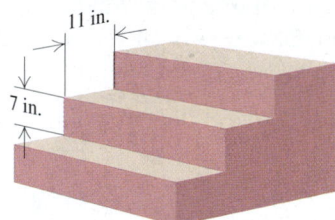

61. Explain why the order in which coordinates are subtracted to find slope does not matter so long as y-coordinates and x-coordinates are subtracted in the same order.

62. If one line has a slope of -3 and another has a slope of 2, which line is steeper? Why?

SKILL MAINTENANCE

Solve.

63. $ax + by = c$, for y

64. $rx - mn = p$, for r

65. $ax - by = c$, for y

66. $rs + nt = q$, for t

Evaluate.

67. $\frac{2}{3}x - 5$, for $x = 12$

68. $\frac{3}{5}x - 7$, for $x = 15$

SYNTHESIS

69. The points $(-4, -3)$, $(1, 4)$, $(4, 2)$, and $(-1, -5)$ are vertices of a quadrilateral. Use slopes to explain why the quadrilateral is a parallelogram.

70. Can the points $(-4, 0)$, $(-1, 5)$, $(6, 2)$, and $(2, -3)$ be vertices of a parallelogram? Why or why not?

71. A line passes through $(4, -7)$ and never enters the first quadrant. What numbers could the line have for its slope?

72. A line passes through $(2, 5)$ and never enters the second quadrant. What numbers could the line have for its slope?

73. *Architecture.* Architects often use the equation $x + y = 18$ to determine the height y, in inches, of the riser of a step when the tread is x inches wide. Express the slope of stairs designed with this equation without using the variable y.

In Exercises 74 and 75, the slope of each line is $-\frac{2}{3}$, but the numbering on one axis is missing. How many units should each tick mark on that unnumbered axis represent?

74.

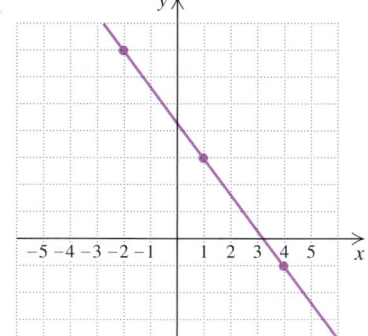

75.

Slope–Intercept Form

If we know the slope and the *y*-intercept of a line, it is possible to graph the line. In this section, we will discover that a line's slope and *y*-intercept can be determined directly from the line's equation, provided the equation is written in a certain form.

Using the *y*-intercept and the Slope to Graph a Line

Let's modify the car production situation that first appeared in Section 3.5. Suppose that as a new workshift begins, 4 cars have already been produced. At the Michigan plant, 3 cars were being produced every 2 hours, a rate of $\frac{3}{2}$ cars per hour. If this rate remains the same regardless of how many cars have already been produced, the table and graph shown here can be made.

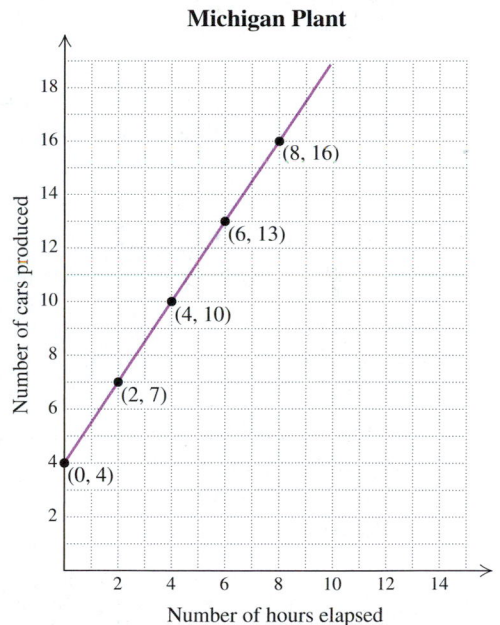

Michigan Plant

Michigan Plant	
Hours Elapsed	**Cars Produced**
0	4
2	7
4	10
6	13
8	16

To confirm that the production rate is still $\frac{3}{2}$, we calculate the slope. Recall that

$$\text{Slope} = \frac{\text{change in } y}{\text{change in } x} = \frac{\text{rise}}{\text{run}} = \frac{y_2 - y_1}{x_2 - x_1},$$

where (x_1, y_1) and (x_2, y_2) are any two points on the graphed line. Here we select $(0, 4)$ and $(2, 7)$:

$$\text{Slope} = \frac{\text{change in } y}{\text{change in } x} = \frac{7 - 4}{2 - 0} = \frac{3}{2}.$$

Knowing that the slope is $\frac{3}{2}$, we could have drawn the graph by plotting $(0, 4)$ and from there moving *up* 3 units and *to the right* 2 units. This would have located the point $(2, 7)$. Using $(0, 4)$ and $(2, 7)$, we can then draw the line. This is the method used in the next example.

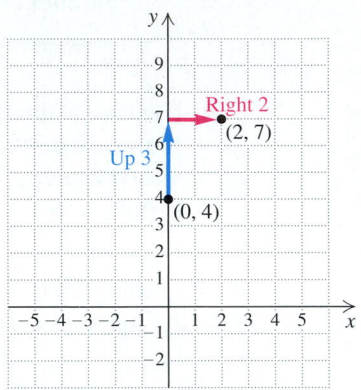

Example 1

Draw a line that has slope $\frac{1}{4}$ and y-intercept $(0, 2)$.

Solution   We plot $(0, 2)$ and from there move *up* 1 unit and *to the right* 4 units. This locates the point $(4, 3)$. We plot $(4, 3)$ and draw a line passing through $(0, 2)$ and $(4, 3)$, as shown on the right below.

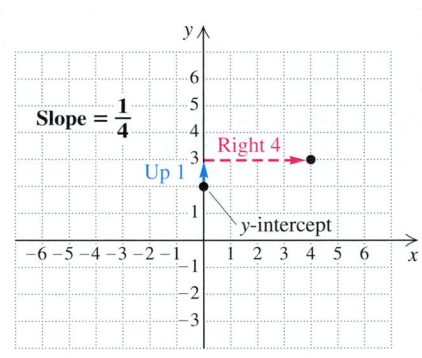
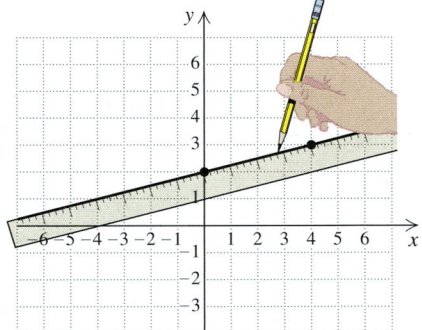

Equations in Slope–Intercept Form

It is not difficult to find a line's slope and y-intercept from its equation. Recall from Section 3.3 that to find the y-intercept of an equation's graph, we replace x with 0 and solve the resulting equation for y. For example, to find the y-intercept of the graph of $y = 2x + 3$, we replace x with 0 and solve as follows:

$$y = 2x + 3$$
$$= 2 \cdot 0 + 3 = 0 + 3 = 3. \qquad \text{The } y\text{-intercept is } (0, 3).$$

The y-intercept of the graph of $y = 2x + 3$ is $(0, 3)$. It can be similarly shown that the graph of $y = mx + b$ has the y-intercept $(0, b)$.

To calculate the slope of the graph of $y = 2x + 3$, we need two ordered pairs that are solutions of the equation. The y-intercept $(0, 3)$ is one pair; a second pair, $(1, 5)$, can be found by substituting 1 for x. We then have

$$\text{Slope} = \frac{\text{change in } y}{\text{change in } x} = \frac{5 - 3}{1 - 0} = \frac{2}{1} = 2.$$

Note that the slope, 2, is also the x-coefficient in $y = 2x + 3$. It can be similarly shown that the graph of any equation of the form $y = mx + b$ has slope m (see Exercise 85).

> ### The Slope–Intercept Equation
>
> The equation $y = mx + b$ is called the *slope–intercept equation*.
> The equation represents a line of slope m with y-intercept $(0, b)$.

The equation of any nonvertical line can be written in this form.

E x a m p l e   2   Find the slope and the y-intercept of each line.

a) $y = \frac{4}{5}x - 8$ **b)** $2x + y = 5$ **c)** $3x + 4y = 7$

Solution

a) We rewrite $y = \frac{4}{5}x - 8$ as $y = \frac{4}{5}x + (-8)$. Now we simply read the slope and the y-intercept from the equation:

$$y = \tfrac{4}{5}x + (-8).$$

The slope is $\frac{4}{5}$. The y-intercept is $(0, -8)$.

b) We first solve for y to find an equivalent equation in the form $y = mx + b$:

$$2x + y = 5$$
$$y = -2x + 5. \qquad \text{Adding } -2x \text{ to both sides}$$

The slope is -2. The y-intercept is $(0, 5)$.

c) We rewrite the equation in the form $y = mx + b$:

$$3x + 4y = 7$$
$$4y = -3x + 7 \qquad \text{Adding } -3x \text{ to both sides}$$
$$y = \tfrac{1}{4}(-3x + 7) \qquad \text{Multiplying both sides by } \tfrac{1}{4}$$
$$y = -\tfrac{3}{4}x + \tfrac{7}{4}. \qquad \text{Using the distributive law}$$

The slope is $-\frac{3}{4}$, or $\frac{-3}{4}$, or $\frac{3}{-4}$. The y-intercept is $\left(0, \frac{7}{4}\right)$.

E x a m p l e   3

A line has slope $-\frac{12}{5}$ and y-intercept $(0, 11)$. Find an equation of the line.

Solution   We use the slope–intercept equation, substituting $-\frac{12}{5}$ for m and 11 for b:

$$y = mx + b = -\frac{12}{5}x + 11.$$

The desired equation is $y = -\frac{12}{5}x + 11$.

E x a m p l e   4

Determine an equation for the graph of car production shown at the beginning of this section.

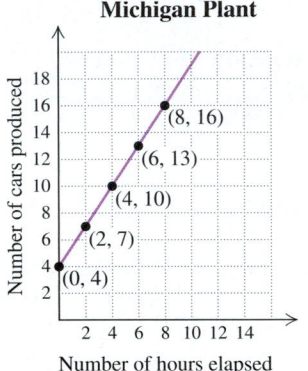

Michigan Plant

Solution   To write an equation for a line, we can use slope–intercept form, provided the slope and the y-intercept are known. Using the coordinates of two points, we already found that the slope, or rate of production, is $\frac{3}{2}$. Since $(0, 4)$ is given, we know the y-intercept as well. The desired equation is

$$y = \frac{3}{2}x + 4, \qquad \text{Using } \tfrac{3}{2} \text{ for } m \text{ and 4 for } b$$

where y is the number of cars produced after x hours.

Graphing and Slope–Intercept Form

In Example 1, we drew a graph, knowing only the slope and the y-intercept. In Example 2, we determined the slope and the y-intercept of a line by examining its equation. We now combine the two procedures to develop a quick way to graph a linear equation.

E x a m p l e   5

Graph: **(a)** $y = \frac{3}{4}x + 5$; **(b)** $2x + 3y = 3$.

Solution

a) From the equation $y = \frac{3}{4}x + 5$, we see that the slope of the graph is $\frac{3}{4}$ and the y-intercept is $(0, 5)$. We plot $(0, 5)$ and then consider the slope, $\frac{3}{4}$. Starting at $(0, 5)$, we plot a second point by moving *up* 3 units (since the numerator is *positive* and corresponds to the change in y) and *to the right* 4 units (since the denominator is *positive* and corresponds to the change in x). We reach a new point, $(4, 8)$.

We can also rewrite the slope as $\frac{-3}{-4}$. We again start at the y-intercept, $(0, 5)$, but move *down* 3 units (since the numerator is *negative* and corresponds to the change in y) and *to the left* 4 units (since the denominator is *negative* and corresponds to the change in x). We reach another point, $(-4, 2)$. Once two or three points have been plotted, the line representing all solutions of $y = \frac{3}{4}x + 5$ can be drawn.

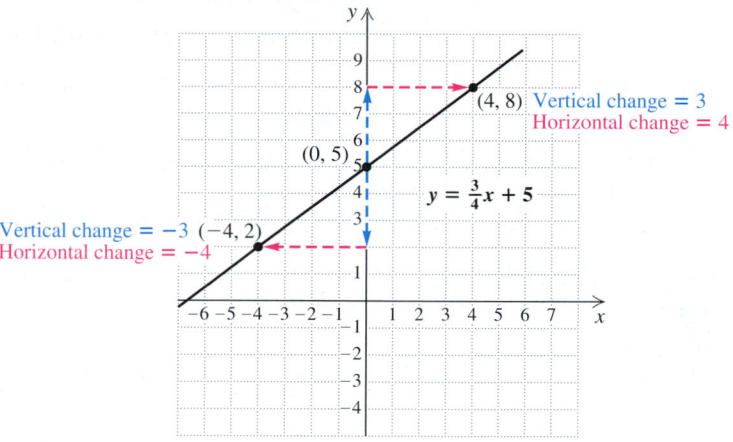

b) To graph $2x + 3y = 3$, we first rewrite it in slope–intercept form:

$$2x + 3y = 3$$

$\quad\quad\quad 3y = -2x + 3$ Adding $-2x$ to both sides

$\quad\quad\quad\ y = \frac{1}{3}(-2x + 3)$ Multiplying both sides by $\frac{1}{3}$

$\quad\quad\quad\ y = -\frac{2}{3}x + 1.$ Using the distributive law

To graph $y = -\frac{2}{3}x + 1$, we first plot the y-intercept, $(0, 1)$. We can think of the slope as $\frac{-2}{3}$. Starting at $(0, 1)$ and using the slope, we find a second point by moving *down* 2 units (since the numerator is *negative*) and *to the right* 3 units (since the denominator is *positive*). We plot the new point, $(3, -1)$. In a similar manner, we can move from the point $(3, -1)$ to locate a third point, $(6, -3)$. The line can then be drawn.

Since $-\frac{2}{3} = \frac{2}{-3}$, an alternative approach is to again plot $(0, 1)$, but this time move *up* 2 units (since the numerator is *positive*) and *to the left* 3 units (since the denominator is *negative*). This leads to another point on the graph, $(-3, 3)$.

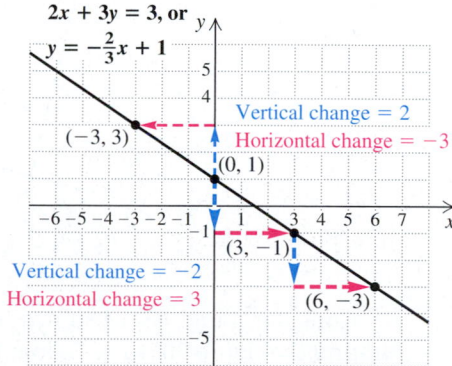

It is important to be able to use both $\frac{2}{-3}$ and $\frac{-2}{3}$ to draw the graph.

Parallel and Perpendicular Lines

Parallel lines extend indefinitely without intersecting. If two lines are vertical, they are parallel. We can tell whether nonvertical lines are parallel by looking at their slopes.

> **Slope and Parallel Lines**
> Two lines are parallel if they have the same slope. Vertical lines are also parallel.

Example 6

technology connection

Using a standard $[-10, 10, -10, 10]$ window, graph the equations $y_1 = \frac{2}{3}x + 1$, $y_2 = \frac{3}{8}x + 1$, $y_3 = \frac{2}{3}x + 5$, and $y_4 = \frac{3}{8}x + 5$. If you can, use your grapher in the MODE that graphs equations *simultaneously*. Once all lines have been drawn, try to decide which equation corresponds to each line. After matching equations with lines, you can check your matches by using TRACE and the up and down arrow keys to move from one line to the next. The number of the equation will appear in a corner of the screen.

1. Graph $y_1 = -\frac{3}{4}x - 2$, $y_2 = -\frac{1}{5}x - 2$, $y_3 = -\frac{3}{4}x - 5$, and $y_4 = -\frac{1}{5}x - 5$ using the SIMULTANEOUS mode. Then match each line with the corresponding equation. Check using TRACE.

Determine whether the graphs of $y = -3x + 4$ and $6x + 2y = -10$ are parallel.

Solution   One of the two equations given,
$$y = -3x + 4,$$
represents a line with slope -3 and y-intercept $(0, 4)$. To find the slope of the other line, we need to rewrite
$$6x + 2y = -10$$
in slope–intercept form:
$$6x + 2y = -10$$
$$2y = -6x - 10 \qquad \text{Adding } -6x \text{ to both sides}$$
$$y = -3x - 5. \qquad \text{The slope is } -3 \text{ and the } y\text{-intercept is } (0, -5).$$

Since both lines have slope -3 but different y-intercepts, (they are not the same line), the graphs are parallel. There is no need for us to actually graph either equation.

If one line is vertical and another is horizontal, they are perpendicular. There are other instances in which two lines are perpendicular.

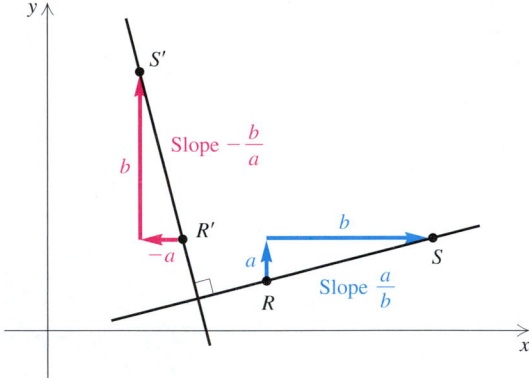

Consider a line $\overleftrightarrow{RS}$, as shown in the graph, with slope a/b. Then think of rotating the figure 90° to get a line $\overleftrightarrow{R'S'}$ perpendicular to $\overleftrightarrow{RS}$. For the new line, the rise and the run are interchanged, but the rise is now negative. Thus the slope of the new line is $-b/a$. Let's multiply the slopes:
$$\frac{a}{b}\left(-\frac{b}{a}\right) = -1.$$

This can help us determine which lines are perpendicular.

> ### Slope and Perpendicular Lines
>
> Two lines are perpendicular if the product of their slopes is -1.
> (If one line has slope m, the slope of a line perpendicular to it is $-1/m$. That is, we take the reciprocal and change the sign.)
>   Lines are also perpendicular if one is vertical and the other is horizontal.

E x a m p l e   7

Determine whether the graphs of $2x + y = 8$ and $y = \frac{1}{2}x + 7$ are perpendicular.

Solution   First, we find the slope of each line. The second equation,

$$y = \tfrac{1}{2}x + 7,$$

is in slope–intercept form. It represents a line with slope $\frac{1}{2}$.
    To find the slope of the other line, we rewrite

$$2x + y = 8$$

in slope–intercept form:

$$2x + y = 8$$
$$y = -2x + 8. \qquad \text{Adding } -2x \text{ to both sides}$$

The slope of the line is -2.
    The lines are perpendicular if the product of their slopes is -1. Since

$$\frac{1}{2}(-2) = -1,$$

the graphs are perpendicular.

E x a m p l e   8

Write a slope–intercept equation for the line whose graph is described.

a) Parallel to the graph of $2x - 3y = 7$, with y-intercept $(0, -1)$
b) Perpendicular to the graph of $2x - 3y = 7$, with y-intercept $(0, -1)$

Solution   We begin by determining the slope of the line represented by $2x - 3y = 7$:

$$2x - 3y = 7$$
$$-3y = -2x + 7 \qquad \text{Adding } -2x \text{ to both sides}$$
$$y = \tfrac{2}{3}x - \tfrac{7}{3}. \qquad \text{Dividing both sides by } -3$$

The slope is $\frac{2}{3}$.

a) A line parallel to the graph of $2x - 3y = 7$ has a slope of $\frac{2}{3}$. Since the y-intercept is $(0, -1)$, the slope–intercept equation is

$$y = \tfrac{2}{3}x - 1.$$

b) A line perpendicular to the graph of $2x - 3y = 7$ has a slope that is the negative reciprocal of $\frac{2}{3}$, or $-\frac{3}{2}$. Since the y-intercept is $(0, -1)$, the slope–intercept equation is

$$y = -\tfrac{3}{2}x - 1.$$

Exercise Set 3.6

FOR EXTRA HELP

Digital Video Tutor CD 2
Videotape 6

InterAct Math

Math Tutor Center

MathXL

MyMathLab.com

Draw a line that has the given slope and y-intercept.

1. Slope $\frac{2}{5}$; y-intercept $(0, 1)$

2. Slope $\frac{3}{5}$; y-intercept $(0, -1)$

3. Slope $\frac{5}{3}$; y-intercept $(0, -2)$

4. Slope $\frac{5}{2}$; y-intercept $(0, 1)$

5. Slope $-\frac{3}{4}$; y-intercept $(0, 5)$

6. Slope $-\frac{4}{5}$; y-intercept $(0, 6)$

7. Slope 2; y-intercept $(0, -4)$

8. Slope -2; y-intercept $(0, -3)$

9. Slope -3; y-intercept $(0, 2)$

10. Slope 3; y-intercept $(0, 4)$

Find the slope and the y-intercept of each line.

11. $y = \frac{3}{7}x + 5$

12. $y = -\frac{3}{8}x + 6$

13. $y = -\frac{5}{6}x + 2$

14. $y = \frac{7}{2}x + 4$

15. $y = \frac{9}{4}x - 7$

16. $y = \frac{2}{9}x - 1$

17. $y = -\frac{2}{5}x$

18. $y = \frac{4}{3}x$

19. $-2x + y = 4$

20. $-5x + y = 5$

21. $3x - 4y = 12$

22. $3x - 2y = 18$

23. $x - 5y = -8$

24. $x - 6y = 9$

Aha! **25.** $y = 4$

26. $y - 3 = 5$

Find the slope–intercept equation for the line with the indicated slope and y-intercept.

27. Slope 3; y-intercept $(0, 7)$

28. Slope -4; y-intercept $(0, -2)$

29. Slope $\frac{7}{8}$; y-intercept $(0, -1)$

30. Slope $\frac{5}{7}$; y-intercept $(0, 4)$

31. Slope $-\frac{5}{3}$; y-intercept $(0, -8)$

32. Slope $\frac{3}{4}$; y-intercept $(0, 23)$

Aha! **33.** Slope 0; y-intercept $(0, 3)$

34. Slope 7; y-intercept $(0, 0)$

Graph.

35. $y = \frac{3}{5}x + 2$

36. $y = -\frac{3}{5}x - 1$

37. $y = -\frac{3}{5}x + 1$

38. $y = \frac{3}{5}x - 2$

39. $y = \frac{5}{3}x + 3$

40. $y = \frac{5}{3}x - 2$

41. $y = -\frac{3}{2}x - 2$

42. $y = -\frac{4}{3}x + 3$

43. $2x + y = 1$

44. $3x + y = 2$

45. $3x - y = 4$

46. $2x - y = 5$

47. $2x + 3y = 9$

48. $4x + 5y = 15$

49. $x - 4y = 12$

50. $x + 5y = 20$

Solve.

51. *Cost of water.* Freda and Phil are keeping track of their water bills. One month, while they were away on vacation, they used no water and were billed $9. The next month they used 70,000 gal and their bill was for $16. Let y represent the size of a monthly bill and x the amount of water used (in 10,000-gal units). Find and graph an equation of the form $y = mx + b$. Then determine the rate that they pay in dollars per 10,000 gallons.

52. *Cost of cable TV.* Allegra and Larry are keeping track of how much their cable TV service costs. When service began, they paid $50 for installation. After one month, their total costs had risen to $70, and after two months they had paid a total of $90 for their cable service. Let y represent the amount paid for x months of service. Find and graph an equation of the form $y = mx + b$. Then determine their monthly rate.

53. *Refrigerator size.* Kitchen designers recommend that a refrigerator be selected on the basis of the number of people in the household. For 1–2 people, a 16 ft^3 model is suggested. For each additional person, an additional 1.5 ft^3 is recommended. If x is the number of residents in excess of 2, find the slope–intercept equation for the recommended size of a refrigerator.

📓 **54.** *Telephone service.* In a recent promotion, AT&T charged a monthly fee of \$4.95 plus 7¢ for each minute of long-distance phone calls. If x is the number of minutes of long-distance calls, find the slope–intercept equation for the monthly bill.

Determine whether each pair of equations represents parallel lines.

55. $y = \frac{2}{3}x + 7,$
$\quad y = \frac{2}{3}x - 5$

56. $y = -\frac{5}{4}x + 1,$
$\quad y = \frac{5}{4}x + 3$

57. $y = 2x - 5,$
$\quad 4x + 2y = 9$

58. $y = -3x + 1,$
$\quad 6x + 2y = 8$

59. $3x + 4y = 8,$
$\quad 7 - 12y = 9x$

60. $3x = 5y - 2,$
$\quad 10y = 4 - 6x$

Determine whether each pair of equations represents perpendicular lines.

61. $y = 4x - 5,$
$\quad 4y = 8 - x$

62. $2x - 5y = -3,$
$\quad 2x + 5y = 4$

63. $x - 2y = 5,$
$\quad 2x + 4y = 8$

64. $y = -x + 7,$
$\quad y - x = 3$

65. $2x + 3y = 1,$
$\quad 3x - 2y = 1$

66. $y = 5 - 3x,$
$\quad 3x - y = 8$

Write a slope–intercept equation of the line whose graph is described.

67. Parallel to the graph of $y = 5x - 7$; y-intercept $(0, 11)$

68. Parallel to the graph of $2x - y = 1$; y-intercept $(0, -3)$

69. Perpendicular to the graph of $2x + y = 0$; y-intercept $(0, 0)$

70. Perpendicular to the graph of $y = \frac{1}{3}x + 7$; y-intercept $(0, 5)$

Aha! **71.** Parallel to the graph of $y = x$; y-intercept $(0, 3)$

Aha! **72.** Perpendicular to the graph of $y = x$; y-intercept $(0, 0)$

73. Perpendicular to the graph of $x + y = 3$; y-intercept $(0, -4)$

74. Parallel to the graph of $3x + 2y = 5$; y-intercept $(0, -1)$

📓 **75.** Can a horizontal line be graphed using the method of Example 5? Why or why not?

📓 **76.** Can a vertical line be graphed using the method of Example 5? Why or why not?

SKILL MAINTENANCE

Solve.

77. $y - k = m(x - h)$, for y

78. $y - 9 = -2(x + 4)$, for y

Simplify.

79. $-5 - (-7)$

80. $7 - (-9)$

81. $-3 - 6$

82. $-2 - 8$

SYNTHESIS

📓 **83.** Explain how it is possible for an incorrect graph to be drawn, even after plotting three points that line up.

📓 **84.** Which would you prefer, and why: graphing an equation of the form $y = mx + b$ or graphing an equation of the form $Ax + By = C$?

85. Show that the slope of the line given by $y = mx + b$ is m. (*Hint*: Substitute both 0 and 1 for x to find two pairs of coordinates. Then use the formula, Slope = change in y/change in x.)

86. Write an equation of the line with the same slope as the line given by $5x + 2y = 8$ and the same y-intercept as the line given by $3x - 7y = 10$.

87. Write an equation of the line parallel to the line given by $2x - 6y = 10$ and having the same y-intercept as the line given by $9x + 6y = 18$.

88. Find an equation of the line parallel to the line given by $3x - 2y = 8$ and having the same y-intercept as the line given by $2y + 3x = -4$.

89. Find an equation of the line perpendicular to the line given by $2x + 5y = 6$ that passes through $(2, 6)$. (*Hint*: Draw a graph.)

📓 **90.** *Aerobic exercise.* The formula $T = -\frac{3}{4}a + 165$ can be used to determine the *target heart rate,* in beats per minute, for a person, a years old, participating in aerobic exercise. Graph the equation and interpret the significance of its slope.

CORNER

Draw the Graph and Match the Math

Focus: Slope–intercept form

Time: 15–20 minutes

Group size: 3

Materials: Graph paper and straightedges

It is important not only to be able to graph equations written in slope–intercept form, but to be able to match a linear graph with an appropriate equation.

ACTIVITY

1. Each group member should select a different one of the following sets of equations:

   A. $y = \frac{4}{3}x - 5$,   **B.** $y = \frac{2}{5}x + 1$,   **C.** $y = \frac{3}{5}x + 2$,

   $y = \frac{4}{3}x + 2$, $y = \frac{3}{4}x + 1$, $y = \frac{3}{5}x - 2$,

   $y = \frac{1}{2}x - 5$, $y = \frac{2}{5}x - 1$, $y = \frac{4}{3}x + 2$,

   $y = \frac{1}{2}x + 2$; $y = \frac{3}{4}x - 1$; $y = \frac{4}{3}x - 2$,

2. Working independently, each group member should graph the four equations he or she

has selected. Do not label the graphs with their corresponding equations, but instead list the four equations (in any random order) across the top of the graph paper.

3. After all group members have completed part (2), the sheets should be passed, clockwise, to the person on the left. This person should then attempt to match each of the four equations listed at the top of the graph paper with the appropriate graph below. If no graph appears to be appropriate, discuss the relevant equation with the group member who drew the graphs. If necessary, turn to the third group member for help in identifying any incorrect graphs.

4. Once all four equations and graphs have been matched, share your answers with the rest of the group. Make sure everyone agrees on all of the matches.

Point–Slope Form

3.7

Writing Equations in Point–Slope Form • Graphing and Point–Slope Form

There are many applications in which a slope—or a rate of change—and an ordered pair are known. When the ordered pair is the *y*-intercept, an equation in slope–intercept form can be easily produced. When the ordered pair represents a point other than the *y*-intercept, a different form, known as *point–slope form*, is more convenient.

Writing Equations in Point–Slope Form

Consider a line with slope 2 passing through the point $(4, 1)$, as shown in the figure. In order for a point (x, y) to be on the line, the coordinates x and y must be solutions of the slope equation

$$\frac{y - 1}{x - 4} = 2.$$

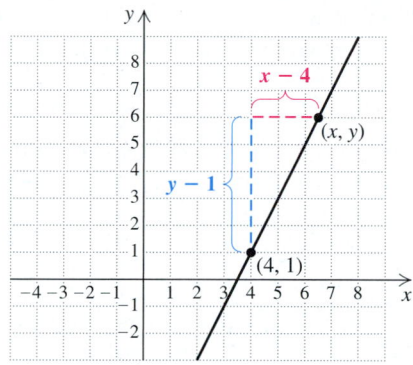

Take a moment to examine this equation. Pairs like $(5, 3)$ and $(3, -1)$ are solutions, since

$$\frac{3 - 1}{5 - 4} = 2 \quad \text{and} \quad \frac{-1 - 1}{3 - 4} = 2.$$

Note, however, that $(4, 1)$ is not itself a solution of the equation:

$$\frac{1 - 1}{4 - 4} \neq 2.$$

To avoid this difficulty, we can use the multiplication principle:

$$(x - 4) \cdot \frac{y - 1}{x - 4} = 2(x - 4) \qquad \text{Multiplying both sides by } x - 4$$

$$y - 1 = 2(x - 4). \qquad \text{Removing a factor equal to 1: } \frac{x - 4}{x - 4} = 1$$

This is considered **point–slope form** for the line shown above. A point–slope equation can be written any time a line's slope and a point on the line are known.

> ### The Point–Slope Equation
>
> The equation $y - y_1 = m(x - x_1)$ is called the *point–slope equation* for the line with slope m that contains the point (x_1, y_1).

Point–slope form is especially useful in more advanced mathematics courses, where problems similar to the following often arise.

E x a m p l e   1   Write a point–slope equation for the line with slope $\frac{1}{5}$ that contains the point $(7, 2)$.

Solution   We substitute $\frac{1}{5}$ for m, 7 for x_1, and 2 for y_1:

$$y - y_1 = m(x - x_1) \qquad \text{Using the point–slope equation}$$
$$y - 2 = \tfrac{1}{5}(x - 7) \qquad \text{Substituting}$$

E x a m p l e   2   Write a point–slope equation for the line with slope $-\frac{4}{3}$ that contains the point $(1, -6)$.

Solution   We substitute $-\frac{4}{3}$ for m, 1 for x_1, and -6 for y_1:

$$y - y_1 = m(x - x_1) \qquad \text{Using the point–slope equation}$$
$$y - (-6) = -\tfrac{4}{3}(x - 1). \qquad \text{Substituting}$$

E x a m p l e   3   Write the slope–intercept equation for the line with slope 3 that contains the point $(1, 9)$.

Solution   There are two parts to this solution. First, we write an equation in point–slope form:

$$y - y_1 = m(x - x_1)$$
$$y - 9 = 3(x - 1). \qquad \text{Substituting}$$

Next, we find an equivalent equation of the form $y = mx + b$:

$$y - 9 = 3(x - 1)$$
$$y - 9 = 3x - 3 \qquad \text{Using the distributive law}$$
$$y = 3x + 6. \qquad \text{Adding 9 to both sides to get slope–intercept form}$$

Graphing and Point–Slope Form

When we know a line's slope and a point that is on the line, we can draw the graph, much as we did in Section 3.6.

E x a m p l e   4   Graph the line with slope 2 that passes through $(-3, 1)$.

Solution   We plot $(-3, 1)$, move *up* 2 and *to the right* 1 (since $2 = \frac{2}{1}$), and draw the line.

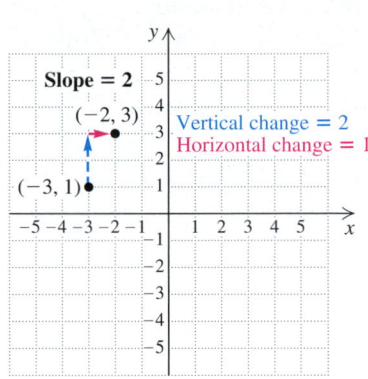

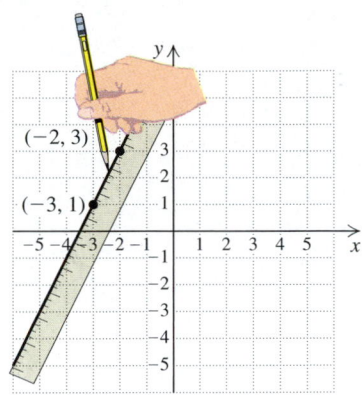

Example 5

Graph: $y - 2 = 3(x - 4)$.

Solution   Since $y - 2 = 3(x - 4)$ is in point–slope form, we know that the line has slope 3, or $\frac{3}{1}$, and passes through the point $(4, 2)$. We plot $(4, 2)$ and then find a second point by moving *up* 3 units and *to the right* 1 unit. The line can then be drawn, as shown below.

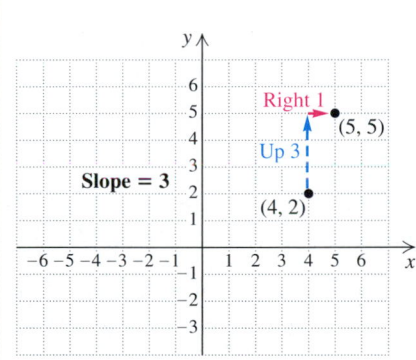

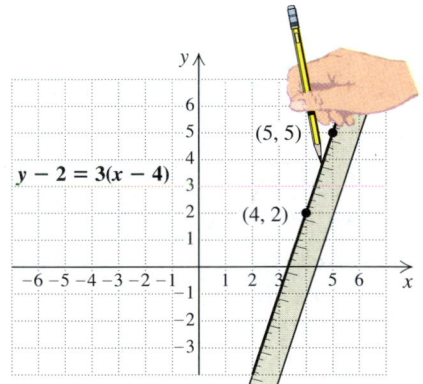

Example 6

Graph: $y + 4 = -\frac{5}{2}(x + 3)$.

Solution   Once we have written the equation in point–slope form, $y - y_1 = m(x - x_1)$, we can proceed much as we did in Example 5. To find an equivalent equation in point–slope form, we subtract opposites instead of adding:

$$y + 4 = -\frac{5}{2}(x + 3)$$

$$y - (-4) = -\frac{5}{2}(x - (-3)).$$

Subtracting a negative instead of adding a positive. This is now in point–slope form.

From this last equation, $y - (-4) = -\frac{5}{2}(x - (-3))$, we see that the line passes through $(-3, -4)$ and has slope $-\frac{5}{2}$, or $\frac{5}{-2}$.

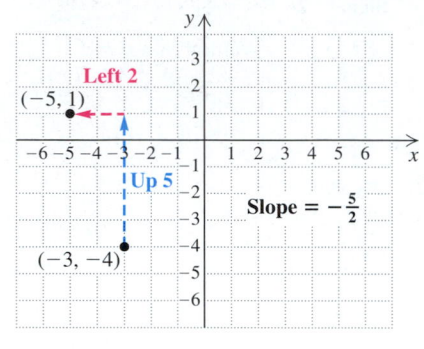

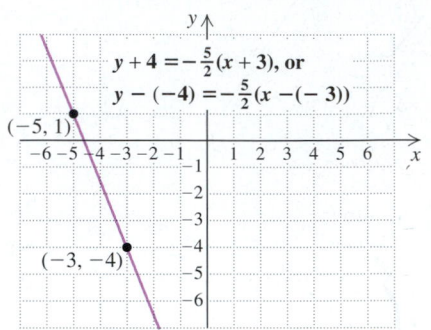

FOR EXTRA HELP

Digital Video Tutor CD 2
Videotape 6   InterAct Math   Math Tutor Center   MathXL   MyMathLab.com

Exercise Set 3.7

Write a point–slope equation for the line with the given slope that contains the given point.

1. $m = 6;\ (2, 7)$ **2.** $m = 4;\ (3, 5)$

3. $m = \frac{3}{5};\ (9, 2)$ **4.** $m = \frac{2}{3};\ (4, 1)$

5. $m = -4;\ (3, 1)$ **6.** $m = -5;\ (6, 2)$

7. $m = \frac{3}{2};\ (5, -4)$ **8.** $m = \frac{4}{3};\ (7, -1)$

9. $m = \frac{5}{4};\ (-2, 6)$ **10.** $m = \frac{7}{2};\ (-3, 4)$

11. $m = -2;\ (-4, -1)$ **12.** $m = -3;\ (-2, -5)$

13. $m = 1;\ (-2, 8)$ **14.** $m = -1;\ (-3, 6)$

Write the slope–intercept equation for the line with the given slope that contains the given point.

15. $m = 2;\ (5, 7)$ **16.** $m = 3;\ (6, 2)$

17. $m = \frac{7}{4};\ (4, -2)$ **18.** $m = \frac{8}{3};\ (3, -4)$

19. $m = -3;\ (1, -5)$ **20.** $m = -2;\ (3, -1)$

21. $m = -4;\ (-2, -1)$ **22.** $m = -5;\ (-1, -4)$

23. $m = \frac{2}{3};\ (6, 5)$ **24.** $m = \frac{3}{2};\ (4, 7)$

25. $m = -\frac{5}{6};\ (3, 2)$ **26.** $m = -\frac{3}{4};\ (2, 5)$

27. Graph the line with slope $\frac{4}{3}$ that passes through the point $(1, 2)$.

28. Graph the line with slope $\frac{2}{5}$ that passes through the point $(3, 4)$.

29. Graph the line with slope $-\frac{3}{4}$ that passes through the point $(2, 5)$.

30. Graph the line with slope $-\frac{3}{2}$ that passes through the point $(1, 4)$.

Graph.

31. $y - 2 = \frac{1}{2}(x - 1)$ **32.** $y - 5 = \frac{1}{3}(x - 2)$

33. $y - 1 = -\frac{1}{2}(x - 3)$ **34.** $y - 1 = -\frac{1}{4}(x - 3)$

35. $y + 2 = \frac{1}{2}(x - 3)$ **36.** $y - 1 = \frac{1}{3}(x + 5)$

37. $y + 4 = 3(x + 1)$ **38.** $y + 3 = 2(x + 1)$

39. $y - 4 = -2(x + 1)$ **40.** $y + 3 = -1(x - 4)$

41. $y + 3 = -(x + 2)$ **42.** $y + 4 = 3(x + 2)$

43. $y + 1 = -\frac{3}{5}(x + 2)$ **44.** $y + 2 = -\frac{2}{3}(x + 1)$

45. $y - 1 = -\frac{7}{2}(x + 5)$ **46.** $y - 3 = -\frac{7}{4}(x + 1)$

47. Can equations for horizontal or vertical lines be written in point–slope form? Why or why not?

48. Describe a situation in which it is easier to graph the equation of a line in point–slope form rather than slope–intercept form.

SKILL MAINTENANCE

Simplify.

49. $(-5)^3$

50. $(-2)^6$

51. $3 \cdot 2^4 - 5 \cdot 2^3$

52. $5 \cdot 3^2 - 7 \cdot 3$

53. $(-2)^3(-3)^2$

54. $(5 - 7)^2(3 - 2 \cdot 2)$

SYNTHESIS

 55. Describe a procedure that can be used to write the slope–intercept equation for any nonvertical line passing through two given points.

56. Any nonvertical line has many equations in point–slope form, but only one in slope–intercept form. Why is this?

Graph.

Aha! **57.** $y - 3 = 0(x - 52)$

58. $y + 4 = 0(x + 93)$

Write two different point–slope equations for the line passing through each pair of points.

59. $(1, 2)$ and $(3, 7)$

60. $(3, 1)$ and $(7, 3)$

61. $(-1, 2)$ and $(3, 8)$

62. $(-3, 1)$ and $(4, 3)$

63. $(-3, 8)$ and $(1, -2)$

64. $(-2, 7)$ and $(4, -3)$

Write the slope–intercept equation for each line shown.

65.

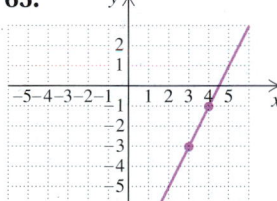

66.

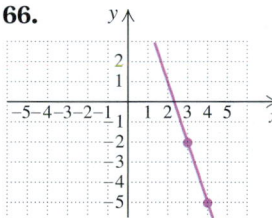

67.

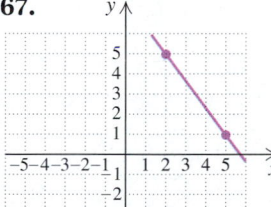

68.

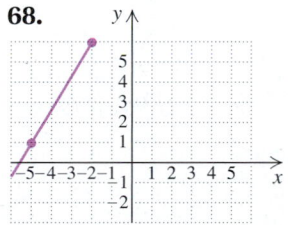

Write the slope–intercept equation for the line containing the given pair of points.

69. $(1, 5)$ and $(4, 2)$

70. $(3, 7)$ and $(4, 8)$

71. $(-3, 1)$ and $(3, 5)$

72. $(-2, 3)$ and $(2, 5)$

73. $(5, 0)$ and $(0, -2)$

74. $(-2, 0)$ and $(0, 3)$

75. $(-2, -4)$ and $(2, -1)$

76. $(-3, 5)$ and $(-1, -3)$

77. Write a point–slope equation of the line passing through $(-4, 7)$ that is parallel to the line given by $2x + 3y = 11$.

78. Write a point–slope equation of the line passing through $(3, -1)$ that is perpendicular to the line given by $4x - 5y = 9$.

Aha! **79.** Write an equation of the line parallel to the line given by $y = 3 - 4x$ that passes through $(0, 7)$.

80. Write the slope–intercept equation of the line that has the same y-intercept as the line $x - 3y = 6$ and contains the point $(5, -1)$.

81. Write the slope–intercept equation of the line that contains the point $(-1, 5)$ and is perpendicular to the line passing through $(2, 7)$ and $(-1, -3)$.

82. Write the slope–intercept equation of the line that has x-intercept $(-2, 0)$ and is parallel to $4x - 8y = 12$.

 83. Why is slope–intercept form more useful than point–slope form when using a grapher? How can point–slope form be modified so that it is more easily used with graphers?

Summary and Review 3

Key Terms

Bar graph, p. 130
Circle graph, p. 131
Pie chart, p. 131
Graph, plot, p. 133
Axis (plural, axes), p. 133
Origin, p. 133
Coordinates, p. 133
Ordered pairs, p. 133
Quadrants, p. 135
Interpolation, p. 135
Extrapolation, p. 135
Graph of equation, p. 143

Linear equation, p. 143
Nonlinear graph, p. 148
y-intercept, p. 153
x-intercept, p. 153
Rate, p. 161
Speed, p. 162
Slope, p. 172
Grade, p. 175
Slope–intercept form, p. 185
Parallel, p. 188
Perpendicular, p. 189
Point–slope form, p. 193

Important Properties and Formulas

To Graph a Linear Equation

1. Select a value for one coordinate and calculate the corresponding value of the other coordinate. Form an ordered pair. This pair is one solution of the equation.
2. Repeat step (1) to find at least one other ordered pair.
3. Plot the ordered pairs and draw a straight line passing through the points. The line represents all solutions of the equation.

To Find Intercepts

To find a y-intercept, let $x = 0$ and solve for y.

To find an x-intercept, let $y = 0$ and solve for x.

$$\text{Slope} = m = \frac{\text{change in } y}{\text{change in } x} = \frac{\text{rise}}{\text{run}} = \frac{y_2 - y_1}{x_2 - x_1}$$

Horizontal line: Slope is 0.
Vertical line: Slope is undefined.
Parallel lines: Slopes are equal or both lines are vertical.
Perpendicular lines: Product of slopes is -1, or one line is horizontal and one is vertical.

Slope–intercept equation: $y = mx + b$
Point–slope equation: $y - y_1 = m(x - x_1)$

Review Exercises

The following circle graph shows a breakdown of the charities to which Americans donated in a recent year (Source: Giving USA 1998/American Association of Fund-Raising Counsel Trust for Philanthropy). Use the graph for Exercises 1 and 2.

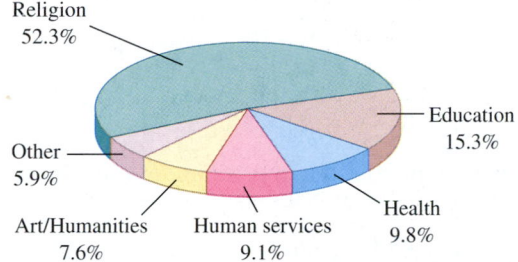

Religion 52.3%
Education 15.3%
Other 5.9%
Art/Humanities 7.6%
Human services 9.1%
Health 9.8%

1. The citizens of Ferrisburg are typical of Americans in general with regard to charitable contributions. If the citizens donated a total of $2 million to charities, how much was given to education?

2. About 5% of the Sophrins' income of $70,000 goes to charity. If their giving habits are typical, approximate the amount that they contribute to human services.

Plot each point.

3. $(2, -3)$ **4.** $(1, 0)$ **5.** $(2, 4)$

In which quadrant is each point located?

6. $(5, -10)$ **7.** $(-16.5, -20.3)$ **8.** $(-14, 7)$

Find the coordinates of each point in the figure.

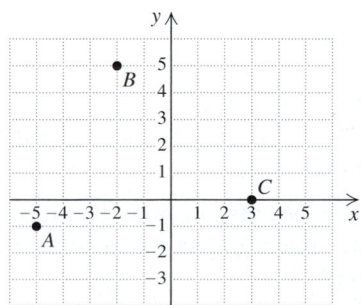

9. A **10.** B **11.** C

Determine whether the equation $y = 2x - 5$ has each ordered pair as a solution.

12. $(-3, 1)$ **13.** $(3, 1)$

14. Show that the ordered pairs $(0, -3)$ and $(2, 1)$ are solutions of the equation $2x - y = 3$. Then use the graph of the two points to determine another solution. Answers may vary.

Graph.

15. $y = x - 5$ **16.** $y = -\frac{1}{4}x$

17. $y = -x + 4$ **18.** $4x + y = 3$

19. $4x + 5 = 3$ **20.** $5x - 2y = 10$

21. *Meal service.* At 8:30 A.M., the Colchester Boy Scouts had served 45 people at their annual pancake breakfast. By 9:15, the total served had reached 65.

a) Find the Boy Scouts' serving rate, in number of meals per minute.

b) Find the Boy Scouts' serving rate, in minutes per meal.

22. *U.S. population.* The following graph shows data for the size of the U.S. population (*Source: Statistical Abstract of the United States*, 1999). At what rate has the population been growing?

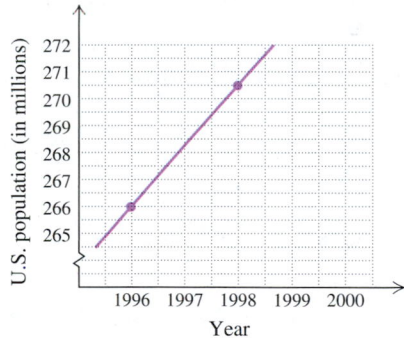

Find the slope of each line.

23.

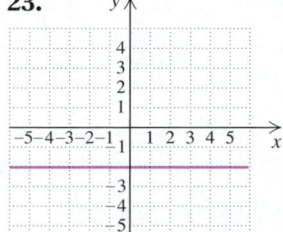

24.

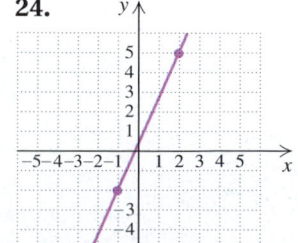

25.

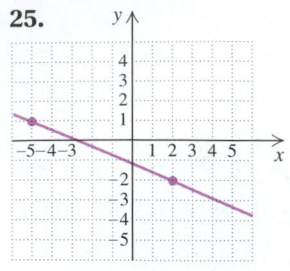

Find the slope of the line containing the given pair of points.

26. $(6, 8)$ and $(-2, -4)$

27. $(5, 1)$ and $(-1, 1)$

28. $(-3, 0)$ and $(-3, 5)$

29. $(-8.3, 4.6)$ and $(-9.9, 1.4)$

30. A road drops 369.6 ft vertically over a horizontal distance of 5280 ft. What is the grade of the road?

31. Find the x-intercept and the y-intercept of the line given by $3x + 2y = 18$.

32. Find the slope and the y-intercept of the line given by $2x + 4y = 20$.

33. Write the slope–intercept equation of the line with slope $-\frac{3}{4}$ and y-intercept $(0, 6)$.

34. Write a point–slope equation for the line with slope $-\frac{1}{2}$ that contains the point $(3, 6)$.

35. Write the slope–intercept equation for the line with slope 4 that contains the point $(-3, -7)$.

Determine without graphing whether each pair of lines is parallel, perpendicular, or neither.

36. $y + 5 = -x$,
$x - y = 2$

37. $3x - 5 = 7y$,
$7y - 3x = 7$

Graph.

38. $y = \frac{2}{3}x - 5$

39. $2x + y = 4$

40. $y = 6$

41. $x = -2$

42. $y + 2 = -\frac{1}{2}(x - 3)$

SYNTHESIS

43. Describe two ways in which a small business might make use of graphs.

44. Explain why the first coordinate of the y-intercept is always 0.

45. Find the value of m in $y = mx + 3$ such that $(-2, 5)$ is on the graph.

46. Find the value of b in $y = -5x + b$ such that $(3, 4)$ is on the graph.

47. Find the area and the perimeter of a rectangle for which $(-2, 2)$, $(7, 2)$, and $(7, -3)$ are three of the vertices.

48. Find three solutions of $y = 4 - |x|$.

Chapter Test 3

Use of tax dollars.   *The following pie chart shows how federal income tax dollars are spent.*

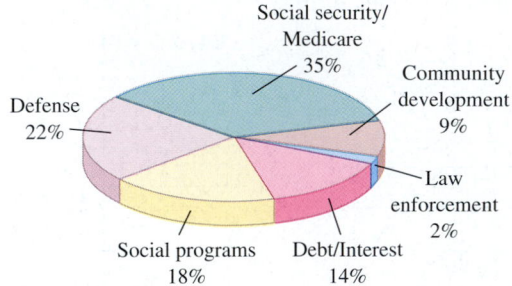

Where Your Tax Dollars Are Spent

Social security/Medicare 35%
Community development 9%
Law enforcement 2%
Debt/Interest 14%
Social programs 18%
Defense 22%

1. Debbie pays 18% of her taxable income of $31,200 in taxes. How much of her income will go to law enforcement?

2. Larry pays 16% of his taxable income of $27,000 in taxes. How much of his income will go to social programs?

In which quadrant is each point located?

3. $\left(-\frac{1}{2}, 7\right)$

4. $(-5, -6)$

Find the coordinates of each point in the figure.

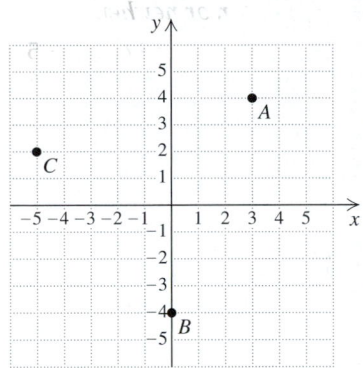

5. A

6. B

7. C

Graph.

8. $y = 2x - 1$

9. $2x - 4y = -8$

10. $y + 1 = 6$

11. $y = \frac{3}{4}x$

12. $y = 7$

13. $2x - y = 3$

14. $x = -1$

Find the x- and y-intercepts. Do not graph.

15. $5x - 3y = 30$

16. $x = 10 - 4y$

Find the slope of the line containing each pair of points.

17. $(4, -1)$ and $(6, 8)$

18. $(-3, -5)$ and $(9, 2)$

19. *Running.* Ted reached the 3-km mark of a race at 2:15 P.M. and the 6-km mark at 2:24 P.M. What is his running rate?

20. Find the slope and the y-intercept of the line given by $y - 3x = 7$.

Determine without graphing whether each pair of lines is parallel, perpendicular, or neither.

21. $4y + 2 = 3x,$
$-3x + 4y = -12$

22. $y = -2x + 5,$
$2y - x = 6$

Graph.

23. $y = \frac{1}{4}x - 2$

24. $y + 4 = -\frac{1}{2}(x - 3)$

25. Write a point–slope equation for the line of slope -3 that contains the point $(6, 8)$.

SYNTHESIS

26. Write an equation of the line that is parallel to the graph of $2x - 5y = 6$ and has the same y-intercept as the graph of $3x + y = 9$.

27. A diagonal of a square connects the points $(-3, -1)$ and $(2, 4)$. Find the area and the perimeter of the square.

Cumulative Review 1–3

1. Evaluate $\dfrac{x}{2y}$ for $x = 60$ and $y = 2$.

2. Multiply: $3(4x - 5y + 7)$.

3. Factor: $15x - 9y + 3$.

4. Find the prime factorization of 42.

5. Find decimal notation: $\frac{9}{20}$.

6. Find the absolute value: $|-4|$.

7. Find the opposite of $-\frac{1}{4}$.

8. Find the reciprocal of $-\frac{1}{4}$.

9. Combine like terms: $2x - 5y + (-3x) + 4y$.

10. Find decimal notation: 78.5%.

Simplify.

11. $\frac{3}{5} - \frac{5}{12}$

12. $3.4 + (-0.8)$

13. $(-2)(-1.4)(2.6)$

14. $\frac{3}{8} \div \left(-\frac{9}{10}\right)$

15. $2 - [32 \div (4 + 2^2)]$

16. $-5 + 16 \div 2 \cdot 4$

17. $y - (3y + 7)$

18. $3(x - 1) - 2[x - (2x + 7)]$

Solve.

19. $1.5 = 2.7 + x$

20. $\frac{2}{7}x = -6$

21. $5x - 9 = 36$

22. $\frac{2}{3} = \frac{-m}{10}$

23. $5.4 - 1.9x = 0.8x$

24. $x - \frac{7}{8} = \frac{3}{4}$

25. $2(2 - 3x) = 3(5x + 7)$

26. $\frac{1}{4}x - \frac{2}{3} = \frac{3}{4} + \frac{1}{3}x$

27. $y + 5 - 3y = 5y - 9$

28. $x - 28 < 20 - 2x$

29. $2(x + 2) \geq 5(2x + 3)$

30. Solve $A = 2\pi rh + \pi r^2$ for h.

31. In which quadrant is the point $(3, -1)$ located?

32. Graph on a number line: $-1 < x \leq 2$.

Graph.

33. $y = -2$ **34.** $2x + 5y = 10$

35. $y = -2x + 1$ **36.** $y = \frac{2}{3}x$

Find the coordinates of the x- and y-intercepts. Do not graph.

37. $2x - 7y = 21$

38. $y = 4x + 5$

Solve.

39. *Donating blood.* Each year 8 million Americans donate blood. This is 5% of those healthy enough to do so (*Source*: *Indianapolis Star*, 10/6/96). How many Americans are eligible to donate blood?

40. *Blood types.* There are 117 million Americans with either O-positive or O-negative blood. Those with O-positive blood outnumber those with O-negative blood by 85.8 million. How many Americans have O-negative blood?

41. Tina paid $126 for a cordless drill, including a 5% sales tax. How much did the drill itself cost?

42. A 143-m wire is cut into three pieces. The second is 3 m longer than the first. The third is four fifths as long as the first. How long is each piece?

43. Cory's contract stipulates that he cannot work more than 40 hr per week. For the first 4 days of one week, he worked 7, 10, 9, and 6 hr. How many hours can he work the fifth day and not violate his contract?

44. *Fund raising.* The graph at the top of the next column shows data from a recent road race held to raise money for Amnesty International. At what rate was money raised?

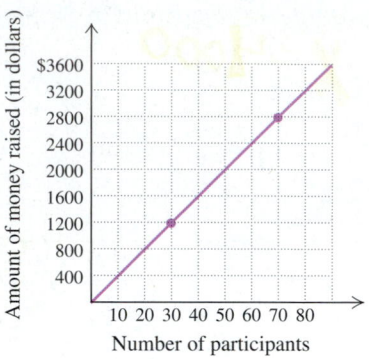

45. Find the slope of the line containing the points $(-4, 1)$ and $(2, -1)$.

46. Write an equation of the line with slope $\frac{2}{7}$ and y-intercept $(0, -4)$.

47. Find the slope and the y-intercept of the line given by $2x + 6y = 18$.

Graph.

48. $y = \frac{4}{3}x - 2$ **49.** $2x + 3y = -12$

50. Write a point–slope equation of the line with slope $-\frac{3}{8}$ that contains the point $(-6, 4)$.

SYNTHESIS

51. *Yearly earnings.* Paula's salary at the end of a year is $26,780. This reflects a 4% salary increase that preceded a 3% cost-of-living adjustment during the year. What was her salary at the beginning of the year?

Solve. If no solution exists, state this.

52. $4|x| - 13 = 3$

53. $4(x + 2) = 9(x - 2) + 16$

54. $2(x + 3) + 4 = 0$

55. $\dfrac{2 + 5x}{4} = \dfrac{11}{28} + \dfrac{8x + 3}{7}$

56. $5(7 + x) = (x + 6)5$

57. Solve $p = \dfrac{2}{m + Q}$ for Q.

58. The points $(-3, 0)$, $(0, 7)$, $(3, 0)$, and $(0, -7)$ are vertices of a parallelogram. Find four equations of lines that intersect to form the parallelogram.

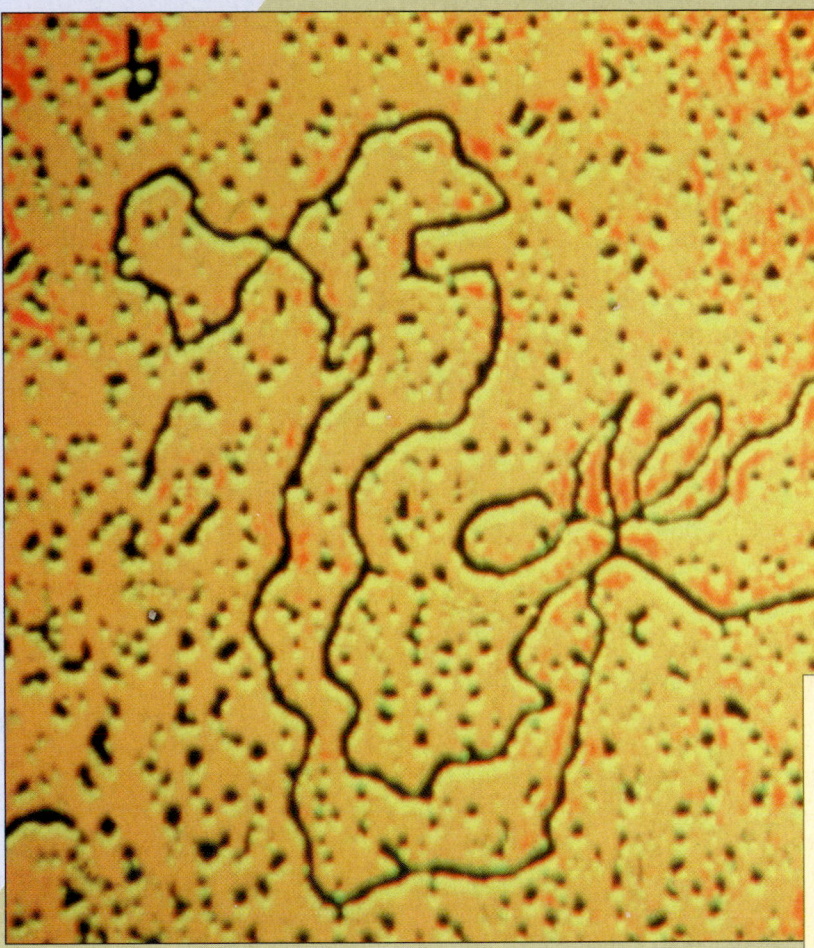

4
Polynomials

AN APPLICATION

A strand of DNA (deoxyribonucleic acid) is about 1.5 m long and 1.3×10^{-10} cm wide (*Source*: Human Genome Project Information). How many times longer is DNA than it is wide?

This problem appears as Exercise 153 in Section 4.8.

A s a biochemist, I cannot conduct an experiment without using math. For example, to investigate the effect of enzyme-to-protein ratios, I must calculate how much protein I have, on the basis of the concentration and volume, and then calculate the total amount of enzyme to add.

KIRA L. FORD
Biochemist
Indianapolis, Indiana

*O*ur work in Chapter 3 concentrated on using graphs to represent solutions of equations in two variables. Here in Chapter 4 we will focus on finding equivalent expressions, not on solving equations.

Algebraic expressions such as $16t^2$, $5a^2 - 3ab$, and $3x^2 - 7x + 5$ are called *polynomials. Polynomials occur frequently in applications and appear in most branches of mathematics. Thus learning to add, subtract, multiply, and divide polynomials is an important part of most courses in elementary algebra and is the focus of this chapter.*

Exponents and Their Properties

4.1

Multiplying Powers with Like Bases • Dividing Powers with Like Bases • Zero as an Exponent • Raising a Power to a Power • Raising a Product or a Quotient to a Power

In Section 4.2, we begin our study of polynomials. Before doing so, however, we must develop some rules for manipulating exponents.

Multiplying Powers with Like Bases

Recall from Section 1.8 that an expression like a^3 means $a \cdot a \cdot a$. We can use this fact to find the product of two expressions that have the same base:

$$a^3 \cdot a^2 = (a \cdot a \cdot a)(a \cdot a) \qquad \text{There are three factors in } a^3\text{; two factors in } a^2.$$

$$= a \cdot a \cdot a \cdot a \cdot a \qquad \text{Using an associative law}$$

$$= a^5.$$

Note that the exponent in a^5 is the sum of the exponents in $a^3 \cdot a^2$. That is, $3 + 2 = 5$. Similarly,

$$b^4 \cdot b^3 = (b \cdot b \cdot b \cdot b)(b \cdot b \cdot b)$$

$$= b^7, \quad \text{where } 4 + 3 = 7.$$

Adding the exponents gives the correct result.

The Product Rule

For any number a and any positive integers m and n,

$$a^m \cdot a^n = a^{m+n}.$$

(To multiply powers with the same base, keep the base and add the exponents.)

E x a m p l e   1

Multiply and simplify each of the following. (Here "simplify" means express the product as one base to a power whenever possible.)

a) $x^2 \cdot x^9$
b) $5 \cdot 5^8 \cdot 5^3$
c) $(r + s)^7 (r + s)^6$
d) $(a^3 b^2)(a^3 b^5)$

Study Tip

When you feel confident in your command of a topic, don't hesitate to help classmates experiencing trouble. Your understanding and retention of a concept will deepen when you explain it to someone else and your classmate will appreciate your help.

Solution

a) $x^2 \cdot x^9 = x^{2+9}$ Adding exponents: $a^m \cdot a^n = a^{m+n}$
$\quad\quad\quad\quad = x^{11}$

b) $5 \cdot 5^8 \cdot 5^3 = 5^1 \cdot 5^8 \cdot 5^3$ Recall that $x^1 = x$ for any number x.
$\quad\quad\quad\quad\quad = 5^{1+8+3}$ Adding exponents
$\quad\quad\quad\quad\quad = 5^{12}$ ——— **Caution!** $5^{12} \neq 5 \cdot 12.$

c) $(r + s)^7 (r + s)^6 = (r + s)^{7+6}$ The base here is $r + s$.
$\quad\quad\quad\quad\quad\quad = (r + s)^{13}$ ——— **Caution!** $(r + s)^{13} \neq r^{13} + s^{13}.$

d) $(a^3 b^2)(a^3 b^5) = a^3 b^2 a^3 b^5$ Using an associative law
$\quad\quad\quad\quad\quad\quad = a^3 a^3 b^2 b^5$ Using a commutative law
$\quad\quad\quad\quad\quad\quad = a^6 b^7$ Adding exponents

Dividing Powers with Like Bases

Recall that any expression that is divided or multiplied by 1 is unchanged. This, together with the fact that anything (besides 0) divided by itself is 1, can lead to a rule for division:

$$\frac{a^5}{a^2} = \frac{a \cdot a \cdot a \cdot a \cdot a}{a \cdot a}$$

$$= \frac{a \cdot a \cdot a}{1} \cdot \frac{a \cdot a}{a \cdot a}$$

$$= \frac{a \cdot a \cdot a}{1} \cdot 1$$

$$= a \cdot a \cdot a = a^3.$$

Note that the exponent in a^3 is the difference of the exponents in a^5/a^2. Similarly,

$$\frac{x^4}{x^3} = \frac{x \cdot x \cdot x \cdot x}{x \cdot x \cdot x} = \frac{x}{1} \cdot \frac{x \cdot x \cdot x}{x \cdot x \cdot x} = \frac{x}{1} \cdot 1 = x^1, \quad \text{or } x.$$

Subtracting the exponents gives the correct result.

> ### The Quotient Rule
>
> For any nonzero number a and any positive integers m and n for which $m > n$,
>
> $$\frac{a^m}{a^n} = a^{m-n}.$$
>
> (To divide powers with the same base, subtract the exponent of the denominator from the exponent of the numerator.)

E x a m p l e   2   Divide and simplify. (Here "simplify" means express the quotient as one base to a power whenever possible.)

a) $\dfrac{x^8}{x^2}$ **b)** $\dfrac{7^9}{7^4}$ **c)** $\dfrac{(5a)^{12}}{(5a)^4}$ **d)** $\dfrac{p^5 q^7}{p^2 q}$

Solution

a) $\dfrac{x^8}{x^2} = x^{8-2}$ Subtracting exponents: $\dfrac{a^m}{a^n} = a^{m-n}$

$\qquad = x^6$

b) $\dfrac{7^9}{7^4} = 7^{9-4}$

$\qquad = 7^5$

c) $\dfrac{(5a)^{12}}{(5a)^4} = (5a)^{12-4} = (5a)^8$ The base here is $5a$.

d) $\dfrac{p^5 q^7}{p^2 q} = \dfrac{p^5}{p^2} \cdot \dfrac{q^7}{q^1} = p^{5-2} \cdot q^{7-1} = p^3 q^6$ Using the quotient rule twice

Zero as an Exponent

The quotient rule can be used to help determine what 0 should mean when it appears as an exponent. Consider a^4/a^4, where a is nonzero. Since the numerator and the denominator are the same,

$$\frac{a^4}{a^4} = 1.$$

On the other hand, using the quotient rule would give us

$$\frac{a^4}{a^4} = a^{4-4} = a^0.$$ Subtracting exponents

Since $a^0 = a^4/a^4 = 1$, this suggests that $a^0 = 1$ for any nonzero value of a.

> ### The Exponent Zero
> For any real number a, $a \neq 0$,
> $$a^0 = 1.$$
> (Any nonzero number raised to the 0 power is 1.)

Note that in the above box, 0^0 is not defined. For this text, we will assume that expressions like a^m do not represent 0^0.

E x a m p l e   3

Simplify: **(a)** 1948^0; **(b)** $(-9)^0$; **(c)** $(3x)^0$; **(d)** $(-1)9^0$.

Solution

a) $1948^0 = 1$ Any nonzero number raised to the 0 power is 1.
b) $(-9)^0 = 1$ Any nonzero number raised to the 0 power is 1. The base here is -9.
c) $(3x)^0 = 1$, for any $x \neq 0$.   The parentheses indicate that the base is $3x$.
d) We have

$$(-1)9^0 = (-1)1 = -1.$$ The base here is 9.

Recall that, unless there are calculations within parentheses, exponents are calculated before multiplication. Since multiplying by -1 is the same as finding the opposite, the expression $(-1)9^0$ could have been written as -9^0. Note that although $-9^0 = -1$, part (b) shows that $(-9)^0 = 1$.

Caution! $-9^0 \neq (-9)^0$, and, in general, $-a^n \neq (-a)^n$.

Raising a Power to a Power

Consider an expression like $(7^2)^4$:

$$(7^2)^4 = (7^2)(7^2)(7^2)(7^2)$$ There are four factors of 7^2.

$$= (7 \cdot 7)(7 \cdot 7)(7 \cdot 7)(7 \cdot 7)$$ We could also use the product rule.

$$= 7 \cdot 7 \cdot 7 \cdot 7 \cdot 7 \cdot 7 \cdot 7 \cdot 7$$ Using an associative law

$$= 7^8.$$

Note that the exponent in 7^8 is the product of the exponents in $(7^2)^4$. Similarly,

$$(y^5)^3 = y^5 \cdot y^5 \cdot y^5$$ There are three factors of y^5.

$$= (y \cdot y \cdot y \cdot y \cdot y)(y \cdot y \cdot y \cdot y \cdot y)(y \cdot y \cdot y \cdot y \cdot y)$$

$$= y^{15}.$$

Once again, we get the same result if we multiply exponents:

$$(y^5)^3 = y^{5 \cdot 3} = y^{15}.$$

> ### The Power Rule
>
> For any number a and any whole numbers m and n,
>
> $$(a^m)^n = a^{mn}.$$
>
> (To raise a power to a power, multiply the exponents and leave the base unchanged.)

Remember that for this text we assume that 0^0 is not considered.

E x a m p l e 4   Simplify: **(a)** $(m^2)^5$; **(b)** $(3^5)^4$.

Solution

a) $(m^2)^5 = m^{2 \cdot 5}$ Multiplying exponents: $(a^m)^n = a^{mn}$

$\qquad = m^{10}$

b) $(3^5)^4 = 3^{5 \cdot 4}$

$\qquad = 3^{20}$

Raising a Product or a Quotient to a Power

When an expression inside parentheses is raised to a power, the inside expression is the base. Let's compare $2a^3$ and $(2a)^3$:

$$2a^3 = 2 \cdot a \cdot a \cdot a; \qquad \text{The base is } a.$$

$$(2a)^3 = (2a)(2a)(2a) \qquad \text{The base is } 2a.$$

$$\qquad = (2 \cdot 2 \cdot 2)(a \cdot a \cdot a) \qquad \text{Using an associative and a commutative law}$$

$$\qquad = 2^3 a^3$$

$$\qquad = 8a^3.$$

We see that $2a^3$ and $(2a)^3$ are *not* equivalent. Note too that $(2a)^3$ can be simplified by cubing each factor. This leads to the following rule for raising a product to a power.

> ### Raising a Product to a Power
>
> For any numbers a and b and any whole number n,
>
> $$(ab)^n = a^n b^n.$$
>
> (To raise a product to a power, raise each factor to that power.)

E x a m p l e 5   Simplify: **(a)** $(4a)^3$; **(b)** $(-5x^4)^2$; **(c)** $(a^7b)^2(a^3b^4)$.

Solution

a) $(4a)^3 = 4^3a^3 = 64a^3$ Raising each factor to the third power and simplifying

b) $(-5x^4)^2 = (-5)^2(x^4)^2$ Raising each factor to the second power. Parentheses are important here.

$\qquad\quad = 25x^8$ Simplifying $(-5)^2$ and using the product rule

c) $(a^7b)^2(a^3b^4) = (a^7)^2b^2a^3b^4$ Raising a product to a power

$\qquad\qquad\qquad = a^{14}b^2a^3b^4$ Multiplying exponents

$\qquad\qquad\qquad = a^{17}b^6$ Adding exponents

Caution!   The rule $(ab)^n = a^nb^n$ applies only to *products* raised to a power, not to sums or differences. For example, $(3 + 4)^2 \neq 3^2 + 4^2$ since $7^2 \neq 9 + 16$.

There is a similar rule for raising a quotient to a power.

Raising a Quotient to a Power

For any numbers a and b, $b \neq 0$, and any whole number n,

$$\left(\frac{a}{b}\right)^n = \frac{a^n}{b^n}.$$

(To raise a quotient to a power, raise the numerator to the power and divide by the denominator to the power.)

E x a m p l e 6   Simplify: **(a)** $\left(\dfrac{x}{5}\right)^2$; **(b)** $\left(\dfrac{5}{a^4}\right)^3$; **(c)** $\left(\dfrac{3a^4}{b^3}\right)^2$.

Solution

a) $\left(\dfrac{x}{5}\right)^2 = \dfrac{x^2}{5^2} = \dfrac{x^2}{25}$ Squaring the numerator and the denominator

b) $\left(\dfrac{5}{a^4}\right)^3 = \dfrac{5^3}{(a^4)^3}$ Raising a quotient to a power

$\qquad\qquad = \dfrac{125}{a^{4 \cdot 3}} = \dfrac{125}{a^{12}}$ Using the power rule and simplifying

c) $\left(\dfrac{3a^4}{b^3}\right)^2 = \dfrac{(3a^4)^2}{(b^3)^2}$ Raising a quotient to a power

$\qquad\qquad = \dfrac{3^2(a^4)^2}{b^{3 \cdot 2}} = \dfrac{9a^8}{b^6}$ Raising a product to a power and using the power rule

In the following summary of definitions and rules, we assume that no denominators are 0 and 0^0 is not considered.

Definitions and Properties of Exponents

For any whole numbers m and n,

1 as an exponent:	$a^1 = a$
0 as an exponent:	$a^0 = 1$
The Product Rule:	$a^m \cdot a^n = a^{m+n}$
The Quotient Rule:	$\dfrac{a^m}{a^n} = a^{m-n}$
The Power Rule:	$(a^m)^n = a^{mn}$
Raising a product to a power:	$(ab)^n = a^n b^n$
Raising a quotient to a power:	$\left(\dfrac{a}{b}\right)^n = \dfrac{a^n}{b^n}$

Exercise Set 4.1

Simplify. Assume that no denominator is zero and 0^0 is not considered.

1. $r^4 \cdot r^6$

2. $8^4 \cdot 8^3$

3. $9^5 \cdot 9^3$

4. $n^3 \cdot n^{20}$

5. $a^6 \cdot a$

6. $y^7 \cdot y^9$

7. $5^7 \cdot 5^8$

8. $t^0 \cdot t^{16}$

9. $(3y)^4(3y)^8$

10. $(2t)^8(2t)^{17}$

11. $(5t)(5t)^6$

12. $(8x)^0(8x)^1$

13. $(a^2b^7)(a^3b^2)$

14. $(m-3)^4(m-3)^5$

15. $(x+1)^5(x+1)^7$

16. $(a^8b^3)(a^4b)$

17. $r^3 \cdot r^7 \cdot r^0$

18. $s^4 \cdot s^5 \cdot s^2$

19. $(xy^4)(xy)^3$

20. $(a^3b)(ab)^4$

21. $\dfrac{7^5}{7^2}$

22. $\dfrac{4^7}{4^3}$

23. $\dfrac{x^{15}}{x^3}$

24. $\dfrac{a^{10}}{a^2}$

25. $\dfrac{t^5}{t}$

26. $\dfrac{x^7}{x}$

27. $\dfrac{(5a)^7}{(5a)^6}$

28. $\dfrac{(3m)^9}{(3m)^8}$

Aha! **29.** $\dfrac{(x+y)^8}{(x+y)^8}$

30. $\dfrac{(a-b)^4}{(a-b)^3}$

31. $\dfrac{18m^5}{6m^2}$

32. $\dfrac{30n^7}{6n^3}$

33. $\dfrac{a^9b^7}{a^2b}$

34. $\dfrac{r^{10}s^7}{r^2s}$

35. $\dfrac{m^9n^8}{m^0n^4}$

36. $\dfrac{a^{10}b^{12}}{a^2b^0}$

Simplify.

37. x^0 when $x = 13$

38. y^0 when $y = 38$

39. $5x^0$ when $x = -4$

40. $7m^0$ when $m = 1.7$

41. $8^0 + 5^0$

42. $(8 + 5)^0$

43. $(-3)^1 - (-3)^0$

44. $(-4)^0 - (-4)^1$

Simplify. Assume that no denominator is zero and 0^0 is not considered.

45. $(x^4)^7$ **46.** $(a^3)^8$ **47.** $(5^8)^2$

48. $(2^5)^3$ **49.** $(m^7)^5$ **50.** $(n^9)^2$

51. $(t^{20})^4$ **52.** $(t^3)^9$ **53.** $(7x)^2$

54. $(5a)^2$ **55.** $(-2a)^3$ **56.** $(-3x)^3$

57. $(4m^3)^2$ **58.** $(5n^4)^2$ **59.** $(a^2b)^7$

60. $(xy^4)^9$ **61.** $(x^3y)^2(x^2y^5)$ **62.** $(a^4b^6)(a^2b)^5$

63. $(2x^5)^3(3x^4)$ **64.** $(5x^3)^2(2x^7)$ **65.** $\left(\dfrac{a}{4}\right)^3$

66. $\left(\dfrac{3}{x}\right)^4$ **67.** $\left(\dfrac{7}{5a}\right)^2$ **68.** $\left(\dfrac{5x}{2}\right)^3$

69. $\left(\dfrac{a^4}{b^3}\right)^5$ **70.** $\left(\dfrac{x^5}{y^2}\right)^7$ **71.** $\left(\dfrac{y^3}{2}\right)^2$

72. $\left(\dfrac{a^5}{2}\right)^3$ **73.** $\left(\dfrac{x^2y}{z^3}\right)^4$ **74.** $\left(\dfrac{x^3}{y^2z}\right)^5$

75. $\left(\dfrac{a^3}{-2b^5}\right)^4$ **76.** $\left(\dfrac{x^5}{-3y^3}\right)^4$ **77.** $\left(\dfrac{5x^7y}{2z^4}\right)^3$

78. $\left(\dfrac{4a^2b}{3c^7}\right)^3$ *Aha!* **79.** $\left(\dfrac{4x^3y^5}{3z^7}\right)^0$ **80.** $\left(\dfrac{5a^7}{2b^5c}\right)^0$

81. Explain in your own words why $-5^2 \neq (-5)^2$.

82. Under what circumstances should exponents be added?

SKILL MAINTENANCE

Factor.

83. $3s - 3r + 3t$ **84.** $-7x + 7y - 7z$

Combine like terms.

85. $9x + 2y - x - 2y$

86. $5a - 7b - 8a + b$

Use the commutative law of addition to write an equivalent expression.

87. $3x + 2y$ **88.** $2xy + 5z$

SYNTHESIS

89. Under what conditions does a^n represent a negative number? Why?

90. Using the quotient rule, explain why 9^0 is 1.

91. Suppose that the width of a square is three times the width of a second square (see the figure at the top of the next column). How do the areas of the squares compare? Why?

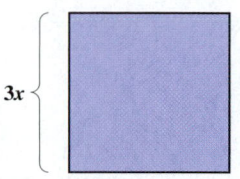
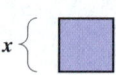

92. Suppose that the width of a cube is twice the width of a second cube. How do the volumes of the cubes compare? Why?

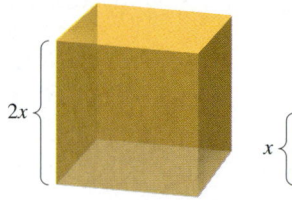

Find a value of the variable that shows that the two expressions are not equivalent. Answers may vary.

93. $(a + 5)^2$; $a^2 + 5^2$ **94.** $3x^2$; $(3x)^2$

95. $\dfrac{a + 7}{7}$; a **96.** $\dfrac{t^6}{t^2}$; t^3

Simplify.

97. $a^{10k} \div a^{2k}$ **98.** $y^{4x} \cdot y^{2x}$

99. $\dfrac{\left(\frac{1}{2}\right)^3\left(\frac{2}{3}\right)^4}{\left(\frac{5}{6}\right)^3}$ **100.** $\dfrac{x^{5t}(x^t)^2}{(x^{3t})^2}$

101. Solve for x:
$$\dfrac{t^{26}}{t^x} = t^x.$$

Replace ▦ *with* $>$, $<$, *or* $=$ *to write a true sentence.*

102. 3^5 ▦ 3^4 **103.** 4^2 ▦ 4^3

104. 4^3 ▦ 5^3 **105.** 4^3 ▦ 3^4

106. 9^7 ▦ 3^{13} **107.** 25^8 ▦ 125^5

In computer science, 1 K of memory refers to 1 kilobyte, or 1×10^3 bytes, of memory. This is really an approximation of 1×2^{10} bytes (since computer memory uses powers of 2). Use the fact that $10^3 \approx 2^{10}$ to estimate each of the following powers of 2. Then compute the power of 2 with a calculator and find the difference between the exact value and the approximation.

108. 2^{14} **109.** 2^{22}

110. 2^{26} **111.** 2^{31}

112. Dana's research project requires 56 K of memory. How many bytes is this?

113. The cash register at Justin's shop has 64 K of memory. How many bytes is this?

4.2

Polynomials

Terms • Types of Polynomials • Degree and Coefficients •
Combining Like Terms • Evaluating Polynomials and Applications

We now examine an important algebraic expression known as a *polynomial*. Certain polynomials have appeared earlier in this text so you already have some experience working with them.

Terms

At this point, we have seen a variety of algebraic expressions like

$$3a^2b^4, \qquad 2l + 2w, \quad \text{and} \quad 5x^2 + x - 2.$$

Of these, $3a^2b^4$, $2l$, $2w$, $5x^2$, x, and -2 are examples of *terms*. A **term** can be a number (like -2), a variable (like x), or a product of numbers and/or variables, which may be raised to powers (like $3a^2b^4$, $2l$, $2w$, or $5x^2$).*

Types of Polynomials

A term that is a product of constants and/or variables is called a **monomial**.† All the terms listed above are monomials. Other examples of monomials are

$$7, \qquad t, \qquad 23x^2y, \quad \text{and} \quad \tfrac{3}{7}a^5.$$

A **polynomial** is a monomial or a sum of monomials. The following are examples of polynomials:

$$4x + 7, \quad \tfrac{2}{3}t^2, \quad 6a + 7, \quad -5n^2 + n - 1, \quad 42r^5, \quad x, \quad \text{and} \quad 0.$$

The following algebraic expressions are *not* polynomials:

$$\textbf{(1)} \ \frac{x + 3}{x - 4}, \qquad \textbf{(2)} \ 5x^3 - 2x^2 + \frac{1}{x}, \qquad \textbf{(3)} \ \frac{1}{x^3 - 2}.$$

Expressions (1) and (3) are not polynomials because they represent quotients, not sums. Expression (2) is not a polynomial because $1/x$ is not a monomial.

When a polynomial is written as a sum of monomials, each monomial is called a *term of the polynomial.*

---

*Later in this text, expressions like $5x^{3/2}$ and $2a^{-7}b$ will be discussed. Such expressions are also considered terms.

†Note that a term, but not a monomial, can include division by a variable.

Example 1   Identify the terms of the polynomial $3t^4 - 5t^6 - 4t + 2$.

Solution   The terms are $3t^4$, $-5t^6$, $-4t$, and 2. We can see this by rewriting all subtractions as additions of opposites:

$$3t^4 - 5t^6 - 4t + 2 = 3t^4 + (-5t^6) + (-4t) + 2.$$

These are the terms of the polynomial.

A polynomial that is composed of two terms is called a **binomial**, whereas those composed of three terms are called **trinomials**. Polynomials with four or more terms have no special name.

Monomials	Binomials	Trinomials	No Special Name
$4x^2$	$2x + 4$	$3t^3 + 4t + 7$	$4x^3 - 5x^2 + xy - 8$
9	$3a^5 + 6bc$	$6x^7 - 8z^2 + 4$	$z^5 + 2z^4 - z^3 + 7z + 3$
$-7a^{19}b^5$	$-9x^7 - 6$	$4x^2 - 6x - \frac{1}{2}$	$4x^6 - 3x^5 + x^4 - x^3 + 2x - 1$

Degree and Coefficients

The **degree of a term** is the number of variable factors in that term. Thus the degree of $7t^2$ is 2 because $7t^2$ has two variable factors: $7t^2 = 7 \cdot t \cdot t$.

Example 2   Determine the degree of each term: **(a)** $8x^4$; **(b)** $3x$; **(c)** 7.

Solution

a) The degree of $8x^4$ is 4. x^4 represents 4 variable factors: $x \cdot x \cdot x \cdot x$.
b) The degree of $3x$ is 1.   There is 1 variable factor.
c) The degree of 7 is 0.   There is no variable factor.

The part of a term that is a constant factor is the **coefficient** of that term. Thus the coefficient of $3x$ is 3, and the coefficient for the term 7 is simply 7.

Example 3   Identify the coefficient of each term in the polynomial

$$4x^3 - 7x^2y + x - 8.$$

Solution

The coefficient of $4x^3$ is 4.

The coefficient of $-7x^2y$ is -7.

The coefficient of the third term is 1, since $x = 1x$.

The coefficient of -8 is simply -8.

The **leading term** of a polynomial is the term of highest degree. Its coefficient is called the **leading coefficient** and its degree is referred to as the **degree of the polynomial**. To see how this terminology is used, consider the polynomial

$$3x^2 - 8x^3 + 5x^4 + 7x - 6.$$

The *terms* are $3x^2$, $-8x^3$, $5x^4$, $7x$,   and -6.
The *coefficients* are   3, -8,   5,   7,   and -6.
The *degree of each term* is   2,   3,   4,   1,   and   0.
The *leading term* is $5x^4$ and the *leading coefficient* is 5.
The *degree of the polynomial* is 4.

Combining Like Terms

Recall from Section 1.8 that *like*, or *similar*, *terms* are either constant terms or terms containing the same variable(s) raised to the same power(s). To simplify certain polynomials, we can often *combine*, or *collect*, like terms.

E x a m p l e   4   Identify the like terms in $4x^3 + 5x - 7x^2 + 2x^3 + x^2$.

Solution

Like terms: $4x^3$ and $2x^3$ Same variable and exponent
Like terms: $-7x^2$ and x^2 Same variable and exponent

E x a m p l e   5   Combine like terms.

a) $2x^3 - 6x^3$ **b)** $5x^2 + 7 + 2x^4 + 4x^2 - 11 - 2x^4$
c) $7a^3 - 5a^2 + 9a^3 + a^2$ **d)** $\frac{2}{3}x^4 - x^3 - \frac{1}{6}x^4 + \frac{2}{5}x^3 - \frac{3}{10}x^3$

Solution

a) $2x^3 - 6x^3 = (2 - 6)x^3$ Using the distributive law
$= -4x^3$

b) $5x^2 + 7 + 2x^4 + 4x^2 - 11 - 2x^4 = 5x^2 + 4x^2 + 2x^4 - 2x^4 + 7 - 11$
$= (5 + 4)x^2 + (2 - 2)x^4 + (7 - 11)$ ⎫ These steps are often
$= 9x^2 + 0x^4 + (-4)$ ⎬ done mentally.
$= 9x^2 - 4$ ⎭

c) $7a^3 - 5a^2 + 9a^3 + a^2 = 7a^3 - 5a^2 + 9a^3 + 1a^2$ When a variable to a power appears without a coefficient, we can write in 1.

$= 16a^3 - 4a^2$

d) $\frac{2}{3}x^4 - x^3 - \frac{1}{6}x^4 + \frac{2}{5}x^3 - \frac{3}{10}x^3 = \left(\frac{2}{3} - \frac{1}{6}\right)x^4 + \left(-1 + \frac{2}{5} - \frac{3}{10}\right)x^3$
$= \left(\frac{4}{6} - \frac{1}{6}\right)x^4 + \left(-\frac{10}{10} + \frac{4}{10} - \frac{3}{10}\right)x^3$
$= \frac{3}{6}x^4 - \frac{9}{10}x^3$
$= \frac{1}{2}x^4 - \frac{9}{10}x^3$

Note in Example 5 that the solutions are written so that the term of highest degree appears first, followed by the term of next highest degree, and so on. This is known as **descending order** and is the form in which answers will normally appear.

Evaluating Polynomials and Applications

When each variable in a polynomial is replaced with a number, the polynomial then represents a number, or *value*, that can be calculated using the rules for order of operations.

E x a m p l e   6

Evaluate $-x^2 + 3x + 9$ for $x = -2$.

Solution   For $x = -2$, we have

$$-x^2 + 3x + 9 = -(-2)^2 + 3(-2) + 9 \qquad \text{\textcolor{magenta}{The negative sign in front of}}$$
$$\textcolor{magenta}{x^2 \text{ remains.}}$$

$$= -4 + (-6) + 9$$
$$= -10 + 9 = -1.$$

E x a m p l e   7

Games in a sports league.   In a sports league of n teams in which each team plays every other team twice, the total number of games to be played is given by the polynomial

$$n^2 - n.$$

A girl's soccer league has 10 teams. How many games are played if each team plays every other team twice?

Solution   We evaluate the polynomial for $n = 10$:

$$n^2 - n = 10^2 - 10$$
$$= 100 - 10$$
$$= 90.$$

The league plays 90 games.

E x a m p l e   8

Medical dosage.   The concentration, in parts per million, of a certain antibiotic in the bloodstream after t hours is given by the polynomial

$$-0.05t^2 + 2t + 2.$$

Find the concentration after 2 hr.

Solution   To find the concentration after 2 hr, we evaluate the polynomial for $t = 2$:

$$-0.05t^2 + 2t + 2 = -0.05(2)^2 + 2(2) + 2$$
$$= -0.05(4) + 2(2) + 2$$
$$= -0.2 + 4 + 2$$
$$= 5.8.$$

The concentration after 2 hr is 5.8 parts per million.

Sometimes, a graph can be used to estimate the value of a polynomial visually.

E x a m p l e   9

Medical dosage.   In the following graph, the polynomial from Example 8 has been graphed by evaluating it for several choices of t. Use the graph to estimate the concentration c of antibiotic in the bloodstream after 14 hr.

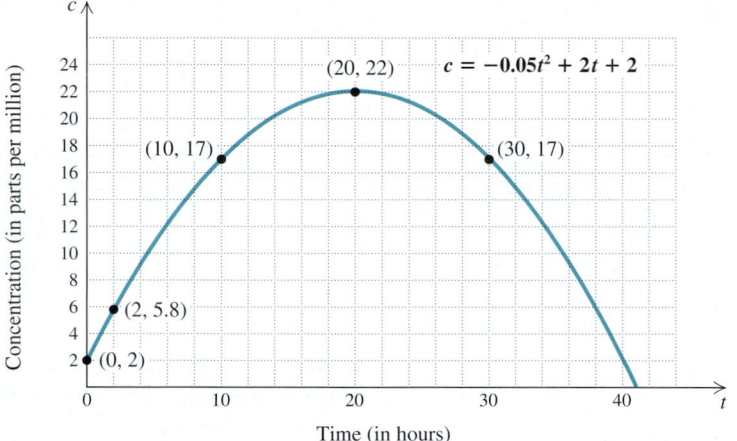

Solution   To estimate the concentration after 14 hr, we locate 14 on the horizontal axis. From there, we move vertically until we meet the curve at some point. From that point, we move horizontally to the c-axis.

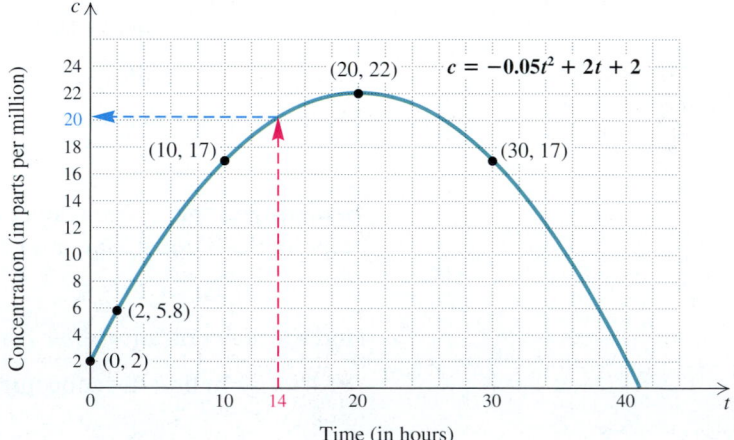

After 14 hr, the concentration of antibiotic in the bloodstream is about 20 parts per million. (For $t = 14$, the value of $-0.05t^2 + 2t + 2$ is approximately 20.)

technology connection

One way to evaluate a polynomial is to use the TRACE key. For example, to evaluate $-0.05x^2 + 2x + 2$ in Example 9, we can use TRACE and then enter the x-value in which we are interested (in this case, 14). The value of the polynomial appears as y, and the cursor automatically appears at $(14, 20.2)$. The Value option of the CALC menu works in a similar way.

1. Use TRACE or CALC Value to find the value of

$$-0.05x^2 + 2x + 2$$

for $x = 25$.

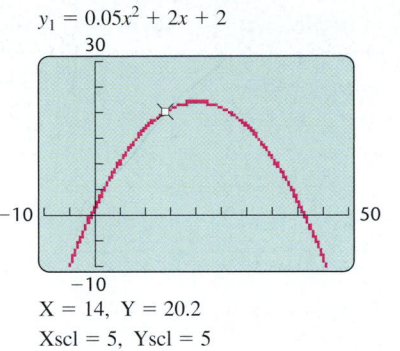

$y_1 = 0.05x^2 + 2x + 2$

X = 14, Y = 20.2
Xscl = 5, Yscl = 5

Exercise Set 4.2

Identify the terms of each polynomial.

1. $7x^4 + x^3 - 5x + 8$

2. $5a^3 + 4a^2 - a - 7$

3. $-t^4 + 7t^3 - 3t^2 + 6$

4. $n^5 - 4n^3 + 2n - 8$

Determine the coefficient and the degree of each term in each polynomial.

5. $4x^5 + 7x$

6. $9a^3 - 4a^2$

7. $9t^2 - 3t + 4$

8. $7x^4 + 5x - 3$

9. $7a^4 + 9a + a^3$

10. $6t^5 - 3t^2 - t$

11. $x^4 - x^3 + 4x - 3$

12. $3a^4 - a^3 + a - 9$

*For each of the following polynomials, **(a)** list the degree of each term; **(b)** determine the leading term and the leading coefficient; and **(c)** determine the degree of the polynomial.*

13. $2a^3 + 7a^5 + a^2$

14. $5x - 9x^2 + 3x^6$

15. $2t + 3 + 4t^2$

16. $3a^2 - 7 + 2a^4$

17. $9x^4 + x^2 + x^7 + 4$

18. $8 + 6x^2 - 3x - x^5$

19. $9a - a^4 + 3 + 2a^3$

20. $-x + 2x^5 - 5x^2 + x^6$

21. Complete the following table for the polynomial
$$7x^2 + 8x^5 - 4x^3 + 6 - \tfrac{1}{2}x^4.$$

Term	Coefficient	Degree of the Term	Degree of the Polynomial
		5	
$-\frac{1}{2}x^4$			
	−4		
		2	
	6		

22. Complete the following table for the polynomial
$$-3x^4 + 6x^3 - 2x^2 + 8x + 7.$$

Term	Coefficient	Degree of the Term	Degree of the Polynomial
	-3		
$6x^3$			
		2	
		1	
	7		

Classify each polynomial as a monomial, binomial, trinomial, or none of these.

23. $x^2 - 23x + 17$ **24.** $-9x^2$

25. $x^3 - 7x^2 + 2x - 4$ **26.** $t^3 + 4$

27. $8t^2 + 5t$ **28.** $4x^2 + 12x + 9$

29. 17

30. $2x^4 - 7x^3 + x^2 + x - 6$

Combine like terms. Write all answers in descending order.

31. $7x^2 + 3x + 4x^2$

32. $5a + 7a^2 + 3a$

33. $3a^4 - 2a + 2a + a^4$

34. $9b^5 + 3b^2 - 2b^5 - 3b^2$

35. $2x^2 - 6x + 3x + 4x^2$

36. $3x^4 - 7x + x^4 - 2x$

37. $9x^3 + 2x - 4x^3 + 5 - 3x$

38. $6x^2 + 2x^4 - 2x^2 - x^4 - 4x^2$

39. $10x^2 + 2x^3 - 3x^3 - 4x^2 - 6x^2 - x^4$

40. $8x^5 - x^4 + 2x^5 + 5x^4 - 4x^4 - x^6$

41. $\frac{1}{5}x^4 + 7 - 2x^2 + 3 - \frac{2}{15}x^4 + 2x^2$

42. $\frac{1}{6}x^3 + 3x^2 - \frac{1}{3}x^3 + 7 + x^2 - 10$

43. $5.9x^2 - 2.1x + 6 + 3.4x - 2.5x^2 - 0.5$

44. $7.4x^3 - 4.9x + 2.9 - 3.5x - 4.3 + 1.9x^3$

45. $6t - 9t^3 + 8t^4 + 4t + 2t^4 + 7t - 3t^3$

46. $5b^2 - 3b + 7b^2 - 4b^3 + 4b - 9b^2 + 10b^3$

Evaluate each polynomial for $x = 3$.

47. $-7x + 5$ **48.** $-5x + 9$

49. $2x^2 - 3x + 7$ **50.** $4x^2 - 6x + 9$

Evaluate each polynomial for $x = -2$.

51. $5x + 7$ **52.** $7 - 3x$

53. $x^2 - 3x + 1$ **54.** $5x - 9 + x^2$

55. $-3x^3 + 7x^2 - 4x - 5$

56. $-2x^3 - 4x^2 + 3x + 1$

Memorizing words. Participants in a psychology experiment were able to memorize an average of M words in t minutes, where $M = -0.001t^3 + 0.1t^2$. Use the following graph for Exercises 57–62.

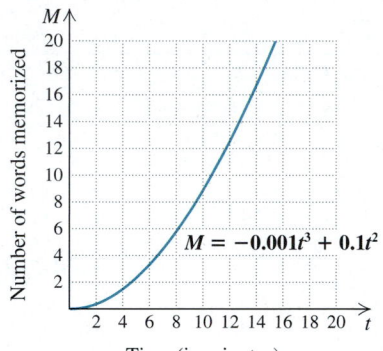

$M = -0.001t^3 + 0.1t^2$

Time (in minutes)

Number of words memorized

57. Estimate the number of words memorized after 10 min.

58. Estimate the number of words memorized after 14 min.

59. Find the approximate value of M for $t = 8$.

60. Find the approximate value of M for $t = 12$.

61. Estimate the value of M when t is 13.

62. Estimate the value of M when t is 7.

63. *Skydiving.* During the first 13 sec of a jump, the number of feet that a skydiver falls in t seconds is approximated by the polynomial
$$11.12t^2.$$

Approximately how far has a skydiver fallen 10 sec after jumping from a plane?

$11.12\,t^2$

64. *Skydiving.* For jumps that exceed 13 sec, the polynomial $173t - 369$ can be used to approximate the distance, in feet, that a skydiver has fallen in t seconds. Approximately how far has a skydiver fallen 20 sec after jumping from a plane?

Daily accidents. *The average number of accidents per day involving drivers of age r can be approximated by the polynomial*

$$0.4r^2 - 40r + 1039.$$

65. Evaluate the polynomial for $r = 18$ to find the daily number of accidents involving 18-year-old drivers.

66. Evaluate the polynomial for $r = 20$ to find the daily number of accidents involving 20-year-old drivers.

Total revenue. *Gigabytes Electronics is selling a new type of computer monitor. Total revenue is the total amount of money taken in. The firm estimates that for the monitor's first year, revenue from the sale of x monitors is*

$$250x - 0.5x^2 \text{ dollars.}$$

67. What is the total revenue from the sale of 40 monitors?

68. What is the total revenue from the sale of 60 monitors?

Total cost. *Gigabytes Electronics estimates that the total cost of producing x monitors is given by*

$$4000 + 0.6x^2 \text{ dollars.}$$

69. What is the total cost of producing 200 monitors?

70. What is the total cost of producing 300 monitors?

Circumference. *The circumference of a circle of radius r is given by the polynomial $2\pi r$, where π is an irrational number. For an approximation of π, use 3.14.*

71. Find the circumference of a circle with radius 10 cm.

72. Find the circumference of a circle with radius 5 ft.

Area of a circle. *The area of a circle of radius r is given by the polynomial πr^2.*

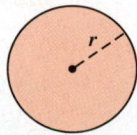

73. Find the area of a circle with radius 7 m.

74. Find the area of a circle with radius 6 ft.

75. Explain how it is possible for a term to not be a monomial.

76. Is it possible to evaluate polynomials without understanding the rules for order of operations? Why or why not?

SKILL MAINTENANCE

Simplify.

77. $-19 + 24$

78. $5 - 14$

Factor.

79. $5x + 15$

80. $7a - 21$

81. A family spent $2011 to drive a car one year, during which the car was driven 14,800 mi. The family spent $972 for insurance and $114 for registration and oil. The only other cost was for gasoline. How much did gasoline cost per mile?

82. The sum of the page numbers on the facing pages of a book is 549. What are the page numbers?

SYNTHESIS

83. Suppose that the coefficients of a polynomial are all integers and the polynomial is evaluated for some integer. Must the value of the polynomial then also be an integer? Why or why not?

84. Is it easier to evaluate a polynomial before or after like terms have been combined? Why?

85. Construct a polynomial in x (meaning that x is the variable) of degree 5 with four terms and coefficients that are integers.

86. Construct a trinomial in y of degree 4 with coefficients that are rational numbers.

87. What is the degree of $(5m^5)^2$?

88. Construct three like terms of degree 4.

Simplify.

89. $\frac{9}{2}x^8 + \frac{1}{9}x^2 + \frac{1}{2}x^9 + \frac{9}{2}x + \frac{9}{2}x^9 + \frac{8}{9}x^2 + \frac{1}{2}x - \frac{1}{2}x^8$

90. $(3x^2)^3 + 4x^2 \cdot 4x^4 - x^4(2x)^2 + ((2x)^2)^3 - 100x^2(x^2)^2$

91. A polynomial in x has degree 3. The coefficient of x^2 is 3 less than the coefficient of x^3. The coefficient of x is three times the coefficient of x^2. The remaining constant is 2 more than the coefficient of x^3. The sum of the coefficients is -4. Find the polynomial.

92. *Path of the Olympic arrow.* The Olympic flame at the 1992 Summer Olympics was lit by a flaming arrow. As the arrow moved d meters horizontally from the archer, its height h, in meters, was approximated by the polynomial

$$-0.0064d^2 + 0.8d + 2.$$

Complete the table for the choices of d given. Then plot the points and draw a graph representing the path of the arrow.

d	$-0.0064d^2 + 0.8d + 2$
0	
30	
60	
90	
120	

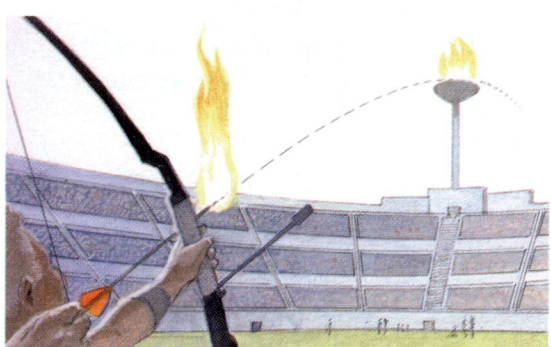

Semester averages. *Professor Kopecki calculates a student's average for her course using*

$$A = 0.3q + 0.4t + 0.2f + 0.1h,$$

with q, t, f, and h representing a student's quiz average, test average, final exam score, and homework average, respectively. In Exercises 93 and 94, find the given student's course average rounded to the nearest tenth.

93. Mary Lou: quizzes: 60, 85, 72, 91; final exam: 84; tests: 89, 93, 90; homework: 88

94. Nigel: quizzes: 95, 99, 72, 79; final exam: 91; tests: 68, 76, 92; homework: 86

95. *Daily accidents.* The average number of accidents per day involving drivers of age r can be approximated by the polynomial

$$0.4r^2 - 40r + 1039.$$

For what age is the number of daily accidents smallest?

In Exercises 96 and 97, complete the table for the given choices of t. Then plot the points and connect them with a smooth curve representing the graph of the polynomial.

96.

t	$-t^2 + 10t - 18$
3	
4	
5	
6	
7	

97.

t	$-t^2 + 6t - 4$
1	
2	
3	
4	
5	

4.3

Addition and Subtraction of Polynomials

Addition of Polynomials • Opposites of Polynomials • Subtraction of Polynomials • Problem Solving

Addition of Polynomials

To add two polynomials, we write a plus sign between them and combine like terms.

E x a m p l e   1   Add.

a) $(-5x^3 + 6x - 1) + (4x^3 + 3x^2 + 2)$

b) $\left(\frac{2}{3}x^4 + 3x^2 - 7x + \frac{1}{2}\right) + \left(-\frac{1}{3}x^4 + 5x^3 - 3x^2 + 3x - \frac{1}{2}\right)$

Solution

a) $(-5x^3 + 6x - 1) + (4x^3 + 3x^2 + 2)$

$\qquad = (-5 + 4)x^3 + 3x^2 + 6x + (-1 + 2)$ Combining like terms; using the distributive law

$\qquad = -x^3 + 3x^2 + 6x + 1$ Note that $-1x^3 = -x^3$.

b) $\left(\frac{2}{3}x^4 + 3x^2 - 7x + \frac{1}{2}\right) + \left(-\frac{1}{3}x^4 + 5x^3 - 3x^2 + 3x - \frac{1}{2}\right)$

$\qquad = \left(\frac{2}{3} - \frac{1}{3}\right)x^4 + 5x^3 + (3 - 3)x^2 + (-7 + 3)x + \left(\frac{1}{2} - \frac{1}{2}\right)$ Combining like terms

$\qquad = \frac{1}{3}x^4 + 5x^3 - 4x$

After some practice, polynomial addition is often performed mentally.

E x a m p l e   2   Add: $(2 - 3x + x^2) + (-5 + 7x - 3x^2 + x^3)$

Solution   We have

$(2 - 3x + x^2) + (-5 + 7x - 3x^2 + x^3)$

$\qquad = (2 - 5) + (-3 + 7)x + (1 - 3)x^2 + x^3$ You might do this step mentally.

$\qquad = -3 + 4x - 2x^2 + x^3.$ Then you would write only this.

The polynomials in the last example are written with the terms arranged according to degree, from least to greatest. Such an arrangement is called *ascending order*. As a rule, answers are written in ascending order when the polynomials being added are given in ascending order. When the polynomials being added are given in descending order, the answer is written in descending order.

We can also add polynomials by writing like terms in columns. Sometimes this makes like terms easier to see.

E x a m p l e   3   Add: $9x^5 - 2x^3 + 6x^2 + 3$ and $5x^4 - 7x^2 + 6$ and $3x^6 - 5x^5 + x^2 + 5.$

Solution   We arrange the polynomials with like terms in columns.

$$
\begin{array}{l}
9x^5 \qquad\quad - 2x^3 + 6x^2 + 3 \\
\qquad\quad 5x^4 \qquad\qquad - 7x^2 + 6 \quad \text{\color{#cc0066}{We leave spaces for missing terms.}} \\
\underline{3x^6 - 5x^5 \qquad\qquad\qquad\quad + 1x^2 + 5} \quad \text{\color{#cc0066}{Writing x^2 as $1x^2$}} \\
3x^6 + 4x^5 + 5x^4 - 2x^3 \qquad\quad + 14 \quad \text{\color{#cc0066}{Adding}}
\end{array}
$$

The answer is $3x^6 + 4x^5 + 5x^4 - 2x^3 + 14.$

CONNECTING THE CONCEPTS

In many ways, polynomials are to algebra as numbers are to arithmetic. Like numbers, polynomials can be added, subtracted, multiplied, and divided. In Sections 4.3–4.7, we examine how these operations are performed. Thus we are again learning how to write equivalent expressions, not how to solve equations.

Note that equations like those in Examples 1–3 are written to show how one expression can be rewritten in an equivalent form. This is very different from solving an equation.

There is much to learn about manipulating polynomials. Not until Section 5.7 will we return to the task of solving equations.

Opposites of Polynomials

In Section 1.8, we used the property of -1 to show that the opposite of a sum is the sum of the opposites. This idea can be extended.

> ### The Opposite of a Polynomial
> To find an equivalent polynomial for the *opposite*, or *additive inverse*, of a polynomial, change the sign of every term. This is the same as multiplying the polynomial by -1.

Example 4   Write the opposite of $4x^5 - 7x^3 - 8x + \frac{5}{6}$ in two different forms.

Solution

i) $-\left(4x^5 - 7x^3 - 8x + \frac{5}{6}\right)$

ii) $-4x^5 + 7x^3 + 8x - \frac{5}{6}$ Changing the sign of every term

Thus, $-\left(4x^5 - 7x^3 - 8x + \frac{5}{6}\right)$ is equivalent to $-4x^5 + 7x^3 + 8x - \frac{5}{6}$. Both expressions represent the opposite of $4x^5 - 7x^3 - 8x + \frac{5}{6}$.

Example 5   Simplify: $-(-7x^4 - \frac{5}{9}x^3 + 8x^2 - x + 67)$.

Solution

$$-\left(-7x^4 - \frac{5}{9}x^3 + 8x^2 - x + 67\right) = 7x^4 + \frac{5}{9}x^3 - 8x^2 + x - 67$$

Subtraction of Polynomials

We can now subtract one polynomial from another by adding the opposite of the polynomial being subtracted.

Example 6

Subtract: $(9x^5 + x^3 - 2x^2 + 4) - (-2x^5 + x^4 - 4x^3 - 3x^2)$.

Solution

$$(9x^5 + x^3 - 2x^2 + 4) - (-2x^5 + x^4 - 4x^3 - 3x^2)$$

$$= 9x^5 + x^3 - 2x^2 + 4 + 2x^5 - x^4 + 4x^3 + 3x^2 \qquad \text{Adding the opposite}$$

$$= 11x^5 - x^4 + 5x^3 + x^2 + 4 \qquad \text{Combining like terms}$$

Example 7

Subtract: $(7x^5 + x^3 - 9x) - (3x^5 - 4x^3 + 5)$.

Solution

$$(7x^5 + x^3 - 9x) - (3x^5 - 4x^3 + 5)$$

$$= 7x^5 + x^3 - 9x + (-3x^5) + 4x^3 - 5 \qquad \text{Adding the opposite}$$

$$= 7x^5 + x^3 - 9x - 3x^5 + 4x^3 - 5 \qquad \text{Try to go directly to this step.}$$

$$= 4x^5 + 5x^3 - 9x - 5 \qquad \text{Combining like terms}$$

To subtract using columns, we first replace the coefficients in the polynomial being subtracted with their opposites. We then add as before.

Example 8

Write in columns and subtract: $(5x^2 - 3x + 6) - (9x^2 - 5x - 3)$.

Solution

i)
$$\begin{array}{r} 5x^2 - 3x + 6 \\ -(9x^2 - 5x - 3) \\ \hline \end{array}$$
Writing similar terms in columns

ii)
$$\begin{array}{r} 5x^2 - 3x + 6 \\ -9x^2 + 5x + 3 \\ \hline \end{array}$$
Changing signs and removing parentheses

iii)
$$\begin{array}{r} 5x^2 - 3x + 6 \\ -9x^2 + 5x + 3 \\ \hline -4x^2 + 2x + 9 \end{array}$$
Adding

If you can do so without error, you can arrange the polynomials in columns, mentally find the opposite of each term being subtracted, and write the answer.

Example 9

Write in columns and subtract: $(x^3 + x^2 + 2x - 12) - (-2x^3 + x^2 - 3x)$.

Solution We have

$$\begin{array}{r} x^3 + x^2 + 2x - 12 \\ -(-2x^3 + x^2 - 3x \qquad) \\ \hline 3x^3 \qquad\quad + 5x - 12. \end{array}$$
Leaving space for the missing term

Problem Solving

E x a m p l e 1 0   Find a polynomial for the sum of the areas of rectangles A, B, C, and D.

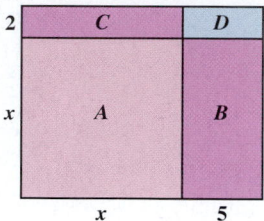

Solution

1. **Familiarize.** Recall that the area of a rectangle is the product of its length and width.

2. **Translate.** We translate the problem to mathematical language. The sum of the areas is a sum of products. We find each product and then add:

$$\underbrace{\text{Area of } A}\quad \text{plus}\quad \underbrace{\text{area of } B}\quad \text{plus}\quad \underbrace{\text{area of } C}\quad \text{plus}\quad \underbrace{\text{area of } D}$$

$$x \cdot x\quad +\quad 5x\quad +\quad 2x\quad +\quad 2 \cdot 5.$$

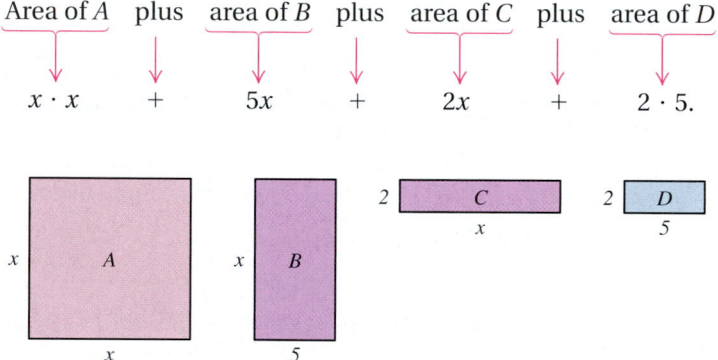

3. **Carry out.** We combine like terms:

$$x^2 + 5x + 2x + 10 = x^2 + 7x + 10.$$

4. **Check.** A partial check is to replace x with a number, say 3. Then we evaluate $x^2 + 7x + 10$ and compare that result with an alternative calculation:

$$3^2 + 7 \cdot 3 + 10 = 9 + 21 + 10 = 40.$$

When we substitute 3 for x and calculate the total area by regarding the figure as one large rectangle, we should also get 40:

$$\text{Total area} = (x + 5)(x + 2) = (3 + 5)(3 + 2) = 8 \cdot 5 = 40.$$

Our check is only partial, since it is possible for an incorrect answer to equal 40 when evaluated for $x = 3$. This would be unlikely, especially if a second choice of x, say $x = 5$, also checks. We leave that check to the student.

5. **State.** A polynomial for the sum of the areas is $x^2 + 7x + 10$.

E x a m p l e   1 1   An 8-ft by 8-ft shed is placed on a square lawn x ft on a side. Find a polynomial for the remaining area.

Solution

1. **Familiarize.**   We make a drawing of the situation as follows.

2. **Translate.**   We reword the problem and translate as follows.

Rewording: Area of lawn − area of shed = area left over

Translating: x ft · x ft   −   8 ft · 8 ft   = Area left over

3. **Carry out.**   We carry out the manipulation by multiplying:

$$x^2 \text{ ft}^2 - 64 \text{ ft}^2 = \text{Area left over.}$$

4. **Check.**   As a partial check, note that the units in the answer are square feet (ft^2), a measure of area, as expected.

5. **State.**   The remaining area in the yard is $(x^2 - 64)$ ft^2.

technology connection

To check polynomial addition or subtraction, we can let y_1 = the expression before the addition or subtraction has been performed and y_2 = the simplified sum or difference. If the addition or subtraction is correct, y_1 will equal y_2 and $y_2 - y_1$ will be 0. We enter $y_2 - y_1$ as y_3, using the VARS key. Below is a check of Example 7 in which

$$y_1 = (7x^5 + x^3 - 9x) - (3x^5 - 4x^3 + 5),$$
$$y_2 = 4x^5 + 5x^3 - 9x - 5,$$

and

$$y_3 = y_2 - y_1.$$

We graph only y_3. If indeed y_1 and y_2 are equivalent, then y_3 should equal 0. This means its graph should coincide with the x-axis. The TRACE or TABLE features can confirm

that y_3 is always 0, or we can select y_3 to be drawn bold at the [**Y=**] window.

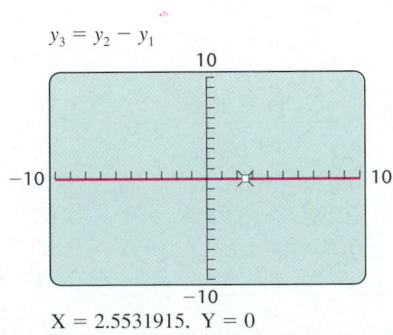

$$y_3 = y_2 - y_1$$

X = 2.5531915. Y = 0

1. Use a grapher to check Examples 1, 2, and 6.

FOR EXTRA HELP

Exercise Set 4.3

 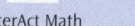
Digital Video Tutor CD 3
Videotape 7

InterAct Math

Math Tutor Center

MathXL

MyMathLab.com

Add.

1. $(2x + 3) + (-7x + 6)$

2. $(5x + 1) + (-9x + 4)$

3. $(-6x + 2) + (x^2 + x - 3)$

4. $(x^2 - 5x + 4) + (8x - 9)$

5. $(7t^2 - 3t + 6) + (2t^2 + 8t - 9)$

6. $(9a^2 + 4a - 5) + (6a^2 - 3a - 1)$

7. $(2m^3 - 4m^2 + m - 7) + (4m^3 + 7m^2 - 4m - 2)$

8. $(5n^3 - n^2 + 4n - 3) + (2n^3 - 4n^2 + 3n - 4)$

9. $(3 + 6a + 7a^2 + 8a^3) + (4 + 7a - a^2 + 6a^3)$

10. $(7 + 4t - 5t^2 + 6t^3) + (2 + t + 6t^2 - 4t^3)$

11. $(9x^8 - 7x^4 + 2x^2 + 5) + (8x^7 + 4x^4 - 2x)$

12. $(4x^5 - 6x^3 - 9x + 1) + (6x^3 + 9x^2 + 9x)$

13. $\left(\frac{1}{4}x^4 + \frac{2}{3}x^3 + \frac{5}{8}x^2 + 7\right) + \left(-\frac{3}{4}x^4 + \frac{3}{8}x^2 - 7\right)$

14. $\left(\frac{1}{3}x^9 + \frac{1}{5}x^5 - \frac{1}{2}x^2 + 7\right) + \left(-\frac{1}{5}x^9 + \frac{1}{4}x^4 - \frac{3}{5}x^5\right)$

15. $(5.3t^2 - 6.4t - 9.1) + (4.2t^3 - 1.8t^2 + 7.3)$

16. $(4.9a^3 + 3.2a^2 - 5.1a) + (2.1a^2 - 3.7a + 4.6)$

17. $-3x^4 + 6x^2 + 2x - 1$
  $ - 3x^2 + 2x + 1$

18. $-4x^3 + 8x^2 + 3x - 2$
  $ - 4x^2 + 3x + 2$

19. $0.15x^4 + 0.10x^3 - 0.9x^2$
  $ - 0.01x^3 + 0.01x^2 + x$
  $1.25x^4 + 0.11x^2 + 0.01$
  $ 0.27x^3 + 0.99$
  $-0.35x^4 + 15x^2 - 0.03$

20. $0.05x^4 + 0.12x^3 - 0.5x^2$
  $ - 0.02x^3 + 0.02x^2 + 2x$
  $1.5x^4 + 0.01x^2 + 0.15$
  $ 0.25x^3 + 0.85$
  $-0.25x^4 + 10x^2 - 0.04$

Write the opposite of each polynomial in two different forms, as in Example 4.

21. $-t^3 + 4t^2 - 9$

22. $-4x^3 - 5x^2 + 2x$

23. $12x^4 - 3x^3 + 3$

24. $5a^3 + 2a - 17$

Simplify.

25. $-(8x - 9)$

26. $-(-6x + 5)$

27. $-(3a^4 - 5a^2 + 9)$

28. $-(-6a^3 + 2a^2 - 7)$

29. $-\left(-4x^4 + 6x^2 + \frac{3}{4}x - 8\right)$

30. $-(-5x^4 + 4x^3 - x^2 + 0.9)$

Subtract.

31. $(7x + 4) - (-2x + 1)$

32. $(5x + 6) - (-2x + 4)$

33. $(-5t + 4) - (t^2 + 2t - 1)$

34. $(a^2 - 5a + 2) - (3a^2 + 2a - 4)$

35. $(6x^4 + 3x^3 - 1) - (4x^2 - 3x + 3)$

36. $(-4x^2 + 2x) - (3x^3 - 5x^2 + 3)$

37. $(1.2x^3 + 4.5x^2 - 3.8x) - (-3.4x^3 - 4.7x^2 + 23)$

38. $(0.5x^4 - 0.6x^2 + 0.7) - (2.3x^4 + 1.8x - 3.9)$

Aha! **39.** $(7x^3 - 2x^2 + 6) - (7x^3 - 2x^2 + 6)$

40. $(8x^5 + 3x^4 + x - 1) - (8x^5 + 3x^4 - 1)$

41. $(6 + 5a + 3a^2 - a^3) - (2 + 3a - 4a^2 + 2a^3)$

42. $(7 + t - 5t^2 + 2t^3) - (1 + 2t - 4t^2 + 5t^3)$

43. $\left(\frac{5}{8}x^3 - \frac{1}{4}x - \frac{1}{3}\right) - \left(-\frac{1}{8}x^3 + \frac{1}{4}x - \frac{1}{3}\right)$

44. $\left(\frac{1}{5}x^3 + 2x^2 - \frac{3}{10}\right) - \left(-\frac{2}{5}x^3 + 2x^2 + \frac{7}{1000}\right)$

45. $(0.07t^3 - 0.03t^2 + 0.01t) - (0.02t^3 + 0.04t^2 - 1)$

46. $(0.9a^3 + 0.2a - 5) - (0.7a^4 - 0.3a - 0.1)$

47. $x^2 + 5x + 6$
  $-(x^2 + 2x + 1)$

48. $x^3 + 3x^2 + 1$
  $-(x^3 + x^2 - 5)$

49. $5x^4 + 6x^3 - 9x^2$
  $-(-6x^4 - 6x^3 + x^2)$

50. $5x^4 - 2x^3 + 6x^2$
$\underline{-(7x^4 + 6x^3 + 7x^2)}$

51. Solve.
   a) Find a polynomial for the sum of the areas of the rectangles shown in the figure.
   b) Find the sum of the areas when $x = 5$ and $x = 7$.

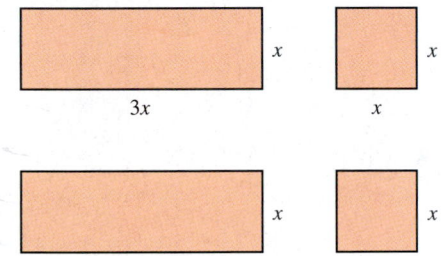

52. Solve.
   a) Find a polynomial for the sum of the areas of the circles shown in the figure.
   b) Find the sum of the areas when $r = 5$ and $r = 11.3$.

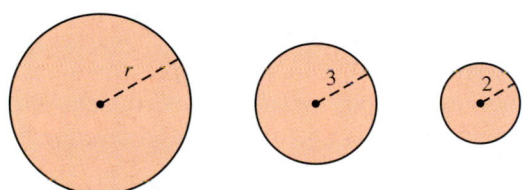

Find a polynomial for the perimeter of each figure in Exercises 53 and 54.

53.

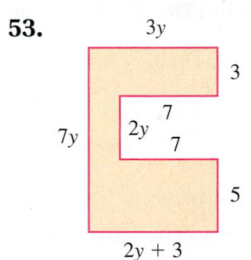

54.

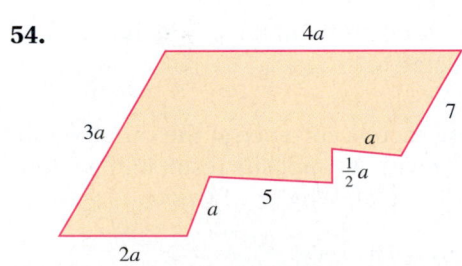

Find two algebraic expressions for the area of each figure. First, regard the figure as one large rectangle,

and then regard the figure as a sum of four smaller rectangles.

55.   **56.**

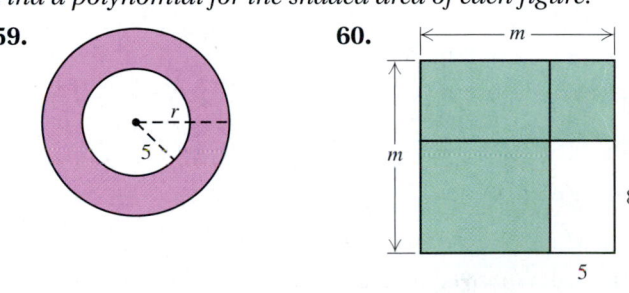

57.   **58.**

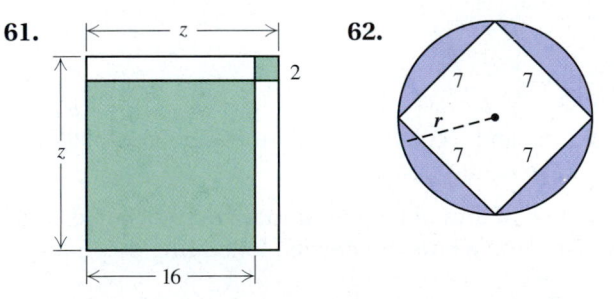

Find a polynomial for the shaded area of each figure.

59.   **60.**

61.   **62.**

63. Find $(y - 2)^2$ by subtracting the white areas from y^2.

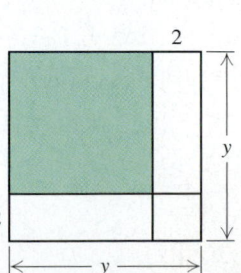

64. Find $(10 - 2x)^2$ by subtracting the white areas from 10^2.

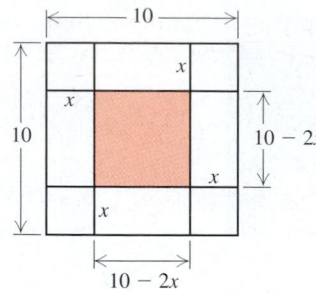

65. Is the sum of two trinomials always a trinomial? Why or why not?

66. What advice would you offer to a student who is successful at adding, but not subtracting, polynomials?

SKILL MAINTENANCE

Simplify.

67. $5(4 + 3) - 5 \cdot 4 - 5 \cdot 3$

68. $7(2 + 6) - 7 \cdot 2 - 7 \cdot 6$

69. $2(5t + 7) + 3t$

70. $3(4t - 5) + 2t$

Solve.

71. $2(x + 3) > 5(x - 3) + 7$

72. $7(x - 8) \le 4(x - 5)$

SYNTHESIS

73. What can be concluded about two polynomials whose sum is zero?

74. Which, if any, of the commutative, associative, and distributive laws are needed for adding polynomials? Why?

Simplify.

75. $(6t^2 - 7t) + (3t^2 - 4t + 5) - (9t - 6)$

76. $(3x^2 - 4x + 6) - (-2x^2 + 4) + (-5x - 3)$

77. $(-8y^2 - 4) - (3y + 6) - (2y^2 - y)$

78. $(5x^3 - 4x^2 + 6) - (2x^3 + x^2 - x) + (x^3 - x)$

79. $(-y^4 - 7y^3 + y^2) + (-2y^4 + 5y - 2) - (-6y^3 + y^2)$

80. $(-4 + x^2 + 2x^3) - (-6 - x + 3x^3) - (-x^2 - 5x^3)$

81. $(345.099x^3 - 6.178x) - (94.508x^3 - 8.99x)$

Find a polynomial for the surface area of the right rectangular solid.

82.

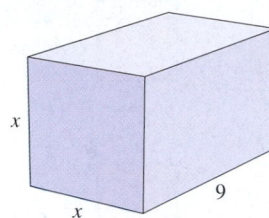

83.

84.

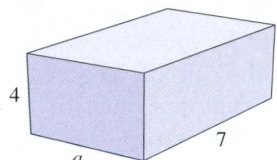

85.

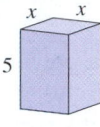

86. *Total profit.* Hadley Electronics is marketing a new kind of stereo. Total revenue is the total amount of money taken in. The firm determines that when it sells x stereos, its total revenue is given by

$$R = 280x - 0.4x^2.$$

Total cost is the total cost of producing x stereos. Hadley Electronics determines that the total cost of producing x stereos is given by

$$C = 5000 + 0.6x^2.$$

The total profit P is

(Total Revenue) − (Total Cost) = $R - C$.

a) Find a polynomial for total profit.

b) What is the total profit on the production and sale of 75 stereos?

c) What is the total profit on the production and sale of 100 stereos?

87. Does replacing each occurrence of the variable x in $4x^7 - 6x^3 + 2x$ with its opposite result in the opposite of the polynomial? Why or why not?

4.4

Multiplying Monomials • Multiplying a Monomial and a Polynomial • Multiplying Any Two Polynomials • Checking by Evaluating

We now multiply polynomials using techniques based largely on the distributive, associative, and commutative laws and the rules for exponents.

Multiplying Monomials

Consider $(3x)(4x)$. We multiply as follows:

$$
\begin{aligned}
(3x)(4x) &= 3 \cdot x \cdot 4 \cdot x && \text{Using an associative law} \\
&= 3 \cdot 4 \cdot x \cdot x && \text{Using a commutative law} \\
&= (3 \cdot 4) \cdot x \cdot x && \text{Using an associative law} \\
&= 12x^2.
\end{aligned}
$$

> **To Multiply Monomials**
>
> To find an equivalent expression for the product of two monomials, multiply the coefficients and then multiply the variables using the product rule for exponents.

Example 1   Multiply: **(a)** $(5x)(6x)$; **(b)** $(3a)(-a)$; **(c)** $(-7x^5)(4x^3)$.

Solution

a) $(5x)(6x) = (5 \cdot 6)(x \cdot x)$ Multiplying the coefficients; multiplying the variables

$\qquad\qquad\quad = 30x^2$ Simplifying

b) $(3a)(-a) = (3a)(-1a)$ Writing $-a$ as $-1a$ can ease calculations.

$\qquad\qquad\quad = (3)(-1)(a \cdot a)$ Using an associative and a commutative law

$\qquad\qquad\quad = -3a^2$

c) $(-7x^5)(4x^3) = (-7 \cdot 4)(x^5 \cdot x^3)$

$\qquad\qquad\quad\ = -28x^{5+3}$ ⎱ Using the product rule

$\qquad\qquad\quad\ = -28x^8$ ⎰ for exponents

After some practice, you can try writing only the answer.

Multiplying a Monomial and a Polynomial

To find an equivalent expression for the product of a monomial, such as $2x$, and a polynomial, such as $5x + 3$, we use the distributive law.

Example 2

Multiply: **(a)** x and $x + 3$; **(b)** $5x(2x^2 - 3x + 4)$.

Solution

a) $x(x + 3) = x \cdot x + x \cdot 3$ Using the distributive law

$\qquad\qquad\quad = x^2 + 3x$

b) $5x(2x^2 - 3x + 4) = (5x)(2x^2) - (5x)(3x) + (5x)(4)$ Using the distributive law

$\qquad\qquad\qquad\qquad = 10x^3 - 15x^2 + 20x$ Performing the three multiplications

The product in Example 2(a) can be visualized as the area of a rectangle with width x and length $x + 3$.

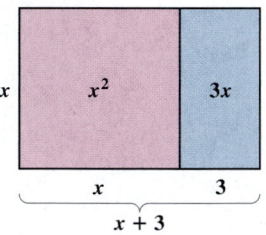

Note that the total area can be expressed as $x(x + 3)$ or, by adding the two smaller areas, $x^2 + 3x$.

The Product of a Monomial and a Polynomial

To multiply a monomial and a polynomial, multiply each term of the polynomial by the monomial.

Try to do this mentally, when possible.

Example 3

Multiply: $2x^2(x^3 - 7x^2 + 10x - 4)$.

Solution

$\qquad\qquad\qquad\qquad$ *Think:* $\underbrace{2x^2 \cdot x^3} - \underbrace{2x^2 \cdot 7x^2} + \underbrace{2x^2 \cdot 10x} - \underbrace{2x^2 \cdot 4}$

$2x^2(x^3 - 7x^2 + 10x - 4) = \quad 2x^5 \quad - \quad 14x^4 \quad + \quad 20x^3 \quad - \quad 8x^2$

Multiplying Any Two Polynomials

Before considering the product of *any* two polynomials, let's look at the product of two binomials.

To find an equivalent expression for the product of two binomials, we again begin by using the distributive law. This time, however, it is a *binomial* rather than a monomial that is being distributed.

E x a m p l e   4

Multiply each of the following.

a) $x + 5$ and $x + 4$

b) $4x - 3$ and $x - 2$

Solution

a) $(x + 5)\ (x + 4) = (x + 5)\ x + (x + 5)\ 4$ — Using the distributive law

$= x(x + 5) + 4(x + 5)$ — Using the commutative law for multiplication

$= x \cdot x + x \cdot 5 + 4 \cdot x + 4 \cdot 5$ — Using the distributive law (twice)

$= x^2 + 5x + 4x + 20$ — Multiplying the monomials

$= x^2 + 9x + 20$ — Combining like terms

b) $(4x - 3)\ (x - 2) = (4x - 3)\ x - (4x - 3)\ 2$ — Using the distributive law

$= x(4x - 3) - 2(4x - 3)$ — Using the commutative law for multiplication. This step is often omitted.

$= x \cdot 4x - x \cdot 3 - 2 \cdot 4x - 2(-3)$ — Using the distributive law (twice)

$= 4x^2 - 3x - 8x + 6$ — Multiplying the monomials

$= 4x^2 - 11x + 6$ — Combining like terms

Study Tip

Try to at least glance at the next section of material that will be covered in class. This will make it easier to concentrate on your instructor's lecture instead of trying to write everything down.

To visualize the product in Example 4(a), consider a rectangle of length $x + 5$ and width $x + 4$.

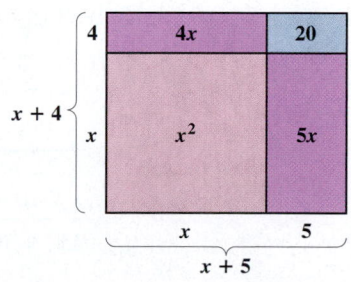

The total area can be expressed as $(x + 5)(x + 4)$ or, by adding the four smaller areas, $x^2 + 5x + 4x + 20$.

Let's consider the product of a binomial and a trinomial. Again we make repeated use of the distributive law.

E x a m p l e   5

Multiply: $(x^2 + 2x - 3)(x + 4)$.

Solution

$$(x^2 + 2x - 3)\ (x + 4)$$

$$= (x^2 + 2x - 3)\ x + (x^2 + 2x - 3)\ 4 \qquad \text{Using the distributive law}$$

$$= x(x^2 + 2x - 3) + 4(x^2 + 2x - 3) \qquad \text{Using the commutative law}$$

$$= x \cdot x^2 + x \cdot 2x - x \cdot 3 + 4 \cdot x^2 + 4 \cdot 2x - 4 \cdot 3 \qquad \text{Using the distributive law (twice)}$$

$$= x^3 + 2x^2 - 3x + 4x^2 + 8x - 12 \qquad \text{Multiplying the monomials}$$

$$= x^3 + 6x^2 + 5x - 12 \qquad \text{Combining like terms}$$

Perhaps you have discovered the following in the preceding examples.

The Product of Two Polynomials

To multiply two polynomials P and Q, select one of the polynomials, say P. Then multiply each term of P by every term of Q and combine like terms.

To use columns for long multiplication, multiply each term in the top row by every term in the bottom row. We write like terms in columns, and then add the results. Such multiplication is like multiplying with whole numbers:

$$
\begin{array}{r}
3\ 2\ 1 \\
\times\quad 1\ 2 \\
\hline
6\ 4\ 2 \\
3\ 2\ 1 \\
\hline
3\ 8\ 5\ 2
\end{array}
\qquad
\begin{array}{r}
300 + 20 + 1 \\
\times \qquad\quad 10 + 2 \\
\hline
600 + 40 + 2 \\
3000 + 200 + 10 \\
\hline
3000 + 800 + 50 + 2
\end{array}
$$

Multiplying the top row by 2
Multiplying the bottom row by 10
Adding

E x a m p l e   6

Multiply: $(4x^3 - 2x^2 + 3x)(x^2 + 2x)$.

Solution

$$
\begin{array}{r}
4x^3 - 2x^2 + 3x \\
x^2 + 2x \\
\hline
8x^4 - 4x^3 + 6x^2 \\
4x^5 - 2x^4 + 3x^3 \\
\hline
4x^5 + 6x^4 - \ x^3 + 6x^2
\end{array}
$$

Multiplying the top row by $2x$
Multiplying the top row by x^2
Combining like terms

Line up like terms in columns.

If a term is missing, it helps to leave space for it so that like terms can be easily aligned.

E x a m p l e   7   Multiply: $(-2x^2 - 3)(5x^3 - 3x + 4)$.

Solution

$$
\begin{array}{r}
5x^3 \quad\quad -3x + \ 4 \\
- 2x^2 \quad\quad\quad - \ 3 \\
\hline
- 15x^3 \quad\quad + 9x - 12 \\
-10x^5 + \ 6x^3 - 8x^2 \quad\quad\quad\quad \\
\hline
-10x^5 - \ 9x^3 - 8x^2 + 9x - 12 \\
\end{array}
$$

Multiplying by -3
Multiplying by $-2x^2$
Combining like terms

With practice, some steps can be skipped. Sometimes we multiply horizontally, while still aligning like terms.

E x a m p l e   8   Multiply: $(2x^3 + 3x^2 - 4x + 6)(3x + 5)$.

Solution

Multiplying by $3x$

$$(2x^3 + 3x^2 - 4x + 6)(3x + 5) = 6x^4 + \ 9x^3 - 12x^2 + 18x$$
$$+ \ 10x^3 + 15x^2 - 20x + 30$$

Multiplying by 5

$$= 6x^4 + 19x^3 + 3x^2 - 2x + 30$$

Checking by Evaluating

How can we be certain that our multiplication (or addition or subtraction) of polynomials is correct? One check is to simply review our calculations. A different type of check, used in Example 10 of Section 4.3, makes use of the fact that equivalent expressions have the same value when evaluated for the same replacement. Thus a quick, partial, check of Example 8 can be made by selecting a convenient replacement for x (say, 1) and comparing the values of the expressions $(2x^3 + 3x^2 - 4x + 6)(3x + 5)$ and $6x^4 + 19x^3 + 3x^2 - 2x + 30$:

$$(2x^3 + 3x^2 - 4x + 6)(3x + 5) = (2 \cdot 1^3 + 3 \cdot 1^2 - 4 \cdot 1 + 6)(3 \cdot 1 + 5)$$
$$= (2 + 3 - 4 + 6)(3 + 5)$$
$$= 7 \cdot 8 = 56;$$

$$6x^4 + 19x^3 + 3x^2 - 2x + 30 = 6 \cdot 1^4 + 19 \cdot 1^3 + 3 \cdot 1^2 - 2 \cdot 1 + 30$$
$$= 6 + 19 + 3 - 2 + 30$$
$$= 28 - 2 + 30 = 56.$$

Since the value of both expressions is 56, the multiplication in Example 8 is very likely correct.

It is possible, by chance, for two expressions that are not equivalent to share the same value when evaluated. For this reason, checking by evaluating is only a partial check. Consult your instructor for the checking approach that he or she prefers.

technology connection

Tables can also be used to check polynomial multiplication. To illustrate, we can check Example 8 by entering $y_1 = (2x^3 + 3x^2 - 4x + 6)(3x + 5)$ and $y_2 = 6x^4 + 19x^3 + 3x^2 - 2x + 30$.

When TABLE is then pressed, we are shown two columns of values—one for y_1 and one for y_2. If our multiplication was correct, the columns of values will match.

X	Y₁	Y₂
−3	36	36
−2	−10	−10
−1	22	22
0	30	30
1	56	56
2	286	286
3	1050	1050
X = −3		

1. Form a table and scroll up and down to check Example 8.
2. Check Example 8 using the method discussed in Section 4.3: Let

$$y_1 = (2x^3 + 3x^2 - 4x + 6)(3x + 5),$$
$$y_2 = 6x^4 + 19x^3 + 3x^2 - 2x + 30,$$

and

$$y_3 = y_2 - y_1.$$

Then check that y_3 is always 0.

Exercise Set 4.4

FOR EXTRA HELP

Digital Video Tutor CD 3
Videotape 7

InterAct Math

Math Tutor Center

MathXL

MyMathLab.com

Multiply.

1. $(5x^4)6$

2. $(4x^3)7$

3. $(-x^2)(-x)$

4. $(-x^3)(x^4)$

5. $(-x^5)(x^3)$

6. $(-x^6)(-x^2)$

7. $(7t^5)(4t^3)$

8. $(10a^2)(3a^2)$

9. $(-0.1x^6)(0.2x^4)$

10. $(0.3x^3)(-0.4x^6)$

11. $\left(-\frac{1}{5}x^3\right)\left(-\frac{1}{3}x\right)$

12. $\left(-\frac{1}{4}x^4\right)\left(\frac{1}{5}x^8\right)$

13. $19t^2 \cdot 0$

14. $(-5n^3)(-1)$

15. $7x^2(-2x^3)(2x^6)$

16. $(-4y^5)(6y^2)(-3y^3)$

17. $3x(-x + 5)$

18. $2x(4x - 6)$

19. $4x(x + 1)$

20. $3x(x + 2)$

21. $(a + 9)3a$

22. $(a - 7)4a$

23. $x^2(x^3 + 1)$

24. $-2x^3(x^2 - 1)$

25. $3x(2x^2 - 6x + 1)$

26. $-4x(2x^3 - 6x^2 - 5x + 1)$

27. $5t^2(3t + 6)$

28. $7t^2(2t + 1)$

29. $-6x^2(x^2 + x)$

30. $-4x^2(x^2 - x)$

31. $\frac{2}{3}a^4\left(6a^5 - 12a^3 - \frac{5}{8}\right)$

32. $\frac{3}{4}t^5\left(8t^6 - 12t^4 + \frac{12}{7}\right)$

33. $(x + 6)(x + 3)$

34. $(x + 5)(x + 2)$

35. $(x + 5)(x - 2)$

36. $(x + 6)(x - 2)$

37. $(a - 6)(a - 7)$

38. $(a - 4)(a - 8)$

39. $(x + 3)(x - 3)$

40. $(x + 6)(x - 6)$

41. $(5 - x)(5 - 2x)$

42. $(3 + x)(6 + 2x)$

43. $\left(t + \frac{3}{2}\right)\left(t + \frac{4}{3}\right)$

44. $\left(a - \frac{2}{5}\right)\left(a + \frac{5}{2}\right)$

45. $\left(\frac{1}{4}a + 2\right)\left(\frac{3}{4}a - 1\right)$

46. $\left(\frac{2}{5}t - 1\right)\left(\frac{3}{5}t + 1\right)$

Draw and label rectangles similar to those following Examples 2 and 4 to illustrate each product.

47. $x(x + 5)$

48. $x(x + 2)$

49. $(x + 1)(x + 2)$

50. $(x + 3)(x + 1)$

51. $(x + 5)(x + 3)$

52. $(x + 4)(x + 6)$

53. $(3x + 2)(3x + 2)$

54. $(5x + 3)(5x + 3)$

Multiply and check.

55. $(x^2 - x + 5)(x + 1)$

56. $(x^2 + x - 7)(x + 2)$

57. $(2a + 5)(a^2 - 3a + 2)$

58. $(3t + 4)(t^2 - 5t + 1)$

59. $(y^2 - 7)(2y^3 + y + 1)$

60. $(a^2 + 4)(5a^3 - 3a - 1)$

61. $(5x^3 - 7x^2 + 1)(x - 3x^2)$

62. $(4x^3 - 5x - 3)(1 + 2x^2)$

63. $(x^2 - 3x + 2)(x^2 + x + 1)$

64. $(x^2 + 5x - 1)(x^2 - x + 3)$

65. $(2t^2 - 5t - 4)(3t^2 - t + 1)$

66. $(5t^2 - t + 1)(2t^2 + t - 3)$

67. $(x + 1)(x^3 + 7x^2 + 5x + 4)$

68. $(x + 2)(x^3 + 5x^2 + 9x + 3)$

69. $\left(x - \frac{1}{2}\right)\left(2x^3 - 4x^2 + 3x - \frac{2}{5}\right)$

70. $\left(x + \frac{1}{3}\right)\left(6x^3 - 12x^2 - 5x + \frac{1}{2}\right)$

71. Is it possible to understand polynomial multiplication without understanding the distributive law? Why or why not?

72. The polynomials

$$(a + b + c + d) \quad \text{and} \quad (r + s + m + p)$$

are multiplied. Without performing the multiplication, determine how many terms the product will contain. Provide a justification for your answer.

SKILL MAINTENANCE

Simplify.

73. $5 - 3 \cdot 2 + 7$

74. $4 + 6 \cdot 5 - 3$

75. $(8 - 2)(8 + 2) + 2^2 - 8^2$

76. $(7 - 3)(7 + 3) + 3^2 - 7^2$

SYNTHESIS

77. Under what conditions will the product of two binomials be a trinomial?

78. How can the following figure be used to show that $(x + 3)^2 \neq x^2 + 9$?

Find a polynomial for the shaded area of each figure.

79.

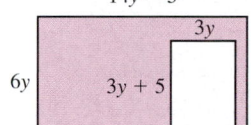

80.

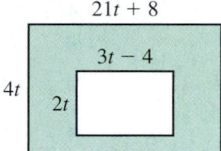

For each figure, determine what the missing number must be in order for the figure to have the given area.

81. Area is $x^2 + 7x + 10$

82. Area is $x^2 + 8x + 15$

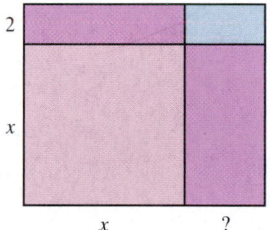

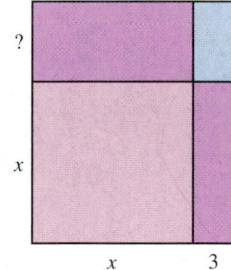

83. A box with a square bottom is to be made from a 12-in.-square piece of cardboard. Squares with side x are cut out of the corners and the sides are folded up. Find the polynomials for the volume and the outside surface area of the box.

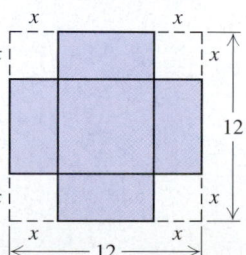

84. An open wooden box is a cube with side x cm. The box, including its bottom, is made of wood that is 1 cm thick. Find a polynomial for the interior volume of the cube.

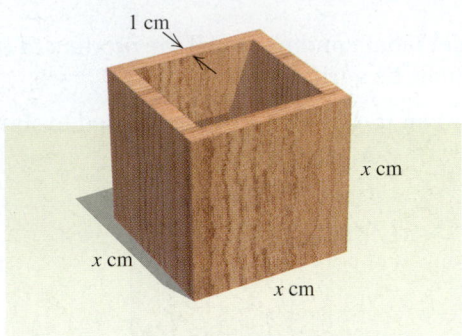

85. Find a polynomial for the volume of the solid shown below.

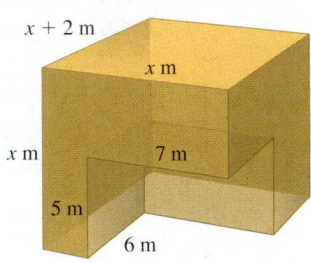

86. A side of a cube is $(x + 2)$ cm long. Find a polynomial for the volume of the cube.

87. A rectangular garden is twice as long as it is wide and is surrounded by a sidewalk that is 4 ft wide (see the figure below). The area of the sidewalk is 256 ft^2. Find the dimensions of the garden.

Compute and simplify.

88. $(x + 3)(x + 6) + (x + 3)(x + 6)$

Aha! **89.** $(x - 2)(x - 7) - (x - 7)(x - 2)$

90. $(x + 5)^2 - (x - 3)^2$

Aha! **91.** Extend the pattern and simplify

$$(x - a)(x - b)(x - c)(x - d) \cdots (x - z).$$

92. Use a grapher to check your answers to Exercises 21, 41, and 61. Use graphs, tables, or both, as directed by your instructor.

CORNER

Slick Tricks with Algebra

Focus: Polynomial multiplication
Time: 15 minutes
Group size: 2

Consider the following dialogue.

Jinny: Cal, let me do a number trick with you. Think of a number between 1 and 7. I'll have you perform some manipulations to this number, you'll tell me the result, and I'll tell me your number.

Cal: OK. I've thought of a number.

Jinny: Good. Write it down so I can't see it. Now double it, and then subtract x from the result.

Cal: Hey, this is algebra!

Jinny: I know. Now square your binomial. After you're through squaring, subtract x^2.

Cal: How did you know I had an x^2? I *thought* this was rigged!

Jinny: It is. Now divide each of the remaining terms by 4 and tell me either your constant term or your x-term. I'll tell you the other term and the number you chose.

Cal: OK. The constant term is 16.

Jinny: Then the other term is $-4x$ and the number you chose was 4.

Cal: You're right! How did you do it?

ACTIVITY

1. Each group member should follow Jinny's instructions. Then determine how Jinny determined Cal's number and the other term.
2. Suppose that, at the end, Cal told Jinny the x-term. How would Jinny have determined Cal's number and the other term?
3. Would Jinny's "trick" work with *any* real number? Why do you think she specified numbers between 1 and 7?

Special Products

4.5

Products of Two Binomials • Multiplying Sums and Differences of Two Terms • Squaring Binomials • Multiplications of Various Types

Certain products of two binomials occur so often that it is helpful to be able to compare them quickly. In this section, we develop methods for computing "special" products more quickly than we were able to in Section 4.4.

Products of Two Binomials

In Section 4.4, we found the product $(x + 5)(x + 4)$ by using the distributive law a total of three times (see p. 231). Note that each term in $x + 5$ is multiplied by each term in $x + 4$. To shorten our work, we can go right to this step:

$$(x + 5)(x + 4) = x \cdot x + x \cdot 4 + 5 \cdot x + 5 \cdot 4$$
$$= x^2 + 4x + 5x + 20$$
$$= x^2 + 9x + 20.$$

Note that the product $x \cdot x$ is found by multiplying the *First* terms of each binomial, $x \cdot 4$ is found by multiplying the *Outer* terms of the two binomials, $5 \cdot x$ is the product of the *Inner* terms of the two binomials, and $5 \cdot 4$ is the product of the *Last* terms of each binomial:

First   Outer   Inner   Last
terms   terms   terms   terms

$$(x + 5)(x + 4) = x \cdot x + 4 \cdot x + 5 \cdot x + 5 \cdot 4.$$

To remember this shortcut for multiplying, we use the initials **FOIL**.

The FOIL Method

To multiply two binomials, $A + B$ and $C + D$, multiply the First terms AC, the Outer terms AD, the Inner terms BC, and then the Last terms BD. Then combine like terms, if possible.

$$(A + B)(C + D) = AC + AD + BC + BD$$

1. Multiply First terms: AC.
2. Multiply Outer terms: AD.
3. Multiply Inner terms: BC.
4. Multiply Last terms: BD.

FOIL

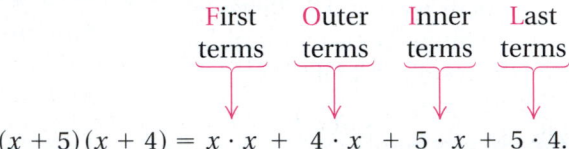

Because addition is commutative, the individual multiplications can be performed in any order. Both FLOI and FIOL yield the same result as FOIL, but FOIL is most easily remembered and most widely used.

Example 1

Multiply: $(x + 8)(x^2 + 5)$.

Solution

$$(x + 8)(x^2 + 5) = x^3 + 5x + 8x^2 + 40 \qquad \text{There are no like terms.}$$
$$= x^3 + 8x^2 + 5x + 40 \qquad \text{Writing in descending order}$$

After multiplying, remember to combine any like terms.

E x a m p l e   2

Multiply.

a) $(x + 7)(x + 4)$
b) $(y + 3)(y - 2)$
c) $(4t^3 + 5t)(3t^2 - 2)$
d) $(3 - 4x)(7 - 5x^3)$

Solution

a) $(x + 7)(x + 4) = x^2 + 4x + 7x + 28$ Using FOIL
$$= x^2 + 11x + 28$$ Combining like terms

b) $(y + 3)(y - 2) = y^2 - 2y + 3y - 6$
$$= y^2 + y - 6$$

c) $(4t^3 + 5t)(3t^2 - 2) = 12t^5 - 8t^3 + 15t^3 - 10t$ Remember to add exponents when multiplying terms with the same base.

$$= 12t^5 + 7t^3 - 10t$$

d) $(3 - 4x)(7 - 5x^3) = 21 - 15x^3 - 28x + 20x^4$
$$= 21 - 28x - 15x^3 + 20x^4$$ Because the original binomials are in *ascending* order, we write the answer that way.

Multiplying Sums and Differences of Two Terms

Consider the product of the sum and difference of the same two terms, such as

$$(x + 5)(x - 5).$$

Since this is the product of two binomials, we can use FOIL. In doing so, we find that the "outer" and "inner" products are opposites:

a) $(x + 5)(x - 5) = x^2 - 5x + 5x - 25$
$$= x^2 - 25;$$

b) $(3a - 2)(3a + 2) = 9a^2 + 6a - 6a - 4$ The "outer" and "inner" terms "drop out." Their sum is zero.
$$= 9a^2 - 4;$$

c) $\left(x^3 + \frac{2}{7}\right)\left(x^3 - \frac{2}{7}\right) = x^6 - \frac{2}{7}x^3 + \frac{2}{7}x^3 - \frac{4}{49}$
$$= x^6 - \frac{4}{49}.$$

Because opposites always add to zero, for products like $(x + 5)(x - 5)$ we can use a shortcut that is faster than FOIL.

> ### The Product of a Sum and Difference
> The product of the sum and difference of the same two terms is the square of the first term minus the square of the second term:
> $$(A + B)(A - B) = \underbrace{A^2 - B^2}.$$
> This is called a *difference of squares*.

Example 3

Multiply.

a) $(x + 4)(x - 4)$
b) $(5 + 2w)(5 - 2w)$
c) $(3a^4 - 5)(3a^4 + 5)$

Solution

$(A + B)(A - B) = A^2 - B^2$ Saying the words can help:

a) $(x + 4)(x - 4) = x^2 - 4^2$ "The square of the first term, x^2, minus the square of the second, 4^2"

$\qquad\qquad\quad = x^2 - 16$ Simplifying

b) $(5 + 2w)(5 - 2w) = 5^2 - (2w)^2$

$\qquad\qquad\qquad = 25 - 4w^2$ Squaring both 5 and $2w$

c) $(3a^4 - 5)(3a^4 + 5) = (3a^4)^2 - 5^2$

$\qquad\qquad\qquad\quad = 9a^8 - 25$ Using the rules for exponents. Remember to multiply exponents when raising a power to a power.

Squaring Binomials

Consider the square of a binomial, such as $(x + 3)^2$. This can be expressed as $(x + 3)(x + 3)$. Since this is the product of two binomials, we can use FOIL. But again, this product occurs so often that a faster method has been developed. Look for a pattern in the following:

a) $(x + 3)^2 = (x + 3)(x + 3)$
$\qquad\quad = x^2 + 3x + 3x + 9$
$\qquad\quad = x^2 + 6x + 9;$

b) $(5 - 3p)^2 = (5 - 3p)(5 - 3p)$
$\qquad\qquad = 25 - 15p - 15p + 9p^2$
$\qquad\qquad = 25 - 30p + 9p^2;$

c) $(a^3 - 7)^2 = (a^3 - 7)(a^3 - 7)$
$\qquad\qquad = a^6 - 7a^3 - 7a^3 + 49$
$\qquad\qquad = a^6 - 14a^3 + 49.$

Perhaps you noticed that in each product the "outer" and "inner" products are identical. The other two terms, the "first" and "last" products, are squares.

The Square of a Binomial

The square of a binomial is the square of the first term, plus twice the product of the two terms, plus the square of the last term:

$$(A + B)^2 = A^2 + 2AB + B^2;$$
$$(A - B)^2 = A^2 - 2AB + B^2.$$

These are called *perfect-square trinomials.**

E x a m p l e   4

Multiply: **(a)** $(x + 7)^2$; **(b)** $(t - 5)^2$; **(c)** $(3a + 0.4)^2$; **(d)** $(5x - 3x^4)^2$.

Solution

$$(A + B)^2 = A^2 + 2 \cdot A \cdot B + B^2$$

Saying the words can help:

a) $(x + 7)^2 = x^2 + 2 \cdot x \cdot 7 + 7^2$

"The square of the first term, x^2, plus twice the product of the terms, $2 \cdot 7x$, plus the square of the second term, 7^2"

$$= x^2 + 14x + 49$$

b) $(t - 5)^2 = t^2 - 2 \cdot t \cdot 5 + 5^2$
$$= t^2 - 10t + 25$$

c) $(3a + 0.4)^2 = (3a)^2 + 2 \cdot 3a \cdot 0.4 + 0.4^2$
$$= 9a^2 + 2.4a + 0.16$$

d) $(5x - 3x^4)^2 = (5x)^2 - 2 \cdot 5x \cdot 3x^4 + (3x^4)^2$
$$= 25x^2 - 30x^5 + 9x^8 \qquad \text{Using the rules for exponents}$$

Caution!   Although the square of a product is the product of the squares, the square of a sum is *not* the sum of the squares. That is, $(AB)^2 = A^2B^2$, but

The term $2AB$ is missing.

$$(A + B)^2 \neq A^2 + B^2.$$

To confirm this inequality, note that

$$(7 + 5)^2 = 12^2 = 144,$$

whereas

$$7^2 + 5^2 = 49 + 25 = 74, \quad \text{and} \quad 74 \neq 144.$$

---

*In some books, these are called *trinomial squares*.

Geometrically, $(A + B)^2$ can be viewed as the area of a square with sides of length $A + B$:

$$(A + B)(A + B) = (A + B)^2.$$

This is equal to the sum of the areas of the four smaller regions:

$$A^2 + AB + AB + B^2 = A^2 + 2AB + B^2.$$

Thus,

$$(A + B)^2 = A^2 + 2AB + B^2.$$

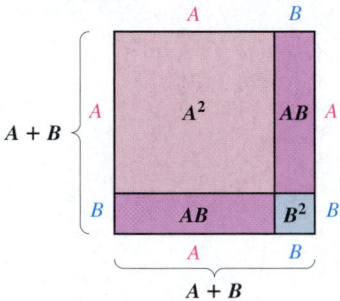

Multiplications of Various Types

Recognizing patterns often helps when new problems are encountered. To simplify a new multiplication problem, always examine what type of product it is so that the best method for finding that product can be used. To do this, ask yourself questions similar to the following.

Multiplying Two Polynomials

1. Is the multiplication the product of a monomial and a polynomial? If so, multiply each term of the polynomial by the monomial.
2. Is the multiplication the product of two binomials? If so:

    a) Is it the product of the sum and difference of the *same* two terms? If so, use the pattern

$$(A + B)(A - B) = A^2 - B^2.$$

    b) Is the product the square of a binomial? If so, use the pattern

$$(A + B)(A + B) = (A + B)^2 = A^2 + 2AB + B^2,$$

    or

$$(A - B)(A - B) = (A - B)^2 = A^2 - 2AB + B^2.$$

    c) If neither (a) nor (b) applies, use FOIL.

3. Is the multiplication the product of two polynomials other than those above? If so, multiply each term of one by every term of the other. Use columns if you wish.

Example 5

Multiply.

a) $(x + 3)(x - 3)$ **b)** $(t + 7)(t - 5)$ **c)** $(x + 7)(x + 7)$
d) $2x^3(9x^2 + x - 7)$ **e)** $(p + 3)(p^2 + 2p - 1)$ **f)** $\left(3x + \frac{1}{4}\right)^2$

Solution

a) $(x + 3)(x - 3) = x^2 - 9$ This is the product of the sum and difference of the same two terms.

b) $(t + 7)(t - 5) = t^2 - 5t + 7t - 35$ Using FOIL
$$= t^2 + 2t - 35$$

c) $(x + 7)(x + 7) = x^2 + 14x + 49$ This is the square of a binomial.

d) $2x^3(9x^2 + x - 7) = 18x^5 + 2x^4 - 14x^3$ Multiplying each term of the trinomial by the monomial

e)
$$
\begin{array}{r}
p^2 + 2p - 1 \\
p + 3 \\
\hline
3p^2 + 6p - 3 \\
p^3 + 2p^2 - p \\
\hline
p^3 + 5p^2 + 5p - 3
\end{array}
$$
Using columns to multiply a binomial and a trinomial

Multiplying by 3
Multiplying by p

f) $\left(3x + \frac{1}{4}\right)^2 = 9x^2 + 2(3x)\left(\frac{1}{4}\right) + \frac{1}{16}$ Squaring a binomial
$$= 9x^2 + \frac{3}{2}x + \frac{1}{16}$$

FOR EXTRA HELP

Exercise Set 4.5

 Digital Video Tutor CD 3 Videotape 8   InterAct Math   Math Tutor Center   MathXL   MyMathLab.com

Multiply.

1. $(x + 4)(x^2 + 3)$
2. $(x^2 - 3)(x - 1)$
3. $(x^3 + 6)(x + 2)$
4. $(x^4 + 2)(x + 12)$
5. $(y + 2)(y - 3)$
6. $(a + 2)(a + 2)$
7. $(3x + 2)(3x + 5)$
8. $(4x + 1)(2x + 7)$
9. $(5x - 6)(x + 2)$
10. $(t - 9)(t + 9)$
11. $(1 + 3t)(2 - 3t)$
12. $(7 - a)(2 + 3a)$
13. $(2x - 7)(x - 1)$
14. $(2x - 1)(3x + 1)$
15. $\left(p - \frac{1}{4}\right)\left(p + \frac{1}{4}\right)$
16. $\left(q + \frac{3}{4}\right)\left(q + \frac{3}{4}\right)$
17. $(x - 0.1)(x + 0.1)$
18. $(x + 0.3)(x - 0.4)$
19. $(2x^2 + 6)(x + 1)$
20. $(2x^2 + 3)(2x - 1)$

21. $(-2x + 1)(x + 6)$
22. $(-x + 4)(2x - 5)$
23. $(a + 9)(a + 9)$
24. $(2y + 7)(2y + 7)$
25. $(1 + 3t)(1 - 5t)$
26. $(1 + 2t)(1 - 3t^2)$
27. $(x^2 + 3)(x^3 - 1)$
28. $(x^4 - 3)(2x + 1)$
29. $(3x^2 - 2)(x^4 - 2)$
30. $(x^{10} + 3)(x^{10} - 3)$
31. $(2t^3 + 5)(2t^3 + 3)$
32. $(5t^2 + 1)(2t^2 + 3)$
33. $(8x^3 + 5)(x^2 + 2)$
34. $(4 - 2x)(5 - 2x^2)$
35. $(4x^2 + 3)(x - 3)$
36. $(7x - 2)(2x - 7)$

Multiply. Try to recognize what type of product each multiplication is before multiplying.

37. $(x + 8)(x - 8)$
38. $(x + 1)(x - 1)$

39. $(2x + 1)(2x - 1)$

40. $(x^2 + 1)(x^2 - 1)$

41. $(5m - 2)(5m + 2)$

42. $(3x^4 + 2)(3x^4 - 2)$

43. $(2x^2 + 3)(2x^2 - 3)$

44. $(6x^5 - 5)(6x^5 + 5)$

45. $(3x^4 - 1)(3x^4 + 1)$

46. $(t^2 - 0.2)(t^2 + 0.2)$

47. $(x^4 + 7)(x^4 - 7)$

48. $(t^3 + 4)(t^3 - 4)$

49. $\left(t - \frac{3}{4}\right)\left(t + \frac{3}{4}\right)$

50. $\left(m - \frac{2}{3}\right)\left(m + \frac{2}{3}\right)$

51. $(x + 2)^2$

52. $(2x - 1)^2$

53. $(3x^5 + 1)^2$

54. $(4x^3 + 1)^2$

55. $\left(a - \frac{2}{5}\right)^2$

56. $\left(t - \frac{1}{5}\right)^2$

57. $(t^3 + 3)^2$

58. $(a^4 + 2)^2$

59. $(2 - 3x^4)^2$

60. $(5 - 2t^3)^2$

61. $(5 + 6t^2)^2$

62. $(3p^2 - p)^2$

63. $(7x - 0.3)^2$

64. $(4a - 0.6)^2$

65. $5a^3(2a^2 - 1)$

66. $9x^3(2x^2 - 5)$

67. $(a - 3)(a^2 + 2a - 4)$

68. $(x^2 - 5)(x^2 + x - 1)$

69. $(3 - 2x^3)^2$

70. $(x - 4x^3)^2$

71. $4x(x^2 + 6x - 3)$

72. $8x(-x^5 + 6x^2 + 9)$

73. $(-t^3 + 1)^2$

74. $(-x^2 + 1)^2$

75. $3t^2(5t^3 - t^2 + t)$

76. $-5x^3(x^2 + 8x - 9)$

77. $(6x^4 - 3)^2$

78. $(8a^3 + 5)^2$

79. $(3x + 2)(4x^2 + 5)$

80. $(2x^2 - 7)(3x^2 + 9)$

81. $(5 - 6x^4)^2$

82. $(3 - 4t^5)^2$

83. $(a + 1)(a^2 - a + 1)$

84. $(x - 5)(x^2 + 5x + 25)$

Find the total area of all shaded rectangles.

85.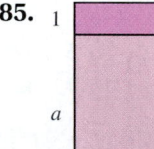

86.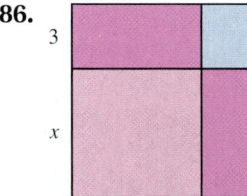

87.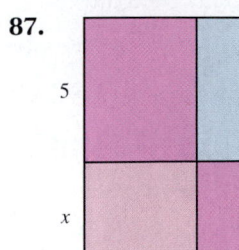

88.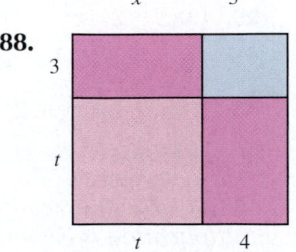

89.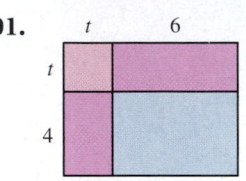

90.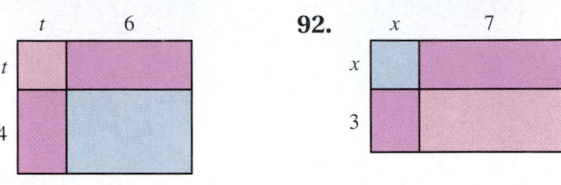

91.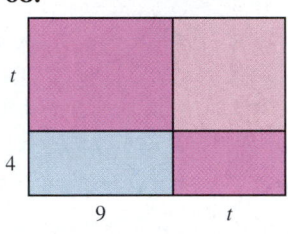

92.

93.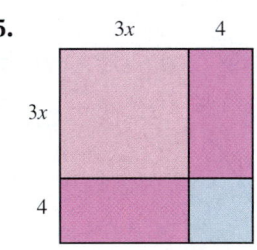

94.

95.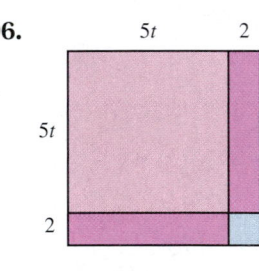

96.

Draw and label rectangles similar to those in Exercises 85–96 to illustrate each of the following.

97. $(x + 5)^2$

98. $(x + 8)^2$

99. $(t + 9)^2$

100. $(a + 12)^2$

101. $(3 + x)^2$

102. $(7 + t)^2$

103. Blair feels that since he can find the product of any two binomials using FOIL, he needn't study the other special products. What advice would you give him?

104. Under what conditions is the product of two binomials a binomial?

SKILL MAINTENANCE

105. *Energy use.*   In an apartment, lamps, an air conditioner, and a television set are all operating at the same time. The lamps take 10 times as many watts as the television set, and the air conditioner takes 40 times as many watts as the television set. The total wattage used in the apartment is 2550 watts. How many watts are used by each appliance?

106. In what quadrant is the point $(-3, 4)$ located?

Solve.

107. $5xy = 8$, for y

108. $3ab = c$, for a

109. $ax - b = c$, for x

110. $st + r = u$, for t

SYNTHESIS

111. Anais claims that by writing $19 \cdot 21$ as $(20 - 1)(20 + 1)$, she can find the product mentally. How is this possible?

112. The product $(A + B)^2$ can be regarded as the sum of the areas of four regions (as shown following Example 4). How might one visually represent $(A + B)^3$? Why?

Multiply.

Aha! **113.** $(4x^2 + 9)(2x + 3)(2x - 3)$

114. $(9a^2 + 1)(3a - 1)(3a + 1)$

Aha! **115.** $(3t - 2)^2(3t + 2)^2$

116. $(5a + 1)^2(5a - 1)^2$

117. $(t^3 - 1)^4(t^3 + 1)^4$

118. $(32.41x + 5.37)^2$

Calculate as the difference of squares.

119. 18×22 [*Hint:* $(20 - 2)(20 + 2)$.]

120. 93×107

Solve.

121. $(x + 2)(x - 5) = (x + 1)(x - 3)$

122. $(2x + 5)(x - 4) = (x + 5)(2x - 4)$

The height of a box is 1 more than its length l, and the length is 1 more than its width w. Find a polynomial for the volume V in terms of the following.

123. The length l

124. The width w

Find a polynomial for the total shaded area in each figure.

125.

126.

127.

128. Find three consecutive integers for which the sum of the squares is 65 more than three times the square of the smallest integer.

129. Use a grapher and the method developed on p. 234 to check your answers to Exercises 17, 47, and 83.

Polynomials in Several Variables

4.6

Evaluating Polynomials • Like Terms and Degree • Addition and Subtraction • Multiplication

Thus far, the polynomials that we have studied have had only one variable. Polynomials such as

$$5x + x^2y - 3y + 7, \qquad 9ab^2c - 2a^3b^2 + 8a^2b^3 + 15, \quad \text{and}$$
$$4m^2 - 9n^2$$

contain two or more variables. In this section, we will add, subtract, multiply, and evaluate such **polynomials in several variables.**

Evaluating Polynomials

To evaluate a polynomial in two or more variables, we substitute numbers for the variables. Then we compute, using the rules for order of operations.

E x a m p l e   1

Evaluate the polynomial $4 + 3x + xy^2 + 8x^3y^3$ for $x = -2$ and $y = 5$.

Solution   We substitute -2 for x and 5 for y:

$$4 + 3x + xy^2 + 8x^3y^3 = 4 + 3(-2) + (-2) \cdot 5^2 + 8(-2)^3 \cdot 5^3$$
$$= 4 - 6 - 50 - 8000 = -8052.$$

E x a m p l e   2

Surface area of a right circular cylinder.   The surface area of a right circular cylinder is given by the polynomial

$$2\pi rh + 2\pi r^2,$$

where h is the height and r is the radius of the base. A 12-oz can has a height of 4.7 in. and a radius of 1.2 in. Approximate its surface area.

Solution   We evaluate the polynomial for $h = 4.7$ in. and $r = 1.2$ in. If 3.14 is used to approximate π, we have

$$2\pi rh + 2\pi r^2 \approx 2(3.14)(1.2 \text{ in.})(4.7 \text{ in.}) + 2(3.14)(1.2 \text{ in.})^2$$
$$\approx 2(3.14)(1.2 \text{ in.})(4.7 \text{ in.}) + 2(3.14)(1.44 \text{ in}^2)$$
$$\approx 35.4192 \text{ in}^2 + 9.0432 \text{ in}^2 \approx 44.4624 \text{ in}^2.$$

If the π key of a calculator is used, we have

$$2\pi rh + 2\pi r^2 \approx 2(3.141592654)(1.2 \text{ in.})(4.7 \text{ in.}) + 2(3.141592654)(1.2 \text{ in.})^2$$
$$\approx 44.48495197 \text{ in}^2.$$

Note that the unit in the answer (square inches) is a unit of area. The surface area is about 44.5 in^2 (square inches).

Like Terms and Degree

Recall that the degree of a term is the number of variable factors in the term. For example, the degree of $5x^2$ is 2 because there are two variable factors in $5 \cdot x \cdot x$. Similarly, the degree of $5a^2b^4$ is 6 because there are 6 variable factors in $5 \cdot a \cdot a \cdot b \cdot b \cdot b \cdot b$. Note that 6 can be found by adding the exponents 2 and 4.

As we learned in Section 4.2, the degree of a polynomial is the degree of the term of highest degree.

E x a m p l e   3

Identify the coefficient and the degree of each term and the degree of the polynomial

$$9x^2y^3 - 14xy^2z^3 + xy + 4y + 5x^2 + 7.$$

Solution

Term	Coefficient	Degree	Degree of the Polynomial
$9x^2y^3$	9	5	
$-14xy^2z^3$	-14	6	6
xy	1	2	
$4y$	4	1	
$5x^2$	5	2	
7	7	0	

Note in Example 3 that although both xy and $5x^2$ have degree 2, they are *not* like terms. *Like*, or *similar*, *terms* either have exactly the same variables with exactly the same exponents or are constants. For example,

$8a^4b^7$ and $5b^7a^4$ are like terms

and

-17 and 3 are like terms,

but

$-2x^2y$ and $9xy^2$ are *not* like terms.

As always, combining like terms is based on the distributive law.

E x a m p l e   4

Combine like terms.

a) $9x^2y + 3xy^2 - 5x^2y - xy^2$

b) $7ab - 5ab^2 + 3ab^2 + 6a^3 + 9ab - 11a^3 + b - 1$

Solution

a) $9x^2y + 3xy^2 - 5x^2y - xy^2 = (9 - 5)x^2y + (3 - 1)xy^2$

$\qquad\qquad\qquad\qquad\qquad\quad = 4x^2y + 2xy^2$ Try to go directly to this step.

b) $7ab - 5ab^2 + 3ab^2 + 6a^3 + 9ab - 11a^3 + b - 1$

$\qquad = -2ab^2 + 16ab - 5a^3 + b - 1$

Addition and Subtraction

The procedure used for adding polynomials in one variable is used to add polynomials in several variables.

Example 5

Add.

a) $(-5x^3 + 3y - 5y^2) + (8x^3 + 4x^2 + 7y^2)$

b) $(5ab^2 - 4a^2b + 5a^3 + 2) + (3ab^2 - 2a^2b + 3a^3b - 5)$

Solution

a) $(-5x^3 + 3y - 5y^2) + (8x^3 + 4x^2 + 7y^2)$

$\qquad = (-5 + 8)x^3 + 4x^2 + 3y + (-5 + 7)y^2$ Try to do this step mentally.

$\qquad = 3x^3 + 4x^2 + 3y + 2y^2$

b) $(5ab^2 - 4a^2b + 5a^3 + 2) + (3ab^2 - 2a^2b + 3a^3b - 5)$

$\qquad = 8ab^2 - 6a^2b + 5a^3 + 3a^3b - 3$

When subtracting a polynomial, remember to find the opposite of each term in that polynomial and then add.

Example 6

Subtract: $(4x^2y + x^3y^2 + 3x^2y^3 + 6y) - (4x^2y - 6x^3y^2 + x^2y^2 - 5y)$.

Solution

$\qquad (4x^2y + x^3y^2 + 3x^2y^3 + 6y) - (4x^2y - 6x^3y^2 + x^2y^2 - 5y)$

$\qquad\quad = 4x^2y + x^3y^2 + 3x^2y^3 + 6y - 4x^2y + 6x^3y^2 - x^2y^2 + 5y$

$\qquad\quad = 7x^3y^2 + 3x^2y^3 - x^2y^2 + 11y$ Combining like terms

Multiplication

To multiply polynomials in several variables, multiply each term of one polynomial by every term of the other, just as we did in Sections 4.4 and 4.5.

Example 7

Multiply: $(3x^2y - 2xy + 3y)(xy + 2y)$.

Solution

$$
\begin{array}{r}
3x^2y - 2xy + 3y \\
xy + 2y \\
\hline
6x^2y^2 - 4xy^2 + 6y^2 \\
3x^3y^2 - 2x^2y^2 + 3xy^2 \\
\hline
3x^3y^2 + 4x^2y^2 - xy^2 + 6y^2
\end{array}
$$

Multiplying by $2y$

Multiplying by xy

Adding

The special products discussed in Section 4.5 can speed up our work.

Example 8

Multiply.

a) $(p + 5q)(2p - 3q)$
b) $(3x + 2y)^2$
c) $(a^3 - 7a^2b)^2$
d) $(3x^2y + 2y)(3x^2y - 2y)$
e) $(-2x^3y^2 + 5t)(2x^3y^2 + 5t)$
f) $(2x + 3 - 2y)(2x + 3 + 2y)$

Solution

$$
\begin{array}{cccc}
\text{F} & \text{O} & \text{I} & \text{L}
\end{array}
$$

a) $(p + 5q)(2p - 3q) = 2p^2 - 3pq + 10pq - 15q^2$

$\qquad\qquad\qquad\quad = 2p^2 + 7pq - 15q^2$ Combining like terms

$$(A + B)^2 = A^2 + 2 \cdot A \cdot B + B^2$$

b) $(3x + 2y)^2 = (3x)^2 + 2(3x)(2y) + (2y)^2$ Squaring a binomial

$\qquad\qquad\quad = 9x^2 + 12xy + 4y^2$

$$(A - B)^2 = A^2 - 2 \cdot A \cdot B + B^2$$

c) $(a^3 - 7a^2b)^2 = (a^3)^2 - 2(a^3)(7a^2b) + (7a^2b)^2$ Squaring a binomial

$\qquad\qquad\qquad = a^6 - 14a^5b + 49a^4b^2$ Using the rules for exponents

$$(A + B)(A - B) = A^2 - B^2$$

d) $(3x^2y + 2y)(3x^2y - 2y) = (3x^2y)^2 - (2y)^2$ Recognizing the pattern

$\qquad\qquad\qquad\qquad\quad = 9x^4y^2 - 4y^2$ Using the rules for exponents

e) $(-2x^3y^2 + 5t)(2x^3y^2 + 5t) = (5t - 2x^3y^2)(5t + 2x^3y^2)$ Using the commutative law for addition twice

$\qquad\qquad\qquad\qquad\qquad = (5t)^2 - (2x^3y^2)^2$ Multiplying the sum and the difference of the same two terms

$\qquad\qquad\qquad\qquad\qquad = 25t^2 - 4x^6y^4$

$$(\quad A \quad - B)(\quad A \quad + B) = \quad A^2 \quad - \quad B^2$$

f) $(2x + 3 - 2y)(2x + 3 + 2y) = (2x + 3)^2 - (2y)^2$ Multiplying a sum and a difference

$$= 4x^2 + 12x + 9 - 4y^2$$ Squaring a binomial

Note that in Example 8 we recognized patterns that might have eluded some students, particularly in parts (e) and (f). In part (e), we could have used FOIL, and in part (f), we could have used long multiplication, but doing so would have been slower. By carefully inspecting a problem before "jumping in," we can often save ourselves considerable work.

technology connection

One way to evaluate the polynomial in Example 1 for $x = -2$ and $y = 5$ is to store -2 to X and 5 to Y and enter the polynomial.

```
-2 → X
              -2
5 → Y
               5
4+3X+XY²+8X^3Y^3
           -8052
■
```

Evaluate.

1. $3x^2 - 2y^2 + 4xy + x$, for $x = -6$ and $y = 2.3$
2. $a^2b^2 - 8c^2 + 4abc + 9a$, for $a = 11$, $b = 15$, and $c = -7$

FOR EXTRA HELP

Exercise Set 4.6

 Digital Video Tutor CD 3 Videotape 8   InterAct Math   Math Tutor Center   MathXL   MyMathLab.com

Evaluate each polynomial for $x = 5$ and $y = -2$.

1. $x^2 - 3y^2 + 2xy$

2. $x^2 + 5y^2 - 4xy$

Evaluate each polynomial for $x = 2$, $y = -3$, and $z = -4$.

3. $xyz^2 - z$

4. $xy - xz + yz$

Lung capacity. The polynomial

$$0.041h - 0.018A - 2.69$$

can be used to estimate the lung capacity, in liters, of a female with height h, in centimeters, and age A, in years.

5. Find the lung capacity of a 50-year-old woman who is 160 cm tall.

6. Find the lung capacity of a 20-year-old woman who is 165 cm tall.

Altitude of a launched object.   *The altitude of an object, in meters, is given by the polynomial*

$$h + vt - 4.9t^2,$$

where h is the height, in meters, at which the launch occurs, v is the initial upward speed (or velocity), in meters per second, and t is the number of seconds for which the object is airborne.

7. A model rocket is launched from atop the Eiffel Tower in Paris, 300 m above the ground. If the initial upward speed is 40 meters per second (m/s), how high above the ground will the rocket be 2 sec after having been launched?

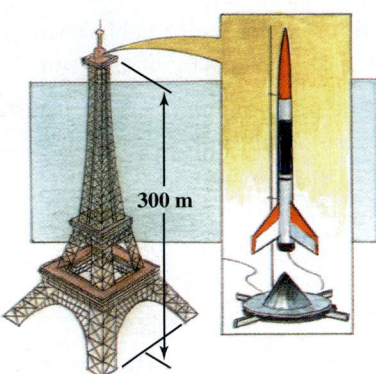

300 m

8. A golf ball is thrown upward with an initial speed of 30 m/s by a golfer atop the Washington Monument, 160 m above the ground. How high above the ground will the ball be after 3 sec?

Surface area of a silo.   *A silo is a structure that is shaped like a right circular cylinder with a half sphere on top. The surface area of a silo of height h and radius r (including the area of the base) is given by the polynomial* $2\pi rh + \pi r^2$.

9. A container of tennis balls is silo-shaped, with a height of $7\frac{1}{2}$ in. and a radius of $1\frac{1}{4}$ in. Find the surface area of the container. Use 3.14 for π.

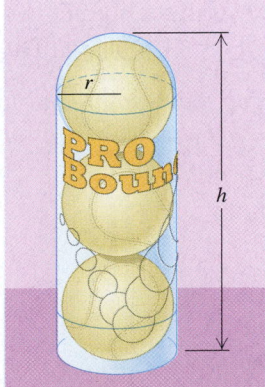

r

h

10. A $1\frac{1}{2}$-oz bottle of roll-on deodorant has a height of 4 in. and a radius of $\frac{3}{4}$ in. Find the surface area of the bottle if the bottle is shaped like a silo. Use 3.14 for π.

Identify the coefficient and the degree of each term of each polynomial. Then find the degree of each polynomial.

11. $x^3y - 2xy + 3x^2 - 5$

12. $xy^2 - y^2 + 9x^2y + 7$

13. $17x^2y^3 - 3x^3yz - 7$

14. $6 - xy + 8x^2y^2 - y^5$

Combine like terms.

15. $7a + b - 4a - 3b$

16. $8r + s - 5r - 4s$

17. $3x^2y - 2xy^2 + x^2 + 5x$

18. $m^3 + 2m^2n - 3m^2 + 3mn^2$

19. $2u^2v - 3uv^2 + 6u^2v - 2uv^2 + 7u^2$

20. $3x^2 + 6xy + 3y^2 - 5x^2 - 10xy$

21. $5a^2c - 2ab^2 + a^2b - 3ab^2 + a^2c - 2ab^2$

22. $3s^2t + r^2t - 4st^2 - s^2t + 3st^2 - 7r^2t$

Add or subtract, as indicated.

23. $(4x^2 - xy + y^2) + (-x^2 - 3xy + 2y^2)$

24. $(2r^3 + 3rs - 5s^2) - (5r^3 + rs + 4s^2)$

25. $(3a^4 - 5ab + 6ab^2) - (9a^4 + 3ab - ab^2)$

26. $(2r^2t - 5rt + rt^2) - (7r^2t + rt - 5rt^2)$

Aha! **27.** $(5r^2 - 4rt + t^2) + (-6r^2 - 5rt - t^2) + (-5r^2 + 4rt - t^2)$

28. $(2x^2 - 3xy + y^2) + (-4x^2 - 6xy - y^2) + (4x^2 + 6xy + y^2)$

29. $(x^3 - y^3) - (-2x^3 + x^2y - xy^2 + 2y^3)$

30. $(a^3 + b^3) - (-5a^3 + 2a^2b - ab^2 + 3b^3)$

31. $(2y^4x^2 - 5y^3x) + (5y^4x^2 - y^3x) + (3y^4x^2 - 2y^3x)$

32. $(5a^2b + 7ab) + (9a^2b - 5ab) + (a^2b - 6ab)$

33. Subtract $7x + 3y$ from the sum of $4x + 5y$ and $-5x + 6y$.

34. Subtract $5a + 2b$ from the sum of $2a + b$ and $3a - 4b$.

Multiply.

35. $(3z - u)(2z + 3u)$ **36.** $(5x + y)(2x - 3y)$

37. $(xy + 7)(xy - 4)$ **38.** $(ab + 3)(ab - 5)$

39. $(2a - b)(2a + b)$ **40.** $(a - 3b)(a + 3b)$

41. $(5rt - 2)(3rt + 1)$ **42.** $(3xy - 1)(4xy + 2)$

43. $(m^3n + 8)(m^3n - 6)$ **44.** $(3 - c^2d^2)(4 + c^2d^2)$

45. $(6x - 2y)(5x - 3y)$ **46.** $(7a - 6b)(5a + 4b)$

47. $(pq + 0.2)(0.4pq - 0.1)$

48. $(ab - 0.6)(0.2ab + 0.3)$

49. $(x + h)^2$ **50.** $(r + t)^2$

51. $(4a + 5b)^2$ **52.** $(3x + 2y)^2$

53. $(c^2 - d)(c^2 + d)$ **54.** $(p^3 - 5q)(p^3 + 5q)$

55. $(ab + cd^2)(ab - cd^2)$ **56.** $(xy + pq)(xy - pq)$

Aha! **57.** $(a + b - c)(a + b + c)$

58. $(x + y + z)(x + y - z)$

59. $[a + b + c][a - (b + c)]$

60. $(a + b + c)(a - b - c)$

Find the total area of each shaded area.

61.

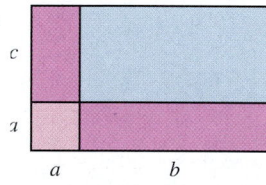

62.

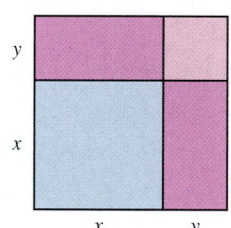

63.

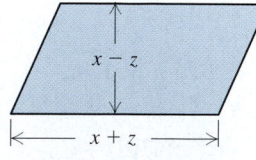

64.

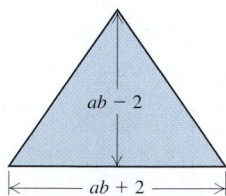

65.

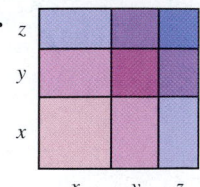

66.

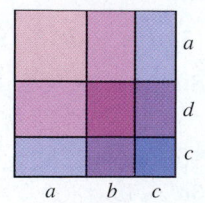

67.

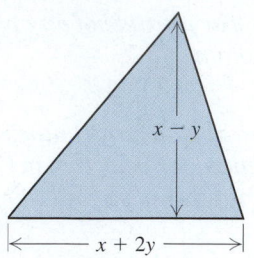

68.

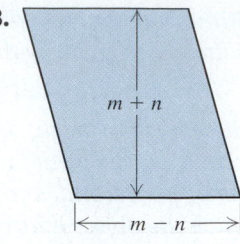

Draw and label rectangles similar to those in Exercises 61, 62, 65, and 66 to illustrate each product.

69. $(r + s)(u + v)$ **70.** $(m + r)(n + v)$

71. $(a + b + c)(a + d + f)$ **72.** $(r + s + t)^2$

73. Is it possible for a polynomial in 4 variables to have a degree less than 4? Why or why not?

74. A fourth-degree polynomial is multiplied by a third-degree polynomial. What is the degree of the product? Explain your reasoning.

SKILL MAINTENANCE

Simplify.

75. $5 + \dfrac{7 + 4 + 2 \cdot 5}{3}$ **76.** $9 - \dfrac{2 + 6 \cdot 3 + 4}{6}$

77. $(4 + 3 \cdot 5 + 8) \div 3 \cdot 3$

78. $(5 + 2 \cdot 7 + 5) \div 2 \cdot 3$

79. $[3 \cdot 5 - 4 \cdot 2 + 7(-3)] \div (-2)$

80. $(7 - 3 \cdot 9 - 2 \cdot 5) \div (-6)$

SYNTHESIS

81. Can the sum of two trinomials in several variables be a binomial in one variable? Why or why not?

82. Can the sum of two trinomials in several variables be a trinomial in one variable? Why or why not?

Find a polynomial for the shaded area. (Leave results in terms of π where appropriate.)

83.

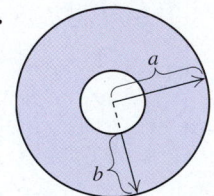

84.

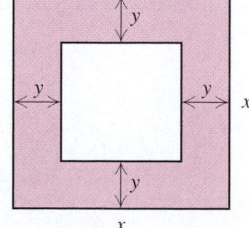

85.

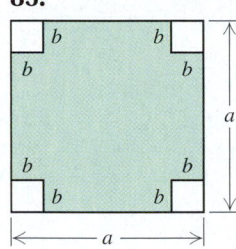

86.

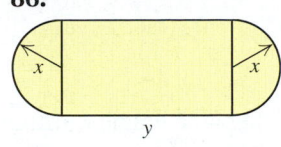

87. Find the shaded area in this figure using each of the approaches given below. Then check that both answers match.

a) Find the shaded area by subtracting the area of the unshaded square from the total area of the figure.

b) Find the shaded area by adding the areas of the three shaded rectangles.

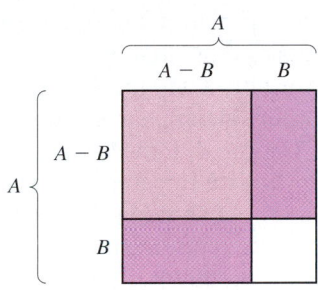

Find a polynomial for the surface area of each solid object shown. (Leave results in terms of π.)

88.

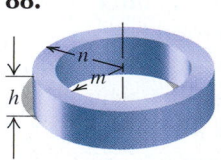

89.

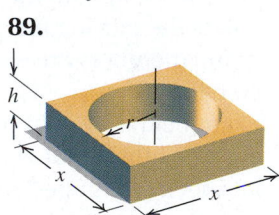

90. The observatory at Danville University is shaped like a silo that is 40 ft high and 30 ft wide (see Exercise 9). The Heavenly Bodies Astronomy Club is to paint the exterior of the observatory using paint that covers 250 ft^2 per gallon. How many gallons should they purchase? Explain your reasoning.

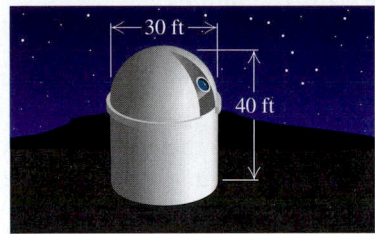

91. Multiply: $(x + a)(x - b)(x - a)(x + b)$.

92. *Interest compounded annually.* An amount of money P that is invested at the yearly interest rate r grows to the amount

$$P(1 + r)^t$$

after t years. Find a polynomial that can be used to determine the amount to which P will grow after 2 yr.

93. *Yearly depreciation.* An investment P that drops in value at the yearly rate r drops in value to

$$P(1 - r)^t$$

after t years. Find a polynomial that can be used to determine the value to which P has dropped after 2 yr.

94. Suppose that $10,400 is invested at 8.5% compounded annually. How much is in the account at the end of 5 yr? (See Exercise 92.)

95. A $90,000 investment in computer hardware is depreciating at a yearly rate of 12.5%. How much is the investment worth after 4 yr? (See Exercise 93.)

CORNER

Finding the Magic Number

Focus: Evaluating polynomials in several variables

Time: 15–25 minutes

Group size: 3

Materials: A coin for each person

When a team nears the end of its schedule in first place, fans begin to discuss the team's "magic number." A team's magic number is the combined number of wins by that team and losses by the second-place team that guarantee the leading team a first-place finish. For example, if the Cubs' magic number is 3 over the Reds, any combination of Cubs wins and Reds losses that totals 3 will guarantee a first-place finish for the Cubs, regardless of how subsequent games are decided. A team's magic number is computed using the polynomial

$$G - P - L + 1,$$

where G is the length of the season, in games, P is the number of games that the leading team has played, and L is the total number of games that the second-place team has lost minus the total number of games that the leading team has lost.

ACTIVITY

1. The standings below are from a fictitious baseball league. Together, the group should calculate the Jaguars' magic number with respect to the Catamounts as well as the Jaguars' magic number with respect to the Wildcats. (Assume that the schedule is 162 games long.)

	W	L
Jaguars	92	64
Catamounts	90	66
Wildcats	89	66

2. Each group member should play the role of one of the teams. To simulate each team's remaining games, coin tosses will be performed. If a group member correctly predicts the side (heads or tails) that comes up, the coin toss represents a win for that team. Should the other side appear, the toss represents a loss. Assume that these games are against other (unlisted) teams in the league. Each group member should perform three coin tosses and then update the standings.

3. Recalculate the two magic numbers, using the updated standings from part (2).

4. Slowly—one coin toss at a time—play out the remainder of the season. Record all wins and losses, update the standings, and recalculate the magic numbers each time all three group members have completed a round of coin tosses.

5. Examine the work in part (4) and explain why a magic number of 0 indicates that a team has been eliminated from contention.

<table>
<tr><td>

Division of Polynomials

</td><td>

4.7

Dividing by a Monomial • Dividing by a Binomial

</td></tr>
</table>

In this section, we study division of polynomials. We will find that polynomial division is similar to division in arithmetic.

Dividing by a Monomial

We first consider division by a monomial. When dividing a monomial by a monomial, we use the quotient rule of Section 4.1 to subtract exponents when bases are the same. For example,

$$\frac{15x^{10}}{3x^4} = 5x^{10-4}$$

$$= 5x^6$$

> **Caution!**   The coefficients are divided but the exponents are subtracted.

and

$$\frac{42a^2b^5}{-3ab^2} = \frac{42}{-3}a^{2-1}b^{5-2}$$

$$= -14ab^3.$$

To divide a polynomial by a monomial, we note that since

$$\frac{A}{C} + \frac{B}{C} = \frac{A+B}{C},$$

it follows that

$$\frac{A+B}{C} = \frac{A}{C} + \frac{B}{C}.$$ Switching the left and right sides of the equation

This is actually how we perform divisions like $86 \div 2$: Although we might simply write

$$\frac{86}{2} = 43,$$

we are really saying

$$\frac{80+6}{2} = \frac{80}{2} + \frac{6}{2} = 40 + 3.$$

Similarly, to divide a polynomial by a monomial, we divide each term by the monomial:

$$\frac{80x^5 + 6x^3}{2x^2} = \frac{80x^5}{2x^2} + \frac{6x^3}{2x^2}$$

$$= \frac{80}{2}x^{5-2} + \frac{6}{2}x^{3-2}$$ Dividing coefficients and subtracting exponents

$$= 40x^3 + 3x.$$

Example 1

Divide $x^4 + 15x^3 - 6x^2$ by $3x$.

Solution   We have

$$\frac{x^4 + 15x^3 - 6x^2}{3x} = \frac{x^4}{3x} + \frac{15x^3}{3x} - \frac{6x^2}{3x}$$

$$= \frac{1}{3}x^{4-1} + \frac{15}{3}x^{3-1} - \frac{6}{3}x^{2-1} \qquad \text{Dividing coefficients and subtracting exponents}$$

$$= \frac{1}{3}x^3 + 5x^2 - 2x. \qquad \text{This is the quotient.}$$

To check, we multiply our answer by $3x$, using the distributive law:

$$3x\left(\frac{1}{3}x^3 + 5x^2 - 2x\right) = 3x \cdot \frac{1}{3}x^3 + 3x \cdot 5x^2 - 3x \cdot 2x$$

$$= x^4 + 15x^3 - 6x^2.$$

This is the polynomial that was being divided, so our answer, $\frac{1}{3}x^3 + 5x^2 - 2x$, checks.

Example 2

Divide and check: $(10a^5b^4 - 2a^3b^2 + 6a^2b) \div 2a^2b$.

Solution   We have

$$\frac{10a^5b^4 - 2a^3b^2 + 6a^2b}{2a^2b} = \frac{10a^5b^4}{2a^2b} - \frac{2a^3b^2}{2a^2b} + \frac{6a^2b}{2a^2b}$$

$$= \frac{10}{2}a^{5-2}b^{4-1} - \frac{2}{2}a^{3-2}b^{2-1} + \frac{6}{2} \qquad \text{Dividing coefficients and subtracting exponents}$$

$$= 5a^3b^3 - ab + 3.$$

Check: $2a^2b(5a^3b^3 - ab + 3) = 2a^2b \cdot 5a^3b^3 - 2a^2b \cdot ab + 2a^2b \cdot 3$

$$= 10a^5b^4 - 2a^3b^2 + 6a^2b$$

Our answer, $5a^3b^3 - ab + 3$, checks.

Dividing by a Binomial

For divisors with more than one term, we use long division, much as we do in arithmetic. Polynomials are written in descending order and any missing terms are written in, using 0 for the coefficients.

E x a m p l e   3

Divide $x^2 + 5x + 6$ by $x + 3$.

Solution   We have

Divide the first term, x^2, by the first term in the divisor: $x^2/x = x$. Ignore the term 3 for the moment.

$$
\begin{array}{r}
x \\
x + 3 \overline{)\, x^2 + 5x + 6} \\
-(x^2 + 3x) \\
\hline
2x
\end{array}
$$

Multiply $x + 3$ by x.

Subtract by mentally changing signs and adding: $x^2 + 5x - (x^2 + 3x) = 2x$.

Now we "bring down" the next term—in this case, 6.

$$
\begin{array}{r}
x \; + 2 \\
x + 3 \overline{)\, x^2 + 5x + 6} \\
-(x^2 + 3x) \\
\hline
2x + 6 \\
-(2x + 6) \\
\hline
0
\end{array}
$$

Divide $2x$ by x: $2x/x = 2$.

Multiply 2 by the divisor, $x + 3$

Subtract: $(2x + 6) - (2x + 6) = 0$.

The quotient is $x + 2$. The remainder is 0, expressed as R 0. A remainder of 0 is generally not listed in an answer.

Check: To check, we multiply the quotient by the divisor and add the remainder, if any, to see if we get the dividend:

Divisor	Quotient		Remainder		Dividend
$(x + 3)$	$(x + 2)$	$+$	0	$=$	$x^2 + 5x + 6.$

Our answer, $x + 2$, checks.

E x a m p l e   4

Divide: $(2x^2 + 5x - 1) \div (2x - 1)$.

Solution   We have

Divide the first term by the first term: $2x^2/(2x) = x$.

$$
\begin{array}{r}
x \\
2x - 1 \overline{)\, 2x^2 + 5x - 1} \\
-(2x^2 - x) \\
\hline
6x
\end{array}
$$

Multiply $2x - 1$ by x.

Subtract by mentally changing signs and adding: $2x^2 + 5x - (2x^2 - x) = 6x$.

Now, we bring down the next term of the dividend, -1.

$$
\begin{array}{r}
x \; + 3 \\
2x - 1 \overline{)\, 2x^2 + 5x - 1} \\
-(2x^2 - x) \\
\hline
6x - 1 \\
-(6x - 3) \\
\hline
2
\end{array}
$$

Divide $6x$ by $2x$: $6x/(2x) = 3$.

Multiply 3 by the divisor, $2x - 1$.

Note that $-1 - (-3) = -1 + 3 = 2$.

The answer is $x + 3$ with R 2.

Another way to write $x + 3$ R 2 is as

$$\underset{\text{Quotient}}{\underline{x + 3}} + \underset{\text{Divisor}}{\underline{\frac{2 \quad \longleftarrow \text{Remainder}}{2x - 1}}}.$$

(This is the way answers will be given at the back of the book.)

Check: To check, we multiply the quotient by the divisor and add the remainder:

$$(2x - 1)(x + 3) + 2 = 2x^2 + 5x - 3 + 2$$
$$= 2x^2 + 5x - 1. \quad \text{Our answer checks.}$$

Our division procedure ends when the degree of the remainder is less than that of the divisor. Check that this was indeed the case in Example 4.

E x a m p l e   5   Divide each of the following.

a) $(x^3 + 1) \div (x + 1)$
b) $(x^4 - 3x^2 + 2x - 3) \div (x^2 - 5)$

Solution

a)
$$\begin{array}{r} x^2 - x + 1 \\ x + 1 \overline{) x^3 + 0x^2 + 0x + 1} \quad \longleftarrow \text{Fill in the missing terms.} \\ \underline{-(x^3 + x^2)} \\ - x^2 + 0x \quad \longleftarrow \text{Subtracting } x^3 + x^2 \text{ from } x^3 + 0x^2 \text{ and} \\ \underline{-(- x^2 - x)} \quad\quad \text{bringing down the } 0x \\ x + 1 \quad \longleftarrow \text{Subtracting } -x^2 - x \text{ from } -x^2 + 0x \text{ and} \\ \underline{-(x + 1)} \quad\quad \text{bringing down the } 1 \\ 0 \end{array}$$

The answer is $x^2 - x + 1$.

Check: $(x + 1)(x^2 - x + 1) = x^3 - x^2 + x + x^2 - x + 1$
$$= x^3 + 1.$$

b)
$$\begin{array}{r} x^2 + 2 \\ x^2 - 5 \overline{) x^4 + 0x^3 - 3x^2 + 2x - 3} \\ \underline{-(x^4 - 5x^2)} \\ 2x^2 + 2x - 3 \\ \underline{-(2x^2 - 10)} \\ 2x + 7 \end{array}$$

Writing in the missing term

Subtracting $x^4 - 5x^2$ from $x^4 - 3x^2$ and bringing down $2x - 3$

$\longleftarrow$ Subtracting $2x^2 - 10$ from $2x^2 + 2x - 3$

Since the remainder, $2x + 7$, is of lower degree than the divisor, the division process stops. The answer is $x^2 + 2$, with R $2x + 7$, or

$$x^2 + 2 + \frac{2x + 7}{x^2 - 5}.$$

Check: $(x^2 - 5)(x^2 + 2) + 2x + 7 = x^4 + 2x^2 - 5x^2 - 10 + 2x + 7$
$$= x^4 - 3x^2 + 2x - 3.$$

Divide and check.

1. $\dfrac{40x^5 - 16x}{8}$

2. $\dfrac{12a^4 - 3a^2}{6}$

3. $\dfrac{u - 2u^2 + u^7}{u}$

4. $\dfrac{50x^5 - 7x^4 + x^2}{x}$

5. $(15t^3 - 24t^2 + 6t) \div (3t)$

6. $(20t^3 - 15t^2 + 30t) \div (5t)$

7. $(25x^6 - 20x^4 - 5x^2) \div (-5x^2)$

8. $(16x^6 + 32x^5 - 8x^2) \div (-8x^2)$

9. $(24t^5 - 40t^4 + 6t^3) \div (4t^3)$

10. $(18t^6 - 27t^5 - 3t^3) \div (9t^3)$

11. $\dfrac{6x^2 - 10x + 1}{2}$

12. $\dfrac{9x^2 + 3x - 2}{3}$

13. $\dfrac{4x^3 + 6x^2 + 4x}{2x^2}$

14. $\dfrac{10x^4 + 15x^3 + 5x}{5x^2}$

15. $\dfrac{9r^2s^2 + 3r^2s - 6rs^2}{-3rs}$

16. $\dfrac{4x^4y - 8x^6y^2 + 12x^8y^6}{4x^4y}$

17. $(x^2 + 4x - 12) \div (x - 2)$

18. $(x^2 - 6x + 8) \div (x - 4)$

19. $(t^2 - 10t - 20) \div (t - 5)$

20. $(t^2 + 8t - 15) \div (t + 4)$

21. $(2x^2 + 11x - 5) \div (x + 6)$

22. $(3x^2 - 2x - 13) \div (x - 2)$

23. $\dfrac{a^3 + 8}{a + 2}$

24. $\dfrac{t^3 + 27}{t + 3}$

25. $\dfrac{t^2 - 15}{t - 4}$

26. $\dfrac{a^2 - 23}{a - 5}$

27. $(3x^2 + 11x - 4) \div (3x - 1)$

28. $(10x^2 + 13x - 3) \div (5x - 1)$

29. $(6a^2 + 17a + 8) \div (2a + 5)$

30. $(10a^2 + 19a + 9) \div (2a + 3)$

31. $\dfrac{2t^3 - 9t^2 + 11t - 3}{2t - 3}$

32. $\dfrac{8t^3 - 22t^2 - 5t + 12}{4t + 3}$

33. $(t^3 - t^2 + t - 1) \div (t + 1)$

34. $(x^3 - x^2 + x - 1) \div (x - 1)$

35. $(t^4 + 4t^2 + 3t - 6) \div (t^2 + 5)$

36. $(t^4 - 2t^2 + 4t - 5) \div (t^2 - 3)$

37. $(4x^4 - 4x^2 - x - 3) \div (2x^2 - 3)$

38. $(6x^4 - 3x^2 + x - 4) \div (2x^2 + 1)$

39. How is the distributive law used when dividing a polynomial by a binomial?

40. On an assignment, Emmy Lou *incorrectly* writes

$$\frac{12x^3 - 6x}{3x} = 4x^2 - 6x.$$

What mistake do you think she is making and how might you convince her that a mistake has been made?

SKILL MAINTENANCE

Simplify.

41. $-4 + (-13)$

42. $-8 + (-15)$

43. $-9 - (-7)$

44. $-2 - (-7)$

45. The perimeter of a rectangle is 640 ft. The length is 15 ft greater than the width. Find the length of the rectangle.

46. Solve: $3(2x - 1) = 7x - 5$.

47. Graph: $3x - 2y = 12$.

48. Plot the points $(4, -1)$, $(0, 5)$, $(-2, 3)$, and $(-3, 0)$.

SYNTHESIS

49. Explain how the quotient of two binomials can have more than two terms.

50. Explain how to form a trinomial for which division by $x - 5$ results in a remainder of 7.

Divide.

51. $(10x^{9k} - 32x^{6k} + 28x^{3k}) \div (2x^{3k})$

52. $(45a^{8k} + 30a^{6k} - 60a^{4k}) \div (3a^{2k})$

53. $(6t^{3h} + 13t^{2h} - 4t^h - 15) \div (2t^h + 3)$

54. $(x^4 + a^2) \div (x + a)$

55. $(5a^3 + 8a^2 - 23a - 1) \div (5a^2 - 7a - 2)$

56. $(15y^3 - 30y + 7 - 19y^2) \div (3y^2 - 2 - 5y)$

57. Divide the sum of $4x^5 - 14x^3 - x^2 + 3$ and $2x^5 + 3x^4 + x^3 - 3x^2 + 5x$ by $3x^3 - 2x - 1$.

58. Divide $5x^7 - 3x^4 + 2x^2 - 10x + 2$ by the sum of $(x - 3)^2$ and $5x - 8$.

If the remainder is 0 when one polynomial is divided by another, the divisor is a factor *of the dividend. Find the value(s) of c for which $x - 1$ is a factor of each polynomial.*

59. $x^2 - 4x + c$

60. $2x^2 - 3cx - 8$

61. $c^2x^2 + 2cx + 1$

Negative Exponents and Scientific Notation

4.8

Negative Integers as Exponents • Scientific Notation • Multiplying and Dividing Using Scientific Notation

We now attach a meaning to negative exponents. Once we understand both positive and negative exponents, we can study a method of writing numbers known as *scientific notation*.

Negative Integers as Exponents

Let's define negative exponents so that the rules that apply to whole-number exponents will hold for all integer exponents. To do so, consider a^{-5} and the rule for adding exponents:

$$a^{-5} = a^{-5} \cdot 1 \qquad \text{Using the identity property of 1}$$

$$= \frac{a^{-5}}{1} \cdot \frac{a^5}{a^5} \qquad \text{Writing 1 as } \frac{a^5}{a^5} \text{ and } a^{-5} \text{ as } \frac{a^{-5}}{1}$$

$$= \frac{a^{-5+5}}{a^5} \qquad \text{Adding exponents}$$

$$= \frac{1}{a^5}. \qquad -5 + 5 = 0 \text{ and } a^0 = 1$$

This leads to our definition of negative exponents.

> **Negative Exponents**
>
> For any real number a that is nonzero and any integer n,
>
> $$a^{-n} = \frac{1}{a^n}.$$
>
> (The numbers a^{-n} and a^n are reciprocals of each other.)

E x a m p l e   1   Express using positive exponents and, if possible, simplify.

a) m^{-3} **b)** 4^{-2} **c)** $(-3)^{-2}$ **d)** ab^{-1}

Solution

a) $m^{-3} = \dfrac{1}{m^3}$ m^{-3} is the reciprocal of m^3.

b) $4^{-2} = \dfrac{1}{4^2} = \dfrac{1}{16}$ 4^{-2} is the reciprocal of 4^2.
                                            Note that $4^{-2} \neq 4(-2)$.

c) $(-3)^{-2} = \dfrac{1}{(-3)^2} = \dfrac{1}{(-3)(-3)} = \dfrac{1}{9}$ $(-3)^{-2}$ is the reciprocal of $(-3)^2$.
                                                 Note that $(-3)^{-2} \neq -\dfrac{1}{3^2}$.

d) $ab^{-1} = a\left(\dfrac{1}{b^1}\right) = a\left(\dfrac{1}{b}\right) = \dfrac{a}{b}$ b^{-1} is the reciprocal of b^1.

Caution!   A negative exponent does not, in itself, indicate that an expression is negative. As shown in Example 1,

$$4^{-2} \neq 4(-2) \quad \text{and} \quad (-3)^{-2} \neq -\frac{1}{3^2}.$$

The following is another way to illustrate why negative exponents are defined as they are.

<table>
<tr><td>On this side, we divide by 5 at each step.</td><td>$125 = 5^3$</td><td>On this side, the exponents decrease by 1.</td></tr>
<tr><td></td><td>$25 = 5^2$</td><td></td></tr>
<tr><td></td><td>$5 = 5^1$</td><td></td></tr>
<tr><td></td><td>$1 = 5^0$</td><td></td></tr>
<tr><td></td><td>$\dfrac{1}{5} = 5^?$</td><td></td></tr>
<tr><td></td><td>$\dfrac{1}{25} = 5^?$</td><td></td></tr>
</table>

To continue the pattern, it follows that

$$\frac{1}{5} = \frac{1}{5^1} = 5^{-1}, \qquad \frac{1}{25} = \frac{1}{5^2} = 5^{-2}, \quad \text{and, in general,} \quad \frac{1}{a^n} = a^{-n}.$$

E x a m p l e   2   Express $\dfrac{1}{x^7}$ using negative exponents.

Solution   We know that $\dfrac{1}{a^n} = a^{-n}$. Thus,

$$\frac{1}{x^7} = x^{-7}.$$

The rules for powers still hold when exponents are negative.

E x a m p l e   3   Simplify. Do not use negative exponents in the answer.

a) $t^5 \cdot t^{-2}$ b) $(y^{-5})^{-7}$ c) $(5x^2y^{-3})^4$

d) $\dfrac{x^{-4}}{x^{-5}}$ e) $\dfrac{1}{t^{-5}}$ f) $\dfrac{s^{-3}}{t^{-5}}$

Solution

a) $t^5 \cdot t^{-2} = t^{5+(-2)} = t^3$ Adding exponents

b) $(y^{-5})^{-7} = y^{(-5)(-7)}$ Multiplying exponents
$\qquad\quad = y^{35}$

c) $(5x^2y^{-3})^4 = 5^4(x^2)^4(y^{-3})^4$ Raising each factor to the fourth power
$$= 625x^8y^{-12} = \frac{625x^8}{y^{12}}$$

d) $\dfrac{x^{-4}}{x^{-5}} = x^{-4-(-5)} = x^1 = x$ We subtract exponents even if the exponent in the denominator is negative.

e) Since $\dfrac{1}{a^n} = a^{-n}$, we have

$$\frac{1}{t^{-5}} = t^{-(-5)} = t^5.$$

f) $\dfrac{s^{-3}}{t^{-5}} = s^{-3} \cdot \dfrac{1}{t^{-5}}$

$\qquad = \dfrac{1}{s^3} \cdot t^5 = \dfrac{t^5}{s^3}$ Using the result from part (e) above

The result from Example 3(f) can be generalized.

Factors and Negative Exponents

For any nonzero real numbers a and b and any integers m and n,

$$\frac{a^{-n}}{b^{-m}} = \frac{b^m}{a^n}.$$

(A factor can be moved to the other side of the fraction bar if the sign of the exponent is changed.)

Example 4

Simplify: $\dfrac{5x^{-7}}{y^2z^{-4}}$.

Solution   We can move the factors x^{-7} and z^{-4} to the other side of the fraction if we change the sign of each exponent:

$$\frac{5x^{-7}}{y^2z^{-4}} = \frac{5z^4}{y^2x^7}, \quad \text{or} \quad \frac{5z^4}{x^7y^2}.$$

Another way to change the sign of the exponent is to take the reciprocal of the base. To understand why this is true, note that

$$\left(\frac{s}{t}\right)^{-5} = \frac{s^{-5}}{t^{-5}} = \frac{t^5}{s^5} = \left(\frac{t}{s}\right)^5.$$

This often provides the easiest way to simplify an expression containing a negative exponent.

> ### Reciprocals and Negative Exponents
>
> For any nonzero real numbers a and b and any integer n,
>
> $$\left(\frac{a}{b}\right)^{-n} = \left(\frac{b}{a}\right)^n.$$
>
> (Any base to a power is equal to the reciprocal of the base raised to the opposite power.)

Example 5

Simplify: $\left(\dfrac{x^4}{2y}\right)^{-3}$.

Solution

$$\left(\frac{x^4}{2y}\right)^{-3} = \left(\frac{2y}{x^4}\right)^3 \qquad \text{Taking the reciprocal of the base and changing the sign of the exponent}$$

$$= \frac{(2y)^3}{(x^4)^3} \qquad \text{Raising a quotient to a power by raising both the numerator and denominator to the power}$$

$$= \frac{2^3y^3}{x^{12}} \qquad \text{Raising a product to a power; using the power rule in the denominator}$$

$$= \frac{8y^3}{x^{12}} \qquad \text{Cubing 2}$$

Definitions and Properties of Exponents

The following summary assumes that no denominators are 0 and that 0^0 is not considered. For any integers m and n,

1 as an exponent:	$a^1 = a$
0 as an exponent:	$a^0 = 1$
Negative exponents:	$a^{-n} = \dfrac{1}{a^n}$,
	$\dfrac{a^{-n}}{b^{-m}} = \dfrac{b^m}{a^n}$,
	$\left(\dfrac{a}{b}\right)^{-n} = \left(\dfrac{b}{a}\right)^n$
The Product Rule:	$a^m \cdot a^n = a^{m+n}$
The Quotient Rule:	$\dfrac{a^m}{a^n} = a^{m-n}$
The Power Rule:	$(a^m)^n = a^{mn}$
Raising a product to a power:	$(ab)^n = a^n b^n$
Raising a quotient to a power:	$\left(\dfrac{a}{b}\right)^n = \dfrac{a^n}{b^n}$

Scientific Notation

When we are working with the very large or very small numbers that frequently occur in science, **scientific notation** provides a useful way of writing numbers. The following are examples of scientific notation.

The mass of the earth:

6.0×10^{24} kilograms (kg) $= 6{,}000{,}000{,}000{,}000{,}000{,}000{,}000{,}000$ kg

The mass of a hydrogen atom:

1.7×10^{-24} g $= 0.0000000000000000000000017$ g

Scientific Notation

Scientific notation for a number is an expression of the type

$$N \times 10^m,$$

where N is at least 1 but less than 10 ($1 \le N < 10$), N is expressed in decimal notation, and m is an integer.

Converting from scientific to decimal notation involves multiplying by a power of 10. Consider the following.

Scientific Notation		
$N \times 10^{m}$	*Multiplication*	*Decimal Notation*
4.52×10^{2}	4.52×100	452.
4.52×10^{1}	4.52×10	45.2
4.52×10^{0}	4.52×1	4.52
4.52×10^{-1}	4.52×0.1	0.452
4.52×10^{-2}	4.52×0.01	0.0452

Note that when m, the power of 10, is positive, the decimal point moves right m places in decimal notation. When m is negative, the decimal point moves left $|m|$ places. We generally try to perform this multiplication mentally.

E x a m p l e   6

Convert to decimal notation: **(a)** 7.893×10^{5}; **(b)** 4.7×10^{-8}.

Solution

a) Since the exponent is positive, the decimal point moves to the right:

7.89300. $7.893 \times 10^{5} = 789,300$ The decimal point moves
5 places   5 places to the right.

b) Since the exponent is negative, the decimal point moves to the left:

0.00000004.7 $4.7 \times 10^{-8} = 0.000000047$ The decimal point
8 places   moves 8 places to
the left.

To convert from decimal to scientific notation, this procedure is reversed.

E x a m p l e   7

Write in scientific notation: **(a)** 83,000; **(b)** 0.0327.

Solution

a) We need to find m such that $83,000 = 8.3 \times 10^{m}$. To change 8.3 to 83,000 requires moving the decimal point 4 places to the right. This can be accomplished by multiplying by 10^{4}. Thus,

$$83,000 = 8.3 \times 10^{4}.   \text{This is scientific notation.}$$

b) We need to find m such that $0.0327 = 3.27 \times 10^{m}$. To change 3.27 to 0.0327 requires moving the decimal point 2 places to the left. This can be accomplished by multiplying by 10^{-2}. Thus,

$$0.0327 = 3.27 \times 10^{-2}.   \text{This is scientific notation.}$$

Conversions to and from scientific notation are often made mentally. Remember that positive exponents are used when representing large numbers and negative exponents are used when representing numbers between 0 and 1.

Multiplying and Dividing Using Scientific Notation

Products and quotients of numbers written in scientific notation are found using the rules for exponents.

E x a m p l e 8

Simplify.

a) $(1.8 \times 10^9) \cdot (2.3 \times 10^{-4})$ **b)** $(3.41 \times 10^5) \div (1.1 \times 10^{-3})$

Solution

a) $(1.8 \times 10^9) \cdot (2.3 \times 10^{-4})$

$\qquad = 1.8 \times 2.3 \times 10^9 \times 10^{-4}$ Using the associative and commutative laws

$\qquad = 4.14 \times 10^{9+(-4)}$ Adding exponents

$\qquad = 4.14 \times 10^5$

b) $(3.41 \times 10^5) \div (1.1 \times 10^{-3})$

$\qquad = \dfrac{3.41 \times 10^5}{1.1 \times 10^{-3}}$

$\qquad = \dfrac{3.41}{1.1} \times \dfrac{10^5}{10^{-3}}$

$\qquad = 3.1 \times 10^{5-(-3)}$ Subtracting exponents

$\qquad = 3.1 \times 10^8$

When a problem is stated using scientific notation, we normally use scientific notation for the answer.

E x a m p l e 9

Simplify.

a) $(3.1 \times 10^5) \cdot (4.5 \times 10^{-3})$ **b)** $(7.2 \times 10^{-7}) \div (8.0 \times 10^6)$

Solution

a) We have

$$(3.1 \times 10^5) \cdot (4.5 \times 10^{-3}) = 3.1 \times 4.5 \times 10^5 \times 10^{-3}$$
$$= 13.95 \times 10^2.$$

Our answer is not yet in scientific notation because 13.95 is not between 1 and 10. We convert to scientific notation as follows:

$$13.95 \times 10^2 = 1.395 \times 10^1 \times 10^2 \qquad \text{Substituting } 1.395 \times 10^1 \text{ for } 13.95$$
$$= 1.395 \times 10^3. \qquad \text{Adding exponents}$$

b) $(7.2 \times 10^{-7}) \div (8.0 \times 10^6) = \dfrac{7.2 \times 10^{-7}}{8.0 \times 10^6} = \dfrac{7.2}{8.0} \times \dfrac{10^{-7}}{10^6}$

$= 0.9 \times 10^{-13}$

$= 9.0 \times 10^{-1} \times 10^{-13}$ Substituting 9.0×10^{-1} for 0.9

$= 9.0 \times 10^{-14}$ Adding exponents

technology connection

A key labeled ⌃ or **EE** is often used to enter scientific notation into a calculator. Sometimes this is a secondary function, meaning that another key—often labeled **SHIFT** or **2nd**—must be pressed first.

To check Example 8(a), we press

1.8 **EE** 9 **×** 2.3 **EE** **(−)** 4.

When we then press **=** or **ENTER**, the result 4.14E5 appears. This represents 4.14×10^5. On some calculators,

the mode SCI must be used in order to display scientific notation.

Calculate each of the following.

1. $(3.8 \times 10^9) \cdot (4.5 \times 10^7)$
2. $(2.9 \times 10^{-8}) \div (5.4 \times 10^6)$
3. $(9.2 \times 10^7) \div (2.5 \times 10^{-9})$

FOR EXTRA HELP

Exercise Set **4.8**

Digital Video Tutor CD 3
Videotape 8   InterAct Math   Math Tutor Center   MathXL   MyMathLab.com

Express using positive exponents. Then, if possible, simplify.

1. 5^{-2}

2. 2^{-4}

3. 10^{-4}

4. 5^{-3}

5. $(-2)^{-6}$

6. $(-3)^{-4}$

7. x^{-8}

8. t^{-5}

9. xy^{-2}

10. $a^{-3}b$

11. $r^{-5}t$

12. xy^{-9}

13. $\dfrac{1}{t^{-7}}$

14. $\dfrac{1}{z^{-9}}$

15. $\dfrac{1}{h^{-8}}$

16. $\dfrac{1}{a^{-12}}$

17. 7^{-1}

18. 3^{-1}

19. $\left(\dfrac{2}{5}\right)^{-2}$

20. $\left(\dfrac{3}{4}\right)^{-2}$

21. $\left(\dfrac{a}{2}\right)^{-3}$

22. $\left(\dfrac{x}{3}\right)^{-4}$

23. $\left(\dfrac{s}{t}\right)^{-7}$

24. $\left(\dfrac{r}{v}\right)^{-5}$

Express using negative exponents.

25. $\dfrac{1}{7^2}$

26. $\dfrac{1}{5^2}$

27. $\dfrac{1}{t^6}$

28. $\dfrac{1}{y^2}$

29. $\dfrac{1}{a^4}$

30. $\dfrac{1}{t^5}$

31. $\dfrac{1}{p^8}$

32. $\dfrac{1}{m^{12}}$

33. $\dfrac{1}{5}$

34. $\dfrac{1}{8}$

35. $\dfrac{1}{t}$

36. $\dfrac{1}{m}$

Simplify. Do not use negative exponents in the answer.

37. $2^{-5} \cdot 2^8$

38. $5^{-8} \cdot 5^9$

39. $x^{-2} \cdot x^{-7}$

40. $x^{-2} \cdot x^{-9}$

41. $t^{-3} \cdot t$

42. $y^{-5} \cdot y$

43. $(a^{-2})^9$

44. $(x^{-5})^6$

45. $(t^{-3})^{-6}$

46. $(a^{-4})^{-7}$

47. $(t^4)^{-3}$

48. $(t^5)^{-2}$

49. $(x^{-2})^{-4}$ **50.** $(t^{-6})^{-5}$ **51.** $(ab)^{-3}$

52. $(xy)^{-6}$ **53.** $(mn)^{-7}$ **54.** $(ab)^{-9}$

55. $(3x^{-4})^2$ **56.** $(2a^{-5})^3$ **57.** $(5r^{-4}t^3)^2$

58. $(4x^5y^{-6})^3$ **59.** $\dfrac{t^7}{t^{-3}}$ **60.** $\dfrac{x^7}{x^{-2}}$

61. $\dfrac{y^{-7}}{y^{-3}}$ **62.** $\dfrac{z^{-6}}{z^{-2}}$ **63.** $\dfrac{y^{-4}}{y^{-9}}$

64. $\dfrac{a^{-6}}{a^{-10}}$ **65.** $\dfrac{x^6}{x}$ **66.** $\dfrac{x}{x^{-1}}$

67. $\dfrac{a^{-7}}{b^{-9}}$ **68.** $\dfrac{x^{-6}}{y^{-10}}$ Aha! **69.** $\dfrac{t^{-7}}{t^{-7}}$

70. $\dfrac{a^{-5}}{b^{-7}}$ **71.** $\dfrac{3x^{-5}}{y^{-6}z^{-2}}$ **72.** $\dfrac{4a^{-6}}{b^{-5}c^{-7}}$

73. $\dfrac{3t^4}{s^{-2}u^{-4}}$ **74.** $\dfrac{5x^{-8}}{y^{-3}z^2}$ **75.** $(x^4y^5)^{-3}$

76. $(t^5x^3)^{-4}$ **77.** $(x^{-6}y^{-2})^{-4}$ **78.** $(x^{-2}y^{-7})^{-5}$

79. $(a^{-5}b^7c^{-2})(a^{-3}b^{-2}c^6)$

80. $(x^3y^{-4}z^{-5})(x^{-4}y^{-2}z^9)$

81. $\left(\dfrac{a^4}{3}\right)^{-2}$ **82.** $\left(\dfrac{y^2}{2}\right)^{-2}$ **83.** $\left(\dfrac{7}{x^{-3}}\right)^2$

84. $\left(\dfrac{3}{a^{-2}}\right)^3$ **85.** $\left(\dfrac{m^{-1}}{n^{-4}}\right)^3$ **86.** $\left(\dfrac{x^2y}{z^{-5}}\right)^3$

87. $\left(\dfrac{2a^2}{3b^4}\right)^{-3}$ **88.** $\left(\dfrac{a^2b}{cd^3}\right)^{-5}$ Aha! **89.** $\left(\dfrac{5x^{-2}}{3y^{-2}z}\right)^0$

90. $\left(\dfrac{4a^3b^{-2}}{5c^{-3}}\right)^1$

Convert to decimal notation.

91. 7.12×10^4 **92.** 8.92×10^2

93. 8.92×10^{-3} **94.** 7.26×10^{-4}

95. 9.04×10^8 **96.** 1.35×10^7

97. 2.764×10^{-10} **98.** 9.043×10^{-3}

99. 4.209×10^7 **100.** 5.029×10^8

Convert to scientific notation.

101. 490,000   **102.** 71,500

103. 0.00583   **104.** 0.0814

105. 78,000,000,000   **106.** 3,700,000,000,000

107. 907,000,000,000,000,000

108. 168,000,000,000,000

109. 0.000000527

110. 0.00000000648

111. 0.000000018

112. 0.00000000002

113. 1,094,000,000,000,000

114. 1,030,200,000,000,000,000

Multiply or divide, and write scientific notation for the result.

115. $(4 \times 10^7)(2 \times 10^5)$

116. $(1.9 \times 10^8)(3.4 \times 10^{-3})$

117. $(3.8 \times 10^9)(6.5 \times 10^{-2})$

118. $(7.1 \times 10^{-7})(8.6 \times 10^{-5})$

119. $(8.7 \times 10^{-12})(4.5 \times 10^{-5})$

120. $(4.7 \times 10^5)(6.2 \times 10^{-12})$

121. $\dfrac{8.5 \times 10^8}{3.4 \times 10^{-5}}$

122. $\dfrac{5.6 \times 10^{-2}}{2.5 \times 10^5}$

123. $(3.0 \times 10^6) \div (6.0 \times 10^9)$

124. $(1.5 \times 10^{-3}) \div (1.6 \times 10^{-6})$

125. $\dfrac{7.5 \times 10^{-9}}{2.5 \times 10^{12}}$

126. $\dfrac{4.0 \times 10^{-3}}{8.0 \times 10^{20}}$

127. Without performing actual computations, explain why 3^{-29} is smaller than 2^{-29}.

128. What is it about scientific notation that makes it so useful?

SKILL MAINTENANCE

Simplify.

129. $(3 - 8)(9 - 12)$ **130.** $(2 - 9)^2$

131. $7 \cdot 2 + 8^2$ **132.** $5 \cdot 6 - 3 \cdot 2 \cdot 4$

133. Plot the points $(-3, 2)$, $(4, -1)$, $(5, 3)$, and $(-5, -2)$.

134. Solve $cx + bt = r$ for t.

SYNTHESIS

135. Explain what requirements must be met in order for x^{-n} to represent a negative integer.

136. Explain why scientific notation cannot be used without an understanding of the rules for exponents.

137. Simplify:
$$\frac{4.2 \times 10^8[(2.5 \times 10^{-5}) \div (5.0 \times 10^{-9})]}{3.0 \times 10^{-12}}.$$

138. Write the reciprocal of 1.25×10^{-6} in scientific notation.

139. Write the reciprocal of 2.5×10^9 in scientific notation.

140. Write $8^{-3} \cdot 32 \div 16^2$ as a power of 2.

141. Write $81^3 \cdot 27 \div 9^2$ as a power of 3.

Simplify.

142. $(7^{-12})^2 \cdot 7^{25}$

Aha! 143. $\dfrac{125^{-4}(25^2)^4}{125}$

144. $\dfrac{27^{-2}(81^2)^3}{9^8}$

145. Determine whether each of the following is true for all pairs of integers m and n and all positive numbers x and y.
    a) $x^m \cdot y^n = (xy)^{mn}$
    b) $x^m \cdot y^m = (xy)^{2m}$
    c) $(x - y)^m = x^m - y^m$

Simplify.

146. $\dfrac{7.4 \times 10^{29}}{(5.4 \times 10^{-6})(2.8 \times 10^8)}$

147. $\dfrac{5.8 \times 10^{17}}{(4.0 \times 10^{-13})(2.3 \times 10^4)}$

148. $\dfrac{(7.8 \times 10^7)(8.4 \times 10^{23})}{2.1 \times 10^{-12}}$

149. $\dfrac{(2.5 \times 10^{-8})(6.1 \times 10^{-11})}{1.28 \times 10^{-3}}$

Write scientific notation for each answer.

150. *Household income.* In 1997, there were about 102.5 million households in the United States. The average income of these households (before taxes) was about \$49,700 (*Source: Statistical Abstract of the United States,* 1999). Find the total income generated by these households.

151. *Computers.* A gigabyte is a measure of a computer's storage capacity. One gigabyte holds about one billion bytes of information. If a firm's computer network contains 2500 gigabytes of memory, how many bytes are in the network?

152. *River discharge.* The average discharge at the mouth of the Amazon River is 4,200,000 cubic feet per second. How much water is discharged from the Amazon River in 1 yr?

153. *Biology.* A strand of DNA (deoxyribonucleic acid) is about 1.5 m long and 1.3×10^{-10} cm wide (*Source:* Human Genome Project Information). How many times longer is DNA than it is wide?

154. *Water contamination.* In the United States, 200 million gal of used motor oil is improperly disposed of each year. One gallon of used oil can contaminate one million gallons of drinking water (*Source: The Macmillan Visual Almanac*). How many gallons of drinking water can 200 million gallons of oil contaminate?

Summary and Review 4

Key Terms

Polynomial, p. 212
Term, p. 212
Monomial, p. 212
Binomial, p. 213
Trinomial, p. 213
Degree of a term, p. 213
Coefficient, p. 213
Leading term, p. 214
Leading coefficient, p. 214
Degree of a polynomial, p. 214

Descending order, p. 215
Ascending order, p. 221
Opposite of a polynomial, p. 222
FOIL, p. 238
Difference of squares, p. 240
Perfect-square trinomial, p. 241
Polynomial in several variables, p. 246
Like terms, p. 247
Scientific notation, p. 260

Important Properties and Formulas

Definitions and Properties of Exponents

Assuming that no denominator is 0 and that 0^0 is not considered, for any integers m and n,

1 as an exponent:	$a^1 = a$
0 as an exponent:	$a^0 = 1$
Negative exponents:	$a^{-n} = \dfrac{1}{a^n},$
	$\dfrac{a^{-n}}{b^{-m}} = \dfrac{b^m}{a^n},$
	$\left(\dfrac{a}{b}\right)^{-n} = \left(\dfrac{b}{a}\right)^n$
The Product Rule:	$a^m \cdot a^n = a^{m+n}$
The Quotient Rule:	$\dfrac{a^m}{a^n} = a^{m-n}$
The Power Rule:	$(a^m)^n = a^{mn}$
Raising a product to a power:	$(ab)^n = a^n b^n$
Raising a quotient to a power:	$\left(\dfrac{a}{b}\right)^n = \dfrac{a^n}{b^n}$

Special Products of Polynomials

$(A + B)(A - B) = A^2 - B^2$
$(A + B)(A + B) = A^2 + 2AB + B^2$
$(A - B)(A - B) = A^2 - 2AB + B^2$

Scientific notation: $N \times 10^m$, where $1 \le N < 10$ and m is an integer

Review Exercises

Simplify.

1. $y^7 \cdot y^3 \cdot y$

2. $(3x)^5 \cdot (3x)^9$

3. $t^8 \cdot t^0$

4. $\dfrac{4^5}{4^2}$

5. $\dfrac{(a+b)^4}{(a+b)^4}$

6. $\left(\dfrac{3t^4}{2s^3}\right)^2$

7. $(-2xy^2)^3$

8. $(2x^3)(-3x)^2$

9. $(a^2b)(ab)^5$

Identify the terms of each polynomial.

10. $3x^2 + 6x + \frac{1}{2}$

11. $-4y^5 + 7y^2 - 3y - 2$

List the coefficients of the terms in each polynomial.

12. $7x^2 - x + 7$

13. $4x^3 + 6x^2 - 5x + \frac{5}{3}$

For each polynomial, (a) list the degree of each term; (b) determine the leading term and the leading coefficient; and (c) determine the degree of the polynomial.

14. $4t^2 + 6 + 15t^5$

15. $-2x^5 + x^4 - 3x^2 + x$

Classify each polynomial as a monomial, a binomial, a trinomial, or none of these.

16. $4x^3 - 1$

17. $4 - 9t^3 - 7t^4 + 10t^2$

18. $7y^2$

Combine like terms and write in descending order.

19. $5x - x^2 + 4x$

20. $\frac{3}{4}x^3 + 4x^2 - x^3 + 7$

21. $-2x^4 + 16 + 2x^4 + 9 - 3x^5$

22. $3x^2 - 2x + 3 - 5x^2 - 1 - x$

23. $-x + \frac{1}{2} + 14x^4 - 7x^2 - 1 - 4x^4$

Evaluate each polynomial for $x = -1$.

24. $7x - 10$

25. $x^2 - 3x + 6$

Add or subtract.

26. $(3x^4 - x^3 + x - 4) + (x^5 + 7x^3 - 3x - 5)$

27. $(3x^4 - 5x^3 + 3x^2) + (4x^5 + 4x^3) + (-5x^5 - 5x^2)$

28. $(5x^2 - 4x + 1) - (3x^2 + 7)$

29. $(3x^5 - 4x^4 + 2x^2 + 3) - (2x^5 - 4x^4 + 3x^3 + 4x^2 - 5)$

30. $\begin{array}{l} -\frac{3}{4}x^4 + \frac{1}{2}x^3 \qquad\qquad\quad + \frac{7}{8} \\ \quad\ - \frac{1}{4}x^3 - \ x^2 - \frac{7}{4}x \\ +\frac{3}{2}x^4 \qquad\quad + \frac{2}{3}x^2 \qquad\quad - \frac{1}{2} \end{array}$

31. $\begin{array}{l} \quad 2x^5 \quad\ - \ x^3 \qquad\ + x + 3 \\ -(3x^5 - x^4 + 4x^3 + 2x^2 - x + 3) \end{array}$

32. The length of a rectangle is 3 m greater than its width.

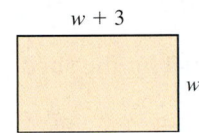

w + 3

w

a) Find a polynomial for the perimeter.
b) Find a polynomial for the area.

Multiply.

33. $3x(-4x^2)$

34. $(7x + 1)^2$

35. $(a - 7)(a + 4)$

36. $(m + 5)(m - 5)$

37. $(4x^2 - 5x + 1)(3x - 2)$

38. $(x - 9)^2$

39. $3t^2(5t^3 - 2t^2 + 4t)$

40. $(a - 7)(a + 7)$

41. $(x - 0.3)(x - 0.75)$

42. $(x^4 - 2x + 3)(x^3 + x - 1)$

43. $(3x - 5)^2$

44. $(2t^2 + 3)(t^2 - 7)$

45. $\left(a - \frac{1}{2}\right)\left(a + \frac{2}{3}\right)$

46. $(3x^2 + 4)(3x^2 - 4)$

47. $(2 - x)(2 + x)$

48. $(2x + 3y)(x - 5y)$

49. Evaluate $2 - 5xy + y^2 - 4xy^3 + x^6$ for $x = -1$ and $y = 2$.

Identify the coefficient and the degree of each term of each polynomial. Then find the degree of each polynomial.

50. $x^5y - 7xy + 9x^2 - 8$

51. $x^2y^5z^9 - y^{40} + x^{13}z^{10}$

Combine like terms.

52. $y + w - 2y + 8w - 5$

53. $6m^3 + 3m^2n + 4mn^2 + m^2n - 5mn^2$

Add or subtract.

54. $(5x^2 - 7xy + y^2) + (-6x^2 - 3xy - y^2)$

55. $(6x^3y^2 - 4x^2y - 6x) - (-5x^3y^2 + 4x^2y + 6x^2 - 6)$

Multiply.

56. $(p - q)(p^2 + pq + q^2)$ **57.** $\left(3a^4 - \frac{1}{3}b^3\right)^2$

58. Find a polynomial for the shaded area.

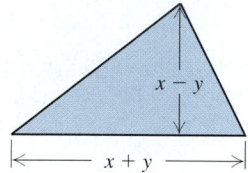

Divide.

59. $(10x^3 - x^2 + 6x) \div 2x$

60. $(6x^3 - 5x^2 - 13x + 13) \div (2x + 3)$

61. $\dfrac{t^4 + t^3 + 2t^2 - t - 3}{t + 1}$

62. Express using a positive exponent: m^{-7}.

63. Express using a negative exponent: $\dfrac{1}{t^8}$.

Simplify.

64. $7^2 \cdot 7^{-4}$ **65.** $\dfrac{a^{-5}b}{a^8b^8}$ **66.** $(x^3)^{-4}$

67. $(2x^{-3}y)^{-2}$ **68.** $\left(\dfrac{2x}{y}\right)^{-3}$

69. Convert to decimal notation: 8.3×10^6.

70. Convert to scientific notation: 0.0000328.

Multiply or divide and write scientific notation for the result.

71. $(3.8 \times 10^4)(5.5 \times 10^{-1})$

72. $\dfrac{1.28 \times 10^{-8}}{2.5 \times 10^{-4}}$

73. *Blood donors.*   Every 4–6 weeks, one of the authors of this text donates 1.14×10^6 cubic millimeters (two pints) of blood platelets to the American Red Cross. In one cubic millimeter of blood, there are about 2×10^5 platelets. Approximate the number of platelets in a typical donation by this author.

SYNTHESIS

74. Explain why $5x^3$ and $(5x)^3$ are not equivalent expressions.

75. If two polynomials of degree n are added, is the sum also of degree n? Why or why not?

76. How many terms are there in each of the following?
   a) $(x - a)(x - b) + (x - a)(x - b)$
   b) $(x + a)(x - b) + (x - a)(x + b)$

77. Combine like terms:

$$-3x^5 \cdot 3x^3 - x^6(2x)^2 + (3x^4)^2 + (2x^2)^4 - 40x^2(x^3)^2.$$

78. A polynomial has degree 4. The x^2-term is missing. The coefficient of x^4 is 2 times the coefficient of x^3. The coefficient of x is 3 less than the coefficient of x^4. The remaining coefficient is 7 less than the coefficient of x. The sum of the coefficients is 15. Find the polynomial.

Aha! **79.** Multiply: $[(x - 5) - 4x^3][(x - 5) + 4x^3]$.

80. Solve: $(x - 7)(x + 10) = (x - 4)(x - 6)$.

Chapter Test 4

Simplify.

1. $t^2 \cdot t^5 \cdot t$

2. $(x + 3)^5(x + 3)^6$

3. $\dfrac{3^5}{3^2}$

4. $\dfrac{(2x)^5}{(2x)^5}$

5. $(x^3)^2$

6. $(-3y^2)^3$

7. $(3x^2)(-2x^5)^3$

8. $(a^3b^2)(ab)^3$

9. Classify the polynomial as a monomial, a binomial, a trinomial, or none of these:

$$6t^2 - 9t.$$

10. Identify the coefficient of each term of the polynomial:

$$\tfrac{1}{3}x^5 - x + 7.$$

11. Determine the degree of each term, the leading term and the leading coefficient, and the degree of the polynomial:

$$2t^3 - t + 7t^5 + 4.$$

12. Evaluate the polynomial $x^2 + 5x - 1$ for $x = -2$.

Combine like terms and write in descending order.

13. $4a^2 - 6 + a^2$

14. $y^2 - 3y - y + \tfrac{3}{4}y^2$

15. $3 - x^2 + 2x^3 + 5x^2 - 6x - 2x + x^5$

Add or subtract.

16. $(3x^5 + 5x^3 - 5x^2 - 3) + (x^5 + x^4 - 3x^2 + 2x - 4)$

17. $\left(x^4 + \tfrac{2}{3}x + 5\right) + \left(4x^4 + 5x^2 + \tfrac{1}{3}x\right)$

18. $(2x^4 + x^3 - 8x^2 - 6x - 3) - (6x^4 - 8x^2 + 2x)$

19. $(x^3 - 0.4x^2 - 12) - (x^5 - 0.3x^3 + 0.4x^2 - 9)$

Multiply.

20. $-3x^2(4x^2 - 3x - 5)$

21. $\left(x - \tfrac{1}{3}\right)^2$

22. $(5t - 7)(5t + 7)$

23. $(3b + 5)(b - 3)$

24. $(x^6 - 4)(x^8 + 4)$

25. $(8 - y)(6 + 5y)$

26. $(2x + 1)(3x^2 - 5x - 3)$

27. $(8a + 3)^2$

28. Combine like terms:

$$x^3y - y^3 + xy^3 + 8 - 6x^3y - x^2y^2 + 11.$$

29. Subtract:

$$(8a^2b^2 - ab + b^3) - (-6ab^2 - 7ab - ab^3 + 5b^3).$$

30. Multiply: $(3x^5 - 4y^5)(3x^5 + 4y^5)$.

Divide.

31. $(12x^4 + 9x^3 - 15x^2) \div 3x^2$

32. $(6x^3 - 8x^2 - 14x + 13) \div (3x + 2)$

33. Express using a positive exponent: 5^{-3}.

34. Express using a negative exponent: $\dfrac{1}{y^8}$.

Simplify.

35. $t^{-4} \cdot t^{-2}$

36. $\dfrac{x^3y^2}{x^8y^{-3}}$

37. $(2a^3b^{-1})^{-4}$

38. $\left(\dfrac{ab}{c}\right)^{-3}$

39. Convert to scientific notation: 3,900,000,000.

40. Convert to decimal notation: 5×10^{-8}.

Multiply or divide and write scientific notation for the result.

41. $\dfrac{5.6 \times 10^6}{3.2 \times 10^{-11}}$

42. $(2.4 \times 10^5)(5.4 \times 10^{16})$

43. A CD-ROM can contain about 600 million pieces of information. How many sound files, each needing 40,000 pieces of information, can a CD-ROM hold? Write scientific notation for the answer.

SYNTHESIS

44. The height of a box is 1 less than its length, and the length is 2 more than its width. Express the volume in terms of the length.

45. Solve: $x^2 + (x - 7)(x + 4) = 2(x - 6)^2$.

5

Polynomials and Factoring

AN APPLICATION

An outdoor-education ropes course includes a cable that slopes downward from a height of 37 ft to a height of 30 ft. The trees that the cable connects are 24 ft apart. How long is the cable?

This problem appears as Exercise 13 in Section 5.8.

O *utdoor education is a part of many of today's physical education programs. Most students probably never think about it, but proper design of a ropes course requires an understanding of both math and physics.*

BETH LOGSDON
Physical Education Teacher
Olathe, Kansas

n Chapter 1, we learned that factoring is multiplying reversed. To factor a polynomial is to find an equivalent expression that is a product. In Sections 5.1–5.6, we factor to find equivalent expressions. In Sections 5.7 and 5.8, we use factoring to solve equations, many of which arise from real-world problems. Factoring polynomials requires a solid command of the multiplication methods studied in Chapter 4.

Introduction to Factoring

5.1

Factoring Monomials • Factoring When Terms Have a Common Factor • Factoring by Grouping • Checking by Evaluating

Just as a number like 15 can be factored as $3 \cdot 5$, a polynomial like $x^2 + 7x$ can be factored as $x(x + 7)$. In both cases, we ask ourselves, "What was multiplied to obtain the given result?" The situation is much like a popular television game show in which an "answer" is given and participants must find a "question" to which the answer corresponds.

Factoring

To *factor* a polynomial is to find an equivalent expression that is a product.

Factoring Monomials

To factor a monomial, we find two monomials whose product is equivalent to the original monomial. For example, $20x^2$ can be factored as $2 \cdot 10x^2$, $4x \cdot 5x$, or $10x \cdot 2x$, as well as several other ways.

Example 1

Find three factorizations of $15x^3$.

Solution

a) $15x^3 = (3 \cdot 5)(x \cdot x^2)$ Thinking of how 15 and x^3 factor

$ = (3x)(5x^2)$ The factors here are $3x$ and $5x^2$.

b) $15x^3 = (3 \cdot 5)(x^2 \cdot x)$

$ = (3x^2)(5x)$ The factors here are $3x^2$ and $5x$.

c) $15x^3 = ((-5)(-3))x^3$

$ = (-5)(-3x^3)$ The factors here are -5 and $-3x^3$.

Recall from Section 1.2 that the word "factor" can be a verb or a noun, depending on the context in which it appears.

Factoring When Terms Have a Common Factor

To multiply a polynomial of two or more terms by a monomial, we multiply each term by the monomial, using the distributive law $a(b + c) = ab + ac$. To factor, we do the reverse. We rewrite a polynomial as a product, using the distributive law with the sides switched: $ab + ac = a(b + c)$. Consider the following:

Multiply   *Factor*

$3x(x^2 + 2x - 4)$ $3x^3 + 6x^2 - 12x$

$\quad = 3x \cdot x^2 + 3x \cdot 2x - 3x \cdot 4$ $\quad = 3x \cdot x^2 + 3x \cdot 2x - 3x \cdot 4$

$\quad = 3x^3 + 6x^2 - 12x;$ $\quad = 3x(x^2 + 2x - 4).$

In the factorization on the right, note that since $3x$ appears as a factor of $3x^3$, $6x^2$, and $-12x$, it is a *common factor* for all the terms of the trinomial $3x^3 + 6x^2 - 12x$.

To factor a polynomial with two or more terms, always try to first find a factor common to all terms. In some cases, there may not be a common factor (other than 1). If a common factor *does* exist, we generally use the common factor with the largest possible coefficient and the largest possible exponent. Such a factor is called the *largest*, or *greatest*, *common factor*.

E x a m p l e   2   Factor: $5x^2 + 15$.

Solution   We have

$$5x^2 + 15 = 5 \cdot x^2 + 5 \cdot 3 \qquad \text{Factoring each term}$$

$$\qquad\qquad = 5(x^2 + 3). \qquad \text{Factoring out the common factor, 5}$$

To check, we multiply: $5(x^2 + 3) = 5 \cdot x^2 + 5 \cdot 3 = 5x^2 + 15$. Since $5x^2 + 15$ is the original polynomial, the factorization $5(x^2 + 3)$ checks.

Caution! $5 \cdot x^2 + 5 \cdot 3$ is a factorization of the *terms* of $5x^2 + 15$, but not of the polynomial itself. The factorization of $5x^2 + 15$ is $5(x^2 + 3)$.

When asked to factor a polynomial in which all terms contain the same variable raised to various powers, we factor out the largest power possible.

E x a m p l e   3 Factor: $24x^3 + 30x^2$.

Solution   The largest factor common to 24 and 30 is 6. The largest power of x common to x^3 and x^2 is x^2. (To see this, think of x^3 as $x^2 \cdot x$.) Thus the largest common factor of $24x^3$ and $30x^2$ is $6x^2$. We factor as follows:

$$24x^3 + 30x^2 = 6x^2 \cdot 4x + 6x^2 \cdot 5 \qquad \text{Factoring each term}$$
$$= 6x^2(4x + 5). \qquad \text{Factoring out } 6x^2$$

Check: $6x^2(4x + 5) = 6x^2 \cdot 4x + 6x^2 \cdot 5 = 24x^3 + 30x^2$, as expected. The factorization $6x^2(4x + 5)$ checks.

Suppose in Example 3 that you did not recognize the *largest* common factor, and removed only part of it, as follows:

$$24x^3 + 30x^2 = 2x^2 \cdot 12x + 2x^2 \cdot 15 \qquad 2x^2 \text{ is a common factor.}$$
$$= 2x^2(12x + 15). \qquad 12x + 15 \text{ itself has a common factor.}$$

Note that $12x + 15$ still has a common factor, 3. To find the largest common factor, continue factoring out common factors, as follows, until no more exist:

$$= 2x^2[3(4x + 5)] \qquad \text{Factoring } 12x + 15. \text{ Remember to rewrite the first common factor, } 2x^2.$$
$$= 6x^2(4x + 5). \qquad \text{Using an associative law}$$

Since $4x + 5$ cannot be factored any further, we say that we have factored *completely.*

E x a m p l e   4 Factor: $15x^5 - 12x^4 + 27x^3 - 3x^2$.

Solution   We have

$$15x^5 - 12x^4 + 27x^3 - 3x^2$$
$$= 3x^2 \cdot 5x^3 - 3x^2 \cdot 4x^2 + 3x^2 \cdot 9x - 3x^2 \cdot 1 \qquad \text{Try to do this mentally.}$$
$$= 3x^2(5x^3 - 4x^2 + 9x - 1). \qquad \text{Factoring out } 3x^2$$

> **Caution!**   Don't forget the term -1. The check below shows why it is essential.

Since $5x^3 - 4x^2 + 9x - 1$ has no common factor, we are finished, except for a check:

$$3x^2(5x^3 - 4x^2 + 9x - 1) = 15x^5 - 12x^4 + 27x^3 - 3x^2. \qquad \text{Our factorization checks.}$$

The factorization is $3x^2(5x^3 - 4x^2 + 9x - 1)$.

If you spot the largest common factor without writing out a factorization of each term, you can write the answer in one step.

E x a m p l e   5

Factor: **(a)** $8m^3 - 16m$; **(b)** $14p^2y^3 - 8py^2 + 2py$.

Solution

a) $8m^3 - 16m = 8m(m^2 - 2)$
b) $14p^2y^3 - 8py^2 + 2py = 2py(7py^2 - 4y + 1)$ Determine the largest common factor by inspection; then carefully fill in the parentheses.

The checks are left to the student.

Tips for Factoring

1. Factor out the largest common factor, if one exists.
2. Factoring can always be checked by multiplying. Multiplication should yield the original polynomial.

Factoring by Grouping

Sometimes algebraic expressions contain a common factor with two or more terms.

E x a m p l e   6

Factor: $x^2(x + 1) + 2(x + 1)$.

Solution   The binomial $x + 1$ is a factor of both $x^2(x + 1)$ and $2(x + 1)$. Thus, $x + 1$ is a common factor:

$$x^2(x + 1) + 2(x + 1) = (x + 1)x^2 + (x + 1)2$$ Using a commutative law twice

$$= (x + 1)(x^2 + 2).$$ Factoring out the common factor, $x + 1$

To check, we could simply reverse the above steps.
The factorization is $(x + 1)(x^2 + 2)$.

In Example 6, the common binomial factor was clearly visible. How do we find such a factor in a polynomial like $5x^3 - x^2 + 15x - 3$? Although there is no factor, other than 1, common to all four terms, $5x^3 - x^2$ and $15x - 3$ can be grouped and factored separately:

$$5x^3 - x^2 = x^2(5x - 1)   \text{and}   15x - 3 = 3(5x - 1).$$

Note that $5x^3 - x^2$ and $15x - 3$ share a common factor of $5x - 1$. This means that the original polynomial, $5x^3 - x^2 + 15x - 3$, can be factored:

$$5x^3 - x^2 + 15x - 3 = (5x^3 - x^2) + (15x - 3) \qquad \text{Using an associative law. This is generally done mentally.}$$

$$= x^2(5x - 1) + 3(5x - 1) \qquad \text{Factoring each binomial}$$

$$= (5x - 1)(x^2 + 3). \qquad \text{Factoring out the common factor, } 5x - 1$$

If a polynomial can be split into groups of terms and the groups share a common factor, then the original polynomial can be factored. This method, known as **factoring by grouping**, can be tried on any polynomial with four or more terms.

Example 7   Factor by grouping.

a) $2x^3 + 8x^2 + x + 4$
b) $8x^4 + 6x - 28x^3 - 21$

Solution

a) $2x^3 + 8x^2 + x + 4 = 2x^2(x + 4) + 1(x + 4) \qquad$ Factoring $2x^3 + 8x^2$ to find a common binomial factor. Writing the 1 helps with the next step.

$$= (x + 4)(2x^2 + 1) \qquad \text{Factoring out the common factor, } x + 4. \text{ The 1 is essential in the factor } 2x^2 + 1.$$

Check: $(x + 4)(2x^2 + 1) = x \cdot 2x^2 + x \cdot 1 + 4 \cdot 2x^2 + 4 \cdot 1 \qquad$ Using FOIL

$$= 2x^3 + x + 8x^2 + 4$$

$$= 2x^3 + 8x^2 + x + 4 \qquad \text{Using a commutative law}$$

The factorization is $(x + 4)(2x^2 + 1)$.

b) We can factor $8x^4 + 6x$ as $2x(4x^3 + 3)$ and $-28x^3 - 21$ as $7(-4x^3 - 3)$ or $-7(4x^3 + 3)$. We use $-7(4x^3 + 3)$ because it shares a common factor, $4x^3 + 3$, with $2x(4x^3 + 3)$:

$$8x^4 + 6x - 28x^3 - 21 = 2x(4x^3 + 3) - 7(4x^3 + 3) \qquad \text{Factoring two binomials. Using } -7 \text{ gives a common binomial factor.}$$

$$= (4x^3 + 3)(2x - 7) \qquad \text{Factoring out the common factor, } 4x^3 + 3$$

Check: $(4x^3 + 3)(2x - 7) = 8x^4 - 28x^3 + 6x - 21$

$\qquad\qquad\qquad\qquad\quad = 8x^4 + 6x - 28x^3 - 21$ This is the original polynomial.

The factorization is $(4x^3 + 3)(2x - 7)$.

Although factoring by grouping can be useful, some polynomials, like $x^3 + x^2 + 2x - 2$, cannot be factored this way. Factoring polynomials of this type is beyond the scope of this text.

Checking by Evaluating

We have seen that one way to check a factorization is to multiply. A second type of check, discussed toward the end of Section 4.4, uses the fact that equivalent expressions have the same value when evaluated for the same replacement. Thus a quick, partial check of Example 7(a) can be made by using a convenient replacement for x (say, 1) and evaluating both $2x^3 + 8x^2 + x + 4$ and $(x + 4)(2x^2 + 1)$:

$$2 \cdot 1^3 + 8 \cdot 1^2 + 1 + 4 = 2 + 8 + 1 + 4$$
$$= 15;$$
$$(1 + 4)(2 \cdot 1^2 + 1) = 5 \cdot 3$$
$$= 15.$$

Since the value of both expressions is the same, the factorization is probably correct.

Keep in mind that it is possible, by chance, for two expressions that are not equivalent to share the same value when evaluated. Because of this, unless several values are used (at least one more than the degree of the polynomial, it turns out), evaluating offers only a partial check. Consult with your instructor before making extensive use of this type of check.

technology connection

We saw in the Technology Connection on p. 234 that a Table of values can be used to check that two expressions are equal. Thus to check Example 7(a), we let $y_1 = 2x^3 + 8x^2 + x + 4$ and $y_2 = (x + 4)(2x^2 + 1)$:

ΔTBL = 1

X	Y₁	Y₂
0	4	4
1	15	15
2	54	54
3	133	133
4	264	264
5	459	459
6	730	730

X = 0

No matter how far up or down we scroll, $y_1 = y_2$. Thus Example 7(a) is correct.

1. Use a Table to check Example 7(b).

FOR EXTRA HELP

Exercise Set **5.1**

 Digital Video Tutor CD 3 Videotape 9   InterAct Math   Math Tutor Center   MathXL   MyMathLab.com

Find three factorizations for each monomial. Answers may vary.

1. $10x^3$

2. $6x^3$

3. $-15a^4$

4. $-8t^5$

5. $26x^5$

6. $25x^4$

Factor. Remember to use the largest common factor and to check by multiplying.

7. $x^2 + 8x$

8. $x^2 + 6x$

9. $10t^2 - 5t$

10. $5a^2 - 15a$

11. $x^3 + 6x^2$

12. $4x^4 + x^2$

13. $8x^4 - 24x^2$

14. $5x^5 + 10x^3$

15. $2x^2 + 2x - 8$

16. $6x^2 + 3x - 15$

17. $7a^6 - 10a^4 - 14a^2$

18. $10t^5 - 15t^4 + 9t^3$

19. $2x^8 + 4x^6 - 8x^4 + 10x^2$

20. $5x^4 - 15x^3 - 25x - 10$

21. $x^5y^5 + x^4y^3 + x^3y^3 - x^2y^2$

22. $x^9y^6 - x^7y^5 + x^4y^4 + x^3y^3$

23. $5a^3b^4 + 10a^2b^3 - 15a^3b^2$

24. $21r^5t^4 - 14r^4t^6 + 21r^3t^6$

Factor.

25. $y(y - 2) + 7(y - 2)$

26. $b(b + 5) + 3(b + 5)$

27. $x^2(x + 3) - 7(x + 3)$

28. $3z^2(2z + 9) + (2z + 9)$

29. $y^2(y + 8) + (y + 8)$

30. $x^2(x - 7) - 3(x - 7)$

Factor by grouping, if possible, and check.

31. $x^3 + 3x^2 + 4x + 12$

32. $6z^3 + 3z^2 + 2z + 1$

33. $3a^3 + 9a^2 + 2a + 6$

34. $3a^3 + 2a^2 + 6a + 4$

35. $9x^3 - 12x^2 + 3x - 4$

36. $10x^3 - 25x^2 + 4x - 10$

37. $4t^3 - 20t^2 + 3t - 15$

38. $6a^3 - 8a^2 + 9a - 12$

39. $7x^3 + 2x^2 - 14x - 4$

40. $5x^3 + 4x^2 - 10x - 8$

41. $6a^3 - 7a^2 + 6a - 7$

42. $7t^3 - 5t^2 + 7t - 5$

43. $x^3 + 8x^2 - 3x - 24$

44. $x^3 + 7x^2 - 2x - 14$

45. $2x^3 + 12x^2 - 5x - 30$

46. $3x^3 + 15x^2 - 5x - 25$

47. $w^3 - 7w^2 + 4w - 28$

48. $p^3 + p^2 - 3p + 10$

49. $x^3 - x^2 - 2x + 5$

50. $y^3 + 8y^2 - 2y - 16$

51. $2x^3 - 8x^2 - 9x + 36$

52. $20g^3 - 4g^2 - 25g + 5$

53. In answering a factoring problem, Taylor says the largest common factor is $-5x^2$ and Natasha says the largest common factor is $5x^2$. Can they both be correct? Why or why not?

54. Write a two-sentence paragraph in which the word "factor" is used at least once as a noun and once as a verb.

SKILL MAINTENANCE

Simplify.

55. $(x + 3)(x + 5)$

56. $(x + 2)(x + 7)$

57. $(a - 7)(a + 3)$

58. $(a + 5)(a - 8)$

59. $(2x + 5)(3x - 4)$

60. $(3t + 2)(4t - 7)$

61. $(3t - 5)^2$

62. $(2t - 9)^2$

SYNTHESIS

63. Marlene recognizes that evaluating provides only a partial check of her factoring. Because of this, she often performs a second check with a different replacement value. Is this a good idea? Why or why not?

64. Josh says that for Exercises 1–52 there is no need to print answers at the back of the book. Is he correct in saying this? Why or why not?

Factor, if possible.

65. $4x^5 + 6x^3 + 6x^2 + 9$

66. $x^6 + x^4 + x^2 + 1$

67. $x^{12} + x^7 + x^5 + 1$

68. $x^3 + x^2 - 2x + 2$

Aha! **69.** $5x^5 - 5x^4 + x^3 - x^2 + 3x - 3$

Aha! **70.** $ax^2 + 2ax + 3a + x^2 + 2x + 3$

71. Write a polynomial of degree 7 for which $3x^2y^3$ is the largest common factor. Answers may vary.

5.2

Factoring Trinomials of the Type $x^2 + bx + c$

Constant Term Positive • Constant Term Negative

We now learn how to factor trinomials like

$$x^2 + 5x + 4 \quad \text{or} \quad x^2 + 3x - 10,$$

for which no common factor exists and the leading coefficient is 1. As preparation for the factoring that follows, compare the following multiplications:

$$
\begin{array}{cccc}
\text{F} & \text{O} & \text{I} & \text{L} \\
\downarrow & \downarrow & \downarrow & \downarrow
\end{array}
$$

$$(x + 2)(x + 5) = x^2 + 5x + 2x + 2 \cdot 5$$
$$= x^2 + 7x + 10;$$

$$(x - 2)(x - 5) = x^2 - 5x - 2x + (-2)(-5)$$
$$= x^2 - 7x + 10;$$

$$(x + 3)(x - 7) = x^2 - 7x + 3x + 3(-7)$$
$$= x^2 - 4x - 21;$$

$$(x - 3)(x + 7) = x^2 + 7x - 3x + (-3)7$$
$$= x^2 + 4x - 21.$$

Note that for all four products:

- The product of the two binomials is a trinomial.
- The coefficient of x in the trinomial is the sum of the constant terms in the binomials.
- The constant term in the trinomial is the product of the constant terms in the binomials.

These observations lead to a method for factoring certain trinomials. The first type we consider has a positive constant term, just as in the first two multiplications above.

Constant Term Positive

To factor a polynomial like $x^2 + 7x + 10$, we think of FOIL in reverse. The x^2 resulted from x times x, which suggests that the first term of each binomial factor is x. Next, we look for numbers p and q such that

$$x^2 + 7x + 10 = (x + p)(x + q).$$

To get the middle term and the last term of the trinomial, we need two numbers p and q whose product is 10 and whose sum is 7. Those numbers are 2 and 5. Thus the factorization is

$$(x + 2)(x + 5). \qquad \textit{Check: } (x + 2)(x + 5) = x^2 + 5x + 2x + 10$$
$$= x^2 + 7x + 10$$

E x a m p l e   1

Factor: $x^2 + 5x + 6$.

Solution   Think of FOIL in reverse. The first term of each factor is x:

$$(x + \quad)(x + \quad).$$

To complete the factorization, we need a constant term for each of these binomial factors. The constants must have a product of 6 and a sum of 5. We list some pairs of numbers that multiply to 6.

Pairs of Factors of 6	Sums of Factors
1, 6	7
2, 3	5 ←
−1, −6	−7
−2, −3	−5

The numbers we seek are 2 and 3.

Since

$$2 \cdot 3 = 6 \quad \text{and} \quad 2 + 3 = 5,$$

the factorization of $x^2 + 5x + 6$ is $(x + 2)(x + 3)$. To check, we simply multiply the two binomials.

Check: $(x + 2)(x + 3) = x^2 + 3x + 2x + 6$
$$= x^2 + 5x + 6. \qquad \text{The product is the original polynomial.}$$

Note that since 5 and 6 are both positive, when factoring $x^2 + 5x + 6$ we need not consider negative factors of 6. Note too that changing the signs of the factors changes only the sign of the sum.

At the beginning of this section, we considered the multiplication $(x - 2)(x - 5)$. For this product, the resulting trinomial, $x^2 - 7x + 10$, has a positive constant term but a negative coefficient of x. This is because the *product* of two negative numbers is always positive, whereas the *sum* of two negative numbers is always negative.

> ### To Factor $x^2 + bx + c$ When c Is Positive
>
> When the constant term of a trinomial is positive, look for two numbers with the same sign. The sign is that of the middle term:
>
> $$x^2 - 7x + 10 = (x - 2)(x - 5);$$
>
> $$x^2 + 7x + 10 = (x + 2)(x + 5).$$

Example 2

Factor: $y^2 - 8y + 12$.

Solution   Since the constant term is positive and the coefficient of the middle term is negative, we look for a factorization of 12 in which both factors are negative. Their sum must be -8.

Pairs of Factors of 12	Sums of Factors
$-1, -12$	-13
$-2, \ -6$	-8 ←
$-3, \ -4$	-7

We need a sum of -8.
The numbers we need are -2 and -6.

The factorization of $y^2 - 8y + 12$ is $(y - 2)(y - 6)$. The check is left to the student.

Constant Term Negative

As we saw in two of the multiplications earlier in this section, the product of two binomials can have a negative constant term:

$$(x + 3)(x - 7) = x^2 - 4x - 21$$

and

$$(x - 3)(x + 7) = x^2 + 4x - 21.$$

Note that when the signs of the constants in the binomials are reversed, only the sign of the middle term in the product changes.

Example 3

Factor: $x^2 - 8x - 20$.

Solution   The constant term, -20, must be expressed as the product of a negative number and a positive number. Since the sum of these two numbers must be negative (specifically, -8), the negative number must have the greater absolute value.

Pairs of Factors of -20	Sums of Factors
$1, -20$	-19
$2, -10$	-8 ←
$4, \ -5$	-1
$5, \ -4$	1
$10, \ -2$	8
$20, \ -1$	19

The numbers we need are 2 and -10.

Because these sums are all positive, for this problem all of the corresponding pairs can be disregarded. Note that in all three pairs, the positive number has the greater absolute value.

The numbers that we are looking for are 2 and -10.

Check: $(x + 2)(x - 10) = x^2 - 10x + 2x - 20$
$$= x^2 - 8x - 20.$$

The factorization is $(x + 2)(x - 10)$.

> ### To Factor $x^2 + bx + c$ When c Is Negative
>
> When the constant term of a trinomial is negative, look for two numbers whose product is negative. One must be positive and the other negative:
>
> $$x^2 - 4x - 21 = (x + 3)(x - 7);$$
>
> $$x^2 + 4x - 21 = (x - 3)(x + 7).$$
>
> Select the two numbers so that the number with the larger absolute value has the same sign as b, the coefficient of the middle term.

E x a m p l e   4

Factor: $t^2 - 24 + 5t$.

Solution   It helps to first write the trinomial in descending order: $t^2 + 5t - 24$. The factorization of the constant term, -24, must have one factor positive and one factor negative. The sum must be 5, so the positive factor must have the larger absolute value. Thus we consider only pairs of factors in which the positive factor has the larger absolute value.

Pairs of Factors of -24	Sums of Factors
$-1, 24$	23
$-2, 12$	10
$-3, \ 8$	5 ← ——— The numbers we need
$-4, \ 6$	2 are -3 and 8.

The factorization is $(t - 3)(t + 8)$. The check is left to the student.

Polynomials in two or more variables, such as $a^2 + 4ab - 21b^2$, are factored in a similar manner.

E x a m p l e   5

Factor: $a^2 + 4ab - 21b^2$.

Solution   It may help to write the trinomial in the equivalent form
$$a^2 + 4ba - 21b^2.$$

This way we think of $-21b^2$ as the "constant" term and $4b$ as the "coefficient" of the middle term. Then we try to express $-21b^2$ as a product of two factors whose sum is $4b$. Those factors are $-3b$ and $7b$.

Check: $(a - 3b)(a + 7b) = a^2 + 7ab - 3ba - 21b^2$
$$= a^2 + 4ab - 21b^2.$$

The factorization is $(a - 3b)(a + 7b)$.

E x a m p l e   6

Factor: $x^2 - x + 5$.

Solution   Since 5 has very few factors, we can easily check all possibilities.

Pairs of Factors of 5	Sums of Factors
5, 1	6
−5, −1	−6

Since there are no factors whose sum is -1, the polynomial is *not* factorable into binomials.

In this text, a polynomial like $x^2 - x + 5$ that cannot be factored further is said to be **prime**. In more advanced courses, polynomials like $x^2 - x + 5$ can be factored and are not considered prime.

Often factoring requires two or more steps. In general, when told to factor, we should *factor completely*. This means that the final factorization should not contain any factors that can be factored further.

E x a m p l e   7

Factor: $2x^3 - 20x^2 + 50x$.

Solution   *Always* look first for a common factor. This time there is one, $2x$, which we factor out first:

$$2x^3 - 20x^2 + 50x = 2x(x^2 - 10x + 25).$$

Now consider $x^2 - 10x + 25$. Since the constant term is positive and the coefficient of the middle term is negative, we look for a factorization of 25 in which both factors are negative. Their sum must be -10.

Pairs of Factors of 25	Sums of Factors	
−25, −1	−26	
−5, −5	−10 ⟵	The numbers we need are −5 and −5.

The factorization of $x^2 - 10x + 25$ is $(x - 5)(x - 5)$, or $(x - 5)^2$.

> **Caution!** When factoring involves more than one step, be careful to write out the *entire* factorization.

Check: $2x(x-5)(x-5) = 2x[x^2 - 10x + 25]$ Multiplying binomials

$\qquad\qquad\qquad\qquad\quad = 2x^3 - 20x^2 + 50x.$ Using the distributive law

The factorization of $2x^3 - 20x^2 + 50x$ is $2x(x-5)(x-5)$, or $2x(x-5)^2$.

Once any common factors have been factored out, the following summary can be used to factor $x^2 + bx + c$.

> ### To Factor $x^2 + bx + c$
>
> **1.** Find a pair of factors that have c as their product and b as their sum.
>
>   **a)** If c is positive, its factors will have the same sign as b.
>
>   **b)** If c is negative, one factor will be positive and the other will be negative. Select the factors such that the factor with the larger absolute value is the factor with the same sign as b.
>
> **2.** Check by multiplying.

FOR EXTRA HELP

Exercise Set **5.2**

 Digital Video Tutor CD 4 Videotape 9   InterAct Math   Math Tutor Center   MathXL   MyMathLab.com

Factor completely. Remember that you can check by multiplying. If a polynomial is prime, state this.

1. $x^2 + 6x + 5$

2. $x^2 + 7x + 6$

3. $x^2 + 7x + 10$

4. $x^2 + 7x + 12$

5. $y^2 + 11y + 28$

6. $x^2 - 6x + 9$

7. $a^2 + 11a + 30$

8. $x^2 + 9x + 14$

9. $x^2 - 5x + 4$

10. $b^2 + 5b + 4$

11. $z^2 - 8z + 7$

12. $a^2 - 4a - 12$

13. $x^2 - 8x + 15$

14. $d^2 - 7d + 10$

15. $y^2 - 11y + 10$

16. $x^2 - 2x - 15$

17. $x^2 + x - 42$

18. $x^2 + 2x - 15$

19. $2x^2 - 14x - 36$

20. $3y^2 - 9y - 84$

21. $x^3 - 6x^2 - 16x$

22. $x^3 - x^2 - 42x$

23. $y^2 + 4y - 45$

24. $x^2 + 7x - 60$

25. $-2x - 99 + x^2$

26. $x^2 - 72 + 6x$

27. $c^4 + c^3 - 56c^2$

28. $5b^2 + 25b - 120$

29. $2a^2 - 4a - 70$

30. $x^5 - x^4 - 2x^3$

31. $x^2 + x + 1$

32. $x^2 + 2x + 3$

33. $7 - 2p + p^2$

34. $11 - 3w + w^2$

35. $x^2 + 20x + 100$

36. $x^2 + 20x + 99$

37. $3x^3 - 63x^2 - 300x$

38. $2x^3 - 40x^2 + 192x$

39. $x^2 - 21x - 72$

40. $4x^2 + 40x + 100$

41. $x^2 - 25x + 144$

42. $y^2 - 21y + 108$

43. $a^4 + a^3 - 132a^2$

44. $a^6 + 9a^5 - 90a^4$

45. $x^2 - \frac{2}{5}x + \frac{1}{25}$

46. $t^2 + \frac{2}{3}t + \frac{1}{9}$

47. $27 + 12y + y^2$

48. $50 + 15x + x^2$

49. $t^2 - 0.3t - 0.10$

50. $y^2 - 0.2y - 0.08$

51. $p^2 + 3pq - 10q^2$

52. $a^2 - 2ab - 3b^2$

53. $m^2 + 5mn + 5n^2$

54. $x^2 - 11xy + 24y^2$

55. $s^2 - 2st - 15t^2$

56. $b^2 + 8bc - 20c^2$

57. $6a^{10} - 30a^9 - 84a^8$

58. $7x^9 - 28x^8 - 35x^7$

59. Marge factors $x^3 - 8x^2 + 15x$ as $(x^2 - 5x)(x - 3)$. Is she wrong? Why or why not? What advice would you offer?

60. Without multiplying $(x - 17)(x - 18)$, explain why it cannot possibly be a factorization of $x^2 + 35x + 306$.

SKILL MAINTENANCE

Solve.

61. $3x - 8 = 0$

62. $2x + 7 = 0$

Multiply.

63. $(x + 6)(3x + 4)$

64. $(7w + 6)^2$

65. In a recent year, 29,090 people were arrested for counterfeiting. This figure was down 1.2% from the year before. How many people were arrested the year before?

66. The first angle of a triangle is four times as large as the second. The measure of the third angle is 30° greater than that of the second. How large are the angles?

SYNTHESIS

67. When searching for a factorization, why do we list pairs of numbers with the correct *product* instead of pairs of numbers with the correct *sum*?

68. What is the advantage of writing out the prime factorization of c when factoring $x^2 + bx + c$ with a large value of c?

69. Find all integers b for which $a^2 + ba - 50$ can be factored.

70. Find all integers m for which $y^2 + my + 50$ can be factored.

Factor each of the following by first factoring out -1.

71. $30 + 7x - x^2$

72. $45 + 4x - x^2$

73. $24 - 10a - a^2$

74. $36 - 9a - a^2$

75. $84 - 8t - t^2$

76. $72 - 6t - t^2$

Factor completely.

77. $x^2 + \frac{1}{4}x - \frac{1}{8}$

78. $x^2 + \frac{1}{2}x - \frac{3}{16}$

79. $\frac{1}{3}a^3 - \frac{1}{3}a^2 - 2a$

80. $a^7 - \frac{25}{7}a^5 - \frac{30}{7}a^6$

81. $x^{2m} + 11x^m + 28$

82. $t^{2n} - 7t^n + 10$

Aha! **83.** $(a + 1)x^2 + (a + 1)3x + (a + 1)2$

84. $ax^2 - 5x^2 + 8ax - 40x - (a - 5)9$
(*Hint*: See Exercise 83.)

Find a polynomial in factored form for the shaded area in each figure. (Leave answers in terms of π.)

85.

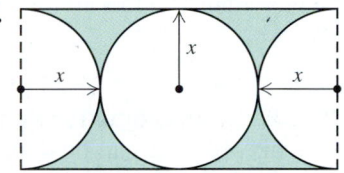

86.

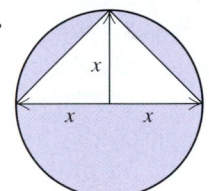

87. Find the volume of a cube if its surface area is $6x^2 + 36x + 54$ square meters.

88. A census taker asks a woman, "How many children do you have?"

"Three," she answers.

"What are their ages?"

She responds, "The product of their ages is 36. The sum of their ages is the house number next door."

The math-savvy census taker walks next door, reads the house number, appears puzzled, and returns to the woman, asking, "Is there something you forgot to tell me?"

"Oh yes," says the woman. "I'm sorry. The oldest child is at the park."

The census taker records the three ages, thanks the woman for her time, and leaves.

How old is each child? Explain how you reached this conclusion. (*Hint*: Consider factorizations.) (*Source*: Adapted from Harnadek, Anita, *Classroom Quickies*. Pacific Grove, CA: Critical Thinking Press and Software)

CORNER

Visualizing Factoring

Focus: Visualizing factoring

Time: 20–30 minutes

Group size: 3

Materials: Graph paper and scissors

The product $(x + 2)(x + 3)$ can be regarded as the area of a rectangle with width $x + 2$ and length $x + 3$. Similarly, factoring a polynomial like $x^2 + 5x + 6$ can be thought of as determining the length and the width of a rectangle that has area $x^2 + 5x + 6$. This is the approach used below.

ACTIVITY

1. **a)** To factor $x^2 + 11x + 10$ geometrically, the group needs to cut out shapes like those below to represent x^2, $11x$, and 10. This can be done by either tracing the figures below or by selecting a value for x, say 4, and using the squares on the graph paper to cut out the following:

   x^2: Using the value selected for x, cut out a square that is x units on each side.

   $11x$: Using the value selected for x, cut out a rectangle that is 1 unit wide and x units long. Repeat this to form 11 such strips.

   10: Cut out two rectangles with whole-number dimensions and an area of 10. One should be 2 units by 5 units and the other 1 unit by 10 units.

   b) The group, working together, should then attempt to use one of the two rectangles with area 10, along with all of the other shapes, to piece together one large rectangle. Only one of the rectangles with area 10 will work.

   c) From the large rectangle formed in part (b), use the length and the width to determine the factorization of $x^2 + 11x + 10$. Where do the dimensions of the rectangle representing 10 appear in the factorization?

2. Repeat step (1) above, but this time use the other rectangle with area 10, and use only 7 of the 11 strips, along with the x^2-shape. Piece together the shapes to form one large rectangle. What factorization do the dimensions of this rectangle suggest?

3. Cut out rectangles with area 12 and use the above approach to factor $x^2 + 8x + 12$. What dimensions should be used for the rectangle with area 12?

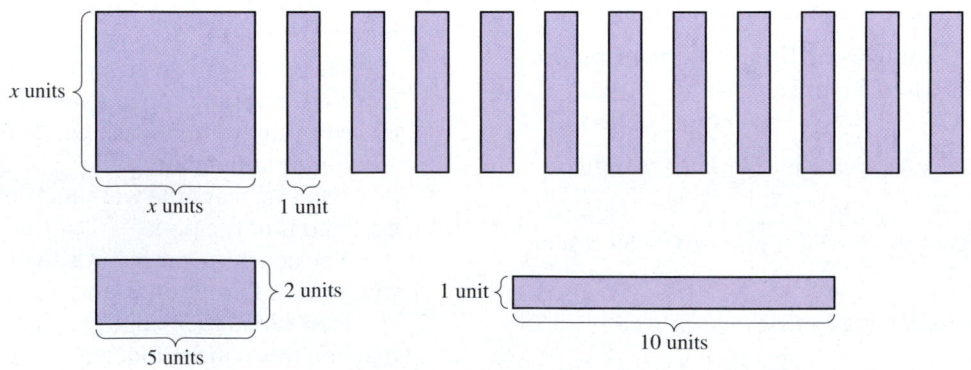

Factoring Trinomials of the Type $ax^2 + bx + c$	**5.3**
	Factoring with FOIL • The Grouping Method

In Section 5.2, we learned a FOIL-based method for factoring trinomials of the type $x^2 + bx + c$. Now we learn to factor trinomials in which the leading, or x^2, coefficient is not 1. First we will use another FOIL-based method and then we will use an alternative method that involves factoring by grouping. Use the method that you prefer or the one selected by your instructor.

Factoring with FOIL

Before factoring trinomials of the type $ax^2 + bx + c$, consider the following:

$$\begin{array}{cccc} \text{F} & \text{O} & \text{I} & \text{L} \end{array}$$
$$(2x + 5)(3x + 4) = 6x^2 + 8x + 15x + 20$$
$$= 6x^2 + 23x + 20.$$

To factor $6x^2 + 23x + 20$, we reverse the multiplication above and look for two binomials whose product is this trinomial. The product of the First terms must be $6x^2$. The product of the Outer terms plus the product of the Inner terms must be $23x$. The product of the Last terms must be 20. This leads us to

$$(2x + 5)(3x + 4). \qquad \text{This factorization is verified above.}$$

How can such a factorization be found without first seeing the corresponding multiplication? Our first approach relies on trial and error and FOIL.

To Factor $ax^2 + bx + c$ Using FOIL

1. Factor out the largest common factor, if one exists.

2. Find two First terms whose product is ax^2:

$$(\boxed{}x +)(\boxed{}x +) = ax^2 + bx + c.$$
$$\text{FOIL}$$

3. Find two Last terms whose product is c:

$$(x + \boxed{})(x + \boxed{}) = ax^2 + bx + c.$$
$$\text{FOIL}$$

4. Repeat steps (2) and (3) until a combination is found for which the sum of the Outer and Inner products is bx:

$$(\boxed{}x + \boxed{})(\boxed{}x + \boxed{}) = ax^2 + bx + c.$$
$$\text{I} \qquad \text{FOIL}$$
$$\text{O}$$

5. Always check by multiplying.

Example 1

Factor: $3x^2 - 10x - 8$.

Solution

1. First, check for a common factor. In this case, there is none (other than 1 or -1).
2. Find two **First** terms whose product is $3x^2$.

   The only possibilities for the **First** terms are $3x$ and x, so any factorization must be of the form

   $$(3x + \quad)(x + \quad).$$

3. Find two **Last** terms whose product is -8.

   Possible factorizations of -8 are

   $$(-8) \cdot 1, \quad 8 \cdot (-1), \quad (-2) \cdot 4, \quad \text{and} \quad 2 \cdot (-4).$$

   Since the **First** terms are not identical, we must also consider

   $$1 \cdot (-8), \quad (-1) \cdot 8, \quad 4 \cdot (-2), \quad \text{and} \quad (-4) \cdot 2.$$

4., 5. Inspect the **O**uter and **I**nner products resulting from steps (2) and (3). Look for a combination in which the sum of the products is the middle term, $-10x$. This may take several tries.

Trial	Product	
$(3x - 8)(x + 1)$	$3x^2 + 3x - 8x - 8$	
	$= 3x^2 - 5x - 8$	Wrong middle term
$(3x + 8)(x - 1)$	$3x^2 - 3x + 8x - 8$	
	$= 3x^2 + 5x - 8$	Wrong middle term
$(3x - 2)(x + 4)$	$3x^2 + 12x - 2x - 8$	
	$= 3x^2 + 10x - 8$	Wrong middle term
$(3x + 2)(x - 4)$	$3x^2 - 12x + 2x - 8$	
	$= 3x^2 - 10x - 8$	Correct middle term!
$(3x + 1)(x - 8)$	$3x^2 - 24x + x - 8$	
	$= 3x^2 - 23x - 8$	Wrong middle term
$(3x - 1)(x + 8)$	$3x^2 + 24x - x - 8$	
	$= 3x^2 + 23x - 8$	Wrong middle term
$(3x + 4)(x - 2)$	$3x^2 - 6x + 4x - 8$	
	$= 3x^2 - 2x - 8$	Wrong middle term
$(3x - 4)(x + 2)$	$3x^2 + 6x - 4x - 8$	
	$= 3x^2 + 2x - 8$	Wrong middle term

The correct factorization is $(3x + 2)(x - 4)$.

Two observations can be made from Example 1. First, we listed all possible trials even though we usually stop after finding the correct factorization. We did this to show that each trial differs only in the middle term of the product. Second, note that as in Section 5.2, only the sign of the middle term changes when the signs in the binomials are reversed.

E x a m p l e   2

Factor: $10x^2 + 37x + 7$.

Solution

1. There is no common factor (other than 1 or -1).
2. Because $10x^2$ factors as $10x \cdot x$ or $5x \cdot 2x$, we have two possibilities:

$$(10x + \quad)(x + \quad) \quad \text{or} \quad (5x + \quad)(2x + \quad).$$

3. There are two pairs of factors of 7 and each can be listed two ways:

$$1, 7 \qquad -1, -7$$

and

$$7, 1 \qquad -7, -1.$$

4., 5. From steps (2) and (3), we see that there are 8 possibilities for factorizations. Look for **O**uter and **I**nner products for which the sum is the middle term. Because all coefficients in $10x^2 + 37x + 7$ are positive, we need consider only positive factors of 7.

Trial	*Product*	
$(10x + 1)(x + 7)$	$10x^2 + 70x + 1x + 7$	
	$= 10x^2 + 71x + 7$	Wrong middle term
$(10x + 7)(x + 1)$	$10x^2 + 10x + 7x + 7$	
	$= 10x^2 + 17x + 7$	Wrong middle term
$(5x + 7)(2x + 1)$	$10x^2 + 5x + 14x + 7$	
	$= 10x^2 + 19x + 7$	Wrong middle term
$(5x + 1)(2x + 7)$	$10x^2 + 35x + 2x + 7$	
	$= 10x^2 + 37x + 7$	Correct middle term!

The correct factorization is $(5x + 1)(2x + 7)$.

E x a m p l e   3

Factor: $24x^2 - 76x + 40$.

Solution

1. First, we factor out the largest common factor, 4:

$$4(6x^2 - 19x + 10).$$

2. Next, we factor $6x^2 - 19x + 10$. Since $6x^2$ can be factored as $3x \cdot 2x$ or $6x \cdot x$, we have two possibilities:

$$(3x + \quad)(2x + \quad) \quad \text{or} \quad (6x + \quad)(x + \quad).$$

3. There are four pairs of factors of 10 and each can be listed two ways:

$$10, 1 \qquad -10, -1 \qquad 5, 2 \qquad -5, -2$$

and

$$1, 10 \qquad -1, -10 \qquad 2, 5 \qquad -2, -5.$$

4., 5. The two possibilities from step (2) and the eight possibilities from step (3) give $2 \cdot 8$, or 16 possibilities for factorizations. We look for **O**uter and **I**nner products resulting from steps (2) and (3) for which the sum is the middle term, $-19x$. Since the sign of the middle term is negative, but the sign of the last term, 10, is positive, the two factors of 10 must both be negative. This means only four pairings from step (3) need be considered. We first try these factors with $(3x + \quad)(2x + \quad)$. If none gives the correct factorization of $6x^2 - 19x + 10$, then we will consider $(6x + \quad)(x + \quad)$.

Trial	*Product*	
$(3x - 10)(2x - 1)$	$6x^2 - 3x - 20x + 10$	
	$= 6x^2 - 23x + 10$	Wrong middle term
$(3x - 1)(2x - 10)$	$6x^2 - 30x - 2x + 10$	
	$= 6x^2 - 32x + 10$	Wrong middle term
$(3x - 5)(2x - 2)$	$6x^2 - 6x - 10x + 10$	
	$= 6x^2 - 16x + 10$	Wrong middle term
$(3x - 2)(2x - 5)$	$6x^2 - 15x - 4x + 10$	
	$= 6x^2 - 19x + 10$	Correct middle term!

Since we have a correct factorization, we need not consider

$$(6x + \quad)(x + \quad).$$

Look again at the possibility $(3x - 5)(2x - 2)$. Without multiplying, we can reject such a possibility. To see why, note that

$$(3x - 5)(2x - 2) = (3x - 5)2(x - 1).$$

The expression $2x - 2$ has a common factor, 2. But we removed the *largest* common factor in step (1). If $2x - 2$ were one of the factors, then 2 would be *another* common factor in addition to the original 4. Thus, $(2x - 2)$ cannot be part of the factorization of $6x^2 - 19x + 10$. Similar reasoning can be used to reject $(3x - 1)(2x - 10)$ as a possible factorization.

Once the largest common factor is factored out, none of the remaining factors can have a common factor.

The factorization of $6x^2 - 19x + 10$ is $(3x - 2)(2x - 5)$, but do not forget the common factor! The factorization of $24x^2 - 76x + 40$ is

$$4(3x - 2)(2x - 5).$$

Tips for Factoring $ax^2 + bx + c$

To factor $ax^2 + bx + c \, (a > 0)$:

- Always factor out the largest common factor, if one exists.
- Once the largest common factor has been factored out of the original trinomial, no binomial factor can contain a common factor (other than 1 or -1).
- If c is positive, then the signs in both binomial factors must match the sign of b.

(continued)

- Reversing the signs in the binomials reverses the sign of the middle term of their product.
- Organize your work so that you can keep track of which possibilities have or have not been checked.
- Always check by multiplying.

E x a m p l e   4

Factor: $10x + 8 - 3x^2$.

Solution   An important problem-solving strategy is to find a way to make new problems look like problems we already know how to solve. The factoring tips above apply only to trinomials of the form $ax^2 + bx + c$, with $a > 0$. This leads us to rewrite $10x + 8 - 3x^2$ in descending order:

$$10x + 8 - 3x^2 = -3x^2 + 10x + 8. \qquad \text{Writing in descending order}$$

Although $-3x^2 + 10x + 8$ looks similar to the trinomials we have factored, the tips above require a positive leading coefficient. This can be attained by factoring out -1:

$$-3x^2 + 10x + 8 = -1(3x^2 - 10x - 8) \qquad \text{Factoring out } -1 \text{ changes the signs of the coefficients.}$$

$$= -1(3x + 2)(x - 4). \qquad \text{Using the result from Example 1}$$

The factorization of $10x + 8 - 3x^2$ is $-1(3x + 2)(x - 4)$.

E x a m p l e   5

Factor: $6p^2 - 13pq - 28q^2$.

Solution   Since no common factor exists, we examine the first term, $6p^2$. There are two possibilities:

$$(2p + \quad)(3p + \quad) \quad \text{or} \quad (6p + \quad)(p + \quad).$$

The last term, $-28q^2$, has the following pairs of factors:

$$28q, -q \qquad 14q, -2q \qquad 7q, -4q,$$

and

$$-28q, \quad q \qquad -14q, \quad 2q \qquad -7q, \quad 4q,$$

as well as each of the pairings reversed.

Some trials, like $(2p + 28q)(3p - q)$ and $(2p + 14q)(3p - 2q)$, cannot be correct because both $(2p + 28q)$ and $(2p + 14q)$ contain a common factor, 2. We try $(2p + 7q)(3p - 4q)$:

$$(2p + 7q)(3p - 4q) = 6p^2 - 8pq + 21pq - 28q^2$$
$$= 6p^2 + 13pq - 28q^2.$$

Our trial is incorrect, but only because of the sign of the middle term. To correctly factor $6p^2 - 13pq - 28q^2$, we simply change the signs in the binomials:

$$(2p - 7q)(3p + 4q) = 6p^2 + 8pq - 21pq - 28q^2$$
$$= 6p^2 - 13pq - 28q^2.$$

The correct factorization is $(2p - 7q)(3p + 4q)$.

The Grouping Method

Another method of factoring trinomials of the type $ax^2 + bx + c$ is known as the *grouping method*. The grouping method relies on rewriting $ax^2 + bx + c$ in the form $ax^2 + px + qx + c$ and then factoring by grouping. To develop this method, consider the following*:

$$(2x + 5)(3x + 4) = 2x \cdot 3x + 2x \cdot 4 + 5 \cdot 3x + 5 \cdot 4 \qquad \text{Using FOIL}$$
$$= 2 \cdot 3 \cdot x^2 + 2 \cdot 4x + 5 \cdot 3x + 5 \cdot 4$$
$$= 2 \cdot 3 \cdot x^2 + (2 \cdot 4 + 5 \cdot 3)x + 5 \cdot 4$$

$$\qquad\qquad\quad \underset{a}{\uparrow\downarrow} \qquad\qquad\quad \underset{b}{\uparrow\downarrow} \qquad\qquad \underset{c}{\uparrow\downarrow}$$

$$= 6x^2 \quad + \quad 23x \quad + \quad 20.$$

Note that reversing these steps shows that $6x^2 + 23x + 20$ can be rewritten as $6x^2 + 8x + 15x + 20$ and then factored by grouping. Note that the numbers that add to b (in this case, $2 \cdot 4$ and $5 \cdot 3$), also multiply to ac (in this case, $2 \cdot 3 \cdot 5 \cdot 4$).

To Factor $ax^2 + bx + c$, Using the Grouping Method

1. Factor out the largest common factor, if one exists.
2. Multiply the leading coefficient a and the constant c.
3. Find a pair of factors of ac whose sum is b.
4. Rewrite the middle term, bx, as a sum or difference using the factors found in step (3).
5. Factor by grouping.
6. Always check by multiplying.

E x a m p l e   6

Factor: $3x^2 - 10x - 8$.

Solution

1. First, we note that there is no common factor (other than 1 or -1).
2. We multiply the leading coefficient, 3, and the constant, -8:

$$3(-8) = -24.$$

---

*This discussion was inspired by a lecture given by Irene Doo at Austin Community College.

3. We next look for a factorization of -24 in which the sum of the factors is the coefficient of the middle term, -10.

Pairs of Factors of -24	Sums of Factors
1, -24	-23
-1, 24	23
2, -12	-10
-2, 12	10
3, -8	-5
-3, 8	5
4, -6	-2
-4, 6	2

$2 + (-12) = -10$

We normally stop listing pairs of factors once we have found the one we are after.

4. Next, we express the middle term as a sum or difference using the factors found in step (3):

$$-10x = 2x - 12x.$$

5. We now factor by grouping as follows:

$$3x^2 - 10x - 8 = 3x^2 + 2x - 12x - 8 \qquad \text{Substituting } 2x - 12x \text{ for } -10x. \text{ We could also use } -12x + 2x.$$

$$= x(3x + 2) - 4(3x + 2) \qquad \text{Factoring by grouping; see Section 5.1}$$

$$= (3x + 2)(x - 4). \qquad \text{Factoring out the common factor, } 3x + 2$$

6. *Check:* $(3x + 2)(x - 4) = 3x^2 - 10x - 8.$

The factorization of $3x^2 - 10x - 8$ is $(3x + 2)(x - 4)$.

E x a m p l e   7   Factor: $8x^3 + 22x^2 - 6x$.

Solution

1. We factor out the largest common factor, $2x$:

$$8x^3 + 22x^2 - 6x = 2x(4x^2 + 11x - 3).$$

2. To factor $4x^2 + 11x - 3$ by grouping, we multiply the leading coefficient, 4, and the constant term, -3:

$$4(-3) = -12.$$

3. We next look for factors of -12 that add to 11.

Pairs of Factors of -12	Sums of Factors
1, -12	-11
-1, 12	11
.	.
.	.
.	.

Since $-1 + 12 = 11$, there is no need to list other pairs of factors.

4. We then rewrite the $11x$ in $4x^2 + 11x - 3$ using

$$11x = -1x + 12x, \quad \text{or} \quad 11x = 12x - 1x.$$

5. Next, we factor by grouping:

$$4x^2 + 11x - 3 = 4x^2 + 12x - 1x - 3 \qquad \begin{array}{l}\text{Rewriting the middle term;} \\ -1x + 12x \text{ could also be} \\ \text{used.}\end{array}$$

$$= 4x(x + 3) - 1(x + 3) \qquad \begin{array}{l}\text{Factoring by grouping.} \\ \text{Removing } -1 \text{ reveals the} \\ \text{common factor, } x + 3.\end{array}$$

$$= (x + 3)(4x - 1). \qquad \begin{array}{l}\text{Factoring out the} \\ \text{common factor}\end{array}$$

6. The factorization of $4x^2 + 11x - 3$ is $(x + 3)(4x - 1)$. But don't forget the common factor, $2x$. The factorization of the original trinomial is

$$2x(x + 3)(4x - 1).$$

FOR EXTRA HELP

Exercise Set 5.3

 Digital Video Tutor CD 4 Videotape 9   InterAct Math   Math Tutor Center   MathXL   MyMathLab.com

Factor completely. If a polynomial is prime, state this.

1. $2x^2 + 7x - 4$

2. $3x^2 + x - 4$

3. $3t^2 + 4t - 15$

4. $5t^2 + t - 18$

5. $6x^2 - 23x + 7$

6. $6x^2 - 13x + 6$

7. $7x^2 + 15x + 2$

8. $3x^2 + 4x + 1$

9. $9a^2 - 6a - 8$

10. $4a^2 - 4a - 15$

11. $3x^2 - 5x - 2$

12. $15x^2 - 19x - 10$

Aha! **13.** $12t^2 - 6t - 6$

14. $18t^2 + 36t - 32$

15. $18t^2 + 3t - 10$

16. $2t^2 + 5t + 2$

17. $15x^2 + 19x + 6$

18. $12x^2 - 31x + 20$

19. $35x^2 + 34x + 8$

20. $28x^2 + 38x - 6$

21. $4 + 6t^2 - 13t$

22. $9 + 8t^2 - 18t$

23. $25x^2 + 40x + 16$

24. $49t^2 + 42t + 9$

25. $16a^2 + 78a + 27$

26. $24x^2 + 47x - 2$

27. $18t^2 + 24t - 10$

28. $35x^2 - 57x - 44$

29. $2x^2 - 15 - x$

30. $2t^2 - 19 - 6t$

31. $6x^2 + 33x + 15$

32. $12x^2 + 28x - 24$

33. $20x^2 - 25x + 5$

34. $30x^2 - 24x - 54$

35. $12x^2 + 68x - 24$

36. $6x^2 + 21x + 15$

37. $4x + 1 + 3x^2$

38. $-9 + 18x^2 + 21x$

Factor. Use factoring by grouping even though it would seem reasonable to first combine like terms.

39. $y^2 + 4y - 2y - 8$

40. $x^2 + 5x - 2x - 10$

41. $8t^2 - 6t - 28t + 21$

42. $35t^2 - 40t + 21t - 24$

43. $6x^2 + 4x + 9x + 6$

44. $3x^2 - 2x + 3x - 2$

45. $2t^2 + 6t - t - 3$

46. $5t^2 + 10t - t - 2$

47. $3a^2 - 12a - a + 4$

48. $2a^2 - 10a - a + 5$

Factor completely. If a polynomial is prime, state this.

49. $9t^2 + 14t + 5$

50. $16t^2 + 23t + 7$

51. $16x^2 + 32x + 7$

52. $9x^2 + 18x + 5$

53. $10a^2 + 25a - 15$

54. $10a^2 - 3a - 18$

55. $2x^2 + 6x - 14$

56. $14x^2 - 35x + 14$

57. $18x^3 + 21x^2 - 9x$

58. $6x^3 - 4x^2 - 10x$

59. $89x + 64 + 25x^2$

60. $47 - 42y + 9y^2$

61. $168x^3 + 45x^2 + 3x$

62. $144x^5 - 168x^4 + 48x^3$

63. $14t^4 - 19t^3 - 3t^2$

64. $70a^4 - 68a^3 + 16a^2$

65. $3x + 45x^2 - 18$

66. $2x + 24x^2 - 40$

67. $9a^2 + 18ab + 8b^2$

68. $3p^2 - 16pq - 12q^2$

69. $35p^2 + 34pq + 8q^2$

70. $10s^2 + 4st - 6t^2$

71. $18x^2 - 6xy - 24y^2$

72. $30a^2 + 87ab + 30b^2$

73. $24a^2 - 34ab + 12b^2$

74. $15a^2 - 5ab - 20b^2$

75. $35x^2 + 34x^3 + 8x^4$

76. $19x^3 - 3x^2 + 14x^4$

77. $18a^7 + 8a^6 + 9a^8$

78. $40a^8 + 16a^7 + 25a^9$

79. Asked to factor $2x^2 - 18x + 36$, Amy *incorrectly* answers

$$2x^2 - 18x + 36 = 2(x^2 + 9x + 18)$$
$$= 2(x + 3)(x + 6).$$

If this were a 10-point quiz question, how many points would you take off? Why?

80. Asked to factor $4x^2 + 28x + 48$, Herb *incorrectly* answers

$$4x^2 + 28x + 48 = (2x + 6)(2x + 8)$$
$$= 2(x + 3)(x + 4).$$

If this were a 10-point quiz question, how many points would you take off? Why?

SKILL MAINTENANCE

81. The earth is a sphere (or ball) that is about 40,000 km in circumference. Find the radius of the earth, in kilometers and in miles. Use 3.14 for π. (*Hint*: 1 km ≈ 0.62 mi.)

82. The second angle of a triangle is 10° less than twice the first. The third angle is 15° more than four times the first. Find the measure of the second angle.

Multiply.

83. $(3x + 1)^2$

84. $(5x - 2)^2$

85. $(4t - 5)^2$

86. $(7a + 1)^2$

87. $(5x - 2)(5x + 2)$

88. $(2x - 3)(2x + 3)$

89. $(2t + 7)(2t - 7)$

90. $(4a + 7)(4a - 7)$

SYNTHESIS

91. Which one of the six tips listed after Example 3 do you find most helpful? Which one do you find least helpful? Explain how you made these determinations.

92. For the trinomial $ax^2 + bx + c$, suppose that a is the product of three different prime factors and c is the product of another two prime factors. How many possible factorizations (like those in Example 1) exist? Explain how you determined your answer.

Factor. If a polynomial is prime, state this.

93. $9a^2b^2 - 15ab - 2$

94. $18x^2y^2 - 3xy - 10$

95. $8x^2y^3 + 10xy^2 + 2y$

96. $9a^2b^3 + 25ab^2 + 16$

97. $9t^{10} + 12t^5 + 4$

98. $16t^{10} - 8t^5 + 1$

99. $-15x^{2m} + 26x^m - 8$

100. $20x^{2n} + 16x^n + 3$

101. $a^{2n+1} - 2a^{n+1} + a$

102. $3a^{6n} - 2a^{3n} - 1$

103. $3(a + 1)^{n+1}(a + 3)^2 - 5(a + 1)^n(a + 3)^3$

104. $7(t - 3)^{2n} + 5(t - 3)^n - 2$

Factoring Perfect-Square Trinomials and Differences of Squares

5.4

Recognizing Perfect-Square Trinomials • Factoring Perfect-Square Trinomials • Recognizing Differences of Squares • Factoring Differences of Squares • Factoring Completely

In Section 4.5, we studied some shortcuts for finding the products of certain binomials. Reversing these procedures provides shortcuts for factoring certain polynomials.

Recognizing Perfect-Square Trinomials

Some trinomials are squares of binomials. For example, $x^2 + 10x + 25$ is the square of the binomial $x + 5$. To see this, we can calculate $(x + 5)^2$. It is $x^2 + 2 \cdot x \cdot 5 + 5^2$, or $x^2 + 10x + 25$. A trinomial that is the square of a binomial is called a **perfect-square trinomial**.

In Section 4.5, we considered squaring binomials as a special-product rule:

$$(A + B)^2 = A^2 + 2AB + B^2;$$
$$(A - B)^2 = A^2 - 2AB + B^2.$$

Reading the right sides first, we see that these equations can be used to factor perfect-square trinomials. Note that in order for a trinomial to be the square of a binomial, it must have the following:

1. Two terms, A^2 and B^2, must be squares, such as

    $4, \quad x^2, \quad 81m^2, \quad 16t^2.$

2. There must be no minus sign before A^2 or B^2.
3. The remaining term is either $2 \cdot A \cdot B$ or $-2 \cdot A \cdot B$, where A and B are the square roots of A^2 and B^2.

E x a m p l e   1   Determine whether each of the following is a perfect-square trinomial.

a) $x^2 + 6x + 9$ b) $t^2 - 8t - 9$ c) $16x^2 + 49 - 56x$

Solution

a) To see if $x^2 + 6x + 9$ is a perfect-square trinomial, note that:

   1. Two terms, x^2 and 9, are squares.
   2. There is no minus sign before x^2 or 9.
   3. The remaining term, $6x$, is $2 \cdot x \cdot 3$, where x and 3 are the square roots of x^2 and 9.

   Thus, $x^2 + 6x + 9$ *is* a perfect-square trinomial.

b) To see if $t^2 - 8t - 9$ is a perfect-square trinomial, note that:

   1. Both t^2 and 9 are squares. But:
   2. Since 9 is being subtracted, $t^2 - 8t - 9$ *is not* a perfect-square trinomial.

c) To see if $16x^2 + 49 - 56x$ is a perfect-square trinomial, it helps to first write it in descending order:

$$16x^2 - 56x + 49.$$

Next, note that:

1. Two terms, $16x^2$ and 49, are squares.
2. There is no minus sign before $16x^2$ or 49.
3. Twice the product of the square roots, $2 \cdot 4x \cdot 7$, is $56x$, the opposite of the remaining term, $-56x$.

Thus, $16x^2 + 49 - 56x$ *is* a perfect-square trinomial.

Factoring Perfect-Square Trinomials

Either of the factoring methods from Section 5.3 can be used to factor perfect-square trinomials, but a faster method is to recognize the following patterns.

> ### Factoring a Perfect-Square Trinomial
> $$A^2 + 2AB + B^2 = (A + B)^2; \qquad A^2 - 2AB + B^2 = (A - B)^2$$

Each factorization uses the square roots of the squared terms and the sign of the remaining term.

E x a m p l e   2

Factor: **(a)** $x^2 + 6x + 9$; **(b)** $x^2 + 49 - 14x$; **(c)** $16x^2 - 40x + 25$.

Solution

a) $x^2 + 6x + 9 = x^2 + 2 \cdot x \cdot 3 + 3^2 = (x + 3)^2$ The sign of the middle term is positive.

$$A^2 + 2 \quad A \quad B + B^2 = (A + B)^2$$

b) $x^2 + 49 - 14x = x^2 - 14x + 49$ Using a commutative law to write in descending order

$$= x^2 - 2 \cdot x \cdot 7 + 7^2 = (x - 7)^2$$

$$A^2 - 2 \quad A \quad B + B^2 = (A - B)^2$$

c) $16x^2 - 40x + 25 = (4x)^2 - 2 \cdot 4x \cdot 5 + 5^2 = (4x - 5)^2$

$$A^2 \quad - 2 \quad A \quad B + B^2 = (A - B)^2$$

With practice, it is possible to spot perfect-square trinomials as they occur and factor them quickly.

E x a m p l e   3

Factor: $4p^2 - 12pq + 9q^2$.

Solution We have

$$4p^2 - 12pq + 9q^2 = (2p)^2 - 2(2p)(3q) + (3q)^2 \qquad \text{Recognizing the perfect-square trinomial}$$

$$= (2p - 3q)^2. \qquad \text{The sign of the middle term is negative.}$$

Check: $(2p - 3q)(2p - 3q) = 4p^2 - 12pq + 9q^2$.

The factorization is $(2p - 3q)^2$.

E x a m p l e   4

Factor: $75m^3 + 60m^2 + 12m$.

Solution *Always* look first for a common factor. This time there is one, $3m$:

$$75m^3 + 60m^2 + 12m = 3m[25m^2 + 20m + 4] \qquad \text{Factoring out the largest common factor}$$

$$= 3m[(5m)^2 + 2(5m)(2) + 2^2] \qquad \text{Recognizing the perfect-square trinomial. Try to do this mentally.}$$

$$= 3m(5m + 2)^2.$$

Check: $3m(5m + 2)^2 = 3m(5m + 2)(5m + 2)$

$$= 3m(25m^2 + 20m + 4)$$

$$= 75m^3 + 60m^2 + 12m.$$

The factorization is $3m(5m + 2)^2$.

Recognizing Differences of Squares

Some binomials represent the difference of two squares. For example, the binomial $16x^2 - 9$ is a difference of two expressions, $16x^2$ and 9, that are squares. To see this, note that $16x^2 = (4x)^2$ and $9 = 3^2$.

Any expression, like $16x^2 - 9$, that can be written in the form $A^2 - B^2$ is called a **difference of squares**. Note that for a binomial to be a difference of squares, it must have the following.

1. There must be two expressions, both squares, such as

$$25x^2, \qquad 9, \qquad 4x^2y^2, \qquad 1, \qquad x^6, \qquad 49y^8.$$

2. The terms in the binomial must have different signs.

Note that in order for a term to be a square, its coefficient must be a perfect square and the power(s) of the variable(s) must be even.

E x a m p l e   5

Determine whether each of the following is a difference of squares.

a) $9x^2 - 64$ **b)** $25 - t^3$ **c)** $-4x^{10} + 36$

Solution

a) To see if $9x^2 - 64$ is a difference of squares, note that:

    1. The first expression is a square: $9x^2 = (3x)^2$.
        The second expression is a square: $64 = 8^2$.
    2. The terms have different signs.

    Thus, $9x^2 - 64$ is a difference of squares, $(3x)^2 - 8^2$.

b) To see if $25 - t^3$ is a difference of squares, note that:

    1. The expression t^3 is not a square.

    Thus, $25 - t^3$ is not a difference of squares.

c) To see if $-4x^{10} + 36$ is a difference of squares, note that:

    1. The expressions $4x^{10}$ and 36 are squares: $4x^{10} = (2x^5)^2$ and $36 = 6^2$.
    2. The terms have different signs.

    Thus, $-4x^{10} + 36$ is a difference of squares, $6^2 - (2x^5)^2$.

Factoring Differences of Squares

To factor a difference of squares, we reverse a pattern from Section 4.5:

> *Factoring a Difference of Squares*
> $A^2 - B^2 = (A + B)(A - B).$

Once we have identified the expressions that are playing the roles of A and B, the factorization can be written directly.

E x a m p l e   6

Factor: **(a)** $x^2 - 4$; **(b)** $m^2 - 9p^2$; **(c)** $9 - 16t^{10}$; **(d)** $50x^2 - 8x^8$.

Solution

a) $x^2 - 4 = x^2 - 2^2 = (x + 2)(x - 2)$

$$A^2 - B^2 = (A + B)(A - B)$$

b) $m^2 - 9p^2 = m^2 - (3p)^2 = (m + 3p)(m - 3p)$

$$A^2 - B^2 = (A + B)(A - B)$$

c) $9 - 16t^{10} = 3^2 - (4t^5)^2$ Using the rules for powers

$$\underset{A^2\ -\ B^2}{}$$

$$= (3 + 4t^5)(3 - 4t^5) \quad\quad \text{Try to go directly to this step.}$$

$$\underset{(A\ +\ B)(A\ -\ B)}{}$$

d) *Always* check first for a common factor. This time there is one, $2x^2$:

$$50x^2 - 8x^8 = 2x^2(25 - 4x^6) \qquad \text{Factoring out the common factor}$$

$$= 2x^2[5^2 - (2x^3)^2] \qquad \text{Recognizing } A^2 - B^2. \text{ Try to do this mentally.}$$

$$= 2x^2(5 + 2x^3)(5 - 2x^3). \qquad \text{Factoring the difference of squares}$$

Check: $\quad 2x^2(5 + 2x^3)(5 - 2x^3) = 2x^2(25 - 4x^6)$

$$= 50x^2 - 8x^8.$$

The factorization is $2x^2(5 + 2x^3)(5 - 2x^3)$.

Caution!   Note in Example 6 that a difference of squares is *not* the square of the difference; that is,

$$A^2 - B^2 \neq (A - B)^2. \qquad \text{To see this, note that}$$
$$(A - B)^2 = A^2 - 2AB + B^2.$$

Factoring Completely

Sometimes, as in Examples 4 and 6(d), a *complete* factorization requires two or more steps. In general, a factorization is complete when no factor can be factored further.

Example 7

Factor: $p^4 - 16$.

Solution   We have

$$p^4 - 16 = (p^2)^2 - 4^2 \qquad \text{Recognizing } A^2 - B^2$$

$$= (p^2 + 4)(p^2 - 4) \qquad \text{Factoring a difference of squares}$$

$$= (p^2 + 4)(p + 2)(p - 2). \qquad \text{Factoring further. The factor } p^2 - 4 \text{ is itself a difference of squares.}$$

Check: $\quad (p^2 + 4)(p + 2)(p - 2) = (p^2 + 4)(p^2 - 4)$

$$= p^4 - 16.$$

The factorization is $(p^2 + 4)(p + 2)(p - 2)$.

Note in Example 7 that the factor $p^2 + 4$ is a *sum* of squares that cannot be factored further.

Caution!   Apart from possibly removing a common factor, you cannot factor a sum of squares. In particular,

$$A^2 + B^2 \neq (A + B)^2.$$

Consider $25x^2 + 100$. Here a sum of squares has a common factor, 25. Factoring, we get $25(x^2 + 4)$, where $x^2 + 4$ is prime.

As you proceed through the exercises, these suggestions may prove helpful.

Tips for Factoring

1. Always look first for a common factor! If there is one, factor it out.
2. Be alert for perfect-square trinomials and differences of squares. Once recognized, they can be factored without trial and error.
3. Always factor completely.
4. Check by multiplying.

FOR EXTRA HELP

Exercise Set 5.4

 Digital Video Tutor CD 4 Videotape 9   InterAct Math   Math Tutor Center   MathXL   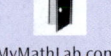 MyMathLab.com

Determine whether each of the following is a perfect-square trinomial.

1. $x^2 - 18x + 81$

2. $x^2 - 16x + 64$

3. $x^2 + 16x - 64$

4. $x^2 - 14x - 49$

5. $x^2 - 3x + 9$

6. $x^2 + 2x + 4$

7. $9x^2 - 36x + 24$

8. $36x^2 - 24x + 16$

Factor completely. Remember to look first for a common factor and to check by multiplying. If a polynomial is prime, state this.

9. $x^2 - 16x + 64$

10. $x^2 - 14x + 49$

11. $x^2 + 14x + 49$

12. $x^2 + 16x + 64$

13. $3x^2 - 6x + 3$

14. $5x^2 - 10x + 5$

15. $4 + 4x + x^2$

16. $4 + x^2 - 4x$

17. $18x^2 - 12x + 2$

18. $25x^2 + 10x + 1$

19. $49 + 56y + 16y^2$

20. $120m + 75 + 48m^2$

21. $x^5 - 18x^4 + 81x^3$

22. $2x^2 - 40x + 200$

23. $2x^3 - 4x^2 + 2x$

24. $x^3 + 24x^2 + 144x$

25. $20x^2 + 100x + 125$

26. $12x^2 + 36x + 27$

27. $49 - 42x + 9x^2$

28. $64 - 112x + 49x^2$

29. $16x^2 + 24x + 9$

30. $2a^2 + 28a + 98$

31. $2 + 20x + 50x^2$

32. $9x^2 + 30x + 25$

33. $4p^2 + 12pq + 9q^2$

34. $25m^2 + 20mn + 4n^2$

35. $a^2 - 12ab + 49b^2$

36. $x^2 - 7xy + 9y^2$

37. $64m^2 + 16mn + n^2$

38 $81p^2 - 18pq + q^2$

39. $32s^2 - 80st + 50t^2$

40. $36a^2 + 96ab + 64b^2$

Determine whether each of the following is a difference of squares.

41. $x^2 - 100$ **42.** $x^2 - 36$ **43.** $x^2 + 36$

44. $x^2 + 4$ **45.** $9t^2 - 32$ **46.** $x^2 - 50y^2$

47. $-25 + 4t^2$ **48.** $-1 + 49t^2$

Factor completely. Remember to look first for a common factor. If a polynomial is prime, state this.

49. $y^2 - 4$ **50.** $x^2 - 36$

51. $p^2 - 9$ **52.** $q^2 + 1$

53. $-49 + t^2$ **54.** $-64 + m^2$

55. $6a^2 - 54$ **56.** $x^2 - 8x + 16$

57. $49x^2 - 14x + 1$ **58.** $3t^2 - 12$

59. $200 - 2t^2$ **60.** $98 - 8w^2$

61. $80a^2 - 45$ **62.** $25x^2 - 4$

63. $5t^2 - 80$ **64.** $4t^2 - 64$

65. $8x^2 - 98$ **66.** $24x^2 - 54$

67. $36x - 49x^3$ **68.** $16x - 81x^3$

69. $49a^4 - 20$ **70.** $25a^4 - 9$

71. $t^4 - 1$ **72.** $x^4 - 16$

73. $3x^3 - 24x^2 + 48x$ **74.** $2a^4 - 36a^3 + 162a^2$

75. $48t^2 - 27$ **76.** $125t^2 - 45$

77. $a^8 - 2a^7 + a^6$ **78.** $x^8 - 8x^7 + 16x^6$

79. $7a^2 - 7b^2$ **80.** $6p^2 - 6q^2$

81. $25x^2 - 4y^2$ **82.** $16a^2 - 9b^2$

83. $1 - a^4b^4$ **84.** $75 - 3m^4n^4$

85. $18t^2 - 8s^2$ **86.** $49x^2 - 16y^2$

87. Explain in your own words how to determine whether a polynomial is a difference of squares.

88. Explain in your own words how to determine if a polynomial is a perfect-square trinomial.

SKILL MAINTENANCE

89. About 5 L of oxygen can be dissolved in 100 L of water at 0°C. This is 1.6 times the amount that can be dissolved in the same volume of water at 20°C. How much oxygen can be dissolved at 20°C?

90. Bonnie is taking an astronomy course. To get an A, she must average at least 90 after four exams. Bonnie scored 96, 98, and 89 on the first three

tests. Determine (in terms of an inequality) what scores on the last test will earn her an A.

Simplify.

91. $(x^3y^5)(x^9y^7)$ **92.** $(5a^2b^3)^2$

Graph.

93. $y = \frac{3}{2}x - 3$ **94.** $3x - 5y = 30$

SYNTHESIS

95. Write directions that would enable someone to construct a polynomial that contains a perfect-square trinomial, a difference of squares, and a common factor.

96. Leon concludes that since $x^2 - 9 = (x - 3)(x + 3)$, it must follow that $x^2 + 9 = (x + 3)(x - 3)$. What mistake(s) is he making?

Factor completely. If a polynomial is prime, state this.

97. $x^8 - 2^8$ **98.** $3x^2 - \frac{1}{3}$

99. $18x^3 - \frac{8}{25}x$ **100.** $0.49p - p^3$

101. $0.64x^2 - 1.21$ **102.** $(x + 3)^4 - 81$

103. $(y - 5)^4 - z^8$ **104.** $x^2 - \left(\frac{1}{x}\right)^2$

105. $a^{2n} - 49b^{2n}$ **106.** $81 - b^{4k}$

107. $x^4 - 8x^2 - 9$ **108.** $9b^{2n} + 12b^n + 4$

109. $16x^4 - 96x^2 + 144$

110. $(y + 3)^2 + 2(y + 3) + 1$

111. $49(x + 1)^2 - 42(x + 1) + 9$

112. $27x^3 - 63x^2 - 147x + 343$

113. Subtract $(x^2 + 1)^2$ from $x^2(x + 1)^2$.

Factor by grouping. Look for a grouping of three terms that is a perfect-square trinomial.

114. $a^2 + 2a + 1 - 9$

115. $y^2 + 6y + 9 - x^2 - 8x - 16$

Find c such that each polynomial is the square of a binomial.

116. $cy^2 + 6y + 1$ **117.** $cy^2 - 24y + 9$

118. Find the value of a if $x^2 + a^2x + a^2$ factors into $(x + a)^2$.

119. Show that the difference of the squares of two consecutive integers is the sum of the integers. (*Hint*: Use x for the smaller number.)

Factoring Sums or Differences of Cubes

5.5

Factoring a Sum of Two Cubes • Factoring a Difference of Two Cubes

Although a sum of two squares cannot be factored unless a common factor exists, a sum of two *cubes* can always be factored. A difference of two cubes can also be factored.

Consider the following products:

$$(A + B)(A^2 - AB + B^2) = A(A^2 - AB + B^2) + B(A^2 - AB + B^2)$$
$$= A^3 - A^2B + AB^2 + A^2B - AB^2 + B^3$$
$$= A^3 + B^3$$

and

$$(A - B)(A^2 + AB + B^2) = A(A^2 + AB + B^2) - B(A^2 + AB + B^2)$$
$$= A^3 + A^2B + AB^2 - A^2B - AB^2 - B^3$$
$$= A^3 - B^3.$$

These equations show how we can factor a sum or a difference of two cubes.

> **To Factor a Sum or Difference of Cubes**
>
> $A^3 + B^3 = (A + B)(A^2 - AB + B^2),$
> $A^3 - B^3 = (A - B)(A^2 + AB + B^2)$

This table of cubes will help in the examples that follow.

N	0.2	0.1	0	1	2	3	4	5	6	7	8
N^3	0.008	0.001	0	1	8	27	64	125	216	343	512

Example 1

Factor: $x^3 - 8$.

Solution   We have

$$x^3 - 8 = x^3 - 2^3 = (x - 2)(x^2 + x \cdot 2 + 2^2).$$
$$A^3 - B^3 = (A - B)(A^2 + A\ B + B^2)$$

This tells us that $x^3 - 8 = (x - 2)(x^2 + 2x + 4)$. Note that we cannot factor $x^2 + 2x + 4$. (It is not a perfect-square trinomial nor can it be factored by trial and error or grouping.) The check is left to the student.

E x a m p l e   2   Factor: $x^3 + 125$.

Solution   We have

$$x^3 + 125 = x^3 + 5^3 = (x + 5)(x^2 - x \cdot 5 + 5^2).$$

$$A^3 + B^3 = (A + B)(A^2 - A\ B + B^2)$$

Thus, $x^3 + 125 = (x + 5)(x^2 - 5x + 25)$. We leave the check to the student.

E x a m p l e   3   Factor: $16a^7b + 54ab^7$.

Solution   We first look for a common factor:

$$16a^7b + 54ab^7 = 2ab[8a^6 + 27b^6]$$
$$= 2ab[(2a^2)^3 + (3b^2)^3] \qquad \text{This is of the form } A^3 + B^3,$$
$$\text{where } A = 2a^2 \text{ and } B = 3b^2.$$
$$= 2ab[(2a^2 + 3b^2)(4a^4 - 6a^2b^2 + 9b^4)].$$

We check using the distributive law:

$$2ab(2a^2 + 3b^2)(4a^4 - 6a^2b^2 + 9b^4)$$
$$= 2ab[2a^2(4a^4 - 6a^2b^2 + 9b^4) + 3b^2(4a^4 - 6a^2b^2 + 9b^4)]$$
$$= 2ab[8a^6 - 12a^4b^2 + 18a^2b^4 + 12b^2a^4 - 18a^2b^4 + 27b^6]$$
$$= 2ab[8a^6 + 27b^6]$$
$$= 16a^7b + 54ab^7.$$

We have a check. The factorization is

$$2ab(2a^2 + 3b^2)(4a^4 - 6a^2b^2 + 9b^4).$$

E x a m p l e   4   Factor: $y^3 - 0.001$.

Solution   Since $0.001 = (0.1)^3$, we have a difference of cubes:

$$y^3 - 0.001 = (y - 0.1)(y^2 + 0.1y + 0.01).$$

The check is left to the student.

E x a m p l e   5   Factor: $r^6 - s^6$.

Solution   We have

$$r^6 - s^6 = (r^3)^2 - (s^3)^2$$
$$= (r^3 + s^3)(r^3 - s^3). \qquad \text{Factoring a difference of two } squares$$

Next, we factor the sum and difference of two cubes:

$$r^6 - s^6 = (r + s)(r^2 - rs + s^2)(r - s)(r^2 + rs + s^2).$$

In Example 5, suppose that we first factored $r^6 - s^6$ as a difference of two cubes:

$$(r^2)^3 - (s^2)^3 = (r^2 - s^2)(r^4 + r^2 s^2 + s^4)$$
$$= (r + s)(r - s)(r^4 + r^2 s^2 + s^4).$$

In this case, we might have missed some factors; $r^4 + r^2 s^2 + s^4$ can be factored as $(r^2 - rs + s^2)(r^2 + rs + s^2)$, but we probably would never have suspected that such a factorization exists.

Remember the following about factoring sums or differences of squares and cubes:

Difference of cubes: $A^3 - B^3 = (A - B)(A^2 + AB + B^2),$

Sum of cubes: $A^3 + B^3 = (A + B)(A^2 - AB + B^2),$

Difference of squares: $A^2 - B^2 = (A + B)(A - B),$

Sum of squares: $A^2 + B^2$ **cannot be factored apart from factoring out a common factor if one exists.**

Exercise Set 5.5

Factor completely.

1. $t^3 + 8$

2. $p^3 + 27$

3. $a^3 - 64$

4. $w^3 - 1$

5. $z^3 + 125$

6. $x^3 + 1$

7. $8a^3 - 1$

8. $27x^3 - 1$

9. $y^3 - 27$

10. $p^3 - 8$

11. $64 + 125x^3$

12. $8 + 27b^3$

13. $125p^3 - 1$

14. $64w^3 - 1$

15. $27m^3 + 64$

16. $8t^3 + 27$

17. $p^3 - q^3$

18. $a^3 + b^3$

19. $x^3 + \frac{1}{8}$

20. $y^3 + \frac{1}{27}$

21. $2y^3 - 128$

22. $3z^3 - 3$

23. $24a^3 + 3$

24. $54x^3 + 2$

25. $rs^3 - 64r$

26. $ab^3 + 125a$

27. $5x^3 + 40z^3$

28. $2y^3 - 54z^3$

29. $x^3 + 0.001$

30. $y^3 + 0.125$

31. $3z^5 - 3z^2$

32. $2y^4 - 128y$

33. $t^6 + 1$

34. $z^6 - 1$

35. $p^6 - q^6$

36. $x^6 - 64y^6$

37. Dino incorrectly believes that
$$a^3 - b^3 = (a - b)(a^2 + b^2).$$
How could you convince him that he is wrong?

38. Is the following statement true or false and why? If A^3 and B^3 have a common factor, then A and B have a common factor.

SKILL MAINTENANCE

Multiply.

39. $(3x + 5)(3x - 5)$ **40.** $(3x + 5)^2$

41. $(x - 7)(x + 4)$ **42.** $(x + 1)(x^2 - x + 1)$

43. In 2005, 55.2 million people will have high-speed Internet access at work. This is 2.3 times the number of people having high-speed access in 2000. (*Source*: Jupiter Media Metrix) How many people had high-speed access in 2000?

44. In January 2001, the Organization of the Petroleum Exporting Countries (OPEC) cut its oil production to 31.3 million barrels per day. This was a decrease of 1.5 million barrels per day. (*Source*: U.S. Department of Energy) What was the daily oil production before the reduction?

SYNTHESIS

45. If $x^3 + c$ is prime, what can you conclude about c? Why?

46. Explain how the geometric model below can be used to verify the formula for factoring $a^3 - b^3$.

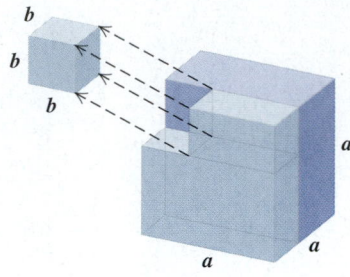

Factor. Assume that variables in exponents represent natural numbers.

47. $125c^6 + 8d^6$ **48.** $64x^6 + 8t^6$

49. $3x^{3a} - 24y^{3b}$ **50.** $\frac{8}{27}x^3 - \frac{1}{64}y^3$

51. $\frac{1}{24}x^3y^3 + \frac{1}{3}z^3$ **52.** $\frac{1}{16}x^{3a} + \frac{1}{2}y^{6a}z^{9b}$

Aha! **53.** $(x + 5)^3 + (x - 5)^3$ **54.** $x^3 - (x + y)^3$

55. $t^4 - 8t^3 - t + 8$ **56.** $5x^3y^6 - \frac{5}{8}$

5.6

Factoring: A General Strategy

Choosing the Right Method

CONNECTING THE CONCEPTS

Thus far, each section in this chapter has examined one or two different methods for factoring polynomials. In practice, when the need for factoring a polynomial arises, we must decide on our own which method to use. As preparation for such a situation, we now encounter polynomials of various types, in random order. Regardless of the polynomial we are faced with, the following guidelines can always be used.

> ### *To Factor a Polynomial*
>
> **A.** Always look for a common factor first. If there is one, factor out the largest common factor. Be sure to include it in your final answer.
>
> **B.** Then look at the number of terms.
>
> *Two terms*: Try factoring as a difference of squares first. Next, try factoring as a sum or a difference of cubes. Do not try to factor a sum of squares: $A^2 + B^2$.
>
> *Three terms*: Determine whether the trinomial is a perfect-square trinomial. If so, factor accordingly. If not, try trial and error, using the standard method or grouping.
>
> *Four terms*: Try factoring by grouping.
>
> **C.** Always *factor completely*. When a factor with more than one term can itself be factored, be sure to factor it.
>
> **D.** Check by multiplying.

Choosing the Right Method

Example 1

Factor: $5t^4 - 80$.

Solution

A. We look for a common factor:

$$5t^4 - 80 = 5(t^4 - 16). \qquad \text{5 is the largest common factor.}$$

B. The factor $t^4 - 16$ is a difference of squares: $(t^2)^2 - 4^2$. We factor it, being careful to rewrite the 5 from step (A):

$$5t^4 - 80 = 5(t^2 + 4)(t^2 - 4). \qquad t^4 - 16 = (t^2 + 4)(t^2 - 4)$$

C. Since $t^2 - 4$ is not prime, we continue factoring:

$$5t^4 - 80 = 5(t^2 + 4)(t^2 - 4) = 5(t^2 + 4)(t - 2)(t + 2)$$

This is a sum of squares with no common factor. It cannot be factored!

D. *Check:* $5(t^2 + 4)(t - 2)(t + 2) = 5(t^2 + 4)(t^2 - 4)$
$$= 5(t^4 - 16) = 5t^4 - 80.$$

The factorization is $5(t^2 + 4)(t - 2)(t + 2)$.

E x a m p l e   2

Factor: $2x^3 + 10x^2 + x + 5$.

Solution

A. We look for a common factor. There is none.

B. Because there are four terms, we try factoring by grouping:

$$2x^3 + 10x^2 + x + 5$$

$$= (2x^3 + 10x^2) + (x + 5) \qquad \text{Separating into two binomials}$$

$$= 2x^2(x + 5) + 1(x + 5) \qquad \text{Factoring out the largest common factor from each binomial. The 1 serves as an aid.}$$

$$= (x + 5)(2x^2 + 1) \qquad \text{Factoring out the common factor, } x + 5$$

C. Nothing can be factored further, so we have factored completely.

D. *Check:* $(x + 5)(2x^2 + 1) = 2x^3 + x + 10x^2 + 5$

$$= 2x^3 + 10x^2 + x + 5.$$

The factorization is $(x + 5)(2x^2 + 1)$.

E x a m p l e   3

Factor: $x^5 - 2x^4 - 35x^3$.

Solution

A. We note that there is a common factor, x^3:

$$x^5 - 2x^4 - 35x^3 = x^3(x^2 - 2x - 35).$$

B. The factor $x^2 - 2x - 35$ is not a perfect-square trinomial. We factor it using trial and error:

$$x^5 - 2x^4 - 35x^3 = x^3(x^2 - 2x - 35)$$

$$= x^3(x - 7)(x + 5).$$

C. Nothing can be factored further, so we have factored completely.

D. *Check:* $x^3(x - 7)(x + 5) = x^3(x^2 - 2x - 35)$

$$= x^5 - 2x^4 - 35x^3.$$

The factorization is $x^3(x - 7)(x + 5)$.

E x a m p l e   4

Factor: $x^2 - 20x + 100$.

Solution

A. We look first for a common factor. There is none.

B. This polynomial is a perfect-square trinomial. We factor it accordingly:

$$x^2 - 20x + 100 = x^2 - 2 \cdot x \cdot 10 + 10^2 \qquad \text{Try to do this step mentally.}$$

$$= (x - 10)^2.$$

C. Nothing can be factored further, so we have factored completely.
D. *Check:* $(x - 10)(x - 10) = x^2 - 20x + 100$.

The factorization is $(x - 10)(x - 10)$, or $(x - 10)^2$.

E x a m p l e   5

Factor: $6x^2y^4 - 21x^3y^5 + 3x^2y^6$.

Solution

A. We first factor out the largest common factor, $3x^2y^4$:

$$6x^2y^4 - 21x^3y^5 + 3x^2y^6 = 3x^2y^4(2 - 7xy + y^2).$$

B. There are three terms in $2 - 7xy + y^2$. Since only y^2 is a square, we do not have a perfect-square trinomial. Can $2 - 7xy + y^2$ be factored by trial and error? A key to the answer is that x appears only in $-7xy$. If $2 - 7xy + y^2$ could be factored into a form like $(1 - y)(2 - y)$, there would be no x in the middle term. Thus, $2 - 7xy + y^2$ cannot be factored.
C. Nothing can be factored further, so we have factored completely.
D. *Check:* $3x^2y^4(2 - 7xy + y^2) = 6x^2y^4 - 21x^3y^5 + 3x^2y^6$.

The factorization is $3x^2y^4(2 - 7xy + y^2)$.

E x a m p l e   6

Factor: $x^6 - 64$.

Solution

A. Look for a common factor. There is none (other than 1 or -1).
B. There are two terms, a difference of squares: $(x^3)^2 - (8)^2$. We factor it:

$$x^6 - 64 = (x^3 + 8)(x^3 - 8). \qquad \text{Note that } x^6 = (x^3)^2.$$

C. One factor is a sum of two cubes, and the other factor is a difference of two cubes. We factor both:

$$x^6 - 64 = (x + 2)(x^2 - 2x + 4)(x - 2)(x^2 + 2x + 4).$$

The factorization is complete because no factor can be factored further.
D. *Check:* $(x + 2)(x^2 - 2x + 4)(x - 2)(x^2 + 2x + 4) = (x^3 + 8)(x^3 - 8)$
$$= x^6 - 64.$$

The factorization is $(x + 2)(x^2 - 2x + 4)(x - 2)(x^2 + 2x + 4)$.

E x a m p l e   7

Factor: $25x^2 + 20xy + 4y^2$.

Solution

A. We look first for a common factor. There is none.

B. There are three terms. Note that the first term and the last term are squares:

$$25x^2 = (5x)^2 \quad \text{and} \quad 4y^2 = (2y)^2.$$

We see that twice the product of $5x$ and $2y$ is the middle term,

$$2 \cdot 5x \cdot 2y = 20xy,$$

so the trinomial is a perfect square.

To factor, we write a binomial squared. The binomial is the sum of the terms being squared:

$$25x^2 + 20xy + 4y^2 = (5x + 2y)^2.$$

C. Nothing can be factored further, so we have factored completely.
D. *Check:* $(5x + 2y)(5x + 2y) = 25x^2 + 20xy + 4y^2$.

The factorization is $(5x + 2y)(5x + 2y)$, or $(5x + 2y)^2$.

E x a m p l e 8

Factor: $p^2q^2 + 7pq + 12$.

Solution

A. We look first for a common factor. There is none.
B. Since only one term is a square, we do not have a perfect-square trinomial. We use trial and error, thinking of the product pq as a single variable:

$$(pq + \quad)(pq + \quad).$$

We factor the last term, 12. All the signs are positive, so we consider only positive factors. Possibilities are 1, 12 and 2, 6 and 3, 4. The pair 3, 4 gives a sum of 7 for the coefficient of the middle term. Thus,

$$p^2q^2 + 7pq + 12 = (pq + 3)(pq + 4).$$

C. Nothing can be factored further, so we have factored completely.
D. *Check:* $(pq + 3)(pq + 4) = p^2q^2 + 7pq + 12$.

The factorization is $(pq + 3)(pq + 4)$.

Compare the variables appearing in Example 7 with those appearing in Example 8. Note that when one variable appears in the leading term and another variable appears in the last term, as in Example 7, each binomial contains two variable terms. When two variables appear in the leading term, as in Example 8, each binomial contains just one variable term.

E x a m p l e 9

Factor: $a^4 - 16b^4$.

Solution

A. We look first for a common factor. There is none.
B. There are two terms. Since $a^4 = (a^2)^2$ and $16b^4 = (4b^2)^2$, we see that we do have a difference of squares. Thus,

$$a^4 - 16b^4 = (a^2 + 4b^2)(a^2 - 4b^2).$$

C. The factor $(a^2 - 4b^2)$ is itself a difference of squares. Thus,

$$a^4 - 16b^4 = (a^2 + 4b^2)\,(a + 2b)\,(a - 2b). \qquad \text{Factoring } a^2 - 4b^2$$

D. *Check:* $(a^2 + 4b^2)(a + 2b)(a - 2b) = (a^2 + 4b^2)(a^2 - 4b^2)$
$$= a^4 - 16b^4.$$

The factorization is $(a^2 + 4b^2)(a + 2b)(a - 2b)$.

Exercise Set 5.6

Factor completely. If a polynomial is prime, state this.

1. $5x^2 - 45$

2. $10a^2 - 640$

3. $a^2 + 25 + 10a$

4. $y^2 + 49 - 14y$

5. $8t^2 - 18t - 5$

6. $2t^2 + 11t + 12$

7. $x^3 - 24x^2 + 144x$

8. $x^3 - 18x^2 + 81x$

9. $x^3 + 3x^2 - 4x - 12$

10. $x^3 - 5x^2 - 25x + 125$

11. $98t^2 - 18$

12. $27t^3 - 3t$

13. $20x^3 - 4x^2 - 72x$

14. $9x^3 + 12x^2 - 45x$

15. $x^2 + 4$

16. $t^2 + 25$

17. $a^4 + 8a^2 + 8a^3 + 64a$

18. $t^4 + 7t^2 - 3t^3 - 21t$

19. $x^5 - 14x^4 + 49x^3$

20. $2x^6 + 8x^5 + 8x^4$

21. $20 - 6x - 2x^2$

22. $45 - 3x - 6x^2$

23. $t^2 - 7t - 6$

24. $t^2 - 8t + 6$

25. $4x^4 - 64$

26. $5x^5 - 80x$

27. $9 + t^8$

28. $t^4 - 9$

29. $x^5 - 4x^4 + 3x^3$

30. $x^6 - 2x^5 + 7x^4$

31. $x^3 - y^3$

32. $8t^3 + 1$

33. $12n^2 + 24n^3$

34. $ax^2 + ay^2$

35. $ab^2 - a^2b$

36. $36mn - 9m^2n^2$

37. $2\pi rh + 2\pi r^2$

38. $4\pi r^2 + 2\pi r$

Aha! **39.** $(a + b)(x - 3) + (a + b)(x + 4)$

40. $5c(a^3 + b) - (a^3 + b)$

41. $n^2 + 2n + np + 2p$

42. $x^2 + x + xy + y$

43. $2x^2 - 4x + xz - 2z$

44. $a^2 - 3a + ay - 3y$

45. $x^2 + y^2 + 2xy$

46. $3x^2 + 13xy - 10y^2$

47. $9c^2 - 6cd + d^2$

48. $4b^2 + a^2 - 4ab$

49. $7p^4 - 7q^4$

50. $a^4b^4 - 16$

51. $25z^2 + 10zy + y^2$

52. $4x^2y^2 + 12xyz + 9z^2$

53. $m^6 - 1$

54. $64t^6 - 1$

55. $a^2 + ab + 2b^2$

56. $4p^2q - pq^2 + 4p^3$

57. $2mn - 360n^2 + m^2$

58. $3b^2 + 17ab - 6a^2$

59. $m^2n^2 - 4mn - 32$

60. $12 + x^2y^2 + 8xy$

61. $a^5b^2 + 3a^4b - 10a^3$

62. $p^2q^2 + 7pq + 6$

63. $54a^4 + 16ab^3$

64. $54x^3y - 250y^4$

65. $2s^6t^2 + 10s^3t^3 + 12t^4$

66. $x^6 + x^5y - 2x^4y^2$

67. $a^2 + 2a^2bc + a^2b^2c^2$

68. $36a^2 - 15a + \frac{25}{16}$

69. $\frac{1}{81}x^2 - \frac{8}{27}x + \frac{16}{9}$

70. $\frac{1}{4}a^2 + \frac{1}{3}ab + \frac{1}{9}b^2$

71. $1 - 16x^{12}y^{12}$

72. $b^4 - 81a^5$

73. Kelly factored $16 - 8x + x^2$ as $(x - 4)^2$, while Tony factored it as $(4 - x)^2$. Are they both correct? Why or why not?

74. Describe in your own words a strategy for factoring polynomials.

SKILL MAINTENANCE

75. Show that the pairs $(-1, 11)$, $(0, 7)$, and $(3, -5)$ are solutions of $y = -4x + 7$.

Solve.

76. $5x - 4 = 0$ **77.** $3x + 7 = 0$

78. $2x + 9 = 0$ **79.** $4x - 9 = 0$

80. Graph: $y = -\frac{1}{2}x + 4$.

SYNTHESIS

 **81.** There are third-degree polynomials in x that we are not yet able to factor, despite the fact that they are not prime. Explain how such a polynomial could be created.

 **82.** Describe a method that could be used to find a binomial of degree 16 that can be expressed as the product of prime binomial factors.

Factor.

83. $-(x^5 + 7x^3 - 18x)$ **84.** $18 + a^3 - 9a - 2a^2$

85. $3a^4 - 15a^2 + 12$ **86.** $x^4 - 7x^2 - 18$

Aha! **87.** $y^2(y + 1) - 4y(y + 1) - 21(y + 1)$

88. $y^2(y - 1) - 2y(y - 1) + (y - 1)$

89. $6(x - 1)^2 + 7y(x - 1) - 3y^2$

90. $(y + 4)^2 + 2x(y + 4) + x^2$

91. $2(a + 3)^4 - (a + 3)^3(b - 2) - (a + 3)^2(b - 2)^2$

92. $5(t - 1)^5 - 6(t - 1)^4(s - 1) + (t - 1)^3(s - 1)^2$

CORNER

Matching Factorizations*

Focus: Factoring

Time: 20 minutes

Group size: Begin with the entire class. The end result is pairs of students. If there is an odd number of students, the instructor should participate.

Materials: Prepared sheets of paper, pins or tape. On half of the sheets, the instructor writes a polynomial. On the remaining sheets, the instructor writes the factorization of those polynomials. The activity is more interesting if the polynomials and factorizations are similar; for example,

$$x^2 - 2x - 8, \quad (x - 2)(x - 4),$$
$$x^2 - 6x + 8, \quad (x - 1)(x - 8),$$
$$x^2 - 9x + 8, \quad (x + 2)(x - 4).$$

ACTIVITY

1. As class members enter the room, the instructor pins or tapes either a polynomial or a factorization to the back of each student. Class members are told only whether their sheet of paper contains a polynomial or a factorization. All students should remain quiet and not tell others what is on their sheet of paper.

2. After all students are wearing a sheet of paper, they should mingle with one another, attempting to match up their factorization with the appropriate polynomial or vice versa. They may ask "yes/no" questions of one another that relate to factoring and polynomials. Answers to the questions should be yes or no. For example, a legitimate question might be "Is my last term negative?", "Do my factors have opposite signs?", or "Does my factorization include the factors of 6?"

3. The game is over when all factorization/polynomial pairs have "found" one another.

---

*Thanks to Jann MacInnes for suggesting this activity.

Solving Polynomial Equations by Factoring

5.7

The Principle of Zero Products • Factoring to Solve Equations • Graphing and Quadratic Equations

---

C O N N E C T I N G   T H E   C O N C E P T S

Chapter 4 and Sections 5.1–5.6 have been devoted to finding equivalent expressions. Whether we are adding, subtracting, multiplying, dividing, or factoring polynomials, the result of our work is an expression that is equivalent to the original expression.

Here in Section 5.7 we return to the task of solving equations. This time, however, the equations will contain a variable raised to a power greater than 1 and will often have more than one solution. Our ability to factor will play a pivotal role in solving such equations.

Whenever two polynomials are set equal to each other, we have a **polynomial equation**. Some examples of polynomial equations are

$$x^2 - x = 6 \quad \text{and} \quad 3y^4 + 2y^2 + 2 = 0.$$

The *degree of a polynomial equation* is the same as the highest degree of any term in the equation. Thus the degree of each equation listed above is 2 and 4. Second-degree equations like $x^2 - x = 6$ and $x^2 + 6x + 5 = 0$ are said to be **quadratic**.

> ### *Quadratic Equation*
> A *quadratic equation* is an equation equivalent to one of the form
> $$ax^2 + bx + c = 0, \quad \text{where } a \neq 0.$$

In order to solve quadratic and other polynomial equations, we need to develop a new principle.

The Principle of Zero Products

Suppose we are told that the product of two numbers is 6. On the basis of this information, it is impossible to know the value of either number—the product could be $2 \cdot 3$, $6 \cdot 1$, $12 \cdot \frac{1}{2}$, and so on. However, if we are told that the product of two numbers is 0, we know that at least one of the two numbers must itself be 0. For example, if $(x + 3)(x - 2) = 0$, we can conclude that either $x + 3$ is 0 or $x - 2$ is 0.

> ### *The Principle of Zero Products*
> An equation $AB = 0$ is true if and only if $A = 0$ or $B = 0$, or both. (A product is 0 if and only if at least one factor is 0.)

Example 1

Solve: $(x + 3)(x - 2) = 0$.

Solution   We are told that the product of $x + 3$ and $x - 2$ is 0. In order for a product to be 0, at least one factor must be 0. We reason that either

$$x + 3 = 0 \quad or \quad x - 2 = 0. \qquad \text{Using the principle of zero products}$$

We solve each equation:

$$x + 3 = 0 \quad or \quad x - 2 = 0$$
$$x = -3 \quad or \quad \qquad x = 2.$$

Both -3 and 2 can be checked in the original equation.

Check: For -3:

$$\frac{(x + 3)(x - 2) = 0}{(-3 + 3)(-3 - 2) \ ? \ 0}$$
$$0(-5) \ \Big| $$
$$0 \ \Big| \ 0 \ \text{TRUE}$$

For 2:

$$\frac{(x + 3)(x - 2) = 0}{(2 + 3)(2 - 2) \ ? \ 0}$$
$$5(0) \ \Big| $$
$$0 \ \Big| \ 0 \ \text{TRUE}$$

The solutions are -3 and 2.

When we are using the principle of zero products, the word "or" is meant to emphasize that any one of the factors could be the one that represents 0.

Example 2

Solve: $(5x + 1)(x - 7) = 0$.

Solution   We have

$$(5x + 1)(x - 7) = 0$$
$$5x + 1 = 0 \quad or \quad x - 7 = 0 \qquad \text{Using the principle of zero products}$$
$$5x = -1 \quad or \quad \qquad x = 7 \qquad \text{Solving the two equations separately}$$
$$x = -\tfrac{1}{5} \quad or \quad \qquad x = 7.$$

Check: For $-\tfrac{1}{5}$:

$$\frac{(5x + 1)(x - 7) = 0}{\left(5\left(-\tfrac{1}{5}\right) + 1\right)\left(-\tfrac{1}{5} - 7\right) \ ? \ 0}$$
$$(-1 + 1)\left(-7\tfrac{1}{5}\right) \ \Big|$$
$$0\left(-7\tfrac{1}{5}\right) \ \Big|$$
$$0 \ \Big| \ 0 \ \text{TRUE}$$

For 7:

$$\frac{(5x + 1)(x - 7) = 0}{(5(7) + 1)(7 - 7) \ ? \ 0}$$
$$(35 + 1)0 \ \Big|$$
$$0 \ \Big| \ 0 \ \text{TRUE}$$

The solutions are $-\tfrac{1}{5}$ and 7.

The principle of zero products can be used whenever a product equals 0—even if a factor has only one term.

Example 3

Solve: $3t(t - 5) = 0$.

Solution We have

$$3t(t - 5) = 0 \qquad \text{The factors are } 3t \text{ and } t - 5.$$
$$3t = 0 \quad or \quad t - 5 = 0 \qquad \text{Using the principle of zero products}$$
$$t = 0 \quad or \qquad t = 5. \qquad \text{Solving the two equations separately}$$

The solutions are 0 and 5. The check is left to the student.

Factoring to Solve Equations

By factoring and using the principle of zero products, we can now solve a variety of quadratic equations.

Example 4

Solve: $x^2 + 5x + 6 = 0$.

Solution This equation differs from those solved in Chapter 2. There are no like terms to combine, and there is a squared term. We first factor the polynomial. Then we use the principle of zero products:

$$x^2 + 5x + 6 = 0$$
$$(x + 2)(x + 3) = 0 \qquad \text{Factoring}$$
$$x + 2 = 0 \quad or \quad x + 3 = 0 \qquad \text{Using the principle of zero products}$$
$$x = -2 \quad or \qquad x = -3.$$

Check: For -2:

$$\frac{x^2 + 5x + 6 = 0}{(-2)^2 + 5(-2) + 6 \;?\; 0}$$
$$4 - 10 + 6$$
$$-6 + 6$$
$$0 \;\big|\; 0 \quad \text{TRUE}$$

For -3:

$$\frac{x^2 + 5x + 6 = 0}{(-3)^2 + 5(-3) + 6 \;?\; 0}$$
$$9 - 15 + 6$$
$$-6 + 6$$
$$0 \;\big|\; 0 \quad \text{TRUE}$$

The solutions are -2 and -3.

The principle of zero products is used even if the factoring consists of only removing a common factor.

Example 5

Solve: $x^2 + 7x = 0$.

Solution Although there is no constant term, the equation is still quadratic. Thus the methods of Chapter 2 are not sufficient. We try factoring instead:

$$x^2 + 7x = 0$$
$$x(x + 7) = 0 \qquad \text{Removing the greatest common factor, } x$$
$$x = 0 \quad or \quad x + 7 = 0$$
$$x = 0 \quad or \qquad x = -7.$$

The solutions are 0 and -7. The check is left to the student.

> **Caution!**   We *must* have 0 on one side of the equation before the principle of zero products can be used. Get all nonzero terms on one side and 0 on the other.

E x a m p l e   6   Solve: **(a)** $x^2 - 8x = -16$; **(b)** $4t^2 = 25$.

Solution

a) We first add 16 to get 0 on one side:

$$x^2 - 8x = -16$$
$$x^2 - 8x + 16 = 0 \qquad \text{Adding 16 to both sides to get 0 on one side}$$
$$(x - 4)(x - 4) = 0 \qquad \text{Factoring}$$
$$x - 4 = 0 \quad or \quad x - 4 = 0 \qquad \text{Using the principle of zero products}$$
$$x = 4 \quad or \qquad\quad x = 4.$$

There is only one solution, 4. The check is left to the student.

b) We have

$$4t^2 = 25$$
$$4t^2 - 25 = 0 \qquad \text{Subtracting 25 from both sides to get 0 on one side}$$
$$(2t - 5)(2t + 5) = 0 \qquad \text{Factoring a difference of squares}$$
$$2t - 5 = 0 \quad or \quad 2t + 5 = 0$$
$$2t = 5 \quad or \qquad 2t = -5 \qquad \text{Solving the two equations separately}$$
$$t = \tfrac{5}{2} \quad or \qquad\quad t = -\tfrac{5}{2}.$$

The solutions are $\frac{5}{2}$ and $-\frac{5}{2}$. The check is left to the student.

When solving quadratic equations by factoring, remember that a factorization is not useful unless 0 is on the other side of the equation.

E x a m p l e   7   Solve: $(x + 3)(2x - 1) = 9$.

Solution   Be careful with an equation like this! Since we need 0 on one side, we multiply out the product on the left and then subtract 9 from both sides:

$$(x + 3)(2x - 1) = 9$$
$$2x^2 + 5x - 3 = 9 \qquad \text{Multiplying on the left}$$
$$2x^2 + 5x - 3 - 9 = 9 - 9 \qquad \text{Subtracting 9 from both sides to get 0 on one side}$$
$$2x^2 + 5x - 12 = 0$$
$$(2x - 3)(x + 4) = 0 \qquad \text{Factoring}$$
$$2x - 3 = 0 \quad or \quad x + 4 = 0 \qquad \text{Using the principle of zero products}$$
$$2x = 3 \quad or \qquad\quad x = -4$$
$$x = \tfrac{3}{2} \quad or \qquad\quad x = -4.$$

Check: For $\frac{3}{2}$:   For -4:

$$\frac{(x+3)(2x-1)=9}{(\frac{3}{2}+3)(2\cdot\frac{3}{2}-1)\ ?\ 9}$$
$$(\frac{9}{2})(2)$$
$$\qquad\qquad 9\ \big|\ 9\quad\text{TRUE}$$

$$\frac{(x+3)(2x-1)=9}{(-4+3)(2(-4)-1)\ ?\ 9}$$
$$(-1)(-9)$$
$$\qquad\qquad 9\ \big|\ 9\quad\text{TRUE}$$

The solutions are $\frac{3}{2}$ and -4.

We can use the principle of zero products to solve polynomials with degree greater than 2, if they can be factored.

Example 8

Solve: $3x^3 - 30x = 9x^2$.

Solution   We have

$$3x^3 - 30x = 9x^2$$

$$3x^3 - 9x^2 - 30x = 0 \qquad\qquad \text{Getting 0 on one side and writing in descending order}$$

$$3x(x^2 - 3x - 10) = 0 \qquad\qquad \text{Factoring out a common factor}$$

$$3x(x+2)(x-5) = 0 \qquad\qquad \text{Factoring the trinomial}$$

$$3x = 0 \quad or \quad x+2 = 0 \quad or \quad x-5 = 0 \qquad \text{Using the principle of zero products}$$

$$x = 0 \quad or \qquad x = -2 \quad or \qquad x = 5.$$

Check:

$$\frac{3x^3 - 30x = 9x^2}{3\cdot 0^3 - 30\cdot 0\ ?\ 9\cdot 0^2}$$
$$0 - 0\ \big|\ 9\cdot 0$$
$$0\ \big|\ 0\quad\text{TRUE}$$

$$\frac{3x^3 - 30x = 9x^2}{3(-2)^3 - 30(-2)\ ?\ 9(-2)^2}$$
$$3(-8) + 60\ \big|\ 9\cdot 4$$
$$-24 + 60\ \big|\ 36$$
$$36\ \big|\ 36\quad\text{TRUE}$$

$$\frac{3x^3 - 30x = 9x^2}{3\cdot 5^3 - 30\cdot 5\ ?\ 9\cdot 5^2}$$
$$3\cdot 125 - 150\ \big|\ 9\cdot 25$$
$$375 - 150\ \big|\ 225$$
$$225\ \big|\ 225\quad\text{TRUE}$$

The solutions are 0, -2, and 5.

Study Tip

The best way to prepare for a final exam is to do so over a period of at least two weeks. First review each chapter, studying the formulas, problems, properties, and procedures in the chapter Summary and Review. Then retake your quizzes and tests. If you miss any questions, spend extra time reviewing the corresponding topics. Watch the videotapes that accompany the text or use the InterAct Math Tutorial Software. Also consider participating in a study group or attending a tutoring or review session.

Graphing and Quadratic Equations

Recall from Chapter 3 that to find the x-intercept of a graph, we replace y with 0 and solve for x. This same procedure is used to find the x-intercepts of the graph of any equation of the form $y = ax^2 + bx + c$. Equations like this are graphed in Chapter 11. Their graphs are shaped like the following curves.

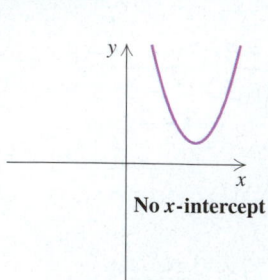

$$y = ax^2 + bx + c$$
$$a \neq 0$$

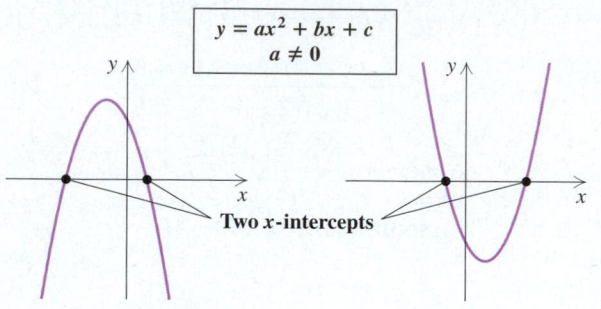

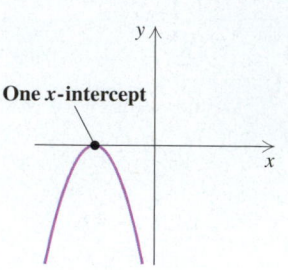

E x a m p l e   9   Find the x-intercepts for the graph of the equation shown. (The grid is intentionally not included.)

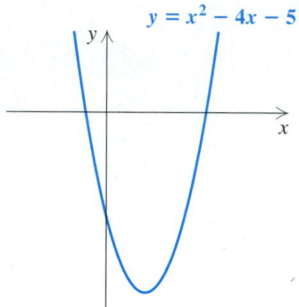

$y = x^2 - 4x - 5$

Solution   To find the x-intercepts, we let $y = 0$ and solve for x:

$$0 = x^2 - 4x - 5 \qquad \text{Substituting 0 for } y$$
$$0 = (x - 5)(x + 1) \qquad \text{Factoring}$$
$$x - 5 = 0 \quad or \quad x + 1 = 0 \qquad \text{Using the principle of zero products}$$
$$x = 5 \quad or \qquad x = -1. \qquad \text{Solving for } x$$

The x-intercepts are $(5, 0)$ and $(-1, 0)$.

A grapher allows us to solve quadratic equations by locating any x-intercepts that might exist. This technique is especially useful when an equation cannot be solved by factoring. As an example, to determine the intercepts in the figure shown at right, we can utilize the ZERO or ROOT option of the CALC menu. This option requires you to enter an x-value slightly less than the x-intercept as a LOWER BOUND. An x-value slightly more than the x-intercept is then entered as an UPPER BOUND. Finally, a GUESS value between the two bounds is entered and the x-intercept, or ZERO or ROOT, is displayed.

Use a grapher to find the solutions, if they exist, accurate to two decimal places.

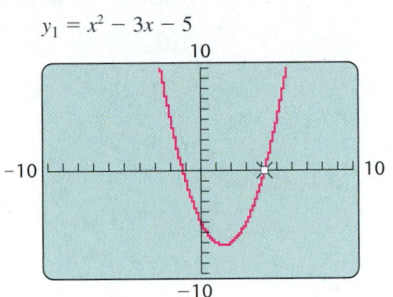

$y_1 = x^2 - 3x - 5$

Zero
X = 4.1925824, Y = 0

1. $x^2 + 4x - 3 = 0$

2. $x^2 - 5x - 2 = 0$

3. $x^2 + 13.54x + 40.95 = 0$

4. $x^2 - 4.43x + 6.32 = 0$

5. $1.235x^2 - 3.409x = 0$

FOR EXTRA HELP

Exercise Set 5.7

Digital Video Tutor CD 4
Videotape 10

InterAct Math

Math Tutor Center

MathXL

MyMathLab.com

Solve using the principle of zero products.

1. $(x + 5)(x + 6) = 0$

2. $(x + 1)(x + 2) = 0$

3. $(x - 3)(x + 7) = 0$

4. $(x + 9)(x - 3) = 0$

5. $(2x - 9)(x + 4) = 0$

6. $(3x - 5)(x + 1) = 0$

7. $(10x - 9)(4x + 7) = 0$

8. $(2x - 7)(3x + 4) = 0$

9. $x(x + 6) = 0$

10. $t(t + 9) = 0$

11. $\left(\frac{2}{3}x - \frac{12}{11}\right)\left(\frac{7}{4}x - \frac{1}{12}\right) = 0$

12. $\left(\frac{1}{9} - 3x\right)\left(\frac{1}{5} + 2x\right) = 0$

13. $5x(2x + 9) = 0$

14. $12x(4x + 5) = 0$

15. $(20 - 0.4x)(7 - 0.1x) = 0$

16. $(1 - 0.05x)(1 - 0.3x) = 0$

17. $(3x - 2)(x + 5)(x - 1) = 0$

18. $(2x + 1)(x + 3)(x - 5) = 0$

Solve by factoring and using the principle of zero products.

19. $x^2 + 7x + 6 = 0$

20. $x^2 - 6x + 5 = 0$

21. $x^2 - 4x - 21 = 0$

22. $x^2 - 7x - 18 = 0$

23. $x^2 - 6x = 0$

24. $x^2 + 8x = 0$

25. $x^3 - 3x^2 + 2x = 0$

26. $t^3 - 6t^2 - 7t = 0$

27. $9x^2 = 4$

28. $4x^2 = 49$

29. $0 = 25 + x^2 + 10x$

30. $0 = 6x + x^2 + 9$

31. $1 + x^2 = 2x$

32. $x^2 + 16 = 8x$

33. $8x^2 = 5x$

34. $3x^2 = 7x$

35. $3x^2 - 7x = 20$

36. $6x^2 - 4x = 10$

37. $2y^2 + 12y = -10$

38. $12y^2 - 5y = 2$

39. $(x - 7)(x + 1) = -16$

40. $(x + 2)(x - 7) = -18$

41. $y(3y + 1) = 2$

42. $t(t - 5) = 14$

43. $81x^2 - 5 = 20$

44. $36m^2 - 9 = 40$

45. $(x - 1)(5x + 4) = 2$

46. $(x + 3)(3x + 5) = 7$

47. $x^2 - 2x = 18 + 5x$

48. $3x^2 - 2x = 9 - 8x$

49. $x^2(2x - 1) = 3x$

50. $x^2 = x(10 - 3x^2)$

51. Use the following graph to solve $x^2 - 3x - 4 = 0$.

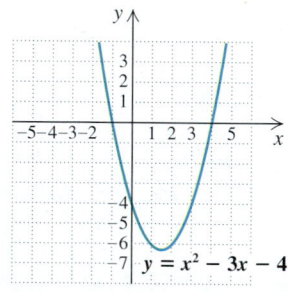

52. Use the following graph to solve $x^2 + x - 6 = 0$.

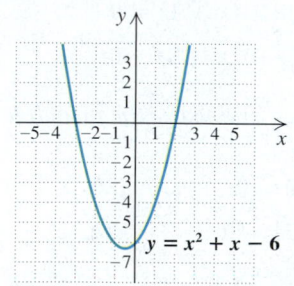

53. Use the following graph to solve $-x^2 + 2x + 3 = 0$.

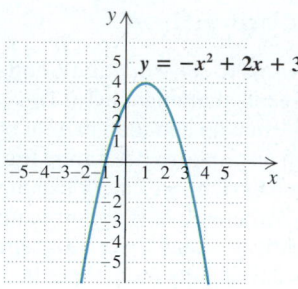

54. Use the following graph to solve $-x^2 - x + 6 = 0$.

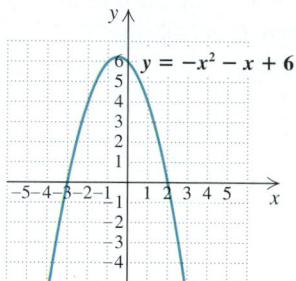

Find the x-intercepts for the graph of each equation. Grids are intentionally not included.

55. $y = x^2 + 3x - 4$

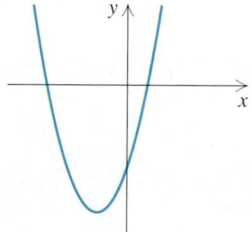

56. $y = x^2 - x - 6$

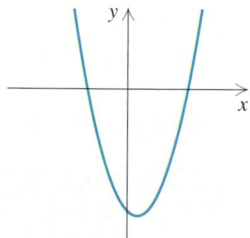

57. $y = x^2 - 2x - 15$

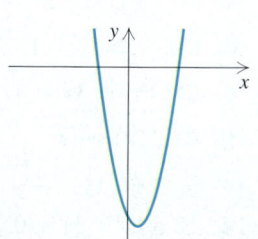

58. $y = x^2 + 2x - 8$

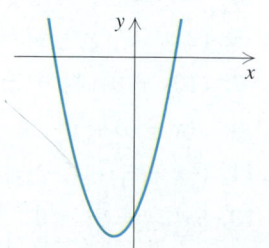

59. $y = 2x^2 + x - 10$

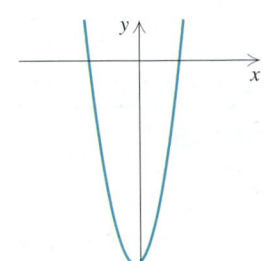

60. $y = 2x^2 + 3x - 9$

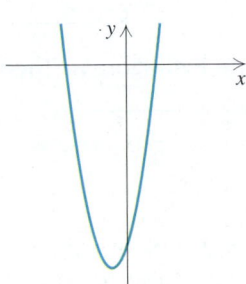

61. The equation $x^2 + 1$ has no real-number solutions. What implications does this have for the graph of $y = x^2 + 1$?

62. What is the difference between a quadratic polynomial and a quadratic equation?

SKILL MAINTENANCE

Translate to an algebraic expression.

63. The square of the sum of a and b

64. The sum of the squares of a and b

65. The sum of two consecutive integers

Translate to an inequality.

66. 5 more than twice a number is less than 19.

67. 7 less than half of a number exceeds 24.

68. 3 less than a number is at least 34.

SYNTHESIS

69. When the principle of zero products is used to solve a quadratic equation, will there always be two solutions? Why or why not?

70. What is wrong with solving $x^2 = 3x$ by dividing both sides of the equation by x?

Solve.

71. $(2x - 5)(x + 7)(3x + 8) = 0$

72. $(4x + 9)(3x - 2)(5x + 1) = 0$

73. Find an equation with integer coefficients that has the given numbers as solutions. For example, 3 and -2 are solutions to $x^2 - x - 6 = 0$.

a) $-4, 5$ **b)** $-1, 7$ **c)** $\frac{1}{4}, 3$
d) $\frac{1}{2}, \frac{1}{3}$ **e)** $\frac{2}{3}, \frac{3}{4}$ **f)** $-1, 2, 3$

Solve.

74. $16(x - 1) = x(x + 8)$

75. $a(9 + a) = 4(2a + 5)$

76. $(t - 5)^2 = 2(5 - t)$

77. $x^2 - \frac{9}{25} = 0$

78. $x^2 - \frac{25}{36} = 0$

Aha! **79.** $(t + 1)^2 = 9$

80. $\frac{27}{25}x^2 = \frac{1}{3}$

81. For each equation on the left, find an equivalent equation on the right.

a) $x^2 + 10x - 2 = 0$ $4x^2 + 8x + 36 = 0$
b) $(x - 6)(x + 3) = 0$ $(2x + 8)(2x - 5) = 0$
c) $5x^2 - 5 = 0$ $9x^2 - 12x + 24 = 0$
d) $(2x - 5)(x + 4) = 0$ $(x + 1)(5x - 5) = 0$
e) $x^2 + 2x + 9 = 0$ $x^2 - 3x - 18 = 0$
f) $3x^2 - 4x + 8 = 0$ $2x^2 + 20x - 4 = 0$

82. Explain how to construct an equation that has seven solutions.

83. Explain how the graph in Exercise 55 can be used to visualize the solutions of
$$x^2 + 3x - 4 = -6.$$

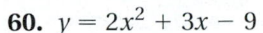 *Use a grapher to find the solutions of each equation. Round solutions to the nearest hundredth.*

84. $x^2 + 1.80x - 5.69 = 0$

85. $x^2 - 9.10x + 15.77 = 0$

86. $-x^2 + 0.63x + 0.22 = 0$

87. $x^2 + 13.74x + 42.00 = 0$

88. $6.4x^2 - 8.45x - 94.06 = 0$

89. $0.84x^2 - 2.30x = 0$

90. $1.23x^2 + 4.63x = 0$

5.8

Solving Applications

Applications • The Pythagorean Theorem

Applications

We can use the five-step problem-solving process and our new methods for solving quadratic equations to solve new types of problems.

Example 1

Page numbers.   The product of the page numbers on two consecutive pages of a book is 156. Find the page numbers.

Solution

1. **Familiarize.**   Consecutive page numbers are one apart, like 49 and 50. Let x = the first page number; then $x + 1$ = the next page number.

2. **Translate.**   We reword the problem before translating:

    Rewording: The first page number times the next page number is 156.

    Translating: x · $(x + 1)$ = 156

3. **Carry out.**   We solve the equation as follows:

    $$x(x + 1) = 156$$
    $$x^2 + x = 156 \qquad \text{Multiplying}$$
    $$x^2 + x - 156 = 0 \qquad \text{Subtracting 156 to get 0 on one side}$$
    $$(x - 12)(x + 13) = 0 \qquad \text{Factoring}$$
    $$x - 12 = 0 \quad or \quad x + 13 = 0 \qquad \text{Using the principle of zero products}$$
    $$x = 12 \quad or \qquad x = -13. \qquad \text{Solving each equation}$$

4. **Check.**   The solutions of the equation are 12 and -13. Since page numbers cannot be negative, -13 must be rejected. On the other hand, if x is 12, then $x + 1$ is 13 and $12 \cdot 13 = 156$. Thus, 12 checks.

5. **State.**   The page numbers are 12 and 13.

Example 2

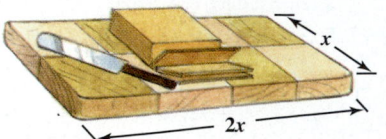

Manufacturing.   Wooden Work, Ltd., builds cutting boards that are twice as long as they are wide. The most popular board that Wooden Work makes has an area of 800 cm². What are the dimensions of the board?

Solution

1. **Familiarize.**   We first make a drawing. Recall that the area of any rectangle is Length · Width. We let x = the width of the board, in centimeters. The length is then $2x$.

2. **Translate.**   We reword and translate as follows:

Rewording: The area of the rectangle is 800 cm².

Translating: $2x \cdot x$ $= \ 800$

3. **Carry out.**   We solve the equation as follows:

$$2x \cdot x = 800$$
$$2x^2 = 800$$

$2x^2 - 800 = 0$ Subtracting 800 to get 0 on one side of
                               the equation

$2(x^2 - 400) = 0$ Removing a common factor of 2

$2(x - 20)(x + 20) = 0$ Factoring a difference of squares

$(x - 20)(x + 20) = 0$ Dividing both sides by 2

$x - 20 = 0$ *or* $x + 20 = 0$ Using the principle of
                                                zero products

$x = 20$ *or* $x = -20.$ Solving each equation

4. **Check.**   The solutions of the equation are 20 and -20. Since the width
   must be positive, -20 cannot be a solution. To check 20 cm, we note that if
   the width is 20 cm, then the length is $2 \cdot 20$ cm $= 40$ cm and the area is
   20 cm $\cdot$ 40 cm $= 800$ cm². Thus the solution 20 checks.

5. **State.**   The cutting board is 20 cm wide and 40 cm long.

E x a m p l e   3

Dimensions of a sail.   The mainsail of Stacey's Lightning-styled sailboat has
an area of 125 ft². If the sail is 15 ft taller than it is wide, find the height and the
width of the sail.

Solution

1. **Familiarize.**   We first make a drawing. The formula for the area of a triangle is Area $= \frac{1}{2} \cdot$ (base) $\cdot$ (height). We let $b =$ the width, in feet, of the
   triangle's base and $b + 15 =$ the height, in feet.

$b + 15$

b

2. **Translate.** We reword and translate as follows:

Rewording: The area of the sail is 125 ft².

Translating: $\frac{1}{2} \cdot b(b + 15)$ = 125.

3. **Carry out.** We solve the equation as follows:

$$\frac{1}{2} \cdot b \cdot (b + 15) = 125$$

$$\frac{1}{2}(b^2 + 15b) = 125 \qquad \text{Multiplying}$$

$$b^2 + 15b = 250 \qquad \text{Multiplying by 2 to clear fractions}$$

$$b^2 + 15b - 250 = 0 \qquad \text{Subtracting 250 to get 0 on one side}$$

$$(b + 25)(b - 10) = 0 \qquad \text{Factoring}$$

$$b + 25 = 0 \quad \textit{or} \quad b - 10 = 0 \quad \text{Using the principle of zero products}$$

$$b = -25 \quad \textit{or} \qquad b = 10.$$

4. **Check.** The width must be positive, so -25 cannot be a solution. Suppose the base is 10 ft. The height would be $10 + 15$, or 25 ft, and the area $\frac{1}{2}(10)(25)$, or 125 ft². These numbers check in the original problem.

5. **State.** Stacey's mainsail is 25 ft tall and 10 ft wide.

Example 4

Games in a league's schedule. In a sports league of x teams in which all teams play each other twice, the total number N of games played is given by

$$x^2 - x = N.$$

The Colchester Youth Soccer League plays a total of 240 games, with all teams playing each other twice. How many teams are in the league?

Solution

1. **Familiarize.** To familiarize yourself with this equation, reread Example 7 in Section 4.2, where we first considered it.

2. **Translate.** We are trying to find the number of teams x in a league in which 240 games are played and all teams play each other twice. We replace N with 240 in the formula above:

$$x^2 - x = 240. \qquad \text{Substituting 240 for } N. \text{ This is now an equation in one variable.}$$

3. **Carry out.** We solve the equation as follows:

$$x^2 - x = 240$$

$$x^2 - x - 240 = 0 \qquad \text{Subtracting 240 to get 0 on one side}$$

$$(x - 16)(x + 15) = 0 \qquad \text{Factoring}$$

$$x - 16 = 0 \quad \textit{or} \quad x + 15 = 0 \quad \text{Using the principle of zero products}$$

$$x = 16 \quad \textit{or} \qquad x = -15.$$

4. **Check.** Since the number of teams must be positive, -15 cannot be a solution. However, 16 checks, since $16^2 - 16 = 256 - 16 = 240$.

5. **State.** There are 16 teams in the league.

The Pythagorean Theorem

The following problems involve the Pythagorean theorem, which relates the lengths of the sides of a *right* triangle. A triangle is a **right triangle** if it has a 90°, or *right*, angle. The side opposite the 90° angle is called the **hypotenuse**. The other sides are called **legs**.

The Pythagorean Theorem

In any right triangle, if a and b are the lengths of the legs and c is the length of the hypotenuse, then

$$a^2 + b^2 = c^2.$$

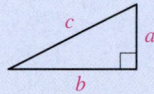

The symbol ⌐ denotes a 90° angle.

E x a m p l e   5

Right triangle geometry.   One leg of a right triangle is 7 m longer than the other. The length of the hypotenuse is 13 m. Find the lengths of the legs.

Solution

1. **Familiarize.** We make a drawing and let $x =$ the length of one leg, in meters. Since the other leg is 7 m longer, we know that $x + 7 =$ the length of the other leg, in meters. The hypotenuse has length 13 m.

2. **Translate.** Applying the Pythagorean theorem, we obtain the following translation:

$$a^2 + b^2 = c^2$$
$$x^2 + (x + 7)^2 = 13^2. \qquad \text{Substituting}$$

3. **Carry out.** We solve the equation as follows:

$x^2 + (x^2 + 14x + 49) = 169$	Squaring the binomial and 13
$2x^2 + 14x + 49 = 169$	Combining like terms
$2x^2 + 14x - 120 = 0$	Subtracting 169 to get 0 on one side
$2(x^2 + 7x - 60) = 0$	Factoring out a common factor
$2(x + 12)(x - 5) = 0$	Factoring
$x + 12 = 0 \quad or \quad x - 5 = 0$	Using the principle of zero products
$x = -12 \quad or \qquad x = 5.$	

4. **Check.** The integer -12 cannot be a length of a side because it is negative. When $x = 5$, $x + 7 = 12$, and $5^2 + 12^2 = 13^2$. So 5 checks.

5. **State.** The lengths of the legs are 5 m and 12 m.

E x a m p l e   6

Roadway design.   Elliott Street is 24 ft wide when it ends at Main Street in Brattleboro, Vermont. A 40-ft long diagonal crosswalk allows pedestrians to cross Main Street to or from either corner of Elliott Street (see the figure). Determine the width of Main Street.

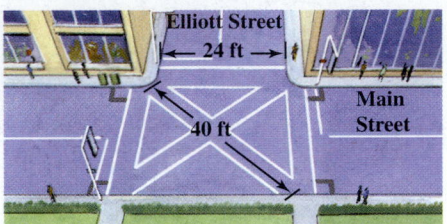

Solution

1. **Familiarize.**   A drawing has already been provided, but we can redraw and label the relevant part.

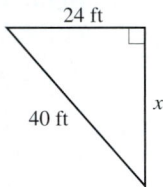

   Note that the two streets intersect at a right angle. We let $x =$ the width of Main Street, in feet.

2. **Translate.**   Since a right triangle is formed, we can use the Pythagorean theorem:

$$a^2 + b^2 = c^2$$
$$x^2 + 24^2 = 40^2. \qquad \text{Substituting}$$

3. **Carry out.**   We solve the equation as follows:

$$x^2 + 576 = 1600 \qquad \text{Squaring 24 and 40}$$
$$x^2 - 1024 = 0 \qquad \text{Subtracting 1600 from both sides}$$
$$(x - 32)(x + 32) = 0 \qquad \text{Note that } 1024 = 32^2. \text{ A calculator might be helpful here.}$$

$$x - 32 = 0 \quad or \quad x + 32 = 0 \qquad \text{Using the principle of zero products}$$
$$x = 32 \quad or \qquad x = -32.$$

4. **Check.**   Since the width of a street must be positive, -32 is not a solution. If the width is 32 ft, we have $32^2 + 24^2 = 1024 + 576 = 1600$, which is 40^2. Thus, 32 checks.

5. **State.**   The width of Main Street is 32 ft.

FOR EXTRA HELP

Digital Video Tutor CD 4
Videotape 10

InterAct Math

Math Tutor Center

MathXL

MyMathLab.com

Exercise Set 5.8

Solve. Use the five-step problem-solving approach.

1. A number is 6 less than its square. Find all such numbers.

2. A number is 2 less than its square. Find all such numbers.

3. One leg of a right triangle is 3 cm longer than the other leg. The length of the hypotenuse is 15 cm. Find the length of each side.

4. One leg of a right triangle is 2 cm shorter than the other leg. The length of the hypotenuse is 10 cm. Find the length of each side.

5. *Page numbers.* The product of the page numbers on two facing pages of a book is 110. Find the page numbers.

6. *Page numbers.* The product of the page numbers on two facing pages of a book is 210. Find the page numbers.

7. The product of two consecutive odd integers is 255. Find the integers.

8. The product of two consecutive even integers is 224. Find the integers.

9. *Framing.* A rectangular picture frame is twice as long as it is wide. If the area of the frame is 288 in², find its dimensions.

10. *Furnishings.* A rectangular table in Arlo's House of Tunes is six times as long as it is wide. If the area of the table is 24 ft², find the length and the width of the table.

11. *Design.* The keypad and viewing window of the TI-83 graphing calculator is rectangular. The length of the rectangle is 2 cm more than twice the width. If the area of the rectangle is 144 cm², find the length and the width.

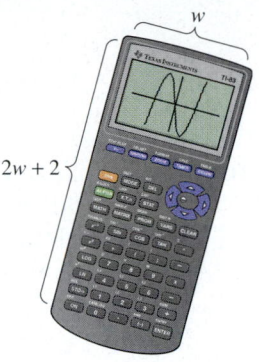

12. *Area of a garden.* The length of a rectangular garden is 4 m greater than the width. The area of the garden is 96 m². Find the length and the width.

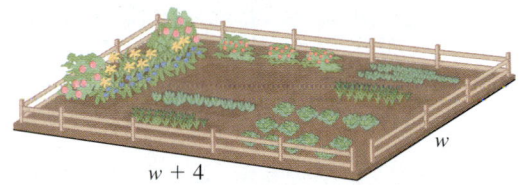

13. *Physical education.* An outdoor-education ropes course includes a cable that slopes downward from a height of 37 ft to a height of 30 ft. The trees that the cable connects are 24 ft apart. How long is the cable?

14. *Aviation.* Engine failure forced Geraldine to pilot her Cessna 150 to an emergency landing. To land, Geraldine's plane glided 17,000 ft over a 15,000-ft stretch of deserted highway. From what altitude did the descent begin?

15. *Dimensions of a triangle.*   A triangle is 10 cm wider than it is tall. The area is 28 cm². Find the height and the base.

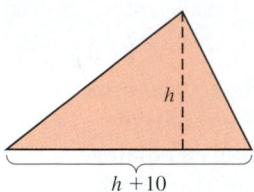

16. *Dimensions of a triangle.*   The height of a triangle is 3 cm less than the length of the base. The area is 35 cm². Find the height and the base.

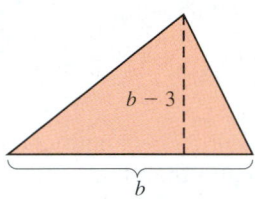

17. *Dimensions of a sail.*   The height of the jib sail on a Lightning sailboat is 5 ft greater than the length of its "foot." If the area of the sail is 42 ft², find the length of the foot and the height of the sail.

18. *Road design.*   A triangular traffic island has a base half as long as its height. Find the base and the height if the island has an area of 64 m².

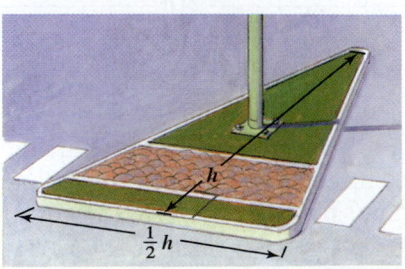

Games in a league.   *Use the formula from Example 4, x² − x = N, for Exercises 19–22. Assume that in each league teams play each other twice.*

19. A women's volleyball league has 20 teams. What is the total number of games to be played?

20. A chess league has 14 teams. What is the total number of games to be played?

21. A women's softball league plays a total of 132 games. How many teams are in the league?

22. A men's basketball league plays a total of 90 games. How many teams are in the league?

23. *Construction.*   The diagonal braces in a lookout tower are 15 ft long and span a distance of 12 ft. How high does each brace reach vertically?

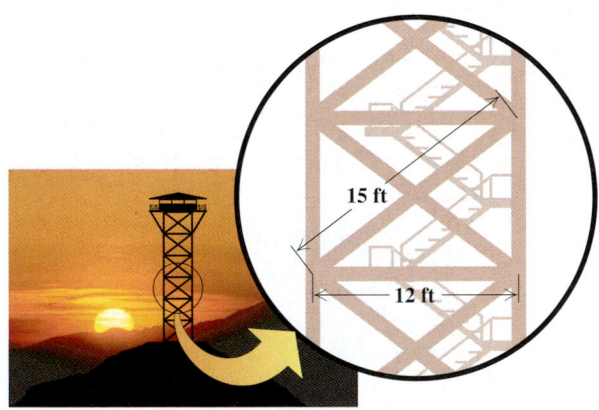

24. *Reach of a ladder.*   Twyla has a 26-ft ladder leaning against her house. If the bottom of the ladder is 10 ft from the base of the house, how high does the ladder reach?

Number of handshakes.   *The number of possible handshakes H within a group of n people is given by H = ½(n² − n). Use this formula for Exercises 25–28.*

25. At a meeting, there are 15 people. How many handshakes are possible?

26. At a party, there are 30 people. How many handshakes are possible?

27. *High-fives.*   After winning the championship, all Los Angeles Laker teammates exchanged "high-fives." Altogether there were 66 high-fives. How many players were there?

28. *Toasting.*   During a toast at a party, there were 190 "clicks" of glasses. How many people took part in the toast?

29. *Architecture.*   An architect has allocated a rectangular space of 264 ft^2 for a square dining room and a 10-ft wide kitchen, as shown in the figure. Find the dimensions of each room.

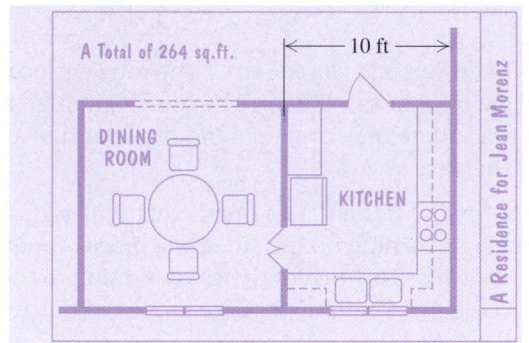

30. *Guy wire.*   The guy wire on a TV antenna is 1 m longer than the height of the antenna. If the guy wire is anchored 3 m from the foot of the antenna, how tall is the antenna?

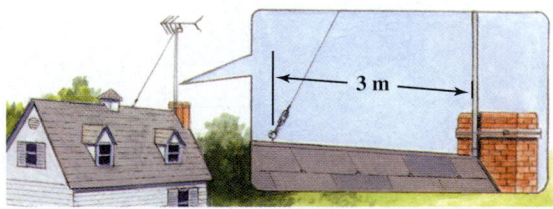

Height of a rocket.   *For Exercises 31–34, assume that a water rocket is launched upward with an initial velocity of 48 ft/sec. Its height h, in feet, after t seconds, is given by* $h = 48t - 16t^2$.

31. Determine the height of the rocket $\frac{1}{2}$ sec after it has been launched.

32. Determine the height 1.5 sec after the rocket has been launched.

33. When will the rocket be exactly 32 ft above the ground?

34. When will the rocket crash into the ground?

35. Write a problem for a classmate to solve such that only one of two solutions of a quadratic equation can be used as an answer.

36. Can we solve any problem that translates to a quadratic equation? Why or why not?

SKILL MAINTENANCE

Simplify.

37. $-\dfrac{2}{3} \cdot \dfrac{4}{7}$

38. $-\dfrac{4}{5} \cdot \dfrac{2}{9}$

39. $\dfrac{5}{6}\left(\dfrac{-7}{9}\right)$

40. $\dfrac{3}{8}\left(-\dfrac{5}{6}\right)$

41. $-\dfrac{2}{3} + \dfrac{4}{7}$

42. $-\dfrac{4}{5} + \dfrac{2}{9}$

43. $\dfrac{5}{6} + \dfrac{-7}{9}$

44. $\dfrac{3}{8} + \left(-\dfrac{5}{6}\right)$

SYNTHESIS

The converse of the Pythagorean theorem is also true. That is, if $a^2 + b^2 = c^2$, *then the triangle is a right triangle (where a and b are the lengths of the legs and c is the length of the hypotenuse). Use this result to answer Exercises 45 and 46.*

45. An archaeologist has measuring sticks of 3 ft, 4 ft, and 5 ft. Explain how she could draw a 7-ft by 9-ft rectangle on a piece of land being excavated.

Aha!

46. Explain how measuring sticks of 5 cm, 12 cm, and 13 cm can be used to draw a right triangle that has two 45° angles.

47. *Telephone service.*   Use the information in the figure below to determine the height of the telephone pole.

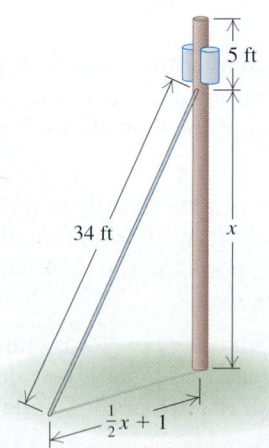

48. *Sailing.* The mainsail of a Lightning sailboat is a right triangle in which the hypotenuse is called the leech. If a 24-ft tall mainsail has a leech length of 26 ft and if Dacron® sailcloth costs $10 per square foot, find the cost of a new mainsail.

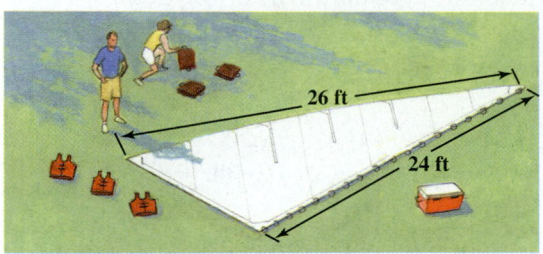

49. *Roofing.* A *square* of shingles covers 100 ft^2 of surface area. How many squares will be needed to reshingle the house shown?

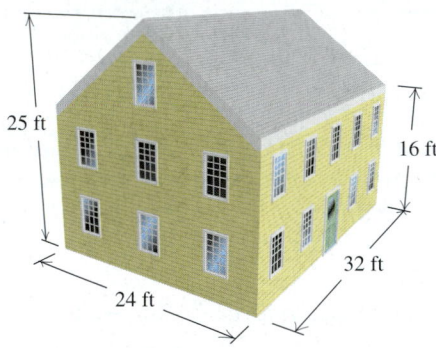

50. Solve for *x*.

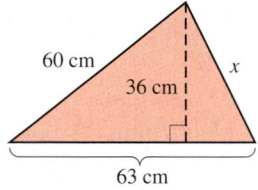

51. The ones digit of a number less than 100 is 4 greater than the tens digit. The sum of the number and the product of the digits is 58. Find the number.

52. *Pool sidewalk.* A cement walk of uniform width is built around a 20-ft by 40-ft rectangular pool. The total area of the pool and the walk is 1500 ft^2. Find the width of the walk.

53. *Dimensions of an open box.* A rectangular piece of cardboard is twice as long as it is wide. A 4-cm square is cut out of each corner, and the sides are

turned up to make a box with an open top. The volume of the box is 616 cm^3. Find the original dimensions of the cardboard.

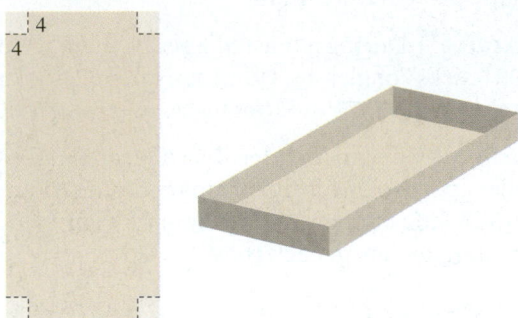

54. *Dimensions of a closed box.* The total surface area of a closed box is 350 m^2. The box is 9 m high and has a square base and lid. Find the length of a side of the base.

55. *Rain-gutter design.* An open rectangular gutter is made by turning up the sides of a piece of metal 20 in. wide. The area of the cross-section of the gutter is 48 in^2. Find the possible depths of the gutter.

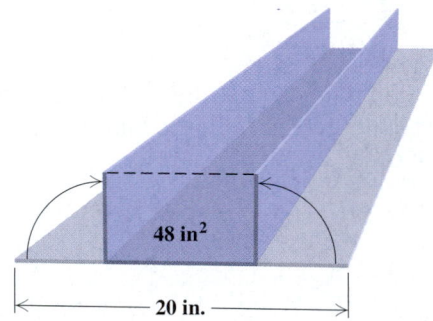

56. The length of each side of a square is increased by 5 cm to form a new square. The area of the new square is $2\frac{1}{4}$ times the area of the original square. Find the area of each square.

57. Find a polynomial for the shaded area in the figure below.

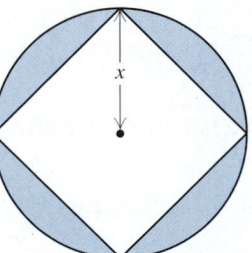

Summary and Review 5

Key Terms

Factor, p. 276

Common factor, p. 277

Factoring by grouping, p. 280

Prime polynomial, p. 287

Factor completely, p. 287

Perfect-square trinomial, p. 300

Difference of squares, p. 302

Sum of cubes, p. 307

Difference of cubes, p. 307

Polynomial equation, p. 317

Degree of a polynomial equation, p. 317

Quadratic equation, p. 317

Right triangle, p. 329

Hypotenuse, p. 329

Leg, p. 329

Important Properties and Formulas

Factoring Formulas

$A^2 + 2AB + B^2 = (A + B)^2;$
$A^2 - 2AB + B^2 = (A - B)^2;$
$A^2 - B^2 = (A + B)(A - B);$
$A^3 + B^3 = (A + B)(A^2 - AB + B^2);$
$A^3 - B^3 = (A - B)(A^2 + AB + B^2)$

To factor a polynomial:

A. Look first for a common factor. If there is one, factor out the largest common factor. Be sure to include it in your final answer.

B. Look at the number of terms.

Two terms: Try factoring as a difference of squares first. Next, try factoring as a sum or a difference of cubes.

Three terms: If you have a perfect-square trinomial, factor accordingly.

If not, try trial and error, using the standard method or grouping.

Four terms: Try factoring by grouping.

C. Always factor completely.

D. Check by multiplying.

The principle of zero products: $AB = 0$ is true if and only if $A = 0$ or $B = 0$, or both.

The Pythagorean theorem: $a^2 + b^2 = c^2$

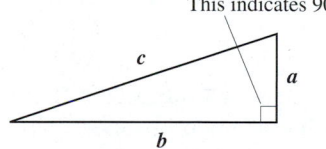

This indicates 90°.

Review Exercises

Find three factorizations of each monomial.

1. $36x^3$
2. $-20x^5$

Factor completely. If a polynomial is prime, state this.

3. $2x^4 + 6x^3$
4. $a^2 - 7a$

5. $4t^2 - 9$
6. $x^2 + 4x - 12$

7. $x^2 + 14x + 49$
8. $12x^3 + 12x^2 + 3x$

9. $6x^3 + 9x^2 + 2x + 3$
10. $6t^2 - 5t - 1$

11. $81a^4 - 1$
12. $9x^3 + 12x^2 - 45x$

13. $2x^2 - 50$
14. $x^4 + 4x^3 - 2x - 8$

15. $a^2b^4 - 36$
16. $8x^6 - 32x^5 + 4x^4$

17. $75 + 12x^2 + 60x$
18. $a^2 + 4$

19. $x^3 - x^2 - 30x$
20. $4x^2 - 25$

21. $9x^2 + 25 - 30x$
22. $6x^2 - 28x - 48$

23. $4t^2 - 13t + 10$
24. $2t^2 - 7t - 4$

25. $18x^2 - 12x + 2$
26. $x^3 - 27$

27. $15 - 8x + x^2$
28. $25x^2 - 20x + 4$

29. $x^2y^2 + xy - 12$
30. $12a^2 + 84ab + 147b^2$

31. $m^2 + 5m + mt + 5t$
32. $250y^3 + 128x^6$

Solve.

33. $(x - 1)(x + 3) = 0$
34. $x^3 + 2x^2 - 35x = 0$

35. $9x^2 = 1$
36. $3x^2 + 2 = 5x$

37. $2x^2 + 5x = 12$
38. $(x + 1)(x - 2) = 4$

39. The square of a number is 12 more than the number. Find all such numbers.

40. Find the x-intercepts for the graph of
$y = 2x^2 - 3x - 5$.

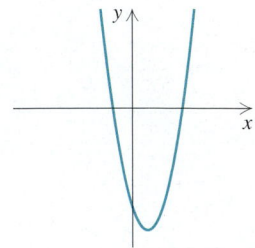

41. A triangular sign is as wide as it is tall. Its area is 800 cm². Find the height and the base.

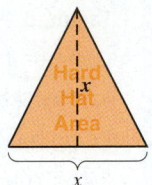

42. A guy wire from a ham radio antenna is 26 m long. It reaches from the top of the antenna to a point on the ground 10 m from the base of the antenna. How tall is the antenna?

SYNTHESIS

43. On a quiz, Edith writes the factorization of $4x^2 - 100$ as $(2x - 10)(2x + 10)$. If this were a 10-point question, how many points would you give Edith? Why?

44. How do the equations solved in this chapter differ from those solved in previous chapters?

Solve.

45. The pages of a book measure 15 cm by 20 cm. Margins of equal width surround the printing on each page and constitute one half of the area of the page. Find the width of the margins.

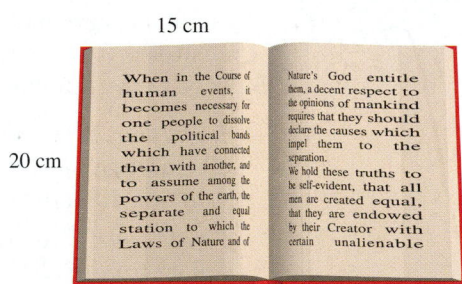

46. The cube of a number is the same as twice the square of the number. Find the number.

47. The length of a rectangle is two times its width. When the length is increased by 20 cm and the width is decreased by 1 cm, the area is 160 cm². Find the original length and width.

Solve.

48. $(x - 2)2x^2 + x(x - 2) - (x - 2)15 = 0$

Aha! **49.** $x^2 + 25 = 0$

Chapter Test 5

1. Find three factorizations of $8x^4$.

Factor completely.

2. $x^2 - 7x + 10$

3. $x^2 + 25 - 10x$

4. $6y^2 - 8y^3 + 4y^4$

5. $x^3 + x^2 + 2x + 2$

6. $x^2 - 5x$

7. $x^3 + 2x^2 - 3x$

8. $3t^3 + 3$

9. $4x^2 - 9$

10. $x^2 - x - 12$

11. $6m^3 + 9m^2 + 3m$

12. $3w^2 - 75$

13. $60x + 45x^2 + 20$

14. $3x^4 - 48$

15. $49x^2 - 84x + 36$

16. $5x^2 - 26x + 5$

17. $x^4 + 2x^3 - 3x - 6$

18. $2m^6 + 16$

19. $4x^2 - 4x - 15$

20. $6t^3 + 9t^2 - 15t$

21. $3m^2 - 9mn - 30n^2$

Solve.

22. $x^2 - x - 20 = 0$

23. $2x^3 + 7x^2 = 15x$

24. $x(x - 3) = 28$

25. Find the x-intercepts for the graph of $y = 3x^2 - 4x - 7$.

26. The length of a rectangle is 2 m more than the width. The area of the rectangle is 48 m². Find the length and the width.

27. A mason wants to be sure she has a right corner in a building's foundation. She marks a point 3 ft from the corner along one wall and another point 4 ft from the corner along the other wall. If the corner is a right angle, what should the distance be between the two marked points?

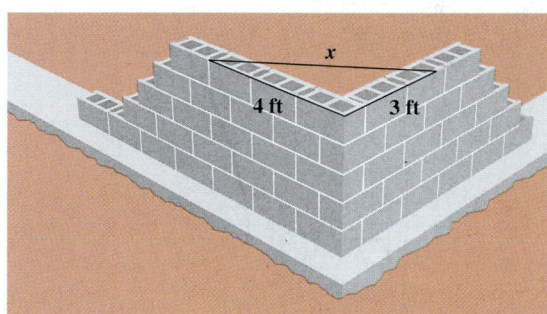

SYNTHESIS

28. The length of a rectangle is five times its width. When the length is decreased by 3 and the width is increased by 2, the area of the new rectangle is 60. Find the original length and width.

29. Factor: $(a + 3)^2 - 2(a + 3) - 35$.

30. Solve: $20x(x + 2)(x - 1) = 5x^3 - 24x - 14x^2$.

6
Rational Expressions and Equations

AN APPLICATION

In South Africa, the design of every woven handbag, or *gipatsi*, is created by repeating two or more patterns around the bag. If a weaver uses a four-strand, a six-strand, and an eight-strand pattern, what is the smallest number of strands needed in order for all three patterns to repeat a whole number of times?

This problem appears as Exercise 80 in Section 6.3.

***W**eavers use math when designing patterns, from calculating the size of the pattern and number of times it repeats to figuring the number of threads per inch and the weight of the thread. I also use math to convert between English and metric units when dyeing warps and woofs.*

ESTELLE CARLSON
**Handweaver and Designer
Los Angeles, California**

*J*ust as fractions are needed to solve certain arithmetic problems, rational expressions *similar to those in the following pages are needed to solve certain algebra problems. We now learn how to simplify, as well as add, subtract, multiply, and divide, rational expressions. These skills will then be used to solve the equations that arise from real-life problems like the one on the preceding page.*

Rational Expressions

6.1

Simplifying Rational Expressions • Factors That Are Opposites

Just as a rational number is a quotient of two integers, a **rational expression** is a quotient of two polynomials. The following are examples of rational expressions:

$$\frac{7}{3}, \qquad \frac{5}{x + 6}, \qquad \frac{t^2 - 5t + 6}{4t^2 - 7}.$$

Rational expressions are examples of *algebraic fractions.* They are also examples of *fractional expressions.*

Because rational expressions indicate division, we must be careful to avoid denominators that are 0. When a variable is replaced with a number that produces a denominator of 0, the rational expression is undefined. For example, in the expression

$$\frac{x + 3}{x - 7},$$

when x is replaced with 7, the denominator is 0, and the expression is undefined:

$$\frac{x + 3}{x - 7} = \frac{7 + 3}{7 - 7} = \frac{10}{0}. \quad \longleftarrow \text{Division by 0 is undefined.}$$

When x is replaced with a number other than 7—say, 6—the expression *is* defined because the denominator is not zero:

$$\frac{x + 3}{x - 7} = \frac{6 + 3}{6 - 7} = \frac{9}{-1} = -9.$$

E x a m p l e   1   Find all numbers for which the rational expression

$$\frac{x + 4}{x^2 - 3x - 10}$$

is undefined.

Solution   The value of the numerator has no bearing on whether or not a rational expression is defined. To determine which numbers make the rational expression undefined, we set the *denominator* equal to 0 and solve:

$$x^2 - 3x - 10 = 0$$

$$(x - 5)(x + 2) = 0 \qquad \text{Factoring}$$

$$x - 5 = 0 \quad or \quad x + 2 = 0 \qquad \text{Using the principle of zero products}$$

$$x = 5 \quad or \qquad x = -2. \qquad \text{Solving each equation}$$

Check:

For $x = 5$:

$$\frac{x + 4}{x^2 - 3x - 10} = \frac{5 + 4}{5^2 - 3 \cdot 5 - 10} \qquad \text{There are no restrictions on the numerator.}$$

$$= \frac{9}{25 - 15 - 10} = \frac{9}{0}. \qquad \text{This expression is undefined.}$$

For $x = -2$:

$$\frac{x + 4}{x^2 - 3x - 10} = \frac{-2 + 4}{(-2)^2 - 3(-2) - 10}$$

$$= \frac{2}{4 + 6 - 10} = \frac{2}{0}. \qquad \text{This expression is undefined.}$$

Thus, $\dfrac{x + 4}{x^2 - 3x - 10}$ is undefined for $x = 5$ and $x = -2$.

---

technology connection

To check Example 1 with a grapher, let $y_1 = x^2 - 3x - 10$ and $y_2 = (x + 4)/(x^2 - 3x - 10)$ or $(x + 4)/y_1$ and use the TABLE feature. Since $x^2 - 3x - 10$ is 0 for $x = -2$, we cannot evaluate y_2 for $x = -2$.

TBL MIN = −2   ΔTBL = 1

X	Y₁	Y₂
−2	0	ERROR
−1	−6	−.5
0	−10	−.4
1	−12	−.4167
2	−12	−.5
3	−10	−.7
4	−6	−1.333

X = −2

Simplifying Rational Expressions

Simplifying rational expressions is similar to simplifying the fractional expressions studied in Section 1.3. We saw, for example, that an expression like $\frac{15}{40}$ can be simplified as follows:

$$\frac{15}{40} = \frac{3 \cdot 5}{8 \cdot 5} \qquad \text{Factoring the numerator and the denominator. Note the common factor, 5.}$$

$$= \frac{3}{8} \cdot \frac{5}{5} \qquad \text{Rewriting as a product of two fractions}$$

$$= \frac{3}{8} \cdot 1 \qquad \frac{5}{5} = 1$$

$$= \frac{3}{8}. \qquad \text{Using the identity property of 1 to remove the factor 1}$$

Similar steps are followed when simplifying rational expressions: We factor and remove a factor equal to 1, using the fact that

$$\frac{ab}{cb} = \frac{a}{c} \cdot \frac{b}{b}.$$

E x a m p l e 2

Simplify: $\dfrac{8x^2}{24x}$.

Solution

$$\frac{8x^2}{24x} = \frac{8 \cdot x \cdot x}{3 \cdot 8 \cdot x}$$ Factoring the numerator and the denominator. Note the common factor of $8 \cdot x$.

$$= \frac{x}{3} \cdot \frac{8x}{8x}$$ Rewriting as a product of two rational expressions

$$= \frac{x}{3} \cdot 1 \qquad \frac{8x}{8x} = 1$$

$$= \frac{x}{3}$$ Removing the factor 1

We say that $8x^2/(24x)$ *simplifies* to $x/3$.* In the work that follows, we assume that all denominators are nonzero.

E x a m p l e 3

Simplify: $\dfrac{5a + 15}{10}$.

Solution

$$\frac{5a + 15}{10} = \frac{5(a + 3)}{5 \cdot 2}$$ Factoring the numerator and the denominator. Note the common factor of 5.

$$= \frac{5}{5} \cdot \frac{a + 3}{2}$$ Rewriting as a product of two rational expressions

$$= 1 \cdot \frac{a + 3}{2} \qquad \frac{5}{5} = 1$$

$$= \frac{a + 3}{2}$$ Removing the factor 1

Sometimes the common factor has two or more terms.

E x a m p l e 4

Simplify.

a) $\dfrac{6x + 12}{7x + 14}$ **b)** $\dfrac{6a^2 + 4a}{2a^2 + 2a}$ **c)** $\dfrac{x^2 - 1}{x^2 + 3x + 2}$

Solution

a) $\dfrac{6x + 12}{7x + 14} = \dfrac{6(x + 2)}{7(x + 2)}$ Factoring the numerator and the denominator

$$= \frac{6}{7} \cdot \frac{x + 2}{x + 2}$$ Rewriting as a product of two rational expressions

---

*In more advanced courses, we would *not* say that $8x^2/(24x)$ simplifies to $x/3$, but would instead say that $8x^2/(24x)$ simplifies to $x/3$ *with the restriction that $x \neq 0$.*

$$\frac{6}{7} \cdot \frac{x+2}{x+2} = \frac{6}{7} \cdot 1 \qquad \frac{x+2}{x+2} = 1$$

$$= \frac{6}{7} \qquad \text{Removing the factor 1}$$

b) $\dfrac{6a^2 + 4a}{2a^2 + 2a} = \dfrac{2a(3a+2)}{2a(a+1)}$ Factoring the numerator and the denominator

$$= \frac{2a}{2a} \cdot \frac{3a+2}{a+1} \qquad \text{Rewriting as a product of two rational expressions}$$

$$= 1 \cdot \frac{3a+2}{a+1} \qquad \frac{2a}{2a} = 1$$

$$= \frac{3a+2}{a+1} \qquad \text{Removing the factor 1. Note in this step that you } cannot \text{ remove the remaining } a\text{'s because } a \text{ is not a factor.}$$

c) $\dfrac{x^2 - 1}{x^2 + 3x + 2} = \dfrac{(x+1)(x-1)}{(x+1)(x+2)}$ Factoring

$$= \frac{x+1}{x+1} \cdot \frac{x-1}{x+2} \qquad \text{Rewriting as a product of two rational expressions}$$

$$= 1 \cdot \frac{x-1}{x+2} \qquad \frac{x+1}{x+1} = 1$$

$$= \frac{x-1}{x+2} \qquad \text{Removing the factor 1}$$

Canceling is a shortcut that can be used—and easily *misused*—when working with rational expressions. As we stated in Section 1.3, canceling must be done with care and understanding. Essentially, canceling streamlines the steps in which we remove a factor equal to 1. Example 4(c) could have been streamlined as follows:

$$\frac{x^2 - 1}{x^2 + 3x + 2} = \frac{(\cancel{x+1})(x-1)}{(\cancel{x+1})(x+2)} \qquad \text{When a factor equal to 1 is noted, it is ``canceled'': } \frac{x+1}{x+1} = 1$$

$$= \frac{x-1}{x+2}. \qquad \text{Simplifying}$$

Caution!   Canceling is often used incorrectly. The following cancellations are *incorrect*:

$$\frac{\cancel{x}+2}{\cancel{x}+3}; \qquad \frac{a^2 - \cancel{5}}{\cancel{5}}; \qquad \frac{6x^2 + 5\cancel{x} + 1}{4x^2 - 3\cancel{x}}.$$

Wrong!   Wrong!   Wrong!

None of the above cancellations removes a factor equal to 1. Factors are parts of products. For example, in $x \cdot 2$, x and 2 are factors, but in $x + 2$, x and 2 are terms, *not* factors. If it is not a factor, then it cannot be canceled.

Example 5   Simplify: $\dfrac{3x^2 - 2x - 1}{x^2 - 3x + 2}$.

Solution   We factor the numerator and the denominator and look for common factors:

$$\frac{3x^2 - 2x - 1}{x^2 - 3x + 2} = \frac{(3x + 1)(x - 1)}{(x - 2)(x - 1)}$$

Try to visualize this as
$$\frac{3x + 1}{x - 2} \cdot \frac{x - 1}{x - 1}.$$

$$= \frac{3x + 1}{x - 2}.$$

Removing a factor equal to 1:
$$\frac{x - 1}{x - 1} = 1$$

When a rational expression is simplified, the result is an equivalent expression. Example 3 says that

$$\frac{5a + 15}{10} \quad \text{is equivalent to} \quad \frac{a + 3}{2}.$$

This result can be partially checked using a value of a. For instance, if $a = 2$, then

$$\frac{5a + 15}{10} = \frac{5 \cdot 2 + 15}{10} = \frac{25}{10} = \frac{5}{2}$$

and

> To see why this check is not foolproof, see Exercise 61.

$$\frac{a + 3}{2} = \frac{2 + 3}{2} = \frac{5}{2}.$$

If evaluating both expressions yields differing results, we know that a mistake has been made. For example, if $(5a + 15)/10$ is incorrectly simplified as $(a + 15)/2$ and we evaluate using $a = 2$, we have

$$\frac{5a + 15}{10} = \frac{5 \cdot 2 + 15}{10} = \frac{5}{2}$$

and

Different results

$$\frac{a + 15}{2} = \frac{2 + 15}{2} = \frac{17}{2},$$

which demonstrates that a mistake has been made.

Factors That Are Opposites

Consider

$$\frac{x - 4}{8 - 2x}, \quad \text{or, equivalently,} \quad \frac{x - 4}{2(4 - x)}.$$

At first glance, the numerator and the denominator do not appear to have any common factors. But $x - 4$ and $4 - x$ are opposites, or additive inverses, of each other. Thus we can find a common factor by factoring out -1 in one expression.

Example 6

Simplify $\dfrac{x-4}{8-2x}$ and check by evaluating.

Solution   We have

$$\dfrac{x-4}{8-2x} = \dfrac{x-4}{2(4-x)} \qquad \text{Factoring}$$

$$= \dfrac{x-4}{2(-1)(x-4)} \qquad \text{Note that } 4-x = -(x-4).$$

$$= \dfrac{x-4}{-2(x-4)} \qquad \begin{array}{l}\text{Had we originally factored out } -2, \text{ we could}\\ \text{have gone directly to this step.}\end{array}$$

$$= \dfrac{1}{-2} \cdot \dfrac{x-4}{x-4} \qquad \begin{array}{l}\text{Rewriting as a product. It is important to}\\ \text{write the 1 in the numerator.}\end{array}$$

$$= -\dfrac{1}{2}. \qquad \begin{array}{l}\text{Removing a factor equal to 1:}\\ (x-4)/(x-4) = 1\end{array}$$

As a partial check, note that for any choice of x other than 4, the value of the rational expression is $-\frac{1}{2}$. For example, if $x = 6$, then

$$\dfrac{x-4}{8-2x} = \dfrac{6-4}{8-2 \cdot 6}$$

$$= \dfrac{2}{-4} = -\dfrac{1}{2}.$$

FOR EXTRA HELP

Exercise Set **6.1**

 Digital Video Tutor CD 4 Videotape 11   InterAct Math   Math Tutor Center   MathXL    MyMathLab.com

List all numbers for which each rational expression is undefined.

1. $\dfrac{25}{-7x}$

2. $\dfrac{14}{-5y}$

3. $\dfrac{t-3}{t+8}$

4. $\dfrac{a-8}{a+7}$

5. $\dfrac{a-4}{3a-12}$

6. $\dfrac{x^2-9}{4x-12}$

7. $\dfrac{x^2-16}{x^2-3x-28}$

8. $\dfrac{p^2-9}{p^2-7p+10}$

9. $\dfrac{m^3-2m}{m^2-25}$

10. $\dfrac{7-3x+x^2}{49-x^2}$

Simplify by removing a factor equal to 1. Show all steps.

11. $\dfrac{60a^2b}{40ab^3}$

12. $\dfrac{45x^3y^2}{9x^5y}$

13. $\dfrac{35x^2y}{14x^3y^5}$

14. $\dfrac{12a^5b^6}{18a^3b}$

15. $\dfrac{9x+15}{12x+20}$

16. $\dfrac{14x-7}{10x-5}$

17. $\dfrac{a^2-9}{a^2+4a+3}$

18. $\dfrac{a^2+5a+6}{a^2-9}$

Simplify, if possible. Then check by evaluating, as in Example 6.

19. $\dfrac{36x^6}{24x^9}$

20. $\dfrac{75a^5}{50a^3}$

21. $\dfrac{-2y + 6}{-8y}$

22. $\dfrac{4x - 12}{6x}$

23. $\dfrac{6a^2 - 3a}{7a^2 - 7a}$

24. $\dfrac{3m^2 + 3m}{6m^2 + 9m}$

25. $\dfrac{t^2 - 16}{t^2 + t - 20}$

26. $\dfrac{a^2 - 4}{a^2 + 5a + 6}$

27. $\dfrac{3a^2 + 9a - 12}{6a^2 - 30a + 24}$

28. $\dfrac{2t^2 - 6t + 4}{4t^2 + 12t - 16}$

29. $\dfrac{x^2 + 8x + 16}{x^2 - 16}$

30. $\dfrac{x^2 - 25}{x^2 - 10x + 25}$

31. $\dfrac{t^2 - 1}{t + 1}$

32. $\dfrac{a^2 - 1}{a - 1}$

33. $\dfrac{y^2 + 4}{y + 2}$

34. $\dfrac{x^2 + 1}{x + 1}$

35. $\dfrac{5x^2 - 20}{10x^2 - 40}$

36. $\dfrac{6x^2 - 54}{4x^2 - 36}$

37. $\dfrac{5y + 5}{y^2 + 7y + 6}$

38. $\dfrac{6t + 12}{t^2 - t - 6}$

39. $\dfrac{y^2 + 3y - 18}{y^2 + 2y - 15}$

40. $\dfrac{a^2 + 10a + 21}{a^2 + 11a + 28}$

41. $\dfrac{(a - 3)^2}{a^2 - 9}$

42. $\dfrac{t^2 - 4}{(t + 2)^2}$

43. $\dfrac{x - 8}{8 - x}$

44. $\dfrac{6 - x}{x - 6}$

45. $\dfrac{7t - 14}{2 - t}$

46. $\dfrac{4a - 12}{3 - a}$

47. $\dfrac{a - b}{3b - 3a}$

48. $\dfrac{q - p}{2p - 2q}$

49. $\dfrac{3x^2 - 3y^2}{2y^2 - 2x^2}$

50. $\dfrac{7a^2 - 7b^2}{3b^2 - 3a^2}$

Aha! **51.** $\dfrac{7s^2 - 28t^2}{28t^2 - 7s^2}$

52. $\dfrac{9m^2 - 4n^2}{4n^2 - 9m^2}$

53. Explain how simplifying is related to the identity property of 1.

54. If a rational expression is undefined for $x = 5$ and $x = -3$, what is the degree of the denominator? Why?

SKILL MAINTENANCE

Simplify.

55. $-\dfrac{2}{3} \cdot \dfrac{6}{7}$

56. $\dfrac{5}{9}\left(\dfrac{-6}{11}\right)$

57. $\dfrac{5}{8} \div \left(-\dfrac{1}{6}\right)$

58. $\dfrac{7}{10} \div \left(-\dfrac{8}{15}\right)$

59. $\dfrac{7}{9} - \dfrac{2}{3} \cdot \dfrac{6}{7}$

60. $\dfrac{2}{3} - \left(\dfrac{3}{4}\right)^2$

SYNTHESIS

61. Terry *incorrectly* simplifies

$$\dfrac{x^2 + x - 2}{x^2 + 3x + 2} \quad \text{as} \quad \dfrac{x - 1}{x + 2}.$$

He then checks his simplification by evaluating both expressions for $x = 1$. Use this situation to explain why evaluating is not a foolproof check.

62. How could you convince someone that $a - b$ and $b - a$ are opposites of each other?

Simplify.

63. $\dfrac{x^4 - y^4}{(y - x)^4}$

64. $\dfrac{16y^4 - x^4}{(x^2 + 4y^2)(x - 2y)}$

65. $\dfrac{(x - 1)(x^4 - 1)(x^2 - 1)}{(x^2 + 1)(x - 1)^2(x^4 - 2x^2 + 1)}$

66. $\dfrac{x^5 - 2x^3 + 4x^2 - 8}{x^7 + 2x^4 - 4x^3 - 8}$

67. $\dfrac{10t^4 - 8t^3 + 15t - 12}{8 - 10t + 12t^2 - 15t^3}$

68. $\dfrac{(t^4 - 1)(t^2 - 9)(t - 9)^2}{(t^4 - 81)(t^2 + 1)(t + 1)^2}$

69. $\dfrac{(t + 2)^3(t^2 + 2t + 1)(t + 1)}{(t + 1)^3(t^2 + 4t + 4)(t + 2)}$

70. $\dfrac{(x^2 - y^2)(x^2 - 2xy + y^2)}{(x + y)^2(x^2 - 4xy - 5y^2)}$

71. Select any number x, multiply by 2, add 5, multiply by 5, subtract 25, and divide by 10. What do you get? Explain how this procedure can be used for a number trick.

Multiplication and Division

6.2

Multiplication • Division

Multiplication and division of rational expressions is similar to multiplication and division with fractions. In this section, we again assume that all denominators are nonzero.

Multiplication

Recall that to multiply fractions, we simply multiply their numerators and multiply their denominators. Rational expressions are multiplied in a similar way.

> **The Product of Two Rational Expressions**
>
> To multiply rational expressions, multiply numerators and multiply denominators:
> $$\frac{A}{B} \cdot \frac{C}{D} = \frac{AC}{BD}.$$
> Then factor and simplify the result if possible.

For example,

$$\frac{3}{5} \cdot \frac{8}{11} = \frac{24}{55} \quad \text{and} \quad \frac{x}{3} \cdot \frac{x+2}{y} = \frac{x(x+2)}{3y}.$$

Fraction bars are grouping symbols, so parentheses are needed when writing some products. Because we generally simplify, we often leave parentheses in the product. There is no need to multiply further.

Example 1   Multiply and simplify.

a) $\dfrac{5a^3}{4} \cdot \dfrac{2}{5a}$

b) $\dfrac{x^2 + 6x + 9}{x^2 - 4} \cdot \dfrac{x - 2}{x + 3}$

c) $\dfrac{x^2 + x - 2}{15} \cdot \dfrac{5}{2x^2 - 3x + 1}$

Solution

a) $\dfrac{5a^3}{4} \cdot \dfrac{2}{5a} = \dfrac{5a^3(2)}{4(5a)}$ Forming the product of the numerators and the product of the denominators

$= \dfrac{2 \cdot 5 \cdot a \cdot a \cdot a}{2 \cdot 2 \cdot 5 \cdot a}$ Factoring the numerator and the denominator

$\left.\begin{array}{l} = \dfrac{2 \cdot 5 \cdot a \cdot a \cdot a}{2 \cdot 2 \cdot 5 \cdot a} \\[2mm] = \dfrac{a^2}{2} \end{array}\right\}$ Removing a factor equal to 1: $\dfrac{2 \cdot 5 \cdot a}{2 \cdot 5 \cdot a} = 1$

b) $\dfrac{x^2 + 6x + 9}{x^2 - 4} \cdot \dfrac{x - 2}{x + 3} = \dfrac{(x^2 + 6x + 9)(x - 2)}{(x^2 - 4)(x + 3)}$ Multiplying the numerators and the denominators

$= \dfrac{(x + 3)(x + 3)(x - 2)}{(x + 2)(x - 2)(x + 3)}$ Factoring the numerator and the denominator

$\left.\begin{array}{l} = \dfrac{(x + 3)(x + 3)(x - 2)}{(x + 2)(x - 2)(x + 3)} \\[2mm] = \dfrac{x + 3}{x + 2} \end{array}\right\}$ Removing a factor equal to 1: $\dfrac{(x + 3)(x - 2)}{(x + 3)(x - 2)} = 1$

c) $\dfrac{x^2 + x - 2}{15} \cdot \dfrac{5}{2x^2 - 3x + 1} = \dfrac{(x^2 + x - 2)5}{15(2x^2 - 3x + 1)}$ Multiplying the numerators and the denominators

$= \dfrac{(x + 2)(x - 1)5}{5(3)(x - 1)(2x - 1)}$ Factoring the numerator and the denominator. Try to go directly to this step.

$\left.\begin{array}{l} = \dfrac{(x + 2)(x - 1)5}{5(3)(x - 1)(2x - 1)} \\[2mm] = \dfrac{x + 2}{3(2x - 1)} \end{array}\right\}$ Removing a factor equal to 1: $\dfrac{5(x - 1)}{5(x - 1)} = 1$

There is no need to multiply out the numerator or the denominator in our results.

Division

As with fractions, reciprocals of rational expressions are found by interchanging the numerator and the denominator. For example,

the reciprocal of $\dfrac{2}{7}$ is $\dfrac{7}{2}$,   and   the reciprocal of $\dfrac{3x}{x + 5}$ is $\dfrac{x + 5}{3x}$.

The Quotient of Two Rational Expressions

To divide by a rational expression, multiply by its reciprocal:

$$\frac{A}{B} \div \frac{C}{D} = \frac{A}{B} \cdot \frac{D}{C} = \frac{AD}{BC}.$$

Then factor and simplify if possible.

E x a m p l e   2

Divide: **(a)** $\dfrac{x}{5} \div \dfrac{7}{y}$; **(b)** $(x + 2) \div \dfrac{x - 1}{x + 3}$.

Solution

a) $\dfrac{x}{5} \div \dfrac{7}{y} = \dfrac{x}{5} \cdot \dfrac{y}{7}$ 　　Multiplying by the reciprocal of the divisor

$\phantom{\dfrac{x}{5} \div \dfrac{7}{y}} = \dfrac{xy}{35}$ 　　Multiplying rational expressions

b) $(x + 2) \div \dfrac{x - 1}{x + 3} = \dfrac{x + 2}{1} \cdot \dfrac{x + 3}{x - 1}$ 　　Multiplying by the reciprocal of the divisor. Writing $x + 2$ as

$\phantom{(x + 2) \div \dfrac{x - 1}{x + 3}} = \dfrac{(x + 2)(x + 3)}{x - 1}$ 　　$\dfrac{x + 2}{1}$ can be helpful.

As usual, we should simplify when possible. Often that will require that we factor one or more polynomials. Our hope is to discover a common factor that appears in both the numerator and the denominator.

E x a m p l e   3

Divide and simplify: $\dfrac{x + 1}{x^2 - 1} \div \dfrac{x + 1}{x^2 - 2x + 1}$.

Solution

$$\frac{x + 1}{x^2 - 1} \div \frac{x + 1}{x^2 - 2x + 1} = \frac{x + 1}{x^2 - 1} \cdot \frac{x^2 - 2x + 1}{x + 1}$$

Multiplying by the reciprocal of the divisor

$$= \frac{(x + 1)(x - 1)(x - 1)}{(x + 1)(x - 1)(x + 1)}$$

Multiplying rational expressions and factoring numerators and denominators

$$= \frac{\cancel{(x + 1)}\,\cancel{(x - 1)}\,(x - 1)}{\cancel{(x + 1)}\,\cancel{(x - 1)}\,(x + 1)}$$

Removing a factor equal to 1: $\dfrac{(x + 1)(x - 1)}{(x + 1)(x - 1)} = 1$

$$= \frac{x - 1}{x + 1}$$

E x a m p l e   4

technology connection

In performing a partial check of Example 4(b), care must be taken in placing parentheses. For example, we enter the original expression in Example 4(b) as $y_1 = ((x^2 - 2x - 3)/(x^2 - 4))/((x + 1)/(x + 5))$ and the simplified expression as $y_2 = ((x - 3)(x + 5))/((x - 2)(x + 2))$. Comparing values of y_1 and y_2, we see that the simplification is probably correct.

X	Y₁	Y₂
−5	ERROR	0
−4	−.5833	−.5833
−3	−2.4	−2.4
−2	ERROR	ERROR
−1	ERROR	5.3333
0	3.75	3.75
1	4	4

X = −5

1. Check Example 4(a).
2. Why are there 3 ERROR messages shown for y_1 on the screen above, and only 1 for y_2?

Divide and, if possible, simplify.

a) $\dfrac{a^2 + 3a + 2}{a^2 + 4} \div (5a^2 + 10a)$

b) $\dfrac{x^2 - 2x - 3}{x^2 - 4} \div \dfrac{x + 1}{x + 5}$

Solution

a) $\dfrac{a^2 + 3a + 2}{a^2 + 4} \div (5a^2 + 10a)$

$= \dfrac{a^2 + 3a + 2}{a^2 + 4} \cdot \dfrac{1}{5a^2 + 10a}$ Multiplying by the reciprocal of the divisor

$= \dfrac{(a + 2)(a + 1)}{(a^2 + 4)5a(a + 2)}$ Multiplying rational expressions and factoring

$= \dfrac{(a + 2)(a + 1)}{(a^2 + 4)5a(a + 2)}$

$= \dfrac{a + 1}{(a^2 + 4)5a}$ Removing a factor equal to 1: $\dfrac{a + 2}{a + 2} = 1$

b) $\dfrac{x^2 - 2x - 3}{x^2 - 4} \div \dfrac{x + 1}{x + 5} = \dfrac{x^2 - 2x - 3}{x^2 - 4} \cdot \dfrac{x + 5}{x + 1}$ Multiplying by the reciprocal of the divisor

$= \dfrac{(x - 3)(x + 1)(x + 5)}{(x - 2)(x + 2)(x + 1)}$ Multiplying rational expressions and factoring

$= \dfrac{(x - 3)(x + 1)(x + 5)}{(x - 2)(x + 2)(x + 1)}$ Removing a factor equal to 1:

$= \dfrac{(x - 3)(x + 5)}{(x - 2)(x + 2)}$ $\dfrac{x + 1}{x + 1} = 1$

Exercise Set 6.2

FOR EXTRA HELP

 Digital Video Tutor CD 4 Videotape 11

 InterAct Math

 Math Tutor Center

 MathXL

 MyMathLab.com

Multiply. Leave each answer in factored form.

1. $\dfrac{9x}{4} \cdot \dfrac{x - 5}{2x + 1}$

2. $\dfrac{3x}{4} \cdot \dfrac{5x + 2}{x - 1}$

3. $\dfrac{a - 4}{a + 6} \cdot \dfrac{a + 2}{a + 6}$

4. $\dfrac{a + 3}{a + 6} \cdot \dfrac{a + 3}{a - 1}$

5. $\dfrac{2x + 3}{4} \cdot \dfrac{x + 1}{x - 5}$

6. $\dfrac{x + 2}{3x - 4} \cdot \dfrac{4}{5x + 6}$

7. $\dfrac{a - 5}{a^2 + 1} \cdot \dfrac{a + 2}{a^2 - 1}$

8. $\dfrac{t + 3}{t^2 - 2} \cdot \dfrac{t + 3}{t^2 - 4}$

9. $\dfrac{x + 4}{2 + x} \cdot \dfrac{x - 1}{x + 1}$

10. $\dfrac{m + 4}{m + 8} \cdot \dfrac{2 + m}{m + 5}$

Multiply and, if possible, simplify.

11. $\dfrac{5a^4}{6a} \cdot \dfrac{2}{a}$

12. $\dfrac{10}{t^7} \cdot \dfrac{3t^2}{25t}$

13. $\dfrac{3c}{d^2} \cdot \dfrac{8d}{6c^3}$

14. $\dfrac{3x^2y}{2} \cdot \dfrac{4}{xy^3}$

15. $\dfrac{x^2 - 3x - 10}{(x - 2)^2} \cdot \dfrac{x - 2}{x - 5}$

16. $\dfrac{t + 2}{t - 2} \cdot \dfrac{t^2 - 5t + 6}{(t + 2)^2}$

17. $\dfrac{a^2 + 25}{a^2 - 4a + 3} \cdot \dfrac{a - 5}{a + 5}$

18. $\dfrac{x + 3}{x^2 + 9} \cdot \dfrac{x^2 + 5x + 4}{x + 9}$

19. $\dfrac{a^2 - 9}{a^2} \cdot \dfrac{5a}{a^2 + a - 12}$

20. $\dfrac{x^2 + 10x - 11}{5x} \cdot \dfrac{x^3}{x + 11}$

21. $\dfrac{4a^2}{3a^2 - 12a + 12} \cdot \dfrac{3a - 6}{2a}$

22. $\dfrac{5v + 5}{v - 2} \cdot \dfrac{2v^2 - 8v + 8}{v^2 - 1}$

23. $\dfrac{t^2 + 2t - 3}{t^2 + 4t - 5} \cdot \dfrac{t^2 - 3t - 10}{t^2 + 5t + 6}$

24. $\dfrac{x^2 + 5x + 4}{x^2 - 6x + 8} \cdot \dfrac{x^2 + 5x - 14}{x^2 + 8x + 7}$

25. $\dfrac{5a^2 - 180}{10a^2 - 10} \cdot \dfrac{20a + 20}{2a - 12}$

26. $\dfrac{2t^2 - 98}{4t^2 - 4} \cdot \dfrac{8t + 8}{16t - 112}$

Aha! **27.** $\dfrac{x^2 + 4x + 4}{(x - 1)^2} \cdot \dfrac{x^2 - 2x + 1}{(x + 2)^2}$

28. $\dfrac{x + 5}{(x + 2)^2} \cdot \dfrac{x^2 + 7x + 10}{(x + 5)^2}$

29. $\dfrac{t^2 + 8t + 16}{(t + 4)^3} \cdot \dfrac{(t + 2)^3}{t^2 + 4t + 4}$

30. $\dfrac{(y - 1)^3}{y^2 - 2y + 1} \cdot \dfrac{y^2 - 4y + 4}{(y - 2)^3}$

Find the reciprocal of each expression.

31. $\dfrac{3x}{7}$

32. $\dfrac{3 - x}{x^2 + 4}$

33. $a^3 - 8a$

34. $\dfrac{7}{a^2 - b^2}$

35. $\dfrac{x^2 + 2x - 5}{x^2 - 4x + 7}$

36. $\dfrac{x^2 - 3xy + y^2}{x^2 + 7xy - y^2}$

Divide and, if possible, simplify.

37. $\dfrac{3}{8} \div \dfrac{5}{2}$

38. $\dfrac{5}{9} \div \dfrac{2}{7}$

39. $\dfrac{x}{4} \div \dfrac{5}{x}$

40. $\dfrac{5}{x} \div \dfrac{x}{12}$

41. $\dfrac{a^5}{b^4} \div \dfrac{a^2}{b}$

42. $\dfrac{x^5}{y^2} \div \dfrac{x^2}{y}$

43. $\dfrac{y + 5}{4} \div \dfrac{y}{2}$

44. $\dfrac{a + 2}{a - 3} \div \dfrac{a - 1}{a + 3}$

45. $\dfrac{4y - 8}{y + 2} \div \dfrac{y - 2}{y^2 - 4}$

46. $\dfrac{x^2 - 1}{x} \div \dfrac{x + 1}{x - 1}$

47. $\dfrac{a}{a - b} \div \dfrac{b}{b - a}$

48. $\dfrac{x - y}{6} \div \dfrac{y - x}{3}$

49. $(y^2 - 9) \div \dfrac{y^2 - 2y - 3}{y^2 + 1}$

50. $(x^2 - 5x - 6) \div \dfrac{x^2 - 1}{x + 6}$

51. $\dfrac{7x - 7}{16} \div \dfrac{x - 1}{6}$

52. $\dfrac{-4 + 2x}{15} \div \dfrac{x - 2}{3}$

53. $\dfrac{-6 + 3x}{5} \div \dfrac{4x - 8}{25}$

54. $\dfrac{-12 + 4x}{12} \div \dfrac{-6 + 2x}{6}$

55. $\dfrac{a + 2}{a - 1} \div \dfrac{3a + 6}{a - 5}$

56. $\dfrac{t - 3}{t + 2} \div \dfrac{4t - 12}{t + 1}$

57. $(2x - 1) \div \dfrac{2x^2 - 11x + 5}{4x^2 - 1}$

58. $(a + 7) \div \dfrac{3a^2 + 14a - 49}{a^2 + 8a + 7}$

59. $\dfrac{x - 5}{x + 5} \div \dfrac{2x^2 - 50}{x^2 + 25}$

60. $\dfrac{3x^2 - 27}{x^2 + 1} \div \dfrac{x + 3}{x - 3}$

61. $\dfrac{a^2 - 10a + 25}{a^2 + 7a + 12} \div \dfrac{a^2 - a - 20}{a^2 + 6a + 9}$

62. $\dfrac{a^2 + 5a + 4}{a^2 - 2a + 1} \div \dfrac{a^2 + 8a + 16}{a^2 - 5a - 6}$

63. $\dfrac{c^2 + 10c + 21}{c^2 - 2c - 15} \div (c^2 + 2c - 35)$

64. $\dfrac{1 - z}{1 + 2z - z^2} \div (1 - z)$

65. $\dfrac{x - y}{x^2 + 2xy + y^2} \div \dfrac{x^2 - y^2}{x^2 - 5xy + 4y^2}$

66. $\dfrac{a^2 - b^2}{a^2 - 4ab + 4b^2} \div \dfrac{a^2 - 3ab + 2b^2}{a - 2b}$

67. A student claims to be able to divide, but not multiply, rational expressions. Why is this claim difficult to believe?

68. Why is it important to insert parentheses when multiplying rational expressions in which the numerators and the denominators contain more than one term?

SKILL MAINTENANCE

Simplify.

69. $\dfrac{3}{4} + \dfrac{5}{6}$

70. $\dfrac{7}{8} + \dfrac{5}{6}$

71. $\dfrac{2}{9} - \dfrac{1}{6}$

72. $\dfrac{3}{10} - \dfrac{7}{15}$

73. $\dfrac{2}{5} - \left(\dfrac{3}{2}\right)^2$

74. $\dfrac{5}{9} + \dfrac{2}{3} \cdot \dfrac{4}{5}$

SYNTHESIS

75. Is the reciprocal of a product the product of the two reciprocals? Why or why not?

76. Explain why the quotient

$$\dfrac{x + 3}{x - 5} \div \dfrac{x - 7}{x + 1}$$

is undefined for $x = 5$, $x = -1$, and $x = 7$.

Simplify.

Aha! **77.** $\dfrac{3x - y}{2x + y} \div \dfrac{3x - y}{2x + y}$

78. $\dfrac{2a^2 - 5ab}{c - 3d} \div (4a^2 - 25b^2)$

79. $(x - 2a) \div \dfrac{a^2x^2 - 4a^4}{a^2x + 2a^3}$

80. $\dfrac{3a^2 - 5ab - 12b^2}{3ab + 4b^2} \div (3b^2 - ab)^2$

81. $\dfrac{3x^2 - 2xy - y^2}{x^2 - y^2} \div (3x^2 + 4xy + y^2)^2$

82. $\dfrac{y^2 - 4xy}{y - x} \div \dfrac{16x^2y^2 - y^4}{4x^2 - 3xy - y^2} \div \dfrac{4}{x^3y^3}$

Aha! **83.** $\dfrac{a^2 - 3b}{a^2 + 2b} \cdot \dfrac{a^2 - 2b}{a^2 + 3b} \cdot \dfrac{a^2 + 2b}{a^2 - 3b}$

84. $\dfrac{z^2 - 8z + 16}{z^2 + 8z + 16} \div \dfrac{(z - 4)^5}{(z + 4)^5} \div \dfrac{3z + 12}{z^2 - 16}$

85. $\dfrac{x^2 - x + xy - y}{x^2 + 6x - 7} \div \dfrac{x^2 + 2xy + y^2}{4x + 4y}$

86. $\dfrac{3x + 3y + 3}{9x} \div \dfrac{x^2 + 2xy + y^2 - 1}{x^4 + x^2}$

87. $\dfrac{(t + 2)^3}{(t + 1)^3} \div \dfrac{t^2 + 4t + 4}{t^2 + 2t + 1} \cdot \dfrac{t + 1}{t + 2}$

88. $\dfrac{3y^3 + 6y^2}{y^2 - y - 12} \div \dfrac{y^2 - y}{y^2 - 2y - 8} \cdot \dfrac{y^2 + 5y + 6}{y^2}$

89. $\dfrac{6y - 4x}{(2x + 3y)^2} \cdot \dfrac{2x - 3y}{x^2 - 9y^2} \div \dfrac{4x^2 - 12xy + 9y^2}{9y^2 + 12xy + 4x^2}$

90. $\dfrac{a^4 - 81b^4}{a^2c - 6abc + 9b^2c} \cdot \dfrac{a + 3b}{a^2 + 9b^2} \div \dfrac{a^2 + 6ab + 9b^2}{(a - 3b)^2}$

COLLABORATIVE

CORNER

Currency Exchange

Focus: Least common multiples and
proportions

Time: 20 minutes

Group size: 2

Travel between different countries usually ne-
cessitates an exchange of currencies. Recently
one Canadian dollar was worth 64 cents in U.S.
funds. Use this exchange rate for the activity that
follows.

ACTIVITY

1. Within each group of two students, one student should play the role of a U.S. citizen
   planning a visit to Canada. The other student
   should play the role of a Canadian planning a
   visit to the United States. Use the exchange
   rate of one Canadian dollar for 64 cents in
   U.S. funds.
2. Determine how much Canadian money the
   U.S. citizen would receive in exchange for $64
   of U.S. funds (this should be easy). Then determine how much U.S. money the Canadian
   would receive in exchange for $64 of Cana-
   dian funds. Finally, determine how much the
   Canadian would receive in exchange for $100

of Canadian funds and how much the U.S.
citizen would receive for $100 of U.S. funds.

3. The answers to part (2) should indicate that
   coins are needed to exchange $64 of Cana-
   dian money for U.S. funds, or to exchange
   $100 of U.S. funds for Canadian money. What
   is the smallest amount of Canadian dollars
   that can be exchanged for U.S. dollars without requiring coins? What is the smallest
   amount of U.S. dollars that can be exchanged
   for Canadian dollars without requiring coins?
   (*Hint*: See part 2.)
4. Use the results from part (3) to find two other
   amounts of U.S. currency that can be exchanged without requiring coins. Answers
   may vary.
5. Find the smallest number *a* for which neither
   conversion—from *a* Canadian dollars to U.S.
   funds or from *a* U.S. dollars to Canadian
   funds—will require coins. (*Hint*: Use LCMs
   and the results of part 2 above.)
6. At one time in 2000, one New Zealand dollar
   was worth about 40 cents in U.S. funds. Find
   the smallest number *a* for which neither conversion—from *a* New Zealand dollars to U.S.
   funds or from *a* U.S. dollars to New Zealand
   funds—will require coins. (*Hint*: See part 5.)

Addition, Subtraction, and Least Common Denominators

6.3

Addition When Denominators Are the Same • Subtraction When Denominators Are the Same • Least Common Multiples and Denominators

Addition When Denominators Are the Same

Recall that to add fractions having the same denominator, like $\frac{2}{7}$ and $\frac{3}{7}$, we add the numerators and keep the common denominator: $\frac{2}{7} + \frac{3}{7} = \frac{5}{7}$. The same procedure is used when rational expressions share a common denominator.

> ### The Sum of Two Rational Expressions
>
> To add when the denominators are the same, add the numerators and keep the common denominator:
>
> $$\frac{A}{B} + \frac{C}{B} = \frac{A + C}{B}.$$

E x a m p l e 1

Add. Simplify the result, if possible.

a) $\dfrac{4}{a} + \dfrac{3 + a}{a}$

b) $\dfrac{3x}{x - 5} + \dfrac{2x + 1}{x - 5}$

c) $\dfrac{2x^2 + 3x - 7}{2x + 1} + \dfrac{x^2 + x - 8}{2x + 1}$

d) $\dfrac{x - 5}{x^2 - 9} + \dfrac{2}{x^2 - 9}$

Solution

a) $\dfrac{4}{a} + \dfrac{3 + a}{a} = \dfrac{7 + a}{a}$ When the denominators are alike, add the numerators and keep the common denominator.

b) $\dfrac{3x}{x - 5} + \dfrac{2x + 1}{x - 5} = \dfrac{5x + 1}{x - 5}$ Adding the numerators and combining like terms

c) $\dfrac{2x^2 + 3x - 7}{2x + 1} + \dfrac{x^2 + x - 8}{2x + 1} = \dfrac{(2x^2 + 3x - 7) + (x^2 + x - 8)}{2x + 1}$

$\qquad\qquad = \dfrac{3x^2 + 4x - 15}{2x + 1}$ Combining like terms in the numerator

d) $\dfrac{x - 5}{x^2 - 9} + \dfrac{2}{x^2 - 9} = \dfrac{x - 3}{x^2 - 9}$ Combining like terms in the numerator: $x - 5 + 2 = x - 3$

$\qquad\qquad = \dfrac{x - 3}{(x - 3)(x + 3)}$ Factoring

$\qquad\qquad = \dfrac{1 \cdot (x - 3)}{(x - 3)(x + 3)}$ Removing a factor equal to 1: $\dfrac{x - 3}{x - 3} = 1$

$\qquad\qquad = \dfrac{1}{x + 3}$

Subtraction When Denominators Are the Same

When two fractions have the same denominator, we subtract one numerator from the other and keep the common denominator: $\frac{5}{7} - \frac{2}{7} = \frac{3}{7}$. The same procedure is used with rational expressions.

> ### *The Difference of Two Rational Expressions*
>
> To subtract when the denominators are the same, subtract the second numerator from the first and keep the common denominator:
>
> $$\frac{A}{B} - \frac{C}{B} = \frac{A - C}{B}.$$

Caution!   A fraction bar is a grouping symbol, just like parentheses, under the numerator. When a numerator is subtracted, be sure to subtract *every* term in that numerator.

Example 2   Subtract: **(a)** $\dfrac{3x}{x + 2} - \dfrac{x - 5}{x + 2}$; **(b)** $\dfrac{x^2}{x - 4} - \dfrac{x + 12}{x - 4}$.

Solution

a) $\dfrac{3x}{x + 2} - \dfrac{x - 5}{x + 2} = \dfrac{3x - (x - 5)}{x + 2}$ The parentheses are needed to make sure that we subtract both terms.

$= \dfrac{3x - x + 5}{x + 2}$ Removing the parentheses and changing signs (using the distributive law)

$= \dfrac{2x + 5}{x + 2}$ Combining like terms

b) $\dfrac{x^2}{x - 4} - \dfrac{x + 12}{x - 4} = \dfrac{x^2 - (x + 12)}{x - 4}$ Remember the parentheses!

$= \dfrac{x^2 - x - 12}{x - 4}$ Removing parentheses (using the distributive law)

$= \dfrac{(x - 4)(x + 3)}{x - 4}$ Factoring, in hopes of simplifying

$= \dfrac{(x - 4)(x + 3)}{x - 4}$ Removing a factor equal to 1: $\dfrac{x - 4}{x - 4} = 1$

$= x + 3$

Least Common Multiples and Denominators

Thus far, every pair of rational expressions that we have added or subtracted shared a common denominator. To add or subtract rational expressions that lack a common denominator, we must first find equivalent rational expressions that *do* have a common denominator.

In algebra, we find a common denominator much as we do in arithmetic. Recall that to add $\frac{1}{12}$ and $\frac{7}{30}$, we first identify the smallest number that contains both 12 and 30 as factors. Such a number, the **least common multiple** (**LCM**) of the denominators, is then used as the **least common denominator** (**LCD**).

Let's find the LCM of 12 and 30 using a method that can also be used with polynomials. We begin by writing the prime factorization of 12:

$$12 = 2 \cdot 2 \cdot 3.$$

Next, we write the prime factorization of 30:

$$30 = 2 \cdot 3 \cdot 5.$$

The LCM must include the factors of each number, so it must include each prime factor the greatest number of times that it appears in either of the factorizations. To find the LCM for 12 and 30, we select one factorization, say

$$2 \cdot 2 \cdot 3,$$

and note that because it lacks a factor of 5, it does not contain the entire factorization of 30. If we multiply $2 \cdot 2 \cdot 3$ by 5, every prime factor occurs just often enough to contain both 12 and 30 as factors.

$$\text{LCM} = 2 \cdot 2 \cdot 3 \cdot 5$$

12 is a factor of the LCM.

30 is a factor of the LCM.

Note that each prime factor—2, 3, and 5—is used the greatest number of times that it appears in either of the individual factorizations. The factor 2 occurs twice and the factors 3 and 5 once each.

> ### To Find the Least Common Denominator (LCD)
>
> 1. Write the prime factorization of each denominator.
> 2. Select one of the factorizations and inspect it to see if it contains the other.
>
>   a) If it does, it represents the LCM of the denominators.
>   b) If it does not, multiply that factorization by any factors of the other denominator that it lacks. The final product is the LCM of the denominators.
>
> The LCD is the LCM of the denominators. It should contain each factor the greatest number of times that it occurs in any of the individual factorizations.

Let's finish adding $\dfrac{1}{12}$ and $\dfrac{7}{30}$:

$$\dfrac{1}{12} + \dfrac{7}{30} = \dfrac{1}{2 \cdot 2 \cdot 3} + \dfrac{7}{2 \cdot 3 \cdot 5}.$$ The least common denominator (LCD) is $2 \cdot 2 \cdot 3 \cdot 5$.

We found above that the LCD is $2 \cdot 2 \cdot 3 \cdot 5$. To get the LCD, we see that the first denominator needs a factor of 5, and the second denominator needs another factor of 2. This is accomplished by multiplying by different forms of 1. We can do this because $a \cdot 1 = a$, for any number a:

$$\dfrac{1}{12} + \dfrac{7}{30} = \dfrac{1}{2 \cdot 2 \cdot 3} \cdot \dfrac{5}{5} + \dfrac{7}{2 \cdot 3 \cdot 5} \cdot \dfrac{2}{2}$$ $\dfrac{5}{5} = 1$ and $\dfrac{2}{2} = 1$

$$= \dfrac{5}{2 \cdot 2 \cdot 3 \cdot 5} + \dfrac{14}{2 \cdot 3 \cdot 5 \cdot 2}$$ The denominators are now the LCD.

$$= \dfrac{19}{60}.$$ Adding the numerators and computing the LCD

Expressions like $\dfrac{5}{36x^2}$ and $\dfrac{7}{24x}$ are added in much the same manner.

E x a m p l e   3   Find the LCD of $\dfrac{5}{36x^2}$ and $\dfrac{7}{24x}$.

Solution

1. We begin by writing the prime factorizations of $36x^2$ and $24x$:

$$36x^2 = 2 \cdot 2 \cdot 3 \cdot 3 \cdot x \cdot x;$$
$$24x = 2 \cdot 2 \cdot 2 \cdot 3 \cdot x.$$

2. Except for a third factor of 2, the factorization of $36x^2$ contains the entire factorization of $24x$. To find the smallest product that contains both $36x^2$ and $24x$ as factors, we multiply $36x^2$ by a third factor of 2:

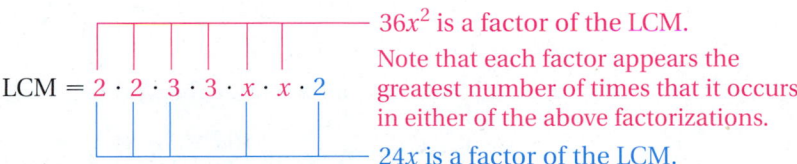

$36x^2$ is a factor of the LCM.

$$\text{LCM} = 2 \cdot 2 \cdot 3 \cdot 3 \cdot x \cdot x \cdot 2$$

Note that each factor appears the greatest number of times that it occurs in either of the above factorizations.

$24x$ is a factor of the LCM.

The LCM is thus $2^3 \cdot 3^2 \cdot x^2$, or $72x^2$, so the LCD is $72x^2$.

We can now add $\dfrac{5}{36x^2}$ and $\dfrac{7}{24x}$:

$$\frac{5}{36x^2} + \frac{7}{24x} = \frac{5}{2 \cdot 2 \cdot 3 \cdot 3 \cdot x \cdot x} + \frac{7}{2 \cdot 2 \cdot 2 \cdot 3 \cdot x}.$$

In Example 3, we found that the LCD is $2 \cdot 2 \cdot 2 \cdot 3 \cdot 3 \cdot x \cdot x$. To obtain equivalent expressions with this LCD, we multiply each expression by 1, using the missing factors of the LCD to write 1:

$$\frac{5}{36x^2} + \frac{7}{24x} = \frac{5}{2 \cdot 2 \cdot 3 \cdot 3 \cdot x \cdot x} \cdot \frac{2}{2} + \frac{7}{2 \cdot 2 \cdot 2 \cdot 3 \cdot x} \cdot \frac{3 \cdot x}{3 \cdot x}$$

The LCD requires another factor of 2.   The LCD requires additional factors of 3 and x.

$$= \frac{10}{2 \cdot 2 \cdot 3 \cdot 3 \cdot x \cdot x \cdot 2} + \frac{21x}{2 \cdot 2 \cdot 2 \cdot 3 \cdot x \cdot 3 \cdot x}$$

Both denominators are now the LCD.

$$= \frac{21x + 10}{72x^2}.$$

You now have the "big" picture of why LCMs are needed when adding rational expressions. For the remainder of this section, we will practice finding LCMs and rewriting rational expressions so that they have the LCD as the denominator. In Section 6.4, we will return to the addition and subtraction of rational expressions.

Example 4

For each pair of polynomials, find the least common multiple.

a) $15a$ and $35b$
b) $21x^3y^6$ and $7x^5y^2$
c) $x^2 + 5x - 6$ and $x^2 - 1$

Solution

a) We write the prime factorizations and then construct the LCM:

$$15a = 3 \cdot 5 \cdot a$$
$$35b = 5 \cdot 7 \cdot b$$

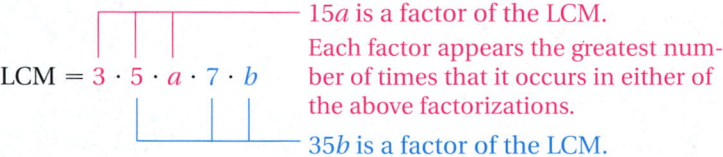

$15a$ is a factor of the LCM.

LCM $= 3 \cdot 5 \cdot a \cdot 7 \cdot b$ Each factor appears the greatest number of times that it occurs in either of the above factorizations.

$35b$ is a factor of the LCM.

The LCM is $3 \cdot 5 \cdot a \cdot 7 \cdot b$, or $105ab$.

b)
$$21x^3y^6 = 3 \cdot 7 \cdot x \cdot x \cdot x \cdot y \cdot y \cdot y \cdot y \cdot y \cdot y$$
$$7x^5y^2 = 7 \cdot x \cdot x \cdot x \cdot x \cdot x \cdot y \cdot y$$

Try to visualize the factors of x and y mentally.

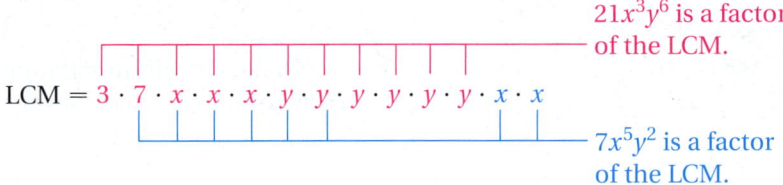

$21x^3y^6$ is a factor of the LCM.

$$\text{LCM} = 3 \cdot 7 \cdot x \cdot x \cdot x \cdot y \cdot y \cdot y \cdot y \cdot y \cdot y \cdot x \cdot x$$

$7x^5y^2$ is a factor of the LCM.

Note that we used the highest power of each factor in $21x^3y^6$ and $7x^5y^2$. The LCM is $21x^5y^6$.

c)
$$x^2 + 5x - 6 = (x - 1)(x + 6)$$
$$x^2 - 1 = (x - 1)(x + 1)$$

$x^2 + 5x - 6$ is a factor of the LCM.

$$\text{LCM} = (x - 1)(x + 6)(x + 1)$$

$x^2 - 1$ is a factor of the LCM.

The LCM is $(x - 1)(x + 6)(x + 1)$. There is no need to multiply this out.

The above procedure can be used to find the LCM of three polynomials as well. We factor each polynomial and then construct the LCM using each factor the greatest number of times that it appears in any one factorization.

E x a m p l e   5

For each group of polynomials, find the LCM.

a) $12x$, $16y$, and $8xyz$
b) $x^2 + 4$, $x + 1$, and 5

Solution

a)
$$12x = 2 \cdot 2 \cdot 3 \cdot x$$
$$16y = 2 \cdot 2 \cdot 2 \cdot 2 \cdot y$$
$$8xyz = 2 \cdot 2 \cdot 2 \cdot x \cdot y \cdot z$$

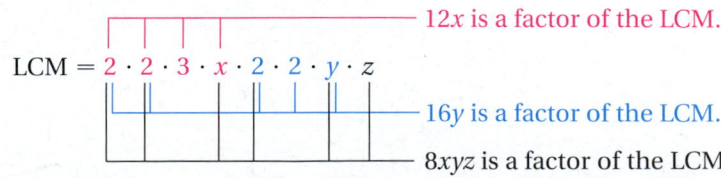

$12x$ is a factor of the LCM.

$$\text{LCM} = 2 \cdot 2 \cdot 3 \cdot x \cdot 2 \cdot 2 \cdot y \cdot z$$

$16y$ is a factor of the LCM.
$8xyz$ is a factor of the LCM.

The LCM is $2^4 \cdot 3 \cdot xyz$, or $48xyz$.

b) Since $x^2 + 4$, $x + 1$, and 5 are not factorable, the LCM is their product: $5(x^2 + 4)(x + 1)$.

To add or subtract rational expressions with different denominators, we must be able to write equivalent expressions that have the LCD.

Example 6   Find equivalent expressions that have the LCD:

$$\frac{x + 3}{x^2 + 5x - 6}, \qquad \frac{x + 7}{x^2 - 1}.$$

Solution   From Example 4(c), we know that the LCD is

$$(x + 6)(x - 1)(x + 1).$$

Since

$$x^2 + 5x - 6 = (x + 6)(x - 1),$$

the factor of the LCD that is missing from the first denominator is $x + 1$. We multiply by 1 using $(x + 1)/(x + 1)$:

$$\left.\begin{aligned}
\frac{x + 3}{x^2 + 5x - 6} &= \frac{x + 3}{(x + 6)(x - 1)} \cdot \frac{x + 1}{x + 1} \\
&= \frac{(x + 3)(x + 1)}{(x + 6)(x - 1)(x + 1)}.
\end{aligned}\right\}$$
Finding an equivalent expression that has the least common denominator

For the second expression, we have $x^2 - 1 = (x + 1)(x - 1)$. The factor of the LCD that is missing is $x + 6$. We multiply by 1 using $(x + 6)/(x + 6)$:

$$\left.\begin{aligned}
\frac{x + 7}{x^2 - 1} &= \frac{x + 7}{(x + 1)(x - 1)} \cdot \frac{x + 6}{x + 6} \\
&= \frac{(x + 7)(x + 6)}{(x + 1)(x - 1)(x + 6)}.
\end{aligned}\right\}$$
Finding an equivalent expression that has the least common denominator

We leave the results in factored form. In Section 6.4, we will carry out the actual addition and subtraction of such rational expressions.

Exercise Set 6.3

FOR EXTRA HELP

Digital Video Tutor CD 4   InterAct Math   Math Tutor Center   MathXL   MyMathLab.com
Videotape 11

Perform the indicated operation. Simplify, if possible.

1. $\dfrac{3}{x} + \dfrac{9}{x}$

2. $\dfrac{4}{a^2} + \dfrac{9}{a^2}$

3. $\dfrac{x}{15} + \dfrac{2x + 5}{15}$

4. $\dfrac{a}{7} + \dfrac{3a - 4}{7}$

5. $\dfrac{4}{a + 3} + \dfrac{5}{a + 3}$

6. $\dfrac{5}{x + 2} + \dfrac{8}{x + 2}$

7. $\dfrac{9}{a + 2} - \dfrac{3}{a + 2}$

8. $\dfrac{8}{x + 7} - \dfrac{2}{x + 7}$

9. $\dfrac{3y + 8}{2y} - \dfrac{y + 1}{2y}$

10. $\dfrac{5 + 3t}{4t} - \dfrac{2t + 1}{4t}$

11. $\dfrac{7x + 8}{x + 1} + \dfrac{4x + 3}{x + 1}$

12. $\dfrac{3a + 13}{a + 4} + \dfrac{2a + 7}{a + 4}$

13. $\dfrac{7x + 8}{x + 1} - \dfrac{4x + 3}{x + 1}$

14. $\dfrac{3a + 13}{a + 4} - \dfrac{2a + 7}{a + 4}$

15. $\dfrac{a^2}{a - 4} + \dfrac{a - 20}{a - 4}$

16. $\dfrac{x^2}{x + 5} + \dfrac{7x + 10}{x + 5}$

17. $\dfrac{x^2}{x - 2} - \dfrac{6x - 8}{x - 2}$

18. $\dfrac{a^2}{a + 3} - \dfrac{2a + 15}{a + 3}$

Aha! **19.** $\dfrac{t^2 - 5t}{t - 1} + \dfrac{5t - t^2}{t - 1}$

20. $\dfrac{y^2 + 6y}{y + 2} + \dfrac{2y + 12}{y + 2}$

21. $\dfrac{x - 4}{x^2 + 5x + 6} + \dfrac{7}{x^2 + 5x + 6}$

22. $\dfrac{x - 5}{x^2 - 4x + 3} + \dfrac{2}{x^2 - 4x + 3}$

23. $\dfrac{3a^2 + 14}{a^2 + 5a - 6} - \dfrac{13a}{a^2 + 5a - 6}$

24. $\dfrac{2a^2 + 15}{a^2 - 7a + 12} - \dfrac{11a}{a^2 - 7a + 12}$

25. $\dfrac{t^2 - 3t}{t^2 + 6t + 9} + \dfrac{2t - 12}{t^2 + 6t + 9}$

26. $\dfrac{y^2 - 7y}{y^2 + 8y + 16} + \dfrac{6y - 20}{y^2 + 8y + 16}$

27. $\dfrac{2x^2 + x}{x^2 - 8x + 12} - \dfrac{x^2 - 2x + 10}{x^2 - 8x + 12}$

28. $\dfrac{2x^2 + 3}{x^2 - 6x + 5} - \dfrac{3 + 2x^2}{x^2 - 6x + 5}$

29. $\dfrac{3 - 2x}{x^2 - 6x + 8} + \dfrac{7 - 3x}{x^2 - 6x + 8}$

30. $\dfrac{1 - 2t}{t^2 - 5t + 4} + \dfrac{4 - 3t}{t^2 - 5t + 4}$

31. $\dfrac{x - 7}{x^2 + 3x - 4} - \dfrac{2x - 3}{x^2 + 3x - 4}$

32. $\dfrac{5 - 3x}{x^2 - 2x + 1} - \dfrac{x + 1}{x^2 - 2x + 1}$

Find the LCM.

33. 15, 27

34. 10, 15

35. 8, 9

36. 12, 15

37. 6, 9, 21

38. 8, 36, 40

Find the LCM.

39. $12x^2,\ 6x^3$

40. $10t^3,\ 5t^4$

41. $15a^4b^7,\ 10a^2b^8$

42. $6a^2b^7,\ 9a^5b^2$

43. $2(y - 3),\ 6(y - 3)$

44. $4(x - 1),\ 8(x - 1)$

45. $x^2 - 4,\ x^2 + 5x + 6$

46. $x^2 + 3x + 2,\ x^2 - 4$

47. $t^3 + 4t^2 + 4t,\ t^2 - 4t$

48. $y^3 - y^2,\ y^4 - y^2$

49. $10x^2y,\ 6y^2z,\ 5xz^3$

50. $8x^3z,\ 12xy^2,\ 4y^5z^2$

51. $a + 1,\ (a - 1)^2,\ a^2 - 1$

52. $x^2 - 9,\ x + 3,\ (x - 3)^2$

53. $m^2 - 5m + 6,\ m^2 - 4m + 4$

54. $2x^2 + 5x + 2,\ 2x^2 - x - 1$

Aha! **55.** $t - 3,\ t + 3,\ (t^2 - 9)^2$

56. $a - 5,\ (a^2 - 10a + 25)^2$

57. $6x^3 - 24x^2 + 18x,\ 4x^5 - 24x^4 + 20x^3$

58. $9x^3 - 9x^2 - 18x,\ 6x^5 - 24x^4 + 24x^3$

Find equivalent expressions that have the LCD.

59. $\dfrac{5}{6x^5},\ \dfrac{y}{12x^3}$

60. $\dfrac{3}{10a^3},\ \dfrac{b}{5a^6}$

61. $\dfrac{3}{2a^2b},\ \dfrac{7}{8ab^2}$

62. $\dfrac{7}{3x^4y^2},\ \dfrac{4}{9xy^3}$

63. $\dfrac{2x}{x^2 - 4}$, $\dfrac{4x}{x^2 + 5x + 6}$

64. $\dfrac{5x}{x^2 - 9}$, $\dfrac{2x}{x^2 + 11x + 24}$

 65. If the LCM of two numbers is their product, what can you conclude about the two numbers?

 66. Explain why the product of two numbers is not always their least common multiple.

SKILL MAINTENANCE

Write each number in two equivalent forms.

67. $\dfrac{7}{-9}$

68. $-\dfrac{3}{2}$

Simplify.

69. $\dfrac{5}{18} - \dfrac{7}{12}$

70. $\dfrac{8}{15} - \dfrac{13}{20}$

Find a polynomial that can represent the shaded area of each figure.

71.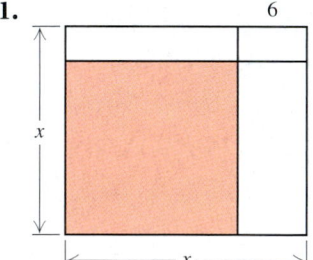

72.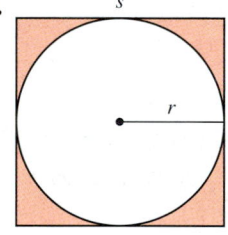

SYNTHESIS

 73. If the LCM of a binomial and a trinomial is the trinomial, what relationship exists between the two expressions?

 74. If the LCM of two third-degree polynomials is a sixth-degree polynomial, what can be concluded about the two polynomials?

Perform the indicated operations. Simplify, if possible.

75. $\dfrac{6x - 1}{x - 1} + \dfrac{3(2x + 5)}{x - 1} + \dfrac{3(2x - 3)}{x - 1}$

76. $\dfrac{2x + 11}{x - 3} \cdot \dfrac{3}{x + 4} + \dfrac{-1}{4 + x} \cdot \dfrac{6x + 3}{x - 3}$

77. $\dfrac{x^2}{3x^2 - 5x - 2} - \dfrac{2x}{3x + 1} \cdot \dfrac{1}{x - 2}$

78. $\dfrac{x + y}{x^2 - y^2} + \dfrac{x - y}{x^2 - y^2} - \dfrac{2x}{x^2 - y^2}$

South African artistry. In South Africa, the design of every woven handbag, or gipatsi *(plural,* sipatsi*) is created by repeating two or more geometric patterns. Each pattern encircles the bag, sharing the strands of fabric with any pattern above or below. The length, or period, of each pattern is the number of strands required to construct the pattern. For a gipatsi to be considered beautiful, each individual pattern must fit a whole number of times around the bag (Source: Gerdes, Paulus,* Women, Art and Geometry in Southern Africa. *Asmara, Eritrea: Africa World Press, Inc., p. 5).*

79. A weaver is using two patterns to create a gipatsi. Pattern A is 10 strands long, and pattern B is 3 strands long. What is the smallest number of strands that can be used to complete the gipatsi?

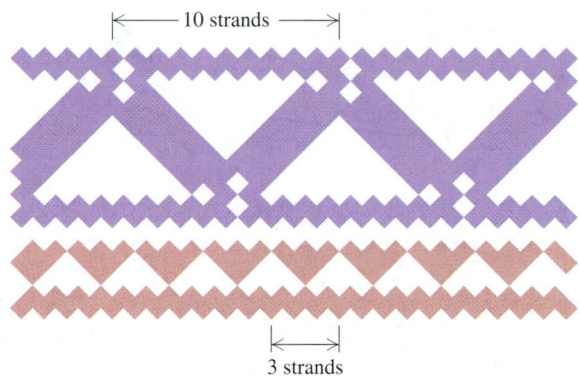

80. A weaver is using a four-strand pattern, a six-strand pattern, and an eight-strand pattern. What is the smallest number of strands that can be used to complete the gipatsi?

81. For technical reasons, the number of strands is generally a multiple of 4. Answer Exercise 79 with this additional requirement in mind.

Find the LCM.

82. 72, 90, 96

83. $8x^2 - 8$, $6x^2 - 12x + 6$, $10 - 10x$

84. $9x^2 - 16$, $6x^2 - x - 12$, $16 - 24x + 9x^2$

85. *Running.* Kim and Jed leave the starting point of a fitness loop at the same time. Kim jogs a lap in 6 min and Jed jogs one in 8 min. Assuming they continue to run at the same pace, when will they next meet at the starting place?

86. *Bus schedules.* Beginning at 5:00 A.M., a hotel shuttle bus leaves Salton Airport every 25 min, and the downtown shuttle bus leaves the airport every 35 min. What time will it be when both shuttles again leave at the same time?

87. *Appliances.* Dishwashers last an average of 10 yr, clothes washers an average of 14 yr, and refrigerators an average of 20 yr (*Source: Energy Savers: Tips on Saving Energy and Money at Home,* produced for the U.S. Department of Energy by the National Renewable Energy Laboratory, 1998). If an apartment house is equipped with new dishwashers, clothes washers, and refrigerators in 2002, in what year will all three appliances need to be replaced at once?

88. Explain how evaluating can be used to perform a partial check on the result of Example 1(d):

$$\frac{x-5}{x^2-9} + \frac{2}{x^2-9} = \frac{1}{x+3}.$$

89. On p. 356, the second step in finding an LCD is to select one of the factorizations of the denominators. Does it matter which one is selected? Why or why not?

Addition and Subtraction with Unlike Denominators

6.4

Adding and Subtracting with LCDs •
When Factors Are Opposites

Adding and Subtracting with LCDs

We now know how to rewrite two rational expressions in an equivalent form that uses the LCD. Once rational expressions share a common denominator, they can be added or subtracted just as in Section 6.3.

> ### To Add or Subtract Rational Expressions Having Different Denominators
>
> 1. Find the LCD.
> 2. Multiply each rational expression by a form of 1 made up of the factors of the LCD missing from that expression's denominator.
> 3. Add or subtract the numerators, as indicated. Write the sum or difference over the LCD.
> 4. Simplify, if possible.

E x a m p l e   1   Add: $\dfrac{5x^2}{8} + \dfrac{7x}{12}$.

Solution

1. First, we find the LCD:

$$\left.\begin{array}{l} 8 = 2 \cdot 2 \cdot 2 \\ 12 = 2 \cdot 2 \cdot 3 \end{array}\right\} \quad \text{LCD} = 2 \cdot 2 \cdot 2 \cdot 3, \text{ or } 24.$$

2. The denominator 8 needs to be multiplied by 3 in order to obtain the LCD. The denominator 12 needs to be multiplied by 2 in order to obtain the LCD. Thus we multiply by $\frac{3}{3}$ and $\frac{2}{2}$ to get the LCD:

$$\begin{aligned} \frac{5x^2}{8} + \frac{7x}{12} &= \frac{5x^2}{2 \cdot 2 \cdot 2} + \frac{7x}{2 \cdot 2 \cdot 3} \\[2mm] &= \frac{5x^2}{2 \cdot 2 \cdot 2} \cdot \frac{3}{3} + \frac{7x}{2 \cdot 2 \cdot 3} \cdot \frac{2}{2} \qquad \text{Multiplying each expression} \\ & \qquad\qquad\qquad\qquad\qquad\qquad\qquad \text{by a form of 1 to get the LCD} \\[2mm] &= \frac{15x^2}{24} + \frac{14x}{24}. \end{aligned}$$

3. Next, we add the numerators:

$$\frac{15x^2}{24} + \frac{14x}{24} = \frac{15x^2 + 14x}{24}.$$

4. Since $15x^2 + 14x$ and 24 have no common factor,

$$\frac{15x^2 + 14x}{24}$$

cannot be simplified any further.

Subtraction is performed in much the same way.

E x a m p l e   2   Subtract: $\dfrac{7}{8x} - \dfrac{5}{12x^2}$.

Solution   We follow, but do not list, the four steps shown above. First, we find the LCD:

$$\left.\begin{array}{l} 8x = 2 \cdot 2 \cdot 2 \cdot x \\ 12x^2 = 2 \cdot 2 \cdot 3 \cdot x \cdot x \end{array}\right\} \quad \text{LCD} = 2 \cdot 2 \cdot 3 \cdot x \cdot x \cdot 2, \text{ or } 24x^2.$$

The denominator $8x$ must be multiplied by $3x$ in order to obtain the LCD. The denominator $12x^2$ must be multiplied by 2 in order to obtain the LCD. Thus we multiply by $\frac{3x}{3x}$ and $\frac{2}{2}$ to get the LCD. Then we subtract and, if possible, simplify.

$$\frac{7}{8x} - \frac{5}{12x^2} = \frac{7}{8x} \cdot \frac{3x}{3x} - \frac{5}{12x^2} \cdot \frac{2}{2}$$

$$= \frac{21x}{24x^2} - \frac{10}{24x^2}$$

> **Caution!** Do not simplify *these* rational expressions or you will lose the LCD.

$$= \frac{21x - 10}{24x^2} \qquad \text{This cannot be simplified, so we are done.}$$

When denominators contain polynomials with two or more terms, the same steps are used.

E x a m p l e   3

Add: $\dfrac{2a}{a^2 - 1} + \dfrac{1}{a^2 + a}.$

Solution   First, we find the LCD:

$$\left. \begin{array}{l} a^2 - 1 = (a - 1)(a + 1) \\ a^2 + a = a(a + 1) \end{array} \right\} \qquad \text{LCD} = (a - 1)(a + 1)a.$$

We multiply by a form of 1 to get the LCD in each expression:

$$\frac{2a}{a^2 - 1} + \frac{1}{a^2 + a} = \frac{2a}{(a - 1)(a + 1)} \cdot \frac{a}{a} + \frac{1}{a(a + 1)} \cdot \frac{a - 1}{a - 1} \qquad \begin{array}{l} \text{Multiplying by} \\ \dfrac{a}{a} \text{ and } \dfrac{a - 1}{a - 1} \\ \text{to get the LCD} \end{array}$$

$$= \frac{2a^2}{(a - 1)(a + 1)a} + \frac{a - 1}{a(a + 1)(a - 1)}$$

$$= \frac{2a^2 + a - 1}{a(a - 1)(a + 1)} \qquad \text{Adding numerators}$$

$$= \left. \begin{array}{c} \dfrac{(2a - 1)(a + 1)}{a(a - 1)(a + 1)} \\[2mm] = \dfrac{2a - 1}{a(a - 1)}. \end{array} \right\} \qquad \begin{array}{l} \text{Simplifying by factoring and} \\ \text{removing a factor equal to 1:} \\ \dfrac{a + 1}{a + 1} = 1 \end{array}$$

E x a m p l e   4

Perform the indicated operations.

a) $\dfrac{x + 4}{x - 2} - \dfrac{x - 7}{x + 5}$

b) $\dfrac{x}{x^2 + 11x + 30} + \dfrac{-5}{x^2 + 9x + 20}$

c) $\dfrac{x}{x^2 + 5x + 6} - \dfrac{2}{x^2 + 3x + 2}$

Solution

a) First, we find the LCD. It is just the product of the denominators:

$$\text{LCD} = (x - 2)(x + 5).$$

We multiply by a form of 1 to get the LCD in each expression. Then we subtract and try to simplify.

$$\frac{x + 4}{x - 2} - \frac{x - 7}{x + 5} = \frac{x + 4}{x - 2} \cdot \frac{x + 5}{x + 5} - \frac{x - 7}{x + 5} \cdot \frac{x - 2}{x - 2}$$

$$= \frac{x^2 + 9x + 20}{(x - 2)(x + 5)} - \frac{x^2 - 9x + 14}{(x - 2)(x + 5)} \qquad \begin{array}{l} \text{Multiplying out} \\ \text{numerators (but} \\ \text{not denominators)} \end{array}$$

$$= \frac{x^2 + 9x + 20 - (x^2 - 9x + 14)}{(x - 2)(x + 5)} \qquad \begin{array}{l} \text{When subtracting} \\ \text{a numerator with} \\ \text{more than one} \\ \text{term, parentheses} \\ \text{are important.} \end{array}$$

$$= \frac{x^2 + 9x + 20 - x^2 + 9x - 14}{(x - 2)(x + 5)} \qquad \begin{array}{l} \text{Removing paren-} \\ \text{theses and sub-} \\ \text{tracting every} \\ \text{term} \end{array}$$

$$= \frac{18x + 6}{(x - 2)(x + 5)}$$

Although $18x + 6$ can be factored as $6(3x + 1)$, doing so will not enable us to simplify our result.

b) $\dfrac{x}{x^2 + 11x + 30} + \dfrac{-5}{x^2 + 9x + 20}$

$$= \frac{x}{(x + 5)(x + 6)} + \frac{-5}{(x + 5)(x + 4)} \qquad \begin{array}{l} \text{Factoring the denominators} \\ \text{in order to find the LCD.} \\ \text{The LCD is } (x + 5)(x + 6)(x + 4). \end{array}$$

$$= \frac{x}{(x + 5)(x + 6)} \cdot \frac{x + 4}{x + 4} + \frac{-5}{(x + 5)(x + 4)} \cdot \frac{x + 6}{x + 6} \qquad \begin{array}{l} \text{Multiplying to} \\ \text{get the LCD} \end{array}$$

$$= \frac{x^2 + 4x}{(x + 5)(x + 6)(x + 4)} + \frac{-5x - 30}{(x + 5)(x + 6)(x + 4)} \qquad \begin{array}{l} \text{Multiplying in} \\ \text{each numerator} \end{array}$$

$$= \frac{x^2 + 4x - 5x - 30}{(x + 5)(x + 6)(x + 4)} \qquad \text{Adding numerators}$$

$$= \frac{x^2 - x - 30}{(x + 5)(x + 6)(x + 4)} \qquad \text{Combining like terms in the numerator}$$

$$= \frac{\cancel{(x + 5)}(x - 6)}{\cancel{(x + 5)}(x + 6)(x + 4)} \left. \begin{array}{l} \\ \\ \\ \end{array} \right\} \qquad \text{Always simplify the result, if possible, by}$$

$$= \frac{x - 6}{(x + 6)(x + 4)} \qquad\qquad\qquad \text{removing a factor equal to 1; here } \frac{x + 5}{x + 5} = 1.$$

c) $\dfrac{x}{x^2 + 5x + 6} - \dfrac{2}{x^2 + 3x + 2}$

$= \dfrac{x}{(x + 2)(x + 3)} - \dfrac{2}{(x + 2)(x + 1)}$ Factoring denominators.
The LCD is $(x + 2)(x + 3)(x + 1)$.

$= \dfrac{x}{(x + 2)(x + 3)} \cdot \dfrac{x + 1}{x + 1} - \dfrac{2}{(x + 2)(x + 1)} \cdot \dfrac{x + 3}{x + 3}$

$= \dfrac{x^2 + x}{(x + 2)(x + 3)(x + 1)} - \dfrac{2x + 6}{(x + 2)(x + 3)(x + 1)}$

$= \dfrac{x^2 + x - (2x + 6)}{(x + 2)(x + 3)(x + 1)}$ Don't forget the parentheses!

$= \dfrac{x^2 + x - 2x - 6}{(x + 2)(x + 3)(x + 1)}$ Remember to subtract each term in $2x + 6$.

$= \dfrac{x^2 - x - 6}{(x + 2)(x + 3)(x + 1)}$ Combining like terms in the numerator

$= \dfrac{\cancel{(x + 2)}(x - 3)}{\cancel{(x + 2)}(x + 3)(x + 1)}$
$= \dfrac{x - 3}{(x + 3)(x + 1)}$ Factoring and simplifying; $\dfrac{x + 2}{x + 2} = 1$

When Factors Are Opposites

Recall from Section 6.1 that expressions of the form $a - b$ and $b - a$ are opposites of each other. When either of these binomials is multiplied by -1, the result is the other binomial:

$-1(a - b) = -a + b = b + (-a) = b - a;$ Multiplication by -1
$-1(b - a) = -b + a = a + (-b) = a - b.$ reverses the order in which subtraction occurs.

E x a m p l e   5   Add: $\dfrac{x}{x - 5} + \dfrac{7}{5 - x}$.

Solution   Since the denominators are opposites of each other, we can find a common denominator by multiplying either rational expression by $-1/-1$. Because polynomials are most often written in descending order, we choose to reverse the subtraction in the second denominator:

$\dfrac{x}{x - 5} + \dfrac{7}{5 - x} = \dfrac{x}{x - 5} + \dfrac{7}{5 - x} \cdot \dfrac{-1}{-1}$ Writing 1 as $-1/-1$ and multiplying to obtain a common denominator

$= \dfrac{x}{x - 5} + \dfrac{-7}{-5 + x}$

$= \dfrac{x}{x - 5} + \dfrac{-7}{x - 5}$ Note that $-5 + x = x + (-5) = x - 5$.

$= \dfrac{x - 7}{x - 5}$.

Sometimes, after factoring to find the LCD, we find a factor in one denominator that is the opposite of a factor in the other denominator. When this happens, multiplication by $-1/-1$ can again be helpful.

Example 6

Perform the indicated operations and simplify.

a) $\dfrac{x}{x^2 - 25} + \dfrac{3}{5 - x}$

b) $\dfrac{x + 9}{x^2 - 4} + \dfrac{6 - x}{4 - x^2} - \dfrac{1 + x}{x^2 - 4}$

Study Tip

It is always best to study for a final exam over a period of at least two weeks. If you have only one or two days of study time, however, begin by studying the formulas, problems, properties, and procedures in each chapter Summary and Review. Then do the exercises in the Cumulative Reviews. Make sure to attend a review session if one is offered.

Solution

a) $\dfrac{x}{x^2 - 25} + \dfrac{3}{5 - x} = \dfrac{x}{(x - 5)(x + 5)} + \dfrac{3}{5 - x}$ Factoring

$\qquad = \dfrac{x}{(x - 5)(x + 5)} + \dfrac{3}{5 - x} \cdot \dfrac{-1}{-1}$ Multiplying by $-1/-1$ changes $5 - x$ to $x - 5$.

$\qquad = \dfrac{x}{(x - 5)(x + 5)} + \dfrac{-3}{x - 5}$ Note that $(5 - x)(-1) = x - 5$.

$\qquad = \dfrac{x}{(x - 5)(x + 5)} + \dfrac{-3}{(x - 5)} \cdot \dfrac{x + 5}{x + 5}$ The LCD is $(x - 5)(x + 5)$.

$\qquad = \dfrac{x}{(x - 5)(x + 5)} + \dfrac{-3x - 15}{(x - 5)(x + 5)}$

$\qquad = \dfrac{-2x - 15}{(x - 5)(x + 5)}$

b) Since $4 - x^2$ is the opposite of $x^2 - 4$, multiplying the second rational expression by $-1/-1$ will lead to a common denominator:

$\dfrac{x + 9}{x^2 - 4} + \dfrac{6 - x}{4 - x^2} - \dfrac{1 + x}{x^2 - 4} = \dfrac{x + 9}{x^2 - 4} + \dfrac{6 - x}{4 - x^2} \cdot \dfrac{-1}{-1} - \dfrac{1 + x}{x^2 - 4}$

$\qquad = \dfrac{x + 9}{x^2 - 4} + \dfrac{x - 6}{x^2 - 4} - \dfrac{1 + x}{x^2 - 4}$

$\qquad = \dfrac{x + 9 + x - 6 - 1 - x}{x^2 - 4}$ Adding and subtracting numerators

$\qquad = \dfrac{x + 2}{x^2 - 4}$

$\qquad = \dfrac{(x + 2) \cdot 1}{(x + 2)(x - 2)}$ Simplifying

$\qquad = \dfrac{1}{x - 2}.$

Exercise Set 6.4

Perform the indicated operation. Simplify, if possible.

1. $\dfrac{3}{x} + \dfrac{7}{x^2}$

2. $\dfrac{5}{x} + \dfrac{6}{x^2}$

3. $\dfrac{1}{6r} - \dfrac{3}{8r}$

4. $\dfrac{4}{9t} - \dfrac{7}{6t}$

5. $\dfrac{4}{xy^2} + \dfrac{2}{x^2y}$

6. $\dfrac{2}{c^2d} + \dfrac{7}{cd^3}$

7. $\dfrac{8}{9t^3} - \dfrac{5}{6t^2}$

8. $\dfrac{-2}{3xy^2} - \dfrac{6}{x^2y^3}$

9. $\dfrac{x+5}{8} + \dfrac{x-3}{12}$

10. $\dfrac{x-4}{9} + \dfrac{x+5}{6}$

11. $\dfrac{a+2}{2} - \dfrac{a-4}{4}$

12. $\dfrac{x-2}{6} - \dfrac{x+1}{3}$

13. $\dfrac{2a-1}{3a^2} + \dfrac{5a+1}{9a}$

14. $\dfrac{a+4}{16a} + \dfrac{3a+4}{4a^2}$

15. $\dfrac{x-1}{4x} - \dfrac{2x+3}{x}$

16. $\dfrac{4z-9}{3z} - \dfrac{3z-8}{4z}$

17. $\dfrac{2c-d}{c^2d} + \dfrac{c+d}{cd^2}$

18. $\dfrac{x+y}{xy^2} + \dfrac{3x+y}{x^2y}$

19. $\dfrac{5x+3y}{2x^2y} - \dfrac{3x+4y}{xy^2}$

20. $\dfrac{4x+2t}{3xt^2} - \dfrac{5x-3t}{x^2t}$

21. $\dfrac{5}{x-1} + \dfrac{5}{x+1}$

22. $\dfrac{3}{x-2} + \dfrac{3}{x+2}$

23. $\dfrac{4}{z-1} - \dfrac{2}{z+1}$

24. $\dfrac{5}{x+5} - \dfrac{3}{x-5}$

25. $\dfrac{2}{x+5} + \dfrac{3}{4x}$

26. $\dfrac{3}{x+1} + \dfrac{2}{3x}$

27. $\dfrac{8}{3t^2-15t} - \dfrac{3}{2t-10}$

28. $\dfrac{3}{2t^2-2t} - \dfrac{5}{2t-2}$

29. $\dfrac{4x}{x^2-25} + \dfrac{x}{x+5}$

30. $\dfrac{2x}{x^2-16} + \dfrac{x}{x-4}$

31. $\dfrac{t}{t-3} - \dfrac{5}{4t-12}$

32. $\dfrac{6}{z+4} - \dfrac{2}{3z+12}$

33. $\dfrac{2}{x+3} + \dfrac{4}{(x+3)^2}$

34. $\dfrac{3}{x-1} + \dfrac{2}{(x-1)^2}$

35. $\dfrac{3}{x+2} - \dfrac{8}{x^2-4}$

36. $\dfrac{2t}{t^2-9} - \dfrac{3}{t-3}$

37. $\dfrac{3a}{4a-20} + \dfrac{9a}{6a-30}$

38. $\dfrac{4a}{5a-10} + \dfrac{3a}{10a-20}$

Aha! **39.** $\dfrac{x}{x-5} + \dfrac{x}{5-x}$

40. $\dfrac{x+4}{x} + \dfrac{x}{x+4}$

41. $\dfrac{7}{a^2+a-2} + \dfrac{5}{a^2-4a+3}$

42. $\dfrac{x}{x^2+2x+1} + \dfrac{1}{x^2+5x+4}$

43. $\dfrac{x}{x^2+9x+20} - \dfrac{4}{x^2+7x+12}$

44. $\dfrac{x}{x^2+5x+6} - \dfrac{2}{x^2+3x+2}$

45. $\dfrac{3z}{z^2-4z+4} + \dfrac{10}{z^2+z-6}$

46. $\dfrac{3}{x^2-9} + \dfrac{2}{x^2-x-6}$

Aha! **47.** $\dfrac{-5}{x^2+17x+16} - \dfrac{0}{x^2+9x+8}$

48. $\dfrac{x}{x^2+15x+56} - \dfrac{1}{x^2+13x+42}$

49. $\dfrac{2x}{5} - \dfrac{x-3}{-5}$

50. $\dfrac{x}{4} - \dfrac{3x-5}{-4}$

51. $\dfrac{y^2}{y-3} + \dfrac{9}{3-y}$

52. $\dfrac{t^2}{t-2} + \dfrac{4}{2-t}$

53. $\dfrac{b-7}{b^2-16} + \dfrac{7-b}{16-b^2}$

54. $\dfrac{a-3}{a^2-25} + \dfrac{a-3}{25-a^2}$

55. $\dfrac{y+2}{y-7} + \dfrac{3-y}{49-y^2}$

56. $\dfrac{4-p}{25-p^2} + \dfrac{p+1}{p-5}$

57. $\dfrac{5x}{x^2-9} - \dfrac{4}{3-x}$

58. $\dfrac{8x}{16-x^2} - \dfrac{5}{x-4}$

59. $\dfrac{3x+2}{3x+6} + \dfrac{x}{4-x^2}$

60. $\dfrac{a}{a^2-1} + \dfrac{2a}{a-a^2}$

61. $\dfrac{4-a^2}{a^2-9} - \dfrac{a-2}{3-a}$

62. $\dfrac{4x}{x^2-y^2} - \dfrac{6}{y-x}$

Perform the indicated operations. Simplify, if possible.

63. $\dfrac{x-3}{2-x} - \dfrac{x+3}{x+2} + \dfrac{x+6}{4-x^2}$

64. $\dfrac{t-5}{1-t} - \dfrac{t+4}{t+1} + \dfrac{t+2}{t^2-1}$

65. $\dfrac{x+5}{x+3} + \dfrac{x+7}{x+2} - \dfrac{7x+19}{(x+3)(x+2)}$

66. $\dfrac{2x+5}{x+1} + \dfrac{x+7}{x+5} - \dfrac{5x+17}{(x+1)(x+5)}$

67. $\dfrac{t}{s+t} - \dfrac{t}{s-t}$

68. $\dfrac{a}{b-a} - \dfrac{b}{b+a}$

69. $\dfrac{1}{x+y} + \dfrac{1}{x-y} - \dfrac{2x}{x^2-y^2}$

70. $\dfrac{2r}{r^2-s^2} + \dfrac{1}{r+s} - \dfrac{1}{r-s}$

71. What is the advantage of using the *least* common denominator—rather than just *any* common denominator—when adding or subtracting rational expressions?

72. Describe a procedure that can be used to add any two rational expressions.

SKILL MAINTENANCE

Simplify.

73. $-\dfrac{3}{7} \div \dfrac{6}{13}$

74. $\dfrac{5}{12} \div \left(-\dfrac{3}{4}\right)$

75. $\dfrac{\frac{2}{9}}{\frac{5}{3}}$

76. $\dfrac{\frac{7}{10}}{\frac{3}{5}}$

Graph.

77. $y = -\dfrac{1}{2}x - 5$

78. $y = \dfrac{1}{2}x - 5$

SYNTHESIS

79. How could you convince someone that

$$\dfrac{1}{3-x} \quad \text{and} \quad \dfrac{1}{x-3}$$

are opposites of each other?

80. Are parentheses as important for adding rational expressions as they are for subtracting rational expressions? Why or why not?

Write expressions for the perimeter and the area of each rectangle.

81.

82.

Perform the indicated operations.

83. $\dfrac{2x+11}{x-3} \cdot \dfrac{3}{x+4} + \dfrac{2x+1}{4+x} \cdot \dfrac{3}{3-x}$

84. $\dfrac{x^2}{3x^2-5x-2} - \dfrac{2x}{3x+1} \cdot \dfrac{1}{x-2}$

Aha! **85.** $\left(\dfrac{x}{x+7} - \dfrac{3}{x+2}\right)\left(\dfrac{x}{x+7} + \dfrac{3}{x+2}\right)$

86. $\dfrac{1}{ay-3a+2xy-6x} - \dfrac{xy+ay}{a^2-4x^2}\left(\dfrac{1}{y-3}\right)^2$

87. $\dfrac{2x^2+5x-3}{2x^2-9x+9} + \dfrac{x+1}{3-2x} + \dfrac{4x^2+8x+3}{x-3} \cdot \dfrac{x+3}{9-4x^2}$

88. $\left(\dfrac{a}{a-b} + \dfrac{b}{a+b}\right)\left(\dfrac{1}{3a+b} + \dfrac{2a+6b}{9a^2-b^2}\right)$

89. Express

$$\dfrac{a-3b}{a-b}$$

as a sum of two rational expressions with denominators that are opposites of each other. Answers may vary.

Complex Rational Expressions

6.5

Using Division to Simplify • Multiplying by the LCD

A **complex rational expression**, or **complex fractional expression**, is a rational expression that has one or more rational expressions within its numerator or denominator. Here are some examples:

$$\frac{1 + \dfrac{2}{x}}{3}, \qquad \frac{\dfrac{x+y}{7}}{\dfrac{2x}{x+1}}, \qquad \frac{\dfrac{4}{3} + \dfrac{1}{5}}{\dfrac{2}{x} - \dfrac{x}{y}}.$$

These are rational expressions within the complex rational expression.

We will consider two methods for simplifying complex rational expressions. Each method offers certain advantages.

Using Division to Simplify (Method 1)

Our first method for simplifying complex rational expressions involves rewriting the expression as a quotient of two rational expressions.

> ### To Simplify a Complex Rational Expression by Dividing
> 1. Add or subtract, as needed, to get a single rational expression in the numerator.
> 2. Add or subtract, as needed, to get a single rational expression in the denominator.
> 3. Divide the numerator by the denominator (invert and multiply).
> 4. If possible, simplify by removing a factor equal to 1.

The key here is to express a complex rational expression as one rational expression divided by another. We can then proceed as in Section 6.2.

E x a m p l e   1

Simplify: $\dfrac{\dfrac{x}{x-3}}{\dfrac{4}{5x-15}}$.

Solution Here the numerator and denominator are already single rational expressions. This allows us to start by dividing (step 3), as in Section 6.2:

$$\frac{\dfrac{x}{x-3}}{\dfrac{4}{5x-15}} = \frac{x}{x-3} \div \frac{4}{5x-15} \qquad \text{Rewriting with a division symbol}$$

$$= \frac{x}{x-3} \cdot \frac{5x-15}{4} \qquad \text{Multiplying by the reciprocal of the divisor (inverting and multiplying)}$$

$$= \frac{x}{x-3} \cdot \frac{5(x-3)}{4} \qquad \text{Factoring and removing a factor equal to 1: } \frac{x-3}{x-3} = 1$$

$$= \frac{5x}{4}.$$

Often we must add or subtract in the numerator and/or denominator before we can divide.

E x a m p l e 2   Simplify.

a) $\dfrac{\dfrac{5}{2a} + \dfrac{1}{a}}{\dfrac{1}{4a} - \dfrac{5}{6}}$

b) $\dfrac{\dfrac{x^2}{y} - \dfrac{5}{x}}{xz}$

Solution

a) $\dfrac{\dfrac{5}{2a} + \dfrac{1}{a}}{\dfrac{1}{4a} - \dfrac{5}{6}} = \dfrac{\dfrac{5}{2a} + \dfrac{1}{a} \cdot \dfrac{2}{2}}{\dfrac{1}{4a} \cdot \dfrac{3}{3} - \dfrac{5}{6} \cdot \dfrac{2a}{2a}}$ ⟵ Multiplying by 1 to get the LCD, $2a$, for the numerator of the complex rational expression

⟵ Multiplying by 1 to get the LCD, $12a$, for the denominator of the complex rational expression

$$= \frac{\dfrac{5}{2a} + \dfrac{2}{2a}}{\dfrac{3}{12a} - \dfrac{10a}{12a}} = \frac{\dfrac{7}{2a}}{\dfrac{3-10a}{12a}} \qquad \substack{\text{⟵ Adding} \\ \text{⟵ Subtracting}}$$

$$= \frac{7}{2a} \div \frac{3-10a}{12a} \qquad \text{Rewriting with a division symbol. This is often done mentally.}$$

$$= \frac{7}{2a} \cdot \frac{12a}{3-10a} \qquad \text{Multiplying by the reciprocal of the divisor (inverting and multiplying)}$$

$$= \frac{7}{2a} \cdot \frac{2a \cdot 6}{3-10a} \qquad \text{Removing a factor equal to 1: } \frac{2a}{2a} = 1$$

$$= \frac{42}{3-10a}.$$

b)
$$\frac{\dfrac{x^2}{y} - \dfrac{5}{x}}{xz} = \frac{\dfrac{x^2}{y} \cdot \dfrac{x}{x} - \dfrac{5}{x} \cdot \dfrac{y}{y}}{xz}$$

← Multiplying by 1 to get the LCD, xy, for the numerator of the complex rational expression

$$= \frac{\dfrac{x^3}{xy} - \dfrac{5y}{xy}}{xz}$$

$$= \frac{\dfrac{x^3 - 5y}{xy}}{xz}$$

← Subtracting

← If you prefer, write xz as $\dfrac{xz}{1}$.

$$= \frac{x^3 - 5y}{xy} \div (xz)$$

Rewriting with a division symbol

$$= \frac{x^3 - 5y}{xy} \cdot \frac{1}{xz}$$

Multiplying by the reciprocal of the divisor (inverting and multiplying)

$$= \frac{x^3 - 5y}{x^2yz}$$

Multiplying by the LCD (Method 2)

A second method for simplifying complex rational expressions relies on multiplying by an expression equal to 1.

> ### To Simplify a Complex Rational Expression by Multiplying by the LCD
>
> 1. Find the LCD of *all* rational expressions within the complex rational expression.
> 2. Multiply the complex rational expression by a factor equal to 1. Write 1 as the LCD over itself (LCD/LCD).
> 3. Distribute and simplify. No fractional expressions should remain within the complex rational expression.
> 4. Factor and, if possible, simplify.

Example 3   Simplify: $\dfrac{\dfrac{1}{2} + \dfrac{3}{4}}{\dfrac{5}{6} - \dfrac{3}{8}}$.

Solution

1. Unlike Method 1, in which $\frac{1}{2} + \frac{3}{4}$ would be treated separately from $\frac{5}{6} - \frac{3}{8}$, here we look for the LCD of *all* four fractions. That LCD is 24.

2. We multiply by a form of 1, using the LCD:

$$\frac{\dfrac{1}{2} + \dfrac{3}{4}}{\dfrac{5}{6} - \dfrac{3}{8}} = \frac{\dfrac{1}{2} + \dfrac{3}{4}}{\dfrac{5}{6} - \dfrac{3}{8}} \cdot \frac{24}{24} \qquad$$ Multiplying by a factor equal to 1, using the LCD: $\dfrac{24}{24} = 1$

3. Using the distributive law, we perform the multiplication:

$$\frac{\dfrac{1}{2} + \dfrac{3}{4}}{\dfrac{5}{6} - \dfrac{3}{8}} \cdot \frac{24}{24} = \frac{\left(\dfrac{1}{2} + \dfrac{3}{4} \right)24}{\left(\dfrac{5}{6} - \dfrac{3}{8} \right)24}$$

Multiplying the numerator by 24

Don't forget the parentheses!

Multiplying the denominator by 24

$$= \frac{\dfrac{1}{2}(24) + \dfrac{3}{4}(24)}{\dfrac{5}{6}(24) - \dfrac{3}{8}(24)} \qquad$$ Using the distributive law

$$= \frac{12 + 18}{20 - 9}, \quad \text{or} \quad \frac{30}{11}. \qquad$$ Simplifying

4. The result, $\frac{30}{11}$, cannot be factored or simplified, so we are done.

Multiplying like this effectively clears fractions in both the top and bottom of the complex rational expression. In Example 4 we follow, but do not list, the same four steps.

E x a m p l e 4

Simplify.

a) $\dfrac{\dfrac{3}{x} + \dfrac{1}{2x}}{\dfrac{1}{3x} - \dfrac{3}{4x}}$ **b)** $\dfrac{1 - \dfrac{1}{x}}{1 - \dfrac{1}{x^2}}$

Solution

a) The denominators within the complex expression are x, $2x$, $3x$, and $4x$, so the LCD is $12x$. We multiply by 1 using $(12x)/(12x)$:

$$\frac{\dfrac{3}{x} + \dfrac{1}{2x}}{\dfrac{1}{3x} - \dfrac{3}{4x}} = \frac{\dfrac{3}{x} + \dfrac{1}{2x}}{\dfrac{1}{3x} - \dfrac{3}{4x}} \cdot \frac{12x}{12x} = \frac{\dfrac{3}{x}(12x) + \dfrac{1}{2x}(12x)}{\dfrac{1}{3x}(12x) - \dfrac{3}{4x}(12x)}. \qquad$$ Using the distributive law

When we multiply by $12x$, all fractions in the numerator and the denominator of the complex rational expression are cleared:

$$\frac{\dfrac{3}{x}(12x) + \dfrac{1}{2x}(12x)}{\dfrac{1}{3x}(12x) - \dfrac{3}{4x}(12x)} = \frac{36 + 6}{4 - 9} = -\frac{42}{5}.$$

b) $\dfrac{1 - \dfrac{1}{x}}{1 - \dfrac{1}{x^2}} = \dfrac{1 - \dfrac{1}{x}}{1 - \dfrac{1}{x^2}} \cdot \dfrac{x^2}{x^2}$ The LCD is x^2 so we multiply by 1 using x^2/x^2.

$= \dfrac{1 \cdot x^2 - \dfrac{1}{x} \cdot x^2}{1 \cdot x^2 - \dfrac{1}{x^2} \cdot x^2}$ Using the distributive law

$= \dfrac{x^2 - x}{x^2 - 1}$ All fractions have been cleared within the complex rational expression.

$= \dfrac{x(x-1)}{(x+1)(x-1)}$

$= \dfrac{x}{x+1}$ Factoring and simplifying: $\dfrac{x-1}{x-1} = 1$

It is important to understand both of the methods studied in this section. Sometimes, as in Example 1, the complex rational expression is either given as—or easily written as—a quotient of two rational expressions. In these cases, Method 1 (using division) is probably the easiest method to use. Other times, as in Example 4(a), it is not difficult to find the LCD of all denominators in the complex rational expression. When this occurs, it is usually easier to use Method 2 (multiplying by the LCD). The more practice you get using both methods, the better you will be at selecting the easier method for any given problem.

technology connection

Be careful to place parentheses properly when entering complex rational expressions into a grapher. Remember to enclose the entire numerator of the complex rational expression in one set of parentheses and the entire denominator in another. For example, we enter the expression in Example 4(a) as

$y_1 = (3/x + 1/(2x))$
$\quad /(1/(3x) - 3/(4x)).$

1. Write Example 4(b) as you would to enter it into a grapher.
2. When must the numerator of a rational expression be enclosed in parentheses? The denominator?

FOR EXTRA HELP

Exercise Set **6.5**

 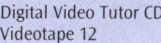 Digital Video Tutor CD 5 Videotape 12   InterAct Math   Math Tutor Center   MathXL   MyMathLab.com

Simplify. Use either method or the method specified by your instructor.

1. $\dfrac{1 + \dfrac{1}{2}}{1 + \dfrac{1}{4}}$

2. $\dfrac{1 + \dfrac{3}{4}}{1 + \dfrac{1}{2}}$

3. $\dfrac{1 + \dfrac{1}{3}}{5 - \dfrac{5}{27}}$

10. $\dfrac{\dfrac{a+4}{a^2}}{\dfrac{a-2}{3a}}$

11. $\dfrac{\dfrac{x}{4} - \dfrac{4}{x}}{\dfrac{1}{4} + \dfrac{1}{x}}$

12. $\dfrac{\dfrac{3}{x} + \dfrac{3}{8}}{\dfrac{x}{8} - \dfrac{3}{x}}$

4. $\dfrac{3 + \dfrac{1}{5}}{1 - \dfrac{3}{5}}$

5. $\dfrac{\dfrac{s}{3} + s}{\dfrac{3}{s} + s}$

6. $\dfrac{\dfrac{1}{x} - 5}{\dfrac{1}{x} + 3}$

13. $\dfrac{\dfrac{1}{5} + \dfrac{1}{x}}{\dfrac{5 + x}{5}}$

14. $\dfrac{\dfrac{1}{3} - \dfrac{1}{a}}{\dfrac{3 - a}{3}}$

15. $\dfrac{\dfrac{1}{t^2} + 1}{\dfrac{1}{t} - 1}$

7. $\dfrac{\dfrac{2}{x}}{\dfrac{3}{x} + \dfrac{1}{x^2}}$

8. $\dfrac{\dfrac{4}{x} - \dfrac{1}{x^2}}{\dfrac{2}{x^2}}$

9. $\dfrac{\dfrac{2a-5}{3a}}{\dfrac{a-1}{6a}}$

16. $\dfrac{2 + \dfrac{1}{x}}{2 - \dfrac{1}{x^2}}$

17. $\dfrac{\dfrac{x^2}{x^2 - y^2}}{\dfrac{x}{x+y}}$

18. $\dfrac{\dfrac{a^2}{a-3}}{\dfrac{2a}{a^2 - 9}}$

19. $\dfrac{\dfrac{2}{a} + \dfrac{4}{a^2}}{\dfrac{5}{a^3} - \dfrac{3}{a}}$

20. $\dfrac{\dfrac{5}{x^3} - \dfrac{1}{x^2}}{\dfrac{2}{x} + \dfrac{3}{x^2}}$

21. $\dfrac{\dfrac{2}{7a^4} - \dfrac{1}{14a}}{\dfrac{3}{5a^2} + \dfrac{2}{15a}}$

22. $\dfrac{\dfrac{5}{4x^3} - \dfrac{3}{8x}}{\dfrac{3}{2x} + \dfrac{3}{4x^3}}$

Aha! **23.** $\dfrac{\dfrac{x}{5y^3} + \dfrac{3}{10y}}{\dfrac{3}{10y} + \dfrac{x}{5y^3}}$

24. $\dfrac{\dfrac{a}{6b^3} + \dfrac{4}{9b^2}}{\dfrac{5}{6b} - \dfrac{1}{9b^3}}$

25. $\dfrac{\dfrac{5}{ab^4} + \dfrac{2}{a^3b}}{\dfrac{5}{a^3b} - \dfrac{3}{ab}}$

26. $\dfrac{\dfrac{2}{x^2y} + \dfrac{3}{xy^2}}{\dfrac{3}{xy^2} + \dfrac{2}{x^2y}}$

27. $\dfrac{2 - \dfrac{3}{x^2}}{2 + \dfrac{3}{x^4}}$

28. $\dfrac{3 - \dfrac{2}{a^4}}{2 + \dfrac{3}{a^3}}$

29. $\dfrac{t - \dfrac{2}{t}}{t + \dfrac{5}{t}}$

30. $\dfrac{x + \dfrac{3}{x}}{x - \dfrac{2}{x}}$

31. $\dfrac{3 + \dfrac{4}{ab^3}}{\dfrac{3 + a}{a^2b}}$

32. $\dfrac{5 + \dfrac{3}{x^2y}}{\dfrac{3 + x}{x^3y}}$

33. $\dfrac{\dfrac{x + 5}{x^2}}{\dfrac{2}{x} - \dfrac{3}{x^2}}$

34. $\dfrac{\dfrac{a + 6}{a^3}}{\dfrac{2}{a^2} + \dfrac{3}{a}}$

35. $\dfrac{x - 3 + \dfrac{2}{x}}{x - 4 + \dfrac{3}{x}}$

36. $\dfrac{x - 2 + \dfrac{1}{x}}{x - 5 + \dfrac{4}{x}}$

37. Is it possible to simplify complex rational expressions without knowing how to divide rational expressions? Why or why not?

38. Why is the distributive law important when simplifying complex rational expressions?

SKILL MAINTENANCE

Solve.

39. $3x - 5 + 2(4x - 1) = 12x - 3$

40. $(x - 1)7 - (x + 1)9 = 4(x + 2)$

41. $\dfrac{3}{4}x - \dfrac{5}{8} = \dfrac{3}{8}x + \dfrac{7}{4}$

42. $\dfrac{5}{9} - \dfrac{2x}{3} = \dfrac{5x}{6} + \dfrac{4}{3}$

43. $x^2 - 7x - 30 = 0$

44. $x^2 + 8x - 20 = 0$

SYNTHESIS

45. Which of the two methods presented would you use to simplify Exercise 18? Why?

46. Which of the two methods presented would you use to simplify Exercise 26? Why?

In Exercises 47–50, find all x-values for which the given expression is undefined.

47. $\dfrac{\dfrac{x - 5}{x - 6}}{\dfrac{x - 7}{x - 8}}$

48. $\dfrac{\dfrac{x + 1}{x + 2}}{\dfrac{x + 3}{x + 4}}$

49. $\dfrac{\dfrac{2x + 3}{5x + 4}}{\dfrac{3}{7} - \dfrac{2x}{9}}$

50. $\dfrac{\dfrac{3x - 5}{2x - 7}}{\dfrac{4x}{5} - \dfrac{5}{6}}$

51. The formula

$$\dfrac{P\left(1 + \dfrac{i}{12}\right)^2}{\dfrac{\left(1 + \dfrac{i}{12}\right)^2 - 1}{\dfrac{i}{12}}},$$

where P is a loan amount and i is an interest rate, arises in certain business situations. Simplify this expression. (*Hint*: Expand the binomials.)

52. Find the simplified form for the reciprocal of

$$\dfrac{2}{x - 1} - \dfrac{1}{3x - 2}.$$

Simplify.

53. $\dfrac{\dfrac{5}{x + 2} - \dfrac{3}{x - 2}}{\dfrac{x}{x - 1} + \dfrac{x}{x + 1}}$

54. $\dfrac{\dfrac{x}{x + 5} + \dfrac{3}{x + 2}}{\dfrac{2}{x + 2} - \dfrac{x}{x + 5}}$

Aha! **55.** $\left[\dfrac{\dfrac{x - 1}{x - 1} - 1}{\dfrac{x + 1}{x - 1} + 1}\right]^5$

56. $1 + \dfrac{1}{1 + \dfrac{1}{1 + \dfrac{1}{x}}}$

57. $\dfrac{\dfrac{z}{1 - \dfrac{z}{2 + 2z}} - 2z}{\dfrac{2z}{5z - 2} - 3}$

58. Under what circumstance(s) will there be no restrictions on the variable appearing in a complex rational expression?

59. Use a grapher to check Example 2(a).

6.6

Solving Rational Equations

Solving a New Type of Equation • A Visual Interpretation

Our study of rational expressions allows us to solve a type of equation that we could not have solved prior to this chapter.

CONNECTING THE CONCEPTS

In Sections 6.1–6.5, we learned how to *simplify expressions* like

$$\frac{x^2 - 1}{x + 2} \cdot \frac{3x}{x + 1} \quad \text{and} \quad \frac{\dfrac{5}{x}}{\dfrac{2}{x^2} + \dfrac{3}{x}}.$$

In this section, we will return to *solving equations*. These equations will look like

the following:

$$x + \frac{6}{x} = -5 \quad \text{and} \quad \frac{3}{x - 5} + \frac{1}{x + 5} = \frac{2}{x^2 - 25}.$$

As always, be careful not to confuse simplifying an expression with solving an equation. When expressions are simplified, the result is an equivalent expression. When equations are solved, the result is a solution.

Solving a New Type of Equation

A **rational**, or **fractional**, **equation** is an equation containing one or more rational expressions, often with the variable in a denominator. Here are some examples:

$$\frac{2}{3} + \frac{5}{6} = \frac{x}{9}, \qquad t + \frac{7}{t} = -5, \qquad \frac{x^2}{x - 1} = \frac{1}{x - 1}.$$

To Solve a Rational Equation

1. List any restrictions that exist. No possible solution can make a denominator equal 0.
2. Clear the equation of fractions by multiplying both sides by the LCD of all rational expressions in the equation.
3. Solve the resulting equation using the addition principle, the multiplication principle, and the principle of zero products, as needed.
4. Check the possible solution(s) in the original equation.

In the examples that follow, we *do not* use the LCD to add or subtract rational expressions. Instead, we use the LCD as a multiplier that will clear fractions.

E x a m p l e   1   Solve: $\dfrac{x}{6} - \dfrac{x}{8} = \dfrac{1}{12}$.

Solution Because no variable appears in a denominator, no restrictions exist. The LCD is 24, so we multiply both sides by 24:

$$24\left(\frac{x}{6} - \frac{x}{8}\right) = 24 \cdot \frac{1}{12}$$

Using the multiplication principle to multiply both sides by the LCD. Parentheses are important!

$$24 \cdot \frac{x}{6} - 24 \cdot \frac{x}{8} = 24 \cdot \frac{1}{12}$$

Using the distributive law

Be sure to multiply *each* term by the LCD.

$$\left.\begin{array}{c} \dfrac{24x}{6} - \dfrac{24x}{8} = \dfrac{24}{12} \\[2mm] 4x - 3x = 2 \\[2mm] x = 2. \end{array}\right\}$$

Simplifying. Note that all fractions have been cleared.

Check:

$$\frac{x}{6} - \frac{x}{8} = \frac{1}{12}$$

$$\begin{array}{c|c} \dfrac{2}{6} - \dfrac{2}{8} & \dfrac{1}{12} \\[2mm] \dfrac{1}{3} - \dfrac{1}{4} & \\[2mm] \dfrac{4}{12} - \dfrac{3}{12} & \\[2mm] \dfrac{1}{12} & \dfrac{1}{12} \quad \text{TRUE} \end{array}$$

This checks, so the solution is 2.

Up to now, the multiplication principle has been used only to multiply both sides of an equation by a nonzero constant. Because rational equations often contain variables in a denominator, clearing fractions may now require us to multiply both sides of an equation by a variable expression. Since a variable expression could represent 0, *multiplying both sides of an equation by a variable expression does not always produce an equivalent equation.* Thus checking in the original equation is very important.

E x a m p l e   2   Solve.

a) $\dfrac{2}{3x} + \dfrac{1}{x} = 10$

b) $x + \dfrac{6}{x} = -5$

c) $1 + \dfrac{3x}{x+2} = \dfrac{-6}{x+2}$

d) $\dfrac{3}{x-5} + \dfrac{1}{x+5} = \dfrac{2}{x^2-25}$

Solution

a) Note that x cannot be 0. The LCD is $3x$, so we multiply both sides by $3x$:

$$\frac{2}{3x} + \frac{1}{x} = 10 \qquad \text{The LCD is } 3x; x \neq 0.$$

$$3x\left(\frac{2}{3x} + \frac{1}{x}\right) = 3x \cdot 10 \qquad \begin{array}{l}\text{Using the multiplication principle to} \\ \text{multiply both sides by the LCD. } \textit{Don't} \\ \textit{forget the parentheses!}\end{array}$$

$$3x \cdot \frac{2}{3x} + 3x \cdot \frac{1}{x} = 3x \cdot 10 \qquad \text{Using the distributive law}$$

$$2 + 3 = 30x \qquad \begin{array}{l}\text{Removing factors equal to 1: } 3x/(3x) = 1 \\ \text{and } x/x = 1. \text{ This clears all fractions.}\end{array}$$

$$5 = 30x$$

$$\frac{5}{30} = x, \quad \text{so } x = \frac{1}{6}. \qquad \begin{array}{l}\text{Since } \frac{1}{6} \neq 0, \text{ which was the} \\ \text{restriction in the first step, this} \\ \textit{should} \text{ check.}\end{array}$$

Check:

$$\frac{2}{3x} + \frac{1}{x} = 10$$

$$\frac{2}{3 \cdot \frac{1}{6}} + \frac{1}{\frac{1}{6}} \; ? \; 10$$

$$\frac{2}{\frac{1}{2}} + \frac{1}{\frac{1}{6}}$$

$$2 \cdot \frac{2}{1} + 1 \cdot \frac{6}{1}$$

$$4 + 6$$

$$10 \; \big| \; 10 \quad \text{TRUE}$$

The solution is $\frac{1}{6}$.

b) Again, note that x cannot be 0. We multiply both sides of the equation by the LCD, x:

$$x + \frac{6}{x} = -5 \qquad \text{We cannot have } x = 0.$$

$$x\left(x + \frac{6}{x}\right) = x(-5) \qquad \begin{array}{l}\text{Multiplying both sides by } x. \\ \textit{Don't forget the parentheses!}\end{array}$$

$$x \cdot x + x \cdot \frac{6}{x} = -5x \qquad \text{Using the distributive law}$$

$$x^2 + 6 = -5x \qquad \begin{array}{l}\text{Removing a factor equal to 1: } x/x = 1. \\ \text{We are left with a quadratic equation.}\end{array}$$

$$x^2 + 5x + 6 = 0 \qquad \begin{array}{l}\text{Using the addition principle to add } 5x \text{ to} \\ \text{both sides}\end{array}$$

$$(x + 3)(x + 2) = 0 \qquad \text{Factoring}$$

$$x + 3 = 0 \quad \textit{or} \quad x + 2 = 0 \qquad \text{Using the principle of zero products}$$

$$x = -3 \quad \textit{or} \quad x = -2 \qquad \begin{array}{l}\text{Since neither solution is 0, the} \\ \text{restriction in the first step, they} \\ \text{should both check.}\end{array}$$

Check: For -3:   For -2:

$$x + \frac{6}{x} = -5$$

$$-3 + \frac{6}{-3} \; ? \; -5$$

$$-3 - 2$$

$$-5 \; \Big| \; -5 \quad \text{TRUE}$$

$$x + \frac{6}{x} = -5$$

$$-2 + \frac{6}{-2} \; ? \; -5$$

$$-2 - 3$$

$$-5 \; \Big| \; -5 \quad \text{TRUE}$$

Both of these check, so there are two solutions, -3 and -2.

c) To avoid division by 0, we must have $x + 2 \neq 0$, or $x \neq -2$. With this restriction in mind, we multiply both sides of the equation by the LCD, $x + 2$:

$$1 + \frac{3x}{x + 2} = \frac{-6}{x + 2} \qquad \text{We cannot have } x = -2.$$

$$(x + 2)\left(1 + \frac{3x}{x + 2}\right) = (x + 2)\frac{-6}{x + 2} \qquad \begin{array}{l}\text{Multiplying both sides} \\ \text{by } x + 2. \text{ Don't forget} \\ \text{the parentheses.}\end{array}$$

$$(x + 2) \cdot 1 + (x + 2)\frac{3x}{x + 2} = (x + 2)\frac{-6}{x + 2} \qquad \begin{array}{l}\text{Using the distributive} \\ \text{law}\end{array}$$

$$x + 2 + 3x = -6 \qquad \begin{array}{l}\text{Removing a factor equal to 1:} \\ (x + 2)/(x + 2) = 1\end{array}$$

$$4x + 2 = -6$$

$$4x = -8$$

$$x = -2 \qquad \text{Above, we stated that } x \neq -2.$$

Because of the above restriction, -2 must be rejected as a solution. The student can confirm that -2 results in division by 0. The equation has no solution.

d)

$$\frac{3}{x - 5} + \frac{1}{x + 5} = \frac{2}{x^2 - 25} \qquad \begin{array}{l}\text{Note that } x \neq 5 \text{ and} \\ x \neq -5. \text{ The LCD is} \\ (x - 5)(x + 5).\end{array}$$

$$(x - 5)(x + 5)\left(\frac{3}{x - 5} + \frac{1}{x + 5}\right) = (x - 5)(x + 5)\frac{2}{(x - 5)(x + 5)}$$

$$\frac{(x - 5)(x + 5)3}{x - 5} + \frac{(x - 5)(x + 5)}{x + 5} = \frac{2(x - 5)(x + 5)}{(x - 5)(x + 5)} \qquad \begin{array}{l}\text{Using the} \\ \text{distributive law}\end{array}$$

$$(x + 5)3 + (x - 5) = 2 \qquad \left\{\begin{array}{l}\text{Removing factors equal to 1:} \\ \frac{x - 5}{x - 5} = 1, \frac{x + 5}{x + 5} = 1, \text{ and} \\ \frac{(x - 5)(x + 5)}{(x - 5)(x + 5)} = 1\end{array}\right.$$

$$3x + 15 + x - 5 = 2 \qquad \text{Using the distributive law}$$

$$4x + 10 = 2$$

$$4x = -8$$

$$x = -2 \qquad \begin{array}{l}-2 \neq 5 \text{ and } -2 \neq -5, \\ \text{so } -2 \textit{ should} \text{ check.}\end{array}$$

We leave it to the student to check that -2 is the solution.

A grapher can be used to check that Example 2(b),

$$x + \frac{6}{x} = -5,$$

has two solutions. To do so, we graph

$$y_1 = x + \frac{6}{x} \quad \text{and}$$

$$y_2 = -5$$

on the same set of axes.

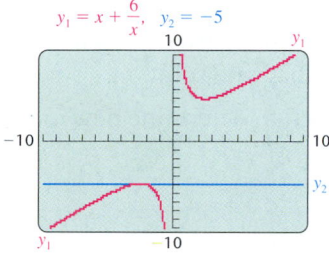

Next, we use the INTERSECT option of the CALC menu to confirm that the points of intersection occur when $x = -3$ and $x = -2$.

A Visual Interpretation

It is possible to solve a rational equation by graphing. The procedure consists of graphing each side of the equation and then determining the first coordinate(s) of any point(s) of intersection. (Since the advent of the graphing calculator, producing such graphs requires little work.) For example, the equation

$$\frac{x}{4} + \frac{x}{2} = 6$$

can be solved by graphing the equations

$$y = \frac{x}{4} + \frac{x}{2} \quad \text{and} \quad y = 6$$

on the same set of axes.

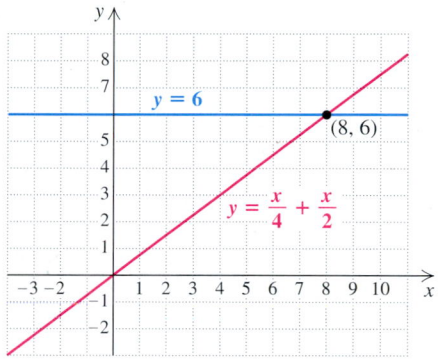

As we can see in the graph above, when $x = 8$, the value of $x/4 + x/2$ is 6. Thus, 8 is the solution of $x/4 + x/2 = 6$. We can check by substitution:

$$\frac{x}{4} + \frac{x}{2} = \frac{8}{4} + \frac{8}{2} = 2 + 4 = 6.$$

Exercise Set 6.6

FOR EXTRA HELP

 Digital Video Tutor CD 5 Videotape 12

 InterAct Math

 Math Tutor Center

 MathXL

 MyMathLab.com

Solve. If no solution exists, state this.

1. $\dfrac{5}{8} - \dfrac{4}{5} = \dfrac{x}{20}$

2. $\dfrac{4}{5} - \dfrac{2}{3} = \dfrac{x}{9}$

3. $\dfrac{1}{3} + \dfrac{5}{6} = \dfrac{1}{x}$

4. $\dfrac{3}{5} + \dfrac{1}{8} = \dfrac{1}{x}$

5. $\dfrac{1}{6} + \dfrac{1}{8} = \dfrac{1}{t}$

6. $\dfrac{1}{8} + \dfrac{1}{10} = \dfrac{1}{t}$

7. $x + \dfrac{5}{x} = -6$

8. $x + \dfrac{6}{x} = -7$

9. $\dfrac{x}{6} - \dfrac{6}{x} = 0$

10. $\dfrac{x}{7} - \dfrac{7}{x} = 0$

11. $\dfrac{5}{x} = \dfrac{6}{x} - \dfrac{1}{3}$

12. $\dfrac{4}{t} = \dfrac{5}{t} - \dfrac{1}{2}$

13. $\dfrac{5}{3t} + \dfrac{3}{t} = 1$

14. $\dfrac{3}{4x} + \dfrac{5}{x} = 1$

15. $\dfrac{x-8}{x+3} = \dfrac{1}{4}$

16. $\dfrac{a-4}{a+5} = \dfrac{3}{8}$

17. $\dfrac{2}{x+1} = \dfrac{1}{x-2}$

18. $\dfrac{5}{x-1} = \dfrac{3}{x+2}$

19. $\dfrac{a}{6} - \dfrac{a}{10} = \dfrac{1}{6}$

20. $\dfrac{t}{8} - \dfrac{t}{12} = \dfrac{1}{8}$

21. $\dfrac{x+1}{3} - 1 = \dfrac{x-1}{2}$

22. $\dfrac{x+2}{5} - 1 = \dfrac{x-2}{4}$

23. $\dfrac{4}{t-5} = \dfrac{t-1}{t-5}$

24. $\dfrac{2}{t-9} = \dfrac{t-7}{t-9}$

25. $\dfrac{3}{x+4} = \dfrac{5}{x}$

26. $\dfrac{2}{x+3} = \dfrac{7}{x}$

27. $\dfrac{a-4}{a-1} = \dfrac{a+2}{a-2}$

28. $\dfrac{x-2}{x-3} = \dfrac{x-1}{x+1}$

29. $\dfrac{4}{x-3} + \dfrac{2x}{x^2-9} = \dfrac{1}{x+3}$

30. $\dfrac{x}{x+4} - \dfrac{4}{x-4} = \dfrac{x^2+16}{x^2-16}$

31. $\dfrac{5}{y-3} - \dfrac{30}{y^2-9} = 1$

32. $\dfrac{1}{x+3} + \dfrac{1}{x-3} = \dfrac{1}{x^2-9}$

33. $\dfrac{4}{8-a} = \dfrac{4-a}{a-8}$

34. $\dfrac{t+10}{7-t} = \dfrac{3}{t-7}$

Aha! **35.** $\dfrac{-2}{x+2} = \dfrac{x}{x+2}$

36. $\dfrac{3}{2x-6} = \dfrac{x}{2x-6}$

37. When solving rational equations, why do we multiply each side by the LCD?

38. Explain the difference between adding rational expressions and solving rational equations.

SKILL MAINTENANCE

39. The sum of two consecutive odd numbers is 276. Find the numbers.

40. The length of a rectangle is 3 yd greater than the width. The area of the rectangle is 10 yd². Find the perimeter.

41. The height of a triangle is 3 cm longer than its base. If the area of the triangle is 54 cm², find the measurements of the base and the height.

42. The product of two consecutive even integers is 48. Find the numbers.

43. *Human physiology.* Between June 9 and June 24, Seth's beard grew 0.9 cm. Find the rate at which Seth's beard grows.

44. *Gardening.* Between July 7 and July 12, Carla's string beans grew 1.4 in. Find the string beans' growth rate.

SYNTHESIS

45. Describe a method that can be used to create rational equations that have no solution.

46. How can a graph be used to determine how many solutions an equation has?

Solve.

47. $1 + \dfrac{x-1}{x-3} = \dfrac{2}{x-3} - x$

48. $\dfrac{4}{y-2} + \dfrac{3}{y^2-4} = \dfrac{5}{y+2} + \dfrac{2y}{y^2-4}$

49. $\dfrac{x}{x^2+3x-4} + \dfrac{x}{x^2+6x+8} = \dfrac{2x}{x^2+x-2} - \dfrac{1}{x^2+6x+8}$

50. $\dfrac{12-6x}{x^2-4} = \dfrac{3x}{x+2} - \dfrac{3-2x}{2-x}$

51. $\dfrac{x^2}{x^2-4} = \dfrac{x}{x+2} - \dfrac{2x}{2-x}$

52. $7 - \dfrac{a-2}{a+3} = \dfrac{a^2-4}{a+3} + 5$

53. $\dfrac{1}{x-1} + x - 5 = \dfrac{5x-4}{x-1} - 6$

54. $\dfrac{5-3a}{a^2+4a+3} - \dfrac{2a+2}{a+3} = \dfrac{3-a}{a+1}$

55. Use a grapher to check the solutions to Examples 1 and 2(c).

56. Use a grapher to check your answers to Exercises 9, 25, and 47.

Applications Using Rational Equations and Proportions

6.7

Problem Solving • Problems Involving Work •
Problems Involving Motion • Problems Involving Proportions

In many areas of study, applications involving rates, proportions, or reciprocals translate to rational equations. By using the five steps for problem solving and the lessons of Section 6.6, we can now solve such problems.

Problem Solving

E x a m p l e   1

A number, plus three times its reciprocal, is -4. Find the number.

Solution

1. **Familiarize.** Let's try to guess the number. Try 2: $2 + 3 \cdot \frac{1}{2} = \frac{7}{2}$. Although $\frac{7}{2} \neq -4$, the guess helps us to better understand how the problem can be translated. We let $x =$ the number for which we are searching.

2. **Translate.** From the *Familiarize* step, we can translate directly:

A number, plus three times its reciprocal, is -4.

$$x \quad + \quad 3 \quad \cdot \quad \frac{1}{x} \quad = -4$$

3. **Carry out.** We solve the equation:

$$x + 3 \cdot \frac{1}{x} = -4 \qquad \text{We note the restriction that } x \text{ cannot equal 0.}$$

$$x\left(x + \frac{3}{x}\right) = x(-4) \qquad \text{Multiplying both sides of the equation by the LCD, } x. \text{ Don't forget the parentheses.}$$

$$x \cdot x + x \cdot \frac{3}{x} = -4x \qquad \text{Using the distributive law}$$

$$x^2 + 3 = -4x \qquad \text{Simplifying}$$

$$\left.\begin{array}{l} x^2 + 4x + 3 = 0 \\ (x + 3)(x + 1) = 0 \\ x + 3 = 0 \quad \text{or} \quad x + 1 = 0 \end{array}\right\} \quad \text{Using the principle of zero products}$$

$$x = -3 \quad \text{or} \qquad x = -1.$$

4. **Check.** Three times the reciprocal of -3 is $3 \cdot \frac{1}{-3}$, or -1. Since $-3 + (-1) = -4$, the number -3 is a solution.
    Three times the reciprocal of -1 is $3 \cdot \frac{1}{-1}$, or -3. Since $-1 + (-3) = -4$, the number -1 is also a solution.

5. **State.** The solutions are -3 and -1.

Problems Involving Work

E x a m p l e   2

Sorting recyclables.   Cecilia and Aaron work as volunteers at a town's recycling depot. Cecilia can sort a day's accumulation of recyclables in 4 hr, while Aaron requires 6 hr to do the same job. How long would it take them, working together, to sort the recyclables?

Solution

1.   **Familiarize.**   We familiarize ourselves with the problem by exploring two common, but *incorrect*, approaches.

   a)   One common incorrect approach is to simply add the two times:

   $$4\,\text{hr} + 6\,\text{hr} = 10\,\text{hr}.$$

   Let's think about this. If Cecilia can do the sorting *alone* in 4 hr, then Cecilia and Aaron *together* should take *less* than 4 hr. Thus we reject 10 hr as a solution and reason that the answer must be less than 4 hr.

   b)   Another incorrect approach is to assume that Cecilia does half the sorting and Aaron does the other half. Then

   Cecilia sorts $\frac{1}{2}$ of the accumulation in $\frac{1}{2}(4\,\text{hr})$, or 2 hr, and
   Aaron sorts $\frac{1}{2}$ of the accumulation in $\frac{1}{2}(6\,\text{hr})$, or 3 hr.

   This would waste time since Cecilia would finish 1 hr earlier than Aaron. In reality, Cecilia would help Aaron after completing her half, so that Aaron would actually sort less than half of the accumulation. This tells us that the entire job will take them between 2 hr and 3 hr.

A correct approach is to consider how much of the sorting is finished in 1 hr, 2 hr, 3 hr, and so on. It takes Cecilia 4 hr to sort the recyclables alone, so her rate is $\frac{1}{4}$ of the job per hour. It takes Aaron 6 hr to do the sorting alone, so his rate is $\frac{1}{6}$ of the job per hour. Working together, they can complete

$$\frac{1}{4} + \frac{1}{6}, \quad \text{or } \frac{5}{12} \text{ of the sorting in 1 hr.} \qquad \textcolor{red}{\text{Together, their rate is } \tfrac{5}{12} \text{ of the job per hour.}}$$

In 2 hr, Cecilia can do $\frac{1}{4} \cdot 2$ of the sorting and Aaron can do $\frac{1}{6} \cdot 2$ of the sorting. Working together, they can complete

$$\frac{1}{4} \cdot 2 + \frac{1}{6} \cdot 2, \quad \text{or } \frac{5}{6} \text{ of the sorting in 2 hr.} \qquad \textcolor{red}{\text{Note that } \tfrac{5}{12} \cdot 2 = \tfrac{5}{6}.}$$

Continuing this reasoning, we can form a table.

Time	Fraction of the Sorting Completed		
	Cecilia	Aaron	Together
1 hr	$\dfrac{1}{4}$	$\dfrac{1}{6}$	$\dfrac{1}{4} + \dfrac{1}{6}$, or $\dfrac{5}{12}$
2 hr	$\dfrac{1}{4} \cdot 2$	$\dfrac{1}{6} \cdot 2$	$\left(\dfrac{1}{4} + \dfrac{1}{6}\right)2$, or $\dfrac{5}{12} \cdot 2$, or $\dfrac{5}{6}$
3 hr	$\dfrac{1}{4} \cdot 3$	$\dfrac{1}{6} \cdot 3$	$\left(\dfrac{1}{4} + \dfrac{1}{6}\right)3$, or $\dfrac{5}{12} \cdot 3$, or $1\dfrac{1}{4}$
t hr	$\dfrac{1}{4} \cdot t$	$\dfrac{1}{6} \cdot t$	$\left(\dfrac{1}{4} + \dfrac{1}{6}\right)t$, or $\dfrac{5}{12} \cdot t$

←— This is too little.

←— This is too much.

From the table, we see that if they work 3 hr, the fraction of the sorting that they complete is $1\frac{1}{4}$, which is more of the job than needs to be done. We need to find a number t for which the fraction of the sorting that is completed in t hours is exactly 1, no more and no less.

2. **Translate.**   From the table, we see that the time we want is some number t for which

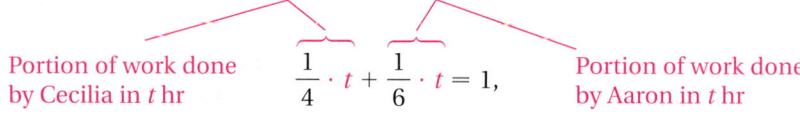

Portion of work done by Cecilia in t hr $\dfrac{1}{4} \cdot t + \dfrac{1}{6} \cdot t = 1$,   Portion of work done by Aaron in t hr

or

$$\left(\dfrac{1}{4} + \dfrac{1}{6}\right)t = 1, \quad \text{or} \quad \dfrac{5}{12} \cdot t = 1.$$

Portion of work done together in t hr

3. **Carry out.**   We can choose any one of the above equations to solve:

$$\dfrac{5}{12} \cdot t = 1$$

$$\dfrac{12}{5} \cdot \dfrac{5}{12} \cdot t = \dfrac{12}{5} \cdot 1 \qquad \text{Multiplying both sides by } \tfrac{12}{5}$$

$$t = \dfrac{12}{5}, \quad \text{or} \quad 2\dfrac{2}{5} \text{ hr.}$$

4. **Check.**   The check can be done following the pattern used in the table of the *Familiarize* step above:

$$\dfrac{1}{4} \cdot \dfrac{12}{5} + \dfrac{1}{6} \cdot \dfrac{12}{5} = \dfrac{3}{5} + \dfrac{2}{5} = \dfrac{5}{5} = 1.$$

A second, partial, check is that (as we predicted in step 1) the answer is between 2 hr and 3 hr.

5. **State.**   Together, it takes Cecilia and Aaron $2\frac{2}{5}$ hr to complete the sorting.

> **The Work Principle**
>
> Suppose that A requires a units of time to complete a task and B requires b units of time to complete the same task. Then
>
> A works at a rate of $\dfrac{1}{a}$ tasks per unit of time,
>
> B works at a rate of $\dfrac{1}{b}$ tasks per unit of time, and
>
> A and B together work at a rate of $\dfrac{1}{a} + \dfrac{1}{b}$ tasks per unit of time.
>
> If A and B, working together, require t units of time to complete the task, then all three of the following equations hold:
>
> $$\frac{1}{a} \cdot t + \frac{1}{b} \cdot t = 1; \qquad \left(\frac{1}{a} + \frac{1}{b}\right)t = 1; \qquad \frac{1}{a} + \frac{1}{b} = \frac{1}{t}.$$

Problems Involving Motion

Problems that deal with distance, speed (or rate), and time are called **motion problems**. Translation of these problems involves the distance formula, $d = r \cdot t$, and/or the equivalent formulas $r = d/t$ and $t = d/r$.

Example 3

Driving speed.   Nancy drives 20 mph faster than her father, Greg. In the same time that Nancy travels 180 mi, her father travels 120 mi. Find their speeds.

Solution

1. **Familiarize.** Suppose that Greg drives 30 mph. Nancy would then be driving at $30 + 20$, or 50 mph. Thus, if r is the speed of Greg's car, in miles per hour, then the speed of Nancy's car is $r + 20$.

   If Greg drove 30 mph, he would drive 120 mi in 120/30, or 4 hr. At 50 mph, Nancy would drive 180 mi in 180/50, or $3\frac{3}{5}$ hr. Because we know that both drivers spend the same amount of time traveling, and because $4 \text{ hr} \neq 3\frac{3}{5}$ hr, we see that our guess of 30 mph is incorrect. We let $t =$ the time, in hours, that is spent traveling and create a table.

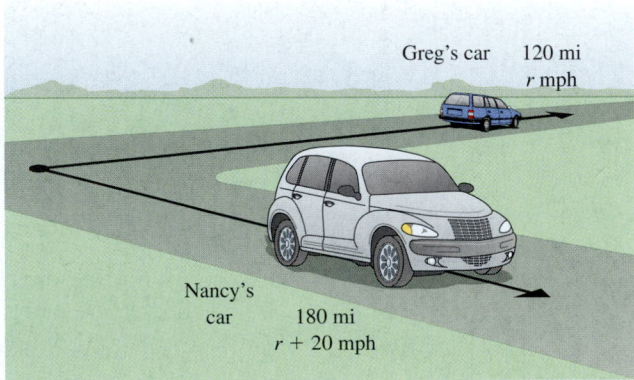

Greg's car   120 mi
r mph

Nancy's
car   180 mi
$r + 20$ mph

$$d \quad = \quad r \quad \cdot \quad t$$

	Distance (in miles)	Speed (in miles per hour)	Time (in hours)
Greg's Car	120	r	t
Nancy's Car	180	$r + 20$	t

2. **Translate.** Examine how we checked our guess. We found, and then compared, the two driving times. The times were found by dividing the distances, 120 mi and 180 mi, by the rates, 30 mph and 50 mph, respectively. Thus the t's in the table above can be replaced, using the formula $t = d/r$. This yields a table that uses only one variable.

	Distance (in miles)	Speed (in miles per hour)	Time (in hours)	
Greg's Car	120	r	$120/r$	← The times must
Nancy's Car	180	$r + 20$	$180/(r + 20)$	← be the same.

Since the times must be the same for both cars, we have the equation

$$\frac{120}{r} = \frac{180}{r + 20}.$$

Note that $\dfrac{\text{mi}}{\text{mph}} = \dfrac{\text{mi}}{\text{mi/hr}} = \text{mi} \cdot \dfrac{\text{hr}}{\text{mi}} = \text{hr}$, so we are indeed comparing two times.

3. **Carry out.** To solve the equation, we first multiply both sides by the LCD, $r(r + 20)$:

$$r(r + 20) \cdot \frac{120}{r} = r(r + 20) \cdot \frac{180}{r + 20}$$

Multiplying both sides by the LCD, $r(r + 20)$. Note that we must have $r \neq 0$ and $r \neq -20$.

$$120(r + 20) = 180r \qquad \text{Simplifying}$$
$$120r + 2400 = 180r \qquad \text{Using the distributive law}$$
$$2400 = 60r \qquad \text{Subtracting } 120r \text{ from both sides}$$
$$40 = r. \qquad \text{Dividing both sides by 60}$$

We now have a possible solution. The speed of Greg's car is 40 mph, and the speed of Nancy's car is $40 + 20$, or 60 mph.

4. **Check.** We first reread the problem to confirm that we were to find the speeds. Note that if Nancy drives 60 mph and Greg drives 40 mph, Nancy is indeed going 20 mph faster than her father. If Nancy travels 180 mi at 60 mph, she drives for 180/60, or 3 hr. If Greg travels 120 mi at 40 mph, he drives for 120/40, or 3 hr. Since the times are the same, the speeds check.

5. **State.** Greg is driving at 40 mph, while Nancy is driving at 60 mph.

Problems Involving Proportions

A **ratio** of two quantities is their quotient. For example, 37% is the ratio of 37 to 100, or $\frac{37}{100}$. A **proportion** is an equation stating that two ratios are equal.

> ### Proportion
>
> An equality of ratios, $A/B = C/D$, is called a *proportion*. The numbers within a proportion are said to be *proportional* to each other.

Proportions can be used to solve a variety of applied problems.

Example 4

Mileage.   In 2000, Honda introduced the Insight, a gasoline–electric car that travels 280 mi on 4 gal of gas. Find the amount of gas required for a 700-mi trip.

Solution   By assuming that the car always burns gas at the same rate, we can form a proportion in which the ratio of miles to gallons is expressed in two ways:

$$\text{Miles} \longrightarrow \frac{280}{4} = \frac{700}{x} \longleftarrow \text{Miles}$$
$$\text{Gallons} \longrightarrow \qquad\qquad \longleftarrow \text{Gallons}$$

To solve for x, we multiply both sides of the equation by the LCD, $4x$:

$$4x \cdot \frac{280}{4} = 4x \cdot \frac{700}{x}$$ We could also simplify first and solve $70 = \frac{700}{x}$.

$$4 \cdot \frac{280x}{4} = x \cdot \frac{4 \cdot 700}{x}$$ Removing factors equal to 1: $\frac{4}{4} = 1$ and $\frac{x}{x} = 1$

$$280x = 4 \cdot 700$$

$$x = \frac{4 \cdot 700}{280}$$ Dividing both sides by 280

$$x = 10.$$ Simplifying

The trip will require 10 gal of gas.

Proportions arise in geometry when we are studying *similar triangles*. If two triangles are **similar**, then their corresponding angles have the same measure and their corresponding sides are proportional. To illustrate, if triangle *ABC* is similar to triangle *RST*, then angles *A* and *R* have the same measure, angles *B* and *S* have the same measure, angles *C* and *T* have the same measure, and

$$\frac{a}{r} = \frac{b}{s} = \frac{c}{t}.$$

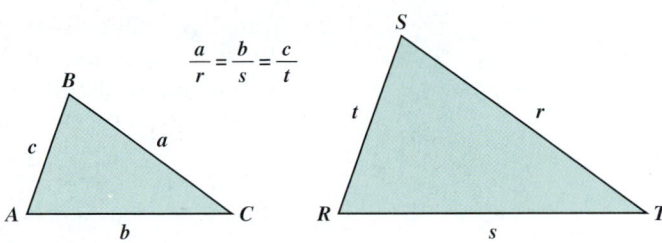

E x a m p l e   5

Similar triangles.   Triangles *ABC* and *XYZ* are similar. Solve for *z* if $x = 10$, $a = 8$, and $c = 5$.

Solution   We make a drawing, write a proportion, and then solve. Note that side *a* is always opposite angle *A*, side *x* is always opposite angle *X*, and so on.

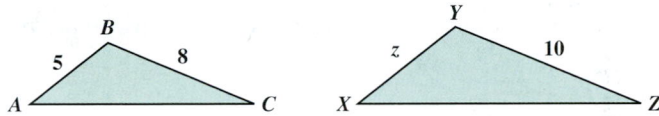

We have

$$\frac{z}{5} = \frac{10}{8}$$ The proportions $\frac{5}{z} = \frac{8}{10}$, $\frac{5}{8} = \frac{z}{10}$, or $\frac{8}{5} = \frac{10}{z}$ could also be used.

$$z = \frac{10}{8} \cdot 5$$ Multiplying both sides by 5

$$z = \frac{50}{8}, \text{ or } 6.25.$$

E x a m p l e   6

Environmental science.   To determine the number of humpback whales in a pod, a marine biologist, using tail markings, identifies 27 members of the pod. Several weeks later, 40 whales from the pod are randomly sighted. Of the 40 sighted, 12 are from the 27 originally identified. Estimate the number of whales in the pod.

Solution

1. **Familiarize.**   If we knew that the 27 whales that were first identified constituted, say, 10% of the pod, we could easily calculate the pod's population from the proportion

    $$\frac{27}{W} = \frac{10}{100},$$

    where *W* is the size of the pod's population. Unfortunately, we are *not* told the percentage of the pod identified. We must reread the problem, looking for numbers that could be used to approximate this percentage.

2. **Translate.** Since 12 of the 40 whales that were later sighted were among those originally identified, the ratio 12/40 estimates the percentage of the pod originally identified. We can then translate to a proportion:

Whales originally identified $\longrightarrow$ $\dfrac{27}{W} = \dfrac{12}{40}$. $\longleftarrow$ Original whales sighted later
Entire pod $\longrightarrow$ $\phantom{\dfrac{27}{W}}$ $\longleftarrow$ Whales sighted later

3. **Carry out.** To solve the proportion, we multiply by the LCD, $40W$:

$$40W \cdot \frac{27}{W} = 40W \cdot \frac{12}{40} \qquad \text{Multiplying both sides by } 40W$$

$$40 \cdot 27 = W \cdot 12 \qquad \text{Removing factors equal to 1: } W/W = 1 \text{ and } 40/40 = 1$$

$$\frac{40 \cdot 27}{12} = W \quad \text{or} \quad W = 90 \qquad \text{Dividing both sides by 12}$$

4. **Check.** The check is left to the student.

5. **State.** There are about 90 whales in the pod.

Exercise Set 6.7

FOR EXTRA HELP

 Digital Video Tutor CD 5 Videotape 12   InterAct Math   Math Tutor Center   MathXL   MyMathLab.com

Solve.

1. A number, minus four times its reciprocal, is 3. Find the number.

2. A number, minus five times its reciprocal, is 4. Find the number.

3. The sum of a number and its reciprocal is 2. Find the number.

4. The sum of a number and five times its reciprocal is 6. Find the number.

5. *Construction.* It takes Fontella 4 hr to put up paneling in a room. Omar takes 5 hr to do the same job. How long would it take them, working together, to panel the room?

6. *Carpentry.* By checking work records, a carpenter finds that Juanita can build a small shed in 12 hr. Anton can do the same job in 16 hr. How long would it take if they worked together?

7. *Shoveling.* Vern can shovel the snow from his driveway in 45 min. Nina can do the same job in 60 min. How long would it take Nina and Vern to shovel the driveway if they worked together?

8. *Raking.* Zoë can rake her yard in 4 hr. Steffi does the same job in 3 hr. How long would it take the two of them, working together, to rake the yard?

9. *Masonry.* By checking work records, a contractor finds that it takes Kenny Dewitt 8 hr to construct a wall of a certain size. It takes Betty Wohnt 6 hr to construct the same wall. How long would it take if they worked together?

10. *Plumbing.* By checking work records, a plumber finds that Raul can plumb a house in 48 hr. Mira can do the same job in 36 hr. How long would it take if they worked together?

11. *Gardening.*   Nicole can weed her vegetable garden in 50 min, while Glen can weed the same garden in 40 min. How long would it take if they worked together?

12. *Harvesting.*   Bobbi can pick a quart of raspberries in 20 min. Blanche can pick a quart in 25 min. How long would it take if Bobbi and Blanche worked together?

13. *Computer printers.*   The HP OfficeJetG85 printer can copy Charlotte's dissertation in 12 min. The HP LaserJet 3200se can copy the same document in 20 min. If the two machines work together, how long would they take to copy the dissertation?

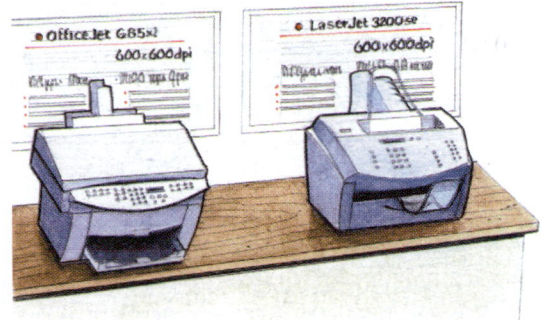

14. *Fax machines.*   The Brother MFC4500® can fax a year-end report in 10 min while the Xerox 850® can fax the same report in 8 min. How long would it take the two machines, working together, to fax the report? (Assume that the recipient has at least two machines for incoming faxes.)

15. *Speed of travel.*   A loaded Roadway truck is moving 40 mph faster than a New York Railways freight train. In the time that it takes the train to travel 150 mi, the truck travels 350 mi. Find their speeds. Complete the following table as part of the familiarization.

| d | = | r | · | t |

	Distance (in miles)	Speed (in miles per hour)	Time (in hours)
Truck	350	r	$\dfrac{350}{r}$
Train	150		

16. *Train speeds.*   A B & M freight train is 14 km/h slower than an AMTRAK passenger train. The B & M train travels 330 km in the same time that it takes the AMTRAK train to travel 400 km. Find their speeds. Complete the following table as part of the familiarization.

| d | = | r | · | t |

	Distance (in km)	Speed (in km/hr)	Time (in hours)
B & M	330		
AMTRAK	400	r	$\dfrac{400}{r}$

17. *Bicycle speed.*   Hank bicycles 5 km/h slower than Kelly. In the time that it takes Hank to bicycle 42 km, Kelly can bicycle 57 km. How fast does each bicyclist travel?

18. *Driving speed.*   Hillary's Lexus travels 30 mph faster than Bill's Harley. In the same time that Bill travels 75 mi, Hillary travels 120 mi. Find their speeds.

19. *Walking speed.*   Bonnie power walks 3 km/h faster than Ralph. In the time that it takes Ralph to walk 7.5 km, Bonnie walks 12 km. Find their speeds.

20. *Cross-country skiing.*   Gerard cross-country skis 4 km/h faster than Sally. In the time that it takes Sally to ski 18 km, Gerard skis 24 km. Find their speeds.

Aha! **21.** *Tractor speed.*   Manley's tractor is just as fast as Caledonia's. It takes Manley 1 hr more than it takes Caledonia to drive to town. If Manley is 20 mi from town and Caledonia is 15 mi from town, how long does it take Caledonia to drive to town?

22. *Boat speed.* Tory and Emilio's motorboats both travel at the same speed. Tory pilots her boat 40 km before docking. Emilio continues for another 2 hr, traveling a total of 100 km before docking. How long did it take Tory to navigate the 40 km?

Geometry. For each pair of similar triangles, find the value of the indicated letter.

23. *b*

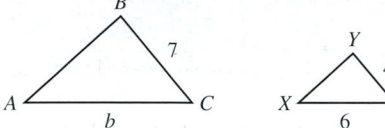

24. *a*

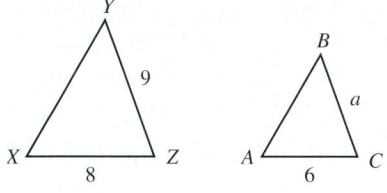

25. *f*

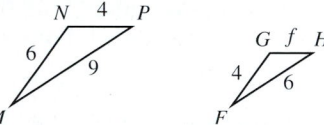

26. *r*

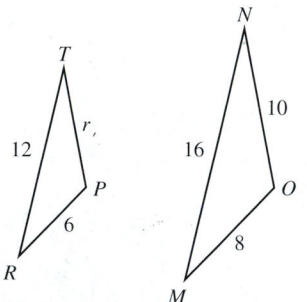

27. *l*

28. *h*

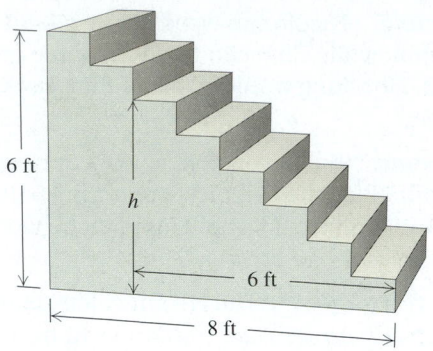

Geometry. When three parallel lines are crossed by two or more lines (transversals), the lengths of corresponding segments are proportional (see the following figure).

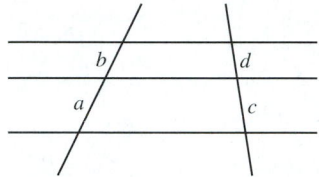

29. If *a* is 8 cm when *b* is 5 cm, find *d* when *c* is 6 cm.

30. If *d* is 7 cm when *c* is 10 cm, find *a* when *b* is 9 cm.

31. If *c* is 2 m longer than *b* and *d* is 2 m shorter than *b*, find all four lengths when *a* is 15 m.

32. If *d* is 2 m shorter than *b* and *c* is 3 m longer than *b*, find *b*, *c*, and *d* when *a* is 18 m.

33. *Coffee harvest.* The coffee beans from 14 trees are needed to produce 7.7 kg of coffee. (This is the average amount that each person in the United States consumes each year.) How many trees are needed to produce 308 kg of coffee?

34. *Walking speed.* Wanda walked 234 km in 14 days. At this rate, how far would she walk in 42 days?

35. *Hemoglobin.* A normal 10-cc specimen of human blood contains 1.2 g of hemoglobin. How much hemoglobin would 16 cc of the same blood contain?

36. *Baking.* In a potato bread recipe, the ratio of milk to flour is $\frac{3}{13}$. If 5 cups of milk are used, how many cups of flour are used?

37. *Wages.* For comparable work, U.S. women earn 77 cents for each dollar earned by a man (*Source: Burlington Free Press,* 12/7/98, p. 3, Business Monday). If a male sales manager earns $42,000, how much would a female earn for comparable work?

Aha! **38.** *Money.* The ratio of the weight of copper to the weight of zinc in a U.S. penny is $\frac{1}{39}$. If 50 kg of zinc is being turned into pennies, how much copper is needed?

39. *Deer population.* To determine the number of deer in the Great Gulf Wilderness, a game warden catches 318 deer, tags them, and lets them loose. Later, 168 deer are caught; 56 of them have tags. Estimate the number of deer in the preserve.

40. *Moose population.* To determine the size of Pine County's moose population, naturalists catch 69 moose, tag them, and then set them free. Months later, 40 moose are caught, of which 15 have tags. Estimate the size of the moose population.

41. *Light bulbs.* A sample of 184 light bulbs contained 6 defective bulbs. How many defective bulbs would you expect in a sample of 1288 bulbs?

42. *Fish population.* To determine the number of trout in a lake, a naturalist catches 112 trout, tags them, and throws them back into the lake. Later, 82 trout are caught; 32 of them have tags. Estimate the number of trout in the lake.

43. *Miles driven.* Emmanuel is allowed to drive his leased car for 45,000 mi in 4 yr without penalty. In the first $1\frac{1}{2}$ yr, Emmanuel has driven 16,000 mi. At this rate will he exceed the mileage allowed for 4 yr?

44. *Firecrackers.* A sample of 144 firecrackers contained 9 "duds." How many duds would you expect in a sample of 320 firecrackers?

45. *Fox population.* To determine the number of foxes in King County, a naturalist catches, tags, and then releases 25 foxes. Later, 36 foxes are caught; 4 of them have tags. Estimate the fox population of the county.

46. *Weight on the moon.* The ratio of the weight of an object on the moon to the weight of that object on Earth is 0.16 to 1.
   a) How much would a 12-ton rocket weigh on the moon?
   b) How much would a 180-lb astronaut weigh on the moon?

47. *Weight on Mars.* The ratio of the weight of an object on Mars to the weight of that object on Earth is 0.4 to 1.
   a) How much would a 12-ton rocket weigh on Mars?
   b) How much would a 120-lb astronaut weigh on Mars?

48. Simplest fractional notation for a rational number is $\frac{9}{17}$. Find an equivalent ratio where the sum of the numerator and the denominator is 104.

49. Is it correct to assume that two workers will complete a task twice as quickly as one person working alone? Why or why not?

50. If two triangles are exactly the same shape and size, are they similar? Why or why not?

SKILL MAINTENANCE

Graph.

51. $y = 2x - 6$

52. $y = -2x + 6$

53. $3x + 2y = 12$

54. $x - 3y = 6$

55. $y = -\frac{3}{4}x + 2$

56. $y = \frac{2}{5}x - 4$

SYNTHESIS

57. Write a problem similar to Example 2 for a classmate to solve. Design the problem so that the translation step is

$$\frac{t}{7} + \frac{t}{5} = 1.$$

58. Write a problem similar to Example 3 for a classmate to solve. Design the problem so that the translation step is

$$\frac{30}{r+4} = \frac{18}{r}.$$

59. *Car cleaning.* Together, Michelle, Sal, and Kristen can wax a car in 1 hr 20 min. To complete the job alone, Michelle needs twice the time that Sal needs and 2 hr more than Kristen. How long would it take each to wax the car working alone?

60. *Programming.* Rosina, Ng, and Oscar can write a computer program in 3 days. Rosina can write the program in 8 days and Ng can do it in 10 days. How many days will it take Oscar to write the program?

61. *Wiring.* Janet can wire a house in 28 hr. Linus can wire a house in 34 hr. How long will it take Janet and Linus, working together, to wire *two* houses?

62. *Quilting.* Ann and Betty work together and sew a quilt in 4 hr. Working alone, Betty would need 6 hr more than Ann to sew a quilt. How long would it take each of them working alone?

63. *Grading.* Alma can grade a batch of placement exams in 3 hr. Kevin can grade a batch in 4 hr. If they work together to grade a batch of exams, what percentage of the exams will have been graded by Alma?

Aha! **64.** *Roofing.* Working alone, Russ can reshingle a roof in 12 hr. When Joan works with Russ, the job takes 6 hr. How long would it take Joan, working alone, to reshingle the roof?

65. *Boating.* The speed of a boat in still water is 10 mph. It travels 24 mi upstream and 24 mi downstream in a total time of 5 hr. What is the speed of the current?

66. *Elections.* Melanie beat her opponent for the presidency of the student senate by a 3-to-2 ratio. If 450 votes were cast, how many votes did Melanie receive?

67. *Commuting.* To reach an appointment 50 mi away, Dr. Wright allowed 1 hr. After driving 30 mi, she realized that her speed would have to be increased 15 mph for the remainder of the trip. What was her speed for the first 30 mi?

68. *Distances.* The shadow from a 40-ft cliff just reaches across a water-filled quarry at the same time that a 6-ft tall diver casts a 10-ft shadow. How wide is the quarry?

69. How soon after 5 o'clock will the hands on a clock first be together?

70. Given that
$$\frac{A}{B} = \frac{C}{D},$$
write three other proportions using A, B, C, and D.

71. If two triangles are similar, are their areas and perimeters proportional? Why or why not?

72. Are the equations
$$\frac{A + B}{B} = \frac{C + D}{D} \quad \text{and} \quad \frac{A}{B} = \frac{C}{D}$$
equivalent? Why or why not?

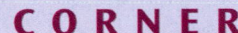

CORNER

Sharing the Workload

COLLABORATIVE

Focus: Modeling, estimation, and work problems

Time: 20–25 minutes

Group size: 3

Materials: Paper, pencils, textbooks, and a watch

Many tasks can be done by two people working together. If both people work at the same rate, each does half the task, and the project is completed in half the time. However, when the work rates differ, the faster worker performs more than half of the task.

ACTIVITY

1. The project is to write down (but not answer) Review Exercises 1–34 from Chapter 6 (pp. 403–404) on a sheet of paper. The problems should be spaced apart and written clearly so that they can be used for studying in the future. Two of the members in each group should write down the exercises, each working at his or her own pace. One worker should begin with Exercise 1 and work forward, while the other worker should begin

with Exercise 34 and work backward. The third group member should record how long it takes for them to reach a point where together the workers have copied all 34 exercises. The time required should be recorded in the appropriate box of the table below.

2. Next, one of the workers from part (1) should be timed copying all 34 exercises working alone. Record that time in the table. The group should then predict how long it will take the second worker, working alone, to copy the exercises.

3. The second worker should then be timed writing down the exercises. Record this time and compare it with the prediction from part (2). How far off was the group's prediction?

4. Let t_1, t_2, and t_3 represent the times required for the first worker, the second worker, and the two workers together, respectively, to complete a task. Then develop a model that can be used to find t_2 when t_1 and t_3 are known.

5. Compare the actual experimental time from part (3) with the time predicted by the model in part (4). List reasons that might account for any discrepancy.

Time Required Working Together	Time Required for One of the Workers, Working Alone	Estimated Time for the Other Worker, Working Alone	Actual Time Required for the Other Worker, Working Alone

Formulas and Equations

6.8

Solving Equations · Solving Formulas

Formulas arise frequently in the natural and social sciences, business, engineering, and health care. In Section 2.3, we saw that the same steps that are used to solve linear equations can be used to solve a formula that appears in this form. Similarly, the steps that are used to solve a rational equation can also be used to solve a formula that appears in this form. Before turning our attention to these formulas, let's briefly review the steps used to solve various types of equations.

Solving Equations

CONNECTING THE CONCEPTS

Below are examples of each type of equation that we have studied so far, along with their solutions.

LINEAR EQUATIONS

$5x + 3 = 2x + 9$

$\quad 3x = 6$ Adding $-2x - 3$ to both sides

$\quad\quad x = 2$ Dividing both sides by 3

The solution is 2.

RATIONAL EQUATIONS

$$\frac{5}{2x} + \frac{4}{3x} = 2 \quad \text{Note that } x \neq 0.$$

$$6x\left(\frac{5}{2x} + \frac{4}{3x}\right) = 6x \cdot 2 \quad \begin{array}{l}\text{Multiplying both}\\ \text{sides by the LCD, } 6x\end{array}$$

$$\frac{6x \cdot 5}{2x} + \frac{6x \cdot 4}{3x} = 12x \quad \begin{array}{l}\text{Using the distributive}\\ \text{law}\end{array}$$

$$15 + 8 = 12x \quad \text{Simplifying}$$

$$23 = 12x$$

$$\frac{23}{12} = x$$

The solution is $\frac{23}{12}$.

POLYNOMIAL EQUATIONS

$$2x^2 + 3 = 5x$$

$$2x^2 - 5x + 3 = 0 \quad \begin{array}{l}\text{Subtracting}\\ 5x \text{ from both}\\ \text{sides}\end{array}$$

$$(2x - 3)(x - 1) = 0 \quad \text{Factoring}$$

$$2x - 3 = 0 \quad or \quad x - 1 = 0 \quad \begin{array}{l}\text{Using the}\\ \text{principle of}\\ \text{zero products}\end{array}$$

$$2x = 3 \quad or \quad\quad x = 1$$

$$x = \frac{3}{2} \quad or \quad\quad x = 1 \quad \text{Solving}$$

The solutions are 1 and $\frac{3}{2}$.

It is always wise to check solutions in the original equation, but this is especially important with rational equations. The checks for the equations above are left to the student.

Example 1   Solve: **(a)** $3x + 17 = 8x$; **(b)** $\dfrac{1}{x} = \dfrac{5}{x}$; **(c)** $\dfrac{8}{x + 2} = \dfrac{x}{x - 1}$.

Solution

a) $3x + 17 = 8x$

$\qquad 17 = 5x \qquad$ Subtracting $3x$ from both sides

$\qquad \dfrac{17}{5} = x \qquad$ Dividing both sides by 5

The solution is $\frac{17}{5}$. The check is left to the student.

b) $\qquad \dfrac{1}{x} = \dfrac{5}{x}$

$x \cdot \dfrac{1}{x} = x \cdot \dfrac{5}{x} \qquad$ Multiplying both sides by the LCD, x

$\dfrac{\cancel{x} \cdot 1}{\cancel{x}} = \dfrac{\cancel{x} \cdot 5}{\cancel{x}} \qquad$ Multiplying

$\qquad 1 = 5 \qquad$ Removing a factor equal to 1: $\dfrac{x}{x} = 1$

There are no replacements for x that can make $1 = 5$ true. The equation has no solution.

c) $\qquad \dfrac{8}{x + 2} = \dfrac{x}{x - 1}$

$(x + 2)(x - 1)\dfrac{8}{x + 2} = (x + 2)(x - 1)\dfrac{x}{x - 1} \qquad$ Multiplying both sides by the LCD, $(x + 2)(x - 1)$

$\dfrac{(x + 2)(x - 1) \cdot 8}{x + 2} = \dfrac{(x + 2)(x - 1) \cdot x}{x - 1} \qquad$ Multiplying

$(x - 1)8 = (x + 2)x \qquad$ $\left\{\begin{array}{l}\text{Removing factors equal to 1:} \\ \dfrac{x + 2}{x + 2} = 1 \text{ and } \dfrac{x - 1}{x - 1} = 1\end{array}\right.$

$8x - 8 = x^2 + 2x \qquad$ Using the distributive law

$0 = x^2 - 6x + 8$

$0 = (x - 2)(x - 4) \qquad$ Factoring

$x - 2 = 0 \quad or \quad x - 4 = 0 \qquad$ Using the principle of zero products

$x = 2 \quad or \qquad x = 4$

Check: For 2:   For 4:

$$\dfrac{8}{x + 2} = \dfrac{x}{x - 1} \qquad\qquad \dfrac{8}{x + 2} = \dfrac{x}{x - 1}$$

$$\dfrac{8}{2 + 2} \overset{?}{\mid} \dfrac{2}{2 - 1} \qquad\qquad \dfrac{8}{4 + 2} \overset{?}{\mid} \dfrac{4}{4 - 1}$$

$$\dfrac{8}{4} \;\mid\; \dfrac{2}{1} \qquad\qquad\qquad \dfrac{8}{6} \;\mid\; \dfrac{4}{3}$$

$$2 \;\mid\; 2 \quad \text{TRUE} \qquad\qquad \dfrac{4}{3} \;\mid\; \dfrac{4}{3} \quad \text{TRUE}$$

Both of these check, so there are two solutions, 2 and 4.

Equations can be classified by the type of expressions they contain, such as *linear, rational,* or *polynomial.* They can also be classified by their solutions. An equation that is never true is a **contradiction**. An **identity** is an equation for which all possible replacements are solutions. If an equation is true for some replacements and false for others, it is a **conditional equation**.

The set of all solutions of an equation is the **solution set** of the equation. The solution set of a contradiction is the empty set, written { } or $\varnothing$. The solution set of an identity is the set of all real numbers, which can be written $\mathbb{R}$.

This new terminology is summarized in the following table.

Type of Equation	Example	Solution Set
Identity	$3x + 1 = 3x + 1$	$\mathbb{R}$
Contradiction	$x + 2 = x + 5$	{ } or $\varnothing$
Conditional	$x + 5 = 3$	$\{-2\}$

E x a m p l e   2   Classify each of the equations in Example 1 as an identity, a contradiction, or a conditional equation.

Solution   The equation in Example 1(a) is true for the replacement $\frac{17}{5}$ and false for any other replacement. Its solution set is $\left\{\frac{17}{5}\right\}$. It is a *conditional equation.*

The equation in Example 1(b) is false for any replacement for x. Its solution set is $\varnothing$. It is a *contradiction.*

The equation in Example 1(c) is true for the replacements 2 and 4. Its solution set is $\{2, 4\}$. It is a *conditional equation.*

Solving Formulas

Formulas can be linear or rational equations, as well as other types of equations. To solve formulas, we use the same steps that we use to solve equations. Probably the greatest difference is that while the solution of an equation is a number, the solution of a formula is generally a variable expression.

E x a m p l e   3   ***Intelligence quotient.***   The formula $Q = \dfrac{100m}{c}$ is used to determine the intelligence quotient, Q, of a person of mental age m and chronological age c. Solve for c.

Solution   We have

$$Q = \frac{100m}{c}$$

$$c \cdot Q = c \cdot \frac{100m}{c} \qquad \text{Multiplying both sides by } c$$

$$cQ = 100m \qquad \text{Removing a factor equal to 1: } \frac{c}{c} = 1$$

$$c = \frac{100m}{Q}. \qquad \text{Dividing both sides by } Q$$

This formula can be used to determine a person's chronological, or actual, age from his or her mental age and intelligence quotient.

E x a m p l e   4

A work formula.   The formula $t/a + t/b = 1$ was used in Section 6.7. Solve this formula for t.

Solution   We have

$$\frac{t}{a} + \frac{t}{b} = 1$$

$$ab\left(\frac{t}{a} + \frac{t}{b}\right) = ab \cdot 1 \qquad \text{Multiplying by the LCD, } ab, \text{ to clear fractions}$$

$$\left.\begin{array}{c} \dfrac{abt}{a} + \dfrac{abt}{b} = ab \\[2mm] bt + at = ab. \end{array}\right\} \quad \begin{array}{l} \text{Multiplying to remove parentheses and} \\[1mm] \text{removing factors equal to 1: } \dfrac{a}{a} = 1 \text{ and } \dfrac{b}{b} = 1 \end{array}$$

If the last equation were $3t + 2t = 6$, we would simply combine like terms ($3t + 2t$ is $5t$) and then divide by the coefficient of t (which would be 5). Since $bt + at$ *cannot* be combined, we factor instead:

$$(b + a)t = ab \qquad \text{Factoring out } t, \text{ the letter for which we are solving}$$

$$t = \frac{ab}{b + a}. \qquad \text{Dividing both sides by } b + a$$

The answer to Example 4 can be used when the times required to do a job independently (a and b) are known and t represents the time required to complete the task working together.

E x a m p l e   5

Young's rule.   Young's rule for determining the size of a particular child's medicine dosage c is

$$c = \frac{a}{a + 12} \cdot d,$$

where a is the child's age and d is the usual adult's dosage (*Source*: Olsen, June Looby, Leon J. Ablon, and Anthony Patrick Giangrasso, *Medical Dosage Calculations*, 6th ed., p. A-31). Solve for a.

Solution   We have

$$c = \frac{a}{a + 12} \cdot d$$

$$(a + 12)c = (a + 12)\frac{a}{a + 12} \cdot d \qquad \text{Multiplying by } a + 12 \text{ to clear the fraction}$$

$$ac + 12c = ad \qquad \text{Simplifying; removing a factor equal to 1: } \frac{a + 12}{a + 12} = 1$$

> **Caution!**   If we next divide by d, we will not isolate a since a would still appear on both sides of the equation.

$$12c = ad - ac \qquad \text{Subtracting } ac \text{ to isolate all } a\text{-terms on one side}$$

$$12c = a(d - c) \qquad \text{Factoring out } a \text{ since it is in more than one term}$$

$$\frac{12c}{d - c} = a. \qquad \text{Dividing by } d - c$$

Exercise Set 6.8

Solve. If the equation is an identity or a contradiction, state this.

1. $3x + 5 = 7x + 1$

2. $x - 9 = 5x + 4$

3. $x^2 = 5x - 6$

4. $3x + 4 = x^2$

5. $2(x - 1) = 2x - 5$

6. $3x - 4(x + 1) = -(x + 4)$

Aha! **7.** $\dfrac{5}{x + 1} = \dfrac{5}{8}$

8. $\dfrac{4}{9} = \dfrac{4}{2 - x}$

9. $4(2x - 1) = 3x - 5$

10. $4x - 7 = 2(5x - 3)$

11. $\dfrac{3}{10x} - \dfrac{4}{5x} = 6$

12. $\dfrac{2}{9x} - \dfrac{5}{6} = 1$

13. $2t^2 - 7t + 3 = 0$

14. $3t^2 - 7t + 2 = 0$

15. $\dfrac{2}{x} - 1 = \dfrac{1 - x}{x}$

16. $\dfrac{3x + 1}{x + 1} = 3$

17. $1 - \dfrac{3}{7n} = \dfrac{5}{14}$

18. $4 - \dfrac{5}{6n} = \dfrac{1}{12}$

Aha! **19.** $\dfrac{x}{x - 2} = \dfrac{2}{x - 2}$

20. $\dfrac{x + 3}{x + 5} = \dfrac{8}{x + 5}$

21. $2\{(x + 2) - 3\} = 3(x + 2) - (x + 8)$

22. $2(x + 5) - 7 = x + 8 - (5 - x)$

23. $\dfrac{x}{x + 1} = \dfrac{x}{x + 2}$

24. $x(x + 2) = x^2 - 3x$

25. $\dfrac{x + 3}{3x + 4} = \dfrac{2}{x + 2}$

26. $\dfrac{x + 2}{1 - x} = \dfrac{1}{x + 3}$

Solve each formula for the specified variable.

27. $s = \dfrac{1}{2}gt^2$, for g

(A physics formula for distance)

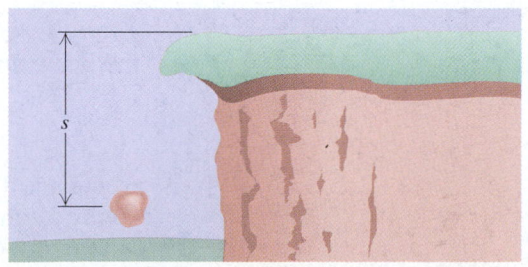

28. $A = \frac{1}{2}bh$, for h
(The area of a triangle)

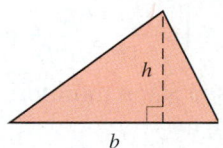

29. $S = 2\pi rh$, for h
(A formula for surface area)

30. $A = P(1 + rt)$, for t
(An interest formula)

31. $A = P + Prt$, for P
(An interest formula)

32. $rl - rS = L$, for r

33. $A = \frac{1}{2}h(b_1 + b_2)$, for h

34. $T = mg + mf$, for m

35. $\frac{1}{180} = \frac{n - 2}{s}$, for n

36. $S = \frac{n}{2}(a + l)$, for a

37. $\frac{m}{n} = p - q$, for n

38. $\frac{M - g}{t} = r + s$, for t

39. $\frac{1}{R} = \frac{1}{r_1} + \frac{1}{r_2}$, for R
(An electricity formula)

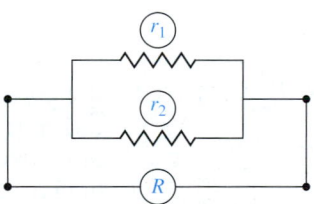

40. $\frac{1}{p} + \frac{1}{q} = \frac{1}{f}$, for f
(An optics formula)

41. $S = 2\pi r(r + h)$, for h
(The surface area of a right circular cylinder)

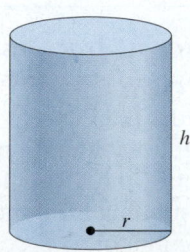

42. $ab = ac + d$, for a

43. $\frac{m}{n} = r$, for n

44. $s + t = \frac{r}{v}$, for v

45. Is it easier to solve

$$\frac{1}{25} + \frac{1}{23} = \frac{1}{x} \text{ for } x,$$

or to solve

$$\frac{1}{p} + \frac{1}{q} = \frac{1}{f} \text{ for } f?$$

Explain why.

46. Explain why someone might want to solve
$A = \frac{1}{2}bh$ for h.
(See Exercise 28.)

SKILL MAINTENANCE

47. Graph: $y = \frac{4}{5}x + 1$.

48. Graph: $y = -\frac{2}{3}x + 1$.

49. Solve: $-\frac{3}{5}x = \frac{9}{10}$.

50. Find the intercepts and graph: $3x + 4y = 24$.

51. Factor: $x^2 - 13x + 30$.

52. Subtract: $(5x^3 - 7x^2 + 9) - (8x^3 - 2x^2 + 4)$.

SYNTHESIS

53. As a step in solving a formula for a certain variable, a student takes the reciprocal on both sides of the equation. Is a mistake being made? Why or why not?

54. Describe a situation in which the result of Example 4,

$$t = \frac{ab}{b + a},$$

would be especially useful.

55. *Health care.* Young's rule for determining the size of a particular child's medicine dosage c is

$$c = \frac{a}{a + 12} \cdot d,$$

where a is the child's age and d is the typical adult dosage (see Example 5). If a child receives 8 mg of antihistamine when the typical adult receives 24 mg, how old is the child?

Solve.

56. $x^2 + px + qx + pq = 0$, for x

57. $\dfrac{a}{b} = \dfrac{b}{a}$, for a

58. $l^2 = lw$, for l

59. $(s + t)^2 = 4st$, for s

60. $\dfrac{n_1}{p_1} + \dfrac{n_2}{p_2} = \dfrac{n_2 - n_1}{R}$, for n_2

61. $u = -F\left(E - \dfrac{P}{T}\right)$, for T

62. *Marine biology.* The formula
$$N = \frac{(b + d)f_1 - v}{(b - v)f_2}$$
is used when monitoring the water in fisheries. Solve for v.

63. *Meteorology.* The formula
$$C = \tfrac{5}{9}(F - 32)$$
is used to convert the Fahrenheit temperature F to the Celsius temperature C. At what temperature are the Fahrenheit and Celsius readings the same?

Summary and Review 6

Key Terms

Rational expression, p. 340

Least Common Multiple, LCM, p. 356

Least Common Denominator, LCD, p. 356

Complex rational expression, p. 371

Rational equation, p. 377

Motion problem, p. 386

Ratio, p. 388

Proportion, p. 388

Similar triangles, p. 388

Contradiction, p. 398

Identity, p. 398

Conditional equation, p. 398

Important Properties and Formulas

To add, subtract, multiply, and divide rational expressions:

$$\frac{A}{B} \cdot \frac{C}{D} = \frac{AC}{BD}; \qquad \frac{A}{B} \div \frac{C}{D} = \frac{A}{B} \cdot \frac{D}{C} = \frac{AD}{BC};$$

$$\frac{A}{B} + \frac{C}{B} = \frac{A + C}{B}; \qquad \frac{A}{B} - \frac{C}{B} = \frac{A - C}{B}.$$

To find the least common denominator (LCD):

1. Write the prime factorization of each denominator.

2. Select one of the factorizations and inspect it to see if it contains the other.

   a) If it does, it represents the LCM of the denominators.

   b) If it does not, multiply that factorization by any factors of the other denominator that it lacks. The final product is the LCM of the denominators.

The LCD is the LCM of the denominators. It contains each factor the greatest number of times that it occurs in any of the individual factorizations.

To add or subtract rational expressions that have different denominators:

1. Find the LCD.

2. Multiply each rational expression by a form of 1 made up of the factors of the LCD missing from that expression's denominator.

3. Add or subtract the numerators, as indicated. Write the sum or difference over the LCD.

4. Simplify, if possible.

To simplify a complex rational expression by dividing:

1. Add or subtract, as needed, to get a single rational expression in the numerator.

2. Add or subtract, as needed, to get a single rational expression in the denominator.

3. Divide the numerator by the denominator (invert and multiply).

4. If possible, simplify by removing a factor equal to 1.

To simplify a complex rational expression by multiplying by the LCD:

1. Find the LCD of all rational expressions *within* the complex rational expression.

2. Multiply the complex rational expression by a factor equal to 1. Write 1 as the LCD over itself (LCD/LCD).

3. Distribute and simplify. No fractional expressions should remain within the complex rational expression.

4. Factor and, if possible, simplify.

To solve a rational equation:

1. List any restrictions that exist. No possible solution can make a denominator equal 0.

2. Clear the equation of fractions by multiplying both sides by the LCD of all rational expressions in the equation.

3. Solve the resulting equation using the addition principle, the multiplication principle, and the principle of zero products, as needed.

4. Check the possible solution(s) in the original equation.

The Work Principle

Suppose that a = the time it takes A to complete a task, b = the time it takes B to complete the task, and t = the time it takes them working together. Then all of the following hold:

$$\frac{1}{a} \cdot t + \frac{1}{b} \cdot t = 1; \left(\frac{1}{a} + \frac{1}{b}\right)t = 1;$$

$$\frac{t}{a} + \frac{t}{b} = 1.$$

Review Exercises

List all numbers for which each expression is undefined.

1. $\dfrac{35}{-x^2}$

2. $\dfrac{9}{a-5}$

3. $\dfrac{x-7}{x^2-36}$

4. $\dfrac{x^2+3x+2}{x^2+x-30}$

5. $\dfrac{-6}{(t+2)^2}$

8. $\dfrac{(y-5)^2}{y^2-25}$

9. $\dfrac{5x^2-20y^2}{2y-x}$

Simplify.

6. $\dfrac{4x^2-8x}{4x^2+4x}$

7. $\dfrac{14x^2-x-3}{2x^2-7x+3}$

Multiply or divide and, if possible, simplify.

10. $\dfrac{a^2-36}{10a} \cdot \dfrac{2a}{a+6}$

11. $\dfrac{8t+8}{2t^2+t-1} \cdot \dfrac{t^2-1}{t^2-2t+1}$

12. $\dfrac{10-5t}{3} \div \dfrac{t-2}{12t}$

13. $\dfrac{4x^4}{x^2 - 1} \div \dfrac{2x^3}{x^2 - 2x + 1}$

14. $\dfrac{x^2 + 1}{x - 2} \cdot \dfrac{2x + 1}{x + 1}$

15. $(t^2 + 3t - 4) \div \dfrac{t^2 - 1}{t + 4}$

Find the LCM.

16. $8a^2b^7,\ 6a^5b^3$ **17.** $x^2 - x,\ x^5 - x^3,\ x^4$

18. $y^2 - y - 2,\ y^2 - 4$

Add or subtract and, if possible, simplify.

19. $\dfrac{x + 8}{x + 7} + \dfrac{10 - 4x}{x + 7}$ **20.** $\dfrac{3}{3x - 9} + \dfrac{x - 2}{3 - x}$

21. $\dfrac{6x - 3}{x^2 - x - 12} - \dfrac{2x - 15}{x^2 - x - 12}$

22. $\dfrac{3x - 1}{2x} - \dfrac{x - 3}{x}$

23. $\dfrac{x + 3}{x - 2} - \dfrac{x}{2 - x}$

24. $\dfrac{2a}{a + 1} - \dfrac{4a}{1 - a^2}$

25. $\dfrac{d^2}{d - c} + \dfrac{c^2}{c - d}$

26. $\dfrac{1}{x^2 - 25} - \dfrac{x - 5}{x^2 - 4x - 5}$

27. $\dfrac{3x}{x + 2} - \dfrac{x}{x - 2} + \dfrac{8}{x^2 - 4}$

28. $\dfrac{2}{5x} + \dfrac{3}{2x + 4}$

Simplify.

29. $\dfrac{\dfrac{1}{z} + 1}{\dfrac{1}{z^2} - 1}$ **30.** $\dfrac{2 + \dfrac{1}{xy^2}}{\dfrac{1 + x}{x^4y}}$ **31.** $\dfrac{\dfrac{c}{d} - \dfrac{d}{c}}{\dfrac{1}{c} + \dfrac{1}{d}}$

Solve.

32. $\dfrac{3}{y} - \dfrac{1}{4} = \dfrac{1}{y}$ **33.** $\dfrac{5}{x + 3} = \dfrac{3}{x + 2}$

34. $\dfrac{15}{x} - \dfrac{15}{x + 2} = 2$

35. Rhetta can polish a helicopter rotor blade in 9 hr. Jason can do the same job in 12 hr. How long would it take if they worked together?

36. The distance by highway between Richmond and Waterbury is 70 km, and the distance by rail is 60 km. A car and a train leave Richmond at the same time and arrive in Waterbury at the same time, the car having traveled 15 km/h faster than the train. Find the speed of the car and the speed of the train.

37. The reciprocal of 1 more than a number is twice the reciprocal of the number itself. What is the number?

38. A sample of 25 doorknobs contained 2 defective doorknobs. How many defective doorknobs would you expect among 375 doorknobs?

39. Triangles ABC and XYZ are similar. Find the value of x.

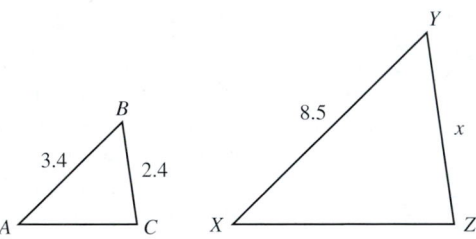

40. A game warden catches, tags, and then releases 15 zebras. A month later, a sample of 20 zebras is caught and 6 of them have tags. Use this information to estimate the size of the zebra population in that area.

Solve. Label each equation as a conditional equation, an identity, or a contradiction.

41. $2(x + 5) + x = 3x + 10$

42. $\dfrac{x}{x - 10} = \dfrac{10}{x - 10}$

43. $2x^2 + 5 = 11x$

Solve for the indicated variable.

44. $2a - b = c$, for a

45. $\dfrac{1}{r} + \dfrac{1}{s} = \dfrac{1}{t}$, for t

46. $F = \dfrac{9C + 160}{5}$, for C

SYNTHESIS

47. For what procedures in this chapter is the LCD used to clear fractions?

48. A student insists on finding a common denominator by always multiplying the denominators of the expressions being added. How could this approach be improved?

Simplify.

49. $\dfrac{2a^2 + 5a - 3}{a^2} \cdot \dfrac{5a^3 + 30a^2}{2a^2 + 7a - 4} \div \dfrac{a^2 + 6a}{a^2 + 7a + 12}$

50. $\dfrac{12a}{(a - b)(b - c)} - \dfrac{2a}{(b - a)(c - b)}$

Aha! 51. $\dfrac{5(x - y)}{(x - y)(x + 2y)} - \dfrac{5(x - 3y)}{(x + 2y)(x - 3y)}$

Chapter Test 6

List all numbers for which each expression is undefined.

1. $\dfrac{8 - x}{3x}$

2. $\dfrac{5}{x + 8}$

3. $\dfrac{x - 7}{x^2 - 49}$

4. $\dfrac{x^2 + x - 30}{x^2 - 3x + 2}$

5. Simplify: $\dfrac{6x^2 + 17x + 7}{2x^2 + 7x + 3}$.

Multiply or divide and, if possible, simplify.

6. $\dfrac{a^2 - 25}{9a} \cdot \dfrac{6a}{5 - a}$

7. $\dfrac{25y^2 - 1}{9y^2 - 6y} \div \dfrac{5y^2 + 9y - 2}{3y^2 + y - 2}$

8. $\dfrac{4x^2 - 1}{x^2 - 2x + 1} \div \dfrac{x - 2}{x^2 + 1}$

9. $(x^2 + 6x + 9) \cdot \dfrac{(x - 3)^2}{x^2 - 9}$

10. Find the LCM:
$$y^2 - 9, \ y^2 + 10y + 21, \ y^2 + 4y - 21.$$

Add or subtract, and, if possible, simplify.

11. $\dfrac{16 + x}{x^3} + \dfrac{7 - 4x}{x^3}$

12. $\dfrac{5 - t}{t^2 + 1} - \dfrac{t - 3}{t^2 + 1}$

13. $\dfrac{x - 4}{x - 3} + \dfrac{x - 1}{3 - x}$

14. $\dfrac{x - 4}{x - 3} - \dfrac{x - 1}{3 - x}$

15. $\dfrac{5}{t - 1} + \dfrac{3}{t}$

16. $\dfrac{1}{x^2 - 16} - \dfrac{x + 4}{x^2 - 3x - 4}$

17. $\dfrac{1}{x - 1} + \dfrac{4}{x^2 - 1} - \dfrac{2}{x^2 - 2x + 1}$

Simplify.

18. $\dfrac{9 - \dfrac{1}{y^2}}{3 - \dfrac{1}{y}}$

19. $\dfrac{\dfrac{3}{a^2 b} - \dfrac{2}{ab^3}}{\dfrac{1}{ab} + \dfrac{2}{a^4 b}}$

Solve.

20. $\dfrac{7}{y} - \dfrac{1}{3} = \dfrac{1}{4}$

21. $\dfrac{15}{x} - \dfrac{15}{x - 2} = -2$

22. Kopy Kwik has 2 copiers. One can copy a year-end report in 20 min. The other can copy the same document in 30 min. How long would it take both machines, working together, to copy the report?

23. A recipe for pizza crust calls for $3\frac{1}{2}$ cups of whole wheat flour and $1\frac{1}{4}$ cups of warm water. If 6 cups of whole wheat flour are used, how much water should be used?

24. Craig drives 20 km/h faster than Marilyn. In the same time that Marilyn drives 225 km, Craig drives 325 km. Find the speed of each car.

Solve. Label each equation as a conditional equation, an identity, or a contradiction.

25. $4x - 2 = 2(x - 1)$

26. $3(x - 1) + 2 = 3x - 1$

Solve for the indicated variable.

27. $d = rt + wt$, for t

28. $\dfrac{1}{ac} = \dfrac{2}{ab} - \dfrac{3}{bc}$, for c

SYNTHESIS

29. Reggie and Rema work together to mulch the flower beds around an office complex in $2\frac{6}{7}$ hr. Working alone, it would take Reggie 6 hr more than

it would take Rema. How long would it take each of them to complete the landscaping working alone?

30. Simplify: $1 - \dfrac{1}{1 - \dfrac{1}{1 - \dfrac{1}{a}}}$.

Cumulative Review 1–6

1. Use the commutative law of addition to write an expression equivalent to $a + 2b$.

2. Write a true sentence using either $<$ or $>$:

$-3.1 \;\blacksquare\; -3.15$.

3. Evaluate $(y - 1)^2$ for $y = -6$.

4. Simplify: $-4[2(x - 3) - 1]$.

Simplify.

5. $-\frac{1}{2} + \frac{3}{8} + (-6) + \frac{3}{4}$

6. $-\frac{72}{108} \div \left(-\frac{2}{3}\right)$

7. $-6.262 \div 1.01$

8. $4 \div (-2) \cdot 2 + 3 \cdot 4$

Solve.

9. $3(x - 2) = 24$

10. $49 = x^2$

11. $-6t = 20$

12. $5x + 7 = -3x - 9$

13. $4(y - 5) = -2(y + 2)$

14. $x^2 + 11x + 10 = 0$

15. $\dfrac{4}{9}t + \dfrac{2}{3} = \dfrac{1}{3}t - \dfrac{2}{9}$

16. $\dfrac{4}{x} + x = 5$

17. $3 - y \geq 2y + 5$

18. $\dfrac{2}{x - 3} = \dfrac{5}{3x + 1}$

19. $2x^2 + 7x = 4$

20. $4(x + 7) < 5(x - 3)$

21. $\dfrac{t^2}{t + 5} = \dfrac{25}{t + 5}$

22. $(2x + 7)(x - 5) = 0$

23. $\dfrac{2}{x^2 - 9} + \dfrac{5}{x - 3} = \dfrac{3}{x + 3}$

Solve each formula.

24. $P = \dfrac{3a}{a + b}$, for a

25. $\frac{3}{4}(x + 2y) = z$, for y

Combine like terms.

26. $x + 2y - 2z + \frac{1}{2}x - z$

27. $2x^3 - 7 + \frac{3}{7}x^2 - 6x^3 - \frac{4}{7}x^2 + 5$

Graph.

28. $y = \frac{3}{4}x + 5$

29. $x = -3$

30. $4x + 5y = 20$

31. $y = 6$

32. Find the slope of the line containing the points $(1, 5)$ and $(2, 3)$.

Simplify.

33. $\dfrac{x^{-5}}{x^{-3}}$

34. $y^2 \cdot y^{-10}$

35. $-(2a^2b^7)^2$

36. Subtract:

$(-8y^2 - y + 2) - (y^3 - 6y^2 + y - 5)$.

Multiply.

37. $4(3x + 4y + z)$

38. $(2x^2 - 1)(x^3 + x - 3)$

39. $(6x - 5y)^2$

40. $(x + 3)(2x - 7)$

41. $(2x^3 + 1)(2x^3 - 1)$

Factor.

42. $6x - 2x^2 - 24x^4$

43. $16x^2 - 81$

44. $t^2 - 10t + 24$

45. $8x^2 + 10x + 3$

46. $6x^2 - 28x + 16$

47. $4t^2 - 36$

48. $25t^2 + 40t + 16$

49. $3x^2 + 10x - 8$

50. $x^4 + 2x^3 - 3x - 6$

Simplify.

51. $\dfrac{y^2 - 36}{2y + 8} \cdot \dfrac{y + 4}{y + 6}$

52. $\dfrac{x^2 - 1}{x^2 - x - 2} \div \dfrac{x - 1}{x - 2}$

53. $\dfrac{5ab}{a^2 - b^2} + \dfrac{a + b}{a - b}$

54. $\dfrac{x + 2}{4 - x} - \dfrac{x + 3}{x - 4}$

55. $\dfrac{1 + \dfrac{2}{x}}{1 - \dfrac{4}{x^2}}$

56. $\dfrac{\dfrac{1}{t} + 2t}{t - \dfrac{2}{t^2}}$

Divide.

57. $\dfrac{15x^4 - 12x^3 + 6x^2 + 2x + 18}{3x^2}$

58. $(15x^4 - 12x^3 + 6x^2 + 2x + 18) \div (x + 3)$

Solve.

59. Linnae has \$36 budgeted for stationery. Engraved stationery costs \$20 for the first 25 sheets and \$0.08 for each additional sheet. How many sheets of stationery can Linnae order and still stay within her budget?

60. The price of a box of cereal increased 15% to \$4.14. What was the price of the cereal before the increase?

61. If the sides of a square are increased by 2 ft, the area of the original square plus the area of the enlarged square is 452 ft². Find the length of a side of the original square.

62. The sum of two consecutive even integers is -554. Find the integers.

63. It takes Dina 50 min to shovel 9 in. of snow from her driveway. It takes Nell 75 min to do the same job. How long would it take if they worked together?

64. Phil's Ford Focus travels 10 km/h faster than Harley's. In the same time that Harley drives 120 km, Phil travels 150 km. Find the speed of each car.

65. A 78-in. board is to be cut into two pieces. One piece must be twice as long as the other. How long should the shorter piece be?

SYNTHESIS

66. Simplify: $(x + 7)(x - 4) - (x + 8)(x - 5)$.

67. Solve: $\frac{1}{3}|n| + 8 = 56$.

Aha! **68.** Multiply: $[4y^3 - (y^2 - 3)][4y^3 + (y^2 - 3)]$.

69. Factor: $2a^{32} - 13{,}122b^{40}$.

70. Solve: $x(x^2 + 3x - 28) - 12(x^2 + 3x - 28) = 0$.

71. Simplify: $-\left|0.875 - \left(-\frac{1}{8}\right) - 8\right|$.

72. Solve: $\dfrac{2}{x - 3} \cdot \dfrac{3}{x + 3} - \dfrac{4}{x^2 - 7x + 12} = 0$.

73. Jesse can peel a bushel of potatoes in 60 min. When Jesse and Priscilla work together, the job takes 20 min. When working together with Jesse, what percentage of the work is performed by Priscilla?

7
Functions and Graphs

AN APPLICATION

A typical adult dosage of an antihistamine is 24 mg. Young's rule for determining the dosage size $c(a)$ for a typical child of age a is

$$c(a) = \frac{24a}{a + 12}.$$

(*Source*: Olsen, June Looby, Leon J. Ablon, and Anthony Patric Giangrasso, *Medical Dosage Calculations*, 6th ed.) What should the dosage be for a typical 8-yr-old child?

This problem appears as Example 5 in Section 7.1.

> N*urses dispense medicine on a regular basis. Since medicine often comes in a dosage different from what a doctor prescribes, I must calculate the correct dosage. Math also helps in critical thinking, such as determining dosage on the basis of body weight.*

RENA J. WEBSTER
Registered Nurse
Indianapolis, Indiana

Introduction to Functions

7.1

Correspondences and Functions • Functions and Graphs •
Function Notation and Equations • Applications

We now develop the idea of a *function*—one of the most important concepts in mathematics. A function is a special kind of correspondence between two sets.

Correspondences and Functions

When forming ordered pairs to graph equations, we often say that the first coordinate of each ordered pair *corresponds* to the second coordinate. In much the same way, a function is a special kind of correspondence between two sets. For example,

To each person in a class	there corresponds	his or her biological mother.
To each item in a shop	there corresponds	its price.
To each real number	there corresponds	the cube of that number.

In each example, the first set is called the **domain**. The second set is called the **range**. For any member of the domain, there is *just one* member of the range to which it corresponds. This kind of correspondence is called a **function**.

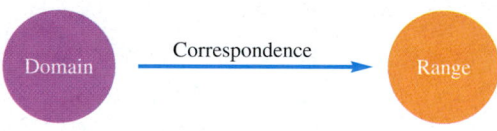

Example 1 Determine whether each correspondence is a function.

a)
$$4 \longrightarrow 2$$
$$1 \nearrow$$
$$-3 \longrightarrow 5$$

b) Ford ⟶ Caravan
Chrysler ⟶ F150 Pickup
General Motors ⟶ Metro
⟶ Camaro

Solution

a) The correspondence *is* a function because each member of the domain corresponds to *just one* member of the range.

b) The correspondence *is not* a function because a member of the domain (General Motors) corresponds to more than one member of the range.

> **Function**
>
> A *function* is a correspondence between a first set, called the *domain,* and a second set, called the *range,* such that each member of the domain corresponds to *exactly one* member of the range.

Example 2

Determine whether each correspondence is a function.

Domain	*Correspondence*	*Range*
a) An elevator full of people	Each person's weight	A set of positive numbers
b) $\{-2, 0, 1, 2\}$	Each number's square	$\{0, 1, 4\}$
c) The books in a college bookstore	Each book's author	A set of people

Solution

a) The correspondence *is* a function, because each person has *only one* weight.

b) The correspondence *is* a function, because every number has *only one* square.

c) The correspondence *is not* a function, because some books have *more than one* author.

Although the correspondence in Example 2(c) is not a function, it is a *relation.*

> **Relation**
>
> A *relation* is a correspondence between a first set, called the *domain,* and a second set, called the *range,* such that each member of the domain corresponds to *at least one* member of the range.

Functions and Graphs

The functions in Examples 1(a) and 2(b) can be expressed as sets of ordered pairs. Example 1(a) can be written $\{(-3, 5), (1, 2), (4, 2)\}$ and Example 2(b) can be written $\{(-2, 4), (0, 0), (1, 1), (2, 4)\}$. We can graph these functions as follows.

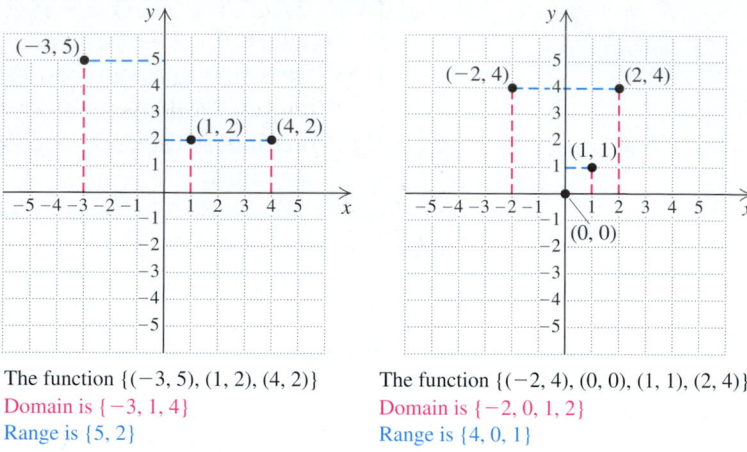

The function $\{(-3, 5), (1, 2), (4, 2)\}$
Domain is $\{-3, 1, 4\}$
Range is $\{5, 2\}$

The function $\{(-2, 4), (0, 0), (1, 1), (2, 4)\}$
Domain is $\{-2, 0, 1, 2\}$
Range is $\{4, 0, 1\}$

When a function is graphed, the domain is generally located on the horizontal axis and the range is located on the vertical axis. Functions are often represented by lower- or upper-case letters.

E x a m p l e   3

For the function f shown here, determine each of the following.

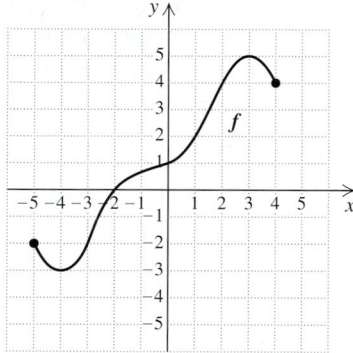

a) What member of the range is paired with 2 (in the domain)

b) What member of the domain is paired with -3 (in the range)

Solution

a) To determine what member of the range is paired with 2, we locate 2 on the horizontal axis (this is where the domain is located). Next, we find the point on the graph of f for which 2 is the first coordinate. From that point, we can look to the vertical axis to find the corresponding y-coordinate, 4. The "input" 2 has the "output" 4.

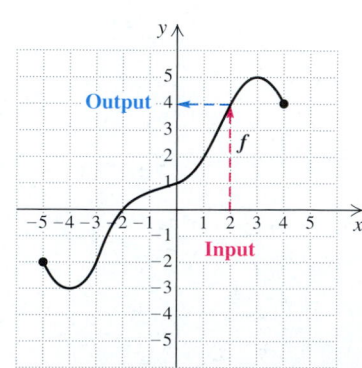

b) To determine what member of the domain is paired with -3, we locate -3 on the vertical axis. (This is where the range is located; see below.) From there we look left and right to the graph of f to find any points for which -3 is the second coordinate. One such point exists, $(-4, -3)$. We observe that -4 is the only element of the domain paired with -3.

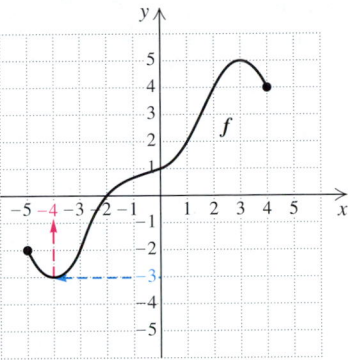

Function Notation and Equations

To understand function notation, it sometimes helps to imagine a "function machine." Think of putting a member of the domain (an *input*) into the machine. The machine is programmed to produce the appropriate member of the range (the *output*).

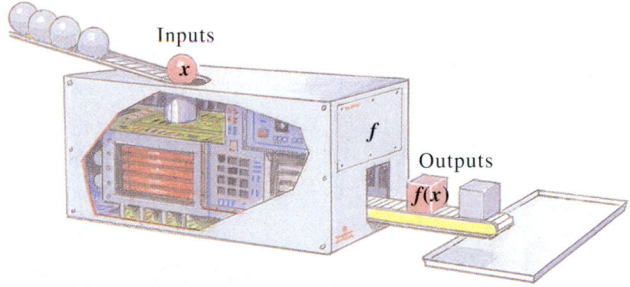

The function pictured has been named f. Here x represents an arbitrary input, and $f(x)$—read "f of x," "f at x," or "the value of f at x"—represents the corresponding output. In Example 3, the input 2 has the output 4. Using function notation, we write $f(2) = 4$, read "f of 2 is 4."

Note that $f(x)$ *does not mean f times x*.

Most functions are described by equations. For example, $f(x) = 2x + 3$ describes the function that takes an input x, multiplies it by 2, and then adds 3.

$$f(x) = 2x + 3$$

Input
Double   Add 3

To calculate the output $f(4)$, we take the input 4, double it, and add 3 to get 11. That is, we substitute 4 into the formula for $f(x)$:

$$f(4) = 2 \cdot 4 + 3$$
$$= 11.$$

Sometimes, in place of $f(x) = 2x + 3$, we write $y = 2x + 3$, where it is understood that the value of y, the *dependent variable*, depends on our choice of x, the *independent variable*. To understand why $f(x)$ notation is so useful, consider two equivalent statements:

a) If $f(x) = 2x + 3$, then $f(4) = 11$.
b) If $y = 2x + 3$, then the value of y is 11 when x is 4.

The notation used in part (a) is far more concise.

Example 4

Find the indicated function value.

a) $f(5)$, for $f(x) = 3x + 2$
b) $g(-2)$, for $g(r) = 5r^2 + 3r$
c) $h(4)$, for $h(x) = 7$
d) $F(a + 1)$, for $F(x) = 3x + 2$

Solution Finding function values is much like evaluating an algebraic expression.

a) $f(5) = 3 \cdot 5 + 2 = 17$

b) $g(-2) = 5(-2)^2 + 3(-2)$
$$= 5 \cdot 4 - 6 = 14$$

c) For the function given by $h(x) = 7$, all inputs share the same output, 7. Therefore, $h(4) = 7$. The function h is an example of a *constant function*.

d) $F(a + 1) = 3(a + 1) + 2$
$$= 3a + 3 + 2 = 3a + 5$$

Note that whether we write $f(x) = 3x + 2$, or $f(t) = 3t + 2$, or $f(\square) = 3\square + 2$, we still have $f(5) = 17$. Thus the independent variable can be thought of as a *dummy variable*. The letter chosen for the dummy variable is not as important as the algebraic manipulations to which it is subjected.

Applications

Function notation is often used in formulas. For example, to emphasize that the area A of a circle is a function of its radius r, instead of

$$A = \pi r^2,$$

we can write

$$A(r) = \pi r^2.$$

E x a m p l e   5   A typical adult dosage of an antihistamine is 24 mg. Young's rule for determining the dosage size $c(a)$ for a typical child of age a is

$$c(a) = \frac{24a}{a + 12}.^*$$

What should the dosage be for a typical 8-yr-old child?

Solution   We find $c(8)$:

$$c(8) = \frac{24(8)}{8 + 12} = \frac{192}{20} = 9.6.$$

The dosage for a typical 8-yr-old child is 9.6 mg.

When a function is given as a graph in a problem-solving situation, we are often asked to determine certain quantities on the basis of the graph. Later in this text, we will develop models that can be used for calculations. For now we simply use the graph to estimate the coordinates of an unknown point by using other points with known coordinates. When the unknown point is *between* the known points, this process is called **interpolation**. If the unknown point extends *beyond* the known points, the process is called **extrapolation**.

E x a m p l e   6   **Working mothers.** More women than ever before are returning to the workforce within a year of giving birth. According to the U.S. Census Bureau, in 1980 38% of women who gave birth returned to work within a year. The figure grew to 43% in 1983, 50% in 1986, 54% in 1992, 55% in 1995, and 59% in 1998. Estimate the percentage of women who returned to work within a year of giving birth in 1989 (a year the Census Bureau did not include) and in 2001.

Solution

1. and **2.   Familiarize** and **Translate.**   The given information enables us to plot and connect six points. We let the horizontal axis represent the year and the vertical axis the percentage of women who gave birth and returned to work within one year. We label the function itself P.

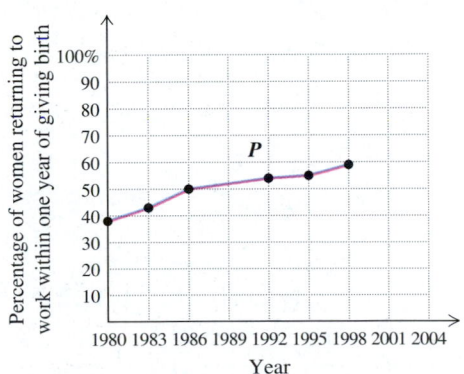

---

*Source: Olsen, June Looby, Leon J. Ablon, and Anthony Patric Giangrasso, *Medical Dosage Calculations*, 6th ed.

3. **Carry out.** To estimate the percentage of women who returned to work within a year of giving birth in 1989, we locate the point directly above the year 1989. We then estimate its second coordinate by moving horizontally from that point to the y-axis. Although our result is not exact, we see that $P(1989) \approx 52$.

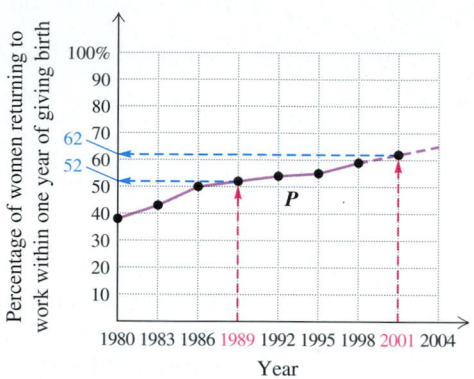

To estimate the percentage of women who returned to work within a year of giving birth in 2001, we extend the graph and extrapolate. It appears that $P(2001) \approx 62$.

4. **Check.** A precise check requires consulting an outside information source. Since 52% is between 50% and 54% and 62% is greater than 59%, our estimates seem plausible.

5. **State.** In 1989, about 52% of all women who gave birth returned to work within a year. By 2001, that percentage would be predicted to be about 62%.

FOR EXTRA HELP

Exercise Set 7.1

 Digital Video Tutor CD 5 Videotape 13

 InterAct Math

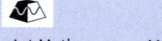 Math Tutor Center

 MathXL

 MyMathLab.com

Determine whether each correspondence is a function.

1.

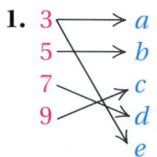

2.

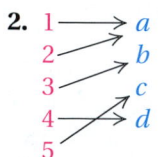

3. *Girl's Age (in months)*   *Average Daily Weight Gain (in grams)*

$2 \longrightarrow 21.8$
$9 \longrightarrow 11.7$
$16 \longrightarrow 8.5$
$23 \longrightarrow 7.0$

Source: American Family Physician, December 1993, p. 1435.

4. *Boy's Age (in months)*   *Average Daily Weight Gain (in grams)*

$2 \longrightarrow 24.3$
$9 \longrightarrow 11.7$
$16 \longrightarrow 8.2$
$23 \longrightarrow 7.0$

Source: American Family Physician, December 1993, p. 1435.

5. *Predator*   *Prey*

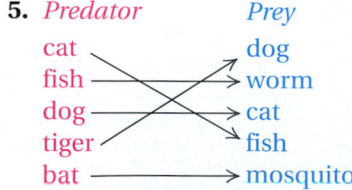

6.

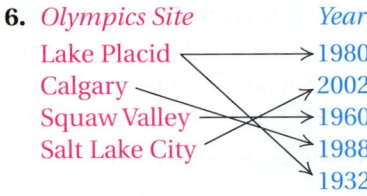

Olympics Site   *Year*
Lake Placid → 1980
Calgary → 2002
Squaw Valley → 1960
Salt Lake City → 1988
1932

Determine whether each of the following is a function. Identify any relations that are not functions.

Domain	Correspondence	*Range*
7. A yard full of Christmas trees	Each tree's price	A set of prices
8. The swordfish stored in a boat	Each fish's weight	A set of weights
9. The members of a rock band	An instrument the person can play	A set of instruments
10. The students in a math class	Each person's seat number	A set of numbers
11. A set of numbers	Square each number and then add 4.	A set of numbers
12. A set of shapes	The area of each shape	A set of numbers

For each graph of a function, determine **(a)** $f(1)$ *and* **(b)** *any x-values for which* $f(x) = 2$.

13.

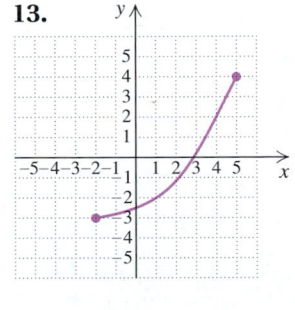

14.

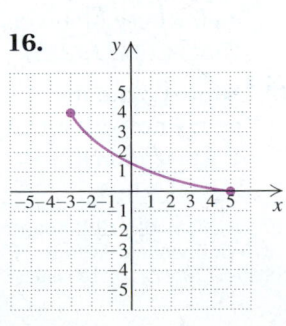

15.

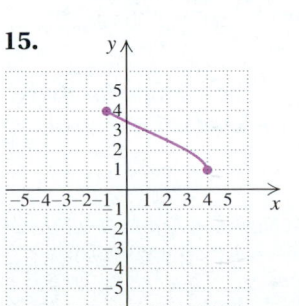

16.

17.

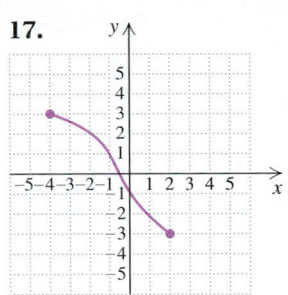

18.

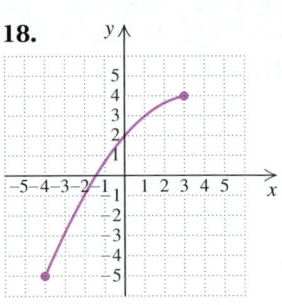

19.

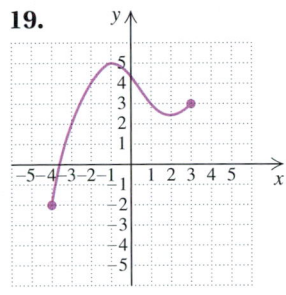

20.

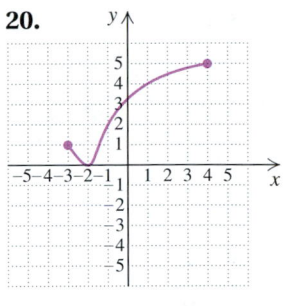

21.

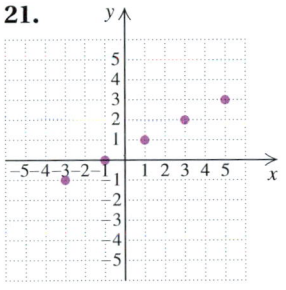

22.

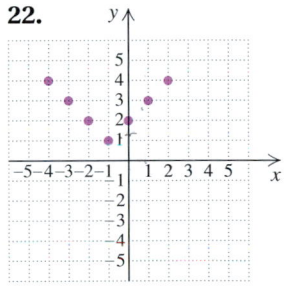

23.

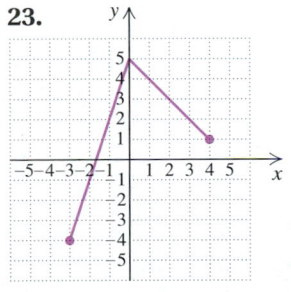

24.

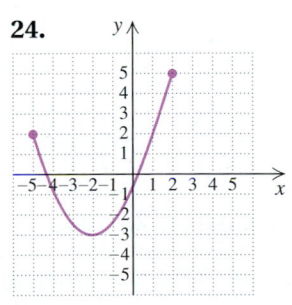

25.

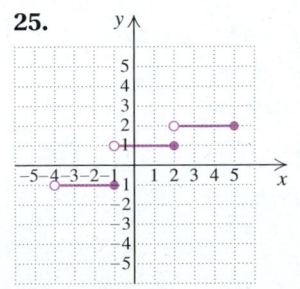

26.

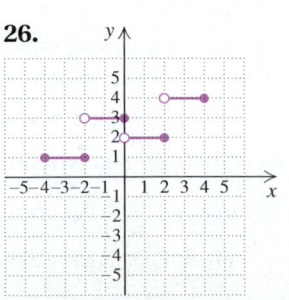

Find the function values.

27. $g(x) = x + 3$

    a) $g(0)$ **b)** $g(-4)$ **c)** $g(-7)$
    d) $g(8)$ **e)** $g(a + 2)$

28. $h(x) = x - 2$

    a) $h(4)$ **b)** $h(8)$ **c)** $h(-3)$
    d) $h(-4)$ **e)** $h(a - 1)$

29. $f(n) = 5n^2 + 4n$

    a) $f(0)$ **b)** $f(-1)$ **c)** $f(3)$
    d) $f(t)$ **e)** $f(2a)$

30. $g(n) = 3n^2 - 2n$

    a) $g(0)$ **b)** $g(-1)$ **c)** $g(3)$
    d) $g(t)$ **e)** $g(2a)$

31. $f(x) = \dfrac{x - 3}{2x - 5}$

    a) $f(0)$ **b)** $f(4)$ **c)** $f(-1)$
    d) $f(3)$ **e)** $f(x + 2)$

32. $s(x) = \dfrac{3x - 4}{2x + 5}$

    a) $s(10)$ **b)** $s(2)$ **c)** $s\left(\frac{1}{2}\right)$
    d) $s(-1)$ **e)** $s(x + 3)$

The function A described by $A(s) = s^2 \dfrac{\sqrt{3}}{4}$ gives the area of an equilateral triangle with side s.

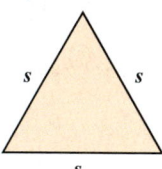

33. Find the area when a side measures 4 cm.

34. Find the area when a side measures 6 in.

The function V described by $V(r) = 4\pi r^2$ gives the surface area of a sphere with radius r.

35. Find the area when the radius is 3 in.

36. Find the area when the radius is 5 cm.

Chemistry.   *The function F described by*

$$F(C) = \tfrac{9}{5}C + 32$$

gives the Fahrenheit temperature corresponding to the Celsius temperature C.

37. Find the Fahrenheit temperature equivalent to $-10°C$.

38. Find the Fahrenheit temperature equivalent to $5°C$.

Archaeology.   *The function H described by*

$$H(x) = 2.75x + 71.48$$

can be used to predict the height, in centimeters, of a woman whose humerus (the bone from the elbow to the shoulder) is x cm long. Predict the height of a woman whose humerus is the length given.

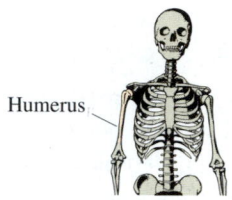

Humerus

39. 32 cm   **40.** 35 cm

Heart attacks and cholesterol.   *For Exercises 41 and 42, use the following graph, which shows the annual heart attack rate per 10,000 men as a function of blood cholesterol level.*[*]

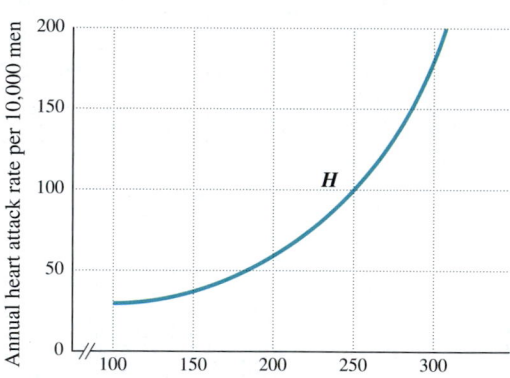

41. Approximate the annual heart attack rate for those men whose blood cholesterol level is 225 mg/dl. That is, find $H(225)$.

42. Approximate the annual heart attack rate for those men whose blood cholesterol level is 275 mg/dl. That is, find $H(275)$.

[*]Copyright 1989, CSPI. Adapted from *Nutrition Action Healthletter* (1875 Connecticut Avenue, N.W., Suite 300, Washington, DC 20009-5728. $24 for 10 issues).

Voting attitudes.   *For Exercises 43 and 44, use this graph, which shows the percentage of people responding yes to the question, "If your (political) party nominated a generally well-qualified person for president who happened to be a woman, would you vote for that person?" (Source: The New York Times, August 13, 2000)*

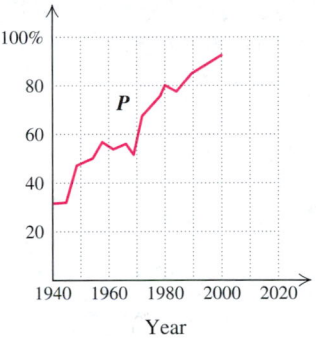

Year

43. Approximate the percentage of Americans willing to vote for a woman for president in 1960. That is, find $P(1960)$.

44. Approximate the percentage of Americans willing to vote for a woman for president in 2000. That is, find $P(2000)$.

Blood alcohol level.   *The following table can be used to predict the number of drinks required for a person of a specified weight to be considered legally intoxicated (blood alcohol level of 0.08 or above). One 12-oz glass of beer, a 5-oz glass of wine, or a cocktail containing 1 oz of a distilled liquor all count as one drink. Assume that all drinks are consumed within one hour.*

12 oz   5 oz   1 oz

Input, Body Weight (in pounds)	Output, Number of Drinks
100	2.5
160	4
180	4.5
200	5

45. Use the data in the table below left to draw a graph and to estimate the number of drinks that a 140-lb person would have to drink to be considered intoxicated.

46. Use the graph from Exercise 45 to estimate the number of drinks a 120-lb person would have to drink to be considered intoxicated.

Incidence of AIDS.   *The following table indicates the number of cases of AIDS reported in each of several years (Source: U.S. Centers for Disease Control and Prevention).*

Input, Year	Output, Number of Cases Reported
1994	77,103
1996	66,497
1998	48,269

47. Use the data in the table above to draw a graph and to estimate the number of cases of AIDS reported in 2000.

48. Use the graph from Exercise 47 to estimate the number of cases of AIDS reported in 1997.

Population growth.   *The town of Falconburg recorded the following dates and populations.*

Input, Year	Output, Population (in tens of thousands)
1995	5.8
1997	6
1999	7
2001	7.5

49. Use the data in the table above to draw a graph of the population as a function of time. Then estimate what the population was in 1998.

50. Use the graph in Exercise 49 to predict Falconburg's population in 2003.

51. *Retailing.*   Shoreside Gifts is experiencing constant growth. They recorded a total of $250,000 in sales in 1996 and $285,000 in 2001. Use a graph that displays the store's total sales as a function of time to predict total sales for 2005.

52. Use the graph in Exercise 51 to estimate what the total sales were in 1999.

Researchers at Yale University have suggested that the following graphs* may represent three different aspects of love.

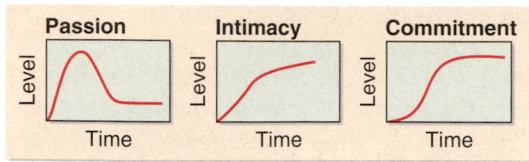

53. In what unit would you measure time if the horizontal length of each graph were ten units? Why?

54. Do you agree with the researchers that these graphs should be shaped as they are? Why or why not?

SKILL MAINTENANCE

Simplify.

55. $\dfrac{10 - 3^2}{9 - 2 \cdot 3}$

56. $\dfrac{2^4 - 10}{6 - 4 \cdot 3}$

57. The surface area of a rectangular solid of length l, width w, and height h is given by $S = 2lh + 2lw + 2wh$. Solve for l.

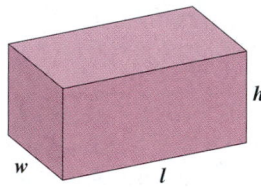

58. Solve the formula in Exercise 57 for w.

Solve for y.

59. $2x + 3y = 6$

60. $5x - 4y = 8$

SYNTHESIS

61. Which would you trust more and why: estimates made using interpolation or those made using extrapolation?

62. Explain in your own words why every function is a relation, but not every relation is a function.

---

*From "A Triangular Theory of Love," by R. J. Sternberg, 1986, *Psychological Review*, **93**(2), 119–135. Copyright 1986 by the American Psychological Association, Inc. Reprinted by permission.

For Exercises 63 and 64, let $f(x) = 3x^2 - 1$ and $g(x) = 2x + 5$.

63. Find $f(g(-4))$ and $g(f(-4))$.

64. Find $f(g(-1))$ and $g(f(-1))$.

Pregnancy. For Exercises 65–68, use the following graph of a woman's "stress test." This graph shows the size of a pregnant woman's contractions as a function of time.

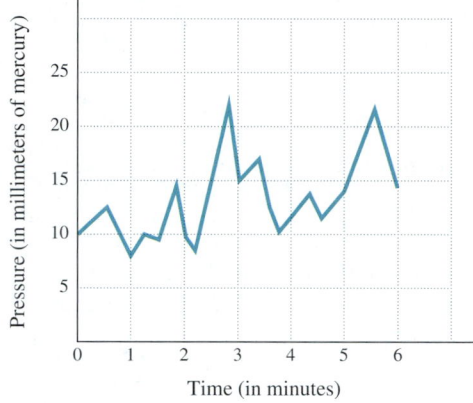

65. How large is the largest contraction that occurred during the test?

66. At what time during the test did the largest contraction occur?

67. On the basis of the information provided, how large a contraction would you expect 60 seconds after the end of the test? Why?

68. What is the frequency of the largest contraction?

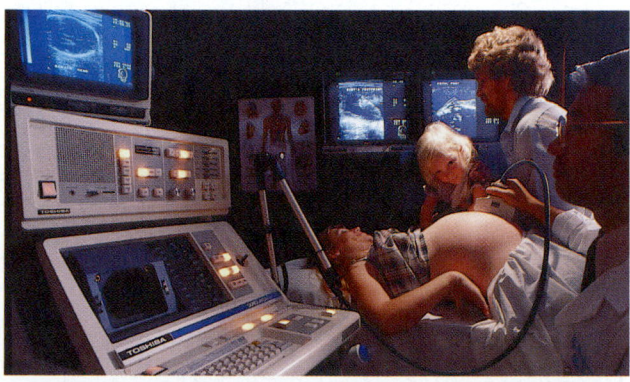

69. The *greatest integer function* $f(x) = [\![x]\!]$ is defined as follows: $[\![x]\!]$ is the greatest integer that is less than or equal to x. For example, if $x = 3.74$, then

$[\![x]\!] = 3$; and if $x = -0.98$, then $[\![x]\!] = -1$. Graph the greatest integer function for $-5 \le x \le 5$. (The notation $f(x) = \text{INT}[x]$ is used in many graphers and computer programs.)

70. Suppose that a function g is such that $g(-1) = -7$ and $g(3) = 8$. Find a formula for g if $g(x)$ is of the form $g(x) = mx + b$, where m and b are constants.

71. *Energy expenditure.*   On the basis of the information given at right, what burns more energy: walking $4\frac{1}{2}$ mph for two hours or bicycling 14 mph for one hour?

Approximate Energy Expenditure by a 150-Pound Person in Various Activities

Activity	Calories per Hour
Walking, $2\frac{1}{2}$ mph	210
Bicycling, $5\frac{1}{2}$ mph	210
Walking, $3\frac{3}{4}$ mph	300
Bicycling, 13 mph	660

Source: Based on material prepared by Robert E. Johnson, M.D., Ph.D., and colleagues, University of Illinois.

CORNER

Calculating License Fees

Focus: Functions

Time: 15–20 minutes

Group size: 3–4

The California Department of Motor Vehicles calculates automobile registration fees (VLF) according to the schedule shown below.

ACTIVITY

1. Determine the original sale price of the oldest vehicle owned by a member of your group. If necessary, use the price and age of a family member's vehicle. Be sure to note the year in which the car was purchased.

2. Use the schedule below to calculate the vehicle license fee (VLF) for the vehicle in part (1) above for each year from the year of purchase to the present. To speed your work, each

group member can find the fee for a few different years.

3. Graph the results from part (2). On the x-axis, plot years beginning with the year of purchase, and on the y-axis, plot $V(x)$, the VLF as a function of year.

4. What is the lowest VLF that the owner of this car will ever have to pay, according to this schedule? Compare your group's answer with other groups' answers.

5. Does your group feel that California's method for calculating registration fees is fair? Why or why not? How could it be improved?

6. Try, as a group, to find an algebraic form for the function $y = V(x)$.

7. *Optional out-of-class extension:* Create a program for a grapher that accepts two inputs (initial value of the vehicle and year of purchase) and produces $V(x)$ as the output.

DMV
A Public Service Agency

VEHICLE LICENSE FEE INFORMATION

The 2% **Vehicle License Fee (VLF)** is in lieu of a personal property tax on vehicles. Most VLF revenue is returned to City and County Local Governments (see reverse side). The license fee charged is based upon the sale price or vehicle value when initially registered in California. The vehicle value is adjusted for any subsequent sale or transfer, that occurred 8/19/91 or later, excluding sales or transfers between specified relatives.

The VLF is calculated by rounding the sale price to the nearest **odd** hundred dollar. That amount is reduced by a percentage utilizing an eleven year schedule (shown to the right), and 2% of that amount is the fee charged. See the accompanying example for a vehicle purchased last year for $9,199. This would be the second registration year following that purchase.

WHERE DO YOUR DMV FEES GO? SEE REVERSE SIDE.

DMV77 8(REV.8/95)   95 30123

PERCENTAGE SCHEDULE
Rev. & Tax. Code Sec. 10753.2
(Trailer coaches have a different schedule)

1st Year	100%	7th Year	40%
2nd Year	90%	8th Year	30%
3rd Year	80%	9th Year	25%
4th Year	70%	10th year	20%
5th Year	60%	11th Year	
6th Year	50%	onward	15%

VLF CALCULATION EXAMPLE

Purchase Price:	$9,199
Rounded to:	$9,100
Times the Percentage:	90%
Equals Fee Basis of:	$8,190
Times 2% Equals:	$163.80
Rounded to:	$164

Domain and Range

7.2

**Determining the Domain and the Range •
Restrictions on Domain • Functions Defined Piecewise**

In Section 7.1, we saw that a function is a correspondence from a set called the *domain* to a set called the *range*. In this section, we look more closely at the concepts of domain and range.

Determining the Domain and the Range

When a function is given as a set of ordered pairs, the domain is the set of all first coordinates and the range is the set of all second coordinates.

Example 1   Find the domain and the range for the function f given by

$$f = \{(2, 0), (-1, 5), (8, 0), (-3, 2)\}.$$

Solution   The first coordinates are $2, -1, 8$, and -3. The second coordinates are $0, 5$, and 2. Thus we have

Domain of $f = \{2, -1, 8, -3\}$ and

Range of $f = \{0, 5, 2\}$.

We can also determine the domain and the range of a function from its graph.

Example 2   Find the domain and the range of the function f shown here.

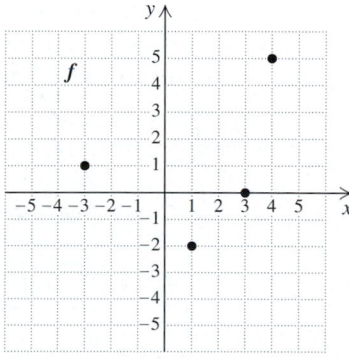

Solution   Here f can be written $\{(-3, 1), (1, -2), (3, 0), (4, 5)\}$. The domain is the set of all first coordinates, $\{-3, 1, 3, 4\}$, and the range is the set of all second coordinates, $\{1, -2, 0, 5\}$.

In Example 2, we could also have found the domain and the range directly, without first writing f, by observing the x- and y-values used in the graph.

E x a m p l e   3   Find the domain and the range of the function f shown here.

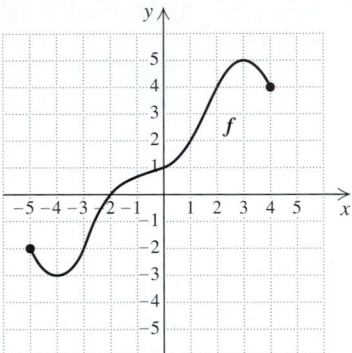

Solution   The domain of the function is the set of all x-values that are in the graph. Because there are no breaks in the graph of f, these extend continuously from -5 to 4 and can be viewed as the curve's shadow, or *projection*, on the x-axis. Thus the domain is $\{x \mid -5 \le x \le 4\}$.

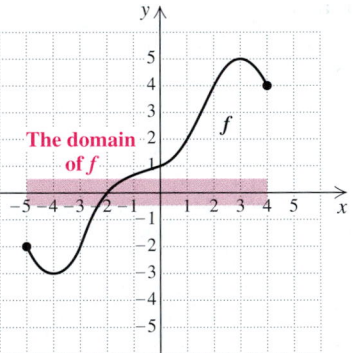

The range of the function is the set of all y-values that are in the graph. These extend continuously from -3 to 5, and can be viewed as the curve's projection on the y-axis. Thus the range is $\{y \mid -3 \le y \le 5\}$.

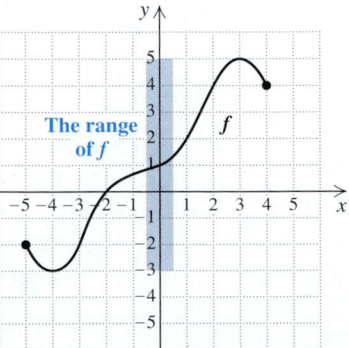

In Example 3, the *endpoints* $(-5, -2)$ and $(4, 4)$ emphasize that the function is not defined for values of x less than -5 or greater than 4.

The graphs of some functions have no endpoints.

E x a m p l e   4   Find the domain and the range of the function f shown here.

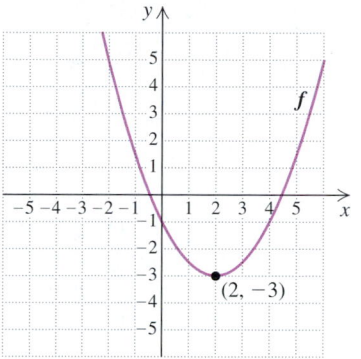

Solution   The domain is the set of all x-values that are in the graph. There are no endpoints, which indicates that the graph extends indefinitely. Only a portion of the graph is shown; in fact, it is impossible to show the entire graph. For any x-value, there is a point on the graph. Thus,

$$\text{Domain of } f = \{x \mid x \text{ is a real number}\}.$$

Again, this can be viewed as the curve's projection on the x-axis.

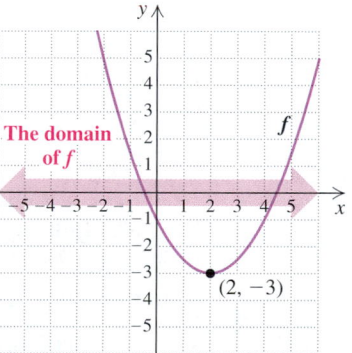

The range of the function is the set of all y-values that are in the graph. This can be viewed as the projection of the curve on the y-axis. The function has no y-values less than -3, and every y-value greater than or equal to -3 corresponds to at least one member of the domain. Thus,

$$\text{Range of } f = \{y \mid y \text{ is a real number } and \; y \geq -3\}.$$

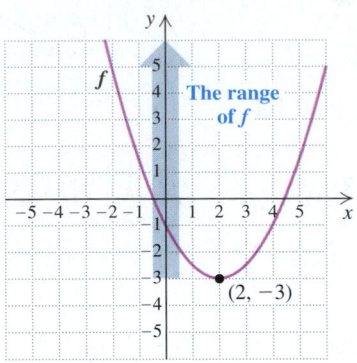

The set of all real numbers is often abbreviated $\mathbb{R}$. Thus, in Example 4, we could write Domain of $f = \mathbb{R}$.

A closed dot, as in Example 3, emphasizes that a particular point is on a graph. An open dot, $\circ$, indicates that a particular point is *not* on a graph.

E x a m p l e   5   Find the domain and the range of the function f shown here.

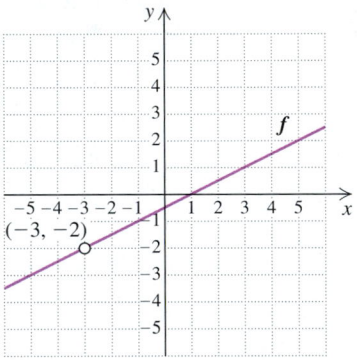

Solution   The domain of f is the set of all x-values that are in the graph. The open dot in the graph at $(-3, -2)$ indicates that there is no y-value that corresponds to $x = -3$; that is, the function is not defined for $x = -3$. Thus, -3 is not in the domain of the function, and

Domain of $f = \{x \mid x$ is a real number $and\ x \neq -3\}$.

There is no function value at $(-3, -2)$, so -2 is not in the range of the function. Thus we have

Range of $f = \{y \mid y$ is a real number $and\ y \neq -2\}$.

When a function is described by an equation, we assume that the domain is the set of all real numbers for which function values can be calculated. If an x-value is not in the domain of a function, the graph of the function will not include any point above or below that x-value.

E x a m p l e   6

For each equation, determine the domain of f.

a) $f(x) = 3x - 7$ **b)** $f(x) = \dfrac{7}{2x - 6}$ **c)** $f(t) = \dfrac{t + 1}{t^2 - 4}$

Solution

a) We ask ourselves, "Is there any number x for which we cannot compute $3x - 7$?" Since we can multiply any number by 3, and subtract 7 from any number, the answer is no. Thus the domain of f is $\mathbb{R}$, the set of all real numbers.

b) Is there any number x for which $\dfrac{7}{2x - 6}$ cannot be computed? Since $\dfrac{7}{2x - 6}$ cannot be computed when $2x - 6$ is 0, the answer is yes. To determine what x-value causes the denominator to be 0, we solve an equation:

$$2x - 6 = 0 \qquad \text{Setting the denominator equal to 0}$$
$$2x = 6 \qquad \text{Adding 6 to both sides}$$
$$x = 3. \qquad \text{Dividing both sides by 2}$$

Thus, 3 is *not* in the domain of f, whereas all other real numbers are. The domain of f is $\{x \mid x \text{ is a real number } and \ x \neq 3\}$.

c) The expression $\dfrac{t + 1}{t^2 - 4}$ is undefined when $t^2 - 4 = 0$:

$$t^2 - 4 = 0 \qquad \text{Setting the denominator equal to 0}$$
$$(t + 2)(t - 2) = 0 \qquad \text{Factoring}$$
$$t + 2 = 0 \quad or \quad t - 2 = 0 \qquad \text{Using the principle of zero products}$$
$$t = -2 \quad or \qquad t = 2. \qquad \text{Solving; these are the values for which } (t + 1)/(t^2 - 4) \text{ is undefined.}$$

Thus we have

Domain of $f = \{t \mid t \text{ is a real number } and \ t \neq -2 \ and \ t \neq 2\}$.

Note that when the numerator, $t + 1$, is zero, the function value is 0 and *is* defined.

technology connection

To visualize Example 6(b), note that the graph of $y_1 = \dfrac{7}{2x - 6}$ has a break at $x = 3$.

$y_1 = 7/(2x - 6)$

X = 3, Y =   DOT mode

Restrictions on Domain

If a function is used as a model for an application, the problem situation may require restrictions on the domain; for example, length and time are generally nonnegative, and a person's age does not increase indefinitely.

E x a m p l e   7

Prize tee shirts.   During intermission at sporting events, it has become common for team mascots to use a powerful slingshot to launch tightly rolled tee shirts into the stands. The height $h(t)$, in feet, of an airborne tee shirt t seconds after being launched can be approximated by

$$h(t) = -15t^2 + 70t + 25.$$

What is the domain of the function?

Solution   The expression $-15t^2 + 70t + 25$ can be evaluated for any number *t*, so any restrictions on the domain will come from the problem situation.

First, we note that *t* cannot be negative, since it represents time from launch, so we have $t \geq 0$. If we make a drawing, we also note that the function will not be defined for values of *t* that make the height negative.

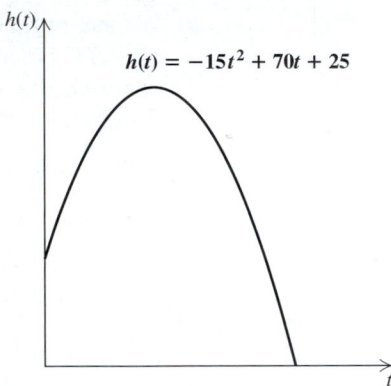

$h(t) = -15t^2 + 70t + 25$

Thus an upper limit for *t* will be the positive value of *t* for which $h(t) = 0$. Solving, we obtain

$$h(t) = 0$$

$$-15t^2 + 70t + 25 = 0 \qquad \text{Substituting } -15t^2 + 70t + 25 \text{ for } h(t)$$

$$\left.\begin{array}{l} -5(3t^2 - 14t - 5) = 0 \\ -5(3t + 1)(t - 5) = 0 \end{array}\right\} \qquad \text{Factoring}$$

$$3t + 1 = 0 \quad or \quad t - 5 = 0 \qquad \text{Using the principle of zero products}$$

$$\left.\begin{array}{ll} 3t = -1 & or \quad t = 5 \\ t = -\dfrac{1}{3} & or \quad t = 5. \end{array}\right\} \qquad \text{Solving for } t$$

We already know that $-\frac{1}{3}$ is not in the domain of the function because of the restriction $t \geq 0$ above.

The tee shirt will hit the ground after 5 sec, so we have $t \leq 5$. Putting the two restrictions together, we have $t \geq 0$ *and* $t \leq 5$, so the

Domain of $h = \{t | t \text{ is a real number } and \ 0 \leq t \leq 5\}$.

Functions Defined Piecewise

Some functions are defined by different equations for various parts of their domains. Such functions are said to be **piecewise** defined. For example, the function given by $f(x) = |x|$ is described by

$$f(x) = \begin{cases} x, & \text{if } x \geq 0, \\ -x, & \text{if } x < 0. \end{cases}$$

To evaluate a piecewise-defined function for an input *a*, we first determine what part of the domain *a* belongs to. Then we use the appropriate formula for that part of the domain.

E x a m p l e   8   Find each function value for the function f given by

$$f(x) = |x| = \begin{cases} x, & \text{if } x \geq 0, \\ -x, & \text{if } x < 0. \end{cases}$$

a) $f(4)$ **b)** $f(-10)$

Solution

a) Since $4 \geq 0$, we use the equation $f(x) = x$. Thus, $f(4) = 4$.
b) Since $-10 < 0$, we use the equation $f(x) = -x$. Thus,
$f(-10) = -(-10) = 10$.

E x a m p l e   9   Find each function value for the function g given by

$$g(x) = \begin{cases} x + 2, & \text{if } x \leq -2, \\ x^2, & \text{if } -2 < x \leq 5, \\ 3x, & \text{if } x > 5. \end{cases}$$

a) $g(-2)$ **b)** $g(3)$ **c)** $g(10)$

Solution

a) Since $-2 \leq -2$, we use the first equation, $g(x) = x + 2$:

$$g(-2) = -2 + 2 = 0.$$

b) Since $-2 < 3 \leq 5$, we use the second equation, $g(x) = x^2$:

$$g(3) = 3^2 = 9.$$

c) Since $10 > 5$, we use the last equation, $g(x) = 3x$:

$$g(10) = 3 \cdot 10 = 30.$$

FOR EXTRA HELP

Exercise Set **7.2**

Digital Video Tutor CD 5   InterAct Math   Math Tutor Center   MathXL   MyMathLab.com
Videotape 13

Find the domain and the range for each function given.

1. $f = \{(2, 8), (9, 3), (-2, 10), (-4, 4)\}$

2. $g = \{(1, 2), (2, 3), (3, 4), (4, 5)\}$

3. $g = \{(0, 0), (4, -2), (-5, 0), (-1, -2)\}$

4. $f = \{(3, 7), (2, 7), (1, 7), (0, 7)\}$

For each graph of a function f, determine the domain and the range of f.

5.

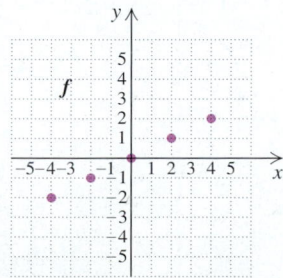

6.

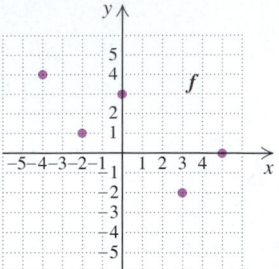

7.

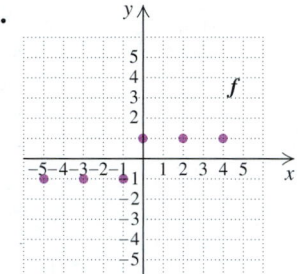

8.

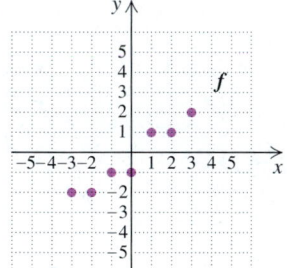

9.

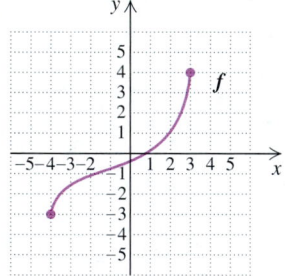

10.

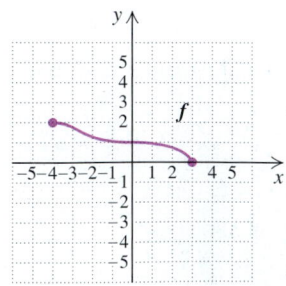

11.

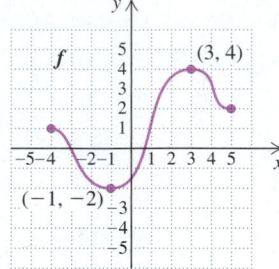

12.

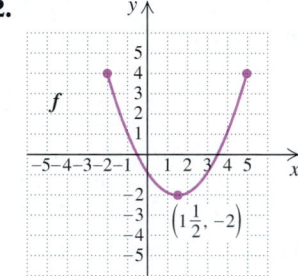

13.

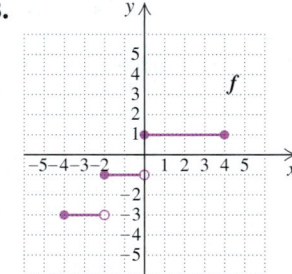

14.

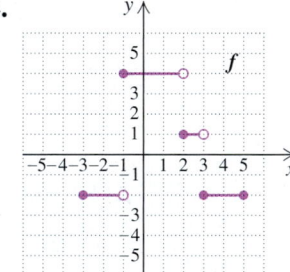

15.

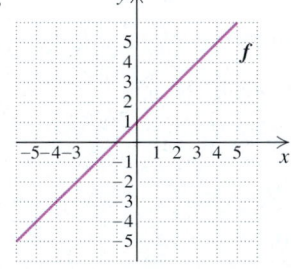

16.

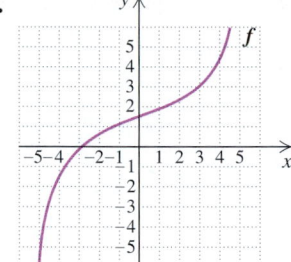

17.

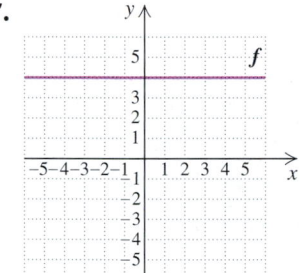

18.

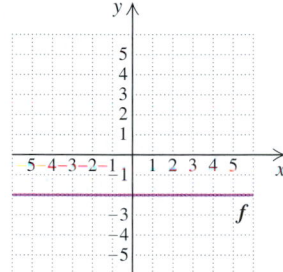

19.

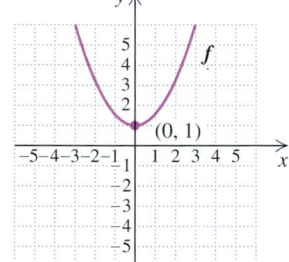

20.

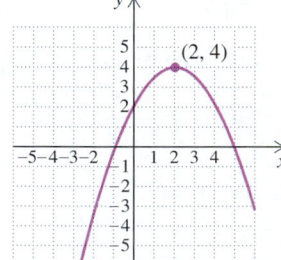

21.

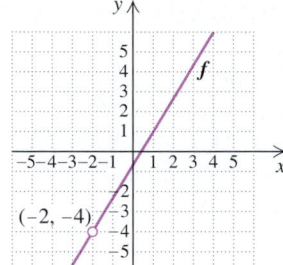

22.

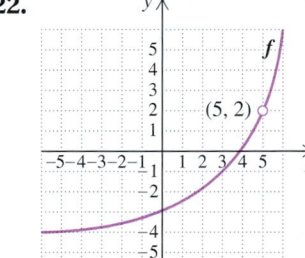

23.

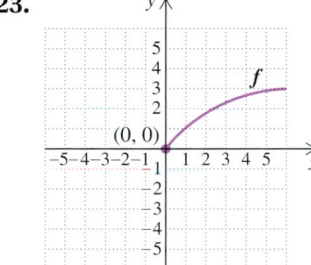

24.

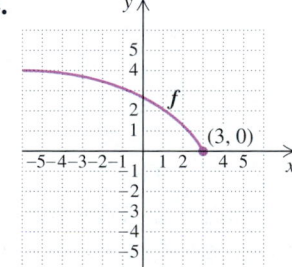

Find the domain of f.

25. $f(x) = \dfrac{5}{x - 3}$

26. $f(x) = \dfrac{7}{6 - x}$

27. $f(x) = \dfrac{3}{2x - 1}$

28. $f(x) = \dfrac{9}{4x + 3}$

29. $f(x) = 2x + 1$

30. $f(x) = x^2 + 3$

31. $f(x) = |5 - x|$

32. $f(x) = |3x - 4|$

33. $f(x) = \dfrac{5}{x^2 - 9}$

34. $f(x) = \dfrac{x}{x^2 - 2x + 1}$

35. $f(x) = x^2 - 9$

36. $f(x) = x^2 - 2x + 1$

37. $f(x) = \dfrac{2x - 7}{x^2 + 8x + 7}$

38. $f(x) = \dfrac{x + 5}{2x^2 - x - 3}$

39. *Records in the 400-m run.*   The record R for the 400-m run t years after 1930 is given by

$$R(t) = 46.8 - 0.075t.$$

What is the domain of the function?

40. *Records in the 1500-m run.*   The record R for the 1500-m run t years after 1930 is given by

$$R(t) = 3.85 - 0.0075t.$$

What is the domain of the function?

41. *Consumer demand.*   The amount A of coffee that consumers are willing to buy at price p is given by

$$A(p) = -2.5p + 26.5.$$

What is the domain of the function?

42. *Seller's supply.*   The amount A of coffee that suppliers are willing to supply at price p is given by

$$A(p) = 2p - 11.$$

What is the domain of the function?

43. *Pressure at sea depth.*   The pressure P, in atmospheres, at a depth d feet beneath the surface of the ocean is given by

$$P(d) = 0.03d + 1.$$

What is the domain of the function?

44. *Perimeter.*   The perimeter P of an equilateral triangle with sides of length s is given by

$$P(s) = 3s.$$

What is the domain of the function?

45. *Fireworks displays.*   The height h, in feet, of a "weeping willow" fireworks display, t seconds after having been launched from an 80-ft high rooftop, is given by

$$h(t) = -16t^2 + 64t + 80.$$

What is the domain of the function?

46. *Safety flares.*   The height h, in feet, of a safety flare, t seconds after having been launched from a height of 224 ft, is given by

$$h(t) = -16t^2 + 80t + 224.$$

What is the domain of the function?

Find the indicated function values for each function.

47. $f(x) = \begin{cases} x, & \text{if } x < 0, \\ 2x + 1, & \text{if } x \geq 0 \end{cases}$

a) $f(-5)$ **b)** $f(0)$ **c)** $f(10)$

48. $g(x) = \begin{cases} x - 5, & \text{if } x \leq 5, \\ 3x, & \text{if } x > 5 \end{cases}$

a) $g(0)$ **b)** $g(5)$ **c)** $g(6)$

49. $G(x) = \begin{cases} x - 5, & \text{if } x < -1, \\ x, & \text{if } -1 \leq x \leq 2, \\ x + 2, & \text{if } x > 2 \end{cases}$

a) $G(0)$ **b)** $G(2)$ **c)** $G(5)$

50. $F(x) = \begin{cases} 2x, & \text{if } x \leq 0, \\ x, & \text{if } 0 < x \leq 3, \\ -5x, & \text{if } x > 3 \end{cases}$

a) $F(-1)$ **b)** $F(3)$ **c)** $F(10)$

51. $f(x) = \begin{cases} x^2 - 10, & \text{if } x < -10, \\ x^2, & \text{if } -10 \leq x \leq 10, \\ x^2 + 10, & \text{if } x > 10 \end{cases}$

a) $f(-10)$ **b)** $f(10)$ **c)** $f(11)$

52. $f(x) = \begin{cases} 2x^2 - 3, & \text{if } x \leq 2, \\ x^2, & \text{if } 2 < x < 4, \\ 5x - 7, & \text{if } x \geq 4 \end{cases}$

a) $f(0)$ **b)** $f(3)$ **c)** $f(6)$

53. Explain why the domain of the function given by $f(x) = \dfrac{x + 3}{2}$ is $\mathbb{R}$, but the domain of the function given by $g(x) = \dfrac{2}{x + 3}$ is not $\mathbb{R}$.

54. Alayna asserts that for a function described by a set of ordered pairs, the range of the function will always have the same number of elements as there are ordered pairs. Is she correct? Why or why not?

SKILL MAINTENANCE

Graph.

55. $y = 2x - 3$

56. $y = x + 5$

Find the slope and the y-intercept of each line.

57. $y = \dfrac{2}{3}x - 4$

58. $y = -\dfrac{1}{4}x + 6$

59. $y = \dfrac{4}{3}x$

60. $y = -5x$

SYNTHESIS

61. Ethan states that $f(x) = \dfrac{x^2}{x}$ and $g(x) = x$ represent the same function. Is he correct? Why or why not?

62. Explain why the domain of a function can be viewed as the projection of its graph on the *x*-axis.

Sketch the graph of a function for which the domain and range are as given. Graphs may vary.

63. Domain: $\mathbb{R}$; range: $\mathbb{R}$

64. Domain: $\{3, 1, 4\}$; range: $\{0, 5\}$

65. Domain: $\{x \mid 1 \le x \le 5\}$; range: $\{y \mid 0 \le y \le 2\}$

66. Domain: $\{x \mid x$ is a real number *and* $x \ne 1\}$; range: $\{y \mid y$ is a real number *and* $y \ne -2\}$

For each graph of a function f, determine the domain and the range of f.

67.

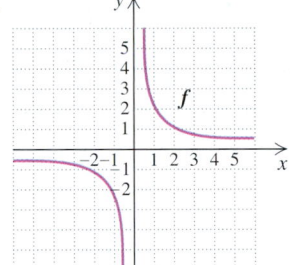

68.

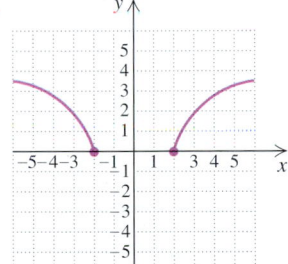

69.

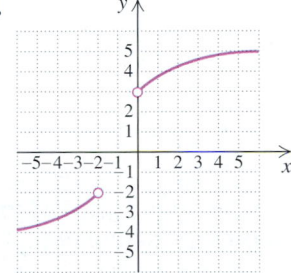

70.

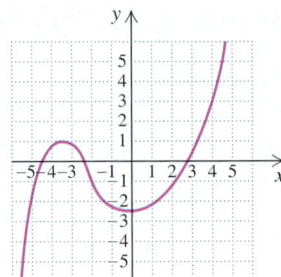

Graph each function on a grapher and estimate its domain and range from the graph.

71. $f(x) = |x - 3|$

72. $f(x) = |x| - 3$

73. $f(x) = \dfrac{3}{x - 2}$

74. $f(x) = \dfrac{-1}{x + 3}$

75. Use a grapher to estimate the range of the function in Exercise 45.

76. Use a grapher to estimate the range of the function in Exercise 46.

77.–82. *For Exercises 77–82, graph the functions given in each of Exercises 47–52, respectively.*

83. A grapher will interpret an expression like $x \ge 1$ as true or false, depending on the value of *x*. If the expression is true, the grapher assigns a value of 1 to the expression. If the expression is false, the grapher assigns a value of 0. To graph a piecewise-defined function using a grapher, multiply each part of the definition by its domain, using the TEST menu to enter the inequality symbol. Thus the function in Exercise 47 is entered as $y_1 = x(x < 0) + (2x + 1)(x \ge 0)$. Use a grapher in DOT mode to check your answers to Exercises 77–82.

Graphs of Functions

The Vertical-Line Test • Linear Functions • Nonlinear Functions

A function can be classified both by the type of equation that is used and by the type of graph it represents. In this section, we will graph a variety of functions that are described by different equations. Before doing so, we examine what types of graphs can represent functions.

The Vertical-Line Test

Recall that a function is a correspondence in which each member of the domain corresponds to *exactly* one member of the range. Thus the correspondence

$$\{(-1, 3), (4, -2), (-4, 3), (-1, 5)\}$$

is not a function because the member −1 of the domain corresponds to the members 3 and 5 of the range. Note that the point $(-1, 5)$ is directly above the point $(-1, 3)$.

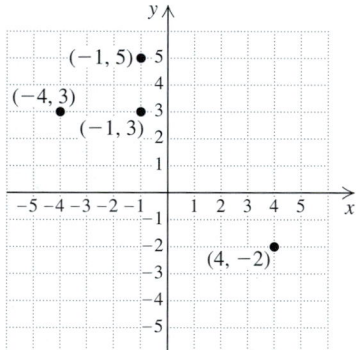

Any time two points, such as $(-1, 3)$ and $(-1, 5)$, lie on the same vertical line, the graph containing those points cannot represent a function. This observation is the basis of the *vertical-line test*.

> ### The Vertical-Line Test
>
> If it is possible for a vertical line to cross a graph more than once, then the graph is not the graph of a function.

Not a function. Three
y-values correspond
to one x-value.

A function

A function

Not a function. Two
y-values correspond
to one x-value.

Linear Functions

In Chapter 3, we graphed *linear equations* like

$$2x + 3y = 6, \qquad y = \frac{1}{2}x - 3, \qquad y = 4, \quad \text{and} \quad x = -2.$$

We can use the vertical-line test to determine which types of linear graphs represent functions. Consider the graphs of the equations listed above.

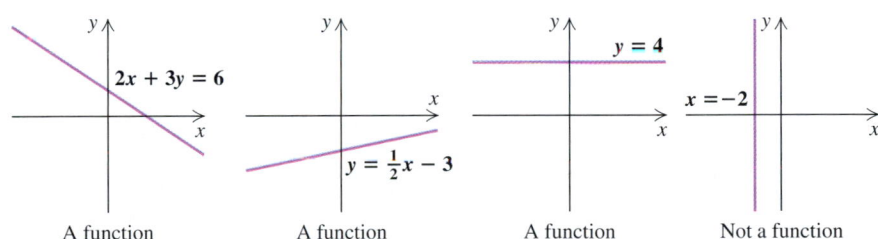

A function

A function

A function

Not a function

Any vertical line that passes through the graphs of $2x + 3y = 6$, $y = \frac{1}{2}x - 3$, and $y = 4$ will cross the graph only once. However, the vertical line through the point $(-2, 0)$ will cross the graph of $x = -2$ at *every* point. In general, *any* straight line that is not vertical is the graph of a function. A **linear function** is a function described by any linear equation whose graph is not vertical. A horizontal line represents a **constant function**.

> ### Linear Function
>
> A function described by an equation of the form $f(x) = mx + b$ is a *linear function*. Its graph is a straight line with slope m and y-intercept $(0, b)$.
>
> When $m = 0$, the function described by $f(x) = b$ is called a *constant function*. Its graph is a horizontal line through $(0, b)$.

E x a m p l e 1

Graph: $f(x) = 3x + 2$.

Solution   The notations

$$f(x) = 3x + 2 \quad \text{and} \quad y = 3x + 2$$

are often used interchangeably. The function notation emphasizes that the second coordinate in each ordered pair is determined by the first coordinate of that pair.

We graph $f(x) = 3x + 2$ in the same way that we would graph $y = 3x + 2$. The vertical axis can be labeled y or $f(x)$. We could use a table of values or, since this is a linear function, use the slope and the y-intercept to graph the function.

Since $f(x) = 3x + 2$ is in the form $f(x) = mx + b$, we can tell from the equation that the slope is 3, or $\frac{3}{1}$, and the y-intercept is $(0, 2)$. We plot $(0, 2)$ and go *up* 3 units and *to the right* 1 unit to determine another point on the line, $(1, 5)$. After we have sketched the line, a third point can be calculated as a check.

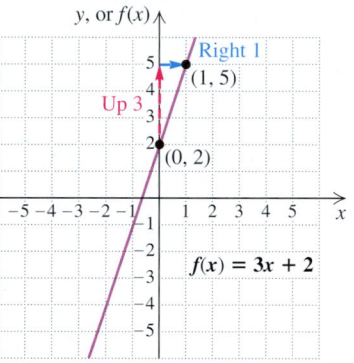

E x a m p l e 2

Graph: $f(x) = -3$.

Solution   This is a constant function. For every input x, the output is -3. The graph is a horizontal line.

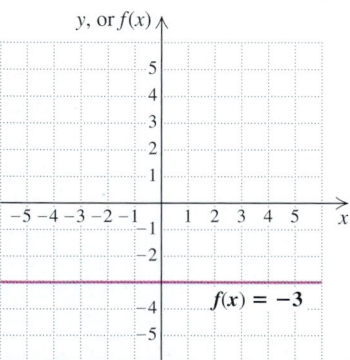

Linear functions are common in today's world.

E x a m p l e   3

Salvage value. Tyline Electric uses the function $S(t) = -700t + 3500$ to determine the *salvage value S(t)*, in dollars, of a color photocopier t years after its purchase.

a) What do the numbers -700 and 3500 signify?
b) How long will it take the copier to *depreciate* completely?
c) What is the domain of S?

Solution   Drawing, or at least visualizing, a graph can be useful here.

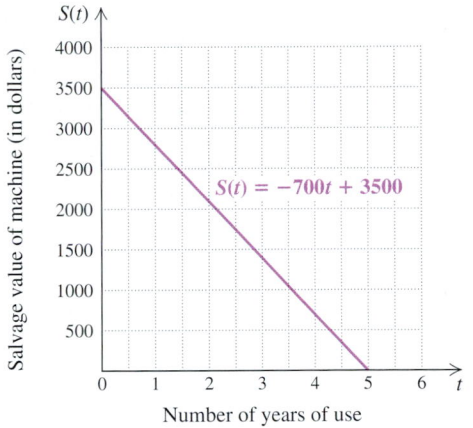

a) At time $t = 0$, we have $S(0) = -700 \cdot 0 + 3500 = 3500$. Thus the number 3500 signifies the original cost of the copier, in dollars.

This function is written in slope–intercept form. Since the output is measured in dollars and the input in years, the number -700 signifies that the value of the copier is declining at a rate of $700 per year.

b) The copier will have depreciated completely when its value drops to 0. To learn when this occurs, we determine when $S(t) = 0$:

$$S(t) = 0 \qquad \text{\color{magenta}A graph is not always available.}$$
$$-700t + 3500 = 0 \qquad \text{\color{magenta}Substituting } -700t + 3500 \text{ for } S(t)$$
$$-700t = -3500 \qquad \text{\color{magenta}Subtracting 3500 from both sides}$$
$$t = 5. \qquad \text{\color{magenta}Dividing both sides by } -700$$

The copier will have depreciated completely in 5 yr.

c) Neither the number of years of service nor the salvage value can be negative. In part (b) we found that after 5 yr the salvage value will have dropped to 0. Thus the domain of S is $\{t \mid 0 \leq t \leq 5\}$. The graph above serves as a visual check of this result.

Often we can formulate a linear function as a model for an application.

E x a m p l e   4

Cost projections. Cleartone Communications charges $50 for a cellular phone and $40 per month for calls made under its Call Anywhere plan. Formulate a mathematical model for the cost. Then use the model to determine the time required for the total cost to reach $250.

Solution

1. **Familiarize.** The problem describes a situation in which a monthly fee is charged after an initial purchase has been made. After 1 month of service, the total cost will be \$50 + \$40 = \$90. After 2 months, the total cost will be \$50 + \$40 · 2 = \$130. This can be generalized in a model if we let $C(t)$ represent the total cost, in dollars, for t months of service.

2. **Translate.** We rephrase and translate as follows:

Rephrasing: The total cost is the cost of the phone plus \$40 per month.

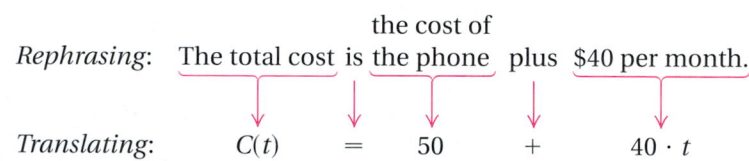

Translating: $C(t) = 50 + 40 \cdot t$

where $t \geq 0$ (since there cannot be a negative number of months).

3. **Carry out.** The model can be written $C(t) = 40t + 50$. To determine the time required for the total cost to reach \$250, we substitute 250 for $C(t)$ and solve for t:

$$C(t) = 40t + 50$$
$$250 = 40t + 50 \qquad \text{Substituting}$$
$$200 = 40t \qquad \text{Subtracting 50 from both sides}$$
$$5 = t. \qquad \text{Dividing both sides by 40}$$

We find that the total cost will reach \$250 in 5 months.

4. **Check.** We evaluate:

$$C(5) = 40 \cdot 5 + 50$$
$$= 200 + 50$$
$$= 250.$$

Our answer checks.

5. **State.** It takes 5 months for the total cost to reach \$250.

E x a m p l e   5

Tattoo removal.   In 1996, an estimated 275,000 Americans visited a doctor for tattoo removal. That figure was expected to grow to 410,000 in 2000 (*Source*: Mike Meyers, staff writer, Star-Tribune Newspaper of the Twin Cities Minneapolis–St. Paul, copyright 2000). Assuming constant growth since 1995, how many people will visit a doctor for tattoo removal in 2005?

Solution

1. **Familiarize.**   Constant growth indicates a constant rate of change, so a linear relationship can be assumed. If we let n represent the number of people, in thousands, who visit a doctor for tattoo removal and t the number of years since 1995, we can form the pairs $(1, 275)$ and $(5, 410)$. After choosing suitable scales on the two axes, we draw the graph.

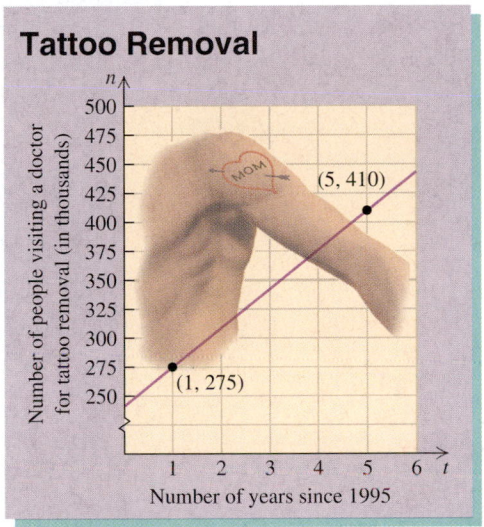

2. **Translate.**   To find an equation relating n and t, we first find the slope of the line. This corresponds to the *growth rate*:

$$m = \frac{410 \text{ thousand people} - 275 \text{ thousand people}}{5 \text{ years} - 1 \text{ year}}$$

$$= \frac{135 \text{ thousand people}}{4 \text{ years}}$$

$$= 33.75 \text{ thousand people per year.}$$

Next, we use the point–slope equation, discussed in Section 3.7, and solve for n:

$$n - 275 = 33.75(t - 1) \qquad \text{Writing point–slope form}$$
$$n - 275 = 33.75t - 33.75 \qquad \text{Using the distributive law}$$
$$n = 33.75t + 241.25. \qquad \text{Adding 275 to both sides}$$

3. **Carry out.**   Using function notation, we have

$$n(t) = 33.75t + 241.25.$$

To predict the number of people who will visit a doctor for tattoo removal in 2005, we find

$$n(10) = 33.75 \cdot 10 + 241.25 \qquad \text{2005 is 10 years from 1995.}$$
$$= 578.75. \qquad\qquad \text{This represents 578,750 people.}$$

4. **Check.** To check, we can repeat our calculations. We could also extend the graph to see if $(10, 578.75)$ appears to be on the line.

5. **State.** Assuming constant growth, there will be about 578,750 people visiting a doctor for tattoo removal in 2005.

Since $f(x) = mx + b$ can be evaluated for any choice of x, the domain of all linear functions is $\mathbb{R}$, the set of all real numbers.

The second coordinate of every ordered pair in a constant function $f(x) = b$ is the number b. The range of a constant function thus consists of one number, b. For a nonconstant linear function, the graph extends indefinitely both up and down, so the range is the set of all real numbers, or $\mathbb{R}$.

> ### Domain and Range of a Linear Function
> The domain of any linear function $f(x) = mx + b$ is
>
> $\{x \mid x \text{ is a real number}\}$, or $\mathbb{R}$.
>
> The range of any linear function $f(x) = mx + b$, $m \neq 0$, is
>
> $\{y \mid y \text{ is a real number}\}$, or $\mathbb{R}$.
>
> The range of any constant function $f(x) = b$ is $\{b\}$.

Example 6   Determine the domain and the range of each of the following functions.

a) f, where $f(x) = 2x - 10$ **b)** g, where $g(x) = 4$

Solution

a) Since $f(x) = 2x - 10$ describes a linear function, but not a constant function,

Domain of $f = \mathbb{R}$ and

Range of $f = \mathbb{R}$.

b) The function described by $g(x) = 4$ is a constant function. Thus,

Domain of $g = \mathbb{R}$ and

Range of $g = \{4\}$.

The graphs of nonlinear functions can get quite complex. We will now define several types of nonlinear functions and discuss some of their characteristics. The detailed study of their graphs appears later in this text or in other courses.

Nonlinear Functions

A function for which the graph is not a straight line is a **nonlinear function**. Some important types of nonlinear functions are described below.

Type of function	Description	Example
Absolute-value function	Described by an absolute-value equation	$f(x) = |x|$
Polynomial function	Described by a polynomial equation	$p(x) = x^3 - 4x^2 + 1$
Quadratic function	Described by a polynomial equation of degree 2 (a quadratic equation)	$q(x) = x^2 + 5x + 2$
Rational function	Described by a rational equation	$r(x) = \dfrac{x+1}{x-2}$

Note that linear and quadratic functions are special kinds of polynomial functions.

Example 7

State whether each equation describes a linear function, an absolute-value function, a general polynomial function, a quadratic function, or a rational function.

a) $f(x) = x^2 - 9$ **b)** $g(x) = \dfrac{3}{x}$

c) $h(x) = \dfrac{1}{4}x - 16$ **d)** $v(x) = 4x^4 - 13$

Solution

a) Since f is described by a polynomial equation of degree 2, f is a *quadratic function*.
b) Since g is described by a rational equation, g is a *rational function*.
c) The function h is described by a linear equation, so h is a *linear function*. Note that although $\frac{1}{4}$ is a fraction, there are no variables in a denominator.
d) Since v is described by a polynomial equation, v is a *polynomial function*.

Since the graphs of nonlinear functions are not straight lines, we usually need to calculate more than two or three points to determine the shape of the graph.

Example 8

Graph the function given by $f(x) = |x|$, and determine the domain and the range of f.

Solution   We calculate function values for several choices of x and list the results in a table.

$f(0) = |0| = 0,$
$f(1) = |1| = 1,$
$f(2) = |2| = 2,$
$f(-1) = |-1| = 1,$
$f(-2) = |-2| = 2$

x	$f(x) = |x|$	$(x, f(x))$
0	0	$(0, 0)$
1	1	$(1, 1)$
2	2	$(2, 2)$
-1	1	$(-1, 1)$
-2	2	$(-2, 2)$

Study Tip

Doing your homework faithfully is an important step to success in this course. Choose a time and place where you can focus on your work. Schedule homework as a part of your routine, doing it as soon as possible after each class. When you write your work, include each step in order to organize your thinking and avoid errors.

To check Example 8, we graph $y_1 = |x|$ (which is entered $y_1 = \text{abs}(x)$, using the NUM option of the MATH menu). Note that the graph appears without interruption for any piece of the x-axis that we examine.

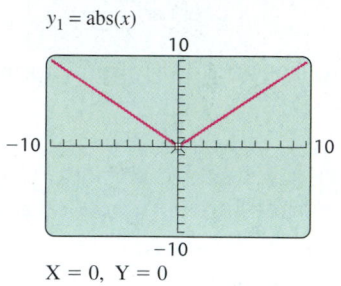

$y_1 = \text{abs}(x)$

$X = 0, \ Y = 0$

Note, too, that only positive y-values are used.

When we plot these points, we observe a pattern. The value of the function is 0 when x is 0. Function values increase both as x increases from 0 and as x decreases from 0. The graph of f is V-shaped, with the "point" of the V at the origin.

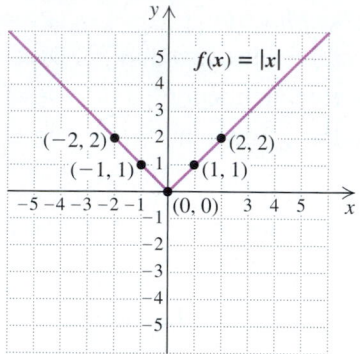

Because we can find the absolute value of any real number, we have

Domain of $f = \mathbb{R}$.

Because the absolute value of a number is never negative, we have

Range of $f = \{y \,|\, y \geq 0\}$.

Graphs of *polynomial functions* generally become more complex as the degree of the polynomial increases. In Chapter 11, we will study in greater detail the graphs of *quadratic functions*, or functions of the form

$$q(x) = ax^2 + bx + c, \, a \neq 0.$$

Since a polynomial can be evaluated for any real number, the domain of a polynomial is the set of all real numbers.

A *rational function* contains a variable in a denominator; thus its domain may be restricted. Division by zero is undefined, so any values of the variable that make a denominator 0 are not in the domain of the function.

Example 9   Determine the domain of f.

a) $f(x) = x^3 + 5x^2 - 4x + 1$ **b)** $f(x) = \dfrac{x^2 - 4}{x + 2}$

Solution

a) $f(x) = x^3 + 5x^2 - 4x + 1$ describes a polynomial function. The domain of any polynomial function is $\mathbb{R}$, so the domain of f is $\mathbb{R}$.

b) $f(x) = \dfrac{x^2 - 4}{x + 2}$ describes a rational function. Note that $f(x)$ is undefined for $x + 2 = 0$, or, equivalently, for $x = -2$. Thus the domain of $f = \{x \,|\, x \text{ is a real number and } x \neq -2\}$.

CONNECTING THE CONCEPTS

In this section, we have defined several types of functions and described some of their characteristics. They are listed together below for easy reference.

LINEAR FUNCTION

$f(x) = mx + b, m \neq 0$
Graph: straight line;
Domain: $\mathbb{R}$;
Range: $\mathbb{R}$

CONSTANT FUNCTION

$f(x) = b$
Graph: horizontal line;
Domain: $\mathbb{R}$;
Range: $\{b\}$

ABSOLUTE-VALUE FUNCTION

Described by an absolute-value equation
Domain: $\mathbb{R}$

QUADRATIC FUNCTION

$p(x) = ax^2 + bx + c, a \neq 0$
Domain: $\mathbb{R}$

POLYNOMIAL FUNCTION

Described by a polynomial equation
Domain: $\mathbb{R}$

RATIONAL FUNCTION

Described by a rational equation
Domain consists of all real numbers except those that make a denominator 0

FOR EXTRA HELP

Exercise Set 7.3

 Digital Video Tutor CD 5 Videotape 13

 InterAct Math

 Math Tutor Center

 MathXL

 MyMathLab.com

Determine whether each of the following is the graph of a function.

1.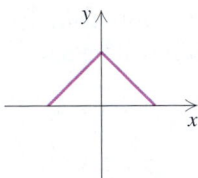

2.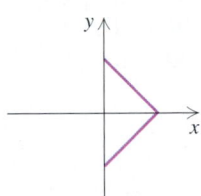

3.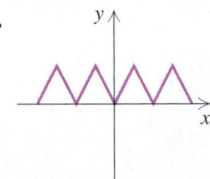

4.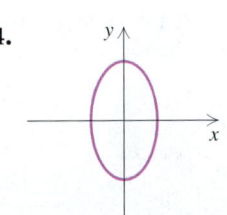

5.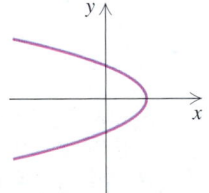

6.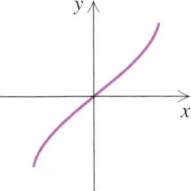

7.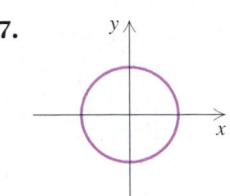

8.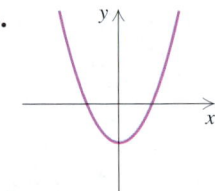

9.

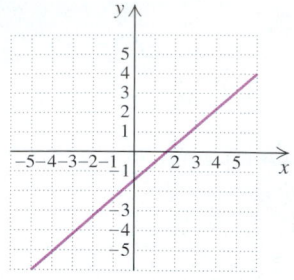

10.

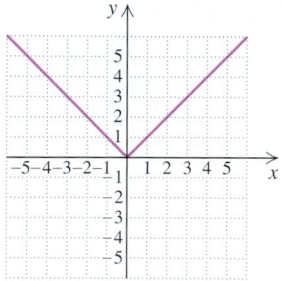

11.

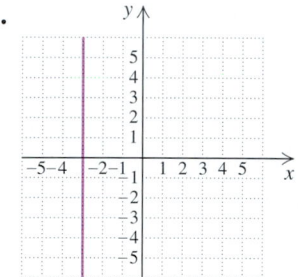

12.

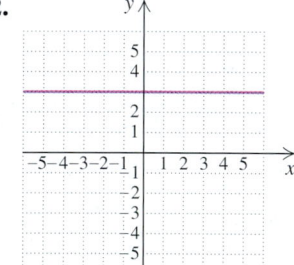

13. *Salvage value.* Green Glass Recycling uses the function given by $F(t) = -5000t + 90,000$ to determine the salvage value $F(t)$, in dollars, of a waste removal truck t years after it has been put into use.

   a) What do the numbers -5000 and $90,000$ signify?

   b) How long will it take the truck to depreciate completely?

   c) What is the domain of F?

14. *Salvage value.* Consolidated Shirt Works uses the function given by $V(t) = -2000t + 15,000$ to determine the salvage value $V(t)$, in dollars, of a color

separator t years after it has been put into use.

   a) What do the numbers -2000 and $15,000$ signify?

   b) How long will it take the machine to depreciate completely?

   c) What is the domain of V?

15. *Trade-in value.* The trade-in value of a Homelite snowblower can be determined using the function given by $v(n) = -150n + 900$. Here $v(n)$ is the trade-in value, in dollars, after n winters of use.

   a) What do the numbers -150 and 900 signify?

   b) When will the trade-in value of the snowblower be $300?

   c) What is the domain of v?

16. *Trade-in value.* The trade-in value of a John Deere riding lawnmower can be determined using the function given by $T(x) = -300x + 2400$. Here $T(x)$ is the trade-in value, in dollars, after x summers of use.

   a) What do the numbers -300 and 2400 signify?

   b) When will the value of the mower be $1200?

   c) What is the domain of T?

17. *Telephone charges.* Skytone Calling charges $50 for a telephone and $25 per month under its economy plan. Formulate a linear function to model the cost, and determine the amount of time required for the total cost to reach $150.

18. *Cellular phone charges.* The Cellular Connection charges $60 for a cellular phone and $40 per month under its economy plan. Determine the amount of time required for the total cost to reach $260.

19. *Hair growth.* In May, Tina had her hair cut in a 1-in.-long "buzz cut." Her hair grows at a rate of $\frac{1}{2}$ in. per month. Formulate a linear function to model the length of Tina's hair t months after the cut, and determine when her hair will be 3 in. long.

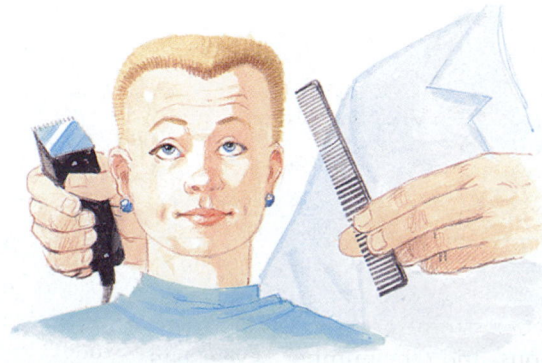

20. *Landscaping.*   On Saturday, Shelby Lawncare cut the lawn at Great Harrington Community College to a height of 2 in. Since then, the grass has grown at a rate of $\frac{1}{8}$ in. per day. Formulate a linear function to model the length of the lawn t days after having been cut, and determine when the grass will be $3\frac{1}{2}$ in. high.

21. *Cost of a taxi ride.*   A taxi ride in Pelham costs $2 plus $0.75 per mile traveled. Formulate a linear function to model the cost of a taxi ride for d miles, and determine the length of a taxi ride that cost $4.25.

22. *Natural gas demand.*   In 1960, the demand for natural gas was 20 quadrillion joules and was growing at a rate of $\frac{1}{5}$ quadrillion joules per year. Formulate a linear function to model the natural gas demand t years after 1960, and determine when the demand was 28 quadrillion joules.

In Exercises 23–28, assume that a constant rate of change exists for each model formed.

23. *Records in the 400-meter run.*   In 1930, the record for the 400-m run was 46.8 sec. In 1970, it was 43.8 sec. Let $R(t)$ represent the record in the 400-m run and t the number of years since 1930.

   a) Find a linear function that fits the data.
   b) Use the function of part (a) to predict the record in 2003; in 2006.
   c) When will the record be 40 sec?

24. *Records in the 1500-meter run.*   In 1930, the record for the 1500-m run was 3.85 min. In 1950, it was 3.70 min. Let $R(t)$ represent the record in the 1500-m run and t the number of years since 1930.

   a) Find a linear function that fits the data.
   b) Use the function of part (a) to predict the record in 2002; in 2006.
   c) When will the record be 3.1 min?

25. *PAC contributions.*   In 1992, Political Action Committees (PACs) contributed $178.6 million to congressional candidates. In 2000, the figure rose to $243.1 million (*Source*: Congressional Research Service and Federal Election Commission). Let $A(t)$ represent the amount of PAC contributions, in millions, and t the number of years since 1992.

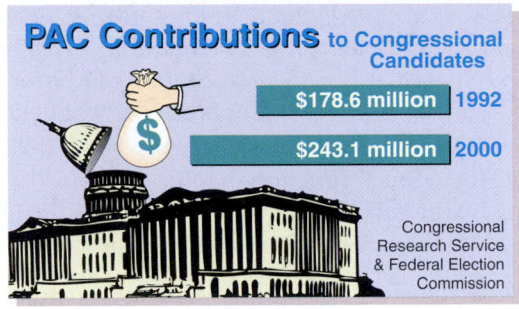

   a) Find a linear function $A(t)$ that fits the data.
   b) Use the function of part (a) to predict the amount of PAC contributions in 2008.

26. *Recycling.*   In 1993, Americans recycled 43.8 million tons of solid waste. In 1997, the figure grew to 60.8 million tons. (*Source*: *Statistical Abstract of the United States,* 1999) Let $N(t)$ represent the number of tons recycled, in millions, and t the number of years since 1993.

   a) Find a linear function that fits the data.
   b) Use the function of part (a) to predict the amount recycled in 2005.

27. *National Park land.*   In 1994, the National Park system consisted of about 74.9 million acres. By 1997, the figure had grown to 77.5 million acres. (*Source*: *Statistical Abstract of the United States,* 1999) Let $A(t)$ represent the amount of land in the National Park system, in millions of acres, t years after 1994.

   a) Find a linear function that fits the data.
   b) Use the function of part (a) to predict the amount of land in the National Park system in 2006.

28. *Life expectancy of females in the United States.* In 1990, the life expectancy of females was 78.8 yr. In 1997, it was 79.2 yr. (*Source: Statistical Abstract of the United States,* 1999) Let $E(t)$ represent life expectancy and t the number of years since 1990.

a) Find a linear function that fits the data.

b) Use the function of part (a) to predict the life expectancy of females in 2008.

Classify each function as a linear function, an absolute-value function, a quadratic function, another polynomial function, or a rational function, and determine the domain of the function.

29. $f(x) = \dfrac{1}{3}x - 7$

30. $g(x) = \dfrac{x}{x + 1}$

31. $p(x) = x^2 + x + 1$

32. $t(x) = |x - 7|$

33. $f(t) = \dfrac{12}{3t + 4}$

34. $g(n) = 15 - 10n$

35. $f(x) = 0.02x^4 - 0.1x + 1.7$

36. $f(a) = 2|a + 3|$

37. $f(x) = \dfrac{x}{2x - 5}$

38. $g(x) = \dfrac{2x}{3x - 4}$

39. $f(n) = \dfrac{4n - 7}{n^2 + 3n + 2}$

40. $h(x) = \dfrac{x - 5}{2x^2 - 2}$

41. $f(n) = 200 - 0.1n$

42. $g(t) = \dfrac{t^2 - 3t + 7}{8}$

Given the graph of each function, determine the range of f.

43.

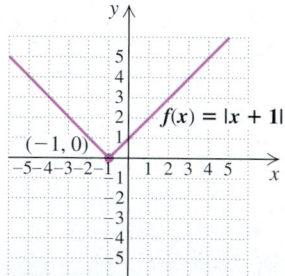

44.

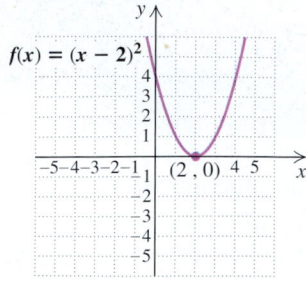

45.

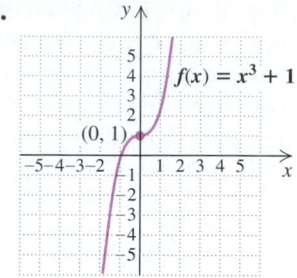

46.

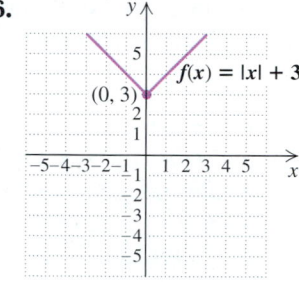

47.

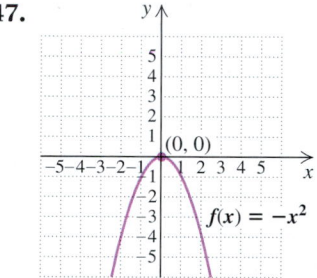

48.

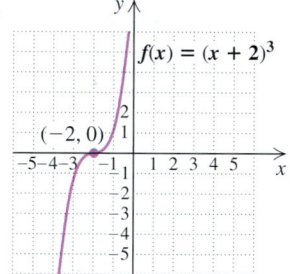

Graph each function and determine its domain and range.

49. $f(x) = x + 3$

50. $f(x) = 2x - 1$

51. $f(x) = -1$

52. $g(x) = 2$

53. $f(x) = |x| + 1$

54. $g(x) = |x - 3|$

55. $g(x) = x^2$

56. $f(x) = x^2 + 2$

57. Explain why the vertical-line test works.

58. Explain why the range of a constant function consists of only one number.

Perform the indicated operations.

59. $(x^2 + 2x + 7) + (3x^2 - 8)$

60. $(3x^3 - x^2 + x) - (x^3 + 2x - 7)$

61. $(2x + 1)(x - 7)$

62. $(x - 3)(x + 4)$

63. $(x^3 + x^2 - 4x + 7) - (3x^2 - x + 2)$

64. $(2x^2 + x - 3) + (x^3 + 7)$

SYNTHESIS

65. Is it possible for the domain of a rational function to be $\mathbb{R}$? Why or why not?

66. *Meteorology.*   Wind chill is a measure of how cold the wind makes you feel. Below are some measurements of wind chill for a 15-mph breeze. How can you tell from the data that a linear function will give an approximate fit?

Temperature	15-mph Wind Chill
30°	9°
25°	2°
20°	−5°
15°	−11°
10°	−18°
5°	−25°
0°	−31°

Source: National Oceanic & Atmospheric Administration, as reported in the *Burlington Free Press*, 17 January 1992.

67. *Engineering.*   Wind friction, or *air resistance,* increases with speed. At the top of the next column are some measurements made in a wind tunnel. Plot the data and explain why a linear function does or does not give an approximate fit.

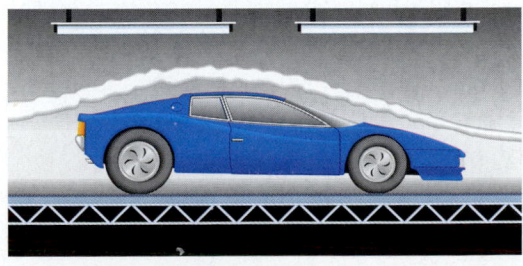

Velocity (in kilometers per hour)	Force of Resistance (in newtons)
10	3
21	4.2
34	6.2
40	7.1
45	15.1
52	29.0

For Exercises 68–71, assume that a linear equation models each situation.

68. *Depreciation of a computer.*   After 6 mos of use, the value of Pearl's computer had dropped to $900. After 8 mos, the value had gone down to $750. How much did the computer cost originally?

69. *Temperature conversion.*   Water freezes at 32° Fahrenheit and at 0° Celsius. Water boils at 212°F and at 100°C. What Celsius temperature corresponds to a room temperature of 70° F?

70. *Operating expenses.*   The total cost for operating Ming's Wings was $7500 after 4 mos and $9250 after 7 mos. Predict the total cost after 10 mos.

71. *Medical insurance.*   In 1993, health insurance companies collected $124.7 billion in premiums and paid out $103.6 billion in benefits. In 1996, they collected $137.1 billion in premiums and paid out $113.8 billion in benefits. What percentage of premiums will be paid out in benefits in 2005?

72. *Cost of a road call.*   Dave's Foreign Auto Village charges $35 for a road call plus $10 for each 15-min unit of time. A 32-min road call with Dave's costs the same as a 44-min road call. Thus a linear graph is not a precise representation of the situation. Draw a graph with a series of "steps" that more accurately reflects the situation.

73. *Parking fees.*   Karla's Parking charges $3.00 to park plus 50¢ for each 15-min unit of time. It costs as much to park at Karla's for 16 min as it does for 29 min. Thus a linear graph is not a precise representation of the situation. Draw a graph with a series of "steps" that more accurately reflects the situation.

Given that $f(x) = mx + b$, classify each of the following as true or false.

74. $f(c + d) = f(c) + f(d)$

75. $f(cd) = f(c)f(d)$

76. $f(kx) = kf(x)$

77. $f(c - d) = f(c) - f(d)$

78. For a linear function g, $g(3) = -5$ and $g(7) = -1$.
  a) Find an equation for g.
  b) Find $g(-2)$.
  c) Find a such that $g(a) = 75$.

 79. When several data points are available and they appear to be nearly collinear, a procedure known as *linear regression* can be used to find an equation for the line that most closely fits the data.
  a) Use a grapher with a LINEAR REGRESSION option and the table that follows to find a linear function that predicts a woman's life expectancy as a function of the year in which she was born. Compare this with the answer to Exercise 28.
  b) Predict the life expectancy in 2008 and compare your answer with the answer to Exercise 28. Which answer seems more reliable? Why?

Life Expectancy of Women

Year, x	Life Expectancy, y (in years)
1920	54.6
1930	61.6
1940	65.2
1950	71.1
1960	73.1
1970	74.7
1980	77.5
1990	78.8
1999	79.2

Sources: Statistical Abstract of the United States and The World Almanac 1999.

80. *Cost of a speeding ticket.*   The following penalty schedule is used to determine the cost of a speeding ticket in certain states. Use this schedule to graph the cost of a speeding ticket as a function of the number of miles per hour over the limit that a driver is going.

The Algebra of Functions

7.4

**The Sum, Difference, Product, or Quotient of Two Functions •
Domains and Graphs**

We now return to the idea of a function as a machine and examine four ways in which functions can be combined.

The Sum, Difference, Product, or Quotient of Two Functions

Suppose that a is in the domain of two functions, f and g. The input a is paired with $f(a)$ by f and with $g(a)$ by g. The outputs can then be added to get $f(a) + g(a)$.

E x a m p l e   1

Let $f(x) = x + 4$ and $g(x) = x^2 + 1$. Find $f(2) + g(2)$.

Solution   We visualize two function machines. Because 2 is in the domain of each function, we can compute $f(2)$ and $g(2)$.

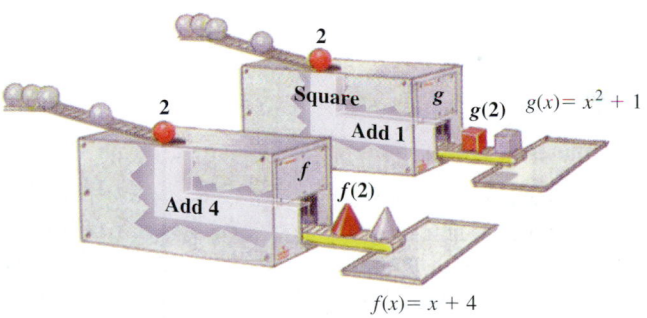

Since

$$f(2) = 2 + 4 = 6 \quad \text{and} \quad g(2) = 2^2 + 1 = 5,$$

we have

$$f(2) + g(2) = 6 + 5 = 11.$$

In Example 1, suppose that we were to write $f(x) + g(x)$ as $(x + 4) + (x^2 + 1)$, or $f(x) + g(x) = x^2 + x + 5$. This could then be regarded as a "new" function. The notation $(f + g)(x)$ is generally used to denote a function formed in this manner. Similar notations exist for subtraction, multiplication, and division of functions.

> **The Algebra of Functions**
>
> If f and g are functions and x is in the domain of both functions, then:
>
> **1.** $(f + g)(x) = f(x) + g(x)$;
> **2.** $(f - g)(x) = f(x) - g(x)$;
> **3.** $(f \cdot g)(x) = f(x) \cdot g(x)$;
> **4.** $(f/g)(x) = f(x)/g(x)$, provided $g(x) \neq 0$.

E x a m p l e   2

For $f(x) = x^2 - x$ and $g(x) = x + 2$, find the following.

a) $(f + g)(3)$ **b)** $(f - g)(x)$ and $(f - g)(-1)$
c) $(f/g)(x)$ and $(f/g)(-4)$ **d)** $(f \cdot g)(3)$

Solution

a) Since $f(3) = 3^2 - 3 = 6$ and $g(3) = 3 + 2 = 5$, we have

$$(f + g)(3) = f(3) + g(3)$$
$$= 6 + 5 \qquad \text{Substituting}$$
$$= 11.$$

Alternatively, we could first find $(f + g)(x)$:

$$(f + g)(x) = f(x) + g(x)$$
$$= x^2 - x + x + 2$$
$$= x^2 + 2. \qquad \text{Combining like terms}$$

Thus,

$$(f + g)(3) = 3^2 + 2 = 11. \qquad \text{Our results match.}$$

b) We have

$$(f - g)(x) = f(x) - g(x)$$
$$= x^2 - x - (x + 2) \qquad \text{Substituting}$$
$$= x^2 - 2x - 2. \qquad \begin{array}{l}\text{Removing parentheses and}\\ \text{combining like terms}\end{array}$$

Thus,

$$(f - g)(-1) = (-1)^2 - 2(-1) - 2 \qquad \begin{array}{l}\text{Using } (f - g)(x) \text{ is faster than}\\ \text{using } f(x) - g(x).\end{array}$$
$$= 1. \qquad \text{Simplifying}$$

c) We have

$$(f/g)(x) = f(x)/g(x)$$
$$= \frac{x^2 - x}{x + 2}. \qquad \text{We assume that } x \neq -2.$$

Thus,

$$(f/g)(-4) = \frac{(-4)^2 - (-4)}{-4 + 2} \qquad \text{Substituting}$$
$$= \frac{20}{-2} = -10.$$

Study Tip

This text was designed to help you study. For example, the examples are solved step-by-step, with substitutions highlighted in red. Also, important concepts are set aside in boxes or in bold print.

d) Using our work in part (a), we have

$$(f \cdot g)(3) = f(3) \cdot g(3)$$
$$= 6 \cdot 5$$
$$= 30.$$

Alternatively, we could first find $(f \cdot g)(x)$:

$$(f \cdot g)(x) = f(x) \cdot g(x)$$
$$= (x^2 - x)(x + 2)$$
$$= x^3 + x^2 - 2x. \qquad \text{Multiplying and combining like terms}$$

Then

$$(f \cdot g)(3) = 3^3 + 3^2 - 2 \cdot 3$$
$$= 27 + 9 - 6$$
$$= 30.$$

Domains and Graphs

Although applications involving products and quotients of functions rarely appear in newspapers, situations involving sums or differences of functions often do appear in print. For example, the following graphs are similar to those published by the California Department of Education to promote breakfast programs in which students eat a balanced meal of fruit or juice, toast or cereal, and 2% or whole milk. The combination of carbohydrate, protein, and fat gives a sustained release of energy, delaying the onset of hunger for several hours.

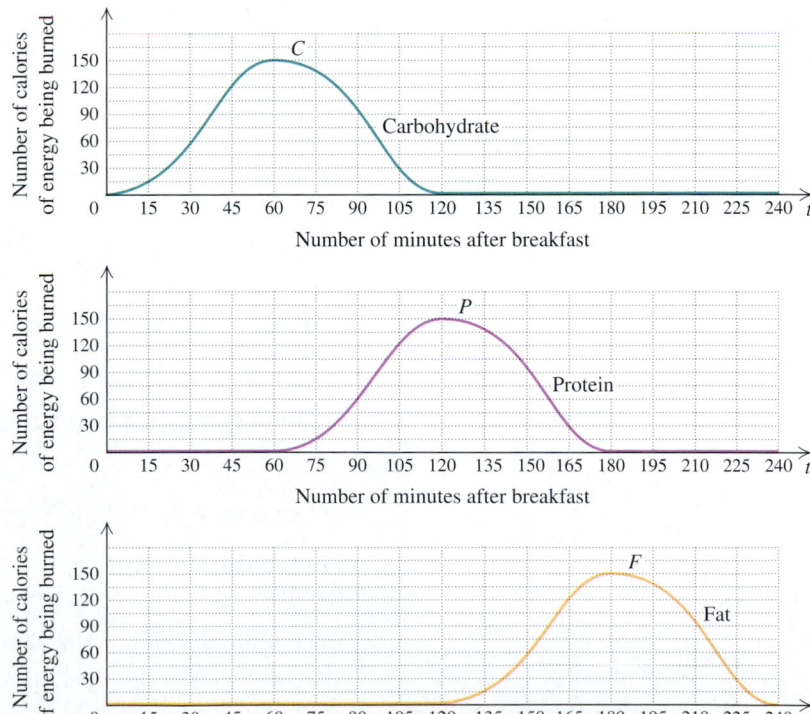

When the three graphs are superimposed, and the calorie expenditures added, it becomes clear that a balanced meal results in a steady, sustained supply of energy.

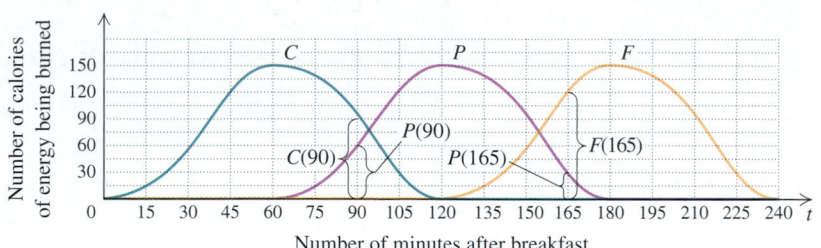

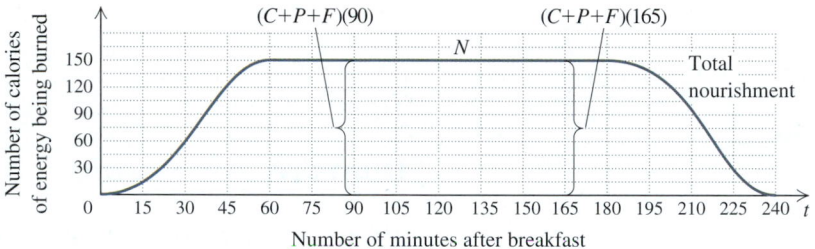

Note that for $t > 120$, we have $C(t) = 0$; for $t < 60$ or $t > 180$, we have $P(t) = 0$; and for $t < 120$, we have $F(t) = 0$. For any point on this last graph, we have

$$N(t) = (C + P + F)(t) = C(t) + P(t) + F(t).$$

To find $(f + g)(a)$, $(f - g)(a)$, $(f \cdot g)(a)$, or $(f/g)(a)$, we must first be able to find $f(a)$ and $g(a)$. Thus we need to ensure that a is in the domain of both f and g.

Example 3   Let

$$f(x) = \frac{5}{x} \quad \text{and} \quad g(x) = \frac{2x - 6}{x + 1}.$$

Find the domain of $f + g$, the domain of $f - g$, and the domain of $f \cdot g$.

Solution   Note that because division by 0 is undefined, we have

Domain of $f = \{x \mid x$ is a real number *and* $x \neq 0\}$

and

Domain of $g = \{x \mid x$ is a real number *and* $x \neq -1\}$.

In order to find $f(a) + g(a)$, $f(a) - g(a)$, or $f(a) \cdot g(a)$, we must know that a is in *both* of the above domains. Thus,

Domain of $f + g =$ Domain of $f - g =$ Domain of $f \cdot g$

$= \{x \mid x$ is a real number *and* $x \neq 0$ *and* $x \neq -1\}$.

Suppose in Example 3 that we want to find $(f/g)(3)$. Finding $f(3)$ and $g(3)$ poses no problem:

$$f(3) = \frac{5}{3} \quad \text{and} \quad g(3) = \frac{2 \cdot 3 - 6}{3 + 1} = 0;$$

but then

$$(f/g)(3) = f(3)/g(3) = \tfrac{5}{3} \; / \; 0 \; . \qquad \textcolor{magenta}{\text{Division by 0 is undefined.}}$$

Thus, although 3 is in the domain of both f and g, it is not in the domain of f/g.

Determining the Domain

The domain of $f + g$, $f - g$, or $f \cdot g$ is the set of all values common to the domains of f and g.

The domain of f/g is the set of all values common to the domains of f and g, excluding any values for which $g(x)$ is 0.

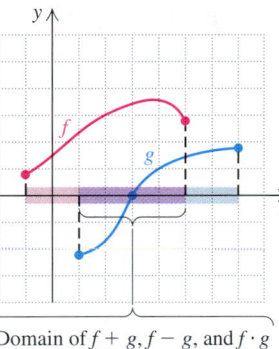

Domain of $f + g$, $f - g$, and $f \cdot g$

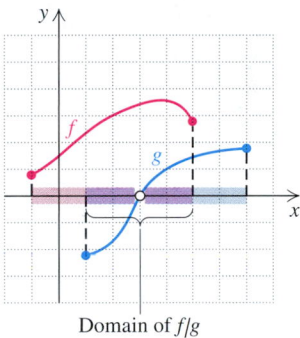

Domain of f/g

Example 4

Given $f(x) = 1/x$ and $g(x) = 2x - 7$, find the domains of $f + g$, $f - g$, $f \cdot g$, and f/g.

Solution   The domain of f is $\{x \mid x \neq 0\}$ or $\{x \mid x$ is a real number *and* $x \neq 0\}$. The domain of g is $\mathbb{R}$. Thus the domains of $f + g$, $f - g$, and $f \cdot g$ are the set of all elements common to both the domain of f and the domain of g. We have

the domain of $f + g =$ the domain of $f - g =$ the domain of $f \cdot g$

$$= \{x \mid x \text{ is a real number } and \ x \neq 0\}.$$

The domain of f/g is $\{x \mid x$ is a real number *and* $x \neq 0\}$, with the additional restriction that $g(x) \neq 0$. To determine what x-values would make $g(x) = 0$, we solve:

$$2x - 7 = 0 \qquad \textcolor{magenta}{\text{Replacing } g(x) \text{ with } 2x - 7}$$
$$x = \tfrac{7}{2}.$$

Since $g(x) = 0$ for $x = \frac{7}{2}$,

the domain of $f/g = \left\{x \mid x \text{ is a real number } and \ x \neq 0 \text{ and } x \neq \tfrac{7}{2}\right\}.$

technology connection

A partial check of Example 4 can be performed by setting up a table so the TABLE MINIMUM is 0 and the increment of change ($\triangle$Tbl) is 0.7. (Other choices, like 0.1, will also work.) Next, we let $y_1 = 1/x$ and $y_2 = 2x - 7$. Using the Y-VARS key to write $y_3 = y_1 + y_2$ and $y_4 = y_1/y_2$, we can create the table of values shown here. Note that when x is 3.5, a value for y_3 can be found, but y_4 is undefined. When setting up these functions, you may wish to enter y_1 and y_2 without "selecting" either. Otherwise, the table's columns must be scrolled to display y_3 and y_4.

X	Y3	Y4
0	ERROR	ERROR
.7	−4.171	−.2551
1.4	−3.486	−.1701
2.1	−2.324	−.1701
2.8	−1.043	−.2551
3.5	.28571	ERROR
4.2	1.6381	.17007
X = 0		

Use a similar approach to partially check Example 3.

Division by 0 is not the only condition that can force restrictions on the domain of a function. In Chapter 10, we will examine functions similar to that given by $f(x) = \sqrt{x}$, for which the concern is taking the square root of a negative number.

Exercise Set 7.4

Let $f(x) = -3x + 1$ and $g(x) = x^2 + 2$. Find the following.

1. $f(2) + g(2)$

2. $f(-1) + g(-1)$

3. $f(5) - g(5)$

4. $f(4) - g(4)$

5. $f(-1) \cdot g(-1)$

6. $f(-2) \cdot g(-2)$

7. $f(-4)/g(-4)$

8. $f(3)/g(3)$

9. $g(1) - f(1)$

10. $g(2)/f(2)$

11. $(f + g)(x)$

12. $(g - f)(x)$

Let $F(x) = x^2 - 2$ and $G(x) = 5 - x$. Find the following.

13. $(F + G)(x)$

14. $(F + G)(a)$

15. $(F + G)(-4)$

16. $(F + G)(-5)$

17. $(F - G)(3)$

18. $(F - G)(2)$

19. $(F \cdot G)(-3)$

20. $(F \cdot G)(-4)$

21. $(F/G)(x)$

22. $(G - F)(x)$

23. $(F/G)(-2)$

24. $(F/G)(-1)$

The following graph shows the number of women, in millions, who had a child in the last year. Here $W(t)$ represents the number of women under 30 who gave birth in year t, $R(t)$ the number of women 30 and older who gave birth in year t, and $N(t)$ the total number of women who gave birth in year t.

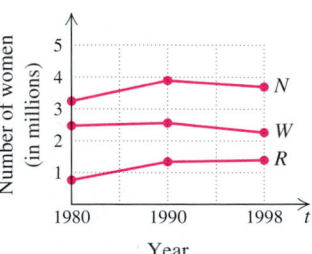

25. Use estimates of $R(1980)$ and $W(1980)$ to estimate $N(1980)$.

26. Use estimates of $R(1990)$ and $W(1990)$ to estimate $N(1990)$.

27. Which group of women was responsible for the drop in the number of births from 1990 to 1998?

28. Which group of women was responsible for the rise in the number of births from 1980 to 1990?

Often function addition is represented by stacking the individual functions directly on top of each other. The graph below indicates how the three major airports servicing New York City have been utilized. The braces indicate the values of the individual functions.

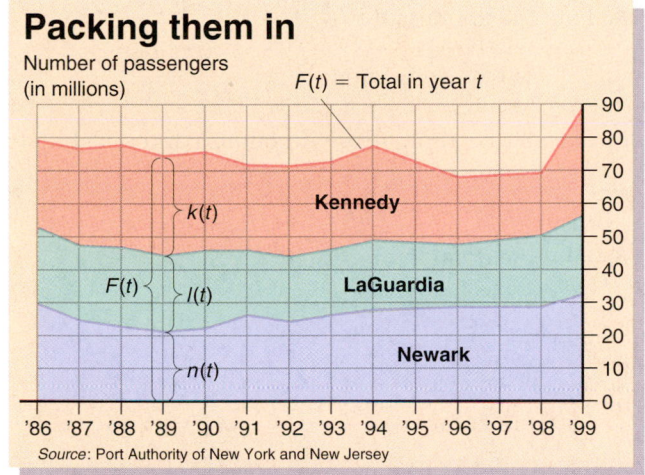

Packing them in

Number of passengers (in millions)

$F(t)$ = Total in year t

Source: Port Authority of New York and New Jersey

29. Estimate $(n + l)('98)$. What does it represent?

30. Estimate $(k + l)('98)$. What does it represent?

31. Estimate $(k - l)('94)$. What does it represent?

32. Estimate $(k - n)('94)$. What does it represent?

33. Estimate $(n + l + k)('99)$. What does it represent?

34. Estimate $(n + l + k)('98)$. What does it represent?

For each pair of functions f and g, determine the domain of the sum, difference, and product of the two functions.

35. $f(x) = x^2$,
$g(x) = 7x - 4$

36. $f(x) = 5x - 1$,
$g(x) = 2x^2$

37. $f(x) = \dfrac{1}{x - 3}$,

$g(x) = 4x^3$

38. $f(x) = 3x^2$,

$g(x) = \dfrac{1}{x - 9}$

39. $f(x) = \dfrac{2}{x}$,
$g(x) = x^2 - 4$

40. $f(x) = x^3 + 1$,
$g(x) = \dfrac{5}{x}$

41. $f(x) = x + \dfrac{2}{x - 1}$,

$g(x) = 3x^3$

42. $f(x) = 9 - x^2$,

$g(x) = \dfrac{3}{x - 6} + 2x$

43. $f(x) = \dfrac{3}{x - 2}$,

$g(x) = \dfrac{5}{4 - x}$

44. $f(x) = \dfrac{5}{x - 3}$,

$g(x) = \dfrac{1}{x - 2}$

For each pair of functions f and g, determine the domain of f/g.

45. $f(x) = x^4$,
$g(x) = x - 3$

46. $f(x) = 2x^3$,
$g(x) = 5 - x$

47. $f(x) = 3x - 2$,
$g(x) = 2x - 8$

48. $f(x) = 5 + x$,
$g(x) = 6 - 2x$

49. $f(x) = \dfrac{3}{x - 4}$,

$g(x) = 5 - x$

50. $f(x) = \dfrac{1}{2 - x}$,

$g(x) = 7 - x$

51. $f(x) = \dfrac{2x}{x + 1}$,

$g(x) = 2x + 5$

52. $f(x) = \dfrac{7x}{x - 2}$,

$g(x) = 3x + 7$

For Exercises 53–60, consider the functions F and G as shown.

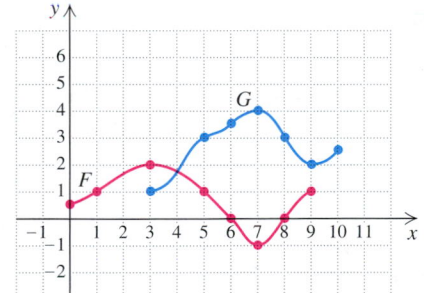

53. Determine $(F + G)(5)$ and $(F + G)(7)$.

54. Determine $(F \cdot G)(6)$ and $(F \cdot G)(9)$.

55. Determine $(G - F)(7)$ and $(G - F)(3)$.

56. Determine $(F/G)(3)$ and $(F/G)(7)$.

57. Find the domains of F, G, $F + G$, and F/G.

58. Find the domains of $F - G$, $F \cdot G$, and G/F.

59. Graph $F + G$.

60. Graph $G - F$.

*In the following graph, W(t) represents the number of gallons of whole milk, L(t) the number of gallons of low-fat milk, and S(t) the number of gallons of skim milk consumed by the average American in year t.**

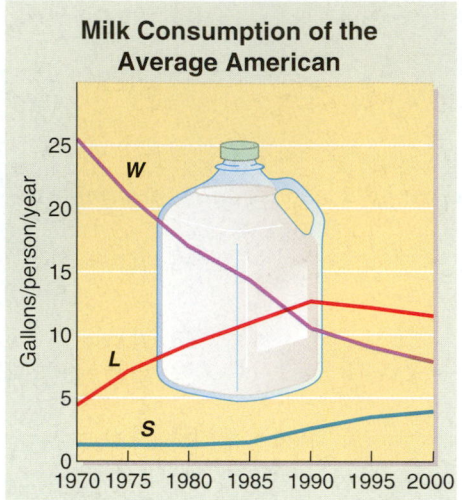

Milk Consumption of the Average American

61. Explain in words what $(W - S)(t)$ represents and what it would mean to have $(W - S)(t) < 0$.

62. Consider $(W + L + S)(t)$ and explain why you feel that total milk consumption per person did or did not change over the years 1970–1998.

SKILL MAINTENANCE

Solve.

63. $4x - 7y = 8$, for x

64. $3x - 8y = 5$, for y

65. $5x + 2y = -3$, for y

66. $6x + 5y = -2$, for x

Translate each of the following. Do not solve.

67. Five more than twice a number is 49.

68. Three less than half of some number is 57.

69. The sum of two consecutive integers is 145.

70. The difference between a number and its opposite is 20.

SYNTHESIS

71. If $f(x) = c$, where c is some positive constant, describe how the graphs of $y = g(x)$ and $y = (f + g)(x)$ will differ.

72. Examine the graphs following Example 2 and explain how they might be modified to represent the absorption of 200 mg of Advil® taken four times a day.

73. Find the domain of f/g, if
$$f(x) = \frac{3x}{2x + 5} \quad \text{and} \quad g(x) = \frac{x^4 - 1}{3x + 9}.$$

74. Find the domain of F/G, if
$$F(x) = \frac{1}{x - 4} \quad \text{and} \quad G(x) = \frac{x^2 - 4}{x - 3}.$$

75. Sketch the graph of two functions f and g such that the domain of f/g is
$$\{x \,|\, -2 \le x \le 3 \text{ and } x \ne 1\}.$$

76. Find the domain of m/n, if
$$m(x) = 3x \text{ for } -1 < x < 5$$
and
$$n(x) = 2x - 3.$$

77. Find the domains of $f + g$, $f - g$, $f \cdot g$, and f/g, if
$$f = \{(-2, 1), (-1, 2), (0, 3), (1, 4), (2, 5)\}$$
and
$$g = \{(-4, 4), (-3, 3), (-2, 4), (-1, 0), (0, 5), (1, 6)\}.$$

78. For f and g as defined in Exercise 77, find $(f + g)(-2)$, $(f \cdot g)(0)$, and $(f/g)(1)$.

79. Write equations for two functions f and g such that the domain of $f + g$ is
$$\{x \,|\, x \text{ is a real number and } x \ne -2 \text{ and } x \ne 5\}.$$

80. Let $y_1 = 2.5x + 1.5$, $y_2 = x - 3$, and $y_3 = y_1/y_2$. Depending on whether the CONNECTED or DOT mode is used, the graph of y_3 appears as follows.

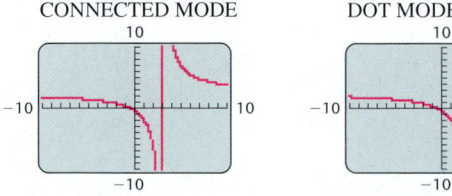
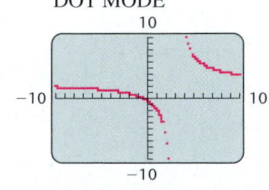

CONNECTED MODE   DOT MODE

Use algebra to determine which graph more accurately represents y_3.

81. Using the window $[-5, 5, -1, 9]$, graph $y_1 = 5$, $y_2 = x + 2$, and $y_3 = \sqrt{x}$. Then predict what shape the graphs of $y_1 + y_2$, $y_1 + y_3$, and $y_2 + y_3$ will take. Use a grapher to check each prediction.

82. Use the TABLE feature on a grapher to check your answers to Exercises 37, 43, 45, and 51. (See the Technology Connection on p. 454.)

C O R N E R

COLLABORATIVE

Time On Your Hands

Focus: The algebra of functions

Time: 10–15 minutes

Group size: 2–3

The graph and data at right chart the average retirement age $R(x)$ and life expectancy $E(x)$ of U.S. citizens in year x.

ACTIVITY

1. Working as a team, perform the appropriate calculations and then graph $E - R$.
2. What does $(E - R)(x)$ represent? In what fields of study or business might the function $E - R$ prove useful?
3. Should E and R really be calculated separately for men and women? Why or why not?

4. What advice would you give to someone considering early retirement?

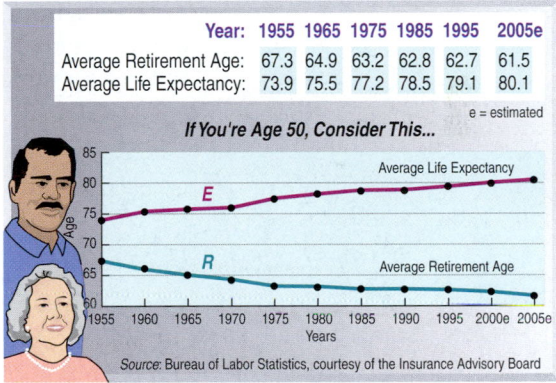

Year:	1955	1965	1975	1985	1995	2005e
Average Retirement Age:	67.3	64.9	63.2	62.8	62.7	61.5
Average Life Expectancy:	73.9	75.5	77.2	78.5	79.1	80.1

e = estimated

If You're Age 50, Consider This...

Source: Bureau of Labor Statistics, courtesy of the Insurance Advisory Board

Variation and Problem Solving

7.5

Direct Variation • Inverse Variation • Joint and Combined Variation

Many problems lead to equations of the form $y = kx$ or $y = k/x$, for some constant k. Such equations are called *equations of variation.*

Direct Variation

A bicycle tour is traveling at a rate of 15 km/h. In 1 hr, it goes 15 km. In 2 hr, it goes 30 km. In 3 hr, it goes 45 km, and so on. In the graph below, we use the number of hours as the first coordinate and the number of kilometers traveled as the second coordinate: $(1, 15), (2, 30), (3, 45), (4, 60)$, and so on. Note that the second coordinate is always 15 times the first.

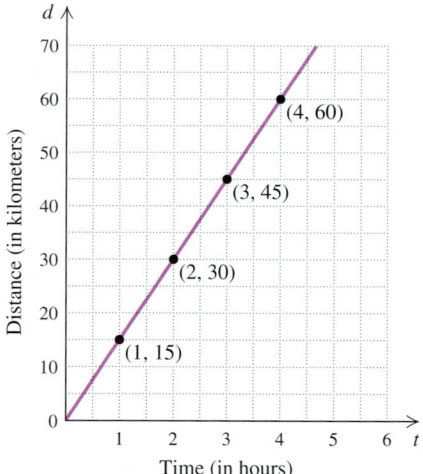

In this example, distance is a constant multiple of time, so we say that there is **direct variation** and that distance **varies directly** as time. The **equation of variation** is $d = 15t$. Using function notation, we write $d(t) = 15t$, to indicate that distance is a function of time.

> ### Direct Variation
>
> When a situation translates to an equation of the form $y = kx$, or $f(x) = kx$, with k a nonzero constant, we say that y *varies directly* as x. The equation $y = kx$ is called an *equation of direct variation.*

In direct variation with $k > 0$, as one variable increases, the other variable increases as well.

The terminologies

"*y* varies as *x*,"

"*y* is directly proportional to *x*," and

"*y* is proportional to *x*"

also imply direct variation and are used in many situations. The constant *k* is called the **constant of proportionality** or the **variation constant**. It can be found if one pair of values of *x* and *y* is known. Once *k* is known, other pairs can be determined.

_E x a m p l e   1_

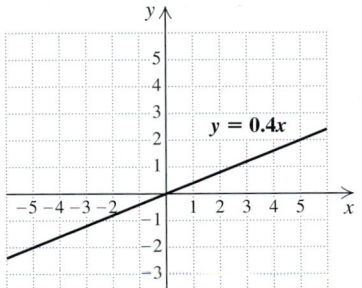

A visualization of Example 1

If *y* varies directly as *x* and $y = 2$ when $x = 5$, find the equation of variation.

Solution   We substitute to find *k*:

$$y = kx$$
$$2 = k \cdot 5 \qquad \text{Substituting to solve for } k$$
$$\tfrac{2}{5} = k, \quad \text{or} \quad k = 0.4. \qquad \text{Dividing both sides by 5}$$

Thus the equation of variation is $y = 0.4x$. A visualization of the situation is shown at left.

From these last two graphs, we see that when *y* varies directly as *x*, the constant of proportionality is also the slope of the associated graph—the rate at which *y* changes with respect to *x*.

_E x a m p l e   2_

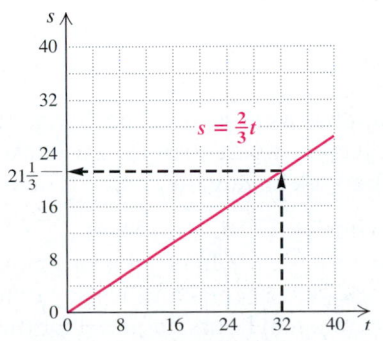

A visualization of Example 2

Find an equation in which *s* varies directly as *t* and $s = 10$ when $t = 15$. Then find the value of *s* when $t = 32$.

Solution   We have

$$s = kt \qquad \text{We know that } s \text{ varies directly as } t.$$
$$10 = k \cdot 15 \qquad \text{Substituting 10 for } s \text{ and 15 for } t$$
$$\tfrac{10}{15} = k, \quad \text{or} \quad k = \tfrac{2}{3}. \qquad \text{Solving for } k$$

Thus the equation of variation is $s = \tfrac{2}{3}t$. When $t = 32$, we have

$$s = \tfrac{2}{3}t$$
$$s = \tfrac{2}{3} \cdot 32 \qquad \text{Substituting 32 for } t \text{ in the equation of variation}$$
$$s = \tfrac{64}{3}, \text{ or } 21\tfrac{1}{3}.$$

The value of *s* is $21\tfrac{1}{3}$ when $t = 32$.

E x a m p l e   3

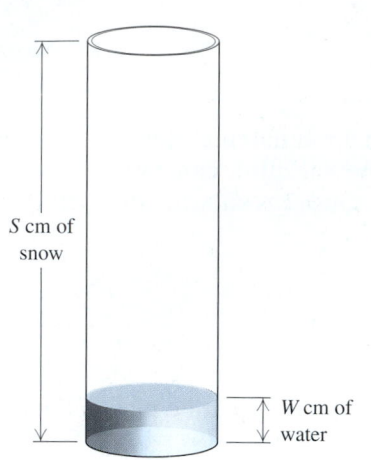

S cm of snow

W cm of water

Water from melting snow.   The number of centimeters *W* of water produced from melting snow varies directly as the number of centimeters *S* of snow. Meteorologists know that under certain conditions, 150 cm of snow will melt to 16.8 cm of water. The average annual snowfall in Alta, Utah, is 500 in. Assuming the above conditions, how much water will replace the 500 in. of snow?

Solution

1. **Familiarize.**   Because of the phrase "*W*... varies directly as ...*S*," we express the amount of water as a function of the amount of snow. Thus, $W(S) = kS$, where *k* is the variation constant. Knowing that 150 cm of snow becomes 16.8 cm of water, we have $W(150) = 16.8$. Because we are using ratios, it does not matter whether we work in inches or centimeters, provided the same units are used for *W* and *S*.

2. **Translate.**   We find the variation constant using the data and then find the equation of variation:

$$W(S) = kS$$

$$W(150) = k \cdot 150 \qquad \text{\color{magenta}Replacing } S \text{ with } 150$$

$$16.8 = k \cdot 150 \qquad \text{\color{magenta}Replacing } W(150) \text{ with } 16.8$$

$$\frac{16.8}{150} = k \qquad \text{\color{magenta}Solving for } k$$

$$0.112 = k. \qquad \text{\color{magenta}This is the variation constant.}$$

The equation of variation is $W(S) = 0.112S$. This is the translation.

3. **Carry out.**   To find how much water 500 in. of snow will become, we compute $W(500)$:

$$W(S) = 0.112S$$

$$W(500) = 0.112(500) \qquad \text{\color{magenta}Substituting 500 for } S$$

$$W = 56.$$

4. **Check.**   To check, we could reexamine all our calculations. Note that our answer seems reasonable since 500/56 and 150/16.8 are equal.

5. **State.**   Alta's 500 in. of snow will be replaced with 56 in. of water.

Inverse Variation

To see what we mean by inverse variation, suppose a car is traveling 20 mi. At 20 mph, the trip will take 1 hr. At 40 mph, it will take $\frac{1}{2}$ hr. At 60 mph, it will take $\frac{1}{3}$ hr, and so on. This gives rise to pairs of numbers, all having the same product:

$$(20, 1), \left(40, \tfrac{1}{2}\right), \left(60, \tfrac{1}{3}\right), \left(80, \tfrac{1}{4}\right), \quad \text{and so on.}$$

Note that the product of each pair of numbers is 20. Whenever a situation gives rise to pairs of numbers for which the product is constant, we say that there is **inverse variation**. Since $r \cdot t = 20$, the time *t*, in hours, required for the car to travel 20 mi at *r* mph is given by

$$t = \frac{20}{r} \quad \text{or, using function notation,} \quad t(r) = \frac{20}{r}.$$

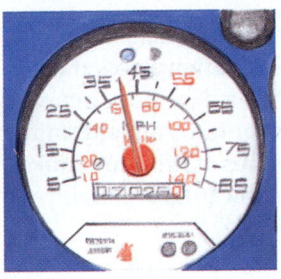

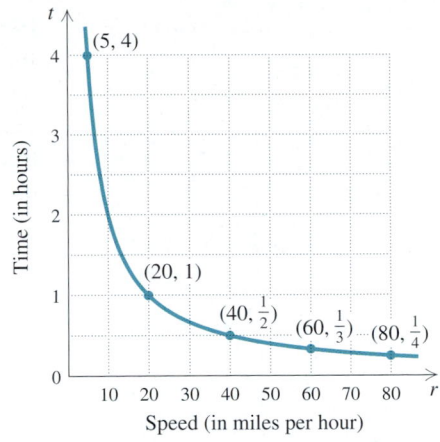

> ### Inverse Variation
>
> When a situation translates to an equation of the form $y = k/x$, or $f(x) = k/x$, with k a nonzero constant, we say that y *varies inversely as* x. The equation $y = k/x$ is called an *equation of inverse variation*.

In inverse variation with $k > 0$, as one variable increases, the other variable decreases.

The terminology

"y is inversely proportional to x"

also implies inverse variation and is used in some situations. The constant k is again called the *constant of proportionality* or the *variation constant*.

Example 4   Find the variation constant and an equation of variation if y varies inversely as x, and $y = 32$ when $x = 0.2$.

Solution   We know that $(0.2, 32)$ is a solution of

$$y = \frac{k}{x}.$$

Therefore,

$$32 = \frac{k}{0.2} \qquad \text{Substituting}$$

$$(0.2)32 = k$$

$$6.4 = k. \qquad \text{Solving for } k$$

The variation constant is 6.4. The equation of variation is

$$y = \frac{6.4}{x}.$$

There are many real-life problems that translate to an equation of inverse variation.

Example 5

Ultraviolet index.   The ultraviolet, or UV, index is a measure issued daily by the National Weather Service that indicates the strength of the sun's rays in a particular locale. For those people whose skin is quite sensitive, a UV rating of 7 will cause sunburn after 10 min (*Source: Los Angeles Times*, 3/24/98). Given that the number of minutes it takes to burn, t, varies inversely as the UV rating u, how long will it take a highly-sensitive person to burn on a day with a UV rating of 2?

Solution

1. **Familiarize.**   Because of the phrase "...varies inversely as the UV index," we express the amount of time needed to burn as a function of the UV rating: $t(u) = k/u$.

2. **Translate.**   We use the given information to solve for k. Then we find the equation of variation.

$$t(u) = \frac{k}{u} \qquad \text{Using function notation}$$

$$t(7) = \frac{k}{7} \qquad \text{Replacing } u \text{ with } 7$$

$$10 = \frac{k}{7} \qquad \text{Replacing } t(7) \text{ with } 10$$

$$70 = k \qquad \text{Solving for } k, \text{ the variation constant}$$

The equation of variation is $t(u) = 70/u$. This is the translation.

3. **Carry out.**   To find how long it would take a highly-sensitive person to burn on a day with a UV index of 2, we calculate $t(2)$:

$$t(2) = \frac{70}{2} = 35. \qquad t = 35 \text{ when } u = 2.$$

4. **Check.**   We could now recheck each step. Note that, as expected, as the UV rating goes *down*, the time it takes to burn goes *up*.

5. **State.**   On a day with a UV rating of 2, a highly-sensitive person will begin to burn after 35 min of exposure.

Joint and Combined Variation

When a variable varies directly with more than one other variable, we say that there is *joint variation*. For example, in the formula for the volume of a right circular cylinder, $V = \pi r^2 h$, we say that V varies *jointly* as h and the square of r.

> ### *Joint Variation*
> y varies *jointly* as x and z if, for some nonzero constant, k, $y = kxz$.

E x a m p l e   6   Find an equation of variation if y varies jointly as x and z, and $y = 30$ when $x = 2$ and $z = 3$.

Solution   We have

$$y = kxz,$$

so

$$30 = k \cdot 2 \cdot 3$$
$$k = 5. \qquad \text{The variation constant is 5.}$$

The equation of variation is $y = 5xz$.

Joint variation is one form of *combined variation*. In general, when a variable varies directly and/or inversely, at the same time, with more than one other variable, there is **combined variation**. Examples 6 and 7 are examples of combined variation.

E x a m p l e   7   Find an equation of variation if y varies jointly as x and z and inversely as the square of w, and $y = 105$ when $x = 3$, $z = 20$, and $w = 2$.

Solution   The equation of variation is of the form

$$y = k \cdot \frac{xz}{w^2},$$

so, substituting, we have

$$105 = k \cdot \frac{3 \cdot 20}{2^2}$$
$$105 = k \cdot 15$$
$$k = 7.$$

Thus,

$$y = 7 \cdot \frac{xz}{w^2}.$$

FOR EXTRA HELP

Exercise Set 7.5

   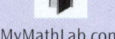

Digital Video Tutor CD 6   InterAct Math   Math Tutor Center   MathXL   MyMathLab.com
Videotape 14

Find the variation constant and an equation of variation if y varies directly as x and the following conditions apply.

1. $y = 28$ when $x = 4$

2. $y = 5$ when $x = 12$

3. $y = 3.4$ when $x = 2$

4. $y = 2$ when $x = 5$

5. $y = 2$ when $x = \frac{1}{3}$

6. $y = 0.9$ when $x = 0.5$

7. *Hooke's law.*   Hooke's law states that the distance d that a spring is stretched by a hanging object varies directly as the mass m of the object. If the distance is 20 cm when the mass is 3 kg, what is the distance when the mass is 5 kg?

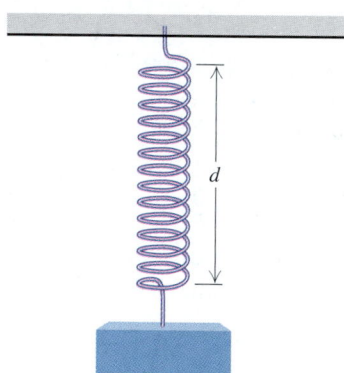

8. *Ohm's law.*   The electric current I, in amperes, in a circuit varies directly as the voltage V. When 15 volts are applied, the current is 5 amperes. What is the current when 18 volts are applied?

9. *Use of aluminum cans.*   The number N of aluminum cans used each year varies directly as the number of people using the cans. If 250 people use 60,000 cans in one year, how many cans are used each year in Dallas, which has a population of 1,008,000?

10. *Weekly allowance.*   According to Fidelity Investments *Investment Vision Magazine,* the average weekly allowance A of children varies directly as

their grade level, G. In a recent year, the average allowance of a 9th-grade student was $9.66 per week. What was the average weekly allowance of a 4th-grade student?

Aha! **11.** *Mass of water in a human.*   The number of kilograms W of water in a human body varies directly as the mass of the body. A 96-kg person contains 64 kg of water. How many kilograms of water are in a 48-kg person?

12. *Weight on Mars.*   The weight M of an object on Mars varies directly as its weight E on Earth. A person who weighs 95 lb on Earth weighs 38 lb on Mars. How much would a 100-lb person weigh on Mars?

13. *Relative aperture.*   The relative aperture, or f-stop, of a 23.5-mm lens is directly proportional to the focal length F of the lens. If a lens with a 150-mm focal length has an f-stop of 6.3, find the f-stop of a 23.5-mm lens with a focal length of 80 mm.

14. *Lead pollution.*   The average U.S. community of population 12,500 released about 385 tons of lead into the environment in a recent year.* How many tons were released nationally? Use 250,000,000 as the U.S. population.

Find the variation constant and an equation of variation in which y varies inversely as x, and the following conditions exist.

15. $y = 3$ when $x = 20$

16. $y = 16$ when $x = 4$

17. $y = 28$ when $x = 4$

18. $y = 9$ when $x = 5$

19. $y = 27$ when $x = \frac{1}{3}$

20. $y = 81$ when $x = \frac{1}{9}$

*Conservation Matters, Autumn 1995 issue. (Boston: Conservation Law Foundation), p. 30.

Solve.

21. *Ultraviolet index.*   At an ultraviolet, or UV, rating of 4, those people who are moderately sensitive to the sun will burn in 70 min (*Source: Los Angeles Times*, 3/24/98). Given that the number of minutes it takes to burn, t, varies inversely with the UV rating, u, how long will it take moderately-sensitive people to burn when the UV rating is 14?

22. *Current and resistance.*   The current I in an electrical conductor varies inversely as the resistance R of the conductor. If the current is $\frac{1}{2}$ ampere when the resistance is 240 ohms, what is the current when the resistance is 540 ohms?

23. *Volume and pressure.*   The volume V of a gas varies inversely as the pressure P upon it. The volume of a gas is 200 cm^3 under a pressure of 32 kg/cm^2. What will be its volume under a pressure of 40 kg/cm^2?

24. *Pumping rate.*   The time t required to empty a tank varies inversely as the rate r of pumping. If a Briggs and Stratton pump can empty a tank in 45 min at the rate of 600 kL/min, how long will it take the pump to empty the tank at 1000 kL/min?

25. *Work rate.*   The time T required to do a job varies inversely as the number of people P working. It takes 5 hr for 7 volunteers to pick up rubbish from 1 mi of roadway. How long would it take 10 volunteers to complete the job?

26. *Wavelength and frequency.*   The wavelength W of a radio wave varies inversely as its frequency F. A wave with a frequency of 1200 kilohertz has a length of 300 meters. What is the length of a wave with a frequency of 800 kilohertz?

Find an equation of variation in which:

27. y varies directly as the square of x, and $y = 6$ when $x = 3$.

28. y varies directly as the square of x, and $y = 0.15$ when $x = 0.1$.

29. y varies inversely as the square of x, and $y = 6$ when $x = 3$.

30. y varies inversely as the square of x, and $y = 0.15$ when $x = 0.1$.

31. y varies jointly as x and the square of z, and $y = 105$ when $x = 14$ and $z = 5$.

32. y varies jointly as x and z and inversely as w, and $y = \frac{3}{2}$ when $x = 2$, $z = 3$, and $w = 4$.

33. y varies jointly as w and the square of x and inversely as z, and $y = 49$ when $w = 3$, $x = 7$, and $z = 12$.

34. y varies directly as x and inversely as w and the square of z, and $y = 4.5$ when $x = 15$, $w = 5$, and $z = 2$.

Solve.

35. *Intensity of light.*   The intensity I of light from a light bulb varies inversely as the square of the distance d from the bulb. Suppose I is 90 W/m^2 (watts per square meter) when the distance is 5 m. What would the intensity be 7.5 m from the bulb?

36. *Stopping distance of a car.*   The stopping distance d of a car after the brakes have been applied varies directly as the square of the speed r. If a car traveling 60 mph can stop in 200 ft, what stopping distance corresponds to a speed of 36 mph?

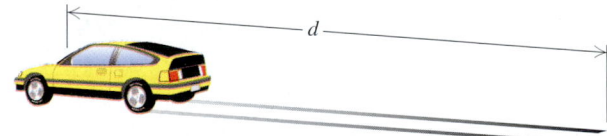

37. *Volume of a gas.*   The volume V of a given mass of a gas varies directly as the temperature T and inversely as the pressure P. If $V = 231$ cm^3 when $T = 42°$ and $P = 20$ kg/cm^2, what is the volume when $T = 30°$ and $P = 15$ kg/cm^2?

38. *Intensity of a signal.*   The intensity I of a television signal varies inversely as the square of the distance d from the transmitter. If the intensity is 25 W/m^2 at a distance of 2 km, what is the intensity 6.25 km from the transmitter?

39. *Atmospheric drag.*   Wind resistance, or atmospheric drag, tends to slow down moving objects. Atmospheric drag W varies jointly as an object's surface area A and velocity v. If a car traveling at a speed of 40 mph with a surface area of 37.8 ft^2 experiences a drag of 222 N (Newtons), how fast must a car with 51 ft^2 of surface area travel in order to experience a drag force of 430 N?

40. *Drag force.* The drag force F on a boat varies jointly as the wetted surface area A and the square of the velocity of the boat. If a boat going 6.5 mph experiences a drag force of 86 N when the wetted surface area is 41.2 ft^2, find the wetted surface area of a boat traveling 8.2 mph with a drag force of 94 N.

State whether each situation represents direct variation, inverse variation, or neither. Give reasons for your answers.

41. The cost of mailing a package in the United States and the distance that it travels

42. A runner's speed in a race and the time it takes to run the race

43. The weight of a turkey and the cooking time

44. The number of plays it takes to go 80 yd for a touchdown and the average gain per play

SKILL MAINTENANCE

Solve.

45. $2x - 5 = 8$

46. $3a + 2 = 7$

47. $\dfrac{1}{x + 1} = \dfrac{3}{x}$

48. $6x^2 = 11x + 35$

49. $3a + 1 = 3(a + 1)$

50. $a + (a - 5) = 2a - 5$

SYNTHESIS

51. Suppose that the number of customer complaints is inversely proportional to the number of employees hired. Will a firm reduce the number of complaints more by expanding from 5 to 10 employees, or from 20 to 25? Explain. Consider using a graph to help justify your answer.

52. If y varies inversely as the cube of x and x is multiplied by 0.5, what is the effect on y?

Write an equation of variation for each situation. Leave k in each equation as the variation constant.

53. *Ecology.* In a stream, the amount of salt S carried varies directly as the sixth power of the speed of the stream v.

54. *Acoustics.* The square of the pitch P of a vibrating string varies directly as the tension t on the string.

55. *Lighting.* The intensity of illumination I from a light source varies inversely as the square of the distance d from the source.

56. *Peanut sales.* The number of bags of peanuts B sold at the circus varies directly as the number of people N in attendance.

57. *Wind energy.* The power P in a windmill varies directly as the cube of the wind speed v.

Write an equation of variation for each situation. Include a value for the variation constant in each equation.

58. *Geometry.* The perimeter P of an equilateral octagon varies directly as the length S of a side.

59. *Geometry.* The circumference C of a circle varies directly as the radius r.

60. *Geometry.* The area of a circle varies directly as the square of the length of the radius.

61. *Geometry.* The volume V of a sphere varies directly as the cube of the radius r.

Describe, in words, the variation given by the equation. Assume k is a constant.

62. $Q = \dfrac{kp^2}{q^3}$

63. $W = \dfrac{km_1M_1}{d^2}$

64. *Tension of a musical string.* The tension T on a string in a musical instrument varies jointly as the string's mass per unit length m, the square of its length l, and the square of its fundamental frequency f. A 2-m–long string of mass 5 gm/m with a fundamental frequency of 80 has a tension of 100 N. How long should the same string be if its tension is going to be changed to 72 N?

65. *Volume and cost.*   A peanut butter jar in the shape of a right circular cylinder is 4 in. high and 3 in. in diameter and sells for \$1.20. If we assume that cost is proportional to volume, how much should a jar 6 in. high and 6 in. in diameter cost?

66. *Golf distance finder.*   A device used in golf to estimate the distance d to a hole measures the size s that the 7-ft pin *appears* to be in a viewfinder. The viewfinder uses the principle, diagrammed here, that s gets bigger when d gets smaller. If $s = 0.56$ in. when $d = 50$ yd, find an equation of variation that expresses d as a function of s. What is d when $s = 0.40$ in.?

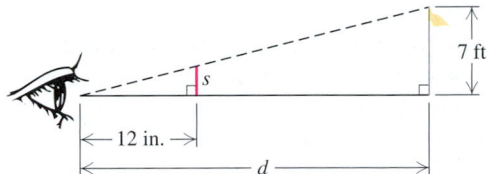

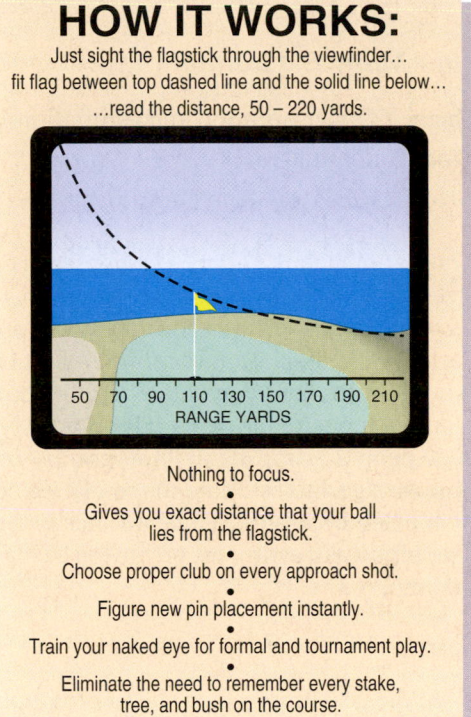

HOW IT WORKS:

Just sight the flagstick through the viewfinder...
fit flag between top dashed line and the solid line below...
...read the distance, 50 – 220 yards.

50   70   90   110   130   150   170   190   210
RANGE YARDS

Nothing to focus.
•
Gives you exact distance that your ball
lies from the flagstick.
•
Choose proper club on every approach shot.
•
Figure new pin placement instantly.
•
Train your naked eye for formal and tournament play.
•
Eliminate the need to remember every stake,
tree, and bush on the course.

COLLABORATIVE

CORNER

How Many Is a Million?

Focus: Direct variation and estimation

Time: 15 minutes

Group size: 2 or 3 and entire class

The National Park Service's estimates of crowd sizes for static (stationary) mass demonstrations vary directly as the area covered by the crowd. Park Service officials have found that at basic "shoulder-to-shoulder" demonstrations, 1 acre of land (about 45,000 ft^2) holds about 9000 people. Using aerial photographs, officials impose a grid to estimate the total area covered by the demonstrators. Once this has been accomplished, estimates of crowd size can be prepared.

ACTIVITY

1. In the grid imposed on the photograph below, each square represents 10,000 ft^2. Esti-
   mate the size of the crowd photographed. Then compare your group's estimate with those of other groups. What might explain discrepancies between estimates? List ways in which your group's estimate could be made more accurate.

2. Park Service officials use an "acceptable margin of error" of no more than 20%. Using all estimates from part (1) above and allowing for error, find a range of values within which you feel certain that the actual crowd size lies.

3. The Million Man March of 1995 was not a static demonstration because of a periodic turnover of people in attendance (many people stayed for only part of the day's festivities). How might you change your methodology to compensate for this complication?

Summary and Review 7

Key Terms

Function, p. 410

Domain, p. 410

Range, p. 410

Relation, p. 411

Input, p. 413

Output, p. 413

Dependent variable, p. 414

Independent variable, p. 414

Constant function, p. 414

Dummy variable, p. 414

Interpolation, p. 415

Extrapolation, p. 415

Projection, p. 424

Endpoints, p. 425

Piecewise-defined function, p. 428

Linear function, p. 435

Salvage value, p. 437

Depreciate, p. 437

Growth rate, p. 439

Nonlinear function, p. 441

Absolute-value function, p. 441

Polynomial function, p. 441

Quadratic function, p. 441

Rational function, p. 441

Equation of variation, p. 458

Direct variation, p. 458

Constant of proportionality, p. 459

Variation constant, p. 459

Inverse variation, p. 460

Joint variation, p. 463

Combined variation, p. 463

Important Properties and Formulas

The Vertical-Line Test

A graph represents a function if it is not possible to draw a vertical line that intersects the graph more than once.

The Domain and the Range of a Linear Function

The domain of any linear function $f(x) = mx + b$ is $\mathbb{R}$.

The range of a linear function $f(x) = mx + b, m \neq 0$, is $\mathbb{R}$.

The range of a constant function $f(x) = b$ is $\{b\}$.

The Algebra of Functions

1. $(f + g)(x) = f(x) + g(x)$
2. $(f - g)(x) = f(x) - g(x)$
3. $(f \cdot g)(x) = f(x) \cdot g(x)$
4. $(f/g)(x) = f(x)/g(x)$, provided $g(x) \neq 0$

Variation

y varies directly as x if there is some nonzero constant k such that $y = kx$.

y varies inversely as x if there is some nonzero constant k such that $y = k/x$.

y varies jointly as x and z if there is some nonzero constant k such that $y = kxz$.

Review Exercises

1. For the following graph of f, determine **(a)** $f(2)$; **(b)** the domain of f; **(c)** any x-values for which $f(x) = 2$; and **(d)** the range of f.

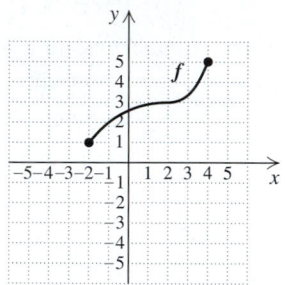

2. Find $g(-3)$ for $g(x) = \dfrac{x}{x + 1}$.

3. Find $f(2a)$ for $f(x) = x^2 + 2x - 3$.

4. The function $A(t) = 0.233t + 5.87$ can be used to estimate the median age of cars in the United States t years after 1990 (*Source*: The Polk Co.). (In this context, a median age of 3 yr means that half the cars are more than 3 yr old and half are less.) Predict the median age of cars in 2010; that is, find $A(20)$.

The following table shows the annual soft-drink production in the United States, in number of 12-oz cans per person (Sources: National Soft Drink Association; Beverage World).

Input, Year	Output, Number of 12-oz Cans per Person
1977	360
1987	480
1997	580

5. Use the data in the table above to draw a graph and to estimate the annual soft-drink production in 1990.

6. Use the graph from Exercise 5 to estimate the annual soft-drink production in 2005.

*For each of the graphs in Exercises 7–10, **(a)** determine whether the graph represents a function and **(b)** if so, determine the domain and the range of the function.*

7.

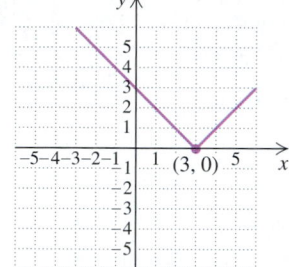

8.

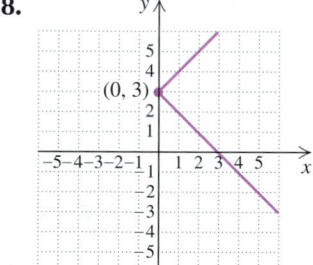

9.

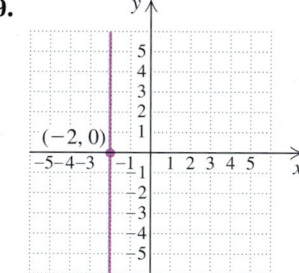

10.

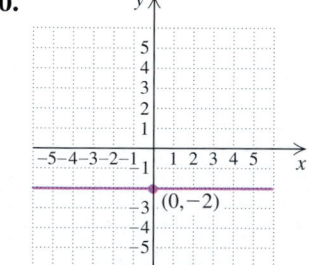

Find the domain of each function.

11. $f(x) = 3x^2 - 7$

12. $g(x) = \dfrac{x^2}{x - 1}$

13. $f(t) = \dfrac{1}{t^2 + 5t + 4}$

14. Each salesperson at Knobby's Furniture is paid $P(x)$ dollars, where $P(x) = 0.05x + 200$ and x is the value of the salesperson's sales for the week. What is the domain of the function?

15. If a used model 8T lawnmower is returned after purchase, the amount that Turner's Hardware will refund is given by the function

$$r(t) = 850 - 25t,$$

where t is the number of weeks since the date of purchase. What is the domain of the function?

16. For the function given by

$$f(x) = \begin{cases} 2 - x, & \text{for } x \le -2, \\ x^2, & \text{for } -2 < x \le 5, \\ x + 10, & \text{for } x > 5, \end{cases}$$

find **(a)** $f(-3)$; **(b)** $f(-2)$; **(c)** $f(4)$; and **(d)** $f(25)$.

17. It costs \$75 plus \$15 a month to join the Family Fitness Center. Formulate a linear function to model the cost for t months of membership, and determine the time required for the cost to reach \$180.

18. In 1955, the U.S. minimum wage was \$0.75, and in 1997, it was \$5.15. Let W represent the minimum wage, in dollars, t years after 1955.
   a) Find a linear function $W(t)$ that fits the data.
   b) Use the function of part (a) to predict the minimum wage in 2005.

Classify each function as a linear function, an absolute-value function, a quadratic function, another polynomial function, or a rational function.

19. $f(x) = |3x - 7|$

20. $g(x) = 4x^5 - 8x^3 + 7$

21. $p(x) = x^2 + x - 10$

22. $h(t) = 4n - 17$

23. $s(t) = \dfrac{t + 1}{t + 2}$

Graph each function and determine its domain and range.

24. $f(x) = 3$ **25.** $f(x) = 2x + 1$

26. $g(x) = |x + 1|$

Let $g(x) = 3x - 6$ and $h(x) = x^2 + 1$. Find the following.

27. $(g \cdot h)(4)$

28. $(g - h)(-2)$

29. $(g/h)(-1)$

30. The domains of $g + h$ and $g \cdot h$

31. The domain of h/g

32. Find an equation of variation in which y varies directly as x, and $y = 30$ when $x = 4$.

33. Find an equation of variation in which y varies inversely as x, and $y = 3$ when $x = \frac{1}{4}$.

34. Find an equation of variation in which y varies jointly as x and the square of w and inversely as z, and $y = 150$ when $x = 6$, $w = 10$, and $z = 2$.

35. The amount of waste generated by a restaurant varies directly as the number of customers served. A typical McDonalds that serves 2000 customers per day generates 238 lb of waste daily (*Source:* Environmental Defense Fund Study, November 1990). How many pounds of waste would be generated daily by a McDonalds that serves 1700 customers a day?

36. A warning dye is used by people in lifeboats to aid searching airplanes. The radius r of the circle formed by the dye varies directly as the square root of the volume V. It is found that 4 L of dye will spread to a circle of radius 5 m. How much dye is needed to form a circle with a 20-m radius?

SYNTHESIS

37. If two functions have the same domain and range, are the functions identical? Why or why not?

38. Jenna believes that 0 is never in the domain of a rational function. Is she correct? Why or why not?

39. Homespun Jellies charges \$2.49 for each jar of preserves. Shipping charges are \$3.75 for handling, plus \$0.60 per jar. Find a linear function for determining the cost of shipping x jars of preserves.

40. Determine the domain and the range of the function graphed below.

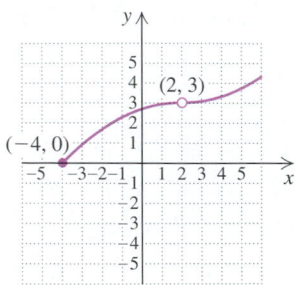

41. Rosewood Graphics recently promised its employees 6% raises each year for the next five years. Amy currently earns $30,000 a year. Can she use a linear function to predict her salary for the next five years? Why or why not?

Chapter Test 7

1. For the following graph of f, determine **(a)** $f(-2)$; **(b)** the domain of f; **(c)** any x-value for which $f(x) = \frac{1}{2}$; and **(d)** the range of f.

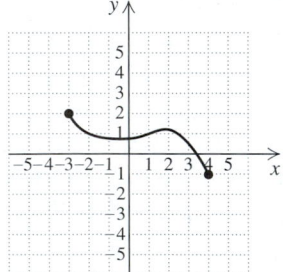

For each of the following graphs, **(a)** *determine whether the graph represents a function and* **(b)** *if so, determine the domain and the range of the function.*

4.

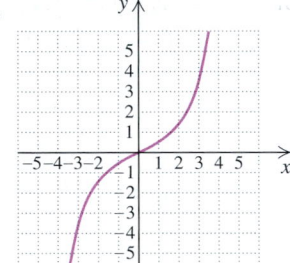

2. The function $S(t) = 1.2t + 21.4$ can be used to estimate the total U.S. sales of books, in billions of dollars, t years after 1992.

   a) Predict the total U.S. sales of books in 2008.

   b) What do the numbers 1.2 and 21.4 signify?

3. There were 43.3 million international visitors to the United States in 1995, and 48.5 million in 1999 (*Source*: Tourism Industries, International Trade Administration, Department of Commerce). Draw a graph and estimate the number of international visitors in 1997.

5.

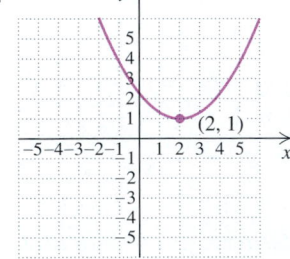

6.

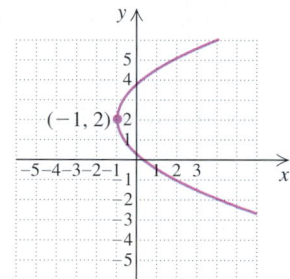

7. The distance d, in miles, that Kerry is from Chicago is given by the function $d(t) = 240 - 60t$, where t is the number of hours since he left Indianapolis. What is the domain of the function?

8. For the function given by
$$f(x) = \begin{cases} x^2, & \text{for } x < 0, \\ 3x - 5, & \text{for } 0 \le x \le 2, \\ x + 7, & \text{for } x > 2, \end{cases}$$
find **(a)** $f(0)$ and **(b)** $f(3)$.

9. For a catered party, Jennette's Catering charges $25 per person plus a $75 setup fee. Formulate a linear function to model the cost of a party for n people, and determine the number of people in a party that costs $500.

10. Jon rented a van for one day and drove it 250 mi at a cost of $100. Josh rented the same van for one day and drove it 300 mi at a cost of $115. Let $C(m)$ represent the cost, in dollars, of driving m miles.
  a) Find a linear function that fits the data.
  b) Use the function to find how much it will cost to rent the van for one day and drive it 500 mi.

Classify each function as a linear function, a quadratic function, another polynomial function, an absolute-value function, or a rational function. Then find the domain of each function.

11. $f(x) = \dfrac{1}{4}x + 7$ **12.** $g(x) = \dfrac{3}{x - 5}$

13. $p(x) = 4x^2 + 7$

Graph each function and determine its domain and range.

14. $f(x) = \dfrac{1}{3}x - 2$ **15.** $g(x) = x^2 - 1$

16. $h(x) = -\dfrac{1}{2}$

Let $g(x) = \dfrac{1}{x + 4}$ and $h(x) = x - 7$. Find the following.

17. $g(0)$ **18.** $h(-3)$

19. $g(2a)$ **20.** $(g \cdot h)(-1)$

21. The domain of g **22.** The domain of h

23. The domain of $g + h$ **24.** The domain of g/h

25. Find an equation of variation in which y varies directly as x, and $y = 10$ when $x = 20$.

26. The number of workers n needed to clean a stadium after a game varies inversely as the amount of time t allowed for the cleanup. If it takes 25 workers to clean the stadium when there are 6 hr allowed for the job, how many workers are needed if the stadium must be cleaned in 5 hr?

27. The surface area of a balloon varies directly as the square of its radius. The area is 3.4 in^2 when the radius is 5 in. What is the area when the radius is 7 in.?

SYNTHESIS

28. The function $f(t) = 5 + 15t$ can be used to determine a bicycle racer's location, in miles from the starting line, measured t hours after passing the 5-mi mark.
  a) How far from the start will the racer be 1 hr and 40 min after passing the 5-mi mark?
  b) Assuming a constant rate, how fast is the racer traveling?

29. Given that $f(x) = 5x^2 + 1$ and $g(x) = 4x - 3$, find an expression for $h(x)$ so that the domain of $f/g/h$ is
$$\{x \mid x \text{ is a real number } and \ x \ne \tfrac{3}{4} \ and \ x \ne \tfrac{2}{7}\}.$$
Answers may vary.

8

Systems of Linear Equations and Problem Solving

AN APPLICATION

Industrial biochemists routinely use a machine to mix a buffer of 10% acetone by adding 100% acetone to water. One day, instead of adding 5 L of acetone to create a vat of buffer, a machine added 10 L. How much additional water was needed to bring the concentration down to 10%?

This problem appears as Exercise 63 in Section 8.3.

STEPHANI HARNICK
Environmental Engineer
McCordsville, IN

*C*hemical engineers often use math to calculate ratios in solutions and to evaluate formulas. As an environmental engineer, I need math to calculate air emissions to ensure acceptable air quality.

*T*he most difficult part of problem solving is almost always translating the problem situation to mathematical language. Once a problem has been translated, the rest is usually straightforward. In this chapter, we study systems of equations *and how to solve them using graphing, substitution, elimination, and matrices. Systems of equations often provide the easiest way to model real-world situations in fields such as psychology, sociology, business, education, engineering, and science.*

Systems of Equations in Two Variables

8.1

Translating • Identifying Solutions • Solving Systems Graphically

Translating

Problems involving two unknown quantities are often solved most easily if we can first translate the situation to two equations in two unknowns.

E x a m p l e   1

Real estate.   Translate the following problem situation to mathematical language, using two equations.

> In 1996, the Simon Property Group and the DeBartolo Realty Corporation merged to form the largest real estate company in the United States, owning 183 shopping centers in 32 states. Prior to merging, Simon owned twice as many properties as DeBartolo. How many properties did each company own before the merger?

> DeBartolo shareholders would receive 0.68 share of Simon common stock for each share of DeBartolo common stock. Simon also would agree to repay $1.5 billion in DeBartolo debt. At Tuesday's closing price of $ 23.625 a share for common stock, the transaction is valued at roughly $3 billion.
>
> Executives say the proposed company, Simon DeBartolo Group, would be the largest real estate company in the United States, worth $7.5 billion. **Not included in the deal:** DeBartolo's ownership stake in the San Francisco 49ers, or the Indiana Pacers, owned separately by the Simon

Solution

1. **Familiarize.**   We have already seen problems in which we need to look up certain formulas or the meaning of certain words. Here we need only observe that the words shopping centers and properties are being used interchangeably.

   Often problems contain information that has no bearing on the situation being discussed. In this case, the number 32 is irrelevant to the question being asked. Instead we focus on the number of properties owned and the phrase "twice as many." Rather than guess and check, let's proceed to the next step, using *x* for the number of properties originally owned by Simon and *y* for the number of properties originally owned by DeBartolo.

2. **Translate.**   There are two statements to translate. First we look at the total number of properties involved:

Rewording: The number of   the number of
Simon properties plus DeBartolo properties total 183.

Translating: x $+$ y $=$ 183

The second statement compares the number of properties that each company held before merging:

Rewording: The number of   twice the number of
Simon properties was DeBartolo properties

Translating: x $=$ $2 \cdot y$

We have now translated the problem to a pair, or **system**, **of equations**:

$$x + y = 183,$$
$$x = 2y.$$

We will complete the solution of this problem in Section 8.3.

Problems like Example 1 *can* be solved using one variable; however, as problems become complicated, you will find that using more than one variable (and more than one equation) is often the preferable approach.

E x a m p l e   2

Purchasing.   Recently the Champlain Valley Community Music Center purchased 120 stamps for $30.30. If the stamps were a combination of 20¢ postcard stamps and 34¢ first-class stamps, how many of each type were bought?

Solution

1. **Familiarize.**   To familiarize ourselves with this problem, let's guess that the music center bought 60 stamps at 20¢ each and 60 stamps at 34¢ each. The total cost would then be

$$60 \cdot \$0.20 + 60 \cdot \$0.34 = \$12.00 + \$20.40, \text{ or } \$32.40.$$

Since $\$32.40 \neq \$30.30$, our guess is incorrect. Rather than guess again, let's see how algebra can be used to translate the problem.

2. **Translate.**   We let p = the number of postcard stamps and f = the number of first-class stamps. The information can be organized in a table, which will help with the translating.

Type of Stamp	Postcard	First-class	Total
Number Sold	p	f	120
Price	$0.20	$0.34	
Amount	$0.20p$	$0.34f$	$30.30

$\rightarrow p + f = 120$

$\rightarrow 0.20p + 0.34f = 30.30$

The first row of the table and the first sentence of the problem indicate that a total of 120 stamps were bought:

$$p + f = 120.$$

Since each postcard stamp cost \$0.20 and p stamps were bought, $0.20p$ represents the amount paid, in dollars, for the postcard stamps. Similarly, $0.34f$ represents the amount paid, in dollars, for the first-class stamps. This leads to a second equation:

$$0.20p + 0.34f = 30.30.$$

Multiplying both sides by 100, we can clear the decimals. This gives the following system of equations as the translation:

$$p + f = 120,$$
$$20p + 34f = 3030.$$

Identifying Solutions

A *solution* of a system of equations in two variables is an ordered pair of numbers that makes *both* equations true.

E x a m p l e 3   Determine whether $(-4, 7)$ is a solution of the system

$$x + y = 3,$$
$$5x - y = -27.$$

Solution   We use alphabetical order of the variables. Thus we replace x with -4 and y with 7:

$$\begin{array}{c|c} x + y = 3 \\ \hline -4 + 7 \; ? \; 3 \\ 3 \;\bigm|\; 3 \quad \text{TRUE} \end{array} \qquad \begin{array}{c|c} 5x - y = -27 \\ \hline 5(-4) - 7 \; ? \; -27 \\ -20 - 7 \\ -27 \;\bigm|\; -27 \quad \text{TRUE} \end{array}$$

The pair $(-4, 7)$ makes both equations true, so it is a solution of the system. We can also describe the solution by writing $x = -4$ and $y = 7$. Set notation can also be used to list the solution set $\{(-4, 7)\}$.

Solving Systems Graphically

Recall that the graph of an equation is a drawing that represents its solution set. If we graph the equations in Example 3, we find that $(-4, 7)$ is the only point common to both lines. Thus one way to solve a system of two equations is to graph both equations and identify any points of intersection. The coordinates of each point of intersection represent a solution of that system.

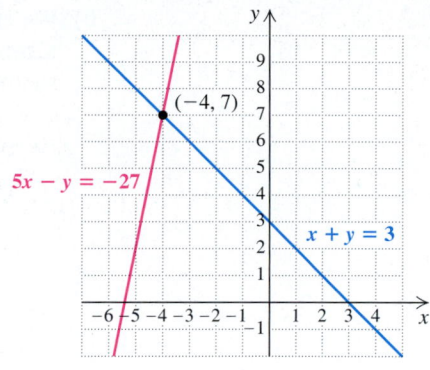

$$x + y = 3,$$
$$5x - y = -27$$

Most pairs of lines have exactly one point in common. We will soon see, however, that this is not always the case.

E x a m p l e   4   Solve each system graphically.

a) $y - x = 1,$
    $y + x = 3$

b) $y = -3x + 5,$
    $y = -3x - 2$

c) $3y - 2x = 6,$
    $-12y + 8x = -24$

Solution

a) We graph each equation using any method studied in Chapter 3. All ordered pairs from line L_1 are solutions of the first equation. All ordered pairs from line L_2 are solutions of the second equation. The point of intersection has coordinates that make *both* equations true. Apparently, $(1, 2)$ is the solution. Graphing is not always accurate, so solving by graphing may yield approximate answers. Our check below shows that $(1, 2)$ is indeed the solution.

$$y - x = 1,$$
$$y + x = 3$$

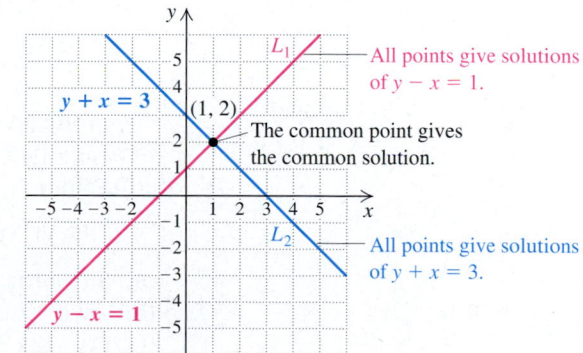

Check: $y - x = 1$ $y + x = 3$
       $\overline{2 - 1\ ?\ 1}$ $\overline{2 + 1\ ?\ 3}$
           $1\ |\ 1$ TRUE $3\ |\ 3$ TRUE

b) We graph the equations. The lines have the same slope, -3, and different y-intercepts, so they are parallel. There is no point at which they cross, so the system has no solution.

$$y = -3x + 5,$$
$$y = -3x - 2$$

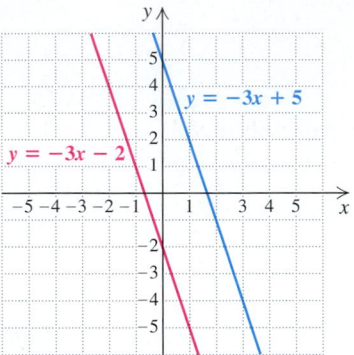

c) We graph the equations and find that the same line is drawn twice. Thus any solution of one equation is a solution of the other. Each equation has an infinite number of solutions, so the system itself has an infinite number of solutions. We check one solution, $(0, 2)$, which is the y-intercept of each equation.

$$3y - 2x = 6,$$
$$-12y + 8x = -24$$

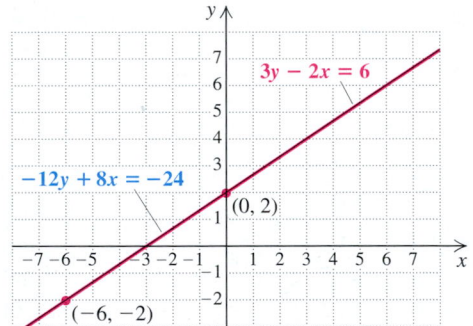

Check:

$$\begin{array}{c|c}
3y - 2x = 6 \\ \hline
3(2) - 2(0) \;?\; 6 \\
6 - 0 \;\bigg| \\
6 \;\bigg|\; 6 \quad \text{TRUE}
\end{array}
\qquad
\begin{array}{c|c}
-12y + 8x = -24 \\ \hline
-12(2) + 8(0) \;?\; -24 \\
-24 + 0 \;\bigg| \\
-24 \;\bigg|\; -24 \quad \text{TRUE}
\end{array}$$

You can check that $(-6, -2)$ is another solution of both equations. In fact, any pair that is a solution of one equation is a solution of the other equation as well. Thus the solution set is

$$\{(x, y) \mid 3y - 2x = 6\}$$

or, in words, "the set of all pairs (x, y) for which $3y - 2x = 6$." Since the two equations are equivalent, we could have written instead $\{(x, y) \mid -12y + 8x = -24\}$.

On most graphers, an INTERSECT option allows you to find the co-ordinates of the intersection directly. This is especially useful when equations contain fractions or decimals or when the coordinates of the intersection are not integers. To illustrate, consider the following system:

$$3.45x + 4.21y = 8.39,$$
$$7.12x - 5.43y = 6.18.$$

Most graphers require equations to have y alone on one side. After solving for y in each equation, we can obtain the graph shown. Using INTERSECT, we see that, to the nearest hundredth, the coordinates of the intersection are $(1.47, 0.79)$.

$$y_1 = (8.39 - 3.45x)/4.21,$$
$$y_2 = (6.18 - 7.12x)/(-5.43)$$

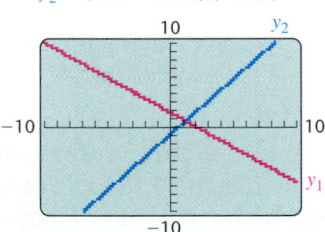

Use a grapher to solve each of the following systems. Make sure that all x- and y-coordinates are correct to the nearest hundredth.

1. $y = -5.43x + 10.89,$
   $y = 6.29x - 7.04$
2. $y = 123.52x + 89.32,$
   $y = -89.22x + 33.76$
3. $2.18x + 7.81y = 13.78,$
   $5.79x - 3.45y = 8.94$
4. $-9.25x - 12.94y = -3.88,$
   $21.83x + 16.33y = 13.69$

When we graph a system of two linear equations in two variables, one of the following three outcomes will occur.

1. The lines have one point in common, and that point is the only solution of the system (see Example 4a). Any system that has at least one solution is said to be **consistent**.
2. The lines are parallel, with no point in common, and the system has no solution (see Example 4b). This type of system is called **inconsistent**.
3. The lines coincide, sharing the same graph. Because every solution of one equation is a solution of the other, the system has an infinite number of solutions (see Example 4c). Since it has a solution, this type of system is consistent.

When one equation in a system can be obtained by multiplying both sides of another equation by a constant, the two equations are said to be **dependent**. Thus the equations in Example 4(c) are dependent, but those in Examples 4(a) and 4(b) are **independent**. For systems of three or more equations, the definitions of dependent and independent must be slightly modified.

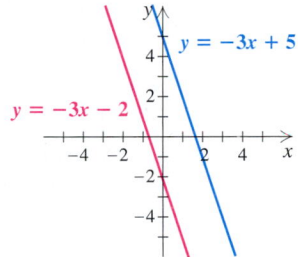

Graphs are parallel.
The system is *inconsistent* because there is no solution. Since the equations are not equivalent, they are *independent*.

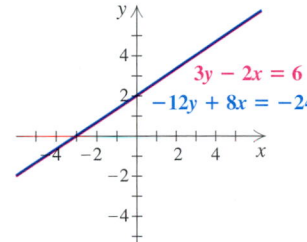

Equations have the same graph.
The system is *consistent* and has an infinite number of solutions. The equations are *dependent* since they are equivalent.

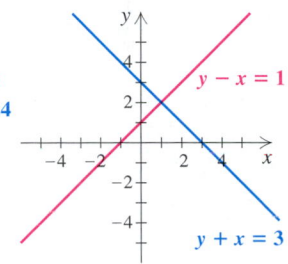

Graphs intersect at one point.
The system is *consistent* and has one solution. Since neither equation is a multiple of the other, they are *independent*.

Graphing is helpful when solving systems because it allows us to "see" the solution. It can also be used on systems of nonlinear equations, and in many applications, it provides a satisfactory answer. However, graphing often lacks precision, especially when fractional or decimal solutions are involved. In Section 8.2, we will develop two algebraic methods of solving systems. Both methods produce exact answers.

FOR EXTRA HELP

Exercise Set 8.1

Digital Video Tutor CD 6   InterAct Math   Math Tutor Center   MathXL   MyMathLab.com
Videotape 15

Determine whether the ordered pair is a solution of the given system of equations. Remember to use alphabetical order of variables.

1. $(1, 2)$; $4x - y = 2$,
   $10x - 3y = 4$

2. $(-1, -2)$; $2x + y = -4$,
   $x - y = 1$

3. $(2, 5)$; $y = 3x - 1$,
   $2x + y = 4$

4. $(-1, -2)$; $x + 3y = -7$,
   $3x - 2y = 12$

5. $(1, 5)$; $x + y = 6$,
   $y = 2x + 3$

6. $(5, 2)$; $a + b = 7$,
   $2a - 8 = b$

Aha! 7. $(3, 1)$; $3x + 4y = 13$,
   $6x + 8y = 26$

8. $(4, -2)$; $-3x - 2y = -8$,
   $8 = 3x + 2y$

Solve each system graphically. Be sure to check your solution. If a system has an infinite number of solutions, use set-builder notation to write the solution set. If a system has no solution, state this.

9. $x - y = 3$,
   $x + y = 5$

10. $x + y = 4$,
    $x - y = 2$

11. $3x + y = 5$,
    $x - 2y = 4$

12. $2x - y = 4$,
    $5x - y = 13$

13. $4y = x + 8$,
    $3x - 2y = 6$

14. $4x - y = 9$,
    $x - 3y = 16$

15. $x = y - 1$,
    $2x = 3y$

16. $a = 1 + b$,
    $b = 5 - 2a$

17. $x = -3$,
    $y = 2$

18. $x = 4$,
    $y = -5$

19. $t + 2s = -1$,
    $s = t + 10$

20. $b + 2a = 2$,
    $a = -3 - b$

21. $2b + a = 11$,
    $a - b = 5$

22. $y = -\frac{1}{3}x - 1$,
    $4x - 3y = 18$

23. $y = -\frac{1}{4}x + 1$,
    $2y = x - 4$

24. $6x - 2y = 2$,
    $9x - 3y = 1$

25. $y - x = 5$,
    $2x - 2y = 10$

26. $y = -x - 1$,
    $4x - 3y = 24$

27. $y = 3 - x$,
    $2x + 2y = 6$

28. $2x - 3y = 6$,
    $3y - 2x = -6$

29. For the systems in the odd-numbered exercises 9–27, which are consistent?

30. For the systems in the even-numbered exercises 10–28, which are consistent?

31. For the systems in the odd-numbered exercises 9–27, which contain dependent equations?

32. For the systems in the even-numbered exercises 10–28, which contain dependent equations?

Translate each problem situation to a system of equations. Do not attempt to solve, but save for later use.

33. The difference between two numbers is 11. Twice the smaller plus three times the larger is 123. What are the numbers?

34. The sum of two numbers is −42. The first number minus the second number is 52. What are the numbers?

35. *Retail sales.* Paint Town sold 45 paintbrushes, one kind at $8.50 each and another at $9.75 each. In all, $398.75 was taken in for the brushes. How many of each kind were sold?

36. *Retail sales.* Mountainside Fleece sold 40 neckwarmers. Polarfleece neckwarmers sold for $9.90 each and wool ones sold for $12.75 each. In all, $421.65 was taken in for the neckwarmers. How many of each type were sold?

37. *Geometry.* Two angles are supplementary.* One angle is 3° less than twice the other. Find the measures of the angles.

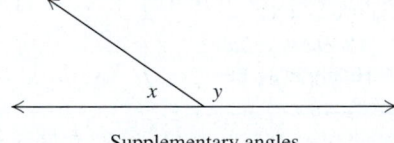

Supplementary angles

---

*The sum of the measures of two supplementary angles is 180°.

38. *Geometry.* Two angles are complementary.* The sum of the measures of the first angle and half the second angle is 64°. Find the measures of the angles.

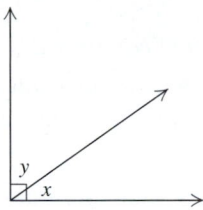

Complementary angles

39. *Basketball scoring.* Wilt Chamberlain once scored 100 points, setting a record for points scored in an NBA game. Chamberlain took only two-point shots and (one-point) foul shots and made a total of 64 shots. How many shots of each type did he make?

40. *Fundraising.* The St. Mark's Community Barbecue served 250 dinners. A child's plate cost $3.50 and an adult's plate cost $7.00. A total of $1347.50 was collected. How many of each type of plate was served?

41. *Sales of pharmaceuticals.* In 2001, the Diabetic Express charged $21.95 for a vial of Humulin insulin and $20.95 for a vial of Novolin insulin. If a total of $1077.50 was collected for 50 vials of insulin, how many vials of each type were sold?

42. *Court dimensions.* The perimeter of a standard basketball court is 288 ft. The length is 44 ft longer than the width. Find the dimensions.

$P = 288$ ft

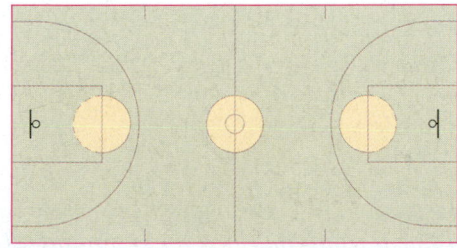

43. *Court dimensions.* The perimeter of a standard tennis court used for doubles is 228 ft. The width is 42 ft less than the length. Find the dimensions.

44. *Basketball scoring.* The Fenton College Cougars made 40 field goals in a recent basketball game, some 2-pointers and the rest 3-pointers. Altogether the 40 baskets counted for 89 points. How many of each type of field goal was made?

*The sum of the measures of two complementary angles is 90°.

45. *Hockey rankings.* Hockey teams receive 2 points for a win and 1 point for a tie. The Wildcats once won a championship with 60 points. They won 9 more games than they tied. How many wins and how many ties did the Wildcats have?

46. *Radio airplay.* Roscoe must play 12 commercials during his 1-hr radio show. Each commercial is either 30 sec or 60 sec long. If the total commercial time during that hour is 10 min, how many commercials of each type does Roscoe play?

47. *Nontoxic floor wax.* A nontoxic floor wax can be made from lemon juice and food-grade linseed oil. The amount of oil should be twice the amount of lemon juice. How much of each ingredient is needed to make 32 oz of floor wax? (The mix should be spread with a rag and buffed when dry.)

48. *Lumber production.* Denison Lumber can convert logs into either lumber or plywood. In a given day, the mill turns out 42 pallets of plywood and lumber. It makes a profit of $25 on a pallet of lumber and $40 on a pallet of plywood. How many pallets of each type must be produced and sold in order to make a profit of $1245?

49. *Video rentals.* J. P.'s Video rents general-interest films for $3.00 each and children's films for $1.50 each. In one day, a total of $213 was taken in from the rental of 77 videos. How many of each type of video was rented?

50. *Airplane seating.* An airplane has a total of 152 seats. The number of coach-class seats is 5 more than six times the number of first-class seats. How many of each type of seat are there on the plane?

51. Write a problem for a classmate to solve that requires writing a system of two equations. Devise the problem so that the solution is "The Lakers made 6 three-point baskets and 31 two-point baskets."

52. Write a problem for a classmate to solve that can be translated into a system of two equations. Devise the problem so that the solution is "Shelly gave 9 haircuts and 5 shampoos."

SKILL MAINTENANCE

Solve.

53. $2(4x - 3) - 7x = 9$

54. $6y - 3(5 - 2y) = 4$

55. $4x - 5x = 8x - 9 + 11x$

56. $8x - 2(5 - x) = 7x + 3$

Solve.

57. $3x + 4y = 7$, for y

58. $2x - 5y = 9$, for y

SYNTHESIS

Technology in U.S. schools. *For Exercises 59–62, consider the following graph.*

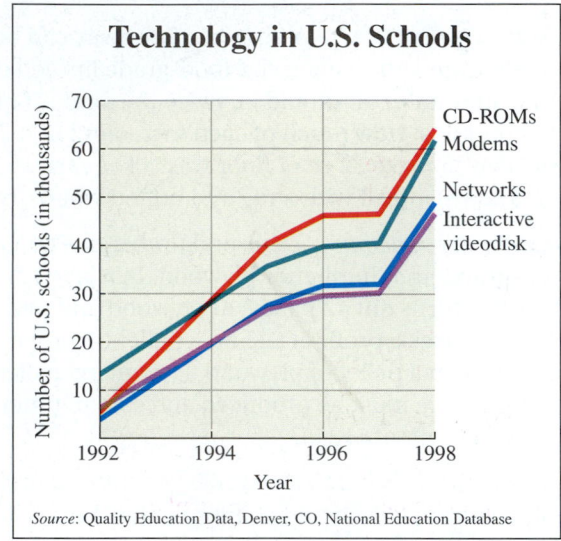

Technology in U.S. Schools

Number of U.S. schools (in thousands)

CD-ROMs
Modems
Networks
Interactive videodisk

Year

Source: Quality Education Data, Denver, CO, National Education Database

59. Is it accurate to state that there have always been more schools with CD-ROMs than with networks? Why or why not?

60. Is it accurate to state that there have always been more schools with networks than with interactive videodisks? Why or why not?

61. During which year did the number of schools with CD-ROMs first exceed the number of schools with modems?

62. During which year did the number of schools owning an interactive videodisk increase the most? At what rate did it increase?

63. For each of the following conditions, write a system of equations.
   a) $(5, 1)$ is a solution.
   b) There is no solution.
   c) There is an infinite number of solutions.

64. A system of linear equations has $(1, -1)$ and $(-2, 3)$ as solutions. Determine:
   a) a third point that is a solution, and
   b) how many solutions there are.

65. The solution of the following system is $(4, -5)$. Find A and B.
$$Ax - 6y = 13,$$
$$x - By = -8.$$

Translate to a system of equations. Do not solve.

66. *Ages.* Burl is twice as old as his son. Ten years ago, Burl was three times as old as his son. How old are they now?

67. *Work experience.* Lou and Juanita are mathematics professors at a state university. Together, they have 46 years of service. Two years ago, Lou had taught 2.5 times as many years as Juanita. How long has each taught at the university?

68. *Design.* A piece of posterboard has a perimeter of 156 in. If you cut 6 in. off the width, the length becomes four times the width. What are the dimensions of the original piece of posterboard?

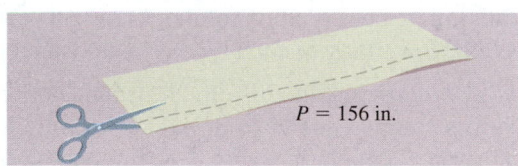

$P = 156$ in.

69. *Nontoxic scouring powder.* A nontoxic scouring powder is made up of 4 parts baking soda and 1 part vinegar. How much of each ingredient is needed for a 16-oz mixture?

Solve graphically.

70. $y = |x|,$
$x + 4y = 15$

71. $x - y = 0,$
$y = x^2$

In Exercises 72–75, use a grapher to solve each system of linear equations for x and y. Round all coordinates to the nearest hundredth.

72. $y = 8.23x + 2.11,$
$y = -9.11x - 4.66$

73. $y = -3.44x - 7.72,$
$y = 4.19x - 8.22$

74. $14.12x + 7.32y = 2.98,$
$21.88x - 6.45y = -7.22$

75. $5.22x - 8.21y = -10.21,$
$-12.67x + 10.34y = 12.84$

Solving by Substitution or Elimination

The Substitution Method • The Elimination Method • Comparing Methods

The Substitution Method

One algebraic (nongraphical) method for solving systems of equations, the *substitution method*, relies on having a variable isolated.

Example 1

Solve the system

$$x + y = 4, \quad (1)$$
$$x = y + 1. \quad (2)$$

For easy reference, we have numbered the equations.

Solution   Equation (2) says that x and $y + 1$ name the same number. Thus we can substitute $y + 1$ for x in equation (1):

$$x + y = 4 \qquad \text{Equation (1)}$$
$$(y + 1) + y = 4. \qquad \text{Substituting } y + 1 \text{ for } x$$

We solve this last equation, using methods learned earlier:

$$(y + 1) + y = 4$$
$$2y + 1 = 4 \qquad \text{Removing parentheses and combining like terms}$$
$$2y = 3 \qquad \text{Subtracting 1 from both sides}$$
$$y = \tfrac{3}{2}. \qquad \text{Dividing by 2}$$

We now return to the original pair of equations and substitute $\tfrac{3}{2}$ for y in either equation so that we can solve for x. For this problem, calculations are slightly easier if we use equation (2):

$$x = y + 1 \qquad \text{Equation (2)}$$
$$= \tfrac{3}{2} + 1 \qquad \text{Substituting } \tfrac{3}{2} \text{ for } y$$
$$= \tfrac{3}{2} + \tfrac{2}{2} = \tfrac{5}{2}.$$

We obtain the ordered pair $\left(\tfrac{5}{2}, \tfrac{3}{2}\right)$. A check ensures that it is a solution:

Check:

$x + y = 4$	$x = y + 1$
$\tfrac{5}{2} + \tfrac{3}{2}$? 4	$\tfrac{5}{2}$? $\tfrac{3}{2} + 1$
$\tfrac{8}{2}$	$\tfrac{3}{2} + \tfrac{2}{2}$
4 | 4 TRUE	$\tfrac{5}{2}$ | $\tfrac{5}{2}$ TRUE

Since $\left(\tfrac{5}{2}, \tfrac{3}{2}\right)$ checks, it is the solution.

A visualization of Example 1.

Note that the coordinates of the intersection are not obvious.

The exact solution to Example 1 is difficult to find graphically because it involves fractions. Despite this, the graph shown does serve as a check and provides a visualization of the problem.

If neither equation in a system has a variable alone on one side, we first isolate a variable in one equation and then substitute.

Example 2

Solve the system

$$2x + y = 6, \quad (1)$$
$$3x + 4y = 4. \quad (2)$$

Solution First, we select an equation and solve for one variable. To isolate y, we can subtract $2x$ from both sides of equation (1):

$$2x + y = 6 \qquad (1)$$
$$y = 6 - 2x. \qquad (3) \qquad \text{Subtracting } 2x \text{ from both sides}$$

Next, we proceed as in Example 1, by substituting:

$$3x + 4(6 - 2x) = 4 \qquad \text{Substituting } 6 - 2x \text{ for } y \text{ in equation (2).}$$
$$\text{Use parentheses!}$$
$$3x + 24 - 8x = 4 \qquad \text{Distributing to remove parentheses}$$
$$3x - 8x = 4 - 24 \qquad \text{Subtracting 24 from both sides}$$
$$-5x = -20$$
$$x = 4. \qquad \text{Dividing both sides by } -5$$

Next, we substitute 4 for x in either equation (1), (2), or (3). It is easiest to use equation (3) because it has already been solved for y:

$$y = 6 - 2x$$
$$= 6 - 2(4)$$
$$= 6 - 8 = -2.$$

The pair $(4, -2)$ appears to be the solution. We check in equations (1) and (2).

Check:

$$\begin{array}{c|c} 2x + y = 6 \\ \hline 2(4) + (-2) \ ? \ 6 \\ 8 - 2 \\ 6 \ | \ 6 \quad \text{TRUE} \end{array} \qquad \begin{array}{c|c} 3x + 4y = 4 \\ \hline 3(4) + 4(-2) \ ? \ 4 \\ 12 - 8 \\ 4 \ | \ 4 \quad \text{TRUE} \end{array}$$

Since $(4, -2)$ checks, it is the solution.

Some systems have no solution, as we saw graphically in Section 8.1. How do we recognize such systems if we are solving by an algebraic method?

Example 3

Solve the system

$$y = -3x + 5, \quad (1)$$
$$y = -3x - 2. \quad (2)$$

Solution We solved this system graphically in Example 4(b) of Section 8.1, and found that the lines are parallel and the system has no solution. Let's now try to solve the system by substitution. Proceeding as in Example 1, we substi-

Study Tip

Begin reviewing for tests and exams early, taking time each week to review important problems, formulas, and properties. Use the chapter summaries and reviews and the cumulative reviews to help you study. Also review the Connecting the Concepts in each chapter.

tute $-3x - 2$ for y in the first equation:

$$-3x - 2 = -3x + 5 \qquad \text{Substituting } -3x - 2 \text{ for } y \text{ in equation (1)}$$

$$-2 = 5. \qquad \text{Adding } 3x \text{ to both sides; } -2 = 5 \text{ is}$$
$$\text{a contradiction.}$$

When we add $3x$ to get the x-terms on one side, the x-terms drop out and we end up with a contradiction—that is, an equation that is always false. When solving algebraically yields a contradiction, the system has no solution.

The Elimination Method

The *elimination method* for solving systems of equations makes use of the *addition principle*: If $a = b$, then $a + c = b + c$. Consider the following system:

$$2x - 3y = 0, \qquad (1)$$
$$-4x + 3y = -1. \qquad (2)$$

To see why the elimination method works well with this system, notice the $-3y$ in one equation and the $3y$ in the other. These terms are opposites. If we add all terms on the left side of the equations, $-3y$ and $3y$ add to 0, and in effect, the variable y is "eliminated."

To use the addition principle for equations, note that according to equation (2), $-4x + 3y$ and -1 are the same number. Thus we can work vertically and add $-4x + 3y$ to the left side of equation (1) and -1 to the right side:

$$2x - 3y = 0 \qquad (1)$$
$$\underline{-4x + 3y = -1} \qquad (2)$$
$$-2x + 0y = -1. \qquad \text{Adding}$$

This eliminates the variable y, and leaves an equation with just one variable, x, for which we solve:

$$-2x = -1$$
$$x = \tfrac{1}{2}.$$

Next, we substitute $\tfrac{1}{2}$ for x in equation (1) and solve for y:

$$2 \cdot \tfrac{1}{2} - 3y = 0 \qquad \text{Substituting. We also could have used equation (2).}$$
$$1 - 3y = 0$$
$$-3y = -1, \text{ so } y = \tfrac{1}{3}.$$

Check:

$$\frac{2x - 3y = 0}{2\left(\tfrac{1}{2}\right) - 3\left(\tfrac{1}{3}\right) \; ? \; 0} \qquad \frac{-4x + 3y = -1}{-4\left(\tfrac{1}{2}\right) + 3\left(\tfrac{1}{3}\right) \; ? \; -1}$$
$$\phantom{2\left(\tfrac{1}{2}\right)} 1 - 1 \phantom{-4\left(\tfrac{1}{2}\right)} -2 + 1$$
$$0 \mid 0 \text{ TRUE} \qquad\qquad -1 \mid -1 \text{ TRUE}$$

Since $\left(\tfrac{1}{2}, \tfrac{1}{3}\right)$ checks, it is the solution.

To eliminate a variable, we must sometimes multiply before adding.

Example 4   Solve the system

$$5x + 4y = 22, \qquad (1)$$
$$-3x + 8y = 18. \qquad (2)$$

Solution   If we add the left sides of the two equations, we will not eliminate a variable. However, if the $4y$ in equation (1) were changed to $-8y$, we would. To accomplish this change, we multiply both sides of equation (1) by -2:

$$-10x - 8y = -44 \qquad \text{Multiplying both sides of equation (1) by } -2$$
$$\underline{-3x + 8y = 18}$$
$$-13x + 0 = -26 \qquad \text{Adding}$$
$$x = 2. \qquad \text{Solving for } x$$

Then

$$-3 \cdot 2 + 8y = 18 \qquad \text{Substituting 2 for } x \text{ in equation (2)}$$
$$-6 + 8y = 18$$
$$\left.\begin{array}{r} 8y = 24 \\ y = 3. \end{array}\right\} \quad \text{Solving for } y$$

We obtain $(2, 3)$, or $x = 2, y = 3$. We leave it to the student to confirm that this checks and is the solution.

Sometimes we must multiply twice in order to make two terms become opposites.

Example 5   Solve the system

$$2x + 3y = 17, \qquad (1)$$
$$5x + 7y = 29. \qquad (2)$$

Solution   We multiply so that the x-terms are eliminated.

$$2x + 3y = 17, \xrightarrow{\substack{\text{Multiplying both} \\ \text{sides by 5}}} 10x + 15y = 85$$
$$5x + 7y = 29 \xrightarrow{\substack{\text{Multiplying both} \\ \text{sides by } -2}} \underline{-10x - 14y = -58}$$
$$0 + y = 27 \qquad \text{Adding}$$
$$y = 27.$$

Next, we substitute to find x:

$$2x + 3 \cdot 27 = 17 \qquad \text{Substituting 27 for } y \text{ in equation (1)}$$
$$2x + 81 = 17$$
$$\left.\begin{array}{r} 2x = -64 \\ x = -32. \end{array}\right\} \quad \text{Solving for } x$$

Check:

$$2x + 3y = 17$$

$$2(-32) + 3(27) \; ? \; 17$$
$$-64 + 81$$
$$17 \; | \; 17 \quad \text{TRUE}$$

$$5x + 7y = 29$$

$$5(-32) + 7(27) \; ? \; 29$$
$$-160 + 189$$
$$29 \; | \; 29 \quad \text{TRUE}$$

We obtain $(-32, 27)$, or $x = -32$, $y = 27$, as the solution.

Example 6

Solve the system

$$3y - 2x = 6, \qquad (1)$$
$$-12y + 8x = -24. \qquad (2)$$

Solution   We graphed this system in Example 4(c) of Section 8.1, and found that the lines coincide and the system has an infinite number of solutions. Suppose we were to solve this system using the elimination method:

$$12y - 8x = \quad 24 \qquad \text{Multiplying both sides of equation (1) by 4}$$
$$\underline{-12y + 8x = -24}$$
$$0 = 0. \qquad \text{We obtain an identity; } 0 = 0 \text{ is always true.}$$

Note that both variables have been eliminated and what remains is an identity—that is, an equation that is always true. Any pair that is a solution of equation (1) is also a solution of equation (2). The equations are dependent and the solution set is infinite:

$$\{(x, y) \,|\, 3y - 2x = 6\}.$$

Special Cases

When solving a system of two linear equations in two variables:

1. If an identity is obtained, such as $0 = 0$, then the system has an infinite number of solutions. The equations are dependent and, since a solution exists, the system is consistent.*
2. If a contradiction is obtained, such as $0 = 7$, then the system has no solution. The system is inconsistent.

Should decimals or fractions appear, it often helps to *clear* before solving.

Example 7

Solve the system

$$0.2x + 0.3y = 1.7,$$
$$\tfrac{1}{7}x + \tfrac{1}{5}y = \tfrac{29}{35}.$$

Solution   We have

$$0.2x + 0.3y = 1.7, \longrightarrow \text{Multiplying both sides by 10} \longrightarrow 2x + 3y = 17$$
$$\tfrac{1}{7}x + \tfrac{1}{5}y = \tfrac{29}{35} \longrightarrow \text{Multiplying both sides by 35} \longrightarrow 5x + 7y = 29.$$

---

*Consistent systems and dependent equations are discussed in greater detail in Section 8.4.

We multiplied both sides of the first equation by 10 to clear the decimals. Multiplication by 35, the least common denominator, clears the fractions in the second equation. The problem is now identical to Example 5. The solution is $(-32, 27)$, or $x = -32$, $y = 27$.

Comparing Methods

The following table is a summary that compares the graphical, substitution, and elimination methods for solving systems of equations.

CONNECTING THE CONCEPTS

We now have three different methods for solving systems of equations. Each method has certain strengths and weaknesses, as outlined below.

Method	Strengths	Weaknesses
Graphical	Solutions are displayed graphically. Works with any system that can be graphed.	Inexact when solutions involve numbers that are not integers. Solution may not appear on the part of the graph drawn.
Substitution	Yields exact solutions. Easy to use when a variable is alone on one side.	Introduces extensive computations with fractions when solving more complicated systems. Solutions are not displayed graphically.
Elimination	Yields exact solutions. Easy to use when fractions or decimals appear in the system. The preferred method for systems of 3 or more equations in 3 or more variables (see Section 8.4).	Solutions are not displayed graphically.

(continued)

Before selecting a method to use, try to remember the strengths and weaknesses of each method. If possible, begin solving the system mentally before settling on the method that seems best suited for that particular system. Selecting the "best" method for a problem is a bit like selecting one of three different saws with which to cut a piece of wood. The "best" choice depends on what kind of wood is being cut and what type of cut is being made, as well as your skill level with each saw.

Note that each of the three methods was introduced using a rather simple example. As the examples became more complicated, additional steps were required in order to "turn" the new problem into a more familiar format. This is a common approach in mathematics: We perform one or more steps to make a "new" problem resemble a problem we already know how to solve.

Exercise Set 8.2

For Exercises 1–48, if a system has an infinite number of solutions, use set-builder notation to write the solution set. If a system has no solution, state this.

Solve using the substitution method.

1. $y = 5 - 4x,$
$2x - 3y = 13$

2. $x = 8 - 4y,$
$3x + 5y = 3$

3. $2y + x = 9,$
$x = 3y - 3$

4. $9x - 2y = 3,$
$3x - 6 = y$

5. $3s - 4t = 14,$
$5s + t = 8$

6. $m - 2n = 16,$
$4m + n = 1$

7. $4x - 2y = 6,$
$2x - 3 = y$

8. $t = 4 - 2s,$
$t + 2s = 6$

9. $-5s + t = 11,$
$4s + 12t = 4$

10. $5x + 6y = 14,$
$-3y + x = 7$

11. $2x + 2y = 2,$
$3x - y = 1$

12. $4p - 2q = 16,$
$5p + 7q = 1$

13. $3a - b = 7,$
$2a + 2b = 5$

14. $5x + 3y = 4,$
$x - 4y = 3$

15. $2x - 3 = y,$
$y - 2x = 1$

16. $a - 2b = 3,$
$3a = 6b + 9$

Solve using the elimination method.

17. $x + 3y = 7,$
$-x + 4y = 7$

18. $x + y = 9,$
$2x - y = -3$

19. $2x + y = 6,$
$x - y = 3$

20. $x - 2y = 6,$
$-x + 3y = -4$

21. $9x + 3y = -3,$
$2x - 3y = -8$

22. $6x - 3y = 18,$
$6x + 3y = -12$

23. $5x + 3y = 19,$
$2x - 5y = 11$

24. $3x + 2y = 3,$
$9x - 8y = -2$

25. $5r - 3s = 24,$
$3r + 5s = 28$

26. $5x - 7y = -16,$
$2x + 8y = 26$

27. $6s + 9t = 12,$
$4s + 6t = 5$

28. $10a + 6b = 8,$
$5a + 3b = 2$

29. $\frac{1}{2}x - \frac{1}{6}y = 3,$
$\frac{2}{5}x + \frac{1}{2}y = 2$

30. $\frac{1}{3}x + \frac{1}{5}y = 7,$
$\frac{1}{6}x - \frac{2}{5}y = -4$

31. $\frac{x}{2} + \frac{y}{3} = \frac{7}{6},$
$\frac{2x}{3} + \frac{3y}{4} = \frac{5}{4}$

32. $\frac{2x}{3} + \frac{3y}{4} = \frac{11}{12},$
$\frac{x}{3} + \frac{7y}{18} = \frac{1}{2}$

Aha! **33.** $12x - 6y = -15,$
$-4x + 2y = 5$

34. $8s + 12t = 16,$
$6s + 9t = 12$

35. $0.2a + 0.3b = 1,$
$0.3a - 0.2b = 4$

36. $-0.4x + 0.7y = 1.3,$
$0.7x - 0.3y = 0.5$

Solve using any appropriate method.

37. $a - 2b = 16,$
$b + 3 = 3a$

38. $5x - 9y = 7,$
$7y - 3x = -5$

39. $10x + y = 306,$
$10y + x = 90$

40. $3(a - b) = 15,$
$4a = b + 1$

41. $3y = x - 2,$
$x = 2 + 3y$

42. $x + 2y = 8,$
$x = 4 - 2y$

43. $3s - 7t = 5,$
$7t - 3s = 8$

44. $2s - 13t = 120,$
$-14s + 91t = -840$

45. $0.05x + 0.25y = 22,$
$0.15x + 0.05y = 24$

46. $2.1x - 0.9y = 15,$
$-1.4x + 0.6y = 10$

47. $13a - 7b = 9,$
$2a - 8b = 6$

48. $3a - 12b = 9,$
$14a - 11b = 5$

49. Describe a procedure that can be used to write an inconsistent system of equations.

50. Describe a procedure that can be used to write a system that has an infinite number of solutions.

SKILL MAINTENANCE

51. The fare for a taxi ride from Johnson Street to Elm Street is $5.20. If the rate of the taxi is $1.00 for the first $\frac{1}{2}$ mi and 30¢ for each additional $\frac{1}{4}$ mi, how far is it from Johnson Street to Elm Street?

52. A student's average after 4 tests is 78.5. What score is needed on the fifth test in order to raise the average to 80?

53. *Home remodeling.* In a recent year, Americans spent $35 billion to remodel bathrooms and kitchens. Twice as much was spent on kitchens as on bathrooms. (*Source: Indianapolis Star*) How much was spent on each?

54. A 480-m wire is cut into three pieces. The second piece is three times as long as the first. The third is four times as long as the second. How long is each piece?

55. *Car rentals.* Badger Rent-A-Car rents a compact car at a daily rate of $34.95 plus 10¢ per mile. A businessperson is allotted $80 for car rental. How many miles can she travel on the $80 budget?

56. *Car rentals.* Badger rents midsized cars at a rate of $43.95 plus 10¢ per mile. A tourist has a car-rental budget of $90. How many miles can he travel on the $90?

SYNTHESIS

57. Some systems are more easily solved by substitution and some are more easily solved by elimination. Write guidelines that could be used to help someone determine which method to use.

58. Explain how it is possible to solve Exercise 33 mentally.

59. If $(1, 2)$ and $(-3, 4)$ are two solutions of $f(x) = mx + b$, find m and b.

60. If $(0, -3)$ and $\left(-\frac{3}{2}, 6\right)$ are two solutions of $px - qy = -1$, find p and q.

61. Determine a and b for which $(-4, -3)$ is a solution of the system

$ax + by = -26,$

$bx - ay = 7.$

62. Solve for x and y in terms of a and b:

$5x + 2y = a,$

$x - y = b.$

Solve.

63. $\dfrac{x+y}{2} - \dfrac{x-y}{5} = 1,$

$\dfrac{x-y}{2} + \dfrac{x+y}{6} = -2$

64. $3.5x - 2.1y = 106.2,$
$4.1x + 16.7y = -106.28$

Each of the following is a system of nonlinear equations. However, each is reducible to linear, since an appropriate substitution (say, u for $1/x$ and v for $1/y$) yields a linear system. Make such a substitution, solve for the new variables, and then solve for the original variables.

65. $\dfrac{2}{x} + \dfrac{1}{y} = 0,$

$\dfrac{5}{x} + \dfrac{2}{y} = -5$

66. $\dfrac{1}{x} - \dfrac{3}{y} = 2,$

$\dfrac{6}{x} + \dfrac{5}{y} = -34$

67. A student solving the system

$17x + 19y = 102,$

$136x + 152y = 826$

graphs both equations on a grapher and gets the following screen.

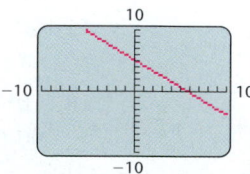

The student then (incorrectly) concludes that the equations are dependent and the solution set is infinite. How can algebra be used to convince the student that a mistake has been made?

CORNER

How Many Two's? How Many Three's?

Focus: Systems of linear equations

Time: 20 minutes

Group size: 3

The box score at right, from a basketball game in 2001 between the Philadelphia 76ers and the Indiana Pacers, contains information on how many field goals and free throws each player attempted and made. For example, the line "Rose 9–19 7–7 26" means that Indiana's Jalen Rose made 9 field goals out of 19 attempts and 7 free throws out of 7 attempts, for a total of 26 points. (Each free throw is worth 1 point and each field goal is worth either 2 or 3 points, depending on how far from the basket it was shot.)

ACTIVITY

1. Work as a group to develop a system of two equations in two unknowns that can be used to determine how many 2-pointers and how many 3-pointers were made by Philadelphia's Allen Iverson.

2. Each group member should solve the system from part (1) in a different way: one person algebraically, one person by making a table and methodically checking all combinations

of 2- and 3-pointers, and one person by guesswork. Compare answers when this has been completed.

3. Determine, as a group, how many 2- and 3-pointers the Philadelphia 76ers made as a team.

■**76ers 104, Pacers 93:** At Philadelphia, Allen Iverson scored 37 points and had seven rebounds as Philadelphia beat Indiana, defeating the team that knocked the 76ers out of the playoffs the past two years. The victory was only the third in nine games for the 76ers, the team with the best record in the East. (*Source:* AP in the *Burlington Free Press,* 4/2/01)

Indiana (93)
Harrington 2-8 0-0 4, O'Neal 2-5 1-1 5, Perkins 0-4 0-0 0, Rose 9-19 7-7 26, R. Miller 3-14 4-4 11, Best 4-5 3-4 11, Croshere 3-9 15-16 21, Edney 2-6 4-4 8, Foster 1-2 1-4 3, Bender 2-2 0-0 4.
Totals 28-74 35-40 93.

Philadelphia (104)
Lynch 2-5 0-0 4, Hill 6-7 3-3 15, Mutombo 0-2 5-6 5, Snow 4-16 2-2 10, Iverson 12-27 10-15 37, Jones 4-11 3-4 11, Geiger 4-5 1-2 9, Ollie 2-4 4-6 8, Buford 1-3 0-0 2, Sanchez 0-0 0-0 0, MacCulloch 1-1 1-2 3.
Totals 36-81 29-40 104.

Indiana	18	24	21	30	—	93
Philadelphia	25	29	23	27	—	104

Solving Applications: Systems of Two Equations

8.3

Total-Value and Mixture Problems • Motion Problems

You are in a much better position to solve problems now that systems of equations can be used. Using systems often makes the translating step easier.

Example 1

Real estate.   In 1996, the Simon Property Group and the DeBartolo Realty Corporation merged to form the largest real estate company in the United States, owning 183 shopping centers in 32 states. Prior to merging, Simon owned twice as many properties as DeBartolo. How many properties did each company own before the merger?

Solution   The *Familiarize* and *Translate* steps have been done in Example 1 of Section 8.1. The resulting system of equations is

$$x + y = 183,$$
$$x = 2y,$$

where x is the number of properties originally owned by Simon and y is the number of properties originally owned by DeBartolo.

3. **Carry out.**   We solve the system of equations. Since one equation already has a variable isolated, let's use the substitution method:

$$x + y = 183$$
$$2y + y = 183 \qquad \text{Substituting } 2y \text{ for } x$$
$$3y = 183 \qquad \text{Combining like terms}$$
$$y = 61.$$

We return to the second equation and substitute 61 for y and compute x:

$$x = 2y = 2 \cdot 61 = 122.$$

Apparently, Simon owned 122 properties and DeBartolo 61.

4. **Check.**   The sum of 122 and 61 is 183, so the total number of properties is correct. Since 122 is twice 61, the numbers check.

5. **State.**   Prior to merging, Simon owned 122 properties and DeBartolo owned 61.

Total-Value and Mixture Problems

Example 2

Purchasing.   Recently the Champlain Valley Community Music Center purchased 120 stamps for \$30.30. If the stamps were a combination of 20¢ postcard stamps and 34¢ first-class stamps, how many of each type were bought?

Solution   The *Familiarize* and *Translate* steps were completed in Example 2 of Section 8.1.

3. **Carry out.**   We are to solve the system of equations

$$p + f = 120, \qquad (1)$$
$$20p + 34f = 3030 \qquad (2) \qquad \text{Working in cents rather than dollars}$$

where p is the number of postcard stamps bought and f is the number of first-class stamps bought. Because both equations are in the form $Ax + By = C$, let's use the elimination method to solve the system. We can eliminate p by multiplying both sides of equation (1) by -20 and adding them to the corresponding sides of equation (2):

$$-20p - 20f = -2400 \qquad \text{Multiplying both sides of equation (1) by } -20$$
$$\underline{20p + 34f = 3030}$$
$$14f = 630 \qquad \text{Adding}$$
$$f = 45. \qquad \text{Solving for } f$$

To find p, we substitute 45 for f in equation (1) and then solve for p:

$$p + f = 120 \qquad \text{Equation (1)}$$
$$p + 45 = 120 \qquad \text{Substituting 45 for } f$$
$$p = 75. \qquad \text{Solving for } p$$

We obtain $(45, 75)$, or $f = 45$ and $p = 75$.

4. **Check.**   We check in the original problem. Recall that f is the number of first-class stamps and p the number of postcard stamps.

Number of stamps: $f + p = 45 + 75 = 120$

Cost of first-class stamps: $\$0.34f = 0.34 \times 45 = \$15.30$

Cost of postcard stamps: $\$0.20p = 0.20 \times 75 = \underline{\$15.00}$

$$\text{Total} = \$30.30$$

The numbers check.

5. **State.**   The music center bought 45 first-class stamps and 75 postcard stamps.

Example 2 involved two types of items (first-class stamps and postcard stamps), the quantity of each type bought, and the total value of the items. We refer to this type of problem as a *total-value problem.*

E x a m p l e   3

Blending teas.   Tara's Tea Terrace sells loose Black tea for 95¢ an ounce and Lapsang Souchong for $1.43 an ounce. Tara wants to make a 1-lb mixture of the two types, called Imperial Blend, that sells for $1.10 an ounce. How much tea of each type should Tara use?

Solution

1. **Familiarize.**   This problem is similar to Example 2. Rather than postcard stamps and first-class stamps, we have ounces of Black tea and ounces of Lapsang Souchong. Instead of a different price for each type of stamp, we have a different price per ounce for each type of tea. Finally, rather than knowing the total cost of the stamps, we know the weight and the price per

ounce of the Imperial Blend. It is important to note that we can find the total value of the blend by multiplying 16 ounces (1 lb) times $1.10 per ounce. Although we could make and check a guess, we proceed to let $b =$ the number of ounces of Black tea and $l =$ the number of ounces of Lapsang Souchong.

2. **Translate.** Since a 16-oz batch is being made, we must have

$$b + l = 16.$$

To find a second equation, note that the total value of the 16-oz blend must match the combined value of the separate ingredients:

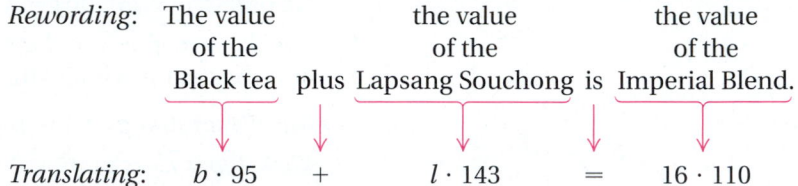

These equations can also be obtained from a table.

	Black Tea	Lapsang Souchong	Imperial Blend	
Number of Ounces	b	l	16	$\rightarrow b + l = 16$
Price per Ounce	95¢	143¢	110¢	
Value of Tea	$95b$	$143l$	$16 \cdot 110$, or 1760¢	$\rightarrow 95b + 143l = 1760$

We have translated to a system of equations:

$$b + \quad l = 16, \qquad (1)$$
$$95b + 143l = 1760. \qquad (2)$$

3. **Carry out.** We can solve using substitution. When equation (1) is solved for b, we have $b = 16 - l$. Substituting $16 - l$ for b in equation (2), we find l:

$$95(16 - l) + 143l = 1760 \qquad \text{Substituting}$$
$$1520 - 95l + 143l = 1760 \qquad \text{Using the distributive law}$$
$$48l = 240 \qquad \text{Combining like terms; subtracting 1520 from both sides}$$
$$l = 5. \qquad \text{Dividing both sides by 48}$$

We have $l = 5$ and, from equation (1) above, $b + l = 16$. Thus, $b = 11$.

4. **Check.** If 11 oz of Black tea and 5 oz of Lapsang Souchong are combined, a 16-oz, or 1-lb, blend will result. The value of 11 oz of Black tea is $11($0.95)$, or $10.45. The value of 5 oz of Lapsang Souchong is $5($1.43)$, or $7.15, so the combined value of the blend is $10.45 + $7.15 = 17.60. A 16-oz batch, priced at $1.10 an ounce, would also be worth $17.60, so our answer checks.

5. State. The Imperial Blend should be made by combining 11 oz of Black tea with 5 oz of Lapsang Souchong.

E x a m p l e   4

Student loans. Dawn's student loans totaled $9600. Part was a Perkins loan made at 5% interest and the rest was a Federal Education Loan made at 8% interest. After one year, Dawn's loans accumulated $633 in interest. What was the original amount of each loan?

Solution

1. Familiarize. We begin with a guess. If $7000 was borrowed at 5% and $2600 was borrowed at 8%, the two loans would total $9600. The interest would then be 0.05($7000), or $350, and 0.08($2600), or $208, for a total of $558 in interest. Our guess was wrong, but checking the guess familiarized us with the problem.

2. Translate. We let $p =$ the amount of the Perkins loan and $f =$ the amount of the Federal Education Loan. We then organize a table in which each column comes from the formula for simple interest:

$$Principal \cdot Rate \cdot Time = Interest.$$

	Perkins Loan	Federal Loan	Total	
Principal	p	f	$9600	$\longrightarrow p + f = 9600$
Rate of Interest	5%	8%		
Time	1 yr	1 yr		
Interest	0.05p	0.08f	$633	$\longrightarrow 0.05p + 0.08f = 633$

The total amount borrowed is found in the first row of the table:

$$p + f = 9600.$$

A second equation, representing the accumulated interest, can be found in the last row:

$$0.05p + 0.08f = 633, \quad \text{or} \quad 5p + 8f = 63{,}300. \qquad \text{Clearing decimals}$$

3. Carry out. The system can be solved by elimination:

$$
\begin{aligned}
p + f &= 9600, \\
5p + 8f &= 63{,}300.
\end{aligned}
\quad\xrightarrow[\text{sides by } -5]{\text{Multiplying both}}\quad
\begin{aligned}
-5p - 5f &= -48{,}000 \\
\underline{5p + 8f} &= \underline{63{,}300} \\
3f &= 15{,}300
\end{aligned}
$$

$$
\begin{aligned}
p + f &= 9600 \quad\longleftarrow\quad f = 5100 \\
p + 5100 &= 9600 \\
p &= 4500.
\end{aligned}
$$

We find that $p = 4500$ and $f = 5100$.

4. **Check.** The total amount borrowed is $4500 + $5100, or $9600. The interest on $4500 at 5% for 1 yr is 0.05($4500), or $225. The interest on $5100 at 8% for 1 yr is 0.08($5100), or $408. The total amount of interest is $225 + $408, or $633, so the numbers check.

5. **State.** The Perkins loan was for $4500 and the Federal Education Loan was for $5100.

Before proceeding to Example 5, briefly scan Examples 2–4 for similarities. Note that in each case, one of the equations in the system is a simple sum while the other equation represents a sum of products. Example 5 continues this pattern with what is commonly called a *mixture problem*.

> ### Problem-Solving Tip
>
> When solving a problem, see if it is patterned or modeled after a problem that you have already solved.

E x a m p l e   5

Mixing fertilizers. Yardbird Gardening, Inc., carries two brands of fertilizer containing nitrogen and water. "Gently Green" is 5% nitrogen and "Sun Saver" is 15% nitrogen. Yardbird Gardening needs to combine the two types of solutions in order to make 100 L of a solution that is 12% nitrogen. How much of each brand should be used?

Solution

1. **Familiarize.** We make a drawing and then make a guess to gain familiarity with the problem.

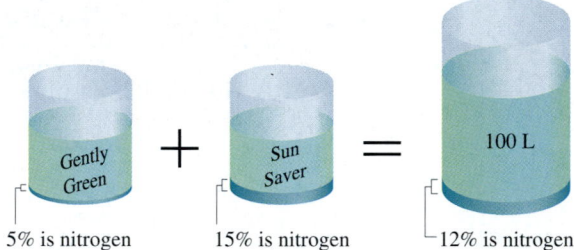

Suppose that 40 L of Gently Green and 60 L of Sun Saver are mixed. The resulting mixture will be the right size, 100 L, but will it be the right strength? To find out, note that 40 L of Gently Green would contribute $0.05(40) = 2$ L of nitrogen to the mixture while 60 L of Sun Saver would contribute $0.15(60) = 9$ L of nitrogen to the mixture. Altogether, 40 L of Gently Green and 60 L of Sun Saver would make 100 L of a mixture that has $2 + 9 = 11$ L of nitrogen. Since this would mean that the final mixture is only 11% nitrogen, our guess of 40 L and 60 L is incorrect. Still, the process of checking our guess has familiarized us with the problem.

2. **Translate.** Let g = the number of liters of Gently Green and s = the number of liters of Sun Saver. The information can be organized in a table.

	Gently Green	Sun Saver	Mixture
Number of Liters	g	s	100
Percent of Nitrogen	5%	15%	12%
Amount of Nitrogen	0.05g	0.15s	0.12×100, or 12 liters

$\longrightarrow g + s = 100$

$\longrightarrow 0.05g + 0.15s = 12$

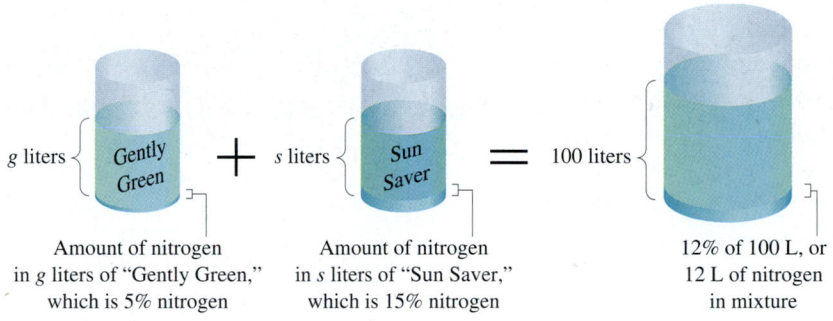

g liters $\{$ Gently Green	$+$	s liters $\{$ Sun Saver	$=$	100 liters $\{$

Amount of nitrogen in g liters of "Gently Green," which is 5% nitrogen

Amount of nitrogen in s liters of "Sun Saver," which is 15% nitrogen

12% of 100 L, or 12 L of nitrogen in mixture

If we add g and s in the first row, we get one equation. It represents the total amount of mixture: $g + s = 100$.

If we add the amounts of nitrogen listed in the third row, we get a second equation. This equation represents the amount of nitrogen in the mixture: $0.05g + 0.15s = 12$.

After clearing decimals, we have translated the problem to the system

$$g + \quad s = 100, \qquad (1)$$
$$5g + 15s = 1200. \qquad (2)$$

3. **Carry out.**   We use the elimination method to solve the system:

$$-5g - \quad 5s = -500 \qquad \text{Multiplying both sides of equation (1) by } -5$$
$$\underline{5g + 15s = \quad 1200}$$
$$10s = \quad 700 \qquad \text{Adding}$$
$$s = \quad 70; \qquad \text{Solving for } s$$

$$g + 70 = 100 \qquad \text{Substituting into equation (1)}$$
$$g = 30. \qquad \text{Solving for } g$$

4. **Check.**   Remember, g is the number of liters of Gently Green and s is the number of liters of Sun Saver.

Total amount of mixture: $\qquad\qquad g + s = 30 + 70 = 100$

Total amount of nitrogen: $\qquad 5\% \text{ of } 30 + 15\% \text{ of } 70 = 1.5 + 10.5 = 12$

Percentage of nitrogen in mixture:
$$\frac{\text{Total amount of nitrogen}}{\text{Total amount of mixture}} = \frac{12}{100} = 12\%$$

The numbers check in the original problem.

5. **State.** Yardbird Gardening should mix 30 L of Gently Green with 70 L of Sun Saver.

Motion Problems

When a problem deals with distance, speed (rate), and time, recall the following.

> ### Distance, Rate, and Time Equations
>
> If r represents rate, t represents time, and d represents distance, then:
>
> $$d = rt, \qquad r = \frac{d}{t}, \quad \text{and} \quad t = \frac{d}{r}.$$

Be sure to remember at least one of these equations. The others can be obtained by using algebraic manipulations as needed.

Example 6

Train travel. A Vermont Railways freight train, loaded with logs, leaves Boston, heading to Washington D.C. at a speed of 60 km/h. Two hours later, an Amtrak® Metroliner leaves Boston, bound for Washington D.C., on a parallel track at 90 km/h. At what point will the Metroliner catch up to the freight train?

Solution

1. **Familiarize.** Let's make a guess—say, 180 km—and check to see if it is correct. The freight train, traveling 60 km/h, would reach a point 180 km from Boston in $\frac{180}{60} = 3$ hr. The Metroliner, traveling 90 km/h, would cover 180 km in $\frac{180}{90} = 2$ hr. Since 3 hr is *not* two hours more than 2 hr, our guess of 180 km is incorrect. Although our guess is wrong, we see that the time that the trains are running and the point at which they meet are both unknown. We let $t =$ the number of hours that the freight train is running before they meet and $d =$ the distance at which the trains meet. Since the freight train has a 2-hr head start, the Metroliner runs for $t - 2$ hours before catching up to the freight train, at which point both trains have traveled the same distance.

60 km/h
d kilometers
t hours

90 km/h
d kilometers
$t - 2$ hours

Trains meet here

2. **Translate.**   We can organize the information in a chart. Each row is determined by the formula *Distance = Rate · Time*.

	Distance	Rate	Time	
Freight Train	d	60	t	⟶ $d = 60t$
Metroliner	d	90	$t - 2$	⟶ $d = 90(t - 2)$

Using *Distance = Rate · Time* twice, we get two equations:

$$d = 60t, \qquad (1)$$
$$d = 90(t - 2). \qquad (2)$$

3. **Carry out.**   We solve the system using substitution:

$$60t = 90(t - 2) \qquad \text{Substituting } 60t \text{ for } d \text{ in equation (2)}$$
$$60t = 90t - 180$$
$$-30t = -180$$
$$t = 6.$$

The time for the freight train is 6 hr, which means that the time for the Metroliner is $6 - 2$, or 4 hr. Remember that it is distance, not time, that the problem asked for. Thus for $t = 6$, we have $d = 60 \cdot 6 = 360$ km.

4. **Check.**   At 60 km/h, the freight train will travel $60 \cdot 6$, or 360 km, in 6 hr. At 90 km/h, the Metroliner will travel $90 \cdot (6 - 2) = 360$ km in 4 hr. The numbers check.

5. **State.**   The freight train will catch up to the Metroliner at a point 360 km from Boston.

E x a m p l e   7

Jet travel.   An F16 jet flies 4 hr west with a 60-mph tailwind. Returning *against* the wind takes 5 hr. Find the speed of the jet with no wind.

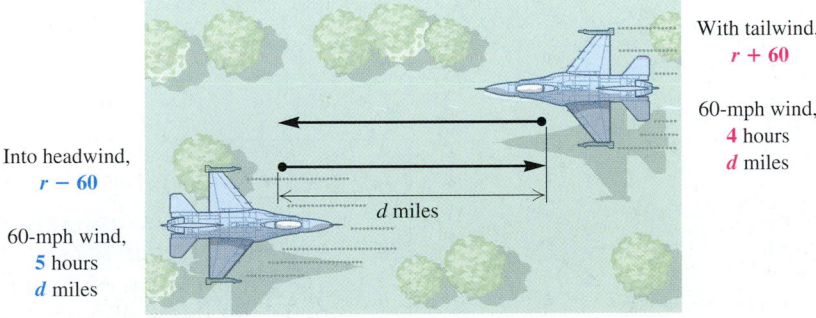

With tailwind,
r + 60

60-mph wind,
4 hours
d miles

Into headwind,
r − 60

60-mph wind,
5 hours
d miles

d miles

Solution

1. **Familiarize.**   We imagine the situation and make a drawing. Note that the wind *speeds up* the jet on the outbound flight, but *slows down* the jet on the return flight. Since the distances traveled each way must be the same, we can check a guess of the jet's speed with no wind. Suppose the speed of the

jet with no wind is 400 mph. The jet would then fly $400 + 60 = 460$ mph with the wind and $400 - 60 = 340$ mph into the wind. In 4 hr, the jet would travel $460 \cdot 4 = 1840$ mi with the wind and $340 \cdot 5 = 1700$ mi against the wind. Since $1840 \neq 1700$, our guess of 400 mph is incorrect. Rather than guess again, let's have $r =$ the speed, in miles per hour, of the jet in still air. Then $r + 60 =$ the jet's speed with the wind and $r - 60 =$ the jet's speed against the wind. We also let $d =$ the distance traveled, in miles.

2. **Translate.**   The information can be organized in a chart. The distances traveled are the same, so we use *Distance = Rate* (or *Speed*) *· Time*. Each row of the chart gives an equation.

	Distance	Rate	Time	
With Wind	d	$r + 60$	4	$\longrightarrow d = (r + 60)4$
Against Wind	d	$r - 60$	5	$\longrightarrow d = (r - 60)5$

The two equations constitute a system:

$$d = (r + 60)4, \qquad (1)$$
$$d = (r - 60)5. \qquad (2)$$

3. **Carry out.**   We solve the system using substitution:

$$(r - 60)5 = (r + 60)4 \qquad \text{Substituting } (r - 60)5 \text{ for } d \text{ in equation (1)}$$
$$5r - 300 = 4r + 240 \qquad \text{Using the distributive law}$$
$$r = 540. \qquad \text{Solving for } r$$

4. **Check.**   When $r = 540$, the speed with the wind is $540 + 60 = 600$ mph, and the speed against the wind is $540 - 60 = 480$ mph. The distance with the wind, $600 \cdot 4 = 2400$ mi, matches the distance into the wind, $480 \cdot 5 = 2400$ mi, so we have a check.

5. **State.**   The speed of the jet with no wind is 540 mph.

Tips for Solving Motion Problems

1. Draw a diagram using an arrow or arrows to represent distance and the direction of each object in motion.
2. Organize the information in a chart.
3. Look for times, distances, or rates that are the same. These often can lead to an equation.
4. Translating to a system of equations allows for the use of two variables.
5. Always make sure that you have answered the question asked.

Exercise Set 8.3

1.–18. *For Exercises 1–18, solve Exercises 33–50 from pp. 482–483.*

19. *Sales.* Staples® recently sold a box of Flair® felt-tip pens for $12 and a four-pack of Sanford® Uni-ball® pens for $8. At the start of a recent fall semester, a combination of 40 boxes and four-packs of these pens was sold for a total of $372. How many of each type were purchased?

20. *Sales.* Staples recently sold a wirebound graph-paper notebook for $2.50 and a college-ruled note-book made of recycled paper for $2.30. At the start of a recent spring semester, a combination of 50 of these notebooks was sold for a total of $118.60. How many of each type were sold?

21. *Blending coffees.* The Coffee Counter charges $9.00 per pound for Kenyan French Roast coffee and $8.00 per pound for Sumatran coffee. How much of each type should be used to make a 20-lb blend that sells for $8.40 per pound?

22. *Mixed nuts.* The Nutty Professor sells cashews for $6.75 per pound and Brazil nuts for $5.00 per pound. How much of each type should be used to make a 50-lb mixture that sells for $5.70 per pound?

23. *Catering.* Casella's Catering is planning a wedding reception. The bride and groom would like to serve a nut mixture containing 25% peanuts. Casella has available mixtures that are either 40% or 10% peanuts. How much of each type should be mixed to get a 10-lb mixture that is 25% peanuts?

24. *Livestock feed.* Soybean meal is 16% protein and corn meal is 9% protein. How many pounds of each should be mixed to get a 350-lb mixture that is 12% protein?

25. *Ink remover.* Etch Clean Graphics uses one cleanser that is 25% acid and a second that is 50% acid. How many liters of each should be mixed to get 10 L of a solution that is 40% acid?

26. *Blending granola.* Deep Thought Granola is 25% nuts and dried fruit. Oat Dream Granola is 10% nuts and dried fruit. How much of Deep Thought and how much of Oat Dream should be mixed to form a 20-lb batch of granola that is 19% nuts and dried fruit?

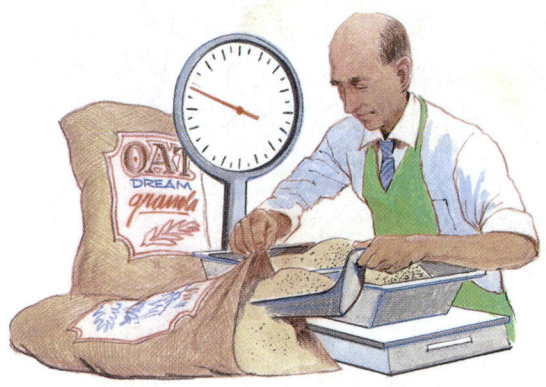

27. *Student loans.* Lomasi's two student loans totaled $12,000. One of her loans was at 6% simple interest and the other at 9%. After one year, Lomasi owed $855 in interest. What was the amount of each loan?

28. *Investments.* An executive nearing retirement made two investments totaling $15,000. In one year, these investments yielded $1432 in simple interest. Part of the money was invested at 9% and the rest at 10%. How much was invested at each rate?

29. *Automotive maintenance.* "Arctic Antifreeze" is 18% alcohol and "Frost No-More" is 10% alcohol. How many liters of each should be mixed to get 20 L of a mixture that is 15% alcohol?

30. *Food science.*   The following bar graph shows the milk fat percentages in three dairy products. How many pounds each of whole milk and cream should be mixed to form 200 lb of milk for cream cheese?

31. *Real estate.*   The perimeter of an oceanfront lot is 190 m. The width is one fourth of the length. Find the dimensions.

32. *Architecture.*   The rectangular ground floor of the John Hancock building has a perimeter of 860 ft. The length is 100 ft more than the width. Find the length and the width.

x

$x + 100$

33. *Making change.*   Cecilia makes a $9.25 purchase at the bookstore with a $20 bill. The store has no bills and gives her the change in quarters and fifty-cent pieces. There are 30 coins in all. How many of each kind are there?

34. *Teller work.*   Ashford goes to a bank and gets change for a $50 bill consisting of all $5 bills and $1 bills. There are 22 bills in all. How many of each kind are there?

35. *Train travel.*   A train leaves Danville Junction and travels north at a speed of 75 km/h. Two hours later, an express train leaves on a parallel track and travels north at 125 km/h. How far from the station will they meet?

36. *Car travel.*   Two cars leave Salt Lake City, traveling in opposite directions. One car travels at a speed of 80 km/h and the other at 96 km/h. In how many hours will they be 528 km apart?

37. *Canoeing.*   Alvin paddled for 4 hr with a 6-km/h current to reach a campsite. The return trip against the same current took 10 hr. Find the speed of Alvin's canoe in still water.

38. *Boating.*   Mia's motorboat took 3 hr to make a trip downstream with a 6-mph current. The return trip against the same current took 5 hr. Find the speed of the boat in still water.

39. *Point of no return.*   A plane flying the 3458-mi trip from New York City to London has a 50-mph tailwind. The flight's *point of no return* is the point at which the flight time required to return to New York is the same as the time required to continue to London. If the speed of the plane in still air is 360 mph, how far is New York from the point of no return?

40. *Point of no return.*   A plane is flying the 2553-mi trip from Los Angeles to Honolulu into a 60-mph headwind. If the speed of the plane in still air is 310 mph, how far from Los Angeles is the plane's point of no return? (See Exercise 39.)

41. Write at least three study tips of your own for someone beginning this exercise set.

42. In what ways are Examples 3 and 4 similar? In what sense are their systems of equations similar?

SKILL MAINTENANCE

Evaluate.

43. $2x - 3y + 12$, for $x = 5$ and $y = 2$

44. $7x - 4y + 9$, for $x = 2$ and $y = 3$

45. $5a - 7b + 3c$, for $a = -2$, $b = 3$, and $c = 1$

46. $3a - 8b - 2c$, for $a = -4$, $b = -1$, and $c = 3$

47. $4 - 2y + 3z$, for $y = \frac{1}{3}$ and $z = \frac{1}{4}$

48. $3 - 5y + 4z$, for $y = \frac{1}{2}$ and $z = \frac{1}{5}$

SYNTHESIS

49. Suppose that in Example 3 you are asked only for the amount of Black tea needed for the Imperial Blend. Would the method of solving the problem change? Why or why not?

50. Write a problem similar to Example 2 for a classmate to solve. Design the problem so that the solution is "The florist sold 14 hanging plants and 9 flats of petunias."

51.–54. *For Exercises 51–54, solve Exercises 66–69 from Exercise Set 8.1.*

55. *Retail.*   Some of the world's best and most expensive coffee is Hawaii's Kona coffee. In order for coffee to be labeled "Kona Blend," it must contain at least 30% Kona beans. Bean Town Roasters has 40 lb of Mexican coffee. How much Kona coffee must they add if they wish to market it as Kona Blend?

56. *Automotive maintenance.*   The radiator in Michelle's car contains 6.3 L of antifreeze and water. This mixture is 30% antifreeze. How much of this mixture should she drain and replace with pure antifreeze so that there will be a mixture of 50% antifreeze?

57. *Exercise.*   Natalie jogs and walks to school each day. She averages 4 km/h walking and 8 km/h jogging. From home to school is 6 km and Natalie makes the trip in 1 hr. How far does she jog in a trip?

58. *Book sales.*   A limited edition of a book published by a historical society was offered for sale to members. The cost was one book for $12 or two books for $20 (maximum of two per member). The society sold 880 books, for a total of $9840. How many members ordered two books?

59. The tens digit of a two-digit positive integer is 2 more than three times the units digit. If the digits are interchanged, the new number is 13 less than half the given number. Find the given integer. (*Hint*: Let x = the tens-place digit and y = the units-place digit; then $10x + y$ is the number.)

60. *Train travel.*   A train leaves Union Station for Central Station, 216 km away, at 9 A.M. One hour later, a train leaves Central Station for Union Station. They meet at noon. If the second train had started at 9 A.M. and the first train at 10:30 A.M., they would still have met at noon. Find the speed of each train.

61. *Wood stains.*   Williams' Custom Flooring has 0.5 gal of stain that is 20% brown and 80% neutral. A customer orders 1.5 gal of a stain that is 60% brown and 40% neutral. How much pure brown stain and how much neutral stain should be added to the original 0.5 gal in order to make up the order?*

62. *Fuel economy.*   Grady's station wagon gets 18 miles per gallon (mpg) in city driving and 24 mpg in highway driving. The car is driven 465 mi on 23 gal of gasoline. How many miles were driven in the city and how many were driven on the highway?

63. *Biochemistry.*   Industrial biochemists routinely use a machine to mix a buffer of 10% acetone by adding 100% acetone to water. One day, instead of adding 5 L of acetone to create a vat of buffer, a machine added 10 L. How much additional water was needed to bring the concentration down to 10%?

64. *Gender.*   Phil and Phyllis are siblings. Phyllis has twice as many brothers as she has sisters. Phil has the same number of brothers as sisters. How many girls and how many boys are in the family?

65. See Exercise 61 above. Let x = the amount of pure brown stain added to the original 0.5 gal. Find a function $P(x)$ that can be used to determine the percentage of brown stain in the 1.5-gal mixture. On a grapher, draw the graph of P and use INTERSECT to confirm the answer to Exercise 61.

---

*This problem was suggested by Professor Chris Burditt of Yountville, California.

Systems of Equations in Three Variables

8.4

Identifying Solutions • Solving Systems in Three Variables • Dependency, Inconsistency, and Geometric Considerations

---

CONNECTING THE CONCEPTS

As often happens in mathematics, once an idea is thoroughly understood, it can be extended to increasingly more complicated problems. This is precisely the situation for the material in Sections 8.4–8.7: We will extend the elimination method of Section 8.2 to systems of three equations in three unknowns. Although we will not do so in this text, the approach that we use can be further extended to systems with four equations in four unknowns, five equations in five unknowns, and so on.

Another common occurrence in mathematics is the streamlining of a sequence of steps that are used repeatedly. In Sections 8.6 and 8.7, we develop different notations that streamline the calculations of Sections 8.2 and 8.4.

Some problems translate directly to two equations. Others more naturally call for a translation to three or more equations. In this section, we learn how to solve systems of three linear equations. Later, we will use such systems in problem-solving situations.

Identifying Solutions

A **linear equation in three variables** is an equation equivalent to one in the form $Ax + By + Cz = D$, where A, B, C, and D are real numbers. We refer to the form $Ax + By + Cz = D$ as *standard form* for a linear equation in three variables.

A solution of a system of three equations in three variables is an ordered triple (x, y, z) that makes *all three* equations true.

Example 1   Determine whether $\left(\frac{3}{2}, -4, 3\right)$ is a solution of the system

$$4x - 2y - 3z = 5,$$
$$-8x - y + z = -5,$$
$$2x + y + 2z = 5.$$

Solution   We substitute $\left(\frac{3}{2}, -4, 3\right)$ into the three equations, using alphabetical order:

$$
\begin{array}{c}
4x - 2y - 3z = 5 \\
\hline
4 \cdot \frac{3}{2} - 2(-4) - 3 \cdot 3 \ ?\ 5 \\
6 + 8 - 9 \\
5 \ \Big|\ 5 \quad \text{TRUE}
\end{array}
\qquad
\begin{array}{c}
-8x - y + z = -5 \\
\hline
-8 \cdot \frac{3}{2} - (-4) + 3 \ ?\ -5 \\
-12 + 4 + 3 \\
-5 \ \Big|\ -5 \quad \text{TRUE}
\end{array}
$$

$$
\begin{array}{c}
2x + y + 2z = 5 \\
\hline
2 \cdot \frac{3}{2} + (-4) + 2 \cdot 3 \ ?\ 5 \\
3 - 4 + 6 \\
5 \ \Big|\ 5 \quad \text{TRUE}
\end{array}
$$

The triple makes all three equations true, so it is a solution.

Solving Systems in Three Variables

Graphical methods for solving linear equations in three variables are problematic, because a three-dimensional coordinate system is required and the graph of a linear equation in three variables is a plane. The substitution method *can* be used but becomes very cumbersome unless one or more of the equations has only two variables. Fortunately, the elimination method allows us to manipulate a system of three equations in three variables so that a simpler system of two equations in two variables is formed. Once that simpler system has been solved, we can substitute into one of the three original equations and solve for the third variable.

Example 2

Solve the following system of equations:

$$
\begin{aligned}
x + y + z &= 4, & (1) \\
x - 2y - z &= 1, & (2) \\
2x - y - 2z &= -1. & (3)
\end{aligned}
$$

Solution   We select *any* two of the three equations and work to get one equation in two variables. Let's add equations (1) and (2):

$$
\begin{array}{ll}
x + y + z = 4 & (1) \\
\underline{x - 2y - z = 1} & (2) \\
2x - y \quad\ = 5. & (4) \qquad \text{Adding to eliminate } z
\end{array}
$$

Next, we select a different pair of equations and eliminate the *same variable* we did above. Let's use equations (1) and (3) to again eliminate z. Be careful here! A common error is to eliminate a different variable in this step.

$$
\begin{aligned}
x + y + z &= 4, \\
2x - y - 2z &= -1
\end{aligned}
\quad \xrightarrow[\text{equation (1) by 2}]{\text{Multiplying both sides of}} \quad
\begin{aligned}
2x + 2y + 2z &= 8 \\
\underline{2x - y - 2z} &= -1 \\
4x + y \quad\ &= 7 \quad (5)
\end{aligned}
$$

Now we solve the resulting system of equations (4) and (5). That solution will give us two of the numbers in the solution of the original system.

$$2x - y = 5 \qquad (4)$$
$$\underline{4x + y = 7} \qquad (5)$$
$$6x = 12 \qquad \text{Adding}$$
$$x = 2$$

Note that we now have two equations in two variables. Had we not eliminated the same variable in both of the above steps, this would not be the case.

We can use either equation (4) or (5) to find y. We choose equation (5):

$$4x + y = 7 \qquad (5)$$
$$4 \cdot 2 + y = 7 \qquad \text{Substituting 2 for } x \text{ in equation (5)}$$
$$8 + y = 7$$
$$y = -1.$$

We now have $x = 2$ and $y = -1$. To find the value for z, we use any of the original three equations and substitute to find the third number, z. Let's use equation (1) and substitute our two numbers in it:

$$x + y + z = 4 \qquad (1)$$
$$2 + (-1) + z = 4 \qquad \text{Substituting 2 for } x \text{ and } -1 \text{ for } y$$
$$1 + z = 4$$
$$z = 3.$$

We have obtained the triple $(2, -1, 3)$. It should check in *all three* equations:

$$\begin{array}{c|c}\underline{x + y + z = 4}\\ 2 + (-1) + 3 ? 4\\ 4 \mid 4 \quad \text{TRUE}\end{array} \qquad \begin{array}{c|c}\underline{x - 2y - z = 1}\\ 2 - 2(-1) - 3 ? 1\\ 1 \mid 1 \quad \text{TRUE}\end{array}$$

$$\begin{array}{c|c}\underline{2x - y - 2z = -1}\\ 2 \cdot 2 - (-1) - 2 \cdot 3 ? -1\\ -1 \mid -1 \quad \text{TRUE}\end{array}$$

The solution is $(2, -1, 3)$.

Solving Systems of Three Linear Equations

To use the elimination method to solve systems of three linear equations:

1. Write all equations in the standard form $Ax + By + Cz = D$.
2. Clear any decimals or fractions.
3. Choose a variable to eliminate. Then select two of the three equations and work to get one equation in two variables.
4. Next, use a different pair of equations and eliminate the same variable that you did in step (3).
5. Solve the system of equations that resulted from steps (3) and (4).
6. Substitute the solution from step (5) into one of the original three equations and solve for the third variable. Then check.

Example 3

Solve the system

$$4x - 2y - 3z = 5, \qquad (1)$$
$$-8x - y + z = -5, \qquad (2)$$
$$2x + y + 2z = 5. \qquad (3)$$

Solution

1., 2. The equations are already in standard form with no fractions or decimals.

3. Next, select a variable to eliminate. We decide on y because the y-terms are opposites of each other in equations (2) and (3). We add:

$$
\begin{array}{ll}
-8x - y + z = -5 & (2) \\
\underline{2x + y + 2z = 5} & (3) \\
-6x + 3z = 0. & (4) \qquad \text{Adding}
\end{array}
$$

4. We use another pair of equations to create a second equation in x and z. That is, we eliminate the same variable, y, as in step (3). We use equations (1) and (3):

$$
\begin{array}{l}
4x - 2y - 3z = 5, \\
2x + y + 2z = 5
\end{array}
\xrightarrow[\text{equation (3) by 2}]{\text{Multiplying both sides of}}
\begin{array}{l}
4x - 2y - 3z = 5 \\
\underline{4x + 2y + 4z = 10} \\
8x + z = 15. \qquad (5)
\end{array}
$$

5. Now we solve the resulting system of equations (4) and (5). That allows us to find two parts of the ordered triple.

$$
\begin{array}{l}
-6x + 3z = 0, \\
8x + z = 15
\end{array}
\xrightarrow[\text{equation (5) by } -3]{\text{Multiplying both sides of}}
\begin{array}{l}
-6x + 3z = 0 \\
\underline{-24x - 3z = -45} \\
-30x = -45 \\
 x = \frac{-45}{-30} = \frac{3}{2}
\end{array}
$$

We use equation (5) to find z:

$$
\begin{array}{ll}
8x + z = 15 & \\
8 \cdot \frac{3}{2} + z = 15 & \text{Substituting } \frac{3}{2} \text{ for } x \\
12 + z = 15 & \\
z = 3. &
\end{array}
$$

6. Finally, we use any of the original equations and substitute to find the third number, y. We choose equation (3):

$$
\begin{array}{ll}
2x + y + 2z = 5 & (3) \\
2 \cdot \frac{3}{2} + y + 2 \cdot 3 = 5 & \text{Substituting } \frac{3}{2} \text{ for } x \text{ and 3 for } z \\
3 + y + 6 = 5 & \\
y + 9 = 5 & \\
y = -4. &
\end{array}
$$

The solution is $\left(\frac{3}{2}, -4, 3\right)$. The check was performed as Example 1.

Sometimes, certain variables are missing at the outset.

Example 4

Solve the system

$$x + y + z = 180, \quad (1)$$
$$x \quad\quad - z = -70, \quad (2)$$
$$2y - z = \quad 0. \quad (3)$$

Solution

1., 2. The equations appear in standard form with no fractions or decimals.

3., 4. Note that there is no y in equation (2). Thus, at the outset, we already have y eliminated from one equation. We need another equation with y eliminated, so we use equations (1) and (3):

$$x + y + z = 180, \xrightarrow{\text{Multiplying both sides of equation (1) by } -2} -2x - 2y - 2z = -360$$
$$2y - z = 0 \qquad\qquad\qquad\qquad\qquad \underline{ 2y - z = \quad 0}$$
$$\qquad\qquad\qquad\qquad\qquad\qquad -2x \qquad - 3z = -360. \quad (4)$$

5., 6. Now we solve the resulting system of equations (2) and (4):

$$x - z = -70, \xrightarrow{\text{Multiplying both sides of equation (2) by } 2} 2x - 2z = -140$$
$$-2x - 3z = -360 \qquad\qquad\qquad\qquad\qquad \underline{-2x - 3z = -360}$$
$$\qquad\qquad\qquad\qquad\qquad\qquad\qquad\qquad -5z = -500$$
$$\qquad\qquad\qquad\qquad\qquad\qquad\qquad\qquad\quad z = \quad 100.$$

Continuing as in Examples 2 and 3, we get the solution $(30, 50, 100)$. The check is left to the student.

Dependency, Inconsistency, and Geometric Considerations

Each equation in Examples 2, 3, and 4 has a graph that is a plane in three dimensions. The solutions are points common to the planes of each system. Since three planes can have an infinite number of points in common or no points at all in common, we need to generalize the concept of *consistency*.

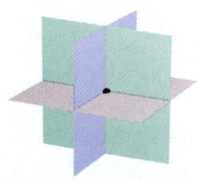   

One solution: planes intersecting in exactly one point. System is consistent.

The planes intersect along a common line. An infinite number of points are common to the three planes. System is consistent.

Three parallel planes. There is no common point of intersection. System is inconsistent.

Planes intersect two at a time, but there is no point common to all three. System is inconsistent.

> **Consistency**
>
> A system of equations that has at least one solution is said to be **consistent.**
>
> A system of equations that has no solution is said to be **inconsistent.**

E x a m p l e   5

Solve:

$$y + 3z = 4, \qquad (1)$$
$$-x - y + 2z = 0, \qquad (2)$$
$$x + 2y + z = 1. \qquad (3)$$

Solution   The variable x is missing in equation (1). By adding equations (2) and (3), we can find a second equation in which x is missing:

$$-x - y + 2z = 0 \qquad (2)$$
$$\underline{x + 2y + z = 1} \qquad (3)$$
$$y + 3z = 1. \qquad (4) \qquad \text{Adding}$$

Equations (1) and (4) form a system in y and z. We solve as before:

$$y + 3z = 4, \quad \xrightarrow[\text{equation (1) by } -1]{\text{Multiplying both sides of}} \quad -y - 3z = -4$$
$$y + 3z = 1 \qquad\qquad\qquad\qquad \underline{y + 3z = 1}$$
$$\text{This is a contradiction.} \longrightarrow 0 = -3. \qquad \text{Adding}$$

Since we end up with a *false* equation, or contradiction, we know that the system has no solution. It is *inconsistent*.

The notion of *dependency* from Section 8.1 can also be extended.

E x a m p l e   6

Solve:

$$2x + y + z = 3, \qquad (1)$$
$$x - 2y - z = 1, \qquad (2)$$
$$3x + 4y + 3z = 5. \qquad (3)$$

Solution   Our plan is to first use equations (1) and (2) to eliminate z. Then we will select another pair of equations and again eliminate z:

$$2x + y + z = 3$$
$$\underline{x - 2y - z = 1}$$
$$3x - y \quad\quad = 4. \quad (4)$$

Next, we use equations (2) and (3) to eliminate z again:

$$x - 2y - z = 1, \xrightarrow{\text{Multiplying both sides of equation (2) by 3}} 3x - 6y - 3z = 3$$
$$3x + 4y + 3z = 5 \qquad\qquad \underline{3x + 4y + 3z = 5}$$
$$6x - 2y \quad\quad = 8. \quad (5)$$

We now try to solve the resulting system of equations (4) and (5):

$$3x - y = 4, \xrightarrow{\text{Multiplying both sides of equation (4) by } -2} -6x + 2y = -8$$
$$6x - 2y = 8 \qquad\qquad \underline{6x - 2y = \quad 8}$$
$$0 = \quad 0. \quad (6)$$

Equation (6), which is an identity, indicates that equations (1), (2), and (3) are *dependent*. This means that the original system of three equations is equivalent to a system of two equations. One way to see this is to observe that two times equation (1), minus equation (2), is equation (3). Thus removing equation (3) from the system does not affect the solution of the system.* In writing an answer to this problem, we simply state that "the equations are dependent."

Recall that when dependent equations appeared in Section 8.1, the solution sets were always infinite in size and were written in set-builder notation. There, all systems of dependent equations were *consistent*. This is not always the case for systems of three or more equations. The following figures illustrate some possibilities geometrically.

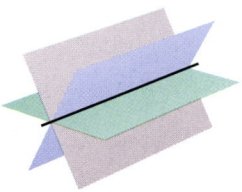

The planes intersect along a common line. The equations are dependent and the system is consistent. There is an infinite number of solutions.

The planes coincide. The equations are dependent and the system is consistent. There is an infinite number of solutions.

Two planes coincide. The third plane is parallel. The equations are dependent and the system is inconsistent. There is no solution.

---

*A set of equations is dependent if at least one equation can be expressed as a sum of multiples of other equations in that set.

Exercise Set **8.4**

1. Determine whether $(2, -1, -2)$ is a solution of the system

$$x + y - 2z = 5,$$
$$2x - y - z = 7,$$
$$-x - 2y + 3z = 6.$$

2. Determine whether $(1, -2, 3)$ is a solution of the system

$$x + y + z = 2,$$
$$x - 2y - z = 2,$$
$$3x + 2y + z = 2.$$

Solve each system. If a system's equations are dependent or if there is no solution, state this.

3. $x + y + z = 6,$
   $2x - y + 3z = 9,$
   $-x + 2y + 2z = 9$

4. $2x - y + z = 10,$
   $4x + 2y - 3z = 10,$
   $x - 3y + 2z = 8$

5. $2x - y - 3z = -1,$
   $2x - y + z = -9,$
   $x + 2y - 4z = 17$

6. $x - y + z = 6,$
   $2x + 3y + 2z = 2,$
   $3x + 5y + 4z = 4$

7. $2x - 3y + z = 5,$
   $x + 3y + 8z = 22,$
   $3x - y + 2z = 12$

8. $6x - 4y + 5z = 31,$
   $5x + 2y + 2z = 13,$
   $x + y + z = 2$

9. $3a - 2b + 7c = 13,$
   $a + 8b - 6c = -47,$
   $7a - 9b - 9c = -3$

10. $x + y + z = 0,$
   $2x + 3y + 2z = -3,$
   $-x + 2y - 3z = -1$

11. $2x + 3y + z = 17,$
   $x - 3y + 2z = -8,$
   $5x - 2y + 3z = 5$

12. $2x + y - 3z = -4,$
   $4x - 2y + z = 9,$
   $3x + 5y - 2z = 5$

13. $2x + y + z = -2,$
   $2x - y + 3z = 6,$
   $3x - 5y + 4z = 7$

14. $2x + y + 2z = 11,$
   $3x + 2y + 2z = 8,$
   $x + 4y + 3z = 0$

15. $x - y + z = 4,$
   $5x + 2y - 3z = 2,$
   $4x + 3y - 4z = -2$

16. $-2x + 8y + 2z = 4,$
   $x + 6y + 3z = 4,$
   $3x - 2y + z = 0$

17. $a + 2b + c = 1,$
   $7a + 3b - c = -2,$
   $a + 5b + 3c = 2$

18. $4x - y - z = 4,$
   $2x + y + z = -1,$
   $6x - 3y - 2z = 3$

19. $5x + 3y + \frac{1}{2}z = \frac{7}{2},$
   $0.5x - 0.9y - 0.2z = 0.3,$
   $3x - 2.4y + 0.4z = -1$

20. $r + \frac{3}{2}s + 6t = 2,$
   $2r - 3s + 3t = 0.5,$
   $r + s + t = 1$

21. $3p + 2r = 11,$
   $q - 7r = 4,$
   $p - 6q = 1$

22. $4a + 9b = 8,$
   $8a + 6c = -1,$
   $6b + 6c = -1$

23. $x + y + z = 105,$
   $10y - z = 11,$
   $2x - 3y = 7$

24. $x + y + z = 57,$
   $-2x + y = 3,$
   $x - z = 6$

25. $2a - 3b = 2,$
   $7a + 4c = \frac{3}{4},$
   $2c - 3b = 1$

26. $a - 3c = 6,$
   $b + 2c = 2,$
   $7a - 3b - 5c = 14$

27. $x + y + z = 182,$ *(Aha!)*
   $y = 2 + 3x,$
   $z = 80 + x$

28. $l + m = 7,$
   $3m + 2n = 9,$
   $4l + n = 5$

29. $x + y = 0,$
   $x + z = 1,$
   $2x + y + z = 2$

30. $x + z = 0,$
   $x + y + 2z = 3,$
   $y + z = 2$

31. $y + z = 1,$
   $x + y + z = 1,$
   $x + 2y + 2z = 2$

32. $x + y + z = 1,$
   $-x + 2y + z = 2,$
   $2x - y = -1$

33. Abbie recommends that a frustrated classmate double- and triple-check each step of work when attempting to solve a system of three equations. Is this good advice? Why or why not?

34. Describe a method for writing an inconsistent system of three equations in three variables.

SKILL MAINTENANCE

Translate each sentence to mathematics.

35. One number is twice another.

36. The sum of two numbers is three times the first number.

37. The sum of three consecutive numbers is 45.

38. One number plus twice another number is 17.

39. The sum of two numbers is five times a third number.

40. The product of two numbers is twice their sum.

SYNTHESIS

41. Is it possible for a system of three equations to have exactly two ordered triples in its solution set? Why or why not?

42. Describe a procedure that could be used to solve a system of four equations in four variables.

Solve.

43. $\dfrac{x+2}{3} - \dfrac{y+4}{2} + \dfrac{z+1}{6} = 0,$
$\dfrac{x-4}{3} + \dfrac{y+1}{4} - \dfrac{z-2}{2} = -1,$
$\dfrac{x+1}{2} + \dfrac{y}{2} + \dfrac{z-1}{4} = \dfrac{3}{4}$

44. $w + x + y + z = 2,$
$w + 2x + 2y + 4z = 1,$
$w - x + y + z = 6,$
$w - 3x - y + z = 2$

45. $w + x - y + z = 0,$
$w - 2x - 2y - z = -5,$
$w - 3x - y + z = 4,$
$2w - x - y + 3z = 7$

For Exercises 46 and 47, let u represent $1/x$, v represent $1/y$, and w represent $1/z$. Solve for u, v, and w, and then solve for x, y, and z.

46. $\dfrac{2}{x} - \dfrac{1}{y} - \dfrac{3}{z} = -1,$
$\dfrac{2}{x} - \dfrac{1}{y} + \dfrac{1}{z} = -9,$
$\dfrac{1}{x} + \dfrac{2}{y} - \dfrac{4}{z} = 17$

47. $\dfrac{2}{x} + \dfrac{2}{y} - \dfrac{3}{z} = 3,$
$\dfrac{1}{x} - \dfrac{2}{y} - \dfrac{3}{z} = 9,$
$\dfrac{7}{x} - \dfrac{2}{y} + \dfrac{9}{z} = -39$

Determine k so that each system is dependent.

48. $x - 3y + 2z = 1,$
$2x + y - z = 3,$
$9x - 6y + 3z = k$

49. $5x - 6y + kz = -5,$
$x + 3y - 2z = 2,$
$2x - y + 4z = -1$

In each case, three solutions of an equation in x, y, and z are given. Find the equation.

50. $Ax + By + Cz = 12;$
$\left(1, \frac{3}{4}, 3\right), \left(\frac{4}{3}, 1, 2\right),$ and $(2, 1, 1)$

51. $z = b - mx - ny;$
$(1, 1, 2), (3, 2, -6),$ and $\left(\frac{3}{2}, 1, 1\right)$

52. Write an inconsistent system of equations that contains dependent equations.

COLLABORATIVE CORNER

Finding the Preferred Approach

Focus: Systems of three linear equations

Time: 10–15 minutes

Group size: 3

Consider the six steps outlined on p. 508 along with the following system:

$$2x + 4y = 3 - 5z,$$
$$0.3x = 0.2y + 0.7z + 1.4,$$
$$0.04x + 0.03y = 0.07 + 0.04z.$$

ACTIVITY

1. Working independently, each group member should solve the system above. One person should begin by eliminating x, one should first eliminate y, and one should first eliminate z. Write neatly so that others can follow your steps.

2. Once all group members have solved the system, compare your answers. If the answers do not check, exchange notebooks and check each other's work. If a mistake is detected, allow the person who made the mistake to make the repair.

3. Decide as a group which of the three approaches above (if any) ranks as easiest and which (if any) ranks as most difficult. Then compare your rankings with the other groups in the class.

Solving Applications: Systems of Three Equations

8.5

Applications of Three Equations in Three Unknowns

Solving systems of three or more equations is important in many applications. Such systems arise in the natural and social sciences, business, and engineering. In mathematics, purely numerical applications also arise.

Example 1

The sum of three numbers is 4. The first number minus twice the second, minus the third is 1. Twice the first number minus the second, minus twice the third is −1. Find the numbers.

Solution

1. **Familiarize.**   There are three statements involving the same three numbers. Let's label these numbers x, y, and z.

2. **Translate.**   We can translate directly as follows.

   The sum of the three numbers is 4.

   $$x + y + z = 4$$

   The first number minus twice the second minus the third is 1.

   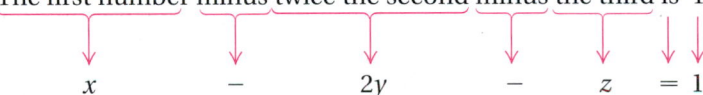

   $$x - 2y - z = 1$$

   Twice the first number minus the second minus twice the third is −1.

   $$2x - y - 2z = -1$$

   We now have a system of three equations:

   $$x + y + z = 4,$$
   $$x - 2y - z = 1,$$
   $$2x - y - 2z = -1.$$

3. **Carry out.**   We need to solve the system of equations. Note that we found the solution, $(2, -1, 3)$, in Example 2 of Section 8.4.

4. **Check.**   The first statement of the problem says that the sum of the three numbers is 4. That checks, because $2 + (-1) + 3 = 4$. The second statement says that the first number minus twice the second, minus the third is 1: $2 - 2(-1) - 3 = 1$. That checks. The check of the third statement is left to the student.

5. **State.**   The three numbers are 2, −1, and 3.

Example 2

Architecture. In a triangular cross section of a roof, the largest angle is 70° greater than the smallest angle. The largest angle is twice as large as the remaining angle. Find the measure of each angle.

Solution

1. **Familiarize.** The first thing we do is make a drawing, or a sketch.

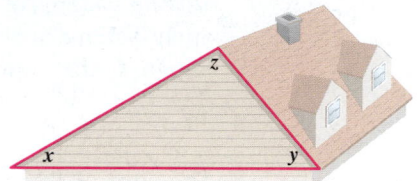

Since we don't know the size of any angle, we use x, y, and z to represent the three measures, from smallest to largest. Recall that the measures of the angles in any triangle add up to 180°.

2. **Translate.** This geometric fact about triangles gives us one equation:

$$x + y + z = 180.$$

Two of the statements can be translated almost directly.

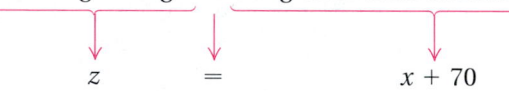

We now have a system of three equations:

$$\begin{aligned} x + y + z &= 180, \\ x + 70 &= z, \\ 2y &= z; \end{aligned} \quad \text{or} \quad \begin{aligned} x + y + z &= 180, \\ x \quad - z &= -70, \\ 2y - z &= 0. \end{aligned}$$

Rewriting in standard form

3. **Carry out.** The system was solved in Example 4 of Section 8.4. The solution is $(30, 50, 100)$.

4. **Check.** The sum of the numbers is 180, so that checks. The measure of the largest angle, 100°, is 70° greater than the measure of the smallest angle, 30°, so that checks. The measure of the largest angle is also twice the measure of the remaining angle, 50°. Thus we have a check.

5. **State.** The angles in the triangle measure 30°, 50°, and 100°.

Example 3

Cholesterol levels. Americans have become very conscious of their cholesterol levels. Recent studies indicate that a child's intake of cholesterol should be no more than 300 mg per day. By eating 1 egg, 1 cupcake, and 1 slice of pizza, a child consumes 302 mg of cholesterol. If the child eats 2 cupcakes and 3 slices of pizza, he or she takes in 65 mg of cholesterol. By eating 2 eggs and 1 cupcake, a child consumes 567 mg of cholesterol. How much cholesterol is in each item?

Solution

1. **Familiarize.**   After we have read the problem a few times, it becomes clear that an egg contains considerably more cholesterol than the other foods. Let's guess that one egg contains 200 mg of cholesterol and one cupcake contains 50 mg. Because of the third sentence in the problem, it would follow that a slice of pizza contains 52 mg of cholesterol since $200 + 50 + 52 = 302$.

   To see if our guess satisfies the other statements in the problem, we find the amount of cholesterol that 2 cupcakes and 3 slices of pizza would contain: $2 \cdot 50 + 3 \cdot 52 = 256$. Since this does not match the 65 mg listed in the fourth sentence of the problem, our guess was incorrect. Rather than guess again, we examine how we checked our guess and let e, c, and $s =$ the number of milligrams of cholesterol in an egg, a cupcake, and a slice of pizza, respectively.

2. **Translate.**   By rewording some of the sentences in the problem, we can translate it into three equations.

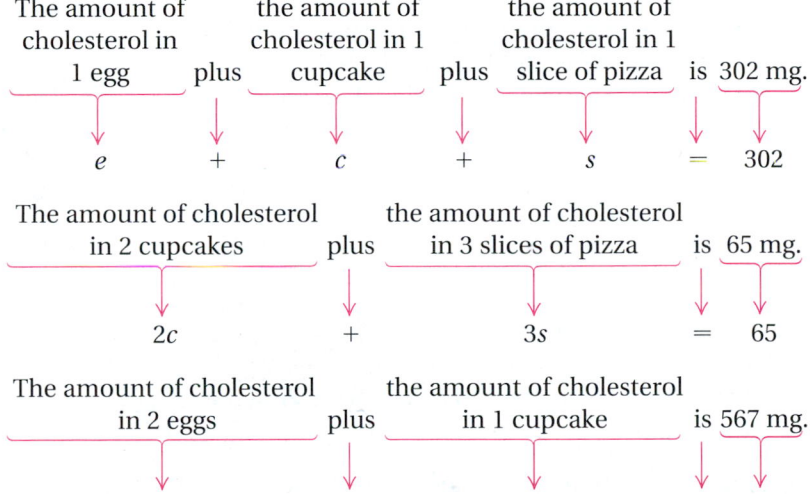

We now have a system of three equations:

$$\begin{aligned} e + c + \ \ s &= 302, \\ 2c + 3s &= 65, \\ 2e + c \qquad &= 567. \end{aligned}$$

3. **Carry out.**   We solve and get $e = 274$, $c = 19$, $s = 9$, or $(274, 19, 9)$.

4. **Check.**   The sum of 274, 19, and 9 is 302 so the total cholesterol in 1 egg, 1 cupcake, and 1 slice of pizza checks. Two cupcakes and three slices of pizza would contain $2 \cdot 19 + 3 \cdot 9 = 65$ mg, while two eggs and one cupcake would contain $2 \cdot 274 + 19 = 567$ mg of cholesterol. The answer checks.

5. **State.**   An egg contains 274 mg of cholesterol, a cupcake contains 19 mg of cholesterol, and a slice of pizza contains 9 mg of cholesterol.

Exercise Set 8.5

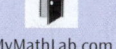
Solve.

1. The sum of three numbers is 57. The second is 3 more than the first. The third is 6 more than the first. Find the numbers.

2. The sum of three numbers is 5. The first number minus the second plus the third is 1. The first minus the third is 3 more than the second. Find the numbers.

3. The sum of three numbers is 26. Twice the first minus the second is 2 less than the third. The third is the second minus three times the first. Find the numbers.

4. The sum of three numbers is 105. The third is 11 less than ten times the second. Twice the first is 7 more than three times the second. Find the numbers.

5. *Geometry.*   In triangle *ABC*, the measure of angle *B* is three times that of angle *A*. The measure of angle *C* is 20° more than that of angle *A*. Find the angle measures.

6. *Geometry.*   In triangle *ABC*, the measure of angle *B* is twice the measure of angle *A*. The measure of angle *C* is 80° more than that of angle *A*. Find the angle measures.

7. *Automobile pricing.*   A recent basic model of a particular automobile had a price of $12,685. The basic model with the added features of automatic transmission and power door locks was $14,070. The basic model with air conditioning (AC) and power door locks was $13,580. The basic model with AC and automatic transmission was $13,925. What was the individual cost of each of the three options?

8. *Lens production.*   When Sight-Rite's three polishing machines, A, B, and C, are all working, 5700 lenses can be polished in one week. When only A and B are working, 3400 lenses can be polished in one week. When only B and C are working, 4200 lenses can be polished in one week. How many lenses can be polished in a week by each machine?

Aha! 9. *Welding rates.*   Elrod, Dot, and Wendy can weld 74 linear feet per hour when working together. Elrod and Dot together can weld 44 linear feet per hour, while Elrod and Wendy can weld 50 linear feet per hour. How many linear feet per hour can each weld alone?

10. *Telemarketing.*   Sven, Tillie, and Isaiah can process 740 telephone orders per day. Sven and Tillie together can process 470 orders, while Tillie and Isaiah together can process 520 orders per day. How many orders can each person process alone?

11. *Restaurant management.*   Kyle works at Dunkin® Donuts, where a 10-oz cup of coffee costs $1.05, a 14-oz cup costs $1.35, and a 20-oz cup costs $1.65. During one busy period, Kyle served 34 cups of coffee, emptying five 96-oz pots while collecting a total of $45. How many cups of each size did Kyle fill?

10 oz
$1.05

14 oz
$1.35

20 oz
$1.65

12. *Advertising.*   In a recent year, companies spent a total of $84.8 billion on newspaper, television, and radio ads. The total amount spent on television and radio ads was only $2.6 billion more than the amount spent on newspaper ads alone. The amount spent on newspaper ads was $5.1 billion more than what was spent on television ads. How much was spent on each form of advertising?

13. *Investments.*   A business class divided an imaginary investment of $80,000 among three mutual funds. The first fund grew by 10%, the second by 6%, and the third by 15%. Total earnings were $8850. The earnings from the first fund were $750 more than the earnings from the third. How much was invested in each fund?

14. *Restaurant management.*   McDonald's® recently sold small soft drinks for 87¢, medium soft drinks for $1.08, and large soft drinks for $1.54. During a lunch-time rush, Chris sold 40 soft drinks for a total of $43.40. The number of small and large drinks, combined, was 10 fewer than the number of medium drinks. How many drinks of each size were sold?

small	medium	large
$0.87	$1.08	$1.54

15. *Nutrition.*   A dietician in a hospital prepares meals under the guidance of a physician. Suppose that for a particular patient a physician prescribes a meal to have 800 calories, 55 g of protein, and 220 mg of vitamin C. The dietician prepares a meal of roast beef, baked potatoes, and broccoli according to the data in the following table.

	Calories	Protein (in grams)	Vitamin C (in milligrams)
Roast Beef, 3 oz	300	20	0
Baked Potato	100	5	20
Broccoli, 156 g	50	5	100

How many servings of each food are needed in order to satisfy the doctor's orders?

16. *Nutrition.*   Repeat Exercise 15 but replace the broccoli with asparagus, for which a 180-g serving contains 50 calories, 5 g of protein, and 44 mg of vitamin C. Which meal would you prefer eating?

17. *Crying rate.*   The sum of the average number of times a man, a woman, and a one-year-old child cry each month is 71.7. A one-year-old cries 46.4 more times than a man. The average number of times a one-year-old cries per month is 28.3 more than the average number of times combined that a man and a woman cry. What is the average number of times per month that each cries?

18. *Obstetrics.*   In the United States, the highest incidence of fraternal twin births occurs among Asian-Americans, then African-Americans, and then Caucasians. Out of every 15,400 births, the total number of fraternal twin births for all three is 739, where there are 185 more for Asian-Americans than African-Americans and 231 more for Asian-Americans than Caucasians. How many births of fraternal twins are there for each group out of every 15,400 births?

19. *Basketball scoring.*   The New York Knicks recently scored a total of 92 points on a combination of 2-point field goals, 3-point field goals, and 1-point foul shots. Altogether, the Knicks made 50 baskets and 19 more 2-pointers than foul shots. How many shots of each kind were made?

20. *History.*   Find the year in which the first U.S. transcontinental railroad was completed. The following are some facts about the number. The sum of the digits in the year is 24. The ones digit is 1 more than the hundreds digit. Both the tens and the ones digits are multiples of 3.

21. Problems like Exercises 11 and 12 could be classified as total-value problems. How do these problems differ from the total-value problems of Section 8.3?

22. Write a problem for a classmate to solve. Design the problem so that it translates to a system of three equations in three variables.

SKILL MAINTENANCE

Simplify.

23. $5(-3) + 7$

24. $-4(-6) + 9$

25. $-6(8) + (-7)$

26. $7(-9) + (-8)$

27. $-7(2x - 3y + 5z)$

28. $-6(4a + 7b - 9c)$

29. $-4(2a + 5b) + 3a + 20b$

30. $3(2x - 7y) + 5x + 21y$

SYNTHESIS

31. Consider Exercise 19. Suppose there were no foul shots made. Would there still be a solution? Why or why not?

32. Consider Exercise 11. Suppose Kyle collected $46. Could the problem still be solved? Why or why not?

33. Find a three-digit positive integer such that the sum of all three digits is 14, the tens digit is 2 more than the ones digit, and if the digits are reversed, the number is unchanged.

34. *Ages.* Tammy's age is the sum of the ages of Carmen and Dennis. Carmen's age is 2 more than the sum of the ages of Dennis and Mark. Dennis's age is four times Mark's age. The sum of all four ages is 42. How old is Tammy?

35. *Ticket revenue.* The Pops concert audience of 100 people consists of adults, students, and children. The ticket prices are $10 for adults, $3 for students, and 50¢ for children. The total amount of money taken in is $100. How many adults, students, and children are in attendance? Does there seem to be some information missing? Do some more careful reasoning.

36. *Sharing raffle tickets.* Hal gives Tom as many raffle tickets as Tom first had and Gary as many as Gary first had. In like manner, Tom then gives Hal and Gary as many tickets as each then has. Similarly, Gary gives Hal and Tom as many tickets as each then has. If each finally has 40 tickets, with how many tickets does Tom begin?

37. Find the sum of the angle measures at the tips of the star in this figure.

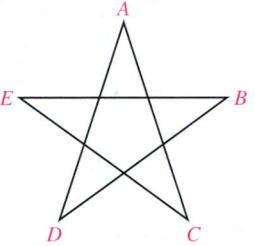

Elimination Using Matrices

8.6

Matrices and Systems • Row-Equivalent Operations

In solving systems of equations, we perform computations with the constants. The variables play no important role until the end. Thus we can simplify writing a system by omitting the variables. For example, the system

$$3x + 4y = 5,$$
$$x - 2y = 1$$

simplifies to

$$\begin{array}{ccc} 3 & 4 & 5 \\ 1 & -2 & 1 \end{array}$$

if we do not write the variables, the operation of addition, and the equals signs.

Matrices and Systems

In the example above, we have written a rectangular array of numbers. Such an array is called a **matrix** (plural, **matrices**). We ordinarily write brackets around matrices. The following are matrices:

$$\begin{bmatrix} -3 & 1 \\ 0 & 5 \end{bmatrix}, \begin{bmatrix} 2 & 0 & -1 & 3 \\ -5 & 2 & 7 & -1 \\ 4 & 5 & 3 & 0 \end{bmatrix}, \begin{bmatrix} 2 & 3 \\ 7 & 15 \\ -2 & 23 \\ 4 & 1 \end{bmatrix}$$ The individual numbers are called *elements* or *entries.*

The **rows** of a matrix are horizontal, and the **columns** are vertical.

$$\begin{bmatrix} 5 & -2 & 2 \\ 1 & 0 & 1 \\ 0 & 1 & 2 \end{bmatrix} \longrightarrow \text{row 1} \\ \longrightarrow \text{row 2} \\ \longrightarrow \text{row 3}$$

column 1   column 2   column 3

Let's see how matrices can be used to solve a system.

E x a m p l e   1

Solve the system

$$5x - 4y = -1,$$
$$-2x + 3y = 2.$$

Solution   We write a matrix using only coefficients and constants, listing x-coefficients in the first column and y-coefficients in the second. Note that in each matrix a dashed line separates the coefficients from the constants:

$$\begin{bmatrix} 5 & -4 & \vdots & -1 \\ -2 & 3 & \vdots & 2 \end{bmatrix}.$$

As an aid for understanding, we list the corresponding system in the margin.

$$5x - 4y = -1,$$
$$-2x + 3y = 2$$

Our goal is to transform

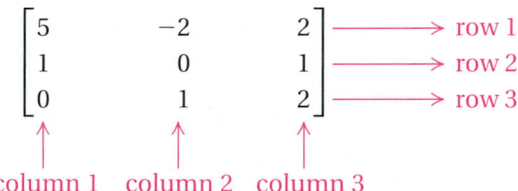

The variables x and y can then be reinserted to form equations from which we can complete the solution.

   We do calculations that are similar to those that we would do if we wrote the entire equations. The first step is to multiply and/or interchange the rows so that each number in the first column below the first number is a multiple of that number. Here that means multiplying Row 2 by 5. This corresponds to multiplying both sides of the second equation by 5.

$$5x - 4y = -1,$$
$$-10x + 15y = 10$$

$$\begin{bmatrix} 5 & -4 & \vdots & -1 \\ -10 & 15 & \vdots & 10 \end{bmatrix}$$ New Row 2 = 5(Row 2 from above)

Next, we multiply the first row by 2, add this to Row 2, and write that result as the "new" Row 2. This corresponds to multiplying the first equation by 2 and adding the result to the second equation in order to eliminate a variable. Write out these computations as necessary—we perform them mentally.

$$5x - 4y = -1,$$
$$7y = 8$$

$$\begin{bmatrix} 5 & -4 & | & -1 \\ 0 & 7 & | & 8 \end{bmatrix}$$

2(5 -4 | -1) = (10 -8 | -2) and
(10 -8 | -2) + (-10 15 | 10) = (0 7 | 8)
New Row 2 = 2(Row 1) + (Row 2)

If we now reinsert the variables, we have

$$5x - 4y = -1, \qquad (1)$$
$$7y = 8. \qquad (2)$$

We can now proceed as before, solving equation (2) for y:

$$7y = 8 \qquad (2)$$
$$y = \tfrac{8}{7}.$$

Next, we substitute $\tfrac{8}{7}$ for y in equation (1):

$$5x - 4y = -1 \qquad (1)$$
$$5x - 4 \cdot \tfrac{8}{7} = -1 \qquad \text{Substituting } \tfrac{8}{7} \text{ for } y \text{ in equation (1)}$$
$$x = \tfrac{5}{7}. \qquad \text{Solving for } x$$

The solution is $\left(\tfrac{5}{7}, \tfrac{8}{7}\right)$. The check is left to the student.

Example 2

Solve the system

$$2x - y + 4z = -3,$$
$$x \quad\;\; - 4z = 5,$$
$$6x - y + 2z = 10.$$

Solution   We first write a matrix, using only the constants. Where there are missing terms, we must write 0's:

$$2x - y + 4z = -3,$$
$$x \quad\;\; - 4z = 5,$$
$$6x - y + 2z = 10$$

$$\begin{bmatrix} 2 & -1 & 4 & | & -3 \\ 1 & 0 & -4 & | & 5 \\ 6 & -1 & 2 & | & 10 \end{bmatrix}$$

Our goal is to transform the matrix to one of the form

$$ax + by + cz = d,$$
$$ey + fz = g,$$
$$hz = i$$

$$\begin{bmatrix} a & b & c & | & d \\ 0 & e & f & | & g \\ 0 & 0 & h & | & i \end{bmatrix}.$$

A matrix of this form can be rewritten as a system of equations that is equivalent to the original system, and from which a solution can be easily found.

The first step is to multiply and/or interchange the rows so that each number in the first column is a multiple of the first number in the first row. In this case, we do so by interchanging Rows 1 and 2:

$$x \quad\;\; - 4z = 5,$$
$$2x - y + 4z = -3,$$
$$6x - y + 2z = 10$$

$$\begin{bmatrix} 1 & 0 & -4 & | & 5 \\ 2 & -1 & 4 & | & -3 \\ 6 & -1 & 2 & | & 10 \end{bmatrix}$$

This corresponds to interchanging the first two equations.

Next, we multiply the first row by -2, add it to the second row, and replace Row 2 with the result:

$$x \quad - \quad 4z = 5,$$
$$-y + 12z = -13,$$
$$6x - y + 2z = 10$$

$$\begin{bmatrix} 1 & 0 & -4 & | & 5 \\ 0 & -1 & 12 & | & -13 \\ 6 & -1 & 2 & | & 10 \end{bmatrix}.$$

$-2(1 \;\; 0 \;\; -4 \;|\; 5) = (-2 \;\; 0 \;\; 8 \;|\; -10)$ and
$(-2 \;\; 0 \;\; 8 \;|\; -10) + (2 \;\; -1 \;\; 4 \;|\; -3) =$
$(0 \;\; -1 \;\; 12 \;|\; -13)$

Now we multiply the first row by -6, add it to the third row, and replace Row 3 with the result:

$$x \quad - \quad 4z = 5,$$
$$-y + 12z = -13,$$
$$-y + 26z = -20$$

$$\begin{bmatrix} 1 & 0 & -4 & | & 5 \\ 0 & -1 & 12 & | & -13 \\ 0 & -1 & 26 & | & -20 \end{bmatrix}.$$

$-6(1 \;\; 0 \;\; -4 \;|\; 5) = (-6 \;\; 0 \;\; 24 \;|\; -30)$ and
$(-6 \;\; 0 \;\; 24 \;|\; -30) + (6 \;\; -1 \;\; 2 \;|\; 10) =$
$(0 \;\; -1 \;\; 26 \;|\; -20)$

Next, we multiply Row 2 by -1, add it to the third row, and replace Row 3 with the result:

$$x \quad - \quad 4z = 5,$$
$$-y + 12z = -13,$$
$$14z = -7$$

$$\begin{bmatrix} 1 & 0 & -4 & | & 5 \\ 0 & -1 & 12 & | & -13 \\ 0 & 0 & 14 & | & -7 \end{bmatrix}.$$

$-1(0 \;\; -1 \;\; 12 \;|\; -13) = (0 \;\; 1 \;\; -12 \;|\; 13)$
and $(0 \;\; 1 \;\; -12 \;|\; 13) + (0 \;\; -1 \;\; 26 \;|\; -20) =$
$(0 \;\; 0 \;\; 14 \;|\; -7)$

Reinserting the variables gives us

$$x \quad - \quad 4z = 5,$$
$$- y + 12z = -13,$$
$$14z = -7.$$

We now solve this last equation for z and get $z = -\frac{1}{2}$. Next, we substitute $-\frac{1}{2}$ for z in the preceding equation and solve for y: $-y + 12\left(-\frac{1}{2}\right) = -13$, so $y = 7$. Since there is no y-term in the first equation of this last system, we need only substitute $-\frac{1}{2}$ for z to solve for x: $x - 4\left(-\frac{1}{2}\right) = 5$, so $x = 3$. The solution is $\left(3, 7, -\frac{1}{2}\right)$. The check is left to the student.

technology connection

Row-equivalent operations can be performed on a grapher. For example, to interchange the first and second rows of the matrix, as in step (1) of Example 2 above, we enter the matrix as matrix **A** and select "rowSwap" from the MATRIX MATH menu. Some graphers will not automatically store the matrix produced using a row-equivalent operation, so when several operations are to be performed in succession, it is helpful to store the result of each operation as it is produced. In the window at right, we see both the matrix produced by the rowSwap operation and the indication that this matrix is stored as matrix **B**.

```
rowSwap([A],1,2)→[B]
[[1  0 -4  5]
 [2 -1  4 -3]
 [6 -1  2 10]]
```

1. Use a grapher to proceed through all the steps in Example 2.

The operations used in the preceding example correspond to those used to produce equivalent systems of equations. We call the matrices **row-equivalent** and the operations that produce them **row-equivalent operations**.

Row-Equivalent Operations

> ### Row-Equivalent Operations
>
> Each of the following row-equivalent operations produces a row-equivalent matrix:
>
> **a)** Interchanging any two rows.
> **b)** Multiplying all elements of a row by a nonzero constant.
> **c)** Replacing a row with the sum of that row and a multiple of another row.

The best overall method for solving systems of equations is by row-equivalent matrices; even computers are programmed to use them. Matrices are part of a branch of mathematics known as linear algebra. They are also studied in many courses in finite mathematics.

FOR EXTRA HELP

Exercise Set 8.6

 Digital Video Tutor CD 6 Videotape 16   InterAct Math   Math Tutor Center   MathXL   MyMathLab.com

Solve using matrices.

1. $9x - 2y = 5,$
$3x - 3y = 11$

2. $4x + y = 7,$
$5x - 3y = 13$

3. $x + 4y = 8,$
$3x + 5y = 3$

4. $x + 4y = 5,$
$-3x + 2y = 13$

5. $6x - 2y = 4,$
$7x + y = 13$

6. $3x + 4y = 7,$
$-5x + 2y = 10$

7. $3x + 2y + 2z = 3,$
$x + 2y - z = 5,$
$2x - 4y + z = 0$

8. $4x - y - 3z = 19,$
$8x + y - z = 11,$
$2x + y + 2z = -7$

9. $p - 2q - 3r = 3,$
$2p - q - 2r = 4,$
$4p + 5q + 6r = 4$

10. $x + 2y - 3z = 9,$
$2x - y + 2z = -8,$
$3x - y - 4z = 3$

11. $3p + 2r = 11,$
$q - 7r = 4,$
$p - 6q = 1$

12. $4a + 9b = 8,$
$8a + 6c = -1,$
$6b + 6c = -1$

13. $2x + 2y - 2z - 2w = -10,$
$w + y + z + x = -5,$
$x - y + 4z + 3w = -2,$
$w - 2y + 2z + 3x = -6$

14. $-w - 3y + z + 2x = -8,$
$x + y - z - w = -4,$
$w + y + z + x = 22,$
$x - y - z - w = -14$

Solve using matrices.

15. *Coin value.* A collection of 34 coins consists of dimes and nickels. The total value is $1.90. How many dimes and how many nickels are there?

16. *Coin value.* A collection of 43 coins consists of dimes and quarters. The total value is $7.60. How many dimes and how many quarters are there?

17. *Mixed granola.* Grace sells two kinds of granola. One is worth $4.05 per pound and the other is worth $2.70 per pound. She wants to blend the two granolas to get a 15-lb mixture worth $3.15 per pound. How much of each kind of granola should be used?

18. *Trail mix.* Phil mixes nuts worth $1.60 per pound with oats worth $1.40 per pound to get 20 lb of trail mix worth $1.54 per pound. How many pounds of nuts and how many pounds of oats should be used?

19. *Investments.* Elena receives $212 per year in simple interest from three investments totaling $2500. Part is invested at 7%, part at 8%, and part at 9%. There is $1100 more invested at 9% than at 8%. Find the amount invested at each rate.

20. *Investments.* Miguel receives $306 per year in simple interest from three investments totaling $3200. Part is invested at 8%, part at 9%, and part at 10%. There is $1900 more invested at 10% than at 9%. Find the amount invested at each rate.

21. Explain how you can recognize dependent equations when solving with matrices.

22. Explain how you can recognize an inconsistent system when solving with matrices.

SKILL MAINTENANCE

Simplify.

23. $5(-3) - (-7)4$ **24.** $8(-5) - (-2)9$

25. $-2(5 \cdot 3 - 4 \cdot 6) - 3(2 \cdot 7 - 15) + 4(3 \cdot 8 - 5 \cdot 4)$

26. $6(2 \cdot 7 - 3(-4)) - 4(3(-8) - 10) + 5(4 \cdot 3 - (-2)7)$

SYNTHESIS

27. If the matrices

$$\begin{bmatrix} a_1 & b_1 & | & c_1 \\ d_1 & e_1 & | & f_1 \end{bmatrix} \quad \text{and} \quad \begin{bmatrix} a_2 & b_2 & | & c_2 \\ d_2 & e_2 & | & f_2 \end{bmatrix}$$

share the same solution, does it follow that the corresponding entries are all equal to each other ($a_1 = a_2$, $b_1 = b_2$, etc.)? Why or why not?

28. Explain how the row-equivalent operations make use of the addition, multiplication, and distributive properties.

29. The sum of the digits in a four-digit number is 10. Twice the sum of the thousands digit and the tens digit is 1 less than the sum of the other two digits. The tens digit is twice the thousands digit. The ones digit equals the sum of the thousands digit and the hundreds digit. Find the four-digit number.

30. Solve for x and y:

$$ax + by = c,$$
$$dx + ey = f.$$

Determinants and Cramer's Rule

8.7

Determinants of 2 × 2 Matrices • Cramer's Rule: 2 × 2 Systems • Cramer's Rule: 3 × 3 Systems

Determinants of 2 × 2 Matrices

When a matrix has *m* rows and *n* columns, it is called an "*m* by *n*" matrix. Thus its *dimensions* are denoted by $m \times n$. If a matrix has the same number of rows and columns, it is called a **square matrix**. Associated with every square matrix is a number called its **determinant**, defined as follows for 2 × 2 matrices.

2 × 2 Determinants

The determinant of a two-by-two matrix $\begin{bmatrix} a & c \\ b & d \end{bmatrix}$ is denoted $\begin{vmatrix} a & c \\ b & d \end{vmatrix}$ and is defined as follows:

$$\begin{vmatrix} a & c \\ b & d \end{vmatrix} = ad - bc.$$

E x a m p l e   1

Evaluate: $\begin{vmatrix} 2 & -5 \\ 6 & 7 \end{vmatrix}$.

Solution   We multiply and subtract as follows:

$$\begin{vmatrix} 2 & -5 \\ 6 & 7 \end{vmatrix} = 2 \cdot 7 - 6 \cdot (-5) = 14 + 30 = 44.$$

Cramer's Rule: 2 × 2 Systems

One of the many uses for determinants is in solving systems of linear equations in which the number of variables is the same as the number of equations and the constants are not all 0. Let's consider a system of two equations:

$$a_1 x + b_1 y = c_1,$$
$$a_2 x + b_2 y = c_2.$$

If we use the elimination method, a series of steps can show that

$$x = \frac{c_1 b_2 - c_2 b_1}{a_1 b_2 - a_2 b_1} \quad \text{and} \quad y = \frac{a_1 c_2 - a_2 c_1}{a_1 b_2 - a_2 b_1}.$$

Determinants can be used in these expressions for x and y.

Cramer's Rule: 2 × 2 Systems

The solution of the system

$$a_1 x + b_1 y = c_1,$$
$$a_2 x + b_2 y = c_2,$$

if it is unique, is given by

$$x = \frac{\begin{vmatrix} c_1 & b_1 \\ c_2 & b_2 \end{vmatrix}}{\begin{vmatrix} a_1 & b_1 \\ a_2 & b_2 \end{vmatrix}}, \quad y = \frac{\begin{vmatrix} a_1 & c_1 \\ a_2 & c_2 \end{vmatrix}}{\begin{vmatrix} a_1 & b_1 \\ a_2 & b_2 \end{vmatrix}}.$$

(continued)

The equations above make sense only if the determinant in the denominator is not 0. If the denominator *is* 0, then one of two things happens.

1. If the denominator is 0 and the other two determinants in the numerators are also 0, then the equations in the system are dependent.

2. If the denominator is 0 and at least one of the other determinants in the numerators is not 0, then the system is inconsistent.

To use Cramer's rule, we find the determinants and compute x and y as shown above. Note that the denominators are identical and the coefficients of x and y appear in the same position as in the original equations. In the numerator of x, the constants c_1 and c_2 replace a_1 and a_2. In the numerator of y, c_1 and c_2 replace b_1 and b_2.

Example 2   Solve using Cramer's rule:

$$2x + 5y = 7,$$
$$5x - 2y = -3.$$

Solution   We have

$$x = \frac{\begin{vmatrix} 7 & 5 \\ -3 & -2 \end{vmatrix}}{\begin{vmatrix} 2 & 5 \\ 5 & -2 \end{vmatrix}} \qquad \text{Using Cramer's rule}$$

$$= \frac{7(-2) - (-3)5}{2(-2) - 5 \cdot 5} = -\frac{1}{29}$$

and

$$y = \frac{\begin{vmatrix} 2 & 7 \\ 5 & -3 \end{vmatrix}}{\begin{vmatrix} 2 & 5 \\ 5 & -2 \end{vmatrix}} \qquad \text{Using Cramer's rule}$$

$$= \frac{2(-3) - 5 \cdot 7}{-29} = \frac{41}{29}. \qquad \text{The denominator is the same as in the expression for } x.$$

The solution is $\left(-\frac{1}{29}, \frac{41}{29}\right)$. The check is left to the student.

Cramer's Rule: 3 × 3 Systems

A similar method has been developed for solving systems of three linear equations in 3 unknowns. However, before stating the rule, we must extend our terminology.

3 × 3 Determinants

The determinant of a three-by-three matrix is defined as follows:

$$\begin{vmatrix} a_1 & b_1 & c_1 \\ a_2 & b_2 & c_2 \\ a_3 & b_3 & c_3 \end{vmatrix} = a_1 \begin{vmatrix} b_2 & c_2 \\ b_3 & c_3 \end{vmatrix} \overset{\text{Subtract.}}{-} a_2 \begin{vmatrix} b_1 & c_1 \\ b_3 & c_3 \end{vmatrix} \overset{\text{Add.}}{+} a_3 \begin{vmatrix} b_1 & c_1 \\ b_2 & c_2 \end{vmatrix}$$

Note that the a's come from the first column. Note too that the 2×2 determinants above can be obtained by crossing out the row and the column in which the a occurs.

For a_1:
$$\begin{vmatrix} a_1 & b_1 & c_1 \\ a_2 & b_2 & c_2 \\ a_3 & b_3 & c_3 \end{vmatrix}$$

For a_2:
$$\begin{vmatrix} a_1 & b_1 & c_1 \\ a_2 & b_2 & c_2 \\ a_3 & b_3 & c_3 \end{vmatrix}$$

For a_3:
$$\begin{vmatrix} a_1 & b_1 & c_1 \\ a_2 & b_2 & c_2 \\ a_3 & b_3 & c_3 \end{vmatrix}$$

Example 3   Evaluate:

$$\begin{vmatrix} -1 & 0 & 1 \\ -5 & 1 & -1 \\ 4 & 8 & 1 \end{vmatrix}.$$

Solution   We have

$$\begin{vmatrix} -1 & 0 & 1 \\ -5 & 1 & -1 \\ 4 & 8 & 1 \end{vmatrix} = -1 \begin{vmatrix} 1 & -1 \\ 8 & 1 \end{vmatrix} \overset{\text{Subtract.}}{-} (-5) \begin{vmatrix} 0 & 1 \\ 8 & 1 \end{vmatrix} \overset{\text{Add}}{+} 4 \begin{vmatrix} 0 & 1 \\ 1 & -1 \end{vmatrix}$$

$$= -1(1 + 8) + 5(0 - 8) + 4(0 - 1) \qquad \text{Evaluating the three determinants}$$

$$= -9 - 40 - 4 = -53.$$

Cramer's Rule: 3 × 3 Systems

The solution of the system

$$a_1x + b_1y + c_1z = d_1,$$
$$a_2x + b_2y + c_2z = d_2,$$
$$a_3x + b_3y + c_3z = d_3$$

is found by considering the following determinants:

$$D = \begin{vmatrix} a_1 & b_1 & c_1 \\ a_2 & b_2 & c_2 \\ a_3 & b_3 & c_3 \end{vmatrix}, \qquad D_x = \begin{vmatrix} d_1 & b_1 & c_1 \\ d_2 & b_2 & c_2 \\ d_3 & b_3 & c_3 \end{vmatrix},$$

D contains only coefficients.

In D_x, the d's replace the a's.

$$D_y = \begin{vmatrix} a_1 & d_1 & c_1 \\ a_2 & d_2 & c_2 \\ a_3 & d_3 & c_3 \end{vmatrix}, \qquad D_z = \begin{vmatrix} a_1 & b_1 & d_1 \\ a_2 & b_2 & d_2 \\ a_3 & b_3 & d_3 \end{vmatrix}.$$

In D_y, the d's replace the b's.

In D_z, the d's replace the c's.

If a unique solution exists, it is given by

$$x = \frac{D_x}{D}, \qquad y = \frac{D_y}{D}, \qquad z = \frac{D_z}{D}.$$

Example 4

Solve using Cramer's rule:

$$x - 3y + 7z = 13,$$
$$x + y + z = 1,$$
$$x - 2y + 3z = 4.$$

Solution   We compute D, D_x, D_y, and D_z:

$$D = \begin{vmatrix} 1 & -3 & 7 \\ 1 & 1 & 1 \\ 1 & -2 & 3 \end{vmatrix} = -10; \qquad D_x = \begin{vmatrix} 13 & -3 & 7 \\ 1 & 1 & 1 \\ 4 & -2 & 3 \end{vmatrix} = 20;$$

$$D_y = \begin{vmatrix} 1 & 13 & 7 \\ 1 & 1 & 1 \\ 1 & 4 & 3 \end{vmatrix} = -6; \qquad D_z = \begin{vmatrix} 1 & -3 & 13 \\ 1 & 1 & 1 \\ 1 & -2 & 4 \end{vmatrix} = -24.$$

Then

$$x = \frac{D_x}{D} = \frac{20}{-10} = -2;$$

$$y = \frac{D_y}{D} = \frac{-6}{-10} = \frac{3}{5};$$

$$z = \frac{D_z}{D} = \frac{-24}{-10} = \frac{12}{5}.$$

The solution is $\left(-2, \frac{3}{5}, \frac{12}{5}\right)$. The check is left to the student.

In Example 4, we need not have evaluated D_z. Once x and y were found, we could have substituted them into one of the equations to find z.

To use Cramer's rule, we divide by D, provided $D \neq 0$. If $D = 0$ and at least one of the other determinants is not 0, then the system is inconsistent. If *all* the determinants are 0, then the equations in the system are dependent.

Exercise Set 8.7

FOR EXTRA HELP

 Digital Video Tutor CD 6 Videotape 16

 InterAct Math

 Math Tutor Center

 MathXL

 MyMathLab.com

Evaluate.

1. $\begin{vmatrix} 5 & 1 \\ 2 & 4 \end{vmatrix}$

2. $\begin{vmatrix} 3 & 2 \\ 2 & -3 \end{vmatrix}$

3. $\begin{vmatrix} 6 & -9 \\ 2 & 3 \end{vmatrix}$

4. $\begin{vmatrix} 3 & 2 \\ -7 & 5 \end{vmatrix}$

5. $\begin{vmatrix} 1 & 4 & 0 \\ 0 & -1 & 2 \\ 3 & -2 & 1 \end{vmatrix}$

6. $\begin{vmatrix} 3 & 0 & -2 \\ 5 & 1 & 2 \\ 2 & 0 & -1 \end{vmatrix}$

7. $\begin{vmatrix} -1 & -2 & -3 \\ 3 & 4 & 2 \\ 0 & 1 & 2 \end{vmatrix}$

8. $\begin{vmatrix} 1 & 2 & 2 \\ 2 & 1 & 0 \\ 3 & 3 & 1 \end{vmatrix}$

9. $\begin{vmatrix} -4 & -2 & 3 \\ -3 & 1 & 2 \\ 3 & 4 & -2 \end{vmatrix}$

10. $\begin{vmatrix} 2 & -1 & 1 \\ 1 & 2 & -1 \\ 3 & 4 & -3 \end{vmatrix}$

Solve using Cramer's rule.

11. $5x + 8y = 1,$
$3x + 7y = 5$

12. $3x - 4y = 6,$
$5x + 9y = 10$

13. $5x - 4y = -3,$
$7x + 2y = 6$

14. $-2x + 4y = 3,$
$3x - 7y = 1$

15. $3x - y + 2z = 1,$
$x - y + 2z = 3,$
$-2x + 3y + z = 1$

16. $3x + 2y - z = 4,$
$3x - 2y + z = 5,$
$4x - 5y - z = -1$

17. $2x - 3y + 5z = 27,$
$x + 2y - z = -4,$
$5x - y + 4z = 27$

18. $x - y + 2z = -3,$
$x + 2y + 3z = 4,$
$2x + y + z = -3$

19. $r - 2s + 3t = 6,$
$2r - s - t = -3,$
$r + s + t = 6$

20. $a \quad - 3c = 6,$
$b + 2c = 2,$
$7a - 3b - 5c = 14$

 **21.** What is it about Cramer's rule that makes it useful?

 **22.** Which version of Cramer's rule do you find more useful: the version for 2×2 systems or the version for 3×3 systems? Why?

SKILL MAINTENANCE

Solve.

23. $0.5x - 2.34 + 2.4x = 7.8x - 9$

24. $5x + 7x = -144$

25. A piece of wire 32.8 ft long is to be cut into two pieces, and those pieces are each to be bent to make a square. The length of a side of one square is to be 2.2 ft greater than the length of a side of the other. How should the wire be cut?

26. *Inventory.* The Freeport College store paid $1728 for an order of 45 calculators. The store paid $9 for each scientific calculator. The others, all graphing calculators, cost the store $58 each. How many of each type of calculator was ordered?

27. *Insulation.* The Mazzas' attic required three and a half times as much insulation as did the Kranepools'. Together, the two attics required 36 rolls of insulation. How much insulation did each attic require?

28. *Sales of food.* High Flyin' Wings charges $12 for a bucket of chicken wings and $7 for a chicken dinner. After filling 28 orders for buckets and dinners, High Flyin' Wings had collected $281. How many buckets and how many dinners did they sell?

SYNTHESIS

29. Cramer's rule states that whenever the equations $a_1x + b_1y = c_1$ and $a_2x + b_2y = c_2$ are dependent, we have

$$\begin{vmatrix} a_1 & b_1 \\ a_2 & b_2 \end{vmatrix} = 0.$$

Explain why this occurs.

30. Under what conditions can a 3×3 system of linear equations be consistent but unable to be solved using Cramer's rule?

Solve.

31. $\begin{vmatrix} y & -2 \\ 4 & 3 \end{vmatrix} = 44$

32. $\begin{vmatrix} 2 & x & -1 \\ -1 & 3 & 2 \\ -2 & 1 & 1 \end{vmatrix} = -12$

33. $\begin{vmatrix} m+1 & -2 \\ m-2 & 1 \end{vmatrix} = 27$

34. Show that an equation of the line through (x_1, y_1) and (x_2, y_2) can be written

$$\begin{vmatrix} x & y & 1 \\ x_1 & y_1 & 1 \\ x_2 & y_2 & 1 \end{vmatrix} = 0.$$

8.8

Business and Economic Applications

Break-Even Analysis • Supply and Demand

Break-Even Analysis

When a company manufactures x units of a product, it spends money. This is **total cost** and can be thought of as a function C, where $C(x)$ is the total cost of producing x units. When the company sells x units of the product, it takes in money. This is **total revenue** and can be thought of as a function R, where $R(x)$ is the total revenue from the sale of x units. **Total profit** is the money taken in less the money spent, or total revenue minus total cost. Total profit from the production and sale of x units is a function P given by

$$\textbf{Profit = Revenue − Cost,} \quad \text{or} \quad P(x) = R(x) − C(x).$$

If $R(x)$ is greater than $C(x)$, the company makes money. If $C(x)$ is greater than $R(x)$, the company has a loss. When $R(x) = C(x)$, the company breaks even.

There are two kinds of costs. First, there are costs like rent, insurance, machinery, and so on. These costs, which must be paid whether a product is produced or not, are called *fixed costs*. When a product is being produced, there are costs for labor, materials, marketing, and so on. These are called *variable costs*, because they vary according to the amount being produced. The sum of the fixed cost and the variable cost gives the *total cost* of producing a product.

> ***Caution!*** Do not confuse "cost" with "price." When we discuss the *cost* of an item, we are referring to what it costs to produce the item. The *price* of an item is what a consumer pays to purchase the item and is used when calculating revenue.

Example 1

Manufacturing radios. Ergs, Inc., is planning to make a new kind of radio. Fixed costs will be $90,000, and it will cost $15 to produce each radio (variable costs). Each radio sells for $26.

a) Find the total cost $C(x)$ of producing x radios.
b) Find the total revenue $R(x)$ from the sale of x radios.
c) Find the total profit $P(x)$ from the production and sale of x radios.
d) What profit or loss will the company realize from the production and sale of 3000 radios? of 14,000 radios?
e) Graph the total-cost, total-revenue, and total-profit functions using the same set of axes. Determine the break-even point.

Solution

a) Total cost is given by

$$C(x) = \text{(Fixed costs)} \ \text{plus} \ \text{(Variable costs)},$$
$$\text{or} \quad C(x) = \quad 90,000 \quad + \quad 15x,$$

where x is the number of radios produced.

b) Total revenue is given by

$$R(x) = 26x.$$ $26 times the number of radios sold. We assume that every radio produced is sold.

c) Total profit is given by

$$P(x) = R(x) - C(x) \qquad \text{Profit is revenue minus cost.}$$
$$= 26x - (90,000 + 15x)$$
$$= 11x - 90,000.$$

d) Profits will be

$$P(3000) = 11 \cdot 3000 - 90,000 = -\$57,000$$

when 3000 radios are produced and sold, and

$$P(14,000) = 11 \cdot 14,000 - 90,000 = \$64,000$$

when 14,000 radios are produced and sold. Thus the company loses money if only 3000 radios are sold, but makes money if 14,000 are sold.

e) The graphs of each of the three functions are shown below:

$R(x) = 26x,$ This represents the revenue function.

$C(x) = 90{,}000 + 15x,$ This represents the cost function.

$P(x) = 11x - 90{,}000.$ This represents the profit function.

$R(x)$, $C(x)$, and $P(x)$ are all in dollars.

The revenue function has a graph that goes through the origin and has a slope of 26. The cost function has an intercept on the \$-axis of 90,000 and has a slope of 15. The profit function has an intercept on the \$-axis of $-90{,}000$ and has a slope of 11. It is shown by the dashed line. The red dashed line shows a "negative" profit, which is a loss. (That is what is known as "being in the red.") The black dashed line shows a "positive" profit, or gain. (That is what is known as "being in the black.")

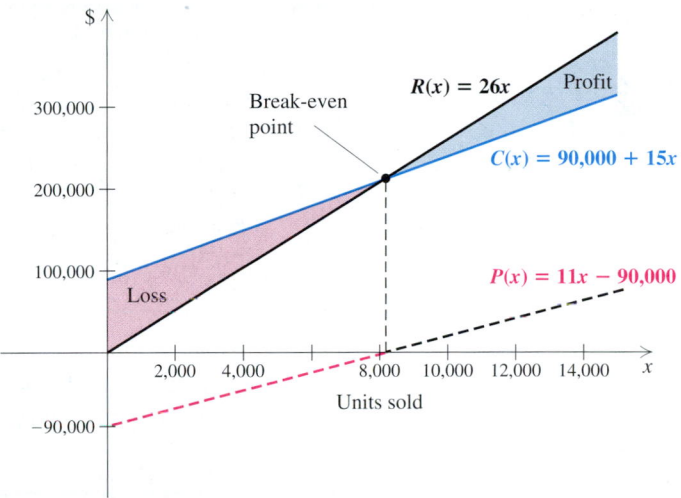

Profits occur where the revenue is greater than the cost. Losses occur where the revenue is less than the cost. The **break-even point** occurs where the graphs of R and C cross. Thus to find the break-even point, we solve a system:

$R(x) = 26x,$

$C(x) = 90{,}000 + 15x.$

Since both revenue and cost are in *dollars* and they are equal at the break-even point, the system can be rewritten as

$d = 26x,$ (1)

$d = 90{,}000 + 15x$ (2)

and solved using substitution:

$26x = 90{,}000 + 15x$ Substituting $26x$ for d in equation (2)

$11x = 90{,}000$

$x \approx 8181.8.$

The firm will break even if it produces and sells about 8182 radios (8181 will yield a tiny loss and 8182 a tiny gain), and takes in a total of $R(8182) = 26 \cdot 8182 = \$212{,}732$ in revenue. Note that the x-coordinate of the break-even point can also be found by solving $P(x) = 0$. The break-even point is (8182 radios, \$212,732).

Supply and Demand

As the price of coffee varies, the amount sold varies. The table and graph below show that consumers will demand less as the price goes up.

Demand Function, D

Price, p, per Kilogram	Quantity, $D(p)$ (in millions of kilograms)
$ 8.00	25
9.00	20
10.00	15
11.00	10
12.00	5

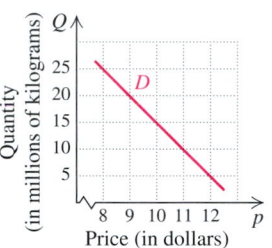

As the price of coffee varies, the amount available varies. The table and graph below show that sellers will supply less as the price goes down.

Supply Function, S

Price, p, per Kilogram	Quantity, $S(p)$ (in millions of kilograms)
$ 9.00	5
9.50	10
10.00	15
10.50	20
11.00	25

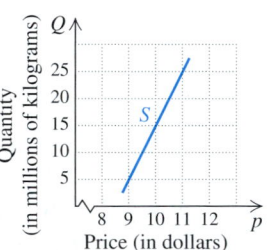

Let's look at the above graphs together. We see that as price increases, demand decreases. As price increases, supply increases. The point of intersection is called the **equilibrium point**. At that price, the amount that the seller will supply is the same amount that the consumer will buy. The situation is analogous to a buyer and a seller negotiating the price of an item. The equilibrium point is the price and quantity that they finally agree on.

Any ordered pair of coordinates from the graph is (price, quantity), because the horizontal axis is the price axis and the vertical axis is the quantity axis. If D is a demand function and S is a supply function, then the equilibrium point is where demand equals supply:

$$D(p) = S(p).$$

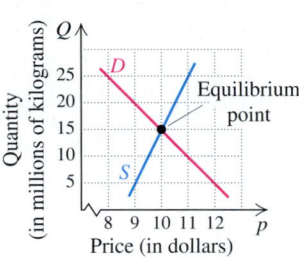

Example 2

Find the equilibrium point for the demand and supply functions given:

$$D(p) = 1000 - 60p, \quad (1)$$
$$S(p) = 200 + 4p. \quad (2)$$

Solution   Since both demand and supply are *quantities* and they are equal at the equilibrium point, we rewrite the system as

$$q = 1000 - 60p, \quad (1)$$
$$q = 200 + 4p. \quad (2)$$

We substitute $200 + 4p$ for q in equation (1) and solve:

$$200 + 4p = 1000 - 60p \qquad \text{Substituting } 200 + 4p \text{ for } q \text{ in equation (1)}$$
$$200 + 64p = 1000 \qquad \text{Adding } 60p \text{ to both sides}$$
$$64p = 800 \qquad \text{Adding } -200 \text{ to both sides}$$
$$p = \tfrac{800}{64} = 12.5.$$

Thus the equilibrium price is $12.50 per unit.

To find the equilibrium quantity, we substitute $12.50 into either $D(p)$ or $S(p)$. We use $S(p)$:

$$S(12.5) = 200 + 4(12.5) = 200 + 50 = 250.$$

Thus the equilibrium quantity is 250 units, and the equilibrium point is ($12.50, 250).

FOR EXTRA HELP

Exercise Set **8.8**

Digital Video Tutor CD 7   InterAct Math   Math Tutor Center   MathXL   MyMathLab.com
Videotape 16

For each of the following pairs of total-cost and total-revenue functions, find **(a)** *the total-profit function and* **(b)** *the break-even point.*

1. $C(x) = 45x + 300,000;$
$R(x) = 65x$

2. $C(x) = 25x + 270,000;$
$R(x) = 70x$

3. $C(x) = 10x + 120,000;$
$R(x) = 60x$

4. $C(x) = 30x + 49{,}500;$
$R(x) = 85x$

5. $C(x) = 40x + 22{,}500;$
$R(x) = 85x$

6. $C(x) = 20x + 10{,}000;$
$R(x) = 100x$

7. $C(x) = 22x + 16{,}000;$
$R(x) = 40x$

8. $C(x) = 15x + 75{,}000;$
$R(x) = 55x$

Aha! **9.** $C(x) = 75x + 100{,}000;$
$R(x) = 125x$

10. $C(x) = 20x + 120{,}000;$
$R(x) = 50x$

Find the equilibrium point for each of the following pairs of demand and supply functions.

11. $D(p) = 1000 - 10p,$
$S(p) = 230 + p$

12. $D(p) = 2000 - 60p,$
$S(p) = 460 + 94p$

13. $D(p) = 760 - 13p,$
$S(p) = 430 + 2p$

14. $D(p) = 800 - 43p,$
$S(p) = 210 + 16p$

15. $D(p) = 7500 - 25p,$
$S(p) = 6000 + 5p$

16. $D(p) = 8800 - 30p,$
$S(p) = 7000 + 15p$

17. $D(p) = 1600 - 53p,$
$S(p) = 320 + 75p$

18. $D(p) = 5500 - 40p,$
$S(p) = 1000 + 85p$

Solve.

19. *Computer manufacturing.*   Biz.com Electronics is planning to introduce a new line of computers. The fixed costs for production are $125,300. The variable costs for producing each computer are $450. The revenue from each computer is $800. Find the following.
  a) The total cost $C(x)$ of producing x computers
  b) The total revenue $R(x)$ from the sale of x computers
  c) The total profit $P(x)$ from the production and sale of x computers
  d) The profit or loss from the production and sale of 100 computers; of 400 computers
  e) The break-even point

20. *Manufacturing lamps.*   City Lights, Inc., is planning to manufacture a new type of lamp. The fixed costs for production are $22,500. The variable costs for producing each lamp are estimated to be $40.

The revenue from each lamp is to be $85. Find the following.
  a) The total cost $C(x)$ of producing x lamps
  b) The total revenue $R(x)$ from the sale of x lamps
  c) The total profit $P(x)$ from the production and sale of x lamps
  d) The profit or loss from the production and sale of 3000 lamps; of 400 lamps
  e) The break-even point

21. *Manufacturing caps.*   Martina's Custom Printing is planning on adding painter's caps to its product line. For the first year, the fixed costs for setting up production are $16,404. The variable costs for producing a dozen caps are $6.00. The revenue on each dozen caps will be $18.00. Find the following.
  a) The total cost $C(x)$ of producing x dozen caps
  b) The total revenue $R(x)$ from the sale of x dozen caps
  c) The total profit $P(x)$ from the production and sale of x dozen caps
  d) The profit or loss from the production and sale of 3000 dozen caps; of 1000 dozen caps
  e) The break-even point

22. *Sport coat production.*   Sarducci's is planning a new line of sport coats. For the first year, the fixed costs for setting up production are $10,000. The variable costs for producing each coat are $30. The revenue from each coat is to be $80. Find the following.
  a) The total cost $C(x)$ of producing x coats
  b) The total revenue $R(x)$ from the sale of x coats
  c) The total profit $P(x)$ from the production and sale of x coats
  d) The profit or loss from the production and sale of 2000 coats; of 50 coats
  e) The break-even point

23. In Example 1, the slope of the line representing Revenue is the sum of the slopes of the other two lines. This is not a coincidence. Explain why.

24. Variable costs and fixed costs are often compared to the slope and the y-intercept, respectively, of an equation for a line. Explain why you feel this analogy is or is not valid.

SKILL MAINTENANCE

Solve.

25. $3x - 9 = 27$

26. $4x - 7 = 53$

27. $4x - 5 = 7x - 13$

28. $2x + 9 = 8x - 15$

29. $7 - 2(x - 8) = 14$

30. $6 - 4(3x - 2) = 10$

SYNTHESIS

31. Ian claims that since his fixed costs are $1000, he need sell only 20 birdbaths at $50 each in order to break even. Does this sound plausible? Why or why not?

32. In this section, we examined supply and demand functions for coffee. Does it seem realistic to you for the graph of D to have a constant slope? Why or why not?

33. *Yo-yo production.* Bing Boing Hobbies is willing to produce 100 yo-yo's at $2.00 each and 500 yo-yo's at $8.00 each. Research indicates that the public will buy 500 yo-yo's at $1.00 each and 100 yo-yo's at $9.00 each. Find the equilibrium point.

34. *Loudspeaker production.* Fidelity Speakers, Inc., has fixed costs of $15,400 and variable costs of $100 for each pair of speakers produced. If the speakers sell for $250 a pair, how many pairs of speakers must be produced (and sold) in order to have enough profit to cover the fixed costs of two additional facilities? Assume that all fixed costs are identical.

Use a grapher to solve.

35. *Dog food production.* Puppy Love, Inc., will soon begin producing a new line of puppy food. The marketing department predicts that the demand function will be $D(p) = -14.97p + 987.35$ and the supply function will be $S(p) = 98.55p - 5.13$.

a) To the nearest cent, what price per unit should be charged in order to have equilibrium between supply and demand?

b) The production of the puppy food involves $87,985 in fixed costs and $5.15 per unit in variable costs. If the price per unit is the value you found in part (a), how many units must be sold in order to break even?

36. *Computer production.* Number Cruncher Computers, Inc., is planning a new line of computers, each of which will sell for $970. The fixed costs in setting up production are $1,235,580 and the variable costs for each computer are $697.

a) What is the break-even point? (Round to the nearest whole number.) (

b) The marketing department at Number Cruncher is not sure that $970 is the best price. Their demand function for the new computers is given by $D(p) = -304.5p + 374,580$ and their supply function is given by $S(p) = 788.7p - 576,504$. To the nearest dollar, what price p would result in equilibrium between supply and demand?

Summary and Review 8

Key Terms

Important Properties and Formulas

When solving a system of two linear equations in two variables:

1. If an identity is obtained, such as $0 = 0$, then the system has an infinite number of solutions. The equations are dependent and, since a solution exists, the system is consistent.
2. If a contradiction is obtained, such as $0 = 7$, then the system has no solution. The system is inconsistent.

To use the elimination method to solve systems of three linear equations:

1. Write all equations in the standard form $Ax + By + Cz = D$.
2. Clear any decimals or fractions.
3. Choose a variable to eliminate. Then select two of the three equations and work to get one equation in two variables.
4. Next, use a different pair of equations and eliminate the same variable that you did in step (3).
5. Solve the system of equations that resulted from steps (3) and (4).
6. Substitute the solution from step (5) into one of the original three equations and solve for the third variable. Then check.

Row-Equivalent Operations

Each of the following row-equivalent operations produces a row-equivalent matrix:

a) Interchanging any two rows.
b) Multiplying all elements of a row by a nonzero constant.
c) Replacing a row by the sum of that row and a multiple of another row.

Determinant of a 2 × 2 Matrix

$$\begin{vmatrix} a & c \\ b & d \end{vmatrix} = ad - bc$$

Determinant of a 3 × 3 Matrix

$$\begin{vmatrix} a_1 & b_1 & c_1 \\ a_2 & b_2 & c_2 \\ a_3 & b_3 & c_3 \end{vmatrix} = a_1 \begin{vmatrix} b_2 & c_2 \\ b_3 & c_3 \end{vmatrix} - a_2 \begin{vmatrix} b_1 & c_1 \\ b_3 & c_3 \end{vmatrix} + a_3 \begin{vmatrix} b_1 & c_1 \\ b_2 & c_2 \end{vmatrix}$$

Cramer's Rule: 2 × 2 Systems

The solution of the system

$$a_1 x + b_1 y = c_1,$$
$$a_2 x + b_2 y = c_2$$

if it is unique, is given by

$$x = \frac{\begin{vmatrix} c_1 & b_1 \\ c_2 & b_2 \end{vmatrix}}{\begin{vmatrix} a_1 & b_1 \\ a_2 & b_2 \end{vmatrix}}, \qquad y = \frac{\begin{vmatrix} a_1 & c_1 \\ a_2 & c_2 \end{vmatrix}}{\begin{vmatrix} a_1 & b_1 \\ a_2 & b_2 \end{vmatrix}}.$$

Cramer's Rule: 3 × 3 Systems

The solution of the system

$$a_1 x + b_1 y + c_1 z = d_1,$$
$$a_2 x + b_2 y + c_2 z = d_2,$$
$$a_3 x + b_3 y + c_3 z = d_3$$

is found by considering the following determinants:

$$D = \begin{vmatrix} a_1 & b_1 & c_1 \\ a_2 & b_2 & c_2 \\ a_3 & b_3 & c_3 \end{vmatrix}, \qquad D_x = \begin{vmatrix} d_1 & b_1 & c_1 \\ d_2 & b_2 & c_2 \\ d_3 & b_3 & c_3 \end{vmatrix},$$

$$D_y = \begin{vmatrix} a_1 & d_1 & c_1 \\ a_2 & d_2 & c_2 \\ a_3 & d_3 & c_3 \end{vmatrix}, \qquad D_z = \begin{vmatrix} a_1 & b_1 & d_1 \\ a_2 & b_2 & d_2 \\ a_3 & b_3 & d_3 \end{vmatrix}.$$

If a unique solution exists, it is given by

$$x = \frac{D_x}{D}, \qquad y = \frac{D_y}{D}, \qquad z = \frac{D_z}{D}.$$

Review Exercises

For Exercises 1–9, if a system has an infinite number of solutions, use set-builder notation to write the solution set. If a system has no solution, state this.

Solve graphically.

1. $3x + 2y = -4,$
$y = 3x + 7$

2. $2x + 3y = 12,$
$4x - y = 10$

Solve using the substitution method.

3. $9x - 6y = 2,$
$x = 4y + 5$

4. $y = x + 2,$
$y - x = 8$

5. $x - 3y = -2,$
$7y - 4x = 6$

Solve using the elimination method.

6. $8x - 2y = 10,$
$-4y - 3x = -17$

7. $4x - 7y = 18,$
$9x + 14y = 40$

8. $3x - 5y = -4,$
$5x - 3y = 4$

9. $1.5x - 3 = -2y,$
$3x + 4y = 6$

Solve.

10. Glynn bought two DVD's and one videocassette for $72. If he had purchased one DVD and two video-cassettes, he would have spent $15 less. What is the price of a DVD? What is the price of a videocassette?

11. A freight train leaves Chicago at midnight traveling south at a speed of 44 mph. One hour later, a passenger train, going 55 mph, travels south from Chicago on a parallel track. How many hours will the passenger train travel before it overtakes the freight train?

12. Yolanda wants 14 L of fruit punch that is 10% juice. At the store, she finds punch that is 15% juice and punch that is 8% juice. How much of each should she purchase?

Solve. If a system's equations are dependent or if there is no solution, state this.

13. $x + 4y + 3z = 2,$
$2x + y + z = 10,$
$-x + y + 2z = 8$

14. $4x + 2y - 6z = 34,$
$2x + y + 3z = 3,$
$6x + 3y - 3z = 37$

15. $2x - 5y - 2z = -4,$
$7x + 2y - 5z = -6,$
$-2x + 3y + 2z = 4$

16. $-5x + 5y = -6,$
$2x - 2y = 4$

17. $3x + y = 2,$
$x + 3y + z = 0,$
$x + z = 2$

Solve.

18. In triangle ABC, the measure of angle A is four times the measure of angle C, and the measure of angle B is 45° more than the measure of angle C. What are the measures of the angles of the triangle?

19. Find the three-digit number in which the sum of the digits is 11, the tens digit is 3 less than the sum of the hundreds and ones digits, and the ones digit is 5 less than the hundreds digit.

20. Lynn has $159 in her purse, consisting of $20, $5, and $1 bills. The number of $20 bills is the same as the total number of $1 and $5 bills. If she has 14 bills in her purse, how many of each denomination does she have?

Solve using matrices. Show your work.

21. $3x + 4y = -13,$
$5x + 6y = 8$

22. $3x - y + z = -1,$
$2x + 3y + z = 4,$
$5x + 4y + 2z = 5$

Evaluate.

23. $\begin{vmatrix} -2 & 4 \\ -3 & 5 \end{vmatrix}$

24. $\begin{vmatrix} 2 & 3 & 0 \\ 1 & 4 & -2 \\ 2 & -1 & 5 \end{vmatrix}$

Solve using Cramer's rule. Show your work.

25. $2x + 3y = 6,$
$x - 4y = 14$

26. $2x + y + z = -2,$
$2x - y + 3z = 6,$
$3x - 5y + 4z = 7$

27. Find the equilibrium point for the demand and supply functions

$$S(p) = 60 + 7p$$

and

$$D(p) = 120 - 13p.$$

28. Robbyn is beginning to produce organic honey. For the first year, the fixed costs for setting up production are $18,000. The variable costs for producing each pint of honey are $1.50. The revenue from each pint of honey is $6. Find the following.

a) The total cost $C(x)$ of producing x pints of honey

b) The total revenue $R(x)$ from the sale of x pints of honey
c) The total profit $P(x)$ from the production and sale of x pints of honey
d) The profit or loss from the production and sale of 1500 pints of honey; of 5000 pints of honey
e) The break-even point

SYNTHESIS

29. How would you go about solving a problem that involves four variables?

30. Explain how a system of equations can be both dependent and inconsistent.

31. Robbyn is quitting a job that pays $27,000 a year to make honey (see Exercise 28). How many pints of honey must she produce and sell in order to make the same amount that she made in the job she left?

32. Solve graphically:
$$y = x + 2,$$
$$y = x^2 + 2.$$

33. The graph of $f(x) = ax^2 + bx + c$ contains the points $(-2, 3)$, $(1, 1)$, and $(0, 3)$. Find a, b, and c and give a formula for the function.

Chapter Test 8

1. Solve graphically:
$$2x + y = 8,$$
$$y - x = 2.$$

Solve, if possible, using the substitution method.

2. $x + 3y = -8,$
$4x - 3y = 23$

3. $2x + 4y = -6,$
$y = 3x - 9$

Solve, if possible, using the elimination method.

4. $4x - 6y = 3,$
$6x - 4y = -3$

5. $4y + 2x = 18,$
$3x + 6y = 26$

6. The perimeter of a rectangle is 96. The length of the rectangle is 6 less than twice the width. Find the dimensions of the rectangle.

7. Between her home mortgage (loan), car loan, and credit card bill (loan), Rema is $75,300 in debt. Rema's credit card bill accumulates 1.5% interest, her car loan 1% interest, and her mortgage 0.6% interest each month. After one month, her total accumulated interest is $460.50. The interest on Rema's credit card bill was $4.50 more than the interest on her car loan. Find the amount of each loan.

Solve. If a system's equations are dependent or if there is no solution, state this.

8. $-3x + y - 2z = 8,$
$-x + 2y - z = 5,$
$2x + y + z = -3$

9. $6x + 2y - 4z = 15,$
$-3x - 4y + 2z = -6,$
$4x - 6y + 3z = 8$

10. $2x + 2y = 0,$
$4x + 4z = 4,$
$2x + y + z = 2$

11. $3x + 3z = 0,$
$2x + 2y = 2,$
$3y + 3z = 3$

Solve using matrices.

12. $7x - 8y = 10,$
$9x + 5y = -2$

13. $x + 3y - 3z = 12,$
$3x - y + 4z = 0,$
$-x + 2y - z = 1$

Evaluate.

14. $\begin{vmatrix} 4 & -2 \\ 3 & 7 \end{vmatrix}$

15. $\begin{vmatrix} 3 & 4 & 2 \\ 2 & -5 & 4 \\ 4 & 5 & -3 \end{vmatrix}$

16. Solve using Cramer's rule:
$$8x - 3y = 5,$$
$$2x + 6y = 3.$$

17. An electrician, a carpenter, and a plumber are hired to work on a house. The electrician earns $21 per hour, the carpenter $19.50 per hour, and the plumber $24 per hour. The first day on the job, they worked a total of 21.5 hr and earned a total of $469.50. If the plumber worked 2 more hours than the carpenter did, how many hours did the electrician work?

18. Find the equilibrium point for the demand and supply functions
$$D(p) = 79 - 8p \quad \text{and} \quad S(p) = 37 + 6p.$$

19. Complete Communications, Inc., is producing a new family radio service model. For the first year, the fixed costs for setting up production are $40,000. The variable costs for producing each radio are $25. The revenue from each radio is $70. Find the following.

a) The total cost $C(x)$ of producing x radios
b) The total revenue $R(x)$ from the sale of x radios
c) The total profit $P(x)$ from the production and sale of x radios
d) The profit or loss from the production and sale of 300 radios; of 900 radios
e) The break-even point

SYNTHESIS

20. The graph of the function $f(x) = mx + b$ contains the points $(-1, 3)$ and $(-2, -4)$. Find m and b.

21. At a county fair, an adult's ticket sold for $5.50, a senior citizen's ticket for $4.00, and a child's ticket for $1.50. On opening day, the number of adults' and senior citizens' tickets sold was 30 more than the number of children's tickets sold. The number of adults' tickets sold was 6 more than four times the number of senior citizens' tickets sold. Total receipts from the ticket sales were $11,219.50. How many of each type of ticket were sold?

9

Inequalities and Problem Solving

AN APPLICATION

Slobberbone receives $750 plus 15% of receipts over $750 for playing a club date. If a club charges a $6 cover charge, how many people must attend in order for the band to receive $1200?

This problem appears as Exercise 47 in Section 9.1.

I use math in my job much more than I had thought I would. Bands are paid using various formulas, and I need to calculate capacity and admission prices to estimate income. I also use math to determine contracts, budgets, and royalty statements.

AMY POJMAN
Musician Management
New York, NY

nequalities are mathematical sentences containing symbols such as < (is less than). In this chapter, we use the principles for solving inequalities developed in Chapter 2 to solve compound inequalities. We also combine our knowledge of inequalities and systems of equations to solve systems of inequalities.

Interval Notation and Problem Solving

9.1

Solving Inequalities • Interval Notation • Problem Solving

Solving Inequalities

Recall from Chapter 1 that an **inequality** is any sentence containing $<, >, \leq, \geq,$ or $\neq$ (see Section 1.4)—for example,

$$-2 < a, \qquad x > 4, \qquad x + 3 \leq 6, \qquad 6 - 7y \geq 10y - 4, \quad \text{and} \quad 5x \neq 10.$$

Any replacement for the variable that makes an inequality true is called a **solution**. The set of all solutions is called the **solution set**. When all solutions of an inequality are found, we say that we have **solved** the inequality.

We can use two principles, developed in Chapter 2, to solve inequalities.

The Addition Principle for Inequalities

For any real numbers a, b, and c:

$$a < b \text{ is equivalent to } a + c < b + c;$$
$$a > b \text{ is equivalent to } a + c > b + c.$$

Similar statements hold for $\leq$ and $\geq$.

The Multiplication Principle for Inequalities

For any real numbers a and b, and for any *positive* number c,

$$a < b \text{ is equivalent to } ac < bc;$$
$$a > b \text{ is equivalent to } ac > bc.$$

For any real numbers a and b, and for any *negative* number c,

$$a < b \text{ is equivalent to } ac > bc;$$
$$a > b \text{ is equivalent to } ac < bc.$$

Similar statements hold for $\leq$ and $\geq$.

The *graph* of an inequality is a drawing that represents its solutions. An inequality in one variable can be graphed on a number line. Inequalities in two variables can be graphed on a coordinate plane, and appear later in this chapter.

The solutions of the inequality $x < 4$ are graphed on the following number line:

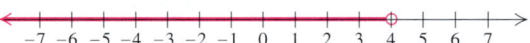

The open dot indicates that 4 is not a solution, and the shading indicates that all real numbers less than 4 are solutions.

We can write the solution set using *set-builder notation* (see Section 2.6):

$\{x \mid x < 4\}$.

This is read

"The set of all x such that x is less than 4."

Interval Notation

Another way to write solutions of an inequality in one variable is to use **interval notation**. Interval notation uses parentheses, ( ), and brackets, [ ].

If a and b are real numbers such that $a < b$, we define the **open interval** **(*a*, *b*)** as the set of all numbers x for which $a < x < b$. Thus,

$(a, b) = \{x \mid a < x < b\}$.   Parentheses are used to exclude endpoints.

Its graph excludes the endpoints:

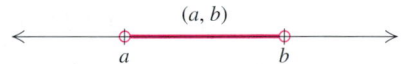

> ***Caution!***   Do not confuse the *interval* (a, b) with the *ordered pair* (a, b). The context in which the notation appears usually makes the meaning clear.

The **closed interval [*a*, *b*]** is defined as the set of all numbers x for which $a \leq x \leq b$. Thus,

$[a, b] = \{x \mid a \leq x \leq b\}$.   Brackets are used to include endpoints.

Its graph includes the endpoints*:

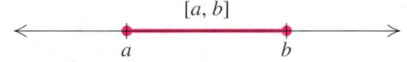

*Some books use the representations ⟵( )⟶ and ⟵[ ]⟶ instead of, respectively,
  ⟵∘——∘⟶ and ⟵•——•⟶ .

There are two kinds of **half-open intervals**, defined as follows:

1. $(a, b] = \{x \mid a < x \le b\}$. This is open on the left. Its graph is as follows:

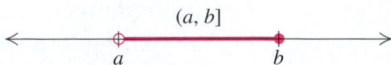

2. $[a, b) = \{x \mid a \le x < b\}$. This is open on the right. Its graph is as follows:

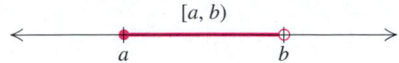

We use the symbols ∞ and $-\infty$ to represent positive and negative infinity, respectively. Thus the notation (a, ∞) represents the set of all real numbers greater than a, and $(-\infty, a)$ represents the set of all real numbers less than a.

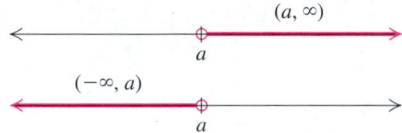

The notations $[a, \infty)$ and $(-\infty, a]$ are used when we want to include the endpoint a.

E x a m p l e   1

Graph $y \ge -2$ on a number line and write the solution set using both set-builder and interval notations.

Solution   Using set-builder notation, we write the solution set as $\{y \mid y \ge -2\}$.
Using interval notation, we write the solution set as $[-2, \infty)$.
To graph the solution, we shade all numbers to the right of -2 and use a solid dot to indicate that -2 is also a solution.

E x a m p l e   2

Solve: $16 - 7y \ge 10y - 4$. Write the solution set in both set-builder and interval notation. **(b)** $-3(x + 8) - 5x > 4x - 9$.

Solution

$$16 - 7y \ge 10y - 4$$

$$-16 + 16 - 7y \ge -16 + 10y - 4 \qquad \text{Adding } -16 \text{ to both sides}$$

$$-7y \ge 10y - 20$$

$$-10y + (-7y) \ge -10y + 10y - 20 \qquad \text{Adding } -10y \text{ to both sides}$$

$$-17y \ge -20$$

The symbol must be reversed.

$$-\tfrac{1}{17} \cdot (-17y) \le -\tfrac{1}{17} \cdot (-20) \qquad \text{Multiplying both sides by } -\tfrac{1}{17}$$
$$\text{or dividing both sides by } -17$$

$$y \le \tfrac{20}{17}$$

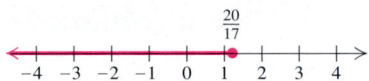

The solution set is $\left\{ y \,\middle|\, y \le \frac{20}{17} \right\}$, or $\left(-\infty, \frac{20}{17} \right]$.

Caution!   Remember that whenever we multiply or divide both sides of an inequality by a negative number, we must reverse the inequality symbol.

We can use interval notation to describe the values for which one function's value is greater than another's.

E x a m p l e   3

To check Example 3, graph $y_1 = -3(x + 8) - 5x$ and $y_2 = 4x - 9$ and identify those x-values for which $y_1 > y_2$.

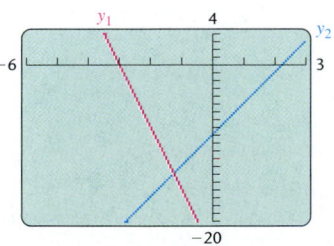

The INTERSECT option helps us find that $y_1 = y_2$ when $x = -1.25$. Note that $y_1 > y_2$ for x-values in the interval $(-\infty, -1.25)$.

On many graphers, the solution can be found by using the Y-VARS and TEST keys to enter $y_3 = y_1 > y_2$. The solution set is then displayed as an interval (shown by a horizontal line 1 unit above the x-axis).

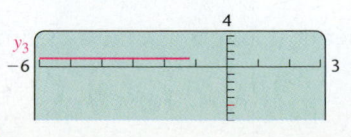

Let $f(x) = -3(x + 8) - 5x$ and $g(x) = 4x - 9$. Find all values of x for which $f(x) > g(x)$.

Solution   We are looking for values of x for which $f(x) > g(x)$:

$$f(x) > g(x)$$

$-3(x + 8) - 5x > 4x - 9$	Replacing $f(x)$ with $-3(x + 8) - 5x$ and $g(x)$ with $4x - 9$
$-3x - 24 - 5x > 4x - 9$	Using the distributive law
$-24 - 8x > 4x - 9$	
$-24 - 8x + 8x > 4x - 9 + 8x$	Adding $8x$ to both sides
$-24 > 12x - 9$	
$-24 + 9 > 12x - 9 + 9$	Adding 9 to both sides
$-15 > 12x$	
	The symbol stays the same.
$-\frac{5}{4} > x$	Dividing by 12 and simplifying

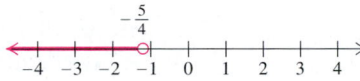

The solution set is $\left\{ x \,\middle|\, -\frac{5}{4} > x \right\}$, or $\left\{ x \,\middle|\, x < -\frac{5}{4} \right\}$, or $\left(-\infty, -\frac{5}{4} \right)$.

Problem Solving

Many problem-solving situations translate to inequalities.

Example 4

Records in the men's 200-m dash.   Michael Johnson set a world record of 19.32 sec in the men's 200-m dash in the 1996 Olympics. If $R(t)$ is given in seconds, then the function given by

$$R(t) = -0.045t + 19.32$$

can be used to predict the world record in the men's 200-m dash t years after 1996. Determine (in terms of an inequality) those years for which the world record will be less than 19.0 sec.

Solution

1. **Familiarize.**   We already have a formula. To become more familiar with it, we might make a substitution for t. Suppose we want to know the record after 20 years, in the year 2016. We substitute 20 for t:

    $$R(20) = -0.045(20) + 19.32 = 18.42 \text{ sec}.$$

    We see that by 2016, the record will be less than 19.0 sec. To predict the exact year in which the 19.0-sec mark will be broken, we could make other guesses that are less than 20. Instead, we proceed to the next step.

2. **Translate.**   The record $R(t)$ is to be *less than* 19.0 sec. Thus we have

    $$R(t) < 19.0.$$

    We replace $R(t)$ with $-0.045t + 19.32$ to find the times t that solve the inequality:

    $$-0.045t + 19.32 < 19.0. \qquad \text{Substituting } -0.045t + 19.32 \text{ for } R(t)$$

3. **Carry out.**   We solve the inequality:

    $$-0.045t + 19.32 < 19.0$$
    $$-0.045t < -0.32 \qquad \text{Adding } -19.32 \text{ to both sides}$$
    $$t > 7.1. \qquad \text{Dividing both sides by } -0.045, \text{ reversing the symbol, and rounding}$$

4. **Check.**   A partial check is to substitute a value for t greater than 7.1. We did that in the *Familiarize* step.

5. **State.**   The record will be less than 19.0 sec for races occurring more than 7.1 years after 1996, or approximately all years after 2003.

Earnings plans.   On a new job, Rose can be paid in one of two ways:

Plan A: A salary of $600 per month, plus a commission of 4% of sales;

Example 5

Plan B: A salary of $800 per month, plus a commission of 6% of sales in excess of $10,000.

For what amount of monthly sales is plan A better than plan B, if we assume that sales are always more than $10,000?

Solution

1. **Familiarize.**   Listing the given information in a table will be helpful.

Plan A: Monthly Income	Plan B: Monthly Income
$600 salary 4% of sales *Total*: $600 + 4% of sales	$800 salary 6% of sales over $10,000 *Total*: $800 + 6% of sales over $10,000

Next, suppose that Rose sold a certain amount—say, $12,000—in one month. Which plan would be better? Under plan A, she would earn $600 plus 4% of $12,000, or

$$600 + 0.04(12,000) = \$1080.$$

Since with plan B commissions are paid only on sales in excess of $10,000, Rose would earn $800 plus 6% of ($12,000 − $10,000), or

$$800 + 0.06(2000) = \$920.$$

This shows that for monthly sales of $12,000, plan A is better. Similar calculations will show that for sales of $30,000 a month, plan B is better. To determine *all* values for which plan A earns more money, we must solve an inequality that is based on the calculations above.

2. **Translate.**   We let S = the amount of monthly sales, in dollars. Examining the calculations in the *Familiarize* step, we see that monthly income from plan A is $600 + 0.04S$ and from plan B is $800 + 0.06(S - 10,000)$. We want to find all values of S for which

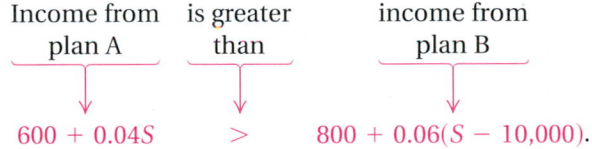

$$600 + 0.04S \quad > \quad 800 + 0.06(S - 10,000).$$

3. **Carry out.**   We solve the inequality:

$$600 + 0.04S > 800 + 0.06(S - 10,000)$$
$$600 + 0.04S > 800 + 0.06S - 600 \qquad \text{Using the distributive law}$$
$$600 + 0.04S > 200 + 0.06S \qquad \text{Combining like terms}$$
$$400 > 0.02S \qquad \text{Subtracting 200 and } 0.04S \text{ from both sides}$$
$$20,000 > S, \text{ or } S < 20,000. \qquad \text{Dividing both sides by 0.02}$$

4. **Check.**   For $S = 20,000$, the income from plan A is

$$600 + 4\% \cdot 20,000, \text{ or } \$1400.$$

The income from plan B is

$$800 + 6\% \cdot (20,000 - 10,000), \text{ or } \$1400.$$

This confirms that for sales totaling $20,000, Rose's pay is the same under either plan.

In the *Familiarize* step, we saw that for sales of $12,000, plan A pays more. Since $12,000 < 20,000$, this is a partial check. Since we cannot check all possible values of S, we will stop here.

5. **State.**   For monthly sales of less than $20,000, plan A is better.

Exercise Set **9.1**

Graph each inequality, and write the solution set using both set-builder and interval notation.

1. $y < 6$

2. $x > 4$

3. $x \geq -4$

4. $t \leq 6$

5. $t > -3$

6. $y < -3$

7. $x \leq -7$

8. $x \geq -6$

Solve. Then graph.

9. $y - 9 > -18$

10. $y - 8 > -14$

11. $y - 20 \leq -6$

12. $x - 11 \leq -2$

13. $9t < -81$

14. $8x \geq 24$

15. $-8y \leq 3.2$

16. $-9x \geq -8.1$

17. $-\frac{5}{6}y \leq -\frac{3}{4}$

18. $-\frac{3}{4}x \geq -\frac{5}{8}$

19. $5y + 13 > 28$

20. $2x + 7 < 19$

21. $-9x + 3x \geq -24$

22. $5y + 2y \leq -21$

23. Let $f(x) = 8x - 9$ and $g(x) = 3x - 11$. Find all values of x for which $f(x) < g(x)$.

24. Let $f(x) = 2x - 7$ and $g(x) = 5x - 9$. Find all values of x for which $f(x) < g(x)$.

25. Let $f(x) = 0.4x + 5$ and $g(x) = 1.2x - 4$. Find all values of x for which $g(x) \geq f(x)$.

26. Let $f(x) = \frac{3}{8} + 2x$ and $g(x) = 3x - \frac{1}{8}$. Find all values of x for which $g(x) \geq f(x)$.

Solve.

27. $4(3y - 2) \geq 9(2y + 5)$

28. $4m + 7 \geq 14(m - 3)$

29. $5(t - 3) + 4t < 2(7 + 2t)$

30. $2(4 + 2x) > 2x + 3(2 - 5x)$

31. $5[3m - (m + 4)] > -2(m - 4)$

32. $8x - 3(3x + 2) - 5 \geq 3(x + 4) - 2x$

33. $19 - (2x + 3) \leq 2(x + 3) + x$

34. $13 - (2c + 2) \geq 2(c + 2) + 3c$

35. $\frac{1}{4}(8y + 4) - 17 < -\frac{1}{2}(4y - 8)$

36. $\frac{1}{3}(6x + 24) - 20 > -\frac{1}{4}(12x - 72)$

37. $2[8 - 4(3 - x)] - 2 \geq 8[2(4x - 3) + 7] - 50$

38. $5[3(7 - t) - 4(8 + 2t)] - 20 \leq -6[2(6 + 3t) - 4]$

Phone rates. In Vermont, Verizon charges customers $13.55 *for monthly service plus* 2.2¢ *per minute for local phone calls between* 9 A.M. *and* 9 P.M. *weekdays. The charge for off-peak local calls is* 0.5¢ *per minute. Calls are free after the total monthly charges reach* $39.40.

39. Assume that only peak local calls were made. For how long must a customer speak on the phone if the $39.40 maximum charge is to apply?

40. Assume that only off-peak calls were made. For how long must a customer speak on the phone if the $39.40 maximum charge is to apply?

41. *Checking-account rates.* The Hudson Bank offers two checking-account plans. Their Anywhere plan charges 20¢ per check whereas their Acu-checking plan costs $2 per month plus 12¢ per check. For what numbers of checks per month will the Acu-checking plan cost less?

42. *Moving costs.* Musclebound Movers charges $85 plus $40 an hour to move households across town. Champion Moving charges $60 an hour for cross-town moves. For what lengths of time is Champion more expensive?

43. *Wages.* Toni can be paid in one of two ways:
Plan A: A salary of $400 per month, plus a commission of 8% of gross sales;
Plan B: A salary of $610 per month, plus a commission of 5% of gross sales.

For what amount of gross sales should Toni select plan A?

44. *Wages.* Branford can be paid for his masonry work in one of two ways:
Plan A: $300 plus $9.00 per hour;
Plan B: Straight $12.50 per hour.

Suppose that the job takes n hours. For what values of n is plan B better for Branford?

45. *Insurance benefits.* Bayside Insurance offers two plans. Under plan A, Giselle would pay the first

$50 of her medical bills and 20% of all bills after that. Under plan B, Giselle would pay the first $250 of bills, but only 10% of the rest. For what amount of medical bills will plan B save Giselle money? (Assume that her bills will exceed $250.)

46. *Wedding costs.*   The Arnold Inn offers two plans for wedding parties. Under plan A, the inn charges $30 for each person in attendance. Under plan B, the inn charges $1300 plus $20 for each person in excess of the first 25 who attend. For what size parties will plan B cost less? (Assume that more than 25 guests will attend.)

47. *Show business.*   Slobberbone receives $750 plus 15% of receipts over $750 for playing a club date. If a club charges a $6 cover charge, how many people must attend in order for the band to receive at least $1200?

48. *Temperature conversion.*   The function

$$C(F) = \tfrac{5}{9}(F - 32)$$

can be used to find the Celsius temperature $C(F)$ that corresponds to $F°$ Fahrenheit.

a) Gold is solid at Celsius temperatures less than 1063°C. Find the Fahrenheit temperatures for which gold is solid.

b) Silver is solid at Celsius temperatures less than 960.8°C. Find the Fahrenheit temperatures for which silver is solid.

49. *Manufacturing.*   Ergs, Inc., is planning to make a new kind of radio. Fixed costs will be $90,000, and variable costs will be $15 for the production of each radio. The total-cost function for x radios is

$$C(x) = 90{,}000 + 15x.$$

The company makes $26 in revenue for each radio sold. The total-revenue function for x radios is

$$R(x) = 26x.$$

(See Section 8.8.)

a) When $R(x) < C(x)$, the company loses money. Find the values of x for which the company loses money.

b) When $R(x) > C(x)$, the company makes a profit. Find the values of x for which the company makes a profit.

50. *Publishing.*   The demand and supply functions for a locally produced poetry book are approximated by

$$D(p) = 2000 - 60p \quad \text{and}$$
$$S(p) = 460 + 94p,$$

where p is the price in dollars (see Section 8.8).

a) Find those values of p for which demand exceeds supply.

b) Find those values of p for which demand is less than supply.

51. Explain in your own words why the inequality symbol must be reversed when both sides of an inequality are multiplied by a negative number.

52. Why isn't roster notation used to write solutions of inequalities?

SKILL MAINTENANCE

Find the domain of f.

53. $f(x) = \dfrac{3}{x - 2}$

54. $f(x) = \dfrac{x - 5}{4x + 12}$

55. $f(x) = \dfrac{5x}{7 - 2x}$

56. $f(x) = \dfrac{x + 3}{9 - 4x}$

Simplify.

57. $9x - 2(x - 5)$

58. $8x + 7(2x - 1)$

SYNTHESIS

59. A Presto photocopier costs $510 and an Exact Image photocopier costs $590. Write a problem that involves the cost of the copiers, the cost per page of photocopies, and the number of copies for which the Presto machine is the more expensive machine to own.

60. Explain how the addition principle can be used to avoid ever needing to multiply or divide both sides of an inequality by a negative number.

Solve. Assume that a, b, c, d, and m are positive constants.

61. $3ax + 2x \geq 5ax - 4$; assume $a > 1$

62. $6by - 4y \leq 7by + 10$

63. $a(by - 2) \geq b(2y + 5)$; assume $a > 2$

64. $c(6x - 4) < d(3 + 2x)$; assume $3c > d$

65. $c(2 - 5x) + dx > m(4 + 2x)$; assume $5c + 2m < d$

66. $a(3 - 4x) + cx < d(5x + 2)$; assume $c > 4a + 5d$

Determine whether each statement is true or false. If false, give an example that shows this.

67. For any real numbers a, b, c, and d, if $a < b$ and $c < d$, then $a - c < b - d$.

68. For all real numbers x and y, if $x < y$, then $x^2 < y^2$.

69. Are the inequalities

$$x < 3 \quad \text{and} \quad x + \frac{1}{x} < 3 + \frac{1}{x}$$

equivalent? Why or why not?

70. Are the inequalities

$$x < 3 \quad \text{and} \quad 0 \cdot x < 0 \cdot 3$$

equivalent? Why or why not?

Solve. Then graph.

71. $x + 5 \leq 5 + x$

72. $x + 8 < 3 + x$

73. $x^2 > 0$

74. Assume that the graphs of $y_1 = -\frac{1}{2}x + 5$, $y_2 = x - 1$, and $y_3 = 2x - 3$ are as shown. Solve each inequality, referring only to the figure below.

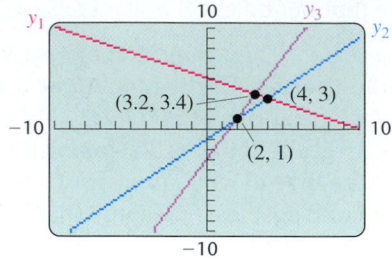

a) $-\frac{1}{2}x + 5 > x - 1$
b) $x - 1 \leq 2x - 3$
c) $2x - 3 \geq -\frac{1}{2}x + 5$

75. Using an approach similar to that in the Technology Connection on p. 547, use a grapher to check your answers to Exercises 9, 25, 41, and 45.

CORNER

Reduce, Reuse, and Recycle

Focus: Inequalities and problem solving

Time: 15–20 minutes

Group size: 2

In the United States, the amount of solid waste (rubbish) being recycled is slowly catching up to the amount being generated. In 1991, each person generated, on average, 4.3 lb of solid waste every day, of which 0.8 lb was recycled. In 2000, each person generated, on average, 4.4 lb of solid waste, of which 1.3 lb was recycled. (*Sources*: U.S. Census 2000 and EPA Municipal Solid Waste Factbook)

ACTIVITY

Assume that the amount of solid waste being generated and the amount recycled are both increasing linearly. One group member should find a linear function w for which $w(t)$ represents the number of pounds of waste generated per person per day t years after 1991. The other group member should find a linear function r for which $r(t)$ represents the number of pounds recycled per person per day t years after 1991. Finally, working together, the group should determine those years for which the amount recycled will meet or exceed the amount generated.

COLLABORATIVE

9.2

Intersections, Unions, and Compound Inequalities

Intersections of Sets and Conjunctions of Sentences • Unions of Sets and Disjunctions of Sentences • Interval Notation and Domains

We now consider **compound inequalities**—that is, sentences like "$-2 < x$ and $x < 1$" or "$x < -3$ or $x > 3$" that are formed using the word *and* or the word *or*.

Intersections of Sets and Conjunctions of Sentences

The **intersection** of two sets A and B is the set of all elements that are common to both A and B. We denote the intersection of sets A and B as

$$A \cap B.$$

The intersection of two sets is often pictured as shown here.

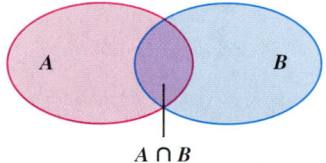

$A \cap B$

Example 1   Find the intersection: $\{1, 2, 3, 4, 5\} \cap \{-2, -1, 0, 1, 2, 3\}$.

Solution   The numbers 1, 2, and 3 are common to both sets, so the intersection is $\{1, 2, 3\}$.

When two or more sentences are joined by the word *and* to make a compound sentence, the new sentence is called a **conjunction** of the sentences. The following is a conjunction of inequalities:

$$-2 < x \quad and \quad x < 1.$$

A number is a solution of a conjunction if it is a solution of both of the separate parts. For example, -1 is a solution because it is a solution of $-2 < x$ as well as $x < 1$.

Below we show the graph of $-2 < x$, followed by the graph of $x < 1$, and finally the graph of the conjunction $-2 < x$ and $x < 1$. *Note that the solution set of a conjunction is the intersection of the solution sets of the individual sentences.*

$\{x \mid -2 < x\}$ (graph on number line from -7 to 7, open circle at -2, shaded right) $(-2, \infty)$

$\{x \mid x < 1\}$ (graph on number line from -7 to 7, open circle at 1, shaded left) $(-\infty, 1)$

$\{x \mid -2 < x\} \cap \{x \mid x < 1\}$ $= \{x \mid -2 < x \text{ and } x < 1\}$ (graph on number line from -7 to 7, open circles at -2 and 1, shaded between) $(-2, 1)$

Because there are numbers that are both greater than -2 and less than 1, the conjunction $-2 < x$ *and* $x < 1$ can be abbreviated by $-2 < x < 1$. Thus the interval $(-2, 1)$ can be represented as $\{x \mid -2 < x < 1\}$, the set of all numbers that are *simultaneously* greater than -2 *and* less than 1. Note that for $a < b$,

$$a < x \quad and \quad x < b \quad \textbf{can be abbreviated} \quad a < x < b;$$

and, equivalently,

$$b > x \quad and \quad x > a \quad \textbf{can be abbreviated} \quad b > x > a.$$

E x a m p l e   2

Solve and graph: $-1 \le 2x + 5 < 13$.

Solution   This inequality is an abbreviation for the conjunction

$$-1 \le 2x + 5 \quad and \quad 2x + 5 < 13.$$

The word *and* corresponds to set *intersection*. To solve the conjunction, we solve each of the two inequalities separately and then find the intersection of the solution sets:

$$-1 \le 2x + 5 \quad and \quad 2x + 5 < 13$$
$$-6 \le 2x \qquad and \qquad 2x < 8 \qquad \text{\color{red}Subtracting 5 from both sides of each inequality}$$
$$-3 \le x \qquad and \qquad x < 4. \qquad \text{\color{red}Dividing both sides of each inequality by 2}$$

We now abbreviate the answer:

$$-3 \le x < 4.$$

The solution set is $\{x \mid -3 \le x < 4\}$, or, in interval notation, $[-3, 4)$. The graph is the intersection of the two separate solution sets.

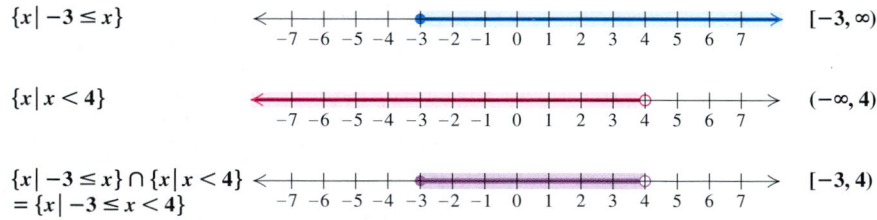

$\{x \mid -3 \le x\}$ $[-3, \infty)$

$\{x \mid x < 4\}$ $(-\infty, 4)$

$\{x \mid -3 \le x\} \cap \{x \mid x < 4\}$ $[-3, 4)$
$= \{x \mid -3 \le x < 4\}$

The steps in Example 2 are sometimes combined as follows:

$$-1 \le 2x + 5 < 13$$
$$-1 - 5 \le 2x + 5 - 5 < 13 - 5$$
$$-6 \le 2x < 8$$
$$-3 \le x < 4.$$

Such an approach saves some writing and will prove useful in Section 9.3.

E x a m p l e 3

Solve and graph: $2x - 5 \geq -3 \ and \ 5x + 2 \geq 17$.

Solution   We first solve each inequality separately:

$$2x - 5 \geq -3 \quad and \quad 5x + 2 \geq 17$$
$$2x \geq 2 \quad and \quad 5x \geq 15$$
$$x \geq 1 \quad and \quad x \geq 3.$$

Next, we find the intersection of the two separate solution sets.

$\{x \,|\, x \geq 1\}$

 $[1, \infty)$

$\{x \,|\, x \geq 3\}$ $[3, \infty)$

$\{x \,|\, x \geq 1\} \cap \{x \,|\, x \geq 3\}$
$= \{x \,|\, x \geq 3\}$ $[3, \infty)$

The numbers common to both sets are greater than or equal to 3. Thus the solution set is $\{x \,|\, x \geq 3\}$, or, in interval notation, $[3, \infty)$. You should check that any number in $[3, \infty)$ satisfies the conjunction whereas numbers outside $[3, \infty)$ do not.

> **Mathematical Use of the Word "and"**
>
> The word "and" corresponds to "intersection" and to the symbol " $\cap$ ". Any solution of a conjunction must make each part of the conjunction true.

Sometimes there is no way to solve both parts of a conjunction at once.

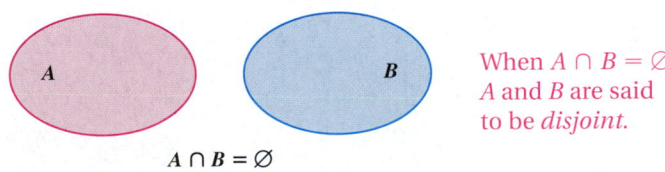

$A \cap B = \emptyset$

When $A \cap B = \emptyset$, A and B are said to be *disjoint*.

E x a m p l e 4

Solve and graph: $2x - 3 > 1 \ and \ 3x - 1 < 2$.

Solution   We solve each inequality separately:

$$2x - 3 > 1 \quad and \quad 3x - 1 < 2$$
$$2x > 4 \quad and \quad 3x < 3$$
$$x > 2 \quad and \quad x < 1.$$

The solution set is the intersection of the individual inequalities.

$\{x \mid x > 2\}$ (2, ∞)

$\{x \mid x < 1\}$ (−∞, 1)

$\{x \mid x > 2\} \cap \{x \mid x < 1\}$
$= \{x \mid x > 2 \text{ and } x < 1\} = \varnothing$ ∅

Since no number is both greater than 2 and less than 1, the solution set is the empty set, ∅.

Unions of Sets and Disjunctions of Sentences

The **union** of two sets A and B is the collection of elements belonging to A and/or B. We denote the union of A and B by

$$A \cup B.$$

The union of two sets is often pictured as shown at left.

Example 5

Find the union: $\{2, 3, 4\} \cup \{3, 5, 7\}$.

Solution   The numbers in either or both sets are 2, 3, 4, 5, and 7, so the union is $\{2, 3, 4, 5, 7\}$.

When two or more sentences are joined by the word *or* to make a compound sentence, the new sentence is called a **disjunction** of the sentences. Here is an example:

$$x < -3 \quad or \quad x > 3.$$

A number is a solution of a disjunction if it is a solution of either of the separate parts. For example, -5 is a solution of this disjunction since -5 is a solution of $x < -3$. Below we show the graph of $x < -3$, followed by the graph of $x > 3$, and finally the graph of the disjunction $x < -3$ or $x > 3$. *Note that the solution set of a disjunction is the union of the solution sets of the individual sentences.*

$\{x \mid x < -3\}$ (−∞, −3)

$\{x \mid x > 3\}$ (3, ∞)

$\{x \mid x < -3\} \cup \{x \mid x > 3\}$
$= \{x \mid x < -3 \text{ or } x > 3\}$ (−∞, −3) ∪ (3, ∞)

The solution set of $x < -3$ or $x > 3$ is $\{x \mid x < -3 \text{ or } x > 3\}$, or in interval notation, $(-\infty, -3) \cup (3, \infty)$. There is no simpler way to write the solution.

> ### *Mathematical Use of the Word "or"*
>
> The word "or" corresponds to "union" and to the symbol " ∪ ". For a number to be a solution of a disjunction, it must be in *at least one* of the solution sets of the individual sentences.

E x a m p l e   6   Solve and graph: $7 + 2x < -1 \text{ or } 13 - 5x \leq 3$.

Solution   We solve each inequality separately, retaining the word *or*:

$$7 + 2x < -1 \quad or \quad 13 - 5x \leq 3$$
$$2x < -8 \quad or \qquad -5x \leq -10$$

——— Reverse the symbol.

$$x < -4 \quad or \qquad x \geq 2.$$

To find the solution set of the disjunction, we consider the individual graphs. We graph $x < -4$ and then $x \geq 2$. Then we take the union of the graphs.

$\{x \mid x < -4\}$

-6 -5 -4 -3 -2 -1 0 1 2 3 4 5 6 $(-\infty, -4)$

$\{x \mid x \geq 2\}$

-6 -5 -4 -3 -2 -1 0 1 2 3 4 5 6 $[2, \infty)$

$\{x \mid x < -4\} \cup \{x \mid x \geq 2\}$
$= \{x \mid x < -4 \text{ or } x \geq 2\}$

-6 -5 -4 -3 -2 -1 0 1 2 3 4 5 6 $(-\infty, -4) \cup [2, \infty)$

The solution set is $\{x \mid x < -4 \text{ or } x \geq 2\}$, or $(-\infty, -4) \cup [2, \infty)$.

Caution!   A compound inequality like

$$x < -4 \quad or \quad x \geq 2,$$

as in Example 6, *cannot* be expressed as $2 \leq x < -4$ because to do so would be to say that x is *simultaneously* less than -4 and greater than or equal to 2. No number is both less than -4 *and* greater than 2, but many are less than -4 *or* greater than 2.

E x a m p l e   7   Solve: $-2x - 5 < -2 \text{ or } x - 3 < -10$.

Solution   We solve the individual inequalities separately, retaining the word *or*:

$$-2x - 5 < -2 \quad or \quad x - 3 < -10$$
$$-2x < 3 \quad or \qquad x < -7$$

Reverse the symbol.

$$x > -\tfrac{3}{2} \quad or \qquad x < -7.$$

——— Keep the word "or."

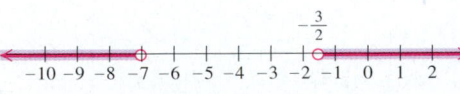

The solution set is $\left\{x \mid x < -7 \text{ or } x > -\frac{3}{2}\right\}$, or $(-\infty, -7) \cup \left(-\frac{3}{2}, \infty\right)$.

Example 8

Solve: $3x - 11 < 4$ *or* $4x + 9 \geq 1$.

Solution   We solve the individual inequalities separately, retaining the word *or*:

$$3x - 11 < 4 \quad or \quad 4x + 9 \geq 1$$
$$3x < 15 \quad or \quad 4x \geq -8$$
$$x < 5 \quad or \quad x \geq -2.$$

⬑————————— Keep the word "or."

To find the solution set, we first look at the individual graphs.

$\{x \,|\, x < 5\}$

$(-\infty, 5)$

$\{x \,|\, x \geq -2\}$

$[-2, \infty)$

$\{x \,|\, x < 5\} \cup \{x \,|\, x \geq -2\}$
$= \{x \,|\, x < 5 \text{ or } x \geq -2\}$

$(-\infty, \infty) = \mathbb{R}$

Since *all* numbers are less than 5 or greater than or equal to -2, the two sets fill the entire number line. Thus the solution set is $\mathbb{R}$, the set of all real numbers.

Interval Notation and Domains

In Section 7.2, we saw that if $g(x) = (5x - 2)/(3x - 7)$, then the domain of $g = \{x \,|\, x \text{ is a real number } and \; x \neq \frac{7}{3}\}$. We can now represent such a set using interval notation:

$$\{x \,|\, x \text{ is a real number } and \; x \neq \tfrac{7}{3}\} = \left(-\infty, \tfrac{7}{3}\right) \cup \left(\tfrac{7}{3}, \infty\right).$$

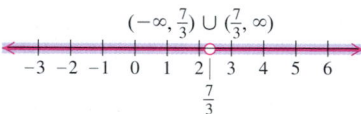

$\left(-\infty, \tfrac{7}{3}\right) \cup \left(\tfrac{7}{3}, \infty\right)$

Example 9

Use interval notation to write the domain of f if $f(x) = \sqrt{x + 2}$.

Solution   The expression $\sqrt{x + 2}$ is not a real number when $x + 2$ is negative. Thus the domain of f is the set of all x-values for which $x + 2 \geq 0$. Since $x + 2 \geq 0$ is equivalent to $x \geq -2$, we have

$$\text{Domain of } f = \{x \,|\, x \geq -2\} = [-2, \infty).$$

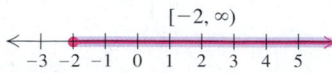

$[-2, \infty)$

technology connection

To visualize the domain of a sum, difference, or product of two functions, we have graphed below $y_1 = \sqrt{3 - x}$ and $y_2 = \sqrt{x + 1}$.

$y_1 = \sqrt{(3 - x)}, \quad y_2 = \sqrt{(x + 1)}$

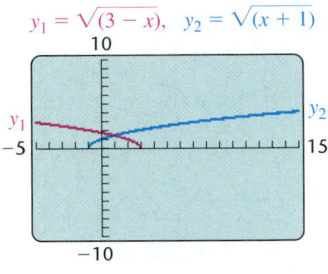

1. Determine algebraically the domains for y_1 and y_2 and TRACE each curve to verify your answer.
2. The Y-VARS option of the $\boxed{\text{VARS}}$ key permits us to access the functions already entered. Use this feature to let $y_3 = y_1 + y_2$, $y_4 = y_1 - y_2$, $y_5 = y_2 - y_1$, and $y_6 = y_1 \cdot y_2$. Determine the domains of $y_1 + y_2$, $y_1 - y_2$, $y_2 - y_1$, and $y_1 \cdot y_2$ algebraically and then check by graphing.

FOR EXTRA HELP

Exercise Set 9.2

Digital Video Tutor CD 7
Videotape 17

 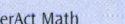
InterAct Math

Math Tutor Center

MathXL

MyMathLab.com

Find each indicated intersection or union.

1. $\{7, 9, 11\} \cap \{9, 11, 13\}$

2. $\{2, 4, 8\} \cup \{8, 9, 10\}$

3. $\{1, 5, 10, 15\} \cup \{5, 15, 20\}$

4. $\{2, 5, 9, 13\} \cap \{5, 8, 10\}$

5. $\{a, b, c, d, e, f\} \cap \{b, d, f\}$

6. $\{a, b, c\} \cup \{a, c\}$

7. $\{r, s, t\} \cup \{r, u, t, s, v\}$

8. $\{m, n, o, p\} \cap \{m, o, p\}$

9. $\{3, 6, 9, 12\} \cap \{5, 10, 15\}$

10. $\{1, 5, 9\} \cup \{4, 6, 8\}$

11. $\{3, 5, 7\} \cup \varnothing$

12. $\{3, 5, 7\} \cap \varnothing$

Graph and write interval notation for each compound inequality.

13. $3 < x < 8$

14. $0 \le y \le 4$

15. $-6 \le y \le -2$

16. $-9 \le x < -5$

17. $x < -2 \text{ or } x > 3$

18. $x < -5 \text{ or } x > 1$

19. $x \le -1 \text{ or } x > 5$

20. $x \le -5 \text{ or } x > 2$

21. $-4 \le -x < 2$

22. $x > -7 \text{ and } x < -2$

23. $x > -2 \text{ and } x < 4$

24. $3 > -x \ge -1$

25. $5 > a \text{ or } a > 7$

26. $t \ge 2 \text{ or } -3 > t$

27. $x \ge 5 \text{ or } -x \ge 4$

28. $-x < 3 \text{ or } x < -6$

29. $4 > y \text{ and } y \ge -6$

30. $6 > -x \ge 0$

31. $x < 7 \text{ and } x \ge 3$

32. $x \ge -3 \text{ and } x < 3$

Aha! 33. $t < 2 \text{ or } t < 5$

34. $t > 4 \text{ or } t > -1$

35. $x > -1 \text{ or } x \le 3$

36. $4 > x \text{ or } x \ge -3$

37. $x \ge 5 \text{ and } x > 7$

38. $x \le -4 \text{ and } x < 1$

Solve and graph each solution set.

39. $-1 < t + 2 < 7$

40. $-3 < t + 1 \le 5$

41. $2 < x + 3 \text{ and } x + 1 \le 5$

42. $-1 < x + 2 \text{ and } x - 4 < 3$

43. $-7 \le 2a - 3 \text{ and } 3a + 1 < 7$

44. $-4 \le 3n + 2 \text{ and } 2n - 3 \le 5$

Aha! **45.** $x + 7 \le -2 \text{ or } x + 7 \ge -3$

46. $x + 5 < -3 \text{ or } x + 5 \ge 4$

47. $2 \le f(x) \le 8$, where $f(x) = 3x - 1$

48. $7 \ge g(x) \ge -2$, where $g(x) = 3x - 5$

49. $-21 \le f(x) < 0$, where $f(x) = -2x - 7$

50. $4 > g(t) \ge 2$, where $g(t) = -3t - 8$

51. $f(x) \le 2 \text{ or } f(x) \ge 8$, where $f(x) = 3x - 1$

52. $g(x) \le -2 \text{ or } g(x) \ge 10$, where $g(x) = 3x - 5$

53. $f(x) < -1 \text{ or } f(x) > 1$, where $f(x) = 2x - 7$

54. $g(x) < -7 \text{ or } g(x) > 7$, where $g(x) = 3x + 5$

55. $6 > 2a - 1 \text{ or } -4 \le -3a + 2$

56. $3a - 7 > -10 \text{ or } 5a + 2 \le 22$

57. $a + 4 < -1 \text{ and } 3a - 5 < 7$

58. $1 - a < -2 \text{ and } 2a + 1 > 9$

59. $3x + 2 < 2 \text{ or } 4 - 2x < 14$

60. $2x - 1 > 5 \text{ or } 3 - 2x \ge 7$

61. $2t - 7 \le 5 \text{ or } 5 - 2t > 3$

62. $5 - 3a \le 8 \text{ or } 2a + 1 > 7$

For f(x) as given, use interval notation to write the domain of f.

63. $f(x) = \dfrac{9}{x + 7}$ **64.** $f(x) = \dfrac{2}{x + 3}$

65. $f(x) = \sqrt{x - 6}$ **66.** $f(x) = \sqrt{x - 2}$

67. $f(x) = \dfrac{x + 3}{2x - 5}$ **68.** $f(x) = \dfrac{x - 1}{3x + 4}$

69. $f(x) = \sqrt{2x + 8}$ **70.** $f(x) = \sqrt{8 - 4x}$

71. $f(x) = \sqrt{8 - 2x}$ **72.** $f(x) = \sqrt{10 - 2x}$

73. Why can the conjunction $2 < x \text{ and } x < 5$ be rewritten as $2 < x < 5$, but the disjunction $2 < x \text{ or } x < 5$ cannot be rewritten as $2 < x < 5$?

74. Can the solution set of a disjunction be empty? Why or why not?

SKILL MAINTENANCE

Graph.

75. $y = 5$ **76.** $y = -2$

77. $f(x) = |x|$ **78.** $g(x) = x - 1$

Solve each system graphically.

79. $y = x - 3$,   **80.** $y = x + 2$,
    $y = 5$ $y = -3$

SYNTHESIS

81. What can you conclude about a, b, c, and d, if $[a, b] \cup [c, d] = [a, d]$? Why?

82. What can you conclude about a, b, c, and d, if $[a, b] \cap [c, d] = [a, b]$? Why?

83. Use the accompanying graph of $f(x) = 2x - 5$ to solve $-7 < 2x - 5 < 7$.

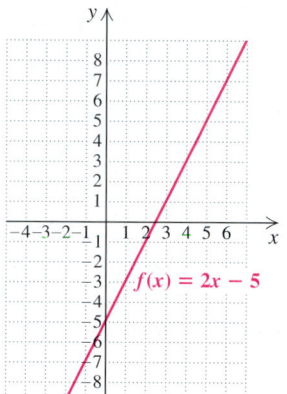

84. Use the accompanying graph of $g(x) = 4 - x$ to solve $4 - x < -2 \text{ or } 4 - x > 7$.

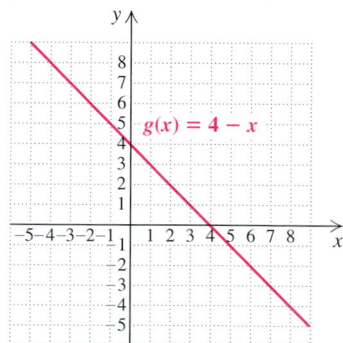

85. *Minimizing tolls.*   A \$3.00 toll is charged to cross the bridge from Sanibel Island to mainland Florida. A six-month pass, costing \$15.00, reduces the toll to \$0.50. A one-year pass, costing \$150, allows for free crossings. How many crossings per year does it take, on average, for the two six-month passes to be the most economical choice? Assume a constant number of trips per month.

86. *Pressure at sea depth.*   The function

$$P(d) = 1 + \frac{d}{33}$$

gives the pressure, in atmospheres (atm), at a depth of d feet in the sea. For what depths d is the pressure at least 1 atm and at most 7 atm?

87. *Converting dress sizes.*   The function

$$f(x) = 2(x + 10)$$

can be used to convert dress sizes x in the United States to dress sizes $f(x)$ in Italy. For what dress sizes in the United States will dress sizes in Italy be between 32 and 46?

88. *Solid-waste generation.*   The function

$$w(t) = 0.01t + 4.3$$

can be used to estimate the number of pounds of solid waste, $w(t)$, produced daily, on average, by each person in the United States, t years after 1991. For what years will waste production range from 4.5 to 4.75 lb per person per day?

89. *Temperatures of liquids.*   The formula

$$C = \tfrac{5}{9}(F - 32)$$

can be used to convert Fahrenheit temperatures F to Celsius temperatures C.

a) Gold is liquid for Celsius temperatures C such that $1063° \leq C < 2660°$. Find a comparable inequality for Fahrenheit temperatures.

b) Silver is liquid for Celsius temperatures C such that $960.8° \leq C < 2180°$. Find a comparable inequality for Fahrenheit temperatures.

90. *Records in the women's 100-m dash.*   Florence Griffith Joyner set a world record of 10.49 sec in the women's 100-m dash in 1988. The function

$$R(t) = -0.0433t + 10.49$$

can be used to predict the world record in the women's 100-m dash t years after 1988. Predict (in terms of an inequality) those years for which the world record was between 11.5 and 10.8 sec. (Measure from the middle of 1988.)

Solve and graph.

91. $4a - 2 \leq a + 1 \leq 3a + 4$

92. $4m - 8 > 6m + 5 \;\; or \;\; 5m - 8 < -2$

93. $x - 10 < 5x + 6 \leq x + 10$

94. $3x < 4 - 5x < 5 + 3x$

Determine whether each sentence is true or false for all real numbers a, b, and c.

95. If $-b < -a$, then $a < b$.

96. If $a \leq c$ and $c \leq b$, then $b > a$.

97. If $a < c$ and $b < c$, then $a < b$.

98. If $-a < c$ and $-c > b$, then $a > b$.

For f(x) as given, use interval notation to write the domain of f.

99. $f(x) = \dfrac{\sqrt{5 + 2x}}{x - 1}$

100. $f(x) = \dfrac{\sqrt{3 - 4x}}{x + 7}$

101. Let $y_1 = -1$, $y_2 = 2x + 5$, and $y_3 = 13$. Then use the graphs of y_1, y_2, and y_3 to check the solution to Example 2.

102. Let $y_1 = -2x - 5$, $y_2 = -2$, $y_3 = x - 3$, and $y_4 = -10$. Then use the graphs of y_1, y_2, y_3, and y_4 to check the solution to Example 7.

103. Use a grapher to check your answers to Exercises 33–36 and Exercises 53–56.

104. On many graphers, the TEST key provides access to inequality symbols, while the LOGIC option of that same key accesses the conjunction *and* and the disjunction *or*. Thus, if $y_1 = x > -2$ and $y_2 = x < 4$, Exercise 23 can be checked by forming the expression $y_3 = y_1$ *and* y_2. As in the Technology Connection on p. 547, the interval(s) in the solution set are shown as a horizontal line 1 unit above the x-axis. (Be careful to "deselect" y_1 and y_2 so that only y_3 is drawn.) Use this approach to check your answers to Exercises 29 and 60.

COLLABORATIVE

CORNER

Saving on Shipping Costs

Focus: Compound inequalities and solution sets

Time: 20–30 minutes

Group size: 2–3

At present (2001), the U.S. Postal Service charges 21 cents per ounce plus an additional 13-cent delivery fee (1 oz or less costs 34 cents; more than 1 oz, but not more than 2 oz, costs 55 cents; and so on). Rapid Delivery charges $1.05 per pound plus an additional $2.50 delivery fee (up to 16 oz costs $3.55; more than 16 oz, but less than or equal to 32 oz, costs $4.60; and so on). Let x be the weight, in ounces, of an item being mailed.*

———————————

*Based on an article by Michael Contino in *Mathematics Teacher,* May 1995.

ACTIVITY

One group member should determine the function p, where $p(x)$ represents the cost, in dollars, of mailing x ounces at a post office. Another group member should determine the function r, where $r(x)$ represents the cost, in dollars, of mailing x ounces with Rapid Delivery. The third group member should graph p and r on the same set of axes. Finally, working together, use the graph to determine those weights for which the Postal Service is less expensive. Express your answer using both set-builder and interval notation.

Absolute-Value Equations and Inequalities

9.3

Equations with Absolute Value • Inequalities with Absolute Value

Equations with Absolute Value

Recall from Section 1.4 that the absolute value of a number a is the number of units that a is from zero. Another definition uses opposites.

> **Absolute Value**
>
> The absolute value of x, denoted $|x|$, is defined as
>
> $$|x| = \begin{cases} x, & \text{if } x \geq 0, \\ -x, & \text{if } x < 0. \end{cases}$$
>
> (When x is nonnegative, the absolute value of x is x. When x is negative, the absolute value of x is the opposite of x.)

To better understand this definition, suppose x is -5. Then $|x| = |-5| = 5$, and 5 is the opposite of -5. This shows that when x represents a negative number, we have $|x| = -x$.

Since distance is always nonnegative, we can think of a number's absolute value as its distance from zero on a number line.

Example 1

Find the solution set: **(a)** $|x| = 4$; **(b)** $|x| = 0$; **(c)** $|x| = -7$.

Solution

a) We interpret $|x| = 4$ to mean that the number x is 4 units from zero on a number line. There are two such numbers, 4 and -4. Thus the solution set is $\{-4, 4\}$.

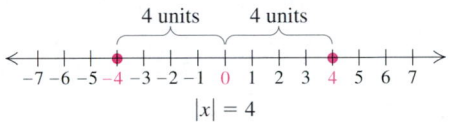

A second way to visualize this problem is to graph $f(x) = |x|$ (see Section 7.3). We also graph $g(x) = 4$. The x-values of the points of intersection are the solutions of $|x| = 4$.

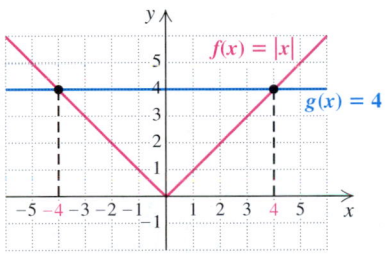

b) We interpret $|x| = 0$ to mean that x is 0 units from zero on a number line. The only number that satisfies this is 0 itself. Thus the solution set is $\{0\}$.

c) Since distance is always nonnegative, it doesn't make sense to talk about a number that is -7 units from zero. Remember: The absolute value of a number is never negative. Thus, $|x| = -7$ has no solution; the solution set is $\varnothing$.

Example 1 leads us to the following principle for solving equations.

Study Tip

The absolute-value principle will be used with a variety of replacements for X. Make sure that the principle, as stated here, makes sense before going further.

The Absolute-Value Principle for Equations

For any positive number p and any algebraic expression X:

a) The solutions of $|X| = p$ are those numbers that satisfy $X = -p$ or $X = p$.

b) The equation $|X| = 0$ is equivalent to the equation $X = 0$.

c) The equation $|X| = -p$ has no solution.

E x a m p l e   2

technology connection

A

To check Example 2(a), we let $y_1 = \text{abs}(2x + 5)$ and $y_2 = 13$. Using the window $[-12, 8, -2, 18]$, we use INTERSECT to find the points of intersection, $(-9, 13)$ and $(4, 13)$. The x-coordinates, -9 and 4, are the solutions.

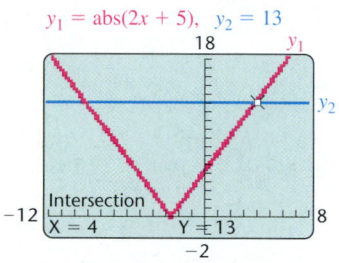

1. Use a grapher to show that Example 2(b) has no solution.

Find the solution set: **(a)** $|2x + 5| = 13$; **(b)** $|4 - 7x| = -8$.

Solution

a) We use the absolute-value principle, replacing X with $2x + 5$ and p with 13:

$$|X| = p$$
$$|2x + 5| = 13$$
$$2x + 5 = -13 \quad or \quad 2x + 5 = 13$$
$$2x = -18 \quad or \qquad\quad 2x = 8$$
$$x = -9 \quad or \qquad\quad\; x = 4.$$

Check: For -9:

$$\begin{array}{c|c} |2x + 5| = 13 \\ \hline |2(-9) + 5| & ?\; 13 \\ |-18 + 5| & \\ |-13| & \\ 13 & 13 \quad \text{TRUE} \end{array}$$

For 4:

$$\begin{array}{c|c} |2x + 5| = 13 \\ \hline |2 \cdot 4 + 5| & ?\; 13 \\ |8 + 5| & \\ |13| & \\ 13 & 13 \quad \text{TRUE} \end{array}$$

The number $2x + 5$ is 13 units from zero if x is replaced with -9 or 4. The solution set is $\{-9, 4\}$.

b) The absolute-value principle reminds us that absolute value is always nonnegative. The equation $|4 - 7x| = -8$ has no solution. The solution set is $\varnothing$.

To use the absolute-value principle, we must be sure that the absolute-value expression is alone on one side of the equation.

E x a m p l e   3

Given that $f(x) = 2|x + 3| + 1$, find all x for which $f(x) = 15$.

Solution   Since we are looking for $f(x) = 15$, we substitute:

$$f(x) = 15$$
$$2|x + 3| + 1 = 15 \qquad\qquad \text{Replacing } f(x) \text{ with } 2|x + 3| + 1$$
$$2|x + 3| = 14 \qquad\qquad \text{Subtracting 1 from both sides}$$
$$|x + 3| = 7 \qquad\qquad \text{Dividing both sides by 2}$$
$$x + 3 = -7 \quad or \quad x + 3 = 7 \qquad \text{Replacing } X \text{ with } x + 3 \text{ and } p \text{ with 7 in the absolute-value principle}$$
$$x = -10 \quad or \qquad\quad x = 4.$$

We leave it to the student to check that $f(-10) = f(4) = 15$. The solution set is $\{-10, 4\}$.

E x a m p l e   4

Solve: $|x - 2| = 3$.

Solution   Because this equation is of the form $|a - b| = c$, it can be solved in two different ways.

Method 1.   We interpret $|x - 2| = 3$ as stating that the number $x - 2$ is 3 units from zero. Using the absolute-value principle, we replace X with $x - 2$ and p with 3:

$$|X| = p$$
$$|x - 2| = 3$$
$$x - 2 = -3 \quad or \quad x - 2 = 3 \qquad \text{Using the absolute-value principle}$$
$$x = -1 \quad or \qquad x = 5.$$

Method 2.   This approach is helpful in calculus. The expressions $|a - b|$ and $|b - a|$ can be used to represent the *distance between* a and b on the number line. For example, the distance between 7 and 8 is given by $|8 - 7|$ or $|7 - 8|$. From this viewpoint, the equation $|x - 2| = 3$ states that the distance between x and 2 is 3 units. We draw a number line and locate all numbers that are 3 units from 2.

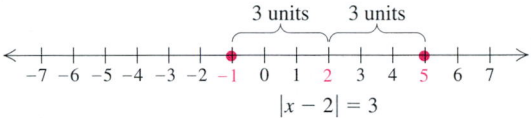

The solutions of $|x - 2| = 3$ are -1 and 5.

Check: The check consists of observing that both methods give the same solutions. The solution set is $\{-1, 5\}$.

Sometimes an equation has two absolute-value expressions. Consider $|a| = |b|$. This means that a and b are the same distance from zero.

If a and b are the same distance from zero, then either they are the same number or they are opposites.

E x a m p l e   5

Solve: $|2x - 3| = |x + 5|$.

Solution   Either $2x - 3 = x + 5$ (they are the same number) or $2x - 3 = -(x + 5)$ (they are opposites). We solve each equation separately:

This assumes these
numbers are the same.   This assumes these
                                                                        numbers are opposites.

$$2x - 3 = x + 5 \quad or \quad 2x - 3 = -(x + 5)$$
$$x - 3 = 5 \quad or \quad 2x - 3 = -x - 5$$
$$x = 8 \quad or \quad 3x - 3 = -5$$
$$3x = -2$$
$$x = -\tfrac{2}{3}.$$

The check is left to the student. The solutions are 8 and $-\tfrac{2}{3}$ and the solution set is $\left\{-\tfrac{2}{3}, 8\right\}$.

Inequalities with Absolute Value

Our methods for solving equations with absolute value can be adapted for solving inequalities. Inequalities of this sort arise regularly in more advanced courses.

E x a m p l e   6

Solve $|x| < 4$. Then graph.

Solution   The solutions of $|x| < 4$ are all numbers whose *distance from zero is less than* 4. By substituting or by looking at the number line, we can see that numbers like $-3, -2, -1, -\frac{1}{2}, -\frac{1}{4}, 0, \frac{1}{4}, \frac{1}{2}, 1, 2,$ and 3 are all solutions. In fact, the solutions are all the numbers between -4 and 4. The solution set is $\{x \,|\, -4 < x < 4\}$. In interval notation, the solution set is $(-4, 4)$. The graph is as follows:

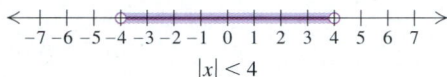

$|x| < 4$

We can also visualize Example 6 by graphing $f(x) = |x|$ and $g(x) = 4$, as in Example 1. The solution set consists of all x-values for which $(x, f(x))$ is below the horizontal line $g(x) = 4$. These x-values comprise the interval $(-4, 4)$.

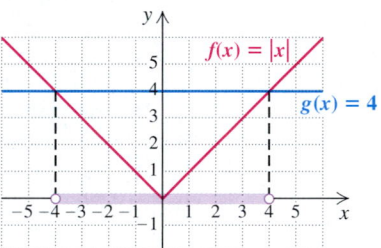

E x a m p l e   7

Solve $|x| \geq 4$. Then graph.

Solution   The solutions of $|x| \geq 4$ are all numbers that are at least 4 units from zero—in other words, those numbers x for which $x \leq -4$ or $4 \leq x$. The solution set is $\{x \,|\, x \leq -4 \ or \ x \geq 4\}$. In interval notation, the solution set is $(-\infty, -4] \cup [4, \infty)$. We can check mentally with numbers like $-4.1, -5, 4.1,$ and 5. The graph is as follows:

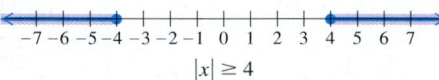

$|x| \geq 4$

As with Examples 1 and 6, Example 7 can be visualized by graphing $f(x) = |x|$ and $g(x) = 4$. The solution set of $|x| \geq 4$ consists of all x-values for which $(x, f(x))$ is on or above the horizontal line $g(x) = 4$. These x-values comprise $(-\infty, -4] \cup [4, \infty)$.

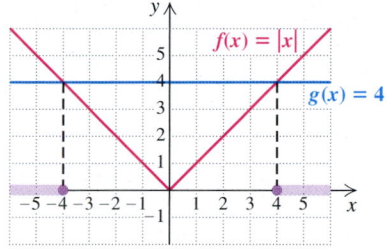

Examples 1, 6, and 7 illustrate three types of problems in which absolute-value symbols appear. The following is a general principle for solving such problems.

> ### Principles for Solving Absolute-Value Problems
>
> For any positive number p and any expression X:
>
> **a)** The solutions of $|X| = p$ are those numbers that satisfy $X = -p \, or \, X = p$.
>
>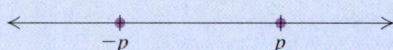
>
> **b)** The solutions of $|X| < p$ are those numbers that satisfy $-p < X < p$.
>
>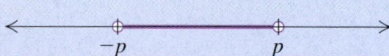
>
> **c)** The solutions of $|X| > p$ are those numbers that satisfy $X < -p \, or \, p < X$.
>
>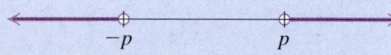

Of course, if p is negative, any value of X will satisfy the inequality $|X| > p$ because absolute value is never negative. By the same reasoning, $|X| < p$ has no solution when p is not positive. Thus, $|2x - 7| > -3$ is true for any real number x, and $|2x - 7| < -3$ has no solution.

Note that an inequality of the form $|X| < p$ corresponds to a *con*junction, whereas an inequality of the form $|X| > p$ corresponds to a *dis*junction.

Example 8

Solve $|3x - 2| < 4$. Then graph.

Solution We use part (b) of the principles listed above. In this case, X is $3x - 2$ and p is 4:

$$|X| < p$$

$|3x - 2| < 4$ Replacing X with $3x - 2$ and p with 4

$-4 < 3x - 2 < 4$ The number $3x - 2$ must be within 4 units of zero.

$-2 < \quad 3x \quad < 6$ Adding 2

$-\frac{2}{3} < \quad x \quad < 2.$ Multiplying by $\frac{1}{3}$

The solution set is $\left\{x \mid -\frac{2}{3} < x < 2\right\}$. In interval notation, the solution set is $\left(-\frac{2}{3}, 2\right)$. The graph is as follows:

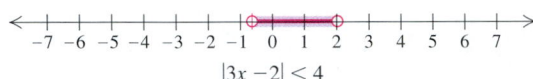

$|3x - 2| < 4$

Example 9

Given that $f(x) = |4x + 2|$, find all x for which $f(x) \geq 6$.

Solution We have

$$f(x) \geq 6,$$

or $|4x + 2| \geq 6.$ Substituting

To solve, we use part (c) of the principles listed above. In this case, X is $4x + 2$ and p is 6:

$$|X| \geq p$$

$|4x + 2| \geq 6$ Replacing X with $4x + 2$ and p with 6

$4x + 2 \leq -6 \quad or \quad 6 \leq 4x + 2$ The number $4x + 2$ must be at least 6 units from zero.

$4x \leq -8 \quad or \quad 4 \leq 4x$ Adding -2

$x \leq -2 \quad or \quad 1 \leq x.$ Multiplying by $\frac{1}{4}$

The solution set is $\{x \mid x \leq -2 \ or \ x \geq 1\}$. In interval notation, the solution is $(-\infty, -2] \cup [1, \infty)$. The graph is as follows:

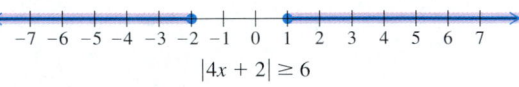

$|4x + 2| \geq 6$

---

technology connection
B

To solve an inequality like $|4x + 2| \geq 6$ with a grapher, graph the equations $y_1 = \text{abs}(4x + 2)$ and $y_2 = 6$. Then use INTERSECT to find where $y_1 = y_2$. The x-values on the graph of $y_1 = |4x + 2|$ that are *on or above* the line $y = 6$ solve the inequality.

$y_1 = \text{abs}(4x + 2), \quad y_2 = 6$

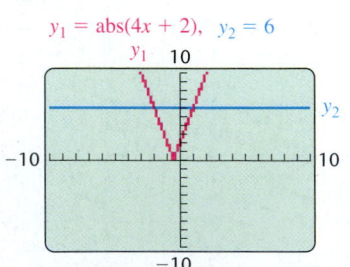

How can the same graph be used to solve the inequality $|4x + 2| < 6$ or the equation $|4x + 2| = 6$? Try using this procedure to solve Example 8 on a grapher.

Exercise Set 9.3

FOR EXTRA HELP

Digital Video Tutor CD 7
Videotape 17

InterAct Math

Math Tutor Center

MathXL

MyMathLab.com

Solve.

1. $|x| = 4$

2. $|x| = 9$

Aha! **3.** $|x| = -5$

4. $|x| = -3$

5. $|y| = 7.3$

6. $|p| = 0$

7. $|m| = 0$

8. $|t| = 5.5$

9. $|5x + 2| = 7$

10. $|2x - 3| = 4$

11. $|7x - 2| = -9$

12. $|3x - 10| = -8$

13. $|x - 3| = 8$

14. $|x - 2| = 6$

15. $|x - 6| = 1$

16. $|x - 5| = 3$

17. $|x - 4| = 5$

18. $|x - 7| = 9$

19. $|2y| - 5 = 13$

20. $|5x| - 3 = 37$

21. $7|z| + 2 = 16$

22. $5|q| - 2 = 9$

23. $\left|\dfrac{4 - 5x}{6}\right| = 3$

24. $\left|\dfrac{2x - 1}{3}\right| = 5$

25. $|t - 7| + 1 = 4$

26. $|m + 5| + 9 = 16$

27. $3|2x - 5| - 7 = -1$

28. $5 - 2|3x - 4| = -5$

29. Let $f(x) = |3x - 4|$. Find all x for which $f(x) = 8$.

30. Let $f(x) = |2x - 7|$. Find all x for which $f(x) = 10$.

31. Let $f(x) = |x| - 2$. Find all x for which $f(x) = 6.3$.

32. Let $f(x) = |x| + 7$. Find all x for which $f(x) = 18$.

33. Let $f(x) = \left|\dfrac{3x - 2}{5}\right|$. Find all x for which $f(x) = 2$.

34. Let $f(x) = \left|\dfrac{1 - 2x}{3}\right|$. Find all x for which $f(x) = 1$.

Solve.

35. $|x + 4| = |2x - 7|$

36. $|3x + 5| = |x - 6|$

37. $|x - 9| = |x + 6|$

38. $|x + 4| = |x - 3|$

39. $|5t + 7| = |4t + 3|$

40. $|3a - 1| = |2a + 4|$

Aha! **41.** $|n - 3| = |3 - n|$

42. $|y - 2| = |2 - y|$

43. $|7 - a| = |a + 5|$

44. $|6 - t| = |t + 7|$

45. $\left|\dfrac{1}{2}x - 5\right| = \left|\dfrac{1}{4}x + 3\right|$

46. $\left|2 - \dfrac{2}{3}x\right| = \left|4 + \dfrac{7}{8}x\right|$

Solve and graph.

47. $|a| \le 7$

48. $|x| < 2$

49. $|x| > 8$

50. $|a| \ge 3$

51. $|t| > 0$

52. $|t| \ge 1.7$

53. $|x - 3| < 5$

54. $|x - 1| < 3$

55. $|x + 2| \le 6$

56. $|x + 4| \le 1$

57. $|x - 3| + 2 > 7$

58. $|x - 4| + 5 > 2$

Aha! **59.** $|2y - 7| > -5$

60. $|3y - 4| > 8$

61. $|3a - 4| + 2 \ge 8$

62. $|2a - 5| + 1 \ge 9$

63. $|y - 3| < 12$

64. $|p - 2| < 3$

65. $9 - |x + 4| \le 5$

66. $12 - |x - 5| \le 9$

67. $|4 - 3y| > 8$

68. $|7 - 2y| < -6$

Aha! **69.** $|3 - 4x| < -5$

70. $7 + |4a - 5| \le 26$

71. $\left|\dfrac{2 - 5x}{4}\right| \ge \dfrac{2}{3}$

72. $\left|\dfrac{1 + 3x}{5}\right| > \dfrac{7}{8}$

73. $|m + 5| + 9 \le 16$

74. $|t - 7| + 3 \ge 4$

75. $25 - 2|a + 3| > 19$

76. $30 - 4|a + 2| > 12$

77. Let $f(x) = |2x - 3|$. Find all x for which $f(x) \le 4$.

78. Let $f(x) = |5x + 2|$. Find all x for which $f(x) \le 3$.

79. Let $f(x) = 2 + |3x - 4|$. Find all x for which $f(x) \ge 13$.

80. Let $f(x) = |2 - 9x|$. Find all x for which $f(x) \ge 25$.

81. Let $f(x) = 7 + |2x - 1|$. Find all x for which $f(x) < 16$.

82. Let $f(x) = 5 + |3x + 2|$. Find all x for which $f(x) < 19$.

83. Explain in your own words why -7 is not a solution of $|x| < 5$.

84. Explain in your own words why $[6, \infty)$ is only part of the solution of $|x| \ge 6$.

SKILL MAINTENANCE

Solve using substitution or elimination.

85. $2x - 3y = 7,$
$3x + 2y = -10$

86. $3x - 5y = 9,$
$4x - 3y = 1$

87. $x = -2 + 3y,$
$x - 2y = 2$

88. $y = 3 - 4x,$
$2x - y = -9$

Solve graphically.

89. $x + 2y = 9,$
$3x - y = -1$

90. $2x + y = 7,$
$-3x - 2y = 10$

SYNTHESIS

91. Is it possible for an equation in x of the form $|ax + b| = c$ to have exactly one solution? Why or why not?

92. Explain why the inequality $|x + 5| \geq 2$ can be interpreted as "the number x is at least 2 units from -5."

93. From the definition of absolute value, $|x| = x$ only when $x \geq 0$. Solve $|3t - 5| = 3t - 5$ using this same reasoning.

Solve.

94. $|x + 2| > x$

95. $2 \leq |x - 1| \leq 5$

96. $|5t - 3| = 2t + 4$

97. $t - 2 \leq |t - 3|$

Find an equivalent inequality with absolute value.

98. $-3 < x < 3$

99. $-5 \leq y \leq 5$

100. $x \leq -6 \text{ or } 6 \leq x$

101. $x < -4 \text{ or } 4 < x$

102. $x < -8 \text{ or } 2 < x$

103. $-5 < x < 1$

104. x is less than 2 units from 7.

105. x is less than 1 unit from 5.

Write an absolute-value inequality for which the interval shown is the solution.

106.
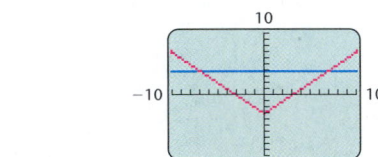
$$\xleftarrow{\hspace{0.5cm}} \underset{-7\ -6\ -5\ -4\ -3\ -2\ -1\ \ 0\ \ 1\ \ 2\ \ 3\ \ 4\ \ 5\ \ 6\ \ 7}{\bullet\!-\!-\!-\!\bullet} \xrightarrow{\hspace{0.5cm}}$$

107.
$$\xleftarrow{\hspace{0.3cm}} \underset{-5\ -4\ -3\ -2\ -1\ \ 0\ \ 1\ \ 2\ \ 3\ \ 4\ \ 5\ \ 6\ \ 7\ \ 8\ \ 9}{\circ\!-\!-\!-\!\circ} \xrightarrow{\hspace{0.3cm}}$$

108.
$$\xleftarrow{\hspace{0.3cm}} \underset{-7\ -6\ -5\ -4\ -3\ -2\ -1\ \ 0\ \ 1\ \ 2\ \ 3\ \ 4\ \ 5\ \ 6\ \ 7}{\circ\!-\!-\!-\!\circ} \xrightarrow{\hspace{0.3cm}}$$

109.
$$\xleftarrow{\hspace{0.3cm}} \underset{0\ \ 1\ \ 2\ \ 3\ \ 4\ \ 5\ \ 6\ \ 7\ \ 8\ \ 9\ 10\ 11\ 12\ 13\ 14}{\bullet\!-\!-\!-\!\bullet} \xrightarrow{\hspace{0.3cm}}$$

110. *Motion of a spring.* A weighted spring is bouncing up and down so that its distance d above the ground satisfies the inequality $|d - 6 \text{ ft}| \leq \frac{1}{2}$ ft (see the figure below). Find all possible distances d.

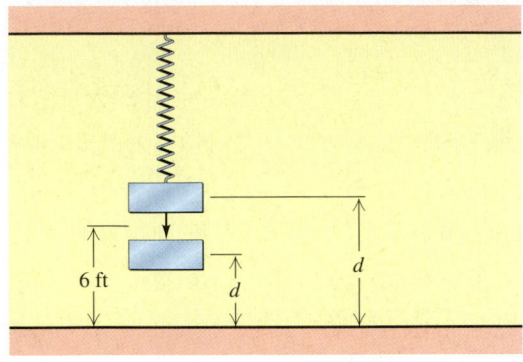

111. Use the accompanying graph of $f(x) = |2x - 6|$ to solve $|2x - 6| \leq 4$.

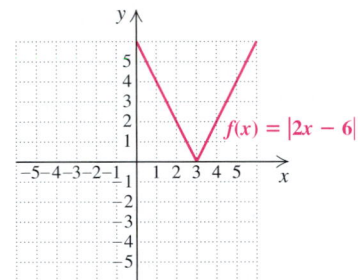

112. Describe a procedure that could be used to solve any equation of the form $g(x) < c$ graphically.

113. Use a grapher to check the solutions to Examples 3 and 5.

114. Use a grapher to check your answers to Exercises 1, 9, 15, 41, 53, 63, 71, and 95.

115. Isabel is using the following graph to solve $|x - 3| < 4$.

How can you tell that a mistake has been made?

Inequalities in Two Variables

9.4

Graphs of Linear Inequalities • Systems of Linear Inequalities

In Section 2.6, we graphed inequalities in one variable on a number line. Now we graph inequalities in two variables on a plane.

Graphs of Linear Inequalities

When the equals sign in a linear equation is replaced with an inequality sign, a **linear inequality** is formed. Solutions of linear inequalities are ordered pairs.

E x a m p l e   1   Determine whether $(-3, 2)$ and $(6, -7)$ are solutions of the inequality $5x - 4y > 13$.

Solution   Below, on the left, we replace x with -3 and y with 2. On the right, we replace x with 6 and y with -7.

$$
\begin{array}{c|c}
5x - 4y > 13 \\
\hline
5(-3) - 4 \cdot 2 \;?\; 13 \\
-15 - 8 \\
-23 \;\big|\; 13 \quad \text{FALSE}
\end{array}
\qquad
\begin{array}{c|c}
5x - 4y > 13 \\
\hline
5(6) - 4(-7) \;?\; 13 \\
30 + 28 \\
58 \;\big|\; 13 \quad \text{TRUE}
\end{array}
$$

Since $-23 > 13$ is false, $(-3, 2)$ is not a solution.

Since $58 > 13$ is true, $(6, -7)$ is a solution.

The graph of a linear equation is a straight line. The graph of a linear inequality is a half-plane, bordered by the graph of the *related equation*. To find an inequality's related equation, we simply replace the inequality sign with an equals sign.

E x a m p l e   2   Graph: $y \leq x$.

Solution   We first graph the related equation $y = x$. Every solution of $y = x$ is an ordered pair, like $(3, 3)$, in which both coordinates are the same. The graph of $y = x$ is shown on the left below. Since the inequality symbol is $\leq$, the line is drawn solid and is part of the graph of $y \leq x$.

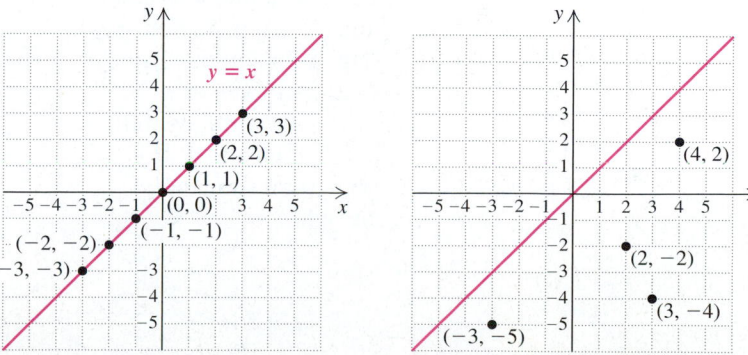

Note that in the graph on the right each ordered pair on the half-plane below $y = x$ contains a y-coordinate that is less than the x-coordinate. All these pairs represent solutions of $y \leq x$. We check one pair, $(4, 2)$, as follows:

$$\frac{y \leq x}{2 \mid 4} \quad \text{TRUE}$$

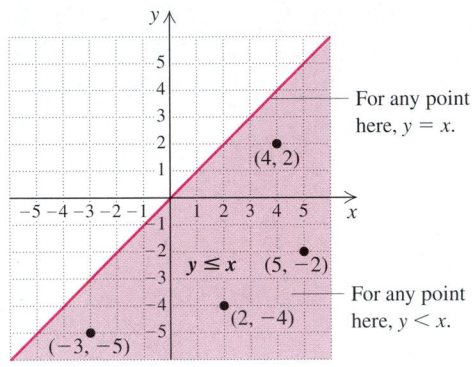

It turns out that *any* point on the same side of $y = x$ as $(4, 2)$ is also a solution. Thus, if one point in a half-plane is a solution, then *all* points in that half-plane are solutions. We complete the drawing of the solution set by shading the half-plane below $y = x$. The complete solution set consists of the shaded half-plane and the line. Note too that for any inequality of the form $y \leq f(x)$ or $y < f(x)$, we shade *below* the graph of $y = f(x)$.

For any point here, $y = x$.

$y \leq x$ $(5, -2)$

For any point here, $y < x$.

E x a m p l e   3

Graph: $8x + 3y > 24$.

Solution   First, we sketch the line $8x + 3y = 24$. Since the inequality sign is $>$, points on this line do not represent solutions of the inequality, so the line is drawn dashed. Points representing solutions of $8x + 3y > 24$ are in either the half-plane above the line or the half-plane below the line. To determine which, we select a point that is not on the line and determine whether it is a solution of $8x + 3y > 24$. We try $(-3, 4)$ as a *test point*:

$$\frac{8x + 3y > 24}{\begin{array}{c|c} 8(-3) + 3 \cdot 4 \;?\; 24 \\ -24 + 12 \\ -12 & 24 \quad \text{FALSE} \end{array}}$$

Since $-12 > 24$ is *false*, $(-3, 4)$ is not a solution. Thus no point in the half-plane containing $(-3, 4)$ is a solution. The points in the other half-plane *are* solutions, so we shade that half-plane and obtain the graph shown at right.

This point is not a solution.

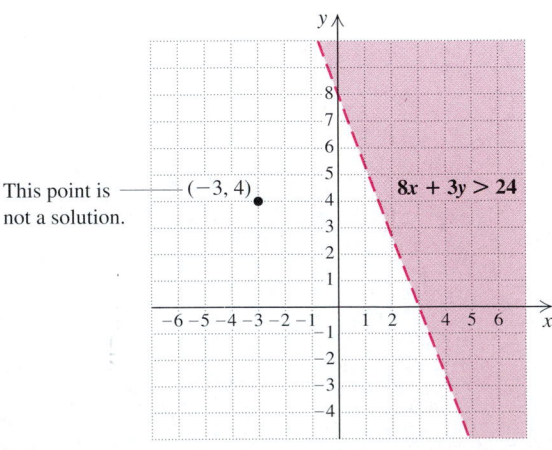

$(-3, 4)$

$8x + 3y > 24$

> **Steps for Graphing Linear Inequalities**
>
> 1. Replace the inequality sign with an equals sign and graph this related equation. If the inequality symbol is $<$ or $>$, draw the line dashed. If the inequality symbol is $\leq$ or $\geq$, draw the line solid.
> 2. The graph consists of a half-plane on one side of the line and, if the line is solid, the line as well.
>
>   a) If the inequality is of the form $y < f(x)$ or $y \leq f(x)$, shade *below* the line. If the inequality is of the form $y > f(x)$ or $y \geq f(x)$, shade *above* the line.
>   b) If y is not isolated, either use algebra to isolate y and proceed as in part (a) or select a point not on the line. If the point represents a solution of the inequality, shade the half-plane containing the point. If it does not, shade the other half-plane.

E x a m p l e   4

Graph: $6x - 2y < 12$.

Solution   We first graph the related equation, $6x - 2y = 12$, as a dashed line. This line passes through the points $(2, 0)$, $(0, -6)$, and $(3, 3)$, and serves as the boundary of the solution set of the inequality. Since y is not isolated, we determine which half-plane to shade by testing a point *not* on the line. The pair $(0, 0)$ is easy to substitute:

$$\frac{6x - 2y < 12}{6 \cdot 0 - 2 \cdot 0 \overset{?}{} 12}$$
$$0 - 0 \mid$$
$$0 \mid 12 \quad \text{TRUE}$$

Since the inequality $0 < 12$ is *true*, the point $(0, 0)$ is a solution, as are all points in the half-plane containing $(0, 0)$. The graph is shown below.

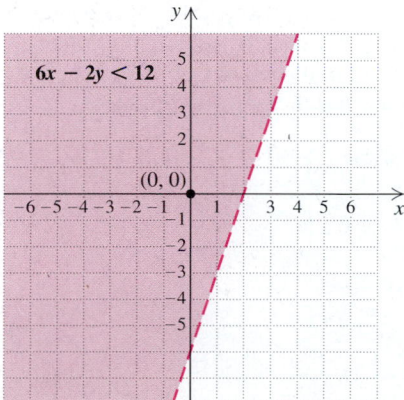

Example 5   Graph $x > -3$ on a plane.

Solution   There is a missing variable in this inequality. If we graph the inequality on a line, its graph is as follows:

However, we can also write this inequality as $x + 0y > -3$ and graph it on a plane. We can use the same technique as in the examples above. First, we graph the related equation $x = -3$ in the plane, using a dashed line. Then we test some point, say, $(2, 5)$:

$$\frac{x + 0y > -3}{2 + 0 \cdot 5 \; ? \; -3}$$
$$2 \; | \; -3 \quad \text{TRUE}$$

Since $(2, 5)$ is a solution, all points in the half-plane containing $(2, 5)$ are solutions. We shade that half-plane. Another approach is to simply note that the solutions of $x > -3$ are all pairs with first coordinates greater than -3.

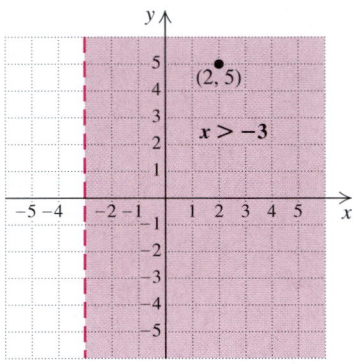

On many graphers, an inequality like $y < \frac{6}{5}x + 3.49$ can be drawn by entering $(6/5)x + 3.49$ as y_1, moving the cursor to the GraphStyle icon just to the left of y_1, pressing ENTER until ◣ appears, and then pressing GRAPH .

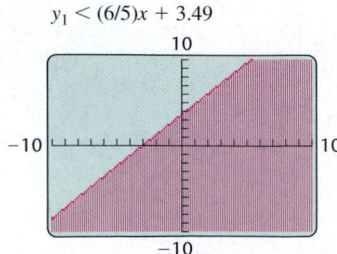

$$y_1 < (6/5)x + 3.49$$

Had we wished to shade above the line, we would have selected ◥ as the GraphStyle. Note that on some graphers a SHADE option must be selected, and on nearly all graphers, the boundary line will always appear solid.

Graph each of the following. Solve for y first if necessary.

1. $y > x + 3.5$
2. $7y \le 2x + 5$
3. $8x - 2y < 11$
4. $11x + 13y + 4 \ge 0$

Example 6

Graph $y \leq 4$ on a plane.

Solution   The inequality is of the form $y \leq f(x)$, so we shade below the solid line representing solutions of $y = f(x)$, or in this case, $y = 4$.

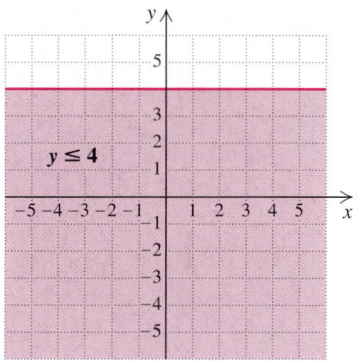

This inequality can also be graphed by drawing $y = 4$ and testing a point above or below the line. The student should check that this results in a graph identical to the one above.

Systems of Linear Inequalities

To graph a system of equations, we graph the individual equations and then find the intersection of the individual graphs. We do the same thing for a system of inequalities, that is, we graph each inequality and find the intersection of the individual graphs.

Example 7

Graph the system

$$x + y \leq 4,$$
$$x - y < 4.$$

Solution   To graph $x + y \leq 4$, we graph $x + y = 4$ using a solid line. Since the test point $(0, 0)$ *is* a solution and $(0, 0)$ is below the line, we shade the half-plane below the line red. The arrows near the ends of the line are another way of indicating the half-plane containing solutions.

Next, we graph $x - y < 4$. We graph $x - y = 4$ using a dashed line and consider $(0, 0)$ as a test point. Again, $(0, 0)$ is a solution, so we shade that side of the line blue. The solution set of the system is the region that is shaded purple (both red and blue) and part of the line $x + y = 4$.

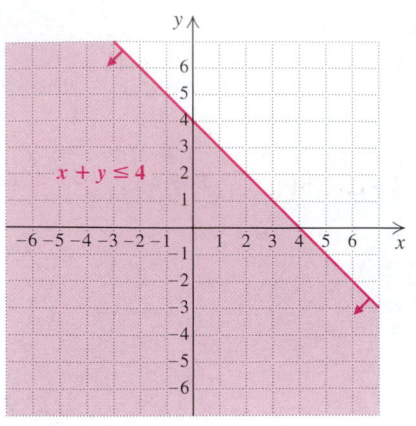
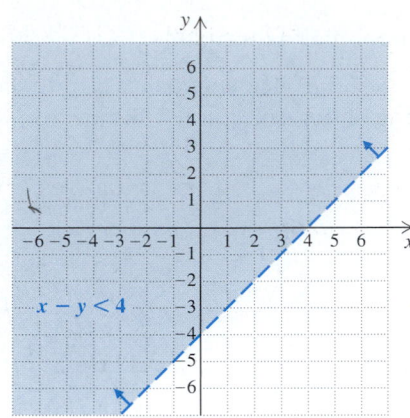
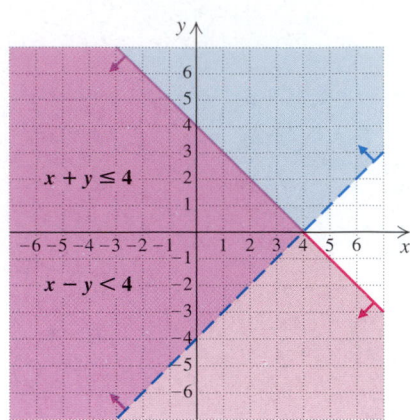

E x a m p l e   8   Graph: $-2 < x \le 3$.

Solution   This is a system of inequalities:

$$-2 < x,$$
$$x \le 3.$$

We graph the equation $-2 = x$, and see that the graph of the first inequality is the half-plane to the right of the line $-2 = x$. It is shaded red.

We graph the second inequality, starting with the line $x = 3$, and find that its graph is the line and also the half-plane to its left. It is shaded blue.

The solution set of the system is the region that is the intersection of the individual graphs. Since it is shaded both blue and red, it appears to be purple. All points in this region have x-coordinates that are greater than -2 but do not exceed 3.

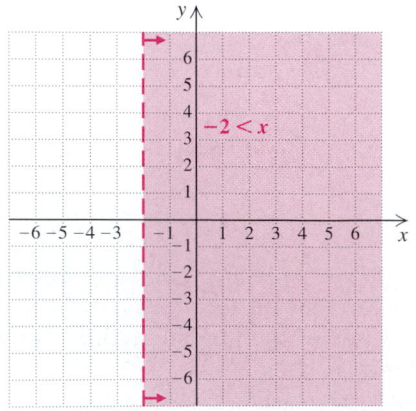
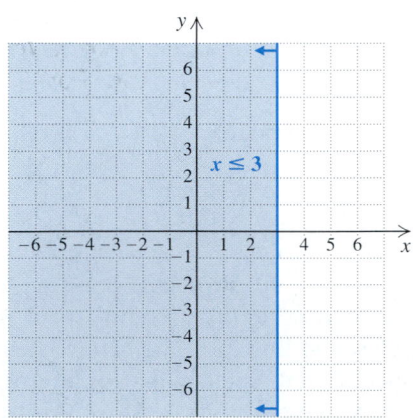
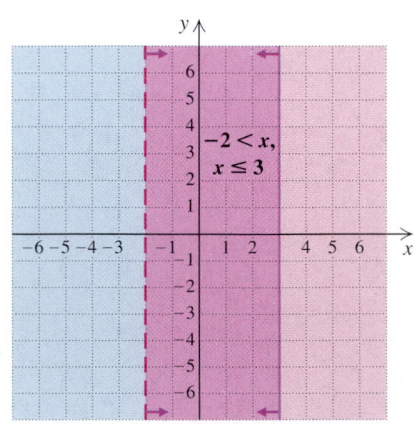

A system of inequalities may have a graph that consists of a polygon and its interior. In Section 9.5, we will have use for the corners, or *vertices* (singular, vertex), of such a graph.

E x a m p l e   9   Graph the system of inequalities. Find the coordinates of any vertices formed.

$$6x - 2y \leq 12, \qquad (1)$$
$$y - 3 \leq 0, \qquad (2)$$
$$x + y \geq 0 \qquad (3)$$

Solution   We graph the lines

$$6x - 2y = 12,$$
$$y - 3 = 0,$$

and $\qquad x + y = 0$

using solid lines. The regions for each inequality are indicated by the arrows near the ends of the lines. We note where the regions overlap and shade the region of solutions purple.

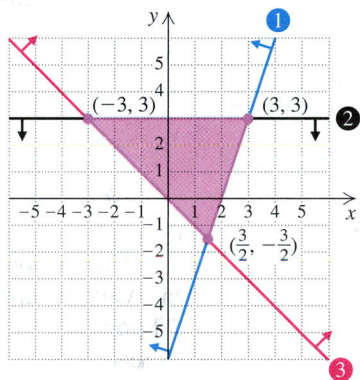

To find the vertices, we solve three different systems of equations. The system of related equations from inequalities (1) and (2) is

$$6x - 2y = 12,$$
$$y - 3 = 0.$$

Solving, we obtain the vertex $(3, 3)$.

The system of related equations from inequalities (1) and (3) is

$$6x - 2y = 12,$$
$$x + y = 0.$$

Solving, we obtain the vertex $\left(\frac{3}{2}, -\frac{3}{2}\right)$.

The system of related equations from inequalities (2) and (3) is

$$y - 3 = 0,$$
$$x + y = 0.$$

Solving, we obtain the vertex $(-3, 3)$.

C O N N E C T I N G   T H E   C O N C E P T S

We have now solved a variety of linear equations, inequalities, systems of equations, and systems of inequalities. In each case, there are different ways to represent the solution. Below is a list of the different types of problems we have solved, along with illustrations of each type.

Type	*Example*	*Solution*	*Graph*
Linear equations in one variable	$2x - 8 = 3(x + 5)$	A number	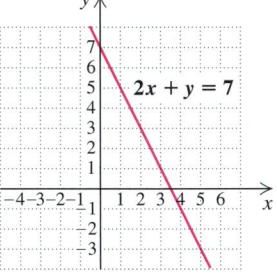
Linear inequalities in one variable	$-3x + 5 > 2$	A set of numbers; an interval	
Linear equations in two variables	$2x + y = 7$	A set of ordered pairs	
Linear inequalities in two variables	$x + y \geq 4$	A set of ordered pairs	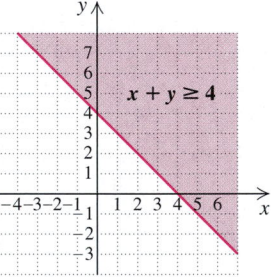
System of equations in two variables	$x + y = 3,$ $5x - y = -27$	An ordered pair or a (possibly empty) set of ordered pairs	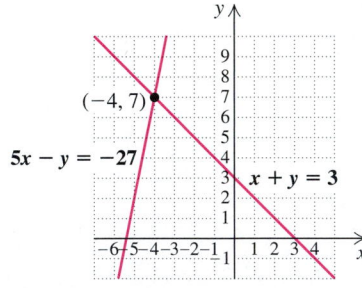
System of inequalities in two variables	$6x - 2y \leq 12,$ $y - 3 \leq 0,$ $x + y \geq 0$	A set of ordered pairs	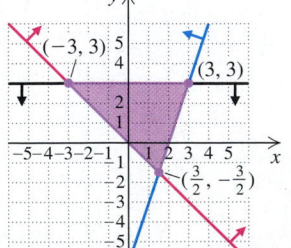

Keeping in mind how these solutions vary and what their graphs look like will help you as you progress further in this book and in mathematics in general.

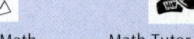

Determine whether each ordered pair is a solution of the given inequality.

1. $(-4, 2)$; $2x + 3y < -1$

2. $(3, -6)$; $4x + 2y \leq -2$

3. $(8, 14)$; $2y - 3x \geq 9$

4. $(7, 20)$; $3x - y > -1$

Graph on a plane.

5. $y > \frac{1}{2}x$

6. $y > 2x$

7. $y \geq x - 3$

8. $y < x + 3$

9. $y \leq x + 4$

10. $y > x - 2$

11. $x - y \leq 5$

12. $x + y < 4$

13. $2x + 3y < 6$

14. $3x + 4y \leq 12$

15. $2y - x \leq 4$

16. $2y - 3x > 6$

17. $2x - 2y \geq 8 + 2y$

18. $3x - 2 \leq 5x + y$

19. $y \geq 2$

20. $x < -5$

21. $x \leq 7$

22. $y > -3$

23. $-2 < y < 6$

24. $-4 < y < -1$

25. $-4 \leq x \leq 5$

26. $-3 \leq y \leq 4$

27. $0 \leq y \leq 3$

28. $0 \leq x \leq 6$

Graph each system.

29. $y > x$,
     $y < -x + 2$

30. $y < x$,
     $y > -x + 1$

31. $y \geq x$,
     $y \leq 2x - 4$

32. $y \geq x$,
     $y \leq -x + 4$

33. $y \leq -3$,
     $x \geq -1$

34. $y \geq -3$,
     $x \geq 1$

35. $x > -4$,
     $y < -2x + 3$

36. $x < 3$,
     $y > -3x + 2$

37. $y \leq 3$,
     $y \geq -x + 2$

38. $y \geq -2$,
     $y \geq x + 3$

39. $x + y \leq 6$,
     $x - y \leq 4$

40. $x + y < 1$,
     $x - y < 2$

41. $y + 3x > 0$,
     $y + 3x < 2$

42. $y - 2x \geq 1$,
     $y - 2x \leq 3$

Graph each system of inequalities. Find the coordinates of any vertices formed.

43. $y \leq 2x - 1$,
     $y \geq -2x + 1$,
     $x \leq 3$

44. $2y - x \leq 2$,
     $y - 3x \geq -4$,
     $y \geq -1$

45. $x + 2y \leq 12$,
     $2x + y \leq 12$,
     $x \geq 0$,
     $y \geq 0$

46. $x - y \leq 2$,
     $x + 2y \geq 8$,
     $y \leq 4$

47. $8x + 5y \leq 40$,
     $x + 2y \leq 8$,
     $x \geq 0$,
     $y \geq 0$

48. $4y - 3x \geq -12$,
     $4y + 3x \geq -36$,
     $y \leq 0$,
     $x \leq 0$

49. $y - x \geq 1$,
     $y - x \leq 3$,
     $2 \leq x \leq 5$

50. $3x + 4y \geq 12$,
     $5x + 6y \leq 30$,
     $1 \leq x \leq 3$

51. In Example 7, is the point $(4, 0)$ part of the solution set? Why or why not?

52. When graphing linear inequalities, Ron makes a habit of always shading above the line when the symbol $\geq$ is used. Is this wise? Why or why not?

SKILL MAINTENANCE

53. *Catering.*   Sandy's Catering needs to provide 10 lb of mixed nuts for a wedding reception. Peanuts cost $2.50 per pound and fancy nuts cost $7 per pound. If $40 has been allocated for nuts, how many pounds of each type should be mixed?

54. *Household waste.*   The Hendersons generate two and a half times as much trash as their neighbors, the Savickis. Together, the two households produce 14 bags of trash each month. How much trash does each household produce?

55. *Paid admissions.*   There were 203 tickets sold for a volleyball game. For activity-card holders the price was $1.25, and for noncard holders the price was $2. The total amount of money collected was $310. How many of each type of ticket were sold?

56. *Paid admissions.*   There were 200 tickets sold for a women's basketball game. Tickets for students were \$2 each and for adults were \$3 each. The total amount collected was \$530. How many of each type of ticket were sold?

57. *Landscaping.* Grass seed is being spread on a triangular traffic island. If the grass seed can cover an area of 200 ft^2 and the island's base is 16 ft long, how tall a triangle can the seed fill?

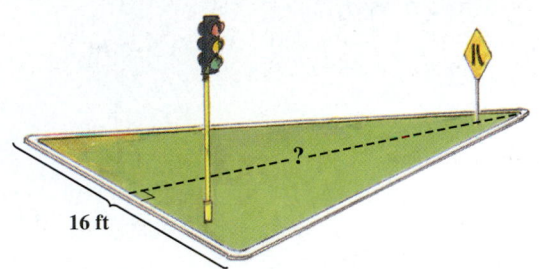

16 ft

58. *Interest rate.* What rate of interest is required in order for a principal of \$320 to earn \$17.60 in half a year?

SYNTHESIS

59. Explain how a system of linear inequalities could have a solution set containing exactly one pair.

60. Do all systems of linear inequalities have solutions? Why or why not?

Graph.

61. $x + y > 8$,
$x + y \leq -2$

62. $\quad x + y \geq 1$,
$-x + y \geq 2$,
$\quad\quad x \leq 4$,
$\quad\quad y \geq 0$,
$\quad\quad y \leq 4$,
$\quad\quad x \leq 2$

63. $\quad x - 2y \leq 0$,
$-2x + \ y \leq 2$,
$\quad\quad x \leq 2$,
$\quad\quad y \leq 2$,
$\quad x + \ y \leq 4$

64. Write four systems of four inequalities that describe a 2-unit by 2-unit square that has $(0, 0)$ as one of the vertices.

65. *Luggage size.* Unless an additional fee is paid, most major airlines will not check any luggage that is more than 62 in. long. The U.S. Postal Service will ship a package only if the sum of the package's length and girth (distance around its midsection) does not exceed 108 in. Concert Productions is ordering several 62-in. long trunks that will be both mailed and checked as luggage. Using w and h for width and height (in inches), respectively, write and graph an inequality that represents all acceptable combinations of width and height.

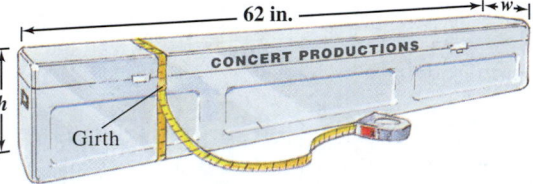

66. *Hockey wins and losses.* The Skating Stars figure that they need at least 60 points for the season in order to make the playoffs. A win is worth 2 points and a tie is worth 1 point. Graph a system of inequalities that describes the situation. (*Hint*: Let w = the number of wins and t = the number of ties.)

67. *Elevators.* Many elevators have a capacity of 1 metric ton (1000 kg). Suppose that c children, each weighing 35 kg, and a adults, each 75 kg, are on an elevator. Graph a system of inequalities that indicates when the elevator is overloaded.

68. *Widths of a basketball floor.* Sizes of basketball floors vary due to building sizes and other constraints such as cost. The length L is to be at most 94 ft and the width W is to be at most 50 ft. Graph a system of inequalities that describes the possible dimensions of a basketball floor.

69. Use a grapher to graph each inequality.
a) $3x + 6y > 2$ **b)** $x - 5y \leq 10$
c) $13x - 25y + 10 \leq 0$ **d)** $2x + 5y > 0$

70. Use a grapher to check your answers to Exercises 29–42. Then use INTERSECT to determine any point(s) of intersection.

CORNER

The Rule of 85

Focus: Linear inequalities

Time: 20–30 minutes

Group size: 3

Under a proposed "Rule of 85," full-time faculty in the California State Teachers Retirement System (kindergarten through community college) who are a years old with y years of service would have the option of retirement if $a + y \geq 85$.

ACTIVITY

1. Decide, as a group, the age range of full-time teachers. Express this age range as an inequality involving a.
2. Decide, as a group, the number of years someone could teach full-time before retiring. Express this answer as a compound inequality involving y.
3. Using the Rule of 85 and the answers to parts (1) and (2) above, write a system of inequalities. Then, using a scale of 5 yr per square, graph the system. To facilitate comparisons with graphs from other groups, plot a on the horizontal axis and y on the vertical axis.
4. Compare the graphs from all groups. Try to reach consensus on the graph that most clearly illustrates what the status would be of someone who would have the option of retirement under the Rule of 85.
5. If your instructor is agreeable to the idea, attempt to represent him or her with a point on your graph.

Applications Using Linear Programming

9.5

Objective Functions and Constraints • Linear Programming

There are many real-world situations in which we need to find a greatest value (a maximum) or a least value (a minimum). For example, most businesses would like to know how to make the *most* profit and how to make their expenses the *least* possible. Some such problems can be solved using systems of inequalities.

Objective Functions and Constraints

Often a quantity we wish to maximize depends on two or more other quantities. For example, a gardener's profits P might depend on the number of shrubs s and the number of trees t that are planted. If the gardener makes a $5 profit

from each shrub and a $9 profit from each tree, the total profit, in dollars, is given by the **objective function**

$$P = 5s + 9t.$$

Thus the gardener might be tempted to simply plant lots of trees since they yield the greater profit. This would be a good idea were it not for the fact that the number of trees and shrubs planted—and thus the total profit—is subject to the demands, or **constraints**, of the situation. For example, to improve drainage, the gardener might be required to plant at least 3 shrubs. Thus the objective function would be subject to the *constraint*

$$s \geq 3.$$

Because of the limited space, the gardener might also be required to plant no more than 10 plants. This would subject the objective function to a *second* constraint:

$$s + t \leq 10.$$

Finally, the gardener might be told to spend no more than $350 on the plants. If the shrubs cost $20 each and the trees cost $50 each, the objective function is subject to a *third* constraint:

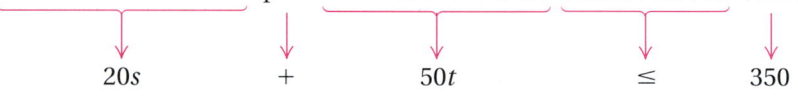

In short, the gardener wishes to maximize the objective function

$$P = 5s + 9t$$

subject to the constraints

$$s + t \leq 10,$$
$$s \geq 3,$$
$$20s + 50t \leq 350,$$
$$s \geq 0,$$
$$t \geq 0.$$

Because the number of trees and shrubs cannot be negative

These constraints form a system of linear inequalities that can be graphed.

Linear Programming

The gardener's problem is "How many shrubs and trees should be planted, subject to the constraints listed, in order to maximize profit?" To solve such a problem, we use an important result from a branch of mathematics known as **linear programming**.

The Corner Principle

Suppose that an objective function $F = ax + by + c$ depends on x and y (with a, b, and c constants). Suppose also that F is subject to constraints on x and y, which form a system of linear inequalities. If F has a minimum or a maximum value, it can then be found as follows:

1. Graph the system of inequalities and find the vertices.
2. Find the value of the objective function at each vertex. The largest and the smallest of those values are the maximum and the minimum of the function, respectively.
3. The ordered pair at which the maximum or minimum occurs indicates the choice of (x, y) for which that maximum or minimum occurs.

This result was proven during World War II, when linear programming was developed to help with shipping troops and supplies to Europe.

Example 1

Solve the gardener's problem discussed above.

Solution   We are asked to maximize $P = 5s + 9t$, subject to the constraints

$$s + t \leq 10,$$
$$s \geq 3,$$
$$20s + 50t \leq 350,$$
$$s \geq 0,$$
$$t \geq 0.$$

We graph the system, using the techniques of Section 9.4. The portion of the graph that is shaded represents all pairs that satisfy the constraints. It is sometimes called the *feasible region*.

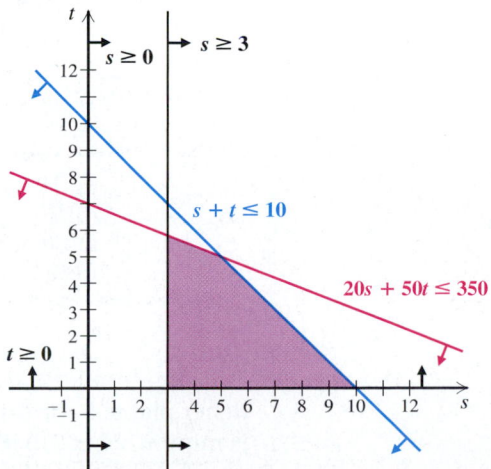

According to the corner principle, P is maximized at one of the vertices of the shaded region. To determine the coordinates of the vertices, we solve the following systems:

$$\left. \begin{array}{r} 20s + 50t = 350, \\ s = 3; \end{array} \right\}$$ The student can verify that the solution of this system is $(3, 5.8)$.

$$\left. \begin{array}{r} s + t = 10, \\ 20s + 50t = 350; \end{array} \right\}$$ The student can verify that the solution of this system is $(5, 5)$.

$$\left. \begin{array}{r} s + t = 10, \\ t = 0; \end{array} \right\}$$ The solution of this system is $(10, 0)$.

$$\left. \begin{array}{r} t = 0, \\ s = 3. \end{array} \right\}$$ The solution of this system is $(3, 0)$.

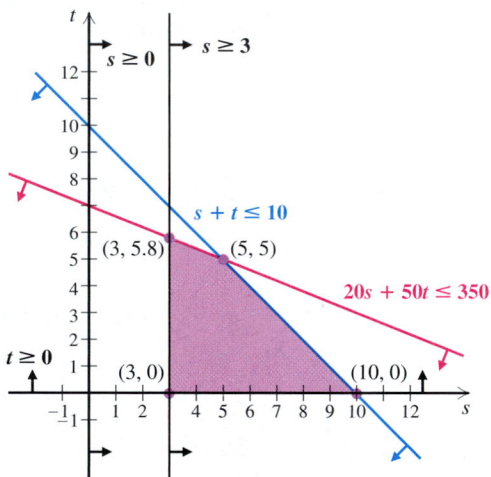

We now find the value of P at each vertex.

Vertex (s, t)	Profit $P = 5s + 9t$	
$(3, 5.8)$	$5(3) + 9(5.8) = 67.2$	
$(5, 5)$	$5(5) + 9(5) = 70$	← Maximum
$(10, 0)$	$5(10) + 9(0) = 50$	
$(3, 0)$	$5(3) + 9(0) = 15$	← Minimum

The largest value of P occurs at $(5, 5)$. Thus profit is maximized at \$70 if the gardener plants 5 shrubs and 5 trees. Incidentally, we have also shown that profit is minimized at \$15 if 3 shrubs and 0 trees are planted.

Example 2   ***Test scores.***   Corinna is taking a test in which multiple-choice questions are worth 10 points each and short-answer questions are worth 15 points each. It takes her 3 min to answer each multiple-choice question and 6 min to answer each short-answer question. The total time allowed is 60 min, and no more than 16 questions can be answered. Assuming that all her answers are correct, how many items of each type should Corinna answer in order to get the best score?

Solution

1. **Familiarize.**   Tabulating information will help us to see the picture.

Type	Number of Points for Each	Time Required for Each	Number Answered
Multiple-choice Short-answer	10 15	3 min 6 min	x y
Total time: 60 min			
Total number of items: 16 or fewer			

Note that we use x to represent the number of multiple-choice questions and y to represent the number of short-answer questions that are answered.

2. **Translate.**   In this case, it helps to extend the table.

Type	Number of Points for Each	Time Required for Each	Number Answered	Total Time for Each Type	Total Points for Each Type
Multiple-choice Short-answer	10 15	3 min 6 min	x y	$3x$ $6y$	$10x$ $15y$
Total			$x + y \leq 16$	$3x + 6y \leq 60$	$10x + 15y$

Because no more than 16 items may be answered

Because the time cannot exceed 60 min

This is what we want to maximize: the total score on the test.

Suppose that the total score on the test is T. We write T as the objective function in terms of x and y:

$$T = 10x + 15y.$$

We wish to maximize T subject to the constraints on x and y listed above:

$$x + \ y \leq 16,$$
$$3x + 6y \leq 60,$$
$$x \geq 0,$$
$$y \geq 0.$$

Because the number of questions answered cannot be negative

3. **Carry out.** The mathematical manipulation consists of graphing the system and evaluating T at each vertex. The graph is as follows:

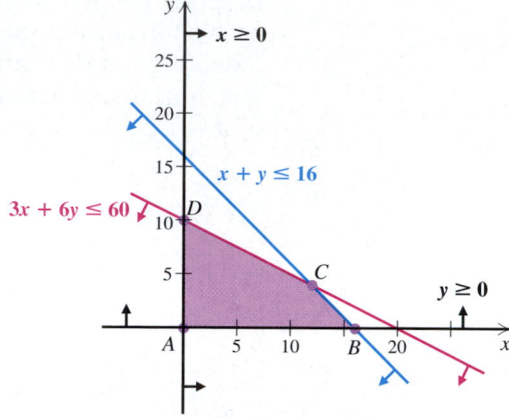

We can find the coordinates of each vertex by solving a system of two linear equations. The coordinates of point A are obviously $(0, 0)$. To find the coordinates of point C, we solve the system

$$x + \quad y = 16, \quad (1)$$
$$3x + 6y = 60, \quad (2)$$

as follows:

$$-3x - 3y = -48 \qquad \text{Multiplying both sides of equation (1) by } -3$$
$$\underline{3x + 6y = 60}$$
$$3y = 12 \qquad \text{Adding}$$
$$y = 4.$$

Then we find that $x = 12$. Thus the coordinates of vertex C are $(12, 4)$. Point B is the x-intercept of the line given by $x + y = 16$, so B is $(16, 0)$. Point D is the y-intercept of $3x + 6y = 60$, so D is $(0, 10)$. Computing the test score for each ordered pair, we obtain the following:

Vertex (x, y)	Score $T = 10x + 15y$
$A\,(0, 0)$	0
$B\,(16, 0)$	160
$C\,(12, 4)$	180
$D\,(0, 10)$	150

The greatest score in the table is 180, obtained when 12 multiple-choice and 4 short-answer questions are answered.

4. **Check.** We can check that $T \le 180$ for any other pair in the shaded region. This is left to the student.

5. **State.** In order to maximize her score, Corinna should answer 12 multiple-choice questions and 4 short-answer questions.

Exercise Set 9.5

Find the maximum and the minimum values of each objective function and the values of x and y at which they occur.

1. $F = 2x + 14y$,
   subject to
   $5x + 3y \leq 34$,
   $3x + 5y \leq 30$,
   $x \geq 0$,
   $y \geq 0$

2. $G = 7x + 8y$,
   subject to
   $3x + 2y \leq 12$,
   $2y - x \leq 4$,
   $x \geq 0$,
   $y \geq 0$

3. $P = 8x - y + 20$,
   subject to
   $6x + 8y \leq 48$,
   $0 \leq y \leq 4$,
   $0 \leq x \leq 7$

4. $Q = 24x - 3y + 52$,
   subject to
   $5x + 4y \leq 20$,
   $0 \leq y \leq 4$,
   $0 \leq x \leq 3$

5. $F = 2y - 3x$,
   subject to
   $y \leq 2x + 1$,
   $y \geq -2x + 3$,
   $x \leq 3$

6. $G = 5x + 2y + 4$,
   subject to
   $y \leq 2x + 1$,
   $y \geq -x + 3$,
   $x \leq 5$

Solve.

7. *Lunch-time profits.*   Elrod's lunch cart sells burritos and chili. To stay in business, Elrod must sell at least 10 orders of chili and 30 burritos each day. Because of limited space, no more than 40 orders of chili or 70 burritos can be made. The total number of orders cannot exceed 90. If profit is $1.65 per chili order and $1.05 per burrito, how many of each item should Elrod sell in order to maximize profit?

8. *Milling.*   Johnson Lumber can convert logs into either lumber or plywood. In a given week, the mill can turn out 400 units of production, of which 100 units of lumber and 150 units of plywood are required by regular customers. The profit on a unit of lumber is $20 and on a unit of plywood is $30. How many units of each type should the mill produce in order to maximize profit?

9. *Cycle production.*   Yawaka manufactures motorcycles and bicycles. To stay in business, the number of bicycles made cannot exceed 3 times the number of motorcycles made. Yawaka lacks the facilities to produce more than 60 motorcycles or more than 120 bicycles. The total production of motorcycles and bicycles cannot exceed 160. The profit on a motorcycle is $1340 and on a bicycle is $200. Find the number of each that should be manufactured in order to maximize profit.

10. *Gas mileage.*   Roschelle owns a car and a moped. She has at most 12 gal of gasoline to be used between the car and the moped. The car's tank holds at most 18 gal and the moped's 3 gal. The mileage for the car is 20 mpg and for the moped is 100 mpg. How many gallons of gasoline should each vehicle use if Roschelle wants to travel as far as possible? What is the maximum number of miles?

11. *Test scores.*   Phil is about to take a test that contains matching questions worth 10 points each and essay questions worth 25 points each. He must do at least 3 matching questions, but time restricts doing more than 12. Phil must do at least 4 essays, but time restricts doing more than 15. If no more than 20 questions can be answered, how many of each type should Phil do in order to maximize his score? What is this maximum score?

12. *Test scores.*   Edy is about to take a test that contains short-answer questions worth 4 points each and word problems worth 7 points each. Edy must do at least 5 short-answer questions, but time restricts doing more than 10. She must do at least 3 word problems, but time restricts doing more than 10. Edy can do no more than 18 questions in total. How many of each type of question must Edy do in order to maximize her score? What is this maximum score?

Aha! **13.** *Investing.*   Rosa is planning to invest up to $40,000 in corporate or municipal bonds, or both. She must invest from $6000 to $22,000 in corporate bonds, and she does not want to invest more than $30,000 in municipal bonds. The interest on corporate bonds is 8% and on municipal bonds is $7\frac{1}{2}$%. This is simple interest for one year. How much should Rosa invest in each type of bond in order to earn the most interest? What is the maximum interest?

14. *Grape growing.* Auggie's vineyard consists of 240 acres upon which he wishes to plant Merlot and Cabernet grapes. Profit per acre of Merlot is $400 and profit per acre of Cabernet is $300. Furthermore, the total number of hours of labor available during the harvest season is 3200. Each acre of Merlot requires 20 hr of labor and each acre of Cabernet requires 10 hr of labor. Determine how the land should be divided between Merlot and Cabernet in order to maximize profit.

15. *Coffee blending.* The Coffee Peddler has 1440 lb of Sumatran coffee and 700 lb of Kona coffee. A batch of Hawaiian Blend requires 8 lb of Kona and 12 lb of Sumatran, and yields a profit of $90. A batch of Classic Blend requires 4 lb of Kona and 16 lb of Sumatran, and yields a $55 profit. How many batches of each kind should be made in order to maximize profit? What is the maximum profit? (*Hint*: Organize the information in a table.)

16. *Investing.* Jamaal is planning to invest up to $22,000 in City Bank or the Southwick Credit Union, or both. He wants to invest at least $2000 but no more than $14,000 in City Bank. The Southwick Credit Union does not insure more than a $15,000 investment, so he will invest no more than that in the Southwick Credit Union. The interest in City Bank is 6% and in the credit union is $6\frac{1}{2}$%. This is simple interest for one year. How much should he invest in each bank in order to earn the most interest? What is the maximum interest?

17. *Textile production.* It takes Cosmic Stitching 2 hr of cutting and 4 hr of sewing to make a knit suit. To make a worsted suit, it takes 4 hr of cutting and 2 hr of sewing. At most 20 hr per day are available for cutting and at most 16 hr per day are available for sewing. The profit on a knit suit is $68 and on a worsted suit is $62. How many of each kind of suit should be made in order to maximize profit?

18. *Biscuit production.* The Hockeypuck Biscuit Factory makes two types of biscuits, Biscuit Jumbos and Mitimite Biscuits. The oven can cook at most 200 biscuits per hour. Jumbos each require 2 oz of flour, Mitimites require 1 oz of flour, and there is at most 1440 oz of flour available. The income from Jumbos is $1.00 and from Mitimites is $0.80. How many of each type of biscuit should be made in an hour in order to maximize income? What is the maximum income?

19. Before a student begins work in this section, what three sections of the text would you suggest he or she study? Why?

20. What does the use of the word "constraint" in this section have in common with the use of the word in everyday speech?

SKILL MAINTENANCE

Evaluate.

21. $5x^3 - 4x^2 - 7x + 2$, for $x = -2$

22. $6t^3 - 3t^2 + 5t$, for $t = 2$

Simplify.

23. $3(2x - 5) + 4(x + 5)$

24. $4(5t - 7) + 6(t + 8)$

25. $6x - 3(x + 2)$

26. $8t - 2(3t - 1)$

SYNTHESIS

27. Explain how Exercises 16 and 18 can be answered by logical reasoning without linear programming.

28. Write a linear programming problem for a classmate to solve. Devise the problem so that profit must be maximized subject to at least two (nontrivial) constraints.

29. *Airplane production.* Alpha Tours has two types of airplanes, the T3 and the S5, and contracts requiring accommodations for a minimum of 2000 first-class, 1500 tourist-class, and 2400 economy-class passengers. The T3 costs $30 per mile to operate and can accommodate 40 first-class, 40 tourist-class, and 120 economy-class passengers, whereas the S5 costs $25 per mile to operate and can accommodate 80 first-class, 30 tourist-class, and 40 economy-class passengers. How many of each type of airplane should be used in order to minimize the operating cost?

$30/mi
T3
ALPHATOURS

$25/mi
S5
ALPHATOURS

30. *Airplane production.*   A new airplane, the T4, is now available, having an operating cost of $37.50

per mile and accommodating 40 first-class, 40 tourist-class, and 80 economy-class passengers. If the T3 of Exercise 29 were replaced with the T4, how many S5's and how many T4's would be needed in order to minimize the operating cost?

31. *Furniture production.*   P. J. Edward Furniture Design produces chairs and sofas. The chairs require 20 ft of wood, 1 lb of foam rubber, and 2 sq yd of fabric. The sofas require 100 ft of wood, 50 lb of foam rubber, and 20 sq yd of fabric. The company has 1900 ft of wood, 500 lb of foam rubber, and 240 sq yd of fabric. The chairs can be sold for $80 each and the sofas for $1200 each. How many of each should be produced in order to maximize income?

Summary and Review 9

Key Terms

Inequality, p. 544
Solution, p. 544
Solution set, p. 544
Set-builder notation, p. 545
Interval notation, p. 545
Open interval, p. 545
Closed interval, p. 545
Half-open interval, p. 546

Compound inequality, p. 553
Intersection, p. 553
Conjunction, p. 553
Union, p. 556
Disjunction, p. 556
Absolute value, p. 562
Linear inequality, p. 571
Half-plane, p. 571

Related equation, p. 571
Test point, p. 572
Vertices (singular, vertex), p. 577
Objective function, p. 582
Constraint, p. 582
Linear programming, p. 582
Feasible region, p. 583

Important Properties and Formulas

The Addition Principle for Inequalities

For any real numbers a, b, and c:

$a < b$ is equivalent to $a + c < b + c$;
$a > b$ is equivalent to $a + c > b + c$.

Similar statements hold for $\leq$ and $\geq$.

The Multiplication Principle for Inequalities

For any real numbers a and b, and for any positive number c,

$a < b$ is equivalent to $ac < bc$;
$a > b$ is equivalent to $ac > bc$.

For any real numbers a and b, and for any *negative* number c,

$a < b$ is equivalent to $ac > bc$;
$a > b$ is equivalent to $ac < bc$.

Similar statements hold for $\leq$ and $\geq$.

Set intersection:

$A \cap B = \{x \mid x$ is in A and x is in $B\}$

Set union:

$A \cup B = \{x \mid x$ is in A or in B, or both$\}$

Intersection corresponds to "and"; union corresponds to "or."

$|x| = x$ if $x \geq 0$;　　$|x| = -x$ if $x < 0$.

The Absolute-Value Principles for Equations and Inequalities

For any positive number p and any algebraic expression X:

a) The solutions of $|X| = p$ are those numbers that satisfy $X = -p$ or $X = p$.

b) The solutions of $|X| < p$ are those numbers that satisfy $-p < X < p$.

c) The solutions of $|X| > p$ are those numbers that satisfy $X < -p$ or $p < X$.

If $|X| = 0$, then $X = 0$. If p is negative, then $|X| = p$ and $|X| < p$ have no solution, and any value of X will satisfy $|X| > p$.

Steps for Graphing Linear Inequalities

1. Graph the related equation. Draw the line *dashed* if the inequality symbol is $<$ or $>$ and *solid* if the inequality symbol is $\leq$ or $\geq$.

2. The graph consists of a half-plane on one side of the line and, if the line is solid, the line as well.

   a) Shade *below* the line if the inequality is of the form $y < f(x)$ or $y \leq f(x)$. Shade *above* the line if the inequality is of the form $y > f(x)$ or $y \geq f(x)$.

b) If y is not isolated, either isolate y and proceed as in part (a) or select a test point not on the line. If the point represents a solution of the inequality, shade the half-plane containing the point. If it does not, shade the other half-plane.

The Corner Principle

Suppose that an objective function $F = ax + by + c$ depends on x and y. Suppose also that F is subject to constraints on x and y, which form a system of linear inequalities. If F has a minimum or a maximum value, it can then be found as follows:

1. Graph the system of inequalities and find the vertices.

2. Find the value of the objective function at each vertex. The largest and the smallest of those values are the maximum and the minimum of the function, respectively.

3. The ordered pair at which the maximum or minimum occurs indicates the choice of (x, y) for which that maximum or minimum occurs.

Review Exercises

Graph each inequality and write the solution set using both set-builder and interval notation.

1. $x \leq -2$

2. $a + 7 \leq -14$

3. $y - 5 \geq -12$

4. $4y > -15$

5. $-0.3y < 9$

6. $-6x - 5 < 4$

7. $-\frac{1}{2}x - \frac{1}{4} > \frac{1}{2} - \frac{1}{4}x$

8. $0.3y - 7 < 2.6y + 15$

9. $-2(x - 5) \geq 6(x + 7) - 12$

10. Let $f(x) = 3x - 5$ and $g(x) = 11 - x$. Find all values of x for which $f(x) \leq g(x)$.

Solve.

11. Jessica can choose between two summer jobs. She can work as a checker in a discount store for $8.40 an hour, or she can mow lawns for $12.00 an hour. In order to mow lawns, she must buy a $450 lawnmower. How many hours of labor will it take Jessica to make more money mowing lawns?

12. Clay is going to invest $4500, part at 6% and the rest at 7%. What is the most he can invest at 6% and still be guaranteed $300 in interest each year?

13. Find the intersection:
$$\{1, 2, 5, 6, 9\} \cap \{1, 3, 5, 9\}.$$

14. Find the union:
$$\{1, 2, 5, 6, 9\} \cup \{1, 3, 5, 9\}.$$

Graph and write interval notation.

15. $x \leq 3 \, and \, x > -5$

16. $x \leq 3 \, or \, x > -5$

Solve and graph each solution set.

17. $-4 < x + 3 \leq 5$

18. $-15 < -4x - 5 < 0$

19. $3x < -9 \, or \, -5x < -5$

20. $2x + 5 < -17 \, or \, -4x + 10 \leq 34$

21. $2x + 7 \leq -5 \, or \, x + 7 \geq 15$

22. $f(x) < -5 \, or \, f(x) > 5$, where $f(x) = 3 - 5x$

For $f(x)$ as given, use interval notation to write the domain of f.

23. $f(x) = \dfrac{x}{x - 3}$

24. $f(x) = \sqrt{x + 3}$

25. $f(x) = \sqrt{8 - 3x}$

Solve.

26. $|x| = 4$

27. $|t| \geq 3.5$

28. $|x - 2| = 7$

29. $|2x + 5| < 12$

30. $|3x - 4| \geq 15$

31. $|2x + 5| = |x - 9|$

32. $|5n + 6| = -8$

33. $\left| \dfrac{x + 4}{8} \right| \leq 1$

34. $2|x - 5| - 7 > 3$

35. Let $f(x) = |3x - 5|$. Find all x for which $f(x) < 0$.

36. Graph $x - 2y \geq 6$ on a plane.

Graph each system of inequalities. Find the coordinates of any vertices formed.

37. $x + 3y > -1$,
   $x + 3y < 4$

38. $x - 3y \leq 3$,
   $x + 3y \geq 9$,
   $y \leq 6$

39. Find the maximum and the minimum values of
$$F = 3x + y + 4$$
subject to
$$y \leq 2x + 1,$$
$$x \leq 7,$$
$$y \geq 3.$$

40. Edsel Computers has two manufacturing plants. The Oregon plant cannot produce more than 60 computers a month, while the Ohio plant cannot produce more than 120 computers a month. The Electronics Outpost sells at least 160 Edsel computers each month. It costs $40 to ship a computer to The Electronics Outpost from the Oregon plant and $25 to ship from the Ohio plant. How many computers should be shipped from each plant in order to minimize cost?

SYNTHESIS

41. Explain in your own words why $|X| = p$ has two solutions when p is positive and no solution when p is negative.

42. Explain why the graph of the solution of a system of linear inequalities is the intersection, not the union, of the individual graphs.

43. Solve: $|2x + 5| \leq |x + 3|$.

44. Classify as true or false: If $x < 3$, then $x^2 < 9$. If false, give an example showing why.

45. Just-For-Fun manufactures marbles with a 1.1-cm diameter and a ± 0.03-cm manufacturing tolerance, or allowable variation in diameter. Write the tolerance as an inequality with absolute value.

Chapter Test 9

Graph each inequality and write the solution set using both set-builder and interval notation.

1. $x - 2 < 10$

2. $-0.6y < 30$

3. $-4y - 3 \geq 5$

4. $3a - 5 \leq -2a + 6$

5. $4(5 - x) < 2x + 5$

6. $-8(2x + 3) + 6(4 - 5x) \geq 2(1 - 7x) - 4(4 + 6x)$

7. Let $f(x) = -5x - 1$ and $g(x) = -9x + 3$. Find all values of x for which $f(x) > g(x)$.

8. Lia can rent a van for either $40 per day with unlimited mileage or $30 per day with 100 free miles and an extra charge of 15¢ for each mile over 100. For what numbers of miles traveled would the unlimited mileage plan save Lia money?

9. A refrigeration repair company charges $40 for the first half-hour of work and $30 for each additional hour. Blue Mountain Camp has budgeted $100 to repair its walk-in cooler. For what lengths of a service call will the budget not be exceeded?

10. Find the intersection:

$$\{1, 3, 5, 7, 9\} \cap \{3, 5, 11, 13\}.$$

11. Find the union:

$$\{1, 3, 5, 7, 9\} \cup \{3, 5, 11, 13\}.$$

12. Write the domain of f using interval notation if $f(x) = \sqrt{7 - x}$.

Solve and graph each solution set.

13. $-3 < x - 2 < 4$

14. $-11 \leq -5t - 2 < 0$

15. $3x - 2 < 7 \ or \ x - 2 > 4$

16. $-3x > 12 \ or \ 4x > -10$

17. $-\frac{1}{3} \leq \frac{1}{6}x - 1 < \frac{1}{4}$

18. $|x| = 9$

19. $|a| > 3$

20. $|4x - 1| < 4.5$

21. $|-5t - 3| \geq 10$

22. $|2 - 5x| = -10$

23. $g(x) < -3 \ or \ g(x) > 3$, where $g(x) = 4 - 2x$

24. Let $f(x) = |x + 10|$ and $g(x) = |x - 12|$. Find all values of x for which $f(x) = g(x)$.

Graph the system of inequalities. Find the coordinates of any vertices formed.

25. $x + y \geq 3,$
$x - y \geq 5$

26. $2y - x \geq -7,$
$2y + 3x \leq 15,$
$y \leq 0,$
$x \leq 0$

27. Find the maximum and the minimum values of

$$F = 5x + 3y$$

subject to

$$x + y \leq 15,$$
$$1 \leq x \leq 6,$$
$$0 \leq y \leq 12.$$

28. Sassy Salon makes $12 on each manicure and $18 on each haircut. A manicure takes 30 minutes and a haircut takes 50 minutes, and there are 5 stylists who each work 6 hours a day. If the salon can schedule 50 appointments a day, how many should be manicures and how many haircuts in order to maximize profit? What is the maximum profit?

SYNTHESIS

Solve. Write the solution set using interval notation.

29. $|2x - 5| \leq 7 \ and \ |x - 2| \geq 2$

30. $7x < 8 - 3x < 6 + 7x$

31. Write an absolute-value inequality for which the interval shown is the solution.

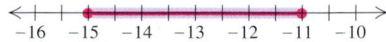

Cumulative Review 1–9

1. Evaluate
$$\frac{2x - y^2}{x + y}$$
for $x = 3$ and $y = -4$.

2. Convert to scientific notation: 5,760,000,000.

3. Determine the slope and the y-intercept for the line given by $7x - 4y = 12$.

4. Find an equation for the line that passes through the points $(-1, 7)$ and $(2, -3)$.

5. Solve the system
$$5x - 2y = -23,$$
$$3x + 4y = 7.$$

6. Solve the system
$$-3x + 4y + z = -5,$$
$$x - 3y - z = 6,$$
$$2x + 3y + 5z = -8.$$

7. Briar Creek Elementary School sold 45 pizzas for a fundraiser. Small pizzas sold for $7.00 each and large pizzas for $10.00 each. The total amount of funds received from the sale was $402. How many of each size pizza were sold?

8. The sum of three numbers is 20. The first number is 3 less than twice the third number. The second number minus the third number is -7. What are the numbers?

9. Trex Company makes decking material from waste wood fibers and reclaimed polyethylene. Its sales rose from $3.5 million in 1993 to $74.3 million in 1999 (*Source: Business Week,* May 29, 2000). Calculate the rate at which sales were rising.

10. In 1989, the average length of a visit to a physician in an HMO was 15.4 min; and in 1998, it was 17.9 min (*Sources:* Rutgers University Study and National Center for Health Statistics). Let V represent the average length of a visit t years after 1989.
   a) Find a linear function $V(t)$ that fits the data.
   b) Use the function of part (a) to predict the average length of a visit in 2005.

11. If
$$f(x) = \frac{x - 2}{x - 5},$$
find **(a)** $f(3)$ and **(b)** the domain of f.

Solve.

12. $8x = 1 + 16x^2$

13. $625 = 49y^2$

14. $20 > 2 - 6x$

15. $\frac{1}{3}x - \frac{1}{5} \geq \frac{1}{5}x - \frac{1}{3}$

16. $-8 < x + 2 < 15$

17. $3x - 2 < -6 \text{ or } x + 3 > 9$

18. $|x| > 6.4$

19. $|4x - 1| \leq 14$

20. $\dfrac{2}{n} - \dfrac{7}{n} = 3$

21. $\dfrac{6}{x - 5} = \dfrac{2}{2x}$

22. $\dfrac{3x}{x - 2} - \dfrac{6}{x + 2} = \dfrac{24}{x^2 - 4}$

23. $\dfrac{3x^2}{x + 2} + \dfrac{5x - 22}{x - 2} = \dfrac{-48}{x^2 - 4}$

24. Let $f(x) = |3x - 5|$. Find all values of x for which $f(x) = 2$.

25. Write the domain of f using interval notation if $f(x) = \sqrt{x - 7}$.

Solve.

26. $5m - 3n = 4m + 12$, for n

27. $P = \dfrac{3a}{a + b}$, for a

Graph on a plane.

28. $4x \geq 5y + 20$ **29.** $y = \frac{1}{3}x - 2$

Perform the indicated operations and simplify.

30. $(2x^2 - 3x + 1) + (6x - 3x^3 + 7x^2 - 4)$

31. $(5x^3y^2)(-3xy^2)$

32. $(3a + b - 2c) - (-4b + 3c - 2a)$

33. $(5x^2 - 2x + 1)(3x^2 + x - 2)$

34. $(2x^2 - y)^2$

35. $(2x^2 - y)(2x^2 + y)$

36. $(-5m^3n^2 - 3mn^3) +$
$\qquad\qquad (-4m^2n^2 + 4m^3n^2) - (2mn^3 - 3m^2n^2)$

37. $\dfrac{y^2 - 36}{2y + 8} \cdot \dfrac{y + 4}{y + 6}$

38. $\dfrac{x^4 - 1}{x^2 - x - 2} \div \dfrac{x^2 + 1}{x - 2}$

39. $\dfrac{5ab}{a^2 - b^2} + \dfrac{a + b}{a - b}$

40. $\dfrac{2}{m + 1} + \dfrac{3}{m - 5} - \dfrac{m^2 - 1}{m^2 - 4m - 5}$

41. $y - \dfrac{2}{3y}$

42. Simplify: $\dfrac{\dfrac{1}{x} - \dfrac{1}{y}}{x + y}$.

43. Divide: $(9x^3 + 5x^2 + 2) \div (x + 2)$.

Factor.

44. $4x^3 + 18x^2$

45. $x^2 + 8x - 84$

46. $16y^2 - 81$

47. $64x^3 + 8$

48. $t^2 - 16t + 64$

49. $x^6 - x^2$

50. $0.027b^3 - 0.008c^3$

51. $20x^2 + 7x - 3$

52. $3x^2 - 17x - 28$

53. $x^5 - x^3y + x^2y - y^2$

54. If $f(x) = x^2 - 4$ and $g(x) = x^2 - 7x + 10$, find the domain of f/g.

55. A digital data circuit can transmit a particular set of data in 4 sec. An analog phone circuit can transmit the same data in 20 sec. How long would it take, working together, for both circuits to transmit the data?

56. The floor area of a rental trailer is rectangular. The length is 3 ft more than the width. A rug of area 54 ft^2 exactly fills the floor of the trailer. Find the perimeter of the trailer.

57. The sum of the squares of three consecutive even integers is equal to 8 more than three times the square of the second number. Find the integers.

58. *Logging.* The volume of wood V in a tree trunk varies jointly as the height h and the square of the girth g (girth is distance around). If the volume is 35 ft^3 when the height is 20 ft and the girth is 5 ft, what is the height when the volume is 85.75 ft^3 and the girth is 7 ft?

SYNTHESIS

59. Multiply: $(x - 4)^3$.

60. Solve: $x^4 + 225 = 34x^2$.

Solve.

61. $4 \le |3 - x| \le 6$

62. $\dfrac{18}{x - 9} + \dfrac{10}{x + 5} = \dfrac{28x}{x^2 - 4x - 45}$

63. $16x^3 = x$

10

Exponents and Radicals

AN APPLICATION

In steel production, the temperature of the molten metal is so great that conventional thermometers melt. Instead, sound is transmitted across the surface of the metal to a receiver on the far side and the speed of sound is measured. The formula

$$S(t) = 1087.7 \sqrt{\frac{9t + 2617}{2457}}$$

gives the speed of sound $S(t)$, in feet per second, at a temperature of t degrees Celsius. Find the temperature of a blast furnace where sound travels 1502.3 ft/sec.

This problem appears as Exercise 59 in Section 10.6.

*M*athematics has transformed the way we produce steel from using the skills of an experienced operator, to an accurate science using highly automated computer control systems.

DANIEL GOLDSTEIN
Research Engineer
Bethlehem Steel
Bethlehem, PA

I *n this chapter, we learn about square roots, cube roots, fourth roots, and so on. These roots are studied in connection with the manipulation of radical expressions and the solution of real-world applications. Fractional exponents are also studied and are used to ease some of our work with radicals. The chapter closes with an examination of the complex-number system.*

Radical Expressions and Functions

10.1

Square Roots and Square Root Functions • Expressions of the Form $\sqrt{a^2}$ • Cube Roots • Odd and Even *n*th Roots

In this section, we consider roots, such as square roots and cube roots. We look at the symbolism that is used and ways in which symbols can be manipulated to get equivalent expressions. All of this will be important in problem solving.

Square Roots and Square Root Functions

When a number is raised to the second power, the number is squared. Often we need to know what number was squared in order to produce some value *a*. If such a number can be found, we call that number a *square root of a*.

> ### *Square Root*
> The number *c* is a *square root* of *a* if $c^2 = a$.

For example,

9 has -3 and 3 as square roots because $(-3)^2 = 9$ and $3^2 = 9$.

25 has -5 and 5 as square roots because $(-5)^2 = 25$ and $5^2 = 25$.

-4 does not have a real-number square root because there is no real number *c* such that $c^2 = -4$.

Note that every positive number has two square roots, whereas 0 has only itself as a square root. Negative numbers do not have real-number square roots, although later in this chapter we will work with a number system in which such square roots do exist.

E x a m p l e   1   Find the two square roots of 64.

Solution   The square roots are 8 and -8, because $8^2 = 64$ and $(-8)^2 = 64$.

Whenever we refer to *the* square root of a number, we mean the nonnegative square root of that number. This is often referred to as the *principal square root* of the number.

> ### Principal Square Root
>
> The *principal square root* of a nonnegative number is its nonnegative square root. The symbol $\sqrt{}$ is called a *radical sign* and is used to indicate the principal square root of the number over which it appears.

E x a m p l e   2

Simplify each of the following.

a) $\sqrt{25}$

b) $\sqrt{\dfrac{25}{64}}$

c) $-\sqrt{64}$

d) $\sqrt{0.0049}$

Solution

a) $\sqrt{25} = 5$ $\sqrt{}$ indicates the principal square root. Note that $\sqrt{25} \neq -5$.

b) $\sqrt{\dfrac{25}{64}} = \dfrac{5}{8}$ Since $\left(\dfrac{5}{8}\right)^2 = \dfrac{25}{64}$

c) $-\sqrt{64} = -8$ Since $\sqrt{64} = 8$, $-\sqrt{64} = -8$.

d) $\sqrt{0.0049} = 0.07$ $(0.07)(0.07) = 0.0049$

In addition to being read as "the principal square root of a," $\sqrt{a}$ is also read as "the square root of a," or simply "root a." Any expression in which a radical sign appears is called a *radical expression*. The following are radical expressions:

$$\sqrt{5}, \qquad \sqrt{a}, \qquad -\sqrt{3x}, \qquad \sqrt{\dfrac{y^2 + 7}{y}}.$$

The expression under the radical sign is called the **radicand**. In the expressions above, the radicands are 5, a, $3x$, and $(y^2 + 7)/y$.

All but the most basic calculators give values for square roots. These values are, for the most part, approximations. For example, on many calculators, if you enter 5 and then press $\boxed{\sqrt{}}$, a number like

2.23606798

appears, depending on how the calculator rounds. (On some calculators, the $\boxed{\sqrt{}}$ key is pressed first.) The exact value of $\sqrt{5}$ is not given by any repeating or terminating decimal. The same is true for the square root of any whole number that is not a perfect square. We discussed such *irrational numbers* in Chapter 1.

The square-root function, given by

$$f(x) = \sqrt{x},$$

has the interval $[0, \infty)$ as its domain. We can draw its graph by selecting convenient values for x and calculating the corresponding outputs. Once these ordered pairs have been graphed, a smooth curve can be drawn.

$f(x) = \sqrt{x}$

x	$\sqrt{x}$	$(x, f(x))$
0	0	$(0, 0)$
1	1	$(1, 1)$
4	2	$(4, 2)$
9	3	$(9, 3)$

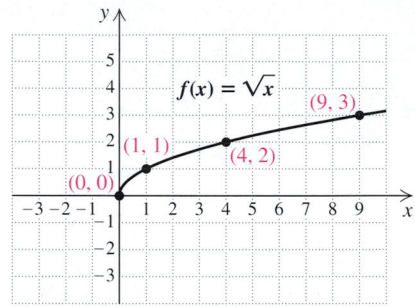

Example 3   For each function, find the indicated function value.

a) $f(x) = \sqrt{3x - 2}$; $f(1)$ **b)** $g(z) = -\sqrt{6z + 4}$; $g(3)$

Solution

a) $f(1) = \sqrt{3 \cdot 1 - 2}$ Substituting

$\quad\quad = \sqrt{1} = 1$ Simplifying

b) $g(3) = -\sqrt{6 \cdot 3 + 4}$ Substituting

$\quad\quad = -\sqrt{22}$ Simplifying

$\quad\quad \approx -4.69041576$ Using a calculator to approximate $\sqrt{22}$

Expressions of the Form $\sqrt{a^2}$

It is tempting to write $\sqrt{a^2} = a$, but the next example shows that, as a rule, this is untrue.

Example 4   Evaluate $\sqrt{x^2}$ for the following values: **(a)** 5; **(b)** 0; **(c)** -5.

Solution

a) $\sqrt{5^2} = \sqrt{25} = 5$
$\quad\quad\quad$ Same

b) $\sqrt{0^2} = \sqrt{0} = 0$
$\quad\quad\quad$ Same

c) $\sqrt{(-5)^2} = \sqrt{25} = 5$
$\quad\quad\quad$ Opposites   Note that $\sqrt{(-5)^2} \neq -5$.

You may have noticed that evaluating $\sqrt{a^2}$ is just like evaluating $|a|$.

> ### *Simplifying $\sqrt{a^2}$*
>
> For any real number a,
>
> $$\sqrt{a^2} = |a|.$$
>
> (The principal square root of a^2 is the absolute value of a.)

When a radicand is the square of a variable expression, like $(x + 5)^2$ or $36t^2$, absolute-value signs are needed when simplifying. We use absolute-value signs unless we know that the expression being squared is nonnegative. This assures that our result is never negative.

E x a m p l e   5

Simplify each expression. Assume that the variable can represent any real number.

a) $\sqrt{(x + 1)^2}$ b) $\sqrt{x^2 - 8x + 16}$

c) $\sqrt{a^8}$ d) $\sqrt{t^6}$

Solution

a) $\sqrt{(x + 1)^2} = |x + 1|$ Since $x + 1$ might be negative (for example, if $x = -3$), absolute-value notation is necessary.

b) $\sqrt{x^2 - 8x + 16} = \sqrt{(x - 4)^2} = |x - 4|$ Since $x - 4$ might be negative, absolute-value notation is necessary.

c) Note that $(a^4)^2 = a^8$ and that a^4 is never negative. Thus,

$$\sqrt{a^8} = a^4.$$ Absolute-value notation is unnecessary here.

d) Note that $(t^3)^2 = t^6$. Thus,

$$\sqrt{t^6} = |t^3|.$$ Since t^3 might be negative, absolute-value notation is necessary.

technology connection A

To see the necessity of the absolute-value signs, let $y_1 = \sqrt{x^2}$, $y_2 = x$, and $y_3 = \text{abs}(x)$. Then use a graph or table to show that $y_1 \ne y_2$ and $y_3 \ne y_2$, but $y_1 = y_3$.

E x a m p l e   6

Simplify each expression. Assume that no radicands were formed by raising negative quantities to even powers.

a) $\sqrt{y^2}$ b) $\sqrt{a^{10}}$ c) $\sqrt{9x^2 - 6x + 1}$

Solution

a) $\sqrt{y^2} = y$ We are assuming that y is nonnegative, so no absolute-value notation is necessary. When y *is* negative, $\sqrt{y^2} \ne y$.

b) $\sqrt{a^{10}} = a^5$ Assuming that a^5 is nonnegative. Note that $(a^5)^2 = a^{10}$.

c) $\sqrt{9x^2 - 6x + 1} = \sqrt{(3x - 1)^2} = 3x - 1$ Assuming that $3x - 1$ is nonnegative

Cube Roots

We often need to know what number was cubed in order to produce a certain value. When such a number is found, we say that we have found a *cube root*. For example,

2 is the cube root of 8 because $2^3 = 2 \cdot 2 \cdot 2 = 8$;

-4 is the cube root of -64 because $(-4)^3 = (-4)(-4)(-4) = -64$.

> ### Cube Root
>
> The number c is the *cube root* of a if $c^3 = a$. In symbols, we write $\sqrt[3]{a}$ to denote the cube root of a.

The cube-root function, given by

$$f(x) = \sqrt[3]{x},$$

has $\mathbb{R}$ as its domain. We can draw its graph by selecting convenient values for x and calculating the corresponding outputs. Once these ordered pairs have been graphed, a smooth curve can be drawn.

$$f(x) = \sqrt[3]{x}$$

x	$\sqrt[3]{x}$	$(x, f(x))$
0	0	$(0, 0)$
1	1	$(1, 1)$
8	2	$(8, 2)$
-1	-1	$(-1, -1)$
-8	-2	$(-8, -2)$

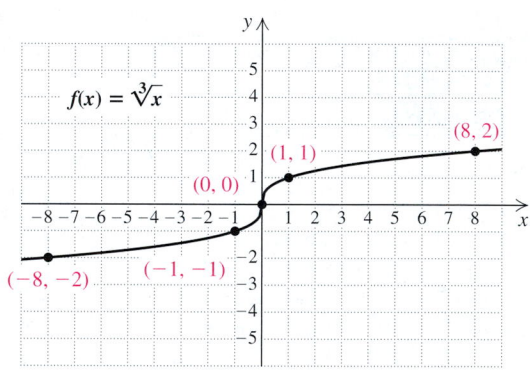

In the real-number system, every number has exactly one cube root. The cube root of a positive number is positive, and the cube root of a negative number is negative. Absolute-value signs are not used when finding cube roots.

Example 7   For each function, find the indicated function value.

a) $f(y) = \sqrt[3]{y}$; $f(125)$ **b)** $g(x) = \sqrt[3]{x - 3}$; $g(-24)$

Solution

a) $f(125) = \sqrt[3]{125} = 5$ Since $5 \cdot 5 \cdot 5 = 125$

b) $g(-24) = \sqrt[3]{-24 - 3}$
$\qquad = \sqrt[3]{-27}$
$\qquad = -3$ Since $(-3)(-3)(-3) = -27$

Example 8

Simplify: $\sqrt[3]{-8y^3}$.

Solution

$$\sqrt[3]{-8y^3} = -2y \qquad \text{Since } (-2y)(-2y)(-2y) = -8y^3$$

Odd and Even *n*th Roots

The fifth root of a number a is the number c for which $c^5 = a$. There are also 6th roots, 7th roots, and so on. We write $\sqrt[n]{a}$ for the nth root. The number n is called the *index* (plural, *indices*). When the index is 2, we do not write it.

Example 9

Find each of the following.

a) $\sqrt[5]{32}$

b) $\sqrt[5]{-32}$

c) $-\sqrt[5]{32}$

d) $-\sqrt[5]{-32}$

Solution

a) $\sqrt[5]{32} = 2$ Since $2^5 = 32$

b) $\sqrt[5]{-32} = -2$ Since $(-2)^5 = -32$

c) $-\sqrt[5]{32} = -2$ Taking the opposite of $\sqrt[5]{32}$

d) $-\sqrt[5]{-32} = -(-2) = 2$ Taking the opposite of $\sqrt[5]{-32}$

Note that every number has just one real root when n is odd. Odd roots of positive numbers are positive and odd roots of negative numbers are negative. Absolute-value signs are not used when finding odd roots.

Example 10

Find each of the following.

a) $\sqrt[7]{x^7}$

b) $\sqrt[9]{(x - 1)^9}$

Solution

a) $\sqrt[7]{x^7} = x$

b) $\sqrt[9]{(x - 1)^9} = x - 1$

When the index n is even, we say that we are taking an *even root*. Every positive real number has two real nth roots when n is even. One root is positive and one is negative. Negative numbers do not have real nth roots when n is even.

When n is even, the notation $\sqrt[n]{a}$ indicates the nonnegative nth root. Thus, when we are finding even nth roots, absolute-value signs are often necessary.

E x a m p l e   1 1

Simplify each expression, if possible. Assume that variables can represent any real number.

a) $\sqrt[4]{16}$

b) $-\sqrt[4]{16}$

c) $\sqrt[4]{-16}$

d) $\sqrt[4]{81x^4}$

e) $\sqrt[6]{(y+7)^6}$

Solution

a) $\sqrt[4]{16} = 2$ Since $2^4 = 16$

b) $-\sqrt[4]{16} = -2$ Taking the opposite of $\sqrt[4]{16}$

c) $\sqrt[4]{-16}$ cannot be simplified. $\sqrt[4]{-16}$ is not a real number.

d) $\sqrt[4]{81x^4} = 3|x|$ Use absolute-value notation since x could represent a negative number.

e) $\sqrt[6]{(y+7)^6} = |y+7|$ Use absolute-value notation since $y + 7$ could be negative.

We summarize as follows.

Simplifying *n*th Roots

n		a	$\sqrt[n]{a}$	$\sqrt[n]{a^n}$
Even		Positive	Positive	$\|a\|$ (or a)
		Negative	Not a real number	$\|a\|$ (or $-a$)
Odd		Positive	Positive	a
		Negative	Negative	a

E x a m p l e   1 2

technology connection

B

Use a grapher to draw the graph of $f(x) = \sqrt{x+2}$. Then check graphically that the domain of f is $[-2, \infty)$.

Determine the domain of $g(x) = \sqrt[6]{7 - 3x}$.

Solution   Since the index is even, the radicand, $7 - 3x$, must be nonnegative. We solve the inequality:

$$7 - 3x \geq 0 \qquad \text{We cannot find the 6th root of a negative number.}$$
$$-3x \geq -7$$
$$x \leq \tfrac{7}{3}. \qquad \text{Multiplying both sides by } -\tfrac{1}{3} \text{ and reversing the inequality}$$

Thus,

$$\text{Domain of } g = \left\{ x \,\middle|\, x \leq \tfrac{7}{3} \right\}$$
$$= \left(-\infty, \tfrac{7}{3} \right].$$

FOR EXTRA HELP

Digital Video Tutor CD 7
Videotape 19

InterAct Math

Math Tutor Center

MathXL

MyMathLab.com

Exercise Set 10.1

For each number, find the square roots.

1. 16 **2.** 49

3. 144 **4.** 9

5. 81 **6.** 400

7. 900 **8.** 225

Simplify.

9. $-\sqrt{\dfrac{49}{36}}$ **10.** $-\sqrt{\dfrac{361}{9}}$ **11.** $\sqrt{441}$

12. $\sqrt{196}$ **13.** $-\sqrt{\dfrac{16}{81}}$ **14.** $-\sqrt{\dfrac{81}{144}}$

15. $\sqrt{0.09}$ **16.** $\sqrt{0.36}$

17. $-\sqrt{0.0049}$ **18.** $\sqrt{0.0144}$

Identify the radicand and the index for each expression.

19. $5\sqrt{p^2 + 4}$ **20.** $-7\sqrt{y^2 - 8}$

21. $x^2 y^3 \sqrt[3]{\dfrac{x}{y+4}}$ **22.** $a^2 b^3 \sqrt[3]{\dfrac{a}{a^2 - b}}$

For each function, find the specified function value, if it exists.

23. $f(t) = \sqrt{5t - 10}$; $f(6), f(2), f(1), f(-1)$

24. $g(x) = \sqrt{x^2 - 25}$; $g(-6), g(3), g(6), g(13)$

25. $t(x) = -\sqrt{2x + 1}$; $t(4), t(0), t(-1), t\left(-\tfrac{1}{2}\right)$

26. $p(z) = \sqrt{2z^2 - 20}$; $p(4), p(3), p(-5), p(0)$

27. $f(t) = \sqrt{t^2 + 1}$; $f(0), f(-1), f(-10)$

28. $g(x) = -\sqrt{(x + 1)^2}$; $g(-3), g(4), g(-5)$

29. $g(x) = \sqrt{x^3 + 9}$; $g(-2), g(-3), g(3)$

30. $f(t) = \sqrt{t^3 - 10}$; $f(2), f(3), f(4)$

Simplify. Remember to use absolute-value notation when necessary.

31. $\sqrt{36x^2}$ **32.** $\sqrt{25t^2}$

33. $\sqrt{(-6b)^2}$ **34.** $\sqrt{(-7c)^2}$

35. $\sqrt{(7 - t)^2}$ **36.** $\sqrt{(a + 1)^2}$

37. $\sqrt{y^2 + 16y + 64}$ **38.** $\sqrt{x^2 - 4x + 4}$

39. $\sqrt{9x^2 - 30x + 25}$ **40.** $\sqrt{4x^2 + 28x + 49}$

41. $-\sqrt[4]{625}$ **42.** $\sqrt[4]{256}$

43. $-\sqrt[5]{3^5}$ **44.** $\sqrt[5]{-1}$

45. $\sqrt[5]{-\dfrac{1}{32}}$ **46.** $\sqrt[5]{-\dfrac{32}{243}}$

47. $\sqrt[8]{y^8}$ **48.** $\sqrt[6]{x^6}$

49. $\sqrt[4]{(7b)^4}$ **50.** $\sqrt[4]{(5a)^4}$

51. $\sqrt[12]{(-10)^{12}}$ **52.** $\sqrt[10]{(-6)^{10}}$

53. $\sqrt[1976]{(2a + b)^{1976}}$ **54.** $\sqrt[414]{(a + b)^{414}}$

55. $\sqrt{x^{10}}$ **56.** $\sqrt{a^{22}}$

57. $\sqrt{a^{14}}$ **58.** $\sqrt{x^{16}}$

Simplify. Assume that no radicands were formed by raising negative quantities to even powers.

59. $\sqrt{25t^2}$ **60.** $\sqrt{16x^2}$

61. $\sqrt{(7c)^2}$ **62.** $\sqrt{(6b)^2}$

63. $\sqrt{(5 + b)^2}$ **64.** $\sqrt{(a + 1)^2}$

65. $\sqrt{9x^2 + 36x + 36}$ **66.** $\sqrt{4x^2 + 8x + 4}$

67. $\sqrt{25t^2 - 20t + 4}$ **68.** $\sqrt{9t^2 - 12t + 4}$

69. $-\sqrt[3]{64}$ **70.** $\sqrt[3]{27}$

71. $\sqrt[4]{81x^4}$ **72.** $\sqrt[4]{16x^4}$

73. $-\sqrt[5]{-100,000}$ **74.** $\sqrt[3]{-216}$

75. $-\sqrt[3]{-64x^3}$ **76.** $-\sqrt[3]{-125y^3}$

77. $\sqrt{a^{14}}$ **78.** $\sqrt{a^{22}}$

79. $\sqrt{(x + 3)^{10}}$ **80.** $\sqrt{(x - 2)^8}$

For each function, find the specified function value, if it exists.

81. $f(x) = \sqrt[3]{x + 1}$; $f(7), f(26), f(-9), f(-65)$

82. $g(x) = -\sqrt[3]{2x - 1}$; $g(0), g(-62), g(-13), g(63)$

83. $g(t) = \sqrt[4]{t - 3}$; $g(19), g(-13), g(1), g(84)$

84. $f(t) = \sqrt[4]{t + 1}$; $f(0), f(15), f(-82), f(80)$

Determine the domain of each function described.

85. $f(x) = \sqrt{x - 5}$ **86.** $g(x) = \sqrt{x + 8}$

87. $g(t) = \sqrt[4]{t + 3}$

88. $f(x) = \sqrt[4]{x - 7}$

89. $g(x) = \sqrt[4]{5 - x}$

90. $g(t) = \sqrt[3]{2t - 5}$

91. $f(t) = \sqrt[5]{2t + 9}$

92. $f(t) = \sqrt[6]{2t + 5}$

93. $h(z) = -\sqrt[6]{5z + 3}$

94. $d(x) = -\sqrt[4]{7x - 5}$

Aha! **95.** $f(t) = 7 + \sqrt[8]{t^8}$

96. $g(t) = 9 + \sqrt[6]{t^6}$

97. Explain how to write the negative square root of a number using radical notation.

98. Does the square root of a number's absolute value always exist? Why or why not?

SKILL MAINTENANCE

Simplify. Do not use negative exponents in your answer.

99. $(a^3 b^2 c^5)^3$

100. $(5a^7 b^8)(2a^3 b)$

101. $(2a^{-2} b^3 c^{-4})^{-3}$

102. $(5x^{-3} y^{-1} z^2)^{-2}$

103. $\dfrac{8x^{-2} y^5}{4x^{-6} z^{-2}}$

104. $\dfrac{10a^{-6} b^{-7}}{2a^{-2} c^{-3}}$

SYNTHESIS

105. Under what conditions does the nth root of x^3 exist? Explain your reasoning.

106. Under what conditions does the nth root of x^2 exist? Explain your reasoning.

107. *Spaces in a parking lot.* A parking lot has attendants to park the cars. The number N of stalls needed for waiting cars before attendants can get to them is given by the formula $N = 2.5\sqrt{A}$, where A is the number of arrivals in peak hours. Find the number of spaces needed for the given number of arrivals in peak hours: **(a)** 25; **(b)** 36; **(c)** 49; **(d)** 64.

Determine the domain of each function described. Then draw the graph of each function.

108. $f(x) = \sqrt{x + 5}$

109. $g(x) = \sqrt{x} + 5$

110. $g(x) = \sqrt{x} - 2$

111. $f(x) = \sqrt{x - 2}$

112. Find the domain of f if
$$f(x) = \frac{\sqrt{x + 3}}{\sqrt[4]{2 - x}}.$$

113. Find the domain of g if
$$g(x) = \frac{\sqrt[4]{5 - x}}{\sqrt[6]{x + 4}}.$$

114. Use a grapher to check your answers to Exercises 31, 39, and 49. On some graphers, a MATH key is needed to enter higher roots.

115. Use a grapher to check your answers to Exercises 112 and 113. (See Exercise 114.)

10.2

Rational Numbers as Exponents

Rational Exponents • Negative Rational Exponents • Laws of Exponents • Simplifying Radical Expressions

In Section 1.8, we considered the natural numbers as exponents. Our discussion of exponents was expanded to include all integers in Section 4.8. In this section, we expand the study still further—to include all rational numbers. This will give meaning to expressions like $a^{1/3}$, $7^{-1/2}$, and $(3x)^{4/5}$. Such notation will help us simplify certain radical expressions.

Rational Exponents

Consider $a^{1/2} \cdot a^{1/2}$. If we still want to add exponents when multiplying, it must follow that $a^{1/2} \cdot a^{1/2} = a^{1/2 + 1/2}$, or a^1. This suggests that $a^{1/2}$ is a square root of a. Similarly, $a^{1/3} \cdot a^{1/3} \cdot a^{1/3} = a^{1/3 + 1/3 + 1/3}$, or a^1, so $a^{1/3}$ should mean $\sqrt[3]{a}$.

$$a^{1/n} = \sqrt[n]{a}$$

$a^{1/n}$ means $\sqrt[n]{a}$. When a is nonnegative, n can be any natural number greater than 1. When a is negative, n must be odd.

Note that the denominator of the exponent becomes the index and the base becomes the radicand.

E x a m p l e   1   Write an equivalent expression using radical notation.

a) $x^{1/2}$ **b)** $(-8)^{1/3}$ **c)** $(abc)^{1/5}$

Solution

a) $x^{1/2} = \sqrt{x}$

b) $(-8)^{1/3} = \sqrt[3]{-8} = -2$ The denominator of the exponent becomes the index. The base becomes the radicand.

c) $(abc)^{1/5} = \sqrt[5]{abc}$

E x a m p l e   2   Write an equivalent expression using exponential notation.

a) $\sqrt[5]{9xy}$ **b)** $\sqrt[7]{\dfrac{x^3 y}{4}}$ **c)** $\sqrt{5x}$

Solution   Parentheses are required to indicate the base.

a) $\sqrt[5]{9xy} = (9xy)^{1/5}$

b) $\sqrt[7]{\dfrac{x^3 y}{4}} = \left(\dfrac{x^3 y}{4}\right)^{1/7}$ The index becomes the denominator of the exponent. The radicand becomes the base.

c) $\sqrt{5x} = (5x)^{1/2}$ For square roots, the index 2 is understood without being written.

How shall we define $a^{2/3}$? If the property for multiplying exponents is to hold, we must have $a^{2/3} = (a^{1/3})^2$ and $a^{2/3} = (a^2)^{1/3}$. This would suggest that $a^{2/3} = (\sqrt[3]{a})^2$ and $a^{2/3} = \sqrt[3]{a^2}$. We make our definition accordingly.

Positive Rational Exponents

For any natural numbers m and n ($n \neq 1$) and any real number a for which $\sqrt[n]{a}$ exists,

$$a^{m/n} \text{ means } (\sqrt[n]{a})^m, \text{ or } \sqrt[n]{a^m}.$$

E x a m p l e   3

Write an equivalent expression using radical notation and simplify.

a) $27^{2/3}$ **b)** $25^{3/2}$

Solution

a) $27^{2/3} = \sqrt[3]{27^2}$, or $(\sqrt[3]{27})^2$ It is easier to simplify using $(\sqrt[3]{27})^2$.
$\qquad\quad = 3^2,\quad$ or 9

b) $25^{3/2} = \sqrt[2]{25^3}$, or $(\sqrt[2]{25})^3$ We normally omit the index 2.
$\qquad\quad = 5^3,\quad$ or 125   Taking the square root and cubing

E x a m p l e   4

Write an equivalent expression using exponential notation.

a) $\sqrt[3]{9^4}$ **b)** $(\sqrt[4]{7xy})^5$

Solution

a) $\sqrt[3]{9^4} = 9^{4/3}$

b) $(\sqrt[4]{7xy})^5 = (7xy)^{5/4}$

$\left.\right\}$ The index becomes the denominator of the fractional exponent.

technology connection
A

To approximate $7^{2/3}$, we enter
7 [^] (2/3).

1. Why are the parentheses needed above?
2. Compare the graphs of $y_1 = x^{1/2}$, $y_2 = x$, and $y_3 = x^{3/2}$.

Negative Rational Exponents

Recall that $x^{-2} = 1/x^2$. Negative rational exponents behave similarly.

> ### Negative Rational Exponents
> For any rational number m/n and any nonzero real number a for which $a^{m/n}$ exists,
>
> $$a^{-m/n} \quad \text{means} \quad \frac{1}{a^{m/n}}.$$

> **Caution!**   A negative exponent does not indicate that the expression in which it appears is negative.

E x a m p l e   5

Write an equivalent expression with positive exponents and, if possible, simplify.

a) $9^{-1/2}$ **b)** $(5xy)^{-4/5}$ **c)** $64^{-2/3}$

d) $4x^{-2/3}y^{1/5}$ **e)** $\left(\dfrac{3r}{7s}\right)^{-5/2}$

Solution

a) $9^{-1/2} = \dfrac{1}{9^{1/2}}$ $9^{-1/2}$ is the reciprocal of $9^{1/2}$.

Since $9^{1/2} = \sqrt{9} = 3$, the answer simplifies to $\dfrac{1}{3}$.

b) $(5xy)^{-4/5} = \dfrac{1}{(5xy)^{4/5}}$ $(5xy)^{-4/5}$ is the reciprocal of $(5xy)^{4/5}$.

c) $64^{-2/3} = \dfrac{1}{64^{2/3}}$ $64^{-2/3}$ is the reciprocal of $64^{2/3}$.

Since $64^{2/3} = (\sqrt[3]{64})^2 = 4^2 = 16$, the answer simplifies to $\dfrac{1}{16}$.

d) $4x^{-2/3}y^{1/5} = 4 \cdot \dfrac{1}{x^{2/3}} \cdot y^{1/5} = \dfrac{4y^{1/5}}{x^{2/3}}$

e) In Section 4.8, we found that $(a/b)^{-n} = (b/a)^n$. This property holds for *any* negative exponent:

$$\left(\frac{3r}{7s}\right)^{-5/2} = \left(\frac{7s}{3r}\right)^{5/2}.$$ Writing the reciprocal of the base and changing the sign of the exponent

Laws of Exponents

The same laws hold for rational exponents as for integer exponents.

Laws of Exponents

For any real numbers a and b and any rational exponents m and n for which a^m, a^n, and b^m are defined:

1. $a^m \cdot a^n = a^{m+n}$ In multiplying, add exponents if the bases are the same.

2. $\dfrac{a^m}{a^n} = a^{m-n}$ In dividing, subtract exponents if the bases are the same. (Assume $a \neq 0$.)

3. $(a^m)^n = a^{m \cdot n}$ To raise a power to a power, multiply the exponents.

4. $(ab)^m = a^m b^m$ To raise a product to a power, raise each factor to the power and multiply.

Example 6   Use the laws of exponents to simplify.

a) $3^{1/5} \cdot 3^{3/5}$ **b)** $a^{1/4}/a^{1/2}$
c) $(7.2^{2/3})^{3/4}$ **d)** $(a^{-1/3}b^{2/5})^{1/2}$

Solution

a) $3^{1/5} \cdot 3^{3/5} = 3^{1/5+3/5} = 3^{4/5}$ Adding exponents

b) $\dfrac{a^{1/4}}{a^{1/2}} = a^{1/4-1/2} = a^{1/4-2/4}$ Subtracting exponents after finding a common denominator

$\qquad\qquad = a^{-1/4}, \text{ or } \dfrac{1}{a^{1/4}}$ $a^{-1/4}$ is the reciprocal of $a^{1/4}$.

c) $(7.2^{2/3})^{3/4} = 7.2^{2/3 \cdot 3/4} = 7.2^{6/12}$ Multiplying exponents

$\qquad\qquad\quad = 7.2^{1/2}$ Using arithmetic to simplify the exponent

d) $(a^{-1/3}b^{2/5})^{1/2} = a^{-1/3 \cdot 1/2} \cdot b^{2/5 \cdot 1/2}$ Raising a product to a power and multiplying exponents

$\qquad\qquad = a^{-1/6}b^{1/5}, \text{ or } \dfrac{b^{1/5}}{a^{1/6}}$

Simplifying Radical Expressions

Many radical expressions can be simplified using rational exponents.

> ### To Simplify Radical Expressions
> 1. Convert radical expressions to exponential expressions.
> 2. Use arithmetic and the laws of exponents to simplify.
> 3. Convert back to radical notation as needed.

Example 7

Use rational exponents to simplify. Do not use fractional exponents in the final answer.

a) $\sqrt[6]{(5x)^3}$ 　　　　　　　　　　　　　　**b)** $\sqrt[5]{t^{20}}$

c) $(\sqrt[3]{ab^2c})^{12}$ 　　　　　　　　　　　**d)** $\sqrt{\sqrt[3]{x}}$

Solution

a) $\sqrt[6]{(5x)^3} = (5x)^{3/6}$ 　　Converting to exponential notation

　　　　　　　$= (5x)^{1/2}$ 　　Simplifying the exponent

　　　　　　　$= \sqrt{5x}$ 　　Returning to radical notation

b) $\sqrt[5]{t^{20}} = t^{20/5}$ 　　Converting to exponential notation

　　　　　$= t^4$ 　　Simplifying the exponent

c) $(\sqrt[3]{ab^2c})^{12} = (ab^2c)^{12/3}$ 　　Converting to exponential notation

　　　　　　　　$= (ab^2c)^4$ 　　Simplifying the exponent

　　　　　　　　$= a^4b^8c^4$ 　　Using the laws of exponents

d) $\sqrt{\sqrt[3]{x}} = \sqrt{x^{1/3}}$ 　　Converting the radicand to exponential notation

　　　　　$= (x^{1/3})^{1/2}$ 　　Try to go directly to this step.

　　　　　$= x^{1/6}$ 　　Using the laws of exponents

　　　　　$= \sqrt[6]{x}$ 　　Returning to radical notation

technology connection
B

One way to check Example 7(a) is to let $y_1 = (5x)^{3/6}$ and $y_2 = \sqrt{5x}$. Then see if the graphs of y_1 and y_2 coincide. An alternative is to let $y_3 = y_2 - y_1$ and see if $y_3 = 0$. Check Example 7(a) using one of the checks just described.

1. Why are rational exponents especially useful when working on a grapher?

FOR EXTRA HELP

Exercise Set 10.2

Digital Video Tutor CD 7
Videotape 19

InterAct Math

Math Tutor Center

MathXL

MyMathLab.com

Note: Assume for all exercises that even roots are of nonnegative quantities and that all denominators are nonzero.

Write an equivalent expression using radical notation and, if possible, simplify.

1. $x^{1/4}$ 　　　　**2.** $y^{1/5}$ 　　　　**3.** $16^{1/2}$

4. $8^{1/3}$ 　　　　**5.** $81^{1/4}$ 　　　　**6.** $64^{1/6}$

7. $9^{1/2}$ 　　　　**8.** $25^{1/2}$ 　　　　**9.** $(xyz)^{1/3}$

10. $(ab)^{1/4}$ 　　**11.** $(a^2b^2)^{1/5}$ 　　**12.** $(x^3y^3)^{1/4}$

13. $a^{2/3}$ 　　　　**14.** $b^{3/2}$ 　　　　**15.** $16^{3/4}$

16. $4^{7/2}$ 　　　　**17.** $49^{3/2}$ 　　　　**18.** $27^{4/3}$

19. $9^{5/2}$ **20.** $81^{3/2}$ **21.** $(81x)^{3/4}$

22. $(125a)^{2/3}$ **23.** $(25x^4)^{3/2}$ **24.** $(9y^6)^{3/2}$

Write an equivalent expression using exponential notation.

25. $\sqrt[3]{20}$ **26.** $\sqrt[3]{19}$ **27.** $\sqrt{17}$

28. $\sqrt{6}$ **29.** $\sqrt{x^3}$ **30.** $\sqrt{a^5}$

31. $\sqrt[5]{m^2}$ **32.** $\sqrt[5]{n^4}$ **33.** $\sqrt[4]{cd}$

34. $\sqrt[5]{xy}$ **35.** $\sqrt[5]{xy^2z}$ **36.** $\sqrt[7]{x^3y^2z^2}$

37. $\left(\sqrt{3mn}\right)^3$ **38.** $\left(\sqrt[3]{7xy}\right)^4$ **39.** $\left(\sqrt[7]{8x^2y}\right)^5$

40. $\left(\sqrt[6]{2a^5b}\right)^7$ **41.** $\dfrac{2x}{\sqrt[3]{z^2}}$ **42.** $\dfrac{3a}{\sqrt[5]{c^2}}$

Write an equivalent expression with positive exponents and, if possible, simplify.

43. $x^{-1/3}$ **44.** $y^{-1/4}$

45. $(2rs)^{-3/4}$ **46.** $(5xy)^{-5/6}$

47. $\left(\dfrac{1}{8}\right)^{-2/3}$ **48.** $\left(\dfrac{1}{16}\right)^{-3/4}$

49. $\dfrac{1}{a^{-5/7}}$ **50.** $\dfrac{1}{a^{-3/5}}$

51. $2a^{3/4}b^{-1/2}c^{2/3}$ **52.** $5x^{-2/3}y^{4/5}z$

53. $2^{-1/3}x^4y^{-2/7}$ **54.** $3^{-5/2}a^3b^{-7/3}$

55. $\left(\dfrac{7x}{8yz}\right)^{-3/5}$ **56.** $\left(\dfrac{2ab}{3c}\right)^{-5/6}$

57. $\dfrac{7x}{\sqrt[3]{z}}$ **58.** $\dfrac{6a}{\sqrt[4]{b}}$

59. $\dfrac{5a}{3c^{-1/2}}$ **60.** $\dfrac{2z}{5x^{-1/3}}$

Use the laws of exponents to simplify. Do not use negative exponents in any answers.

61. $5^{3/4} \cdot 5^{1/8}$ **62.** $11^{2/3} \cdot 11^{1/2}$

63. $\dfrac{3^{5/8}}{3^{-1/8}}$ **64.** $\dfrac{8^{7/11}}{8^{-2/11}}$

65. $\dfrac{4.1^{-1/6}}{4.1^{-2/3}}$ **66.** $\dfrac{2.3^{-3/10}}{2.3^{-1/5}}$

67. $(10^{3/5})^{2/5}$ **68.** $(5^{5/4})^{3/7}$

69. $a^{2/3} \cdot a^{5/4}$ **70.** $x^{3/4} \cdot x^{2/3}$

Aha! **71.** $(64^{3/4})^{4/3}$ **72.** $(27^{-2/3})^{3/2}$

73. $(m^{2/3}n^{-1/4})^{1/2}$ **74.** $(x^{-1/3}y^{2/5})^{1/4}$

Use rational exponents to simplify. Do not use fractional exponents in the final answer.

75. $\sqrt[6]{a^2}$ **76.** $\sqrt[6]{t^4}$

77. $\sqrt[3]{x^{15}}$ **78.** $\sqrt[4]{a^{12}}$

79. $\sqrt[6]{x^{18}}$ **80.** $\sqrt[5]{a^{10}}$

81. $\left(\sqrt[3]{ab}\right)^{15}$ **82.** $\left(\sqrt[7]{xy}\right)^{14}$

83. $\sqrt[8]{(3x)^2}$ **84.** $\sqrt[4]{(7a)^2}$

85. $\left(\sqrt[10]{3a}\right)^5$ **86.** $\left(\sqrt[8]{2x}\right)^6$

87. $\sqrt[4]{\sqrt{x}}$ **88.** $\sqrt[3]{\sqrt[6]{m}}$

89. $\sqrt{(ab)^6}$ **90.** $\sqrt[4]{(xy)^{12}}$

91. $\left(\sqrt[3]{x^2y^5}\right)^{12}$ **92.** $\left(\sqrt[5]{a^2b^4}\right)^{15}$

93. $\sqrt[3]{\sqrt[4]{xy}}$ **94.** $\sqrt[5]{\sqrt{2a}}$

95. If $f(x) = (x + 5)^{1/2}(x + 7)^{-1/2}$, find the domain of f. Explain how you found your answer.

96. Explain why $\sqrt[3]{x^6} = x^2$ for any value of x, whereas $\sqrt[2]{x^6} = x^3$ only when $x \geq 0$.

SKILL MAINTENANCE

Simplify.

97. $3x(x^3 - 2x^2) + 4x^2(2x^2 + 5x)$

98. $5t^3(2t^2 - 4t) - 3t^4(t^2 - 6t)$

99. $(3a - 4b)(5a + 3b)$

100. $(7x - y)^2$

101. *Real estate taxes.* For homes under \$100,000, the real-estate transfer tax in Vermont is 0.5% of the selling price. Find the selling price of a home that had a transfer tax of \$467.50.

102. What numbers are their own squares?

SYNTHESIS

103. Let $f(x) = 5x^{-1/3}$. Under what condition will we have $f(x) > 0$? Why?

104. If $g(x) = x^{3/n}$, in what way does the domain of g depend on whether n is odd or even?

Use rational exponents to simplify.

105. $\sqrt[5]{x^2y\sqrt{xy}}$

106. $\sqrt{x\sqrt[3]{x^2}}$

107. $\sqrt[4]{\sqrt[3]{8x^3y^6}}$

108. $\sqrt[12]{p^2 + 2pq + q^2}$

Music. *The function* $f(x) = k2^{x/12}$ *can be used to determine the frequency, in cycles per second, of a musical note that is x half-steps above a note with frequency k.**

109. The frequency of middle C on a piano is 262 cycles per second. Find the frequency of the C that is one octave (12 half-steps) higher.

110. The frequency of concert A for a trumpet is 440 cycles per second. Find the frequency of the A that is two octaves (24 half-steps) above concert A (few trumpeters can reach this note.)

111. Show that the G that is 7 half-steps (a "perfect fifth") above middle C (see Exercise 109) has a frequency that is about 1.5 times that of middle C.

112. Show that the C sharp that is 4 half-steps (a "major third") above concert A (see Exercise 110) has a frequency that is about 25% greater than that of concert A.

113. *Road pavement messages.* In a psychological study, it was determined that the proper length L of the letters of a word printed on pavement is given by

$$L = \frac{0.000169 d^{2.27}}{h},$$

where d is the distance of a car from the lettering and h is the height of the eye above the surface of the road. All units are in meters. This formula says that if a person is h meters above the surface of the road and is to be able to recognize a message d meters away, that message will be the most recognizable if the length of the letters is L. Find L to the nearest tenth of a meter, given d and h.

a) $h = 1$ m, $d = 60$ m
b) $h = 0.9906$ m, $d = 75$ m
c) $h = 2.4$ m, $d = 80$ m
d) $h = 1.1$ m, $d = 100$ m

114. *Dating fossils.* The function $r(t) = 10^{-12} 2^{-t/5700}$ expresses the ratio of carbon isotopes to carbon atoms in a fossil that is t years old. What ratio of carbon isotopes to carbon atoms would be present in a 1900-year-old bone?

115. *Physics.* The equation $m = m_0(1 - v^2 c^{-2})^{-1/2}$, developed by Albert Einstein, is used to determine the mass m of an object that is moving v meters per second and has mass m_0 before the motion begins. The constant c is the speed of light, approximately 3×10^8 m/sec. Suppose that a particle with mass 8 mg is accelerated to a speed of $\frac{9}{5} \times 10^8$ m/sec. Without using a calculator, find the new mass of the particle.

116. Use a grapher in the SIMULTANEOUS mode with the LABEL OFF format to graph

$$y_1 = x^{1/2}, \qquad y_2 = 3x^{2/5},$$
$$y_3 = x^{4/7}, \quad \text{and} \quad y_4 = \tfrac{1}{5}x^{3/4}.$$

Then, looking only at coordinates, match each graph with its equation.

*This application was inspired by information provided by Dr. Homer B. Tilton of Pima Community College East.

COLLABORATIVE

CORNER

Are Equivalent Fractions Equivalent Exponents?

Focus: Functions and rational exponents

Time: 10–20 minutes

Group size: 3

Materials: Graph paper

In arithmetic, we have seen that $\frac{1}{3}, \frac{1}{6} \cdot 2$, and $2 \cdot \frac{1}{6}$ all represent the same number. Interestingly,

$$f(x) = x^{1/3},$$
$$g(x) = (x^{1/6})^2, \quad \text{and}$$
$$h(x) = (x^2)^{1/6}$$

represent three *different* functions.

ACTIVITY

1. Selecting a variety of values for x and using the definition of positive rational exponents, one group member should graph f, a second group member should graph g, and a third group member should graph h. Be sure to check whether negative x-values are in the domain of the function.

2. Compare the three graphs and check each other's work. How and why do the graphs differ?

3. Decide as a group which graph, if any, would best represent the graph of $k(x) = x^{2/6}$. Then be prepared to explain your reasoning to the entire class. (*Hint*: Study the definition of $a^{m/n}$ on p. 605 carefully.)

Multiplying Radical Expressions

10.3

Multiplying Radical Expressions • Simplifying by Factoring • Multiplying and Simplifying

Multiplying Radical Expressions

Note that $\sqrt{4}\,\sqrt{25} = 2 \cdot 5 = 10$. Also $\sqrt{4 \cdot 25} = \sqrt{100} = 10$. Likewise,

$$\sqrt[3]{27}\,\sqrt[3]{8} = 3 \cdot 2 = 6 \quad \text{and} \quad \sqrt[3]{27 \cdot 8} = \sqrt[3]{216} = 6.$$

These examples suggest the following.

> **The Product Rule for Radicals**
>
> For any real numbers $\sqrt[n]{a}$ and $\sqrt[n]{b}$,
>
> $$\sqrt[n]{a} \cdot \sqrt[n]{b} = \sqrt[n]{a \cdot b}.$$
>
> (To multiply, when the indices match, multiply the radicands.)

Fractional exponents can be used to derive this rule:

$$\sqrt[n]{a} \cdot \sqrt[n]{b} = a^{1/n} \cdot b^{1/n} = (a \cdot b)^{1/n} = \sqrt[n]{a \cdot b}.$$

E x a m p l e 1

Multiply.

a) $\sqrt{3} \cdot \sqrt{5}$

b) $\sqrt{x+3}\,\sqrt{x-3}$

c) $\sqrt[3]{4} \cdot \sqrt[3]{5}$

d) $\sqrt[4]{\dfrac{y}{5}} \cdot \sqrt[4]{\dfrac{7}{x}}$

Solution

a) When no index is written, roots are understood to be square roots with an unwritten index of two. We apply the product rule:

$$\sqrt{3} \cdot \sqrt{5} = \sqrt{3 \cdot 5}$$
$$= \sqrt{15}.$$

b) $\sqrt{x+3}\,\sqrt{x-3} = \sqrt{(x+3)(x-3)}$ 　　The product of two square roots
　　　　　　　　　　　$= \sqrt{x^2 - 9}$ 　　　　　is the square root of the product.

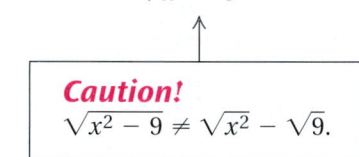

Caution!
$$\sqrt{x^2 - 9} \neq \sqrt{x^2} - \sqrt{9}.$$

c) Both $\sqrt[3]{4}$ and $\sqrt[3]{5}$ have indices of three, so to multiply we can use the product rule:

$$\sqrt[3]{4} \cdot \sqrt[3]{5} = \sqrt[3]{4 \cdot 5} = \sqrt[3]{20}.$$

d) $\sqrt[4]{\dfrac{y}{5}} \cdot \sqrt[4]{\dfrac{7}{x}} = \sqrt[4]{\dfrac{y}{5} \cdot \dfrac{7}{x}} = \sqrt[4]{\dfrac{7y}{5x}}$ 　　In Section 10.4, we discuss other ways to write answers like this.

Important: The product rule for radicals applies only when radicals have the same index.

Simplifying by Factoring

An integer p is a *perfect square* if there exists a rational number q for which $q^2 = p$. We say that p is a *perfect cube* if $q^3 = p$ for some rational number q. In general, p is a *perfect nth power* if $q^n = p$ for some rational number q. The product rule allows us to simplify $\sqrt[n]{ab}$ when a or b is a perfect nth power.

Using the Product Rule to Simplify
$$\sqrt[n]{ab} = \sqrt[n]{a} \cdot \sqrt[n]{b}.$$

---

technology connection A

To check Example 1(b), let $y_1 = \sqrt{x+3}\,\sqrt{x-3}$ and $y_2 = \sqrt{x^2 - 9}$ and compare.

1. What should the graph of

$$y = \sqrt{x^2 - 9}$$
$$\quad - \sqrt{x+3}\,\sqrt{x-3}$$

look like?

To illustrate, suppose we wish to simplify $\sqrt{20}$. Since this is a *square* root, we check to see if there is a factor of 20 that is a perfect square. There is one, 4, so we express 20 as $4 \cdot 5$ and use the product rule:

$$\sqrt{20} = \sqrt{4 \cdot 5} \qquad \text{Factoring the radicand (4 is a perfect square)}$$
$$= \sqrt{4} \cdot \sqrt{5} \qquad \text{Factoring into two radicals}$$
$$= 2\sqrt{5}. \qquad \text{Taking the square root of 4}$$

> ### To Simplify a Radical Expression with Index n by Factoring
>
> 1. Express the radicand as a product in which one factor is the largest perfect nth power possible.
> 2. Take the nth root of each factor.
> 3. Simplification is complete when no radicand has a factor that is a perfect nth power.

Example 2   Simplify by factoring: **(a)** $\sqrt{200}$; **(b)** $\sqrt[3]{32}$; **(c)** $\sqrt[4]{48}$; **(d)** $\sqrt{18x^2y}$.

Solution

a) $\sqrt{200} = \sqrt{100 \cdot 2} = \sqrt{100} \cdot \sqrt{2} = 10\sqrt{2}$ This is the largest perfect-square factor of 200.

b) $\sqrt[3]{32} = \sqrt[3]{8 \cdot 4} = \sqrt[3]{8} \cdot \sqrt[3]{4} = 2\sqrt[3]{4}$ This is the largest perfect-cube (third-power) factor of 32.

c) $\sqrt[4]{48} = \sqrt[4]{16 \cdot 3} = \sqrt[4]{16} \cdot \sqrt[4]{3} = 2\sqrt[4]{3}$ This is the largest fourth-power factor of 48.

d) $\sqrt{18x^2y} = \sqrt{9x^2 \cdot 2y}$ $9x^2$ is a perfect square.
$$= \sqrt{9x^2} \cdot \sqrt{2y} \qquad \text{Factoring into two radicals}$$
$$= |3x|\sqrt{2y}, \text{ or } 3|x|\sqrt{2y} \qquad \text{Taking the square root of } 9x^2$$

Example 3   If $f(x) = \sqrt{3x^2 - 6x + 3}$, find a simplified form for $f(x)$.

Solution

$$f(x) = \sqrt{3x^2 - 6x + 3}$$
$$= \sqrt{3(x^2 - 2x + 1)} \Bigg\} \qquad \text{Factoring the radicand; } x^2 - 2x + 1 \text{ is a perfect square.}$$
$$= \sqrt{(x - 1)^2 \cdot 3}$$
$$= \sqrt{(x - 1)^2} \cdot \sqrt{3} \qquad \text{Factoring into two radicals}$$
$$= |x - 1|\sqrt{3} \qquad \text{Taking the square root of } (x - 1)^2$$

In many situations that do not involve functions, it is safe to assume that no radicands were formed by raising negative quantities to even powers. We now make this assumption and thus discontinue the use of absolute-value notation when taking even roots unless functions are involved.

E x a m p l e   4

Simplify: **(a)** $\sqrt{x^7 y^{11} z^9}$; **(b)** $\sqrt[3]{16a^7 b^{14}}$.

Solution

a) There are many ways to factor $x^7 y^{11} z^9$. Because of the square root (index of 2), we identify the largest exponents that are multiples of 2:

$$\sqrt{x^7 y^{11} z^9} = \sqrt{x^6 \cdot x \cdot y^{10} \cdot y \cdot z^8 \cdot z} \qquad \text{Using the largest even powers of } x, y, \text{ and } z$$

$$= \sqrt{x^6}\, \sqrt{y^{10}}\, \sqrt{z^8}\, \sqrt{xyz} \qquad \text{Factoring into several radicals}$$

$$= x^{6/2}\, y^{10/2}\, z^{8/2} \sqrt{xyz} \qquad \text{Converting to fractional exponents}$$

$$= x^3 y^5 z^4 \sqrt{xyz}.$$

Check: $(x^3 y^5 z^4 \sqrt{xyz})^2 = (x^3)^2 (y^5)^2 (z^4)^2 (\sqrt{xyz})^2$
$$= x^6 \cdot y^{10} \cdot z^8 \cdot xyz = x^7 y^{11} z^9$$

Our check shows that $x^3 y^5 z^4 \sqrt{xyz}$ is the square root of $x^7 y^{11} z^9$.

b) There are many ways to factor $16a^7 b^{14}$. Because of the cube root (index of 3), we identify factors with the largest exponents that are multiples of 3:

$$\sqrt[3]{16a^7 b^{14}} = \sqrt[3]{8 \cdot 2 \cdot a^6 \cdot a \cdot b^{12} \cdot b^2} \qquad \text{Using the largest perfect-cube factors}$$

$$= \sqrt[3]{8}\, \sqrt[3]{a^6}\, \sqrt[3]{b^{12}}\, \sqrt[3]{2ab^2} \qquad \text{Factoring into several radicals}$$

$$= 2\ a^{6/3}\, b^{12/3} \sqrt[3]{2ab^2} \qquad \text{Converting to fractional exponents}$$

$$= 2a^2 b^4 \sqrt[3]{2ab^2}$$

Check: $(2a^2 b^4 \sqrt[3]{2ab^2})^3 = 2^3 (a^2)^3 (b^4)^3 (\sqrt[3]{2ab^2})^3$
$$= 8 \cdot a^6 \cdot b^{12} \cdot 2ab^2 = 16a^7 b^{14}$$

We see that $2a^2 b^4 \sqrt[3]{2ab^2}$ is the cube root of $16a^7 b^{14}$.

Example 4 demonstrates the following.

> To simplify an *n*th root, identify factors in the radicand with exponents that are multiples of *n*.

Multiplying and Simplifying

We have used the product rule for radicals to find products and also to simplify radical expressions. For some radical expressions, it is possible to do both: First find a product and then simplify.

E x a m p l e   5   Multiply and simplify.

a) $\sqrt{15}\,\sqrt{6}$ **b)** $3\sqrt[3]{25}\cdot 2\sqrt[3]{5}$ **c)** $\sqrt[4]{8x^3y^5}\,\sqrt[4]{4x^2y^3}$

Solution

a) $\sqrt{15}\,\sqrt{6} = \sqrt{15\cdot 6}$ Multiplying radicands

$\phantom{\sqrt{15}\,\sqrt{6}} = \sqrt{90} = \sqrt{9\cdot 10}$ 9 is a perfect square.

$\phantom{\sqrt{15}\,\sqrt{6}} = 3\sqrt{10}$

b) $3\sqrt[3]{25}\cdot 2\sqrt[3]{5} = 3\cdot 2\cdot\sqrt[3]{25\cdot 5}$ Using a commutative law; multiplying radicands

$\phantom{3\sqrt[3]{25}\cdot 2\sqrt[3]{5}} = 6\cdot\sqrt[3]{125}$ 125 is a perfect cube.

$\phantom{3\sqrt[3]{25}\cdot 2\sqrt[3]{5}} = 6\cdot 5, \text{ or } 30$

c) $\sqrt[4]{8x^3y^5}\,\sqrt[4]{4x^2y^3} = \sqrt[4]{32x^5y^8}$ Multiplying radicands

$\phantom{\sqrt[4]{8x^3y^5}\,\sqrt[4]{4x^2y^3}} = \sqrt[4]{16x^4y^8\cdot 2x}$ Identifying perfect fourth-power factors

$\phantom{\sqrt[4]{8x^3y^5}\,\sqrt[4]{4x^2y^3}} = \sqrt[4]{16}\,\sqrt[4]{x^4}\,\sqrt[4]{y^8}\,\sqrt[4]{2x}$ Factoring into radicals

$\phantom{\sqrt[4]{8x^3y^5}\,\sqrt[4]{4x^2y^3}} = 2xy^2\sqrt[4]{2x}$ Finding the fourth roots; assume $x \geq 0$.

The checks are left to the student.

Exercise Set **10.3**

Multiply.

1. $\sqrt{10}\,\sqrt{7}$ **2.** $\sqrt{5}\,\sqrt{7}$

3. $\sqrt[3]{2}\,\sqrt[3]{5}$ **4.** $\sqrt[3]{7}\,\sqrt[3]{2}$

5. $\sqrt[4]{8}\,\sqrt[4]{9}$ **6.** $\sqrt[4]{6}\,\sqrt[4]{3}$

7. $\sqrt{5a}\,\sqrt{6b}$ **8.** $\sqrt{2x}\,\sqrt{13y}$

9. $\sqrt[5]{9t^2}\,\sqrt[5]{2t}$ **10.** $\sqrt[5]{8y^3}\,\sqrt[5]{10y}$

11. $\sqrt{x-a}\,\sqrt{x+a}$ **12.** $\sqrt{y-b}\,\sqrt{y+b}$

13. $\sqrt[3]{0.5x}\,\sqrt[3]{0.2x}$ **14.** $\sqrt[3]{0.7y}\,\sqrt[3]{0.3y}$

15. $\sqrt[4]{x-1}\,\sqrt[4]{x^2+x+1}$ **16.** $\sqrt[5]{x-2}\,\sqrt[5]{(x-2)^2}$

17. $\sqrt{\dfrac{x}{6}}\,\sqrt{\dfrac{7}{y}}$ **18.** $\sqrt{\dfrac{7}{t}}\,\sqrt{\dfrac{s}{11}}$

19. $\sqrt[7]{\dfrac{x-3}{4}}\,\sqrt[7]{\dfrac{5}{x+2}}$ **20.** $\sqrt[6]{\dfrac{a}{b-2}}\,\sqrt[6]{\dfrac{3}{b+2}}$

Simplify by factoring.

21. $\sqrt{50}$ **22.** $\sqrt{27}$ **23.** $\sqrt{28}$

24. $\sqrt{45}$ **25.** $\sqrt{8}$ **26.** $\sqrt{18}$

27. $\sqrt{198}$ **28.** $\sqrt{325}$ **29.** $\sqrt{36a^4b}$

30. $\sqrt{175y^8}$ **31.** $\sqrt[3]{8x^3y^2}$ **32.** $\sqrt[3]{27ab^6}$

33. $\sqrt[3]{-16x^6}$ **34.** $\sqrt[3]{-32a^6}$

Find a simplified form of $f(x)$. Assume that x can be any real number.

35. $f(x) = \sqrt[3]{125x^5}$    **36.** $f(x) = \sqrt[3]{16x^6}$

37. $f(x) = \sqrt{49(x-3)^2}$ **38.** $f(x) = \sqrt{81(x-1)^2}$

39. $f(x) = \sqrt{5x^2-10x+5}$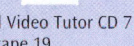

40. $f(x) = \sqrt{2x^2+8x+8}$

Simplify. Assume that no radicands were formed by raising negative numbers to even powers.

41. $\sqrt{a^3b^4}$ **42.** $\sqrt{x^6y^9}$

43. $\sqrt[3]{x^5y^6z^{10}}$ **44.** $\sqrt[3]{a^6b^7c^{13}}$

45. $\sqrt[5]{-32a^7b^{11}}$ **46.** $\sqrt[4]{16x^5y^{11}}$

47. $\sqrt[5]{a^6b^8c^9}$

48. $\sqrt[5]{x^{13}y^8z^{17}}$

49. $\sqrt[4]{810x^9}$

50. $\sqrt[3]{-80a^{14}}$

Multiply and simplify.

51. $\sqrt{15}\,\sqrt{5}$

52. $\sqrt{6}\,\sqrt{3}$

53. $\sqrt{10}\,\sqrt{14}$

54. $\sqrt{15}\,\sqrt{21}$

55. $\sqrt[3]{2}\,\sqrt[3]{4}$

56. $\sqrt[3]{9}\,\sqrt[3]{3}$

Aha! **57.** $\sqrt{18a^3}\,\sqrt{18a^3}$

58. $\sqrt{75x^7}\,\sqrt{75x^7}$

59. $\sqrt[3]{5a^2}\,\sqrt[3]{2a}$

60. $\sqrt[3]{7x}\,\sqrt[3]{3x^2}$

61. $\sqrt{3x^5}\,\sqrt{15x^2}$

62. $\sqrt{5a^7}\,\sqrt{15a^3}$

63. $\sqrt[3]{s^2t^4}\,\sqrt[3]{s^4t^6}$

64. $\sqrt[3]{x^2y^4}\,\sqrt[3]{x^2y^6}$

65. $\sqrt[3]{(x+5)^2}\,\sqrt[3]{(x+5)^4}$

66. $\sqrt[3]{(a-b)^5}\,\sqrt[3]{(a-b)^7}$

67. $\sqrt[4]{12a^3b^7}\,\sqrt[4]{4a^2b^5}$

68. $\sqrt[4]{9x^7y^2}\,\sqrt[4]{9x^2y^9}$

69. $\sqrt[5]{x^3(y+z)^4}\,\sqrt[5]{x^3(y+z)^6}$

70. $\sqrt[5]{a^3(b-c)^4}\,\sqrt[5]{a^7(b-c)^4}$

71. Why do we need to know how to multiply radical expressions before learning how to simplify radical expressions?

72. Why is it incorrect to say that, in general, $\sqrt{x^2}=x$?

SKILL MAINTENANCE

Perform the indicated operation and, if possible, simplify.

73. $\dfrac{3x}{16y}+\dfrac{5y}{64x}$

74. $\dfrac{2}{a^3b^4}+\dfrac{6}{a^4b}$

75. $\dfrac{4}{x^2-9}-\dfrac{7}{2x-6}$

76. $\dfrac{8}{x^2-25}-\dfrac{3}{2x-10}$

Simplify.

77. $\dfrac{9a^4b^7}{3a^2b^5}$

78. $\dfrac{12a^2b^7}{4ab^2}$

SYNTHESIS

79. Explain why it is true that
$$\sqrt[n]{ab}=\sqrt[n]{a}\cdot\sqrt[n]{b}.$$

80. Is the equation $\sqrt{(2x+3)^8}=(2x+3)^4$ always, sometimes, or never true? Why?

81. *Speed of a skidding car.*   Police can estimate the speed at which a car was traveling by measuring its skid marks. The function
$$r(L)=2\sqrt{5L}$$
can be used, where L is the length of a skid mark, in feet, and $r(L)$ is the speed, in miles per hour. Find the exact speed and an estimate (to the nearest tenth mile per hour) for the speed of a car that left skid marks **(a)** 20 ft long; **(b)** 70 ft long; **(c)** 90 ft long.

82. *Wind chill temperature.*   When the temperature is T degrees Celsius and the wind speed is v meters per second, the *wind chill temperature*, T_w, is the temperature (with no wind) that it feels like. Here is a formula for finding wind chill temperature:
$$T_w=33-\frac{(10.45+10\sqrt{v}-v)(33-T)}{22}.$$

Estimate the wind chill temperature (to the nearest tenth of a degree) for the given actual temperatures and wind speeds.

a) $T=7°C$, $v=8$ m/sec
b) $T=0°C$, $v=12$ m/sec
c) $T=-5°C$, $v=14$ m/sec
d) $T=-23°C$, $v=15$ m/sec

Simplify. Assume that all variables are nonnegative.

83. $\left(\sqrt{r^3t}\right)^7$

84. $\left(\sqrt[3]{25x^4}\right)^4$

85. $\left(\sqrt[3]{a^2b^4}\right)^5$

86. $\left(\sqrt{a^3b^5}\right)^7$

Draw and compare the graphs of each group of equations.

87. $f(x)=\sqrt{x^2-2x+1}$,
$g(x)=x-1$,
$h(x)=|x-1|$

88. $f(x)=\sqrt{x^2+2x+1}$,
$g(x)=x+1$,
$h(x)=|x+1|$

89. If $f(t) = \sqrt{t^2 - 3t - 4}$, what is the domain of f?

90. What is the domain of g, if $g(x) = \sqrt{x^2 - 6x + 8}$?

Solve.

91. $\sqrt[3]{5x^{k+1}} \sqrt[3]{25x^k} = 5x^7$, for k

92. $\sqrt[5]{4a^{3k+2}} \sqrt[5]{8a^{6-k}} = 2a^4$, for k

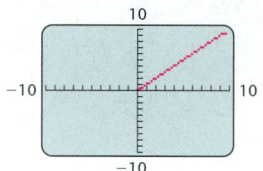 **93.** Use a grapher to check your answers to Exercises 15, 35, and 59.

94. Rony is puzzled. When he uses a grapher to graph $y = \sqrt{x} \cdot \sqrt{x}$, he gets the following screen. Explain why Rony did not get the complete line $y = x$.

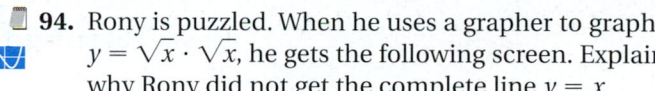

Dividing Radical Expressions

10.4

Dividing and Simplifying • Rationalizing Denominators and Numerators

Dividing and Simplifying

Just as the root of a product can be expressed as the product of two roots, the root of a quotient can be expressed as the quotient of two roots. For example,

$$\sqrt[3]{\frac{27}{8}} = \frac{3}{2} \quad \text{and} \quad \frac{\sqrt[3]{27}}{\sqrt[3]{8}} = \frac{3}{2}.$$

This example suggests the following.

> **The Quotient Rule for Radicals**
> For any real numbers $\sqrt[n]{a}$ and $\sqrt[n]{b}$, $b \neq 0$,
> $$\sqrt[n]{\frac{a}{b}} = \frac{\sqrt[n]{a}}{\sqrt[n]{b}}.$$

Remember that an nth root is simplified when its radicand has no factors that are perfect nth powers. Recall too that we assume that no radicands represent negative quantities raised to an even power.

Example 1

Simplify by taking the roots of the numerator and the denominator.

a) $\sqrt[3]{\dfrac{27}{125}}$ **b)** $\sqrt{\dfrac{25}{y^2}}$

Solution

a) $\sqrt[3]{\dfrac{27}{125}} = \dfrac{\sqrt[3]{27}}{\sqrt[3]{125}} = \dfrac{3}{5}$ Taking the cube roots of the numerator and the denominator

b) $\sqrt{\dfrac{25}{y^2}} = \dfrac{\sqrt{25}}{\sqrt{y^2}} = \dfrac{5}{y}$ Taking the square roots of the numerator and the denominator. Assume $y > 0$.

As in Section 10.3, any radical expressions appearing in the answers should be simplified as much as possible.

Example 2

Simplify: **(a)** $\sqrt{\dfrac{16x^3}{y^8}}$; **(b)** $\sqrt[3]{\dfrac{27y^{14}}{8x^3}}$.

Solution

a) $\sqrt{\dfrac{16x^3}{y^8}} = \dfrac{\sqrt{16x^3}}{\sqrt{y^8}}$

$= \dfrac{\sqrt{16x^2 \cdot x}}{\sqrt{y^8}} = \dfrac{4x\sqrt{x}}{y^4}$ Simplifying the numerator and the denominator

b) $\sqrt[3]{\dfrac{27y^{14}}{8x^3}} = \dfrac{\sqrt[3]{27y^{14}}}{\sqrt[3]{8x^3}}$

$= \dfrac{\sqrt[3]{27y^{12}y^2}}{\sqrt[3]{8x^3}} = \dfrac{\sqrt[3]{27y^{12}}\,\sqrt[3]{y^2}}{\sqrt[3]{8x^3}} = \dfrac{3y^4\sqrt[3]{y^2}}{2x}$ Simplifying the numerator and the denominator

If we read from right to left, the quotient rule tells us that to divide two radical expressions that have the same index, we can divide the radicands.

Example 3

Divide and, if possible, simplify.

a) $\dfrac{\sqrt{80}}{\sqrt{5}}$ **b)** $\dfrac{5\sqrt[3]{32}}{\sqrt[3]{2}}$ **c)** $\dfrac{\sqrt{72xy}}{2\sqrt{2}}$ **d)** $\dfrac{\sqrt[4]{18a^9b^5}}{\sqrt[4]{3b}}$

Solution

a) $\dfrac{\sqrt{80}}{\sqrt{5}} = \sqrt{\dfrac{80}{5}} = \sqrt{16} = 4$ Because the indices match, we can divide the radicands.

b) $\dfrac{5\sqrt[3]{32}}{\sqrt[3]{2}} = 5\sqrt[3]{\dfrac{32}{2}} = 5\sqrt[3]{16}$

$= 5\sqrt[3]{8 \cdot 2}$

$= 5\sqrt[3]{8}\,\sqrt[3]{2} = 5 \cdot 2\sqrt[3]{2}$

$= 10\sqrt[3]{2}$

c) $\dfrac{\sqrt{72xy}}{2\sqrt{2}} = \dfrac{1}{2}\dfrac{\sqrt{72xy}}{\sqrt{2}} = \dfrac{1}{2}\sqrt{\dfrac{72xy}{2}} = \dfrac{1}{2}\sqrt{36xy}$

$\qquad = \dfrac{1}{2}\sqrt{36}\sqrt{xy} = \dfrac{1}{2}\cdot 6\sqrt{xy}$

$\qquad = 3\sqrt{xy}$

Because the indices match, we can divide the radicands.

d) $\dfrac{\sqrt[4]{18a^9b^5}}{\sqrt[4]{3b}} = \sqrt[4]{\dfrac{18a^9b^5}{3b}}$

$\qquad = \sqrt[4]{6a^9b^4} = \sqrt[4]{a^8b^4}\sqrt[4]{6a}$

$\qquad = a^2b\sqrt[4]{6a}$

Note that 8 is the largest power less than 9 that is a multiple of the index 4.

Partial check: $(a^2b)^4 = a^8b^4$

Rationalizing Denominators and Numerators

When a radical expression appears in a denominator, it can be useful to find an equivalent expression in which the denominator no longer contains a radical.* The procedure for finding such an expression is called **rationalizing the denominator**. We carry this out by multiplying by 1 in either of two ways.

One way is to multiply by 1 *under* the radical to make the denominator of the radicand a perfect power.

E x a m p l e   4

Rationalize each denominator.

a) $\sqrt{\dfrac{7}{3}}$ 　　　　　　　　　　**b)** $\sqrt[3]{\dfrac{5}{16}}$

Solution

a) We multiply by 1 under the radical, using $\frac{3}{3}$. We do this so that the denominator of the radicand will be a perfect square:

$\sqrt{\dfrac{7}{3}} = \sqrt{\dfrac{7}{3}\cdot\dfrac{3}{3}}$ 　　Multiplying by 1 under the radical

$\qquad = \sqrt{\dfrac{21}{9}}$ 　　The denominator, 9, is now a perfect square.

$\qquad = \dfrac{\sqrt{21}}{\sqrt{9}}$ 　　Using the quotient rule for radicals

$\qquad = \dfrac{\sqrt{21}}{3}.$

*See Exercise 65 on p. 622.

b) Note that $16 = 4^2$. Thus, to make the denominator a perfect cube, we multiply under the radical by $\frac{4}{4}$:

$$\sqrt[3]{\frac{5}{16}} = \sqrt[3]{\frac{5}{4 \cdot 4} \cdot \frac{4}{4}} \qquad \text{Since the index is 3, we need 3 identical factors in the denominator.}$$

$$= \sqrt[3]{\frac{20}{4^3}} \qquad \text{The denominator is now a perfect cube.}$$

$$= \frac{\sqrt[3]{20}}{\sqrt[3]{4^3}}$$

$$= \frac{\sqrt[3]{20}}{4}.$$

Another way to rationalize a denominator is to multiply by 1 *outside* the radical.

Example 5   Rationalize each denominator.

a) $\sqrt{\dfrac{4}{5b}}$ **b)** $\dfrac{\sqrt[3]{a}}{\sqrt[3]{9x}}$ **c)** $\dfrac{3x}{\sqrt[5]{2x^2y^3}}$

Solution

a) We rewrite the expression as a quotient of two radicals. Then we simplify and multiply by 1:

$$\sqrt{\frac{4}{5b}} = \frac{\sqrt{4}}{\sqrt{5b}} = \frac{2}{\sqrt{5b}} \qquad \text{We assume } b > 0.$$

$$= \frac{2}{\sqrt{5b}} \cdot \frac{\sqrt{5b}}{\sqrt{5b}} \qquad \text{Multiplying by 1}$$

$$= \frac{2\sqrt{5b}}{(\sqrt{5b})^2} \qquad \text{Try to do this step mentally.}$$

$$= \frac{2\sqrt{5b}}{5b}.$$

b) To rationalize the denominator $\sqrt[3]{9x}$, note that $9x$ is $3 \cdot 3 \cdot x$. In order for this radicand to be a cube, we need another factor of 3 and two more factors of x. Thus we multiply by 1, using $\sqrt[3]{3x^2}/\sqrt[3]{3x^2}$:

$$\frac{\sqrt[3]{a}}{\sqrt[3]{9x}} = \frac{\sqrt[3]{a}}{\sqrt[3]{9x}} \cdot \frac{\sqrt[3]{3x^2}}{\sqrt[3]{3x^2}} \qquad \text{Multiplying by 1}$$

$$= \frac{\sqrt[3]{3ax^2}}{\sqrt[3]{27x^3}} \longleftarrow \text{This radicand is now a perfect cube.}$$

$$= \frac{\sqrt[3]{3ax^2}}{3x}.$$

c) To change the radicand $2x^2y^3$ into a perfect fifth power, we need four more factors of 2, three more factors of x, and two more factors of y. Thus we multiply by 1, using $\sqrt[5]{2^4x^3y^2}/\sqrt[5]{2^4x^3y^2}$, or $\sqrt[5]{16x^3y^2}/\sqrt[5]{16x^3y^2}$:

$$\frac{3x}{\sqrt[5]{2x^2y^3}} = \frac{3x}{\sqrt[5]{2x^2y^3}} \cdot \frac{\sqrt[5]{16x^3y^2}}{\sqrt[5]{16x^3y^2}} \qquad \text{\color{magenta}Multiplying by 1}$$

$$= \frac{3x\sqrt[5]{16x^3y^2}}{\sqrt[5]{32x^5y^5}} \longleftarrow \quad \text{\color{magenta}This radicand is now a perfect fifth power.}$$

$$= \frac{3x\sqrt[5]{16x^3y^2}}{2xy} = \frac{3\sqrt[5]{16x^3y^2}}{2y}. \qquad \text{\color{magenta}Always simplify if possible.}$$

Sometimes in calculus it is necessary to rationalize a numerator. To do so, we multiply by 1 to make the radicand in the *numerator* a perfect power.

E x a m p l e   6   Rationalize each numerator: **(a)** $\sqrt{\dfrac{7}{5}}$; **(b)** $\dfrac{\sqrt[3]{4a^2}}{\sqrt[3]{5b}}$.

Solution

a) $\sqrt{\dfrac{7}{5}} = \sqrt{\dfrac{7}{5} \cdot \dfrac{7}{7}}$ {\color{magenta}Multiplying by 1 under the radical. We also could have multiplied by $\sqrt{7}/\sqrt{7}$ outside the radical.}

$= \sqrt{\dfrac{49}{35}}$ {\color{magenta}The numerator is now a perfect square.}

$= \dfrac{\sqrt{49}}{\sqrt{35}}$ {\color{magenta}Using the quotient rule for radicals}

$= \dfrac{7}{\sqrt{35}}$

b) $\dfrac{\sqrt[3]{4a^2}}{\sqrt[3]{5b}} = \dfrac{\sqrt[3]{4a^2}}{\sqrt[3]{5b}} \cdot \dfrac{\sqrt[3]{2a}}{\sqrt[3]{2a}}$ {\color{magenta}Multiplying by 1}

$= \dfrac{\sqrt[3]{8a^3}}{\sqrt[3]{10ba}} \longleftarrow$ {\color{magenta}This radicand is now a perfect cube.}

$= \dfrac{2a}{\sqrt[3]{10ab}}$

In Section 10.5, we will discuss rationalizing denominators and numerators in which two terms appear.

Exercise Set 10.4

Simplify by taking the roots of the numerator and the denominator. Assume all variables represent positive numbers.

1. $\sqrt{\dfrac{25}{36}}$

2. $\sqrt{\dfrac{100}{81}}$

3. $\sqrt[3]{\dfrac{64}{27}}$

4. $\sqrt[3]{\dfrac{343}{1000}}$

5. $\sqrt{\dfrac{49}{y^2}}$

6. $\sqrt{\dfrac{121}{x^2}}$

7. $\sqrt{\dfrac{25y^3}{x^4}}$

8. $\sqrt{\dfrac{36a^5}{b^6}}$

9. $\sqrt[3]{\dfrac{27a^4}{8b^3}}$

10. $\sqrt[3]{\dfrac{64x^7}{216y^6}}$

11. $\sqrt[4]{\dfrac{16a^4}{b^4c^8}}$

12. $\sqrt[4]{\dfrac{81x^4}{y^8z^4}}$

13. $\sqrt[4]{\dfrac{a^5b^8}{c^{10}}}$

14. $\sqrt[4]{\dfrac{x^9y^{12}}{z^6}}$

15. $\sqrt[5]{\dfrac{32x^6}{y^{11}}}$

16. $\sqrt[5]{\dfrac{243a^9}{b^{13}}}$

17. $\sqrt[6]{\dfrac{x^6y^8}{z^{15}}}$

18. $\sqrt[6]{\dfrac{a^9b^{12}}{c^{13}}}$

Divide and, if possible, simplify. Assume all variables represent positive numbers.

19. $\dfrac{\sqrt{35x}}{\sqrt{7x}}$

20. $\dfrac{\sqrt{28y}}{\sqrt{4y}}$

21. $\dfrac{\sqrt[3]{270}}{\sqrt[3]{10}}$

22. $\dfrac{\sqrt[3]{40}}{\sqrt[3]{5}}$

23. $\dfrac{\sqrt{40xy^3}}{\sqrt{8x}}$

24. $\dfrac{\sqrt{56ab^3}}{\sqrt{7a}}$

25. $\dfrac{\sqrt[3]{96a^4b^2}}{\sqrt[3]{12a^2b}}$

26. $\dfrac{\sqrt[3]{189x^5y^7}}{\sqrt[3]{7x^2y^2}}$

27. $\dfrac{\sqrt{100ab}}{5\sqrt{2}}$

28. $\dfrac{\sqrt{75ab}}{3\sqrt{3}}$

29. $\dfrac{\sqrt[4]{48x^9y^{13}}}{\sqrt[4]{3xy^{-2}}}$

30. $\dfrac{\sqrt[5]{64a^{11}b^{28}}}{\sqrt[5]{2ab^{-2}}}$

31. $\dfrac{\sqrt[3]{x^3-y^3}}{\sqrt[3]{x-y}}$

32. $\dfrac{\sqrt[3]{r^3+s^3}}{\sqrt[3]{r+s}}$

Hint: Factor and then simplify.

Rationalize each denominator. Assume all variables represent positive numbers.

33. $\sqrt{\dfrac{5}{7}}$

34. $\sqrt{\dfrac{11}{6}}$

35. $\dfrac{6\sqrt{5}}{5\sqrt{3}}$

36. $\dfrac{4\sqrt{5}}{3\sqrt{2}}$

37. $\sqrt[3]{\dfrac{16}{9}}$

38. $\sqrt[3]{\dfrac{2}{9}}$

39. $\dfrac{\sqrt[3]{3a}}{\sqrt[3]{5c}}$

40. $\dfrac{\sqrt[3]{7x}}{\sqrt[3]{3y}}$

41. $\dfrac{\sqrt[3]{5y^4}}{\sqrt[3]{6x^4}}$

42. $\dfrac{\sqrt[3]{3a^4}}{\sqrt[3]{7b^2}}$

43. $\sqrt[3]{\dfrac{2}{x^2y}}$

44. $\sqrt[3]{\dfrac{5}{ab^2}}$

45. $\sqrt{\dfrac{7a}{18}}$

46. $\sqrt{\dfrac{3x}{10}}$

47. $\sqrt{\dfrac{9}{20x^2y}}$

48. $\sqrt{\dfrac{7}{32a^2b}}$ *Aha!* **49.** $\sqrt{\dfrac{10ab^2}{72a^3b}}$

50. $\sqrt{\dfrac{21x^2y}{75xy^5}}$

Rationalize each numerator. Assume all variables represent positive numbers.

51. $\dfrac{\sqrt{5}}{\sqrt{7x}}$

52. $\dfrac{\sqrt{10}}{\sqrt{3x}}$

53. $\sqrt{\dfrac{14}{21}}$

54. $\sqrt{\dfrac{12}{15}}$

55. $\dfrac{4\sqrt{13}}{3\sqrt{7}}$

56. $\dfrac{5\sqrt{21}}{2\sqrt{5}}$

57. $\dfrac{\sqrt[3]{7}}{\sqrt[3]{2}}$

58. $\dfrac{\sqrt[3]{5}}{\sqrt[3]{4}}$

59. $\sqrt{\dfrac{7x}{3y}}$

60. $\sqrt{\dfrac{6a}{5b}}$

61. $\sqrt[3]{\dfrac{2a^5}{5b}}$

62. $\sqrt[3]{\dfrac{2a^4}{7b}}$

63. $\sqrt{\dfrac{x^3y}{2}}$

64. $\sqrt{\dfrac{ab^5}{3}}$

65. Explain why it is easier to approximate

$$\dfrac{\sqrt{2}}{2} \quad \text{than} \quad \dfrac{1}{\sqrt{2}}$$

if no calculator is available and $\sqrt{2} \approx 1.414213562$.

66. A student *incorrectly* claims that

$$\dfrac{5+\sqrt{2}}{\sqrt{18}} = \dfrac{5+\sqrt{1}}{\sqrt{9}} = \dfrac{5+1}{3}.$$

How could you convince the student that a mistake has been made? How would you explain the correct way of rationalizing the denominator?

SKILL MAINTENANCE

Multiply.

67. $\dfrac{3}{x-5} \cdot \dfrac{x-1}{x+5}$

68. $\dfrac{7}{x+4} \cdot \dfrac{x-2}{x-4}$

Simplify.

69. $\dfrac{a^2 - 8a + 7}{a^2 - 49}$

70. $\dfrac{t^2 + 9t - 22}{t^2 - 4}$

71. $(5a^3b^4)^3$

72. $(3x^4)^2(5xy^3)^2$

SYNTHESIS

73. Is it possible to understand how to rationalize a denominator without knowing how to multiply rational expressions? Why or why not?

74. Is the quotient of two irrational numbers always an irrational number? Why or why not?

75. *Pendulums.* The *period* of a pendulum is the time it takes to complete one cycle, swinging to and fro. For a pendulum that is L centimeters long, the period T is given by the formula

$$T = 2\pi\sqrt{\dfrac{L}{980}},$$

where T is in seconds. Find, to the nearest hundredth of a second, the period of a pendulum of length **(a)** 65 cm; **(b)** 98 cm; **(c)** 120 cm. Use a calculator's $\boxed{\pi}$ key if possible.

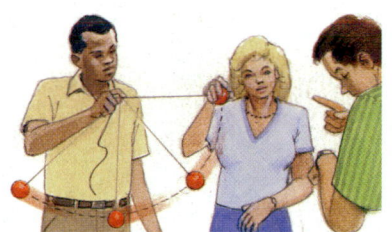

Perform the indicated operations.

76. $\dfrac{7\sqrt{a^2b}\,\sqrt{25xy}}{5\sqrt{a^{-4}b^{-1}}\,\sqrt{49x^{-1}y^{-3}}}$

77. $\dfrac{(\sqrt[3]{81mn^2})^2}{(\sqrt[3]{mn})^2}$

78. $\dfrac{\sqrt{44x^2y^9z}\,\sqrt{22y^9z^6}}{(\sqrt{11xy^8z^2})^2}$

79. $\sqrt{a^2 - 3} - \dfrac{a^2}{\sqrt{a^2 - 3}}$

80. $5\sqrt{\dfrac{x}{y}} + 4\sqrt{\dfrac{y}{x}} - \dfrac{3}{\sqrt{xy}}$

81. Provide a reason for each step in the following derivation of the quotient rule:

$$\sqrt[n]{\dfrac{a}{b}} = \left(\dfrac{a}{b}\right)^{1/n} \qquad \underline{\hspace{3cm}}$$

$$= \dfrac{a^{1/n}}{b^{1/n}} \qquad \underline{\hspace{3cm}}$$

$$= \dfrac{\sqrt[n]{a}}{\sqrt[n]{b}} \qquad \underline{\hspace{3cm}}.$$

82. Show that $\dfrac{\sqrt[n]{a}}{\sqrt[n]{b}}$ is the nth root of $\dfrac{a}{b}$ by raising it to the nth power and simplifying.

83. Let $f(x) = \sqrt{18x^3}$ and $g(x) = \sqrt{2x}$. Find $(f/g)(x)$ and specify the domain of f/g.

84. Let $f(t) = \sqrt{2t}$ and $g(t) = \sqrt{50t^3}$. Find $(f/g)(t)$ and specify the domain of f/g.

85. Let $f(x) = \sqrt{x^2 - 9}$ and $g(x) = \sqrt{x - 3}$. Find $(f/g)(x)$ and specify the domain of f/g.

Expressions Containing Several Radical Terms

10.5

Adding and Subtracting Radical Expressions • Products and Quotients of Two or More Radical Terms • Terms with Differing Indices

Radical expressions like $6\sqrt{7} + 4\sqrt{7}$ or $(\sqrt{a} + \sqrt{b})(\sqrt{a} - \sqrt{b})$ contain more than one *radical term* and can sometimes be simplified.

Adding and Subtracting Radical Expressions

When two radical expressions have the same indices and radicands, they are said to be **like radicals**. Like radicals can be combined (added or subtracted) in much the same way that we combined like terms earlier in this text.

E x a m p l e 1

Simplify by combining like radical terms.

a) $6\sqrt{7} + 4\sqrt{7}$

b) $\sqrt[3]{2} - 7x\sqrt[3]{2} + 5\sqrt[3]{2}$

c) $6\sqrt[5]{4x} + 3\sqrt[5]{4x} - \sqrt[3]{4x}$

Solution

a) $6\sqrt{7} + 4\sqrt{7} = (6 + 4)\sqrt{7}$ Using the distributive law (factoring out $\sqrt{7}$)

$\qquad\qquad\qquad = 10\sqrt{7}$ You can think: 6 square roots of 7 plus 4 square roots of 7 results in 10 square roots of 7.

b) $\sqrt[3]{2} - 7x\sqrt[3]{2} + 5\sqrt[3]{2} = (1 - 7x + 5)\sqrt[3]{2}$ Factoring out $\sqrt[3]{2}$

$\qquad\qquad\qquad\qquad = (6 - 7x)\sqrt[3]{2}$ These parentheses are important!

c) $6\sqrt[5]{4x} + 3\sqrt[5]{4x} - \sqrt[3]{4x} = (6 + 3)\sqrt[5]{4x} - \sqrt[3]{4x}$ Try to do this step mentally.

$\qquad\qquad\qquad\qquad = 9\sqrt[5]{4x} - \sqrt[3]{4x}$ Because the indices differ, we are done.

Our ability to simplify radical expressions can help us to find like radicals even when, at first, it may appear that none exists.

E x a m p l e 2

Simplify by combining like radical terms, if possible.

a) $3\sqrt{8} - 5\sqrt{2}$

b) $9\sqrt{5} - 4\sqrt{3}$

c) $\sqrt[3]{2x^6y^4} + 7\sqrt[3]{2y}$

Solution

a) $3\sqrt{8} - 5\sqrt{2} = 3\sqrt{4 \cdot 2} - 5\sqrt{2}$

$\qquad\qquad = 3\sqrt{4} \cdot \sqrt{2} - 5\sqrt{2}$ Simplifying $\sqrt{8}$

$\qquad\qquad = 3 \cdot 2 \cdot \sqrt{2} - 5\sqrt{2}$

$\qquad\qquad = 6\sqrt{2} - 5\sqrt{2}$

$\qquad\qquad = \sqrt{2}$ Combining like radicals

b) $9\sqrt{5} - 4\sqrt{3}$ cannot be simplified.

c) $\sqrt[3]{2x^6y^4} + 7\sqrt[3]{2y} = \sqrt[3]{x^6y^3 \cdot 2y} + 7\sqrt[3]{2y}$

$\qquad\qquad = \sqrt[3]{x^6y^3} \cdot \sqrt[3]{2y} + 7\sqrt[3]{2y}$ Simplifying $\sqrt[3]{2x^6y^4}$

$\qquad\qquad = x^2y \cdot \sqrt[3]{2y} + 7\sqrt[3]{2y}$

$\qquad\qquad = (x^2y + 7)\sqrt[3]{2y}$ Factoring to combine like radical terms

Products and Quotients of Two or More Radical Terms

Radical expressions often contain factors that have more than one term. The procedure for multiplying out such expressions is similar to finding products of polynomials. Some products will yield like radical terms, which we can now combine.

Example 3

Multiply.

a) $\sqrt{3}\left(x - \sqrt{5}\right)$

b) $\sqrt[3]{y}\left(\sqrt[3]{y^2} + \sqrt[3]{2}\right)$

c) $\left(4\sqrt{3} + \sqrt{2}\right)\left(\sqrt{3} - 5\sqrt{2}\right)$

d) $\left(\sqrt{a} + \sqrt{b}\right)\left(\sqrt{a} - \sqrt{b}\right)$

Solution

a) $\sqrt{3}\left(x - \sqrt{5}\right) = \sqrt{3} \cdot x - \sqrt{3} \cdot \sqrt{5}$ Using the distributive law

$\qquad\qquad\qquad = x\sqrt{3} - \sqrt{15}$ Multiplying radicals

b) $\sqrt[3]{y}\left(\sqrt[3]{y^2} + \sqrt[3]{2}\right) = \sqrt[3]{y} \cdot \sqrt[3]{y^2} + \sqrt[3]{y} \cdot \sqrt[3]{2}$ Using the distributive law

$\qquad\qquad\qquad\quad = \sqrt[3]{y^3} + \sqrt[3]{2y}$ Multiplying radicals

$\qquad\qquad\qquad\quad = y + \sqrt[3]{2y}$ Simplifying $\sqrt[3]{y^3}$

c) $\left(4\sqrt{3} + \sqrt{2}\right)\left(\sqrt{3} - 5\sqrt{2}\right) = \overset{\text{F}}{4(\sqrt{3})^2} - \overset{\text{O}}{20\sqrt{3} \cdot \sqrt{2}} + \overset{\text{I}}{\sqrt{2} \cdot \sqrt{3}} - \overset{\text{L}}{5(\sqrt{2})^2}$

$\qquad\qquad\qquad\qquad\qquad = 4 \cdot 3 - 20\sqrt{6} + \sqrt{6} - 5 \cdot 2$ Multiplying radicals

$\qquad\qquad\qquad\qquad\qquad = 12 - 20\sqrt{6} + \sqrt{6} - 10$

$\qquad\qquad\qquad\qquad\qquad = 2 - 19\sqrt{6}$ Combining like terms

d) $\left(\sqrt{a} + \sqrt{b}\right)\left(\sqrt{a} - \sqrt{b}\right) = (\sqrt{a})^2 - \sqrt{a}\,\sqrt{b} + \sqrt{a}\,\sqrt{b} - (\sqrt{b})^2$ Using FOIL

$\qquad\qquad\qquad\qquad\qquad = a - b$ Combining like terms

In Example 3(d) above, you may have noticed that since the outer and inner products in FOIL are opposites, the result, $a - b$, is not itself a radical expression. Pairs of radical terms, like $\sqrt{a} + \sqrt{b}$ and $\sqrt{a} - \sqrt{b}$, are called **conjugates**. The use of conjugates allows us to rationalize denominators or numerators with two terms.

Example 4

Rationalize each denominator: **(a)** $\dfrac{4}{\sqrt{3} + x}$; **(b)** $\dfrac{4 + \sqrt{2}}{\sqrt{5} - \sqrt{2}}$.

Solution

a) $\dfrac{4}{\sqrt{3}+x} = \dfrac{4}{\sqrt{3}+x} \cdot \dfrac{\sqrt{3}-x}{\sqrt{3}-x}$ Multiplying by 1, using the conjugate of $\sqrt{3}+x$, which is $\sqrt{3}-x$

$\phantom{\dfrac{4}{\sqrt{3}+x}} = \dfrac{4(\sqrt{3}-x)}{(\sqrt{3}+x)(\sqrt{3}-x)}$ Multiplying numerators and denominators

$\phantom{\dfrac{4}{\sqrt{3}+x}} = \dfrac{4(\sqrt{3}-x)}{(\sqrt{3})^2 - x^2}$ Using FOIL in the denominator

$\phantom{\dfrac{4}{\sqrt{3}+x}} = \dfrac{4\sqrt{3}-4x}{3 - x^2}$ Simplifying

b) $\dfrac{4+\sqrt{2}}{\sqrt{5}-\sqrt{2}} = \dfrac{4+\sqrt{2}}{\sqrt{5}-\sqrt{2}} \cdot \dfrac{\sqrt{5}+\sqrt{2}}{\sqrt{5}+\sqrt{2}}$ Multiplying by 1, using the conjugate of $\sqrt{5}-\sqrt{2}$, which is $\sqrt{5}+\sqrt{2}$

$\phantom{\dfrac{4+\sqrt{2}}{\sqrt{5}-\sqrt{2}}} = \dfrac{(4+\sqrt{2})(\sqrt{5}+\sqrt{2})}{(\sqrt{5}-\sqrt{2})(\sqrt{5}+\sqrt{2})}$ Multiplying numerators and denominators

$\phantom{\dfrac{4+\sqrt{2}}{\sqrt{5}-\sqrt{2}}} = \dfrac{4\sqrt{5}+4\sqrt{2}+\sqrt{2}\sqrt{5}+(\sqrt{2})^2}{(\sqrt{5})^2-(\sqrt{2})^2}$ Using FOIL

$\phantom{\dfrac{4+\sqrt{2}}{\sqrt{5}-\sqrt{2}}} = \dfrac{4\sqrt{5}+4\sqrt{2}+\sqrt{10}+2}{5-2}$ Squaring in the denominator and the numerator

$\phantom{\dfrac{4+\sqrt{2}}{\sqrt{5}-\sqrt{2}}} = \dfrac{4\sqrt{5}+4\sqrt{2}+\sqrt{10}+2}{3}$

To rationalize a numerator with more than one term, we use the conjugate of the numerator.

E x a m p l e 5   Rationalize the numerator: $\dfrac{4+\sqrt{2}}{\sqrt{5}-\sqrt{2}}$.

Solution

$\dfrac{4+\sqrt{2}}{\sqrt{5}-\sqrt{2}} = \dfrac{4+\sqrt{2}}{\sqrt{5}-\sqrt{2}} \cdot \dfrac{4-\sqrt{2}}{4-\sqrt{2}}$ Multiplying by 1, using the conjugate of $4+\sqrt{2}$, which is $4-\sqrt{2}$

$\phantom{\dfrac{4+\sqrt{2}}{\sqrt{5}-\sqrt{2}}} = \dfrac{16-(\sqrt{2})^2}{4\sqrt{5}-\sqrt{5}\sqrt{2}-4\sqrt{2}+(\sqrt{2})^2}$

$\phantom{\dfrac{4+\sqrt{2}}{\sqrt{5}-\sqrt{2}}} = \dfrac{14}{4\sqrt{5}-\sqrt{10}-4\sqrt{2}+2}$

Terms with Differing Indices

Sometimes it is necessary to determine products or quotients involving radical terms with indices that differ from each other. When this occurs, we can convert to exponential notation, use the rules for exponents, and then convert back to radical notation.

E x a m p l e 6 Divide and, if possible, simplify: $\dfrac{\sqrt[4]{(x+y)^3}}{\sqrt{x+y}}$.

Solution

$$\frac{\sqrt[4]{(x+y)^3}}{\sqrt{x+y}} = \frac{(x+y)^{3/4}}{(x+y)^{1/2}}$$ Converting to exponential notation

$$= (x+y)^{3/4-1/2}$$ Since the bases are identical, we can subtract exponents: $\frac{3}{4} - \frac{1}{2} = \frac{3}{4} - \frac{2}{4} = \frac{1}{4}$.

$$\left.\begin{array}{l} = (x+y)^{1/4} \\ = \sqrt[4]{x+y} \end{array}\right\}$$ Converting back to radical notation

The steps used in Example 6 can be used in a variety of situations.

> ### To Simplify Products or Quotients with Differing Indices
> 1. Convert all radical expressions to exponential notation.
> 2. When the bases are identical, subtract exponents to divide and add exponents to multiply. This may require finding a common denominator.
> 3. Convert back to radical notation and, if possible, simplify.

E x a m p l e 7 Multiply and simplify: $\sqrt{x^3}\,\sqrt[3]{x}$.

Solution

$$\sqrt{x^3}\,\sqrt[3]{x} = x^{3/2} \cdot x^{1/3}$$ Converting to exponential notation

$$= x^{11/6}$$ Adding exponents: $\frac{3}{2} + \frac{1}{3} = \frac{9}{6} + \frac{2}{6}$

$$= \sqrt[6]{x^{11}}$$ Converting back to radical notation

$$\left.\begin{array}{l} = \sqrt[6]{x^6}\,\sqrt[6]{x^5} \\ = x\sqrt[6]{x^5} \end{array}\right\}$$ Simplifying

E x a m p l e 8 If $f(x) = \sqrt[3]{x^2}$ and $g(x) = \sqrt{x} + \sqrt[4]{x}$, find $(f \cdot g)(x)$.

Solution Recall from Section 7.4 that $(f \cdot g)(x) = f(x) \cdot g(x)$. Thus,

$$(f \cdot g)(x) = \sqrt[3]{x^2}\left(\sqrt{x} + \sqrt[4]{x}\right)$$ x is assumed to be nonnegative.

$$= x^{2/3}(x^{1/2} + x^{1/4})$$ Converting to exponential notation

$$= x^{2/3} \cdot x^{1/2} + x^{2/3} \cdot x^{1/4}$$ Using the distributive law

$$= x^{2/3+1/2} + x^{2/3+1/4}$$ Adding exponents

$$= x^{7/6} + x^{11/12}$$ $\frac{2}{3} + \frac{1}{2} = \frac{4}{6} + \frac{3}{6}; \frac{2}{3} + \frac{1}{4} = \frac{8}{12} + \frac{3}{12}$

$$= \sqrt[6]{x^7} + \sqrt[12]{x^{11}}$$ Converting back to radical notation

$$\left.\begin{array}{l} = \sqrt[6]{x^6}\,\sqrt[6]{x} + \sqrt[12]{x^{11}} \\ = x\sqrt[6]{x} + \sqrt[12]{x^{11}} \end{array}\right\}$$ Simplifying

If factors are raised to powers that share a common denominator, we can write the final result as a single radical expression.

E x a m p l e 9

Divide and, if possible, simplify: $\dfrac{\sqrt[3]{a^2b^4}}{\sqrt{ab}}$.

Solution

$$\frac{\sqrt[3]{a^2b^4}}{\sqrt{ab}} = \frac{(a^2b^4)^{1/3}}{(ab)^{1/2}} \qquad \text{Converting to exponential notation}$$

$$= \frac{a^{2/3}b^{4/3}}{a^{1/2}b^{1/2}} \qquad \text{Using the product and power rules}$$

$$= a^{2/3-1/2}b^{4/3-1/2} \qquad \text{Subtracting exponents}$$

$$= a^{1/6}b^{5/6}$$

$$= \sqrt[6]{a}\,\sqrt[6]{b^5} \qquad \text{Converting to radical notation}$$

$$= \sqrt[6]{ab^5} \qquad \text{Using the product rule for radicals}$$

FOR EXTRA HELP

Exercise Set 10.5

 Digital Video Tutor CD 8 Videotape 20   InterAct Math   Math Tutor Center   MathXL   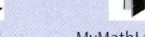 MyMathLab.com

Add or subtract. Simplify by combining like radical terms, if possible. Assume that all variables and radicands represent nonnegative real numbers.

1. $3\sqrt{7} + 2\sqrt{7}$

2. $8\sqrt{5} + 9\sqrt{5}$

3. $9\sqrt[3]{5} - 6\sqrt[3]{5}$

4. $14\sqrt[5]{2} - 6\sqrt[5]{2}$

5. $4\sqrt[3]{y} + 9\sqrt[3]{y}$

6. $9\sqrt[4]{t} - 3\sqrt[4]{t}$

7. $8\sqrt{2} - 6\sqrt{2} + 5\sqrt{2}$

8. $2\sqrt{6} + 8\sqrt{6} - 3\sqrt{6}$

9. $9\sqrt[3]{7} - \sqrt{3} + 4\sqrt[3]{7} + 2\sqrt{3}$

10. $5\sqrt{7} - 8\sqrt[4]{11} + \sqrt{7} + 9\sqrt[4]{11}$

11. $8\sqrt{27} - 3\sqrt{3}$

12. $9\sqrt{50} - 4\sqrt{2}$

13. $3\sqrt{45} + 7\sqrt{20}$

14. $5\sqrt{12} + 16\sqrt{27}$

15. $3\sqrt[3]{16} + \sqrt[3]{54}$

16. $\sqrt[3]{27} - 5\sqrt[3]{8}$

17. $\sqrt{5a} + 2\sqrt{45a^3}$

18. $4\sqrt{3x^3} - \sqrt{12x}$

19. $\sqrt[3]{6x^4} + \sqrt[3]{48x}$

20. $\sqrt[3]{54x} - \sqrt[3]{2x^4}$

21. $\sqrt{4a - 4} + \sqrt{a - 1}$

22. $\sqrt{9y + 27} + \sqrt{y + 3}$

23. $\sqrt{x^3 - x^2} + \sqrt{9x - 9}$

24. $\sqrt{4x - 4} - \sqrt{x^3 - x^2}$

Multiply. Assume all variables represent nonnegative real numbers.

25. $\sqrt{7}(3 - \sqrt{7})$

26. $\sqrt{3}(4 + \sqrt{3})$

27. $4\sqrt{2}(\sqrt{3} - \sqrt{5})$

28. $3\sqrt{5}(\sqrt{5} - \sqrt{2})$

29. $\sqrt{3}(2\sqrt{5} - 3\sqrt{4})$

30. $\sqrt{2}(3\sqrt{10} - 2\sqrt{2})$

31. $\sqrt[3]{2}(\sqrt[3]{4} - 2\sqrt[3]{32})$

32. $\sqrt[3]{3}(\sqrt[3]{9} - 4\sqrt[3]{21})$

33. $\sqrt[3]{a}\left(\sqrt[3]{a^2} + \sqrt[3]{24a^2}\right)$ **34.** $\sqrt[3]{x}\left(\sqrt[3]{3x^2} - \sqrt[3]{81x^2}\right)$

35. $\left(5 + \sqrt{6}\right)\left(5 - \sqrt{6}\right)$ **36.** $\left(2 - \sqrt{5}\right)\left(2 + \sqrt{5}\right)$

37. $\left(3 - 2\sqrt{7}\right)\left(3 + 2\sqrt{7}\right)$

38. $\left(4 + 3\sqrt{2}\right)\left(4 - 3\sqrt{2}\right)$

39. $\left(3 + \sqrt{5}\right)^2$ **40.** $\left(7 + \sqrt{3}\right)^2$

41. $\left(2\sqrt{7} - 4\sqrt{2}\right)\left(3\sqrt{7} + 6\sqrt{2}\right)$

42. $\left(4\sqrt{5} + 3\sqrt{3}\right)\left(3\sqrt{5} - 4\sqrt{3}\right)$

43. $\left(2\sqrt[3]{3} - \sqrt[3]{2}\right)\left(\sqrt[3]{3} + 2\sqrt[3]{2}\right)$

44. $\left(3\sqrt[4]{7} + \sqrt[4]{6}\right)\left(2\sqrt[4]{9} - 3\sqrt[4]{6}\right)$

45. $\left(\sqrt{3x} + \sqrt{y}\right)^2$

46. $\left(\sqrt{t} - \sqrt{2r}\right)^2$

Rationalize each denominator.

47. $\dfrac{2}{3 + \sqrt{5}}$ **48.** $\dfrac{3}{4 - \sqrt{7}}$

49. $\dfrac{2 + \sqrt{5}}{6 - \sqrt{3}}$ **50.** $\dfrac{1 + \sqrt{2}}{3 + \sqrt{5}}$

51. $\dfrac{\sqrt{a}}{\sqrt{a} + \sqrt{b}}$ **52.** $\dfrac{\sqrt{z}}{\sqrt{x} - \sqrt{z}}$

Aha! **53.** $\dfrac{\sqrt{7} - \sqrt{3}}{\sqrt{3} - \sqrt{7}}$ **54.** $\dfrac{\sqrt{7} + \sqrt{5}}{\sqrt{5} + \sqrt{2}}$

55. $\dfrac{3\sqrt{2} - \sqrt{7}}{4\sqrt{2} + \sqrt{5}}$ **56.** $\dfrac{5\sqrt{3} - \sqrt{11}}{2\sqrt{3} - 5\sqrt{2}}$

57. $\dfrac{5\sqrt{3} - 3\sqrt{2}}{3\sqrt{2} - 2\sqrt{3}}$ **58.** $\dfrac{7\sqrt{2} + 4\sqrt{3}}{4\sqrt{3} - 3\sqrt{2}}$

Rationalize each numerator.

59. $\dfrac{\sqrt{7} + 2}{5}$ **60.** $\dfrac{\sqrt{3} + 1}{4}$

61. $\dfrac{\sqrt{6} - 2}{\sqrt{3} + 7}$ **62.** $\dfrac{\sqrt{10} + 4}{\sqrt{2} - 3}$

63. $\dfrac{\sqrt{x} - \sqrt{y}}{\sqrt{x} + \sqrt{y}}$ **64.** $\dfrac{\sqrt{a} + \sqrt{b}}{\sqrt{a} - \sqrt{b}}$

Perform the indicated operation and simplify. Assume all variables represent nonnegative real numbers.

65. $\sqrt{a}\,\sqrt[4]{a^3}$ **66.** $\sqrt[3]{x^2}\,\sqrt[6]{x^5}$

67. $\sqrt[5]{b^2}\,\sqrt{b^3}$ **68.** $\sqrt[4]{a^3}\,\sqrt[3]{a^2}$

69. $\sqrt{xy^3}\,\sqrt[3]{x^2y}$ **70.** $\sqrt[5]{a^3b}\,\sqrt{ab}$

71. $\sqrt[4]{9ab^3}\,\sqrt{3a^4b}$ **72.** $\sqrt{2x^3y^3}\,\sqrt[3]{4xy^2}$

73. $\sqrt[3]{xy^2z}\,\sqrt{x^3yz^2}$ **74.** $\sqrt{a^4b^3c^4}\,\sqrt[3]{ab^2c}$

75. $\dfrac{\sqrt[3]{x^2}}{\sqrt[5]{x}}$ **76.** $\dfrac{\sqrt[3]{a^2}}{\sqrt[4]{a}}$

77. $\dfrac{\sqrt[5]{a^4b}}{\sqrt[3]{ab}}$ **78.** $\dfrac{\sqrt[4]{x^2y^3}}{\sqrt[3]{xy}}$

79. $\dfrac{\sqrt[5]{x^3y^4}}{\sqrt{xy}}$ **80.** $\dfrac{\sqrt{ab^3}}{\sqrt[5]{a^2b^3}}$

81. $\dfrac{\sqrt[3]{(2 + 5x)^2}}{\sqrt[4]{2 + 5x}}$ **82.** $\dfrac{\sqrt[4]{(3x - 1)^3}}{\sqrt[5]{(3x - 1)^3}}$

83. $\dfrac{\sqrt[4]{(5 + 3x)^3}}{\sqrt[3]{(5 + 3x)^2}}$ **84.** $\dfrac{\sqrt[3]{(2x + 1)^2}}{\sqrt[5]{(2x + 1)^2}}$

85. $\sqrt[3]{x^2y}\left(\sqrt{xy} - \sqrt[5]{xy^3}\right)$

86. $\sqrt[4]{a^2b}\left(\sqrt[3]{a^2b} - \sqrt[5]{a^2b^2}\right)$

87. $\left(m + \sqrt[3]{n^2}\right)\left(2m + \sqrt[4]{n}\right)$

88. $\left(r - \sqrt[4]{s^3}\right)\left(3r - \sqrt[5]{s}\right)$

In Exercises 89–92, $f(x)$ and $g(x)$ are as given. Find $(f \cdot g)(x)$. Assume all variables represent nonnegative real numbers.

89. $f(x) = \sqrt[4]{x}$, $g(x) = \sqrt[4]{2x} - \sqrt[4]{x^{11}}$

90. $f(x) = \sqrt[4]{x^7} + \sqrt[4]{3x^2}$, $g(x) = \sqrt[4]{x}$

91. $f(x) = x + \sqrt{7}$, $g(x) = x - \sqrt{7}$

92. $f(x) = x - \sqrt{2}$, $g(x) = x + \sqrt{6}$

Let $f(x) = x^2$. Find each of the following.

93. $f\left(5 - \sqrt{2}\right)$ **94.** $f\left(7 + \sqrt{3}\right)$

95. $f\left(\sqrt{3} + \sqrt{5}\right)$ **96.** $f\left(\sqrt{6} - \sqrt{3}\right)$

97. Why do we need to know how to multiply radical expressions before learning how to add them?

98. In what way(s) is combining like radical terms the same as combining like terms that are monomials?

SKILL MAINTENANCE

Solve.

99. $\dfrac{12x}{x - 4} - \dfrac{3x^2}{x + 4} = \dfrac{384}{x^2 - 16}$

100. $\dfrac{2}{3} + \dfrac{1}{t} = \dfrac{4}{5}$

101. The width of a rectangle is one-fourth the length. The area is twice the perimeter. Find the dimensions of the rectangle.

102. The sum of a number and its square is 20. Find the number.

103. $5x^2 - 6x + 1 = 0$

104. $7t^2 - 8t + 1 = 0$

SYNTHESIS

105. Ramon *incorrectly* writes

$$\sqrt[5]{x^2} \cdot \sqrt{x^3} = x^{2/5} \cdot x^{3/2} = \sqrt[5]{x^3}.$$

What mistake do you suspect he is making?

106. After examining the expression $\sqrt[4]{25xy^3} \sqrt{5x^4y}$ Dyan (correctly) concludes that x and y are both nonnegative. Explain how she could reach this conclusion.

For Exercises 107–110, fill in the blanks by selecting from the following words:

radicands, indices, bases, denominators.

Words can be used more than once.

107. To add radical expressions, the _____ and the _____ must be the same.

108. To multiply radical expressions, the _____ must be the same.

109. To add rational expressions, the _____ must be the same.

110. To find a product by adding exponents, the _____ must be the same.

Find a simplified form for $f(x)$. Assume $x \geq 0$.

111. $f(x) = \sqrt{20x^2 + 4x^3} - 3x\sqrt{45 + 9x} + \sqrt{5x^2 + x^3}$

112. $f(x) = \sqrt{x^3 - x^2} + \sqrt{9x^3 - 9x^2} - \sqrt{4x^3 - 4x^2}$

113. $f(x) = \sqrt[4]{x^5 - x^4} + 3\sqrt[4]{x^9 - x^8}$

114. $f(x) = \sqrt[4]{16x^4 + 16x^5} - 2\sqrt[4]{x^8 + x^9}$

Simplify.

115. $\frac{1}{2}\sqrt{36a^5bc^4} - \frac{1}{2}\sqrt[3]{64a^4bc^6} + \frac{1}{6}\sqrt{144a^3bc^6}$

116. $7x\sqrt{(x+y)^3} - 5xy\sqrt{x+y} - 2y\sqrt{(x+y)^3}$

117. $\sqrt{27a^5(b+1)}\sqrt[3]{81a(b+1)^4}$

118. $\sqrt{8x(y+z)^5}\sqrt[3]{4x^2(y+z)^2}$

119. $\dfrac{\dfrac{1}{\sqrt{w}} - \sqrt{w}}{\dfrac{\sqrt{w}+1}{\sqrt{w}}}$

120. $\dfrac{1}{4+\sqrt{3}} + \dfrac{1}{\sqrt{3}} + \dfrac{1}{\sqrt{3}-4}$

Express each of the following as the product of two radical expressions.

121. $x - 5$ **122.** $y - 7$ **123.** $x - a$

Multiply.

124. $\sqrt{9 + 3\sqrt{5}}\sqrt{9 - 3\sqrt{5}}$

125. $\left(\sqrt{x+2} - \sqrt{x-2}\right)^2$

For Exercises 126–129, assume that all radicands are positive and that no denominator is 0.

Rationalize each denominator.

126. $\dfrac{a - \sqrt{a+b}}{\sqrt{a+b} - b}$ **127.** $\dfrac{b + \sqrt{b}}{1 + b + \sqrt{b}}$

Rationalize each numerator.

128. $\dfrac{\sqrt{y+18} - \sqrt{y}}{18}$ **129.** $\dfrac{\sqrt{x+6} - 5}{\sqrt{x+6} + 5}$

 130. Use a grapher to check your answers to Exercises 19, 33, and 75.

Solving Radical Equations

10.6

The Principle of Powers • Equations with Two Radical Terms

CONNECTING THE CONCEPTS

In Sections 10.1–10.5, we learned how to manipulate radical expressions as well as expressions containing rational exponents. We performed this work to find *equivalent expressions*.

Now that we know how to work with radicals and rational exponents, we can learn how to solve a new type of equation. As in our earlier work with equations, finding *equivalent equations* will be part of our strategy. What is different, however, is that now we will use a step that does not always produce equivalent equations. Checking solutions will therefore be more important than ever.

The Principle of Powers

A **radical equation** is an equation in which the variable appears in a radicand. Examples are

$$\sqrt[3]{2x} + 1 = 5, \quad \sqrt{a} + \sqrt{a - 2} = 7, \quad \text{and} \quad 4 - \sqrt{3x + 1} = \sqrt{6 - x}.$$

To solve such equations, we need a new principle. Suppose an equation $a = b$ is true. If we square both sides, we get another true equation: $a^2 = b^2$. This can be generalized.

The Principle of Powers

If $a = b$, then $a^n = b^n$ for any exponent n.

Note that the principle of powers is an "if–then" statement. The statement obtained by interchanging the two parts of the sentence—"if $a^n = b^n$ for some exponent n, then $a = b$"—is *not always true*. For example, $3^2 = (-3)^2$ *is* true, but $3 = -3$ *is not* true. More generally, $3^n = (-3)^n$ is true for any even number n, whereas $3 = -3$ is false. For this reason, when we raise both sides of an equation to an even power, it will be essential for us to check the answer in the original equation.

E x a m p l e   1

Solve: $\sqrt{x} - 3 = 4$.

Solution

$$\sqrt{x} - 3 = 4$$
$$\sqrt{x} = 7 \qquad \text{Adding 3 to both sides to isolate the radical}$$
$$(\sqrt{x})^2 = 7^2 \qquad \text{Using the principle of powers}$$
$$x = 49$$

Check:
$$\sqrt{x} - 3 = 4$$
$$\overline{\sqrt{49} - 3 \;?\; 4}$$
$$7 - 3 \quad \Big|$$
$$4 \;\Big|\; 4 \qquad \text{TRUE}$$

The solution is 49.

E x a m p l e   2

Solve: $\sqrt{x} - 5 = -7$.

Solution

$$\sqrt{x} - 5 = -7$$
$$\sqrt{x} = -2 \qquad \text{Adding 5 to both sides to isolate the radical}$$

The equation $\sqrt{x} = -2$ has no solution because the principal square root of a number is never negative. We continue as in Example 1 for comparison.

$$(\sqrt{x})^2 = (-2)^2 \qquad \text{Using the principle of powers}$$
$$x = 4$$

Check:
$$\sqrt{x} - 5 = -7$$
$$\overline{\sqrt{4} - 5 \;?\; -7}$$
$$2 - 5 \quad \Big|$$
$$-3 \;\Big|\; -7 \qquad \text{FALSE}$$

The number 4 does not check. Thus the equation $\sqrt{x} - 5 = -7$ has no real-number solution.

> **Caution!** Raising both sides of an equation to an even power may not produce an equivalent equation. In this case, a check is essential.

Note in Example 2 that $x = 4$ has solution 4, but that $\sqrt{x} - 5 = -7$ has *no* solution. Thus the equations $x = 4$ and $\sqrt{x} - 5 = -7$ are *not* equivalent.

To Solve an Equation with a Radical Term

1. Isolate the radical term on one side of the equation.
2. Use the principle of powers and solve the resulting equation.
3. Check any possible solution in the original equation.

E x a m p l e   3

technology connection

To solve Example 3 with a grapher, graph the curves $y_1 = x$ and $y_2 = (x + 7)^{1/2} + 5$ on the same set of axes.

$$y_1 = x, \ y_2 = (x + 7)^{1/2} + 5$$

Using the INTERSECT option of the CALC menu, determine the point of intersection. The intersection should appear to occur when $x = 9$. Note that there is no intersection when $x = 2$, as predicted in the check of Example 3.

1. Use a grapher to solve Examples 1, 2, 4, 5, and 6. Compare your answers with those found using the algebraic methods shown.

Solve: $x = \sqrt{x + 7} + 5$.

Solution

$$x = \sqrt{x + 7} + 5$$

$$x - 5 = \sqrt{x + 7}$$ Subtracting 5 from both sides. This isolates the radical term.

$$\left.\begin{array}{c}(x - 5)^2 = \left(\sqrt{x + 7}\right)^2 \\ x^2 - 10x + 25 = x + 7\end{array}\right\}$$ Using the principle of powers; squaring both sides

$$x^2 - 11x + 18 = 0$$ Adding $-x - 7$ to both sides to write the quadratic equation in standard form

$$(x - 9)(x - 2) = 0$$ Factoring

$$x = 9 \quad or \quad x = 2$$ Using the principle of zero products

The possible solutions are 9 and 2. Let's check.

Check: For 9:

$$\begin{array}{c|c} \multicolumn{2}{c}{x = \sqrt{x + 7} + 5} \\ \hline 9 \ ? \ \sqrt{9 + 7} + 5 & \\ 9 \ | \ 9 & \text{TRUE} \end{array}$$

For 2:

$$\begin{array}{c|c} \multicolumn{2}{c}{x = \sqrt{x + 7} + 5} \\ \hline 2 \ ? \ \sqrt{2 + 7} + 5 & \\ 2 \ | \ 8 & \text{FALSE} \end{array}$$

Since 9 checks but 2 does not, the solution is 9.

It is important to isolate a radical term before using the principle of powers. Suppose in Example 3 that both sides of the equation were squared *before* isolating the radical. We then would have had the expression $\left(\sqrt{x + 7} + 5\right)^2$ or $x + 7 + 10\sqrt{x + 7} + 25$ on the right side, and the radical would have remained in the problem.

E x a m p l e   4

Solve: $(2x + 1)^{1/3} + 5 = 0$.

Solution   We need not use radical notation to solve:

$$(2x + 1)^{1/3} + 5 = 0$$

$$(2x + 1)^{1/3} = -5$$ Subtracting 5 from both sides

$$[(2x + 1)^{1/3}]^3 = (-5)^3$$ Cubing both sides

$$(2x + 1)^1 = (-5)^3$$ Multiplying exponents. Try to do this mentally.

$$2x + 1 = -125$$

$$2x = -126$$ Subtracting 1 from both sides

$$x = -63.$$

Because both sides were raised to an *odd* power, it is not essential that we check the answer. The student can show that -63 checks and is the solution.

Equations with Two Radical Terms

A strategy for solving equations with two or more radical terms is as follows.

To Solve an Equation with Two or More Radical Terms

1. Isolate one of the radical terms.
2. Use the principle of powers.
3. If a radical remains, perform steps (1) and (2) again.
4. Solve the resulting equation.
5. Check possible solutions in the original equation.

Example 5

Solve: $\sqrt{2x-5} = 1 + \sqrt{x-3}$.

Solution

$$\sqrt{2x-5} = 1 + \sqrt{x-3}$$

$$\left(\sqrt{2x-5}\right)^2 = \left(1 + \sqrt{x-3}\right)^2 \qquad \text{One radical is already isolated.}$$
$$\text{We square both sides.}$$

> This is like squaring a binomial. We square 1, then find twice the product of 1 and $\sqrt{x-3}$ and then the square of $\sqrt{x-3}$.

$$2x - 5 = 1 + 2\sqrt{x-3} + \left(\sqrt{x-3}\right)^2$$

$$2x - 5 = 1 + 2\sqrt{x-3} + (x-3)$$

$$x - 3 = 2\sqrt{x-3} \qquad \text{Isolating the remaining radical term}$$

$$(x-3)^2 = \left(2\sqrt{x-3}\right)^2 \qquad \text{Squaring both sides}$$

$$x^2 - 6x + 9 = 4(x-3) \qquad \text{Remember to square both the 2 and the } \sqrt{x-3} \text{ on the right side.}$$

$$x^2 - 6x + 9 = 4x - 12$$

$$x^2 - 10x + 21 = 0$$

$$(x-7)(x-3) = 0 \qquad \text{Factoring}$$

$$x = 7 \quad or \quad x = 3 \qquad \text{Using the principle of zero products}$$

We leave it to the student to show that 7 and 3 both check and are the solutions.

> **Caution!** A common error in solving equations like
>
> $$\sqrt{2x - 5} = 1 + \sqrt{x - 3}$$
>
> is to obtain $1 + (x - 3)$ as the square of the right side. This is wrong because $(A + B)^2 \neq A^2 + B^2$. For example,
>
> $$(1 + 2)^2 \neq 1^2 + 2^2$$
> $$3^2 \neq 1 + 4$$
> $$9 \neq 5.$$

E x a m p l e   6

Let $f(x) = \sqrt{x + 5} - \sqrt{x - 7}$. Find all x-values for which $f(x) = 2$.

Solution   We must have $f(x) = 2$, or

$$\sqrt{x + 5} - \sqrt{x - 7} = 2. \qquad \text{Substituting for } f(x)$$

To solve, we isolate one radical term and square both sides:

$$\sqrt{x + 5} = 2 + \sqrt{x - 7} \qquad \begin{array}{l}\text{Adding } \sqrt{x - 7} \text{ to both sides.}\\ \text{This isolates one of the}\\ \text{radical terms.}\end{array}$$

$$\left(\sqrt{x + 5}\right)^2 = \left(2 + \sqrt{x - 7}\right)^2 \qquad \begin{array}{l}\text{Using the principle of powers}\\ \text{(squaring both sides)}\end{array}$$

$$x + 5 = 4 + 4\sqrt{x - 7} + (x - 7) \qquad \begin{array}{l}\text{Using}\\ (A + B)^2 = A^2 + 2AB + B^2\end{array}$$

$$5 = 4\sqrt{x - 7} - 3 \qquad \begin{array}{l}\text{Adding } -x \text{ to both sides and}\\ \text{combining like terms}\end{array}$$

$$8 = 4\sqrt{x - 7} \qquad \begin{array}{l}\text{Isolating the remaining}\\ \text{radical term}\end{array}$$

$$2 = \sqrt{x - 7}$$

$$2^2 = \left(\sqrt{x - 7}\right)^2 \qquad \text{Squaring both sides}$$

$$4 = x - 7$$

$$11 = x.$$

Check: $f(11) = \sqrt{11 + 5} - \sqrt{11 - 7}$

$\qquad\qquad = \sqrt{16} - \sqrt{4}$

$\qquad\qquad = 4 - 2 = 2.$

We will have $f(x) = 2$ when $x = 11$.

FOR EXTRA HELP

Exercise Set 10.6

Digital Video Tutor CD 8
Videotape 20

InterAct Math

Math Tutor Center

MathXL

MyMathLab.com

Solve.

1. $\sqrt{x+3}=5$

2. $\sqrt{5x+1}=8$

3. $\sqrt{2x-1}=2$

4. $\sqrt{3x+1}=6$

5. $\sqrt{x-2}-7=-4$

6. $\sqrt{y+1}-5=8$

7. $\sqrt{y+4}+6=7$

8. $\sqrt{x-7}+3=10$

9. $\sqrt[3]{x-2}=3$

10. $\sqrt[3]{x+5}=2$

11. $\sqrt[4]{x+3}=2$

12. $\sqrt[4]{y-1}=3$

13. $8\sqrt{y}=y$

14. $3\sqrt{x}=x$

15. $3x^{1/2}+12=9$

16. $2y^{1/2}-7=9$

17. $\sqrt[3]{y}=-4$

18. $\sqrt[3]{x}=-3$

19. $x^{1/4}-2=1$

20. $t^{1/3}-2=3$

Aha! **21.** $(y-3)^{1/2}=-2$

22. $(x+2)^{1/2}=-4$

23. $\sqrt[4]{3x+1}-4=-1$

24. $\sqrt[4]{2x+3}-5=-2$

25. $(x+7)^{1/3}=4$

26. $(y-7)^{1/4}=3$

27. $\sqrt[3]{3y+6}+7=8$

28. $\sqrt[3]{6x+9}+5=2$

29. $\sqrt{3t+4}=\sqrt{4t+3}$

30. $\sqrt{2t-7}=\sqrt{3t-12}$

31. $3(4-t)^{1/4}=6^{1/4}$

32. $2(1-x)^{1/3}=4^{1/3}$

33. $3+\sqrt{5-x}=x$

34. $x=\sqrt{x-1}+3$

35. $\sqrt{4x-3}=2+\sqrt{2x-5}$

36. $3+\sqrt{z-6}=\sqrt{z+9}$

37. $\sqrt{20-x}+8=\sqrt{9-x}+11$

38. $4+\sqrt{10-x}=6+\sqrt{4-x}$

39. $\sqrt{x+2}+\sqrt{3x+4}=2$

40. $\sqrt{6x+7}-\sqrt{3x+3}=1$

41. If $f(x)=\sqrt{x}+\sqrt{x-9}$, find x such that $f(x)=1$.

42. If $g(x)=\sqrt{x}+\sqrt{x-5}$, find x such that $g(x)=5$.

43. If $f(x)=\sqrt{x-2}-\sqrt{4x+1}$, find a such that $f(a)=-3$.

44. If $g(x)=\sqrt{2x+7}-\sqrt{x+15}$, find a such that $g(a)=-1$.

45. If $f(x)=\sqrt{2x-3}$ and $g(x)=\sqrt{x+7}-2$, find x such that $f(x)=g(x)$.

46. If $f(x)=2\sqrt{3x+6}$ and $g(x)=5+\sqrt{4x+9}$, find x such that $f(x)=g(x)$.

47. If $f(t)=4-\sqrt{t-3}$ and $g(t)=(t+5)^{1/2}$, find a such that $f(a)=g(a)$.

48. If $f(t)=7+\sqrt{2t-5}$ and $g(t)=3(t+1)^{1/2}$, find a such that $f(a)=g(a)$.

49. Explain in your own words why it is important to check your answers when using the principle of powers.

50. Describe a procedure that could be used to write radical equations that have no solution.

SKILL MAINTENANCE

51. The base of a triangle is 2 in. longer than the height. The area is $31\frac{1}{2}$ in². Find the height and the base.

52. During a one-hour television show, there were 12 commercials. Some of the commercials were 30 sec long and the others were 60 sec long. If the number of 30-sec commercials was 6 less than the total number of minutes of commercial time during the show, how many 60-sec commercials were used?

53. Elaine can sew a quilt in 6 fewer hours than Gonzalo can. When they work together, it takes them 4 hr. How long would it take each of them alone to sew the quilt?

54. Jackie can paint an apartment in 5.5 hr. Her partner, Grant, can paint the same apartment in 7.5 hr. How long would it take the two of them, working together, to paint the apartment?

Graph.

55. $y>3x+5$

56. $f(x)=\frac{2}{3}x-7$

SYNTHESIS

57. The principle of powers is an "if–then" statement that becomes false when the sentence parts are interchanged. Give an example of another such if–then statement.

58. Is checking essential when the principle of powers is used with an odd power n? Why or why not?

Steel manufacturing.   *In the production of steel and other metals, the temperature of the molten metal is so great that conventional thermometers melt. Instead, sound is transmitted across the surface of the metal to a receiver on the far side and the speed of the sound is measured. The formula*

$$S(t) = 1087.7\sqrt{\dfrac{9t + 2617}{2457}}$$

gives the speed of sound S(t), in feet per second, at a temperature of t degrees Celsius.

59. Find the temperature of a blast furnace where sound travels 1502.3 ft/sec.

60. Find the temperature of a blast furnace where sound travels 1880 ft/sec.

61. Solve the above equation for *t*.

Automotive repair.   *For an engine with a displacement of* 2.8 L, *the function given by*

$$d(n) = 0.75\sqrt{2.8n}$$

can be used to determine the diameter size of the carburetor's opening, in millimeters. Here n is the number of rpm's at which the engine achieves peak performance. (Source: macdizzy.com)

62. If a carburetor's opening is 81 mm, for what number of rpm's will the engine produce peak power?

63. If a carburetor's opening is 84 mm, for what number of rpm's will the engine produce peak power?

Escape velocity.   *A formula for the escape velocity v of a satellite is*

$$v = \sqrt{2gr}\sqrt{\dfrac{h}{r + h}},$$

where g is the force of gravity, r is the planet or star's radius, and h is the height of the satellite above the planet or star's surface.

64. Solve for *h*.   **65.** Solve for *r*.

Sighting to the horizon.   *The function $D(h) = 1.2\sqrt{h}$ can be used to approximate the distance D, in miles, that a person can see to the horizon from a height h, in feet.*

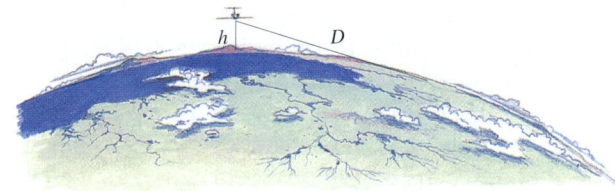

66. How far above sea level must a pilot fly in order to see a horizon that is 180 mi away?

67. How high above sea level must a sailor climb in order to see 10.2 mi out to sea?

Solve.

68. $\dfrac{x + \sqrt{x + 1}}{x - \sqrt{x + 1}} = \dfrac{5}{11}$ **69.** $\left(\dfrac{z}{4} - 5\right)^{2/3} = \dfrac{1}{25}$

70. $(z^2 + 17)^{3/4} = 27$ **71.** $\sqrt{\sqrt{y} + 49} = 7$

72. $x^2 - 5x - \sqrt{x^2 - 5x - 2} = 4$
(*Hint*: Let $u = x^2 - 5x - 2$.)

73. $\sqrt{8 - b} = b\sqrt{8 - b}$

Without graphing, determine the x-intercepts of the graphs given by each of the following.

74. $f(x) = \sqrt{x - 2} - \sqrt{x + 2} + 2$

75. $g(x) = 6x^{1/2} + 6x^{-1/2} - 37$

76. $f(x) = (x^2 + 30x)^{1/2} - x - (5x)^{1/2}$

77. Use a grapher to check your answers to Exercises 4, 10, and 26.

78. Saul is trying to solve Exercise 67 using a grapher. Without resorting to trial and error, how can he determine a suitable viewing window for finding the solution?

79. Use a grapher to check your answers to Exercises 21, 29, and 35.

C O R N E R

Tailgater Alert

Focus: Radical equations and problem solving

Time: 15–25 minutes

Group size: 2–3

Materials: Calculators or square-root tables

The faster a car is traveling, the more distance it needs to stop. Thus it is important for drivers to allow sufficient space between their vehicle and the vehicle in front of them. Police recommend that for each 10 mph of speed, a driver allow 1 car length. Thus a driver going 30 mph should have at least 3 car lengths between his or her vehicle and the one in front.

In Exercise Set 10.3, the function $r(L) = 2\sqrt{5L}$ was used to find the speed, in miles per hour, that a car was traveling when it left skid marks L feet long.

ACTIVITY

1. Each group member should estimate the length of a car in which he or she frequently travels. (Each should use a different length, if possible.)
2. Using a calculator as needed, each group member should complete the table below.

Column 1 gives a car's speed s, column 2 lists the minimum amount of space between cars traveling s miles per hour, as recommended by police. Column 3 is the speed that a vehicle *could* travel were it forced to stop in the distance listed in column 2, using the above function.

Column 1 s (in miles per hour)	Column 2 L(s) (in feet)	Column 3 r(L) (in miles per hour)
20		
30		
40		
50		
60		
70		

3. Determine whether there are any speeds at which the "1 car length per 10 mph" guideline might not suffice. On what reasoning do you base your answer? Compare tables to determine how car length affects the results. What recommendations would your group make to a new driver?

Geometric Applications

10.7

Using the Pythagorean Theorem • Two Special Triangles

Using the Pythagorean Theorem

There are many kinds of problems that involve powers and roots. Many also involve right triangles and the Pythagorean theorem, which we studied in Section 5.8 and restate here.

The Pythagorean Theorem*

In any right triangle, if a and b are the lengths of the legs and c is the length of the hypotenuse, then

$$a^2 + b^2 = c^2.$$

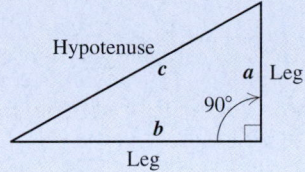

In using the Pythagorean theorem, we often make use of the following principle.

The Principle of Square Roots

For any nonnegative real number n,

If $x^2 = n$, then $x = \sqrt{n}$ or $x = -\sqrt{n}$.

E x a m p l e   1

Baseball. A baseball diamond is actually a square 90 ft on a side. Suppose a catcher fields a ball along the third-base line 10 ft from home plate. How far would the catcher's throw to first base be? Give an exact answer and an approximation to three decimal places.

Solution We first make a drawing and let $d =$ the distance, in feet, to first base. Note that a right triangle is formed in which the length of the leg from home to first base is 90 ft. The length of the leg from home to where the catcher fields the ball is 10 ft.

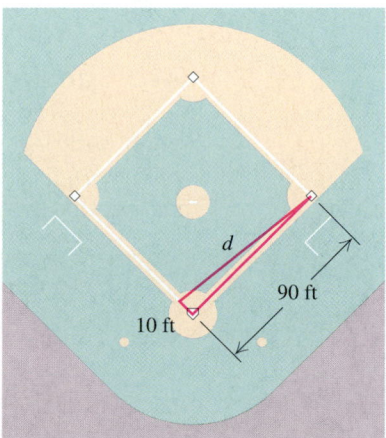

*The converse of the Pythagorean theorem also holds. That is, if a, b, and c are the lengths of the sides of a triangle and $a^2 + b^2 = c^2$, then the triangle is a right triangle.

We substitute these values into the Pythagorean theorem to find d:

$$d^2 = 90^2 + 10^2$$
$$d^2 = 8100 + 100$$
$$d^2 = 8200.$$

We now use the principle of square roots: If $d^2 = 8200$, then $d = \sqrt{8200}$ or $d = -\sqrt{8200}$. In this case, since d is a length, it follows that d is the positive square root of 8200:

$d = \sqrt{8200}$ ft   This is an exact answer.

$d \approx 90.6$ ft.   Using a calculator for an approximation

E x a m p l e   2

Guy wires.   The base of a 40-ft-long guy wire is located 15 ft from the telephone pole that it is anchoring. How high up the pole does the guy wire reach? Give an exact answer and an approximation to three decimal places.

Solution   We make a drawing and let $h =$ the height on the pole that the guy wire reaches. A right triangle is formed in which the length of one leg is 15 ft and the length of the hypotenuse is 40 ft. Using the Pythagorean theorem, we have

$$h^2 + 15^2 = 40^2$$
$$h^2 + 225 = 1600$$
$$h^2 = 1375$$
$$h = \sqrt{1375}.$$

Exact answer:
$$h = \sqrt{1375} \text{ ft}$$

Approximation:
$$h \approx 37.081 \text{ ft}$$ Using a calculator

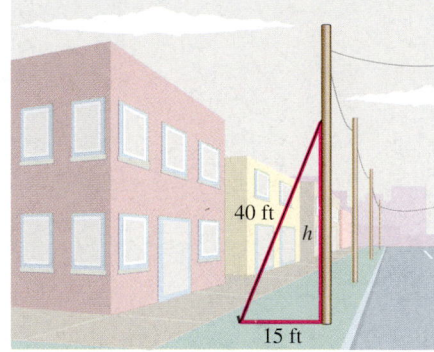

Two Special Triangles

When both legs of a right triangle are the same size, we call the triangle an *isosceles right triangle,* as shown at left. If one leg of an isosceles right triangle has length a, we can find a formula for the length of the hypotenuse as follows:

$$c^2 = a^2 + b^2$$
$c^2 = a^2 + a^2$ Because the triangle is isosceles, both legs are the same size: $a = b$.
$c^2 = 2a^2$.   Combining like terms

Next, we use the principle of square roots. Because a, b, and c are lengths, there is no need to consider negative square roots or absolute values. Thus,

$c = \sqrt{2a^2}$ Using the principle of square roots
$$c = \sqrt{a^2 \cdot 2} = a\sqrt{2}.$$

E x a m p l e   3

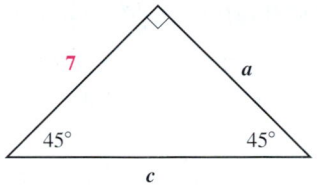

One leg of an isosceles right triangle measures 7 cm. Find the length of the hypotenuse. Give an exact answer and an approximation to three decimal places.

Solution   We substitute:

$$c = a\sqrt{2}$$ This equation is worth memorizing.

$$c = 7\sqrt{2}.$$

Exact answer: $c = 7\sqrt{2}$ cm

Approximation: $c \approx 9.899$ cm   Using a calculator

When the hypotenuse of an isosceles right triangle is known, the lengths of the legs can be found.

E x a m p l e   4

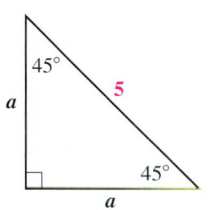

The hypotenuse of an isosceles right triangle is 5 ft long. Find the length of a leg. Give an exact answer and an approximation to three decimal places.

Solution   We replace c with 5 and solve for a:

$$5 = a\sqrt{2}$$ Substituting 5 for c in $c = a\sqrt{2}$

$$\frac{5}{\sqrt{2}} = a$$ Dividing both sides by $\sqrt{2}$

$$\frac{5\sqrt{2}}{2} = a.$$ Rationalize the denominator if desired.

Exact answer: $a = \dfrac{5}{\sqrt{2}}$ ft, or $\dfrac{5\sqrt{2}}{2}$ ft

Approximation: $a \approx 3.536$ ft   Using a calculator

A second special triangle is known as a $30°\text{–}60°\text{–}90°$ right triangle, so named because of the measures of its angles. Note that in an equilateral triangle, all sides have the same length and all angles are $60°$. An altitude, drawn dashed in the figure, bisects, or splits in half, one angle and one side. Two $30°\text{–}60°\text{–}90°$ right triangles are thus formed.

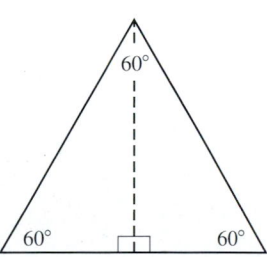

Because of the way in which the altitude is drawn, if a represents the length of the shorter leg in a $30°\text{–}60°\text{–}90°$ right triangle, then $2a$ represents the length of the hypotenuse. We have

$$a^2 + b^2 = (2a)^2$$ Using the Pythagorean theorem

$$a^2 + b^2 = 4a^2$$

$$b^2 = 3a^2$$ Adding $-a^2$ to both sides

$$b = \sqrt{3a^2}$$

$$= \sqrt{a^2 \cdot 3} = a\sqrt{3}.$$

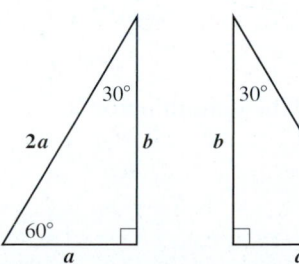

E x a m p l e   5

The shorter leg of a 30°–60°–90° right triangle measures 8 in. Find the lengths of the other sides. Give exact answers and, where appropriate, an approximation to three decimal places.

Solution   The hypotenuse is twice as long as the shorter leg, so we have

$$c = 2a \qquad \text{This relationship is worth memorizing.}$$
$$= 2 \cdot 8 = 16 \text{ in.}$$

The length of the longer leg is the length of the shorter leg times $\sqrt{3}$. This gives us

$$b = a\sqrt{3} \qquad \text{This is also worth memorizing.}$$
$$= 8\sqrt{3} \text{ in.}$$

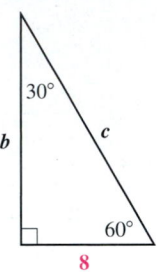

Exact answer: $c = 16$ in., $b = 8\sqrt{3}$ in.

Approximation: $b \approx 13.856$ in.

E x a m p l e   6

The length of the longer leg of a 30°–60°–90° right triangle is 14 cm. Find the length of the hypotenuse. Give an exact answer and an approximation to three decimal places.

Solution   The length of the hypotenuse is twice the length of the shorter leg. We first find a, the length of the shorter leg, by using the length of the longer leg:

$$14 = a\sqrt{3} \qquad \text{Substituting 14 for } b \text{ in } b = a\sqrt{3}$$
$$\frac{14}{\sqrt{3}} = a. \qquad \text{Dividing by } \sqrt{3}$$

Since the hypotenuse is twice as long as the shorter leg, we have

$$c = 2a$$
$$= 2 \cdot \frac{14}{\sqrt{3}} \qquad \text{Substituting}$$
$$= \frac{28}{\sqrt{3}} \text{ cm.}$$

Exact answer: $c = \dfrac{28}{\sqrt{3}}$ cm, or $\dfrac{28\sqrt{3}}{3}$ cm if the denominator is rationalized.

Approximation: $c \approx 16.166$ cm

Lengths Within Isosceles and 30°–60°–90° Right Triangles

The length of the hypotenuse in an isosceles right triangle is the length of a leg times $\sqrt{2}$.

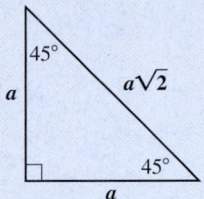

The length of the longer leg in a 30°–60°–90° right triangle is the length of the shorter leg times $\sqrt{3}$. The hypotenuse is twice as long as the shorter leg.

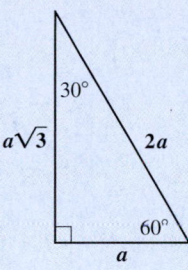

FOR EXTRA HELP

Exercise Set **10.7**

Digital Video Tutor CD 8
Videotape 20

InterAct Math

Math Tutor Center

MathXL

MyMathLab.com

In a right triangle, find the length of the side not given. Give an exact answer and, where appropriate, an approximation to three decimal places.

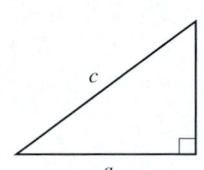

1. $a = 5, b = 3$

2. $a = 8, b = 10$

Aha! **3.** $a = 9, b = 9$

4. $a = 10, b = 10$

5. $b = 12, c = 13$

6. $a = 5, c = 12$

7. $c = 6, a = \sqrt{5}$

8. $c = 8, a = 4\sqrt{3}$

9. $b = 2, c = \sqrt{15}$

10. $a = 1, c = \sqrt{20}$

Aha! **11.** $a = 1, c = \sqrt{2}$

12. $c = 2, a = 1$

In Exercises 13–20, give an exact answer and, where appropriate, an approximation to three decimal places.

13. *Guy wire.*   How long is a guy wire if it reaches from the top of a 15-ft pole to a point on the ground 10 ft from the pole?

14. *Softball.*   A slow-pitch softball diamond is actually a square 65 ft on a side. How far is it from home to second base?

15. *Baseball.* Suppose the catcher in Example 1 makes a throw to second base from the same location. How far is that throw?

16. *Television sets.* What does it mean to refer to a 20-in. TV set or a 25-in. TV set? Such units refer to the diagonal of the screen. A 20-in. TV set has a width of 16 in. What is its height?

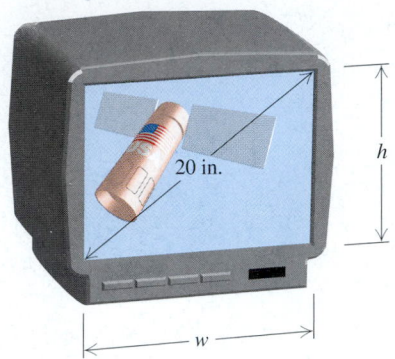

17. *Television sets.* A 25-in. TV set has a screen with a height of 15 in. What is its width? (See Exercise 16.)

18. *Speaker placement.* A stereo receiver is in a corner of a 12-ft by 14-ft room. Speaker wire will run under a rug, diagonally, to a speaker in the far corner. If 4 ft of slack is required on each end, how long a piece of wire should be purchased?

19. *Distance over water.* To determine the width of a pond, a surveyor locates two stakes at either end of the pond and uses instrumentation to place a third stake so that the distance across the pond is the length of a hypotenuse. If the third stake is 90 m from one stake and 70 m from the other, how wide is the pond?

20. *Vegetable garden.* Benito and Dominique are planting a 30-ft by 40-ft vegetable garden and are laying it out using string. They would like to know the length of a diagonal to make sure that right angles are formed. Find the length of a diagonal.

For each triangle, find the missing length(s). Give an exact answer and, where appropriate, an approximation to three decimal places.

21.
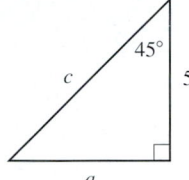

22.
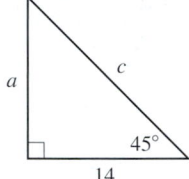

23.
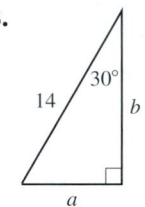

24.
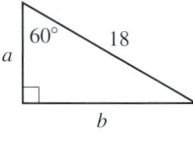

25.
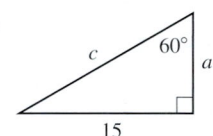

26.
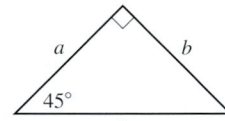

27.
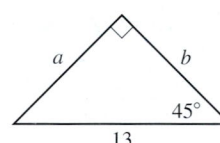

28.
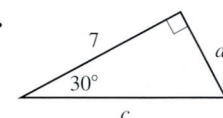

29.
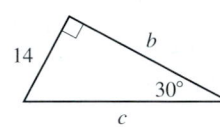

30.
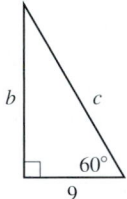

31.
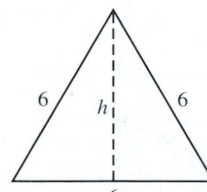

32.
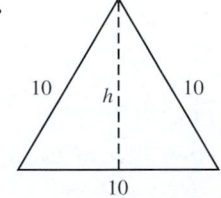

33.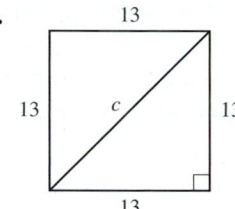

34.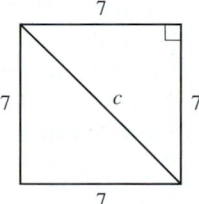

35.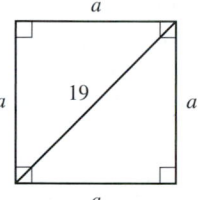

36.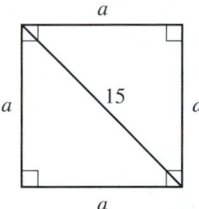

In Exercises 37–42, give an exact answer and, where appropriate, an approximation to three decimal places.

37. *Bridge expansion.*   During the summer heat, a 2-mi bridge expands 2 ft in length. If we assume that the bulge occurs straight up the middle, how high is the bulge? (The answer may surprise you. Most bridges have expansion spaces to avoid such buckling.)

38. Triangle *ABC* has sides of lengths 25 ft, 25 ft, and 30 ft. Triangle *PQR* has sides of lengths 25 ft, 25 ft, and 40 ft. Which triangle has the greater area and by how much?

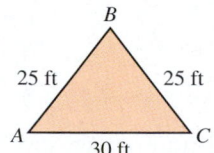 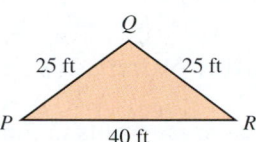

39. *Camping tent.*   The entrance to a pup tent is the shape of an equilateral triangle. If the base of the tent is 4 ft wide, how tall is the tent?

40. Each side of a regular octagon has length *s*. Find a formula for the distance *d* between the parallel sides of the octagon.

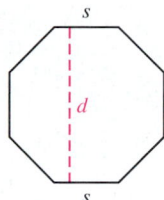

41. The diagonal of a square has length $8\sqrt{2}$ ft. Find the length of a side of the square.

42. The length and the width of a rectangle are given by consecutive integers. The area of the rectangle is 90 cm². Find the length of a diagonal of the rectangle.

43. Find all points on the *y*-axis of a Cartesian coordinate system that are 5 units from the point (3, 0).

44. Find all points on the *x*-axis of a Cartesian coordinate system that are 5 units from the point (0, 4).

45. Write a problem for a classmate to solve in which the solution is: "The height of the tepee is $5\sqrt{3}$ yd."

46. Write a problem for a classmate to solve in which the solution is: "The height of the window is $15\sqrt{3}$ ft."

SKILL MAINTENANCE

Simplify.

47. $47(-1)^{19}$

48. $(-5)(-1)^{13}$

Factor.

49. $x^3 - 9x$

50. $7a^3 - 28a$

Solve.

51. $|3x - 5| = 7$

52. $|2x - 3| = |x + 7|$

SYNTHESIS

53. Are there any right triangles, other than those with sides measuring 3, 4, and 5, that have consecutive numbers for the lengths of the sides? Why or why not?

54. If a 30°–60°–90° triangle and an isosceles right triangle have the same perimeter, which will have the greater area? Why?

55. A cube measures 5 cm on each side. How long is the diagonal that connects two opposite corners of the cube? Give an exact answer.

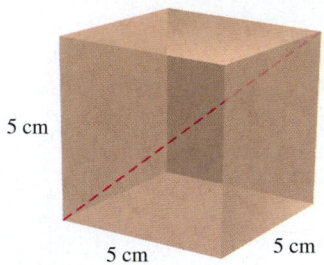

56. *Roofing.* Kit's cottage, which is 24 ft wide and 32 ft long, needs a new roof. By counting clapboards that are 4 in. apart, Kit determines that the peak of the roof is 6 ft higher than the sides. If one packet of shingles covers 100 square feet, how many packets will the job require?

57. *Painting.*   (Refer to Exercise 56.) A gallon of paint covers about 275 square feet. If Kit's first floor is 10 ft high, how many gallons of paint should be bought to paint the house? What assumption(s) is made in your answer?

58. *Contracting.*   Oxford Builders has an extension cord on their generator that permits them to work, with electricity, anywhere in a circular area of 3850 ft². Find the dimensions of the largest square room they could work on without having to relocate the generator to reach each corner of the floor plan.

59. *Contracting.*   Cerrelli Construction has an extension cord on their generator that permits them to work, with electricity, anywhere in a circular area of 6160 ft². Find the dimensions of the largest cube-shaped room they could work on without having to relocate the generator to reach the corners of the ceiling. Assume that the generator sits on the floor.

The Complex Numbers

10.8

Imaginary and Complex Numbers • Addition and Subtraction • Multiplication • Conjugates and Division • Powers of *i*

Imaginary and Complex Numbers

Negative numbers do not have square roots in the real-number system. However, a larger number system that contains the real-number system is designed so that negative numbers *do* have square roots. That system is called the **complex-number system**, and it makes use of a number that is a square root of -1. We call this new number *i*.

> ### The Number i
>
> We define the number i such that $i = \sqrt{-1}$ and $i^2 = -1$.

To express roots of negative numbers in terms of i, we can use the fact that in the complex numbers, $\sqrt{-p} = \sqrt{-1}\,\sqrt{p} = i\sqrt{p}$ or $\sqrt{p}i$, for any positive number p.

E x a m p l e   1   Express in terms of i: **(a)** $\sqrt{-7}$; **(b)** $\sqrt{-16}$; **(c)** $-\sqrt{-13}$; **(d)** $-\sqrt{-50}$.

Solution

a) $\sqrt{-7} = \sqrt{-1 \cdot 7} = \sqrt{-1} \cdot \sqrt{7} = i\sqrt{7}$, or $\sqrt{7}i$ i is *not* under the radical.

b) $\sqrt{-16} = \sqrt{-1 \cdot 16} = \sqrt{-1} \cdot \sqrt{16} = i \cdot 4 = 4i$

c) $-\sqrt{-13} = -\sqrt{-1 \cdot 13} = -\sqrt{-1} \cdot \sqrt{13} = -i\sqrt{13}$, or $-\sqrt{13}i$

d) $-\sqrt{-50} = -\sqrt{-1} \cdot \sqrt{25} \cdot \sqrt{2} = -i \cdot 5 \cdot \sqrt{2} = -5i\sqrt{2}$, or $-5\sqrt{2}i$

> ### Imaginary Numbers
>
> An *imaginary number* is a number that can be written in the form $a + bi$, where a and b are real numbers and $b \neq 0$.

Don't let the name "imaginary" fool you. Imaginary numbers appear in fields such as engineering and the physical sciences. The following are examples of imaginary numbers:

$5 + 4i$,   Here $a = 5, b = 4$.

$\sqrt{5} - \pi i$,   Here $a = \sqrt{5}, b = -\pi$.

$17i$.   Here $a = 0, b = 17$.

When a and b are real numbers and b is allowed to be 0, the number $a + bi$ is said to be **complex**.

> ### Complex Numbers
>
> A *complex number* is any number that can be written in the form $a + bi$, where a and b are real numbers. (Note that a and b both can be 0.)

The following are examples of complex numbers:

$7 + 3i$ (here $a \neq 0, b \neq 0$); $4i$ (here $a = 0, b \neq 0$);

8 (here $a \neq 0, b = 0$); 0 (here $a = 0, b = 0$).

Complex numbers like $17i$ or $4i$, in which $a = 0$ and $b \neq 0$, are imaginary numbers with no real part. Such numbers are called *pure imaginary numbers*.

Note that when $b = 0$, we have $a + 0i = a$, so every real number is a complex number. The relationships among various real and complex numbers are shown below.

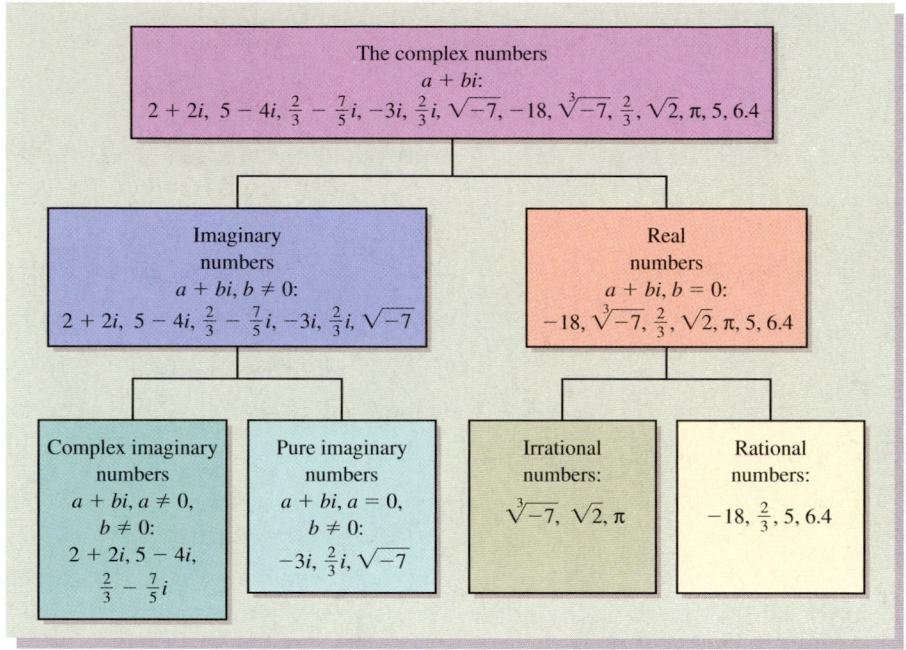

Note that although $\sqrt{-7}$ and $\sqrt[3]{-7}$ are both complex numbers, $\sqrt{-7}$ is imaginary whereas $\sqrt[3]{-7}$ is real.

Addition and Subtraction

The complex numbers obey the commutative, associative, and distributive laws. Thus we can add and subtract them as we do binomials.

E x a m p l e   2

Add or subtract and simplify.

a) $(8 + 6i) + (3 + 2i)$ **b)** $(4 + 5i) - (6 - 3i)$

Solution

a) $(8 + 6i) + (3 + 2i) = (8 + 3) + (6i + 2i)$ Combining the real parts and the imaginary parts

$$= 11 + (6 + 2)i = 11 + 8i$$

b) $(4 + 5i) - (6 - 3i) = (4 - 6) + [5i - (-3i)]$ Note that the 6 and the $-3i$ are both being subtracted.

$$= -2 + 8i$$

Multiplication

For complex numbers, the property $\sqrt{a}\,\sqrt{b} = \sqrt{ab}$ does *not* hold in general, but it does hold when $a = -1$ and b is nonnegative. To multiply square roots of negative real numbers, we first express them in terms of i. For example,

$$\sqrt{-2} \cdot \sqrt{-5} = \sqrt{-1} \cdot \sqrt{2} \cdot \sqrt{-1} \cdot \sqrt{5}$$
$$= i \cdot \sqrt{2} \cdot i \cdot \sqrt{5}$$
$$= i^2\sqrt{10}$$
$$= -1\sqrt{10} = -\sqrt{10} \text{ is correct!}$$

Caution!   With complex numbers, simply multiplying radicands is *incorrect*: $\sqrt{-2} \cdot \sqrt{-5} \ne \sqrt{10}$.

With this in mind, we can now multiply complex numbers.

E x a m p l e   3

Multiply and simplify. When possible, write answers in the form $a + bi$.

a) $\sqrt{-16} \cdot \sqrt{-25}$ **b)** $\sqrt{-5} \cdot \sqrt{-7}$ **c)** $-3i \cdot 8i$

d) $-4i(3 - 5i)$ **e)** $(1 + 2i)(1 + 3i)$

Solution

a) $\sqrt{-16} \cdot \sqrt{-25} = \sqrt{-1} \cdot \sqrt{16} \cdot \sqrt{-1} \cdot \sqrt{25}$
$$= i \cdot 4 \cdot i \cdot 5$$
$$= i^2 \cdot 20$$
$$= -1 \cdot 20 \qquad i^2 = -1$$
$$= -20$$

b) $\sqrt{-5} \cdot \sqrt{-7} = \sqrt{-1} \cdot \sqrt{5} \cdot \sqrt{-1} \cdot \sqrt{7}$ Try to do this step mentally.
$$= i \cdot \sqrt{5} \cdot i \cdot \sqrt{7}$$
$$= i^2 \cdot \sqrt{35}$$
$$= -1 \cdot \sqrt{35} \qquad i^2 = -1$$
$$= -\sqrt{35}$$

c) $-3i \cdot 8i = -24 \cdot i^2$
$$= -24 \cdot (-1) \qquad i^2 = -1$$
$$= 24$$

d) $-4i(3 - 5i) = -4i \cdot 3 + (-4i)(-5i)$ Using the distributive law
$$= -12i + 20i^2$$
$$= -12i - 20 \qquad\qquad i^2 = -1$$
$$= -20 - 12i \qquad\qquad \text{Writing in the form } a + bi$$

e) $(1 + 2i)(1 + 3i) = 1 + 3i + 2i + 6i^2$ Multiplying every term of one number by every term of the other (FOIL)
$$= 1 + 3i + 2i - 6 \qquad i^2 = -1$$
$$= -5 + 5i \qquad\qquad\qquad \text{Combining like terms}$$

Conjugates and Division

Conjugates of complex numbers are defined as follows.

> ### Conjugate of a Complex Number
>
> The *conjugate* of a complex number $a + bi$ is $a - bi$, and the *conjugate* of $a - bi$ is $a + bi$.

E x a m p l e 4

Find the conjugate.

a) $-3 + 7i$

b) $14 - 5i$

c) $4i$

Solution

a) $-3 + 7i$ The conjugate is $-3 - 7i$.

b) $14 - 5i$ The conjugate is $14 + 5i$.

c) $4i$ The conjugate is $-4i$. Note that $4i = 0 + 4i$.

The product of a complex number and its conjugate is a real number.

E x a m p l e 5

Multiply: $(5 + 7i)(5 - 7i)$.

Solution

$$(5 + 7i)(5 - 7i) = 5^2 - (7i)^2 \qquad \text{Using } (A + B)(A - B) = A^2 - B^2$$
$$= 25 - 49i^2$$
$$= 25 - 49(-1) \qquad i^2 = -1$$
$$= 25 + 49 = 74$$

Conjugates are used when dividing complex numbers. The procedure is much like that used to rationalize denominators in Section 10.5.

E x a m p l e 6

Divide and simplify to the form $a + bi$.

a) $\dfrac{-5 + 9i}{1 - 2i}$

b) $\dfrac{7 + 3i}{5i}$

Solution

a) To divide and simplify $(-5 + 9i)/(1 - 2i)$, we multiply by 1, using the conjugate of the denominator to form 1:

$$\frac{-5 + 9i}{1 - 2i} = \frac{-5 + 9i}{1 - 2i} \cdot \frac{1 + 2i}{1 + 2i}$$ Multiplying by 1 using the conjugate of the denominator in the symbol for 1

$$= \frac{(-5 + 9i)(1 + 2i)}{(1 - 2i)(1 + 2i)}$$ Multiplying numerators; multiplying denominators

$$= \frac{-5 - 10i + 9i + 18i^2}{1^2 - 4i^2}$$ Using FOIL

$$= \frac{-5 - i - 18}{1 - 4(-1)}$$ $i^2 = -1$

$$= \frac{-23 - i}{5}$$

$$= -\frac{23}{5} - \frac{1}{5}i$$ Writing in the form $a + bi$

b) When the denominator is a pure imaginary number, it is easiest if we multiply by i/i:

$$\frac{7 + 3i}{5i} = \frac{7 + 3i}{5i} \cdot \frac{i}{i}$$ Multiplying by 1 using i/i. We can also use the conjugate of $5i$ to write $-5i/(-5i)$.

$$= \frac{7i + 3i^2}{5i^2}$$ Multiplying

$$= \frac{7i + 3(-1)}{5(-1)}$$ $i^2 = -1$

$$= \frac{7i - 3}{-5}$$

$$= \frac{-3}{-5} + \frac{7}{-5}i, \text{ or } \frac{3}{5} - \frac{7}{5}i.$$

Powers of *i*

Answers to problems involving complex numbers are generally written in the form $a + bi$. In the following discussion, we show why there is no need to use powers of i (other than 1) when writing answers.

Recall that -1 raised to an *even* power is 1, and -1 raised to an *odd* power is -1. Simplifying powers of i can then be done by using the fact that $i^2 = -1$ and expressing the given power of i in terms of i^2. Consider the following:

$$i, \text{ or } \sqrt{-1},$$
$$i^2 = -1,$$
$$i^3 = i^2 \cdot i = (-1)i = -i,$$
$$i^4 = (i^2)^2 = (-1)^2 = 1,$$
$$i^5 = i^4 \cdot i = (i^2)^2 \cdot i = (-1)^2 \cdot i = i,$$
$$i^6 = (i^2)^3 = (-1)^3 = -1.$$

The pattern is now repeating.

Note that the powers of i cycle themselves through the values i, -1, $-i$, and 1 and that even powers of i are -1 or 1 whereas odd powers of i are i or $-i$.

Example 7   Simplify: **(a)** i^{18}; **(b)** i^{24}; **(c)** i^{29}; **(d)** i^{75}.

Solution

a) $i^{18} = (i^2)^9$ Using the power rule

$\quad = (-1)^9 = -1$ -1 to an odd power is -1

b) $i^{24} = (i^2)^{12}$ Using the power rule

$\quad = (-1)^{12} = 1$ -1 to an even power is 1

c) $i^{29} = i^{28}i^1$ Using the product rule. This is a key step when i is raised to an odd power.

$\quad = (i^2)^{14}i$ Using the power rule

$\quad = (-1)^{14}i$

$\quad = 1 \cdot i = i$

d) $i^{75} = i^{74}i^1$ Using the product rule

$\quad = (i^2)^{37}i$ Using the power rule

$\quad = (-1)^{37}i$

$\quad = -1 \cdot i = -i$

FOR EXTRA HELP

Exercise Set 10.8

 Digital Video Tutor CD 8 Videotape 20   InterAct Math   Math Tutor Center   MathXL   MyMathLab.com

Express in terms of i.

1. $\sqrt{-25}$
2. $\sqrt{-36}$
3. $\sqrt{-13}$

4. $\sqrt{-19}$
5. $\sqrt{-18}$
6. $\sqrt{-98}$

7. $\sqrt{-3}$
8. $\sqrt{-4}$
9. $\sqrt{-81}$

10. $\sqrt{-27}$
11. $\sqrt{-300}$
12. $-\sqrt{-75}$

13. $-\sqrt{-49}$
14. $-\sqrt{-125}$

15. $4 - \sqrt{-60}$
16. $6 - \sqrt{-84}$

17. $\sqrt{-4} + \sqrt{-12}$
18. $-\sqrt{-76} + \sqrt{-125}$

19. $\sqrt{-72} - \sqrt{-25}$
20. $\sqrt{-18} - \sqrt{-100}$

Perform the indicated operation and simplify. Write each answer in the form a + bi.

21. $(7 + 8i) + (5 + 3i)$
22. $(4 - 5i) + (3 + 9i)$

23. $(9 + 8i) - (5 + 3i)$
24. $(9 + 7i) - (2 + 4i)$

25. $(5 - 3i) - (9 + 2i)$
26. $(7 - 4i) - (5 - 3i)$

27. $(-2 + 6i) - (-7 + i)$
28. $(-5 - i) - (7 + 4i)$

29. $6i \cdot 9i$
30. $7i \cdot 6i$

31. $7i \cdot (-8i)$
32. $(-4i)(-6i)$

33. $\sqrt{-49}\sqrt{-25}$
34. $\sqrt{-36}\sqrt{-9}$

35. $\sqrt{-6}\sqrt{-7}$
36. $\sqrt{-5}\sqrt{-2}$

37. $\sqrt{-15}\sqrt{-10}$
38. $\sqrt{-6}\sqrt{-21}$

39. $2i(7 + 3i)$
40. $5i(2 + 6i)$

41. $-4i(6 - 5i)$
42. $-7i(3 - 4i)$

43. $(1 + 5i)(4 + 3i)$
44. $(1 + i)(3 + 2i)$

45. $(5 - 6i)(2 + 5i)$
46. $(6 - 5i)(3 + 4i)$

47. $(-4 + 5i)(3 - 4i)$
48. $(7 - 2i)(2 - 6i)$

49. $(7 - 3i)(4 - 7i)$
50. $(5 - 3i)(4 - 5i)$

51. $(-3 + 6i)(-3 + 4i)$

52. $(-2 + 3i)(-2 + 5i)$

53. $(2 + 9i)(-3 - 5i)$

54. $(-5 - 4i)(3 + 7i)$

55. $(1 - 2i)^2$

56. $(4 - 2i)^2$

57. $(3 + 2i)^2$

58. $(2 + 3i)^2$

59. $(-5 - 2i)^2$

60. $(-2 + 3i)^2$

61. $\dfrac{3}{2 - i}$

62. $\dfrac{4}{3 + i}$

63. $\dfrac{3i}{5 + 2i}$

64. $\dfrac{4i}{5 - 3i}$

65. $\dfrac{7}{9i}$

66. $\dfrac{5}{8i}$

67. $\dfrac{5 - 3i}{4i}$

68. $\dfrac{2 + 7i}{5i}$

Aha! **69.** $\dfrac{7i + 14}{7i}$

70. $\dfrac{6i + 3}{3i}$

71. $\dfrac{4 + 5i}{3 - 7i}$

72. $\dfrac{5 + 3i}{7 - 4i}$

73. $\dfrac{3 - 2i}{4 + 3i}$

74. $\dfrac{5 - 2i}{3 + 6i}$

Simplify.

75. i^7

76. i^{11}

77. i^{24}

78. i^{35}

79. i^{42}

80. i^{64}

81. i^9

82. $(-i)^{71}$

83. $(-i)^6$

84. $(-i)^4$

85. $(5i)^3$

86. $(-3i)^5$

87. $i^2 + i^4$

88. $5i^5 + 4i^3$

89. Is the product of two imaginary numbers always an imaginary number? Why or why not?

90. In what way(s) are conjugates of complex numbers similar to the conjugates used in Section 10.5?

SKILL MAINTENANCE

For Exercises 91–94, let
$$f(x) = x^2 - 3x \quad \text{and} \quad g(x) = 2x - 5.$$

91. Find $(f + g)(-2)$.

92. Find $(f - g)(4)$.

93. Find $(f \cdot g)(5)$.

94. Find $(f/g)(3)$.

Solve.

95. $28 = 3x^2 - 17x$

96. $|3x + 7| < 22$

SYNTHESIS

97. Is the set of real numbers a subset of the complex numbers? Why or why not?

98. Is the union of the set of imaginary numbers and the set of real numbers the set of complex numbers? Why or why not?

A function g is given by
$$g(z) = \frac{z^4 - z^2}{z - 1}.$$

99. Find $g(3i)$.

100. Find $g(1 + i)$.

101. Find $g(5i - 1)$.

102. Find $g(2 - 3i)$.

103. Evaluate
$$\frac{1}{w - w^2} \quad \text{for} \quad w = \frac{1 - i}{10}.$$

Simplify.

104. $\dfrac{i^5 + i^6 + i^7 + i^8}{(1 - i)^4}$

105. $(1 - i)^3(1 + i)^3$

106. $\dfrac{5 - \sqrt{5}i}{\sqrt{5}i}$

107. $\dfrac{6}{1 + \dfrac{3}{i}}$

108. $\left(\dfrac{1}{2} - \dfrac{1}{3}i\right)^2 - \left(\dfrac{1}{2} + \dfrac{1}{3}i\right)^2$

109. $\dfrac{i - i^{38}}{1 + i}$

Summary and Review 10

Key Terms

Square root, p. 596
Principal square root, p. 597
Radical sign, p. 597
Radical expression, p. 597
Radicand, p. 597
Square-root function, p. 598
Cube root, p. 600
nth root, p. 601
Index (plural, indices), p. 601
Even root, p. 601

Rational exponent, p. 604
Perfect square, p. 612
Perfect cube, p. 612
Perfect nth power, p. 612
Rationalizing, p. 619
Radical term, p. 623
Like radicals, p. 624
Conjugates, p. 625
Radical equation, p. 631
Isosceles right triangle, p. 640

$30°$–$60°$–$90°$ right triangle, p. 641
Complex-number system, p. 646
Imaginary number, p. 647
Complex number, p. 647
Pure imaginary number, p. 648
Conjugate of a complex number, p. 650

Important Properties and Formulas

The number c is a square root of a if $c^2 = a$.

The number c is the cube root of a if $c^3 = a$.

For any real number a:

a) $\sqrt[n]{a^n} = |a|$ when n is even. Unless a is known to be nonnegative, absolute-value notation is needed when n is even.

b) $\sqrt[n]{a^n} = a$ when n is odd. Absolute-value notation is not used when n is odd.

$a^{1/n}$ means $\sqrt[n]{a}$. When a is nonnegative, n can be any natural number greater than 1. When a is negative, n must be odd.

For any natural numbers m and n ($n \neq 1$), and any real number a,

$a^{m/n}$ means $(\sqrt[n]{a})^m$ or $\sqrt[n]{a^m}$.

When a is negative, n must be odd.

For any rational number m/n and any nonzero real number a for which $a^{m/n}$ exists,

$a^{-m/n}$ means $\dfrac{1}{a^{m/n}}$.

For any real numbers a and b and any rational exponents m and n for which a^m, a^n, and b^m are defined:

1. $a^m \cdot a^n = a^{m+n}$ In multiplying, add exponents if the bases are the same.

2. $\dfrac{a^m}{a^n} = a^{m-n}$ In dividing, subtract exponents if the bases are the same. (Assume $a \neq 0$.)

3. $(a^m)^n = a^{m \cdot n}$ To raise a power to a power, multiply the exponents.

4. $(ab)^m = a^m b^m$ To raise a product to a power, raise each factor to the power and multiply.

The Product Rule for Radicals

For any real numbers $\sqrt[n]{a}$ and $\sqrt[n]{b}$,

$$\sqrt[n]{a}\,\sqrt[n]{b} = \sqrt[n]{a \cdot b}.$$

The Quotient Rule for Radicals

For any real numbers $\sqrt[n]{a}$ and $\sqrt[n]{b}$, $b \neq 0$,

$$\sqrt[n]{\frac{a}{b}} = \frac{\sqrt[n]{a}}{\sqrt[n]{b}}.$$

Some Ways to Simplify Radical Expressions

1. *Simplifying by factoring.* Factor the radicand and look for factors raised to powers that are divisible by the index.

   Example: $\sqrt[3]{a^6 b} = \sqrt[3]{a^6}\,\sqrt[3]{b} = a^2\sqrt[3]{b}$

2. *Using rational exponents to simplify.* Convert to exponential notation and then use arithmetic and the laws of exponents to simplify the exponents. Then convert back to radical notation as needed.

   Example: $\sqrt[3]{p} \cdot \sqrt[4]{q^3} = p^{1/3} \cdot q^{3/4}$
   $$= p^{4/12} \cdot q^{9/12}$$
   $$= \sqrt[12]{p^4 q^9}$$

3. *Combining like radical terms.*

   Example:
   $$\sqrt{8} + 3\sqrt{2} = \sqrt{4} \cdot \sqrt{2} + 3\sqrt{2}$$
   $$= 2\sqrt{2} + 3\sqrt{2} = 5\sqrt{2}$$

The Principle of Powers

If $a = b$, then $a^n = b^n$ for any exponent n.

To solve an equation with a radical term:

1. Isolate the radical term on one side of the equation.
2. Use the principle of powers and solve the resulting equation.
3. Check any possible solution in the original equation.

To solve an equation with two or more radical terms:

1. Isolate one of the radical terms.
2. Use the principle of powers.
3. If a radical remains, repeat steps (1) and (2).
4. Solve the resulting equation.
5. Check possible solutions in the original equation.

The Pythagorean Theorem

$a^2 + b^2 = c^2$

The Principle of Square Roots

If $x^2 = n$, then $x = \sqrt{n}$ or $x = -\sqrt{n}$.

Special Triangles

The length of the hypotenuse in an isosceles right triangle is the length of a leg times $\sqrt{2}$.

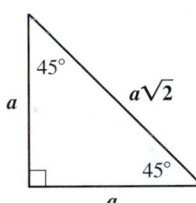

The length of the longer leg in a 30°–60°–90° right triangle is the length of the shorter leg times $\sqrt{3}$. The hypotenuse is twice as long as the shorter leg.

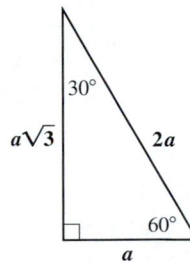

A complex number is any number that can be written in the form $a + bi$, where a and b are real numbers and $i = \sqrt{-1}$.

Review Exercises

Simplify.

1. $\sqrt{\dfrac{49}{36}}$

2. $-\sqrt{0.25}$

Let $f(x) = \sqrt{2x - 7}$. Find the following.

3. $f(16)$

4. The domain of f

Simplify. Assume that each variable can represent any real number.

5. $\sqrt{49a^2}$

6. $\sqrt{(c + 8)^2}$

7. $\sqrt{x^2 - 6x + 9}$

8. $\sqrt{4x^2 + 4x + 1}$

9. $\sqrt[5]{-32}$

10. $\sqrt[3]{-\dfrac{64x^6}{27}}$

11. $\sqrt[4]{x^{12}y^8}$

12. $\sqrt[6]{64x^{12}}$

13. Write an equivalent expression using exponential notation: $\left(\sqrt[3]{5ab}\right)^4$.

14. Write an equivalent expression using radical notation: $(16a^6)^{3/4}$.

Use rational exponents to simplify. Assume $x, y \geq 0$.

15. $\sqrt{x^6 y^{10}}$

16. $\left(\sqrt[6]{x^2 y}\right)^2$

Simplify. Do not use negative exponents in the answers.

17. $(x^{-2/3})^{3/5}$

18. $\dfrac{7^{-1/3}}{7^{-1/2}}$

19. If $f(x) = \sqrt{25(x - 3)^2}$, find a simplified form for $f(x)$.

Perform the indicated operation and, if possible, simplify. Write all answers using radical notation.

20. $\sqrt{5x}\,\sqrt{3y}$

21. $\sqrt[3]{a^5 b}\,\sqrt[3]{27b}$

22. $\sqrt[3]{-24x^{10}y^8}\,\sqrt[3]{18x^7 y^4}$

23. $\dfrac{\sqrt[3]{60xy^3}}{\sqrt[3]{10x}}$

24. $\dfrac{\sqrt{75x}}{2\sqrt{3}}$

25. $\sqrt[4]{\dfrac{48a^{11}}{c^8}}$

26. $5\sqrt[3]{x} + 2\sqrt[3]{x}$

27. $2\sqrt{75} - 7\sqrt{3}$

28. $\sqrt[3]{8x^4} + \sqrt[3]{xy^6}$

29. $\sqrt{50} + 2\sqrt{18} + \sqrt{32}$

30. $\left(\sqrt{5} - 3\sqrt{8}\right)\left(\sqrt{5} + 2\sqrt{8}\right)$

31. $\sqrt[4]{x}\,\sqrt{x}$

32. $\dfrac{\sqrt[3]{x^2}}{\sqrt[4]{x}}$

33. If $f(x) = x^2$, find $f\left(a - \sqrt{2}\right)$.

34. Rationalize the denominator:
$$\dfrac{2\sqrt{3}}{\sqrt{2} + \sqrt{3}}.$$

35. Rationalize the numerator of the expression in Exercise 34.

Solve.

36. $\sqrt{y + 4} - 2 = 3$

37. $(x + 1)^{1/3} = -5$

38. $1 + \sqrt{x} = \sqrt{3x - 3}$

39. If $f(x) = \sqrt[4]{x + 2}$, find a such that $f(a) = 2$.

Solve. Give an exact answer and, where appropriate, an approximation to three decimal places.

40. The diagonal of a square has length 10 cm. Find the length of a side of the square.

41. A bookcase is 5 ft tall and has a 7-ft diagonal brace, as shown. How wide is the bookcase?

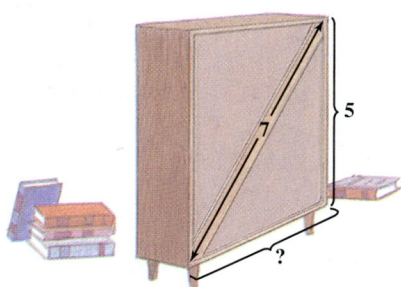

42. Find the missing lengths. Give exact answers and, where appropriate, an approximation to three decimal places.

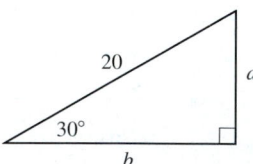

43. Express in terms of i and simplify: $-\sqrt{-8}$.

44. Add: $(-4 + 3i) + (2 - 12i)$.

45. Subtract: $(4 - 7i) - (3 - 8i)$.

Multiply.

46. $(2 + 5i)(2 - 5i)$

47. i^{13}

48. $(6 - 3i)(2 - i)$

49. Divide and simplify to the form $a + bi$:

$$\frac{7 - 2i}{3 + 4i}.$$

SYNTHESIS

50. Explain why $\sqrt[n]{x^n} = |x|$ when n is even, but $\sqrt[n]{x^n} = x$ when n is odd.

51. What is the difference between real numbers and complex numbers?

52. Solve:

$$\sqrt{11x + \sqrt{6 + x}} = 6.$$

53. Simplify:

$$\frac{2}{1 - 3i} - \frac{3}{4 + 2i}.$$

Chapter Test 10

Simplify. Assume that variables can represent any real number.

1. $\sqrt{75}$

2. $\sqrt[3]{-\dfrac{8}{x^6}}$

3. $\sqrt{100a^2}$

4. $\sqrt{x^2 - 8x + 16}$

5. $\sqrt[5]{x^{12}y^8}$

6. $\sqrt{\dfrac{25x^2}{36y^4}}$

7. $\sqrt[3]{2x}\,\sqrt[3]{5y^2}$

8. $\dfrac{\sqrt[5]{x^3y^4}}{\sqrt[5]{xy^2}}$

9. $\sqrt[4]{x^3y^2}\,\sqrt{xy}$

10. $\dfrac{\sqrt[5]{a^2}}{\sqrt[4]{a}}$

11. $7\sqrt{2} - 2\sqrt{2}$

12. $\sqrt{x^4y} + \sqrt{9y^3}$

13. $\left(7 + \sqrt{x}\right)\left(2 - 3\sqrt{x}\right)$

14. Write an equivalent expression using radical notation: $(2a^3b)^{5/6}$.

15. Write an equivalent expression using exponential notation: $\sqrt{7xy}$.

16. If $f(x) = \sqrt{8 - 4x}$, determine the domain of f.

17. If $f(x) = x^2$, find $f\left(5 + \sqrt{2}\right)$.

18. Rationalize the denominator:

$$\frac{\sqrt{3}}{1 + \sqrt{2}}.$$

Solve.

19. $x = \sqrt{2x - 5} + 4$

20. $\sqrt{x} = \sqrt{x + 1} - 5$

Solve. Give exact answers and, where appropriate, approximations to three decimal places.

21. One leg of an isosceles right triangle is 7 cm long. Find the lengths of the other sides.

22. A referee jogs diagonally from one corner of a 50-ft by 90-ft basketball court to the far corner. How far does she jog? Give an exact answer and an approximation to three decimal places.

23. Express in terms of i and simplify: $\sqrt{-50}$.

24. Subtract: $(7 + 8i) - (-3 + 6i)$.

25. Multiply: $\sqrt{-16}\,\sqrt{-36}$.

26. Multiply. Write the answer in the form $a + bi$.

$$(4 - i)^2$$

27. Divide and simplify to the form $a + bi$:

$$\frac{-3 + i}{2 - 7i}.$$

28. Simplify: i^{37}.

SYNTHESIS

29. Solve:

$$\sqrt{2x - 2} + \sqrt{7x + 4} = \sqrt{13x + 10}.$$

30. Simplify:

$$\frac{1 - 4i}{4i(1 + 4i)^{-1}}.$$

11

Quadratic Functions and Equations

AN APPLICATION

The number of pounds of milk per day recommended for a calf that is x weeks old can be approximated by $p(x)$, where $p(x) = -0.2x^2 + 1.3x + 6.2$ (*Source*: C. Chaloux, University of Vermont, 1998). When is the milk consumption of a calf greatest and how much milk does the calf consume at that time?

This problem appears as Example 1 in Section 11.8.

*D*etermining an animal's nutritional requirements and adjusting dosages of medicine require quick calculations and an understanding of the appropriate formulas.

KAREN ANDERSON
Doctor of Veterinary Medicine
Waitsfield, VT

n translating problem situations to mathematics, we often obtain a function or equation containing a second-degree polynomial in one variable. Such functions or equations are said to be quadratic. *In this chapter, we will study a variety of equations, inequalities, and applications for which we will need to solve quadratic equations or graph quadratic functions.*

Quadratic Equations

11.1

The Principle of Square Roots • Completing the Square

In Section 5.7, we solved quadratic equations like $3x^2 = 2 - x$ by factoring. Let's review that procedure.

Example 1   Solve: $3x^2 = 2 - x$.

Solution   To use the principle of zero products, we must first have zero on one side of the equation. We then factor:

$$3x^2 = 2 - x$$

$$3x^2 + x - 2 = 0 \qquad \text{Adding } -2 + x \text{ to both sides to obtain standard form}$$

$$(3x - 2)(x + 1) = 0 \qquad \text{Factoring}$$

$$3x - 2 = 0 \quad or \quad x + 1 = 0 \qquad \text{Using the principle of zero products}$$

$$3x = 2 \quad or \qquad x = -1$$

$$x = \tfrac{2}{3} \quad or \qquad x = -1.$$

Check: For $\tfrac{2}{3}$:

$$\begin{array}{c|c} \multicolumn{2}{c}{3x^2 = 2 - x} \\ \hline 3\left(\tfrac{2}{3}\right)^2 \;?\; 2 - \tfrac{2}{3} & \\ 3 \cdot \tfrac{4}{9} & \tfrac{6}{3} - \tfrac{2}{3} \\ \tfrac{4}{3} & \tfrac{4}{3} \quad \text{TRUE} \end{array}$$

For -1:

$$\begin{array}{c|c} \multicolumn{2}{c}{3x^2 = 2 - x} \\ \hline 3(-1)^2 \;?\; 2 - (-1) & \\ 3 \cdot 1 & 2 + 1 \\ 3 & 3 \quad \text{TRUE} \end{array}$$

The solutions are -1 and $\tfrac{2}{3}$.

Example 2   Solve: $x^2 = 25$.

Solution   We have

$$x^2 = 25$$

$$x^2 - 25 = 0 \qquad \text{Writing in standard form}$$

$$(x - 5)(x + 5) = 0 \qquad \text{Factoring}$$

$$x - 5 = 0 \quad or \quad x + 5 = 0 \qquad \text{Using the principle of zero products}$$

$$x = 5 \quad or \qquad x = -5.$$

The solutions are 5 and -5. The checks are left to the student.

The Principle of Square Roots

Consider the equation $x^2 = 25$ again. We know from Chapter 10 that the number 25 has two real-number square roots, namely, 5 and -5. Note that these are the solutions of the equation in Example 2. Thus square roots can provide a quick method for solving equations of the type $x^2 = k$.

> **The Principle of Square Roots**
>
> For any real number k, if $x^2 = k$, then $x = \sqrt{k}$ or $x = -\sqrt{k}$.

Example 3

Solve: $3x^2 = 6$.

Solution We have

$$3x^2 = 6$$
$$x^2 = 2 \qquad \text{Multiplying by } \tfrac{1}{3}$$
$$x = \sqrt{2} \quad or \quad x = -\sqrt{2}. \qquad \text{Using the principle of square roots}$$

We often use the symbol $\pm\sqrt{2}$ to represent the two numbers $\sqrt{2}$ and $-\sqrt{2}$. We check as follows.

Check: For $\sqrt{2}$:

$$\frac{3x^2 = 6}{3(\sqrt{2})^2 \;?\; 6}$$
$$3 \cdot 2 \;\big|$$
$$6 \;\big|\; 6 \quad \text{TRUE}$$

For $-\sqrt{2}$:

$$\frac{3x^2 = 6}{3(-\sqrt{2})^2 \;?\; 6}$$
$$3 \cdot 2 \;\big|$$
$$6 \;\big|\; 6 \quad \text{TRUE}$$

The solutions are $\sqrt{2}$ and $-\sqrt{2}$, or $\pm\sqrt{2}$.

Sometimes we rationalize denominators to simplify answers, although this is not as common as it once was.

Example 4

Solve: $-5x^2 + 2 = 0$.

Solution We have

$$-5x^2 + 2 = 0$$
$$x^2 = \frac{2}{5} \qquad \text{Isolating } x^2$$
$$x = \sqrt{\frac{2}{5}} \quad or \quad x = -\sqrt{\frac{2}{5}}. \qquad \text{Using the principle of square roots}$$

The solutions are $\sqrt{\dfrac{2}{5}}$ and $-\sqrt{\dfrac{2}{5}}$. This can also be written as $\pm\sqrt{\dfrac{2}{5}}$, or, if we rationalize the denominator, $\pm\dfrac{\sqrt{10}}{5}$. The checks are left to the student.

Sometimes we get solutions that are imaginary numbers.

E x a m p l e   5

Solve: $4x^2 + 9 = 0$.

Solution   We have

$$4x^2 + 9 = 0$$
$$x^2 = -\frac{9}{4} \qquad \text{Isolating } x^2$$
$$x = \sqrt{-\frac{9}{4}} \quad or \quad x = -\sqrt{-\frac{9}{4}} \qquad \text{Using the principle of square roots}$$
$$x = \sqrt{\frac{9}{4}}\sqrt{-1} \quad or \quad x = -\sqrt{\frac{9}{4}}\sqrt{-1}$$
$$x = \frac{3}{2}i \qquad\quad or \quad x = -\frac{3}{2}i.$$

Check: Since the solutions are opposites and the equation has an x^2-term and no x-term, we can check both solutions at once.

$$\frac{4x^2 + 9 = 0}{}$$
$$4\left(\pm\frac{3}{2}i\right)^2 + 9 \ ? \ 0$$
$$4 \cdot \frac{9}{4} \cdot i^2 + 9$$
$$9(-1) + 9$$
$$0 \ \big| \ 0 \ \text{TRUE}$$

The solutions are $\frac{3}{2}i$ and $-\frac{3}{2}i$, or $\pm\frac{3}{2}i$.

The principle of square roots can be restated in a more general form that pertains to more complicated algebraic expressions than just x.

> **The Principle of Square Roots (*Generalized Form*)**
>
> For any real number k and any algebraic expression X,
>
> $$\text{If} \quad X^2 = k, \quad \text{then} \quad X = \sqrt{k} \quad \text{or} \quad X = -\sqrt{k}.$$

E x a m p l e   6

Let $f(x) = (x - 2)^2$. Find all x-values for which $f(x) = 7$.

Solution   We are asked to find all x-values for which

$$f(x) = 7,$$

or

$$(x - 2)^2 = 7. \qquad \text{Substituting } (x - 2)^2 \text{ for } f(x)$$

The generalized principle of square roots gives us

$$x - 2 = \sqrt{7} \quad or \quad x - 2 = -\sqrt{7} \qquad \text{Replacing } X \text{ with } x - 2$$
$$x = 2 + \sqrt{7} \quad or \qquad x = 2 - \sqrt{7}.$$

Check: $f(2 + \sqrt{7}) = (2 + \sqrt{7} - 2)^2 = (\sqrt{7})^2 = 7.$

Similarly,

$$f(2 - \sqrt{7}) = (2 - \sqrt{7} - 2)^2 = (-\sqrt{7})^2 = 7.$$

The solutions are $2 + \sqrt{7}$ and $2 - \sqrt{7}$, or $2 \pm \sqrt{7}$.

In Example 6, one side of the equation is the square of a binomial and the other side is a constant. Sometimes an equation must be factored in order to appear in this form.

Example 7

Solve: $x^2 + 6x + 9 = 2.$

Solution We have

$$x^2 + 6x + 9 = 2 \qquad\qquad \text{The left side is the square of a binomial.}$$

$$(x + 3)^2 = 2 \qquad\qquad \text{Factoring}$$

$$x + 3 = \sqrt{2} \qquad or \quad x + 3 = -\sqrt{2} \qquad \text{Using the principle of square roots}$$

$$x = -3 + \sqrt{2} \quad or \qquad x = -3 - \sqrt{2}. \qquad \text{Adding } -3 \text{ to both sides}$$

The solutions are $-3 + \sqrt{2}$ and $-3 - \sqrt{2}$, or $-3 \pm \sqrt{2}$. The checks are left to the student.

Completing the Square

By using a method called *completing the square*, we can use the principle of square roots to solve *any* quadratic equation.

Example 8

Solve: $x^2 + 6x + 4 = 0.$

Solution We have

$$x^2 + 6x + 4 = 0$$

$$x^2 + 6x \quad = -4 \qquad\qquad \text{Subtracting 4 from both sides}$$

$$x^2 + 6x + 9 = -4 + 9 \qquad\qquad \text{Adding 9 to both sides. We explain this shortly.}$$

$$(x + 3)^2 = 5 \qquad\qquad \text{Factoring the perfect-square trinomial}$$

$$x + 3 = \pm\sqrt{5} \qquad\qquad \text{Using the principle of square roots. Remember that } \pm\sqrt{5} \text{ represents two numbers.}$$

$$x = -3 \pm \sqrt{5}. \qquad\qquad \text{Adding } -3 \text{ to both sides}$$

Check: For $-3 + \sqrt{5}$:

$$x^2 + 6x + 4 = 0$$

$$\frac{(-3 + \sqrt{5})^2 + 6(-3 + \sqrt{5}) + 4 \stackrel{?}{\;\;\;} 0}{}$$

$$9 - 6\sqrt{5} + 5 - 18 + 6\sqrt{5} + 4$$

$$9 + 5 - 18 + 4 - 6\sqrt{5} + 6\sqrt{5}$$

$$0 \mid 0 \;\; \text{TRUE}$$

For $-3 - \sqrt{5}$:

$$x^2 + 6x + 4 = 0$$

$$\frac{(-3 - \sqrt{5})^2 + 6(-3 - \sqrt{5}) + 4 \stackrel{?}{\;\;\;} 0}{}$$

$$9 + 6\sqrt{5} + 5 - 18 - 6\sqrt{5} + 4$$

$$9 + 5 - 18 + 4 + 6\sqrt{5} - 6\sqrt{5}$$

$$0 \mid 0 \;\; \text{TRUE}$$

The solutions are $-3 + \sqrt{5}$ and $-3 - \sqrt{5}$, or $-3 \pm \sqrt{5}$.

Let's examine how the above solutions were found. The decision to add 9 to both sides in Example 8 was made because it creates a perfect-square trinomial on the left side. The 9 was determined by taking half of the coefficient of x and squaring it—that is,

$$\left(\tfrac{1}{2} \cdot 6\right)^2 = 3^2, \quad \text{or} \quad 9.$$

To help see why this procedure works, examine the following drawings.

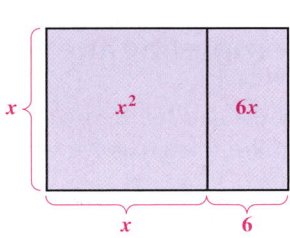
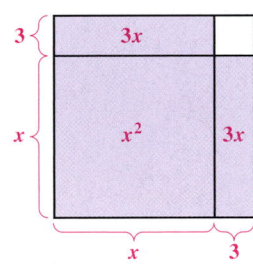

Note that the shaded areas in both figures represent the same area, $x^2 + 6x$. However, only the figure on the right, in which the $6x$ is halved, can be converted into a square with the addition of a constant term. The constant term, 9, can be interpreted as the "missing" piece of the diagram on the right. It *completes* the square.

To complete the square for $x^2 + bx$, we add $(b/2)^2$.

Example 9, which follows, is one of the few examples in this text for which we are neither solving an equation nor writing an equivalent expression. Instead we are simply gaining practice in finding numbers that complete the square. The trinomial that we create is *not* equivalent to the original binomial.

E x a m p l e   9

Complete the square. Then write the trinomial in factored form.

a) $x^2 + 14x$
b) $x^2 - 5x$
c) $x^2 + \frac{3}{4}x$

Solution

a) We take half of the coefficient of x and square it.

$$x^2 + 14x$$

Half of 14 is 7, and $7^2 = 49$. We add 49.

Thus, $x^2 + 14x + 49$ is a perfect-square trinomial. It is equivalent to $(x + 7)^2$. We must add 49 in order for $x^2 + 14x$ to become a perfect-square trinomial.

b) We take half of the coefficient of x and square it:

$$x^2 - 5x$$

$\frac{1}{2} \cdot (-5) = -\frac{5}{2}$,   and $\left(-\frac{5}{2}\right)^2 = \frac{25}{4}$.

Thus, $x^2 - 5x + \frac{25}{4}$ is a perfect-square trinomial. It is equivalent to $\left(x - \frac{5}{2}\right)^2$. Note that for purposes of factoring, it is best *not* to convert $\frac{25}{4}$ to decimal notation.

c) We take half of the coefficient of x and square it:

$$x^2 + \frac{3}{4}x$$

$\frac{1}{2} \cdot \frac{3}{4} = \frac{3}{8}$,   and $\left(\frac{3}{8}\right)^2 = \frac{9}{64}$.

Thus, $x^2 + \frac{3}{4}x + \frac{9}{64}$ is a perfect-square trinomial. It is equivalent to $\left(x + \frac{3}{8}\right)^2$.

We can now use the method of completing the square to solve equations similar to Example 8.

E x a m p l e   1 0

Solve: **(a)** $x^2 - 8x - 7 = 0$; **(b)** $x^2 + 5x - 3 = 0$.

Solution

a) $x^2 - 8x - 7 = 0$

$x^2 - 8x = 7$ Adding 7 to both sides. We can now complete the square on the left side.

$x^2 - 8x + 16 = 7 + 16$ Adding 16 to both sides to complete the square: $\frac{1}{2}(-8) = -4$, and $(-4)^2 = 16$

$(x - 4)^2 = 23$ Factoring

$x - 4 = \pm\sqrt{23}$ Using the principle of square roots

$x = 4 \pm \sqrt{23}$ Adding 4 to both sides

The solutions are $4 - \sqrt{23}$ and $4 + \sqrt{23}$, or $4 \pm \sqrt{23}$. The checks are left to the student.

b) $x^2 + 5x - 3 = 0$

$x^2 + 5x = 3$ — Adding 3 to both sides

$x^2 + 5x + \dfrac{25}{4} = 3 + \dfrac{25}{4}$ — Completing the square: $\frac{1}{2} \cdot 5 = \frac{5}{2}$, and $\left(\frac{5}{2}\right)^2 = \frac{25}{4}$

$\left(x + \dfrac{5}{2}\right)^2 = \dfrac{37}{4}$ — Factoring and simplifying

$x + \dfrac{5}{2} = \pm\dfrac{\sqrt{37}}{2}$ — Using the principle of square roots and the quotient rule for radicals

$x = \dfrac{-5 \pm \sqrt{37}}{2}$ — Adding $-\frac{5}{2}$ to both sides

The checks are left to the student. The solutions are $\left(-5 - \sqrt{37}\right)/2$ and $\left(-5 + \sqrt{37}\right)/2$, or $\left(-5 \pm \sqrt{37}\right)/2$.

Before we can complete the square, the coefficient of x^2 must be 1. When it is not 1, we divide both sides of the equation by whatever that coefficient may be.

Example 11

Solve: $3x^2 + 7x - 2 = 0$.

Solution   We have

$3x^2 + 7x - 2 = 0$

$3x^2 + 7x = 2$ — Adding 2 to both sides

$x^2 + \dfrac{7}{3}x = \dfrac{2}{3}$ — Dividing both sides by 3

$x^2 + \dfrac{7}{3}x + \dfrac{49}{36} = \dfrac{2}{3} + \dfrac{49}{36}$ — Completing the square: $\left(\frac{1}{2} \cdot \frac{7}{3}\right)^2 = \frac{49}{36}$

$\left(x + \dfrac{7}{6}\right)^2 = \dfrac{73}{36}$ — Factoring and simplifying

$x + \dfrac{7}{6} = \pm\dfrac{\sqrt{73}}{6}$ — Using the principle of square roots and the quotient rule for radicals

$x = \dfrac{-7 \pm \sqrt{73}}{6}.$ — Adding $-\frac{7}{6}$ to both sides

The solutions are

$\dfrac{-7 - \sqrt{73}}{6}$ and $\dfrac{-7 + \sqrt{73}}{6}$,   or $\dfrac{-7 \pm \sqrt{73}}{6}.$

The checks are left to the student.

The procedure used in Example 11 is important because it can be used to solve *any* quadratic equation.

> ### To Solve a Quadratic Equation in x by Completing the Square
>
> 1. Isolate the terms with variables on one side of the equation, and arrange them in descending order.
> 2. Divide both sides by the coefficient of x^2 if that coefficient is not 1.
> 3. Complete the square by taking half of the coefficient of x and adding its square to both sides.
> 4. Express one side as the square of a binomial and simplify the other side.
> 5. Use the principle of square roots.
> 6. Solve for x by adding or subtracting on both sides.

Problem Solving

If you put money in a savings account, the bank will pay you interest. As interest is paid into your account, the bank will start paying you interest on both the original amount and the interest already earned. This is called **compounding interest**. If interest is paid yearly, we say that it is **compounded annually**.

> ### The Compound-Interest Formula
>
> If an amount of money P is invested at interest rate r, compounded annually, then in t years, it will grow to the amount A given by
>
> $$A = P(1 + r)^t.$$

We can use quadratic equations to solve certain interest problems.

Example 12

Investment growth.   Rosa invested $4000 at interest rate r, compounded annually. In 2 yr, it grew to $4410. What was the interest rate?

Solution

1. **Familiarize.**   We are already familiar with the compound-interest formula. If we were not, we would need to consult an outside source.

2. **Translate.**   The translation consists of substituting into the formula:

$$A = P(1 + r)^t$$

$4410 = 4000(1 + r)^2.$ Substituting

3. **Carry out.**   We solve for r:

$$4410 = 4000(1 + r)^2$$

$$\frac{4410}{4000} = (1 + r)^2 \qquad \text{Dividing both sides by 4000}$$

$$\frac{441}{400} = (1 + r)^2 \qquad \text{Simplifying}$$

$$\pm\sqrt{\frac{441}{400}} = 1 + r \qquad \text{Using the principle of square roots}$$

$$\pm\frac{21}{20} = 1 + r \qquad \text{Simplifying}$$

$$-\frac{20}{20} \pm \frac{21}{20} = r \qquad \text{Adding } -1, \text{ or } -\frac{20}{20}, \text{ to both sides}$$

$$\frac{1}{20} = r \quad or \quad -\frac{41}{20} = r.$$

4. **Check.**   Since the interest rate cannot be negative, we need only check $\frac{1}{20}$, or 5%. If $4000 were invested at 5% interest, compounded annually, then in 2 yr it would grow to $4000(1.05)^2$, or $4410. The number 5% checks.

5. **State.**   The interest rate was 5%.

Example 13   ***Free-falling objects.***   The formula $s = 16t^2$ is used to approximate the distance s, in feet, that an object falls freely from rest in t seconds. The RCA Building in New York City is 850 ft tall. How long will it take an object to fall from the top?

Solution

1. **Familiarize.**   We make a drawing to help visualize the problem.

2. **Translate.**   We substitute into the formula:

$$s = 16t^2$$
$$850 = 16t^2.$$

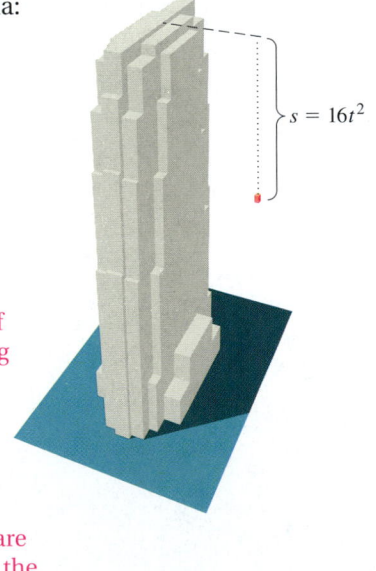

$$s = 16t^2$$

3. **Carry out.**   We solve for t:

$$850 = 16t^2$$

$$\frac{850}{16} = t^2$$

$$53.125 = t^2$$

$$\sqrt{53.125} = t \qquad \text{Using the principle of square roots; rejecting the negative square root since } t \text{ cannot be negative in this problem}$$

$$7.3 \approx t. \qquad \text{Using a calculator to approximate the square root and rounding to the nearest tenth}$$

4. **Check.**   Since $16(7.3)^2 = 852.64 \approx 850$, our answer checks.

5. **State.**   It takes about 7.3 sec for an object to fall freely from the top of the RCA Building.

technology connection B

As we saw in Section 5.7, a grapher can be used to find approximate solutions of any quadratic equation that has real-number solutions.

To check Example 10(a), we graph $y = x^2 - 8x - 7$ and use the ZERO or ROOT option of the CALC menu. When asked for a Left and Right Bound, we enter cursor positions to the left of and to the right of the root. A Guess between the bounds is entered and a value for the root then appears.

1. Use a grapher to check the second solution of Example 10(a).
2. Use a grapher to confirm the solutions in Examples 8 and 10(b).
3. Can a grapher be used to find *exact* solutions in Example 11? Why or why not?

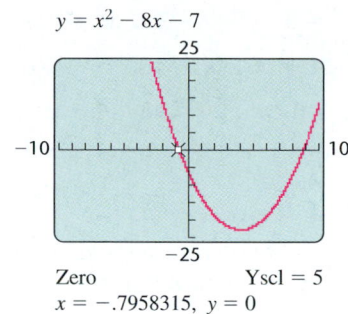

$y = x^2 - 8x - 7$

Zero   Yscl = 5
$x = -.7958315, \ y = 0$

4. Use a grapher to confirm that there are no real-number solutions in Example 5.

Exercise Set 11.1

FOR EXTRA HELP

 Digital Video Tutor CD 8 Videotape 21   InterAct Math   Math Tutor Center    MathXL   MyMathLab.com

Solve.

1. $7x^2 = 21$
2. $4x^2 = 20$
3. $25x^2 + 4 = 0$
4. $9x^2 + 16 = 0$
5. $3t^2 - 2 = 0$
6. $5t^2 - 7 = 0$
7. $(x + 2)^2 = 25$
8. $(x - 1)^2 = 49$
9. $(a + 5)^2 = 8$
10. $(a - 13)^2 = 18$
11. $(x - 1)^2 = -49$
12. $(x + 1)^2 = -9$
13. $\left(t + \frac{3}{2}\right)^2 = \frac{7}{2}$
14. $\left(y + \frac{3}{4}\right)^2 = \frac{17}{16}$
15. $x^2 - 6x + 9 = 100$
16. $x^2 - 10x + 25 = 64$
17. Let $f(x) = (x - 5)^2$. Find x such that $f(x) = 16$.
18. Let $g(x) = (x - 2)^2$. Find x such that $g(x) = 25$.
19. Let $F(t) = (t + 4)^2$. Find t such that $F(t) = 13$.
20. Let $f(t) = (t + 6)^2$. Find t such that $f(t) = 15$.

Aha! 21. Let $g(x) = x^2 + 14x + 49$. Find x such that $g(x) = 49$.

22. Let $F(x) = x^2 + 8x + 16$. Find x such that $F(x) = 9$.

Complete the square. Then write the perfect-square trinomial in factored form.

23. $x^2 + 8x$
24. $x^2 + 16x$
25. $x^2 - 6x$
26. $x^2 - 10x$
27. $x^2 - 24x$
28. $x^2 - 18x$
29. $t^2 + 9t$
30. $t^2 + 3t$
31. $x^2 - 3x$
32. $x^2 - 7x$
33. $x^2 + \frac{2}{3}x$
34. $x^2 + \frac{2}{5}x$
35. $t^2 - \frac{5}{3}t$
36. $t^2 - \frac{5}{6}t$
37. $x^2 + \frac{9}{5}x$
38. $x^2 + \frac{9}{4}x$

Solve by completing the square. Show your work.

39. $x^2 + 6x = 7$ **40.** $x^2 + 8x = 9$

41. $x^2 - 10x = 22$ **42.** $x^2 - 4x = -9$

43. $x^2 + 8x + 7 = 0$ **44.** $x^2 + 10x + 9 = 0$

45. $x^2 - 10x + 21 = 0$ **46.** $x^2 - 10x + 24 = 0$

47. $t^2 + 5t + 3 = 0$ **48.** $t^2 + 6t + 7 = 0$

49. $x^2 + 10 = 6x$ **50.** $x^2 + 23 = 10x$

51. $s^2 + 4s + 13 = 0$ **52.** $t^2 + 12t + 25 = 0$

Solve by completing the square. Remember to first divide, as in Example 11, to make sure that the coefficient of x^2 is 1.

53. $2x^2 - 5x - 3 = 0$ **54.** $3x^2 + 5x - 2 = 0$

55. $4x^2 + 8x + 3 = 0$ **56.** $9x^2 + 18x + 8 = 0$

57. $6x^2 - x = 15$ **58.** $6x^2 - x = 2$

59. $2x^2 + 4x + 1 = 0$ **60.** $2x^2 + 5x + 2 = 0$

61. $3x^2 - 5x - 3 = 0$ **62.** $4x^2 - 6x - 1 = 0$

Interest. *Use $A = P(1 + r)^t$ to find the interest rate in Exercises 63–68. Refer to Example 12.*

63. $2000 grows to $2420 in 2 yr

64. $2560 grows to $2890 in 2 yr

65. $1280 grows to $1805 in 2 yr

66. $1000 grows to $1440 in 2 yr

67. $6250 grows to $6760 in 2 yr

68. $6250 grows to $7290 in 2 yr

Free-falling objects. *Use $s = 16t^2$ for Exercises 69–72. Refer to Example 13.*

69. The CN Tower in Toronto, at 1815 ft, is the world's tallest self-supporting tower (no guy wires) (*Source: The Guinness Book of Records*). How long would it take an object to fall freely from the top?

70. Reaching 745 ft above the water, the towers of California's Golden Gate Bridge are the world's tallest bridge towers (*Source: The Guinness Book of Records*). How long would it take an object to fall freely from the top?

71. The Gateway Arch in St. Louis is 640 ft high. How long would it take an object to fall freely from the top?

72. The Sears Tower in Chicago is 1454 ft tall. How long would it take an object to fall freely from the top?

73. Explain in your own words a sequence of steps that can be used to solve any quadratic equation in the quickest way.

74. Write an interest-rate problem for a classmate to solve. Devise the problem so that the solution is "The loan was made at 7% interest."

SKILL MAINTENANCE

Evaluate.

75. $at^2 - bt$, for $a = 3$, $b = 5$, and $t = 4$

76. $mn^2 - mp$, for $m = -2$, $n = 7$, and $p = 3$

Simplify.

77. $\sqrt[3]{270}$ **78.** $\sqrt{80}$

Let $f(x) = \sqrt{3x - 5}$.

79. Find $f(10)$.   **80.** Find $f(18)$.

SYNTHESIS

81. What would be better: to receive 3% interest every 6 months, or to receive 6% interest every 12 months? Why?

82. Write a problem involving a free-falling object for a classmate to solve (see Example 13). Devise the problem so that the solution is "The object takes about 4.5 sec to fall freely from the top of the structure."

Find b such that each trinomial is a square.

83. $x^2 + bx + 81$ **84.** $x^2 + bx + 49$

85. If $f(x) = 2x^5 - 9x^4 - 66x^3 + 45x^2 + 280x$ and $x^2 - 5$ is a factor of $f(x)$, find all a for which $f(a) = 0$.

86. If $f(x) = \left(x - \frac{1}{3}\right)(x^2 + 6)$ and $g(x) = \left(x - \frac{1}{3}\right)\left(x^2 - \frac{2}{3}\right)$, find all a for which $(f + g)(a) = 0$.

87. *Boating.* A barge and a fishing boat leave a dock at the same time, traveling at a right angle to each other. The barge travels 7 km/h slower than the fishing boat. After 4 hr, the boats are 68 km apart. Find the speed of each boat.

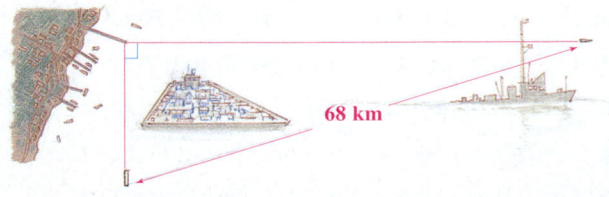

68 km

88. Find three consecutive integers such that the square of the first plus the product of the other two is 67.

89. Exercises 17, 21, and 41 can be solved on a grapher without first rewriting in standard form. Simply let y_1 represent the left side of the equation and y_2 the right side. Then use a grapher to determine the x-coordinate of any point of intersection. Use a grapher to solve Exercises 17, 21, and 41 in this manner.

90. Use a grapher to check your answers to Exercises 5, 45, 59, and 61.

91. Example 12 can be solved with a grapher by graphing each side of

$$4410 = 4000(1 + r)^2.$$

How could you determine, from a reading of the problem, a suitable viewing window? What might that window be?

The Quadratic Formula

11.2

Solving Using the Quadratic Formula • Approximating Solutions

There are at least two reasons for learning to complete the square. One is to enhance your ability to graph certain equations that appear later in this and other chapters. Another is to develop a general formula for solving quadratic equations.

Solving Using the Quadratic Formula

Each time you solve by completing the square, the procedure is the same. In mathematics, when a procedure is repeated many times, a formula is often developed to speed up our work.

We begin with a quadratic equation in standard form,

$$ax^2 + bx + c = 0,$$

with $a > 0$. For $a < 0$, a slightly different derivation is needed (see Exercise 52), but the result is the same. Let's solve by completing the square. As the steps are performed, compare them with Example 11 on p. 666.

$$ax^2 + bx = -c \qquad \text{Adding } -c \text{ to both sides}$$

$$x^2 + \frac{b}{a}x = -\frac{c}{a} \qquad \text{Dividing both sides by } a$$

Half of $\frac{b}{a}$ is $\frac{b}{2a}$ and $\left(\frac{b}{2a}\right)^2$ is $\frac{b^2}{4a^2}$. We add $\frac{b^2}{4a^2}$ to both sides:

$$x^2 + \frac{b}{a}x + \frac{b^2}{4a^2} = -\frac{c}{a} + \frac{b^2}{4a^2}$$ Adding $\frac{b^2}{4a^2}$ to complete the square

$$\left(x + \frac{b}{2a}\right)^2 = -\frac{4ac}{4a^2} + \frac{b^2}{4a^2}$$

Factoring on the left side; finding a common denominator on the right side

$$\left(x + \frac{b}{2a}\right)^2 = \frac{b^2 - 4ac}{4a^2}$$

$$x + \frac{b}{2a} = \pm\frac{\sqrt{b^2 - 4ac}}{2a}$$

Using the principle of square roots and the quotient rule for radicals; since $a > 0$, $\sqrt{4a^2} = 2a$

$$x = \frac{-b \pm \sqrt{b^2 - 4ac}}{2a}.$$ Adding $-\frac{b}{2a}$ to both sides

It is important that you remember the quadratic formula and know how to use it.

The Quadratic Formula

The solutions of $ax^2 + bx + c = 0$, $a \neq 0$, are given by

$$x = \frac{-b \pm \sqrt{b^2 - 4ac}}{2a}.$$

Example 1

Solve $5x^2 + 8x = -3$ using the quadratic formula.

Solution We first find standard form and determine a, b, and c:

$$5x^2 + 8x + 3 = 0;$$ Adding 3 to both sides to get 0 on one side

$$a = 5, \quad b = 8, \quad c = 3.$$

Next, we use the quadratic formula:

$$x = \frac{-b \pm \sqrt{b^2 - 4ac}}{2a}$$

$$x = \frac{-8 \pm \sqrt{8^2 - 4 \cdot 5 \cdot 3}}{2 \cdot 5}$$ Substituting

$$x = \frac{-8 \pm \sqrt{64 - 60}}{10}$$ Be sure to write the fraction bar all the way across.

$$x = \frac{-8 \pm \sqrt{4}}{10} = \frac{-8 \pm 2}{10}$$

$$x = \frac{-8 + 2}{10} \quad or \quad x = \frac{-8 - 2}{10}$$

$$x = \frac{-6}{10} \qquad or \quad x = \frac{-10}{10}$$

$$x = -\frac{3}{5} \qquad or \quad x = -1.$$

The solutions are $-\frac{3}{5}$ and -1. The checks are left to the student.

Because $5x^2 + 8x + 3$ can be factored as $(5x + 3)(x + 1)$, the quadratic formula may not have been the fastest way of solving Example 1. However, because the quadratic formula works for *any* quadratic equation, we need not spend too much time struggling to solve a quadratic equation by factoring.

To Solve a Quadratic Equation

1. If the equation can be easily written in the form $ax^2 = p$ or $(x + k)^2 = d$, use the principle of square roots as in Section 11.1.
2. If step (1) does not apply, write the equation in the form $ax^2 + bx + c = 0$.
3. Try factoring and using the principle of zero products.
4. If factoring seems to be difficult or impossible, use the quadratic formula.

The solutions of a quadratic equation can always be found using the quadratic formula. They cannot always be found by factoring.

Recall that a second-degree polynomial in one variable is said to be quadratic. Similarly, a second-degree polynomial function in one variable is said to be a **quadratic function**.

Example 2   For the quadratic function given by $f(x) = 5x^2 - 8x - 3$, find all x for which $f(x) = 0$.

Solution   We substitute and solve for x:

$$f(x) = 0$$
$$5x^2 - 8x - 3 = 0 \qquad \text{Substituting. This cannot be solved by factoring.}$$
$$a = 5, \quad b = -8, \quad c = -3.$$

We then substitute into the quadratic formula:

$$x = \frac{-(-8) \pm \sqrt{(-8)^2 - 4 \cdot 5 \cdot (-3)}}{2 \cdot 5}$$

$$= \frac{8 \pm \sqrt{64 + 60}}{10}$$

$$= \frac{8 \pm \sqrt{124}}{10}. \qquad \text{Note that 4 is a perfect-square factor of 124.}$$

technology connection

A

On many graphers, it is possible to check Example 2 by graphing $y_1 = 5x^2 - 8x - 3$, pressing TRACE , and entering $(4 + \sqrt{31})/5$. A rational approximation for x and the y-value 0 appear.

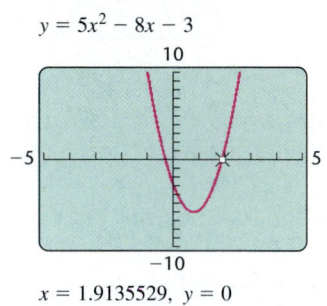

$y = 5x^2 - 8x - 3$

$x = 1.9135529, \ y = 0$

Use this approach to check the other solution of Example 2.

Thus,

$$x = \frac{8 \pm \sqrt{4 \cdot 31}}{10} \qquad \textcolor{red}{124 = 4 \cdot 31}$$

$$= \frac{8 \pm 2\sqrt{31}}{10} \qquad \textcolor{red}{\sqrt{4} = 2}$$

$$= \frac{2(4 \pm \sqrt{31})}{2 \cdot 5} = \frac{4 \pm \sqrt{31}}{5}. \qquad \textcolor{red}{\text{Removing a factor equal to 1: } \tfrac{2}{2} = 1}$$

> **Caution!** To avoid a common error, *factor the numerator and the denominator* when removing a factor equal to 1.

The solutions are

$$\frac{4 + \sqrt{31}}{5} \quad \text{and} \quad \frac{4 - \sqrt{31}}{5}.$$

The checks are left to the student.

Some quadratic equations have solutions that are imaginary numbers.

Example 3

Solve: $x^2 + 2 = -x$.

Solution We first find standard form:

$$x^2 + x + 2 = 0. \qquad \textcolor{red}{\text{Adding } x \text{ to both sides}}$$

Since we cannot factor $x^2 + x + 2$, we use the quadratic formula with $a = 1$, $b = 1$, and $c = 2$:

$$x = \frac{-1 \pm \sqrt{1^2 - 4 \cdot 1 \cdot 2}}{2 \cdot 1} \qquad \textcolor{red}{\text{Substituting}}$$

$$= \frac{-1 \pm \sqrt{1 - 8}}{2}$$

$$= \frac{-1 \pm \sqrt{-7}}{2}$$

$$= \frac{-1 \pm i\sqrt{7}}{2}, \text{ or } -\frac{1}{2} \pm \frac{\sqrt{7}}{2}i.$$

The solutions are $-\dfrac{1}{2} - \dfrac{\sqrt{7}}{2}i$ and $-\dfrac{1}{2} + \dfrac{\sqrt{7}}{2}i$. The checks are left to the student.

The quadratic formula is sometimes used to solve equations that do not originally appear to be quadratic.

E x a m p l e   4

If $f(x) = 2 + \dfrac{7}{x}$ and $g(x) = \dfrac{4}{x^2}$, find all x for which $f(x) = g(x)$.

Solution   We set $f(x)$ equal to $g(x)$ and solve:

$$f(x) = g(x)$$

$$2 + \frac{7}{x} = \frac{4}{x^2}. \qquad \text{Substituting. Note that } x \neq 0.$$

This is a rational equation similar to those in Section 6.6. To solve, we multiply both sides by the LCD, x^2:

$$x^2\left(2 + \frac{7}{x}\right) = x^2 \cdot \frac{4}{x^2}$$

$$2x^2 + 7x = 4 \qquad \text{Simplifying}$$

$$2x^2 + 7x - 4 = 0. \qquad \text{Subtracting 4 from both sides}$$

We have

$$a = 2, \quad b = 7, \quad \text{and} \quad c = -4.$$

Substituting then gives us

$$x = \frac{-7 \pm \sqrt{7^2 - 4 \cdot 2 \cdot (-4)}}{2 \cdot 2}$$

$$= \frac{-7 \pm \sqrt{49 + 32}}{4}$$

$$= \frac{-7 \pm \sqrt{81}}{4}$$

$$= \frac{-7 \pm 9}{4}$$

$$x = \frac{-7 + 9}{4} = \frac{1}{2} \quad or \quad x = \frac{-7 - 9}{4} = -4. \qquad \begin{array}{l}\text{Both answers should} \\ \text{check since } x \neq 0.\end{array}$$

You can confirm that $f\left(\tfrac{1}{2}\right) = g\left(\tfrac{1}{2}\right)$ and $f(-4) = g(-4)$. The solutions are $\tfrac{1}{2}$ and -4.

technology connection

B

We saw in Sections 5.7 and 11.1 how graphers can solve quadratic equations. To determine whether quadratic equations are solved more quickly on a grapher or by using the quadratic formula, solve Examples 2 and 4 both ways. Which method is faster? Which method is more precise? Why?

Checking the solutions of Examples 2 and 3 can be cumbersome. Fortunately, when the quadratic formula is used to solve a quadratic equation, the results will always check in that equation provided the formula has been properly used. Thus checking for computational errors is usually sufficient.

Approximating Solutions

When the solution of an equation is irrational, a rational-number approximation is often useful. This is often the case in real-world applications similar to those found in Section 11.3.

E x a m p l e   5   Use a calculator to approximate the solutions of Example 2.

Solution   On most calculators, one of the following sequences of keystrokes can be used to approximate $(4 + \sqrt{31})/5$:

$$\boxed{(}\; \boxed{4} \; \boxed{+} \; \boxed{\sqrt{}} \; 31 \; \boxed{)} \; \boxed{)} \; \boxed{\div} \; 5 \; \boxed{\text{ENTER}}; \quad \text{or}$$

$$31 \; \boxed{\sqrt{}} \; \boxed{+} \; \boxed{4} \; \boxed{=} \; \boxed{\div} \; \boxed{5} \; \boxed{=}.$$

Similar keystrokes can be used to approximate $(4 - \sqrt{31})/5$.

The solutions of Example 2 are approximately 1.913552873 and -0.3135528726.

FOR EXTRA HELP

Exercise Set 11.2

Digital Video Tutor CD 8
Videotape 21

InterAct Math

Math Tutor Center

MathXL

MyMathLab.com

Solve.

1. $x^2 + 7x - 3 = 0$

2. $x^2 - 7x + 4 = 0$

3. $3p^2 = 18p - 6$

4. $3u^2 = 8u - 5$

5. $x^2 + x + 2 = 0$

6. $x^2 + x + 1 = 0$

7. $x^2 + 13 = 4x$

8. $x^2 + 13 = 6x$

9. $h^2 + 4 = 6h$

10. $r^2 + 3r = 8$

11. $3 + \dfrac{8}{x} = \dfrac{1}{x^2}$

12. $2 + \dfrac{5}{x^2} = \dfrac{9}{x}$

13. $3x + x(x - 2) = 4$

14. $4x + x(x - 3) = 5$

15. $12t^2 + 9t = 1$

16. $15t^2 + 7t = 2$

17. $25x^2 - 20x + 4 = 0$

18. $36x^2 + 84x + 49 = 0$

19. $7x(x + 2) + 5 = 3x(x + 1)$

20. $5x(x - 1) - 7 = 4x(x - 2)$

21. $14(x - 4) - (x + 2) = (x + 2)(x - 4)$

22. $11(x - 2) + (x - 5) = (x + 2)(x - 6)$

23. $5x^2 = 13x + 17$

24. $25x = 3x^2 + 28$

25. $x^2 + 9 = 4x$

26. $x^2 + 7 = 3x$

27. $x^3 - 8 = 0$ (*Hint*: Factor the difference of cubes. Then use the quadratic formula.)

28. $x^3 + 1 = 0$

29. Let $f(x) = 3x^2 - 5x - 1$. Find x such that $f(x) = 0$.

30. Let $g(x) = 4x^2 - 2x - 3$. Find x such that $g(x) = 0$.

31. Let

$$f(x) = \frac{7}{x} + \frac{7}{x + 4}.$$

Find all x for which $f(x) = 1$.

32. Let

$$g(x) = \frac{2}{x} + \frac{2}{x + 3}.$$

Find all x for which $g(x) = 1$.

33. Let

$$F(x) = \frac{x + 3}{x} \quad \text{and} \quad G(x) = \frac{x - 4}{3}.$$

Find all x for which $F(x) = G(x)$.

34. Let

$$f(x) = \frac{3 - x}{4} \quad \text{and} \quad g(x) = \frac{1}{4x}.$$

Find all x for which $f(x) = g(x)$.

35. Let

$$f(x) = \frac{15 - 2x}{6} \quad \text{and} \quad g(x) = \frac{3}{x}.$$

Find all x for which $f(x) = g(x)$.

36. Let
$$f(x) = x + 5 \quad \text{and} \quad g(x) = \frac{3}{x - 5}.$$
Find all x for which $f(x) = g(x)$.

Solve. Use a calculator to approximate solutions as rational numbers.

37. $x^2 + 4x - 7 = 0$

38. $x^2 + 6x + 4 = 0$

39. $x^2 - 6x + 4 = 0$

40. $x^2 - 4x + 1 = 0$

41. $2x^2 - 3x - 7 = 0$

42. $3x^2 - 3x - 2 = 0$

43. Are there any equations that can be solved by the quadratic formula but not by completing the square? Why or why not?

44. The list on p. 673 does not mention completing the square as a method of solving quadratic equations. Why not?

SKILL MAINTENANCE

45. *Coffee beans.*   Twin Cities Roasters has Kenyan coffee for which they pay $6.75 a pound and Peruvian coffee for which they pay $11.25 a pound. How much of each kind should be mixed in order to obtain a 50-lb mixture that is worth $8.55 a pound?

46. *Donuts.*   South Street Bakers charges $1.10 for a cream-filled donut and 85¢ for a glazed donut. On a recent Sunday, a total of 90 glazed and cream-filled donuts were sold for $88.00. How many of each type were sold?

Simplify.

47. $\sqrt{27a^2b^5} \cdot \sqrt{6a^3b}$

48. $\sqrt{8a^3b} \cdot \sqrt{12ab^5}$

49. $\dfrac{\dfrac{3}{x - 1}}{\dfrac{1}{x + 1} + \dfrac{2}{x - 1}}$

50. $\dfrac{\dfrac{4}{a^2b}}{\dfrac{3}{a} - \dfrac{4}{b^2}}$

SYNTHESIS

51. Suppose you had a large number of quadratic equations to solve and none of the equations had a constant term. Would you use factoring or the quadratic formula to solve these equations? Why?

52. If $a < 0$ and $ax^2 + bx + c = 0$, then $-a$ is positive and the equivalent equation, $-ax^2 - bx - c = 0$, can be solved using the quadratic formula.
   a) Find this solution, replacing a, b, and c in the formula with $-a$, $-b$, and $-c$ from the equation.
   b) How does the result of part (a) indicate that the quadratic formula "works" regardless of the sign of a?

For Exercises 53–55, let
$$f(x) = \frac{x^2}{x - 2} + 1 \quad and \quad g(x) = \frac{4x - 2}{x - 2} + \frac{x + 4}{2}.$$

53. Find the x-intercepts of the graph of f.

54. Find the x-intercepts of the graph of g.

55. Find all x for which $f(x) = g(x)$.

Solve.

56. $x^2 - 0.75x - 0.5 = 0$

57. $z^2 + 0.84z - 0.4 = 0$

58. $\left(1 + \sqrt{3}\right)x^2 - \left(3 + 2\sqrt{3}\right)x + 3 = 0$

59. $\sqrt{2}x^2 + 5x + \sqrt{2} = 0$

60. $ix^2 - 2x + 1 = 0$

61. One solution of $kx^2 + 3x - k = 0$ is -2. Find the other.

62. Use a grapher to solve Exercises 3, 17, and 37.

63. Use a grapher to solve Exercises 9, 25, and 33. Use the method of graphing each side of the equation.

64. Can a grapher be used to solve *any* quadratic equation? Why or why not?

Applications Involving Quadratic Equations

11.3

Solving Problems • Solving Formulas

Solving Problems

As we found in Section 6.7, some problems translate to rational equations. The solution of such rational equations can involve quadratic equations.

Example 1

Motorcycle travel.   Makita rode her motorcycle 300 mi at a certain average speed. Had she averaged 10 mph more, the trip would have taken 1 hr less. Find the average speed of the motorcycle.

Solution

1. **Familiarize.**   We make a drawing, labeling it with the known and unknown information. As in Section 6.7, we can organize the information in a table. We let r represent the rate, in miles per hour, and t the time, in hours, for Makita's trip.

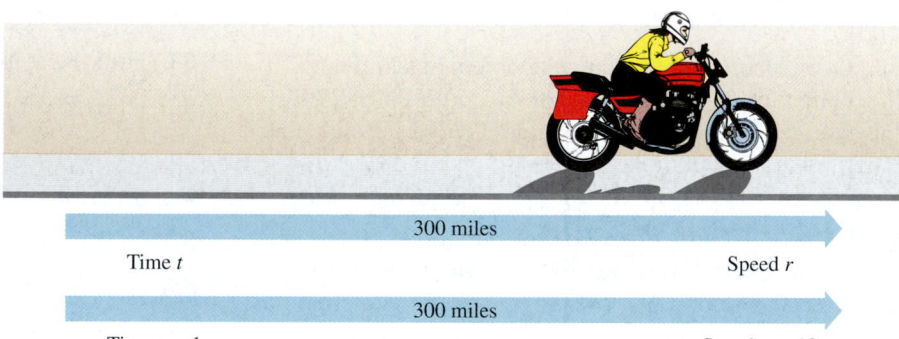

Distance	Speed	Time
300	r	t
300	$r + 10$	$t - 1$

$\longrightarrow r = \dfrac{300}{t}$

$\longrightarrow r + 10 = \dfrac{300}{t - 1}$

Recall that the definition of speed, $r = d/t$, relates the three quantities.

2. **Translate.**   From the first two lines of the table, we obtain

$$r = \frac{300}{t} \quad \text{and} \quad r + 10 = \frac{300}{t - 1}.$$

3. **Carry out.**   A system of equations has been formed. We substitute for r from the first equation into the second and solve the resulting equation:

$$\frac{300}{t} + 10 = \frac{300}{t-1} \qquad \text{Substituting } 300/t \text{ for } r$$

$$t(t-1) \cdot \left[\frac{300}{t} + 10\right] = t(t-1) \cdot \frac{300}{t-1} \qquad \begin{array}{l}\text{Multiplying by the} \\ \text{LCD}\end{array}$$

$$\cancel{t}(t-1) \cdot \frac{300}{\cancel{t}} + t(t-1) \cdot 10 = t\cancel{(t-1)} \cdot \frac{300}{\cancel{t-1}} \qquad \begin{array}{l}\text{Using the distributive} \\ \text{law and removing} \\ \text{factors that equal 1:} \\ \dfrac{t}{t} = 1; \dfrac{t-1}{t-1} = 1\end{array}$$

$$\left.\begin{array}{l}300(t-1) + 10(t^2 - t) = 300t \\ 300t - 300 + 10t^2 - 10t = 300t \\ 10t^2 - 10t - 300 = 0\end{array}\right\} \qquad \begin{array}{l}\text{Rewriting in} \\ \text{standard form}\end{array}$$

$$t^2 - t - 30 = 0 \qquad \begin{array}{l}\text{Multiplying by } \frac{1}{10} \text{ or} \\ \text{dividing by 10}\end{array}$$

$$(t-6)(t+5) = 0 \qquad \text{Factoring}$$

$$t = 6 \quad or \quad t = -5. \qquad \begin{array}{l}\text{Principle of zero} \\ \text{products}\end{array}$$

4. **Check.**   Note that we have solved for t, not r as required. Since negative time has no meaning here, we disregard the -5 and use 6 hr to find r:

$$r = \frac{300 \text{ mi}}{6 \text{ hr}} = 50 \text{ mph}.$$

> **_Caution!_**   Always make sure that you find the quantity asked for in the problem.

To see if 50 mph checks, we increase the speed 10 mph to 60 mph and see how long the trip would have taken at that speed:

$$t = \frac{d}{r} = \frac{300 \text{ mi}}{60 \text{ mph}} = 5 \text{ hr}. \qquad \text{Note that mi/mph} = \text{mi} \div \frac{\text{mi}}{\text{hr}} =$$

$$\cancel{\text{mi}} \cdot \frac{\text{hr}}{\cancel{\text{mi}}} = \text{hr}.$$

This is 1 hr less than the trip actually took, so the answer checks.

5. **State.**   Makita's motorcycle traveled at an average speed of 50 mph.

Solving Formulas

Recall that to solve a formula for a certain letter, we use the principles for solving equations to get that letter alone on one side.

E x a m p l e   2

Period of a pendulum.   The time T required for a pendulum of length l to swing back and forth (complete one period) is given by the formula $T = 2\pi\sqrt{l/g}$, where g is the earth's gravitational constant. Solve for l.

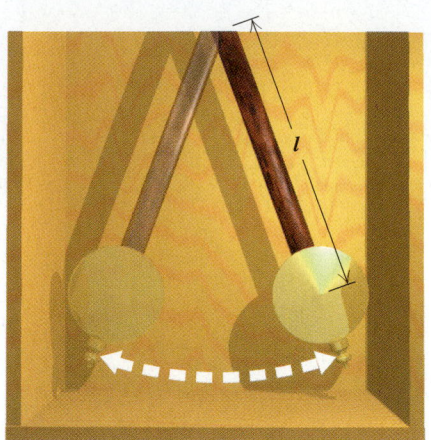

Solution   We have

$$T = 2\pi\sqrt{\dfrac{l}{g}} \qquad \text{This is a radical equation (see Section 10.6).}$$

$$T^2 = \left(2\pi\sqrt{\dfrac{l}{g}}\right)^2 \qquad \text{Principle of powers (squaring both sides)}$$

$$T^2 = 2^2\pi^2\dfrac{l}{g}$$

$$gT^2 = 4\pi^2 l \qquad \text{Multiplying both sides by } g \text{ to clear fractions}$$

$$\dfrac{gT^2}{4\pi^2} = l. \qquad \text{Dividing both sides by } 4\pi^2$$

We now have l alone on one side and l does not appear on the other side, so the formula is solved for l.

In most formulas, variables represent nonnegative numbers, so we do not need to use absolute-value signs when taking square roots.

E x a m p l e   3

Hang time.*   An athlete's *hang time* is the amount of time that the athlete can remain airborne when jumping. A formula relating an athlete's vertical leap V, in inches, to hang time T, in seconds, is $V = 48T^2$. Solve for T.

---

*This formula is taken from an article by Peter Brancazio, "The Mechanics of a Slam Dunk," *Popular Mechanics*, November 1991. Courtesy of Professor Peter Brancazio, Brooklyn College.

Solution

$$48T^2 = V$$

$$T^2 = \frac{V}{48} \qquad \text{Dividing by 48 to get } T^2 \text{ alone}$$

$$T = \frac{\sqrt{V}}{\sqrt{48}} \qquad \text{Using the principle of square roots and} \\ \text{the quotient rule for radicals}$$

$$\left. \begin{array}{l} = \dfrac{\sqrt{V}}{\sqrt{16}\,\sqrt{3}} = \dfrac{\sqrt{V}}{4\sqrt{3}} \\[2ex] = \dfrac{\sqrt{V}}{4\sqrt{3}} \cdot \dfrac{\sqrt{3}}{\sqrt{3}} = \dfrac{\sqrt{3V}}{12}. \end{array} \right\} \quad \text{Rationalizing the denominator}$$

E x a m p l e   4

Falling distance.   An object tossed downward with an initial speed (velocity) of v_0 will travel a distance of s meters, where $s = 4.9t^2 + v_0 t$ and t is measured in seconds. Solve for t.

Solution   Since t is squared in one term and raised to the first power in the other term, the equation is quadratic in t.

$$4.9t^2 + v_0 t = s$$

$$4.9t^2 + v_0 t - s = 0 \qquad \text{Writing standard form}$$

$$a = 4.9, \quad b = v_0, \quad c = -s$$

$$t = \frac{-v_0 \pm \sqrt{v_0^2 - 4(4.9)(-s)}}{2(4.9)} \qquad \text{Using the quadratic formula}$$

Since the negative square root would yield a negative value for t, we use only the positive root:

$$t = \frac{-v_0 + \sqrt{v_0^2 + 19.6s}}{9.8}.$$

The following list of steps should help you when solving formulas for a given letter. Try to remember that when solving a formula, you use the same approach that you would to solve an equation.

> ### *To Solve a Formula for a Letter—Say, b*
>
> 1. Clear fractions and use the principle of powers, as needed. (In some cases, you may clear the fractions first, and in some cases, you may use the principle of powers first.) Perform these steps until radicals containing b are gone and b is not in any denominator.
> 2. Combine all terms with b^2 in them. Also combine all terms with b in them.
> 3. If b^2 does not appear, you can solve by using just the addition and multiplication principles as in Sections 2.3 and 6.8.
> 4. If b^2 appears but b does not, solve the equation for b^2. Then use the principle of square roots to solve for b.
> 5. If there are terms containing both b and b^2, put the equation in standard form and use the quadratic formula.

FOR EXTRA HELP

Exercise Set 11.3

 Digital Video Tutor CD 8 Videotape 21   InterAct Math   Math Tutor Center   MathXL   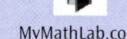 MyMathLab.com

Solve.

1. *Canoeing.* During the first part of a canoe trip, Tim covered 60 km at a certain speed. He then traveled 24 km at a speed that was 4 km/h slower. If the total time for the trip was 8 hr, what was the speed on each part of the trip?

2. *Car trips.* During the first part of a trip, Meira's Honda traveled 120 mi at a certain speed. Meira then drove another 100 mi at a speed that was 10 mph slower. If Meira's total trip time was 4 hr, what was her speed on each part of the trip?

3. *Car trips.* Sandi's Subaru travels 280 mi averaging a certain speed. If the car had gone 5 mph faster, the trip would have taken 1 hr less. Find Sandi's average speed.

4. *Car trips.* Petra's Plymouth travels 200 mi averaging a certain speed. If the car had gone 10 mph faster, the trip would have taken 1 hr less. Find Petra's average speed.

5. *Air travel.* A Cessna flies 600 mi at a certain speed. A Beechcraft flies 1000 mi at a speed that is 50 mph faster, but takes 1 hr longer. Find the speed of each plane.

6. *Air travel.* A turbo-jet flies 50 mph faster than a super-prop plane. If a turbo-jet goes 2000 mi in 3 hr less time than it takes the super-prop to go 2800 mi, find the speed of each plane.

7. *Bicycling.* Naoki bikes the 40 mi to Hillsboro averaging a certain speed. The return trip is made at a speed that is 6 mph slower. Total time for the round trip is 14 hr. Find Naoki's average speed on each part of the trip.

8. *Car speed.* On a sales trip, Gail drives the 600 mi to Richmond averaging a certain speed. The return trip is made at an average speed that is 10 mph slower. Total time for the round trip is 22 hr. Find Gail's average speed on each part of the trip.

9. *Navigation.* The current in a typical Mississippi River shipping route flows at a rate of 4 mph. In order for a barge to travel 24 mi upriver and then return in a total of 5 hr, approximately how fast must the barge be able to travel in still water?

10. *Navigation.* The Hudson River flows at a rate of 3 mph. A patrol boat travels 60 mi upriver and returns in a total time of 9 hr. What is the speed of the boat in still water?

11. *Filling a pool.*   Two wells are used to fill a swimming pool. Working together, they can fill the pool in 4 hr. One well, working alone, can fill the pool in 6 hr less time than the other. How long would the smaller one take, working alone, to fill the pool?

12. *Filling a tank.*   Two pipes are connected to the same tank. Working together, they can fill the tank in 2 hr. The larger pipe, working alone, can fill the tank in 3 hr less time than the smaller one. How long would the smaller one take, working alone, to fill the tank?

13. *Paddleboats.*   Ellen paddles 1 mi upstream and 1 mi back in a total time of 1 hr. The speed of the river is 2 mph. Find the speed of Ellen's paddleboat in still water.

14. *Rowing.*   Dan rows 10 km upstream and 10 km back in a total time of 3 hr. The speed of the river is 5 km/h. Find Dan's speed in still water.

Solve each formula for the indicated letter. Assume that all variables represent nonnegative numbers.

15. $A = 4\pi r^2$, for r
(Surface area of a sphere)

16. $A = 6s^2$, for s
(Surface area of a cube)

17. $A = 2\pi r^2 + 2\pi rh$, for r
(Surface area of a right cylindrical solid)

18. $F = \dfrac{Gm_1 m_2}{r^2}$, for r
(Law of gravity)

19. $N = \dfrac{kQ_1 Q_2}{s^2}$, for s
(Number of phone calls between two cities)

20. $A = \pi r^2$, for r
(Area of a circle)

21. $T = 2\pi \sqrt{\dfrac{l}{g}}$, for g
(A pendulum formula)

22. $a^2 + b^2 = c^2$, for b
(Pythagorean formula in two dimensions)

23. $a^2 + b^2 + c^2 = d^2$, for c
(Pythagorean formula in three dimensions)

24. $N = \dfrac{k^2 - 3k}{2}$, for k
(Number of diagonals of a polygon)

25. $s = v_0 t + \dfrac{gt^2}{2}$, for t
(A motion formula)

26. $A = \pi r^2 + \pi rs$, for r
(Surface area of a cone)

27. $N = \frac{1}{2}(n^2 - n)$, for n
(Number of games if n teams play each other once)

28. $A = A_0(1 - r)^2$, for r
(A business formula)

29. $V = 3.5\sqrt{h}$, for h
(Distance to horizon from a height)

30. $W = \sqrt{\dfrac{1}{LC}}$, for L
(An electricity formula)

Aha! **31.** $at^2 + bt + c = 0$, for t
(An algebraic formula)

32. $A = P_1(1 + r)^2 + P_2(1 + r)$, for r
(An investment formula)

Solve. Refer to Exercises 15–32 and Examples 2–4 for the appropriate formula.

33. *Falling distance.*
   a) An object is dropped 500 m from an airplane. How long does it take the object to reach the ground?
   b) An object is thrown downward 500 m from the plane at an initial velocity of 30 m/sec. How long does it take the object to reach the ground?
   c) How far will an object fall in 5 sec, when thrown downward at an initial velocity of 30 m/sec?

34. *Falling distance.*
   a) An object is dropped 75 m from an airplane. How long does it take the object to reach the ground?
   b) An object is thrown downward with an initial velocity of 30 m/sec from a plane 75 m above the ground. How long does it take the object to reach the ground?
   c) How far will an object fall in 2 sec, if thrown downward at an initial velocity of 30 m/sec?

35. *Bungee jumping.* Jesse is tied to one end of a 40-m elasticized (bungee) cord. The other end of the cord is tied to the middle of a train trestle. If Jesse jumps off the bridge, for how long will he fall before the cord begins to stretch? (See Example 4 and let $v_0 = 0$.)

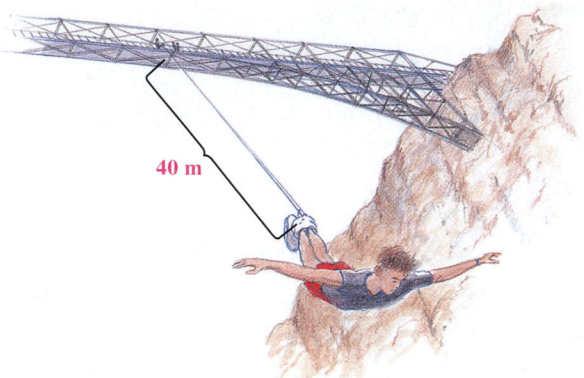

40 m

36. *Bungee jumping.* Sheila is tied to a bungee cord (see Exercise 35) and falls for 2.5 sec before her cord begins to stretch. How long is the bungee cord?

37. *Hang time.* The NBA's Vince Carter has a vertical leap of about 36 in. What is his hang time?

38. *League schedules.* In a volleyball league, each team plays each of the other teams once. If a total of 66 games is played, how many teams are in the league?

39. *Downward speed.* An object thrown downward from a 100-m cliff travels 51.6 m in 3 sec. What was the initial velocity of the object?

40. *Downward speed.* An object thrown downward from a 200-m cliff travels 91.2 m in 4 sec. What was the initial velocity of the object?

41. *Compound interest.* A firm invests $3000 in a savings account for 2 yr. At the beginning of the second year, an additional $1700 is invested. If a total of $5253.70 is in the account at the end of the second year, what is the annual interest rate? (*Hint*: See Exercise 32.)

42. *Compound interest.* A business invests $10,000 in a savings account for 2 yr. At the beginning of the second year, an additional $3500 is invested. If a total of $15,569.75 is in the account at the end of the second year, what is the annual interest rate? (*Hint*: See Exercise 32.)

43. Marti is tied to a bungee cord that is twice as long as the cord tied to Pedro. Will Marti's fall take twice as long as Pedro's before their cords begin to stretch? Why or why not? (See Exercises 35 and 36.)

44. Under what circumstances would a negative value for t, time, have meaning?

SKILL MAINTENANCE

Evaluate.

45. $b^2 - 4ac$, for $a = 5$, $b = 6$, and $c = 7$

46. $\sqrt{b^2 - 4ac}$, for $a = 3$, $b = 4$, and $c = 5$

Simplify.

47. $\dfrac{x^2 + xy}{2x}$

48. $\dfrac{a^3 - ab^2}{ab}$

49. $\dfrac{3 + \sqrt{45}}{6}$

50. $\dfrac{2 - \sqrt{28}}{10}$

SYNTHESIS

51. In what ways do the motion problems of this section (like Example 1) differ from the motion problems in Chapter 6 (see p. 386)?

52. Write a problem for a classmate to solve. Devise the problem so that **(a)** the solution is found after solving a rational equation and **(b)** the solution is "The express train travels 90 mph."

53. *Biochemistry.*   The equation

$$A = 6.5 - \frac{20.4t}{t^2 + 36}$$

is used to calculate the acid level A in a person's blood t minutes after sugar is consumed. Solve for t.

54. *Special relativity.*   Einstein found that an object of mass m_0, traveling velocity v, has its mass become

$$m = \frac{m_0}{\sqrt{1 - \dfrac{v_2}{c^2}}},$$

where c is the speed of light. Solve the formula for c.

55. *The Golden Rectangle.*   For over 2000 yr, the proportions of a "golden" rectangle have been considered visually appealing. A rectangle of width w and length l is considered "golden" if

$$\frac{w}{l} = \frac{l}{w + l}.$$

Solve for l.

56. *Diagonal of a cube.*   Find a formula that expresses the length of the three-dimensional diagonal of a cube as a function of the cube's surface area.

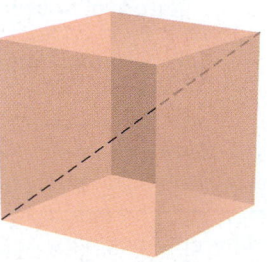

57. Find a number for which the reciprocal of 1 less than the number is the same as 1 more than the number.

58. *Purchasing.*   A discount store bought a quantity of beach towels for $250 and sold all but 15 at a profit of $3.50 per towel. With the total amount received, the manager could buy 4 more than twice as many as were bought before. Find the cost per towel.

59. Solve for n:

$$mn^4 - r^2pm^3 - r^2n^2 + p = 0.$$

60. *Surface area.*   Find a formula that expresses the diameter of a right cylindrical solid as a function of its surface area and its height.

61. A sphere is inscribed in a cube as shown in the figure below. Express the surface area of the sphere as a function of the surface area S of the cube.

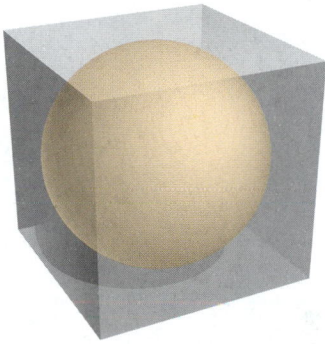

62. Explain how Exercises 1–14 can be solved without factoring, completing the square, or using the quadratic formula.

Studying Solutions of Quadratic Equations

11.4

The Discriminant • Writing Equations from Solutions

The Discriminant

Sometimes in mathematics it is enough to know what *type* of number a solution will be, without actually solving the equation. To illustrate, suppose we want to know if the equation $4x^2 + 7x - 15 = 0$ has rational solutions (and thus can be solved by factoring). Using the quadratic formula, we would have

$$x = \frac{-b \pm \sqrt{b^2 - 4ac}}{2a} = \frac{-7 \pm \sqrt{7^2 - 4 \cdot 4 \cdot (-15)}}{2 \cdot 4}.$$

Note that the radicand, $7^2 - 4 \cdot 4 \cdot (-15)$, determines what type of number the solutions will be. Since $7^2 - 4 \cdot 4 \cdot (-15) = 49 - 16(-15) = 289$, and since 289 is a perfect square $(\sqrt{289} = 17)$, we know that the solutions of the equation will be two rational numbers. This means that $4x^2 + 7x - 15 = 0$ *can* be solved by factoring.

It is the expression $b^2 - 4ac$, known as the **discriminant**, that determines what type of number the solutions of a quadratic equation will be:

- When $b^2 - 4ac$ simplifies to 0, it doesn't matter if we use $+\sqrt{b^2 - 4ac}$ or $-\sqrt{b^2 - 4ac}$; we get the same solution twice. Thus, when the discriminant is 0, there is one *repeated* solution and it will be rational.
- When the discriminant is positive, there are two different real-number solutions. As we saw above, when $b^2 - 4ac$ is a perfect square, these solutions are rational numbers. When $b^2 - 4ac$ is positive but not a perfect square, there are two irrational solutions and they will be conjugates of each other (see p. 625).
- When the discriminant is negative, there are two imaginary-number solutions and they will be complex conjugates of each other.

Discriminant $b^2 - 4ac$	Nature of Solutions
0	One solution; a rational number

Positive
  Perfect square
  Not a perfect square
	Two different real-number solutions
	  Solutions are rational.
	  Solutions are irrational conjugates.

Negative	Two different imaginary-number solutions (complex conjugates)

Example 1

For each equation, determine what type of number the solutions will be and how many solutions exist.

a) $9x^2 - 12x + 4 = 0$ **b)** $x^2 + 5x + 8 = 0$ **c)** $2x^2 + 7x - 3 = 0$

technology connection

Recall that the real-number solutions of $ax^2 + bx + c = 0$ are the x-intercepts of the graph of $y = ax^2 + bx + c$. Use a grapher to confirm that part (a) of Example 1 has one real solution, part (b) has no real solution, and part (c) has two real solutions.

Solution

a) For $9x^2 - 12x + 4 = 0$, we have

$$a = 9, \quad b = -12, \quad c = 4.$$

We substitute and compute the discriminant:

$$b^2 - 4ac = (-12)^2 - 4 \cdot 9 \cdot 4$$
$$= 144 - 144 = 0.$$

There is just one solution, and it is rational. This tells us that $9x^2 - 12x + 4 = 0$ can be solved by factoring.

b) For $x^2 + 5x + 8 = 0$, we have

$$a = 1, \quad b = 5, \quad c = 8.$$

We substitute and compute the discriminant:

$$b^2 - 4ac = 5^2 - 4 \cdot 1 \cdot 8$$
$$= 25 - 32 = -7.$$

Since the discriminant is negative, there are two imaginary-number solutions that are complex conjugates of each other.

c) For $2x^2 + 7x - 3 = 0$, we have

$$a = 2, \quad b = 7, \quad c = -3;$$
$$b^2 - 4ac = 7^2 - 4 \cdot 2(-3)$$
$$= 49 - (-24) = 73.$$

The discriminant is a positive number that is not a perfect square. Thus there are two irrational solutions that are conjugates of each other.

Discriminants can also be used to determine the number of real-number solutions of $ax^2 + bx + c = 0$. This can be used as an aid in graphing.

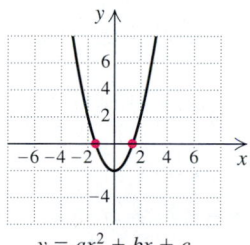

$y = ax^2 + bx + c$
$b^2 - 4ac > 0$
Two real solutions
of $ax^2 + bx + c = 0$
Two x-intercepts

$y = ax^2 + bx + c$
$b^2 - 4ac = 0$
One real solution
of $ax^2 + bx + c = 0$
One x-intercept

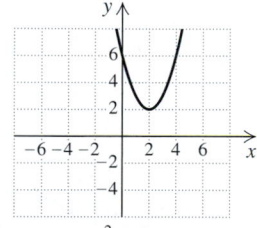

$y = ax^2 + bx + c$
$b^2 - 4ac < 0$
No real solutions
of $ax^2 + bx + c = 0$
No x-intercept

Writing Equations from Solutions

We know by the principle of zero products that $(x - 2)(x + 3) = 0$ has solutions 2 and -3. If we know the solutions of an equation, we can write an equation, using the principle in reverse.

E x a m p l e   2 Find an equation for which the given numbers are solutions.

a) 3 and $-\frac{2}{5}$ **b)** $2i$ and $-2i$

c) $5\sqrt{7}$ and $-5\sqrt{7}$ **d)** $-4, 0,$ and 1

Solution

a)
$$x = 3 \quad or \quad x = -\tfrac{2}{5}$$
$$x - 3 = 0 \quad or \quad x + \tfrac{2}{5} = 0 \qquad \text{Getting 0's on one side}$$
$$(x - 3)\left(x + \tfrac{2}{5}\right) = 0 \qquad \text{Using the principle of zero products (multiplying)}$$
$$x^2 + \tfrac{2}{5}x - 3x - 3 \cdot \tfrac{2}{5} = 0 \qquad \text{Multiplying}$$
$$x^2 - \tfrac{13}{5}x - \tfrac{6}{5} = 0 \qquad \text{Combining like terms}$$
$$5x^2 - 13x - 6 = 0 \qquad \text{Multiplying both sides by 5 to clear fractions}$$

Note that multiplying both sides by the LCD, 5, clears the equation of fractions. Had we preferred, we could have multiplied $x + \frac{2}{5} = 0$ by 5, thus clearing fractions *before* using the principle of zero products.

b)
$$x = 2i \quad or \quad x = -2i$$
$$x - 2i = 0 \quad or \quad x + 2i = 0 \qquad \text{Getting 0's on one side}$$
$$(x - 2i)(x + 2i) = 0 \qquad \text{Using the principle of zero products (multiplying)}$$
$$x^2 - (2i)^2 = 0 \qquad \text{Finding the product of a sum and difference}$$
$$x^2 - 4i^2 = 0$$
$$x^2 + 4 = 0 \qquad\qquad i^2 = -1$$

c)
$$x = 5\sqrt{7} \quad or \quad x = -5\sqrt{7}$$
$$x - 5\sqrt{7} = 0 \quad or \quad x + 5\sqrt{7} = 0 \qquad \text{Getting 0's on one side}$$
$$\left(x - 5\sqrt{7}\right)\left(x + 5\sqrt{7}\right) = 0 \qquad \text{Using the principle of zero products}$$
$$x^2 - \left(5\sqrt{7}\right)^2 = 0 \qquad \text{Finding the product of a sum and difference}$$
$$x^2 - 25 \cdot 7 = 0$$
$$x^2 - 175 = 0$$

d)
$$x = -4 \quad or \quad x = 0 \quad or \quad x = 1$$
$$x + 4 = 0 \quad or \quad x = 0 \quad or \quad x - 1 = 0 \qquad \text{Getting 0's on one side}$$
$$(x + 4)x(x - 1) = 0 \qquad \text{Using the principle of zero products}$$
$$x(x^2 + 3x - 4) = 0 \qquad \text{Multiplying}$$
$$x^3 + 3x^2 - 4x = 0$$

To check any of these equations, we can simply substitute one or more of the given solutions. For example, in Example 2(d) above,

$$(-4)^3 + 3(-4)^2 - 4(-4) = -64 + 3 \cdot 16 + 16$$
$$= -64 + 48 + 16 = 0.$$

The other checks are left to the student.

Exercise Set 11.4

For each equation, determine what type of number the solutions are and how many solutions exist.

1. $x^2 - 5x + 3 = 0$

2. $x^2 - 7x + 5 = 0$

3. $x^2 + 5 = 0$

4. $x^2 + 3 = 0$

5. $x^2 - 3 = 0$

6. $x^2 - 5 = 0$

7. $4x^2 - 12x + 9 = 0$

8. $4x^2 + 8x - 5 = 0$

9. $x^2 - 2x + 4 = 0$

10. $x^2 + 4x + 6 = 0$

11. $6t^2 - 19t - 20 = 0$

12. $9t^2 - 48t + 64 = 0$

13. $6x^2 + 5x - 4 = 0$

14. $10x^2 - x - 2 = 0$

 **15.** $9t^2 - 3t = 0$

16. $4m^2 + 7m = 0$

17. $x^2 + 4x = 8$

18. $x^2 + 5x = 9$

19. $2a^2 - 3a = -5$

20. $3a^2 + 5 = 7a$

21. $y^2 + \frac{9}{4} = 4y$

22. $x^2 = \frac{1}{2}x - \frac{3}{5}$

Write a quadratic equation having the given numbers as solutions.

23. $-7, 3$

24. $-6, 4$

25. 3, only solution (*Hint*: It must be a repeated solution.)

26. -5, only solution

27. $-2, -5$

28. $-1, -3$

29. $4, \frac{2}{3}$

30. $5, \frac{3}{4}$

31. $\frac{1}{2}, \frac{1}{3}$

32. $-\frac{1}{4}, -\frac{1}{2}$

33. $-0.6, 1.4$

34. $2.4, -0.4$

35. $-\sqrt{7}, \sqrt{7}$

36. $-\sqrt{3}, \sqrt{3}$

37. $3\sqrt{2}, -3\sqrt{2}$

38. $2\sqrt{5}, -2\sqrt{5}$

39. $3i, -3i$

40. $4i, -4i$

41. $5 - 2i, 5 + 2i$

42. $2 - 7i, 2 + 7i$

43. $2 - \sqrt{10}, 2 + \sqrt{10}$

44. $3 - \sqrt{14}, 3 + \sqrt{14}$

Write a third-degree equation having the given numbers as solutions.

45. $-2, 1, 5$

46. $-5, 0, 2$

47. $-1, 0, 3$

48. $-2, 2, 3$

49. Under what condition(s) is the discriminant *not* the fastest way to determine how many and what type of solutions exist?

50. Describe a procedure that could be used to write an equation having the first 7 natural numbers as solutions.

SKILL MAINTENANCE

Simplify.

51. $(3a^2)^4$

52. $(4x^3)^2$

Find the x-intercepts of the graph of each equation.

53. $y = x^2 - 7x - 8$

54. $y = x^2 - 6x + 8$

55. During a one-hour television show, there were 12 commercials. Some of the commercials were 30 sec long and the others were 60 sec long. The amount of time for 30-sec commercials was 6 min less than the total number of minutes of commercial time during the show. How many 30-sec commercials were used?

56. Graph: $y = -\frac{3}{7}x + 4$.

SYNTHESIS

57. If we assume that a quadratic equation has integers for coefficients, will the product of the solutions always be a real number? Why or why not?

58. Can a fourth-degree equation have three irrational solutions? Why or why not?

59. The graph of an equation of the form

$$y = ax^2 + bx + c$$

is a curve similar to the one shown below. Determine a, b, and c from the information given.

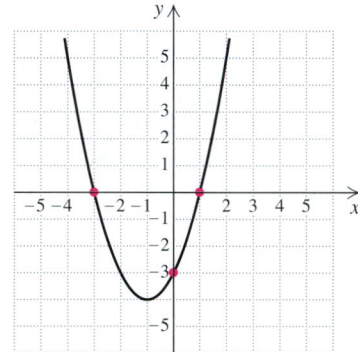

60. Show that the product of the solutions of $ax^2 + bx + c = 0$ is c/a.

For each equation under the given condition, **(a)** *find k and* **(b)** *find the other solution.*

61. $kx^2 - 2x + k = 0$; one solution is -3

62. $x^2 - kx + 2 = 0$; one solution is $1 + i$

63. $x^2 - (6 + 3i)x + k = 0$; one solution is 3

64. Show that the sum of the solutions of $ax^2 + bx + c = 0$ is $-b/a$.

65. Show that whenever there is just one solution of $ax^2 + bx + c = 0$, that solution is of the form $-b/2a$.

66. Find h and k, where $3x^2 - hx + 4k = 0$, the sum of the solutions is -12, and the product of the solutions is 20. (*Hint*: See Exercises 60 and 64.)

67. Suppose that $f(x) = ax^2 + bx + c$, with $f(-3) = 0$, $f(\frac{1}{2}) = 0$, and $f(0) = -12$. Find a, b, and c.

68. Find an equation for which $2 - \sqrt{3}$, $2 + \sqrt{3}$, $5 - 2i$, and $5 + 2i$ are solutions.

Aha! **69.** Find an equation for which $1 - \sqrt{5}$ and $3 + 2i$ are two of the solutions.

70. A discriminant that is a perfect square indicates that factoring can be used to solve the quadratic equation. Why?

71. While solving a quadratic equation of the form $ax^2 + bx + c = 0$ with a grapher, Shawn-Marie gets the following screen.

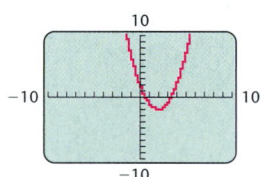

How could the sign of the discriminant help her check the graph?

Equations Reducible to Quadratic	**11.5**

Recognizing Equations in Quadratic Form • Radical and Rational Equations

Recognizing Equations in Quadratic Form

Certain equations that are not really quadratic can be thought of in such a way that they can be solved as quadratic. For example, because the square of x^2 is

x^4, the equation $x^4 - 9x^2 + 8 = 0$ is said to be "quadratic in x^2":

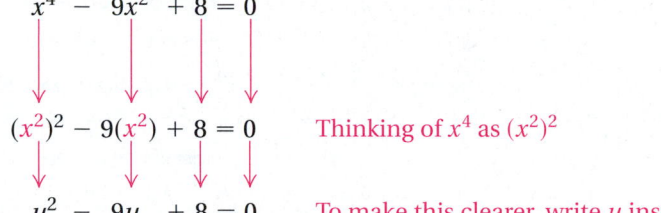

$$x^4 \; - \; 9x^2 \; + 8 = 0$$

$$(x^2)^2 \; - \; 9(x^2) \; + 8 = 0 \qquad \text{Thinking of } x^4 \text{ as } (x^2)^2$$

$$u^2 \; - \; 9u \; + 8 = 0. \qquad \text{To make this clearer, write } u \text{ instead of } x^2.$$

The equation $u^2 - 9u + 8 = 0$ can be solved by factoring or by the quadratic formula. Then, remembering that $u = x^2$, we can solve for x. Equations that can be solved like this are said to be *reducible to quadratic*, or *in quadratic form*.

E x a m p l e   1

Solve: $x^4 - 9x^2 + 8 = 0$.

Solution   Let $u = x^2$. Then we solve by substituting u for x^2 and u^2 for x^4:

$$u^2 - 9u + 8 = 0$$
$$(u - 8)(u - 1) = 0 \qquad\qquad \text{Factoring}$$
$$u - 8 = 0 \quad or \quad u - 1 = 0 \qquad \text{Principle of zero products}$$
$$u = 8 \quad or \qquad u = 1.$$

Caution!   A common error is to solve for u and then forget to solve for x. Remember that you must find values for the *original* variable!

We replace u with x^2 and solve these equations:

$$x^2 = 8 \qquad or \quad x^2 = 1$$
$$x = \pm\sqrt{8} \quad or \quad x = \pm 1$$
$$x = \pm 2\sqrt{2} \quad or \quad x = \pm 1.$$

To check, note that for $x = 2\sqrt{2}$, we have $x^2 = 8$ and $x^4 = 64$. Similarly, for $x = -2\sqrt{2}$, we have $x^2 = 8$ and $x^4 = 64$. When $x = 1$, we have $x^2 = 1$ and $x^4 = 1$, and when $x = -1$, we have $x^2 = 1$ and $x^4 = 1$. Thus instead of making four checks, we need make only two.

Check:

For $\pm 2\sqrt{2}$:

$$\frac{x^4 - 9x^2 + 8 = 0}{(\pm 2\sqrt{2})^4 - 9(\pm 2\sqrt{2})^2 + 8 \; ? \; 0}$$
$$64 - 9 \cdot 8 + 8 \quad\Big|$$
$$0 \;\Big|\; 0 \quad \text{TRUE}$$

For ± 1:

$$\frac{x^4 - 9x^2 + 8 = 0}{(\pm 1)^4 - 9(\pm 1)^2 + 8 \; ? \; 0}$$
$$1 - 9 + 8 \quad\Big|$$
$$0 \;\Big|\; 0 \quad \text{TRUE}$$

The solutions are $1, -1, 2\sqrt{2}$, and $-2\sqrt{2}$.

Example 1 can be solved directly by factoring:

$$x^4 - 9x^2 + 8 = 0$$
$$(x^2 - 1)(x^2 - 8) = 0$$
$$x^2 - 1 = 0 \quad or \quad x^2 - 8 = 0$$
$$x^2 = 1 \quad or \quad x^2 = 8$$
$$x = \pm 1 \quad or \quad x = \pm 2\sqrt{2}.$$

There is nothing wrong with this approach. However, in the examples that follow, you will note that it becomes increasingly difficult to solve the equation without first making a substitution.

Radical and Rational Equations

Sometimes rational equations, radical equations, or equations containing fractional exponents are reducible to quadratic. It is especially important that answers to these equations be checked in the original equation.

E x a m p l e   2

Solve: $x - 3\sqrt{x} - 4 = 0$.

Solution   This radical equation could be solved using the method discussed in Section 10.6. However, if we note that the square of $\sqrt{x}$ is x, we can regard the equation as "quadratic in $\sqrt{x}$."

We let $u = \sqrt{x}$ and consequently $u^2 = x$:

$$x - 3\sqrt{x} - 4 = 0$$
$$u^2 - 3u - 4 = 0 \qquad \text{Substituting}$$
$$(u - 4)(u + 1) = 0$$
$$u = 4 \quad or \quad u = -1. \qquad \text{Using the principle of zero products}$$

Next, we replace u with $\sqrt{x}$ and solve these equations:

$$\sqrt{x} = 4 \quad or \quad \sqrt{x} = -1.$$

Squaring gives us $x = 16$ or $x = 1$ and also makes checking essential.

Check: For 16:

$$\begin{array}{c|c} x - 3\sqrt{x} - 4 = 0 \\ \hline 16 - 3\sqrt{16} - 4 \; ? \; 0 \\ 16 - 3 \cdot 4 - 4 \\ 0 & 0 \quad \text{TRUE} \end{array}$$

For 1:

$$\begin{array}{c|c} x - 3\sqrt{x} - 4 = 0 \\ \hline 1 - 3\sqrt{1} - 4 \; ? \; 0 \\ 1 - 3 \cdot 1 - 4 \\ -6 & 0 \quad \text{FALSE} \end{array}$$

The number 16 checks, but 1 does not. Had we noticed that $\sqrt{x} = -1$ has no solution (since principal roots are never negative), we could have solved only the equation $\sqrt{x} = 4$. The solution is 16.

Example 3

technology
connection

Check Example 3 with a grapher.
Use the ZERO, ROOT, or
INTERSECT option, if possible.

Find the x-intercepts of the graph of $f(x) = (x^2 - 1)^2 - (x^2 - 1) - 2$.

Solution   The x-intercepts occur where $f(x) = 0$ so we must have

$$(x^2 - 1)^2 - (x^2 - 1) - 2 = 0. \qquad \text{Setting } f(x) \text{ equal to 0}$$

This equation is quadratic in $x^2 - 1$, so we let $u = x^2 - 1$ and $u^2 = (x^2 - 1)^2$:

$$u^2 - u - 2 = 0 \qquad \text{Substituting in } (x^2 - 1)^2 - (x^2 - 1) - 2 = 0$$

$$(u - 2)(u + 1) = 0$$

$$u = 2 \qquad or \qquad u = -1. \qquad \text{Using the principle of zero products}$$

Next, we replace u with $x^2 - 1$ and solve these equations:

$$x^2 - 1 = 2 \qquad or \quad x^2 - 1 = -1$$

$$x^2 = 3 \qquad or \qquad x^2 = 0 \qquad \text{Adding 1 to both sides}$$

$$x = \pm\sqrt{3} \quad or \qquad x = 0. \qquad \text{Using the principle of square roots}$$

The x-intercepts occur at $\left(-\sqrt{3}, 0\right)$, $(0, 0)$, and $\left(\sqrt{3}, 0\right)$.

     Sometimes great care must be taken in deciding what substitution to make.

Example 4

Solve: $m^{-2} - 6m^{-1} + 4 = 0$.

Solution   Note that the square of m^{-1} is $(m^{-1})^2$, or m^{-2}. This allows us to regard the equation as quadratic in m^{-1}.
     We let $u = m^{-1}$ and $u^2 = m^{-2}$:

$$u^2 - 6u + 4 = 0 \qquad\qquad\qquad \text{Substituting}$$

$$u = \frac{-(-6) \pm \sqrt{(-6)^2 - 4 \cdot 1 \cdot 4}}{2 \cdot 1} \qquad \begin{array}{l}\text{Using the quadratic}\\\text{formula}\end{array}$$

$$u = \frac{6 \pm \sqrt{20}}{2} = \frac{2 \cdot 3 \pm 2\sqrt{5}}{2} \left.\vphantom{\begin{array}{c}a\\b\\c\end{array}}\right\} \quad \text{Simplifying}$$

$$u = 3 \pm \sqrt{5}.$$

Next, we replace u with m^{-1} and solve:

$$m^{-1} = 3 \pm \sqrt{5}$$

$$\frac{1}{m} = 3 \pm \sqrt{5} \qquad\qquad \text{Recall that } m^{-1} = \frac{1}{m}.$$

$$1 = m(3 \pm \sqrt{5}) \qquad\qquad \text{Multiplying both sides by } m$$

$$\frac{1}{3 \pm \sqrt{5}} = m. \qquad\qquad \text{Dividing both sides by } 3 \pm \sqrt{5}$$

We can check both solutions as follows.

Check:

For $1/(3 - \sqrt{5})$:

$$\frac{m^{-2} - 6m^{-1} + 4 = 0}{\left(\dfrac{1}{3 - \sqrt{5}}\right)^{-2} - 6\left(\dfrac{1}{3 - \sqrt{5}}\right)^{-1} + 4 \; ? \; 0}$$

$$(3 - \sqrt{5})^2 - 6(3 - \sqrt{5}) + 4$$

$$9 - 6\sqrt{5} + 5 - 18 + 6\sqrt{5} + 4$$

$$0 \;\bigm|\; 0 \quad \text{TRUE}$$

For $1/(3 + \sqrt{5})$:

$$\frac{m^{-2} - 6m^{-1} + 4 = 0}{\left(\dfrac{1}{3 + \sqrt{5}}\right)^{-2} - 6\left(\dfrac{1}{3 + \sqrt{5}}\right)^{-1} + 4 \; ? \; 0}$$

$$(3 + \sqrt{5})^2 - 6(3 + \sqrt{5}) + 4$$

$$9 + 6\sqrt{5} + 5 - 18 - 6\sqrt{5} + 4$$

$$0 \;\bigm|\; 0 \quad \text{TRUE}$$

Both numbers check. The solutions are $1/(3 - \sqrt{5})$ and $1/(3 + \sqrt{5})$, or approximately 1.309016994 and 0.1909830056.

Example 5

Solve: $t^{2/5} - t^{1/5} - 2 = 0$.

Solution Note that the square of $t^{1/5}$ is $(t^{1/5})^2$, or $t^{2/5}$. The equation is therefore quadratic in $t^{1/5}$, so we let $u = t^{1/5}$ and $u^2 = t^{2/5}$:

$$u^2 - u - 2 = 0 \qquad \text{Substituting}$$

$$(u - 2)(u + 1) = 0$$

$$u = 2 \quad or \quad u = -1. \qquad \text{Using the principle of zero products}$$

Now we replace u with $t^{1/5}$ and solve:

$$t^{1/5} = 2 \quad or \quad t^{1/5} = -1$$

$$t = 32 \quad or \quad t = -1. \qquad \text{Principle of powers; raising to the 5th power}$$

Check:

For 32:

$$\frac{t^{2/5} - t^{1/5} - 2 = 0}{32^{2/5} - 32^{1/5} - 2 \; ? \; 0}$$

$$(32^{1/5})^2 - 32^{1/5} - 2$$

$$2^2 - 2 - 2$$

$$0 \;\bigm|\; 0 \quad \text{TRUE}$$

For -1:

$$\frac{t^{2/5} - t^{1/5} - 2 = 0}{(-1)^{2/5} - (-1)^{1/5} - 2 \; ? \; 0}$$

$$[(-1)^{1/5}]^2 - (-1)^{1/5} - 2$$

$$(-1)^2 - (-1) - 2$$

$$0 \;\bigm|\; 0 \quad \text{TRUE}$$

Both numbers check. The solutions are 32 and -1.

The following tips may prove useful.

To Solve an Equation That Is Reducible to Quadratic

1. The equation is quadratic in form if the variable factor in one term is the square of the variable factor in the other variable term.
2. Write down any substitutions that you are making.
3. Whenever you make a substitution, be sure to solve for the variable that is used in the original equation.
4. Check possible answers in the original equation.

FOR EXTRA HELP

Exercise Set **11.5**

Digital Video Tutor CD 9   InterAct Math   Math Tutor Center   MathXL   MyMathLab.com
Videotape 21

Solve.

1. $x^4 - 10x^2 + 9 = 0$

2. $x^4 - 5x^2 + 4 = 0$

3. $x^4 - 12x^2 + 27 = 0$

4. $x^4 - 9x^2 + 20 = 0$

5. $9x^4 - 14x^2 + 5 = 0$

6. $4x^4 - 19x^2 + 12 = 0$

7. $x - 4\sqrt{x} - 1 = 0$

8. $x - 2\sqrt{x} - 6 = 0$

9. $(x^2 - 7)^2 - 3(x^2 - 7) + 2 = 0$

10. $(x^2 - 1)^2 - 5(x^2 - 1) + 6 = 0$

11. $(1 + \sqrt{x})^2 + 5(1 + \sqrt{x}) + 6 = 0$

12. $(3 + \sqrt{x})^2 + 3(3 + \sqrt{x}) - 10 = 0$

13. $x^{-2} - x^{-1} - 6 = 0$

14. $2x^{-2} - x^{-1} - 1 = 0$

15. $4x^{-2} + x^{-1} - 5 = 0$

16. $m^{-2} + 9m^{-1} - 10 = 0$

17. $t^{2/3} + t^{1/3} - 6 = 0$

18. $w^{2/3} - 2w^{1/3} - 8 = 0$

19. $y^{1/3} - y^{1/6} - 6 = 0$

20. $t^{1/2} + 3t^{1/4} + 2 = 0$

21. $t^{1/3} + 2t^{1/6} = 3$

22. $m^{1/2} + 6 = 5m^{1/4}$

23. $(3 - \sqrt{x})^2 - 10(3 - \sqrt{x}) + 23 = 0$

24. $(5 + \sqrt{x})^2 - 12(5 + \sqrt{x}) + 33 = 0$

25. $16\left(\dfrac{x - 1}{x - 8}\right)^2 + 8\left(\dfrac{x - 1}{x - 8}\right) + 1 = 0$

26. $9\left(\dfrac{x + 2}{x + 3}\right)^2 - 6\left(\dfrac{x + 2}{x + 3}\right) + 1 = 0$

Find all x-intercepts of the given function f. If none exist, state this.

27. $f(x) = 5x + 13\sqrt{x} - 6$

28. $f(x) = 3x + 10\sqrt{x} - 8$

29. $f(x) = (x^2 - 3x)^2 - 10(x^2 - 3x) + 24$

30. $f(x) = (x^2 - 6x)^2 - 2(x^2 - 6x) - 35$

31. $f(x) = x^{2/5} + x^{1/5} - 6$

32. $f(x) = x^{1/2} - x^{1/4} - 6$

Aha! 33. $f(x) = \left(\dfrac{x^2 + 2}{x}\right)^4 + 7\left(\dfrac{x^2 + 2}{x}\right)^2 + 5$

34. $f(x) = \left(\dfrac{x^2 + 1}{x}\right)^4 + 4\left(\dfrac{x^2 + 1}{x}\right)^2 + 12$

35. To solve $25x^6 - 10x^3 + 1 = 0$, Don lets $u = 5x^3$ and Robin lets $u = x^3$. Can they both be correct? Why or why not?

36. Can the examples and exercises of this section be understood without knowing the rules for exponents? Why or why not?

SKILL MAINTENANCE

Graph.

37. $f(x) = \frac{3}{2}x$

38. $f(x) = -\frac{2}{3}x$

39. $f(x) = \dfrac{2}{x}$

40. $f(x) = \dfrac{3}{x}$

41. Solution A is 18% alcohol and solution B is 45% alcohol. How much of each should be mixed together to get 12 L of a solution that is 36% alcohol?

42. If $g(x) = x^2 - x$, find $g(a + 1)$.

SYNTHESIS

43. Describe a procedure that could be used to solve any equation of the form $ax^4 + bx^2 + c = 0$.

44. Describe a procedure that could be used to write an equation that is quadratic in $3x^2 - 1$. Then explain how the procedure could be adjusted to write equations that are quadratic in $3x^2 - 1$ and have no real-number solution.

Solve.

45. $5x^4 - 7x^2 + 1 = 0$

46. $3x^4 + 5x^2 - 1 = 0$

47. $(x^2 - 4x - 2)^2 - 13(x^2 - 4x - 2) + 30 = 0$

48. $(x^2 - 5x - 1)^2 - 18(x^2 - 5x - 1) + 65 = 0$

49. $\dfrac{x}{x-1} - 6\sqrt{\dfrac{x}{x-1}} - 40 = 0$

50. $\left(\sqrt{\dfrac{x}{x-3}}\right)^2 - 24 = 10\sqrt{\dfrac{x}{x-3}}$

51. $a^5(a^2 - 25) + 13a^3(25 - a^2) + 36a(a^2 - 25) = 0$

52. $a^3 - 26a^{3/2} - 27 = 0$

53. $x^6 - 28x^3 + 27 = 0$

54. $x^6 + 7x^3 - 8 = 0$

55. Use a grapher to check your answers to Exercises 1, 3, 29, and 47.

56. Use a grapher to solve
$$x^4 - x^3 - 13x^2 + x + 12 = 0.$$

57. While trying to solve $0.05x^4 - 0.8 = 0$ with a grapher, Murray gets the following screen.

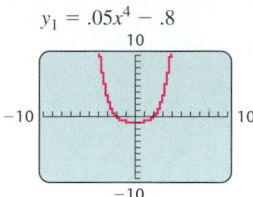

$y_1 = .05x^4 - .8$

Can Murray solve this equation with a grapher? Why or why not?

Quadratic Functions and Their Graphs

11.6

The Graph of $f(x) = ax^2$ • The Graph of $f(x) = a(x - h)^2$ • The Graph of $f(x) = a(x - h)^2 + k$

We have already used quadratic functions when we solved equations earlier in this chapter. In this section and the next, we learn to graph such functions.

The Graph of $f(x) = ax^2$

The most basic quadratic function is $f(x) = x^2$.

Example 1 Graph: $f(x) = x^2$.

Solution We choose some values for x and compute $f(x)$ for each. Then we plot the ordered pairs and connect them with a smooth curve.

technology connection

A

To examine the effect of a when graphing $f(x) = ax^2$, first graph $y_1 = x^2$ in a $[-5, 5, -10, 10]$ window. Without erasing this curve, graph $y_2 = 3x^2$. How do the graphs compare? Now include the graph of $y_3 = \frac{1}{3}x^2$. Describe the effect of multiplying x^2 by a, for $a > 1$ or $0 < a < 1$.

Clear the display and graph $y_1 = x^2$ again. Now include the graph of $y_2 = -x^2$. Note how it differs from the graph of $y_1 = x^2$. Next, graph $y_3 = \frac{2}{3}x^2$ and $y_4 = -\frac{2}{3}x^2$ and compare them. Describe the effect of multiplying x^2 by a, for $a < -1$ or $-1 < a < 0$.

x	$f(x) = x^2$	$(x, f(x))$
-3	9	$(-3, 9)$
-2	4	$(-2, 4)$
-1	1	$(-1, 1)$
0	0	$(0, 0)$
1	1	$(1, 1)$
2	4	$(2, 4)$
3	9	$(3, 9)$

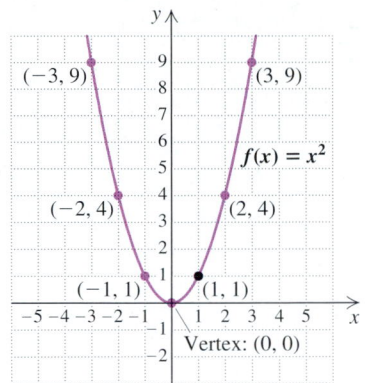

All quadratic functions have graphs similar to the one in Example 1. Such curves are called *parabolas*. They are cup-shaped and symmetric with respect to a vertical line known as the parabola's *axis of symmetry*. For the graph of $f(x) = x^2$, the y-axis (or the line $x = 0$) is the axis of symmetry. Were the paper folded on this line, the two halves of the curve would match. The point $(0, 0)$ is known as the *vertex* of this parabola.

By plotting points, we can compare the graphs of $g(x) = \frac{1}{2}x^2$ and $h(x) = 2x^2$ with the graph of $f(x) = x^2$.

x	$h(x) = 2x^2$
-3	18
-2	8
-1	2
0	0
1	2
2	8
3	18

x	$g(x) = \frac{1}{2}x^2$
-3	$\frac{9}{2}$
-2	2
-1	$\frac{1}{2}$
0	0
1	$\frac{1}{2}$
2	2
3	$\frac{9}{2}$

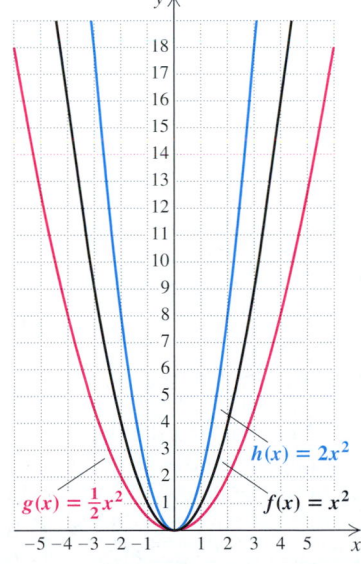

Note that the graph of $g(x) = \frac{1}{2}x^2$ is a wider parabola than the graph of $f(x) = x^2$, and the graph of $h(x) = 2x^2$ is narrower. The vertex and the axis of symmetry, however, remain $(0, 0)$ and $x = 0$, respectively.

When we consider the graph of $k(x) = -\frac{1}{2}x^2$, we see that the parabola is the same shape as the graph of $g(x) = \frac{1}{2}x^2$, but opens downward. We say that the graphs of k and g are *reflections* of each other across the x-axis.

x	$k(x) = -\frac{1}{2}x^2$
-3	$-\frac{9}{2}$
-2	-2
-1	$-\frac{1}{2}$
0	0
1	$-\frac{1}{2}$
2	-2
3	$-\frac{9}{2}$

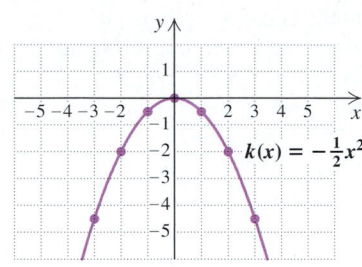

To investigate the effect of h on the graph of $f(x) = a(x - h)^2$, let $y_1 = 7x^2$ and $y_2 = 7(x - 1)^2$. Graph both y_1 and y_2 in the window $[-5, 5, -5, 5]$ and compare. Using the TABLE feature, compare y-values, beginning at $x = 1$ and increasing x by one unit at a time. On many graphers, the G-T or HORIZ modes can be used to view a split screen showing both the graph and the table.

Next, let $y_3 = 7(x - 2)^2$ and compare its graph and y-values with those of y_1 and y_2.

Finally, replace y_2 and y_3 with $y_2 = 7(x + 1)^2$ and $y_3 = 7(x + 2)^2$. Compare graphs and y-values and describe the effect of h on the graph of $f(x) = a(x - h)^2$.

> ### Graphing $f(x) = ax^2$
>
> The graph of $f(x) = ax^2$ is a parabola with $x = 0$ as its axis of symmetry. Its vertex is the origin.
>
> For $a > 0$, the parabola opens upward. For $a < 0$, the parabola opens downward.
>
> If $|a|$ is greater than 1, the parabola is narrower than $y = x^2$.
>
> If $|a|$ is between 0 and 1, the parabola is wider than $y = x^2$.

The Graph of $f(x) = a(x - h)^2$

Why not now consider graphs of

$$f(x) = ax^2 + bx + c,$$

where b and c are not both 0? In effect, we will do that, but in a disguised form. It turns out to be convenient to first graph $f(x) = a(x - h)^2$, where h is some constant. This allows us to observe similarities to the graphs drawn above.

E x a m p l e   2   Graph: $f(x) = (x - 3)^2$.

Solution   We choose some values for x and compute $f(x)$. Note that when an input here is 3 more than an input for Example 1, the outputs match. We plot the points and draw the curve.

x	$f(x) = (x - 3)^2$
-1	16
0	9
1	4
2	1
3	0
4	1
5	4
6	9

← Vertex

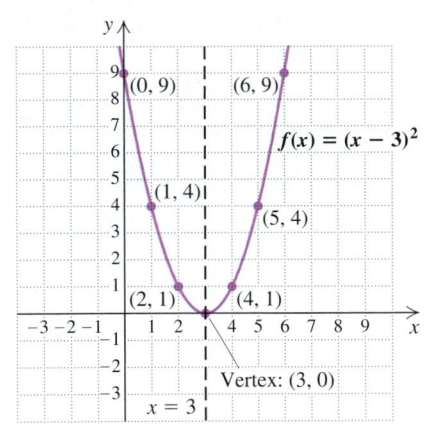

Note that $f(x)$ is smallest when $x - 3$ is 0, that is, for $x = 3$. Thus the line $x = 3$ is now the axis of symmetry and the point $(3, 0)$ is the vertex. Had we recognized earlier that $x = 3$ is the axis of symmetry, we could have computed some values on one side, such as $(4, 1)$, $(5, 4)$, and $(6, 9)$, and then used symmetry to get their mirror images $(2, 1)$, $(1, 4)$, and $(0, 9)$ without further computation.

E x a m p l e   3

Graph: $g(x) = -2(x + 3)^2$.

Solution   We choose some values for x and compute $g(x)$. Note that $g(x)$ is greatest when $x + 3$ is 0, that is, for $x = -3$. Thus the line given by $x = -3$ is the axis of symmetry and the point $(-3, 0)$ is the vertex. We plot some points and draw the curve.

x	$g(x) = -2(x + 3)^2$
-5	-8
-4	-2
-3	0
-2	-2
-1	-8

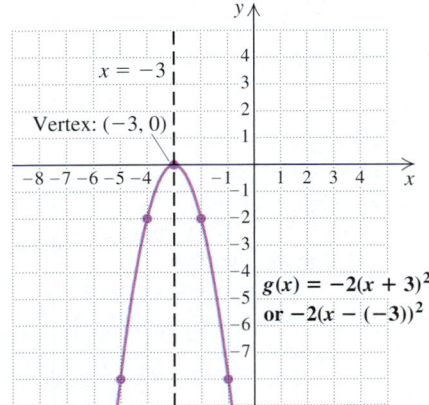

In Example 2, we found that the graph of $f(x) = (x - 3)^2$ looks just like the graph of $y = x^2$, except that it is moved, or *translated*, 3 units to the right. In Example 3, we found that the graph of $g(x) = -2(x + 3)^2$ looks like the graph of $y = -2x^2$, except that it is shifted 3 units to the left. These results can be generalized as follows.

Graphing $f(x) = a(x - h)^2$

The graph of $f(x) = a(x - h)^2$ has the same shape as the graph of $y = ax^2$.

If h is positive, the graph of $y = ax^2$ is shifted h units to the right.

If h is negative, the graph of $y = ax^2$ is shifted $|h|$ units to the left.

The vertex is $(h, 0)$ and the axis of symmetry is $x = h$.

The Graph of $f(x) = a(x - h)^2 + k$

Given a graph of $f(x) = a(x - h)^2$, what happens if we add a constant k? Suppose that we add 2. This increases $f(x)$ by 2, so the curve is moved up. If k is negative, the curve is moved down. The axis of symmetry for the parabola remains $x = h$, but the vertex will be at (h, k), or, equivalently, $(h, f(h))$.

Note that if a parabola opens upward $(a > 0)$, the function value, or y-value, at the vertex is a least, or *minimum*, value. That is, it is less than the y-value at any other point on the graph. If the parabola opens downward $(a < 0)$, the function value at the vertex is a greatest, or *maximum*, value.

technology connection
C

To study the effect of k on the graph of $f(x) = a(x - h)^2 + k$, let $y_1 = 7(x - 1)^2$ and $y_2 = 7(x - 1)^2 + 2$. Graph both y_1 and y_2 in the window $[-5, 5, -5, 5]$ and use TRACE or a TABLE to compare the y-values for any given x-value.

Next, let $y_3 = 7(x - 1)^2 - 4$ and compare its graph and y-values with those of y_1 and y_2. Try other values of k, including decimals and fractions. Describe the effect of adding k to the right side of $f(x) = a(x - h)^2$.

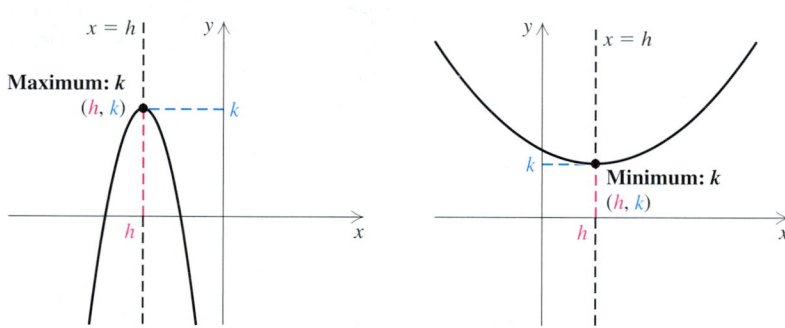

Graphing $f(x) = a(x - h)^2 + k$

The graph of $f(x) = a(x - h)^2 + k$ has the same shape as the graph of $y = a(x - h)^2$.

If k is positive, the graph of $y = a(x - h)^2$ is shifted k units up.

If k is negative, the graph of $y = a(x - h)^2$ is shifted $|k|$ units down.

The vertex is (h, k), and the axis of symmetry is $x = h$.

For $a > 0$, k is the minimum function value. For $a < 0$, k is the maximum function value.

E x a m p l e   4

Graph $g(x) = (x - 3)^2 - 5$, and find the minimum function value.

Solution   The graph will look like that of $f(x) = (x - 3)^2$ (see Example 2) but shifted 5 units down. You can confirm this by plotting some points. For instance, $g(4) = (4 - 3)^2 - 5 = -4$, whereas in Example 2, $f(4) = (4 - 3)^2 = 1$. The vertex is now $(3, -5)$, and the minimum function value is -5.

x	$g(x) = (x - 3)^2 - 5$
0	4
1	-1
2	-4
3	-5
4	-4
5	-1
6	4

←— Vertex

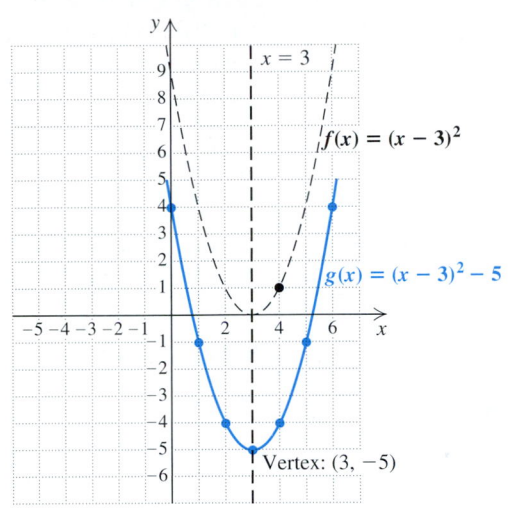

E x a m p l e   5

Graph $h(x) = \frac{1}{2}(x - 3)^2 + 5$, and find the minimum function value.

Solution   The graph looks just like that of $f(x) = \frac{1}{2}x^2$ but moved 3 units to the right and 5 units up. The vertex is $(3, 5)$, and the axis of symmetry is $x = 3$. We draw $f(x) = \frac{1}{2}x^2$ and then shift the curve over and up. The minimum function value is 5. By plotting some points, we have a check.

x	$h(x) = \frac{1}{2}(x - 3)^2 + 5$
0	$9\frac{1}{2}$
1	7
3	5
5	7
6	$9\frac{1}{2}$

←— Vertex

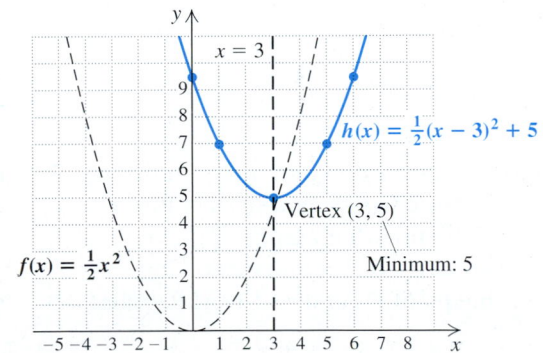

E x a m p l e 6

Graph $y = -2(x + 3)^2 + 5$. Find the vertex, the axis of symmetry, and the maximum or minimum value.

Solution   We first express the equation in the equivalent form

$$y = -2[x - (-3)]^2 + 5.$$

The graph looks like that of $y = -2x^2$ translated 3 units to the left and 5 units up. The vertex is $(-3, 5)$, and the axis of symmetry is $x = -3$. Since -2 is negative, we know that 5, the second coordinate of the vertex, is the maximum y-value.

We compute a few points as needed, selecting convenient x-values on either side of the vertex. The graph is shown here.

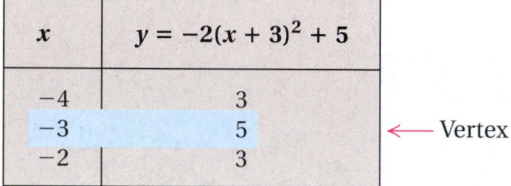

x	$y = -2(x + 3)^2 + 5$
-4	3
-3	5
-2	3

← Vertex

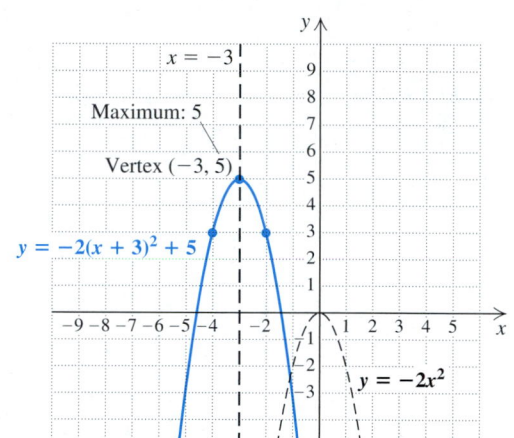

CONNECTING THE CONCEPTS

The ability to graph a function is an important skill. Later in this chapter, as well as in future courses, you will find that graphs of polynomial functions can be used as a tool for solving equations, inequalities, and real-world applications. In the process of learning how to graph quadratic functions, we have developed the ability to reflect or shift (translate) a graph. This skill will prove useful not only in future courses, but in Chapters 12 and 13 as well.

Exercise Set 11.6

Graph.

1. $f(x) = x^2$

2. $f(x) = -x^2$

3. $f(x) = -2x^2$

4. $f(x) = -3x^2$

5. $g(x) = \frac{1}{3}x^2$

6. $g(x) = \frac{1}{4}x^2$

Aha! **7.** $h(x) = -\frac{1}{3}x^2$

8. $h(x) = -\frac{1}{4}x^2$

9. $f(x) = \frac{5}{2}x^2$

10. $f(x) = \frac{3}{2}x^2$

For each of the following, graph the function, label the vertex, and draw the axis of symmetry.

11. $g(x) = (x + 4)^2$

12. $g(x) = (x + 1)^2$

13. $f(x) = (x - 1)^2$

14. $f(x) = (x - 2)^2$

15. $h(x) = (x - 3)^2$

16. $h(x) = (x - 4)^2$

17. $f(x) = -(x + 4)^2$

18. $f(x) = -(x - 2)^2$

19. $g(x) = -(x - 1)^2$

20. $g(x) = -(x + 1)^2$

21. $f(x) = 2(x + 1)^2$

22. $f(x) = 2(x + 4)^2$

23. $h(x) = -\frac{1}{2}(x - 4)^2$

24. $h(x) = -\frac{3}{2}(x - 2)^2$

25. $f(x) = \frac{1}{2}(x - 1)^2$

26. $f(x) = \frac{1}{3}(x + 2)^2$

27. $f(x) = -2(x + 5)^2$

28. $f(x) = 2(x + 7)^2$

29. $h(x) = -3\left(x - \frac{1}{2}\right)^2$

30. $h(x) = -2\left(x + \frac{1}{2}\right)^2$

For each of the following, graph the function and find the vertex, the axis of symmetry, and the maximum value or the minimum value.

31. $f(x) = (x - 5)^2 + 2$

32. $f(x) = (x + 3)^2 - 2$

33. $f(x) = (x + 1)^2 - 3$

34. $f(x) = (x - 1)^2 + 2$

35. $g(x) = (x + 4)^2 + 1$

36. $g(x) = -(x - 2)^2 - 4$

37. $h(x) = -2(x - 1)^2 - 3$

38. $h(x) = -2(x + 1)^2 + 4$

39. $f(x) = 2(x + 4)^2 + 1$

40. $f(x) = 2(x - 5)^2 - 3$

41. $g(x) = -\frac{3}{2}(x - 1)^2 + 4$

42. $g(x) = \frac{3}{2}(x + 2)^2 - 3$

Without graphing, find the vertex, the axis of symmetry, and the maximum value or the minimum value.

43. $f(x) = 8(x - 9)^2 + 7$

44. $f(x) = 10(x + 5)^2 - 6$

45. $h(x) = -\frac{2}{7}(x + 6)^2 + 11$

46. $h(x) = -\frac{3}{11}(x - 7)^2 - 9$

47. $f(x) = 5\left(x + \frac{1}{4}\right)^2 - 13$

48. $f(x) = 6\left(x - \frac{1}{4}\right)^2 + 15$

49. $f(x) = \sqrt{2}(x + 4.58)^2 + 65\pi$

50. $f(x) = 4\pi(x - 38.2)^2 - \sqrt{34}$

51. Explain, without plotting points, why the graph of $y = x^2 - 4$ looks like the graph of $y = x^2$ translated 4 units down.

52. Explain, without plotting points, why the graph of $y = (x + 2)^2$ looks like the graph of $y = x^2$ translated 2 units to the left.

SKILL MAINTENANCE

Graph using intercepts.

53. $2x - 7y = 28$

54. $6x - 3y = 36$

Solve each system.

55. $3x + 4y = -19,$
$7x - 6y = -29$

56. $5x + 7y = 9,$
$3x - 4y = -11$

Complete the square.

57. $x^2 + 5x +$ _____

58. $x^2 - 9x +$ _____

SYNTHESIS

59. Before graphing a quadratic function, Sophie always plots five points. First, she calculates and plots the coordinates of the vertex. Then she plots *four* more points after calculating *two* more ordered pairs. How is this possible?

60. If the graphs of $f(x) = a_1(x - h_1)^2 + k_1$ and $g(x) = a_2(x - h_2)^2 + k_2$ have the same shape, what, if anything, can you conclude about the a's, the h's, and the k's? Why?

Write an equation for a function having a graph with the same shape as the graph of $f(x) = \frac{3}{5}x^2$, but with the given point as the vertex.

61. $(4, 1)$ **62.** $(2, 6)$ **63.** $(3, -1)$

64. $(5, -6)$ **65.** $(-2, -5)$ **66.** $(-4, -2)$

For each of the following, write the equation of the parabola that has the shape of $f(x) = 2x^2$ or $g(x) = -2x^2$ and has a maximum or minimum value at the specified point.

67. Maximum: $(5, 0)$ **68.** Minimum: $(2, 0)$

69. Minimum: $(-4, 0)$ **70.** Maximum: $(0, 3)$

71. Maximum: $(3, 8)$ **72.** Minimum: $(-2, 3)$

Find an equation for a quadratic function F that satisfies the following conditions.

73. The graph of F is the same shape as the graph of f, where $f(x) = 3(x + 2)^2 + 7$, and $F(x)$ is a minimum at the same point that $g(x) = -2(x - 5)^2 + 1$ is a maximum.

74. The graph of F is the same shape as the graph of f, where $f(x) = -\frac{1}{3}(x - 2)^2 + 7$, and $F(x)$ is a maximum at the same point that $g(x) = 2(x + 4)^2 - 6$ is a minimum.

Functions other than parabolas can be translated. When calculating $f(x)$, if we replace x with $x - h$, where h is a constant, the graph will be moved horizontally. If we replace $f(x)$ with $f(x) + k$, the graph will be moved vertically.

Use the graph below for Exercises 75–80.

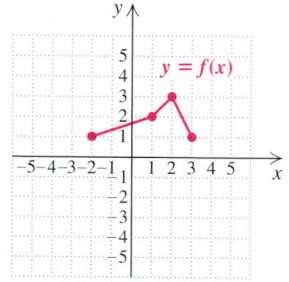

Draw a graph of each of the following.

75. $y = f(x - 1)$

76. $y = f(x + 2)$

77. $y = f(x) + 2$

78. $y = f(x) - 3$

79. $y = f(x + 3) - 2$

80. $y = f(x - 3) + 1$

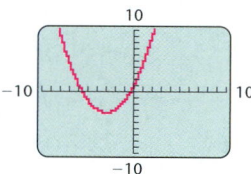 **81.** Use the TRACE and/or TABLE features of a grapher to confirm the maximum and minimum values given as answers to Exercises 43, 45, and 47. Be sure to adjust the window appropriately. On some graphers, a maximum or minimum option may be available by using a CALC key.

82. Use a grapher to check your graphs for Exercises 10, 20, and 40.

83. While trying to graph $y = -\frac{1}{2}x^2 + 3x + 1$, Omar gets the following screen.

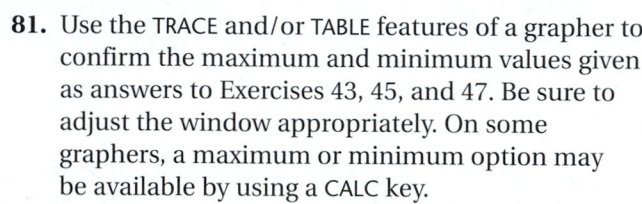

How can Omar tell at a glance that a mistake has been made?

CORNER

Match the Graph

COLLABORATIVE

Focus: Graphing quadratic functions

Time: 15–20 minutes

Group size: 6

Materials: Index cards

ACTIVITY

1. On each of six index cards, write one of the following equations:

$$y = \tfrac{1}{2}(x - 3)^2 + 1; \qquad y = \tfrac{1}{2}(x - 1)^2 + 3;$$
$$y = \tfrac{1}{2}(x + 1)^2 - 3; \qquad y = \tfrac{1}{2}(x + 3)^2 + 1;$$
$$y = \tfrac{1}{2}(x + 3)^2 - 1; \qquad y = \tfrac{1}{2}(x + 1)^2 + 3.$$

2. Fold each index card and mix up the six cards in a hat or bag. Then, one by one, each group member should select one of the equations. Do not let anyone see your equation.

3. Each group member should carefully graph the equation selected. Make the graph large enough so that when it is finished, it can be easily viewed by the rest of the group. Be sure to scale the axes and label the vertex, but **do not label the graph with the equation used.**

4. When all group members have drawn a graph, place the graphs in a pile. The group should then match and agree on the correct equation for each graph *with no help from the person who drew the graph.* If a mistake has been made and a graph has no match, determine what its equation *should* be.

5. Compare your group's labeled graphs with those of other groups to reach consensus within the class on the correct label for each graph.

More About Graphing Quadratic Functions

11.7

Completing the Square • Finding Intercepts

Completing the Square

By *completing the square* (see Section 11.1), we can rewrite any polynomial $ax^2 + bx + c$ in the form $a(x - h)^2 + k$. Once that has been done, the procedures discussed in Section 11.6 will enable us to graph any quadratic function.

E x a m p l e   1   Graph: $g(x) = x^2 - 6x + 4$.

Solution   We have

$$g(x) = x^2 - 6x + 4$$
$$= (x^2 - 6x) + 4.$$

To complete the square inside the parentheses, we take half the x-coefficient, $\frac{1}{2} \cdot (-6) = -3$, and square it to get $(-3)^2 = 9$. Then we add $9 - 9$ inside the parentheses:

$$g(x) = (x^2 - 6x + 9 - 9) + 4 \qquad \text{The effect is of adding 0.}$$
$$= (x^2 - 6x + 9) + (-9 + 4) \qquad \text{Using the associative law of addition to regroup}$$
$$= (x - 3)^2 - 5. \qquad \text{Factoring and simplifying}$$

This equation was graphed in Example 4 of Section 11.6. The graph is that of $f(x) = x^2$ translated right 3 units and down 5 units. The vertex is $(3, -5)$.

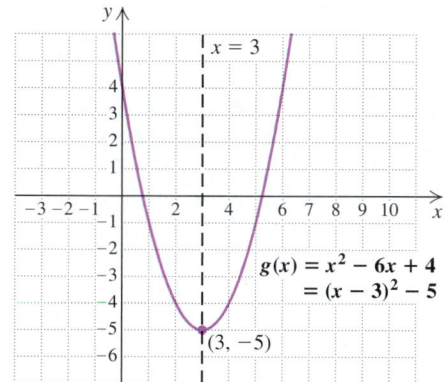

When the leading coefficient is not 1, we factor out that number from the first two terms. Then we complete the square.

E x a m p l e   2   Graph: $f(x) = 3x^2 + 12x + 13$.

Solution   Since the coefficient of x^2 is not 1, we need to factor out that number—in this case, 3—from the first two terms. Remember that we want the form $f(x) = a(x - h)^2 + k$:

$$f(x) = 3x^2 + 12x + 13$$
$$= 3(x^2 + 4x) + 13.$$

Now we complete the square as before. We take half of the x-coefficient, $\frac{1}{2} \cdot 4 = 2$, and square it: $2^2 = 4$. Then we add $4 - 4$ inside the parentheses:

$$f(x) = 3(x^2 + 4x + 4 - 4) + 13. \qquad \text{Adding } 4 - 4, \text{ or 0, inside the parentheses}$$

The distributive law allows us to separate the -4 from the perfect-square trinomial so long as it is multiplied by 3:

$$f(x) = 3(x^2 + 4x + 4) + 3(-4) + 13 \qquad \text{This leaves a perfect-square trinomial inside the parentheses.}$$
$$= 3(x + 2)^2 + 1. \qquad \text{Factoring and simplifying}$$

The vertex is $(-2, 1)$, and the axis of symmetry is $x = -2$. The coefficient of x^2 is 3, so the graph is narrow and opens upward. We choose a few x-values on either side of the vertex, compute *y*-values, and then graph the parabola.

x	$f(x) = 3(x + 2)^2 + 1$
−2	1
−3	4
−1	4

← Vertex

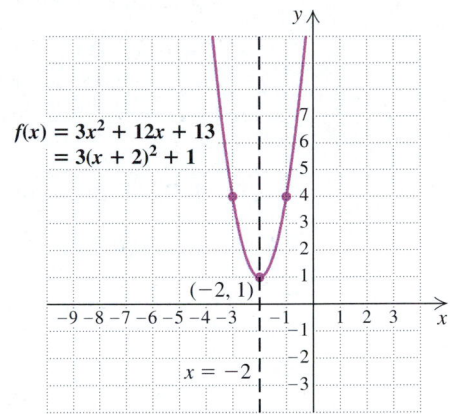

$f(x) = 3x^2 + 12x + 13$
$= 3(x + 2)^2 + 1$

$(-2, 1)$

$x = -2$

E x a m p l e   3

Graph: $f(x) = -2x^2 + 10x - 7$.

Solution   We first find the vertex by completing the square. To do so, we factor out -2 from the first two terms of the expression. This makes the coefficient of x^2 inside the parentheses 1:

$$f(x) = -2x^2 + 10x - 7$$
$$= -2(x^2 - 5x) - 7.$$

Now we complete the square as before. We take half of the *x*-coefficient and square it to get $\frac{25}{4}$. Then we add $\frac{25}{4} - \frac{25}{4}$ inside the parentheses:

$$f(x) = -2\left(x^2 - 5x + \frac{25}{4} - \frac{25}{4}\right) - 7$$
$$= -2\left(x^2 - 5x + \frac{25}{4}\right) + (-2)\left(-\frac{25}{4}\right) - 7 \qquad \text{Multiplying by } -2, \text{ using the distributive law, and regrouping}$$
$$= -2\left(x - \frac{5}{2}\right)^2 + \frac{11}{2}. \qquad \text{Factoring and simplifying}$$

The vertex is $\left(\frac{5}{2}, \frac{11}{2}\right)$, and the axis of symmetry is $x = \frac{5}{2}$. The coefficient of x^2, -2, is negative, so the graph opens downward. We plot a few points on either side of the vertex, including the *y*-intercept, $f(0)$, and graph the parabola.

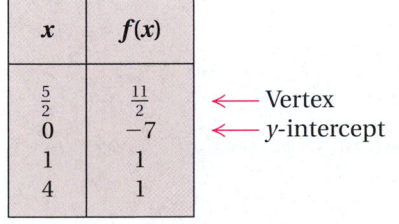

x	$f(x)$
$\frac{5}{2}$	$\frac{11}{2}$
0	-7
1	1
4	1

← Vertex
← *y*-intercept

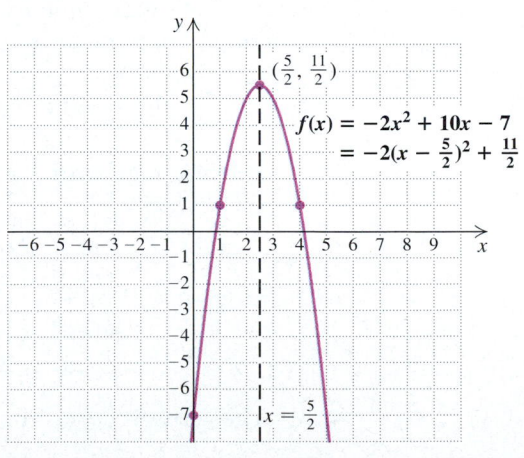

$\left(\frac{5}{2}, \frac{11}{2}\right)$

$f(x) = -2x^2 + 10x - 7$
$= -2\left(x - \frac{5}{2}\right)^2 + \frac{11}{2}$

$x = \frac{5}{2}$

The method used in Examples 1–3 can be generalized to find a formula for locating the vertex. We complete the square as follows:

$$f(x) = ax^2 + bx + c$$

$$= a\left(x^2 + \frac{b}{a}x\right) + c. \qquad \text{Factoring } a \text{ out of the first two terms.}$$
Check by multiplying.

Half of the x-coefficient, $\dfrac{b}{a}$, is $\dfrac{b}{2a}$. We square it to get $\dfrac{b^2}{4a^2}$ and add $\dfrac{b^2}{4a^2} - \dfrac{b^2}{4a^2}$ inside the parentheses. Then we distribute the a and regroup terms:

$$f(x) = a\left(x^2 + \frac{b}{a}x + \frac{b^2}{4a^2} - \frac{b^2}{4a^2}\right) + c$$

$$= a\left(x^2 + \frac{b}{a}x + \frac{b^2}{4a^2}\right) + a\left(-\frac{b^2}{4a^2}\right) + c \qquad \text{Using the distributive law}$$

$$= a\left(x + \frac{b}{2a}\right)^2 + \frac{-b^2}{4a} + \frac{4ac}{4a} \qquad \text{Factoring and finding a common denominator}$$

$$= a\left[x - \left(-\frac{b}{2a}\right)\right]^2 + \frac{4ac - b^2}{4a}.$$

Thus we have the following.

> ### The Vertex of a Parabola
>
> The vertex of the parabola given by $f(x) = ax^2 + bx + c$ is
>
> $$\left(-\frac{b}{2a}, f\left(-\frac{b}{2a}\right)\right) \quad \text{or} \quad \left(-\frac{b}{2a}, \frac{4ac - b^2}{4a}\right).$$
>
> The x-coordinate of the vertex is $-b/(2a)$. The axis of symmetry is $x = -b/(2a)$. The second coordinate of the vertex is most commonly found by computing $f\left(-\dfrac{b}{2a}\right)$.

Let's reexamine Example 3 to see how we could have found the vertex directly. From the formula above,

$$\text{the } x\text{-coordinate of the vertex is } -\frac{b}{2a} = -\frac{10}{2(-2)} = \frac{5}{2}.$$

Substituting $\frac{5}{2}$ into $f(x) = -2x^2 + 10x - 7$, we find the second coordinate of the vertex:

$$f\left(\tfrac{5}{2}\right) = -2\left(\tfrac{5}{2}\right)^2 + 10\left(\tfrac{5}{2}\right) - 7$$
$$= -2\left(\tfrac{25}{4}\right) + 25 - 7$$
$$= -\tfrac{25}{2} + 18$$
$$= -\tfrac{25}{2} + \tfrac{36}{2} = \tfrac{11}{2}.$$

The vertex is $\left(\tfrac{5}{2}, \tfrac{11}{2}\right)$. The axis of symmetry is $x = \tfrac{5}{2}$.

We have actually developed two methods for finding the vertex. One is by completing the square and the other is by using a formula. You should check with your instructor about which method to use.

Finding Intercepts

The points at which a graph crosses an axis are called intercepts. For $f(x) = ax^2 + bx + c$, the y-intercept is simply $(0, c)$. To find x-intercepts, we look for points where $y = 0$ or $f(x) = 0$. To find the x-intercepts of the quadratic function given by $f(x) = ax^2 + bx + c$, we solve the equation

$$0 = ax^2 + bx + c.$$

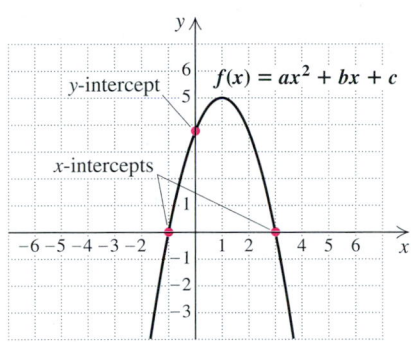

E x a m p l e   4   Find the x- and y-intercepts of the graph of $f(x) = x^2 - 2x - 2$.

Solution   The y-intercept is simply $(0, f(0))$, or $(0, -2)$. To find the x-intercepts, we solve the equation

$$0 = x^2 - 2x - 2.$$

We are unable to factor $x^2 - 2x - 2$, so we use the quadratic formula and get $x = 1 \pm \sqrt{3}$. Thus the x-intercepts are $\left(1 - \sqrt{3}, 0\right)$ and $\left(1 + \sqrt{3}, 0\right)$.

If graphing, we would approximate, to get $(-0.7, 0)$ and $(2.7, 0)$.

For each quadratic function, (a) find the vertex and the axis of symmetry and (b) graph the function.

1. $f(x) = x^2 + 4x + 5$

2. $f(x) = x^2 + 2x - 5$

3. $g(x) = x^2 - 6x + 13$

4. $g(x) = x^2 - 4x + 5$

5. $f(x) = x^2 + 8x + 20$

6. $f(x) = x^2 - 10x + 21$

7. $h(x) = 2x^2 - 16x + 25$

8. $h(x) = 2x^2 + 16x + 23$

9. $f(x) = -x^2 + 2x + 5$

10. $f(x) = -x^2 - 2x + 7$

11. $g(x) = x^2 + 3x - 10$

12. $g(x) = x^2 + 5x + 4$

13. $f(x) = 3x^2 - 24x + 50$

14. $f(x) = 4x^2 + 8x - 3$

15. $h(x) = x^2 + 7x$

16. $h(x) = x^2 - 5x$

17. $f(x) = -2x^2 - 4x - 6$

18. $f(x) = -3x^2 + 6x + 2$

19. $g(x) = 2x^2 - 8x + 3$

20. $g(x) = 2x^2 + 5x - 1$

21. $f(x) = -3x^2 + 5x - 2$

22. $f(x) = -3x^2 - 7x + 2$

23. $h(x) = \frac{1}{2}x^2 + 4x + \frac{19}{3}$

24. $h(x) = \frac{1}{2}x^2 - 3x + 2$

Find the x- and y-intercepts. If no x-intercepts exist, state this.

25. $f(x) = x^2 - 6x + 3$

26. $f(x) = x^2 + 5x + 2$

27. $g(x) = -x^2 + 2x + 3$

28. $g(x) = x^2 - 6x + 9$

Aha! **29.** $f(x) = x^2 - 9x$

30. $f(x) = x^2 - 7x$

31. $h(x) = -x^2 + 4x - 4$

32. $h(x) = 4x^2 - 12x + 3$

33. $f(x) = 2x^2 - 4x + 6$

34. $f(x) = x^2 - x + 2$

35. Does the graph of every quadratic function have a *y*-intercept? Why or why not?

36. Is it possible for the graph of a quadratic function to have only one *x*-intercept if the vertex is off the *x*-axis? Why or why not?

SKILL MAINTENANCE

Solve each system.

37. $5x - 3y = 16,$
$4x + 2y = 4$

38. $2x - 5y = 9,$
$5x - 15y = 20$

39. $4a - 5b + c = 3,$
$3a - 4b + 2c = 3,$
$a + b - 7c = -2$

40. $2a - 7b + c = 25,$
$a + 5b - 2c = -18,$
$3a - b + 4c = 14$

Solve.

41. $\sqrt{4x - 4} = \sqrt{x + 4} + 1$

42. $\sqrt{5x - 4} + \sqrt{13 - x} = 7$

SYNTHESIS

43. If the graphs of two quadratic functions have the same *x*-intercepts, will they also have the same vertex? Why or why not?

44. Suppose that the graph of $f(x) = ax^2 + bx + c$ has $(x_1, 0)$ and $(x_2, 0)$ as *x*-intercepts. Explain why the graph of $g(x) = -ax^2 - bx - c$ will also have $(x_1, 0)$ and $(x_2, 0)$ as *x*-intercepts.

For each quadratic function, find (a) the maximum or minimum value and (b) the x- and y-intercepts.

45. $f(x) = 2.31x^2 - 3.135x - 5.89$

46. $f(x) = -18.8x^2 + 7.92x + 6.18$

47. Graph the function
$$f(x) = x^2 - x - 6.$$
Then use the graph to approximate solutions to each of the following equations.

a) $x^2 - x - 6 = 2$

b) $x^2 - x - 6 = -3$

48. Graph the function

$$f(x) = \frac{x^2}{2} + x - \frac{3}{2}.$$

Then use the graph to approximate solutions to each of the following equations.

a) $\dfrac{x^2}{2} + x - \dfrac{3}{2} = 0$

b) $\dfrac{x^2}{2} + x - \dfrac{3}{2} = 1$

c) $\dfrac{x^2}{2} + x - \dfrac{3}{2} = 2$

Find an equivalent equation of the type

$$f(x) = a(x - h)^2 + k.$$

49. $f(x) = mx^2 - nx + p$

50. $f(x) = 3x^2 + mx + m^2$

51. A quadratic function has $(-1, 0)$ as one of its intercepts and $(3, -5)$ as its vertex. Find an equation for the function.

52. A quadratic function has $(4, 0)$ as one of its intercepts and $(-1, 7)$ as its vertex. Find an equation for the function.

Graph.

53. $f(x) = |x^2 - 1|$

54. $f(x) = |x^2 - 3x - 4|$

55. $f(x) = |2(x - 3)^2 - 5|$

 56. Use a grapher to check your answers to Exercises 9, 23, 33, 45, and 47.

Problem Solving and Quadratic Functions

11.8

Maximum and Minimum Problems • Fitting Quadratic Functions to Data

Let's look now at some of the many situations in which quadratic functions are used for problem solving.

Maximum and Minimum Problems

We have seen that for any quadratic function f, the value of $f(x)$ at the vertex is either a maximum or a minimum. Thus problems in which a quantity must be maximized or minimized can often be solved by finding the coordinates of a vertex. This assumes that the problem can be modeled with a quadratic function.

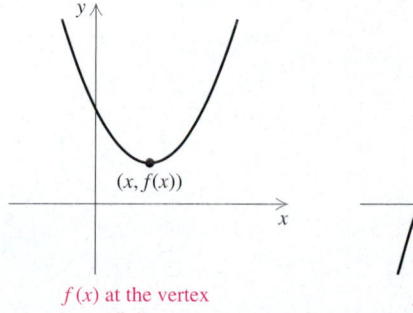

$f(x)$ at the vertex
a minimum

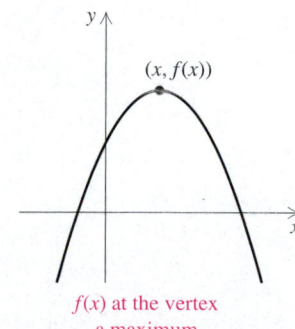

$f(x)$ at the vertex
a maximum

Example 1

Newborn calves. The number of pounds of milk per day recommended for a calf that is x weeks old can be approximated by $p(x)$, where $p(x) = -0.2x^2 + 1.3x + 6.2$ (*Source*: C. Chaloux, University of Vermont, 1998). When is a calf's milk consumption greatest and how much milk does it consume at that time?

Solution

1., 2. Familiarize and **Translate.** We are given the function for milk consumption by a calf. Note that it is a quadratic function of x, the calf's age in weeks. If it is not difficult to do so, we can generate a graph of the function. The graph (shown at left) indicates that the calf's consumption increases and then decreases.

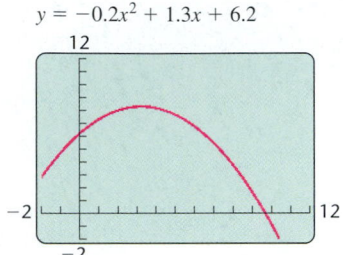

$y = -0.2x^2 + 1.3x + 6.2$

A visualization for Example 1

3. Carry out. We can either complete the square,

$$p(x) = -0.2x^2 + 1.3x + 6.2$$
$$= -0.2(x^2 - 6.5x) + 6.2$$
$$= -0.2(x^2 - 6.5x + 3.25^2 - 3.25^2) + 6.2 \qquad \text{Completing the square; } \frac{6.5}{2} = 3.25$$
$$= -0.2(x^2 - 6.5x + 3.25^2) + (-0.2)(-3.25^2) + 6.2$$
$$= -0.2(x - 3.25)^2 + 8.3125 \qquad \text{Factoring and simplifying}$$

or we can use $-b/(2a) = -1.3/(-0.4) = 3.25$. Using a calculator, we find that

$$p(3.25) = -0.2(3.25)^2 + 1.3(3.25) + 6.2 = 8.3125.$$

4. Check. Both of the approaches in step (3) indicate that a maximum occurs when $x = 3.25$, or $3\frac{1}{4}$. The graph also serves as a check.

5. State. A calf's milk consumption is greatest when the calf is $3\frac{1}{4}$ weeks old. At that time, it drinks about 8.3 lb of milk per day.

Example 2

Fenced-in land. What are the dimensions of the largest rectangular pen that a farmer can enclose with 64 m of electric fence?

Solution

1. Familiarize. We make a drawing and label it. Recall these important formulas:

Perimeter: $2w + 2l$;

Area: $l \cdot w$.

To get a better feel for the problem, we can look at some possible dimensions for a rectangular pen that can be enclosed with 64 m of fence. All possibilities are chosen so that $2w + 2l = 64$.

l	w	Perimeter	Area
22 m	10 m	64 m	220 m^2
20 m	12 m	64 m	240 m^2
18 m	14 m	64 m	252 m^2
.	.		.
.	.		.
.	.		.

What choice of l and w will maximize A?

2. **Translate.** We have two equations: One guarantees that the perimeter is 64 m; the other expresses area in terms of length and width.

$$2w + 2l = 64$$
$$A = l \cdot w$$

3. **Carry out.** We need to express A as a function of l or w but not both. To do so, we solve for l in the first equation to obtain $l = 32 - w$. Substituting for l in the second equation, we get a quadratic function:

$$A = (32 - w)w \qquad \text{Substituting for } l$$
$$= -w^2 + 32w. \qquad \text{This is a parabola opening downward, so a maximum exists.}$$

Completing the square, we get

$$A = -(w^2 - 32w + 256 - 256)$$
$$= -(w - 16)^2 + 256.$$

The maximum function value, 256 m^2, occurs when $w = 16$ m and $l = 32 - 16$, or 16 m.

4. **Check.** Note that 256 m^2 is greater than any of the values for A found in the *Familiarize* step. To be more certain, we could check values other than those used in that step. For example, if $w = 15$ m, then $l = 32 - 15 = 17$ m, and $A = 15 \cdot 17 = 255$ m^2. The same area results if $w = 17$ m and $l = 15$ m. Since 256 m^2 is greater than 255 m^2, it looks as though we have a maximum.

5. **State.** The largest rectangular pen that can be enclosed is 16 m by 16 m.

technology connection

A

To generate a table of values on your grapher, let x represent the width of the pen, in meters. If l represents the length, in meters, we must have $64 = 2x + 2l$. Next, solve for l and use that expression for y_1. Then let $y_2 = x$ and $y_3 = y_1 \cdot y_2$ and create a table.

Fitting Quadratic Functions to Data

Whenever a certain quadratic function fits a situation, that function can be determined if three inputs and their outputs are known. Each of the given ordered pairs is called a *data point*.

E x a m p l e   3

Hydrology.   The drawing below shows the cross section of a river. Typically rivers are deepest in the middle, with the depth decreasing to 0 at the edges. A hydrologist measures the depths D, in feet, of a river at distances x, in feet, from one bank. The results are listed in the table below.

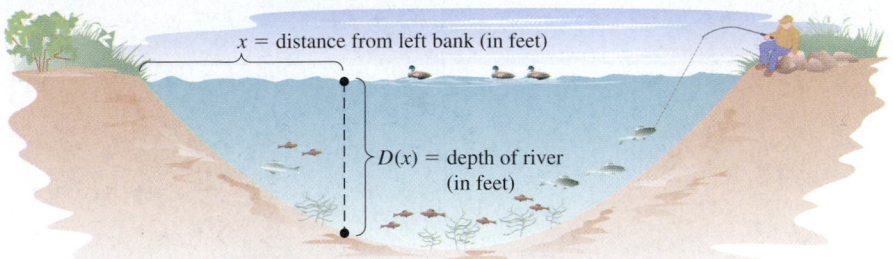

x = distance from left bank (in feet)

$D(x)$ = depth of river (in feet)

Distance x, from the Left Bank (in feet)	Depth, D, of the River (in feet)
0	0
15	10.2
25	17
50	20
90	7.2
100	0

a) Plot the data and decide whether the data seem to fit a quadratic function.

b) Use the data points $(0, 0)$, $(50, 20)$, and $(100, 0)$ to find a quadratic function that fits the data.

c) Use the function to estimate the depth of the river 70 ft from the left bank.

Solution

a)   We plot the data points as follows.

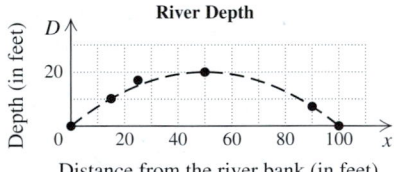

River Depth

Depth (in feet)

Distance from the river bank (in feet)

The data seem to rise and fall in a manner similar to a quadratic function. The dashed curve represents a quadratic function that closely fits the data. Note that it need not pass through each point.

b) We are looking for a quadratic function of the form

$$D(x) = ax^2 + bx + c.$$

To determine the constants a, b, and c, we use the points $(0, 0)$, $(50, 20)$, and $(100, 0)$ and substitute:

$$0 = a \cdot 0^2 + b \cdot 0 + c,$$
$$20 = a \cdot 50^2 + b \cdot 50 + c,$$
$$0 = a \cdot 100^2 + b \cdot 100 + c.$$

technology connection

B

Enter the data from the table in Example 3 using the EDIT option of the STAT menu. Then use the CALC option of the STAT menu to find the quadratic function that most closely fits the data. Finally, graph the function just developed.

After simplifying, we see that we need to solve the system

$$0 = c,$$
$$20 = 2{,}500a + 50b + c,$$
$$0 = 10{,}000a + 100b + c.$$

Since $c = 0$, the system reduces to a system of two equations in two variables:

$$20 = 2{,}500a + 50b, \qquad (1)$$
$$0 = 10{,}000a + 100b. \qquad (2)$$

We multiply both sides of equation (1) by -2, add, and solve for a:

$$-40 = -5{,}000a - 100b,$$
$$\underline{0 = 10{,}000a + 100b}$$
$$-40 = 5000a \qquad \text{Adding}$$

$$\frac{-40}{5000} = a \qquad \text{Solving for } a$$

$$-0.008 = a.$$

Next, we substitute -0.008 for a in equation (2) and solve for b:

$$0 = 10{,}000(-0.008) + 100b$$
$$0 = -80 + 100b$$
$$80 - 100b$$
$$0.8 = b.$$

We can now rewrite $D(x) = ax^2 + bx + c$ as

$$D(x) = -0.008x^2 + 0.8x.$$

c) To find the depth 70 ft from the riverbank, we substitute:

$$D(70) = -0.008(70)^2 + 0.8(70) = 16.8.$$

At a distance of 70 ft from the riverbank, the river is 16.8 ft deep.

Exercise Set **11.8**

Solve.

1. *Stock prices.*   The value of a share of R. P. Mugahti can be represented by $V(x) = x^2 - 6x + 13$, where x is the number of months after January 2001. What is the lowest value $V(x)$ will reach, and when will that occur?

2. *Minimizing cost.*   Aki's Bicycle Designs has determined that when x hundred bicycles are built, the average cost per bicycle is given by

    $$C(x) = 0.1x^2 - 0.7x + 2.425,$$

    where $C(x)$ is in hundreds of dollars. What is the minimum average cost per bicycle and how many bicycles should be built to achieve that minimum?

3. *Ticket sales.*   The number of tickets sold each day for an upcoming performance of Handel's Messiah is given by

    $$N(x) = -0.4x^2 + 9x + 11,$$

    where x is the number of days since the concert was first announced. When will daily ticket sales peak and how many tickets will be sold that day?

4. *Maximizing profit.*   Recall that total profit P is the difference between total revenue R and total cost C. Given $R(x) = 1000x - x^2$ and $C(x) = 3000 + 20x$, find the total profit, the maximum value of the total profit, and the value of x at which it occurs.

5. *Architecture.*   An architect is designing an atrium for a hotel. The atrium is to be rectangular with a perimeter of 720 ft of brass piping. What dimensions will maximize the area of the atrium?

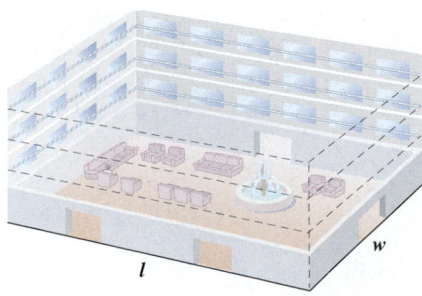

6. *Stained-glass window design.*   An artist is designing a rectangular stained-glass window with a perimeter of 84 in. What dimensions will yield the maximum area?

7. *Garden design.*   A farmer decides to enclose a rectangular garden, using the side of a barn as one side of the rectangle. What is the maximum area that the farmer can enclose with 40 ft of fence? What should the dimensions of the garden be in order to yield this area?

8. *Patio design.*   A stone mason has enough stones to enclose a rectangular patio with 60 ft of perimeter, assuming that the attached house forms one side of the rectangle. What is the maximum area that the mason can enclose? What should the dimensions of the patio be in order to yield this area?

9. *Molding plastics.*   Economite Plastics plans to produce a one-compartment vertical file by bending the long side of an 8-in. by 14-in. sheet of plastic along two lines to form a U shape. How tall should the file be in order to maximize the volume that the file can hold?

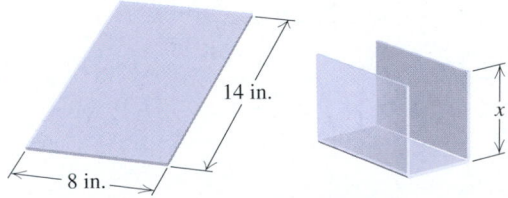

14 in.

8 in.

x

10. *Composting.*   A rectangular compost container is to be formed in a corner of a fenced yard, with 8 ft

of chicken wire completing the other two sides of the rectangle. If the chicken wire is 3 ft high, what dimensions of the base will maximize the container's volume?

11. What is the maximum product of two numbers that add to 18? What numbers yield this product?

12. What is the maximum product of two numbers that add to 26? What numbers yield this product?

13. What is the minimum product of two numbers that differ by 8? What are the numbers?

14. What is the minimum product of two numbers that differ by 7? What are the numbers?

Aha! **15.** What is the maximum product of two numbers that add to −10? What numbers yield this product?

16. What is the maximum product of two numbers that add to −12? What numbers yield this product?

For Exercises 17–24, state whether the graph appears to represent a quadratic function.

17.

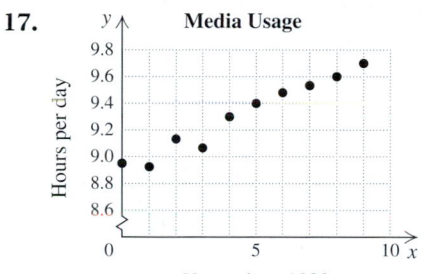

Media Usage

18.

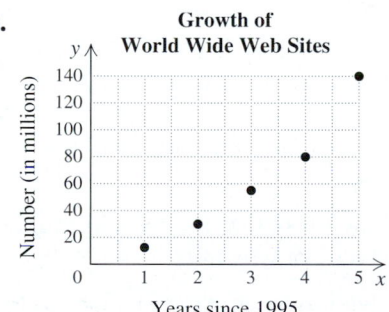

Growth of World Wide Web Sites

19.

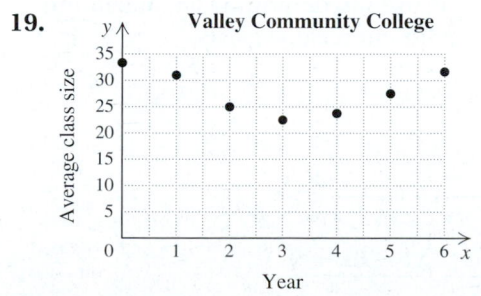

Valley Community College

20.

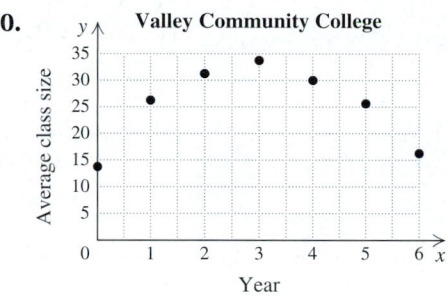

Valley Community College

21.

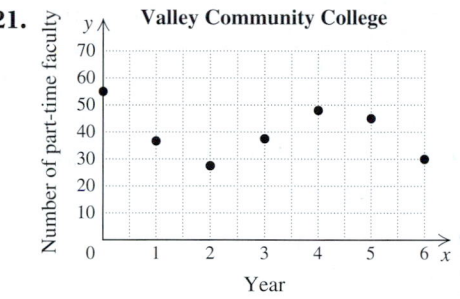

Valley Community College

22.

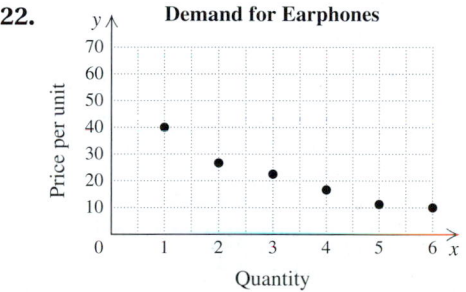

Demand for Earphones

23.

U.S. Trade Deficit with Japan

24.

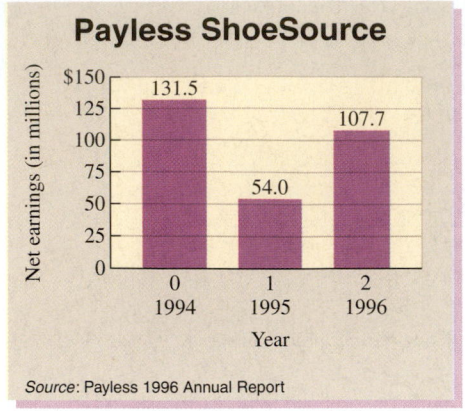

Payless ShoeSource

Net earnings (in millions)

Year	
0 / 1994	131.5
1 / 1995	54.0
2 / 1996	107.7

Source: Payless 1996 Annual Report

Find a quadratic function that fits the set of data points.

25. $(1, 4), (-1, -2), (2, 13)$

26. $(1, 4), (-1, 6), (-2, 16)$

27. $(2, 0), (4, 3), (12, -5)$

28. $(-3, -30), (3, 0), (6, 6)$

29. a) Find a quadratic function that fits the following data.

Travel Speed (in kilometers per hour)	Number of Nighttime Accidents (for every 200 million kilometers driven)
60	400
80	250
100	250

b) Use the function to estimate the number of nighttime accidents that occur at 50 km/h.

30. a) Find a quadratic function that fits the following data.

Travel Speed (in kilometers per hour)	Number of Daytime Accidents (for every 200 million kilometers driven)
60	100
80	130
100	200

b) Use the function to estimate the number of daytime accidents that occur at 50 km/h.

31. *Archery.* The Olympic flame tower at the 1992 Summer Olympics was lit at a height of about 27 m by a flaming arrow that was launched about 63 m from the base of the tower. If the arrow landed about 63 m beyond the tower, find a quadratic function that expresses the height h of the arrow as a function of the distance d that it traveled horizontally.

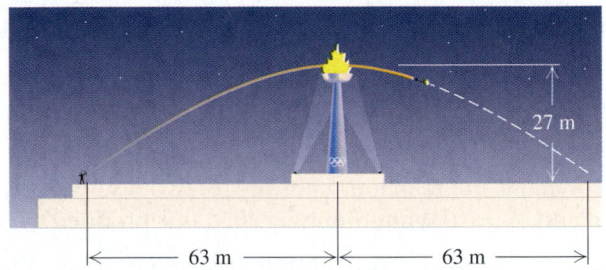

32. *Pizza prices.* Pizza Unlimited has the following prices for pizzas.

Diameter	Price
8 in.	$ 6.00
12 in.	$ 8.50
16 in.	$11.50

Is price a quadratic function of diameter? It probably should be, because the price should be proportional to the area, and the area is a quadratic function of the diameter. (The area of a circular region is given by $A = \pi r^2$ or $(\pi/4) \cdot d^2$.)

a) Express price as a quadratic function of diameter using the data points $(8, 6), (12, 8.50)$, and $(16, 11.50)$.

b) Use the function to find the price of a 14-in. pizza.

33. Does every nonlinear function have a minimum or maximum value? Why or why not?

34. Explain how the leading coefficient of a quadratic function can be used to determine if a maximum or a minimum function value exists.

SKILL MAINTENANCE

Simplify.

35. $\dfrac{x}{x^2 + 17x + 72} - \dfrac{8}{x^2 + 15x + 56}$

36. $\dfrac{x^2 - 9}{x^2 - 8x + 7} \div \dfrac{x^2 + 6x + 9}{x^2 - 1}$

37. $\dfrac{t^2 - 4}{t^2 - 7t - 8} \cdot \dfrac{t^2 - 64}{t^2 - 5t + 6}$

38. $\dfrac{t}{t^2 - 10t + 21} + \dfrac{t}{t^2 - 49}$

Solve.

39. $5x - 9 < 31$ **40.** $3x - 8 \geq 22$

SYNTHESIS

41. Write a problem for a classmate to solve. Design the problem so that its solution requires finding the minimum or maximum value.

42. Explain what restrictions should be placed on the quadratic functions developed in Exercises 29 and 32 and why such restrictions are needed.

43. *Norman window.*   A *Norman window* is a rectangle with a semicircle on top. Big Sky Windows is designing a Norman window that will require 24 ft of trim. What dimensions will allow the maximum amount of light to enter a house?

44. *Minimizing area.*   A 36-in. piece of string is cut into two pieces. One piece is used to form a circle while the other is used to form a square. How should the string be cut so that the sum of the areas is a minimum?

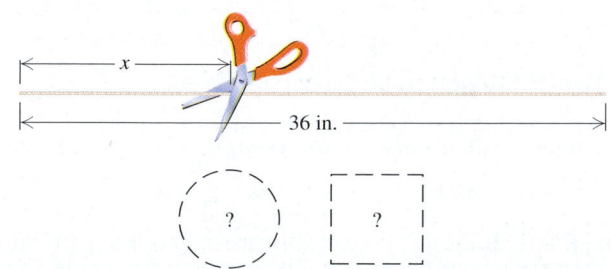

45. *Crop yield.*   An orange grower finds that she gets an average yield of 40 bushels (bu) per tree when she plants 20 trees on an acre of ground. Each time she adds a tree to an acre, the yield per tree decreases by 1 bu, due to congestion. How many trees per acre should she plant for maximum yield?

46. *Cover charges.*   When the owner of Sweet Sounds charges a $10 cover charge, an average of 80 people will attend a show. For each 25¢ increase in admission price, the average number attending decreases by 1. What should the owner charge in order to make the most money?

47. *Trajectory of a launched object.*   The height above the ground of a launched object is a quadratic function of the time that it is in the air. Suppose that a flare is launched from a cliff 64 ft above sea level. If 3 sec after being launched the flare is again level with the cliff, and if 2 sec after that it lands in the sea, what is the maximum height that the flare will reach?

48. *Bridge design.*   The cables supporting a straight-line suspension bridge are nearly parabolic in shape. Suppose that a suspension bridge is being designed with concrete supports 160 ft apart and with vertical cables 30 ft above road level at the midpoint of the bridge and 80 ft above road level at a point 50 ft from the midpoint of the bridge. How long are the longest vertical cables?

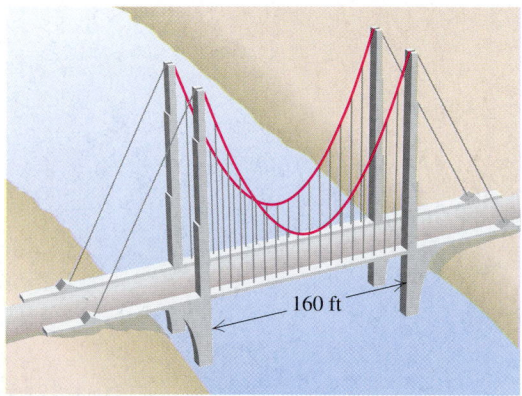

160 ft

49. Many graphers have a QUADREG option as part of the CALC option of the STAT menu. Such a feature can quickly fit a quadratic function to a LIST of ordered pairs. Use the QUADREG feature to check your answers to Exercises 25, 27, and 29.

CORNER

Quadratic Counter Settings

Focus: Modeling quadratic functions

Time: 20–30 minutes

Group size: 3 or 4

Materials: Graphers are optional.

The Panasonic Portable Stereo System RX-DT680® has a counter for finding locations on an audio cassette. When a fully wound cassette with 45 min of music on a side begins to play, the counter is at 0. After 15 min of music has played, the counter reads 250 and after 35 min, it reads 487. When the 45-min side is finished playing, the counter reads 590.

ACTIVITY

1. The paragraph above describes four ordered pairs of the form (counter number, minutes played). Three pairs are enough to find a function of the form

$$T(n) = an^2 + bn + c,$$

where $T(n)$ represents the time, in minutes, that the tape has run at counter reading n hundred. Each group member should select a different set of three points from the four given and then fit a quadratic function to the data.

2. Of the 3 or 4 functions found in part (1) above, which fits the data "best"? One way to answer this is to see how well each function predicts other pairs. The same counter used above reads 432 after a 45-min tape has played for 30 min. Which function comes closest to predicting this?

3. If a grapher is available with a QUADREG option (see Exercise 49), what function does it fit to the four pairs originally listed?

4. If a class member has access to a Panasonic System RX-DT680, see how well the functions developed above predict the counter readings for a tape that has played for 5 or 10 min.

Polynomial and Rational Inequalities

11.9

Quadratic and Other Polynomial Inequalities • Rational Inequalities

Quadratic and Other Polynomial Inequalities

Inequalities like the following are called *polynomial inequalities*:

$$x^3 - 5x > x^2 + 7, \qquad 4x - 3 < 9, \qquad 5x^2 - 3x + 2 \geq 0.$$

Second-degree polynomial inequalities in one variable are called *quadratic inequalities*. To solve polynomial inequalities, we often focus attention on where the outputs of a polynomial function are positive and where they are negative.

E x a m p l e   1

Solve: $x^2 + 3x - 10 > 0$.

Solution   Consider the "related" function $f(x) = x^2 + 3x - 10$ and its graph. Its graph opens upward since the leading coefficient is positive.

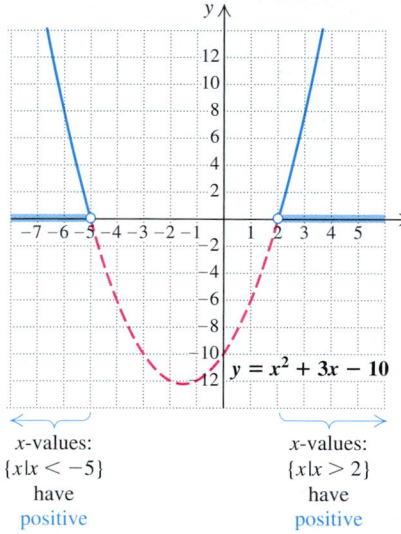

x-values:
$\{x|x < -5\}$
have
positive
y-values.

x-values:
$\{x|x > 2\}$
have
positive
y-values.

Values of y will be positive to the left and right of the x-intercepts, as shown. To find the intercepts, we set the polynomial equal to 0 and solve:

$$x^2 + 3x - 10 = 0$$
$$(x + 5)(x - 2) = 0$$
$$x + 5 = 0 \quad or \quad x - 2 = 0$$
$$x = -5 \quad or \quad x = 2.$$

Thus the solution set of the inequality is

$$\{x | x < -5 \, or \, x > 2\}, \quad or \quad (-\infty, -5) \cup (2, \infty).$$

Any inequality with 0 on one side can be solved by considering a graph of the related function and finding intercepts as in Example 1. Sometimes the quadratic formula is needed to find the intercepts.

E x a m p l e   2

Solve: $x^2 - 2x \leq 2$.

Solution   We first find standard form with 0 on one side:

$$x^2 - 2x - 2 \leq 0. \qquad \text{This is equivalent to the original inequality.}$$

The graph of $f(x) = x^2 - 2x - 2$ is a parabola opening upward. Values of $f(x)$ are negative for x-values between the x-intercepts. We find the x-intercepts by solving $f(x) = 0$:

$$x = \frac{-b \pm \sqrt{b^2 - 4ac}}{2a}$$

$$= \frac{-(-2) \pm \sqrt{(-2)^2 - 4 \cdot 1(-2)}}{2 \cdot 1}$$

$$= \frac{2 \pm \sqrt{12}}{2} = \frac{2 \pm 2\sqrt{3}}{2} = 1 \pm \sqrt{3}.$$

Inputs in this interval have negative or 0 outputs.

At the x-intercepts, $1 - \sqrt{3}$ and $1 + \sqrt{3}$, the value of $f(x)$ is 0. Thus the solution set of the inequality is

$$\left[1 - \sqrt{3}, 1 + \sqrt{3}\right], \quad \text{or} \quad \left\{x \,|\, 1 - \sqrt{3} \le x \le 1 + \sqrt{3}\right\}.$$

Note that in Example 2 it was not essential to actually draw the graph. The important information came from the location of the x-intercepts and the sign of $f(x)$ on each side of those intercepts.

In the next example, we solve a third-degree polynomial inequality, without graphing, by locating the x-intercepts, or **zeros**, of f and then using *test points* to determine the sign of $f(x)$ over each interval of the x-axis.

E x a m p l e   3   Solve: $5x^3 + 10x^2 - 15x > 0$.

Solution   We first solve the related equation:

$$5x^3 + 10x^2 - 15x = 0$$

$$5x(x^2 + 2x - 3) = 0$$

$$5x(x + 3)(x - 1) = 0$$

$$5x = 0 \quad or \quad x + 3 = 0 \quad or \quad x - 1 = 0$$

$$x = 0 \quad or \qquad\quad x = -3 \quad or \qquad\quad x = 1.$$

We see that if $f(x) = 5x^3 + 10x^2 - 15x$, then the zeros of f are $-3, 0$, and 1. These zeros divide the number line, or x-axis, into four intervals: A, B, C, and D.

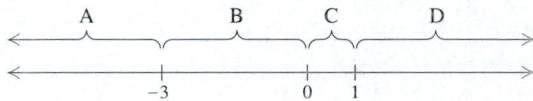

Next, using one convenient test value from each interval, we determine the sign of $f(x)$ for that interval. We know that, within each interval, the sign of $f(x)$ cannot change. If it did, there would need to be another zero in that interval. Using the factored form of $f(x)$ eases the computations:

$$f(x) = 5x(x + 3)(x - 1).$$

For interval A,

$$f(-4) = 5(-4)((-4) + 3)((-4) - 1)$$

-4 is a convenient value in interval A.

$$= -20(-1)(-5)$$

$$= -100.$$

$f(-4)$ is negative.

For interval B,

$$f(-1) = 5(-1)((-1) + 3)((-1) - 1)$$

-1 is a convenient value in interval B.

$$= -5(2)(-2)$$

$$= 20.$$

$f(-1)$ is positive.

For interval C,

$$f\left(\tfrac{1}{2}\right) = \underbrace{5 \cdot \tfrac{1}{2}}_{\text{Positive}} \cdot \underbrace{\left(\tfrac{1}{2} + 3\right)}_{\text{Positive}} \cdot \underbrace{\left(\tfrac{1}{2} - 1\right)}_{\text{Negative}}.$$
$$\underbrace{\phantom{5 \cdot \tfrac{1}{2} \cdot \left(\tfrac{1}{2} + 3\right) \cdot \left(\tfrac{1}{2} - 1\right)}}_{\text{Negative}}$$

$\tfrac{1}{2}$ is a convenient value in interval C.
Only the sign is important. The product is negative, so $f\left(\tfrac{1}{2}\right)$ is negative.

For interval D,

$$f(2) = \underbrace{5 \cdot 2}_{\text{Positive}} \cdot \underbrace{(2 + 3)}_{\text{Positive}} \cdot \underbrace{(2 - 1)}_{\text{Positive}}.$$

2 is a convenient value in interval D.
$f(2)$ is positive.

Recall that we are looking for all x for which $5x^3 + 10x^2 - 15x > 0$. The calculations above indicate that $f(x)$ is positive for any number in intervals B and D. The solution set of the original inequality is

$$(-3, 0) \cup (1, \infty), \quad \text{or} \quad \{x \mid -3 < x < 0 \ or \ x > 1\}.$$

Note that the calculations in Example 3 were made simpler by using the factored form of the polynomial. The process was simplified further when, for intervals C and D, we concentrated on only the *sign* of $f(x)$. In the next example, we determine the sign of a polynomial function over each interval by tracking the sign of each factor. By looking at how many positive or negative factors are being multiplied, we will be able to determine the sign of the polynomial function.

Example 4

To solve $2.3x^2 \leq 9.11 - 2.94x$, we first rewrite the inequality in the form $2.3x^2 + 2.94x - 9.11 \leq 0$ and graph the function $f(x) = 2.3x^2 + 2.94x - 9.11$.

$y = 2.3x^2 + 2.94x - 9.11$

To find the values of x for which $f(x) \leq 0$, we focus on the region in which the graph lies *on or below* the x-axis. From this graph, it appears that this region begins somewhere between -3 and -2, and continues to somewhere between 1 and 2. Using the ZERO or ROOT option of CALC, we can find the endpoints of this region. To two decimal places, the endpoints are -2.73 and 1.45. The solution set is approximately $\{x \mid -2.73 \leq x \leq 1.45\}$.

Had the inequality been $2.3x^2 > 9.11 - 2.94x$, we would look for portions of the graph that lie *above* the x-axis. An approximate solution set of such an inequality would be $\{x \mid x < -2.73 \text{ or } x > 1.45\}$.

Use a grapher to solve each inequality. Round the values of the endpoints to the nearest hundredth.

1. $4.32x^2 - 3.54x - 5.34 \leq 0$
2. $7.34x^2 - 16.55x - 3.89 \geq 0$
3. $10.85x^2 + 4.28x + 4.44 > 7.91x^2 + 7.43x + 13.03$
4. $5.79x^3 - 5.68x^2 + 10.68x > 2.11x^3 + 16.90x - 11.69$

Solve: $4x^3 - 4x \leq 0$.

Solution   We first solve the related equation:

$$4x^3 - 4x = 0$$
$$4x(x^2 - 1) = 0$$
$$4x(x + 1)(x - 1) = 0$$
$$4x = 0 \quad or \quad x + 1 = 0 \quad or \quad x - 1 = 0$$
$$x = 0 \quad or \quad \quad x = -1 \quad or \quad \quad x = 1.$$

The function $f(x) = 4x^3 - 4x$ has zeros at $-1, 0$, and 1. We could now use test values, as in Example 3. Instead, let's use the factorization $f(x) = 4x(x + 1)(x - 1)$. The product $4x(x + 1)(x - 1)$ is positive or negative, depending on the signs of the factors $4x$, $x + 1$, and $x - 1$. This is easily determined using a chart.

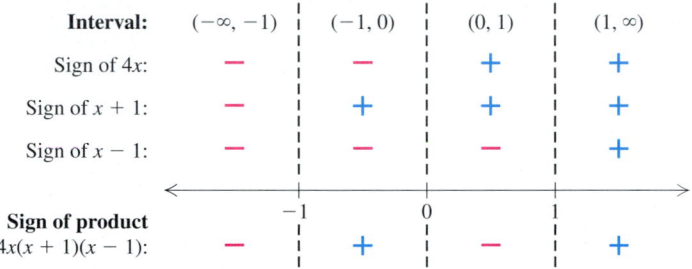

A product is negative when it has an odd number of negative factors. Since the $\leq$ sign allows for equality, the endpoints $-1, 0$, and 1 are solutions. From the chart, we see that the solution set is

$$(-\infty, -1] \cup [0, 1], \quad or \quad \{x \mid x \leq -1 \text{ or } 0 \leq x \leq 1\}.$$

To Solve a Polynomial Inequality Using Factors

1. Get 0 on one side and solve the related polynomial equation by factoring.
2. Use the numbers found in step (1) to divide the number line into intervals.
3. Using a test value from each interval, determine the sign of each factor over that interval.
4. Determine the sign of the product of the factors over each interval. Remember that the product of an odd number of negative numbers is negative.
5. Select the interval(s) for which the inequality is satisfied and write set-builder notation or interval notation for the solution set. Include the endpoints of the intervals when $\leq$ or $\geq$ is used.

Rational Inequalities

Inequalities involving rational expressions are called **rational inequalities**. Like polynomial inequalities, rational inequalities can be solved using test values. Unlike polynomials, however, rational expressions often have values for which the expression is undefined.

E x a m p l e   5

Solve: $\dfrac{x-3}{x+4} \geq 2$.

Solution   We write the related equation by changing the $\geq$ symbol to $=$:

$$\frac{x-3}{x+4} = 2.$$

Next, we solve this related equation:

$$(x+4) \cdot \frac{x-3}{x+4} = (x+4) \cdot 2 \quad \text{Multiplying both sides by the LCD, } x+4$$
$$x - 3 = 2x + 8$$
$$-11 = x. \qquad\qquad \text{Solving for } x$$

In the case of rational inequalities, we also need to find any values that make the denominator 0. We set the denominator equal to 0 and solve:

$$\left.\begin{array}{l} x + 4 = 0 \\ x = -4. \end{array}\right\} \quad \begin{array}{l} \text{This tells us that } -4 \text{ is not in the} \\ \text{domain of } f \text{ if } f(x) = \dfrac{x-3}{x+4}. \end{array}$$

Now we use -11 and -4 to divide the number line into intervals:

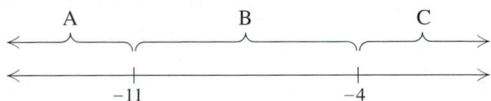

We test a number in each interval to see where the original inequality is satisfied:

$$\frac{x-3}{x+4} \geq 2.$$

A: Test -15, $\dfrac{-15-3}{-15+4} = \dfrac{-18}{-11}$

$$= \frac{18}{11} \ngeq 2 \quad \begin{array}{l} -15 \text{ } \textit{is not} \text{ a solution, so interval A is} \\ \text{not part of the solution set.} \end{array}$$

B: Test -8, $\dfrac{-8-3}{-8+4} = \dfrac{-11}{-4}$

$$= \frac{11}{4} \geq 2 \quad \begin{array}{l} -8 \text{ } \textit{is} \text{ a solution, so interval B is part} \\ \text{of the solution set.} \end{array}$$

C: Test 1, $\dfrac{1 - 3}{1 + 4} = \dfrac{-2}{5}$

$= -\dfrac{2}{5} \not\geq 2$ *1 is not* a solution, so interval C is not part of the solution set.

The solution set includes the interval B. The endpoint -11 is included because the inequality symbol is $\geq$ and -11 is a solution of the related equation. The number -4 is *not* included because $(x - 3)/(x + 4)$ is undefined for $x = -4$. Thus the solution set of the original inequality is

$$[-11, -4), \quad \text{or} \quad \{x \mid -11 \leq x < -4\}.$$

There is an interesting visual interpretation of Example 5. If we graph the function $f(x) = (x - 3)/(x + 4)$, we see that the solutions of the inequality $(x - 3)/(x + 4) \geq 2$ can be found by inspection. We simply sketch the line $y = 2$ and locate all x-values for which $f(x) \geq 2$.

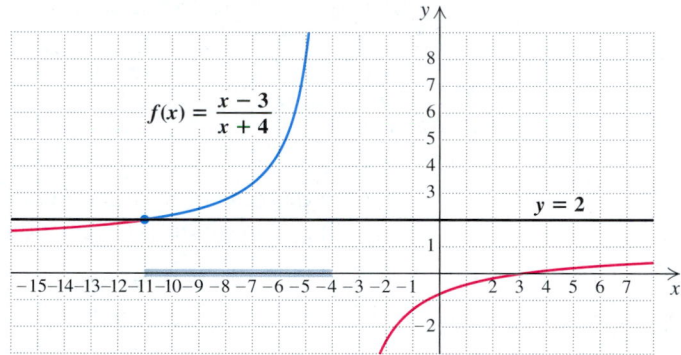

Because graphing rational functions can be very time-consuming, we generally just use test values.

> ### To Solve a Rational Inequality
>
> 1. Change the inequality symbol to an equals sign and solve the related equation.
> 2. Find any replacements for which the rational expression is undefined.
> 3. Use the numbers found in steps (1) and (2) to divide the number line into intervals.
> 4. Substitute a test value from each interval into the inequality. If the number is a solution, then the interval to which it belongs is part of the solution set.
> 5. Select the interval(s) and any endpoints for which the inequality is satisfied and write set-builder or interval notation for the solution set. If the inequality symbol is $\leq$ or $\geq$, then the solutions from step (1) are also included in the solution set. Those numbers found in step (2) should be excluded from the solution set, even if they are solutions from step (1).

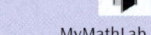
Exercise Set 11.9

Solve.

1. $(x + 4)(x - 3) < 0$

2. $(x - 5)(x + 2) > 0$

3. $(x + 7)(x - 2) \geq 0$

4. $(x - 1)(x + 4) \leq 0$

5. $x^2 - x - 2 < 0$

6. $x^2 + x - 2 < 0$

7. $25 - x^2 \geq 0$

8. $4 - x^2 \geq 0$

Aha! **9.** $x^2 + 4x + 4 < 0$

10. $x^2 + 6x + 9 < 0$

11. $x^2 - 4x < 12$

12. $x^2 + 6x > -8$

13. $3x(x + 2)(x - 2) < 0$

14. $5x(x + 1)(x - 1) > 0$

15. $(x + 3)(x - 2)(x + 1) > 0$

16. $(x - 1)(x + 2)(x - 4) < 0$

17. $(x + 3)(x + 2)(x - 1) < 0$

18. $(x - 2)(x - 3)(x + 1) < 0$

19. $\dfrac{1}{x + 3} < 0$

20. $\dfrac{1}{x + 4} > 0$

21. $\dfrac{x + 1}{x - 5} \geq 0$

22. $\dfrac{x - 2}{x + 5} \leq 0$

23. $\dfrac{3x + 2}{2x - 4} \leq 0$

24. $\dfrac{5 - 2x}{4x + 3} \leq 0$

25. $\dfrac{x + 1}{x + 6} > 1$

26. $\dfrac{x - 1}{x - 2} < 1$

27. $\dfrac{(x - 2)(x + 1)}{x - 5} \leq 0$

28. $\dfrac{(x + 4)(x - 1)}{x + 3} \geq 0$

29. $\dfrac{x}{x + 3} \geq 0$

30. $\dfrac{x - 2}{x} \leq 0$

31. $\dfrac{x - 5}{x} < 1$

32. $\dfrac{x}{x - 1} > 2$

33. $\dfrac{x - 1}{(x - 3)(x + 4)} \leq 0$

34. $\dfrac{x + 2}{(x - 2)(x + 7)} \geq 0$

35. $4 < \dfrac{1}{x}$

36. $\dfrac{1}{x} \leq 5$

37. Explain how any quadratic inequality can be solved by examining a parabola.

38. Describe a method for creating a quadratic inequality for which there is no solution.

SKILL MAINTENANCE

Simplify.

39. $(2a^3b^2c^4)^3$

40. $(5a^4b^7)^2$

41. 2^{-5}

42. 3^{-4}

43. If $f(x) = 3x^2$, find $f(a + 1)$.

44. If $g(x) = 5x - 3$, find $g(a + 2)$.

SYNTHESIS

45. Step (5) on p. 726 states that even when the inequality symbol is $\leq$ or $\geq$, the solutions from step (1) are not always part of the solution set. Why?

46. Describe a method that could be used to create quadratic inequalities that have $(-\infty, a] \cup [b, \infty)$ as the solution set.

Find each solution set.

47. $x^2 + 2x < 5$

48. $x^4 + 2x^2 \geq 0$

49. $x^4 + 3x^2 \leq 0$

50. $\left| \dfrac{x + 2}{x - 1} \right| \leq 3$

51. *Total profit.*   Derex, Inc., determines that its total-profit function is given by

$$P(x) = -3x^2 + 630x - 6000.$$

a) Find all values of x for which Derex makes a profit.

b) Find all values of x for which Derex loses money.

52. *Height of a thrown object.*   The function

$$S(t) = -16t^2 + 32t + 1920$$

gives the height S, in feet, of an object thrown from a cliff that is 1920 ft high. Here t is the time, in seconds, that the object is in the air.

a) For what times does the height exceed 1920 ft?

b) For what times is the height less than 640 ft?

53. *Number of handshakes.* There are n people in a room. The number N of possible handshakes by the people is given by the function

$$N(n) = \frac{n(n-1)}{2}.$$

For what number of people n is $66 \le N \le 300$?

54. *Number of diagonals.* A polygon with n sides has D diagonals, where D is given by the function

$$D(n) = \frac{n(n-3)}{2}.$$

Find the number of sides n if

$$27 \le D \le 230.$$

Use a grapher to graph each function and find solutions of $f(x) = 0$. Then solve the inequalities $f(x) < 0$ and $f(x) > 0$.

55. $f(x) = x^3 - 2x^2 - 5x + 6$

56. $f(x) = \frac{1}{3}x^3 - x + \frac{2}{3}$

57. $f(x) = x + \frac{1}{x}$

58. $f(x) = x - \sqrt{x},\, x \ge 0$

59. $f(x) = \frac{x^3 - x^2 - 2x}{x^2 + x - 6}$

60. $f(x) = x^4 - 4x^3 - x^2 + 16x - 12$

61. Use a grapher to solve Exercises 11, 25, and 35 by drawing two curves, one for each side of the inequality.

Summary and Review 11

Key Terms

Quadratic equation, p. 660

Principle of square roots, p. 661

Completing the square, p. 663

Compounding interest annually, p. 667

The quadratic formula, p. 672

Quadratic function, p. 673

Discriminant, p. 686

Reducible to quadratic, p. 691

Parabola, p. 697

Axis of symmetry, p. 697

Vertex, p. 697

Reflection, p. 698

Translated, p. 700

Minimum value, p. 700

Maximum value, p. 700

Data point, p. 713

Polynomial inequality, p. 720

Quadratic inequality, p. 720

Zero, p. 722

Rational inequality, p. 725

Important Properties and Formulas

The Principle of Square Roots

For any real number k, if $x^2 = k$, then $x = \sqrt{k}$ or $x = -\sqrt{k}$.

For any real number k and any algebraic expression X, if $X^2 = k$, then $X = \sqrt{k}$ or $X = -\sqrt{k}$.

To complete the square for $x^2 + bx$, add $(b/2)^2$.

To solve a quadratic equation in x by completing the square:

1. Isolate the terms with variables on one side of the equation, and arrange them in descending order.
2. Divide both sides by the coefficient of x^2 if that coefficient is not 1.
3. Complete the square by taking half of the coefficient of x and adding its square to both sides.

4. Express one side as the square of a binomial and simplify the other side.
5. Use the principle of square roots.
6. Solve for x by adding or subtracting on both sides.

The Quadratic Formula

The solutions of $ax^2 + bx + c = 0$, $a \neq 0$, are given by

$$x = \frac{-b \pm \sqrt{b^2 - 4ac}}{2a}.$$

To solve a quadratic equation:

1. If the equation can be easily written in the form $ax^2 = p$ or $(x + k)^2 = d$, use the principle of square roots.
2. If step (1) does not apply, write the equation in $ax^2 + bx + c = 0$ form.
3. Try factoring and using the principle of zero products.
4. If factoring seems to be difficult or impossible, use the quadratic formula.

The solutions of a quadratic equation can always be found using the quadratic formula. They cannot always be found by factoring.

To solve a formula for a letter—say, b:

1. Clear fractions and use the principle of powers, as needed. (In some cases you may clear the fractions first, and in some cases you may use the principle of powers first.) Perform these steps until radicals containing b are gone and b is not in any denominator.
2. Combine all terms with b^2 in them. Also combine all terms with b in them.
3. If b^2 does not appear, you can solve by using just the addition and multiplication principles as in Sections 2.3 and 6.8.
4. If b^2 appears but b does not, solve the equation for b^2. Then use the principle of square roots to solve for b.

5. If there are terms containing both b and b^2, put the equation in standard form and use the quadratic formula.

Discriminant $b^2 - 4ac$	Nature of Solutions
0	One solution; a rational number
Positive	Two different real-number solutions
Perfect square	Solutions are rational.
Not a perfect square	Solutions are irrational conjugates.
Negative	Two different imaginary-number solutions (complex conjugates)

The graph of $g(x) = ax^2$ is a parabola with $x = 0$ as its axis of symmetry; its vertex is the origin.

For $a > 0$, the parabola opens upward. For $a < 0$, the parabola opens downward.

If $|a|$ is greater than 1, the parabola is narrower than $f(x) = x^2$.

If $|a|$ is between 0 and 1, the parabola is wider than $f(x) = x^2$.

The graph of $f(x) = a(x - h)^2$ has the same shape as the graph of $y = ax^2$.

If h is positive, the graph of $y = ax^2$ is shifted h units to the right.

If h is negative, the graph of $y = ax^2$ is shifted $|h|$ units to the left.

The vertex is $(h, 0)$, and the axis of symmetry is $x = h$.

The graph of $f(x) = a(x - h)^2 + k$ has the same shape as the graph of $y = a(x - h)^2$.

If k is positive, the graph of $y = a(x - h)^2$ is shifted k units up.

If k is negative, the graph of $y = a(x - h)^2$ is shifted $|k|$ units down.

The vertex is (h, k), and the axis of symmetry is $x = h$.

For $a > 0$, k is the minimum function value. For $a < 0$, k is the maximum function value.

Formulas

Compound interest: $\quad A = P(1 + r)^t$
Free-fall distance,
in feet: $\qquad\qquad s = 16t^2$

The vertex of the parabola given by $f(x) = ax^2 + bx + c$ is

$$\left(-\frac{b}{2a}, f\left(-\frac{b}{2a}\right)\right),$$

or

$$\left(-\frac{b}{2a}, \frac{4ac - b^2}{4a}\right).$$

The x-coordinate of the vertex is $-b/(2a)$. The axis of symmetry is $x = -b/(2a)$.

To solve a polynomial inequality using factors:

1. Get 0 on one side and solve the related polynomial equation by factoring.
2. Use the numbers found in step (1) to divide the number line into intervals.
3. Using a test value from each interval, determine the sign of each factor over that interval.
4. Determine the sign of the product of the factors over each interval. Remember that the product of an odd number of negative numbers is negative.
5. Select the interval(s) for which the inequality is satisfied and write set-builder or interval notation for the solution set. Include the endpoints of the intervals when $\leq$ or $\geq$ is used.

To solve a rational inequality:

1. Change the inequality symbol to an equals sign and solve the related equation.
2. Find any replacements for which the rational expression is undefined.
3. Use the numbers found in steps (1) and (2) to divide the number line into intervals.
4. Substitute a test value from each interval into the inequality. If the number is a solution, then the interval to which it belongs is part of the solution set.
5. Select the interval(s) and any endpoints for which the inequality is satisfied and write set-builder or interval notation for the solution set. If the inequality symbol is $\leq$ or $\geq$, then the solutions to step (1) are also included in the solution set. Those numbers found in step (2) should be excluded from the solution set, even if they are solutions from step (1).

Review Exercises

Solve.

1. $2x^2 - 7 = 0$

2. $14x^2 + 5x = 0$

3. $x^2 - 12x + 36 = 9$

4. $x^2 - 5x + 9 = 0$

5. $x(3x + 4) = 4x(x - 1) + 15$

6. $x^2 + 9x = 1$

7. $x^2 - 5x - 2 = 0$. Use a calculator to approximate the solutions with rational numbers.

8. Let $f(x) = 4x^2 - 3x - 1$. Find x such that $f(x) = 0$.

Complete the square. Then write the perfect-square trinomial in factored form.

9. $x^2 - 12x$

10. $x^2 + \frac{3}{5}x$

11. Solve by completing the square. Show your work.
$$x^2 - 6x + 1 = 0$$

12. \$2500 grows to \$3025 in 2 yr. Use the formula $A = P(1 + r)^t$ to find the interest rate.

13. The Peachtree Center Plaza in Atlanta, Georgia, is 723 ft tall. Use the formula $s = 16t^2$ to approximate how long it would take an object to fall from the top.

Solve.

14. A corporate pilot must fly from company headquarters to a manufacturing plant and back in 4 hr. The distance between headquarters and the plant is 300 mi. If there is a 20-mph headwind going and a 20-mph tailwind returning, how fast must the plane be able to travel in still air?

15. Working together, Erica and Shawna can answer a day's worth of technical support questions in 4 hr. Working alone, Erica takes 6 hr longer than Shawna. How long would it take Shawna to answer the questions alone?

For each equation, determine what type of number the solutions will be.

16. $x^2 + 3x - 6 = 0$

17. $x^2 + 2x + 5 = 0$

18. Write a quadratic equation having the solutions $\sqrt{5}$ and $-\sqrt{5}$.

19. Write a quadratic equation having -4 as its only solution.

20. Find all x-intercepts of the graph of
$$f(x) = x^4 - 13x^2 + 36.$$

Solve.

21. $15x^{-2} - 2x^{-1} - 1 = 0$

22. $(x^2 - 4)^2 - (x^2 - 4) - 6 = 0$

23. a) Graph: $f(x) = -3(x + 2)^2 + 4$.
b) Label the vertex.
c) Draw the axis of symmetry.
d) Find the maximum or the minimum value.

24. For the function $f(x) = 2x^2 - 12x + 23$:
a) find the vertex and the axis of symmetry;
b) graph the function.

25. Find the x- and y-intercepts of
$$f(x) = x^2 - 9x + 14.$$

26. Solve $N = 3\pi\sqrt{1/p}$ for p.

27. Solve $2A + T = 3T^2$ for T.

State whether the graph appears to represent a quadratic function.

28.

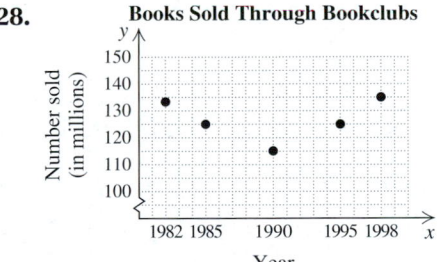

Books Sold Through Bookclubs

Source: Statistical Abstract of the United States, 2000

29.

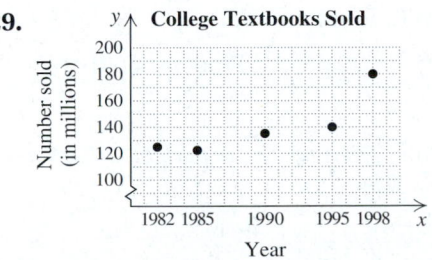

College Textbooks Sold

Source: Statistical Abstract of the United States, 2000

30. Eastgate Consignments wants to build an area for children to play in while their parents shop. They have 30 ft of low fencing. What is the maximum area they can enclose? What dimensions will yield this area?

31. The following table lists the number of books (in millions) sold through bookclubs x years after 1980. (See Exercise 28.)

Years Since 1980	Number of Books Sold Through Bookclubs (in millions)
5	130
10	108
15	123

a) Find the quadratic function that fits the data.
b) Use the function to estimate the number of books sold through bookclubs in 2000.

Solve.

32. $x^3 - 3x > 2x^2$

33. $\dfrac{x-5}{x+3} \le 0$

SYNTHESIS

34. Explain how the x-intercepts of a quadratic function can be used to help find the maximum or minimum value of the function.

35. Suppose that the quadratic formula is used to solve a quadratic equation. If the discriminant is a perfect square, could factoring have been used to solve the equation? Why or why not?

36. What is the greatest number of solutions that an equation of the form $ax^4 + bx^2 + c = 0$ can have? Why?

37. Discuss two ways in which completing the square was used in this chapter.

38. A quadratic function has x-intercepts at -3 and 5. If the y-intercept is at -7, find an equation for the function.

39. Find h and k if, for $3x^2 - hx + 4k = 0$, the sum of the solutions is 20 and the product is 80.

40. The average of two positive integers is 171. One of the numbers is the square root of the other. Find the integers.

Chapter Test 11

Solve.

1. $3x^2 - 16 = 0$

2. $4x(x-2) - 3x(x+1) = -18$

3. $x^2 + x + 1 = 0$

4. $2x + 5 = x^2$

5. $x^{-2} - x^{-1} = \dfrac{3}{4}$

6. $x^2 + 3x = 5$. Use a calculator to approximate the solutions with rational numbers.

7. Let $f(x) = 12x^2 - 19x - 21$. Find x such that $f(x) = 0$.

Complete the square. Then write the perfect-square trinomial in factored form.

8. $x^2 + 14x$

9. $x^2 - \dfrac{2}{7}x$

10. Solve by completing the square. Show your work.
$$x^2 + 10x + 15 = 0$$

Solve.

11. The Connecticut River flows at a rate of 4 km/h for the length of a popular scenic route. In order for a cruiser to travel 60 km upriver and then return in a total of 8 hr, how fast must the boat be able to travel in still water?

12. Brock and Ian can assemble a swing set in $1\frac{1}{2}$ hr. Working alone, it takes Ian 4 hr longer than Brock to assemble the swing set. How long would it take Brock, working alone, to assemble the swing set?

13. Determine the type of number that the solutions of $x^2 + 5x + 17 = 0$ will be.

14. Write a quadratic equation having solutions -2 and $\frac{1}{3}$.

15. Find all x-intercepts of the graph of
$$f(x) = (x^2 + 4x)^2 + 2(x^2 + 4x) - 3.$$

16. a) Graph: $f(x) = 4(x - 3)^2 + 5$.
b) Label the vertex.
c) Draw the axis of symmetry.
d) Find the maximum or the minimum function value.

17. For the function $f(x) = 2x^2 + 4x - 6$:
a) find the vertex and the axis of symmetry;
b) graph the function.

18. Find the x- and y-intercepts of
$$f(x) = x^2 - x - 6.$$

19. Solve $V = \frac{1}{3}\pi(R^2 + r^2)$ for r.

20. State whether the graph appears to represent a quadratic function.

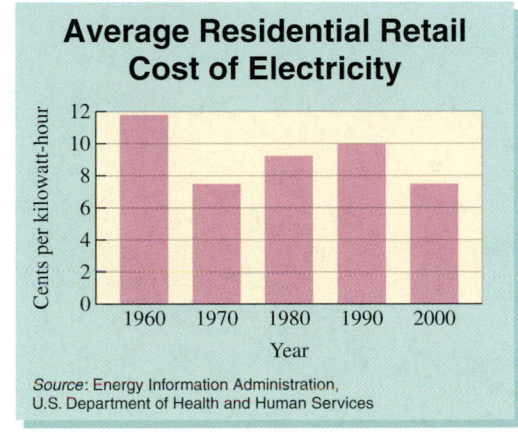

Average Residential Retail Cost of Electricity

Cents per kilowatt-hour

Year

Source: Energy Information Administration, U.S. Department of Health and Human Services

21. Jay's Custom Pickups has determined that when x hundred truck caps are built, the average cost per cap is given by
$$C(x) = 0.2x^2 - 1.3x + 3.4025,$$
where $C(x)$ is in hundreds of dollars. What is the minimum cost per truck cap and how many caps should be built to achieve that minimum?

22. Find the quadratic function that fits the data points $(0, 0)$, $(3, 0)$, and $(5, 2)$.

Solve.

23. $x^2 + 5x \leq 6$

24. $x - \dfrac{1}{x} > 0$

SYNTHESIS

25. One solution of $kx^2 + 3x - k = 0$ is -2. Find the other solution.

26. Find a fourth-degree polynomial equation, with integer coefficients, for which $2 - \sqrt{3}$ and $5 - i$ are solutions.

27. Find a polynomial equation, with integer coefficients, for which 5 is a repeated root and $\sqrt{2}$ and $\sqrt{3}$ are solutions.

12

Exponential and Logarithmic Functions

AN APPLICATION

The number of computers infected by a virus *t* days after it first appears usually increases exponentially. In 2000, the "Love Bug" virus spread from 100 computers to about 1,000,000 computers in 2 hr (120 min). Assuming exponential growth, estimate how long it took the Love Bug virus to infect 80,000 computers.

This problem appears as Example 5 in Section 12.7.

I use algebra to calculate bandwidth availability and network capacity and to create security profiles. I also use matrices in data storage.

CHRISTOPHER KENLY
Chief Consultant
TOS Consulting
Dallas, TX

*T*he functions that we consider in this chapter are interesting not only from a purely intellectual point of view, but also for their rich applications to many fields. We will look at applications such as compound interest and population growth, to name just two.

The basis of the theory centers on functions having variable exponents (exponential functions). Results follow from those functions and their properties.

Composite and Inverse Functions

12.1

Composite Functions • Inverses and One-to-One Functions • Finding Formulas for Inverses • Graphing Functions and Their Inverses • Inverse Functions and Composition

Composite Functions

In the real world, functions frequently occur in which some quantity depends on a variable that, in turn, depends on another variable. For instance, a firm's profits may depend on the number of items the firm produces, which may in turn depend on the number of employees hired. Functions like this are called **composite functions**.

For example, the function g that gives a correspondence between women's shoe sizes in the United States and those in Italy is given by $g(x) = 2x + 24$, where x is the U.S. size and $g(x)$ is the Italian size. Thus a U.S. size 4 corresponds to a shoe size of $g(4) = 2 \cdot 4 + 24$, or 32, in Italy.

There is also a function that gives a correspondence between women's shoe sizes in Italy and those in Britain. This particular function is given by $f(x) = \frac{1}{2}x - 14$, where x is the Italian size and $f(x)$ is the corresponding British size. Thus an Italian size 32 corresponds to a British size $f(32) = \frac{1}{2} \cdot 32 - 14$, or 2.

It seems reasonable to conclude that a shoe size of 4 in the United States corresponds to a size of 2 in Britain and that some function h describes this correspondence. Can we find a formula for h? If we look at the following tables, we might guess that such a formula is $h(x) = x - 2$, and that is indeed correct. But, for more complicated formulas, we would need to use algebra.

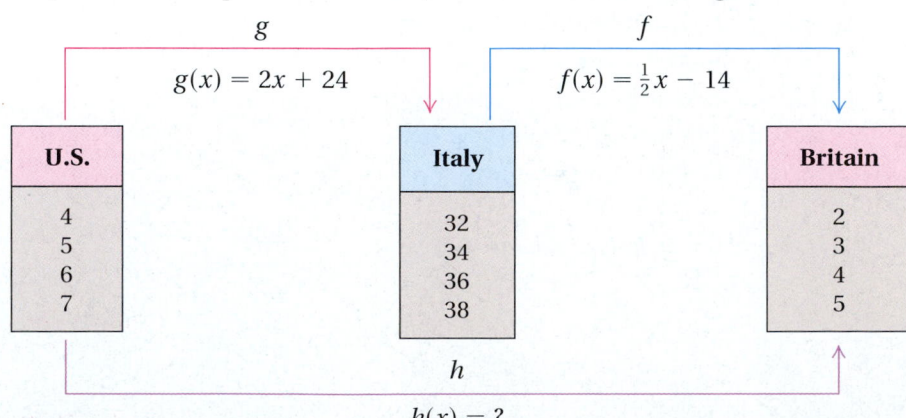

Size x shoes in the United States correspond to size $g(x)$ shoes in Italy, where

$$g(x) = 2x + 24.$$

Size n shoes in Italy correspond to size $f(n)$ shoes in Britain. Thus size $g(x)$ shoes in Italy correspond to size $f(g(x))$ shoes in Britain. Since the x in the expression $f(g(x))$ represents a U.S. shoe size, we can find the British shoe size that corresponds to a U.S. size x as follows:

$$f(g(x)) = f(2x + 24) = \tfrac{1}{2} \cdot (2x + 24) - 14 \qquad \text{Using } g(x) \text{ as an input}$$
$$= x + 12 - 14 = x - 2.$$

This gives a formula for h: $h(x) = x - 2$. Thus a shoe size of 4 in the United States corresponds to a shoe size of $h(4) = 4 - 2$, or 2, in Britain. The function h is called the *composition* of f and g and is denoted $f \circ g$ (read "the composition of f and g," "f composed with g," or "f circle g").

> ### Composition of Functions
>
> The *composite function* $f \circ g$, the *composition* of f and g, is defined as
>
> $$(f \circ g)(x) = f(g(x)).$$

We can visualize the composition of functions as follows.

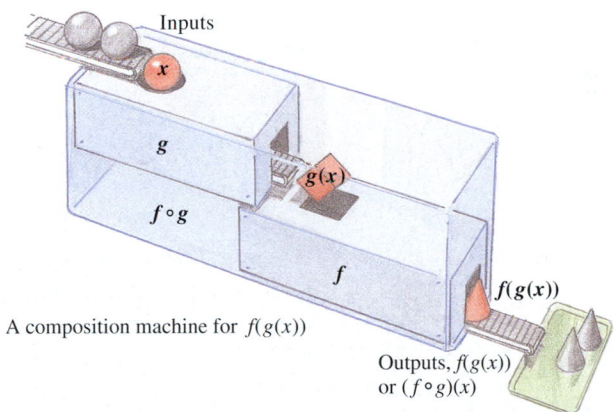

A composition machine for $f(g(x))$

Outputs, $f(g(x))$ or $(f \circ g)(x)$

E x a m p l e   1   Given $f(x) = 3x$ and $g(x) = 1 + x^2$:

a) Find $(f \circ g)(5)$ and $(g \circ f)(5)$.
b) Find $(f \circ g)(x)$ and $(g \circ f)(x)$.

Solution   Consider each function separately:

$$f(x) = 3x \qquad \text{This function multiplies each input by 3.}$$

and

$$g(x) = 1 + x^2. \qquad \text{This function adds 1 to the square of each input.}$$

In Example 2, we will see that if $g(x) = x - 1$ and $f(x) = \sqrt{x}$, then $f(g(x)) = \sqrt{x - 1}$. One way to show this on a grapher is to let $y_1 = x - 1$ and $y_2 = \sqrt{y_1}$ (this is accomplished by using the Y-VARS option of the VARS key's menu to enter y_1). We then let $y_3 = \sqrt{x - 1}$ and use graphs or a table to show that $y_2 = y_3$.

Another approach is to let $y_1 = x - 1$ and $y_2 = \sqrt{x}$ and have $y_4 = y_2(y_1)$. If we again let $y_3 = \sqrt{x - 1}$, we complete the check by showing that $y_3 = y_4$.

1. Check Example 3 by using one of the above approaches.

a) To find $(f \circ g)(5)$, we first find $g(5)$ by substituting in the formula for g: Square 5 and add 1, to get 26. We then use 26 as an input for f:

$$(f \circ g)(5) = f(g(5)) = f(1 + 5^2) \qquad \text{Using } g(x) = 1 + x^2$$
$$= f(26) = 3 \cdot 26 = 78. \qquad \text{Using } f(x) = 3x$$

To find $(g \circ f)(5)$, we first find $f(5)$ by substituting into the formula for f: Multiply 5 by 3, to get 15. We then use 15 as an input for g:

$$(g \circ f)(5) = g(f(5)) = g(3 \cdot 5) \qquad \text{Note that } f(5) = 3 \cdot 5 = 15.$$
$$= g(15) = 1 + 15^2 = 1 + 225 = 226.$$

b) We find $(f \circ g)(x)$ by substituting $g(x)$ for x in the equation for $f(x)$:

$$(f \circ g)(x) = f(g(x)) = f(1 + x^2) \qquad \text{Using } g(x) = 1 + x^2$$
$$= 3(1 + x^2) = 3 + 3x^2. \qquad \text{Using } f(x) = 3x. \textit{ These parentheses indicate multiplication.}$$

To find $(g \circ f)(x)$, we substitute $f(x)$ for x in the equation for $g(x)$:

$$(g \circ f)(x) = g(f(x)) = g(3x) \qquad \text{Substituting } 3x \text{ for } f(x)$$
$$= 1 + (3x)^2 = 1 + 9x^2.$$

As a check, note that $(g \circ f)(5) = 1 + 9 \cdot 5^2 = 1 + 9 \cdot 25 = 226$, as expected from part (a) above.

Example 1 shows that, in general, $(f \circ g)(5) \neq (g \circ f)(5)$ and $(f \circ g)(x) \neq (g \circ f)(x)$.

E x a m p l e   2   Given $f(x) = \sqrt{x}$ and $g(x) = x - 1$, find $(f \circ g)(x)$ and $(g \circ f)(x)$.

Solution

$$(f \circ g)(x) = f(g(x)) = f(x - 1) = \sqrt{x - 1}$$
$$(g \circ f)(x) = g(f(x)) = g(\sqrt{x}) = \sqrt{x} - 1$$

In fields ranging from chemistry to geology and economics, one needs to recognize how a function can be regarded as the composition of two "simpler" functions. This is sometimes called *de*composition.

E x a m p l e   3   If $h(x) = (7x + 3)^2$, find $f(x)$ and $g(x)$ such that $h(x) = (f \circ g)(x)$.

Solution   To find $h(x)$, we can think of first forming $7x + 3$ and then squaring. This suggests that $g(x) = 7x + 3$ and $f(x) = x^2$. We check by forming the composition:

$$h(x) = (f \circ g)(x) = f(g(x))$$
$$= f(7x + 3) = (7x + 3)^2.$$

This is probably the most "obvious" answer to the question. There can be other less obvious answers. For example, if

$$f(x) = (x - 1)^2$$

and

$$g(x) = 7x + 4,$$

then

$$h(x) = (f \circ g)(x) = f(g(x)) = f(7x + 4)$$
$$= (7x + 4 - 1)^2 = (7x + 3)^2.$$

Inverses and One-to-One Functions

Let's consider the following two functions. We think of them as relations, or correspondences.

Professions and Their Median Yearly Salary in 1998 Dollars*

Domain (Set of Inputs)	Range (Set of Outputs)
Teacher	$39,300
Registered nurse	$40,690
Emergency medical technician	$20,290
Computer programmer	$47,550
Veterinarian	$50,950
Accountant	$37,860

U.S. Senators and Their States

Domain (Set of Inputs)	Range (Set of Outputs)
Clinton	New York
Schumer	
McCain	Arizona
Kyl	
Feinstein	California
Boxer	

Suppose we reverse the arrows. We obtain what is called the **inverse relation**. Are these inverse relations functions?

Professions and Their Median Yearly Salary in 1998 Dollars

Range (Set of Outputs)	Domain (Set of Inputs)
Teacher	$39,300
Registered nurse	$40,690
Emergency medical technician	$20,290
Computer programmer	$47,550
Veterinarian	$50,950
Accountant	$37,860

U.S. Senators and Their States

Range (Set of Outputs)	Domain (Set of Inputs)
Clinton	New York
Schumer	
McCain	Arizona
Kyl	
Feinstein	California
Boxer	

Recall that for each input, a function provides exactly one output. However, a function can have the same output for two or more different inputs. Thus it is possible for different inputs to correspond to the same output. Only when this

*Source: U.S. Bureau of Labor Statistics, 2001

possibility is *excluded* will the inverse be a function. For the functions listed above, this means the inverse of the "Profession" correspondence is a function, but the inverse of the "U.S. Senator" correspondence is not.

In the Profession function, different inputs have different outputs. It is an example of a **one-to-one function**. In the U.S. Senator function, *Clinton* and *Schumer* are both paired with *New York*. Thus the U.S. Senator function is not one-to-one.

> ### One-To-One Function
>
> A function f is *one-to-one* if different inputs have different outputs. That is, if for any $a \neq b$, we have $f(a) \neq f(b)$, the function f is one-to-one. If a function is one-to-one, then its inverse correspondence is also a function.

How can we tell graphically whether a function is one-to-one?

E x a m p l e   4   Shown here is the graph of a function similar to those we will study in Section 12.2. Determine whether the function is one-to-one and thus has an inverse that is a function.

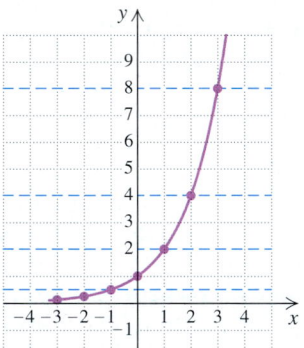

Solution   A function is one-to-one if different inputs have different outputs—that is, if no two x-values have the same y-value. For this function, we cannot find two x-values that have the same y-value. Note that this means that no horizontal line can be drawn so that it crosses the graph more than once. The function is one-to-one so its inverse is a function.

The graph of every function must pass the vertical-line test. In order for a function to have an inverse that is a function, it must pass the *horizontal-line test* as well.

> ### The Horizontal-Line Test
>
> A function is one-to-one, and thus has an inverse that is a function, if it is impossible to draw a horizontal line that intersects its graph more than once.

Example 5   Determine whether the function $f(x) = x^2$ is one-to-one and thus has an inverse that is a function.

Solution   The graph of $f(x) = x^2$ is shown here. Many horizontal lines cross the graph more than once—in particular, the line $y = 4$. Note that where the line crosses, the first coordinates are -2 and 2. Although these are different inputs, they have the same output. That is, $-2 \neq 2$, but

$$f(-2) = (-2)^2 = 4 = 2^2 = f(2).$$

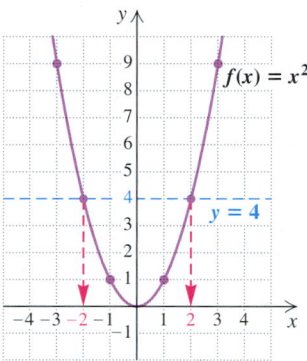

Thus the function is not one-to-one and no inverse function exists.

Finding Formulas for Inverses

When the inverse of f is also a function, it is denoted f^{-1} (read "f-inverse").

> **Caution!**   The -1 in f^{-1} is *not* an exponent!

Suppose a function is described by a formula. If it has an inverse that is a function, how do we find a formula for the inverse? For any equation in two variables, if we interchange the variables, we obtain an equation of the inverse correspondence. If it is a function, we proceed as follows to find a formula for f^{-1}.

> ### To Find a Formula for f^{-1}
>
> First check a graph to make sure that f is one-to-one. Then:
>
> **1.** Replace $f(x)$ with y.
> **2.** Interchange x and y. (This gives the inverse function.)
> **3.** Solve for y.
> **4.** Replace y with $f^{-1}(x)$. (This is inverse function notation.)

E x a m p l e 6   Determine if each function is one-to-one and if it is, find a formula for $f^{-1}(x)$.

a) $f(x) = x + 2$

b) $f(x) = 2x - 3$

Solution

a) The graph of $f(x) = x + 2$ is shown below. It passes the horizontal-line test, so it is one-to-one. Thus its inverse is a function.

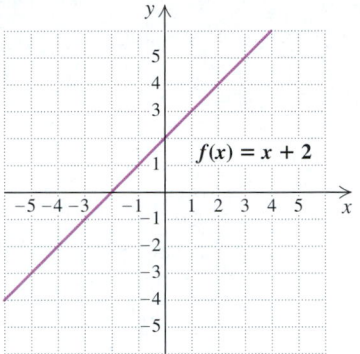

1. Replace $f(x)$ with y: $y = x + 2$.
2. Interchange x and y: $x = y + 2$.   This gives the inverse function.
3. Solve for y: $x - 2 = y$.
4. Replace y with $f^{-1}(x)$: $f^{-1}(x) = x - 2$.   We also "reversed" the equation.

In this case, the function f added 2 to all inputs. Thus, to "undo" f, the function f^{-1} must subtract 2 from its inputs.

b) The function $f(x) = 2x - 3$ is also linear. Any linear function that is not constant will pass the horizontal-line test. Thus, f is one-to-one.

1. Replace $f(x)$ with y: $y = 2x - 3$.
2. Interchange x and y: $x = 2y - 3$.
3. Solve for y: $x + 3 = 2y$

$$\frac{x + 3}{2} = y.$$

4. Replace y with $f^{-1}(x)$: $f^{-1}(x) = \dfrac{x + 3}{2}.$

Graphing Functions and Their Inverses

How do the graphs of a function and its inverse compare?

E x a m p l e 7   Graph $f(x) = 2x - 3$ and $f^{-1}(x) = (x + 3)/2$ on the same set of axes. Then compare.

Solution   The graph of each function follows. Note that the graph of f^{-1} can be drawn by reflecting the graph of f across the line $y = x$. That is, if we graph $f(x) = 2x - 3$ in wet ink and fold the paper along the line $y = x$, the graph of $f^{-1}(x) = (x + 3)/2$ will appear as the impression made by f.

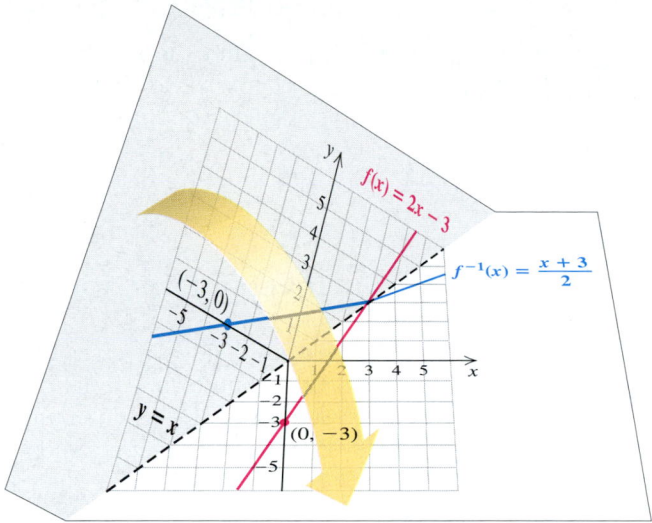

When x and y are interchanged to find a formula for the inverse, we are, in effect, reflecting or flipping the graph of $f(x) = 2x - 3$ across the line $y = x$. For example, when the coordinates of the y-intercept of the graph of f, $(0, -3)$, are reversed, we get the x-intercept of the graph of f^{-1}, $(-3, 0)$.

Visualizing Inverses

The graph of f^{-1} is a reflection of the graph of f across the line $y = x$.

E x a m p l e   8

Consider $g(x) = x^3 + 2$.

a) Determine whether the function is one-to-one.

b) If it is one-to-one, find a formula for its inverse.

c) Graph the inverse, if it exists.

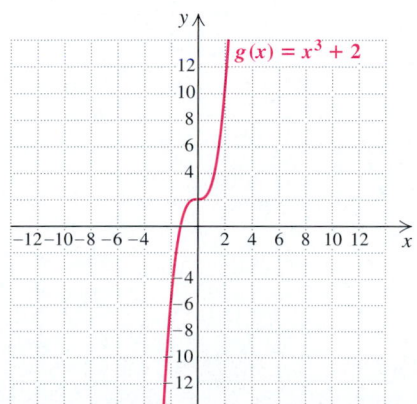

Solution

a) The graph of $g(x) = x^3 + 2$ is shown at right. It passes the horizontal-line test and thus has an inverse.

b) 1. Replace $g(x)$ with y: $y = x^3 + 2.$ Using $g(x) = x^3 + 2$

2. Interchange x and y: $x = y^3 + 2.$

3. Solve for y: $x - 2 = y^3$

$\sqrt[3]{x - 2} = y.$ Since a number has only one cube root, we can solve for y.

4. Replace y with $g^{-1}(x)$: $g^{-1}(x) = \sqrt[3]{x - 2}.$

c) To find the graph, we reflect the graph of $g(x) = x^3 + 2$ across the line $y = x$, as we did in Example 7. We can also substitute into $g^{-1}(x) = \sqrt[3]{x - 2}$ and plot points. The graphs of g and g^{-1} are shown together at left.

Inverse Functions and Composition

Let's consider inverses of functions in terms of a function machine. Suppose that a one-to-one function f is programmed into a machine. If the machine has a reverse switch, when the switch is thrown, the machine performs the inverse function f^{-1}. Inputs then enter at the opposite end, and the entire process is reversed.

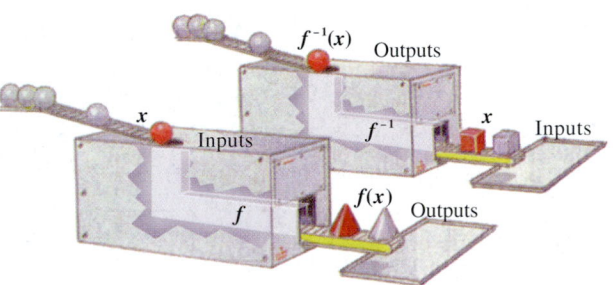

Consider $g(x) = x^3 + 2$ and $g^{-1}(x) = \sqrt[3]{x - 2}$ from Example 8. For the input 3,

$$g(3) = 3^3 + 2 = 27 + 2 = 29.$$

The output is 29. Now we use 29 for the input in the inverse:

$$g^{-1}(29) = \sqrt[3]{29 - 2} = \sqrt[3]{27} = 3.$$

The function g takes 3 to 29. The inverse function g^{-1} takes the number 29 back to 3.

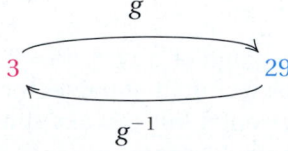

In general, for any output $f(x)$, the function f^{-1} takes that output back to x. Similarly, for any output $f^{-1}(x)$, the function f takes that output back to x.

> ### Composition and Inverses
>
> If a function f is one-to-one, then f^{-1} is the unique function for which
>
> $$(f^{-1} \circ f)(x) = x \quad \text{and} \quad (f \circ f^{-1})(x) = x.$$

E x a m p l e 9 Let $f(x) = 2x + 1$. Show that

$$f^{-1}(x) = \frac{x - 1}{2}.$$

Solution We find $(f^{-1} \circ f)(x)$ and $(f \circ f^{-1})(x)$ and check to see that each is x.

$$(f^{-1} \circ f)(x) = f^{-1}(f(x)) = f^{-1}(2x + 1)$$

$$= \frac{(2x + 1) - 1}{2}$$

$$= \frac{2x}{2} = x$$

$$(f \circ f^{-1})(x) = f(f^{-1}(x)) = f\left(\frac{x - 1}{2}\right)$$

$$= 2 \cdot \frac{x - 1}{2} + 1$$

$$= x - 1 + 1 = x$$

---

technology connection B

To determine whether $y_1 = 2x + 6$ and $y_2 = \frac{1}{2}x - 3$ might be inverses of each other, we have drawn both functions, along with the line $y = x$, on a "squared" set of axes. It *appears* that y_1 and y_2 are inverses of each other. For further verification, we can examine a table of values in which $y_1 = 2x + 6$ and $y_2 = \frac{1}{2} \cdot y_1 - 3$. Note that y_2 "undoes" what y_1 "does."

TBL MIN $= -3$ ΔTBL $= 1$ $y_2 = \frac{1}{2}y_1 - 3$

X	Y₁	Y₂
-3	0	-3
-2	2	-2
-1	4	-1
0	6	0
1	8	1
2	10	2
3	12	3

X = 3

$y_1 = 2x + 6$, $y_2 = \frac{1}{2}x - 3$

A final, visual, check can be made by graphing

$$y_1 = 2x + 6 \quad \text{and} \quad y_2 = \frac{1}{2}x - 3$$

and then pressing ⟨ **DRAW** ⟩ and selecting the DRAWINV option. Once this has been selected, we use VARS to enter y_1. The resulting graph of the inverse of y_1 should coincide with y_2.

1. Use a grapher to check Examples 7, 8, and 9.
2. Will DRAWINV work for *any* choice of y_1? Why or why not?

Exercise Set 12.1

FOR EXTRA HELP

Digital Video Tutor CD 9
Videotape 23

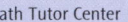
InterAct Math

Math Tutor Center

MathXL

MyMathLab.com

Find $(f \circ g)(1)$, $(g \circ f)(1)$, $(f \circ g)(x)$, and $(g \circ f)(x)$.

1. $f(x) = x^2 + 3$; $g(x) = 2x + 1$

2. $f(x) = 2x + 1$; $g(x) = x^2 - 5$

3. $f(x) = 3x - 1$; $g(x) = 5x^2 + 2$

4. $f(x) = 3x^2 + 4$; $g(x) = 4x - 1$

5. $f(x) = x + 7$; $g(x) = 1/x^2$

6. $f(x) = 1/x^2$; $g(x) = x + 2$

Find $f(x)$ and $g(x)$ such that $h(x) = (f \circ g)(x)$. Answers may vary.

7. $h(x) = (7 + 5x)^2$

8. $h(x) = (3x - 1)^2$

9. $h(x) = \sqrt{2x + 7}$

10. $h(x) = \sqrt{5x + 2}$

11. $h(x) = \dfrac{2}{x - 3}$

12. $h(x) = \dfrac{3}{x} + 4$

13. $h(x) = \dfrac{1}{\sqrt{7x + 2}}$

14. $h(x) = \sqrt{x - 7} - 3$

15. $h(x) = \dfrac{1}{\sqrt{3x}} + \sqrt{3x}$

16. $h(x) = \dfrac{1}{\sqrt{2x}} - \sqrt{2x}$

Determine whether each function is one-to-one.

17. $f(x) = x - 5$

18. $f(x) = 5 - 2x$

Aha! **19.** $f(x) = x^2 + 1$

20. $f(x) = 1 - x^2$

21. $g(x) = x^3$

22. $g(x) = \sqrt{x} + 1$

23. $g(x) = |x|$

24. $h(x) = |x| - 1$

*For each function, **(a)** determine whether it is one-to-one; **(b)** if it is one-to-one, find a formula for the inverse.*

25. $f(x) = x - 4$

26. $f(x) = x - 2$

27. $f(x) = 3 + x$

28. $f(x) = 9 + x$

29. $g(x) = x + 5$

30. $g(x) = x + 8$

31. $f(x) = 4x$

32. $f(x) = 7x$

33. $g(x) = 4x - 1$

34. $g(x) = 4x - 6$

35. $h(x) = 5$

36. $h(x) = -2$

Aha! **37.** $f(x) = \dfrac{1}{x}$

38. $f(x) = \dfrac{3}{x}$

39. $f(x) = \dfrac{2x + 1}{3}$

40. $f(x) = \dfrac{3x + 2}{5}$

41. $f(x) = x^3 - 5$

42. $f(x) = x^3 + 2$

43. $g(x) = (x - 2)^3$

44. $g(x) = (x + 7)^3$

45. $f(x) = \sqrt{x}$

46. $f(x) = \sqrt{x - 1}$

47. $f(x) = 2x^2 + 1, x \geq 0$

48. $f(x) = 3x^2 - 2, x \geq 0$

Graph each function and its inverse using the same set of axes.

49. $f(x) = \frac{1}{3}x - 2$

50. $g(x) = x + 4$

51. $f(x) = x^3$

52. $f(x) = x^3 - 1$

53. $g(x) = -2x + 3$

54. $g(x) = \sqrt{x}$

55. $F(x) = -\sqrt{x}$

56. $f(x) = -\frac{1}{2}x + 1$

57. $f(x) = 3 - x^2, x \geq 0$

58. $f(x) = x^2 - 1, x \leq 0$

59. Let $f(x) = \frac{4}{5}x$. Show that
$$f^{-1}(x) = \tfrac{5}{4}x.$$

60. Let $f(x) = (x + 7)/3$. Show that
$$f^{-1}(x) = 3x - 7.$$

61. Let $f(x) = (1 - x)/x$. Show that
$$f^{-1}(x) = \dfrac{1}{x + 1}.$$

62. Let $f(x) = x^3 - 5$. Show that
$$f^{-1}(x) = \sqrt[3]{x + 5}.$$

63. *Dress sizes in the United States and France.* A size-6 dress in the United States is size 38 in France. A function that converts dress sizes in the United States to those in France is
$$f(x) = x + 32.$$

a) Find the dress sizes in France that correspond to sizes 8, 10, 14, and 18 in the United States.

b) Determine whether this function has an inverse that is a function. If so, find a formula for the inverse.

c) Use the inverse function to find dress sizes in the United States that correspond to sizes 40, 42, 46, and 50 in France.

64. *Dress sizes in the United States and Italy.*   A size-6 dress in the United States is size 36 in Italy. A function that converts dress sizes in the United States to those in Italy is

$$f(x) = 2(x + 12).$$

a) Find the dress sizes in Italy that correspond to sizes 8, 10, 14, and 18 in the United States.

b) Determine whether this function has an inverse that is a function. If so, find a formula for the inverse.

c) Use the inverse function to find dress sizes in the United States that correspond to sizes 40, 44, 52, and 60 in Italy.

65. Is there a one-to-one relationship between the numbers and letters on the keypad of a telephone? Why or why not?

66. Mathematicians usually try to select "logical" words when forming definitions. Does the term "one-to-one" seem logical? Why or why not?

SKILL MAINTENANCE

Simplify.

67. $(a^5b^4)^2(a^3b^5)$

68. $(x^3y^5)^2(x^4y^2)$

69. $27^{4/3}$

70. $25^{3/2}$

Solve.

71. $x = \frac{2}{3}y - 7$, for y

72. $x = 10 - 3y$, for y

SYNTHESIS

73. The function $V(t) = 750(1.2)^t$ is used to predict the value, $V(t)$, of a certain rare stamp t years from 2001. Do not calculate $V^{-1}(t)$, but explain how V^{-1} could be used.

74. An organization determines that the cost per person of chartering a bus is given by the function

$$C(x) = \frac{100 + 5x}{x},$$

where x is the number of people in the group and $C(x)$ is in dollars. Determine $C^{-1}(x)$ and explain how this inverse function could be used.

For Exercises 75 and 76, graph the inverse of f.

75.

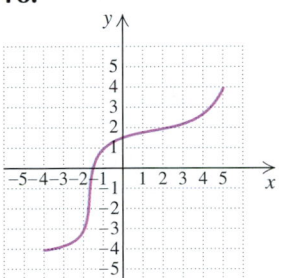

76.

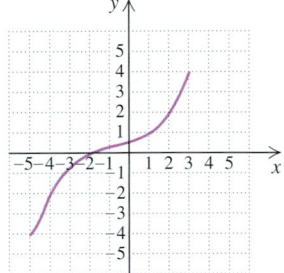

77. *Dress sizes in France and Italy.*   Use the information in Exercises 63 and 64 to find a function for the French dress size that corresponds to a size x dress in Italy.

78. *Dress sizes in Italy and France.*   Use the information in Exercises 63 and 64 to find a function for the Italian dress size that corresponds to a size x dress in France.

79. What relationship exists between the answers to Exercises 77 and 78? Explain how you determined this.

80. Show that function composition is associative by showing that $((f \circ g) \circ h)(x) = (f \circ (g \circ h))(x)$.

81. Show that if $h(x) = (f \circ g)(x)$, then $h^{-1}(x) = (g^{-1} \circ f^{-1})(x)$. (*Hint:* Use Exercise 80.)

Determine whether or not the given functions are inverses of each other.

82. $f(x) = 0.75x^2 + 2$; $g(x) = \sqrt{\dfrac{4(x - 2)}{3}}$

83. $f(x) = 1.4x^3 + 3.2$; $g(x) = \sqrt[3]{\dfrac{x - 3.2}{1.4}}$

84. $f(x) = \sqrt{2.5x + 9.25}$; $g(x) = 0.4x^2 - 3.7, x \geq 0$

85. $f(x) = 0.8x^{1/2} + 5.23$; $g(x) = 1.25(x^2 - 5.23), x \geq 0$

86. $f(x) = 2.5(x^3 - 7.1)$;
$g(x) = \sqrt[3]{0.4x + 7.1}$

87. Match each function in Column A with its inverse from Column B.

Column A

(1) $y = 5x^3 + 10$

(2) $y = (5x + 10)^3$

(3) $y = 5(x + 10)^3$

(4) $y = (5x)^3 + 10$

Column B

A. $y = \dfrac{\sqrt[3]{x} - 10}{5}$

B. $y = \sqrt[3]{\dfrac{x}{5}} - 10$

C. $y = \sqrt[3]{\dfrac{x - 10}{5}}$

D. $y = \dfrac{\sqrt[3]{x - 10}}{5}$

88. How could a grapher be used to determine whether a function is one-to-one?

89. Examine the following table. Does it appear that f and g could be inverses of each other? Why or why not?

x	$f(x)$	$g(x)$
6	6	6
7	6.5	8
8	7	10
9	7.5	12
10	8	14
11	8.5	16
12	9	18

90. Assume in Exercise 89 that f and g are both linear functions. Find equations for $f(x)$ and $g(x)$. Are f and g inverses of each other?

12.2

Exponential Functions

Graphing Exponential Functions • Equations with x and y Interchanged • Applications of Exponential Functions

CONNECTING THE CONCEPTS

Composite and inverse functions, as shown in Section 12.1, are very useful in and of themselves. The reason they are included in this chapter, however, is because they are needed in order to understand the logarithmic functions that appear in Section 12.3. Here in Section 12.2, we make no reference to composite or inverse functions. Instead, we introduce a new type of function, the *exponential function*, so that we can study both it and its inverse in Sections 12.3–12.7.

Consider the graph below. The rapidly rising curve approximates the graph of an *exponential function*. We now consider such functions and some of their applications.

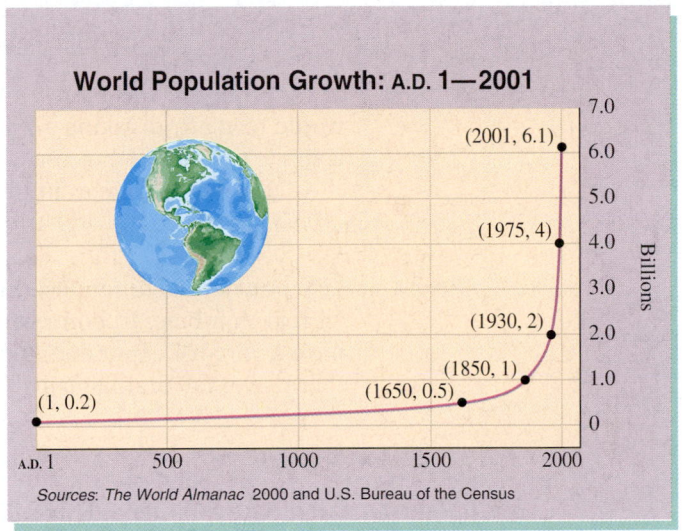

World Population Growth: A.D. 1—2001

Sources: *The World Almanac* 2000 and U.S. Bureau of the Census

Graphing Exponential Functions

In Chapter 10, we studied exponential expressions with rational-number exponents, such as

$$5^{1/4}, \qquad 3^{-3/4}, \qquad 7^{2.34}, \qquad 5^{1.73}.$$

For example, $5^{1.73}$, or $5^{173/100}$, represents the 100th root of 5 raised to the 173rd power. What about expressions with irrational exponents, such as $5^{\sqrt{3}}$ or $7^{-\pi}$? To attach meaning to $5^{\sqrt{3}}$, consider a rational approximation, r, of $\sqrt{3}$. As r gets closer to $\sqrt{3}$, the value of 5^r gets closer to some real number p.

r closes in on $\sqrt{3}$.	5^r closes in on some real number p.
$1.7 < r < 1.8$	$15.426 \approx 5^{1.7} < p < 5^{1.8} \approx 18.119$
$1.73 < r < 1.74$	$16.189 \approx 5^{1.73} < p < 5^{1.74} \approx 16.452$
$1.732 < r < 1.733$	$16.241 \approx 5^{1.732} < p < 5^{1.733} \approx 16.267$

We define $5^{\sqrt{3}}$ to be the number p. To eight decimal places,

$$5^{\sqrt{3}} \approx 16.24245082.$$

Any positive irrational exponent can be defined in a similar way. Negative irrational exponents are then defined using reciprocals. Thus, so long as a is positive, a^x has meaning for *any* real number x. All of the laws of exponents still hold, but we will not prove that here. We now define an *exponential function*.

> **Exponential Function**
>
> The function $f(x) = a^x$, where a is a positive constant, $a \neq 1$, is called the *exponential function*, base a.

We require the base a to be positive to avoid imaginary numbers that would result from taking even roots of negative numbers. The restriction $a \neq 1$ is made to exclude the constant function $f(x) = 1^x$, or $f(x) = 1$.

The following are examples of exponential functions:

$$f(x) = 2^x, \qquad f(x) = \left(\tfrac{1}{3}\right)^x, \qquad f(x) = 5^{-3x}. \qquad \text{Note that } 5^{-3x} = (5^{-3})^x.$$

Like polynomial functions, the domain of an exponential function is the set of all real numbers. In contrast to polynomial functions, exponential functions have a variable exponent. Because of this, graphs of exponential functions either rise or fall dramatically.

Example 1

Graph the exponential function $y = f(x) = 2^x$.

Solution   We compute some function values, thinking of y as $f(x)$, and list the results in a table. It is a good idea to start by letting $x = 0$.

$$f(0) = 2^0 = 1; \qquad f(-1) = 2^{-1} = \frac{1}{2^1} = \frac{1}{2};$$
$$f(1) = 2^1 = 2;$$
$$f(2) = 2^2 = 4; \qquad f(-2) = 2^{-2} = \frac{1}{2^2} = \frac{1}{4};$$
$$f(3) = 2^3 = 8;$$
$$f(-3) = 2^{-3} = \frac{1}{2^3} = \frac{1}{8}$$

Next, we plot these points and connect them with a smooth curve.

x	y, or $f(x)$
0	1
1	2
2	4
3	8
-1	$\frac{1}{2}$
-2	$\frac{1}{4}$
-3	$\frac{1}{8}$

The curve comes very close to the x-axis, but does not touch or cross it.

$$y = f(x) = 2^x$$

Be sure to plot enough points to determine how steeply the curve rises.

Note that as x increases, the function values increase without bound. As x decreases, the function values decrease, getting very close to 0. The x-axis, or the line $y = 0$, is a horizontal *asymptote*, meaning that the curve gets closer and closer to this line the further we move to the left.

E x a m p l e   2

Graph the exponential function $y = f(x) = \left(\frac{1}{2}\right)^x$.

Solution   We compute some function values, thinking of y as $f(x)$, and list the results in a table. Before we do this, note that

$$y = f(x) = \left(\frac{1}{2}\right)^x = (2^{-1})^x = 2^{-x}.$$

Then we have

$$f(0) = 2^{-0} = 1;$$

$$f(1) = 2^{-1} = \frac{1}{2^1} = \frac{1}{2};$$

$$f(2) = 2^{-2} = \frac{1}{2^2} = \frac{1}{4};$$

$$f(3) = 2^{-3} = \frac{1}{2^3} = \frac{1}{8};$$

$$f(-1) = 2^{-(-1)} = 2^1 = 2;$$

$$f(-2) = 2^{-(-2)} = 2^2 = 4;$$

$$f(-3) = 2^{-(-3)} = 2^3 = 8.$$

x	y, or $f(x)$
0	1
1	$\frac{1}{2}$
2	$\frac{1}{4}$
3	$\frac{1}{8}$
-1	2
-2	4
-3	8

Next, we plot these points and connect them with a smooth curve. Note that this curve is a mirror image, or *reflection*, of the above graph of $y = 2^x$ across the y-axis. The line $y = 0$ is again an asymptote.

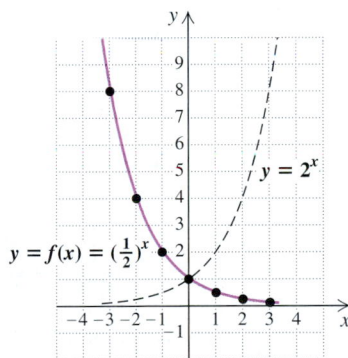

From Examples 1 and 2, we can make the following observations.

A. For $a > 1$, the graph of $f(x) = a^x$ increases from left to right. The greater the value of a, the steeper the curve. (See the figure below.)

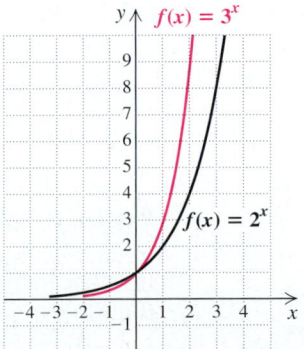

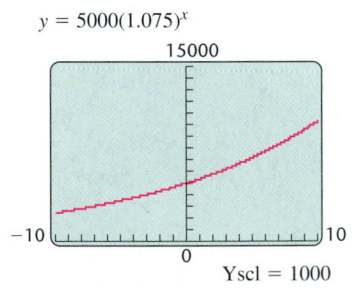

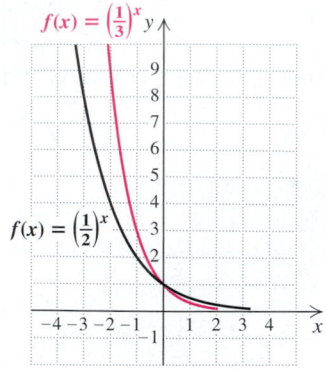

B. For $0 < a < 1$, the graph of $f(x) = a^x$ decreases from left to right. For smaller values of a, the curve becomes steeper. (See the figure at left.)

C. All graphs of $f(x) = a^x$ go through the y-intercept $(0, 1)$.

D. If $f(x) = a^x$, with $a > 0$, $a \neq 1$, the domain of f is all real numbers, and the range of f is all positive real numbers.

E. For $a > 0$, $a \neq 1$, the function given by $f(x) = a^x$ is one-to-one. Its graph passes the horizontal-line test.

E x a m p l e   3

Graph: $y = f(x) = 2^{x-2}$.

Solution   We construct a table of values. Then we plot the points and connect them with a smooth curve. Here $x - 2$ is the *exponent*.

$$f(0) = 2^{0-2} = 2^{-2} = \frac{1}{4}; \qquad\qquad f(-1) = 2^{-1-2} = 2^{-3} = \frac{1}{8};$$

$$f(1) = 2^{1-2} = 2^{-1} = \frac{1}{2}; \qquad\qquad f(-2) = 2^{-2-2} = 2^{-4} = \frac{1}{16}$$

$$f(2) = 2^{2-2} = 2^0 = 1;$$

$$f(3) = 2^{3-2} = 2^1 = 2;$$

$$f(4) = 2^{4-2} = 2^2 = 4;$$

x	y, or $f(x)$
0	$\frac{1}{4}$
1	$\frac{1}{2}$
2	1
3	2
4	4
-1	$\frac{1}{8}$
-2	$\frac{1}{16}$

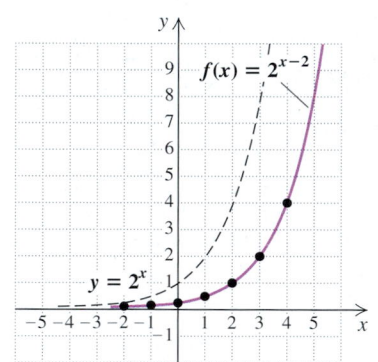

The graph looks just like the graph of $y = 2^x$, but it is translated 2 units to the right. The y-intercept of $y = 2^x$ is $(0, 1)$. The y-intercept of $y = 2^{x-2}$ is $\left(0, \frac{1}{4}\right)$. The line $y = 0$ is again the asymptote.

Equations with x and y Interchanged

It will be helpful in later work to be able to graph an equation in which the x and the y in $y = a^x$ are interchanged.

E x a m p l e   4   Graph: $x = 2^y$.

Solution   Note that x is alone on one side of the equation. To find ordered pairs that are solutions, we choose values for y and then compute values for x:

For $y = 0$, $x = 2^0 = 1$.
For $y = 1$, $x = 2^1 = 2$.
For $y = 2$, $x = 2^2 = 4$.
For $y = 3$, $x = 2^3 = 8$.

For $y = -1$, $x = 2^{-1} = \dfrac{1}{2}$.

For $y = -2$, $x = 2^{-2} = \dfrac{1}{4}$.

For $y = -3$, $x = 2^{-3} = \dfrac{1}{8}$.

x	y
1	0
2	1
4	2
8	3
$\frac{1}{2}$	-1
$\frac{1}{4}$	-2
$\frac{1}{8}$	-3

(1) Choose values for y.
(2) Compute values for x.

We plot the points and connect them with a smooth curve.

This curve does not touch or cross the y-axis, which serves as a vertical asymptote.

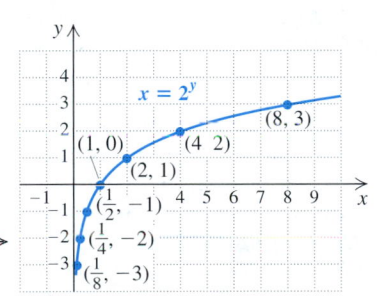

Note too that this curve looks just like the graph of $y = 2^x$, except that it is reflected across the line $y = x$, as shown here.

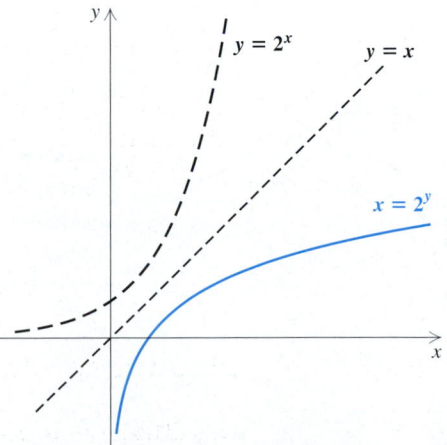

Applications of Exponential Functions

E x a m p l e   5

Interest compounded annually.   The amount of money A that a principal P will be worth after t years at interest rate i, compounded annually, is given by the formula

$$A = P(1 + i)^t.$$ You might review Example 12 in Section 11.1.

Suppose that \$100,000 is invested at 8% interest, compounded annually.

a) Find a function for the amount in the account after t years.
b) Find the amount of money in the account at $t = 0$, $t = 4$, $t = 8$, and $t = 10$.
c) Graph the function.

Solution

a) If $P = \$100{,}000$ and $i = 8\% = 0.08$, we can substitute these values and form the following function:

$$A(t) = \$100{,}000(1 + 0.08)^t$$ Using $A = P(1 + i)^t$
$$= \$100{,}000(1.08)^t.$$

b) To find the function values, a calculator with a power key is helpful.

$$A(0) = \$100{,}000(1.08)^0$$
$$= \$100{,}000(1)$$
$$= \$100{,}000$$

$$A(4) = \$100{,}000(1.08)^4$$
$$= \$100{,}000(1.36048896)$$
$$\approx \$136{,}048.90$$

$$A(8) = \$100{,}000(1.08)^8$$
$$\approx \$100{,}000(1.85093021)$$
$$\approx \$185{,}093.02$$

$$A(10) = \$100{,}000(1.08)^{10}$$
$$\approx \$100{,}000(2.158924997)$$
$$\approx \$215{,}892.50$$

c) We use the function values computed in part (b), and others if we wish, to draw the graph as follows. Note that the axes are scaled differently because of the large numbers.

technology
connection
B

Graphers can quickly find many function values at the touch of a few keys. To see this, let $y_1 = 100{,}000(1.08)^x$. Then use the TABLE feature to check Example 5(b).

Exercise Set 12.2

FOR EXTRA HELP

Digital Video Tutor CD 9
Videotape 23

InterAct Math

Math Tutor Center

MathXL

MyMathLab.com

Graph.

1. $y = f(x) = 2^x$

2. $y = f(x) = 3^x$

3. $y = 5^x$

4. $y = 6^x$

5. $y = 2^x + 3$

6. $y = 2^x + 1$

7. $y = 3^x - 1$

8. $y = 3^x - 2$

9. $y = 2^x - 4$

10. $y = 2^x - 5$

11. $y = 2^{x-1}$

12. $y = 2^{x-2}$

13. $y = 2^{x+3}$

14. $y = 2^{x+1}$

15. $y = \left(\frac{1}{5}\right)^x$

16. $y = \left(\frac{1}{4}\right)^x$

17. $y = \left(\frac{1}{2}\right)^x$

18. $y = \left(\frac{1}{3}\right)^x$

19. $y = 2^{x-3} - 1$

20. $y = 2^{x+1} - 3$

21. $x = 3^y$

22. $x = 6^y$

23. $x = 2^{-y}$

24. $x = 3^{-y}$

25. $x = 5^y$

26. $x = 4^y$

27. $x = \left(\frac{3}{2}\right)^y$

28. $x = \left(\frac{4}{3}\right)^y$

Graph both equations using the same set of axes.

29. $y = 3^x,\ x = 3^y$

30. $y = 2^x,\ x = 2^y$

31. $y = \left(\frac{1}{2}\right)^x,\ x = \left(\frac{1}{2}\right)^y$

32. $y = \left(\frac{1}{4}\right)^x,\ x = \left(\frac{1}{4}\right)^y$

Solve.

33. *Population growth.* The world population $P(t)$, in billions, t years after 1975 can be approximated by

$$P(t) = 4(1.0164)^t$$

(*Sources:* Data from *The World Almanac* 2000 and U.S. Bureau of the Census).

a) What will the world population be in 2004? in 2008? in 2012?

b) Graph the function.

34. *Growth of bacteria.* The bacteria *Escherichi coli* are commonly found in the human bladder. Suppose that 3000 of the bacteria are present at time $t = 0$. Then t minutes later, the number of bacteria present will be

$$N(t) = 3000(2)^{t/20}.$$

a) How many bacteria will be present after 10 min? 20 min? 30 min? 40 min? 60 min?

b) Graph the function.

35. *Marine biology.* Due to excessive whaling prior to the mid 1970s, the humpback whale is considered an endangered species. The worldwide population of humpbacks, $P(t)$, in thousands, t years after 1900 ($t < 70$) can be approximated by*

$$P(t) = 150(0.960)^t.$$

a) How many humpback whales were alive in 1930? in 1960?

b) Graph the function.

36. *Marine biology.* As a result of preservation efforts in most countries in which whaling was common, the humpback whale population has grown since the 1970s. The worldwide population of humpbacks, $P(t)$, in thousands, t years after 1982 can be approximated by*

$$P(t) = 5.5(1.047)^t.$$

a) How many humpback whales were alive in 1992? in 2001?

b) Graph the function.

37. *Recycling aluminum cans.* It is estimated that $\frac{2}{3}$ of all aluminum cans distributed will be recycled each year. A beverage company distributes 250,000 cans. The number still in use after time t, in years, is given by the exponential function

$$N(t) = 250{,}000\left(\tfrac{2}{3}\right)^t.$$

a) How many cans are still in use after 0 yr? 1 yr? 4 yr? 10 yr?

b) Graph the function.

38. *Salvage value.* A photocopier is purchased for $5200. Its value each year is about 80% of the value of the preceding year. Its value, in dollars, after t years is given by the exponential function

$$V(t) = 5200(0.8)^t.$$

a) Find the value of the machine after 0 yr, 1 yr, 2 yr, 5 yr, and 10 yr.

b) Graph the function.

---

*Based on information from the American Cetacean Society, 2001, and the ASK Archive, 1998.

39. *Cellular phones.* The number of cellular phones in use in the United States is increasing exponentially. The number N, in millions, in use is given by the exponential function

$$N(t) = 0.3(1.4477)^t,$$

where t is the number of years after 1985 (*Source*: Cellular Telecommunications and Internet Association).

a) Find the number of cellular phones in use in 1985, 1995, 2005, and 2010.

b) Graph the function.

40. *Spread of zebra mussels.* Beginning in 1988, infestations of zebra mussels started spreading throughout North American waters.* These mussels spread with such speed that water treatment facilities, power plants, and entire ecosystems can become threatened. The function

$$A(t) = 10 \cdot 34^t$$

can be used to estimate the number of square centimeters of lake bottom that will be covered with mussels t years after an infestation covering 10 cm² first occurs.

a) How many square centimeters of lake bottom will be covered with mussels 5 years after an infestation covering 10 cm² first appears? 7 years after the infestation first appears?

b) Graph the function.

41. Without using a calculator, explain why 2^π must be greater than 8 but less than 16.

42. Suppose that $1000 is invested for 5 yr at 7% interest, compounded annually. In what year will the most interest be earned? Why?

SKILL MAINTENANCE

43. 5^{-2}

44. 2^{-5}

45. $1000^{2/3}$

46. $25^{-3/2}$

47. $\dfrac{10a^8b^7}{2a^2b^4}$

48. $\dfrac{24x^6y^4}{4x^2y^3}$

SYNTHESIS

49. Examine Exercise 39. Do you believe that the equation for the number of cellular phones in use in the

United States will be accurate 20 yr from now? Why or why not?

50. Why was it necessary to discuss irrational exponents before graphing exponential functions?

Determine which of the two numbers is larger. Do not use a calculator.

51. $\pi^{1.3}$ or $\pi^{2.4}$

52. $\sqrt{8^3}$ or $8^{\sqrt{3}}$

Graph.

53. $f(x) = 3.8^x$

54. $f(x) = 2.3^x$

55. $y = 2^x + 2^{-x}$

56. $y = \left|\left(\frac{1}{2}\right)^x - 1\right|$

57. $y = |2^x - 2|$

58. $y = 2^{-(x-1)^2}$

59. $y = \left|2^{x^2} - 1\right|$

60. $y = 3^x + 3^{-x}$

Graph both equations using the same set of axes.

61. $y = 3^{-(x-1)}, \ x = 3^{-(y-1)}$

62. $y = 1^x, \ x = 1^y$

63. *Sales of DVD players.* As prices of DVD players continue to drop, sales have grown from $171 million in 1997 to $421 million in 1998 and $1099 million in 1999 (*Source*: *Statistical Abstract of the United States*, 2000). Use the REGRESSION feature in the STAT CALC menu to find an exponential function that models the total sales of DVD players t years after 1997. Then use that function to predict the total sales in 2005.

64. *Spread of AIDS.* In 1985, a total of 8249 cases of AIDS was reported in the United States; in 1988, a total of 31,001 cases; in 1989, a total of 33,722 cases; and in 1990, a total of 41,595 cases. Use the STAT REGRESSION feature of a grapher to find a model for $N(t)$, the number of AIDS cases in the United States t years after 1985. Then estimate the number of cases reported in 1987 and in 1997.

65. *Keyboarding speed.* Ali is studying keyboarding. After he has studied for t hours, Ali's speed, in words per minute, is given by the exponential function

$$S(t) = 200[1 - (0.99)^t].$$

Use a graph and/or table of values to predict Ali's speed after studying for 10 hr, 40 hr, and 80 hr.

66. Consider any exponential function of the form $f(x) = a^x$ with $a > 1$. Will it always follow that $f(3) - f(2) > f(2) - f(1)$, and, in general, $f(n + 2) - f(n + 1) > f(n + 1) - f(n)$? Why or why not? (*Hint*: Think graphically.)

*Many thanks to Dr. Gerald Mackie of the Department of Zoology at the University of Guelph in Ontario for the background information for this exercise.

COLLABORATIVE

CORNER

The True Cost of a New Car

Focus: Car loans and exponential functions

Time: 30 minutes

Group size: 2

Materials: Calculators with exponentiation keys

The formula

$$M = \frac{Pr}{1 - (1 + r)^{-n}}$$

is used to determine the payment size, M, when a loan of P dollars is to be repaid in n equally sized monthly payments. Here r represents the monthly interest rate. Loans repaid in this fashion are said to be *amortized* (spread out equally) over a period of n months.

ACTIVITY

1. Suppose one group member is selling the other a car for $2600, financed at 1% interest per month for 24 months. What should be the size of each monthly payment?

2. Suppose both group members are shopping for the same model new car. To save time, each group member visits a different dealer. One dealer offers the car for $13,000 at 10.5% interest (0.00875 monthly interest) for 60 months (no down payment). The other dealer offers the same car for $12,000, but at 12% interest (0.01 monthly interest) for 48 months (no down payment).

   a) Determine the monthly payment size for each offer (remember to use the *monthly* interest rates). Then determine the total amount paid for the car under each offer. How much of each total is interest?

   b) Work together to find the annual interest rate for which the total cost of 60 monthly payments for the $13,000 car would equal the total amount paid for the $12,000 car (as found in part a above).

Logarithmic Functions

12.3

Graphs of Logarithmic Functions • Converting Exponential and Logarithmic Equations • Solving Certain Logarithmic Equations

We are now ready to study inverses of exponential functions. These functions have many applications and are referred to as *logarithm*, or *logarithmic*, *functions*.

Graphs of Logarithmic Functions

Consider the exponential function $f(x) = 2^x$. Like all exponential functions, f is one-to-one. Can a formula for f^{-1} be found? To answer this, we use the method of Section 12.1:

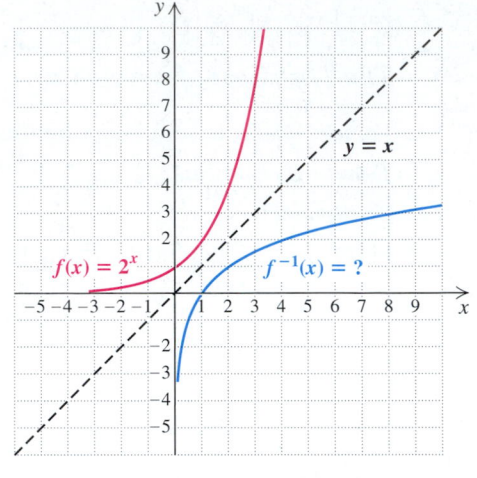

1. Replace $f(x)$ with y:

$$y = 2^x.$$

2. Interchange x and y:

$$x = 2^y.$$

3. Solve for y: y = the power to which we raise 2 to get x.

4. Replace y with $f^{-1}(x)$: $f^{-1}(x)$ = the power to which we raise 2 to get x.

We now define a new symbol to replace the words "the power to which we raise 2 to get x":

> $\log_2 x$, read **"the logarithm, base 2, of x"**, or **"log, base 2, of x,"** means **"the power to which we raise 2 to get x."**

Thus if $f(x) = 2^x$, then $f^{-1}(x) = \log_2 x$. Note that $f^{-1}(8) = \log_2 8 = 3$, because 3 is *the power to which we raise* 2 *to get* 8.

E x a m p l e   1

Simplify: **(a)** $\log_2 32$; **(b)** $\log_2 1$; **(c)** $\log_2 \frac{1}{8}$.

Solution

a) Think of the meaning of $\log_2 32$. It is the exponent to which we raise 2 to get 32. That exponent is 5. Therefore, $\log_2 32 = 5$.

b) We ask ourselves: "To what power do we raise 2 in order to get 1?" That power is 0 (recall that $2^0 = 1$). Thus, $\log_2 1 = 0$.

c) To what power do we raise 2 in order to get $\frac{1}{8}$? Since $2^{-3} = \frac{1}{8}$, we have $\log_2 \frac{1}{8} = -3$.

Although expressions like $\log_2 13$ can only be approximated, we must remember that $\log_2 13$ represents *the power to which we raise* 2 *to get* 13. That is, $2^{\log_2 13} = 13$. A calculator can be used to show that $\log_2 13 \approx 3.7$ and $2^{3.7} \approx 13$. Later in this chapter, we will discuss how a calculator can be used to find such approximations.

For any exponential function $f(x) = a^x$, the inverse is called a **logarithmic function, base a.** The graph of the inverse can, of course, be drawn by reflecting the graph of $f(x) = a^x$ across the line $y = x$. It will be helpful to remember that the inverse of $f(x) = a^x$ is given by $f^{-1}(x) = \log_a x$.

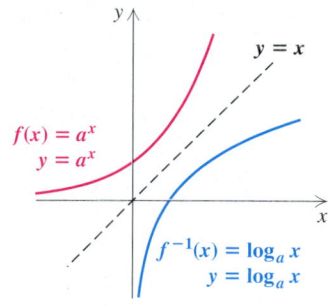

> ### *The Meaning of* $\log_a x$
>
> For $x > 0$ and a a positive constant other than 1, $\log_a x$ is the power to which a must be raised in order to get x. Thus,
>
> $$a^{\log_a x} = x \qquad \text{or equivalently,} \qquad \text{if } y = \log_a x, \text{ then } a^y = x.$$

It is important to remember that *a logarithm is an exponent.* It might help to repeat several times: "The logarithm, base a, of a number x is the power to which a must be raised in order to get x."

E x a m p l e   2

Simplify: $7^{\log_7 85}$.

Solution   Remember that $\log_7 85$ is the power to which 7 is raised to get 85. Raising 7 to that power, we have

$$7^{\log_7 85} = 85.$$

Because logarithmic and exponential functions are inverses of each other, the result in Example 2 should come as no surprise: If $f(x) = \log_7 x$, then

$$\text{for} \quad f(x) = \log_7 x, \text{ we have } f^{-1}(x) = 7^x$$

$$\text{and} \quad f^{-1}(f(x)) = f^{-1}(\log_7 x) = 7^{\log_7 x} = x.$$

Thus, $f^{-1}(f(85)) = 7^{\log_7 85} = 85$.

The following is a comparison of exponential and logarithmic functions.

technology connection

To see that $f(x) = 10^x$ and $g(x) = \log_{10} x$ are inverses of each other, let $y_1 = 10^x$ and $y_2 = \log_{10} x = \log x$. Then, using a squared window, compare both graphs. Finally, let $y_3 = y_1(y_2)$ and $y_4 = y_2(y_1)$ to show, using a table or graphs, that $y_3 = y_4 = x$.

Exponential Function	Logarithmic Function
$y = a^x$	$x = a^y$
$f(x) = a^x$	$f(x) = \log_a x$
$a > 0, a \neq 1$	$a > 0, a \neq 1$
The domain is $\mathbb{R}$.	The range is $\mathbb{R}$.
$y > 0$ (Outputs are positive.)	$x > 0$ (Inputs are positive.)
$f^{-1}(x) = \log_a x$	$f^{-1}(x) = a^x$

E x a m p l e   3   Graph: $y = f(x) = \log_5 x$.

Solution   If $y = \log_5 x$, then $5^y = x$. We can find ordered pairs that are solutions by choosing values for y and computing the x-values.

For $y = 0$, $x = 5^0 = 1$.

For $y = 1$, $x = 5^1 = 5$.   (1) Select y.

For $y = 2$, $x = 5^2 = 25$.   (2) Compute x.

For $y = -1$, $x = 5^{-1} = \frac{1}{5}$.

For $y = -2$, $x = 5^{-2} = \frac{1}{25}$.

x, or 5^y	y
1	0
5	1
25	2
$\frac{1}{5}$	-1
$\frac{1}{25}$	-2

This table shows the following:

$\log_5 1 = 0$;

$\log_5 5 = 1$;

$\log_5 25 = 2$;   These can all be checked

$\log_5 \frac{1}{5} = -1$;   using the equations above.

$\log_5 \frac{1}{25} = -2$.

We plot the set of ordered pairs and connect the points with a smooth curve. The graph of $y = 5^x$ is shown only for reference.

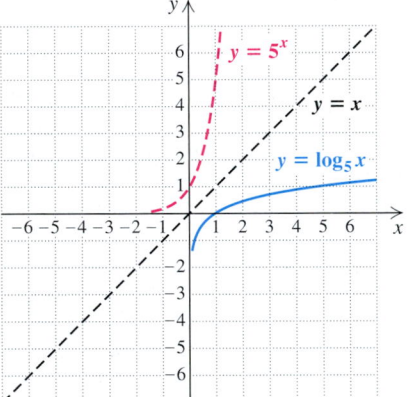

Converting Exponential and Logarithmic Equations

We use the definition of logarithm to convert from *exponential equations* to *logarithmic equations*:

$$y = \log_a x \quad \text{is equivalent to} \quad a^y = x.$$

Caution!   **Do not forget this relationship!** It is probably the most important definition in the chapter. Many times this definition will be used to justify a property we are considering.

E x a m p l e   4

Convert each to a logarithmic equation: **(a)** $8 = 2^x$; **(b)** $y^{-1} = 4$; **(c)** $a^b = c$.

Solution

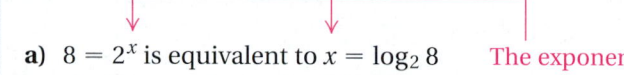

a) $8 = 2^x$ is equivalent to $x = \log_2 8$ The exponent is the logarithm.

The base remains the same.

b) $y^{-1} = 4$ is equivalent to $-1 = \log_y 4$

c) $a^b = c$ is equivalent to $b = \log_a c$

We also use the definition of logarithm to convert from logarithmic equations to exponential equations.

E x a m p l e   5

Convert each to an exponential equation: **(a)** $y = \log_3 5$; **(b)** $-2 = \log_a 7$; **(c)** $a = \log_b d$.

Solution

a) $y = \log_3 5$ is equivalent to $3^y = 5$ The logarithm is the exponent.

The base remains the same.

b) $-2 = \log_a 7$ is equivalent to $a^{-2} = 7$

c) $a = \log_b d$ is equivalent to $b^a = d$

Solving Certain Logarithmic Equations

Some logarithmic equations can be solved by converting to exponential equations.

E x a m p l e   6

Solve: **(a)** $\log_2 x = -3$; **(b)** $\log_x 16 = 2$.

Solution

a) $\log_2 x = -3$

$\qquad 2^{-3} = x$ Converting to an exponential equation

$\qquad \frac{1}{8} = x$ Computing 2^{-3}

Check: $\log_2 \frac{1}{8}$ is the power to which 2 is raised to get $\frac{1}{8}$. Since that power is -3, we have a check. The solution is $\frac{1}{8}$.

b) $\log_x 16 = 2$

$\qquad x^2 = 16$ Converting to an exponential equation

$\quad x = 4 \quad or \quad x = -4$ Principle of square roots

Check: $\log_4 16 = 2$ because $4^2 = 16$. Thus, 4 is a solution of $\log_x 16 = 2$. Because all logarithmic bases must be positive, -4 cannot be a solution. Logarithmic bases must be positive because logarithms are defined using exponential functions that require positive bases. The solution is 4.

One method for solving certain logarithmic and exponential equations relies on the following property, which results from the fact that exponential functions are one-to-one.

> ### The Principle of Exponential Equality
>
> For any real number b, where $b \neq -1, 0$, or 1,
>
> $$b^x = b^y \quad \text{is equivalent to} \quad x = y.$$
>
> (Powers of the same base are equal if and only if the exponents are equal.)

Example 7

Solve: **(a)** $\log_{10} 1000 = x$; **(b)** $\log_4 1 = t$.

Solution

a) We convert $\log_{10} 1000 = x$ to exponential form and solve:

$$10^x = 1000 \qquad \text{Converting to an exponential equation}$$
$$10^x = 10^3 \qquad \text{Writing 1000 as a power of 10}$$
$$x = 3. \qquad \text{Equating exponents}$$

Check: This equation can also be solved directly by determining the power to which we raise 10 in order to get 1000. In both cases we find that $\log_{10} 1000 = 3$, so we have a check. The solution is 3.

b) We convert $\log_4 1 = t$ to exponential form and solve:

$$4^t = 1 \qquad \text{Converting to an exponential equation}$$
$$4^t = 4^0 \qquad \text{Writing 1 as a power of 4. This can be done mentally.}$$
$$t = 0. \qquad \text{Equating exponents}$$

Check: As in part (a), this equation can be solved directly by determining the power to which we raise 4 in order to get 1. In both cases we find that $\log_4 1 = 0$, so we have a check. The solution is 0.

Example 7 illustrates an important property of logarithms.

> ### $\log_a 1$
>
> The logarithm, base a, of 1 is always 0: $\log_a 1 = 0$.

This follows from the fact that $a^0 = 1$ is equivalent to the logarithmic equation $\log_a 1 = 0$. Thus, $\log_{10} 1 = 0$, $\log_7 1 = 0$, and so on.

Another property results from the fact that $a^1 = a$. This is equivalent to the equation $\log_a a = 1$.

$\log_a a$

The logarithm, base a, of a is always 1: $\log_a a = 1$.

Thus, $\log_{10} 10 = 1$, $\log_8 8 = 1$, and so on.

Exercise Set 12.3

Simplify.

1. $\log_{10} 100$

2. $\log_{10} 1000$

3. $\log_2 8$

4. $\log_2 16$

5. $\log_3 81$

6. $\log_3 27$

7. $\log_4 \frac{1}{16}$

8. $\log_4 \frac{1}{4}$

9. $\log_7 \frac{1}{7}$

10. $\log_7 \frac{1}{49}$

11. $\log_5 625$

12. $\log_5 125$

13. $\log_6 6$

14. $\log_7 1$

15. $\log_8 1$

16. $\log_8 8$

Aha! 17. $\log_9 9^7$

18. $\log_9 9^{10}$

19. $\log_{10} 0.1$

20. $\log_{10} 0.01$

21. $\log_9 3$

22. $\log_{16} 4$

23. $\log_9 27$

24. $\log_{16} 64$

25. $\log_{1000} 100$

26. $\log_{27} 9$

27. $5^{\log_5 7}$

28. $6^{\log_6 13}$

Graph.

29. $y = \log_{10} x$

30. $y = \log_2 x$

31. $y = \log_3 x$

32. $y = \log_7 x$

33. $f(x) = \log_6 x$

34. $f(x) = \log_4 x$

35. $f(x) = \log_{2.5} x$

36. $f(x) = \log_{1/2} x$

Graph both functions using the same set of axes.

37. $f(x) = 3^x,\ f^{-1}(x) = \log_3 x$

38. $f(x) = 4^x,\ f^{-1}(x) = \log_4 x$

Convert to logarithmic equations.

39. $10^2 = 100$

40. $10^4 = 10{,}000$

41. $4^{-5} = \frac{1}{1024}$

42. $5^{-3} = \frac{1}{125}$

43. $16^{3/4} = 8$

44. $8^{1/3} = 2$

45. $10^{0.4771} = 3$

46. $10^{0.3010} = 2$

47. $p^k = 3$

48. $m^n = r$

49. $p^m = V$

50. $Q^t = x$

51. $e^3 = 20.0855$

52. $e^2 = 7.3891$

53. $e^{-4} = 0.0183$

54. $e^{-2} = 0.1353$

Convert to exponential equations.

55. $t = \log_3 8$

56. $h = \log_7 10$

57. $\log_5 25 = 2$

58. $\log_6 6 = 1$

59. $\log_{10} 0.1 = -1$

60. $\log_{10} 0.01 = -2$

61. $\log_{10} 7 = 0.845$

62. $\log_{10} 3 = 0.4771$

63. $\log_c m = 8$

64. $\log_b n = 23$

65. $\log_t Q = r$

66. $\log_m P = a$

67. $\log_e 0.25 = -1.3863$

68. $\log_e 0.989 = -0.0111$

69. $\log_r T = -x$

70. $\log_c M = -w$

Solve.

71. $\log_3 x = 2$

72. $\log_4 x = 3$

73. $\log_x 64 = 3$

74. $\log_x 125 = 3$

75. $\log_5 25 = x$

76. $\log_2 16 = x$

77. $\log_4 16 = x$

78. $\log_3 27 = x$

79. $\log_x 7 = 1$

80. $\log_x 8 = 1$

81. $\log_9 x = 1$

82. $\log_6 x = 0$

83. $\log_3 x = -2$

84. $\log_2 x = -1$

85. $\log_{32} x = \frac{2}{5}$

86. $\log_8 x = \frac{2}{3}$

87. Express in words what number is represented by $\log_b c$.

88. Is it true that $2 = b^{\log_b 2}$? Why or why not?

SKILL MAINTENANCE

Simplify.

89. $\dfrac{x^{12}}{x^4}$

90. $\dfrac{a^{15}}{a^3}$

91. $(a^4 b^6)(a^3 b^2)$

92. $(x^3 y^5)(x^2 y^7)$

93. $\dfrac{\dfrac{3}{x} - \dfrac{2}{xy}}{\dfrac{2}{x^2} + \dfrac{1}{xy}}$

94. $\dfrac{\dfrac{4 + x}{x^2 + 2x + 1}}{\dfrac{3}{x + 1} - \dfrac{2}{x + 2}}$

SYNTHESIS

95. Would a manufacturer be pleased or unhappy if sales of a product grew logarithmically? Why?

96. Explain why the number $\log_2 13$ must be between 3 and 4.

97. Graph both equations using the same set of axes:
$$y = \left(\tfrac{3}{2}\right)^x, \qquad y = \log_{3/2} x.$$

Graph.

98. $y = \log_2 (x - 1)$

99. $y = \log_3 |x + 1|$

Solve.

100. $|\log_3 x| = 2$

101. $\log_{125} x = \frac{2}{3}$

102. $\log_4 (3x - 2) = 2$

103. $\log_8 (2x + 1) = -1$

104. $\log_{10} (x^2 + 21x) = 2$

Simplify.

105. $\log_{1/4} \frac{1}{64}$

106. $\log_{1/5} 25$

107. $\log_{81} 3 \cdot \log_3 81$

108. $\log_{10} (\log_4 (\log_3 81))$

109. $\log_2 (\log_2 (\log_4 256))$

110. Show that $b^x = b^y$ is *not* equivalent to $x = y$ for $b = 0$ or $b = 1$.

111. If $\log_b a = x$, does it follow that $\log_a b = 1/x$? Why or why not?

12.4

Properties of Logarithmic Functions

Logarithms of Products • Logarithms of Powers • Logarithms of Quotients • Using the Properties Together

Logarithmic functions are important in many applications and in more advanced mathematics. We now establish some basic properties that are useful in manipulating expressions involving logarithms. As their proofs reveal, the properties of logarithms are related to the properties of exponents.

Logarithms of Products

The first property we discuss is related to the product rule for exponents: $a^m \cdot a^n = a^{m+n}$.

> ### *The Product Rule for Logarithms*
>
> For any positive numbers M, N, and a ($a \neq 1$),
>
> $$\log_a MN = \log_a M + \log_a N.$$
>
> (The logarithm of a product is the sum of the logarithms of the factors.)

Example 1

Express as a sum of logarithms: $\log_2 (4 \cdot 16)$.

Solution   We have

$$\log_2 (4 \cdot 16) = \log_2 4 + \log_2 16. \qquad \text{Using the product rule for logarithms}$$

As a check, note that

$$\log_2 (4 \cdot 16) = \log_2 64 = 6 \qquad 2^6 = 64$$

and that

$$\log_2 4 + \log_2 16 = 2 + 4 = 6. \qquad 2^2 = 4 \text{ and } 2^4 = 16$$

Example 2

Express as a single logarithm: $\log_b 7 + \log_b 5$.

Solution   We have

$$\log_b 7 + \log_b 5 = \log_b (7 \cdot 5) \qquad \text{Using the product rule for logarithms}$$
$$= \log_b 35.$$

The check is left to the student.

A Proof of the Product Rule.   Let $\log_a M = x$ and $\log_a N = y$. Converting to exponential equations, we have $a^x = M$ and $a^y = N$.

Now we multiply the last two equations, to obtain

$$MN = a^x \cdot a^y, \quad \text{or} \quad MN = a^{x+y}.$$

Converting back to a logarithmic equation, we get

$$\log_a MN = x + y.$$

Recalling what x and y represent, we get

$$\log_a MN = \log_a M + \log_a N.$$

Logarithms of Powers

The second basic property is related to the power rule for exponents: $(a^m)^n = a^{mn}$.

> ### The Power Rule for Logarithms
>
> For any positive numbers M and a ($a \neq 1$), and any real number p,
>
> $$\log_a M^p = p \cdot \log_a M.$$
>
> (The logarithm of a power of M is the exponent times the logarithm of M.)

To better understand the power rule, note that

$$\log_a M^3 = \log_a (M \cdot M \cdot M) = \log_a M + \log_a M + \log_a M = 3 \log_a M.$$

E x a m p l e 3 Express as a product: **(a)** $\log_a 9^{-5}$; **(b)** $\log_7 \sqrt[3]{x}$.

Solution

a) $\log_a 9^{-5} = -5 \log_a 9$ Using the power rule for logarithms

b) $\log_7 \sqrt[3]{x} = \log_7 x^{1/3}$ Writing exponential notation

$\phantom{\log_7 \sqrt[3]{x}} = \frac{1}{3} \log_7 x$ Using the power rule for logarithms

A Proof of the Power Rule. Let $x = \log_a M$. We then write the equivalent exponential equation, $a^x = M$. Raising both sides to the pth power, we get

$$(a^x)^p = M^p, \quad \text{or} \quad a^{xp} = M^p.   \text{Multiplying exponents}$$

Converting back to a logarithmic equation gives us

$$\log_a M^p = xp.$$

But $x = \log_a M$, so substituting, we have

$$\log_a M^p = (\log_a M)p = p \cdot \log_a M.$$

Logarithms of Quotients

The third property that we study is similar to the quotient rule for exponents: $a^m/a^n = a^{m-n}$.

> ### The Quotient Rule for Logarithms
>
> For any positive numbers M, N, and a ($a \neq 1$),
>
> $$\log_a \frac{M}{N} = \log_a M - \log_a N.$$
>
> (The logarithm of a quotient is the logarithm of the dividend minus the logarithm of the divisor.)

To better understand the quotient rule, note that

$$\log_a\left(\frac{b^5}{b^3}\right) = \log_a b^2 = 2\log_a b = 5\log_a b - 3\log_a b$$

$$= \log_a b^5 - \log_a b^3.$$

E x a m p l e   4

Express as a difference of logarithms: $\log_t(6/U)$.

Solution

$$\log_t\frac{6}{U} = \log_t 6 - \log_t U \qquad \text{Using the quotient rule for logarithms}$$

E x a m p l e   5

Express as a single logarithm: $\log_b 17 - \log_b 27$.

Solution

$$\log_b 17 - \log_b 27 = \log_b\frac{17}{27} \qquad \text{Using the quotient rule for logarithms "in reverse"}$$

A Proof of the Quotient Rule.   Our proof uses both the product and power rules:

$$\log_a\frac{M}{N} = \log_a MN^{-1} \qquad \text{Rewriting } \frac{M}{N} \text{ with a negative exponent}$$

$$= \log_a M + \log_a N^{-1} \qquad \text{Using the product rule for logarithms}$$

$$= \log_a M + (-1)\log_a N \qquad \text{Using the power rule for logarithms}$$

$$= \log_a M - \log_a N.$$

Using the Properties Together

E x a m p l e   6

Express in terms of the individual logarithms of x, y, and z.

a) $\log_b\dfrac{x^3}{yz}$

b) $\log_a\sqrt[4]{\dfrac{xy}{z^3}}$

Solution

a) $\log_b\dfrac{x^3}{yz} = \log_b x^3 - \log_b yz$ Using the quotient rule for logarithms

$$= 3\log_b x - \log_b yz \qquad \text{Using the power rule for logarithms}$$

$$= 3\log_b x - (\log_b y + \log_b z) \qquad \text{Using the product rule for logarithms. Because of the subtraction, parentheses are essential.}$$

$$= 3\log_b x - \log_b y - \log_b z \qquad \text{Using the distributive law}$$

b) $\log_a \sqrt[4]{\dfrac{xy}{z^3}} = \log_a\left(\dfrac{xy}{z^3}\right)^{1/4}$ Writing exponential notation

$= \dfrac{1}{4} \cdot \log_a \dfrac{xy}{z^3}$ Using the power rule for logarithms

$= \dfrac{1}{4}(\log_a xy - \log_a z^3)$ Using the quotient rule for logarithms. Parentheses are important.

$= \dfrac{1}{4}(\log_a x + \log_a y - 3\log_a z)$ Using the product and power rules for logarithms

---

> **Caution!** When subtraction or multiplication precedes use of the product or quotient rule, parentheses are needed, as in Example 6.

E x a m p l e 7 Express as a single logarithm.

a) $\dfrac{1}{2}\log_a x - 7\log_a y + \log_a z$ **b)** $\log_a \dfrac{b}{\sqrt{x}} + \log_a \sqrt{bx}$

Solution

a) $\dfrac{1}{2}\log_a x - 7\log_a y + \log_a z$

$= \log_a x^{1/2} - \log_a y^7 + \log_a z$ Using the power rule for logarithms

$= \left(\log_a \sqrt{x} - \log_a y^7\right) + \log_a z$ Using parentheses to emphasize the order of operations; $x^{1/2} = \sqrt{x}$

$= \log_a \dfrac{\sqrt{x}}{y^7} + \log_a z$ Using the quotient rule for logarithms

$= \log_a \dfrac{z\sqrt{x}}{y^7}$ Using the product rule for logarithms

b) $\log_a \dfrac{b}{\sqrt{x}} + \log_a \sqrt{bx} = \log_a \dfrac{b \cdot \sqrt{bx}}{\sqrt{x}}$ Using the product rule for logarithms

$= \log_a b\sqrt{b}$ Removing a factor equal to 1: $\dfrac{\sqrt{x}}{\sqrt{x}} = 1$

$= \log_a b^{3/2}$, or $\dfrac{3}{2}\log_a b$ Since $b\sqrt{b} = b^1 \cdot b^{1/2}$

If we know the logarithms of two different numbers (to the same base), the properties allow us to calculate other logarithms.

E x a m p l e 8 Given $\log_a 2 = 0.431$ and $\log_a 3 = 0.683$, find each of the following.

a) $\log_a 6$ **b)** $\log_a \frac{2}{3}$ **c)** $\log_a 81$

d) $\log_a \frac{1}{3}$ **e)** $\log_a 2a$ **f)** $\log_a 5$

Solution

a) $\log_a 6 = \log_a (2 \cdot 3) = \log_a 2 + \log_a 3$ Using the product rule for logarithms

$\qquad\qquad = 0.431 + 0.683 = 1.114$

Check: $a^{1.114} = a^{0.431} \cdot a^{0.683} = 2 \cdot 3 = 6$

b) $\log_a \frac{2}{3} = \log_a 2 - \log_a 3$ Using the quotient rule for logarithms

$\qquad\qquad = 0.431 - 0.683 = -0.252$

c) $\log_a 81 = \log_a 3^4 = 4 \log_a 3$ Using the power rule for logarithms

$\qquad\qquad = 4(0.683) = 2.732$

d) $\log_a \frac{1}{3} = \log_a 1 - \log_a 3$ Using the quotient rule for logarithms

$\qquad\qquad = 0 - 0.683 = -0.683$

e) $\log_a 2a = \log_a 2 + \log_a a$ Using the product rule for logarithms

$\qquad\qquad = 0.431 + 1 = 1.431$

f) $\log_a 5$ *cannot be found using these properties.* $(\log_a 5 \neq \log_a 2 + \log_a 3)$

A final property follows from the product rule: Since $\log_a a^k = k \log_a a$, and $\log_a a = 1$, we have $\log_a a^k = k$.

The Logarithm of the Base to a Power

For any base a,

$$\log_a a^k = k.$$

(The logarithm, base a, of a to a power is the power.)

This property also follows from the definition of logarithm: k is the power to which you raise a in order to get a^k.

E x a m p l e   9

Simplify: **(a)** $\log_3 3^7$; **(b)** $\log_{10} 10^{-5.2}$.

Solution

a) $\log_3 3^7 = 7$ 7 is the power to which you raise 3 in order to get 3^7.

b) $\log_{10} 10^{-5.2} = -5.2$

We summarize the properties covered in this section as follows.

For any positive numbers M, N, and a $(a \neq 1)$:

$$\log_a MN = \log_a M + \log_a N; \qquad \log_a M^p = p \cdot \log_a M;$$

$$\log_a \frac{M}{N} = \log_a M - \log_a N; \qquad \log_a a^k = k.$$

> **Caution!** Keep in mind that, in general,
>
> $\log_a (M + N) \neq \log_a M + \log_a N,$ $\log_a MN \neq (\log_a M)(\log_a N),$
>
> $\log_a (M - N) \neq \log_a M - \log_a N,$ $\log_a \dfrac{M}{N} \neq \dfrac{\log_a M}{\log_a N}.$

Exercise Set 12.4

FOR EXTRA HELP

Digital Video Tutor CD 9
Videotape 23

InterAct Math

Math Tutor Center

MathXL

MyMathLab.com

Express as a sum of logarithms.

1. $\log_3 (81 \cdot 27)$ **2.** $\log_2 (16 \cdot 32)$

3. $\log_4 (64 \cdot 16)$ **4.** $\log_5 (25 \cdot 125)$

5. $\log_c rst$ **6.** $\log_t 3ab$

Express as a single logarithm.

7. $\log_a 5 + \log_a 14$ **8.** $\log_b 65 + \log_b 2$

9. $\log_c t + \log_c y$ **10.** $\log_t H + \log_t M$

Express as a product.

11. $\log_a r^8$ **12.** $\log_b t^5$

13. $\log_c y^6$ **14.** $\log_{10} y^7$

15. $\log_b C^{-3}$ **16.** $\log_c M^{-5}$

Express as a difference of logarithms.

17. $\log_2 \dfrac{53}{17}$ **18.** $\log_3 \dfrac{23}{9}$

19. $\log_b \dfrac{m}{n}$ **20.** $\log_a \dfrac{y}{x}$

Express as a single logarithm.

21. $\log_a 15 - \log_a 3$ **22.** $\log_b 42 - \log_b 7$

23. $\log_b 36 - \log_b 4$ **24.** $\log_a 26 - \log_a 2$

25. $\log_a 7 - \log_a 18$ **26.** $\log_b 5 - \log_b 13$

Express in terms of the individual logarithms of w, x, y, and z.

27. $\log_a x^5 y^7 z^6$ **28.** $\log_a xy^4 z^3$

29. $\log_b \dfrac{xy^2}{z^3}$ **30.** $\log_b \dfrac{x^2 y^5}{w^4 z^7}$

31. $\log_a \dfrac{x^4}{y^3 z}$ **32.** $\log_a \dfrac{x^4}{yz^2}$

33. $\log_b \dfrac{xy^2}{wz^3}$ **34.** $\log_b \dfrac{w^2 x}{y^3 z}$

35. $\log_a \sqrt{\dfrac{x^7}{y^5 z^8}}$ **36.** $\log_c \sqrt[3]{\dfrac{x^4}{y^3 z^2}}$

37. $\log_a \sqrt[3]{\dfrac{x^6 y^3}{a^2 z^7}}$ **38.** $\log_a \sqrt[4]{\dfrac{x^8 y^{12}}{a^3 z^5}}$

Express as a single logarithm and, if possible, simplify.

39. $7 \log_a x + 3 \log_a z$

40. $2 \log_b m + \frac{1}{2} \log_b n$

41. $\log_a x^2 - 2 \log_a \sqrt{x}$

42. $\log_a \dfrac{a}{\sqrt{x}} - \log_a \sqrt{ax}$

43. $\frac{1}{2} \log_a x + 5 \log_a y - 2 \log_a x$

44. $\log_a 2x + 3(\log_a x - \log_a y)$

45. $\log_a (x^2 - 4) - \log_a (x + 2)$

46. $\log_a (2x + 10) - \log_a (x^2 - 25)$

Given $\log_b 3 = 0.792$ and $\log_b 5 = 1.161$. If possible, find each of the following.

47. $\log_b 15$ **48.** $\log_b \frac{5}{3}$

49. $\log_b \frac{3}{5}$ **50.** $\log_b \frac{1}{3}$

51. $\log_b \frac{1}{5}$ **52.** $\log_b \sqrt{b}$

53. $\log_b \sqrt{b^3}$ **54.** $\log_b 3b$

55. $\log_b 8$ **56.** $\log_b 45$

57. $\log_b 75$ **58.** $\log_b 20$

Simplify.

Aha! **59.** $\log_t t^9$ **60.** $\log_p p^4$

61. $\log_e e^m$ **62.** $\log_Q Q^{-2}$

63. A student *incorrectly* reasons that

$$\log_b \frac{1}{x} = \log_b \frac{x}{xx}$$

$$= \log_b x - \log_b x + \log_b x = \log_b x.$$

What mistake has the student made?

64. How could you convince someone that

$$\log_a c \neq \log_c a?$$

SKILL MAINTENANCE

Graph.

65. $f(x) = \sqrt{x} - 3$ **66.** $g(x) = \sqrt{x} + 2$

67. $g(x) = \sqrt[3]{x} + 1$ **68.** $f(x) = \sqrt[3]{x} - 1$

Simplify.

69. $(a^3 b^2)^5 (a^2 b^7)$ **70.** $(x^5 y^3 z^2)(x^2 y z^2)^3$

SYNTHESIS

71. Is it possible to express $\log_b \dfrac{x}{5}$ as a difference of two logarithms without using the quotient rule? Why or why not?

72. Is it true that $\log_a x + \log_b x = \log_{ab} x$? Why or why not?

Express as a single logarithm and, if possible, simplify.

73. $\log_a (x^8 - y^8) - \log_a (x^2 + y^2)$

74. $\log_a (x + y) + \log_a (x^2 - xy + y^2)$

Express as a sum or difference of logarithms.

75. $\log_a \sqrt{1 - s^2}$ **76.** $\log_a \dfrac{c - d}{\sqrt{c^2 - d^2}}$

77. If $\log_a x = 2$, $\log_a y = 3$, and $\log_a z = 4$, what is

$$\log_a \frac{\sqrt[3]{x^2 z}}{\sqrt[3]{y^2 z^{-2}}}?$$

78. If $\log_a x = 2$, what is $\log_a (1/x)$?

79. If $\log_a x = 2$, what is $\log_{1/a} x$?

Classify each of the following as true or false. Assume a, x, P, and Q > 0.

80. $\log_a \left(\dfrac{P}{Q}\right)^x = x \log_a P - \log_a Q$

81. $\log_a (Q + Q^2) = \log_a Q + \log_a (Q + 1)$

82. Use graphs to show that

$$\log x^2 \neq \log x \cdot \log x. \quad (\textit{Note}: \log \text{ means } \log_{10}.)$$

12.5

Common and Natural Logarithms

> Common Logarithms on a Calculator • The Base e and Natural Logarithms on a Calculator • Changing Logarithmic Bases • Graphs of Exponential and Logarithmic Functions, Base e

Any positive number other than 1 can serve as the base of a logarithmic function. However, some numbers are easier to use than others, and there are logarithmic bases that fit into certain applications more naturally than others.

Base-10 logarithms, called **common logarithms**, are useful because they have the same base as our "commonly" used decimal system. Before calculators became so widely available, common logarithms were helpful when performing tedious calculations. In fact, that is why logarithms were invented.

Another logarithmic base widely used today is an irrational number named e. We will consider e and base e, or *natural*, logarithms later in this section. First we examine common logarithms.

Common Logarithms on a Calculator

Before the advent of calculators, tables were developed to list common logarithms. Today we find common logarithms using calculators.

Here, and in most books, the abbreviation **log**, with no base written, is understood to mean logarithm base 10, or a common logarithm. Thus,

$$\log 17 \quad \text{means} \quad \log_{10} 17. \qquad \text{\color{magenta}It is important to remember this abbreviation.}$$

On scientific calculators, the key for common logarithms is usually marked $\boxed{\log}$. To find the common logarithm of a number, we enter that number and press the $\boxed{\log}$ key. On most graphing calculators, we press $\boxed{\log}$, the number, and then $\boxed{\text{ENTER}}$.

Example 1

Use a calculator to find each number: **(a)** $\log 53{,}128$; **(b)** $\dfrac{\log 6500}{\log 0.007}$.

Solution

a) We enter 53,128 and then press $\boxed{\log}$. We find that

$$\log 53{,}128 \approx 4.7253. \qquad \text{\color{magenta}Rounded to four decimal places}$$

b) We enter 6500 and then press $\boxed{\log}$. Next, we press $\boxed{\div}$, enter 0.007, press $\boxed{\log}$ $\boxed{=}$. Be careful not to round until the end:

$$\frac{\log 6500}{\log 0.007} \approx -1.7694. \qquad \text{\color{magenta}Rounded to four decimal places}$$

The inverse of a logarithmic function is an exponential function. Because of this, on many calculators the $\boxed{\log}$ key doubles as the $\boxed{10^x}$ key after a $\boxed{\text{2nd}}$ or $\boxed{\text{SHIFT}}$ key is pressed.

Example 2

Use a calculator to find $10^{3.417}$.

Solution We enter 3.417 and then press $\boxed{10^x}$. On some calculators, $\boxed{10^x}$ is pressed first, followed by 3.417 and $\boxed{\text{ENTER}}$. Since $10^{3.417}$ is irrational, our answer is approximate:

$$10^{3.417} \approx 2612.161354.$$

On calculators without a $\boxed{10^x}$ key, an exponential key, labeled $\boxed{x^y}$, $\boxed{a^x}$, or $\boxed{\wedge}$ may be available. Such a key can raise any positive real number to any real-numbered power.

The Base *e* and Natural Logarithms on a Calculator

When interest is compounded n times a year, the compound interest formula is

$$A = P\left(1 + \frac{r}{n}\right)^{nt},$$

where A is the amount that an initial investment P will be worth after t years at interest rate r. Suppose that $1 is invested at 100% interest for 1 year (no bank would pay this). The preceding formula becomes a function A defined in terms of the number of compounding periods n:

$$A(n) = \left(1 + \frac{1}{n}\right)^n.$$

Let's find some function values. We round to six decimal places, using a calculator.

n	$A(n) = \left(1 + \dfrac{1}{n}\right)^n$
1 (compounded annually)	$2.00
2 (compounded semiannually)	$2.25
3	$2.370370
4 (compounded quarterly)	$2.441406
5	$2.488320
100	$2.704814
365 (compounded daily)	$2.714567
8760 (compounded hourly)	$2.718127

The numbers in this table approach a very important number in mathematics, called e. Because e is irrational, its decimal representation does not terminate or repeat.

The Number *e*

$e \approx 2.7182818284\ldots$

Logarithms base e are called **natural logarithms**, or **Napierian logarithms**, in honor of John Napier (1550–1617), who first "discovered" logarithms.

The abbreviation "ln" is generally used with natural logarithms. Thus,

$\ln 53$ means $\log_e 53$.   It is important to remember this abbreviation.

On most scientific calculators, to find the natural logarithm of a number, we enter that number and press $\boxed{\text{ln}}$. On most graphing calculators, we press $\boxed{\text{ln}}$, the number, and then $\boxed{\text{ENTER}}$.

E x a m p l e   3

Use a calculator to find ln 4568.

Solution   We enter 4568 and then press ⟨ ln ⟩. We find that

$$\ln 4568 \approx 8.4268. \qquad \text{Rounded to four decimal places}$$

On many calculators, the ⟨ ln ⟩ key doubles as the ⟨ e^x ⟩ key after a ⟨ 2nd ⟩ or ⟨ SHIFT ⟩ key has been pressed.

E x a m p l e   4

Use a calculator to find $e^{-1.524}$.

Solution   We enter -1.524 and then press ⟨ e^x ⟩. On some calculators, ⟨ e^x ⟩ is pressed first, followed by -1.524 and ⟨ ENTER ⟩. Since $e^{-1.524}$ is irrational, our answer is approximate:

$$e^{-1.524} \approx 0.2178387868.$$

Changing Logarithmic Bases

Most calculators can find both common logarithms and natural logarithms. To find a logarithm with some other base, a conversion formula is needed.

> ### The Change-of-Base Formula
>
> For any logarithmic bases a and b, and any positive number M,
>
> $$\log_b M = \frac{\log_a M}{\log_a b}.$$
>
> (To find the log, base b, of some number M, find the log of M using another base—usually 10 or e—and divide by the log of b to that same base.)

Proof.   Let $x = \log_b M$. Then,

$$b^x = M \qquad \text{Rewriting } x = \log_b M \text{ in exponential form}$$

$$\log_a b^x = \log_a M \qquad \text{Taking the logarithm, base } a, \text{ on both sides}$$

$$x \log_a b = \log_a M \qquad \text{Using the power rule for logarithms}$$

$$x = \frac{\log_a M}{\log_a b}. \qquad \text{Dividing both sides by } \log_a b$$

But at the outset we stated that $x = \log_b M$. Thus, by substitution, we have

$$\log_b M = \frac{\log_a M}{\log_a b},$$

which is the change-of-base formula.

Example 5

Find $\log_5 8$ using the change-of-base formula.

Solution Because calculators can find base 10 logarithms, we use the change-of-base formula with $a = 10$, $b = 5$, and $M = 8$:

$$\log_5 8 = \frac{\log_{10} 8}{\log_{10} 5} \qquad \text{Substituting into } \log_b M = \frac{\log_a M}{\log_a b}$$

$$\approx \frac{0.903089987}{0.6989700043} \qquad \text{Using } \boxed{\text{log}} \text{ twice}$$

$$\approx 1.2920. \qquad \text{When using a calculator, it is best not to round before dividing.}$$

To check, note that $\ln 8 / \ln 5 \approx 1.2920$. We can also use a calculator to verify that $5^{1.2920} \approx 8$.

Example 6

Find $\log_4 31$.

Solution As shown in the check of Example 5, base e can also be used in the change-of-base formula.

$$\log_4 31 = \frac{\log_e 31}{\log_e 4} \qquad \text{Substituting into } \log_b M = \frac{\log_a M}{\log_a b}$$

$$= \frac{\ln 31}{\ln 4} \approx \frac{3.433987204}{1.386294361} \qquad \text{Using } \boxed{\text{ln}} \text{ twice}$$

$$\approx 2.4771. \qquad \textit{Check: } 4^{2.4771} \approx 31$$

Graphs of Exponential and Logarithmic Functions, Base e

Example 7

Graph $f(x) = e^x$ and $g(x) = e^{-x}$ and state the domain and the range of f and g.

Solution We use a calculator with an $\boxed{e^x}$ key to find approximate values of e^x and e^{-x}. Using these values, we can graph the functions.

x	e^x	e^{-x}
0	1	1
1	2.7	0.4
2	7.4	0.1
−1	0.4	2.7
−2	0.1	7.4

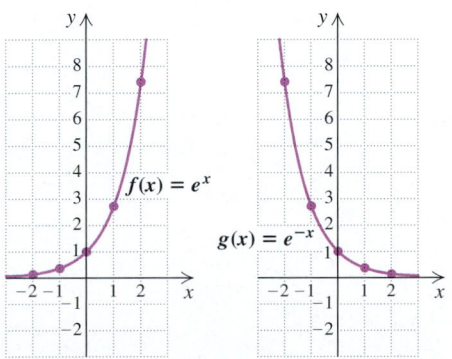

The domain of each function is $\mathbb{R}$ and the range of each function is $(0, \infty)$.

E x a m p l e   8

Graph $f(x) = e^{-0.5x} + 1$ and state the domain and the range of f.

Solution   We find some solutions
with a calculator, plot them, and
then draw the graph. For example,
$f(2) = e^{-0.5(2)} + 1 = e^{-1} + 1 \approx 1.4$.

x	$e^{-0.5x} + 1$
0	2
1	1.6
2	1.4
3	1.2
-1	2.6
-2	3.7
-3	5.5

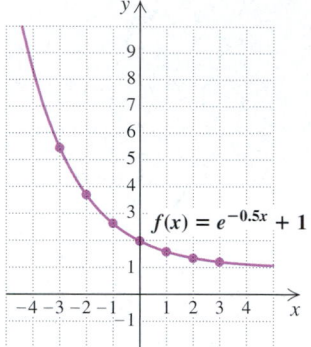

The domain of f is $\mathbb{R}$ and the range is $(1, \infty)$.

E x a m p l e   9

Graph and state the domain and the range of each function.

a) $g(x) = \ln x$ **b)** $f(x) = \ln (x + 3)$

Solution

a) We find some solutions with a calculator and then draw the graph. As expected, the graph is a reflection across the line $y = x$ of the graph of $y = e^x$.

x	$\ln x$
1	0
4	1.4
7	1.9
0.5	-0.7

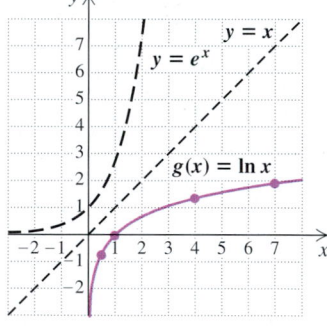

The domain of g is $(0, \infty)$ and the range is $\mathbb{R}$.

b) We find some solutions with a calculator, plot them, and draw the graph.

x	$\ln (x + 3)$
0	1.1
1	1.4
2	1.6
3	1.8
4	1.9
-1	0.7
-2	0
-2.5	-0.7

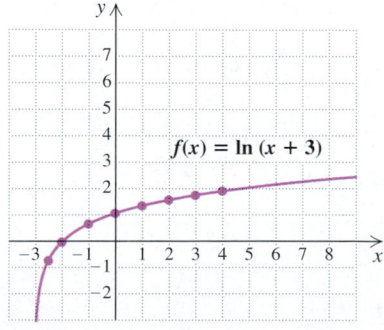

**technology
connection**

Logarithmic functions with
bases other than 10 or e can be
easily drawn on a grapher, provided the change-of-base formula is used.

1. Graph $y = \log_5 x$.
2. Graph $y = \log_7 x$.
3. Graph $y = \log_5 (x + 2)$.
4. Graph $y = \log_7 x + 2$.

The graph of $y = \ln(x + 3)$ is the graph of $y = \ln x$ translated 3 units to the left. The domain is $(-3, \infty)$ and the range is $\mathbb{R}$.

Exercise Set 12.5

Use a calculator to find each of the following to the nearest ten-thousandth.

1. $\log 6$ **2.** $\log 5$ **3.** $\log 72.8$

4. $\log 73.9$ *Aha!* **5.** $\log 1000$ **6.** $\log 100$

7. $\log 0.527$ **8.** $\log 0.493$ **9.** $\dfrac{\log 8200}{\log 150}$

10. $\dfrac{\log 5700}{\log 90}$ **11.** $10^{2.3}$ **12.** $10^{3.4}$

13. $10^{0.173}$ **14.** $10^{0.247}$ **15.** $10^{-2.9523}$

16. $10^{4.8982}$ **17.** $\ln 5$ **18.** $\ln 2$

19. $\ln 57$ **20.** $\ln 30$ **21.** $\ln 0.0062$

22. $\ln 0.00073$ **23.** $\dfrac{\ln 2300}{0.08}$ **24.** $\dfrac{\ln 1900}{0.07}$

25. $e^{2.71}$ **26.** $e^{3.06}$ **27.** $e^{-3.49}$

28. $e^{-2.64}$ **29.** $e^{4.7}$ **30.** $e^{1.23}$

Find each of the following logarithms using the change-of-base formula. Round answers to the nearest ten-thousandth.

31. $\log_6 92$ **32.** $\log_3 78$ **33.** $\log_2 100$

34. $\log_7 100$ **35.** $\log_7 65$ **36.** $\log_5 42$

37. $\log_{0.5} 5$ **38.** $\log_{0.1} 3$ **39.** $\log_2 0.2$

40. $\log_2 0.08$ **41.** $\log_\pi 58$ **42.** $\log_\pi 200$

Graph and state the domain and the range of each function.

43. $f(x) = e^x$ **44.** $f(x) = e^{0.5x}$

45. $f(x) = e^{-0.4x}$ **46.** $f(x) = e^{-x}$

47. $f(x) = e^x + 1$ **48.** $f(x) = e^x + 2$

49. $f(x) = e^x - 2$ **50.** $f(x) = e^x - 3$

51. $f(x) = 0.5e^x$ **52.** $f(x) = 2e^x$

53. $f(x) = 2e^{-0.5x}$ **54.** $f(x) = 0.5e^{2x}$

55. $f(x) = e^{x-2}$ **56.** $f(x) = e^{x-3}$

57. $f(x) = e^{x+3}$ **58.** $f(x) = e^{x+2}$

59. $g(x) = 2 \ln x$ **60.** $g(x) = 3 \ln x$

61. $g(x) = 0.5 \ln x$ **62.** $g(x) = 0.4 \ln x$

63. $g(x) = \ln x + 3$ **64.** $g(x) = \ln x + 2$

65. $g(x) = \ln x - 2$ **66.** $g(x) = \ln x - 3$

67. $g(x) = \ln(x + 1)$ **68.** $g(x) = \ln(x + 2)$

69. $g(x) = \ln(x - 3)$ **70.** $g(x) = \ln(x - 1)$

71. Using a calculator, Zeno *incorrectly* says that $\log 79$ is between 4 and 5. How could you convince him, without using a calculator, that he is mistaken?

72. Examine Exercise 71. What mistake do you believe Zeno made?

SKILL MAINTENANCE

Solve.

73. $4x^2 - 25 = 0$ **74.** $5x^2 - 7x = 0$

75. $17x - 15 = 0$ **76.** $9 - 13x = 0$

77. $x^{1/2} - 6x^{1/4} + 8 = 0$ **78.** $2y - 7\sqrt{y} + 3 = 0$

SYNTHESIS

79. Explain how the graph of $f(x) = e^x$ could be used to graph the function given by $g(x) = 1 + \ln x$.

80. Explain how the graph of $f(x) = \ln x$ could be used to graph the function given by $g(x) = e^{x-1}$.

Given that $\log 2 \approx 0.301$ *and* $\log 3 \approx 0.477$, *find each of the following.*

81. $\log_6 81$ **82.** $\log_9 16$ **83.** $\log_{12} 36$

84. Find a formula for converting common logarithms to natural logarithms.

85. Find a formula for converting natural logarithms to common logarithms.

▦ *Solve for x.*

86. $\log (275x^2) = 38$

87. $\log (492x) = 5.728$

88. $\dfrac{3.01}{\ln x} = \dfrac{28}{4.31}$

89. $\log 692 + \log x = \log 3450$

🗠 *For each function given below, (a) determine the domain and the range, (b) set an appropriate window, and (c) draw the graph. Graphs may vary, depending on the scale used.*

90. $f(x) = 7.4e^x \ln x$

91. $f(x) = 3.4 \ln x - 0.25e^x$

92. $f(x) = x \ln (x - 2.1)$

93. $f(x) = 2x^3 \ln x$

🗠 **94.** Use a grapher to check your answers to Exercises 45, 53, and 59.

🗠 **95.** Use a grapher to check your answers to Exercises 44, 52, and 60.

🗒 **96.** In an attempt to solve $\ln x = 1.5$, Emma gets the
🗠 following graph.

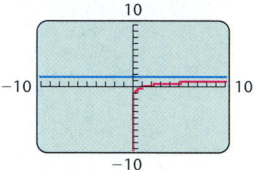

How can Emma tell at a glance that she has made a mistake?

<table>
<tr><td>

Solving Exponential and Logarithmic Equations

</td><td>

12.6

Solving Exponential Equations • Solving Logarithmic Equations

</td></tr>
</table>

Solving Exponential Equations

Equations with variables in exponents, such as $5^x = 12$ and $2^{7x} = 64$, are called **exponential equations**. In Section 12.3, we solved certain exponential equations by using the principle of exponential equality. We restate that principle below.

> **The Principle of Exponential Equality**
>
> For any real number b, where $b \neq -1, 0,$ or 1,
>
> $$b^x = b^y \quad \text{is equivalent to} \quad x = y.$$
>
> (Powers of the same base are equal if and only if the exponents are equal.)

E x a m p l e   1   Solve: $4^{3x-5} = 16$.

Solution   Note that $16 = 4^2$. Thus we can write each side as a power of the same number:

$$4^{3x-5} = 4^2.$$

Since the base is the same, 4, the exponents must be the same. Thus,

$$3x - 5 = 2 \qquad \text{Equating exponents}$$
$$3x = 7$$
$$x = \tfrac{7}{3}.$$

Check:
$$\begin{array}{c|c} 4^{3x-5} = 16 \\ \hline 4^{3 \cdot 7/3 - 5} \;?\; 16 \\ 4^{7-5} \\ 4^2 & 16 \quad \text{TRUE} \end{array}$$

The solution is $\tfrac{7}{3}$.

When it does not seem possible to write both sides of an equation as powers of the same base, we can use the following principle along with the properties developed in Section 12.4.

> ### The Principle of Logarithmic Equality
>
> For any logarithmic base a, and for $x, y > 0$,
>
> $$x = y \quad \text{is equivalent to} \quad \log_a x = \log_a y.$$
>
> (Two expressions are equal if and only if the logarithms of those expressions are equal.)

Because calculators can generally find only common or natural logarithms (without resorting to the change-of-base formula), we usually take the common or natural logarithm on both sides of the equation.

The principle of logarithmic equality is useful anytime a variable appears as an exponent.

Example 2

Solve: $5^x = 12$.

Solution   We have

$$5^x = 12$$
$$\log 5^x = \log 12 \qquad \text{Using the principle of logarithmic equality to take the common logarithm on both sides. Natural logarithms also would work.}$$
$$x \log 5 = \log 12 \qquad \text{Using the power rule for logarithms}$$
$$x = \frac{\log 12}{\log 5} \quad \longleftarrow \boxed{\textbf{\textit{Caution!}}\quad \text{This is } not \log 12 - \log 5.}$$
$$\approx 1.544. \qquad \text{Using a calculator and rounding to three decimal places}$$

Since $5^{1.544} \approx 12$, we have a check. The solution is $\log 12/\log 5$, or approximately 1.544.

Example 3

Solve: $e^{0.06t} = 1500$.

Solution   Since one side is a power of e, we take the *natural logarithm* on both sides:

$$\ln e^{0.06t} = \ln 1500 \qquad \text{Taking the natural logarithm on both sides}$$

$$0.06t = \ln 1500 \qquad \text{Finding the logarithm of the base to a power: } \log_a a^k = k$$

$$t = \frac{\ln 1500}{0.06} \qquad \text{Dividing both sides by 0.06}$$

$$\approx 121.887. \qquad \text{Using a calculator and rounding to three decimal places}$$

Solving Logarithmic Equations

Equations containing logarithmic expressions are called **logarithmic equations**. We saw in Section 12.3 that certain logarithmic equations can be solved by writing an equivalent exponential equation.

Example 4

Solve: $\log_4 (8x - 6) = 3$.

Solution   We write an equivalent exponential equation:

$$4^3 = 8x - 6 \qquad \text{Remember: } \log_a X = y \text{ is equivalent to } a^y = X.$$

$$64 = 8x - 6$$

$$70 = 8x \qquad \text{Adding 6 to both sides}$$

$$x = \tfrac{70}{8}, \text{ or } \tfrac{35}{4}.$$

The check is left to the student. The solution is $\frac{35}{4}$.

Often the properties for logarithms are needed. The goal is to first write an equivalent equation in which the variable appears in just one logarithmic expression. We then isolate that term and solve as in Example 4.

Example 5

Solve.

a) $\log x + \log (x - 3) = 1$
b) $\log_2 (x + 7) - \log_2 (x - 7) = 3$
c) $\log_7 (x + 1) + \log_7 (x - 1) = \log_7 8$

Solution

a) As an aid in solving, we write in the base, 10.

$$\log_{10} x + \log_{10} (x - 3) = 1$$

$$\log_{10} [x(x - 3)] = 1 \qquad \text{Using the product rule for logarithms to obtain a single logarithm}$$

$$x(x - 3) = 10^1 \qquad \text{Writing an equivalent exponential equation}$$

$$x^2 - 3x = 10$$

$$x^2 - 3x - 10 = 0$$

$$(x + 2)(x - 5) = 0 \qquad \text{Factoring}$$

$$x + 2 = 0 \quad \text{or} \quad x - 5 = 0 \qquad \text{Using the principle of zero products}$$

$$x = -2 \quad \text{or} \qquad x = 5$$

Check:

For -2:

$$\log x + \log (x - 3) = 1$$

$$\overline{\log (-2) + \log (-2 - 3)} \; ? \; 1 \quad \text{FALSE}$$

For 5:

$$\log x + \log (x - 3) = 1$$

$$\overline{\log 5 + \log (5 - 3)} \; ? \; 1$$

$$\log 5 + \log 2 \; \big|$$

$$\log 10 \; \big|$$

$$1 \; \big| \; 1 \quad \text{TRUE}$$

The number -2 *does not check* because the logarithm of a negative number is undefined. The solution is 5.

b) We have

$$\log_2 (x + 7) - \log_2 (x - 7) = 3$$

$$\log_2 \frac{x + 7}{x - 7} = 3 \qquad \text{Using the quotient rule for logarithms to obtain a single logarithm}$$

$$\frac{x + 7}{x - 7} = 2^3 \qquad \text{Writing an equivalent exponential equation}$$

$$\frac{x + 7}{x - 7} = 8$$

$$x + 7 = 8(x - 7) \qquad \text{Multiplying by the LCD, } x - 7$$

$$x + 7 = 8x - 56 \qquad \text{Using the distributive law}$$

$$63 = 7x$$

$$9 = x. \qquad \text{Dividing by 7}$$

Check:

$$\log_2 (x + 7) - \log_2 (x - 7) = 3$$

$$\overline{\log_2 (9 + 7) - \log_2 (9 - 7)} \; ? \; 3$$

$$\log_2 16 - \log_2 2 \; \big|$$

$$4 - 1 \; \big|$$

$$3 \; \big| \; 3 \quad \text{TRUE}$$

The solution is 9.

c) We have

$$\log_7 (x + 1) + \log_7 (x - 1) = \log_7 8$$

$$\log_7 [(x + 1)(x - 1)] = \log_7 8 \quad \text{Using the product rule for logarithms}$$

$$\log_7 (x^2 - 1) = \log_7 8 \quad \text{Multiplying}$$

$$x^2 - 1 = 8 \quad \text{Using the principle of logarithmic equality. Study this step carefully.}$$

$$x^2 - 9 = 0$$

$$(x - 3)(x + 3) = 0 \quad \text{Solving the quadratic equation}$$

$$x = 3 \quad or \quad x = -3.$$

We leave it to the student to show that 3 checks but -3 does not. The solution is 3.

technology connection

Exponential and logarithmic equations can be solved by graphing each side of the equation and using INTERSECT to determine the x-coordinate at each intersection.

For example, to solve $e^{0.5x} - 7 = 2x + 6$, we graph $y_1 = e^{0.5x} - 7$ and $y_2 = 2x + 6$ as shown at right. We then use the INTERSECT option of the CALC menu. The x-coordinates at the intersections are approximately -6.48 and 6.52.

Use a grapher to find solutions, accurate to the nearest hundredth, for each of the following equations.

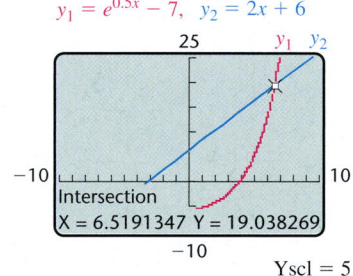
$y_1 = e^{0.5x} - 7, \quad y_2 = 2x + 6$

Intersection
X = 6.5191347 Y = 19.038269
Yscl = 5

1. $e^{7x} = 14$ **2.** $8e^{0.5x} = 3$

3. $xe^{3x-1} = 5$ **4.** $4 \ln (x + 3.4) = 2.5$

5. $\ln 3x = 0.5x - 1$ **6.** $\ln x^2 = -x^2$

Exercise Set **12.6**

Solve. Where appropriate, include approximations to the nearest thousandth.

1. $2^x = 16$

2. $2^x = 8$

3. $3^x = 27$

4. $5^x = 125$

5. $2^{x+3} = 32$

6. $4^{x-2} = 64$

7. $5^{3x} = 625$

8. $3^{2x} = 27$

Aha! **9.** $7^{4x} = 1$

10. $8^{5x} = 1$

11. $4^{2x-1} = 64$

12. $5^{2x-3} = 25$

13. $3^{x^2} \cdot 3^{3x} = 81$

14. $3^{4x} \cdot 3^{x^2} = \frac{1}{27}$

15. $2^x = 15$

16. $2^x = 19$

17. $4^{x+1} = 13$

18. $8^{x-1} = 17$

19. $e^t = 100$

20. $e^t = 1000$

21. $e^{-0.07t} + 3 = 3.08$

22. $e^{0.03t} + 2 = 7$

23. $2^x = 3^{x-1}$

24. $5^x = 3^{x+1}$

25. $4^{x+1} = 5^x$

26. $2^{x+3} = 7^x$

27. $7.2^x - 65 = 0$

28. $4.9^x - 87 = 0$

29. $\log_5 x = 3$

30. $\log_3 x = 4$

31. $\log_4 x = \frac{1}{2}$

32. $\log_2 x = -3$

33. $\log x = 3$

34. $\log x = 1$

35. $2 \log x = -8$

36. $4 \log x = -16$

Aha! **37.** $\ln x = 1$

38. $\ln x = 2$

39. $5 \ln x = -15$

40. $3 \ln x = -3$

41. $\log_2 (8 - 6x) = 5$

42. $\log_5 (2x - 7) = 3$

43. $\log (x - 9) + \log x = 1$

44. $\log (x + 9) + \log x = 1$

45. $\log x - \log (x + 3) = 1$

46. $\log x - \log (x + 7) = -1$

47. $\log_4 (x + 3) - \log_4 (x - 5) = 2$

48. $\log_2 (x + 3) + \log_2 (x - 3) = 4$

49. $\log_7 (x + 1) + \log_7 (x + 2) = \log_7 6$

50. $\log_6 (x + 3) + \log_6 (x + 2) = \log_6 20$

51. $\log_3 (x + 4) + \log_3 (x - 4) = 2$

52. $\log_{14} (x + 3) + \log_{14} (x - 2) = 1$

53. $\log_{12} (x + 5) - \log_{12} (x - 4) = \log_{12} 3$

54. $\log_6 (x + 7) - \log_6 (x - 2) = \log_6 5$

55. $\log_2 (x - 2) + \log_2 x = 3$

56. $\log_4 (x + 6) - \log_4 x = 2$

57. Could Example 2 have been solved by taking the natural logarithm on both sides? Why or why not?

58. Christina finds that the solution of $\log_3 (x + 4) = 1$ is -1, but rejects -1 as an answer. What mistake is she making?

SKILL MAINTENANCE

59. Find an equation of variation if y varies directly as x, and $y = 7.2$ when $x = 0.8$.

60. Find an equation of variation if y varies inversely as x, and $y = 3.5$ when $x = 6.1$.

Solve.

61. $T = 2\pi\sqrt{L/32}$, for L

62. $E = mc^2$, for c
(Assume $E, m, c > 0$.)

63. Joni can key in a musical score in 2 hr. Miles takes 3 hr to key in the same score. How long would it take them, working together, to key in the score?

64. The side exit at the Flynn Theater can empty a capacity crowd in 25 min. The main exit can empty a capacity crowd in 15 min. How long will it take to empty a capacity crowd when both exits are in use?

SYNTHESIS

65. Can the principle of logarithmic equality be expanded to include all functions? That is, is the statement "$m = n$ is equivalent to $f(m) = f(n)$" true for any function f? Why or why not?

66. Explain how Exercises 33–36 could be solved using the graph of $f(x) = \log x$.

Solve.

67. $100^{3x} = 1000^{2x+1}$

68. $27^x = 81^{2x-3}$

69. $8^x = 16^{3x+9}$

70. $\log_x (\log_3 27) = 3$

71. $\log_6 (\log_2 x) = 0$

72. $x \log \frac{1}{8} = \log 8$

73. $\log_5 \sqrt{x^2 - 9} = 1$

74. $2^{x^2+4x} = \frac{1}{8}$

75. $\log (\log x) = 5$

76. $\log_5 |x| = 4$

77. $\log x^2 = (\log x)^2$

78. $\log \sqrt{2x} = \sqrt{\log 2x}$

79. $\log x^{\log x} = 25$

80. $3^{2x} - 8 \cdot 3^x + 15 = 0$

81. $(81^{x-2})(27^{x+1}) = 9^{2x-3}$

82. $3^{2x} - 3^{2x-1} = 18$

83. Given that $2^y = 16^{x-3}$ and $3^{y+2} = 27^x$, find the value of $x + y$.

84. If $x = (\log_{125} 5)^{\log_5 125}$, what is the value of $\log_3 x$?

85. Find the value of x for which the natural logarithm is the same as the common logarithm.

86. Use a grapher to check your answers to Exercises 3, 21, 25, 37, and 77.

Applications of Exponential and Logarithmic Functions

12.7

Applications of Logarithmic Functions • Applications of Exponential Functions

We now consider applications of exponential and logarithmic functions.

Applications of Logarithmic Functions

Example 1

Sound levels. To measure the volume, or "loudness," of a sound, the *decibel* scale is used. The loudness L, in decibels (dB), of a sound is given by

$$L = 10 \cdot \log \frac{I}{I_0},$$

where I is the intensity of the sound, in watts per square meter (W/m^2), and $I_0 = 10^{-12} \, W/m^2$. (I_0 is approximately the intensity of the softest sound that can be heard by the human ear.)

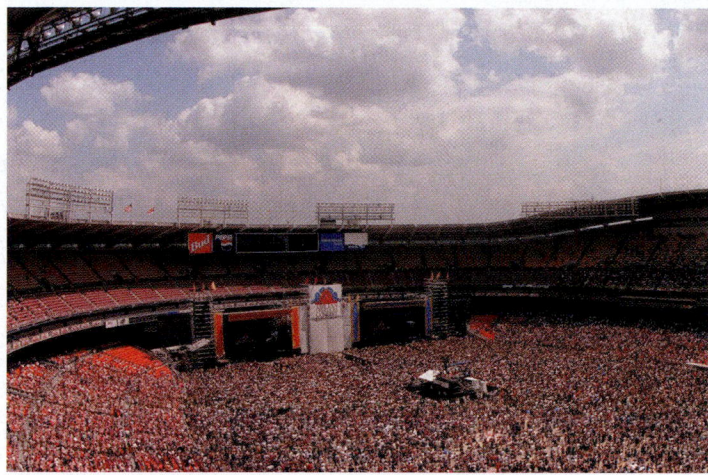

a) It is common for the intensity of sound at live performances of rock music to reach 10^{-1} W/m^2 (even higher close to the stage). How loud, in decibels, is the sound level?

b) The Occupational Safety and Health Administration (OSHA) considers sound levels of 85 dB and above unsafe. What is the intensity of such sounds?

Solution

a) To find the loudness, in decibels, we use the above formula:

$$L = 10 \cdot \log \frac{I}{I_0}$$

$$= 10 \cdot \log \frac{10^{-1}}{10^{-12}} \qquad \text{Substituting}$$

$$= 10 \cdot \log 10^{11} \qquad \text{Subtracting exponents}$$

$$= 10 \cdot 11 \qquad \log 10^a = a$$

$$= 110.$$

The volume of the music is 110 decibels.

b) We substitute and solve for I:

$$L = 10 \cdot \log \frac{I}{I_0}$$

$$85 = 10 \cdot \log \frac{I}{10^{-12}} \qquad \text{Substituting}$$

$$8.5 = \log \frac{I}{10^{-12}} \qquad \text{Dividing both sides by 10}$$

$$8.5 = \log I - \log 10^{-12} \qquad \text{Using the quotient rule for logarithms}$$

$$8.5 = \log I - (-12) \qquad \log 10^a = a$$

$$-3.5 = \log I \qquad \text{Adding } -12 \text{ to both sides}$$

$$10^{-3.5} = I. \qquad \text{Converting to an exponential equation}$$

Earplugs would be recommended for sounds with intensities exceeding $10^{-3.5}$ W/m^2.

E x a m p l e   2   ***Chemistry: pH of liquids.*** In chemistry, the pH of a liquid is a measure of its acidity. We calculate pH as follows:

$$pH = -\log[H^+],$$

where $[H^+]$ is the hydrogen ion concentration in moles per liter.

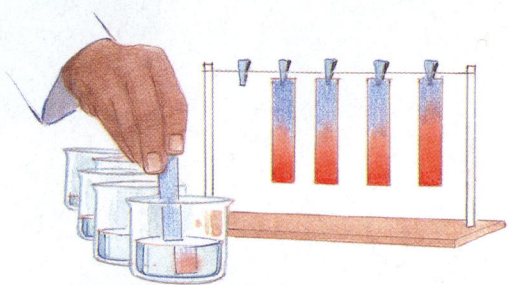

a) The hydrogen ion concentration of human blood is normally about 3.98×10^{-8} moles per liter. Find the pH.
b) The pH of seawater is about 8.3. Find the hydrogen ion concentration.

Solution

a) To find the pH of blood, we use the above formula:

$$
\begin{aligned}
pH &= -\log[H^+] \\
   &= -\log[3.98 \times 10^{-8}] \\
   &\approx -(-7.400117) \qquad \text{Using a calculator} \\
   &\approx 7.4.
\end{aligned}
$$

The pH of human blood is normally about 7.4.

b) We substitute and solve for $[H^+]$:

$$
\begin{aligned}
8.3 &= -\log[H^+] & &\text{Using } pH = -\log[H^+] \\
-8.3 &= \log[H^+] & &\text{Dividing both sides by } -1 \\
10^{-8.3} &= [H^+] & &\text{Converting to an exponential equation} \\
5.01 \times 10^{-9} &\approx [H^+]. & &\text{Using a calculator; writing scientific notation}
\end{aligned}
$$

The hydrogen ion concentration of seawater is about 5.01×10^{-9} moles per liter.

Applications of Exponential Functions

E x a m p l e   3

Interest compounded annually. Suppose that $30,000 is invested at 8% interest, compounded annually. In *t* years, it will grow to the amount *A* given by the function

$$A(t) = 30{,}000(1.08)^t.$$

(See Example 5 in Section 12.2.)

a) How long will it take to accumulate $150,000 in the account?
b) Find the amount of time it takes for the $30,000 to double itself.

Solution

a) We set $A(t) = 150{,}000$ and solve for *t*:

$$150{,}000 = 30{,}000(1.08)^t$$

$$\frac{150{,}000}{30{,}000} = 1.08^t \qquad \text{Dividing both sides by 30,000}$$

$$5 = 1.08^t$$

$$\log 5 = \log 1.08^t \qquad \text{Taking the common logarithm on both sides}$$

$$\log 5 = t \log 1.08 \qquad \text{Using the power rule for logarithms}$$

$$\frac{\log 5}{\log 1.08} = t \qquad \text{Dividing both sides by log 1.08}$$

$$20.9 \approx t. \qquad \text{Using a calculator}$$

Remember that when doing a calculation like this on a calculator, it is best to wait until the end to round off. At an interest rate of 8% per year, it will take about 20.9 yr for $30,000 to grow to $150,000.

b) To find the *doubling time*, we replace $A(t)$ with 60,000 and solve for *t*:

$$60{,}000 = 30{,}000(1.08)^t$$

$$2 = (1.08)^t \qquad \text{Dividing both sides by 30,000}$$

$$\log 2 = \log (1.08)^t \qquad \text{Taking the common logarithm on both sides}$$

$$\log 2 = t \log 1.08 \qquad \text{Using the power rule for logarithms}$$

$$t = \frac{\log 2}{\log 1.08} \approx 9.0. \qquad \text{Dividing both sides by log 1.08 and using a calculator}$$

At an interest rate of 8% per year, the doubling time is about 9 yr.

Like investments, populations often grow exponentially.

> ### *Exponential Growth*
>
> An **exponential growth model** is a function of the form
>
> $$P(t) = P_0 e^{kt}, \quad k > 0,$$
>
> where P_0 is the population at time 0, $P(t)$ is the population at time t, and k is the **exponential growth rate** for the situation. The **doubling time** is the amount of time necessary for the population to double in size.

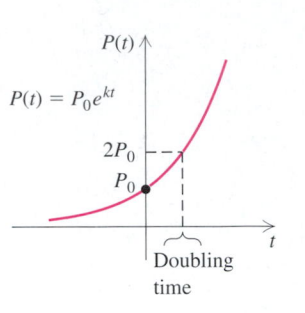

The exponential growth rate is the rate of growth of a population at any *instant* in time. Since the population is continually growing, the percent of total growth after one year will exceed the exponential growth rate.

Example 4

Growth of zebra mussel populations. Zebra mussels, inadvertently imported from Europe, began fouling North American waters in 1988. These mussels are so prolific that lake and river bottoms, as well as water intake pipes, can become blanketed with them, altering an entire ecosystem. In 2000, a portion of the Hudson River contained an average of 10 zebra mussels per square mile. The exponential growth rate was 340% per year.

a) Find the exponential growth model.
b) Predict the number of mussels per square mile in 2003.

Solution

a) In 2000, at $t = 0$, the population was $10/\text{mi}^2$. We substitute 10 for P_0 and 340%, or 3.4, for k. This gives the exponential growth function

$$P(t) = 10e^{3.4t}.$$

b) In 2003, we have $t = 3$ (since 3 yr have passed since 2000). To find the population in 2003, we compute $P(3)$:

$$P(3) = 10e^{3.4(3)} \qquad \text{Using } P(t) = 10e^{3.4t} \text{ from part (a)}$$
$$= 10e^{10.2}$$
$$\approx 269,000. \qquad \text{Using a calculator}$$

The population of zebra mussels in the specified portion of the Hudson River will reach approximately 269,000 per square mile in 2003.

E x a m p l e   5

Spread of a computer virus. The number of computers infected by a virus t days after it first appears usually increases exponentially. In 2000, the "Love Bug" virus spread from 100 computers to about 1,000,000 computers in 2 hr (120 min).

a) Find the exponential growth rate and the exponential growth function.
b) Assuming exponential growth, estimate how long it took the Love Bug virus to infect 80,000 computers.

Solution

a) We use $N(t) = N_0 e^{kt}$, where t is the number of minutes since the first 100 computers were infected. Substituting 100 for N_0 gives

$$N(t) = 100e^{kt}.$$

To find the exponential growth rate, k, note that after 120 min, 1,000,000 computers were infected:

$$\left.\begin{array}{l} N(120) = 100e^{k \cdot 120} \\ 1,000,000 = 100e^{120k} \end{array}\right\} \qquad \text{Substituting}$$

$$10,000 = e^{120k} \qquad \text{Dividing both sides by 100}$$
$$\ln 10,000 = \ln e^{120k} \qquad \text{Taking the natural logarithm on both sides}$$
$$\ln 10,000 = 120k \qquad \ln e^a = a$$
$$\frac{\ln 10,000}{120} = k \qquad \text{Dividing both sides by 120}$$
$$0.077 \approx k. \qquad \text{Using a calculator and rounding}$$

The exponential growth function is given by $N(t) = 100e^{0.077t}$.

b) To estimate how long it took for 80,000 computers to be infected, we replace $N(t)$ with 80,000 and solve for t:

$$80{,}000 = 100e^{0.077t}$$

$$800 = e^{0.077t} \qquad \text{Dividing both sides by 100}$$

$$\ln 800 = \ln e^{0.077t} \qquad \text{Taking the natural logarithm on both sides}$$

$$\ln 800 = 0.077t \qquad \ln e^a = a$$

$$\frac{\ln 800}{0.077} = t \qquad \text{Dividing both sides by 0.077}$$

$$86.8 \approx t. \qquad \text{Using a calculator}$$

Rounding up to 87, we see that, according to this model, it took about 87 min, or 1 hr 27 min, for 80,000 computers to be infected.

E x a m p l e   6

Interest compounded continuously. Suppose that an amount of money P_0 is invested in a savings account at interest rate k, compounded continuously. That is, suppose that interest is computed every "instant" and added to the amount in the account. The balance $P(t)$, after t years, is given by the exponential growth model

$$P(t) = P_0 e^{kt}.$$

a) Suppose that \$30,000 is invested and grows to \$44,754.75 in 5 yr. Find the exponential growth function.

b) What is the doubling time?

Solution

a) We have $P(0) = 30{,}000$. Thus the exponential growth function is

$$P(t) = 30{,}000 e^{kt}, \quad \text{where } k \text{ must still be determined.}$$

Knowing that for $t = 5$ we have $P(5) = 44{,}754.75$, it is possible to solve for k:

$$44{,}754.75 = 30{,}000 e^{k(5)} = 30{,}000 e^{5k}$$

$$\frac{44{,}754.75}{30{,}000} = e^{5k} \qquad \text{Dividing both sides by 30,000}$$

$$1.491825 = e^{5k}$$

$$\ln 1.491825 = \ln e^{5k} \qquad \text{Taking the natural logarithm on both sides}$$

$$\ln 1.491825 = 5k \qquad \ln e^a = a$$

$$\frac{\ln 1.491825}{5} = k \qquad \text{Dividing both sides by 5}$$

$$0.08 \approx k. \qquad \text{Using a calculator and rounding}$$

The interest rate is about 0.08, or 8%, compounded continuously. Note that since interest is being compounded continuously, the interest earned each year is more than 8%. The exponential growth function is

$$P(t) = 30{,}000 e^{0.08t}.$$

b) To find the doubling time T, we replace $P(T)$ with 60,000 and solve for T:

$$60{,}000 = 30{,}000e^{0.08T}$$

$\qquad 2 = e^{0.08T}$ Dividing both sides by 30,000

$\qquad \ln 2 = \ln e^{0.08T}$ Taking the natural logarithm on both sides

$\qquad \ln 2 = 0.08T$ $\ln e^a = a$

$\qquad \dfrac{\ln 2}{0.08} = T$ Dividing both sides by 0.08

$\qquad 8.7 \approx T.$ Using a calculator and rounding

Thus the original investment of \$30,000 will double in about 8.7 yr.

As the results of Examples 3(b) and 6(b) imply, for any specified interest rate, continuous compounding gives the highest yield and the shortest doubling time.

In some real-life situations, a quantity or population is *decreasing* or *decaying* exponentially.

Exponential Decay

An **exponential decay model** is a function of the form

$$P(t) = P_0 e^{-kt}, \quad k > 0,$$

where P_0 is the quantity present at time 0, $P(t)$ is the amount present at time t, and k is the **decay rate**. The **half-life** is the amount of time necessary for half of the quantity to decay.

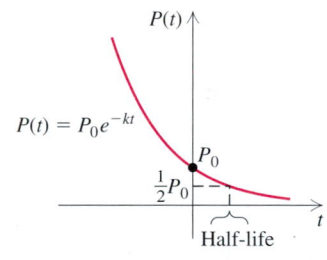

Example 7

Carbon dating. The radioactive element carbon-14 has a half-life of 5750 yr. The percentage of carbon-14 present in the remains of organic matter can be used to determine the age of that organic matter. Recently, while digging in Chaco Canyon, New Mexico, archaeologists found corn pollen that had lost 38.1% of its carbon-14. The age of this corn pollen was evidence that Indians had been cultivating crops in the Southwest centuries earlier than scientists had thought. What was the age of the pollen? (*Source: American Anthropologist*)

Solution   We first find k. To do so, we use the concept of half-life. When $t = 5750$ (the half-life), $P(t)$ will be half of P_0. Then

$$0.5P_0 = P_0 e^{-k(5750)}$$ Substituting in $P(t) = P_0 e^{-kt}$

$$0.5 = e^{-5750k}$$ Dividing both sides by P_0

$$\ln 0.5 = \ln e^{-5750k}$$ Taking the natural logarithm on both sides

$$\ln 0.5 = -5750k$$ $\ln e^a = a$

$$\frac{\ln 0.5}{-5750} = k$$ Dividing

$$0.00012 \approx k.$$ Using a calculator and rounding

Now we have a function for the decay of carbon-14:

$$P(t) = P_0 e^{-0.00012t}.$$ This completes the first part of our solution.

(*Note*: This equation can be used for any subsequent carbon-dating problem.) If the corn pollen has lost 38.1% of its carbon-14 from an initial amount P_0, then 100% − 38.1%, or 61.9%, of P_0 is still present. To find the age t of the pollen, we solve this equation for t:

$$0.619P_0 = P_0 e^{-0.00012t}$$ We want to find t for which $P(t) = 0.619P_0$.

$$0.619 = e^{-0.00012t}$$ Dividing both sides by P_0

$$\ln 0.619 = \ln e^{-0.00012t}$$ Taking the natural logarithm on both sides

$$\ln 0.619 = -0.00012t$$ $\ln e^a = a$

$$\frac{\ln 0.619}{-0.00012} = t$$ Dividing

$$4000 \approx t.$$ Using a calculator

The pollen is about 4000 yr old.

FOR EXTRA HELP

Exercise Set 12.7

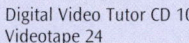
Digital Video Tutor CD 10
Videotape 24

InterAct Math

Math Tutor Center

MathXL

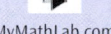
MyMathLab.com

Solve.

1. *Cellular phones.*   The number of cellular phones in use in the United States t years after May 2001 can be predicted by

$$N(t) = 112(1.37)^t,$$

where $N(t)$ is the number of cellular phones in use, in millions.

a) Determine the year in which 200 million cellular phones would be in use.

b) What is the doubling time for the number of cellular phones in use?

2. *Compact discs.*   The number of compact discs N purchased each year, in millions, can be approximated by

$$N(t) = 384(1.13)^t,$$

where t is the number of years after 1991.*

a) After what amount of time will two billion compact discs be sold in a year?

b) What is the doubling time for the number of compact discs sold in one year?

*Based on data from the Recording Industry Association, Washington, DC.

3. *Student loan repayment.* A college loan of $29,000 is made at 8% interest, compounded annually. After t years, the amount due, A, is given by the function

$$A(t) = 29{,}000(1.08)^t.$$

a) After what amount of time will the amount due reach $40,000?

b) Find the doubling time.

4. *Spread of a rumor.* The number of people who have heard a rumor increases exponentially. If all who hear a rumor repeat it to two people a day, and if 20 people start the rumor, the number of people N who have heard the rumor after t days is given by

$$N(t) = 20(3)^t.$$

a) After what amount of time will 1000 people have heard the rumor?

b) What is the doubling time for the number of people who have heard the rumor?

5. *Skateboarding.* The number of skateboarders of age x (with $x \geq 16$), in thousands, can be approximated by

$$N(x) = 600(0.873)^{x-16}$$

(*Sources*: Based on figures from the U.S. Bureau of the Census and *Statistical Abstract of the United States*, 2000).

a) Estimate the number of 41-yr-old skateboarders.

b) At what age are there only 2000 skateboarders?

6. *Recycling aluminum cans.* Approximately two thirds of all aluminum cans distributed will be recycled each year. A beverage company distributes 250,000 cans. The number still in use after t years is given by the function

$$N(t) = 250{,}000\left(\tfrac{2}{3}\right)^t.$$

a) After how many years will 60,000 cans still be in use?

b) After what amount of time will only 1000 cans still be in use?

Use the pH formula given in Example 2 for Exercises 7–10.

7. *Chemistry.* The hydrogen ion concentration of fresh-brewed coffee is about 1.3×10^{-5} moles per liter. Find the pH.

8. *Chemistry.* The hydrogen ion concentration of milk is about 1.6×10^{-7} moles per liter. Find the pH.

9. *Medicine.* When the pH of a patient's blood drops below 7.4, a condition called *acidosis* sets in. Acidosis can be deadly when the patient's pH reaches 7.0. What would the hydrogen ion concentration of the patient's blood be at that point?

10. *Medicine.* When the pH of a patient's blood rises above 7.4, a condition called *alkalosis* sets in. Alkalosis can be deadly when the patient's pH reaches 7.8. What would the hydrogen ion concentration of the patient's blood be at that point?

Use the decibel formula given in Example 1 for Exercises 11–14.

11. *Audiology.* The intensity of sound in normal conversation is about 3.2×10^{-6} W/m². How loud in decibels is this sound level?

12. *Audiology.* The intensity of a riveter at work is about 3.2×10^{-3} W/m². How loud in decibels is this sound level?

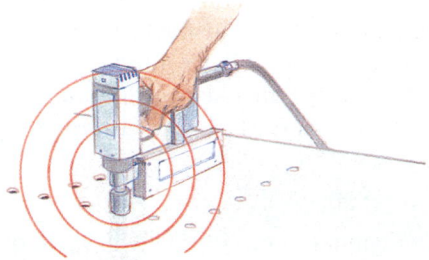

13. *Music.* The band U2 recently performed and sound measurements of 105 dB were recorded. What is the intensity of such sounds?

14. *Music.* The band Strange Folk recently performed in Burlington, VT, and reached sound levels of 111 dB (*Source*: Melissa Garrido, *Burlington Free Press*). What is the intensity of such sounds?

Use the compound-interest formula in Example 6 for Exercises 15 and 16.

15. *Interest compounded continuously.* Suppose that P_0 is invested in a savings account where interest is compounded continuously at 6% per year.

a) Express $P(t)$ in terms of P_0 and 0.06.
b) Suppose that $5000 is invested. What is the balance after 1 yr? after 2 yr?
c) When will an investment of $5000 double itself?

16. *Interest compounded continuously.* Suppose that P_0 is invested in a savings account where interest is compounded continuously at 5% per year.

a) Express $P(t)$ in terms of P_0 and 0.05.
b) Suppose that $1000 is invested. What is the balance after 1 yr? after 2 yr?
c) When will an investment of $1000 double itself?

17. *Population growth.* In 2000, the population of the United States was 283.75 million and the exponential growth rate was 1.3% per year (*Source*: U.S. Bureau of the Census).

a) Find the exponential growth function.
b) Predict the U.S. population in 2005.
c) When will the U.S. population reach 325 million?

18. *World population growth.* In 2001, the world population was 6.1 billion and the exponential growth rate was 1.4% per year (*Source*: U.S. Bureau of the Census).

a) Find the exponential growth function.
b) Predict the world population in 2005.
c) When will the world population be 8.0 billion?

19. *Growth of bacteria.* The bacteria *Escherichi coli* are commonly found in the human bladder. Suppose that 3000 of the bacteria are present at time $t = 0$. Then t minutes later, the number of bacteria present is

$$N(t) = 3000(2)^{t/20}.$$

a) After what amount of time will there be 60,000 bacteria?
b) If 100,000,000 bacteria accumulate, a bladder infection can occur. What amount of time would have to pass in order for a possible infection to occur?
c) What is the doubling time?

20. *Population growth.* The exponential growth rate of the population of Central America is 3.5% per

year (one of the highest in the world). What is the doubling time?

21. *Advertising.* A model for advertising response is given by

$$N(a) = 2000 + 500 \log a, \quad a \ge 1,$$

where $N(a)$ is the number of units sold and a is the amount spent on advertising, in thousands of dollars.

a) How many units were sold after spending $1000 ($a = 1$) on advertising?
b) How many units were sold after spending $8000?
c) Graph the function.
d) How much would have to be spent in order to sell 5000 units?

22. *Forgetting.* Students in an English class took a final exam. They took equivalent forms of the exam at monthly intervals thereafter. The average score $S(t)$, in percent, after t months was found to be given by

$$S(t) = 68 - 20 \log (t + 1), \quad t \ge 0.$$

a) What was the average score when they initially took the test, $t = 0$?
b) What was the average score after 4 months? after 24 months?
c) Graph the function.
d) After what time t was the average score 50?

23. *Public health.* In 1995, an outbreak of Herpes infected 17 people in a large community. By 1996, the number of those infected had grown to 29.

a) Find an exponential growth function that fits the data.
b) Predict the number of people who will be infected in 2001.

24. *Heart transplants.* In 1967, Dr. Christiaan Barnard of South Africa stunned the world by performing the first heart transplant. There were 1418 heart transplants in 1987, and 2185 such transplants in 1999.

a) Find an exponential growth function that fits the data from 1987 and 1999.
b) Use the function to predict the number of heart transplants in 2012.

25. *Oil demand.* The exponential growth rate of the demand for oil in the United States is 10% per year. In what year will the demand be double that of May 1995?

26. *Coal demand.* The exponential growth rate of the demand for coal in the world is 4% per year. When will the demand be double that of 1995?

27. *Decline of discarded yard waste.* The amount of discarded yard waste has declined considerably in recent years because of increased recycling and composting. In 1996, 17.5 million tons were discarded, but by 1998 the figure dropped to 14.5 million tons (*Source*: *Statistical Abstract of the United States*, 2000). Assume the amount of discarded yard waste is decreasing according to the exponential decay model.
   a) Find the value k, and write an exponential function that describes the amount of yard waste discarded t years after 1996.
   b) Estimate the amount of discarded yard waste in 2006.
   c) In what year (theoretically) will only 1 ton of yard waste be discarded?

28. *Decline in cases of mumps.* The number of cases of mumps has dropped exponentially from 5300 in 1990 to 800 in 1996 (*Source*: *Statistical Abstract of the United States*, 2000).
   a) Find the value k, and write an exponential function that can be used to estimate the number of cases t years after 1990.
   b) Estimate the number of cases of mumps in 2004.
   c) In what year (theoretically) will there be only 1 case of mumps?

29. *Archaeology.* When archaeologists found the Dead Sea scrolls, they determined that the linen wrapping had lost 22.3% of its carbon-14. How old is the linen wrapping? (See Example 7.)

30. *Archaeology.* In 1996, researchers found an ivory tusk that had lost 18% of its carbon-14. How old was the tusk? (See Example 7.)

31. *Chemistry.* The exponential decay rate of iodine-131 is 9.6% per day. What is its half-life?

32. *Chemistry.* The decay rate of krypton-85 is 6.3% per year. What is its half-life?

33. *Home construction.* The chemical urea formaldehyde was found in some insulation used in houses built during the mid to late 60s. Unknown at the time was the fact that urea formaldehyde emitted toxic fumes as it decayed. The half-life of urea formaldehyde is 1 yr. What is its decay rate?

34. *Plumbing.* Lead pipes and solder are often found in older buildings. Unfortunately, as lead decays, toxic chemicals can get in the water resting in the pipes. The half-life of lead is 22 yr. What is its decay rate?

35. *Value of a sports card.* Legend has it that because he objected to smoking, and because his first baseball card was issued in cigarette packs, the great shortstop Honus Wagner halted production of his card before many were produced. One of these cards was sold in 1996 for $640,500 and again in 2000 for $1.1 million. For the following questions, assume that the card's value increases exponentially, as it has for many years.

WAGNER, PITTSBURG

   a) Find the exponential growth rate k, and determine an exponential function V that can be used to estimate the dollar value, $V(t)$, of the card t years after 1996.
   b) Predict the value of the card in 2006.
   c) What is the doubling time for the value of the card?
   d) In what year will the value of the card first exceed $2,000,000?

36. *Portrait of Dr. Gachet.*   As of May 2001, the most ever paid for a painting is $82.5 million, paid in 1990 for Vincent Van Gogh's *Portrait of Dr. Gachet.* The same painting sold for $58 million in 1987. Assume that the growth in the value V of the painting is exponential.

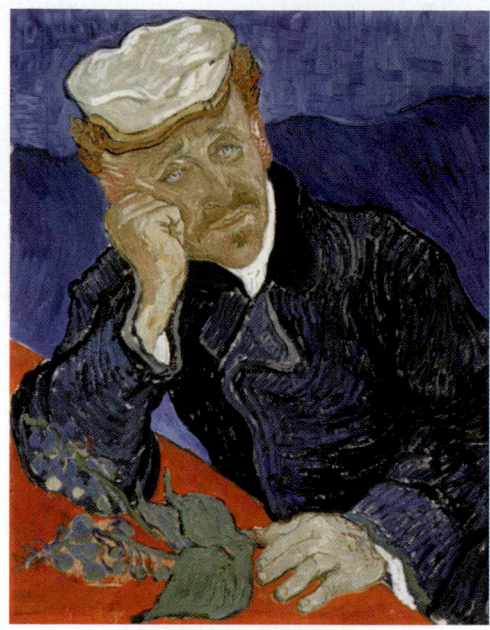

Van Gogh's *Portrait of Dr. Gachet*, oil on canvas.

a) Find the exponential growth rate k, and determine the exponential growth function V, for which $V(t)$ is the painting's value, in millions of dollars, t years after 1987.
b) Estimate the value of the painting in 2007.
c) What is the doubling time for the value of the painting?
d) How long after 1987 will the value of the painting be $1 billion?

37. Write a problem for a classmate to solve in which information is provided and the classmate is asked to find an exponential growth function. Make the problem as realistic as possible.

38. Examine the restriction on t in Exercise 22.
a) What upper limit might be placed on t?
b) In practice, would this upper limit ever be enforced? Why or why not?

SKILL MAINTENANCE

Graph.

39. $y = x^2 - 8x$ **40.** $y = x^2 - 5x - 6$

41. $f(x) = 3x^2 - 5x - 1$ **42.** $g(x) = 2x^2 - 6x + 3$

Solve by completing the square.

43. $x^2 - 8x = 7$ **44.** $x^2 + 10x = 6$

SYNTHESIS

45. Will the model used to predict the number of cellular phones in Exercise 1 still be realistic in 2020? Why or why not?

46. *Atmospheric pressure.*   Atmospheric pressure P at altitude a is given by

$$P = P_0 e^{-0.00005a},$$

where P_0 is the pressure at sea level $\approx 14.7 \text{ lb/in}^2$ (pounds per square inch). Explain how a barometer, or some other device for measuring atmospheric pressure, can be used to find the height of a skyscraper.

47. *Sports salaries.*   In 2001, Derek Jeter of the New York Yankees signed a $189 million 10-yr contract that will pay him $21 million in 2010. How much would Yankee owner George Steinbrenner need to invest in 2001 at 5% interest compounded continuously, in order to have the $21 million for Jeter in 2010? (This is much like finding what $21 million in 2010 will be worth in 2001 dollars.)

48. *Supply and demand.*   The supply and demand for the sale of stereos by Sound Ideas are given by

$$S(x) = e^x \quad \text{and} \quad D(x) = 162{,}755e^{-x},$$

where $S(x)$ is the price at which the company is willing to supply x stereos and $D(x)$ is the demand price for a quantity of x stereos. Find the equilibrium point. (For reference, see Section 8.8.)

49. Use Exercises 1 and 17 to form a model for the percentage of U.S. residents owning a cellular phone t years after 2001.

50. Use the model developed in Exercise 49 to predict the percentage of U.S. residents who will own cellular phones in 2010. Does your prediction seem plausible? Why or why not?

51. *Nuclear energy.*   Plutonium-239 (Pu-239) is used in nuclear energy plants. The half-life of Pu-239 is 24,360 yr (*Source*: *Microsoft Encarta 97 Encyclopedia*). How long will it take for a fuel rod of Pu-239 to lose 90% of its radioactivity?

COLLABORATIVE

CORNER

Investments in Collectibles

Focus: Exponential-growth models

Time: 30 minutes

Group size: 6

Prepaid calling cards have become a big business, not only as a convenient way to place a telephone call, but also as collectibles. Some telephone cards are worth far more than their original cost, particularly if they are unused.

Collectors often estimate the future value of their collections by examining the value's growth in the past. The value of some collectibles grows exponentially.

ACTIVITY

1. Suppose that in 2001, each group member needed to invest $1200 in the telephone cards listed in the following table. Looking only at the approximate value in 2001, each student should select $1200 worth of cards to buy. More than one card of each type can be selected. The card or cards chosen will become that student's portfolio.

2. Each group member should be assigned one of the cards. That person should then form an exponential growth model for the value of that card using the year of issue and the original cost of the card.

3. Using the models developed above, each group member should predict the value of his or her portfolio in 2005. Compare the values. Why, when buying a collector's item, is it important to consider its previous worth?

4. Look at the predicted value of each card in the year 2008. Does an exponential growth model seem appropriate? If possible, find the current value of some of the cards and compare those values with the predicted values.

Card	Approximate Value in 2001	Year of Issue	Original Cost
World Rowing Championships, Spec Set	$200	1994	$16
AT&T, America's Cup	$1200	1992	$50
McDonald's Hamburgers, Proof	$60	1996	$2
Sprint, Coca-Cola collection	$150	1995	$17
New York Telephone, 1992 Democratic National Convention	$350	1992	$1
McDonald's Extra-Value Meal Promotion	$75	1995	$3

Source: Moneycard.com, June 2001

Summary and Review 12

Key Terms

Important Properties and Formulas

To Find a Formula for the Inverse of a Function

First check a graph to make sure that the function f is one-to-one. Then:

1. Replace $f(x)$ with y.
2. Interchange x and y.
3. Solve for y.
4. Replace y with $f^{-1}(x)$.

Composition of f and g: $(f \circ g)(x) = f(g(x))$
Composition and inverses: $(f^{-1} \circ f)(x) = (f \circ f^{-1})(x) = x$
Exponential function: $f(x) = a^x, \quad a > 0, \quad a \neq 1$
Interest compounded annually: $A = P(1 + i)^t$

For $x > 0$ and a a positive constant other than 1, $\log_a x$ is the power to which a must be raised in order to get x. Thus, $a^{\log_a x} = x$, or equivalently, if $y = \log_a x$, then $a^y = x$. *A logarithm is an exponent.*

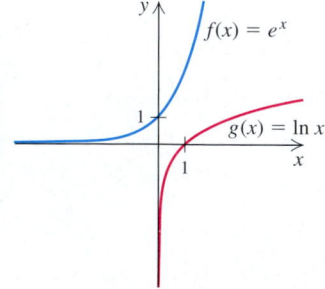

The Principle of Exponential Equality

For any real number b, $b \neq -1, 0$, or 1:

$b^x = b^y$ is equivalent to $x = y$.

The Principle of Logarithmic Equality

For any logarithmic base a, and for $x, y > 0$;

$x = y$ is equivalent to $\log_a x = \log_a y$.

Properties of Logarithms

$\log_a MN = \log_a M + \log_a N$,

$\log_a \dfrac{M}{N} = \log_a M - \log_a N$,

$\log_a M^p = p \cdot \log_a M$,

$\log_a 1 = 0$,

$\log_a a = 1$,

$\log_a a^k = k$,

$\log M = \log_{10} M$,

$\ln M = \log_e M$,

$\log_b M = \dfrac{\log_a M}{\log_a b}$

$e \approx 2.7182818284\ldots$

Loudness of sound:	$L = 10 \cdot \log \dfrac{I}{I_0}$
pH:	$\text{pH} = -\log[H^+]$
Exponential growth:	$P(t) = P_0 e^{kt}, \, k > 0$
Exponential decay:	$P(t) = P_0 e^{-kt}, \, k > 0$
Interest compounded continuously:	$P(t) = P_0 e^{kt}$, where P_0 is the principal invested for t years at interest rate k
Carbon dating:	$P(t) = P_0 e^{-0.00012t}$

Review Exercises

1. Find $(f \circ g)(x)$ and $(g \circ f)(x)$ if $f(x) = x^2 + 1$ and $g(x) = 2x - 3$.

2. If $h(x) = \sqrt{3 - x}$, find $f(x)$ and $g(x)$ such that $h(x) = (f \circ g)(x)$. Answers may vary.

3. Determine whether $f(x) = 4 - x^2$ is one-to-one.

Find a formula for the inverse of each function.

4. $f(x) = x - 3$

5. $g(x) = \dfrac{3x + 1}{2}$

6. $f(x) = 27x^3$

Graph.

7. $f(x) = 3^x + 1$

8. $x = \left(\frac{1}{4}\right)^y$

9. $y = \log_5 x$

Simplify.

10. $\log_3 9$

11. $\log_{10} \frac{1}{10}$

12. $\log_{25} 5$

13. $\log_3 3^{12}$

Convert to logarithmic equations.

14. $10^{-2} = \frac{1}{100}$

15. $25^{1/2} = 5$

Convert to exponential equations.

16. $\log_4 16 = x$

17. $\log_8 1 = 0$

Express in terms of logarithms of x, y, and z.

18. $\log_a x^4 y^2 z^3$

19. $\log_a \dfrac{x^3}{yz^2}$

20. $\log \sqrt[4]{\dfrac{z^2}{x^3 y}}$

Express as a single logarithm and, if possible, simplify.

21. $\log_a 8 + \log_a 15$

22. $\log_a 72 - \log_a 12$

23. $\frac{1}{2} \log a - \log b - 2 \log c$

24. $\frac{1}{3}[\log_a x - 2 \log_a y]$

Simplify.

25. $\log_m m$

26. $\log_m 1$

27. $\log_m m^{17}$

Given $\log_a 2 = 1.8301$ and $\log_a 7 = 5.0999$, find each of the following.

28. $\log_a 14$

29. $\log_a \frac{2}{7}$

30. $\log_a 28$

31. $\log_a 3.5$

32. $\log_a \sqrt{7}$

33. $\log_a \frac{1}{4}$

Use a calculator to find each of the following to the nearest ten-thousandth.

34. $\log 82$

35. $10^{1.789}$

36. $\ln 0.05$

37. $e^{-0.98}$

Find each of the following logarithms using the change-of-base formula. Round answers to the nearest ten-thousandth.

38. $\log_5 2$

39. $\log_{12} 70$

Graph and state the domain and the range of each function.

40. $f(x) = e^x - 1$

41. $g(x) = 0.6 \ln x$

Solve. Where appropriate, include approximations to the nearest ten-thousandth.

42. $2^x = 32$

43. $3^x = \frac{1}{9}$

44. $\log_3 x = -2$

45. $\log_x 32 = 5$

46. $\log x = -4$

47. $3 \ln x = -6$

48. $4^{2x-5} = 16$

49. $2^{x^2} \cdot 2^{4x} = 32$

50. $4^x = 8.3$

51. $e^{-0.1t} = 0.03$

52. $2 \ln x = -6$

53. $\log_3 (2x - 5) = 1$

54. $\log_4 x + \log_4 (x - 6) = 2$

55. $\log x + \log (x - 15) = 2$

56. $\log_3 (x - 4) = 3 - \log_3 (x + 4)$

57. In a business class, students were tested at the end of the course with a final exam. They were tested again after 6 months. The forgetting formula was determined to be

$$S(t) = 62 - 18 \log (t + 1),$$

where t is the time, in months, after taking the first test.

a) Determine the average score when they first took the test (when $t = 0$).

b) What was the average score after 6 months?

c) After what time was the average score 34?

58. A color photocopier is purchased for $5200. Its value each year is about 80% of its value in the preceding year. Its value in dollars after t years is given by the exponential function

$$V(t) = 5200(0.8)^t.$$

  a) After what amount of time will the salvage value be $1200?

  b) After what amount of time will the salvage value be half the original value?

59. The number of customer service complaints against U.S. airlines has grown exponentially from 667 in 1995 to 3664 in 1999 (*Sources*: U.S. Bureau of the Census and *Statistical Abstract of the United States*, 2000).

  a) Find the value k, and write an exponential function that describes the number of customer service complaints t years after 1995.

  b) Predict the number of complaints in 2005.

  c) In what year will there be 10,000 customer service complaints?

60. The value of Jose's stock market portfolio doubled in 3 yr. What was the exponential growth rate?

61. How long will it take $7600 to double itself if it is invested at 8.4%, compounded continuously?

62. How old is a skeleton that has lost 34% of its carbon-14? (Use $P(t) = P_0 e^{-0.00012t}$.)

63. What is the pH of a substance if its hydrogen ion concentration is 2.3×10^{-7} moles per liter? (Use pH $= -\log [\text{H}^+]$.)

64. The intensity of the sound of water at the foot of the Niagara Falls is about 10^{-3} W/m². * How loud in decibels is this sound level?

$$\left(\text{Use } L = 10 \cdot \log \frac{I}{10^{-12}}. \right)$$

SYNTHESIS

65. Explain why negative numbers do not have logarithms.

66. Explain why taking the natural or common logarithm on each side of an equation produces an equivalent equation.

Solve.

67. $\ln (\ln x) = 3$

68. $2^{x^2+4x} = \frac{1}{8}$

69. $5^{x+y} = 25,$
    $2^{2x-y} = 64$

Sound and Hearing, Life Science Library. (New York: Time Incorporated, 1965), p. 173.

Chapter Test 12

1. Find $(f \circ g)(x)$ and $(g \circ f)(x)$ if $f(x) = x + x^2$ and $g(x) = 2x + 1$.

2. If

$$h(x) = \frac{1}{2x^2 + 1},$$

  find $f(x)$ and $g(x)$ such that $h(x) = (f \circ g)(x)$. Answers may vary.

3. Determine whether $f(x) = |x + 1|$ is one-to-one.

Find a formula for the inverse of each function.

4. $f(x) = 4x - 3$

5. $g(x) = (x + 1)^3$

Graph.

6. $f(x) = 2^x - 3$

7. $g(x) = \log_7 x$

Simplify.

8. $\log_5 125$

9. $\log_{100} 10$

10. $3^{\log_3 18}$

Convert to logarithmic equations.

11. $4^{-3} = \frac{1}{64}$

12. $256^{1/2} = 16$

Convert to exponential equations.

13. $m = \log_7 49$

14. $\log_3 81 = 4$

15. Express in terms of logarithms of a, b, and c:

$$\log \frac{a^3 b^{1/2}}{c^2}.$$

16. Express as a single logarithm:

$$\tfrac{1}{3} \log_a x + 2 \log_a z.$$

Simplify.

17. $\log_p p$

18. $\log_t t^{23}$

19. $\log_c 1$

Given $\log_a 2 = 0.301$, $\log_a 6 = 0.778$, and $\log_a 7 = 0.845$, find each of the following.

20. $\log_a \frac{2}{7}$

21. $\log_a 12$

22. $\log_a 16$

Use a calculator to find each of the following to the nearest ten-thousandth.

23. $\log 12.3$

24. $10^{-0.8}$

25. $\ln 0.035$

26. $e^{4.8}$

27. Find $\log_3 10$ using the change-of-base formula. Round to the nearest ten-thousandth.

Graph and state the domain and the range of each function.

28. $f(x) = e^x + 3$

29. $g(x) = \ln(x - 4)$

Solve. Where appropriate, include approximations to the nearest ten-thousandth.

30. $2^x = \frac{1}{32}$

31. $\log_x 25 = 2$

32. $\log_4 x = \frac{1}{2}$

33. $\log x = 4$

34. $5^{4-3x} = 125$

35. $7^x = 1.2$

36. $\ln x = \frac{1}{4}$

37. $\log(x - 3) + \log(x + 1) = \log 5$

38. The average walking speed R of people living in a city of population P, in thousands, is given by $R = 0.37 \ln P + 0.05$, where R is in feet per second.
  a) The population of Albuquerque, New Mexico, is 679,000. Find the average walking speed.
  b) A city has an average walking speed of 2.6 ft/sec. Find the population.

39. The population of Kenya was 30 million in 2000, and the exponential growth rate was 1.5% per year.
  a) Write an exponential function describing the population of Kenya.
  b) What will the population be in 2003? in 2010?
  c) When will the population be 50 million?
  d) What is the doubling time?

40. The U.S. Consumer Price Index, a method of comparing prices of basic items, grew exponentially from 60.0 in 1920 to 511.5 in 2000 (*Sources*: U.S. Bureau of Labor Statistics and U.S. Department of Labor).
  a) Find the value k, and write an exponential function that approximates the Consumer Price Index t years after 1920.
  b) Predict the Consumer Price Index in 2010.
  c) In what year will the Consumer Price Index be 1000?

41. An investment with interest compounded continuously doubled itself in 15 yr. What is the interest rate?

42. How old is an animal bone that has lost 43% of its carbon-14? (Use $P(t) = P_0 e^{-0.00012t}$.)

43. The sound of traffic at a busy intersection averages 75 dB. What is the intensity of such a sound?
$$\left(\text{Use } L = 10 \cdot \log \frac{I}{I_0}.\right)$$

44. The hydrogen ion concentration of water is 1.0×10^{-7} moles per liter. What is the pH? (Use $\text{pH} = -\log[\text{H}^+]$.)

SYNTHESIS

45. Solve: $\log_5 |2x - 7| = 4$.

46. If $\log_a x = 2$, $\log_a y = 3$, and $\log_a z = 4$, find
$$\log_a \frac{\sqrt[3]{x^2 z}}{\sqrt[3]{y^2 z^{-1}}}.$$

Cumulative Review 1–12

1. Evaluate $\dfrac{x^0 + y}{-z}$ for $x = 6$, $y = 9$, and $z = -5$.

Simplify.

2. $\left|-\frac{5}{2} + \left(-\frac{7}{2}\right)\right|$

3. $(-2x^2y^{-3})^{-4}$

4. $(-5x^4y^{-3}z^2)(-4x^2y^2)$

5. $\dfrac{3x^4y^6z^{-2}}{-9x^4y^2z^3}$

6. $2x - 3 - 2[5 - 3(2 - x)]$

7. $3^3 + 2^2 - (32 \div 4 - 16 \div 8)$

Solve.

8. $8(2x - 3) = 6 - 4(2 - 3x)$

9. $4x - 3y = 15,$
     $3x + 5y = 4$

10. $x + y - 3z = -1,$
         $2x - y + z = 4,$
         $-x - y + z = 1$

11. $x(x - 3) = 10$

12. $\dfrac{7}{x^2 - 5x} - \dfrac{2}{x - 5} = \dfrac{4}{x}$

13. $\dfrac{8}{x + 1} + \dfrac{11}{x^2 - x + 1} = \dfrac{24}{x^3 + 1}$

14. $\sqrt{4 - 5x} = 2x - 1$

15. $\sqrt[3]{2x} = 1$

16. $3x^2 + 75 = 0$

17. $x - 8\sqrt{x} + 15 = 0$

18. $x^4 - 13x^2 + 36 = 0$

19. $\log_8 x = 1$

20. $\log_x 49 = 2$

21. $9^x = 27$

22. $3^{5x} = 7$

23. $\log x - \log(x - 8) = 1$

24. $x^2 + 4x > 5$

25. If $f(x) = x^2 + 6x$, find a such that $f(a) = 11$.

26. If $f(x) = |2x - 3|$, find all x for which $f(x) \geq 9$.

Solve.

27. $D = \dfrac{ab}{b + a}$, for a

28. $\dfrac{1}{p} + \dfrac{1}{q} = \dfrac{1}{f}$, for q

29. $M = \dfrac{2}{3}(A + B)$, for B

Evaluate.

30. $\begin{vmatrix} 6 & -5 \\ 4 & -3 \end{vmatrix}$

31. $\begin{vmatrix} 7 & -6 & 0 \\ -2 & 1 & 2 \\ -1 & 1 & -1 \end{vmatrix}$

32. Find the domain of the function f given by
$$f(x) = \dfrac{-4}{3x^2 - 5x - 2}.$$

Solve.

33. The number of Americans filing taxes on the Internet (e-filers) in January and February increased from 2.4 million in 2000 to 3.4 million in 2001 (*Source*: USA Today, April 2, 2001).

   a) At what rate was the number of January and February e-filers increasing?
   b) Find a linear function $E(t)$ that fits the data. Let t represent the number of years since 2000.
   c) Use the function of part (b) to predict the number of e-filers in January and February of 2005.

34. The perimeter of a rectangular garden is 112 m. The length is 16 m more than the width. Find the length and the width.

35. In triangle ABC, the measure of angle B is three times the measure of angle A. The measure of angle C is 105° greater than the measure of angle A. Find the angle measures.

36. Good's Candies makes all their chocolates by hand. It takes Anne 10 min to coat a tray of candies in chocolate. It takes Clay 12 min to coat a tray of candies. How long would it take Anne and Clay, working together, to coat the candies?

37. Joe's Thick and Tasty salad dressing gets 45% of its calories from fat. The Light and Lean dressing gets 20% of its calories from fat. How many ounces of each should be mixed in order to get 15 oz of dressing that gets 30% of its calories from fat?

38. A fishing boat with a trolling motor can move at a speed of 5 km/h in still water. The boat travels 42 km downstream in the same time that it takes to travel 12 km upstream. What is the speed of the stream?

39. What is the minimum product of two numbers whose difference is 14? What are the numbers that yield this product?

📖 *Students in a biology class just took a final exam. A formula for determining what the average exam grade will be t months later is*

$$S(t) = 78 - 15 \log (t + 1).$$

40. The average score when the students first took the test occurs when $t = 0$. Find the students' average score on the final exam.

41. What would the average score be on a retest after 4 months?

The population of Mozambique was 19 million in 2000, and the exponential growth rate was 1.5% per year.

42. Write an exponential function describing the growth of the population of Mozambique.

📖 **43.** Predict what the population will be in 2005 and in 2012.

44. What is the doubling time of the population?

45. y varies directly as the square of x and inversely as z, and $y = 2$ when $x = 5$ and $z = 100$. What is y when $x = 3$ and $z = 4$?

Perform the indicated operations and simplify.

46. $(5p^2q^3 + 6pq - p^2 + p) + (2p^2q^3 + p^2 - 5pq - 9)$

47. $(11x^2 - 6x - 3) - (3x^2 + 5x - 2)$

48. $(3x^2 - 2y)^2$

49. $(5a + 3b)(2a - 3b)$

50. $\dfrac{x^2 + 8x + 16}{2x + 6} \div \dfrac{x^2 + 3x - 4}{x^2 - 9}$

51. $\dfrac{1 + \dfrac{3}{x}}{x - 1 - \dfrac{12}{x}}$

52. $\dfrac{a^2 - a - 6}{a^3 - 27} \cdot \dfrac{a^2 + 3a + 9}{6}$

53. $\dfrac{3}{x + 6} - \dfrac{2}{x^2 - 36} + \dfrac{4}{x - 6}$

Factor.

54. $xy - 2xz + xw$

55. $1 - 125x^3$

56. $6x^2 + 8xy - 8y^2$

57. $x^4 - 4x^3 + 7x - 28$

58. $2m^2 + 12mn + 18n^2$

59. $x^4 - 16y^4$

60. For the function described by
$$h(x) = -3x^2 + 4x + 8,$$
find $h(-2)$.

61. Divide: $(x^4 - 5x^3 + 2x^2 - 6) \div (x - 3)$.

62. Multiply $(5.2 \times 10^4)(3.5 \times 10^{-6})$. Write scientific notation for the answer.

For the radical expressions that follow, assume that all variables represent positive numbers.

63. Divide and simplify:
$$\dfrac{\sqrt[3]{40xy^8}}{\sqrt[3]{5xy}}.$$

64. Multiply and simplify: $\sqrt{7xy^3} \cdot \sqrt{28x^2y}$.

65. Rewrite without rational exponents: $(27a^6b)^{4/3}$.

66. Rationalize the denominator:
$$\dfrac{3 - \sqrt{y}}{2 - \sqrt{y}}.$$

67. Divide and simplify:
$$\dfrac{\sqrt{x + 5}}{\sqrt[5]{x + 5}}.$$

68. Multiply these complex numbers:
$$\left(1 + i\sqrt{3}\right)\left(6 - 2i\sqrt{3}\right).$$

69. Add: $(3 - 2i) + (5 + 3i)$.

70. Find the inverse of f if $f(x) = 7 - 2x$.

71. Find a linear equation with a graph that contains the points $(0, -3)$ and $(-1, 2)$.

72. Find an equation of the line whose graph has a y-intercept of $(0, 7)$ and is perpendicular to the line given by $2x + y = 6$.

Graph.

73. $5x = 15 + 3y$

74. $y = 2x^2 - 4x - 1$

75. $y = \log_3 x$

76. $y = 3^x$

77. $-2x - 3y \leq 6$

78. Graph: $f(x) = 2(x + 3)^2 + 1$.

   a) Label the vertex.

   b) Draw the axis of symmetry.

   c) Find the maximum or minimum value.

79. Graph $f(x) = 2e^x$ and determine the domain and the range.

80. Express in terms of logarithms of a, b, and c:

$$\log\left(\frac{a^2 c^3}{b}\right).$$

81. Express as a single logarithm:

$$3 \log x - \tfrac{1}{2} \log y - 2 \log z.$$

82. Convert to an exponential equation: $\log_a 5 = x$.

83. Convert to a logarithmic equation: $x^3 = t$.

Find each of the following using a calculator. Round to the nearest ten-thousandth.

84. $\log 0.05566$

85. $10^{2.89}$

86. $\ln 12.78$

87. $e^{-1.4}$

SYNTHESIS

Solve.

88. $\dfrac{5}{3x - 3} + \dfrac{10}{3x + 6} = \dfrac{5x}{x^2 + x - 2}$

89. $\log \sqrt{3x} = \sqrt{\log 3x}$

90. A train travels 280 mi at a certain speed. If the speed had been increased by 5 mph, the trip could have been made in 1 hr less time. Find the actual speed.

13

Conic Sections

AN APPLICATION

Each side edge of the Burton® Twin 53 snowboard is an arc of a circle with a "running length" of 1150 mm and a "sidecut depth" of 19.5 mm. What radius is used to manufacture the edge of this board?

This problem, along with a diagram, appears as Exercise 97 in Section 13.1.

A snowboard's shape is designed by complex 3-D geometry, from the side-cut to the running length to the transition zones. By using these geometrical tools, we can design boards to perform like no others.

PETER BERGENDAHL
Board Design Engineer
Burlington, VT

*T*he arcs described in the chapter opening are parts of a circle. A circle is one example of a conic section, meaning that it can be regarded as a cross section of a cone. This chapter presents a variety of equations with graphs that are conic sections. We have already worked with two conic sections, lines and parabolas, in Chapters 3 and 11. There are many applications involving conics, and we will consider some in this chapter.

Conic Sections: Parabolas and Circles

13.1

Parabolas • The Distance and Midpoint Formulas • Circles

This section and the next two examine curves formed by cross sections of cones. These curves are graphs of second-degree equations in two variables. Some are shown below:

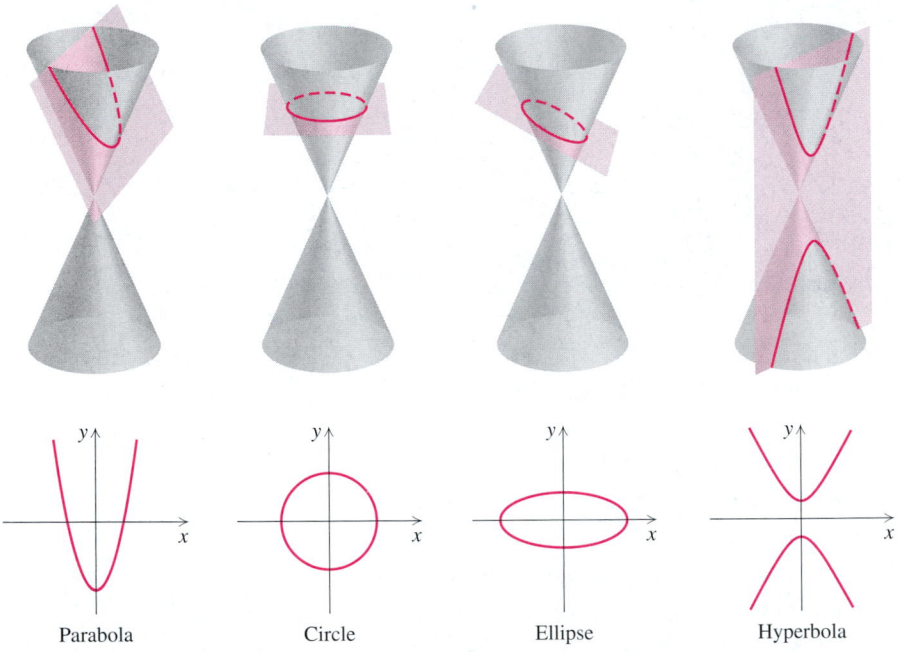

| Parabola | Circle | Ellipse | Hyperbola |

Parabolas

When a cone is cut as shown in the first figure above, the conic section formed is a **parabola**. Parabolas have many applications in electricity, mechanics, and optics. A cross section of a contact lens or satellite dish is a parabola, and arches that support certain bridges are parabolas.

> ### *Equation of a Parabola*
>
> A parabola with a vertical axis of symmetry opens upward or downward and has an equation that can be written in the form
>
> $$y = ax^2 + bx + c.$$
>
> A parabola with a horizontal axis of symmetry opens to the right or left and has an equation that can be written in the form
>
> $$x = ay^2 + by + c.$$

Parabolas with equations of the form $f(x) = ax^2 + bx + c$ were graphed in Chapter 11.

E x a m p l e   1

Graph: $y = x^2 - 4x + 9$.

Solution   To locate the vertex, we can use either of two approaches. One way is to complete the square:

$$y = (x^2 - 4x) + 9 \qquad \text{Note that half of } -4 \text{ is } -2, \text{ and } (-2)^2 = 4.$$
$$= (x^2 - 4x + 4 - 4) + 9 \qquad \text{Adding and subtracting 4}$$
$$= (x^2 - 4x + 4) + (-4 + 9) \qquad \text{Regrouping}$$
$$= (x - 2)^2 + 5. \qquad \text{Factoring and simplifying}$$

The vertex is $(2, 5)$.

A second way to find the vertex is to recall that the x-coordinate of the vertex of the parabola given by $y = ax^2 + bx + c$ is $-b/(2a)$:

$$x = -\frac{b}{2a} = -\frac{-4}{2(1)} = 2.$$

To find the y-coordinate of the vertex, we substitute 2 for x:

$$y = x^2 - 4x + 9 = 2^2 - 4(2) + 9 = 5.$$

Either way, the vertex is $(2, 5)$. Next, we calculate and plot some points on each side of the vertex. Since the x^2-coefficient, 1, is positive, the graph opens upward.

x	y	
2	5	← Vertex
0	9	← y-intercept
1	6	
3	6	
4	9	

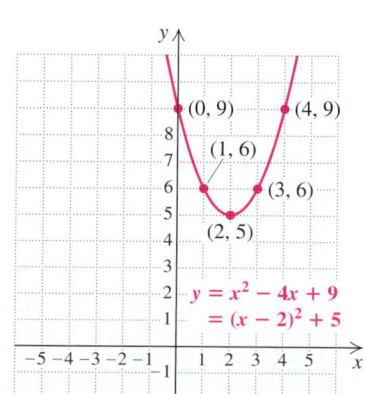

> ### To Graph an Equation of the Form $y = ax^2 + bx + c$
>
> 1. Find the vertex (h, k) either by completing the square to find an equivalent equation $y = a(x - h)^2 + k$, or by using $-b/(2a)$ to find the x-coordinate and substituting to find the y-coordinate.
> 2. Choose other values for x on each side of the vertex, and compute the corresponding y-values.
> 3. The graph opens upward for $a > 0$ and downward for $a < 0$.

Equations of the form $x = ay^2 + by + c$ represent horizontal parabolas. These parabolas open to the right for $a > 0$, open to the left for $a < 0$, and have axes of symmetry parallel to the x-axis.

E x a m p l e 2   Graph: $x = y^2 - 4y + 9$.

Solution   This equation is like that in Example 1 except that x and y are interchanged. The vertex is $(5, 2)$ instead of $(2, 5)$. To find ordered pairs, we choose values for y on each side of the vertex. Then we compute values for x. Note that the x- and y-values of the table in Example 1 are now switched. You should confirm that, by completing the square, we get $x = (y - 2)^2 + 5$.

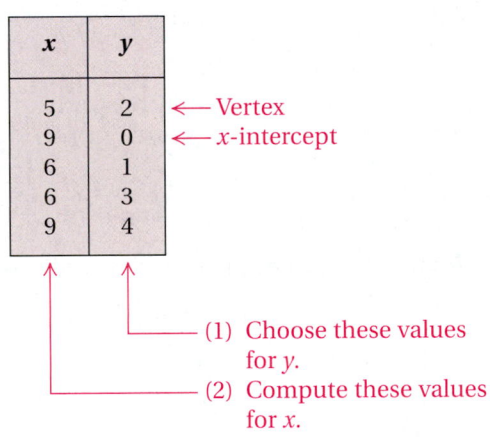

x	y	
5	2	← Vertex
9	0	← x-intercept
6	1	
6	3	
9	4	

(1) Choose these values for y.

(2) Compute these values for x.

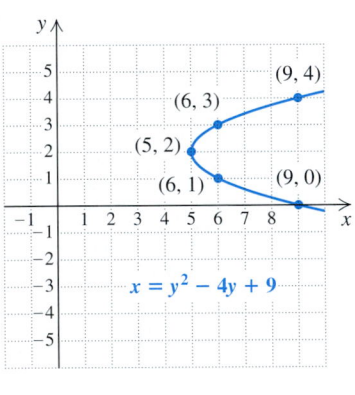

> ### To Graph an Equation of the Form $x = ay^2 + by + c$
>
> 1. Find the vertex (h, k) either by completing the square to find an equivalent equation
>
> $$x = a(y - k)^2 + h,$$
>
>   or by using $-b/(2a)$ to find the y-coordinate and substituting to find the x-coordinate.
> 2. Choose other values for y that are above and below the vertex, and compute the corresponding x-values.
> 3. The graph opens to the right if $a > 0$ and to the left if $a < 0$.

E x a m p l e   3

Graph: $x = -2y^2 + 10y - 7$.

Solution   We use the method of completing the square:

$$x = -2y^2 + 10y - 7$$
$$= -2(y^2 - 5y \qquad) - 7$$
$$= -2\left(y^2 - 5y + \tfrac{25}{4}\right) - 7 - (-2)\tfrac{25}{4} \qquad \tfrac{1}{2}(-5) = \tfrac{-5}{2}; \left(\tfrac{-5}{2}\right)^2 = \tfrac{25}{4}; \text{we add}$$
$$\text{and subtract } (-2)\tfrac{25}{4}.$$
$$= -2\left(y - \tfrac{5}{2}\right)^2 + \tfrac{11}{2}. \qquad \text{Factoring and simplifying}$$

The vertex is $\left(\tfrac{11}{2}, \tfrac{5}{2}\right)$.

For practice, we also find the vertex by first computing its y-coordinate, $-b/(2a)$, and then substituting to find the x-coordinate:

$$y = -\frac{b}{2a} = -\frac{10}{2(-2)} = \frac{5}{2}$$

$$x = -2y^2 + 10y - 7 = -2\left(\tfrac{5}{2}\right)^2 + 10\left(\tfrac{5}{2}\right) - 7$$
$$= \tfrac{11}{2}.$$

To find ordered pairs, we choose values for y on each side of the vertex and then compute values for x. A table is shown below, together with the graph. The graph opens to the left because the y^2-coefficient, -2, is negative.

x	y	
$\frac{11}{2}$	$\frac{5}{2}$	← Vertex
-7	0	← x-intercept
5	2	
5	3	
1	1	
1	4	
-7	5	

(1) Choose these values for y.

(2) Compute these values for x.

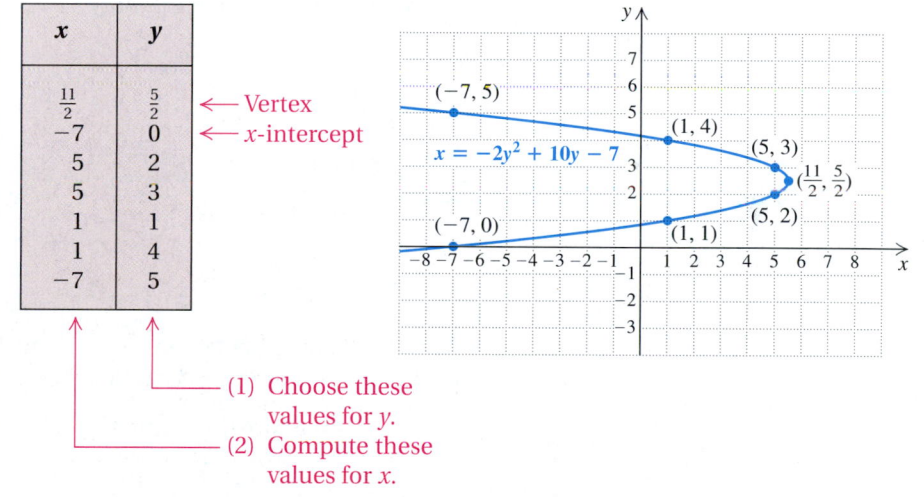

The Distance and Midpoint Formulas

Suppose that two points are on a horizontal line, and thus have the same second coordinate. We can find the distance between them by subtracting their first coordinates. This difference may be negative, depending on the order in which we subtract. So, to make sure we get a positive number, we take the absolute value of this difference. The distance between the points (x_1, y_1) and (x_2, y_1) on a horizontal line is thus $|x_2 - x_1|$. Similarly, the distance between the points (x_2, y_1) and (x_2, y_2) on a vertical line is $|y_2 - y_1|$.

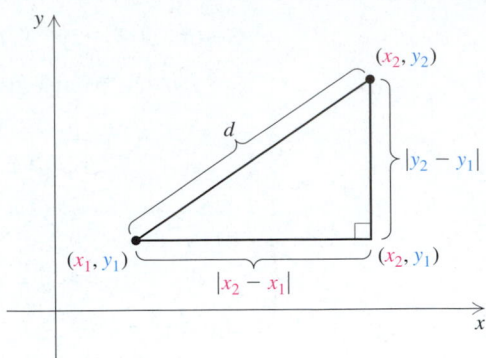

Now consider *any* two points (x_1, y_1) and (x_2, y_2). If $x_1 \neq x_2$ and $y_1 \neq y_2$, these points, along with the point (x_2, y_1), describe a right triangle. The lengths of the legs are $|x_2 - x_1|$ and $|y_2 - y_1|$. We find d, the length of the hypotenuse, by using the Pythagorean theorem:

$$d^2 = |x_2 - x_1|^2 + |y_2 - y_1|^2.$$

Since the square of a number is the same as the square of its opposite, we don't really need these absolute-value signs. Thus,

$$d^2 = (x_2 - x_1)^2 + (y_2 - y_1)^2.$$

Taking the principal square root, we obtain the distance between two points.

The Distance Formula

The distance d between any two points (x_1, y_1) and (x_2, y_2) is given by

$$d = \sqrt{(x_2 - x_1)^2 + (y_2 - y_1)^2}.$$

E x a m p l e   4

Find the distance between $(5, -1)$ and $(-4, 6)$. Find an exact answer and an approximation to three decimal places.

Solution   We substitute into the distance formula:

$$d = \sqrt{(-4 - 5)^2 + [6 - (-1)]^2} \qquad \text{Substituting}$$
$$= \sqrt{(-9)^2 + 7^2}$$
$$= \sqrt{130} \qquad \text{This is exact.}$$
$$\approx 11.402. \qquad \text{Using a calculator for an approximation}$$

The distance formula is needed to develop the formula for a circle, which follows, and to verify certain properties of conic sections. It is also needed to verify a formula for the coordinates of the *midpoint* of a segment connecting two points. We state the midpoint formula and leave its proof to the exercises. Note that although the distance formula involves both subtraction and addition, the midpoint formula uses only addition.

The Midpoint Formula

If the endpoints of a segment are (x_1, y_1) and (x_2, y_2), then the coordinates of the midpoint are

$$\left(\frac{x_1 + x_2}{2}, \frac{y_1 + y_2}{2} \right).$$

(To locate the midpoint, average the x-coordinates and average the y-coordinates.)

E x a m p l e 5   Find the midpoint of the segment with endpoints $(-2, 3)$ and $(4, -6)$.

Solution   Using the midpoint formula, we obtain

$$\left(\frac{-2 + 4}{2}, \frac{3 + (-6)}{2} \right), \quad \text{or} \quad \left(\frac{2}{2}, \frac{-3}{2} \right), \quad \text{or} \quad \left(1, -\frac{3}{2} \right).$$

Circles

One conic section, the **circle**, is a set of points in a plane that are a fixed distance r, called the **radius** (plural, **radii**), from a fixed point (h, k), called the **center**. Note that the word radius can mean either any segment connecting a point on a circle and the center of the circle or the length of such a segment. If a point (x, y) is on the circle, then by the definition of a circle and the distance formula, it follows that

$$r = \sqrt{(x - h)^2 + (y - k)^2}.$$

Squaring both sides gives the equation of a circle in standard form.

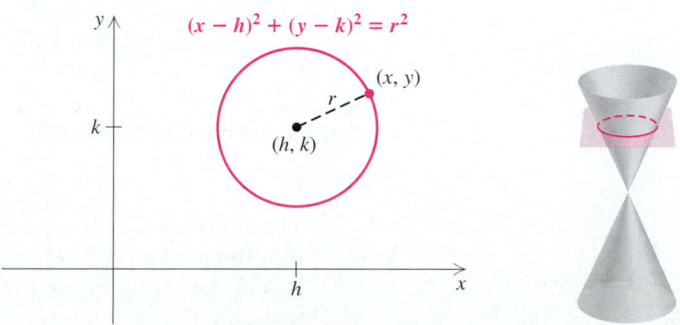

Equation of a Circle

The equation of a circle, centered at (h, k), with radius r, is given by

$$(x - h)^2 + (y - k)^2 = r^2.$$

Note that when $h = 0$ and $k = 0$, the circle is centered at the origin. Otherwise, the circle is translated $|h|$ units horizontally and $|k|$ units vertically.

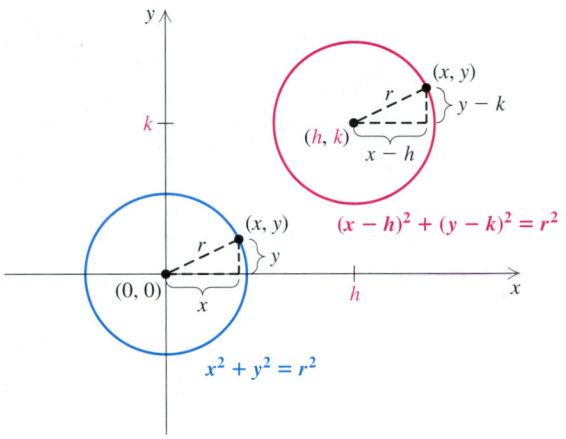

E x a m p l e 6

Find an equation of the circle having center $(4, 5)$ and radius 6.

Solution Using the standard form, we obtain

$$(x - 4)^2 + (y - 5)^2 = 6^2, \qquad \text{Using } (x - h)^2 + (y - k)^2 = r^2$$

or

$$(x - 4)^2 + (y - 5)^2 = 36.$$

E x a m p l e 7

Find the center and the radius and then graph each circle.

a) $(x - 2)^2 + (y + 3)^2 = 4^2$
b) $x^2 + y^2 + 8x - 2y + 15 = 0$

Solution

a) We write standard form:

$$(x - 2)^2 + [y - (-3)]^2 = 4^2.$$

The center is $(2, -3)$ and the radius is 4. To graph, we can use a compass or plot the points $(2, 1)$, $(2, -7)$, $(-2, -3)$, and $(6, -3)$, which are 4 units above, below, left, and right of $(2, -3)$, respectively, and then sketch a circle by hand.

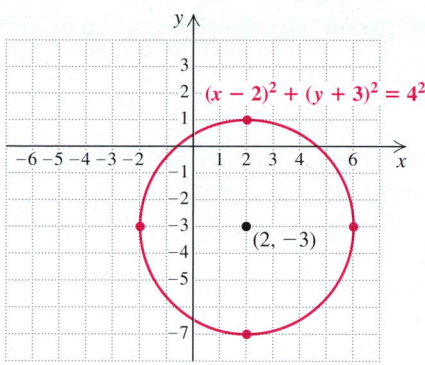

b) To write the equation $x^2 + y^2 + 8x - 2y + 15 = 0$ in standard form, we complete the square twice, once with $x^2 + 8x$ and once with $y^2 - 2y$:

$$x^2 + y^2 + 8x - 2y + 15 = 0$$

$$x^2 + 8x \qquad + y^2 - 2y \qquad = -15 \qquad \text{Grouping the } x\text{-terms and the } y\text{-terms; adding } -15 \text{ to both sides}$$

$$x^2 + 8x + 16 + y^2 - 2y + 1 = -15 + 16 + 1 \qquad \text{Adding } \left(\tfrac{8}{2}\right)^2, \text{ or 16, and } \left(-\tfrac{2}{2}\right)^2, \text{ or 1, to both sides}$$

$$(x + 4)^2 + (y - 1)^2 = 2 \qquad \text{Factoring}$$

$$[x - (-4)]^2 + (y - 1)^2 = \left(\sqrt{2}\right)^2. \qquad \text{Writing standard form}$$

The center is $(-4, 1)$ and the radius is $\sqrt{2}$.

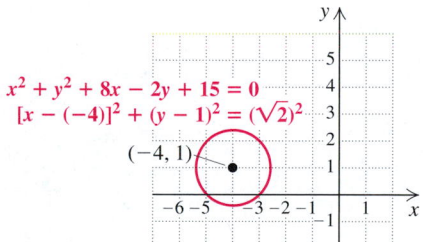

technology connection

Because most graphers can graph only functions, graphing the equation of a circle usually requires two steps:

1. Solve the equation for y. The result will include a $\pm$ sign in front of a radical.
2. Graph two functions, one for the $+$ sign and the other for the $-$ sign, on the same set of axes.

For example, to graph $(x - 3)^2 + (y + 1)^2 = 16$, solve for $y + 1$ and then y:

$$(y + 1)^2 = 16 - (x - 3)^2$$
$$y + 1 = \pm\sqrt{16 - (x - 3)^2}$$
$$y = -1 \pm \sqrt{16 - (x - 3)^2},$$
or
$$y_1 = -1 + \sqrt{16 - (x - 3)^2}$$
and
$$y_2 = -1 - \sqrt{16 - (x - 3)^2}.$$

When both functions are graphed (in a "squared" window to eliminate distortion), the result is as follows.

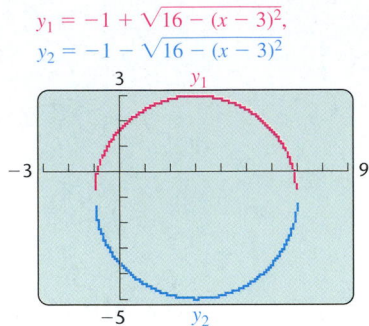

Circles can also be drawn using the CIRCLE option of the DRAW menu, but such graphs cannot be traced.

Use a grapher to graph each of the following equations.

1. $x^2 + y^2 - 16 = 0$
2. $4x^2 + 4y^2 = 100$
3. $x^2 + y^2 + 14x - 16y + 54 = 0$
4. $x^2 + y^2 - 10x - 11 = 0$

FOR EXTRA HELP

Exercise Set **13.1**

 Digital Video Tutor CD 10 Videotape 25   InterAct Math   Math Tutor Center   MathXL    MyMathLab.com

Graph. Be sure to label each vertex.

1. $y = -x^2$
2. $y = 2x^2$
3. $y = -x^2 + 4x - 5$
4. $x = 4 - 3y - y^2$
5. $x = y^2 - 4y + 1$
6. $y = x^2 + 2x + 3$
7. $x = y^2 + 1$
8. $x = 2y^2$
9. $x = -\frac{1}{2}y^2$
10. $x = y^2 - 1$
11. $x = -y^2 - 4y$
12. $x = y^2 + y - 6$
13. $x = 8 - y - y^2$
14. $y = x^2 + 2x + 1$
15. $y = x^2 - 2x + 1$
16. $y = -\frac{1}{2}x^2$
17. $x = -y^2 + 2y - 1$
18. $x = -y^2 - 2y + 3$
19. $x = -2y^2 - 4y + 1$
20. $x = 2y^2 + 4y - 1$

Find the distance between each pair of points. Where appropriate, find an approximation to three decimal places.

21. $(1, 6)$ and $(5, 9)$
22. $(1, 10)$ and $(7, 2)$
23. $(0, -7)$ and $(3, -4)$
24. $(6, 2)$ and $(6, -8)$
25. $(-4, 4)$ and $(6, -6)$
26. $(5, 21)$ and $(-3, 1)$
27. $(8.6, -3.4)$ and $(-9.2, -3.4)$
28. $(5.9, 2)$ and $(3.7, -7.7)$
29. $\left(\frac{5}{7}, \frac{1}{14}\right)$ and $\left(\frac{1}{7}, \frac{11}{14}\right)$

30. $\left(0, \sqrt{7}\right)$ and $\left(\sqrt{6}, 0\right)$

31. $\left(-\sqrt{6}, \sqrt{2}\right)$ and $(0, 0)$

32. $\left(\sqrt{5}, -\sqrt{3}\right)$ and $(0, 0)$

33. $\left(\sqrt{2}, -\sqrt{3}\right)$ and $\left(-\sqrt{7}, \sqrt{5}\right)$

34. $\left(\sqrt{8}, \sqrt{3}\right)$ and $\left(-\sqrt{5}, -\sqrt{6}\right)$

35. $(0, 0)$ and (s, t)

36. (p, q) and $(0, 0)$

Find the midpoint of each segment with the given endpoints.

37. $(-7, 6)$ and $(9, 2)$

38. $(6, 7)$ and $(7, -9)$

39. $(2, -1)$ and $(5, 8)$

40. $(-1, 2)$ and $(1, -3)$

41. $(-8, -5)$ and $(6, -1)$

42. $(8, -2)$ and $(-3, 4)$

43. $(-3.4, 8.1)$ and $(2.9, -8.7)$

44. $(4.1, 6.9)$ and $(5.2, -6.9)$

45. $\left(\frac{1}{6}, -\frac{3}{4}\right)$ and $\left(-\frac{1}{3}, \frac{5}{6}\right)$

46. $\left(-\frac{4}{5}, -\frac{2}{3}\right)$ and $\left(\frac{1}{8}, \frac{3}{4}\right)$

47. $\left(\sqrt{2}, -1\right)$ and $\left(\sqrt{3}, 4\right)$

48. $\left(9, 2\sqrt{3}\right)$ and $\left(-4, 5\sqrt{3}\right)$

Find an equation of the circle satisfying the given conditions.

49. Center $(0, 0)$, radius 6

50. Center $(0, 0)$, radius 5

51. Center $(7, 3)$, radius $\sqrt{5}$

52. Center $(5, 6)$, radius $\sqrt{2}$

53. Center $(-4, 3)$, radius $4\sqrt{3}$

54. Center $(-2, 7)$, radius $2\sqrt{5}$

55. Center $(-7, -2)$, radius $5\sqrt{2}$

56. Center $(-5, -8)$, radius $3\sqrt{2}$

57. Center $(0, 0)$, passing through $(-3, 4)$

58. Center $(3, -2)$, passing through $(11, -2)$

59. Center $(-4, 1)$, passing through $(-2, 5)$

60. Center $(-1, -3)$, passing through $(-4, 2)$

Find the center and the radius of each circle. Then graph the circle.

61. $x^2 + y^2 = 49$ **62.** $x^2 + y^2 = 36$

63. $(x + 1)^2 + (y + 3)^2 = 4$

64. $(x - 2)^2 + (y + 3)^2 = 1$

65. $(x - 4)^2 + (y + 3)^2 = 10$

66. $(x + 5)^2 + (y - 1)^2 = 15$

67. $x^2 + y^2 = 7$

68. $x^2 + y^2 = 8$

69. $(x - 5)^2 + y^2 = \frac{1}{4}$

70. $x^2 + (y - 1)^2 = \frac{1}{25}$

71. $x^2 + y^2 + 8x - 6y - 15 = 0$

72. $x^2 + y^2 + 6x - 4y - 15 = 0$

73. $x^2 + y^2 - 8x + 2y + 13 = 0$

74. $x^2 + y^2 + 6x + 4y + 12 = 0$

75. $x^2 + y^2 + 10y - 75 = 0$

76. $x^2 + y^2 - 8x - 84 = 0$

77. $x^2 + y^2 + 7x - 3y - 10 = 0$

78. $x^2 + y^2 - 21x - 33y + 17 = 0$

79. $36x^2 + 36y^2 = 1$

80. $4x^2 + 4y^2 = 1$

81. Describe a procedure that would use the distance formula to determine whether three points, (x_1, y_1), (x_2, y_2), and (x_3, y_3), are vertices of a right triangle.

82. Does the graph of an equation of a circle include the point that is the center? Why or why not?

SKILL MAINTENANCE

Solve.

83. $\dfrac{x}{4} + \dfrac{5}{6} = \dfrac{2}{3}$ **84.** $\dfrac{t}{6} - \dfrac{1}{9} = \dfrac{7}{12}$

85. A rectangle 10 in. long and 6 in. wide is bordered by a strip of uniform width. If the perimeter of the larger rectangle is twice that of the smaller rectangle, what is the width of the border?

86. One airplane flies 60 mph faster than another. To fly a certain distance, the faster plane takes 4 hr and the slower plane takes 4 hr and 24 min. What is the distance?

Solve each system.

87. $3x - 8y = 5,$
$2x + 6y = 5$

88. $4x - 5y = 9,$
$12x - 10y = 18$

SYNTHESIS

89. Outline a procedure that would use the distance formula to determine whether three points, (x_1, y_1), (x_2, y_2), and (x_3, y_3), are collinear (lie on the same line).

90. Why does the discussion of the distance formula precede the discussion of circles?

Find an equation of a circle satisfying the given conditions.

91. Center $(3, -5)$ and tangent to (touching at one point) the y-axis

92. Center $(-7, -4)$ and tangent to the x-axis

93. The endpoints of a diameter are $(7, 3)$ and $(-1, -3)$.

94. Center $(-3, 5)$ with a circumference of 8π units

95. Find the point on the y-axis that is equidistant from $(2, 10)$ and $(6, 2)$.

96. Find the point on the x-axis that is equidistant from $(-1, 3)$ and $(-8, -4)$.

97. *Snowboarding.* Each side edge of the Burton® Twin 53 snowboard is an arc of a circle with a "running length" of 1150 mm and a "sidecut depth" of 19.5 mm (see the figure below).

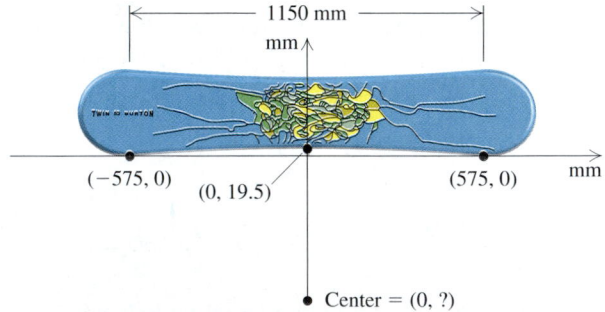

a) Using the coordinates shown, locate the center of the circle. (*Hint*: Equate distances.)
b) What radius is used for the edge of the board?

98. *Snowboarding.* The Burton® Twin 44 snowboard has a running length of 1070 mm and a sidecut depth of 17.5 mm (see Exercise 97). What radius is used for the edge of this snowboard?

99. *Skiing.* The Rossignol® Cut 10.4 ski, when lying flat and viewed from above, has edges that are arcs of a circle. (Actually, each edge is made of two arcs of slightly different radii. The arc for the rear half of the ski edge has a slightly larger radius.)

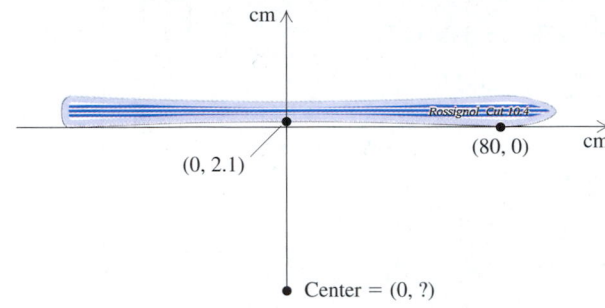

a) Using the coordinates shown, locate the center of the circle. (*Hint*: Equate distances.)
b) What radius is used for the arc passing through $(0, 2.1)$ and $(80, 0)$?

100. *Doorway construction.* Ace Carpentry needs to cut an arch for the top of an entranceway. The arch needs to be 8 ft wide and 2 ft high. To draw the arch, the carpenters will use a stretched string with chalk attached at an end as a compass.

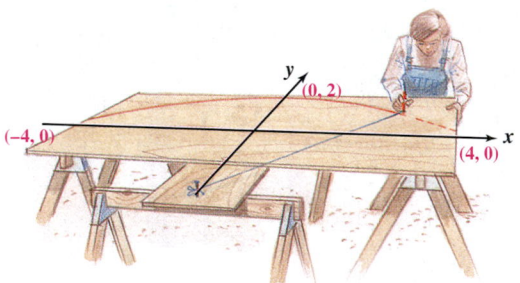

a) Using a coordinate system, locate the center of the circle.
b) What radius should the carpenters use to draw the arch?

101. *Archaeology.* During an archaeological dig, Martina finds the bowl fragment shown below. What was the original diameter of the bowl?

102. *Ferris wheel design.* A ferris wheel has a radius of 24.3 ft. Assuming that the center is 30.6 ft off the ground and that the origin is below the center, as in the following figure, find an equation of the circle.

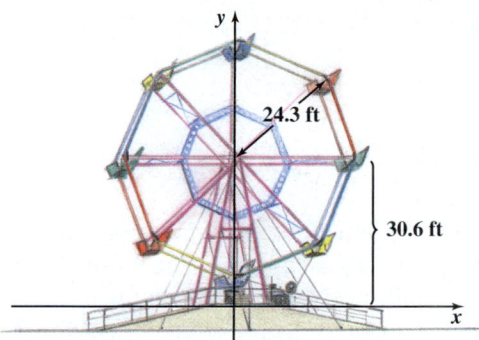

103. Use a graph of the equation $x = y^2 - y - 6$ to approximate to the nearest tenth the solutions of each of the following equations.

a) $y^2 - y - 6 = 2$ (*Hint*: Graph $x = 2$ on the same set of axes as the graph of $x = y^2 - y - 6$.)

b) $y^2 - y - 6 = -3$

104. *Power of a motor.* The horsepower of a certain kind of engine is given by the formula

$$H = \frac{D^2 N}{2.5},$$

where N is the number of cylinders and D is the diameter, in inches, of each piston. Graph this equation, assuming that $N = 6$ (a six-cylinder engine). Let D run from 2.5 to 8.

105. Prove the midpoint formula by showing that

i) the distance from (x_1, y_1) to
$$\left(\frac{x_1 + x_2}{2}, \frac{y_1 + y_2}{2} \right)$$
equals the distance from (x_2, y_2) to
$$\left(\frac{x_1 + x_2}{2}, \frac{y_1 + y_2}{2} \right);$$
and

ii) the points
$$(x_1, y_1), \left(\frac{x_1 + x_2}{2}, \frac{y_1 + y_2}{2} \right),$$
and
$$(x_2, y_2)$$
lie on the same line (see Exercise 89).

106. If the equation $x^2 + y^2 - 6x + 2y - 6 = 0$ is written as $y^2 + 2y + (x^2 - 6x - 6) = 0$, it can be regarded as quadratic in y.

a) Use the quadratic formula to solve for y.

b) Show that the graph of your answer to part (a) coincides with the graph in the Technology Connection on p. 816.

107. How could a grapher best be used to help you sketch the graph of an equation of the form $x = ay^2 + by + c$?

108. Why should a grapher's window be "squared" before graphing a circle?

Conic Sections: Ellipses

13.2

Ellipses Centered at $(0, 0)$ • Ellipses Centered at (h, k)

When a cone is cut at an angle, as shown on the following page, the conic section formed is an *ellipse*. To draw an ellipse, stick two tacks in a piece of cardboard. Then tie a string to the tacks, place a pencil as shown, and draw an oval by moving the pencil while keeping the string taut.

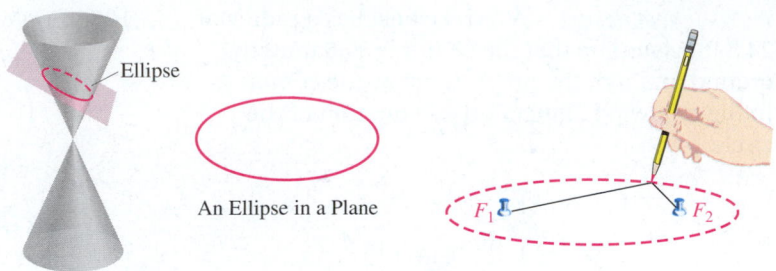

An Ellipse in a Plane

Ellipses Centered at (0, 0)

An **ellipse** is defined as the set of all points in a plane for which the *sum* of the distances from two fixed points F_1 and F_2 is constant. The points F_1 and F_2 are called **foci** (pronounced fō-sī), the plural of focus. In the figure above, the tacks are at the foci and the length of the string is the constant sum of the distances. The midpoint of the segment F_1F_2 is the **center**. The equation of an ellipse is as follows. Its derivation is left to the exercises.

> ### Equation of an Ellipse Centered at the Origin
>
> The equation of an ellipse centered at the origin and symmetric with respect to both axes is
>
> $$\frac{x^2}{a^2} + \frac{y^2}{b^2} = 1, \quad a, b > 0. \qquad \text{(Standard form)}$$

To graph an ellipse centered at the origin, it helps to first find the intercepts. If we replace x with 0, we can find the y-intercepts:

$$\frac{0^2}{a^2} + \frac{y^2}{b^2} = 1$$

$$\frac{y^2}{b^2} = 1$$

$$y^2 = b^2 \quad \text{or} \quad y = \pm b.$$

Thus the y-intercepts are $(0, b)$ and $(0, -b)$. Similarly, the x-intercepts are $(a, 0)$ and $(-a, 0)$. If $a > b$, the ellipse is said to be horizontal and $(-a, 0)$ and $(a, 0)$ are referred to as the **vertices** (singular, **vertex**). If $b > a$, the ellipse is said to be vertical and $(0, -b)$ and $(0, b)$ are then the vertices.

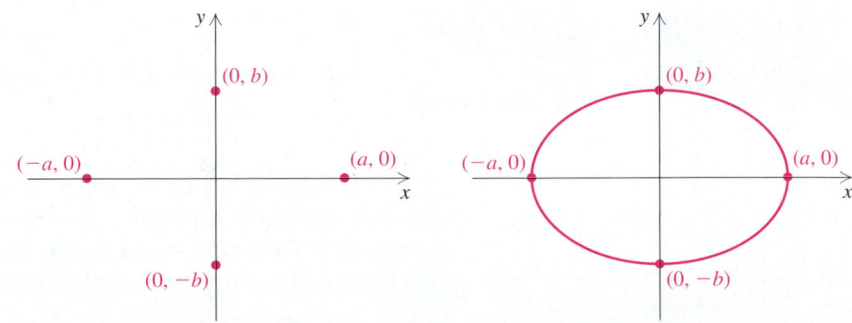

Plotting these four points and drawing an oval-shaped curve, we get a graph of the ellipse. If a more precise graph is desired, we can plot more points.

Using a and b to Graph an Ellipse

For the ellipse

$$\frac{x^2}{a^2} + \frac{y^2}{b^2} = 1,$$

the x-intercepts are $(-a, 0)$ and $(a, 0)$. The y-intercepts are $(0, -b)$ and $(0, b)$.

E x a m p l e   1   Graph the ellipse

$$\frac{x^2}{4} + \frac{y^2}{9} = 1.$$

Solution   Note that

$$\frac{x^2}{4} + \frac{y^2}{9} = \frac{x^2}{2^2} + \frac{y^2}{3^2}.$$ Identifying a and b. Since $b > a$, the ellipse is vertical.

Thus the x-intercepts are $(-2, 0)$ and $(2, 0)$, and the y-intercepts are $(0, -3)$ and $(0, 3)$. We plot these points and connect them with an oval-shaped curve. To plot some other points, we let $x = 1$ and solve for y:

$$\frac{1^2}{4} + \frac{y^2}{9} = 1$$

$$36\left(\frac{1}{4} + \frac{y^2}{9}\right) = 36 \cdot 1$$

$$36 \cdot \frac{1}{4} + 36 \cdot \frac{y^2}{9} = 36$$

$$9 + 4y^2 = 36$$

$$4y^2 = 27$$

$$y^2 = \frac{27}{4}$$

$$y = \pm\sqrt{\frac{27}{4}}$$

$$y \approx \pm 2.6.$$

Thus, $(1, 2.6)$ and $(1, -2.6)$ can also be used to draw the graph. Similarly, the points $(-1, 2.6)$ and $(-1, -2.6)$ can also be computed and plotted.

Example 2

technology connection

Graphing an ellipse on a grapher is much like graphing a circle: We graph it in two pieces after solving for y. To illustrate, let's check Example 2:

$$4x^2 + 25y^2 = 100$$
$$25y^2 = 100 - 4x^2$$
$$y^2 = 4 - \frac{4}{25}x^2$$
$$y = \pm\sqrt{4 - \frac{4}{25}x^2}.$$

Using a squared window, we have our check:

$$y_1 = -\sqrt{4 - \frac{4}{25}x^2}, \quad y_2 = \sqrt{4 - \frac{4}{25}x^2}$$

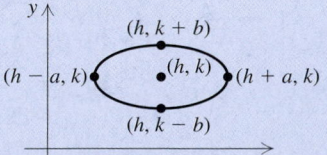

Graph: $4x^2 + 25y^2 = 100$.

Solution   To write the equation in standard form, we divide both sides by 100 to get 1 on the right side:

$$\frac{4x^2 + 25y^2}{100} = \frac{100}{100} \qquad \text{\textcolor{red}{Dividing by 100 to get 1 on the right side}}$$

$$\left.\begin{array}{c}\dfrac{4x^2}{100} + \dfrac{25y^2}{100} = 1 \\[2mm] \dfrac{x^2}{25} + \dfrac{y^2}{4} = 1\end{array}\right\} \qquad \text{\textcolor{red}{Simplifying}}$$

$$\frac{x^2}{5^2} + \frac{y^2}{2^2} = 1. \qquad \textcolor{red}{a = 5, b = 2}$$

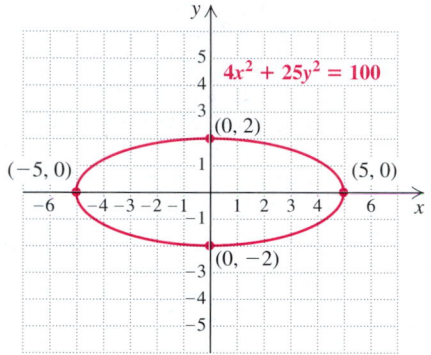

The x-intercepts are $(-5, 0)$ and $(5, 0)$, and the y-intercepts are $(0, -2)$ and $(0, 2)$. We plot the intercepts and connect them with an oval-shaped curve. Other points can also be computed and plotted.

Ellipses Centered at (h, k)

Horizontal and vertical translations, similar to those used in Chapter 11, can be used to graph ellipses that are not centered at the origin.

> ### Equation of an Ellipse Centered at (h, k)
>
> The standard form of a horizontal or vertical ellipse centered at (h, k) is
>
> $$\frac{(x - h)^2}{a^2} + \frac{(y - k)^2}{b^2} = 1.$$
>
> The vertices are $(h + a, k)$ and $(h - a, k)$ if horizontal; $(h, k + b)$ and $(h, k - b)$ if vertical.

Example 3   Graph the ellipse

$$\frac{(x-1)^2}{4} + \frac{(y+5)^2}{9} = 1.$$

Solution   Note that

$$\frac{(x-1)^2}{4} + \frac{(y+5)^2}{9} = \frac{(x-1)^2}{2^2} + \frac{(y+5)^2}{3^2}.$$

Thus, $a = 2$ and $b = 3$. To determine the center of the ellipse, (h, k), note that

$$\frac{(x-1)^2}{2^2} + \frac{(y+5)^2}{3^2} = \frac{(x-1)^2}{2^2} + \frac{(y-(-5))^2}{3^2}.$$

Thus the center is $(1, -5)$. We plot the points 2 units to the left and right of center, as well as the points 3 units above and below center. These are the points $(-1, -5)$, $(-3, -5)$, $(1, -2)$, and $(1, -8)$.

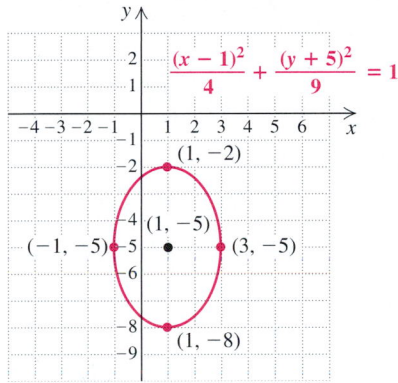

Note that this ellipse is the same as the ellipse in Example 1 but translated 1 unit to the right and 5 units down.

Ellipses have many applications. Communications satellites move in elliptical orbits with the earth as a focus while the earth itself follows an elliptical path around the sun. A medical instrument, the lithotripter, uses shock waves originating at one focus to crush a kidney stone located at the other focus.

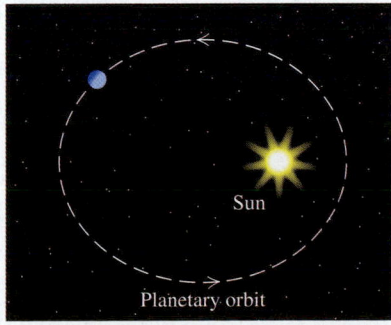

Planetary orbit

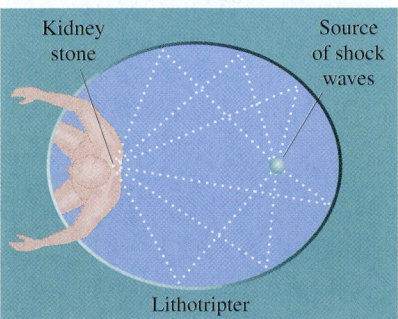

Kidney stone   Source of shock waves

Lithotripter

In some buildings, an ellipsoidal ceiling creates a "whispering gallery" in which a person at one focus can whisper and still be heard clearly at the other focus. This happens because sound waves coming from one focus are all reflected to the other focus. Similarly, light waves bouncing off an ellipsoidal mirror are used in a dentist's or surgeon's reflector light. The light source is located at one focus while the patient's mouth is at the other.

FOR EXTRA HELP

Exercise Set **13.2**

 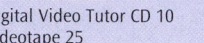
Digital Video Tutor CD 10
Videotape 25

InterAct Math

Math Tutor Center

MathXL

MyMathLab.com

Graph each of the following equations.

1. $\dfrac{x^2}{1} + \dfrac{y^2}{4} = 1$

2. $\dfrac{x^2}{4} + \dfrac{y^2}{1} = 1$

3. $\dfrac{x^2}{25} + \dfrac{y^2}{9} = 1$

4. $\dfrac{x^2}{16} + \dfrac{y^2}{25} = 1$

5. $4x^2 + 9y^2 = 36$

6. $9x^2 + 4y^2 = 36$

7. $16x^2 + 9y^2 = 144$

8. $9x^2 + 16y^2 = 144$

9. $2x^2 + 3y^2 = 6$

10. $5x^2 + 7y^2 = 35$

Aha! **11.** $5x^2 + 5y^2 = 125$

12. $8x^2 + 5y^2 = 80$

13. $3x^2 + 7y^2 - 63 = 0$

14. $3x^2 + 8y^2 - 72 = 0$

15. $8x^2 = 96 - 3y^2$

16. $6y^2 = 24 - 8x^2$

17. $16x^2 + 25y^2 = 1$

18. $9x^2 + 4y^2 = 1$

19. $\dfrac{(x-2)^2}{9} + \dfrac{(y-1)^2}{25} = 1$

20. $\dfrac{(x-3)^2}{25} + \dfrac{(y-4)^2}{9} = 1$

21. $\dfrac{(x+4)^2}{16} + \dfrac{(y-3)^2}{49} = 1$

22. $\dfrac{(x+5)^2}{4} + \dfrac{(y-2)^2}{36} = 1$

23. $12(x-1)^2 + 3(y+4)^2 = 48$
(*Hint:* Divide both sides by 48.)

24. $4(x-5)^2 + 9(y+2)^2 = 36$

Aha! **25.** $4(x+3)^2 + 4(y+1)^2 - 10 = 90$

26. $9(x+6)^2 + (y+2)^2 - 20 = 61$

27. Is the center of an ellipse part of the ellipse itself? Why or why not?

28. Can an ellipse ever be the graph of a function? Why or why not?

SKILL MAINTENANCE

Solve.

29. $\dfrac{3}{x-2} - \dfrac{5}{x-2} = 9$

30. $\dfrac{7}{x+3} - \dfrac{2}{x+3} = 8$

31. $\dfrac{x}{x-4} - \dfrac{3}{x-5} = \dfrac{2}{x-4}$

32. $\dfrac{7}{x-3} - \dfrac{x}{x-2} = \dfrac{4}{x-2}$

33. $9 - \sqrt{2x+1} = 7$

34. $5 - \sqrt{x+3} = 9$

SYNTHESIS

35. An eccentric person builds a pool table in the shape of an ellipse with a hole at one focus and a tiny dot at the other. Guests are amazed at how many bank shots the owner of the pool table makes. Explain why this occurs.

36. Can a circle be considered a special type of ellipse? Why or why not?

Find an equation of an ellipse that contains the following points.

37. $(-9, 0), (9, 0), (0, -11),$ and $(0, 11)$

38. $(-7, 0), (7, 0), (0, -5),$ and $(0, 5)$

39. $(-2, -1), (6, -1), (2, -4),$ and $(2, 2)$

40. $(-6, 3), (4, 3), (-1, 7),$ and $(-1, -1)$

41. *Astronomy.* The maximum distance of the planet Mars from the sun is 2.48×10^8 mi. The minimum distance is 3.46×10^7 mi. The sun is at one focus of the elliptical orbit. Find the distance from the sun to the other focus.

42. Let $(-c, 0)$ and $(c, 0)$ be the foci of an ellipse. Any point $P(x, y)$ is on the ellipse if the sum of the distances from the foci to P is some constant. Use $2a$ to represent this constant.

   a) Show that an equation for the ellipse is given by

   $$\frac{x^2}{a^2} + \frac{y^2}{a^2 - c^2} = 1.$$

   b) Substitute b^2 for $a^2 - c^2$ to get standard form.

43. *President's office.* The Oval Office of the President of the United States is an ellipse 31 ft wide and 38 ft long. Show in a sketch precisely where the President and an adviser could sit to best hear each other using the room's acoustics. (*Hint:* See Exercise 42(b) and the discussion following Example 3.)

44. *Dentistry.* The light source in a dental lamp shines against a reflector that is shaped like a portion of an ellipse in which the light source is one focus of the ellipse. Reflected light enters a patient's mouth at the other focus of the ellipse. If the ellipse from which the reflector was formed is 2 ft wide and 6 ft long, how far should the patient's mouth be from the light source? (*Hint:* See Exercise 42(b).)

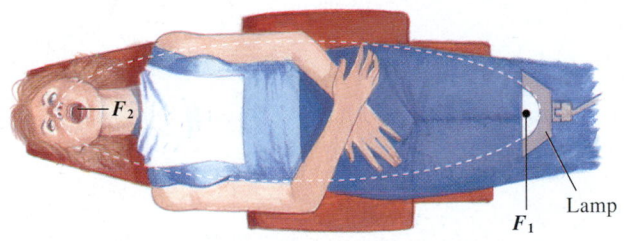

45. *Firefighting.* The size and shape of certain forest fires can be approximated as the union of two "half-ellipses." For the blaze modeled below, the equation of the smaller ellipse—the part of the fire moving *into* the wind—is

$$\frac{x^2}{40,000} + \frac{y^2}{10,000} = 1.$$

The equation of the other ellipse—the part moving *with* the wind—is

$$\frac{x^2}{250,000} + \frac{y^2}{10,000} = 1.$$

(*Source for figure*: "Predicting Wind-Driven Wild Land Fire Size and Shape," Hal E. Anderson, Research Paper INT-305, U.S. Department of Agriculture, Forest Service, February 1983).

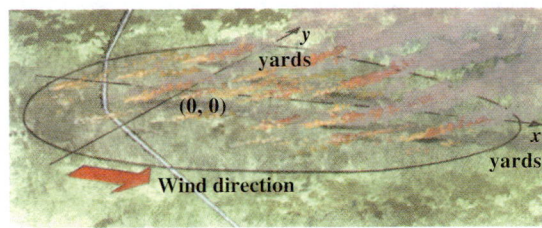

Determine the width and the length of the fire.

For each of the following equations, complete the square as needed and find an equivalent equation in standard form. Then graph the ellipse.

46. $x^2 - 4x + 4y^2 + 8y - 8 = 0$

47. $4x^2 + 24x + y^2 - 2y - 63 = 0$

48. Use a grapher to check your answers to Exercises 3, 17, 21, and 25.

COLLABORATIVE

CORNER

A Cosmic Path

Focus: Ellipses

Time: 20–30 minutes

Group size: 2

Materials: Scientific calculators

In March 1996, the comet Hyakutake came within 21 million mi of the sun, and closer to Earth than any comet in over 500 yr (*Source*: Associated Press newspaper story, 3/20/96). Hyakutake is traveling in an elliptical orbit with the sun at one focus. The comet's average speed is about 100,000 mph (it actually goes much faster near its foci and slower as it gets further from the foci) and one orbit takes about 15,000 yr. (Astronomers estimate the time at 10,000–20,000 yr.)

ACTIVITY

1. The elliptical orbit of Hyakutake is so elongated that the distance traveled in one orbit can be estimated by $4a$ (see the following figure). Use the information above to estimate the distance, in millions of miles, traveled in one orbit. Then determine a.

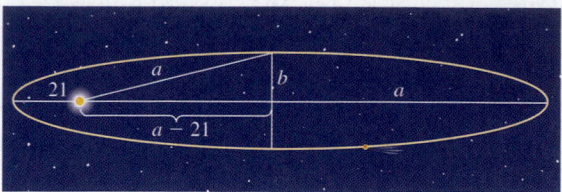

Units are millions of miles.

2. Using the figure above, express b^2 as a function of a. Then solve for b using the value found for a in part (1).

3. Approximately how far will Hyakutake be from the sun at the most distant part of its orbit?

4. Repeat parts (1)–(3), with one group member using the lower estimate of orbit time (10,000 yr) and the other using the upper estimate of orbit time (20,000 yr). By how much do the three answers to part (3) vary?

Conic Sections: Hyperbolas

13.3

Hyperbolas • Hyperbolas (Nonstandard Form) •
Classifying Graphs of Equations

Hyperbolas

A **hyperbola** looks like a pair of parabolas, but the shapes are actually different. A hyperbola has two **vertices** and the line through the vertices is known as an **axis**. The point halfway between the vertices is called the **center**. The two curves that comprise a hyperbola are called **branches**.

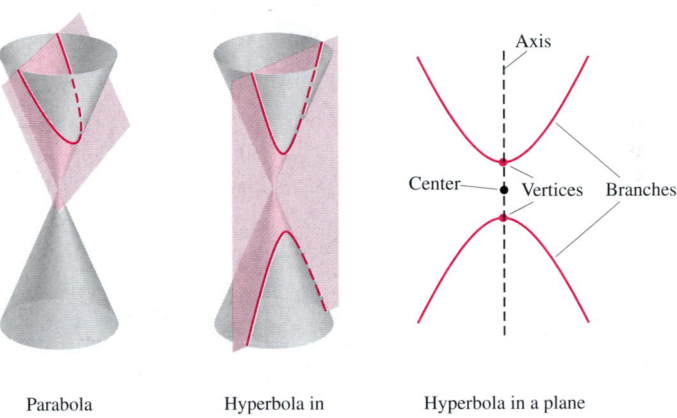

Parabola

Hyperbola in
three dimensions

Hyperbola in a plane

Equation of a Hyperbola Centered at the Origin

Hyperbolas with their centers at the origin* have equations as follows:

$$\frac{x^2}{a^2} - \frac{y^2}{b^2} = 1 \qquad \text{(Axis horizontal);}$$

$$\frac{y^2}{b^2} - \frac{x^2}{a^2} = 1 \qquad \text{(Axis vertical).}$$

Note that both equations have a 1 on the right-hand side and a subtraction symbol between the terms. For the discussion that follows, we assume $a, b > 0$.

To graph a hyperbola, it helps to begin by graphing two lines called **asymptotes**. Although the asymptotes themselves are not part of the graph, they serve as guidelines for an accurate sketch.

---

*Hyperbolas with horizontal or vertical axes and centers *not* at the origin are discussed in Exercises 51–56.

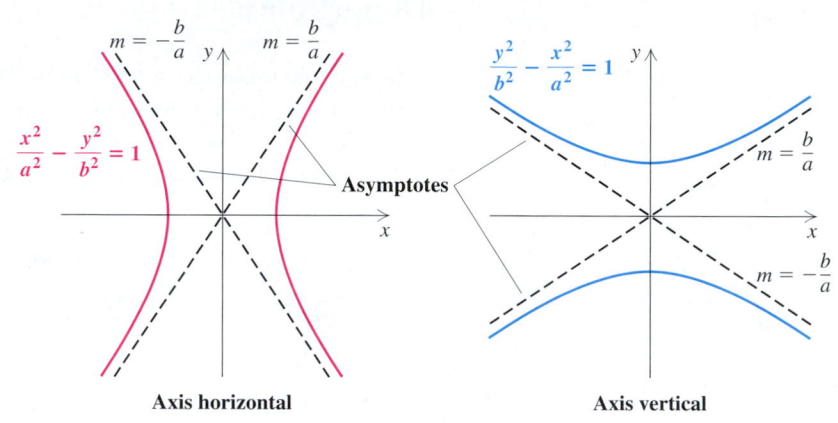

Asymptotes of a Hyperbola

For hyperbolas with equations as shown below, the asymptotes are the lines

$$y = \frac{b}{a}x \quad \text{and} \quad y = -\frac{b}{a}x.$$

Axis horizontal

Axis vertical

As a hyperbola gets farther away from the origin, it gets closer and closer to its asymptotes. The larger $|x|$ gets, the closer the graph gets to an asymptote. The asymptotes act to "constrain" the graph of a hyperbola. Parabolas are *not* constrained by any asymptotes.

In Section 13.2, we found that a and b can be used to determine the width and the length of an ellipse. For hyperbolas, a and b can be used to determine the base and the height of a rectangle that can be used as an aid in sketching asymptotes and locating vertices. This is illustrated in the following example.

E x a m p l e 1

Graph: $\dfrac{x^2}{4} - \dfrac{y^2}{9} = 1.$

Solution Note that

$$\frac{x^2}{4} - \frac{y^2}{9} = \frac{x^2}{2^2} - \frac{y^2}{3^2}, \qquad \text{\textcolor{magenta}{Identifying } } a \text{ \textcolor{magenta}{and} } b$$

so $a = 2$ and $b = 3$. The asymptotes are thus

$$y = \frac{3}{2}x \quad \text{and} \quad y = -\frac{3}{2}x.$$

To help us sketch asymptotes and locate vertices, we use a and b—in this case, 2 and 3—to form the pairs $(-2, 3)$, $(2, 3)$, $(2, -3)$, and $(-2, -3)$. We plot these pairs and lightly sketch a rectangle. The asymptotes pass through the corners and, since this is a horizontal hyperbola, the vertices are where the rectangle intersects the x-axis. Finally, we draw the hyperbola, as shown on the next page.

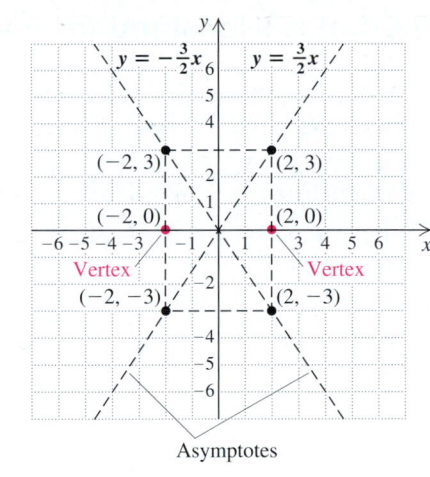

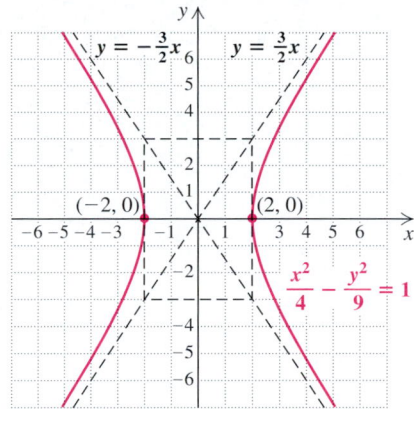

Example 2   Graph: $\dfrac{y^2}{36} - \dfrac{x^2}{4} = 1$.

Solution   Note that

$$\dfrac{y^2}{36} - \dfrac{x^2}{4} = \dfrac{y^2}{6^2} - \dfrac{x^2}{2^2} = 1.$$

> Whether the hyperbola is horizontal or vertical is determined by the nonnegative term. Here there is a y in this term, so the hyperbola is vertical.

Using ± 2 as x-coordinates and ± 6 as y-coordinates, we plot $(2, 6)$, $(2, -6)$, $(-2, 6)$, and $(-2, -6)$, and lightly sketch a rectangle through them. The asymptotes pass through the corners (see the figure on the left below). Since the hyperbola is vertical, its vertices are $(0, 6)$ and $(0, -6)$. Finally, we draw curves through the vertices toward the asymptotes, as shown below.

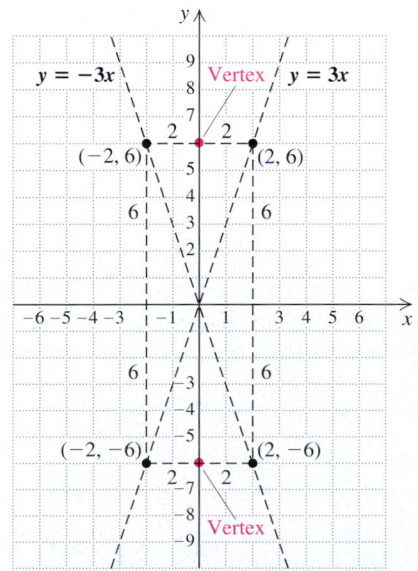

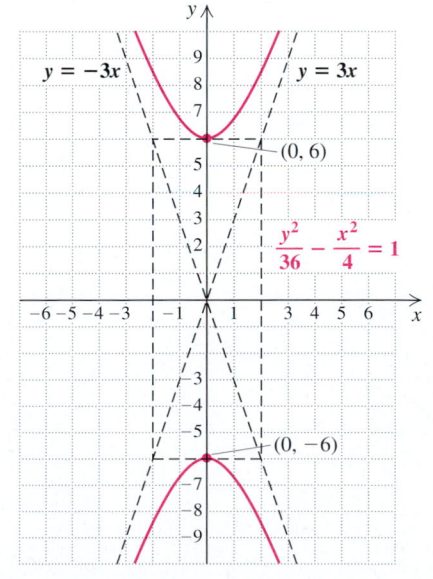

Hyperbolas (Nonstandard Form)

The equations for hyperbolas just examined are the standard ones, but there are other hyperbolas. We consider some of them.

> ### Equation of a Hyperbola in Nonstandard Form
>
> Hyperbolas having the x- and y-axes as asymptotes have equations as follows:
>
> $$xy = c, \quad \text{where } c \text{ is a nonzero constant.}$$

E x a m p l e 3

Graph: $xy = -8$.

Solution We first solve for y:

$$y = -\frac{8}{x}. \qquad \text{Dividing both sides by } x. \text{ Note that } x \neq 0.$$

Next, we find some solutions, keeping the results in a table. Note that x cannot be 0 and that for large values of $|x|$, y will be close to 0. Thus the x- and y-axes serve as asymptotes. We plot the points and draw two curves.

x	y
2	−4
−2	4
4	−2
−4	2
1	−8
−1	8
8	−1
−8	1

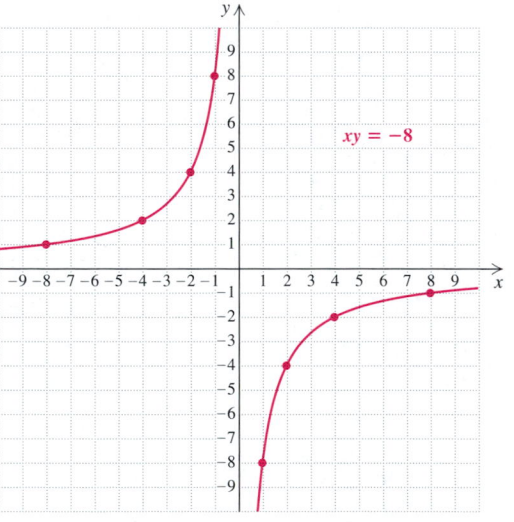

Hyperbolas have many applications. A jet breaking the sound barrier creates a sonic boom with a wave front the shape of a cone. The intersection of the cone with the ground is one branch of a hyperbola. Some comets travel in hyperbolic orbits, and a cross section of many lenses is hyperbolic in shape.

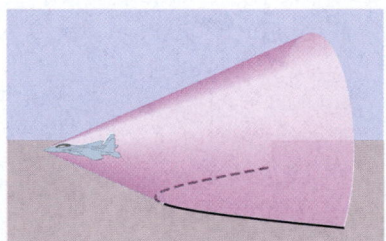

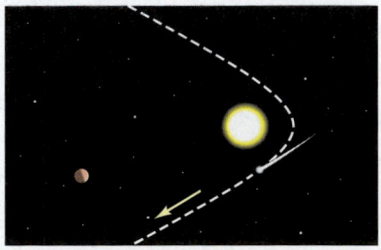

Classifying Graphs of Equations

CONNECTING THE CONCEPTS

Recall that the vertical-line test tells us that circles, ellipses, and hyperbolas in standard form do not represent functions. Of the graphs examined in this chapter, only vertical parabolas and hyperbolas similar to the one in Example 3 can represent functions. Because functions are so important, circles and ellipses generally appear in applications of a purely geometric nature, whereas vertical parabolas

and nonstandard hyperbolas can be used in applications involving geometry or functions.

In Section 13.4, we return to the challenge of solving real-world problems that translate to a system of equations. There we will find that knowing the general shape of the graph of an equation can help us determine how many solutions, if any, may exist.

---

technology connection

The procedure used to graph a hyperbola in standard form on a grapher is similar to that used to draw a circle or ellipse. Consider the graph of the hyperbola given by the equation.

$$\frac{x^2}{25} - \frac{y^2}{49} = 1.$$

The student should confirm that solving for y yields

$$y_1 = \frac{\sqrt{49x^2 - 1225}}{5}$$

$$= \frac{7}{5}\sqrt{x^2 - 25}$$

and $$y_2 = \frac{-\sqrt{49x^2 - 1225}}{5}$$

$$= -\frac{7}{5}\sqrt{x^2 - 25},$$

or $$y_2 = -y_1.$$

When the two pieces are drawn on the same squared window, the result is as shown. Note the problem that the grapher has at points where the graph is nearly vertical.

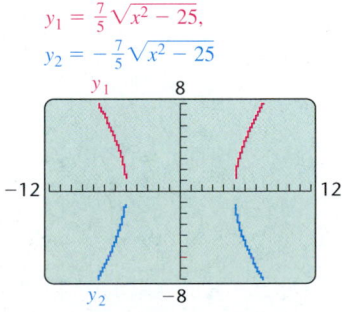

$$y_1 = \frac{7}{5}\sqrt{x^2 - 25},$$
$$y_2 = -\frac{7}{5}\sqrt{x^2 - 25}$$

Use a grapher to draw the graph of each hyperbola. Use a squared window so that the shapes are not distorted.

1. $\dfrac{x^2}{16} - \dfrac{y^2}{60} = 1$ 2. $16x^2 - 3y^2 = 64$

3. $\dfrac{y^2}{20} - \dfrac{x^2}{64} = 1$ 4. $45y^2 - 9x^2 = 441$

We summarize the equations and the graphs of the conic sections studied. We resume the examples with Example 4 on p. 834.

Parabola

$y = ax^2 + bx + c, \quad a > 0$
$\quad = a(x - h)^2 + k$

$y = ax^2 + bx + c, \quad a < 0$
$\quad = a(x - h)^2 + k$

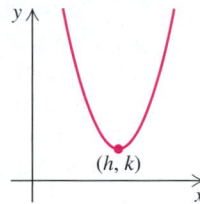

$x = ay^2 + by + c, \quad a > 0$
$\quad = a(y - k)^2 + h$

$x = ay^2 + by + c, \quad a < 0$
$\quad = a(y - k)^2 + h$

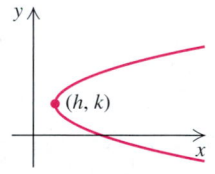
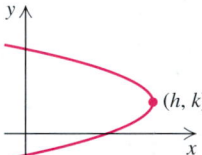

Circle

Center at the origin:

$$x^2 + y^2 = r^2$$

Center at (h, k):

$$(x - h)^2 + (y - k)^2 = r^2$$

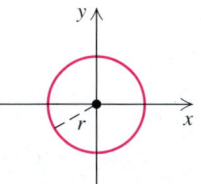
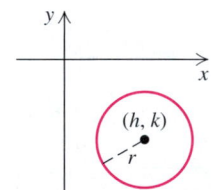

Hyperbola

Center at the origin:

$$\frac{x^2}{a^2} - \frac{y^2}{b^2} = 1 \qquad\qquad \frac{y^2}{b^2} - \frac{x^2}{a^2} = 1$$

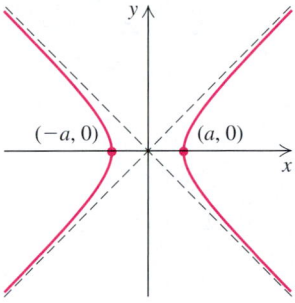

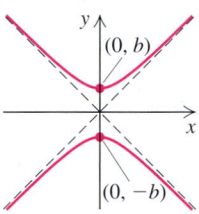

$$xy = c, \quad c > 0 \qquad\qquad xy = c, \quad c < 0$$

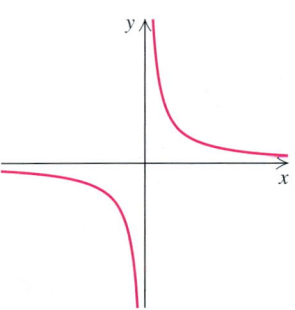

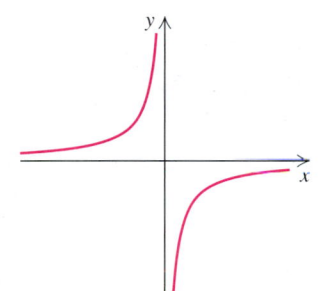

Center at (h, k)*:

$$\frac{(x - h)^2}{a^2} - \frac{(y - k)^2}{b^2} = 1 \qquad\qquad \frac{(y - k)^2}{b^2} - \frac{(x - h)^2}{a^2} = 1$$

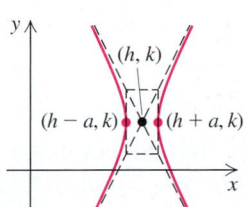

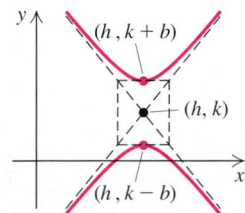

*See Exercises 51–56.

Ellipse

Center at the origin:

$$\frac{x^2}{a^2} + \frac{y^2}{b^2} = 1$$

Center at (h, k):

$$\frac{(x - h)^2}{a^2} + \frac{(y - k)^2}{b^2} = 1$$

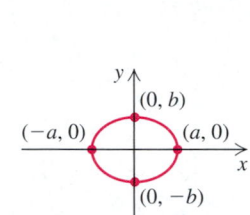

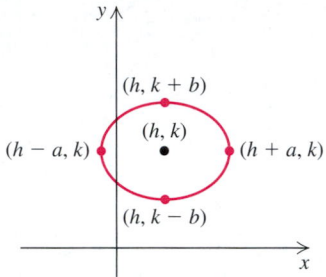

Algebraic manipulations may be needed to express an equation in one of the preceding forms.

Example 4

Classify the graph of each equation as a circle, an ellipse, a parabola, or a hyperbola.

a) $5x^2 = 20 - 5y^2$
b) $x + 3 + 8y = y^2$
c) $x^2 = y^2 + 4$
d) $x^2 = 16 - 4y^2$

Solution

a) We get the terms with variables on one side by adding $5y^2$:

$$5x^2 + 5y^2 = 20.$$

Since x and y are *both* squared, we do not have a parabola. The fact that the squared terms are *added* tells us that we do not have a hyperbola. Do we have a circle? To find out, we need to get $x^2 + y^2$ by itself. We can do that by factoring the 5 out of both terms on the left and then dividing by 5:

$$5(x^2 + y^2) = 20 \qquad \text{Factoring out 5}$$
$$x^2 + y^2 = 4 \qquad \text{Dividing both sides by 5}$$
$$x^2 + y^2 = 2^2. \qquad \text{This is an equation for a circle.}$$

We can see that the graph is a circle with center at the origin and radius 2.

b) The equation $x + 3 + 8y = y^2$ has only one variable squared, so we solve for the other variable:

$$x = y^2 - 8y - 3. \qquad \text{This is an equation for a parabola.}$$

The graph is a horizontal parabola that opens to the right.

c) In $x^2 = y^2 + 4$, both variables are squared, so the graph is not a parabola. We subtract y^2 on both sides and divide by 4 to obtain

$$\frac{x^2}{2^2} - \frac{y^2}{2^2} = 1. \qquad \text{This is an equation for a hyperbola.}$$

The minus sign here indicates that the graph of this equation is a hyperbola. Because it is the x^2-term that is nonnegative, the hyperbola is horizontal.

d) In $x^2 = 16 - 4y^2$, both variables are squared, so the graph cannot be a parabola. We obtain the following equivalent equation:

$$x^2 + 4y^2 = 16.$$

If the coefficients of the terms were the same, we would have the graph of a circle, as in part (a), but they are not. Dividing both sides by 16 yields

$$\frac{x^2}{16} + \frac{y^2}{4} = 1. \qquad \text{This is an equation for an ellipse.}$$

The graph of this equation is a horizontal ellipse.

FOR EXTRA HELP

Exercise Set 13.3

 Digital Video Tutor CD 10 Videotape 25   InterAct Math   Math Tutor Center   MathXL   MyMathLab.com

Graph each hyperbola. Label all vertices and sketch all asymptotes.

1. $\dfrac{y^2}{9} - \dfrac{x^2}{9} = 1$

2. $\dfrac{x^2}{16} - \dfrac{y^2}{16} = 1$

3. $\dfrac{x^2}{4} - \dfrac{y^2}{25} = 1$

4. $\dfrac{y^2}{16} - \dfrac{x^2}{9} = 1$

5. $\dfrac{y^2}{36} - \dfrac{x^2}{9} = 1$

6. $\dfrac{x^2}{25} - \dfrac{y^2}{36} = 1$

7. $y^2 - x^2 = 25$

8. $x^2 - y^2 = 4$

9. $25x^2 - 16y^2 = 400$

10. $4y^2 - 9x^2 = 36$

Graph.

11. $xy = -6$

12. $xy = 6$

13. $xy = 4$

14. $xy = -9$

15. $xy = -2$

16. $xy = -1$

17. $xy = 1$

18. $xy = 2$

Classify each of the following as the equation of a circle, an ellipse, a parabola, or a hyperbola.

19. $x^2 + y^2 - 10x + 8y - 40 = 0$

20. $y + 7 = 3x^2$

21. $9x^2 + 4y^2 - 36 = 0$

22. $1 + 3y = 2y^2 - x$

23. $4x^2 - 9y^2 - 72 = 0$

24. $y^2 + x^2 = 8$

25. $x^2 + y^2 = 2x + 4y + 4$

26. $2y + 13 + x^2 = 8x - y^2$

27. $4x^2 = 64 - y^2$

28. $y = \dfrac{2}{x}$

29. $x - \dfrac{3}{y} = 0$

30. $x - 4 = y^2 - 3y$

31. $y + 6x = x^2 + 5$

32. $x^2 = 16 + y^2$

33. $9y^2 = 36 + 4x^2$

34. $3x^2 + 5y^2 + x^2 = y^2 + 49$

35. $3x^2 + y^2 - x = 2x^2 - 9x + 10y + 40$

36. $4y^2 + 20x^2 + 1 = 8y - 5x^2$

37. $16x^2 + 5y^2 - 12x^2 + 8y^2 - 3x + 4y = 568$

38. $56x^2 - 17y^2 = 234 - 13x^2 - 38y^2$

39. What does graphing hyperbolas have in common with graphing ellipses?

40. Is it possible for a hyperbola to represent the graph of a function? Why or why not?

SKILL MAINTENANCE

Solve each system.

41. $5x + 6y = -12,$
$3x + 9y = 15$

42. $2x + 6y = -6,$
$3x + 5y = 7$

Solve.

43. $y^2 - 3 = 6$

44. $x^2 + 3 = 4$

45. The price of a radio, including 5% sales tax, is $36.75. Find the price of the radio before the tax was added.

46. A basketball team increases its score by 7 points in each of three consecutive games. If the team scored a total of 228 points in all three games, what was its score in the first game?

SYNTHESIS

47. What is it in the equation of a hyperbola that controls how wide open the branches are? Explain your reasoning.

48. If, in

$$\frac{x^2}{a^2} - \frac{y^2}{b^2} = 1,$$

$a = b$, what are the asymptotes of the graph? Why?

Find an equation of a hyperbola satisfying the given conditions.

49. Having intercepts $(0, 6)$ and $(0, -6)$ and asymptotes $y = 3x$ and $y = -3x$

50. Having intercepts $(8, 0)$ and $(-8, 0)$ and asymptotes $y = 4x$ and $y = -4x$

The standard equations for horizontal or vertical hyperbolas centered at (h, k) are as follows:

$$\frac{(x - h)^2}{a^2} - \frac{(y - k)^2}{b^2} = 1$$

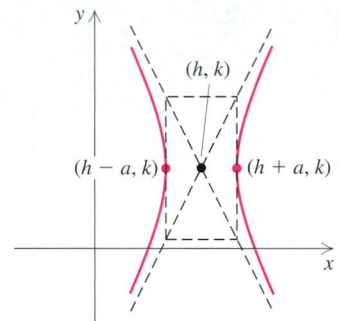

$$\frac{(y - k)^2}{b^2} - \frac{(x - h)^2}{a^2} = 1$$

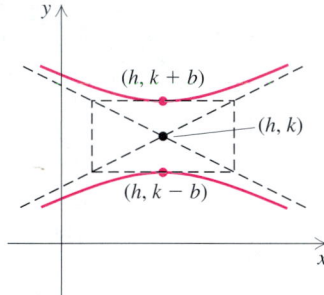

The vertices are as labeled and the asymptotes are

$$y - k = \frac{b}{a}(x - h) \quad and \quad y - k = -\frac{b}{a}(x - h).$$

For each of the following equations of hyperbolas, complete the square, if necessary, and write in standard form. Find the center, the vertices, and the asymptotes. Then graph the hyperbola.

51. $\dfrac{(x - 5)^2}{36} - \dfrac{(y - 2)^2}{25} = 1$

52. $\dfrac{(x - 2)^2}{9} - \dfrac{(y - 1)^2}{4} = 1$

53. $8(y + 3)^2 - 2(x - 4)^2 = 32$

54. $25(x - 4)^2 - 4(y + 5)^2 = 100$

55. $4x^2 - y^2 + 24x + 4y + 28 = 0$

56. $4y^2 - 25x^2 - 8y - 100x - 196 = 0$

57. Use a grapher to check your answers to Exercises 5, 17, 23, and 51.

Nonlinear Systems of Equations

13.4

Systems Involving One Nonlinear Equation • Systems of Two Nonlinear Equations • Problem Solving

The equations appearing in systems of two equations have thus far all been linear. We now consider systems of two equations in which at least one equation is nonlinear.

Systems Involving One Nonlinear Equation

Suppose that a system consists of an equation of a circle and an equation of a line. In what ways can the circle and the line intersect? The figures below represent three ways in which the situation can occur. We see that such a system will have 0, 1, or 2 real solutions.

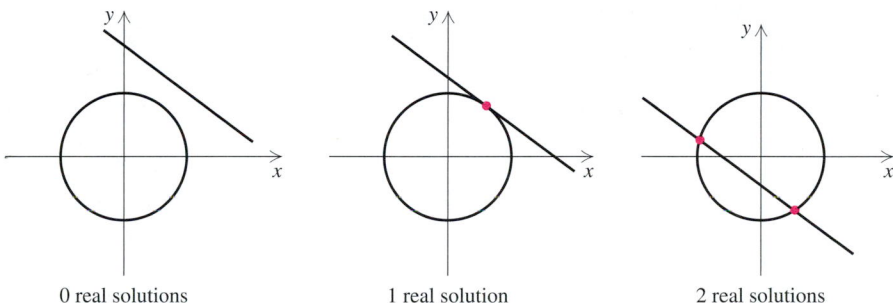

0 real solutions   1 real solution   2 real solutions

Recall that, in addition to graphing, we used both *elimination* and *substitution* to solve systems of linear equations. When solving systems in which one equation is of first degree and one is of second degree, it is preferable to use the *substitution* method.

E x a m p l e   1   Solve the system

$$x^2 + y^2 = 25, \qquad (1) \qquad \text{(The graph is a circle.)}$$
$$3x - 4y = 0. \qquad (2) \qquad \text{(The graph is a line.)}$$

Solution   First, we solve the linear equation, (2), for x:

$$x = \tfrac{4}{3}y. \qquad (3) \qquad \text{We could have solved for } y \text{ instead.}$$

Then we substitute $\tfrac{4}{3}y$ for x in equation (1) and solve for y:

$$\left(\tfrac{4}{3}y\right)^2 + y^2 = 25$$
$$\tfrac{16}{9}y^2 + y^2 = 25$$
$$\tfrac{25}{9}y^2 = 25$$
$$y^2 = 9 \qquad \text{Multiplying both sides by } \tfrac{9}{25}$$
$$y = \pm 3. \qquad \text{Using the principle of square roots}$$

Now we substitute these numbers for y in equation (3) and solve for x:

for $y = 3$, $x = \frac{4}{3}(3) = 4$;

for $y = -3$, $x = \frac{4}{3}(-3) = -4$.

Check: For $(4, 3)$:

$$\frac{x^2 + y^2 = 25}{4^2 + 3^2 \ ? \ 25}$$
$$16 + 9$$
$$25 \ \Big| \ 25 \ \text{TRUE}$$

$$\frac{3x - 4y = 0}{3(4) - 4(3) \ ? \ 0}$$
$$12 - 12$$
$$0 \ \Big| \ 0 \ \text{TRUE}$$

It is left to the student to confirm that $(-4, -3)$ also checks in both equations.

The pairs $(4, 3)$ and $(-4, -3)$ check, so they are solutions. We can see the solutions in the graph. Intersections occur at $(4, 3)$ and $(-4, -3)$.

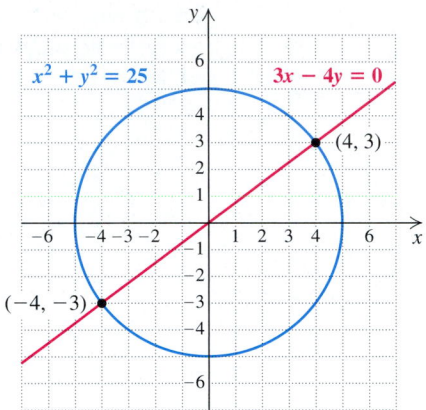

Although we may not know what the graph of each equation in a system looks like, the algebraic approach of Example 1 can still be used.

E x a m p l e   2   Solve the system

$$y + 3 = 2x, \qquad (1)$$
$$x^2 + 2xy = -1. \qquad (2)$$

Solution   First, we solve the linear equation (1) for y:

$$y = 2x - 3. \quad (3)$$

Then we substitute $2x - 3$ for y in equation (2) and solve for x:

$$x^2 + 2x(2x - 3) = -1$$
$$x^2 + 4x^2 - 6x = -1$$
$$5x^2 - 6x + 1 = 0$$
$$(5x - 1)(x - 1) = 0 \qquad \text{Factoring}$$
$$5x - 1 = 0 \quad or \quad x - 1 = 0 \qquad \text{Using the principle of zero products}$$
$$x = \tfrac{1}{5} \quad or \qquad x = 1.$$

Now we substitute these numbers for x in equation (3) and solve for y:

$$\text{for } x = \tfrac{1}{5}, \quad y = 2\left(\tfrac{1}{5}\right) - 3 = -\tfrac{13}{5};$$
$$\text{for } x = 1, \quad y = 2(1) - 3 = -1.$$

You can confirm that $\left(\tfrac{1}{5}, -\tfrac{13}{5}\right)$ and $(1, -1)$ check, so they are both solutions.

Example 3

technology connection A

Systems of equations offer a fine opportunity to use the INTERSECT feature of a grapher, although most graphers will restrict solutions to real numbers.

To solve Example 2,

$$y + 3 = 2x,$$
$$x^2 + 2xy = -1,$$

we solve each equation for y and then graph:

$$\left. \begin{array}{l} y_1 = 2x - 3, \\ y_2 = \dfrac{-1 - x^2}{2x}. \end{array} \right\} \quad \begin{array}{l} \text{Note that} \\ x, y \neq 0. \end{array}$$

$$y_1 = 2x - 3, \quad y_2 = \dfrac{-1 - x^2}{2x}$$

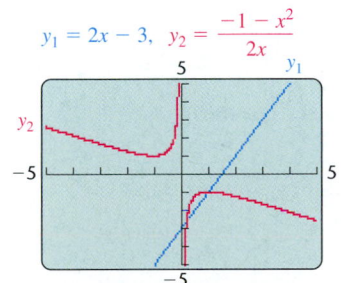

Using INTERSECT, we find the solutions to be $(0.2, -2.6)$ and $(1, -1)$.

Use a grapher to solve each system. Round all values to two decimal places.

1. $4xy - 7 = 0,$
  $x - 3y - 2 = 0$
2. $x^2 + y^2 = 14,$
  $16x + 7y^2 = 0$

Solve the system

$$x + y = 5, \qquad (1) \qquad \text{(The graph is a line.)}$$
$$y = 3 - x^2. \qquad (2) \qquad \text{(The graph is a parabola.)}$$

Solution   We substitute $3 - x^2$ for y in the first equation:

$$x + 3 - x^2 = 5$$
$$-x^2 + x - 2 = 0 \qquad \text{Adding } -5 \text{ to both sides and rearranging}$$
$$x^2 - x + 2 = 0. \qquad \text{Multiplying both sides by } -1$$

Since $x^2 - x + 2$ does not factor, we need the quadratic formula:

$$x = \frac{-b \pm \sqrt{b^2 - 4ac}}{2a}$$
$$= \frac{-(-1) \pm \sqrt{(-1)^2 - 4 \cdot 1 \cdot 2}}{2(1)} \qquad \text{Substituting}$$
$$= \frac{1 \pm \sqrt{1 - 8}}{2} = \frac{1 \pm \sqrt{-7}}{2} = \frac{1}{2} \pm \frac{\sqrt{7}}{2}i.$$

Solving equation (1) for y gives us $y = 5 - x$. Substituting values for x gives

$$y = 5 - \left(\frac{1}{2} + \frac{\sqrt{7}}{2}i\right) = \frac{9}{2} - \frac{\sqrt{7}}{2}i \quad \text{and}$$
$$y = 5 - \left(\frac{1}{2} - \frac{\sqrt{7}}{2}i\right) = \frac{9}{2} + \frac{\sqrt{7}}{2}i.$$

The solutions are

$$\left(\frac{1}{2} + \frac{\sqrt{7}}{2}i, \frac{9}{2} - \frac{\sqrt{7}}{2}i\right) \quad \text{and} \quad \left(\frac{1}{2} - \frac{\sqrt{7}}{2}i, \frac{9}{2} + \frac{\sqrt{7}}{2}i\right).$$

There are no real-number solutions. Note in the figure at right that the graphs do not intersect. Getting only nonreal solutions tells us that the graphs do not intersect.

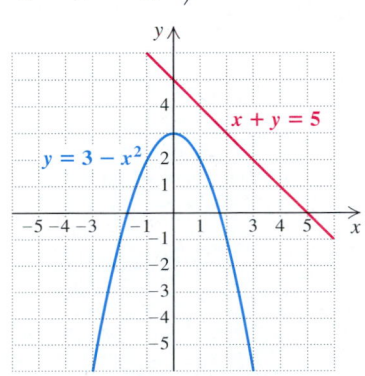

Systems of Two Nonlinear Equations

We now consider systems of two second-degree equations. Graphs of such systems can involve any two conic sections. The following figure shows some ways in which a circle and a hyperbola can intersect.

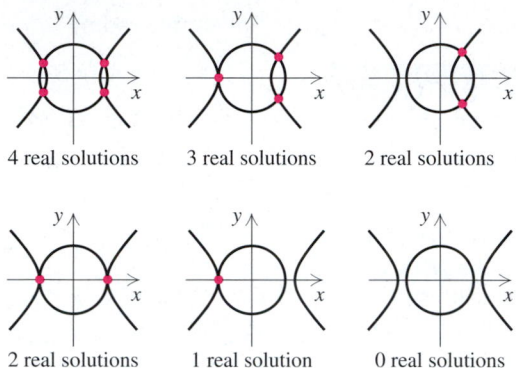

4 real solutions   3 real solutions   2 real solutions

2 real solutions   1 real solution   0 real solutions

To solve systems of two second-degree equations, we can use either substitution or elimination. The elimination method is generally better when both equations are of the form $Ax^2 + By^2 = C$. Then we can eliminate an x^2- or y^2-term in a manner similar to the procedure used in Chapter 8.

E x a m p l e 4   Solve the system

$$2x^2 + 5y^2 = 22, \qquad (1)$$
$$3x^2 - y^2 = -1. \qquad (2)$$

Solution   Here we multiply equation (2) by 5 and then add:

$$\begin{array}{rl}
2x^2 + 5y^2 = & 22 \\
\underline{15x^2 - 5y^2 = -5} & \qquad \text{Multiplying both sides of equation (2) by 5} \\
17x^2 = & 17 \qquad \text{Adding} \\
x^2 = & 1 \\
x = & \pm 1.
\end{array}$$

There is no x-term, and whether x is -1 or 1, we have $x^2 = 1$. Thus we can simultaneously substitute 1 and -1 for x in equation (2):

$$\left.\begin{array}{r}
3 \cdot (\pm 1)^2 - y^2 = -1 \\
3 - y^2 = -1 \\
-y^2 = -4
\end{array}\right\} \quad \begin{array}{l} \text{Since } (-1)^2 = 1^2, \text{ we can evaluate for} \\ x = -1 \text{ and } x = 1 \text{ simultaneously.} \end{array}$$

$$y^2 = 4 \quad \text{or} \quad y = \pm 2.$$

Thus, if $x = 1$, then $y = 2$ or $y = -2$; and if $x = -1$, then $y = 2$ or $y = -2$. The four possible solutions are $(1, 2)$, $(1, -2)$, $(-1, 2)$, and $(-1, -2)$.

Check: Since $(2)^2 = (-2)^2$ and $(1)^2 = (-1)^2$, we can check all four pairs at once.

$$\frac{2x^2 + 5y^2 = 22}{2(\pm 1)^2 + 5(\pm 2)^2 \ ? \ 22}$$
$$2 + 20$$
$$22 \ | \ 22 \quad \text{TRUE}$$

$$\frac{3x^2 - y^2 = -1}{3(\pm 1)^2 - (\pm 2)^2 \ ? \ -1}$$
$$3 - 4$$
$$-1 \ | \ -1 \quad \text{TRUE}$$

The solutions are $(1, 2)$, $(1, -2)$, $(-1, 2)$, and $(-1, -2)$.

When a product of variables is in one equation and the other equation is of the form $Ax^2 + By^2 = C$, we often solve for a variable in the equation with the product and then use substitution.

Example 5

Solve the system

$$x^2 + 4y^2 = 20, \quad (1)$$
$$xy = 4. \quad (2)$$

Solution   First, we solve equation (2) for y:

$$y = \frac{4}{x}. \qquad \text{Dividing both sides by } x. \text{ Note that } x \neq 0.$$

Then we substitute $4/x$ for y in equation (1) and solve for x:

$$x^2 + 4\left(\frac{4}{x}\right)^2 = 20$$

$$x^2 + \frac{64}{x^2} = 20$$

$$x^4 + 64 = 20x^2 \qquad \text{Multiplying by } x^2$$

$$x^4 - 20x^2 + 64 = 0 \qquad \begin{array}{l}\text{Obtaining standard form.} \\ \text{This equation is reducible} \\ \text{to quadratic.}\end{array}$$

$$(x^2 - 4)(x^2 - 16) = 0 \qquad \begin{array}{l}\text{Factoring. If you prefer, let} \\ u = x^2 \text{ and substitute.}\end{array}$$

$$(x - 2)(x + 2)(x - 4)(x + 4) = 0 \qquad \text{Factoring}$$

$$x = 2 \quad or \quad x = -2 \quad or \quad x = 4 \quad or \quad x = -4. \qquad \begin{array}{l}\text{Using the principle} \\ \text{of zero products}\end{array}$$

Since $y = 4/x$, for $x = 2$, we have $y = 4/2$, or 2. Thus, $(2, 2)$ is a solution. Similarly, $(-2, -2)$, $(4, 1)$, and $(-4, -1)$ are solutions. You can show that all four pairs check.

Problem Solving

We now consider applications that can be modeled by a system of equations in which at least one equation is not linear.

technology connection

B

Before Example 4 can be checked with a grapher, each equation must first be solved for y. When this has been done, we have $y_1 = \sqrt{(22 - 2x^2)/5}$ and $y_2 = -\sqrt{(22 - 2x^2)/5}$ for equation (1) and $y_3 = \sqrt{3x^2 + 1}$ and $y_4 = -\sqrt{3x^2 + 1}$ for equation (2). The graph verifies the solutions found algebraically.

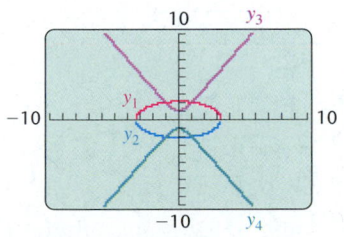

1. Use a grapher to provide a visual check for Example 5.

E x a m p l e   6   **_Architecture._**   For a college gymnasium, an architect wants to lay out a rectangular piece of land that has a perimeter of 204 m and an area of 2565 m². Find the dimensions of the piece of land.

Solution

1. **Familiarize.**   We draw and label a sketch, letting l = the length and w = the width, both in meters.

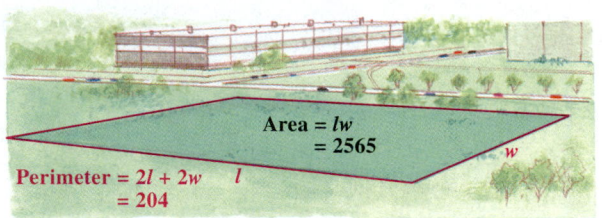

Area = lw
= 2565

Perimeter = $2l + 2w$ l
= 204

w

2. **Translate.**   We then have the following translation:

    Perimeter: $2w + 2l = 204;$

    Area: $lw = 2565.$

3. **Carry out.**   We solve the system

    $$2w + 2l = 204,$$
    $$lw = 2565.$$

    Solving the second equation for l gives us $l = 2565/w$. Then we substitute $2565/w$ for l in the first equation and solve for w:

    $$2w + 2\left(\frac{2565}{w}\right) = 204$$

    $$2w^2 + 2(2565) = 204w \qquad \text{Multiplying both sides by } w$$

    $$2w^2 - 204w + 2(2565) = 0 \qquad \text{Standard form}$$

    $$w^2 - 102w + 2565 = 0 \qquad \text{Multiplying by } \tfrac{1}{2}$$

    > Factoring could be used instead of the quadratic formula, but the numbers are quite large.

    $$w = \frac{-(-102) \pm \sqrt{(-102)^2 - 4 \cdot 1 \cdot 2565}}{2 \cdot 1}$$

    $$w = \frac{102 \pm \sqrt{144}}{2} = \frac{102 \pm 12}{2}$$

    $$w = 57 \quad or \quad w = 45.$$

    If $w = 57$, then $l = 2565/w = 2565/57 = 45$. If $w = 45$, then $l = 2565/w = 2565/45 = 57$. Since length is usually considered to be longer than width, we have the solution $l = 57$ and $w = 45$, or $(57, 45)$.

4. **Check.**   If $l = 57$ and $w = 45$, the perimeter is $2 \cdot 57 + 2 \cdot 45$, or 204. The area is $57 \cdot 45$, or 2565. The numbers check.

5. **State.**   The length is 57 m and the width is 45 m.

E x a m p l e   7   **HDTV dimensions.**   High-definition television (HDTV) offers greater clarity than conventional television. The Kaplans' new HDTV screen has an area of 1296 in^2 and has a $\sqrt{3033}$-in. (about 55-in.) diagonal screen. Find the width and the length of the screen.

Solution

1. **Familiarize.**   We make a drawing and label it. Note that there is a right triangle in the figure. We let l = the length and w = the width, both in inches.

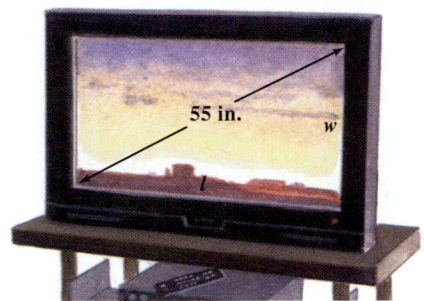

2. **Translate.**   We translate to a system of equations:

   $l^2 + w^2 = \sqrt{3033}^2$ Using the Pythagorean theorem

   $lw = 1296.$ Using the formula for the area of a rectangle

3. **Carry out.**   We solve the system

   $\left.\begin{array}{l} l^2 + w^2 = 3033, \\ lw = 1296 \end{array}\right\}$ You should complete the solution of this system.

   to get $(48, 27)$, $(27, 48)$, $(-48, -27)$, and $(-27, -48)$.

4. **Check.**   Measurements cannot be negative and length is usually greater than width, so we check only $(48, 27)$. In the right triangle, $48^2 + 27^2 = 2304 + 729 = 3033$ or $\sqrt{3033}^2$. The area is $48 \cdot 27 = 1296$, so our answer checks.

5. **State.**   The length is 48 in. and the width is 27 in.

FOR EXTRA HELP

Exercise Set   13.4

Digital Video Tutor CD 10
Videotape 25

InterAct Math

Math Tutor Center

MathXL

MyMathLab.com

Solve. Remember that graphs can be used to confirm all real solutions.

1. $x^2 + y^2 = 25,$
   $y - x = 1$

2. $x^2 + y^2 = 100,$
   $y - x = 2$

3. $9x^2 + 4y^2 = 36,$
   $3x + 2y = 6$

4. $4x^2 + 9y^2 = 36,$
   $3y + 2x = 6$

5. $y = x^2,$
   $3x = y + 2$

6. $y^2 = x + 3,$
   $2y = x + 4$

7. $2y^2 + xy + x^2 = 7,$
$x - 2y = 5$

8. $x^2 - xy + 3y^2 = 27,$
$x - y = 2$

9. $x^2 - y^2 = 16,$
$x - 2y = 1$

10. $x^2 + 4y^2 = 25,$
$x + 2y = 7$

11. $m^2 + 3n^2 = 10,$
$m - n = 2$

12. $x^2 - xy + 3y^2 = 5,$
$x - y = 2$

13. $2y^2 + xy = 5,$
$4y + x = 7$

14. $3x + y = 7,$
$4x^2 + 5y = 24$

15. $p + q = -6,$
$pq = -7$

16. $a + b = 7,$
$ab = 4$

17. $4x^2 + 9y^2 = 36,$
$x + 3y = 3$

18. $2a + b = 1,$
$b = 4 - a^2$

19. $xy = 4,$
$x + y = 5$

20. $a^2 + b^2 = 89,$
$a - b = 3$

Aha! **21.** $y = x^2,$
$x = y^2$

22. $x^2 + y^2 = 25,$
$y^2 = x + 5$

23. $x^2 + y^2 = 9,$
$x^2 - y^2 = 9$

24. $y^2 - 4x^2 = 4,$
$4x^2 + y^2 = 4$

25. $x^2 + y^2 = 25,$
$xy = 12$

26. $x^2 - y^2 = 16,$
$x + y^2 = 4$

27. $x^2 + y^2 = 4,$
$9x^2 + 16y^2 = 144$

28. $x^2 + y^2 = 9,$
$25x^2 + 16y^2 = 400$

29. $x^2 + y^2 = 16,$
$y^2 - 2x^2 = 10$

30. $x^2 + y^2 = 14,$
$x^2 - y^2 = 4$

31. $x^2 + y^2 = 5,$
$xy = 2$

32. $x^2 + y^2 = 20,$
$xy = 8$

33. $x^2 + y^2 = 13,$
$xy = 6$

34. $x^2 + 4y^2 = 20,$
$xy = 4$

35. $3xy + x^2 = 34,$
$2xy - 3x^2 = 8$

36. $2xy + 3y^2 = 7,$
$3xy - 2y^2 = 4$

37. $xy - y^2 = 2,$
$2xy - 3y^2 = 0$

38. $4a^2 - 25b^2 = 0,$
$2a^2 - 10b^2 = 3b + 4$

39. $x^2 - y = 5,$
$x^2 + y^2 = 25$

40. $ab - b^2 = -4,$
$ab - 2b^2 = -6$

Solve.

41. *Computer parts.* Dataport Electronics needs a rectangular memory board that has a perimeter of 28 cm and a diagonal of length 10 cm. What should the dimensions of the board be?

42. *Geometry.* A rectangle has an area of 2 yd² and a perimeter of 6 yd. Find its dimensions.

43. *Geometry.* A rectangle has an area of 20 in² and a perimeter of 18 in. Find its dimensions.

44. *Tile design.* The New World tile company wants to make a new rectangular tile that has a perimeter of 6 in. and a diagonal of length $\sqrt{5}$ in. What should the dimensions of the tile be?

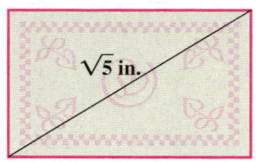

45. *Design of a van.* The cargo area of a delivery van must be 60 ft², and the length of a diagonal must accommodate a 13-ft board. Find the dimensions of the cargo area.

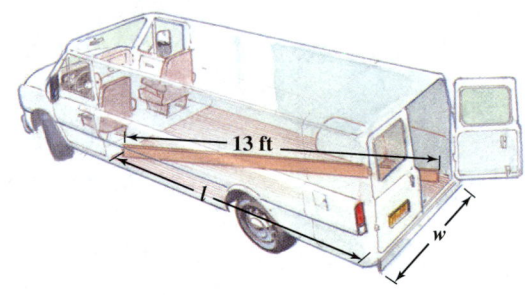

46. *Dimensions of a rug.* The diagonal of a Persian rug is 25 ft. The area of the rug is 300 ft². Find the length and the width of the rug.

47. The product of the lengths of the legs of a right triangle is 156. The hypotenuse has length $\sqrt{313}$. Find the lengths of the legs.

48. The product of two numbers is 60. The sum of their squares is 136. Find the numbers.

49. *Investments.* A certain amount of money saved for 1 yr at a certain interest rate yielded $225 in interest. If $750 more had been invested and the rate had been 1% less, the interest would have been the same. Find the principal and the rate.

50. *Garden design.* A garden contains two square peanut beds. Find the length of each bed if the sum of their areas is 832 ft² and the difference of their areas is 320 ft².

51. The area of a rectangle is $\sqrt{3}$ m², and the length of a diagonal is 2 m. Find the dimensions.

52. The area of a rectangle is $\sqrt{2}$ m², and the length of a diagonal is $\sqrt{3}$ m. Find the dimensions.

53. How can an understanding of conic sections be helpful when a system of nonlinear equations is being solved algebraically?

54. Suppose a system of equations is comprised of one linear and one nonlinear equation. Is it possible for such a system to have three solutions? Why or why not?

SKILL MAINTENANCE

Simplify.

55. $(-1)^9(-2)^4$

56. $(-1)^{10}(-2)^5$

Evaluate each of the following.

57. $\dfrac{(-1)^k}{k-5}$, for $k = 6$

58. $\dfrac{(-1)^k}{k-5}$, for $k = 9$

59. $\dfrac{n}{2}(3+n)$, for $n = 8$

60. $\dfrac{7(1-r^2)}{1-r}$, for $r = 3$

SYNTHESIS

61. Write a problem that translates to a system of two equations. Design the problem so that at least one equation is nonlinear and so that no real solution exists.

62. Write a problem for a classmate to solve. Devise the problem so that a system of two nonlinear equations with exactly one real solution is solved.

63. A piece of wire 100 cm long is to be cut into two pieces and those pieces are each to be bent to make a square. The area of one square is to be 144 cm² greater than that of the other. How should the wire be cut?

64. Find the equation of a circle that passes through $(-2, 3)$ and $(-4, 1)$ and whose center is on the line $5x + 8y = -2$.

65. Find the equation of an ellipse centered at the origin that passes through the points $(2, -3)$ and $(1, \sqrt{13})$.

Solve.

66. $p^2 + q^2 = 13$,

$\dfrac{1}{pq} = -\dfrac{1}{6}$

67. $a + b = \dfrac{5}{6}$,

$\dfrac{a}{b} + \dfrac{b}{a} = \dfrac{13}{6}$

68. *Box design.* Four squares with sides 5 in. long are cut from the corners of a rectangular metal sheet that has an area of 340 in². The edges are bent up to form an open box with a volume of 350 in³. Find the dimensions of the box.

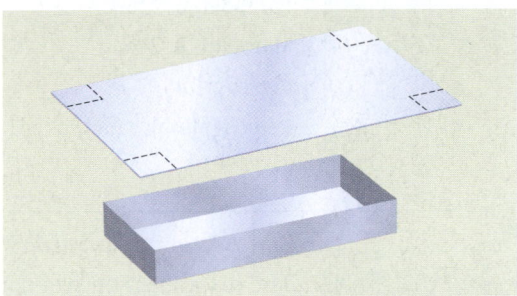

69. *Computer screens.* The ratio of the length to the height of the screen on a computer monitor is 4 to 3. An IBM Thinkpad laptop has a 31-cm diagonal screen. Find the dimensions of the screen.

70. *HDTV screens.* The ratio of the length to the height of an HDTV screen (see Example 7) is 16 to 9. The Remton Lounge has an HDTV screen with a $\sqrt{4901}$-in. (about 70-in.) diagonal screen. Find the dimensions of the screen.

71. *Railing sales.* Fireside Castings finds that the total revenue R from the sale of x units of railing is given by

$$R = 100x + x^2.$$

Fireside also finds that the total cost C of producing x units of the same product is given by

$$C = 80x + 1500.$$

A break-even point is a value of x for which total revenue is the same as total cost; that is, $R = C$. How many units must be sold to break even?

72. Use a grapher to check your answers to Exercises 5, 17, and 39.

Summary and Review 13

Key Terms

Conic section, p. 736

Parabola, p. 736

Midpoint, p. 740

Circle, p. 741

Radius (plural, radii), p. 741

Center, p. 741

Ellipse, p. 748

Foci (singular, focus), p. 748

Vertices (singular, vertex),
p. 748

Hyperbola, p. 755

Axis, p. 755

Branches, p. 755

Asymptote, p. 755

Important Properties and Formulas

The Distance Formula

The distance d between any two points (x_1, y_1) and (x_2, y_2) is given by

$$d = \sqrt{(x_2 - x_1)^2 + (y_2 - y_1)^2}.$$

The Midpoint Formula

If the endpoints of a segment are (x_1, y_1) and (x_2, y_2), then the coordinates of the midpoint are

$$\left(\frac{x_1 + x_2}{2}, \frac{y_1 + y_2}{2} \right).$$

Graphs of conic sections: See the summary of graphs on pp. 832–834.

Parabola

Vertical with vertex at (h, k):

$$y = ax^2 + bx + c$$
$$= a(x - h)^2 + k$$

Horizontal with vertex at (h, k):

$$x = ay^2 + by + c$$
$$= a(y - k)^2 + h$$

Circle

Center at the origin:

$$x^2 + y^2 = r^2$$

Center at (h, k):

$$(x - h)^2 + (y - k)^2 = r^2$$

Ellipse

Center at the origin:

$$\frac{x^2}{a^2} + \frac{y^2}{b^2} = 1$$

Center at (h, k):

$$\frac{(x - h)^2}{a^2} + \frac{(y - k)^2}{b^2} = 1$$

Hyperbola

Center at the origin:

Axis horizontal: $\dfrac{x^2}{a^2} - \dfrac{y^2}{b^2} = 1$ Axis vertical: $\dfrac{y^2}{b^2} - \dfrac{x^2}{a^2} = 1$

With x- and y-axes as asymptotes: $xy = c$

Review Exercises

Find the distance between each pair of points. Where appropriate, find an approximation to three decimal places.

1. $(2, 6)$ and $(6, 6)$

2. $(-1, 1)$ and $(-5, 4)$

3. $(1.4, 3.6)$ and $(4.7, -5.3)$

4. $(2, 3a)$ and $(-1, a)$

Find the midpoint of the segment with the given endpoints.

5. $(1, 6)$ and $(7, 6)$

6. $(-1, 1)$ and $(-5, 4)$

7. $\left(1, \sqrt{3}\right)$ and $\left(\frac{1}{2}, -\sqrt{2}\right)$

8. $(2, 3a)$ and $(-1, a)$

Find the center and the radius of each circle.

9. $(x + 2)^2 + (y - 3)^2 = 2$

10. $(x - 5)^2 + y^2 = 49$

11. $x^2 + y^2 - 6x - 2y + 1 = 0$

12. $x^2 + y^2 + 8x - 6y = 10$

13. Find an equation of the circle with center $(-4, 3)$ and radius $4\sqrt{3}$.

14. Find an equation of the circle with center $(7, -2)$ and radius $2\sqrt{5}$.

Classify each equation as a circle, an ellipse, a parabola, or a hyperbola. Then graph.

15. $4x^2 + 4y^2 = 100$

16. $9x^2 + 2y^2 = 18$

17. $y = -x^2 + 2x - 3$

18. $\dfrac{y^2}{9} - \dfrac{x^2}{4} = 1$

19. $xy = 9$

20. $x = y^2 + 2y - 2$

21. $\dfrac{(x + 1)^2}{3} + (y - 3)^2 = 1$

22. $x^2 + y^2 + 6x - 8y - 39 = 0$

Solve.

23. $x^2 - y^2 = 33,$
     $x + y = 11$

24. $x^2 - 2x + 2y^2 = 8,$
     $2x + y = 6$

25. $x^2 - y = 3,$
     $2x - y = 3$

26. $x^2 + y^2 = 25,$
     $x^2 - y^2 = 7$

27. $x^2 - y^2 = 3,$
     $y = x^2 - 3$

28. $x^2 + y^2 = 18,$
     $2x + y = 3$

29. $x^2 + y^2 = 100,$
     $2x^2 - 3y^2 = -120$

30. $x^2 + 2y^2 = 12,$
     $xy = 4$

31. A rectangular garden has a perimeter of 38 m and an area of 84 m². What are the dimensions of the garden?

32. One type of carton used by table products.com exactly fits both a folded napkin of area 108 in² and a candle of length 15 in., laid diagonally on the bottom of the carton. What are the dimensions of the carton?

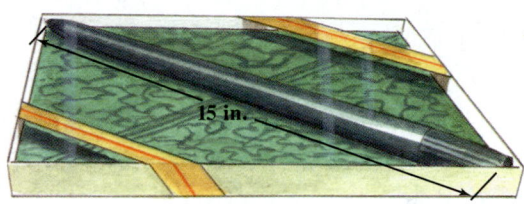

33. The perimeter of a square is 12 cm more than the perimeter of another square. Its area exceeds the area of the other by 39 cm². Find the perimeter of each square.

34. The sum of the areas of two circles is 130π ft². The difference of the circumferences is 16π ft. Find the radius of each circle.

35. How does the graph of a hyperbola differ from the graph of a parabola?

36. Explain why function notation is not used in this chapter, and list the graphs discussed for which function notation could be used.

37. Solve:

$$4x^2 - x - 3y^2 = 9,$$
$$-x^2 + x + \ y^2 = 2.$$

38. Find the points whose distance from $(8, 0)$ and from $(-8, 0)$ is 10.

39. Find an equation of the circle that passes through $(-2, -4)$, $(5, -5)$, and $(6, 2)$.

40. Find an equation of the ellipse with the following intercepts: $(-7, 0)$, $(7, 0)$, $(0, -3)$, and $(0, 3)$.

41. Find the point on the x-axis that is equidistant from $(-3, 4)$ and $(5, 6)$.

Chapter Test 13

Find the distance between each pair of points. Where appropriate, find an approximation to three decimal places.

1. $(4, -1)$ and $(-5, 8)$

2. $(3, -a)$ and $(-3, a)$

Find the midpoint of the segment with the given endpoints.

3. $(4, -1)$ and $(-5, 8)$

4. $(3, -a)$ and $(-3, a)$

Find the center and the radius of each circle.

5. $(x + 2)^2 + (y - 3)^2 = 64$

6. $x^2 + y^2 + 4x - 6y + 4 = 0$

Classify the equation as a circle, an ellipse, a parabola, or a hyperbola. Then graph.

7. $y = x^2 - 4x - 1$

8. $x^2 + y^2 + 2x + 6y + 6 = 0$

9. $\dfrac{x^2}{9} - \dfrac{y^2}{4} = 1$

10. $16x^2 + 4y^2 = 64$

11. $xy = -5$

12. $x = -y^2 + 4y$

Solve.

13. $\dfrac{x^2}{16} + \dfrac{y^2}{9} = 1,$
$\quad 3x + 4y = 12$

14. $x^2 + y^2 = 16,$
$\quad \dfrac{x^2}{16} - \dfrac{y^2}{9} = 1$

15. $x^2 - 2y^2 = 1.$
$\quad xy = 6$

16. $x^2 + y^2 = 10,$
$\quad x^2 = y^2 + 2$

17. A rectangle with diagonal of length $5\sqrt{5}$ has an area of 22. Find the dimensions of the rectangle.

18. Two squares are such that the sum of their areas is 8 m^2 and the difference of their areas is 2 m^2. Find the length of a side of each square.

19. A rectangle has a diagonal of length 20 ft and a perimeter of 56 ft. Find the dimensions of the rectangle.

20. Nikki invested a certain amount of money for 1 yr and earned $72 in interest. Erin invested $240 more than Nikki at an interest rate that was 83% of the rate given to Nikki, but she earned the same amount of interest. Find the principal and interest rate of Nikki's investment.

SYNTHESIS

21. Find an equation of the ellipse passing through $(6, 0)$ and $(6, 6)$ with vertices at $(1, 3)$ and $(11, 3)$.

22. Find the point on the y-axis that is equidistant from $(-3, -5)$ and $(4, -7)$.

23. The sum of two numbers is 36, and the product is 4. Find the sum of the reciprocals of the numbers.

14

Sequences, Series, and Combinatorics

AN APPLICATION

At one point in a recent season, Derek Jeter of the New York Yankees had a batting average of .325. At that time, if someone were to randomly select five of his "at-bats," the probability of his getting exactly 3 hits would be the 3rd term of the binomial expansion of $(0.325 + 0.675)^5$. Find that term and use a calculator to estimate the probability.

This problem appears as Exercise 37 in Section 14.6.

I use math daily, especially to figure percentages such as batting averages, earned run averages, slugging percentages, and on-base percentages. Understanding math also helps me to apply the percentages with meaning.

CHARLIE SCOGGINS
Baseball Writer
The Lowell Sun
Lowell, MA

T he first three sections of this chapter are devoted to sequences *and* series. *A sequence is simply an ordered list. When the members of a sequence are numbers, we can discuss their sum. Such a sum is called a* series.

Section 14.6 presents the binomial theorem, *which is used to expand expressions of the form* $(a + b)^n$. *Such an expansion is itself a series.*

The chapter concludes with a discussion of combinatorics *and* probability.

Sequences and Series

14.1

Sequences • Finding the General Term • Sums and Series • Sigma Notation

Sequences

Suppose that $1000 is invested at 8%, compounded annually. The amounts to which the money grows after 1 year, 2 years, 3 years, and so on, are as follows:

$1080.00, $1166.40, $1259.71, $1360.49, . . .

We can regard this as a function that pairs 1 with $1080.00, 2 with $1166.40, 3 with $1259.71, and so on. A **sequence** (or **progression**) is thus a function, where the domain is a set of consecutive positive integers beginning with 1, and the range varies from sequence to sequence.

If we continue computing the amounts in the account forever, we obtain an **infinite sequence**, with function values

$1080.00, $1166.40, $1259.71, $1360.49, $1469.33, $1586.87, . . .

The three dots at the end indicate that the sequence goes on without stopping. If we stop after a certain number of years, we obtain a **finite sequence:**

$1080.00, $1166.40, $1259.71, $1360.49

> ### Sequences
>
> An *infinite sequence* is a function having for its domain the set of natural numbers: $\{1, 2, 3, 4, 5, \ldots\}$.
>
> A *finite sequence* is a function having for its domain a set of natural numbers: $\{1, 2, 3, 4, 5, \ldots, n\}$, for some natural number n.

As another example, consider the sequence given by

$$a(n) = 2^n, \quad \text{or} \quad a_n = 2^n.$$

The notation a_n means the same as $a(n)$ but is used more commonly with sequences. Some function values (also called *terms* of the sequence) follow:

$$a_1 = 2^1 = 2,$$
$$a_2 = 2^2 = 4,$$
$$a_3 = 2^3 = 8,$$
$$a_6 = 2^6 = 64.$$

The first term of the sequence is a_1, the fifth term is a_5, and the nth term, or **general term**, is a_n. This sequence can also be denoted in the following ways:

$$2, 4, 8, \ldots;$$

or $2, 4, 8, \ldots, 2^n, \ldots.$ The 2^n emphasizes that the nth term of this sequence is found by raising 2 to the nth power.

E x a m p l e 1

Find the first four terms and the 57th term of the sequence for which the general term is given by $a_n = (-1)^n/(n + 1)$.

Solution   We have

$$a_1 = \frac{(-1)^1}{1 + 1} = -\frac{1}{2},$$

$$a_2 = \frac{(-1)^2}{2 + 1} = \frac{1}{3},$$

$$a_3 = \frac{(-1)^3}{3 + 1} = -\frac{1}{4},$$

$$a_4 = \frac{(-1)^4}{4 + 1} = \frac{1}{5},$$

$$a_{57} = \frac{(-1)^{57}}{57 + 1} = -\frac{1}{58}.$$

Note that the expression $(-1)^n$ causes the signs of the terms to alternate between positive and negative, depending on whether n is even or odd.

technology connection

Sequences are entered and graphed much like functions. The difference is that the SEQUENCE MODE must be selected. You can then enter U_n or V_n using n as the variable. Use this approach to check Example 1 with a table of values for the sequence.

Finding the General Term

When only the first few terms of a sequence are known, it is impossible to be certain what the general term is, but a prediction can be made by looking for a pattern.

E x a m p l e 2

For each sequence, predict the general term.

a) $1, 4, 9, 16, 25, \ldots$ **b)** $-1, 2, -4, 8, -16, \ldots$
c) $2, 4, 8, \ldots$

Solution

a) $1, 4, 9, 16, 25, \ldots$

These are squares of consecutive positive integers, so the general term could be n^2.

b) $-1, 2, -4, 8, -16, \ldots$

These are powers of 2 with alternating signs, so the general term may be $(-1)^n[2^{n-1}]$. To check, note that 8 is the fourth term, and

$$(-1)^4[2^{4-1}] = 1 \cdot 2^3$$
$$= 8.$$

c) $2, 4, 8, \ldots$

We regard the pattern as powers of 2, in which case 16 would be the next term and 2^n the general term. The sequence could then be written with more terms as

$$2, 4, 8, 16, 32, 64, 128, \ldots$$

In part (c) above, suppose that the second term is found by adding 2, the third term by adding 4, the next term by adding 6, and so on. In this case, 14 would be the next term and the sequence would be

$$2, 4, 8, 14, 22, 32, 44, 58, \ldots$$

This illustrates that the fewer terms we are given, the greater the uncertainty about the nth term.

Sums and Series

> **Series**
>
> Given the infinite sequence
>
> $$a_1, \ a_2, \ a_3, \ a_4, \ \ldots, \ a_n, \ldots,$$
>
> the sum of the terms
>
> $$a_1 + a_2 + a_3 + \cdots + a_n + \cdots$$
>
> is called an *infinite series*. A *partial sum* is the sum of the first n terms:
>
> $$a_1 + a_2 + a_3 + \cdots + a_n.$$
>
> A partial sum is also called a *finite series* and is denoted S_n.

Example 3 For the sequence $-2, 4, -6, 8, -10, 12, -14$, find: **(a)** S_2; **(b)** S_3; **(c)** S_7.

Solution

a) $S_2 = -2 + 4 = 2$ This is the sum of the first 2 terms.

b) $S_3 = -2 + 4 + (-6) = -4$ This is the sum of the first 3 terms.

c) $S_7 = -2 + 4 + (-6) + 8 + (-10) + 12 + (-14) = -8$ This is the sum of the first 7 terms.

Sigma Notation

When the general term of a sequence is known, the Greek letter Σ (capital sigma) can be used to write a series. For example, the sum of the first four terms of the sequence 3, 5, 7, 9, 11, ..., $2k + 1$, ... can be named as follows, using *sigma notation*, or *summation notation*:

$$\sum_{k=1}^{4} (2k + 1).$$ This represents $(2 \cdot 1 + 1) + (2 \cdot 2 + 1) + (2 \cdot 3 + 1) + (2 \cdot 4 + 1).$

This is read "the sum as k goes from 1 to 4 of $(2k + 1)$." The letter k is called the *index of summation*. The index of summation need not start at 1.

Example 4

Write out and evaluate each sum.

a) $\displaystyle\sum_{k=1}^{5} k^2$ **b)** $\displaystyle\sum_{k=4}^{6} (-1)^k(2k)$ **c)** $\displaystyle\sum_{k=0}^{3} (2^k + 5)$

Solution

a) $\displaystyle\sum_{k=1}^{5} k^2 = 1^2 + 2^2 + 3^2 + 4^2 + 5^2 = 1 + 4 + 9 + 16 + 25 = 55$

Evaluate k^2 for all integers from 1 through 5. Then add.

b) $\displaystyle\sum_{k=4}^{6} (-1)^k(2k) = (-1)^4(2 \cdot 4) + (-1)^5(2 \cdot 5) + (-1)^6(2 \cdot 6)$

$= 8 - 10 + 12 = 10$

c) $\displaystyle\sum_{k=0}^{3} (2^k + 5) = (2^0 + 5) + (2^1 + 5) + (2^2 + 5) + (2^3 + 5)$

$= 6 + 7 + 9 + 13 = 35$

Example 5

Write sigma notation for each sum.

a) $1 + 4 + 9 + 16 + 25$ **b)** $-1 + 3 - 5 + 7$

c) $3 + 9 + 27 + 81 + \cdots$

Solution

a) $1 + 4 + 9 + 16 + 25$

Note that this is a sum of squares, $1^2 + 2^2 + 3^2 + 4^2 + 5^2$, so the general term is k^2. Sigma notation is

$$\sum_{k=1}^{5} k^2.$$ The sum starts with 1^2 and ends with 5^2.

Answers may vary here. For example, another—perhaps less obvious—way of writing $1 + 4 + 9 + 16 + 25$ is

$$\sum_{k=2}^{6} (k - 1)^2.$$

b) $-1 + 3 - 5 + 7$

Except for the alternating signs, this is the sum of the first four positive odd numbers. Note that $2k - 1$ is a formula for the kth positive odd number. It is also important to note that since $(-1)^k = 1$ when k is even and $(-1)^k = -1$ when k is odd, the factor $(-1)^k$ can be used to create the alternating signs. The general term is thus $(-1)^k(2k - 1)$, beginning with $k = 1$. Sigma notation is

$$\sum_{k=1}^{4} (-1)^k(2k - 1).$$

To check, we can evaluate $(-1)^k(2k - 1)$ using 1, 2, 3, and 4. Then we can write the sum of the four terms. We leave this to the student.

c) $3 + 9 + 27 + 81 + \cdots$

This is a sum of powers of 3, and it is also an infinite series. We use the symbol ∞ for infinity and write the series using sigma notation:

$$\sum_{k=1}^{\infty} 3^k.$$

FOR EXTRA HELP

Exercise Set **14.1**

 Digital Video Tutor CD 11 Videotape 26   InterAct Math   Math Tutor Center   MathXL   MyMathLab.com

In each of the following, the nth term of a sequence is given. In each case, find the first 4 terms; the 10th term, a_{10}; and the 15th term, a_{15}.

1. $a_n = 5n - 2$

2. $a_n = 2n + 3$

3. $a_n = \dfrac{n}{n + 1}$

4. $a_n = n^2 + 2$

5. $a_n = n^2 - 2n$

6. $a_n = \dfrac{n^2 - 1}{n^2 + 1}$

7. $a_n = n + \dfrac{1}{n}$

8. $a_n = \left(-\dfrac{1}{2}\right)^{n-1}$

9. $a_n = (-1)^n n^2$

10. $a_n = (-1)^n(n + 3)$

11. $a_n = (-1)^{n+1}(3n - 5)$

12. $a_n = (-1)^n(n^3 - 1)$

Find the indicated term of each sequence.

13. $a_n = 2n - 5$; a_7

14. $a_n = 3n + 2$; a_8

15. $a_n = (3n + 1)(2n - 5)$; a_9

16. $a_n = (3n + 2)^2$; a_6

17. $a_n = (-1)^{n-1}(3.4n - 17.3)$; a_{12}

18. $a_n = (-2)^{n-2}(45.68 - 1.2n)$; a_{23}

19. $a_n = 3n^2(9n - 100)$; a_{11}

20. $a_n = 4n^2(2n - 39)$; a_{22}

21. $a_n = \left(1 + \dfrac{1}{n}\right)^2$; a_{20}

22. $a_n = \left(1 - \dfrac{1}{n}\right)^3$; a_{15}

Look for a pattern and then predict the general term, or nth term, a_n, of each sequence. Answers may vary.

23. $1, 3, 5, 7, 9, \ldots$

24. $2, 4, 6, 8, \ldots$

25. $1, -1, 1, -1, \ldots$

26. $-1, 1, -1, 1, \ldots$

27. $-1, 2, -3, 4, \ldots$

28. $1, -2, 3, -4, \ldots$

29. $-2, 6, -18, 54, \ldots$

30. $-2, 3, 8, 13, 18, \ldots$

31. $\frac{1}{2}, \frac{2}{3}, \frac{3}{4}, \frac{4}{5}, \frac{5}{6}, \ldots$

32. $1 \cdot 2, 2 \cdot 3, 3 \cdot 4, 4 \cdot 5, \ldots$

33. $5, 25, 125, 625, \ldots$

34. $4, 16, 64, 256, \ldots$

35. $-1, 4, -9, 16, \ldots$

36. $1, -4, 9, -16, \ldots$

Find the indicated partial sum for each sequence.

37. $1, -2, 3, -4, 5, -6, \ldots;\ S_7$

38. $1, -3, 5, -7, 9, -11, \ldots;\ S_8$

39. $2, 4, 6, 8, \ldots;\ S_5$

40. $1, \frac{1}{4}, \frac{1}{9}, \frac{1}{16}, \frac{1}{25}, \ldots;\ S_5$

Write out and evaluate each sum.

41. $\displaystyle\sum_{k=1}^{5} \frac{1}{2k}$

42. $\displaystyle\sum_{k=1}^{6} \frac{1}{2k-1}$

43. $\displaystyle\sum_{k=0}^{4} 3^k$

44. $\displaystyle\sum_{k=4}^{7} \sqrt{2k+1}$

45. $\displaystyle\sum_{k=1}^{8} \frac{k}{k+1}$

46. $\displaystyle\sum_{k=1}^{4} \frac{k-2}{k+3}$

47. $\displaystyle\sum_{k=1}^{8} (-1)^{k+1} 2^k$

48. $\displaystyle\sum_{k=1}^{7} (-1)^k 4^{k+1}$

49. $\displaystyle\sum_{k=0}^{5} (k^2 - 2k + 3)$

50. $\displaystyle\sum_{k=0}^{5} (k^2 - 3k + 4)$

51. $\displaystyle\sum_{k=3}^{5} \frac{(-1)^k}{k(k+1)}$

52. $\displaystyle\sum_{k=3}^{7} \frac{k}{2^k}$

Rewrite each sum using sigma notation. Answers may vary.

53. $\frac{2}{3} + \frac{3}{4} + \frac{4}{5} + \frac{5}{6} + \frac{6}{7}$

54. $3 + 6 + 9 + 12 + 15$

55. $1 + 4 + 9 + 16 + 25 + 36$

56. $\frac{1}{1^2} + \frac{1}{2^2} + \frac{1}{3^2} + \frac{1}{4^2} + \frac{1}{5^2}$

57. $4 - 9 + 16 - 25 + \cdots + (-1)^n n^2$

58. $9 - 16 + 25 + \cdots + (-1)^{n+1} n^2$

59. $5 + 10 + 15 + 20 + 25 + \cdots$

60. $7 + 14 + 21 + 28 + 35 + \cdots$

61. $\frac{1}{1 \cdot 2} + \frac{1}{2 \cdot 3} + \frac{1}{3 \cdot 4} + \frac{1}{4 \cdot 5} + \cdots$

62. $\frac{1}{1 \cdot 2^2} + \frac{1}{2 \cdot 3^2} + \frac{1}{3 \cdot 4^2} + \frac{1}{4 \cdot 5^2} + \cdots$

63. The sequence $1, 4, 9, 16, \ldots$ can be written as $f(x) = x^2$ with the domain the set of all positive integers. Explain how the graph of f would compare with the graph of $y = x^2$.

64. Eric says he expects he will prefer sequences to functions because he dislikes fractions. Will his expectations prove correct? Why or why not?

SKILL MAINTENANCE

Evaluate.

65. $\frac{7}{2}(a_1 + a_7)$, for $a_1 = 8$ and $a_7 = 14$

66. $a_1 + (n - 1)d$, for $a_1 = 3$, $n = 6$, and $d = 4$

Multiply.

67. $(x + y)^3$

68. $(a - b)^3$

69. $(2a - b)^3$

70. $(2x + y)^3$

SYNTHESIS

71. Explain why the equation
$$\sum_{k=1}^{n} (a_k + b_k) = \sum_{k=1}^{n} a_k + \sum_{k=1}^{n} b_k$$
is true for any positive integer n. What laws are used to justify this result?

72. Consider the sums
$$\sum_{k=1}^{5} 3k^2 \quad \text{and} \quad 3\sum_{k=1}^{5} k^2.$$
a) Which is easier to evaluate and why?
b) Is it true that
$$\sum_{k=1}^{n} ca_k = c \sum_{k=1}^{n} a_k?$$
Why or why not?

Some sequences are given by a recursive definition. The value of the first term, a_1, is given, and then we are told how to find any subsequent term from the term preceding it. Find the first six terms of each of the following recursively defined sequences.

73. $a_1 = 1,\ a_{n+1} = 5a_n - 2$

74. $a_1 = 0,\ a_{n+1} = a_n^2 + 3$

75. *Cell biology.* A single cell of bacterium divides into two every 15 min. Suppose that the same rate of division is maintained for 4 hr. Give a sequence that lists the number of cells after successive 15-min periods.

76. *Value of a copier.* The value of a color photocopier is $5200. Its scrap value each year is 75% of its value the year before. Give a sequence that lists the scrap value of the machine at the start of each year for a 10-yr period.

77. Find S_{100} and S_{101} for the sequence in which $a_n = (-1)^n$.

Find the first five terms of each sequence; then find S_5.

78. $a_n = \dfrac{1}{2^n} \log 1000^n$ **79.** $a_n = i^n, i = \sqrt{-1}$

80. Find all values for x that solve the following:
$$\sum_{k=1}^{x} i^k = -1.$$

81. The nth term of a sequence is given by
$$a_n = n^5 - 14n^4 + 6n^3 + 416n^2 - 655n - 1050.$$
Use a grapher with a TABLE feature to determine what term in the sequence is 6144.

82. To define a sequence recursively on a grapher (see Exercises 73 and 74), the SEQ MODE is used. The general term U_n or V_n can often be expressed in terms of U_{n-1} or V_{n-1} by pressing [2nd] [7] or [2nd] [8] . The starting values of U_n, V_n, and n are set as one of the WINDOW variables.

Use recursion to determine how many handshakes will occur if a group of 50 people shake hands with one another. To develop the recursion formula, begin with a group of 2 and determine how many additional handshakes occur with the arrival of each new group member.

Arithmetic Sequences and Series

14.2

Arithmetic Sequences • Sum of the First *n* Terms of an Arithmetic Sequence • Problem Solving

In this section, we concentrate on sequences and series that are said to be arithmetic (pronounced ar-ith-MET-ik).

Arithmetic Sequences

In an **arithmetic sequence** (or **progression**), any term (other than the first) can be found by adding the same number to its preceding term. For example, the sequence 2, 5, 8, 11, 14, 17, . . . is arithmetic because adding 3 to any term produces the next term.

Arithmetic Sequence

A sequence is *arithmetic* if there exists a number d, called the *common difference*, such that $a_{n+1} = a_n + d$ for any integer $n \geq 1$.

Example 1

For each arithmetic sequence, identify the first term, a_1, and the common difference, d.

a) $4, 9, 14, 19, 24, \ldots$ **b)** $27, 20, 13, 6, -1, -8, \ldots$

Solution To find a_1, we simply use the first term listed. To find d, we choose any term beyond the first and subtract the preceding term from it.

Sequence	*First Term, a_1*	*Common Difference, d*
a) $4, 9, 14, 19, 24, \ldots$	4	$5 \longleftarrow 9 - 4 = 5$
b) $27, 20, 13, 6, -1, -8, \ldots$	27	$-7 \longleftarrow 20 - 27 = -7$

To find the common difference, we subtracted a_1 from a_2. Had we subtracted a_2 from a_3 or a_3 from a_4, we would have found the same values for d.

Check: As a check, note that when d is added to each term, the result is the next term in the sequence.

a) $4 + 5 = 9, \quad 9 + 5 - 14, \quad 14 + 5 = 19, \quad 19 + 5 = 24$
b) $27 + (-7) = 20, \quad 20 + (-7) = 13, \quad 13 + (-7) = 6, \quad 6 + (-7) = -1,$
$\quad -1 + (-7) = -8$

To find a formula for the general, or nth, term of any arithmetic sequence, we denote the common difference by d and write out the first few terms:

$a_1,$
$a_2 = a_1 + d,$
$a_3 = a_2 + d = (a_1 + d) + d = a_1 + 2d,$ Substituting $a_1 + d$ for a_2
$a_4 = a_3 + d = (a_1 + 2d) + d = a_1 + 3d.$ Substituting $a_1 + 2d$ for a_3

Note that the coefficient of d in each case is 1 less than the subscript.

Generalizing, we obtain the following formula.

To Find a_n for an Arithmetic Sequence

The nth term of an arithmetic sequence with common difference d is

$$a_n = a_1 + (n - 1)d, \quad \text{for any integer } n \geq 1.$$

E x a m p l e   2

Find the 14th term of the arithmetic sequence 6, 9, 12, 15,...

Solution   First we note that $a_1 = 6$, $d = 3$, and $n = 14$. Using the formula for the nth term of an arithmetic sequence, we have

$$a_n = a_1 + (n - 1)d$$
$$a_{14} = 6 + (14 - 1) \cdot 3 = 6 + 13 \cdot 3 = 6 + 39 = 45.$$

The 14th term is 45.

E x a m p l e   3

For the sequence in Example 2, which term is 300? That is, find n if $a_n = 300$.

Solution   We substitute into the formula for the nth term of an arithmetic sequence and solve for n:

$$a_n = a_1 + (n - 1)d$$
$$300 = 6 + (n - 1) \cdot 3$$
$$300 = 6 + 3n - 3$$
$$297 = 3n$$
$$99 = n.$$

The term 300 is the 99th term of the sequence.

Given two terms and their places in an arithmetic sequence, we can construct the sequence.

E x a m p l e   4

The 3rd term of an arithmetic sequence is 14, and the 16th term is 79. Find a_1 and d and construct the sequence.

Solution   We know that $a_3 = 14$ and $a_{16} = 79$. Thus we would have to add d 13 times to get from 14 to 79. That is,

$$14 + 13d = 79. \qquad a_3 \text{ and } a_{16} \text{ are 13 terms apart; } 16 - 3 = 13$$

Solving $14 + 13d = 79$, we obtain

$$13d = 65 \qquad \text{Subtracting 14 from both sides}$$
$$d = 5. \qquad \text{Dividing both sides by 13}$$

We subtract d twice from a_3 to get to a_1. Thus,

$$a_1 = 14 - 2 \cdot 5 = 4. \qquad a_1 \text{ and } a_3 \text{ are 2 terms apart; } 3 - 1 = 2$$

The sequence is 4, 9, 14, 19, ... Note that we could have subtracted d 15 times from a_{16} in order to find a_1.

In general, d should be subtracted $(n - 1)$ times from a_n in order to find a_1.

Sum of the First *n* Terms of an Arithmetic Sequence

When the terms of an arithmetic sequence are added, an **arithmetic series** is formed. To find a formula for computing S_n when the series is arithmetic, we denote the first *n* terms as follows:

This is the next-to-last term. If you add *d* to this term, the result is a_n.

$$a_1, (a_1 + d), (a_1 + 2d), \ldots, (a_n - 2d), (a_n - d), a_n$$

This term is two terms back from the end. If you add *d* to this term, you get the next-to-last term, $a_n - d$.

Thus, S_n is given by

$$S_n = a_1 + (a_1 + d) + (a_1 + 2d) + \cdots + (a_n - 2d) + (a_n - d) + a_n.$$

Using a commutative law, we have a second equation:

$$S_n = a_n + (a_n - d) + (a_n - 2d) + \cdots + (a_1 + 2d) + (a_1 + d) + a_1.$$

Adding corresponding terms on each side of the above equations, we get

$$2S_n = [a_1 + a_n] + [(a_1 + d) + (a_n - d)] + [(a_1 + 2d) + (a_n - 2d)]$$
$$+ \cdots + [(a_n - 2d) + (a_1 + 2d)] + [(a_n - d) + (a_1 + d)]$$
$$+ [a_n + a_1].$$

This simplifies to

$$2S_n = [a_1 + a_n] + [a_1 + a_n] + [a_1 + a_n]$$
$$+ \cdots + [a_n + a_1] + [a_n + a_1] + [a_n + a_1].$$

There are *n* bracketed sums.

Since $[a_1 + a_n]$ is being added *n* times, it follows that

$$2S_n = n[a_1 + a_n].$$

Dividing both sides by 2 leads to the following formula.

To Find S_n for an Arithmetic Sequence

The sum of the first *n* terms of an arithmetic sequence is given by

$$S_n = \frac{n}{2}(a_1 + a_n).$$

E x a m p l e   5   Find the sum of the first 100 positive even numbers.

Solution   The sum is

$$2 + 4 + 6 + \cdots + 198 + 200.$$

This is the sum of the first 100 terms of the arithmetic sequence for which

$$a_1 = 2, \quad n = 100, \quad \text{and} \quad a_n = 200.$$

Substituting in the formula

$$S_n = \frac{n}{2}(a_1 + a_n),$$

we get

$$S_{100} = \frac{100}{2}(2 + 200)$$

$$= 50(202) = 10{,}100.$$

The above formula is useful when we know the first and last terms, a_1 and a_n. To find S_n when a_n is unknown, but a_1, n, and d are known, we can use the formula $a_n = a_1 + (n - 1)d$ to calculate a_n and then proceed as in Example 5.

E x a m p l e   6   Find the sum of the first 15 terms of the arithmetic sequence 4, 7, 10, 13, ...

Solution   Note that

$$a_1 = 4, \quad n = 15, \quad \text{and} \quad d = 3.$$

Before using the formula for S_n, we find a_{15}:

$$a_{15} = 4 + (15 - 1)3 \qquad \text{Substituting into the formula for } a_n$$

$$= 4 + 14 \cdot 3 = 46.$$

Thus, knowing that $a_{15} = 46$, we have

$$S_{15} = \frac{15}{2}(4 + 46) \qquad \text{Using the formula for } S_n$$

$$= \frac{15}{2}(50) = 375.$$

Problem Solving

For some problem-solving situations, the translation may involve sequences or series. In Examples 7 and 8, the calculations and translations can be done in a number of ways. There is often a variety of ways in which a problem can be solved. You should use the one that is best or easiest for you. In this chapter, however, we will try to emphasize sequences and series and their related formulas.

E x a m p l e   7

Hourly wages.   Chris accepts a job managing a CD shop, starting with an hourly wage of $14.25, and is promised a raise of 15¢ per hour every 2 months for 5 years. After 5 years of work, what will be Chris's hourly wage?

Solution

1. **Familiarize.**   It helps to write down the hourly wage for several two-month time periods.

   Beginning: 14.25,

   After two months: 14.40,

   After four months: 14.55,

   and so on.

   What appears is a sequence of numbers: 14.25, 14.40, 14.55, ... Since the same amount is added each time, the sequence is arithmetic.
      We list what we know about arithmetic sequences. The pertinent formulas are

   $$a_n = a_1 + (n - 1)d$$

   and

   $$S_n = \frac{n}{2}(a_1 + a_n).$$

   In this case, we are not looking for a sum, so it is probably the first formula that will give us our answer. We want to determine the last term in a sequence. To do so, we need to know a_1, n, and d. From our list above, we see that

   $$a_1 = 14.25 \quad \text{and} \quad d = 0.15.$$

   What is n? That is, how many terms are in the sequence? After 1 year, there have been 6 raises, since Chris gets a raise every 2 months. There are 5 years, so the total number of raises will be $5 \cdot 6$, or 30. Altogether, there will be 31 terms: the original wage and 30 increased rates.

2. **Translate.**   We want to find a_n for the arithmetic sequence in which $a_1 = 14.25$, $n = 31$, and $d = 0.15$.

3. **Carry out.**   Substituting in the formula for a_n gives us

   $$a_{31} = 14.25 + (31 - 1) \cdot 0.15$$
   $$= 18.75.$$

4. **Check.**   We can check by redoing the calculations or we can calculate in a slightly different way for another check. For example, at the end of a year, there will be 6 raises, for a total raise of $0.90. At the end of 5 years, the total raise will be $5 \times 0.90, or $4.50. If we add that to the original wage of $14.25, we obtain $18.75. The answer checks.

5. **State.**   After 5 years, Chris's hourly wage will be $18.75.

E x a m p l e   8   ***Telephone pole storage.***   A stack of telephone poles has 30 poles in the bottom row. There are 29 poles in the second row, 28 in the next row, and so on. How many poles are in the stack if there are 5 poles in the top row?

Solution

1. **Familiarize.**   A picture will help in this case. The following figure shows the ends of the poles and the way in which they stack. There are 30 poles on the bottom, and we see that there will be one fewer in each succeeding row. How many rows will there be?

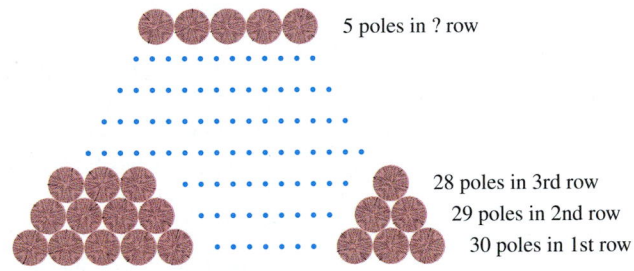

5 poles in ? row

28 poles in 3rd row
29 poles in 2nd row
30 poles in 1st row

   Note that there are $30 - 1 = 29$ poles in the 2nd row, $30 - 2 = 28$ poles in the 3rd row, $30 - 3 = 27$ poles in the 4th row, and so on. The pattern leads to $30 - 25 = 5$ poles in the 26th row.
   The situation is represented by the equation

   $30 + 29 + 28 + \cdots + 5.$ There are 26 terms in this series.

   Thus we have an arithmetic series. We recall the formula

   $$S_n = \frac{n}{2}(a_1 + a_n).$$

2. **Translate.**   We want to find the sum of the first 26 terms of an arithmetic sequence in which $a_1 = 30$ and $a_{26} = 5$.

3. **Carry out.** Substituting into the above formula gives us

$$S_{26} = \frac{26}{2}(30 + 5)$$
$$= 13 \cdot 35 = 455.$$

4. **Check.** In this case, we can check the calculations by doing them again. A longer, harder way would be to do the entire addition:

$$30 + 29 + 28 + \cdots + 5.$$

5. **State.** There are 455 poles in the stack.

FOR EXTRA HELP

Exercise Set **14.2**

Digital Video Tutor CD 11   InterAct Math   Math Tutor Center   MathXL   MyMathLab.com
Videotape 26

Find the first term and the common difference.

1. $2, 6, 10, 14, \ldots$

2. $1.06, 1.12, 1.18, 1.24, \ldots$

3. $6, 2, -2, -6, \ldots$

4. $-9, -6, -3, 0, \ldots$

5. $\frac{3}{2}, \frac{9}{4}, 3, \frac{15}{4}, \ldots$

6. $\frac{3}{5}, \frac{1}{10}, -\frac{2}{5}, \ldots$

7. $\$5.12, \$5.24, \$5.36, \$5.48, \ldots$

8. $\$214, \$211, \$208, \$205, \ldots$

9. Find the 12th term of the arithmetic sequence $3, 7, 11, \ldots$.

10. Find the 11th term of the arithmetic sequence $0.07, 0.12, 0.17, \ldots$.

11. Find the 17th term of the arithmetic sequence $7, 4, 1, \ldots$.

12. Find the 14th term of the arithmetic sequence $3, \frac{7}{3}, \frac{5}{3}, \ldots$.

13. Find the 13th term of the arithmetic sequence $\$1200, \$964.32, \$728.64, \ldots$.

14. Find the 10th term of the arithmetic sequence $\$2345.78, \$2967.54, \$3589.30, \ldots$.

15. In the sequence of Exercise 9, what term is 107?

16. In the sequence of Exercise 10, what term is 1.67?

17. In the sequence of Exercise 11, what term is -296?

18. In the sequence of Exercise 12, what term is -27?

19. Find a_{17} when $a_1 = 2$ and $d = 5$.

20. Find a_{20} when $a_1 = 14$ and $d = -3$.

21. Find a_1 when $d = 4$ and $a_8 = 33$.

22. Find a_1 when $d = 8$ and $a_{11} = 26$.

23. Find n when $a_1 = 5$, $d = -3$, and $a_n = -76$.

24. Find n when $a_1 = 25$, $d = -14$, and $a_n = -507$.

25. For an arithmetic sequence in which $a_{17} = -40$ and $a_{28} = -73$, find a_1 and d. Write the first five terms of the sequence.

26. In an arithmetic sequence, $a_{17} = \frac{25}{3}$ and $a_{32} = \frac{95}{6}$. Find a_1 and d. Write the first five terms of the sequence.

Aha! 27. Find a_1 and d if $a_{13} = 13$ and $a_{54} = 54$.

28. Find a_1 and d if $a_{12} = 24$ and $a_{25} = 50$.

29. Find the sum of the first 20 terms of the arithmetic series $1 + 5 + 9 + 13 + \cdots$.

30. Find the sum of the first 14 terms of the arithmetic series $11 + 7 + 3 + \cdots$.

31. Find the sum of the first 250 natural numbers.

32. Find the sum of the first 400 natural numbers.

33. Find the sum of the even numbers from 2 to 100, inclusive.

34. Find the sum of the odd numbers from 1 to 99, inclusive.

35. Find the sum of all multiples of 6 from 6 to 102, inclusive.

36. Find the sum of all multiples of 4 that are between 15 and 521.

37. An arithmetic series has $a_1 = 4$ and $d = 5$. Find S_{20}.

38. An arithmetic series has $a_1 = 9$ and $d = -3$. Find S_{32}.

Solve.

39. *Band formations.*   The Duxbury marching band has 14 marchers in the front row, 16 in the second row, 18 in the third row, and so on, for 15 rows. How many marchers are in the last row? How many marchers are there altogether?

40. *Gardening.*   A gardener is planting bulbs near an entrance to a college. She has 39 plants in the front row, 35 in the second row, 31 in the third row, and so on. If the pattern is consistent, how many plants will be in the last row? How many plants will there be altogether?

41. *Telephone pole piles.*   How many poles will be in a pile of telephone poles if there are 50 in the first layer, 49 in the second, and so on, until there are 6 in the last layer?

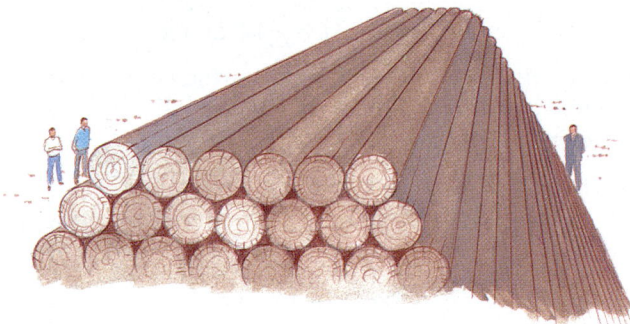

42. *Accumulated savings.*   If 10¢ is saved on October 1, another 20¢ on October 2, another 30¢ on October 3, and so, how much is saved during October? (October has 31 days.)

43. *Accumulated savings.*   Renata saves money in an arithmetic sequence: $600 for the first year, another $700 the second, and so on, for 20 yr. How much does she save in all (disregarding interest)?

44. *Spending.*   Jacob spent $30 on August 1, $50 on August 2, $70 on August 3, and so on. How much did Jacob spend in August? (August has 31 days.)

45. *Auditorium design.*   Theaters are often built with more seats per row as the rows move toward the

back. The Sanders Amphitheater has 20 seats in the first row, 22 in the second, 24 in the third, and so on, for 19 rows. How many seats are in the amphitheater?

46. *Accumulated savings.*   Shirley sets up an investment such that it will return $5000 the first year, $6125 the second year, $7250 the third year, and so on, for 25 yr. How much in all is received from the investment?

47. It is said that as a young child, the mathematician Karl F. Gauss (1777–1855) was able to compute the sum $1 + 2 + 3 + \cdots + 100$ very quickly in his head. Explain how Gauss might have done this and present a formula for the sum of the first n natural numbers. (*Hint*: $1 + 99 = 100$.)

48. If every number in a sequence is doubled and then added, is the result the same as if the numbers were first added and the sum then doubled? Why or why not?

SKILL MAINTENANCE

Simplify.

49. $\dfrac{3}{10x} + \dfrac{2}{15x}$

50. $\dfrac{2}{9t} + \dfrac{5}{12t}$

Convert to an exponential equation.

51. $\log_a P = k$

52. $\ln t = a$

Find an equation of the circle satisfying the given conditions.

53. Center $(0, 0)$, radius 9

54. Center $(-2, 5)$, radius $3\sqrt{2}$

SYNTHESIS

55. Write a problem for a classmate to solve. Devise the problem so that its solution requires computing S_{17} for an arithmetic sequence.

56. The sum of the first n terms of an arithmetic sequence is also given by

$$S_n = \frac{n}{2}[2a_1 + (n-1)d].$$

Use the earlier formulas for a_n and S_n to explain how this equation was developed.

57. Find a formula for the sum of the first n consecutive odd numbers starting with 1:

$$1 + 3 + 5 + \cdots + (2n - 1).$$

58. Find three numbers in an arithmetic sequence for which the sum of the first and third is 10 and the product of the first and second is 15.

59. In an arithmetic sequence, $a_1 = \$8760$ and $d = -\$798.23$. Find the first 10 terms of the sequence.

60. Find the sum of the first 10 terms of the sequence given in Exercise 59.

61. Prove that if p, m, and q are consecutive terms in an arithmetic sequence, then

$$m = \frac{p + q}{2}.$$

62. *Straight-line depreciation.* A company buys a color copier for $5200 on January 1 of a given year. The machine is expected to last for 8 yr, at the end of which time its *trade-in*, or *salvage, value* will be $1100. If the company figures the decline in value to be the same each year, then the trade-in values, after t years, $0 \le t \le 8$, form an arithmetic sequence given by

$$a_t = C - t\left(\frac{C - S}{N}\right),$$

where C is the original cost of the item, N the years of expected life, and S the salvage value.

a) Find the formula for a_t for the straight-line depreciation of the copier.
b) Find the salvage value after 0 yr, 1 yr, 2 yr, 3 yr, 4 yr, 7 yr, and 8 yr.
c) Find a formula that expresses a_t recursively.

63. Use your answer to Exercise 31 to find the sum of all integers from 501 through 750.

Geometric Sequences and Series

14.3

Geometric Sequences • Sum of the First n Terms of a Geometric Sequence • Infinite Geometric Series • Problem Solving

In an arithmetic sequence, a certain number is added to each term to get the next term. When each term in a sequence is *multiplied* by a certain number to get the next term, the sequence is **geometric**. In this section, we examine geometric sequences (or progressions) and *geometric series*.

Geometric Sequences

Consider the sequence

$$2, 6, 18, 54, 162, \ldots.$$

If we multiply each term by 3, we obtain the next term. The multiplier is called the *common ratio* because it is found by dividing any term by the preceding term.

> ### Geometric Sequence
>
> A sequence is *geometric* if there exists a number r, called the *common ratio*, for which
>
> $$\frac{a_{n+1}}{a_n} = r, \quad \text{or} \quad a_{n+1} = a_n \cdot r \quad \text{for any integer } n \geq 1.$$

E x a m p l e   1   For each geometric sequence, find the common ratio.

a) $3, 6, 12, 24, 48, \ldots$ b) $3, -6, 12, -24, 48, -96, \ldots$

c) $\$5200, \$3900, \$2925, \$2193.75, \ldots$

Solution

Sequence	Common Ratio	
a) $3, 6, 12, 24, 48, \ldots$	2	$\frac{6}{3} = 2, \frac{12}{6} = 2$, and so on
b) $3, -6, 12, -24, 48, -96, \ldots$	-2	$\frac{-6}{3} = -2, \frac{12}{-6} = -2$, and so on
c) $\$5200, \$3900, \$2925, \$2193.75, \ldots$	0.75	$\frac{\$3900}{\$5200} = 0.75, \frac{\$2925}{\$3900} = 0.75$

To develop a formula for the general, or nth, term of a geometric sequence, let a_1 be the first term and let r be the common ratio. We write out the first few terms as follows:

$a_1,$

$a_2 = a_1 r,$

$a_3 = a_2 r = (a_1 r)r = a_1 r^2,$ Substituting $a_1 r$ for a_2

$a_4 = a_3 r = (a_1 r^2)r = a_1 r^3.$ Substituting $a_1 r^2$ for a_3

Note that the exponent is 1 less than the subscript.

Generalizing, we obtain the following.

> ### To Find a_n for a Geometric Sequence
>
> The nth term of a geometric sequence with common ratio r is given by
>
> $$a_n = a_1 r^{n-1}, \quad \text{for any integer } n \geq 1.$$

E x a m p l e   2   Find the 7th term of the geometric sequence $4, 20, 100, \ldots$.

Solution   First we note that

$$a_1 = 4 \quad \text{and} \quad n = 7.$$

To find the common ratio, we can divide any term (other than the first) by the term preceding it. Since the second term is 20 and the first is 4,

$$r = \frac{20}{4}, \quad \text{or } 5.$$

The formula

$$a_n = a_1 r^{n-1}$$

gives us

$$a_7 = 4 \cdot 5^{7-1} = 4 \cdot 5^6 = 4 \cdot 15{,}625 = 62{,}500.$$

E x a m p l e   3   Find the 10th term of the geometric sequence

$$64, -32, 16, -8, \ldots.$$

Solution   First, we note that

$$a_1 = 64, \qquad n = 10, \quad \text{and} \quad r = \frac{-32}{64} = -\frac{1}{2}.$$

Then, using the formula for the nth term of a geometric sequence, we have

$$a_{10} = 64 \cdot \left(-\frac{1}{2}\right)^{10-1} = 64 \cdot \left(-\frac{1}{2}\right)^9 = 2^6 \cdot \left(-\frac{1}{2^9}\right) = -\frac{1}{2^3} = -\frac{1}{8}.$$

The 10th term is $-\frac{1}{8}$.

Sum of the First n Terms of a Geometric Sequence

We next develop a formula for S_n when a sequence is geometric:

$$a_1, a_1 r, a_1 r^2, a_1 r^3, \ldots, a_1 r^{n-1}, \ldots.$$

The **geometric series** S_n is given by

$$S_n = a_1 + a_1 r + a_1 r^2 + \cdots + a_1 r^{n-2} + a_1 r^{n-1}. \tag{1}$$

Multiplying both sides by r gives us

$$rS_n = a_1 r + a_1 r^2 + a_1 r^3 + \cdots + a_1 r^{n-1} + a_1 r^n. \tag{2}$$

When we subtract corresponding sides of equation (2) from equation (1), the color terms drop out, leaving

$$S_n - rS_n = a_1 - a_1 r^n$$
$$S_n(1 - r) = a_1(1 - r^n), \qquad \text{Factoring}$$

or

$$S_n = \frac{a_1(1 - r^n)}{1 - r}. \qquad \text{Dividing both sides by } 1 - r$$

> **To Find S_n for a Geometric Sequence**
>
> The sum of the first n terms of a geometric sequence with common ratio r is given by
>
> $$S_n = \frac{a_1(1 - r^n)}{1 - r}, \quad \text{for any } r \neq 1.$$

Example 4 Find the sum of the first 7 terms of the geometric sequence 3, 15, 75, 375, ...

Solution First, we note that

$$a_1 = 3, \quad n = 7, \quad \text{and} \quad r = \frac{15}{3} = 5.$$

Then, substituting in the formula $S_n = \dfrac{a_1(1 - r^n)}{1 - r}$, we have

$$S_7 = \frac{3(1 - 5^7)}{1 - 5} = \frac{3(1 - 78{,}125)}{-4}$$

$$= \frac{3(-78{,}124)}{-4}$$

$$= 58{,}593.$$

Infinite Geometric Series

Suppose we consider the sum of the terms of an infinite geometric sequence, such as 2, 4, 8, 16, 32, ... We get what is called an **infinite geometric series**:

$$2 + 4 + 8 + 16 + 32 + \cdots.$$

Here, as n grows larger and larger, the sum of the first n terms, S_n, becomes larger and larger without bound. There are also infinite series that get closer and closer to some specific number. Here is an example:

$$\frac{1}{2} + \frac{1}{4} + \frac{1}{8} + \frac{1}{16} + \cdots + \frac{1}{2^n} + \cdots.$$

Let's consider S_n for the first four values of n:

$$S_1 = \tfrac{1}{2} \qquad\qquad\quad = \tfrac{1}{2} = 0.5,$$
$$S_2 = \tfrac{1}{2} + \tfrac{1}{4} \qquad\quad = \tfrac{3}{4} = 0.75,$$
$$S_3 = \tfrac{1}{2} + \tfrac{1}{4} + \tfrac{1}{8} \qquad = \tfrac{7}{8} = 0.875,$$
$$S_4 = \tfrac{1}{2} + \tfrac{1}{4} + \tfrac{1}{8} + \tfrac{1}{16} = \tfrac{15}{16} = 0.9375.$$

$\uparrow$

The denominator of the sum is 2^n, where n is the subscript of S. The numerator is $2^n - 1$.

Thus, for this particular series, we have

$$S_n = \frac{2^n - 1}{2^n} = \frac{2^n}{2^n} - \frac{1}{2^n} = 1 - \frac{1}{2^n}.$$

Note that the value of S_n is less than 1 for any value of n, but as n gets larger and larger, the values of $1/2^n$ get closer to 0 and the values of S_n get closer to 1. We say that 1 is the *limit* of S_n and that 1 is the sum of this infinite geometric sequence. An infinite geometric series is denoted S_∞. It can be shown (but we will not do it here) that the sum of the terms of an infinite geometric sequence exists if and only if $|r| < 1$ (that is, the absolute value of the common ratio is less than 1).

To find a formula for the sum of an infinite geometric sequence, we first consider the sum of the first n terms:

$$S_n = \frac{a_1(1 - r^n)}{1 - r} = \frac{a_1 - a_1r^n}{1 - r}. \qquad \text{\textcolor{red}{Using the distributive law}}$$

For $|r| < 1$, it follows that values of r^n get closer to 0 as n gets larger. (Check this by selecting a number between -1 and 1 and finding larger and larger powers on a calculator.) As r^n gets closer to 0, so does a_1r^n. Thus, S_n gets closer to $a_1/(1 - r)$.

The Limit of an Infinite Geometric Series

When $|r| < 1$, the limit of an infinite geometric series is given by

$$S_\infty = \frac{a_1}{1 - r}. \qquad \text{(For } |r| \geq 1\text{, no limit exists.)}$$

Example 5 Determine whether each series has a limit. If a limit exists, find it.

a) $1 + 3 + 9 + 27 + \cdots$
b) $-2 + 1 - \frac{1}{2} + \frac{1}{4} - \frac{1}{8} + \cdots$

Solution

a) Here $r = 3$, so $|r| = |3| = 3$. Since $|r| \not< 1$, the series does *not* have a limit.

b) Here $r = -\frac{1}{2}$, so $|r| = \left|-\frac{1}{2}\right| = \frac{1}{2}$. Since $|r| < 1$, the series *does* have a limit. We find the limit by substituting into the formula for S_∞:

$$S_\infty = \frac{-2}{1 - \left(-\frac{1}{2}\right)} = \frac{-2}{\frac{3}{2}} = -2 \cdot \frac{2}{3} = -\frac{4}{3}.$$

Example 6 Find fractional notation for $0.63636363\ldots$.

Solution We can express this as

$$0.63 + 0.0063 + 0.000063 + \cdots.$$

This is an infinite geometric series, where $a_1 = 0.63$ and $r = 0.01$. Since $|r| < 1$, this series has a limit:

$$S_\infty = \frac{a_1}{1 - r} = \frac{0.63}{1 - 0.01} = \frac{0.63}{0.99} = \frac{63}{99}.$$

Thus fractional notation for $0.63636363\ldots$ is $\frac{63}{99}$, or $\frac{7}{11}$.

Problem Solving

For some problem-solving situations, the translation may involve geometric sequences or series.

E x a m p l e   7

Daily wages.   Suppose someone offered you a job for the month of September (30 days) under the following conditions. You will be paid $0.01 for the first day, $0.02 for the second, $0.04 for the third, and so on, doubling your previous day's salary each day. How much would you earn? (Would you take the job? Make a guess before reading further.)

Solution

1. **Familiarize.**   You earn $0.01 the first day, $0.01(2) the second day, $0.01(2)(2) the third day, and so on. Since each day's wages are a constant multiple of the previous day's wages, a geometric sequence is formed.

2. **Translate.**   The amount earned is the geometric series

$$\$0.01 + \$0.01(2) + \$0.01(2^2) + \$0.01(2^3) + \cdots + \$0.01(2^{29}),$$

where

$$a_1 = \$0.01, \quad n = 30, \quad \text{and} \quad r = 2.$$

3. **Carry out.**   Using the formula

$$S_n = \frac{a_1(1 - r^n)}{1 - r},$$

we have

$$S_{30} = \frac{\$0.01(1 - 2^{30})}{1 - 2}$$

$$= \frac{\$0.01(-1,073,741,823)}{-1} \qquad \text{Using a calculator}$$

$$= \$10,737,418.23.$$

4. **Check.**   The calculations can be repeated as a check.

5. **State.**   The pay exceeds $10.7 million for the month. Most people would probably take the job!

E x a m p l e   8

Loan repayment.   Francine's student loan is in the amount of $6000. Interest is to be 9% compounded annually, and the entire amount is to be paid after 10 yr. How much is to be paid back?

Solution

1. **Familiarize.**   Suppose we let P represent any principal amount. At the end of one year, the amount owed will be $P + 0.09P$, or $1.09P$. That amount will be the principal for the second year. The amount owed at the end of the

second year will be $1.09 \times$ New principal $= 1.09(1.09P)$, or 1.09^2P. Thus the amount owed at the beginning of successive years is as follows:

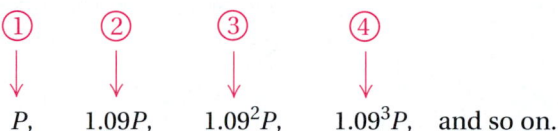

$$P, \quad 1.09P, \quad 1.09^2P, \quad 1.09^3P, \quad \text{and so on.}$$

We have a geometric sequence. The amount owed at the beginning of the 11th year will be the amount owed at the end of the 10th year.

2. **Translate.**   We have a geometric sequence with $a_1 = 6000$, $r = 1.09$, and $n = 11$. The appropriate formula is

$$a_n = a_1 r^{n-1}.$$

3. **Carry out.**   We substitute and calculate:

$$a_{11} = \$6000(1.09)^{11-1} = \$6000(1.09)^{10}$$

$$\approx \$14{,}204.18. \qquad \textcolor{red}{\text{Using a calculator and rounding to the nearest hundredth}}$$

4. **Check.**   A check, by repeating the calculations, is left to the student.

5. **State.**   Francine will owe \$14,204.18 at the end of 10 yr.

E x a m p l e   9

Bungee jumping.   A bungee jumper rebounds 60% of the height jumped. A bungee jump is made using a cord that stretches to 200 ft.

a) After jumping and then rebounding 9 times, how far has a bungee jumper traveled upward (the total rebound distance)?

b) Approximately how far will a jumper have traveled upward (bounced) before coming to rest?

Solution

1. **Familiarize.**   Let's do some calculations and look for a pattern.

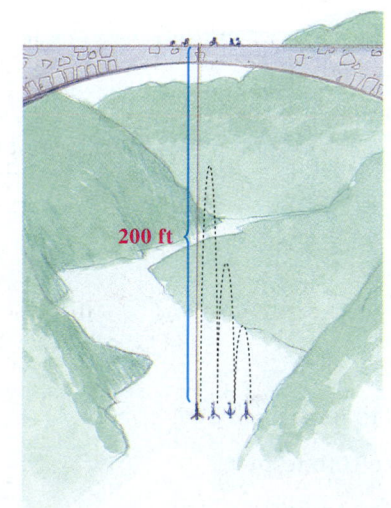

200 ft

First fall:	200 ft
First rebound:	0.6×200, or 120 ft
Second fall:	120 ft, or 0.6×200
Second rebound:	0.6×120, or $0.6(0.6 \times 200)$, which is 72 ft
Third fall:	72 ft, or $0.6(0.6 \times 200)$
Third rebound:	0.6×72, or $0.6(0.6(0.6 \times 200))$, which is 43.2 ft

The rebound distances form a geometric sequence:

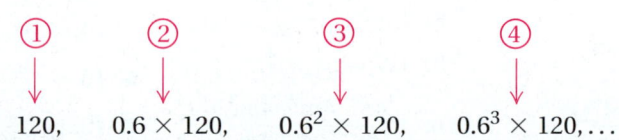

$$120, \quad 0.6 \times 120, \quad 0.6^2 \times 120, \quad 0.6^3 \times 120, \dots .$$

2. Translate.

a) The total rebound distance after 9 bounces is the sum of a geometric sequence. The first term is 120 and the common ratio is 0.6. There will be 9 terms, so we can use the formula

$$S_n = \frac{a_1(1 - r^n)}{1 - r}.$$

b) Theoretically, the jumper will never stop bouncing. Realistically, the bouncing will eventually stop. To approximate the actual distance bounced, we consider an infinite number of bounces and use the formula

$$S_\infty = \frac{a_1}{1 - r}. \qquad \text{Since } r = 0.6 \text{ and } |0.6| < 1, \text{ we know that } S_\infty \text{ exists.}$$

3. Carry out.

a) We substitute into the formula and calculate:

$$S_9 = \frac{120[1 - (0.6)^9]}{1 - 0.6} \approx 297. \qquad \text{Using a calculator}$$

b) We substitute and calculate:

$$S_\infty = \frac{120}{1 - 0.6} = 300.$$

4. Check. We can do the calculations again.

5. State.

a) In 9 bounces, the bungee jumper will have traveled upward a total distance of about 297 ft.

b) The jumper will have traveled upward a total of about 300 ft before coming to rest.

Exercise Set 14.3

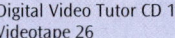

Find the common ratio for each geometric sequence.

1. $7, 14, 28, 56, \ldots$

2. $2, 6, 18, 54, \ldots$

3. $5, -5, 5, -5, \ldots$

4. $-5, -0.5, -0.05, -0.005, \ldots$

5. $\frac{1}{2}, -\frac{1}{4}, \frac{1}{8}, -\frac{1}{16}, \ldots$

6. $\frac{2}{3}, -\frac{4}{3}, \frac{8}{3}, -\frac{16}{3}, \ldots$

7. $75, 15, 3, \frac{3}{5}, \ldots$

8. $12, -4, \frac{4}{3}, -\frac{4}{9}, \ldots$

9. $\frac{1}{m}, \frac{3}{m^2}, \frac{9}{m^3}, \frac{27}{m^4}, \ldots$

10. $4, \frac{4m}{5}, \frac{4m^2}{25}, \frac{4m^3}{125}, \ldots$

Find the indicated term for each geometric sequence.

11. $3, 6, 12, \ldots$; the 7th term

12. $2, 8, 32, \ldots$; the 9th term

13. $5, 5\sqrt{2}, 10, \ldots$; the 9th term

14. $4, 4\sqrt{3}, 12, \ldots$; the 8th term

15. $-\frac{8}{243}, \frac{8}{81}, -\frac{8}{27}, \ldots$; the 10th term

16. $\frac{7}{625}, \frac{-7}{125}, \frac{7}{25}, \ldots$; the 13th term

17. $1000, $1080, $1166.40, \ldots$; the 12th term

18. $1000, $1070, $1144.90, \ldots$; the 11th term

Find the nth, or general, term for each geometric sequence.

19. $1, 3, 9, \ldots$

20. $25, 5, 1, \ldots$

21. $1, -1, 1, -1, \ldots$

22. $2, 4, 8, \ldots$

23. $\dfrac{1}{x}, \dfrac{1}{x^2}, \dfrac{1}{x^3}, \ldots$

24. $5, \dfrac{5m}{2}, \dfrac{5m^2}{4}, \ldots$

For Exercises 25–32, use the formula for S_n to find the indicated sum.

25. S_7 for the geometric series $6 + 12 + 24 + \cdots$

26. S_6 for the geometric series $16 - 8 + 4 - \cdots$

27. S_7 for the geometric series $\frac{1}{18} - \frac{1}{6} + \frac{1}{2} - \cdots$

Aha! **28.** S_5 for the geometric series $7 + 0.7 + 0.07 + \cdots$

29. S_8 for the series $1 + x + x^2 + x^3 + \cdots$

30. S_{10} for the series $1 + x^2 + x^4 + x^6 + \cdots$

31. S_{16} for the geometric sequence
$$200, \$200(1.06), \$200(1.06)^2, \ldots$$

32. S_{23} for the geometric sequence
$$1000, \$1000(1.08), \$1000(1.08)^2, \ldots$$

Determine whether each infinite geometric series has a limit. If a limit exists, find it.

33. $16 + 4 + 1 + \cdots$

34. $8 + 4 + 2 + \cdots$

35. $7 + 3 + \frac{9}{7} + \cdots$

36. $12 + 9 + \frac{27}{4} + \cdots$

37. $3 + 15 + 75 + \cdots$

38. $2 + 3 + \frac{9}{2} + \cdots$

39. $4 - 6 + 9 - \frac{27}{2} + \cdots$

40. $-6 + 3 - \frac{3}{2} + \frac{3}{4} - \cdots$

41. $0.43 + 0.0043 + 0.000043 + \cdots$

42. $0.37 + 0.0037 + 0.000037 + \cdots$

43. $500(1.02)^{-1} + \$500(1.02)^{-2} + \$500(1.02)^{-3} + \cdots$

44. $1000(1.08)^{-1} + \$1000(1.08)^{-2} + \$1000(1.08)^{-3} + \cdots$

Find fractional notation for each infinite sum. (These are geometric series.)

45. $0.7777\ldots$

46. $0.2222\ldots$

47. $8.3838\ldots$

48. $7.4747\ldots$

49. $0.15151515\ldots$

50. $0.12121212\ldots$

Solve. Use a calculator as needed for evaluating formulas.

51. *Rebound distance.* A ping-pong ball is dropped from a height of 20 ft and always rebounds one fourth of the distance fallen. How high does it rebound the 6th time?

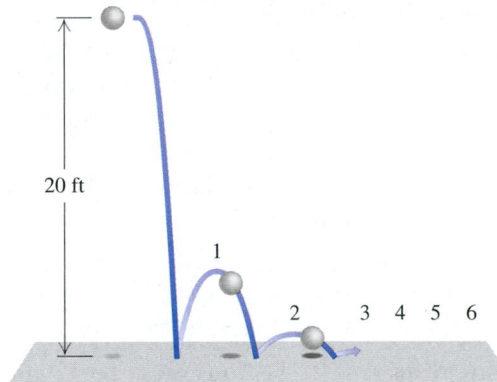

52. *Rebound distance.* Approximate the total of the rebound heights of the ball in Exercise 51.

53. *Population growth.* Yorktown has a current population of 100,000, and the population is increasing by 3% each year. What will the population be in 15 yr?

54. *Doubling time.* How long will it take for the population of Yorktown to double? (See Exercise 53.)

55. *Amount owed.* Gilberto borrows $15,000. The loan is to be repaid in 13 yr at 8.5% interest, compounded annually. How much will be repaid at the end of 13 yr?

56. *Shrinking population.* A population of 5000 fruit flies is dying off at a rate of 4% per minute. How many flies will be alive after 15 min?

57. *Shrinking population.* For the population of fruit flies in Exercise 56, how long will it take for only 1800 fruit flies to remain alive? (See Exercise 56 and use logarithms.) Round to the nearest minute.

58. *Investing.* Leslie is saving money in a retirement account. At the beginning of each year, she invests $1000 at 7%, compounded annually. How much will be in the retirement fund at the end of 40 yr?

59. *Rebound distance.* A superball dropped from the top of the Washington Monument (556 ft high) rebounds three fourths of the distance fallen. How far (up and down) will the ball have traveled when it hits the ground for the 6th time?

60. *Rebound distance.* Approximate the total distance that the ball of Exercise 59 will have traveled when it comes to rest.

61. *Stacking paper.* Construction paper is about 0.02 in. thick. Beginning with just one piece, a stack is doubled again and again 10 times. Find the height of the final stack.

62. *Monthly earnings.* Suppose you accepted a job for the month of February (28 days) under the following conditions. You will be paid $0.01 the first day, $0.02 the second, $0.04 the third, and so on, doubling your previous day's salary each day. How much would you earn?

Aha! **63.** Under what circumstances is it possible for the 5th term of a geometric sequence to be greater than the 4th term but less than the 7th term?

64. When r is negative, a series is said to be *alternating*. Why do you suppose this terminology is used?

SKILL MAINTENANCE

Multiply.

65. $(x + y)(x^2 + 2xy + y^2)$

66. $(a - b)(a^2 - 2ab + b^2)$

Solve the system.

67. $5x - 2y = -3,$
$2x + 5y = -24$

68. $x - 2y + 3z = 4,$
$2x - y + z = -1,$
$4x + y + z = 1$

SYNTHESIS

69. Write a problem for a classmate to solve. Devise the problem so that a geometric series is involved and the solution is "The total amount in the bank is $900(1.08)^{40}$, or about $19,550."

70. The infinite series

$$S_\infty = 2 + \frac{1}{2} + \frac{1}{2 \cdot 3} + \frac{1}{2 \cdot 3 \cdot 4} + \frac{1}{2 \cdot 3 \cdot 4 \cdot 5}$$

$$+ \frac{1}{2 \cdot 3 \cdot 4 \cdot 5 \cdot 6} + \cdots$$

is not geometric, but it does have a sum. Using S_1, S_2, S_3, S_4, S_5, and S_6, make a conjecture about the value of S_∞ and explain your reasoning.

71. Find the sum of the first n terms of
$$x^2 - x^3 + x^4 - x^5 + \cdots.$$

72. Find the sum of the first n terms of
$$1 + x + x^2 + x^3 + \cdots.$$

73. The sides of a square are each 16 cm long. A second square is inscribed by joining the midpoints of the sides, successively. In the second square we repeat the process, inscribing a third square. If this process is continued indefinitely, what is the sum of all of the areas of all the squares? (*Hint:* Use an infinite geometric series.)

74. Show that $0.999\ldots$ is 1.

75. Using Example 5 and Exercises 33–44, explain how the graph of a geometric sequence can be used to determine whether a geometric series has a limit.

76. To compare the *graphs* of an arithmetic and a geometric sequence, we plot n on the horizontal axis and a_n on the vertical axis. Graph Example 1(a) of Section 14.2 and Example 1(a) of Section 14.3 on the same set of axes. How do the graphs of geometric sequences differ from the graphs of arithmetic sequences?

CORNER

Bargaining for a Used Car

Focus: Geometric series

Time: 30 minutes

Group size: 2

Materials: Graphing calculators are optional.

ACTIVITY*

1. One group member ("the seller") has a car for sale and is asking $3500. The second ("the buyer") offers $1500. The seller splits the difference ($2000 ÷ 2 = $1000) and lowers the price to $2500. The buyer then splits the difference again ($1000 ÷ 2 = $500) and counters with $2000. Continue in this manner and stop when you are able to agree on the car's selling price to the nearest penny.

2. What should the buyer's initial offer be in order to achieve a purchase price of $2000? (Check several guesses to find the appropriate initial offer.)

*This activity is based on the article "Bargaining Theory, or Zeno's Used Cars," by James C. Kirby, *The College Mathematics Journal*, **27**(4), September 1996.

3. The seller's price in the bargaining above can be modeled recursively (see Exercises 73, 74, and 82 in Section 14.1) by the sequence

$$a_1 = 3500, \qquad a_n = a_{n-1} - \frac{d}{2^{2n-3}},$$

where d is the difference between the initial price and the first offer. Use this recursively defined sequence to solve parts (1) and (2) above either manually or by using the SEQ MODE and the TABLE feature of a grapher.

4. The first four terms in the sequence in part (3) can be written as

$$a_1, \quad a_1 - \frac{d}{2}, \quad a_1 - \frac{d}{2} - \frac{d}{8},$$

$$a_1 - \frac{d}{2} - \frac{d}{8} - \frac{d}{32}.$$

Use the formula for the limit of an infinite geometric series to find a simple algebraic formula for the eventual sale price, P, when the bargaining process from above is followed. Verify the formula by using it to solve parts (1) and (2) above.

Combinatorics: Permutations

14.4

Fundamental Counting Principle • Permutations • Factorial Notation • Permutations of *n* Objects Taken *r* at a Time

To study probability, it is first necessary to be able to determine the number of ways in which objects in a set can be arranged or combined, certain objects can be chosen, or a succession of events can occur. This study of the theory of counting is called **combinatorics**.

Fundamental Counting Principle

E x a m p l e   1   How many 3-letter code symbols can be formed with the letters A, B, and C without repetition?

Solution   Examples of such symbols are ABC, CBA, ACB, and so on. Consider placing the letters in these frames.

We can select any of the 3 letters for the first letter in the symbol. Once this letter has been selected, the second must be selected from the 2 remaining letters. The third letter is already determined, since only 1 possibility is left. All the possibilities can be found using a **tree diagram**.

Tree Diagram   *Outcomes*

A ⟨ B ——— C   ABC
    C ——— B   ACB

B ⟨ A ——— C   BAC
    C ——— A   BCA

C ⟨ A ——— B   CAB
    B ——— A   CBA

1st pick   2nd pick   3rd pick

There are 3 · 2 · 1, or 6, possibilities. The set of all of the possibilities is:

{ABC, ACB, BAC, BCA, CAB, CBA}.

Suppose we perform an experiment such as selecting letters (as in the preceding example), flipping a coin, or drawing a card. The results are called **outcomes**. An **event** is a set of outcomes. The following principle concerns the counting of events.

> ### *Fundamental Counting Principle*
>
> Given a combined action, or event, in which the first action can be performed in n_1 ways, the second action can be performed n_2 ways, and so on, the total number of ways in which the combined action can be performed is the product
>
> $$n_1 \cdot n_2 \cdot n_3 \cdot \cdots \cdot n_k.$$

In Example 1, the first letter can be chosen in 3 ways, the second in 2 ways, and the third in 1 way. So, by the fundamental counting principle, the total number of ways in which all 3 letters can be selected is $3 \cdot 2 \cdot 1 = 6$.

E x a m p l e 2   How many 3-letter code symbols can be formed with the letters A, B, and C with repetition allowed?

Solution   There are 3 choices for the first letter and, since we allow repetition, 3 choices for the second and 3 for the third. Thus by the fundamental counting principle, there are $3 \cdot 3 \cdot 3$, or 27, 3-letter codes.

E x a m p l e 3   ***New car choices.***   Tyrone bought a 2001 Volkswagen New Beetle family coupe. When he ordered the car, there were 5 series, 22 color combinations, 2 engines, and 2 wheel styles from which to choose. In how many different ways could the New Beetle be ordered?

Solution   There were 5 series, 22 colors, 2 engines, and 2 wheel styles from which to choose. By the fundamental counting principle there were

$$5 \cdot 22 \cdot 2 \cdot 2 = 440$$

different ways in which the New Beetle could be ordered.

Permutations

We now turn our attention to the part of combinatorics that deals with the study of *permutations*. The study of permutations involves *order* and *arrangement*.

> ### *Permutation*
>
> A *permutation* of a set of n objects is an ordered arrangement of all n objects.

In Example 1, we found that there are $3 \cdot 2 \cdot 1$, or 6 *ordered arrangements* of the letters in the set {A, B, C}: ABC, ACB, BAC, BCA, CAB, CBA. Each of these arrangements is called a *permutation* of the letters A, B, and C.

We can find a formula for the total number of permutations of all objects in a set of n objects. We have n choices for the first selection, $n - 1$ for the second, $n - 2$ for the third, and so on. For the nth selection, there is only 1 choice.

> ### Permutations of n Objects
>
> The total number of permutations of a set of n objects, denoted $_nP_n$, is given by
>
> $$_nP_n = n(n - 1)(n - 2) \cdots (3)(2)(1).$$

E x a m p l e   4   Find **(a)** $_4P_4$ and **(b)** $_7P_7$.

Solution

a) Start with 4.

$$_4P_4 = 4 \cdot 3 \cdot 2 \cdot 1 = 24$$

4 factors

b) $_7P_7 = 7 \cdot 6 \cdot 5 \cdot 4 \cdot 3 \cdot 2 \cdot 1 = 5040$

E x a m p l e   5   *Interior design.* Alyssa is arranging a collection of 9 different vases on a mantel. In how many different ways can the vases be arranged?

Solution   Each placement of the vases is an ordered arrangement, or permutation, of a set of 9 objects. The number of permutations is

$$_9P_9 = 9 \cdot 8 \cdot 7 \cdot 6 \cdot 5 \cdot 4 \cdot 3 \cdot 2 \cdot 1 = 362{,}880.$$

E x a m p l e   6   *Batting order.* A baseball manager arranges the batting order as follows: The 4 infielders will bat first, and then the other 5 players will follow. How many different batting orders are possible?

Solution   The infielders can bat in $_4P_4$ different ways; the rest in $_5P_5$ different ways. By the fundamental counting principle, we have $_4P_4 \cdot {_5P_5}$, or 2880, possible batting orders.

Factorial Notation

We will use products of successive natural numbers, such as

$$7 \cdot 6 \cdot 5 \cdot 4 \cdot 3 \cdot 2 \cdot 1,$$

so often that it is convenient to adopt a notation for them.

For the product $7 \cdot 6 \cdot 5 \cdot 4 \cdot 3 \cdot 2 \cdot 1$, we write 7!, read "7-factorial."

> **Factorial Notation**
>
> For any natural number n,
>
> $$n! = n(n-1)(n-2) \cdots (3)(2)(1).$$
>
> For the number 0,
>
> $$0! = 1.$$

We define $0!$ to be 1 so that certain formulas and theorems can be stated concisely and with a consistent pattern.

Here are some examples of factorial notation.

$$7! = 7 \cdot 6 \cdot 5 \cdot 4 \cdot 3 \cdot 2 \cdot 1 = 5040$$
$$6! = \qquad 6 \cdot 5 \cdot 4 \cdot 3 \cdot 2 \cdot 1 = 720$$
$$5! = \qquad\quad 5 \cdot 4 \cdot 3 \cdot 2 \cdot 1 = 120$$
$$4! = \qquad\qquad 4 \cdot 3 \cdot 2 \cdot 1 = 24$$
$$3! = \qquad\qquad\quad 3 \cdot 2 \cdot 1 = 6$$
$$2! = \qquad\qquad\qquad 2 \cdot 1 = 2$$
$$1! = \qquad\qquad\qquad\quad 1 = 1$$
$$0! = \qquad\qquad\qquad\quad 1 = 1$$

We can restate the permutation formula using factorial notation:

$$_nP_n = n!.$$

We will often need to manipulate factorial notation. For example, note that

$$8! = 8 \cdot 7 \cdot 6 \cdot 5 \cdot 4 \cdot 3 \cdot 2 \cdot 1$$
$$= 8 \cdot (7 \cdot 6 \cdot 5 \cdot 4 \cdot 3 \cdot 2 \cdot 1)$$
$$= 8 \cdot 7!$$

Generalizing, we get the following.

For any natural number n, $n! = n(n-1)!$.

By using this result repeatedly, we can further manipulate factorial notation.

Example 7

Rewrite $7!$ with a factor of $5!$.

Solution

$$7! = 7 \cdot 6 \cdot 5!$$

Thus far, we have considered ordered arrangements of all the members of a set. Now we look at permutations of only some of the objects in a set.

Permutations of *n* Objects Taken *r* at a Time

Consider a set of 6 objects, say {A, B, C, D, E, F}. How many ordered arrangements are there of 3 members of the set? We can select the first object in 6 ways.

There are then 5 choices for the second and then 4 choices for the third. By the fundamental counting principle, there are $6 \cdot 5 \cdot 4$, or 120 permutations of a set of 6 objects taken 3 at a time. Note that if we multiply by 1 we have

$$6 \cdot 5 \cdot 4 = \frac{6 \cdot 5 \cdot 4 \cdot 3 \cdot 2 \cdot 1}{3 \cdot 2 \cdot 1}, \quad \text{or} \quad \frac{6!}{3!}.$$

Now consider a set of n objects from which an ordered arrangement of r objects is selected. Since the arrangements are ordered, each is a *permutation* of the set of n objects taken r at a time. The first object can be selected in n ways. The second can be selected in $n - 1$ ways, and so on. The rth can be selected in $n - (r - 1)$ ways. By the fundamental counting principle, the total number of permutations is

$$n(n - 1)(n - 2) \cdots [n - (r - 1)], \quad \text{or}$$
$$n(n - 1)(n - 2) \cdots (n - r + 1).$$

We now multiply by 1:

$$n(n - 1)(n - 2) \cdots (n - r + 1)\frac{(n - r)!}{(n - r)!}$$
$$= \frac{n(n - 1)(n - 2)(n - 3) \cdots (n - r + 1)(n - r)!}{(n - r)!}$$
$$= \frac{n!}{(n - r)!}.$$

This gives us the following.

Permutations of n Objects Taken r at a Time

The number of permutations of a set of n objects taken r at a time, denoted $_nP_r$, is given by

$$_nP_r = n(n - 1)(n - 2) \cdots (n - r + 1) \tag{1}$$

or

$$_nP_r = \frac{n!}{(n - r)!}. \tag{2}$$

Form (1) is most useful in application, but form (2) will be important later in this chapter.

Example 8   Compute $_6P_4$ using both of the above formulas.

Solution   Using form (1), we have

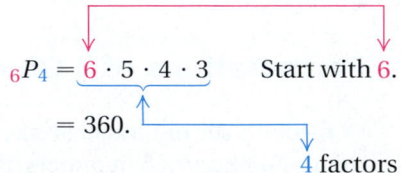

$$_6P_4 = 6 \cdot 5 \cdot 4 \cdot 3 \qquad \text{Start with } 6.$$
$$= 360.$$

4 factors

Most calculators can evaluate permutation notation directly. The $_nP_r$ notation may be on the face of the calculator or under a MATH menu. In general, you must press the value of n, then the $_nP_r$ key or option, and then the value for r. Use a calculator to check Examples 4, 6, 8, and 9.

Using form (2), we have

$$_6P_4 = \frac{6!}{(6-4)!} = \frac{6!}{2!}$$
$$= \frac{6 \cdot 5 \cdot 4 \cdot 3 \cdot 2 \cdot 1}{2 \cdot 1}$$
$$= 6 \cdot 5 \cdot 4 \cdot 3 = 360.$$

Caution! $\dfrac{6!}{2!} \neq 3!$

To see this, note that

$$\frac{6!}{2!} = \frac{6 \cdot 5 \cdot 4 \cdot 3 \cdot 2 \cdot 1}{2 \cdot 1} = 6 \cdot 5 \cdot 4 \cdot 3.$$

E x a m p l e   9

Flags of nations.   The flags of many nations consist of three horizontal stripes, similar to the flag of Hungary, shown at left. The top stripe of this flag is red, the middle one white, and the bottom one green.

Suppose that the following 7 colors are available:

{green, yellow, red, white, black, blue, orange}.

How many different flags of 3 colors can be made without repetition of colors within a flag?

Solution   The order in which the colors are chosen is important in this application. For example, the flag of Bulgaria also consists of three horizontal stripes of the same colors as the Hungarian flag, but they are arranged in a different order. So we need to determine the number of permutations of 7 objects (colors) taken 3 at a time. Using form (1), we get

$$_7P_3 = 7 \cdot 6 \cdot 5 = 210.$$

There are 210 different 3-color flags.

Exercise Set   14.4

FOR EXTRA HELP

 Digital Video Tutor CD 11 Videotape 26   InterAct Math   Math Tutor Center   MathXL   MyMathLab.com

Evaluate.

1. 9!   **2.** 10!   **3.** 11!   **10.** $\frac{10!}{7!}$ **11.** $(8-3)!$ **12.** $(9-5)!$

4. 12!   *Aha!* **5.** 0!   **6.** 1!   **13.** $8! - 3!$ **14.** $9! - 5!$ **15.** $_6P_6$

7. $\frac{7!}{4!}$ **8.** $\frac{8!}{6!}$ **9.** $\frac{9!}{5!}$ **16.** $_5P_5$ **17.** $_4P_3$ **18.** $_7P_5$

19. $_{10}P_7$ **20.** $_{10}P_3$ *Aha!* **21.** $_6P_1$

22. $_{12}P_1$ **23.** $_6P_5$ **24.** $_{12}P_{11}$

Answer each of the following exercises using permutation notation, factorial notation, or other operations. Then evaluate.

25. *Shipping.*   When mailing candles, the Glass Gazebo must choose 1 of 3 types of boxes, 1 of 10 styles of wrapping paper, 1 of 5 possible shipping companies, and 1 of 3 insurance options. In how many different ways can the candles be packed and shipped?

26. *Truck options.*   Mariah is ordering a new F-150 regular cab pickup truck. She must choose 1 of 4 series, 1 of 33 color combinations, 1 of 5 engines, and either two-wheel drive or four-wheel drive. In how many different ways can she select the truck?

How many permutations are there of the letters in each of the following words, if all of the letters are used without repetition?

27. OWL   **28.** WE

29. QUALIFY   **30.** TIMES

31. How many permutations are there of the letters of the word QUALIFY if the letters are taken 4 at a time?

32. How many permutations are there of the letters of the word TIMES if the letters are taken 3 at a time?

33. How many 4-digit numbers can be named using the digits 6, 7, 8, and 9 without repetition? with repetition?

34. How many 5-digit numbers can be named using the digits 2, 3, 4, 5, and 8 without repetition? with repetition?

35. In how many ways can 5 students be arranged in a straight line?

36. In how many ways can 7 athletes be arranged in a straight line?

37. *Phone numbers.*   How many 7-digit phone numbers can be formed with the digits 0, 1, 2, 3, 4, 5, 6, 7, 8, and 9, assuming that no digit is used more than once and the first digit is not 0 or 1?

38. *Choosing officers.*   How many ways can a president, vice president, secretary, and treasurer be chosen from a committee of 12 people?

39. *Coin arrangements.*   A penny, nickel, dime, quarter, and half-dollar are arranged in a straight line.
   a) Considering just the denominations, in how many ways can they be lined up?
   b) Considering the denominations and heads and tails, in how many ways can they be lined up?

40. *Coin arrangements.*   A penny, nickel, dime, and quarter are arranged in a straight line.
   a) Considering just the denominations, in how many ways can they be lined up?
   b) Considering the denominations and heads and tails, in how many ways can they be lined up?

41. Compute $_{52}P_4$.

42. Compute $_{50}P_5$.

43. A state forms its license plates by first listing a number that corresponds to the county in which the car owner lives. Then the plate lists a letter of the alphabet, and this is followed by a number from 1 to 9999. How many such plates are possible if there are 80 counties?

44. *Zip codes.*   Zip codes in Canada are a series of numerals and letters. A zip code for Montreal, Quebec, is H2N 1M5. It consists of a letter for the first, third, and fifth places, and a number from 0 to 9 in the second, fourth, and sixth places.
   a) How many such zip codes are possible?
   b) There are about 31 million people in Canada. Can each person have his or her own zip code?

45. *Zip codes.*   Zip codes in the United States are 5-digit numbers. A zip code in Dallas, Texas, is 75247.
   a) How many zip codes are possible if any of the digits 0 to 9 can be used?
   b) If each post office has its own zip code, how many possible post offices can there be?

46. *Zip codes.*   Zip codes are sometimes given using a 9-digit number like 75247-5456.
   a) How many 9-digit zip codes are possible?
   b) There are about 284 million people in the United States. If each person had a private zip code, would there be enough 9-digit zip codes?

47. *Social security numbers.* A social security number is a 9-digit number like 293-36-0391.

    a) How many social security numbers can there be?

    b) There are about 284 million people in the United States. Can each person have a unique social security number?

48. How "long" is 15!? You own 15 different books and decide to actually make up all possible ordered arrangements of the books on a shelf. About how long, in years, would it take if you can make one arrangement per second?

49. Explain why more code words can be formed from a set of letters if repetition is allowed than if it is not.

50. Explain why the fundamental counting principle involves multiplication, not addition.

SKILL MAINTENANCE

Given $\log_b 7 = 1.946$ and $\log_b 5 = 1.609$, *find each of the following.*

51. $\log_b 35$ **52.** $\log_b \frac{5}{7}$ **53.** $\log_b 49$

Simplify.

54. $\log_p p^{10}$ **55.** $\log_b b^{-5}$ **56.** $\log_Q Q^m$

SYNTHESIS

57. Write a problem for which the solution requires the use of both the fundamental counting principle and permutations.

58. Explain why $_nP_r$ is generally computed using form (1) on p. 880.

Solve for n.

59. $_nP_5 = 7 \cdot {}_nP_4$ **60.** $_nP_4 = 8 \cdot {}_{n-1}P_3$

61. $_nP_5 = 9 \cdot {}_{n-1}P_4$ **62.** $_nP_4 = 8 \cdot {}_nP_3$

63. In how many ways can 3 adults and 3 children be seated in a row of 6 seats:

    a) with no seating restrictions?

    b) if adults and children must alternate?

    c) if a particular adult and child must sit together?

    d) if a particular adult and child must not sit together?

64. A car holds 3 people in the front and 3 people in the back. In how many ways can 6 people be seated in the car?

    a) with no seating restrictions?

    b) if two people must sit together?

    c) if two particular couples must each sit together?

65. One method for factoring a trinomial $ax^2 + bx + c$ is the FOIL method. Factorizations of the form $(px + q)(rx + s)$ are formed, where $pr = a$ and $qs = c$. The number of possible factorizations thus depends on the number of ways in which a and c can be factored. How many possible trial factorizations are there of $6x^2 + 73x + 12$?

66. With reference to Exercise 65, how many possible trial factorizations are there of $5x^2 + 12x + 7$?

Combinatorics: Combinations

14.5

Combinations • Combinations of *n* Objects Taken *r* at a Time

If you play cards, you know that in most situations the *order* in which you hold cards *is not important*! It is just the contents of the hand, or set, of cards. We may sometimes make selections from a set *without regard to order*. Such selections are called **combinations**.

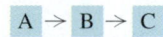

Permutation:
Order considered!

Combination:
Order *not* considered!

Combinations

E x a m p l e 1   Find all the combinations of 3 elements taken from the set of 5 elements $\{A, B, C, D, E\}$. How many are there?

Solution   The combinations are

$$\{A, B, C\}, \quad \{A, B, D\}, \quad \{A, B, E\}, \quad \{A, C, D\}, \quad \{A, C, E\}$$
$$\{A, D, E\}, \quad \{B, C, D\}, \quad \{B, C, E\}, \quad \{B, D, E\}, \quad \{C, D, E\}.$$

There are 10 combinations of 5 objects taken 3 at a time.

When we find all the combinations of 5 objects taken 3 at a time, we are finding all the 3-element subsets.* When we are naming a set, the order of the listing is *not* important. Thus,

$$\{A, C, B\} \quad \text{names the same set as} \quad \{A, B, C\}.$$

> ### *Combination*
>
> A *combination* of *r* objects chosen from a set of *n* objects is a subset of the set of *n* objects.

Because the elements of a set may be listed in any order, it is important to remember that when thinking of *combinations*, we do not think about *order*.

E x a m p l e 2   Find all the subsets of the set $\{A, B, C\}$. Identify these as combinations. How many subsets are there in all?

Solution   There will be subsets containing 0 elements, 1 element, 2 elements, and 3 elements.

a) The set with 0 elements is the empty set, denoted $\emptyset$. The empty set is a subset of every set. In this case, it is the combination of 3 objects taken 0 at a time. There is 1 such combination, $\emptyset$.

b) The following are all the 1-element subsets of $\{A, B, C\}$:

$$\{A\}, \quad \{B\}, \quad \{C\}.$$

These are the combinations of 3 objects taken 1 at a time. There are 3 such combinations.

c) The following are all the 2-element subsets of $\{A, B, C\}$:

$$\{A, B\}, \quad \{A, C\}, \quad \{B, C\}.$$

These are the combinations of 3 objects taken 2 at a time. There are 3 such combinations.

---

*The set A is a subset of B if every element of A is an element of B.

d) The following are all the 3-element subsets of {A, B, C}:

{A, B, C}.

This is the combination of 3 objects taken 3 at a time. There is only 1 such combination. A set is always a subset of itself.

The total number of subsets is $1 + 3 + 3 + 1$, or 8.

Combinations of n Objects Taken r at a Time

We want to develop a formula for computing the number of combinations of n objects taken r at a time, denoted $_nC_r$, without actually listing the combinations, or subsets. In Example 2, we saw that

$$_3C_0 = 1, \qquad _3C_1 = 3, \qquad _3C_2 = 3, \quad \text{and} \quad _3C_3 = 1.$$

Before we derive a general formula for $_nC_r$, let's return to Example 1 and compare the number of combinations with the number of permutations.

Combinations			*Permutations*			
{A, B, C} ⟶ ABC	BCA	CAB	CBA	BAC	ACB
{A, B, D} ⟶ ABD	BDA	DAB	DBA	BAD	ADB
{A, B, E} ⟶ ABE	BEA	EAB	EBA	BAE	AEB
{A, C, D} ⟶ ACD	CDA	DAC	DCA	CAD	ADC
{A, C, E} ⟶ ACE	CEA	EAC	ECA	CAE	AEC
{A, D, E} ⟶ ADE	DEA	EAD	EDA	DAE	AED
{B, C, D} ⟶ BCD	CDB	DBC	DCB	CBD	BDC
{B, C, E} ⟶ BCE	CEB	EBC	ECB	CBE	BEC
{B, D, E} ⟶ BDE	DEB	EBD	EDB	DBE	BED
{C, D, E} ⟶ CDE	DEC	ECD	EDC	DCE	CED

Note that each combination of 3 objects, say {A, C, E}, yields $3!$, or 6, permutations, as shown. It follows that

$$3! \cdot {}_5C_3 = {}_5P_3 = 5 \cdot 4 \cdot 3 = 60,$$

so

$$_5C_3 = \frac{_5P_3}{3!} = \frac{5 \cdot 4 \cdot 3}{3 \cdot 2 \cdot 1} = 10. \qquad \text{Dividing by 3!}$$

In general, for $r \le n$, the number of combinations of n objects taken r at a time, $_nC_r$, times the number of permutations of these r objects, $r!$, must equal the number of permutations of n objects taken r at at time:

$$r! \cdot {}_nC_r = {}_nP_r.$$

Dividing both sides by $r!$, we have

$$_nC_r = \frac{_nP_r}{r!} = {}_nP_r \cdot \frac{1}{r!} = \frac{n!}{(n-r)!} \cdot \frac{1}{r!} = \frac{n!}{(n-r)!\,r!}.$$

Combinations of n Objects Taken r at a Time

The total number of combinations of n objects taken r at a time, denoted $_nC_r$, is given by

$$_nC_r = \frac{n!}{(n-r)!\,r!}.$$

We can make some general observations. First, it is always true that $_nC_n = 1$ because a set of n objects has only 1 subset with n objects, the set itself. Second, $_nC_1 = n$ because a set with n objects has n subsets with 1 element each. Finally, $_nC_0 = 1$ because a set with n objects has only one subset with 0 elements, namely, the empty set $\varnothing$.

There is another kind of notation that is also used for $_nC_r$. It is called **binomial coefficient** notation. The reason for such terminology will be be seen later.

Binomial Coefficient

$$\binom{n}{r} = {_nC_r}$$

It is important to remember that $\binom{n}{r}$ does not mean $n \div r$, or $\dfrac{n}{r}$. The notation $\binom{n}{r}$ is often read "n choose r."

E x a m p l e   3

Simplify: **(a)** $\binom{7}{5}$; **(b)** $\binom{7}{2}$; **(c)** $\binom{6}{6}$

Solution

a) $\binom{7}{5} = \dfrac{7!}{(7-5)!\,5!}$

$= \dfrac{7!}{2!\,5!} = \dfrac{7 \cdot 6 \cdot 5!}{2 \cdot 1 \cdot 5!} = \dfrac{7 \cdot 6}{2 \cdot 1} = \dfrac{7 \cdot 3 \cdot 2}{2} = 21$

b) $\binom{7}{2} = \dfrac{7!}{5!\,2!} = \dfrac{7 \cdot 6 \cdot 5!}{5! \cdot 2 \cdot 1} = \dfrac{7 \cdot 6}{2} = 7 \cdot 3 = 21$

c) $\binom{6}{6} = \dfrac{6!}{0!\,6!} = \dfrac{6!}{1 \cdot 6!}$ Since $0! = 1$

$= \dfrac{6!}{6!} = 1$

technology connection

Combination notation on calculators is generally written $_nC_r$, rather than $\binom{n}{r}$. Use a calculator to check Examples 3, 4, and 5.

In Example 3, we saw that

$$\binom{7}{5} = \binom{7}{2}.$$

This says that the number of 5-element subsets of a set of 7 objects is the same as the number of 2-element subsets of a set of 7 objects. Each choice of a 5-element subset leaves a 2-element subset that has not been chosen. This result can be generalized as follows.

Subsets of Size r and of Size n − r

For any whole numbers r and n, $r \leq n$,

$$\binom{n}{r} = \binom{n}{n-r} \quad \text{and} \quad {}_nC_r = {}_nC_{n-r}.$$

The number of subsets of size r of a set with n objects is the same as the number of subsets of size $n - r$. The number of combinations of n objects taken r at a time is the same as the number of combinations of n objects taken $n - r$ at a time.

Example 4

State lotto.   A state runs a 6-out-of-44-number lotto twice a week that pays at least $1.5 million. Kim purchases a card for $1 and picks any 6 numbers from 1 to 44. How many possible 6-number combinations are there for drawing?

Solution   No order is implied here. Kim can choose any 6 numbers from 1 to 44. Thus the number of combinations is

$$_{44}C_6 = \binom{44}{6}$$

$$= \frac{44!}{38!\,6!}$$

$$= \frac{44 \cdot 43 \cdot 42 \cdot 41 \cdot 40 \cdot 39 \cdot 38!}{38! \cdot 6 \cdot 5 \cdot 4 \cdot 3 \cdot 2 \cdot 1}$$

$$= 7{,}059{,}052.$$

Example 5

A medical office employs 5 office staff and 7 nurses. On any given day, there are 3 office personnel and 4 nurses present. How many different staffing situations are possible?

Solution First we determine the number of ways in which 3 office personnel can be chosen from the group of 5. Since the order in which they are chosen is not important, the office staff can be selected in $_5C_3$ ways.

Similarly, the 4 nurses can be chosen in $_7C_4$ ways.

If we use the fundamental counting principle, it follows that the number of possible staffing situations is

$$_5C_3 \cdot {}_7C_4 = \frac{5!}{2!\,3!} \cdot \frac{7!}{3!\,4!}$$
$$= 10 \cdot 35 = 350.$$

Compare the number of batting orders in Example 6 in Section 14.4 and the number of staffing situations in Example 5 above. In choosing a batting order, order was important, so permutations were used. The medical staffs chosen were subsets, in which order was not important, so combinations were used.

Exercise Set 14.5

FOR EXTRA HELP

 Digital Video Tutor CD 11 Videotape 27   InterAct Math    Math Tutor Center   MathXL    MyMathLab.com

Evaluate.

1. $_{13}C_2$

2. $_9C_6$

Aha! **3.** $\binom{13}{11}$

4. $\binom{9}{3}$

5. $\binom{7}{1}$

6. $\binom{8}{8}$

7. $\dfrac{_5P_3}{3!}$

8. $\dfrac{_{10}P_5}{5!}$

9. $\binom{6}{0}$

10. $\binom{6}{3}$

Aha! **11.** $_{12}C_{11}$

12. $_{12}C_{10}$

13. $_{20}C_{18}$

14. $_{30}C_3$

15. $\binom{35}{2}$

16. $\binom{40}{38}$

17. $_{10}C_5$

18. $_{15}C_{11}$

Answer each of the following exercises using permutation notation, combination notation, factorial notation, or other operations. Then evaluate.

19. *Business club officers.* There are 23 students in a business club. How many sets of 4 officers can be selected?

20. *Basketball league games.* How many basketball games can be played in a 9-team league if each team plays all other teams once? twice?

21. *Test options.* On a test, a student is to select 6 out of 10 questions. In how many ways can the student do this?

22. *Test options.* Of the first 10 questions on a test, a student must answer 7. On the next 5 questions, the student must answer 3. In how many ways can this be done?

23. *Determining lines and triangles.*   How many lines are determined by 8 points, no 3 of which are collinear? How many triangles are determined by the same points?

24. *Determining lines and triangles.*   How many lines are determined by 7 points, no 3 of which are collinear? How many triangles are determined by the same points?

25. *Senate committees.*   Suppose that the Senate of the United States consists of 58 Democrats and 42 Republicans. How many committees made up of 6 Democrats and 4 Republicans can be formed? You need not simplify the expression.

26. *Senate committees.*   Suppose that the Senate of the United States consists of 63 Republicans and 37 Democrats. How many committees made up of 8 Republicans and 12 Democrats can be formed? You need not simplify the expression.

27. *Dice.*   Two 6-sided dice are rolled, one a blue die and one red. How many different ways can the two fall?

28. *Buffet choices.*   A restaurant's Sunday buffet offers 6 seafood entrees, 9 vegetables, and 7 sauces. How many possible combinations of seafood, vegetable, and sauce are there?

29. *Poker hands.*   How many different 5-card poker hands are possible with a 52-card deck? (See Section 14.7 for a description of a 52-card deck.) You need not simplify the expression.

30. *Bridge hands.*   How many different 13-card bridge hands can be dealt from a standard deck of 52 cards? You need not simplify the expression.

31. *Course selection.*   A university offers 5 science courses, 6 humanity courses, and 3 literature courses. In how many ways can a student choose

2 science courses, 3 humanity courses, and 1 literature course?

32. *Compact disc distribution.*   In how many ways can 8 different compact discs be distributed among 3 students if the first gets 2, the second gets 5, and the third gets 1?

33. *Pizza choices.*   Pizza Shack has the following toppings for pizzas:

> extra cheese, pepperoni, sausage, mushroom, onion, green pepper, beef, Canadian bacon, black olives, ham.

How many different kinds of pizza can Pizza Shack serve (excluding size and thickness of pizza)?

34. *Pizza choices.*   Pizza Shack serves round pizzas in two sizes—10-in. and 16-in.—and three thicknesses—original, thin, and pan. Using the toppings listed in Exercise 33 and considering size and thickness, how many different kinds of pizza can Pizza Shack serve?

35. *Poker hands.*   How many 5-card poker hands consisting of 3 aces and 2 cards that are not aces are possible with a 52-card deck?

36. *Poker hands.*   How many 5-card poker hands consisting of 2 kings and 3 cards that are not kings are possible with a 52-card deck?

37. *Ice cream cones.*   Bresler's Ice Cream, a national firm, sells ice cream in 33 flavors.

   a) How many 3-dip cones are possible if order of flavors is to be considered and no flavor is repeated?
   b) How many 3-dip cones are possible if order is to be considered and flavors can be repeated?
   c) How many 3-dip cones are possible if order is not considered and no flavor is repeated?

38. *Ice cream cones.*   Baskin-Robbins Ice Cream, a national firm, sells ice cream in 31 flavors.

   a) How many 2-dip cones are possible if order of flavors is to be considered and no flavor is repeated?
   b) How many 2-dip cones are possible if order is to be considered and flavors can be repeated?
   c) How many 2-dip cones are possible if order is not considered and no flavor is repeated?

39. Explain why a "combination" lock should really be called a "permutation" lock.

40. Explain why there are more permutations of a set of n objects taken r at a time than there are combinations of a set of n objects taken r at a time.

SKILL MAINTENANCE

Solve.

41. $2^x = \frac{1}{4}$

42. $3^{2x} = 27$

43. $\log_5(x + 1) = 2$

44. $\log x + \log(x - 3) = 1$

45. $\log_x 5 = 1$

46. $5^x = 20$

SYNTHESIS

47. Explain why $_mC_0 = 1$ and $_mC_m = 1$.

48. A restaurant that allows its customers to top their own hamburgers advertises that with 28 toppings to choose from, there are 268,435,456 combinations. Describe a procedure for determining this number using the formulas developed in this section.

Simplify.

49. $\begin{pmatrix} m \\ 1 \end{pmatrix}$ **50.** $\begin{pmatrix} m \\ m - 1 \end{pmatrix}$

51. $\begin{pmatrix} m \\ 0 \end{pmatrix}$ **52.** $\begin{pmatrix} m \\ m - 2 \end{pmatrix}$

Solve for n.

53. $\begin{pmatrix} n + 1 \\ 3 \end{pmatrix} = 2 \cdot \begin{pmatrix} n \\ 2 \end{pmatrix}$ **54.** $\begin{pmatrix} n \\ n - 2 \end{pmatrix} = 6$

55. $\begin{pmatrix} n + 2 \\ 4 \end{pmatrix} = 6 \cdot \begin{pmatrix} n \\ 2 \end{pmatrix}$ **56.** $\begin{pmatrix} n \\ 3 \end{pmatrix} = 2 \cdot \begin{pmatrix} n - 1 \\ 2 \end{pmatrix}$

57. *Single-elimination tournaments.* In a single-elimination sports tournament consisting of n teams, a team is eliminated when it loses one game. How many games are required to complete the tournament?

58. *Double-elimination tournaments.* In a double-elimination softball tournament consisting of n teams, a team is eliminated when it loses two games. At most, how many games are required to complete the tournament?

59. *Triangles inscribed in a circle.* There are m points on a circle. How many triangles can be inscribed with these points as vertices?

60. *Parallelograms.* A set of m parallel lines crosses another set of n parallel lines. How many parallelograms are formed?

61. Prove that
$$\begin{pmatrix} n \\ r \end{pmatrix} = \begin{pmatrix} n \\ n - r \end{pmatrix}$$
for any whole numbers n and r.

The Binomial Theorem

14.6

Binomial Expansion Using Pascal's Triangle • Binomial Expansion Using Factorial Notation

CONNECTING THE CONCEPTS

Sequences and series occur in many settings, some of which were mentioned in Sections 14.1–14.3. Although you may not have viewed it this way before, the expression $(x + y)^2$ can be regarded as a series: $x^2 + 2xy + y^2$.

In Chapter 4, we found that the expansion of $(x + y)^n$, for powers greater than 2, can be quite time-consuming. The reason for this extends all the way back to the rules for the order of operations and the properties of exponents: $(x + y)^n \neq x^n + y^n$. Since the terms in the expansion of $(x + y)^n$ have many uses, we devote this section to two methods that streamline the expansion of this important algebraic expression.

Binomial Expansion Using Pascal's Triangle

Consider the following expanded powers of $(a + b)^n$:

$$(a + b)^0 = 1$$
$$(a + b)^1 = a + b$$
$$(a + b)^2 = a^2 + 2a^1b^1 + b^2$$
$$(a + b)^3 = a^3 + 3a^2b^1 + 3a^1b^2 + b^3$$
$$(a + b)^4 = a^4 + 4a^3b^1 + 6a^2b^2 + 4a^1b^3 + b^4$$
$$(a + b)^5 = a^5 + 5a^4b^1 + 10a^3b^2 + 10a^2b^3 + 5a^1b^4 + b^5.$$

Each expansion is a polynomial. There are some patterns to be noted:

1. There is one more term than the power of the binomial, n. That is, there are $n + 1$ terms in the expansion of $(a + b)^n$.
2. In each term, the sum of the exponents is the power to which the binomial is raised.
3. The exponents of a start with n, the power of the binomial, and decrease to 0 (since $a^0 = 1$, the last term has no factor of a). The first term has no factor of b, so powers of b start with 0 and increase to n.
4. The coefficients start at 1, increase through certain values, and then decrease through these same values back to 1. Let's study the coefficients further.

Suppose we wish to expand $(a + b)^8$. The patterns we noticed above indicate 9 terms in the expansion:

$$a^8 + c_1a^7b + c_2a^6b^2 + c_3a^5b^3 + c_4a^4b^4 + c_5a^3b^5 + c_6a^2b^6 + c_7ab^7 + b^8.$$

How can we determine the values for the c's? One method involves writing down the coefficients in a triangular array as follows. We form what is known as **Pascal's triangle**:

$$
\begin{array}{c}
(a+b)^0: \qquad\qquad\qquad 1 \\
(a+b)^1: \qquad\qquad\quad 1 \quad 1 \\
(a+b)^2: \qquad\qquad 1 \quad 2 \quad 1 \\
(a+b)^3: \qquad\quad 1 \quad 3 \quad 3 \quad 1 \\
(a+b)^4: \qquad 1 \quad 4 \quad 6 \quad 4 \quad 1 \\
(a+b)^5: \quad 1 \quad 5 \quad 10 \quad 10 \quad 5 \quad 1
\end{array}
$$

There are many patterns in the triangle. Find as many as you can.

Perhaps you discovered a way to write the next row of numbers, given the numbers in the row above it. There are always 1's on the outside. Each remaining number is the sum of the two numbers above:

$$
\begin{array}{c}
1 \\
1 \quad 1 \\
1 \quad 2 \quad 1 \\
1 \quad 3 \quad 3 \quad 1 \\
1 \quad 4 \quad 6 \quad 4 \quad 1 \\
1 \quad 5 \quad 10 \quad 10 \quad 5 \quad 1 \\
1 \quad 6 \quad 15 \quad 20 \quad 15 \quad 6 \quad 1
\end{array}
$$

We see that in the bottom (seventh) row

the 1st and last numbers are 1;

the 2nd number is $1 + 5$, or 6:

the 3rd number is $5 + 10$, or 15;

the 4th number is $10 + 10$, or 20;

the 5th number is $10 + 5$, or 15; and

the 6th number is $5 + 1$, or 6.

Thus the expansion of $(a+b)^6$ is

$$(a+b)^6 = 1a^6 + 6a^5b + 15a^4b^2 + 20a^3b^3 + 15a^2b^4 + 6ab^5 + 1b^6.$$

To expand $(a+b)^8$, we complete two more rows of Pascal's triangle:

$$
\begin{array}{c}
1 \\
1 \quad 1 \\
1 \quad 2 \quad 1 \\
1 \quad 3 \quad 3 \quad 1 \\
1 \quad 4 \quad 6 \quad 4 \quad 1 \\
1 \quad 5 \quad 10 \quad 10 \quad 5 \quad 1 \\
1 \quad 6 \quad 15 \quad 20 \quad 15 \quad 6 \quad 1 \\
1 \quad 7 \quad 21 \quad 35 \quad 35 \quad 21 \quad 7 \quad 1 \\
1 \quad 8 \quad 28 \quad 56 \quad 70 \quad 56 \quad 28 \quad 8 \quad 1
\end{array}
$$

Thus the expansion of $(a + b)^8$ has coefficients found in the 9th row above:

$$(a + b)^8 = 1a^8 + 8a^7b + 28a^6b^2 + 56a^5b^3 + 70a^4b^4 + 56a^3b^5 + 28a^2b^6 + 8ab^7 + 1b^8.$$

We can generalize our results as follows:

> ### The Binomial Theorem (Form 1)
>
> For any binomial $a + b$ and any natural number n,
>
> $$(a + b)^n = c_0a^nb^0 + c_1a^{n-1}b^1 + c_2a^{n-2}b^2 + \cdots + c_{n-1}a^1b^{n-1} + c_na^0b^n,$$
>
> where the numbers $c_0, c_1, c_2, \ldots, c_n$ are from the $(n + 1)$st row of Pascal's triangle.

Example 1

Expand: $(u - v)^5$.

Solution   Using the binomial theorem, we have $a = u$, $b = -v$, and $n = 5$. We use the 6th row of Pascal's triangle: 1   5   10   10   5   1. Thus,

$$\begin{aligned}
(u - v)^5 &= [u + (-v)]^5 \qquad \text{Rewriting } u - v \text{ as a sum} \\
&= 1(u)^5 + 5(u)^4(-v)^1 + 10(u)^3(-v)^2 + 10(u)^2(-v)^3 \\
&\quad + 5(u)^1(-v)^4 + 1(-v)^5 \\
&= u^5 - 5u^4v + 10u^3v^2 - 10u^2v^3 + 5uv^4 - v^5.
\end{aligned}$$

Note that the signs of the terms alternate between $+$ and $-$. When $-v$ is raised to an odd power, the sign is $-$.

Example 2

Expand: $\left(2t + \dfrac{3}{t}\right)^6$.

Solution   Note that $a = 2t$, $b = 3/t$, and $n = 6$. We use the 7th row of Pascal's triangle: 1   6   15   20   15   6   1. Thus,

$$\begin{aligned}
\left(2t + \frac{3}{t}\right)^6 &= 1(2t)^6 + 6(2t)^5\left(\frac{3}{t}\right)^1 + 15(2t)^4\left(\frac{3}{t}\right)^2 + 20(2t)^3\left(\frac{3}{t}\right)^3 \\
&\quad + 15(2t)^2\left(\frac{3}{t}\right)^4 + 6(2t)^1\left(\frac{3}{t}\right)^5 + 1\left(\frac{3}{t}\right)^6 \\
&= 64t^6 + 6(32t^5)\left(\frac{3}{t}\right) + 15(16t^4)\left(\frac{9}{t^2}\right) + 20(8t^3)\left(\frac{27}{t^3}\right) \\
&\quad + 15(4t^2)\left(\frac{81}{t^4}\right) + 6(2t)\left(\frac{243}{t^5}\right) + \frac{729}{t^6} \\
&= 64t^6 + 576t^4 + 2160t^2 + 4320 + 4860t^{-2} + 2916t^{-4} \\
&\quad + 729t^{-6}.
\end{aligned}$$

Binomial Expansion Using Factorial Notation

The drawback to using Pascal's triangle is that we must compute all the preceding rows in the table to obtain the row needed for the expansion in which we are interested. The following method avoids this difficulty. It will also enable us to find a specific term— say, the 8th term—without computing all the other terms in the expansion. This method is useful in such courses as finite mathematics, calculus, and statistics, and uses the *binomial coefficient* notation

$$\binom{n}{r}$$

developed in Section 14.5.

> ### *The Binomial Theorem (Form 2)*
> For any binomial $a + b$ and any natural number n,
> $$(a + b)^n = \binom{n}{0}a^n + \binom{n}{1}a^{n-1}b + \binom{n}{2}a^{n-2}b^2 + \cdots + \binom{n}{n}b^n.$$

E x a m p l e   3   Expand: $(3x + y)^4$.

Solution   We use the binomial theorem (Form 2) with $a = 3x$, $b = y$, and $n = 4$:

$$(3x + y)^4 = \binom{4}{0}(3x)^4 + \binom{4}{1}(3x)^3y + \binom{4}{2}(3x)^2y^2 + \binom{4}{3}(3x)y^3 + \binom{4}{4}y^4$$

$$= \frac{4!}{4!\,0!}3^4x^4 + \frac{4!}{3!\,1!}3^3x^3y + \frac{4!}{2!\,2!}3^2x^2y^2 + \frac{4!}{1!\,3!}3xy^3 + \frac{4!}{0!\,4!}y^4$$

$$= 81x^4 + 108x^3y + 54x^2y^2 + 12xy^3 + y^4. \qquad \text{Simplifying}$$

E x a m p l e   4   Expand: $(x^2 - 2y)^5$.

Solution   In this case, $a = x^2$, $b = -2y$, and $n = 5$:

$$(x^2 - 2y)^5 = \binom{5}{0}(x^2)^5 + \binom{5}{1}(x^2)^4(-2y) + \binom{5}{2}(x^2)^3(-2y)^2$$

$$+ \binom{5}{3}(x^2)^2(-2y)^3 + \binom{5}{4}(x^2)(-2y)^4 + \binom{5}{5}(-2y)^5$$

$$= \frac{5!}{5!\,0!}x^{10} + \frac{5!}{4!\,1!}x^8(-2y) + \frac{5!}{3!\,2!}x^6(-2y)^2 + \frac{5!}{2!\,3!}x^4(-2y)^3$$

$$+ \frac{5!}{1!\,4!}x^2(-2y)^4 + \frac{5!}{0!\,5!}(-2y)^5$$

$$= x^{10} - 10x^8y + 40x^6y^2 - 80x^4y^3 + 80x^2y^4 - 32y^5.$$

Note that in the binomial theorem (Form 2), $\binom{n}{0}a^n b^0$ gives us the first term, $\binom{n}{1}a^{n-1}b^1$ gives us the second term, $\binom{n}{2}a^{n-2}b^2$ gives us the third term, and so on. This can be generalized to give a method for finding a specific term without writing the entire expansion.

> ### Finding a Specific Term
> The $(r + 1)$st term of $(a + b)^n$ is
> $$\binom{n}{r}a^{n-r}b^r.$$

E x a m p l e   5   Find the 5th term in the expansion of $(2x - 3y)^7$.

Solution   First, we note that $5 = 4 + 1$. Thus, $r = 4$, $a = 2x$, $b = -3y$, and $n = 7$. Then the 5th term of the expansion is

$$\binom{7}{4}(2x)^{7-4}(-3y)^4, \quad \text{or} \quad \frac{7!}{3!\,4!}(2x)^3(-3y)^4, \quad \text{or} \quad 22{,}680x^3y^4.$$

It is because of the binomial theorem that $\binom{n}{r}$ is called a *binomial coefficient*. We can now explain why 0! is defined to be 1. In the binomial expansion, we want $\binom{n}{0}$ to equal 1 and we also want the definition

$$\binom{n}{r} = \frac{n!}{(n - r)!\,r!}$$

to hold for all whole numbers n and r. Thus we must have

$$\binom{n}{0} = \frac{n!}{(n - 0)!\,0!} = \frac{n!}{n!\,0!} = 1.$$

This is satisfied only if 0! is defined to be 1.

FOR EXTRA HELP

Exercise Set 14.6

 Digital Video Tutor CD 11 Videotape 27   InterAct Math   Math Tutor Center   MathXL    MyMathLab.com

Expand. Use both of the methods shown in this section.

1. $(m + n)^5$

2. $(a - b)^4$

3. $(x - y)^6$

4. $(p + q)^7$

5. $(x^2 - 3y)^5$

6. $(3c - d)^7$

7. $(3c - d)^6$

8. $(t^{-2} + 2)^6$

9. $(x - y)^3$

10. $(x - y)^5$

11. $\left(x + \dfrac{2}{y}\right)^9$

12. $\left(3s + \dfrac{1}{t}\right)^9$

13. $(a^2 - b^3)^5$

14. $(x^3 - 2y)^5$

15. $(\sqrt{3} - t)^4$

16. $(\sqrt{5} + t)^6$

17. $(x^{-2} + x^2)^4$

18. $\left(\dfrac{1}{\sqrt{x}} - \sqrt{x}\right)^6$

Find the indicated term for each binomial expression.

19. 3rd, $(a + b)^6$

20. 6th, $(x + y)^7$

21. 12th, $(a - 3)^{14}$

22. 11th, $(x - 2)^{12}$

23. 5th, $(2x^3 + \sqrt{y})^8$

24. 4th, $\left(\dfrac{1}{b^2} + c\right)^7$

25. Middle, $(2u - 3v^2)^{10}$

26. Middle two, $(\sqrt{x} + \sqrt{3})^5$

Aha! **27.** 9th, $(x - y)^8$

28. 10th, $(a - b)^9$

29. Maya claims that she can calculate mentally the first two and the last two terms of the expansion of $(a + b)^n$ for any whole number n. How do you think she does this?

30. Without performing any calculations, how can you tell if the expansions of $(x - y)^8$ and $(y - x)^8$ are equal?

SKILL MAINTENANCE

Solve.

31. $\log_2 x + \log_2(x - 2) = 3$

32. $\log_3(x + 2) - \log_3(x - 2) = 2$

33. $e^t = 280$

34. $\log_5 x^2 = 2$

SYNTHESIS

35. Devise two problems requiring the use of the binomial theorem. Design the problems so that one is solved more easily using Form 1 and the other is solved more easily using Form 2. Then explain what makes one form easier to use than the other in each case.

36. Explain how someone can determine the x^2-term of the expansion of $\left(x - \dfrac{3}{x}\right)^{10}$ without calculating any other terms.

37. *Baseball.* At one point in a recent season. Derek Jeter of the New York Yankees had a batting average of 0.325. At that time, if someone were to randomly select 5 of his "at-bats," the probability of his getting exactly 3 hits would be the 3rd term of the binomial expansion of $(0.325 + 0.675)^5$. Find that term and use a calculator to estimate the probability.

38. *Widows or divorcees.* The probability that a woman will be either widowed or divorced is 85%. If 8 women are randomly selected, the probability that exactly 5 of them will be either widowed or divorced is the 6th term of the binomial expansion of $(0.15 + 0.85)^8$. Use a calculator to estimate that probability.

39. *Baseball.* In reference to Exercise 37, the probability that Jeter will get *at most* 3 hits is found by adding the last 4 terms of the binomial expansion of $(0.325 + 0.675)^5$. Find these terms and use a calculator to estimate the probability.

40. *Widows or divorcees.* In reference to Exercise 38, the probability that *at least* 6 of the women will be widowed or divorced is found by adding the last three terms of the binomial expansion of $(0.15 + 0.85)^8$. Find these terms and use a calculator to estimate the probability.

41. Find the term of
$$\left(\dfrac{3x^2}{2} - \dfrac{1}{3x}\right)^{12}$$
that does not contain x.

42. Find the middle term of $(x^2 - 6y^{3/2})^6$.

43. Find the ratio of the 4th term of

$$\left(p^2 - \frac{1}{2}p\sqrt[3]{q}\right)^5$$

to the 3rd term.

44. Find the term containing $\dfrac{1}{x^{1/6}}$ of

$$\left(\sqrt[3]{x} - \frac{1}{\sqrt{x}}\right)^7.$$

45. What is the degree of $(x^2 + 3)^4$?

46. Multiply: $(x^2 + 2xy + y^2)(x^2 + 2xy + y^2)^2(x + y)$.

Introduction to Probability

14.7

**Experimental and Theoretical Probability •
Experimental Probabilities • Theoretical Probabilities •
Origin and Use of Probability**

We reason that when a coin is tossed, the chances that it will fall heads are 1 out of 2, or the **probability** that it will fall heads is $\frac{1}{2}$. Of course this does not mean that if a coin is tossed ten times, it will necessarily fall heads exactly five times. If the coin is a "fair" coin and is tossed a great many times, however, it will fall heads very nearly half of them.

Experimental and Theoretical Probability

If we toss a coin a great number of times, say 1000, and count the number of times that it falls heads, we can determine the probability of it falling heads. If there are 503 heads, we would calculate the probability of the coin falling heads to be

$$\frac{503}{1000}, \quad \text{or} \quad 0.503.$$

This is an **experimental** determination of probability. Such a determination of probability is quite common. Here, for example, are some probabilities that have been determined *experimentally*.

1. If you kiss someone who has a cold, the probability of your catching a cold is 0.07.

2. A person just released from prison has an 80% probability of returning.

3. The probability that a woman will get breast cancer is $\frac{1}{11}$.

If we consider a coin and *reason* that it is just as likely to fall heads as tails, we would calculate the probability to be $\frac{1}{2}$. This is a **theoretical** determination of probability. Here, for example, are some probabilities that have been determined *theoretically*, using mathematics.

1. If there are 30 people in a room, the probability that two of them have the same birthday (excluding year of birth) is 0.706.
2. If a deck of 52 playing cards is thoroughly shuffled and a card is selected, the probability that the card is a jack is $\frac{1}{13}$, or about 0.077.

Experimental Probabilities

We first consider experimental determination of probability. The basic principle we use to compute such probabilities is as follows.

> ### Principle P (Experimental)
>
> An experiment is performed in which n observations are made. If a situation E, or event, occurs m times out of the n observations, then we say that the *experimental probability* of that event, $P(E)$, is given by
>
> $$P(E) = \frac{m}{n}.$$

E x a m p l e   1

TV game shows. A contestant on the game show "Who Wants to be a Millionaire?" can win one of 11 different prize amounts. The number of contestants in the year 2000 who won each of those amounts is shown in the following graph.

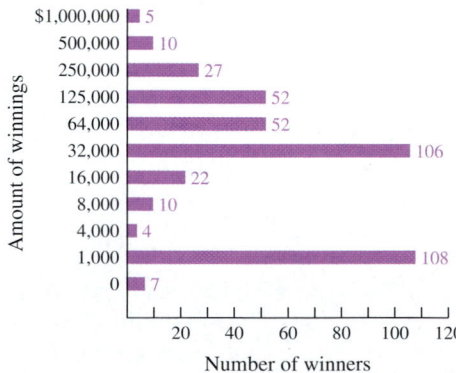

a) Determine the probability that a given contestant will win $1000.
b) Determine the probability that a given contestant will win $250,000 or more.

Solution

a) The total number of contestants, found by adding the number who won each prize amount, is 403. The number of contestants who won $1000 is 108. Thus the probability that a contestant would win $1000 is P, where

$$P = \frac{108}{403}, \quad \text{or} \quad \text{approximately 27\%.}$$

b) There are 27 contestants who have won $250,000, 10 who have won $500,000, and 5 who have won $1,000,000. The number of contestants who have won $250,000 or more is thus $27 + 10 + 5$, or 42. Thus the probability that a contestant will win $250,000 or more is P, where

$$P = \frac{42}{403}, \quad \text{or} \quad \text{approximately 10\%.}$$

A common use of experimental probability is in opinion polls, which are used for various purposes from advertising campaigns to predicting the outcome of an election. The opinions of a relatively small group of people, or *sample*, are used to describe the opinions of an entire population. Interestingly, if the sample is chosen correctly, the results are very accurate.

Example 2

Opinion poll. A typical sample size for a U.S. opinion poll is 1000 (*Source*: www.gallup.com). In a recent poll, participants were asked whether or not they approved of the President's policy on global warming. The results are listed in the following table.

Opinion	Approve of Policy	Disapprove of Policy	Not Sure
Number of Responses	392	415	193

What is the probability that a given person in the United States approves of the President's policy on global warming?

Solution The total number of responses was 1000. There were 392 who approved of the policy. Thus the probability of a given person approving of the policy is P, where

$$P = \frac{392}{1000} = 0.392 = 39.2\%.$$

Theoretical Probabilities

Suppose that we perform an experiment such as flipping a coin, throwing a dart, drawing a card from a deck, or checking an item off an assembly line for quality. The results of an experiment are called **outcomes**. The set of all possible outcomes is called the **sample space**. An **event** is a set of outcomes, that is, a subset of the sample space.

Example 3

List the outcomes and the sample space for each of the following experiments.

a) Tossing a coin
b) Rolling a die
c) Throwing a dart at the dartboard shown at left below

Solution

a) A coin can fall either heads or tails. Thus the outcomes are *falling heads* and *falling tails*. To simplify the writing, we let H stand for the outcome *falling heads* and T stand for *falling tails*. Then the sample space is $\{H, T\}$.

b) A die (pl., dice) is a cube. Each of the six faces of a die contains a pattern of dots representing one of the numbers from 1 to 6. The outcomes are 1, 2, 3, 4, 5, and 6. The sample space is $\{1, 2, 3, 4, 5, 6\}$.

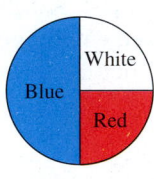

c) If we assume that the dart must hit the board somewhere, the outcomes are *hitting white* (W), *hitting blue* (B), and *hitting red* (R). The sample space is $\{W, B, R\}$.

We denote the probability that an event E occurs as $P(E)$. For example, if "a coin falling heads" is denoted by H, then $P(H)$ represents the probability of the coin falling heads. When all the outcomes of an experiment have the same probability of occurring, we say that they are *equally likely*. If the coin and the die in Example 3 have not been weighted to make one outcome appear more often, they are "fair," and the outcomes are equally likely.

The outcomes in Example 3(c) are *not* equally likely. As shown on the dartboard, labeled below as board A, the probability of hitting blue is greater than the probability of hitting either red or white. The sample space for the experiment of throwing a dart at board B is the same as that for board A, but for board B the outcomes are equally likely.

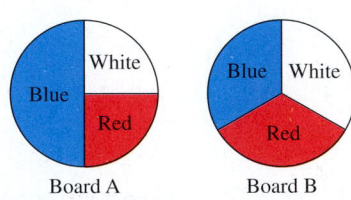

Board A   Board B

A sample space that is comprised of equally likely outcomes allows us to calculate probabilities of other events.

> **Principle P (Theoretical)**
>
> If an event E can occur m ways out of n possible equally likely outcomes of a sample space S, then the *theoretical probability* of that event is given by
>
> $$P(E) = \frac{m}{n}.$$

E x a m p l e   4   What is the probability of rolling a 3 on the die?

Solution   On a fair die, there are 6 equally likely outcomes and there is 1 way to get a 3. By Principle P, $P(3) = \frac{1}{6}$.

E x a m p l e   5   What is the probability of rolling an even number on a die?

Solution   The event is getting an even number. It can occur in 3 ways (getting 2, 4, or 6). The number of equally likely outcomes is 6. By Principle P, $P(\text{even}) = \frac{3}{6}$, or $\frac{1}{2}$.

E x a m p l e   6   Suppose we select, without looking, one marble from a bag containing 3 red marbles and 4 green marbles. What is the probability of selecting a red marble?

Solution   There are 7 equally likely ways of selecting any marble, and since the number of ways of getting a red marble is 3,

$$P(\text{selecting a red marble}) = \frac{3}{7}.$$

We now use a number of examples related to a standard bridge deck of 52 cards. Such a deck is made up as shown in the following figure.

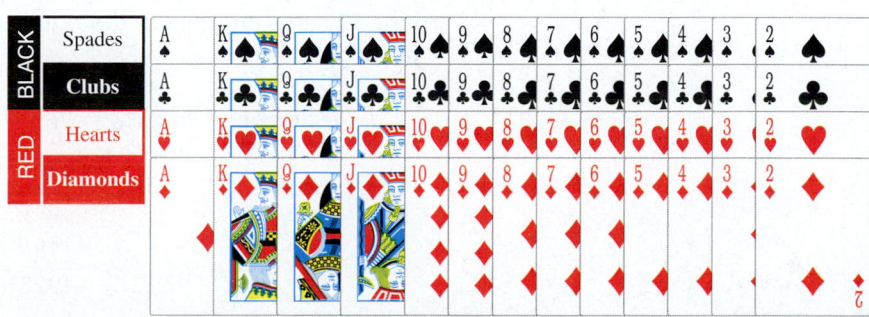

E x a m p l e 7 What is the probability of drawing an ace from a well-shuffled deck of 52 cards?

Solution Since there are 52 outcomes (cards in the deck) and they are equally likely (from a well-shuffled deck) and there are 4 ways to obtain an ace, by Principle *P* we have

$$P(\text{drawing an ace}) = \frac{4}{52}, \quad \text{or} \quad \frac{1}{13}.$$

The following are some results that follow from Principle *P*.

Probability Properties

a) If an event *E* cannot occur, then $P(E) = 0$.
b) If an event *E* is certain to occur (that is, every trial is a success), then $P(E) = 1$.
c) The probability that an event *E* will occur is a number from 0 to 1: $0 \leq P(E) \leq 1$.

For example, in coin tossing, the event that a coin will land on its edge has probability 0. The event that a coin falls either heads or tails has probability 1.

In the following examples, we use the combinations that we studied in Section 14.5 to calculate theoretical probabilities.

E x a m p l e 8 Suppose that 2 cards are drawn from a well-shuffled deck of 52 cards. What is the probability that both of them are spades?

Solution The probability will be

$$P(\text{drawing 2 spades}) = \frac{m}{n}$$

$$= \frac{\text{The number of ways of drawing 2 spades}}{\text{The number of ways of drawing any 2 cards}}.$$

The number of ways *n* of drawing 2 cards from a deck of 52 is $_{52}C_2$. Since 13 of the 52 cards are spades, the number of ways *m* of drawing 2 spades is $_{13}C_2$. Thus,

$$P(\text{drawing 2 spades}) = \frac{m}{n}$$

$$= \frac{_{13}C_2}{_{52}C_2}$$

$$= \frac{78}{1326} = \frac{1}{17}.$$

Example 9

Suppose that 2 students are selected at random from a group that consists of 6 freshmen and 4 sophomores. What is the probability that both of them are sophomores?

Solution   We need to know how many ways n any 2 students can be selected, and also how many ways m 2 sophomores can be selected. The number of ways of selecting 2 students from a group of 10 is $_{10}C_2$. The number of ways of selecting 2 sophomores from a group of 4 is $_4C_2$. Thus the probability of selecting 2 sophomores from the group of 10 is P, where

$$P = \frac{_4C_2}{_{10}C_2} = \frac{6}{45} = \frac{2}{15}.$$

Example 10

Suppose that 3 people are selected at random from a group that consists of 6 students and 4 professors. What is the probability that 1 student and 2 professors are selected?

Solution   The number of ways of selecting 3 people from a group of 10 is $_{10}C_3$. One student can be selected in $_6C_1$ ways, and 2 professors can be selected in $_4C_2$ ways. By the fundamental counting principle, the number of ways of selecting 1 student and 2 professors is $_6C_1 \cdot {_4C_2}$. Thus the probability is

$$P = \frac{_6C_1 \cdot {_4C_2}}{_{10}C_3}, \quad \text{or} \quad \frac{3}{10}.$$

Example 11

What is the probability of getting a total of 8 on a roll of a pair of dice? (Assume that the dice are different—say, one blue and one black.)

Solution   On each die, there are 6 possible outcomes. The outcomes are paired so there are $6 \cdot 6$, or 36, possible ways in which the two can fall.

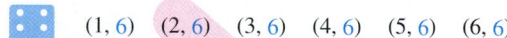

$$
\begin{array}{llllll}
(1,6) & (2,6) & (3,6) & (4,6) & (5,6) & (6,6) \\
(1,5) & (2,5) & (3,5) & (4,5) & (5,5) & (6,5) \\
(1,4) & (2,4) & (3,4) & (4,4) & (5,4) & (6,4) \\
(1,3) & (2,3) & (3,3) & (4,3) & (5,3) & (6,3) \\
(1,2) & (2,2) & (3,2) & (4,2) & (5,2) & (6,2) \\
(1,1) & (2,1) & (3,1) & (4,1) & (5,1) & (6,1)
\end{array}
$$

The pairs that total 8 are shown. Thus there are 5 possible ways of getting a total of 8, so the probability is $\frac{5}{36}$.

Origin and Use of Probability

A desire to calculate odds in games of chance gave rise to the theory of probability. Today the theory of probability and its closely related field, mathematical statistics, have many applications, most of them not related to games of chance. Opinion polls, with such uses as predicting elections, are a familiar example. Quality control, in which a prediction about the percentage of faulty items manufactured is made, is an important application, among many, in business. Still other applications are in the areas of DNA blood testing for genetics and crime investigation, other areas of medicine, and the kinetic theory of gases.

---

1. *Corrective lenses.* In a survey conducted by a statistics student, 100 people were polled to determine the probability of a person wearing either glasses or contact lenses. Of those polled, 57 wore either glasses or contacts. What is the probability that a person wears either glasses or contacts? What is the probability that a person wears neither?

2. *Favorite number.* In another survey, 100 people were polled and asked to select a number from 1 to 5. The results are shown in the following table.

Number Choice	1	2	3	4	5
Number of People Who Chose That Number	18	24	23	23	12

a) What is the probability that the number selected is 1? 2? 3? 4? 5?

b) What general conclusion might a psychologist make from this experiment?

Linguistics. *The number of occurrences of each letter of the English alphabet on the front page of a particular newspaper is listed in the following table. In all, there were 9136 letters on the page.*

Letter	Number of Occurrences
A	853
B	136
C	273
D	286
E	1229
F	173
G	190
H	399
I	539
J	21
K	57
L	417
M	231
N	597
O	705
P	238
Q	4
R	609
S	745
T	789
U	240
V	113
W	127
X	20
Y	124
Z	21

Round answers in Exercises 3–6 to three decimal places.

3. What is the probability of the occurrence of the letter A? E? I? O? U?

4. What is the probability of a vowel occurring?

5. What is the probability of a consonant occurring?

6. Which letter has the least probability of occurring? What is the probability of this letter not occurring?

7. On the popular game show "Wheel of Fortune," players guess letters in order to spell out a phrase, a person, or a thing.

   a) Which consonant has the greatest probability of occurring?

   b) Which vowel has the greatest probability of occurring?

   c) At one time on the show, contestants playing for a grand prize were allowed to guess 5 consonants and 1 vowel in order to discover the secret wording. On the basis of the results of this experiment alone, which 5 consonants and which vowel would you advise a contestant to choose?

Drawing a card. *Suppose that we draw a card from a well-shuffled deck of 52 cards.*

8. How many equally likely outcomes are there?

9. What is the probability of drawing a queen?

10. What is the probability of drawing a heart?

11. What is the probability of drawing a spade or a club?

12. What is the probability of drawing a red card?

13. What is the probability of drawing a 9 or a king?

14. What is the probability of drawing a black ace?

Drawing marbles. *Suppose we select, without looking, one marble from a bag containing 4 red marbles and 10 green marbles.*

15. What is the probability of selecting a red marble?

16. What is the probability of selecting a green marble?

17. What is the probability of selecting a purple marble?

18. What is the probability of selecting a white marble?

Drawing cards. *Suppose that 4 cards are drawn from a well-shuffled deck of 52 cards.*

19. What is the probability that all 4 are spades?

20. What is the probability that all 4 are hearts?

21. *Drawing coins.* From a bag containing 6 nickels, 10 dimes, and 4 quarters, 6 coins are drawn at random, all at once. What is the probability of getting 3 nickels, 2 dimes, and 1 quarter?

22. *Choosing a committee.* From a group of 8 men and 7 women, a committee of 4 is chosen. What is the probability that 2 men and 2 women will be chosen?

23. *Rolling dice.* What is the probability of getting a total of 6 on a roll of a pair of dice?

24. *Rolling dice.* What is the probability of getting snake eyes (a total of 2) on a roll of a pair of dice?

25. *Monopoly®.* In the board game Monopoly, a player can "get out of jail" by rolling doubles (both numbers the same) on a pair of dice. What is the probability of rolling doubles?

26. *Parcheesi®.*   Moves in the board game Parcheesi are based on the roll of a pair of dice. To begin moving around the board, a player must roll a 5 on one or both dice *or* a total of 5 on both dice. What is the probability that a player will roll a 5 or a total of 5?

27. Tracy takes one pill each of vitamins A, C, E, and B-12 every day. He fills a bottle with 7 days' worth of all the vitamins needed. The next day, Tracy selects 4 vitamins at random, all at once, from the bottle. What is the probability that the vitamins will be the 4 desired?

28. Toni is given a box of canned goods and told that it contains 4 cans of meat, 10 cans of vegetables, 6 cans of fruit, and 5 cans of soup. However, the labels have been removed from the cans and they all appear to be identical. Toni selects 4 cans at random, all at once, and opens them for dinner. What is the probability that she will open 1 can each of meat, fruit, vegetables, and soup?

Roulette.   A roulette wheel contains slots numbered 00, 0, 1, 2, 3, . . . , 35, 36. Eighteen of the slots numbered 1 through 36 are colored red and eighteen are colored black. The 00 and 0 slots are uncolored. The wheel is spun, and a ball is rolled around the rim until it falls into a slot. What is the probability that the ball falls in:

29. a black slot?

30. a red or black slot?

31. the 0 slot?

32. either the 00 or 0 slot? (In this case, the house always wins.)

Throwing a dart.   A dart is thrown at the dartboard shown here.

Find each of the following probabilities assuming the dartboard is hit every time.

33. $P(\text{red})$

34. $P(\text{yellow})$

35. $P(\text{blue or yellow})$

36. $P(\text{red or blue})$

37. $P(\text{red or blue or yellow})$

38. $P(\text{green})$

39. Use Principle P to explain why the probability of an event cannot be greater than 1.

40. Use Principle P to explain why the probability of an event cannot be less than 0.

SKILL MAINTENANCE

Solve.

41. $2x + 5y = 7,$
$3x + 2y = 16$

42. $2^{x^2 + 2x} = \frac{1}{2}$

43. Express in terms of logarithms of x, y, and z:
$$\log_a \frac{x^2 y}{z^3}.$$

44. Find an equation of the circle with center $(0, -3)$ and radius $2\sqrt{3}$.

45. Convert to an exponential equation:
$$\log_3 x = 4.$$

46. Convert to a logarithmic equation:
$$4^y = 10.$$

SYNTHESIS

47. Suppose that a coin is tossed repeatedly and that it falls heads 10 times in a row. Is the probability that it will fall tails on the next toss greater than $\frac{1}{2}$? Explain your answer.

48. Explain the difference between experimental and theoretical probability.

Five-card poker hands.   For Exercises 49–58, give a reasoned expression as well as the answer. Read all the exercises before beginning.

49. How many 5-card poker hands can be dealt from a standard 52-card deck?

50. A *royal flush* consists of a 5-card hand with A-K-Q-J-10 of the same suit.
   a) How many royal flushes are there?
   b) What is the probability of getting a royal flush?

51. A *straight flush* consists of 5 cards in sequence in the same suit, but excludes royal flushes. An ace can be used low, before a two.
   a) How many straight flushes are there?
   b) What is the probability of getting a straight flush?

52. *Four of a kind* is a 5-card hand in which exactly 4 of the cards are of the same denomination, such as J-J-J-J-6, 7-7-7-7-A, or 2-2-2-2-5.
   a) How many are there?
   b) What is the probability of getting four of a kind?

53. A *full house* consists of a pair and a three of a kind, such as Q-Q-Q-4-4.
   a) How many are there?
   b) What is the probability of getting a full house?

54. A *pair* is a 5-card hand in which just 2 of the cards are of the same denomination, such as Q-Q-8-A-3.
   a) How many are there?
   b) What is the probability of getting a pair?

55. *Three of a kind* is a 5-card hand in which exactly 3 of the cards are of the same denomination and the other 2 are *not* of the same denomination, such as Q-Q-Q-10-7.
   a) How many are there?
   b) What is the probability of getting three of a kind?

56. A *flush* is a 5-card hand in which all of the cards are of the same suit, but not all in sequence (not a straight flush or royal flush).
   a) How many are there?
   b) What is the probability of getting a flush?

57. *Two pairs* is a hand like Q-Q-3-3-A.
   a) How many are there?
   b) What is the probability of getting two pairs?

58. A *straight* is any 5 cards in sequence, but not of the same suit—for example, 4 of spades, 5 of spades, 6 of diamonds, 7 of hearts, and 8 of clubs.
   a) How many straights are there?
   b) What is the probability of getting a straight?

Summary and Review 14

Key Terms

Important Properties and Formulas

Arithmetic sequence: $a_{n+1} = a_n + d$

nth term of an arithmetic sequence: $a_n = a_1 + (n - 1)d$

Sum of the first n terms of an arithmetic sequence:

$$S_n = \frac{n}{2}(a_1 + a_n)$$

Geometric sequence: $a_{n+1} = a_n \cdot r$

nth term of a geometric sequence: $a_n = a_1 r^{n-1}$

Sum of the first n terms of a geometric sequence:

$$S_n = \frac{a_1(1 - r^n)}{1 - r}$$

Limit of an infinite geometric series: $S_\infty = \dfrac{a_1}{1 - r}, \quad |r| < 1$

Factorial notation: $n! = n(n - 1)(n - 2) \cdots 3 \cdot 2 \cdot 1$

Permutations of n objects taken r at a time: $_nP_r = n(n - 1)(n - 2) \cdots [n - (r - 1)] = \dfrac{n!}{(n - r)!}$

Combinations of n objects taken r at a time:

$$_nC_r = \frac{n!}{(n - r)!\, r!}$$

Binomial coefficient:

$$\binom{n}{r} = \frac{n!}{(n - r)!\, r!}$$

Binomial theorem:
$$(a + b)^n = \binom{n}{0}a^n + \binom{n}{1}a^{n-1}b$$
$$+ \binom{n}{2}a^{n-2}b^2 + \cdots + \binom{n}{n}b^n$$

$(r + 1)$st term of $(a + b)^n$:
$$\binom{n}{r}a^{n-r}b^r$$

Fundamental Counting Principle

Given a combined action, or event, in which the first action can be performed in n_1 ways, the second action can be performed in n_2 ways, and so on, then the total number of ways in which the combined action can be performed is the product

$n_1 \cdot n_2 \cdot n_3 \cdot \cdots \cdot n_k.$

Principle P (Experimental)

An experiment is performed in which n observations are made. If a situation E, or event, occurs m times out of the n observations, then we say that the *experimental probability* of that event, $P(E)$, is given by

$$P(E) = \frac{m}{n}.$$

Principle P (Theoretical)

If an event E can occur m ways out of n possible equally likely outcomes of a sample space S, then the *theoretical probability* of that event is given by

$$P(E) = \frac{m}{n}.$$

If an event E cannot occur, then $P(E) = 0$.

If an event E is certain to occur, then $P(E) = 1$.

The probability that an event E will occur is a number from 0 to 1:

$0 \le P(E) \le 1.$

Review Exercises

Find the first four terms; the 8th term, a_8; and the 12th term, a_{12}.

1. $a_n = 4n - 3$

2. $a_n = \dfrac{n-1}{n^2+1}$

Predict the general term. Answers may vary.

3. $-2, -4, -6, -8, -10, \ldots$

4. $-1, 3, -5, 7, -9, \ldots$

Write out and evaluate each sum.

5. $\displaystyle\sum_{k=1}^{5} (-2)^k$

6. $\displaystyle\sum_{k=2}^{7} (1-2k)$

Rewrite using sigma notation.

7. $4 + 8 + 12 + 16 + 20$

8. $\dfrac{-1}{2} + \dfrac{1}{4} + \dfrac{-1}{8} + \dfrac{1}{16} + \dfrac{-1}{32}$

9. Find the 14th term of the arithmetic sequence $-6, 1, 8, \ldots$.

10. Find d when $a_1 = 11$ and $a_{10} = 35$. Assume an arithmetic sequence.

11. Find a_1 and d when $a_{12} = 25$ and $a_{24} = 40$. Assume an arithmetic sequence.

12. Find the sum of the first 17 terms of the arithmetic series $-8 + (-11) + (-14) + \cdots$.

13. Find the sum of all the multiples of 6 from 12 to 318, inclusive.

14. Find the 20th term of the geometric sequence $2, 2\sqrt{2}, 4, \ldots$.

15. Find the common ratio of the geometric sequence $2, \frac{4}{3}, \frac{8}{9}, \ldots$.

16. Find the nth term of the geometric sequence $-2, 2, -2, \ldots$.

17. Find the nth term of the geometric sequence $3, \frac{3}{4}x, \frac{3}{16}x^2, \ldots$.

18. Find S_6 for the geometric series $3 + 12 + 48 + \cdots$.

19. Find S_{12} for the geometric series $3x - 6x + 12x - \cdots$.

Determine whether each infinite geometric series has a limit. If a limit exists, find it.

20. $6 + 3 + 1.5 + 0.75 + \cdots$

21. $7 - 4 + \frac{16}{7} - \cdots$

22. $2 + (-2) + 2 + (-2) + \cdots$

23. $0.04 + 0.08 + 0.16 + 0.32 + \cdots$

24. $\$2000 + \$1900 + \$1805 + \$1714.75 + \cdots$

25. Find fractional notation for $0.555555\ldots$.

26. Find fractional notation for $1.39393939\ldots$.

Solve.

27. Adam took a telemarketing job, starting with an hourly wage of $11.40. He was promised a raise of 20¢ per hour every 3 mos for 8 yr. At the end of 8 yr, what will be his hourly wage?

28. A stack of poles has 42 poles in the bottom row. There are 41 poles in the second row, 40 poles in the third row, and so on, ending with 1 pole in the top row. How many poles are in the stack?

29. A student loan is in the amount of $10,000. Interest is 7%, compounded annually, and the amount is to be paid off in 12 yr. How much is to be paid back?

30. Find the total rebound distance of a ball, given that it is dropped from a height of 12 m and each rebound is one-third of the preceding one.

Simplify.

31. $7!$

32. $\dbinom{8}{3}$

33. $_6P_2$

34. $\dfrac{9!}{3!}$

35. $_7C_4$

36. $_{10}P_1$

37. If 9 different signal flags are available, how many different displays are possible using 4 flags in a row?

38. The Greek alphabet contains 24 letters. How many fraternity or sorority names can be formed using 3 different letters?

39. In how many different ways can 6 books be arranged on a shelf?

40. The winner of a contest can choose any 8 of

15 prizes. How many different selections can be made?

41. All the houses in a community of patio homes have the same floor plan. The developer achieves variety by offering 3 different trim colors, 2 different locations for the garage, and 4 different types of entrance. Find the number of different homes that can be built in the community.

42. Find the 3rd term of $(a + b)^{20}$.

43. Expand: $(x - 2y)^4$.

44. Before an election, a poll was conducted to see which candidate was favored. Three people were running for a particular office. During the polling, 86 favored A, 97 favored B, and 23 favored C. Assuming that the poll is a valid indicator of the election, what is the probability that a person will vote for A? for B? for C?

45. What is the probability of rolling a 10 on a roll of a pair of dice? on a roll of one die?

46. From a deck of 52 cards, 1 card is drawn. What is the probability that it is a club?

47. From a deck of 52 cards, 3 are drawn at random without replacement. What is the probability that 2 are aces and 1 is a king?

SYNTHESIS

48. What happens to the terms of a geometric sequence with $|r| < 1$ as n gets larger? Why?

49. Compare the two forms of the binomial theorem given in the text. Under what circumstances would one be more useful than the other?

50. Find the sum of the first n terms of the geometric series $1 - x + x^2 - x^3 + \cdots$.

51. Expand: $(x^{-3} + x^3)^5$.

Chapter Test 14

1. Find the first five terms and the 16th term of a sequence with general term $a_n = 6n - 5$.

2. Predict the general term of the sequence
$$\frac{4}{3}, \frac{4}{9}, \frac{4}{27}, \ldots.$$

3. Write out and evaluate:
$$\sum_{k=1}^{5} (3 - 2^k).$$

4. Rewrite using sigma notation:
$$1 + (-8) + 27 + (-64) + 125.$$

5. Find the 12th term, a_{12}, of the arithmetic sequence $9, 4, -1, \ldots$.

Assume arithmetic sequences for Questions 6 and 7.

6. Find the common difference d when $a_1 = 9$ and $a_7 = 11\frac{1}{4}$.

7. Find a_1 and d when $a_5 = 16$ and $a_{10} = -3$.

8. Find the sum of all the multiples of 12 from 24 to 240, inclusive.

9. Find the 6th term of the geometric sequence $72, 18, 4\frac{1}{2}, \ldots$.

10. Find the common ratio of the geometric sequence $22\frac{1}{2}, 15, 10, \ldots$.

11. Find the nth term of the geometric sequence $3, -9, 27, \ldots$.

12. Find the sum of the first nine terms of the geometric series
$$(1 + x) + (2 + 2x) + (4 + 4x) + \cdots.$$

Determine whether each infinite geometric series has a limit. If a limit exists, find it.

13. $0.5 + 0.25 + 0.125 + \cdots$

14. $0.5 + 1 + 2 + 4 + \cdots$

15. $\$1000 + \$80 + \$6.40 + \cdots$

16. Find fractional notation for $0.85858585\ldots$.

17. An auditorium has 31 seats in the first row, 33 seats in the second row, 35 seats in the third row, and so on, for 18 rows. How many seats are in the 17th row?

18. Lindsay's Uncle Ken gave her $100 for her first birthday, $200 for her second birthday, and so on, until her eighteenth birthday. How much did he give her in all?

19. Each week the price of a $15,000 boat will be reduced 5% of the previous week's price. If we assume that it is not sold, what will be the price after 10 weeks?

20. Find the total rebound distance of a ball that is dropped from a height of 18 m, with each rebound two thirds of the preceding one.

21. Simplify: $\begin{pmatrix} 13 \\ 11 \end{pmatrix}$.

22. In how many different ways can 6 people be seated in a row?

23. On a test, a student must answer 4 out of 7 questions. In how many ways can this be done?

24. From a group of 20 seniors and 14 juniors, how many committees consisting of 3 seniors and 2 juniors are possible?

25. Ben and Jerry's Homemade, Inc., an international firm, sells ice cream in 20 flavors. How many 3-dip cones are possible if order of flavors is to be considered and no flavor is repeated?

26. Expand: $(x^2 - 3y)^5$.

27. Find the 4th term in the expansion of $(a + x)^{12}$.

28. What is the probability of getting a total of 7 on a roll of a pair of dice?

29. From a deck of 52 cards, 1 card is drawn. What is the probability of drawing an ace?

30. If 3 marbles are drawn at random all at once from a bag containing 5 green marbles, 7 red marbles, and 4 white marbles, what is the probability that 2 will be green and 1 will be white?

SYNTHESIS

31. Find a formula for the sum of the first n even natural numbers:
$$2 + 4 + 6 + \cdots + 2n.$$

32. Find the sum of the first n terms of
$$1 + \frac{1}{x} + \frac{1}{x^2} + \frac{1}{x^3} + \cdots.$$

Cumulative Review 1–14

Simplify.

1. $(-9x^2y^3)(5x^4y^{-7})$

2. $|-3.5 + 9.8|$

3. $2y - [3 - 4(5 - 2y) - 3y]$

4. $(10 \cdot 8 - 9 \cdot 7)^2 - 54 \div 9 - 3$

5. Evaluate
$$\frac{ab - ac}{bc}$$
for $a = -2$, $b = 3$, and $c = -4$.

Perform the indicated operations and simplify.

6. $(5a^2 - 3ab - 7b^2) - (2a^2 + 5ab + 8b^2)$

7. $(-3x^2 + 4x^3 - 5x - 1) + (9x^3 - 4x^2 + 7 - x)$

8. $(2a - 1)(3a + 5)$

9. $(3a^2 - 5y)^2$

10. $\dfrac{1}{x - 2} - \dfrac{4}{x^2 - 4} + \dfrac{3}{x + 2}$

11. $\dfrac{x^2 - 6x + 8}{3x + 9} \cdot \dfrac{x + 3}{x^2 - 4}$

12. $\dfrac{3x + 3y}{5x - 5y} \div \dfrac{3x^2 + 3y^2}{5x^3 - 5y^3}$

13. $\dfrac{x - \dfrac{a^2}{x}}{1 + \dfrac{a}{x}}$

Factor.

14. $4x^2 - 12x + 9$

15. $27a^3 - 8$

16. $a^3 + 3a^2 - ab - 3b$

17. $15y^4 + 33y^2 - 36$

18. For the function described by

$$f(x) = 3x^2 - 4x,$$

find $f(-2)$.

19. Divide:

$$(7x^4 - 5x^3 + x^2 - 4) \div (x - 2).$$

Solve.

20. $9(x - 1) - 3(x - 2) = 1$

21. $\dfrac{6}{x} + \dfrac{6}{x + 2} = \dfrac{5}{2}$

22. $2x + 1 > 5$ *or* $x - 7 \le 3$

23. $5x + 3y = 2,$
  $3x + 5y = -2$

24. $x + y - z = 0,$
  $3x + y + z = 6,$
  $x - y + 2z = 5$

25. $3\sqrt{x - 1} = 5 - x$

26. $x^4 - 29x^2 + 100 = 0$

27. $x^2 + y^2 = 8,$
  $x^2 - y^2 = 2$

28. $5^x = 8$

29. $\log(x^2 - 25) - \log(x + 5) = 3$

30. $\log_4 x = -2$

31. $7^{2x+3} = 49$

32. $|2x - 1| \le 5$

33. $7x^2 + 14 = 0$

34. $x^2 + 4x = 3$

35. $y^2 + 3y > 10$

36. Let $f(x) = x^2 - 2x$. Find a such that $f(a) \le 48$.

37. If $f(x) = \sqrt{x + 1}$ and $g(x) = \sqrt{x - 2} + 3$, find a such that $f(a) = g(a)$.

Solve.

38. The perimeter of a rectangle is 34 ft. The length of a diagonal is 13 ft. Find the dimensions of the rectangle.

39. A music club offers two types of membership. Limited members pay a fee of $10 a year and can buy CDs for $10 each. Preferred members pay $20 a year and can buy CDs for $7.50 each. For what numbers of annual CD purchases would it be less expensive to be a preferred member?

40. Find three consecutive integers whose sum is 198.

41. A pentagon with all five sides the same size has a perimeter equal to that of an octagon in which all eight sides are the same size. One side of the pentagon is 2 less than three times one side of the octagon. What is the perimeter of each figure?

42. Mark's Natural Foods mixes herbs that cost $2.68 an ounce with herbs that cost $4.60 an ounce to create a seasoning that costs $3.80 an ounce. How many ounces of each herb should be mixed together to make 24 oz of the seasoning?

43. An airplane can fly 190 mi with the wind in the same time it takes to fly 160 mi against the wind. The speed of the wind is 30 mph. How fast can the plane fly in still air?

44. Bianca can tap the sugar maple trees in Southway Park in 21 hr. Delia can tap the trees in 14 hr. How long would it take them, working together, to tap the trees?

45. The centripetal force F of an object moving in a circle varies directly as the square of the velocity v and inversely as the radius r of the circle. If $F = 8$ when $v = 1$ and $r = 10$, what is F when $v = 2$ and $r = 16$?

46. A farmer wants to fence in a rectangular area next to a river. (Note that no fence will be needed along the river.) What is the area of the largest region that can be fenced in with 100 ft of fencing?

Graph.

47. $3x - y = 6$

48. $\dfrac{x^2}{25} + \dfrac{y^2}{4} = 1$

49. $y = \log_2 x$

50. $2x - 3y < -6$

51. Graph: $f(x) = -2(x - 3)^2 + 1$.
  a) Label the vertex.
  b) Draw the axis of symmetry.
  c) Find the maximum or minimum value.

52. Solve $V = P - Prt$ for r.

53. Solve $I = \dfrac{R}{R + r}$ for R.

54. Find a linear equation whose graph has a y-intercept of $(0, -3)$ and is parallel to the line whose equation is $3x - y = 6$.

Find the domain of each function.

55. $f(x) = \sqrt{5 - 3x}$

56. $g(x) = \dfrac{x - 4}{x^2 - 2x + 1}$

57. Multiply $(8.9 \times 10^{-17})(7.6 \times 10^4)$. Write scientific notation for the answer.

58. Multiply and simplify: $\sqrt{8x}\,\sqrt{8x^3y}$.

59. Simplify: $(25x^{4/3}y^{1/2})^{3/2}$.

60. Divide and simplify:

$$\frac{\sqrt[3]{15x}}{\sqrt[3]{3y^2}}.$$

61. Rationalize the denominator:

$$\frac{1 - \sqrt{x}}{1 + \sqrt{x}}.$$

62. Multiply these complex numbers:

$$(3 + 2i)(4 - 7i).$$

63. Write a quadratic equation whose solutions are $5\sqrt{2}$ and $-5\sqrt{2}$.

64. Find the center and the radius of the circle

$$x^2 + y^2 - 4x + 6y - 23 = 0.$$

65. Express as a single logarithm:

$$\tfrac{2}{3}\log_a x - \tfrac{1}{2}\log_a y + 5\log_a z.$$

66. Convert to an exponential equation: $\log_a c = 5$.

Find each of the following using a calculator.

67. $\log 5677.2$ **68.** $10^{-3.587}$

69. $\ln 5677.2$ **70.** $e^{-3.587}$

71. The number of personal computers in Mexico has grown exponentially from 0.12 million in 1985 to 6.0 million in 2000.

    a) Find the exponential growth rate k to three decimal places and write an exponential function describing the number of personal computers in Mexico t years after 1985.

    b) Predict the number of personal computers in Mexico in 2008.

72. Find the distance between the points $(-1, -5)$ and $(2, -1)$.

73. Find the 21st term of the arithmetic sequence $19, 12, 5, \ldots$.

74. Find the sum of the first 25 terms of the arithmetic series $-1 + 2 + 5 + \cdots$.

75. Find the general term of the geometric sequence $16, 4, 1, \ldots$.

76. Find the 7th term of $(a - 2b)^{10}$.

77. Find the sum of the first nine terms of the geometric series $x + 1.5x + 2.25x + \cdots$.

78. On Elyse's 9th birthday, her grandmother opened a savings account for her with \$100. The account draws 6% interest, compounded annually. If Elyse neither adds to nor withdraws any money from the bank, how much will be in the account on her 18th birthday?

79. How many code words can be formed using 4 out of the 5 letters of the word "video" if the letters are not to be repeated?

80. What is the probability of drawing a heart from a well-shuffled deck of 52 cards?

SYNTHESIS

Solve.

81. $\dfrac{9}{x} - \dfrac{9}{x + 12} = \dfrac{108}{x^2 + 12x}$

82. $\log_2 (\log_3 x) = 2$

83. y varies directly as the cube of x and x is multiplied by 0.5. What is the effect on y?

84. Divide these complex numbers:

$$\frac{2\sqrt{6} + 4\sqrt{5}i}{2\sqrt{6} - 4\sqrt{5}i}.$$

85. Diaphantos, a famous mathematician, spent $\tfrac{1}{6}$ of his life as a child, $\tfrac{1}{12}$ as an adolescent, and $\tfrac{1}{7}$ as a bachelor. Five years after he was married, he had a son who died 4 years before his father at half his father's final age. How long did Diaphantos live?

R

Elementary Algebra Review

*T*his chapter is a review of the first six chapters of this text. Each section corresponds to a chapter of the text. For further explanation of the topics in this chapter, refer to the sections or pages referenced in the margin.

Introduction to Algebraic Expressions

R.1

The Real Numbers • Operations on Real Numbers • Algebraic Expressions

The study of algebra requires a thorough understanding of how numbers are manipulated.

The Real Numbers

Numbers can be represented by points on a number line.

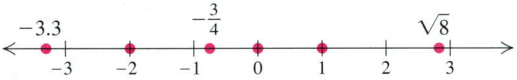

Sets, p. 26

Some sets of numbers are given specific names.

> ### Sets of Numbers
> *Natural numbers*: $\{1, 2, 3, \ldots\}$
>
> *Whole numbers*: $\{0, 1, 2, 3, \ldots\}$
>
> *Integers*: $\{\ldots, -3, -2, -1, 0, 1, 2, 3, \ldots\}$
>
> *Rational numbers*: $\left\{ \dfrac{a}{b} \mid a \text{ and } b \text{ are integers and } b \neq 0 \right\}$
>
> *Real numbers*: $\{x \mid x \text{ is a number corresponding to a point on the number line}\}$
>
> *Irrational numbers*: $\{x \mid x \text{ is a real number and } x \text{ is not a rational number}\}$

Sets of real numbers,
Section 1.4

For example, -5 is an integer, a rational number (because it can be written $-5/1$), and a real number. Since 0.569 can be written 569/1000, it is a rational number and a real number. The number $\sqrt{2}$ is an irrational number and a real number.

Order, p. 30

We can compare, or order, real numbers by their graphs on the number line. For any two numbers, the one to the left is less than the one to the right.

Equation, p. 5

Sentences like $\frac{1}{4} = 0.25$, containing an equals sign, are called **equations**. An **inequality** is a sentence containing $>$ (is greater than), $<$ (is less than), $\geq$ (is

Inequality, p. 32

greater than or equal to), or $\leq$ (is less than or equal to). Equations and inequalities can be true or false.

E x a m p l e   1

Write true or false for each equation or inequality.

a) $-2\frac{1}{3} = -\frac{7}{3}$ **b)** $1 = -1$ **c)** $-5 < -2$
d) $-3 \geq 2$ **e)** $1.1 \leq 1.1$

Solution

a) $-2\frac{1}{3} = -\frac{7}{3}$ is *true* because $-2\frac{1}{3}$ and $-\frac{7}{3}$ represent the same number.
b) $1 = -1$ is a *false* equation.
c) $-5 < -2$ is *true* because -5 is to the left of -2 on the number line.
d) $-3 \geq 2$ is *false* because neither $-3 > 2$ nor $-3 = 2$ is true.
e) $1.1 \leq 1.1$ is *true* because $1.1 = 1.1$ is true.

Absolute value, p. 32

The distance of a number from 0 is called the **absolute value** of the number. The notation $|-4|$ represents the absolute value of -4. The absolute value of a number is never negative.

E x a m p l e   2

Find the absolute value: **(a)** $|-4|$; **(b)** $\left|\frac{11}{3}\right|$; **(c)** $|0|$.

Solution

a) $|-4| = 4$ since -4 is 4 units from 0.
b) $\left|\frac{11}{3}\right| = \frac{11}{3}$ since $\frac{11}{3}$ is $\frac{11}{3}$ units from 0.
c) $|0| = 0$ since 0 is 0 units from itself.

Operations on Real Numbers

Addition, subtraction, multiplication, and division of real numbers are defined using absolute values.

> ### *Rules for Addition of Real Numbers*
>
> **1.** *Positive numbers*: Add as usual. The answer is positive.
> **2.** *Negative numbers*: Add absolute values and make the answer negative.
> **3.** *A positive and a negative number*: Subtract absolute values. Then:
>
> > **a)** If the positive number has the greater absolute value, the answer is positive.
> > **b)** If the negative number has the greater absolute value, the answer is negative.
> > **c)** If the numbers have the same absolute value, the answer is 0.
>
> **4.** *One number is zero*: The sum is the other number.

E x a m p l e   3

Addition, Section 1.5

Addition of fractions, p. 22

Add: **(a)** $-13 + (-9)$; **(b)** $2.7 + (-1.4)$; **(c)** $-\frac{4}{5} + \frac{1}{10}$.

Solution

a) $-13 + (-9) = -22$

Two negatives. *Think*: Add the absolute values, 13 and 9, to get 22. Make the answer *negative*, -22.

b) $2.7 + (-1.4) = 1.3$

A negative and a positive. *Think*: The difference of absolute values is $2.7 - 1.4$, or 1.3. The positive number has the larger absolute value, so the answer is *positive*, 1.3.

c) $-\frac{4}{5} + \frac{1}{10} = -\frac{8}{10} + \frac{1}{10} = -\frac{7}{10}$

A negative and a positive. *Think*: The difference of absolute values is $\frac{8}{10} - \frac{1}{10}$, or $\frac{7}{10}$. The negative number has the larger absolute value, so the answer is *negative*, $-\frac{7}{10}$.

Opposite, p. 41

Every real number has an **opposite**. The opposite of -6 is 6, the opposite of 3.7 is -3.7, and the opposite of 0 is itself. When opposites are added, the result is 0. Finding the opposite of a number is often called "changing its sign."

Subtraction of real numbers is defined in terms of addition and opposites.

> ### Subtraction of Real Numbers
> To subtract, add the opposite of the number being subtracted.

E x a m p l e   4

Subtraction, Section 1.6

Subtract: **(a)** $-6 - (-7.3)$; **(b)** $-20 - 5$.

Solution

a) $-6 - (-7.3) = -6 + 7.3 = 1.3$

Change the subtraction to addition and add the opposite.

b) $-20 - 5 = -20 + (-5) = -25$

Adding the opposite. We can check by addition: $-25 + 5 = -20$.

The rules for multiplication of real numbers are similar to the rules for division.

> ### Rules for Multiplication and Division
> To multiply or divide two real numbers:
> **1.** Using the absolute values, multiply or divide, as indicated.
> **2.** If the signs are the same, the answer is positive.
> **3.** If the signs are different, the answer is negative.

E x a m p l e   5

Multiply or divide, as indicated.

a) $3(-1.5)$

b) $\left(-\frac{1}{3}\right)\left(-\frac{6}{8}\right)$

c) $\frac{-18}{3}$

d) $\left(-\frac{4}{9}\right) \div \left(-\frac{2}{5}\right)$

Solution

a) $3(-1.5) = -4.5$

Think: $3(1.5) = 4.5$. The signs are different, so the answer is negative.

b) $\left(-\frac{1}{3}\right)\left(-\frac{6}{8}\right) = \frac{6}{24} = \frac{1}{4} \cdot \frac{6}{6} = \frac{1}{4}$

Removing a factor of 1: $\frac{6}{6} = 1$. The signs are the same, so the answer is positive.

c) $\frac{-18}{3} = -18 \div 3 = -6$

Think: $18 \div 3 = 6$. The signs are different, so the answer is negative.

d) $\left(-\frac{4}{9}\right) \div \left(-\frac{2}{5}\right) = \left(-\frac{4}{9}\right) \cdot \left(-\frac{5}{2}\right)$

$= \frac{20}{18} = \frac{10}{9} \cdot \frac{2}{2} = \frac{10}{9}$

Multiplying by the reciprocal. The answer is positive.

Multiplication and division, Section 1.7

Multiplication and division of fractions, pp. 19–20

Reciprocal, p. 20

Division by 0, p. 53

Exponential notation, p. 56

Addition, subtraction, and multiplication are defined for all real numbers, but we cannot **divide by 0**.

A product like $2 \cdot 2 \cdot 2 \cdot 2$, in which the factors are the same, is called a **power**. Powers are often written using **exponential notation**:

$$2 \cdot 2 \cdot 2 \cdot 2 = 2^4.$$

There are 4 factors; 4 is the *exponent*.

2 is the *base*.

A number raised to the power of 1 is the number itself; for example, $3^1 = 3$.

An expression containing a series of operations is not necessarily evaluated from left to right. Instead, we perform the operations according to the following rules.

> **Rules for Order of Operations**
>
> **1.** Calculate within the innermost grouping symbols.
> **2.** Simplify all exponential expressions.
> **3.** Perform all multiplication and division, working from left to right.
> **4.** Perform all addition and subtraction, working from left to right.

Example 6

Simplify: $3 - [(4 \times 5) + 12 \div 2^3 \times 6] + 5$.

Solution

$$3 - [(4 \times 5) + 12 \div 2^3 \times 6] + 5$$
$$= 3 - [20 + 12 \div 2^3 \times 6] + 5 \qquad \text{Doing the calculations in the innermost parentheses first}$$
$$= 3 - [20 + 12 \div 8 \times 6] + 5 \qquad \text{Working inside the brackets; evaluating } 2^3$$
$$= 3 - [20 + 1.5 \times 6] + 5 \qquad 12 \div 8 \text{ is the first multiplication or division working from left to right.}$$
$$= 3 - [20 + 9] + 5 \qquad \text{Multiplying}$$
$$= 3 - 29 + 5 \qquad \text{Completing the calculations within the brackets}$$
$$\left.\begin{array}{l} = -26 + 5 \\ = -21 \end{array}\right\} \qquad \text{Adding and subtracting from left to right}$$

Absolute-value symbols and fraction bars are also grouping symbols.

Algebraic Expressions

Algebraic expressions, p. 3

Constant, p. 2

Variable, p. 2

The expressions we have evaluated so far have consisted of numbers, operation signs, and/or grouping symbols. **Algebraic expressions** can also contain letters such as x and t. A number whose value does not vary, such as 100, is called a **constant**. A letter that can represent different numbers is called a **variable**.

Algebraic expressions containing variables can be evaluated by substituting a number for each variable in the expression and following the rules for order of operations.

Example 7

The perimeter P of a rectangle of length l and width w is given by the formula $P = 2l + 2w$. Find the perimeter when l is 16 in. and w is 7.5 in.

Solution   We evaluate, substituting 16 in. for l and 7.5 in. for w and carrying out the operations:

$$P = 2l + 2w$$
$$= 2 \cdot 16 + 2 \cdot 7.5$$
$$= 32 + 15$$
$$= 47 \text{ in.}$$

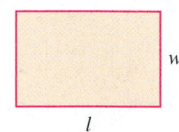

Expressions that represent the same number are said to be **equivalent**. The laws that follow provide methods for writing equivalent expressions.

> ### The Commutative, Associative, and Distributive Laws
>
> #### The Commutative Laws
>
> For any numbers a and b,
>
> $$a + b = b + a \quad \text{and} \quad ab = ba.$$
>
> (Changing the order when adding or multiplying does not affect the answer.)
>
> #### The Associative Laws
>
> For any numbers a, b, and c,
>
> $$a + (b + c) = (a + b) + c \quad \text{and} \quad a \cdot (b \cdot c) = (a \cdot b) \cdot c.$$
>
> (Numbers can be grouped in any manner for addition or multiplication.)
>
> #### The Distributive Law
>
> For any numbers a, b, and c,
>
> $$a(b + c) = ab + ac.$$
>
> (The product of a number and a sum can be written as the sum of two products.)

Example 8

Multiply: $-2(5x - 3)$.

Solution

$$
\begin{aligned}
-2(5x - 3) &= -2(5x + (-3)) && \text{Adding the opposite} \\
&= -2 \cdot 5x + (-2) \cdot (-3) && \text{Using the distributive law} \\
&= (-2 \cdot 5)x + 6 && \text{Using the associative law for} \\
& && \text{multiplication} \\
&= -10x + 6
\end{aligned}
$$

When we reverse the statement of the distributive law to write a sum as an equivalent product, we are **factoring** the expression. To **factor** an expression means to write an equivalent expression that is a product.

Example 9

Factor: $5x + 10y + 5$.

Solution

$$
\begin{aligned}
5x + 10y + 5 &= 5 \cdot x + 5 \cdot 2y + 5 \cdot 1 && \text{The common factor is 5.} \\
&= 5(x + 2y + 1) && \text{Using the distributive law}
\end{aligned}
$$

The **terms** of an algebraic expression are separated by plus signs. When two terms have variable factors that are exactly the same, the terms are called **like**, or **similar terms**. The distributive law enables us to **combine**, or **collect**, **like terms**.

E x a m p l e   1 0

Combine like terms: $-5m + 3n - 4n + 10m$.

Solution

$$-5m + 3n - 4n + 10m$$
$$= -5m + 3n + (-4n) + 10m \qquad \text{Rewriting as addition}$$
$$= -5m + 10m + 3n + (-4n) \qquad \text{Using the commutative law of addition}$$
$$= (-5 + 10)m + (3 + (-4))n \qquad \text{Using the distributive law}$$
$$= 5m + (-n)$$
$$= 5m - n \qquad \text{Rewriting as subtraction}$$

We can also use the distributive law to help simplify algebraic expressions containing parentheses. When there is a subtraction sign before parentheses, we use the fact that $-1 \cdot a = -a$ to remove the parentheses.

> **The Opposite of a Sum**
>
> For any real numbers a and b,
>
> $$-(a + b) = -a + (-b).$$
>
> (The opposite of a sum is the sum of the opposites.)

E x a m p l e   1 1

Simplify: **(a)** $4x - (y - 2x)$; **(b)** $3(t + 2) - 6(t - 1)$.

Solution

a) $4x - (y - 2x) = 4x - y + 2x \qquad$ Removing parentheses and changing the sign of every term

$$\qquad\qquad\qquad = 6x - y \qquad \text{Combining like terms}$$

b) $3(t + 2) - 6(t - 1) = 3t + 6 - 6t + 6 \qquad$ Multiplying each term of $t + 2$ by 3 and each term of $t - 1$ by -6

$$\qquad\qquad\qquad\qquad = -3t + 12 \qquad \text{Combining like terms}$$

We have seen that algebraic expressions can be evaluated for specified numbers. If the expressions on each side of an equation have the same value for a given number, then that number is a **solution** of the equation.

E x a m p l e   1 2

Determine whether each number is a solution of $x - 2 = -5$.

a) 3   **b)** -3

Solution

a) We have

$$x - 2 = -5 \qquad \text{Writing the equation}$$
$$\overline{3 - 2 \; ? \; -5} \qquad \text{Substituting 3 for } x$$
$$1 \; | \; -5 \qquad 1 = -5 \text{ is FALSE}$$

Since $3 - 2 = -5$ is false, 3 is not a solution of $x - 2 = -5$.

b) We have

$$x - 2 = -5$$
$$\overline{-3 - 2 \; ? \; -5}$$
$$-5 \; | \; -5 \quad \text{TRUE}$$

Since $-3 - 2 = -5$ is true, -3 is a solution of $x - 2 = -5$.

Translating to algebraic
expressions, p. 4

Translating to equations, p. 5

Certain word phrases can be translated to algebraic expressions. These in turn can often be used to translate problems to equations.

E x a m p l e   1 3

Time usage.   Translate the following problem to an equation.

The average adult spends 145.6 hr a year shopping for clothes. This is 16 times as many hours as is spent planning for retirement (*Source*: *Perspective*, September 1996). How many hours a year does the average adult spend planning for retirement?

Solution   We let r represent the number of hours spent planning for retirement. We then reword the problem to make the translation more direct.

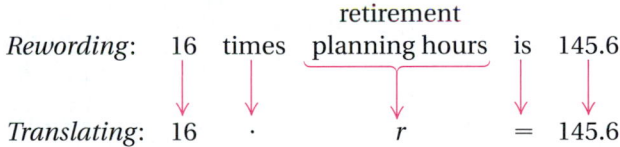

		retirement		
Rewording: 16	times	planning hours	is	145.6.
Translating: 16	$\cdot$	r	$=$	145.6

FOR EXTRA HELP

Exercise Set   R.1

Digital Video Tutor: none
Videotape: none

Math Tutor Center

MathXL

MyMathLab.com

Classify each equation or inequality as true or false.

1. $2.3 = 2.31$
2. $-3 \geq -3$
3. $-10 < -1$

4. $0 \leq -1$
5. $5 > 0$
6. $\frac{1}{10} = 0.1$

Find each absolute value.

7. $|4|$
8. $\left|\frac{11}{4}\right|$

9. $|-1.3|$
10. $|-105|$

Simplify.

11. $(-13) + (-12)$
12. $3 - (-2)$

13. $-\frac{1}{3} - \frac{2}{5}$
14. $\frac{3}{8} \div \frac{3}{5}$

15. $4.2 - 10.7$
16. $(-1.3)(2.8)$

17. $-15 + 0$
18. $\left(-\frac{1}{2}\right) + \frac{1}{8}$

19. $0 \div (-10)$
20. $0 - 32$

21. $\left(-\frac{3}{10}\right) + \left(-\frac{1}{5}\right)$

22. $\left(-\frac{4}{7}\right)\left(\frac{7}{4}\right)$

23. $-3.8 + 9.6$

24. $-0.01 + 1$

25. $(-12) \div 4$

26. $(-3)(30)$

27. $32 - (-7)$

28. $-100 + 35$

29. $(-10)(-17.5)$

30. $-10 - 2.68$

31. $(-68) + 36$

32. $175 \div (-25)$

33. $2 + (-3) + 7 + 10$

34. $-5 + (-15) + 13 + (-1)$

35. $3 \cdot (-2) \cdot (-1) \cdot (-1)$

36. $(-6) \cdot (-5) \cdot (-4) \cdot (-3) \cdot (-2) \cdot (-1)$

37. $(-1)^4 + 2^3$

38. $(-1)^5 + 2^4$

39. $2 \times 6 - 3 \times 5$

40. $12 \div 4 + 15 \div 3$

41. $3 - (2 \cdot 4 + 11)$

42. $3 - 2 \cdot 4 + 11$

43. $4 \cdot 5^2$

44. $7 \cdot 2^3$

45. $25 - 8 \times 3 + 1$

46. $12 - 16 \times 5 + 4$

47. $2 - (3^3 + 16 \div (-2)^3)$

48. $-7 - (8 + 10 \times 2^2)$

49. $|6(-3)| + |(-2)(-9)|$

50. $3 - |2 - 7 + 4|$

51. $\dfrac{7000 + (-10)^3}{10^2 \times (2 + 4)}$

52. $\dfrac{3 - 2 \times 6 - 5}{2(3 + 7)^2}$

53. $2 + 8 \div 2 \times 2$

54. $2 + 8 \div (2 \times 2)$

Evaluate.

55. $x - y$, for $x = 10$ and $y = 3$

56. $2m - n$, for $m = 6$ and $n = 11$

57. $-3 - x^2 + 12x$, for $x = 5$

58. $14 + (y - 5)^2 - 12 \div y$, for $y = -2$

59. The area of a parallelogram with base b and height h is bh. Find the area of the parallelogram when the height is 3.5 cm and the base is 8 cm.

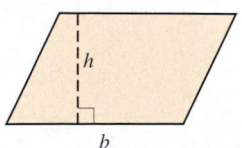

60. The area of a triangle with base b and height h is $\frac{1}{2}bh$. Find the area of the triangle when the height is 2 in. and the base is 6.2 in.

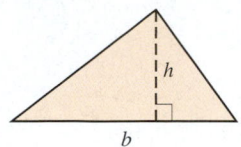

Multiply.

61. $4(2x + 7)$

62. $3(5y + 1)$

63. $5(x - 10)$

64. $4(3x - 2)$

65. $-2(15 - 3x)$

66. $-7(3x - 5)$

67. $2(4a + 6b - 3c)$

68. $5(8p + q - 5r)$

Factor.

69. $8x + 6y$

70. $7p + 14q$

71. $3 + 3w$

72. $4x + 4y$

73. $10x + 50y + 100$

74. $81p + 27q + 36$

Combine like terms.

75. $3p - 2p$

76. $4x + 3x$

77. $4m + 10 - 5m + 12$

78. $3a - 4b - 5b - 6a$

79. $-6x + 7 + 9x$

80. $16r + (-7r) + 3s$

Remove parentheses and simplify.

81. $2p - (7 - 4p)$

82. $4r - (3r + 5)$

83. $2x + 5y - 7(x - y)$

84. $14m - 6(2n - 3m) + n$

85. $6[2a + 4(a - 2b)]$

86. $2[2a + 1 - (3a - 6)]$

87. $3 - 2[5(x - 10y) - (3 + 2y)]$

88. $7 - 4[2(3 - 2x) - 5(4x - 3)]$

Determine whether the given number is a solution of the given equation.

89. $4; 3x - 2 = 10$

90. $12; 100 = 4x + 50$

91. $-3; 4 - x = 1$

92. $-1; 2 = 5 + 3x$

93. $4.6; \dfrac{x}{2} = 2.3$

94. $144; \dfrac{x}{9} = 16$

Translate each problem to an equation. Do not solve.

95. Three times what number is 348?

96. What number added to 256 is 113?

97. *Fast-food calories.* A McDonald's Big Mac® contains 500 calories. This is 69 more calories than a Taco Bell Beef Burrito® provides. How many calories are in a Taco Bell Beef Burrito?

98. *Coca-Cola® consumption.* The average U.S. citizen consumes 296 servings of Coca-Cola each year. This is 7.4 times the international average.

What is the international average per capita consumption of Coke?

99. *Vegetable production.* It takes 42 gal of water to produce 1 lb of broccoli. This is twice the amount of water used to produce 1 lb of lettuce. How many gallons of water does it take to produce 1 lb of lettuce?

100. *Sports costs.* The average annual cost for scuba diving is $470. This is $458 more than the average annual cost to play badminton. What is the average annual cost to play badminton?

Equations, Inequalities, and Problem Solving

R.2

Solving Equations and Formulas • Solving Inequalities • Problem Solving

In this section, we develop a problem-solving approach that can be used to solve many types of problems. Before doing so, we will study some principles used to solve equations and inequalities.

Solving Equations and Formulas

Solution, p. 70

Any replacement for the variable in an equation that makes the equation true is called a *solution* of the equation. To **solve** an equation means to find all of its solutions.

Although the solution of the equation

$$x = 10$$

can be seen to be 10, the solution of an equation like

$$-10 = x + 14$$

is less obvious. For more complicated equations, the following principles are useful.

> ### The Addition and Multiplication Principles
>
> #### The Addition Principle
> For any real numbers a, b, and c,
> $$a = b \quad \text{is equivalent to} \quad a + c = b + c.$$
>
> #### The Multiplication Principle
> For any real numbers a, b, and c, with $c \neq 0$,
> $$a = b \quad \text{is equivalent to} \quad a \cdot c = b \cdot c.$$

These principles can be used individually or together to solve equations. They allow us to write *equivalent equations,* or equations with the same solutions. We can start with one equation and end up with an equivalent equation similar to $x = 10$.

E x a m p l e   1   Solve: $-10 = x + 14$.

Solution   We have

$$-10 = x + 14$$

$$-10 + (-14) = x + 14 + (-14) \qquad \text{Using the addition principle:}$$
Adding -14 to both sides of the equation

$$-24 = x + 0 \qquad \text{Simplifying; } 14 + (-14) = 0$$

$$-24 = x.$$

To check the answer, we substitute -24 for x in the original equation.

Check:
$$\begin{array}{c|c} -10 = x + 14 \\ \hline -10 \;?\; (-24) + 14 \\ -10 \;\big|\; -10 \qquad \text{TRUE} \end{array}$$

The solution of the original equation is -24.

In Example 1, we added the *opposite,* or *additive inverse,* of 14 to get x alone on one side. To solve $x + a = b$ for x, we add $-a$ to (or subtract a from) both sides.

E x a m p l e   2   Solve: $\frac{2}{3}x = \frac{5}{9}$.

Solution

$$\frac{2}{3}x = \frac{5}{9}$$

$$\frac{3}{2} \cdot \frac{2}{3}x = \frac{3}{2} \cdot \frac{5}{9} \qquad \text{Using the multiplication principle: Multiplying both sides of the equation by } \frac{3}{2}$$

$$1 \cdot x = \frac{5}{6} \qquad \text{Simplifying. Note that } \frac{3}{2} \text{ is the reciprocal of } \frac{2}{3}, \text{ so } \frac{3}{2} \cdot \frac{2}{3} = 1.$$

$$x = \frac{5}{6}$$

Check: $\dfrac{2}{3}x = \dfrac{5}{9}$

$\dfrac{2}{3}\left(\dfrac{5}{6}\right)$? $\dfrac{5}{9}$

$\dfrac{5}{9}$ $\bigg|$ $\dfrac{5}{9}$ TRUE

The solution is $\dfrac{5}{6}$.

In Example 2, we multiplied by the *reciprocal*, or *multiplicative inverse*, of $\dfrac{2}{3}$ to get x alone on one side. To solve $ax = b$ for x, we multiply both sides by $\dfrac{1}{a}$ (or divide both sides by a).

To solve an equation like $-3x - 10 = 14$, we first isolate the variable term, $-3x$, using the addition principle. Then we use the multiplication principle to get the variable by itself.

E x a m p l e   3

Solve: $-3x - 10 = 14$.

Solution

$$-3x - 10 = 14$$

$$-3x - 10 + 10 = 14 + 10 \qquad \text{Using the addition principle:}$$
Adding 10 to both sides

> First isolate the x-term.

$$-3x = 24 \qquad \text{Simplifying}$$

$$\dfrac{-3x}{-3} = \dfrac{24}{-3} \qquad \text{Dividing both sides by } -3$$

> Then isolate x.

$$x = -8 \qquad \text{Simplifying}$$

Check: $-3x - 10 = 14$

$-3(-8) - 10$? 14

$24 - 10$ $\bigg|$

14 $\bigg|$ 14 TRUE

The solution is -8.

Equations are generally easier to solve when they do not contain fractions. The easiest way to clear an equation of fractions is to multiply *every term on both sides* of the equation by the least common denominator.

Clearing fractions, p. 81

E x a m p l e   4

Solve: $\dfrac{5}{2} - \dfrac{1}{6}t = \dfrac{2}{3}$.

Solution   The number 6 is the least common denominator, so we multiply both sides by 6.

$$6\left(\dfrac{5}{2} - \dfrac{1}{6}t\right) = 6 \cdot \dfrac{2}{3} \qquad \text{Multiplying both sides by 6}$$

$$6 \cdot \dfrac{5}{2} - 6 \cdot \dfrac{1}{6}t = 6 \cdot \dfrac{2}{3} \qquad \text{Using the distributive law. Be sure to multiply every term by 6.}$$

$$15 - t = 4 \qquad \text{The fractions are cleared.}$$

$$15 - t - 15 = 4 - 15 \qquad \text{Subtracting 15 from both sides}$$

$$-t = -11 \qquad \qquad 15 - t - 15 = 15 + (-t) + (-15)$$
$$= -t + 15 + (-15) = -t$$

$$(-1)(-t) = (-1)(-11) \qquad \text{Multiplying both sides by } -1 \text{ to change the sign}$$

$$t = 11$$

Check: $\dfrac{5}{2} - \dfrac{1}{6}t = \dfrac{2}{3}$

$$\begin{array}{c|c} \dfrac{5}{2} - \dfrac{1}{6}(11) \ ? \ \dfrac{2}{3} & \\ \dfrac{5}{2} - \dfrac{11}{6} & \\ \dfrac{15}{6} - \dfrac{11}{6} & \\ \dfrac{2}{3} & \dfrac{2}{3} \quad \text{TRUE} \end{array}$$

The solution is 11.

To solve equations that contain parentheses, we can use the distributive law to first remove the parentheses. If like terms appear in an equation, we combine them and then solve.

Example   5

Solve: $1 - 3(4 - x) = 2(x + 5) - 3x$.

Solution

$$1 - 3(4 - x) = 2(x + 5) - 3x$$

$$1 - 12 + 3x = 2x + 10 - 3x \qquad \text{Using the distributive law}$$

$$-11 + 3x = -x + 10 \qquad \text{Combining like terms;} \\ 1 - 12 = -11 \text{ and } 2x - 3x = -x$$

$$-11 + 3x + x = 10 \qquad \text{Adding } x \text{ to both sides to get all } x\text{-terms on one side}$$

$$-11 + 4x = 10 \qquad \text{Combining like terms}$$

$$4x = 10 + 11 \qquad \text{Adding 11 to both sides to isolate the } x\text{-term}$$

$$4x = 21 \qquad \text{Simplifying}$$

$$x = \frac{21}{4} \qquad \text{Dividing both sides by 4}$$

Check: $1 - 3(4 - x) = 2(x + 5) - 3x$

$$\begin{array}{c|c} 1 - 3\left(4 - \frac{21}{4}\right) \ ? \ 2\left(\frac{21}{4} + 5\right) - 3\left(\frac{21}{4}\right) & \\ 1 - 3\left(-\frac{5}{4}\right) & 2\left(\frac{41}{4}\right) - \frac{63}{4} \\ 1 + \frac{15}{4} & \frac{82}{4} - \frac{63}{4} \\ \frac{19}{4} & \frac{19}{4} \qquad \text{TRUE} \end{array}$$

The solution is $\frac{21}{4}$.

Formulas, Section 2.3

A **formula** is an equation using two or more letters that represents a relationship between two or more quantities. A formula can be solved for a specified letter using the principles for solving equations.

E x a m p l e   6

The formula

$$A = \frac{a + b + c + d}{4}$$

gives the average A of four test scores a, b, c, and d. Solve for d.

Solution   We have

$$A = \frac{a + b + c + d}{4} \qquad \text{We want the letter } d \text{ alone.}$$

$$4A = a + b + c + d \qquad \text{Multiplying by 4 to clear the fraction}$$

$$4A - a - b - c = d. \qquad \text{Subtracting } a + b + c \text{ from (or adding } -a - b - c \text{ to) both sides. The letter } d \text{ is now isolated.}$$

We can also write this as $d = 4A - a - b - c$. This formula can be used to determine the test score needed to obtain a specified average if three tests have already been taken.

Solving Inequalities

Solutions of inequalities, p. 109

A solution of an inequality is a replacement of the variable that makes the inequality true.

E x a m p l e   7

Determine whether the given number is a solution of $x < -1$: **(a)** 0; **(b)** -10.

Solution

a) Since $0 < -1$ is false, 0 is *not* a solution.

b) Since $-10 < -1$ is true, -10 *is* a solution.

Graphs of inequalities, p. 109

The solutions of an inequality in one variable can be *graphed*, or represented by a drawing, on a number line. All points that are solutions are shaded, and dots are used at the endpoints. An open dot indicates an endpoint that is not a solution and a closed dot indicates an endpoint that is a solution.

E x a m p l e   8

Graph each inequality: **(a)** $m \leq 2$; **(b)** $-1 \leq x < 4$.

Solution

a) The solutions of $m \leq 2$ are shown on the number line by shading points to the left of 2 as well as the point at 2. The closed dot at 2 indicates that 2 is a part of the graph (that is, it is a solution of $m \leq 2$).

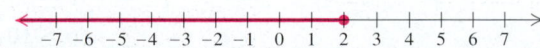

b) In order to be a solution of the inequality $-1 \le x < 4$, a number must be a solution of both $-1 \le x$ and $x < 4$. The solutions are shaded on the number line, with an open dot indicating that 4 is not a solution and a closed dot indicating that -1 is a solution.

Equivalent inequalities, p. 110

As with equations, our goal when solving inequalities is to isolate the variable on one side. We use principles that enable us to write *equivalent inequalities*—inequalities having the same solution set. The addition principle is similar to the addition principle for equations; the multiplication principle contains an important difference.

The addition principle for inequalities, p. 111

The multiplication principle for inequalities, p. 113

> ## The Addition and Multiplication Principles for Inequalities
>
> ### The Addition Principle
>
> For any real numbers a, b, and c,
>
> $$a < b \quad \text{is equivalent to} \quad a + c < b + c, \quad \text{and}$$
> $$a > b \quad \text{is equivalent to} \quad a + c > b + c.$$
>
> ### The Multiplication Principle
>
> For any real numbers a and b, and for any *positive* number c,
>
> $$a < b \quad \text{is equivalent to} \quad ac < bc, \quad \text{and}$$
> $$a > b \quad \text{is equivalent to} \quad ac > bc.$$
>
> For any real numbers a and b, and for any *negative* number c,
>
> $$a < b \quad \text{is equivalent to} \quad ac > bc, \quad \text{and}$$
> $$a > b \quad \text{is equivalent to} \quad ac < bc.$$
>
> Similar statements hold for $\le$ and $\ge$.

Note that when we multiply both sides of an inequality by a negative number, we must reverse the direction of the inequality symbol in order to have an equivalent inequality.

Example 9 Solve and graph each inequality: **(a)** $x - 3 > 7$; **(b)** $-2x \ge 5$.

Solution

a) We have

$$x - 3 > 7$$
$$x - 3 + 3 > 7 + 3 \qquad \text{\color{red}Using the addition principle: Adding 3 to both sides}$$
$$x > 10.$$

Any number greater than 10 is a solution of $x > 10$ and thus a solution of $x - 3 > 7$. The graph is as follows:

We cannot check all the solutions, but a partial check can be made using one of the possible solutions. We can substitute any number greater than 10—say, 15—into the original inequality.

Check: $\dfrac{x - 3 > 7}{15 - 3 \; ? \; 7}$

$\qquad\qquad$ $12 \;\big|\; 7$ Since $12 > 7$, this statement is TRUE.

Had 15 made the inequality false, we would know that we had not found the solution set. Thus any number greater than 10 is a solution.

b) We have

$$-2x \geq 5$$

$$\frac{-2x}{-2} \leq \frac{5}{-2} \qquad \text{Multiplying by } -\frac{1}{2} \text{ or dividing by } -2$$

$\qquad\qquad\qquad$ The symbol must be reversed!

$$x \leq -\frac{5}{2}.$$

Any number less than or equal to $-\frac{5}{2}$ is a solution. The graph is as follows:

Set-builder notation, p. 111

In Example 9, note that $x > 10$ and $x \leq -\frac{5}{2}$ are inequalities that describe the set of all solutions. Since it is impossible to list all the solutions, we use **set-builder notation**. The solution set of Example 9(a) is written

$$\{x \,|\, x > 10\},$$

read "the set of all x such that x is greater than 10." The solution set of Example 9(b) is written

$$\left\{x \,\middle|\, x \leq -\tfrac{5}{2}\right\}.$$

We can use the addition and multiplication principles together to solve inequalities. We can also combine like terms, remove parentheses, and clear fractions and decimals.

E x a m p l e   1 0

Solve: $2 - 3(x + 5) > 4 - 6(x - 1)$.

Solution   We have

$$2 - 3(x + 5) > 4 - 6(x - 1)$$

$$2 - 3x - 15 > 4 - 6x + 6 \qquad \text{Using the distributive law to remove parentheses}$$

$$-3x - 13 > -6x + 10 \qquad \text{Simplifying}$$

$$-3x + 6x > 10 + 13 \qquad \text{Adding } 6x \text{ and also 13, to get all } x\text{-terms on one side and all other terms on the other side}$$

$$3x > 23 \qquad \text{Combining like terms}$$

$$x > \tfrac{23}{3}. \qquad \text{Multiplying by } \tfrac{1}{3}. \text{ The inequality symbol stays the same because } \tfrac{1}{3} \text{ is positive.}$$

The solution set is $\left\{x \mid x > \tfrac{23}{3}\right\}$.

Problem Solving

Problem solving, Section 2.5

One of the most important uses of algebra is as a tool for problem solving. The following five steps can be used to help solve problems of many types.

> ### Five Steps for Problem Solving in Algebra
> 1. *Familiarize* yourself with the problem.
> 2. *Translate* to mathematical language. (This often means write an equation.)
> 3. *Carry out* some mathematical manipulation. (This often means *solve* an equation.)
> 4. *Check* your possible answer in the original problem.
> 5. *State* the answer clearly.

E x a m p l e   1 1

Familiarization step, pp. 98–99

Percent, Section 2.4

Kitchen cabinets.   Cherry kitchen cabinets cost 10% more than oak cabinets. Shelby Custom Cabinets designs a kitchen using $7480 worth of cherry cabinets. How much would the same kitchen cost using oak cabinets?

Solution

1. **Familiarize.**   The *Familiarize* step is often the most important of the five steps, and may require a significant amount of time. Sometimes it helps to make a drawing or a table, make a guess and check it, or look up further information. For this problem, we could review percent notation. We could also make a guess. Let's suppose that the oak cabinets cost $6500. Then the cherry cabinets would cost 10% more, or an additional $(0.10)(\$6500) = \$650$. Altogether the cherry cabinets would cost $6500 + $650 = $7150. Since $7150 \neq $7480, our guess is incorrect, but we see that 10% of the price of the oak cabinets must be added to the price of the oak cabinets to get the price of the cherry cabinets. We let c = the cost of the oak cabinets.

2. **Translate.**   What we learned in the *Familiarize* step leads to the translation of the problem to an equation.

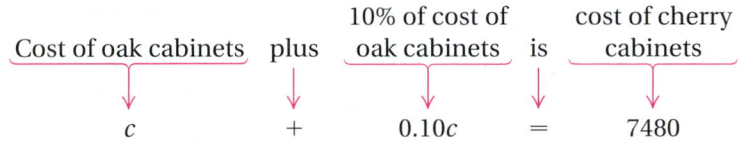

$$c + 0.10c = 7480$$

3. **Carry out.**   We solve the equation:

$$c + 0.10c = 7480$$

$$1c + 0.10c = 7480 \qquad \text{Writing } c \text{ as } 1c \text{ before combining terms}$$

$$1.10c = 7480 \qquad \text{Combining like terms}$$

$$c = \frac{7480}{1.10} \qquad \text{Dividing by 1.10}$$

$$c = 6800.$$

4. **Check.**   We check in the wording of the stated problem: Cherry cabinets cost 10% more, so the additional cost is

$$10\% \text{ of } \$6800 = 0.10(\$6800) = \$680.$$

The total cost of the cherry cabinets is then

$$\$6800 + \$680 = \$7480,$$

which is the amount stated in the problem.

5. **State.**   The oak cabinets would cost $6800.

Sometimes the translation of a problem is an inequality.

E x a m p l e 1 2

Long-distance telephone usage.   Elyse pays a flat rate of 15¢ per minute for long-distance telephone calls. The monthly charge for her local calls is $21.50. How many minutes can she spend calling long distance in a month and not exceed her telephone budget of $50?

Solution

1. **Familiarize.**   Suppose that Elyse spends 5 hr, or 300 min, making long-distance calls one month. Then her bill would be the local service charge plus the long-distance charges, or

$$\$21.50 + \$0.15(300) = \$66.50.$$

This exceeds $50, so we know that the number of long-distance minutes must be less than 300. We let m = the number of minutes of long-distance calls in a month.

2. **Translate.**   The *Familiarize* step helps us reword and translate.

Rewording:	The local service charge	plus	the long-distance charges	cannot exceed	$50.
Translating:	21.50	+	0.15m	≤	50

Solving applications with inequalities, Section 2.7

3. **Carry out.** We solve the inequality:

$$21.50 + 0.15m \le 50$$
$$0.15m \le 28.50 \qquad \text{Subtracting 21.50 from both sides}$$
$$m \le 190. \qquad \text{Dividing by 0.15. The inequality symbol stays the same.}$$

4. **Check.** As a partial check, note that the telephone bill for 190 min of long-distance charges is

$$\$21.50 + \$0.15(190) = \$50.$$

Since this does not exceed the $50 budget, and fewer minutes will cost even less, our answer checks. We also note that 190 is less than 300 min, as noted in the *Familiarize* step.

5. **State.** Elyse will not exceed her budget if she talks long distance for no more than 190 min.

FOR EXTRA HELP

Exercise Set R.2

Digital Video Tutor CD: none
Videotape: none

Math Tutor Center   MathXL   MyMathLab.com

Solve.

1. $y + 5 = 13$

2. $y + 7 = -3$

3. $-3 + m = 9$

4. $-11 = 4 + x$

5. $t + \frac{1}{3} = \frac{1}{4}$

6. $-\frac{2}{3} + p = \frac{1}{6}$

7. $-1.9 = x - 1.1$

8. $x + 4.6 = 1.7$

9. $3y = 13$

10. $2x = 18$

11. $-x = \frac{5}{3}$

12. $-y = -\frac{2}{5}$

13. $-\frac{2}{7}x = -12$

14. $-\frac{1}{4}x = 3$

15. $\frac{-t}{5} = 1$

16. $\frac{2}{3} = -\frac{z}{8}$

17. $3x + 7 = 13$

18. $4x + 3 = -1$

19. $3y - 10 = 15$

20. $12 = 5y - 18$

21. $4x + 7 = 3 - 5x$

22. $2x = 5 + 7x$

23. $2x - 7 = 5x + 1 - x$

24. $a + 7 - 2a = 14 + 7a - 10$

25. $\frac{2}{5} + \frac{1}{3}t = 5$

26. $-\frac{5}{6} + t = \frac{1}{2}$

27. $x + 0.45 = 2.6x$

28. $1.8x + 0.16 = 4.2 - 0.05x$

29. $8(3 - m) + 7 = 47$

30. $2(5 - m) = 5(6 + m)$

31. $4 - (6 + x) = 13$

32. $18 = 9 - (3 - x)$

33. $2 + 3(4 + c) = 1 - 5(6 - c)$

34. $b + (b + 5) - 2(b - 5) = 18 + b$

35. $0.1(a - 0.2) = 1.2 + 2.4a$

36. $\frac{2}{3}\left(\frac{1}{2} - x\right) + \frac{5}{6} = \frac{3}{2}\left(\frac{2}{3}x + 1\right)$

37. $A = lw$, for l

38. $A = lw$, for w

39. $p = 30q$, for q

40. $d = 20t$, for t

41. $I = \dfrac{P}{V}$, for P

42. $b = \dfrac{A}{h}$, for A

43. $q = \dfrac{p + r}{2}$, for p

44. $q = \dfrac{p - r}{2}$, for r

45. $A = \pi r^2 + \pi r^2 h$, for π

46. $ax + by = c$, for a

Determine whether each number is a solution of the given inequality.

47. $x \le -5$

   a) 5   **b)** -5

   c) 0   **d)** -10

48. $y > 0$

    a) -1 **b)** 1

    c) 0 **d)** 100

Solve and graph. Write each answer in set-builder notation.

49. $x + 3 \leq 15$ **50.** $y + 7 < -10$

51. $m - 17 > -5$ **52.** $x + 9 \geq -8$

53. $2x \geq -3$ **54.** $-\frac{1}{2}n \leq 4$

55. $-5t > 15$ **56.** $3x > 10$

Solve. Write each answer in set-builder notation.

57. $2y - 7 > 13$ **58.** $2 - 6y \leq 18$

59. $6 - 5a \leq a$ **60.** $4b + 7 > 2 - b$

61. $2(3 + 5x) \geq 7(10 - x)$ **62.** $2(x + 5) < 8 - 3x$

63. $\frac{2}{3}(6 - x) < \frac{1}{4}(x + 3)$ **64.** $\frac{2}{3}t + \frac{8}{9} \geq \frac{4}{6} - \frac{1}{4}t$

65. $0.7(2 + x) \geq 1.1x + 5.75$

66. $0.4x + 5.7 \leq 2.6 - 3(1.2x - 7)$

Solve. Use the five-step problem-solving process.

67. Three less than the sum of 2 and some number is 6. What is the number?

68. Five times some number is 10 less than the number. What is the number?

69. The sum of two consecutive even integers is 34. Find the numbers.

70. The sum of three consecutive integers is 195. Find the numbers.

71. *Reading.* Leisa is reading a 500-page book. She has twice as many pages to read as she has already finished. How many pages has she already read?

72. *Mowing.* It takes Caleb 50 min to mow his lawn. It will take him three times as many minutes to finish as he has already spent mowing. How long has he already spent mowing?

73. *Perimeter of a rectangle.* The perimeter of a rectangle is 28 cm. The width is 5 cm less than the length. Find the width and the length.

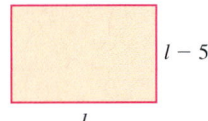

$l - 5$

l

74. *Triangles.* The second angle of a triangle is one third as large as the first. The third angle is 5° more than the first. Find the measure of the second angle.

75. *Water usage.* Rural Water Company charges a monthly service fee of $9.70 plus a volume charge of $2.60 for every hundred cubic feet of water used. How much water was used if the monthly bill is $33.10?

76. *Telephone bills.* Brandon pays $4.95 a month for a long-distance telephone service that offers a flat rate of 7¢ per minute. One month his total long-distance telephone bill was $10.69. How many minutes of long-distance telephone calls were made that month?

77. *Sales prices.* A can of tomatoes is on sale at 20% off for 64¢. What is the normal selling price of the tomatoes?

78. *Plywood.* The price of a piece of plywood rose 5% to $42. What was the original price of the plywood?

R.3

Introduction to Graphing

Points and Ordered Pairs • Graphing Equations • Linear Equations and Slope

The graph of an equation is a drawing representing the solutions of that equation. Every point on the graph is a solution, and every solution is represented by a point.

Points and Ordered Pairs

We can represent, or graph, pairs of numbers such as $(2, -5)$ on a plane. To do so, we use two perpendicular number lines called **axes**. The axes cross at a point called the **origin**. Arrows on the axes show the positive directions.

The order of the **coordinates**, or numbers in a pair, is important. The **first coordinate** indicates horizontal position and the **second coordinate** indicates vertical position. Such pairs of numbers are called **ordered pairs**.

The axes divide the plane into four regions, or **quadrants**, as indicated by Roman numerals in the figure at right. Points on the axes are not considered to be in any quadrant.

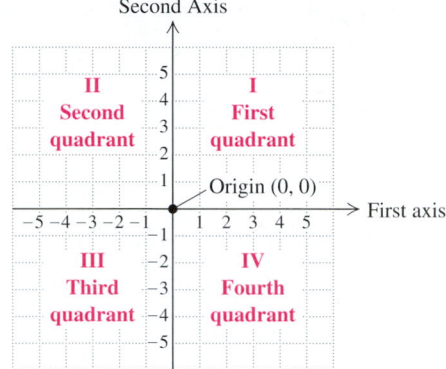

Example 1

List the coordinates of points A, B, C, D, and E, and tell in which quadrant each point is located.

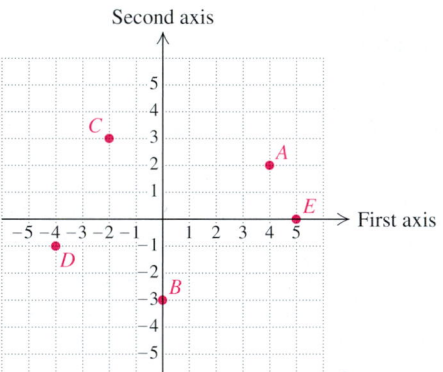

Solution The coordinates and quadrant of each point are as follows:

A: $(4, 2)$, I; B: $(0, -3)$, none; C: $(-2, 3)$, II;

D: $(-4, -1)$, III; E: $(5, 0)$, none

Graphing Equations

When an equation contains two variables, solutions must be ordered pairs. Unless stated otherwise, the first number in each pair replaces the variable that occurs first alphabetically.

E x a m p l e   2

Determine whether each of the following pairs is a solution of $y - x = 3$:
(a) $(1, 4)$; **(b)** $(4, 1)$.

Solution

a) We substitute 1 for x and 4 for y since x occurs first alphabetically:

$$\frac{y - x = 3}{4 - 1 \;\overset{?}{}\; 3}$$
$$3 \;\Big|\; 3 \quad \text{\small TRUE}$$

Since $3 = 3$ is true, the pair $(1, 4)$ *is* a solution.

b) In this case, we replace x with 4 and y with 1:

$$\frac{y - x = 3}{1 - 4 \;\overset{?}{}\; 3}$$
$$-3 \;\Big|\; 3 \quad \text{\small FALSE}$$

Since $-3 = 3$ is false, the pair $(4, 1)$ is *not* a solution.

A curve or line that represents all the solutions of an equation is called its **graph**.

E x a m p l e   3

Graph: $y = -2x + 1$.

Solution   We select a value for x, calculate the corresponding value of y, and form an ordered pair.

If $x = 0$, then $y = -2 \cdot 0 + 1 = 1$, and $(0, 1)$ is a solution. Repeating this step, we find other ordered pairs and list the results in a table. We then plot the points corresponding to the pairs. They appear to form a straight line, so we draw a line through the points.

$$y = -2x + 1$$

x	y	(x, y)
0	1	$(0, 1)$
-1	3	$(-1, 3)$
3	-5	$(3, -5)$
1	-1	$(1, -1)$

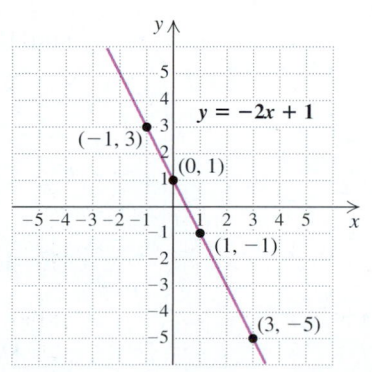

Linear Equations and Slope

The graph in Example 3 was a straight line. An equation whose graph is a straight line is a **linear equation**.

Any equation that can be written in the **standard form** $Ax + By = C$ is linear. Linear equations can also be written in other forms.

Forms of Linear Equations

Standard form: $Ax + By = C$

Slope-intercept form: $y = mx + b$

Point-slope form: $y - y_1 = m(x - x_1)$

The *rate of change* of y with respect to x is called the **slope** of a graph. A line has a constant slope. It can be found using any two points on a line or read from the slope-intercept form of the line's equation.

Slope

The *slope* of the line containing points (x_1, y_1) and (x_2, y_2) is given by

$$m = \frac{\text{change in } y}{\text{change in } x} = \frac{\text{rise}}{\text{run}} = \frac{y_2 - y_1}{x_2 - x_1}.$$

The slope of the line given by the equation $y = mx + b$ is m.

Example 4

Find the slope of the line containing the points $(-2, 1)$ and $(3, -4)$.

Solution From $(-2, 1)$ to $(3, -4)$, the change in y, or the rise, is $-4 - 1$, or -5. The change in x, or the run, is $3 - (-2)$, or 5. Thus

$$\text{Slope} = \frac{\text{change in } y}{\text{change in } x} = \frac{\text{rise}}{\text{run}} = \frac{-4 - 1}{3 - (-2)} = \frac{-5}{5} = -1.$$

Example 5

Find the slope of the line given by the equation $4x - 3y = 9$.

Solution We write the equation in slope-intercept form $y = mx + b$:

$$\begin{aligned}
4x - 3y &= 9 & &\text{We must solve for } y \\
-3y &= -4x + 9 & &\text{Adding } -4x \text{ to both sides} \\
y &= \tfrac{4}{3}x - 3. & &\text{Dividing both sides by } -3
\end{aligned}$$

The slope is $\frac{4}{3}$.

The slope of a line indicates the direction and steepness of its slant. The larger the absolute value of the slope, the steeper the line. The direction of the slant is indicated by the sign of the slope, as shown in the figures at the top of the next page.

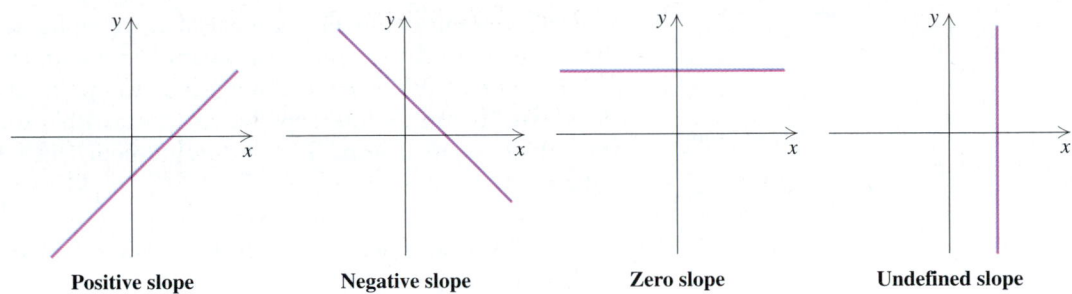

| Positive slope | Negative slope | Zero slope | Undefined slope |

We can also tell from the slopes of two lines whether they are parallel or perpendicular.

Parallel lines, p. 188

Perpendicular lines, p. 189

> ### Parallel and Perpendicular Lines
>
> Two lines are parallel if they have the same slope.
> Two lines are perpendicular if the product of the slopes is -1.

E x a m p l e   6   Tell whether the graphs of each pair of lines are parallel, perpendicular, or neither.

a) $2x - y = 7$,
$y = 2x + 3$

b) $4x - y = 8$,
$x + 4y = 8$

Solution

a) The slope of $y = 2x + 3$ is 2.
To find the slope of $2x - y = 7$, we solve for y:

$$2x - y = 7$$
$$-y = -2x + 7$$
$$y = 2x - 7.$$

The slope of $2x - y = 7$ is also 2. Since the slopes are equal, the lines are parallel.

b) We solve both equations for y in order to determine the slopes of the lines:

$$4x - y = 8$$
$$-y = -4x + 8$$
$$y = 4x - 8.$$

The slope of $4x - y = 8$ is 4.
For the second line, we have

$$x + 4y = 8$$
$$4y = -x + 8$$
$$y = -\tfrac{1}{4}x + 2.$$

The slope of $x + 4y = 8$ is $-\tfrac{1}{4}$. Since $4 \cdot \left(-\tfrac{1}{4}\right) = -1$, the lines are perpendicular.

The *x-intercept* of a line, if it exists, is the point at which the graph crosses the *x*-axis. To find an *x*-intercept, we replace *y* with 0 and calculate *x*.

The *y-intercept* of a line, if it exists, is the point at which the graph crosses the *y*-axis. To find a *y*-intercept, we replace *x* with 0 and calculate *y*. In an equation of the form $y = mx + b$, the *y*-intercept can be read directly from the equation.

The graph of any equation $y = mx + b$ passes through the *y*-intercept $(0, b)$.

If we know an equation is a straight line, we can plot two points on the line and draw the line through those points. The intercepts are often convenient points to use.

E x a m p l e 7

Graph $2x - 5y = 10$ using intercepts.

Solution   To find the *x*-intercept, we let $y = 0$ and solve for *x*:

$$2x - 5 \cdot 0 = 10 \qquad \text{Replacing } y \text{ with } 0$$
$$2x = 10$$
$$x = 5.$$

To find the *y*-intercept, we let $x = 0$ and solve for *y*:

$$2 \cdot 0 - 5y = 10 \qquad \text{Replacing } x \text{ with } 0$$
$$-5y = 10$$
$$y = -2.$$

Thus the *x*-intercept is $(5, 0)$ and the *y*-intercept is $(0, -2)$. The graph is a line, since $2x - 5y = 10$ is in the form $Ax + By = C$. It passes through these two points.

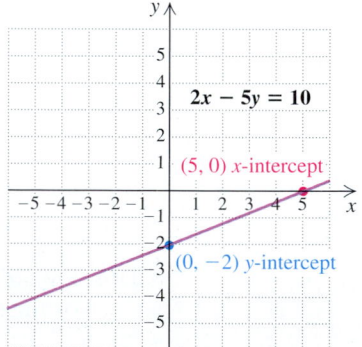

Alternatively, if we know a point on the line and its slope, we can plot the point and "count off" its slope to locate another point on the line.

Example 8

Graph: $y = -\dfrac{1}{2}x + 3$.

Solution   The equation is in slope-intercept form, so we can read the slope and y-intercept directly from the equation.

$$\text{Slope: } -\frac{1}{2}$$

y-intercept: $(0, 3)$

We plot the y-intercept and use the slope to find another point.

Another way to write the slope is $\dfrac{-1}{2}$. This means for a run of 2 units, there is a negative rise, or a fall, of 1 unit. Starting at $(0, 3)$, we move 2 units in the positive horizontal direction and then 1 unit down, to locate the point $(2, 2)$. Then we draw the graph. A third point can be calculated and plotted as a check.

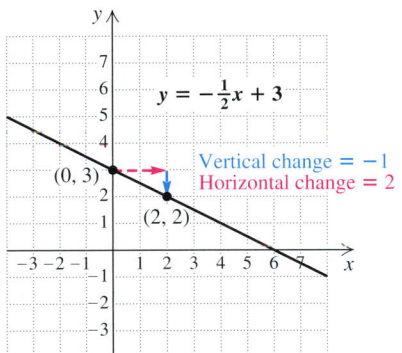

Horizontal and vertical lines intersect only one axis.

Horizontal and Vertical Lines

Horizontal Line	Vertical Line
$y = b$	$x = a$
y-intercept $(0, b)$	x-intercept $(a, 0)$
Slope is 0	Undefined slope

E x a m p l e   9

Graph: $x = 2$.

Solution   The graph is a vertical line, with x-intercept $(2, 0)$.

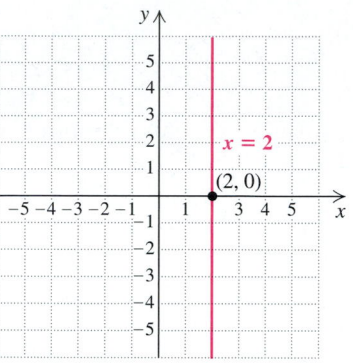

E x a m p l e   1 0

Graph: $y = -5$.

Solution   The graph is a horizontal line, with y-intercept $(0, -5)$.

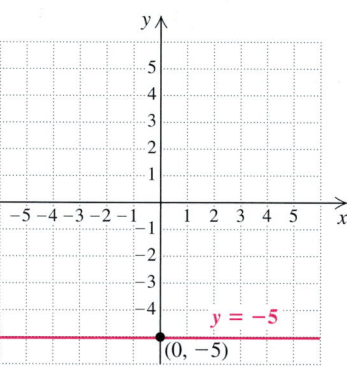

If we know the slope of a line and the coordinates of a point on the line, we can find an equation of the line, using either the slope-intercept equation $y = mx + b$, or the point-slope equation $y - y_1 = m(x - x_1)$.

E x a m p l e   1 1

Find the slope-intercept equation of a line given the following:

a) The slope is 2, and the y-intercept is $(0, -5)$.
b) The graph contains the points $(-2, 1)$ and $(3, -4)$.

Solution

a) Since the slope and the y-intercept are given, we use the slope-intercept equation:

$$y = mx + b$$

$$y = 2x - 5. \qquad \text{Substituting 2 for } m \text{ and } -5 \text{ for } b$$

b) To use the point-slope equation, we need a point on the line and its slope. The slope can be found from the points given:

$$m = \frac{1 - (-4)}{-2 - 3} = \frac{5}{-5} = -1.$$

Either point can be used for (x_1, y_1). Using $(-2, 1)$, we have

$$y - y_1 = m(x - x_1)$$
$$y - 1 = -1(x - (-2)) \qquad \text{Substituting } -2 \text{ for } x_1, 1 \text{ for } y_1, \text{ and } -1 \text{ for } m$$
$$y - 1 = -(x + 2)$$
$$y - 1 = -x - 2$$
$$y = -x - 1. \qquad \text{This is in slope-intercept form}$$

FOR EXTRA HELP

Exercise Set R.3

Digital Video Tutor CD: none
Videotape: none

    Math Tutor Center    MathXL   MyMathLab.com

1. Plot these points.

$(2, -3), (5, 1), (0, 2), (-1, 0),$
$(0, 0), (-2, -5), (-1, 1), (1, -1)$

2. Plot these points.

$(0, -4), (-4, 0), (5, -2), (2, 5),$
$(3, 3), (-3, -1), (-1, 4), (0, 1)$

In which quadrant is each point located?

3. $(2, 1)$

4. $(-2, 5)$

5. $(3, -2.6)$

6. $(-1.7, -5.9)$

7. First coordinates are positive in quadrants _____ and _____ .

8. Second coordinates are negative in quadrants _____ and _____ .

Determine whether each equation has the given ordered pair as a solution.

9. $y = 2x - 5$; $(1, 3)$

10. $4x + 3y = 8$; $(-1, 4)$

11. $a - 5b = -3$; $(2, 1)$

12. $c = d + 1$; $(1, 2)$

Graph.

13. $y = \frac{1}{3}x + 3$

14. $y = -x - 2$

15. $y = -4x$

16. $y = \frac{3}{4}x + 1$

Find the slope of the line containing each given pair of points.

17. $(3, 6)$ and $(2, 5)$

18. $(-1, 7)$ and $(-3, 4)$

19. $\left(-2, -\frac{1}{2}\right)$ and $\left(5, -\frac{1}{2}\right)$

20. $(6.8, 7.5)$ and $(6.8, -3.2)$

Find the slope and the y-intercept of each equation.

21. $y = 2x - 5$

22. $y = 4x - 8$

23. $2x + y = 1$

24. $x - 2y = 3$

Determine whether each pair of lines is parallel, perpendicular, or neither.

25. $x + y = 5,$
$x - y = 1$

26. $2x + y = 3,$
$y = 4 - 2x$

27. $2x + 3y = 1,$
$2x - 3y = 5$

28. $y = \frac{1}{3}x - 7,$
$y + 3x = 1$

Find the intercepts. Then graph.

29. $3 - y = 2x$

30. $2x + 5y = 10$

31. $y = 3x + 5$

32. $y = -x + 7$

33. $3x - 2y = 6$

34. $2y + 1 = x$

Determine the coordinates of the y-intercept of each equation. Then graph the equation.

35. $y = 2x - 5$

36. $y = -\frac{5}{4}x - 3$

37. $2y + 4x = 6$

38. $3y + x = 4$

Find the slope of each line, and graph.

39. $y = 4$

40. $x = -5$

41. $x = 3$

42. $y = -1$

Find the slope–intercept equation of a line given the conditions.

43. The slope is $\frac{1}{3}$ and the y-intercept is $(0, 1)$.

44. The slope is -1 and the y-intercept is $(0, -5)$.

45. The graph contains the points $(0, 3)$ and $(-1, 4)$.

46. The graph contains the points $(5, 1)$ and $(8, 0)$.

R.4

Polynomials

Exponents • Polynomials • Addition and Subtraction of Polynomials • Multiplication of Polynomials • Division of Polynomials

In this section, we define polynomials and learn to manipulate them. Before doing so, however, we must extend our knowledge of exponents.

Exponents

We know that x^4 means $x \cdot x \cdot x \cdot x$ and that x^1 means x. Exponential notation is also defined for zero and negative exponents.

> **Zero and Negative Exponents**
> For any real number a, $a \neq 0$,
> $$a^0 = 1 \quad \text{and} \quad a^{-n} = \frac{1}{a^n}.$$

E x a m p l e   1

Simplify: **(a)** $(-36)^0$; **(b)** $(-2x)^0$.

Solution

a) $(-36)^0 = 1$ since any number (other than 0 itself) raised to the 0 power is 1.

The exponent zero, p. 207

b) $(-2x)^0 = 1$ for any $x \neq 0$.

E x a m p l e   2

Write an equivalent expression using positive exponents:

a) x^{-2}

b) $\dfrac{1}{x^{-2}}$

c) xy^{-1}

Solution

Negative exponents, p. 261

a) $x^{-2} = \dfrac{1}{x^2}$ x^{-2} is the reciprocal of x^2.

b) $\dfrac{1}{x^{-2}} = x^{-(-2)} = x^2$ The reciprocal of x^{-2} is $x^{-(-2)}$, or x^2.

c) $xy^{-1} = x\left(\dfrac{1}{y^1}\right) = \dfrac{x}{y}$ y^{-1} is the reciprocal of y^1.

The following properties hold for any integers m and n and any real numbers a and b, provided no denominators are 0 and 0^0 is not considered.

Properties of Exponents

The Product Rule: $a^m \cdot a^n = a^{m+n}$

The Quotient Rule: $\dfrac{a^m}{a^n} = a^{m-n}$

The Power Rule: $(a^m)^n = a^{mn}$

Raising a product to a power: $(ab)^n = a^n b^n$

Raising a quotient to a power: $\left(\dfrac{a}{b}\right)^n = \dfrac{a^n}{b^n}$

These properties are often used to simplify exponential expressions.

E x a m p l e   3

Simplify.

a) $(x^2 y^{-1})(xy^{-3})$

b) $\dfrac{(3p)^3}{(3p)^{-2}}$

c) $(4x^3 y^{-2})^{-2}$

d) $\left(\dfrac{ab^2}{3c^3}\right)^{-4}$

Solution

a) $(x^2y^{-1})(xy^{-3}) = x^2y^{-1}xy^{-3}$ Using an associative law

$\qquad\qquad\qquad = x^2x^1y^{-1}y^{-3}$ Using a commutative law; $x = x^1$

$\qquad\qquad\qquad = x^{2+1}y^{-1+(-3)}$ Using the product rule: Adding exponents

$\qquad\qquad\qquad = x^3y^{-4}$, or $\dfrac{x^3}{y^4}$

b) $\dfrac{(3p)^3}{(3p)^{-2}} = (3p)^{3-(-2)}$ Using the quotient rule: Subtracting exponents

$\qquad\quad = (3p)^5$

$\qquad\quad = 3^5p^5$ Raising each factor to the fifth power

$\qquad\quad = 243p^5$

c) $(4x^3y^{-2})^{-2} = 4^{-2}(x^3)^{-2}(y^{-2})^{-2}$ Raising each factor to the power -2

$\qquad\qquad\quad = 4^{-2}x^{-6}y^4$ Using the power rule: Multiplying exponents

$\qquad\qquad\quad = \dfrac{1}{4^2}x^{-6}y^4$, or $\dfrac{y^4}{16x^6}$

d) $\left(\dfrac{ab^2}{3c^3}\right)^{-4} = \dfrac{(ab^2)^{-4}}{(3c^3)^{-4}}$ Raising the numerator and the denominator to the -4 power

$\qquad\qquad\quad = \dfrac{a^{-4}(b^2)^{-4}}{3^{-4}(c^3)^{-4}}$ Raising each factor to the -4 power

$\qquad\qquad\quad = \dfrac{a^{-4}b^{-8}}{3^{-4}c^{-12}}$ Multiplying exponents

$\qquad\qquad\quad = \dfrac{3^4c^{12}}{a^4b^8}$, or $\dfrac{81c^{12}}{a^4b^8}$ Rewriting without negative exponents

Polynomials

Algebraic expressions like

$$2x^3 + 3x - 5, \qquad 4x, \qquad -7, \quad \text{and} \quad 2a^3b^2 + ab^3$$

are all examples of **polynomials**. All variables in a polynomial are raised to whole-number powers, and there are no variables in a denominator. The **terms** of a polynomial are separated by addition signs.

A polynomial with one term is called a **monomial**. A polynomial with two terms is called a **binomial**, and one with three terms is called a **trinomial**. The **degree of a term** is the number of variable factors in that term. The **leading term** of a polynomial is the term of highest degree. The **degree of a polynomial** is the degree of the leading term. A polynomial is written in *descending order* when the leading term appears first, followed by the term of next highest degree, and so on.

The number -2 in the term $-2y^3$ is called the **coefficient** of that term. The coefficient of the leading term is the **leading coefficient** of the polynomial. To illustrate this terminology, consider the polynomial

$$4y^2 - 8y^5 + y^3 - 6y + 7.$$

The *terms* are $4y^2$, $-8y^5$, y^3, $-6y$, and 7.

The *coefficients* are 4, -8, 1, -6, and 7.

The *degree of each term* is 2, 5, 3, 1, and 0.

The *leading term* is $-8y^5$ and the *leading coefficient* is -8.

The *degree of the polynomial* is 5.

Like, or *similar*, *terms* are either constant terms or terms containing the same variable(s) raised to the same power(s). Polynomials containing like terms can be simplified by *combining* those terms.

E x a m p l e   4   Combine like terms: $4x^2y + 2xy - x^2y + xy^2$.

Solution   The like terms are $4x^2y$ and $-x^2y$. Thus we have

$$4x^2y + 2xy - x^2y + xy^2 = 4x^2y - x^2y + 2xy + xy^2$$
$$= 3x^2y + 2xy + xy^2.$$

A polynomial can be evaluated by replacing the variable or variables with a number or numbers.

E x a m p l e   5   Evaluate $-a^2 + 2ab + 5b^2$ for $a = -1$ and $b = 3$.

Solution   We replace a with -1 and b with 3 and calculate the value using the rules for the order of operations:

$$-a^2 + 2ab + 5b^2 = -(-1)^2 + 2 \cdot (-1) \cdot 3 + 5 \cdot 3^2$$
$$= -1 - 6 + 45 = 38.$$

Evaluating a polynomial,
p. 215

Polynomials can be added, subtracted, multiplied, and divided.

Addition and Subtraction of Polynomials

Addition of polynomials,
Section 4.3

To add two polynomials, we write a plus sign between them and combine like terms.

E x a m p l e   6   Add: $(4x^3 + 3x^2 + 2x - 7) + (-5x^2 + x - 10)$.

Solution

$$(4x^3 + 3x^2 + 2x - 7) + (-5x^2 + x - 10)$$
$$= 4x^3 + (3 - 5)x^2 + (2 + 1)x + (-7 - 10)$$
$$= 4x^3 - 2x^2 + 3x - 17$$

Opposite of a polynomial,
p. 222

In order to subtract polynomials, we must be able to find the *opposite* of a polynomial. To find the opposite of a polynomial, we replace each term with its opposite. This process is also called *changing the sign* of each term. For example, the opposite of

$$3y^4 - 7y^2 - \tfrac{1}{3}y + 17$$

is

$$-\left(3y^4 - 7y^2 - \tfrac{1}{3}y + 17\right) = -3y^4 + 7y^2 + \tfrac{1}{3}y - 17.$$

To subtract polynomials, we add the opposite of the polynomial being subtracted.

Subtraction of polynomials,

Section 4.3

Example 7

Subtract: $(3a^4 - 2a + 7) - (-a^3 + 5a - 1)$.

Solution

$$(3a^4 - 2a + 7) - (-a^3 + 5a - 1)$$
$$= 3a^4 - 2a + 7 + a^3 - 5a + 1 \qquad \text{Adding the opposite}$$
$$= 3a^4 + a^3 - 7a + 8 \qquad \text{Combining like terms}$$

Multiplication of Polynomials

Multiplication of polynomials,

Section 4.4

We first consider the product of two monomials. To multiply two monomials, we multiply coefficients and then multiply variables using the product rule for exponents.

Example 8

Multiply: $(2x^2y)(-3xy^3)$.

Solution

$$(2x^2y)(-3xy^3) = (2 \cdot (-3))(x^2 \cdot x \cdot y \cdot y^3) \qquad \text{Multiplying coefficients;}$$
$$\text{multiplying variables}$$

$$= -6x^3y^4 \qquad \text{Using the product rule}$$

To multiply a monomial and a polynomial, we multiply each term of the polynomial by the monomial, using the distributive property.

Example 9

Multiply: $4x^3(3x^4 - 2x^3 + 7x - 5)$.

Solution

$$\textit{Think:} \quad \overbrace{4x^3 \cdot 3x^4} - \overbrace{4x^3 \cdot 2x^3} + \overbrace{4x^3 \cdot 7x} - \overbrace{4x^3 \cdot 5}$$
$$4x^3(3x^4 - 2x^3 + 7x - 5) = 12x^7 \quad - \quad 8x^6 \quad + \quad 28x^4 \quad - \quad 20x^3$$

To multiply any two polynomials P and Q, we select one of the polynomials—say, P. We then multiply each term of P by every term of Q and combine like terms.

E x a m p l e   1 0

Multiply: $(2a^3 + 3a - 1)(a^2 - 4a)$.

Solution   It is often helpful to use columns for a long multiplication. We multiply each term at the top by every term at the bottom, write like terms in columns, and add the results.

$$
\begin{array}{r}
2a^3 \quad + 3a \; - 1 \\
a^2 - 4a \\
\hline
-8a^4 \qquad - 12a^2 + 4a \\
2a^5 \qquad + 3a^3 \; - a^2 \\
\hline
2a^5 - 8a^4 + 3a^3 - 13a^2 + 4a
\end{array}
$$

Multiplying the top row by $-4a$

Multiplying the top row by a^2

Combining like terms. Be sure that like terms are lined up in columns.

We could multiply two binomials in the same manner in which we multiplied the polynomials in Example 10. However, by observing the pattern of the products formed, we can develop a method of multiplying two binomials more efficiently.

The FOIL Method

To multiply two binomials, $A + B$ and $C + D$, multiply the First terms AC, the Outer terms AD, the Inner terms BC, and then the Last terms BD. Then combine like terms, if possible.

$$(A + B)(C + D) = AC + AD + BC + BD$$

1. Multiply First terms: AC.
2. Multiply Outer terms: AD.
3. Multiply Inner terms: BC.
4. Multiply Last terms: BD.

FOIL

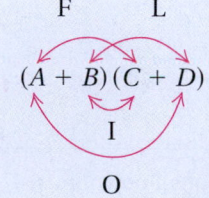

E x a m p l e   1 1

Multiply: $(3x + 4)(x - 2)$.

Solution

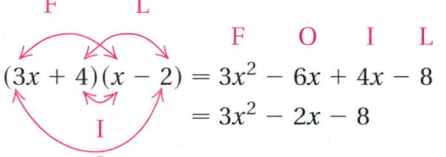

$$
\begin{aligned}
(3x + 4)(x - 2) &= 3x^2 - 6x + 4x - 8 \\
&= 3x^2 - 2x - 8
\end{aligned}
$$

Combining like terms

FOIL, p. 238

Two types of products of binomials occur so often that specific formulas or methods for computing them have been developed. Products of this type are called *special products*.

Multiplying sums and
differences of two terms, p. 240

Squaring binomials, p. 241

> **Special Products**
>
> The product of a sum and difference of the same two terms:
>
> $$(A + B)(A - B) = \underbrace{A^2 - B^2}$$
>
> This is called a *difference of squares.*
>
> The square of a binomial:
>
> $$(A + B)^2 = A^2 + 2AB + B^2$$
> $$(A - B)^2 = A^2 - 2AB + B^2$$

E x a m p l e   1 2

Multiply: **(a)** $(x + 3y)(x - 3y)$; **(b)** $(x^3 + 2)^2$.

Solution

$$(A + B)(A - B) = A^2 - B^2$$

a) $(x + 3y)(x - 3y) = x^2 - (3y)^2$ $A = x$ and $B = 3y$

$$= x^2 - 9y^2$$

$$(A + B)^2 = A^2 + 2 \cdot A \cdot B + B^2$$

b) $(x^3 + 2)^2 = (x^3)^2 + 2 \cdot x^3 \cdot 2 + 2^2$ $A = x^3$ and $B = 2$

$$= x^6 + 4x^3 + 4$$

Division of Polynomials

Division of polynomials,
Section 4.7

Polynomial division is similar to division in arithmetic. First, let's consider division by a monomial. To divide a polynomial by a monomial, we divide each term by the monomial.

E x a m p l e   1 3

Divide: $(3x^5 + 8x^3 - 12x) \div 4x$.

Solution   This division can be written

$$\frac{3x^5 + 8x^3 - 12x}{4x} = \frac{3x^5}{4x} + \frac{8x^3}{4x} - \frac{12x}{4x}$$ Dividing each term by $4x$

$$= \frac{3}{4}x^{5-1} + \frac{8}{4}x^{3-1} - \frac{12}{4}x^{1-1}$$ Dividing coefficients and subtracting exponents

$$= \frac{3}{4}x^4 + 2x^2 - 3.$$

To check, we multiply the quotient by $4x$:

$$\left(\tfrac{3}{4}x^4 + 2x^2 - 3\right)4x = 3x^5 + 8x^3 - 12x.$$ The answer checks.

To use long division, we write polynomials in descending order, including terms with 0 coefficients for missing terms. As shown below in Example 14, the procedure ends when the degree of the remainder is less than the degree of the divisor.

E x a m p l e   1 4

Divide: $(4x^3 - 7x + 1) \div (2x + 1)$.

Solution   The polynomials are already written in descending order, but there is no x^2-term in the dividend. We fill in $0x^2$ for that term.

$$
\begin{array}{r}
2x^2 \\
2x + 1 \overline{)\,4x^3 + 0x^2 - 7x + 1} \\
\underline{4x^3 + 2x^2} \\
-2x^2
\end{array}
$$

Divide the first term of the dividend, $4x^3$, by the first term in the divisor, $2x$: $4x^3/(2x) = 2x^2$.
Multiply $2x^2$ by the divisor, $2x + 1$.
Subtract: $(4x^3 + 0x^2) - (4x^3 + 2x^2) = -2x^2$.

Then we bring down the next term of the dividend, $-7x$.

$$
\begin{array}{r}
2x^2 - x \\
2x + 1 \overline{)\,4x^3 + 0x^2 - 7x + 1} \\
\underline{4x^3 + 2x^2} \\
-2x^2 - 7x \\
\underline{-2x^2 - x} \\
-6x
\end{array}
$$

Divide the first term of $-2x^2 - 7x$ by the first term in the divisor: $-2x^2/(2x) = -x$.
The $-7x$ has been "brought down."
Multiply $-x$ by the divisor, $2x + 1$.
Subtract: $(-2x^2 - 7x) - (-2x^2 - x) = -6x$.

Since the degree of the remainder, $-6x$, is *not* less than the degree of the divisor, we must continue dividing.

$$
\begin{array}{r}
2x^2 - x - 3 \\
2x + 1 \overline{)\,4x^3 + 0x^2 - 7x + 1} \\
\underline{4x^3 + 2x^2} \\
-2x^2 - 7x \\
\underline{-2x^2 - x} \\
- 6x + 1 \\
\underline{- 6x - 3} \\
4
\end{array}
$$

Divide the first term of $-6x + 1$ by the first term in the divisor: $-6x/(2x) = -3$.
The 1 has been "brought down."
Multiply -3 by $2x + 1$.
Subtract.

The answer is $2x^2 - x - 3$ with R4, or

$$\text{Quotient} \longrightarrow 2x^2 - x - 3 + \frac{4}{2x + 1}. \quad \begin{array}{l} \longleftarrow \text{Remainder} \\ \longleftarrow \text{Divisor} \end{array}$$

To check, we can multiply by the divisor and add the remainder:

$$
\begin{aligned}
(2x + 1)(2x^2 - x - 3) + 4 &= 4x^3 - 7x - 3 + 4 \\
&= 4x^3 - 7x + 1.
\end{aligned}
$$

Exercise Set R.4

FOR EXTRA HELP

Digital Video Tutor CD: none
Videotape: none

Math Tutor Center

MathXL

MyMathLab.com

Simplify.

1. a^0, for $a = -25$

2. y^0, for $y = 6.97$

3. $4^0 - 4^1$

4. $8^1 - 8^0$

Write an equivalent expression using positive exponents. Then, if possible, simplify.

5. 8^{-2}

6. 2^{-5}

7. $(-2)^{-3}$

8. $(-3)^{-2}$

9. $(ab)^{-2}$

10. ab^{-2}

11. $\dfrac{1}{y^{-10}}$

12. $\dfrac{1}{x^{-t}}$

Write an equivalent expression using negative exponents.

13. $\dfrac{1}{y^4}$

14. $\dfrac{1}{a^2 b^3}$

15. $\dfrac{1}{x^t}$

16. $\dfrac{1}{n}$

Simplify.

17. $x^5 \cdot x^8$

18. $a^4 \cdot a^{-2}$

19. $\dfrac{a}{a^{-5}}$

20. $\dfrac{p^{-3}}{p^{-8}}$

21. $\dfrac{(4x)^{10}}{(4x)^2}$

22. $\dfrac{a^2 b^9}{a^9 b^2}$

23. $(7^8)^5$

24. $(x^3)^{-7}$

25. $(x^{-2} y^{-3})^{-4}$

26. $(-2a^2)^3$

27. $\left(\dfrac{y^2}{4}\right)^3$

28. $\left(\dfrac{ab^2}{c^3}\right)^4$

29. $\left(\dfrac{2p^3}{3q^4}\right)^{-2}$

30. $\left(\dfrac{2}{x}\right)^{-5}$

Identify the terms of each polynomial.

31. $8x^3 - 6x^2 + x - 7$

32. $-a^2 b + 4a^2 - 8b + 17$

Determine the coefficient and the degree of each term in each polynomial. Then find the degree of each polynomial.

33. $18x^3 + 36x^9 - 7x + 3$

34. $-8y^7 + y + 19$

35. $-x^2 y + 4y^3 - 2xy$

36. $8 - x^2 y^4 + y^7$

Determine the leading term and the leading coefficient of each polynomial.

37. $-p^2 + 4 + 8p^4 - 7p$

38. $13 + 20t - 30t^2 - t^3$

Combine like terms. Write each answer in descending order.

39. $3x^3 - x^2 + x^4 + x^2$　　**40.** $5t - 8t^2 + 4t^2$

41. $3 - 2t^2 + 8t - 3t - 5t^2 + 7$

42. $8x^5 - \frac{1}{3} + \frac{4}{5}x + 1 - \frac{1}{2}x$

Evaluate each polynomial for the given replacements of the variables.

43. $3x^2 - 7x + 10$, for $x = -2$

44. $-y + 3y^2 + 2y^3$, for $y = 3$

45. $a^2 b^3 + 2b^2 - 6a$, for $a = 2$ and $b = -1$

46. $2pq^3 - 5q^2 + 8p$, for $p = -4$ and $q = -2$

The distance s, in feet, traveled by a body falling freely from rest in t seconds is approximated by

$$s = 16t^2.$$

47. A pebble is dropped into a well and takes 3 sec to hit the water. How far down is the surface of the water?

48. An acorn falls from the top of an oak tree and takes 2 sec to hit the ground. How high is the tree?

Add or subtract, as indicated.

49. $(3x^3 + 2x^2 + 8x) + (x^3 - 5x^2 + 7)$

50. $(-6x^4 + 3x^2 - 16) + (4x^2 + 4x - 7)$

51. $(8y^2 - 2y - 3) - (9y^2 - 7y - 1)$

52. $(4t^2 + 6t - 7) - (t + 5)$

53. $(-x^2 y + 2y^2 + y) - (3y^2 + 2x^2 y - 7y)$

54. $(ab + x^2 y^2) + (2ab - x^2 y^2)$

Multiply.

55. $4x^2(3x^3 - 7x + 7)$

56. $a^2b(a^3 + b^2 - ab - 2b)$

57. $(2a + y)(4a + b)$

58. $(x + 7y)(y - 3x)$

59. $(x + 7)(x^2 - 3x + 1)$

60. $(2x - 3)(x^2 - x - 1)$

61. $(x + 7)(x - 7)$

62. $(2x + 1)^2$

63. $(x + y)^2$

64. $(xy + 1)(xy - 1)$

65. $(2x^2 + 7)(3x^2 - 2)$

66. $(x^2 + 2)^2$

67. $(a - 3b)^2$

68. $(1.1x^2 + 5)(0.1x^2 - 2)$

69. $(6a - 5y)(7a + 3y)$

70. $(3p^2 - q^3)^2$

Divide and check.

71. $(3t^5 + 9t^3 - 6t^2 + 15t) \div (-3t)$

72. $(4x^5 + 10x^4 - 16x^2) \div (4x^2)$

73. $(15x^2 - 16x - 15) \div (3x - 5)$

74. $(x^3 - 2x^2 - 14x + 1) \div (x - 5)$

75. $(2x^3 - x^2 + 1) \div (x + 1)$

76. $(2x^3 + 3x^2 - 50) \div (2x - 5)$

77. $(5x^3 + 3x^2 - 5x) \div (x^2 - 1)$

78. $(2x^3 + 3x^2 + 6x + 10) \div (x^2 + 3)$

Polynomials and Factoring

R.5

Common Factors and Factoring by Grouping •
Factoring Trinomials • Factoring Special Forms •
Solving Polynomial Equations by Factoring

Factor, p. 276

The reverse of multiplication is factoring. To *factor* a polynomial is to find an equivalent expression that is a product. To factor a monomial, we find two monomials whose product is equivalent to the original monomial. Many monomials have multiple factorizations. For example, three factorizations of $50x^6$ are $5 \cdot 10x^6$, $5x^3 \cdot 10x^3$, and $2x \cdot 25x^5$.

Common Factors and Factoring by Grouping

Common factor, p. 277

If all the terms in a polynomial share a common factor, that factor can be "factored out" of the polynomial. Whenever you are factoring a polynomial with two or more terms, try to first find the largest common factor of the terms, if one exists.

Example 1

Factor: $3x^6 + 15x^4 - 9x^3$.

Solution   The largest factor common to 3, 15, and -9 is 3. The largest power of x common to x^6, x^4, and x^3 is x^3. Thus the largest common factor of the terms of the polynomial is $3x^3$. We factor as follows:

$$3x^6 + 15x^4 - 9x^3 = 3x^3 \cdot x^3 + 3x^3 \cdot 5x - 3x^3 \cdot 3 \qquad \text{Factoring each term}$$

$$= 3x^3(x^3 + 5x - 3). \qquad \text{Factoring out } 3x^3$$

Factorizations can always be checked by multiplying:

$$3x^3(x^3 + 5x - 3) = 3x^6 + 15x^4 - 9x^3.$$

A polynomial with two or more terms can be a common factor.

Example 2

Factor: $3x^2(x - 2) + 5(x - 2)$.

Solution   The binomial $x - 2$ is a factor of both $3x^2(x - 2)$ and $5(x - 2)$. Thus we have

$$3x^2(x - 2) + 5(x - 2) = (x - 2)(3x^2 + 5). \qquad \text{Factoring out the common factor, } x - 2$$

Factoring by grouping, p. 280

If a polynomial with four terms can be split into two groups of terms, and both groups share a common binomial factor, the polynomial can be factored. This method is known as **factoring by grouping**.

Example 3

Factor by grouping: $2x^3 + 6x^2 - x - 3$.

Solution   First, we consider the polynomial as two groups of terms, $2x^3 + 6x^2$ and $-x - 3$. Then we factor each group separately:

$$2x^3 + 6x^2 - x - 3 = 2x^2(x + 3) - 1(x + 3) \qquad \text{Factoring out } 2x^2 \text{ and } -1 \text{ to give the common binomial factor, } x + 3$$

$$= (x + 3)(2x^2 - 1).$$

The check is left to the student.

Prime polynomial, p. 287

Not every polynomial with four terms is factorable by grouping. A polynomial that is not factorable is said to be *prime*.

Factoring Trinomials

Factoring trinomials of the type $x^2 + bx + c$, Section 5.2

Many trinomials that have no common factor can be written as the product of two binomials. We look first at trinomials of the form

$$x^2 + bx + c,$$

for which the leading coefficient is 1.

Factoring trinomials involves a trial-and-error process. In order for the product of two binomials to be $x^2 + bx + c$, the binomials must look like

$$(x + p)(x + q),$$

where p and q are constants that must be determined. For example, to factor $x^2 + 10x + 16$, we must have

$$x^2 + 10x + 16 = (x + p)(x + q)$$
$$= x^2 + (p + q)x + pq. \qquad \text{Using FOIL}$$

Therefore,

$$p + q = 10,$$
$$pq = 16.$$

Thus we look for two numbers whose product is 16 and whose sum is 10.

E x a m p l e 4

Factor.

a) $x^2 + 10x + 16$ **b)** $x^2 - 8x + 15$
c) $x^2 + x - 6$ **d)** $x^2 - 2x - 24$

Constant term positive, p. 283

Solution

a) The factorization is of the form

$$(x + \quad)(x + \quad).$$

To find the constant terms, we need a pair of factors whose product is 16 and whose sum is 10. Since 16 is positive, its factors will have the same sign as 10—that is, we need consider only positive factors of 16.
    We list the possible factorizations in a table and calculate the sum of each pair of factors.

Pairs of Factors of 16	Sums of Factors
1, 16	17
2, 8	10 ←
4, 4	8

The numbers we seek are 2 and 8.

The factorization of $x^2 + 10x + 16$ is $(x + 2)(x + 8)$. To check, we multiply.

Check: $(x + 2)(x + 8) = x^2 + 8x + 2x + 16 = x^2 + 10x + 16.$

b) For $x^2 - 8x + 15$, c is positive and b is negative. Therefore, the factors of 15 will be negative. Again, we list the possible factorizations in a table.

Pairs of Factors of 15	Sums of Factors
−1, −15	−16
−3, −5	−8 ←

The numbers we need are −3 and −5.

The factorization is $(x - 3)(x - 5)$. The check is left to the student.

Constant term negative, p. 285

c) The constant term in $x^2 + x - 6$ is negative, so one factor will be positive and the other will be negative. The coefficient of x, which is 1, is positive. Thus the positive factor must have the larger absolute value.

Pairs of Factors of -6	Sums of Factors
$-1, 6$	5
$-2, 3$	1

The numbers we need are -2 and 3.

The factorization is $(x - 2)(x + 3)$.

Check: $(x - 2)(x + 3) = x^2 + 3x - 2x - 6$
$= x^2 + x - 6.$

d) For $x^2 - 2x - 24$, c is negative, so one factor of -24 will be negative and one will be positive. Since b is also negative, the negative factor must have the larger absolute value.

Pairs of Factors of -24	Sums of Factors
$1, -24$	-23
$2, -12$	-10
$3, -8$	-5
$4, -6$	-2

The numbers we need are 4 and -6.

The factorization is $(x + 4)(x - 6)$. The check is left to the student.

Always look first for a common factor.

E x a m p l e   5

Factor: $3t^2 - 33st + 84s^2$.

Solution There is a common factor, 3, which we factor out first:

$$3t^2 - 33st + 84s^2 = 3(t^2 - 11st + 28s^2).$$

Now we consider $t^2 - 11st + 28s^2$. Think of $28s^2$ as the "constant" term c and $-11s$ as the "coefficient" b of the middle term. We try to express $28s^2$ as the product of two factors whose sum is $-11s$. These factors are $-4s$ and $-7s$. Thus the factorization of $t^2 - 11st + 28s^2$ is

$(t - 4s)(t - 7s).$ This is not the entire factorization of $3t^2 - 33st + 84s^2$.

We now include the common factor, 3, and write

$$3t^2 - 33st + 84s^2 = 3(t - 4s)(t - 7s).$$ This is the factorization.

The check is left to the student.

Factoring trinomials of the
type $ax^2 + bx + c$, Section 5.3

When the leading coefficient of a trinomial is not 1, the number of trials needed to find a factorization can increase dramatically. We will consider two methods for factoring trinomials of the type $ax^2 + bx + c$: factoring with FOIL and the grouping method.

To Factor $ax^2 + bx + c$ Using FOIL

1. Factor out the largest common factor, if one exists. Here we assume none does.
2. Find two First terms whose product is ax^2:

$$(\quad x + \quad)(\quad x + \quad) = ax^2 + bx + c.$$

FOIL

3. Find two Last terms whose product is c:

$$(\quad x + \quad)(\quad x + \quad) = ax^2 + bx + c.$$

FOIL

4. Repeat steps (2) and (3) until a combination is found for which the sum of the Outer and Inner products is bx:

$$(\quad x + \quad)(\quad x + \quad) = ax^2 + bx + c.$$

I

O

FOIL

5. Always check by multiplying.

E x a m p l e   6

Factor: $20x^3 - 22x^2 - 12x$.

Solution

1) First, we factor out the largest common factor, $2x$:

$$20x^3 - 22x^2 - 12x = 2x(10x^2 - 11x - 6).$$

Factoring with FOIL, p. 291

2) Next, in order to factor the trinomial $10x^2 - 11x - 6$, we search for two terms whose product is $10x^2$. The possibilities are

$$(x + \quad)(10x + \quad) \quad \text{or} \quad (2x + \quad)(5x + \quad).$$

3) There are four pairs of factors of -6. Since the first terms of the binomials are different, the order of the factors is important. So there are eight possibilities for the last terms:

$$1, -6 \qquad -1, 6 \qquad 2, -3 \qquad -2, 3$$

and

$$-6, 1 \qquad 6, -1 \qquad -3, 2 \qquad 3, -2.$$

4) Since each of the eight possibilities from step (3) could be used in either of the two possibilities from step (2), there are $2 \cdot 8$, or 16, possible factorizations. We check the possibilities systematically until we find one that gives the correct factorization. Let's first try factors with $(2x + \quad)(5x + \quad)$.

Trial	*Product*	
$(2x + 1)(5x - 6)$	$10x^2 - 7x - 6$	⟵ Wrong middle term
$(2x - 1)(5x + 6)$	$10x^2 + 7x - 6$	⟵ Wrong middle term. Note that changing the signs in the binomials changed the sign of middle term in the product.
$(2x + 2)(5x - 3)$	$10x^2 + 4x - 6$	⟵ Wrong middle term. We need not consider $(2x - 2)(5x + 3)$.
$(2x - 6)(5x + 1)$	$10x^2 - 28x - 6$	⟵ Wrong middle term. We need not consider $(2x + 6)(5x - 1)$.
$(2x - 3)(5x + 2)$	$10x^2 - 11x - 6$	⟵ Correct middle term

We can stop when we find a correct factorization. Including the common factor $2x$, we now have

$$20x^3 - 22x^2 - 12x = 2x(2x - 3)(5x + 2).$$

This can be checked by multiplying.

The grouping method, p. 296

With practice, some of the trials can be skipped or performed mentally.

The second method of factoring trinomials of the type $ax^2 + bx + c$ involves factoring by grouping.

> ### To Factor $ax^2 + bx + c$, Using the Grouping Method
>
> **1.** Factor out the largest common factor, if one exists.
> **2.** Multiply the leading coefficient a and the constant c.
> **3.** Find a pair of factors of ac whose sum is b.
> **4.** Rewrite the middle term, bx, as a sum or difference using the factors found in step (3).
> **5.** Factor by grouping.
> **6.** Always check by multiplying.

Example 7

Factor: $7x^2 + 31x + 12$.

Solution

1. There is no common factor (other than 1 or -1).
2. We multiply the leading coefficient, 7, and the constant, 12:

$$7 \cdot 12 = 84.$$

3. We look for a pair of factors of 84 whose sum is 31. Since both 84 and 31 are positive, we need consider only positive factors.

Pairs of Factors of 84	Sums of Factors
1, 84	85
2, 42	44
3, 28	31 ← $\quad\quad$ 3 + 28 = 31

4. Next, we rewrite $31x$ using the factors 3 and 28:

$$31x = 3x + 28x.$$

5. We now factor by grouping:

$$7x^2 + 31x + 12 = 7x^2 + 3x + 28x + 12 \qquad \text{Substituting } 3x + 28x \text{ for } 31x$$

$$= x(7x + 3) + 4(7x + 3)$$

$$= (7x + 3)(x + 4). \qquad \text{Factoring out the common factor, } 7x + 3$$

6. *Check:* $(7x + 3)(x + 4) = 7x^2 + 31x + 12.$

Factoring Special Forms

We can factor certain types of polynomials directly, without using trial and error.

> ### *Factoring Formulas*
>
> Perfect-square trinomial: $A^2 + 2AB + B^2 = (A + B)^2,$
> $\qquad\qquad\qquad\qquad\qquad A^2 - 2AB + B^2 = (A - B)^2$
> Difference of squares: $A^2 - B^2 = (A + B)(A - B)$
> Sum of cubes: $A^3 + B^3 = (A + B)(A^2 - AB + B^2)$
> Difference of cubes: $A^3 - B^3 = (A - B)(A^2 + AB + B^2)$

Before using the factoring formulas, it is important to check carefully that the expression being factored is indeed in one of the forms listed. Note that there is no factoring formula for the sum of two squares—unless it has a common factor, we cannot factor the sum of two squares.

E x a m p l e   8

Recognizing and factoring
differences of squares,
pp. 302–304

Factor: $2x^2 - 2$.

Solution   We first factor out a common factor, 2:

$$2x^2 - 2 = 2(x^2 - 1).$$

Looking at $x^2 - 1$, we see that it is a difference of squares, with $A = x$ and $B = 1$. The factorization is thus

$$2x^2 - 2 = 2(x^2 - 1) = 2(x + 1)(x - 1).$$
$$\underset{A^2 - B^2}{\uparrow\quad\uparrow}\qquad\underset{(A + B)(A - B)}{\uparrow\quad\uparrow\quad\uparrow\quad\uparrow}$$

E x a m p l e   9

Recognizing and factoring
perfect-square trinomials,
pp. 300–302

Factor: $x^2y^2 + 20xy + 100$.

Solution   First, we check for a common factor; there is none. The polynomial is a perfect-square trinomial, since x^2y^2 and 100 are squares; there is no minus sign before either square; and $20xy$ is $2 \cdot xy \cdot 10$, where xy and 10 are square roots of x^2y^2 and 100, respectively. The factorization is thus

$$x^2y^2 + 20xy + 100 = (xy)^2 + 2 \cdot xy \cdot 10 + 10^2 = (xy + 10)^2.$$
$$\underset{A^2}{\uparrow}\quad\underset{+ 2 \cdot A \cdot B + B^2}{\uparrow\quad\uparrow\quad\uparrow\quad\uparrow}\quad\underset{= (A + B)^2}{\uparrow\quad\uparrow}$$

E x a m p l e   1 0

Factoring sums or differences
of cubes, Section 5.5

Factor: **(a)** $p^3 - 64$; **(b)** $3y^2 + 27$.

Solution

a) This is a difference of cubes, with $A = p$ and $B = 4$:

$$p^3 - 64 = (p)^3 - (4)^3$$
$$= (p - 4)(p^2 + 4p + 16).$$

b) We factor out the common factor, 3:

$$3y^2 + 27 = 3(y^2 + 9).$$

Since $y^2 + 9$ is a sum of squares, no further factorization is possible.

Factoring completely, p. 304

A polynomial is said to be *factored completely* when no factor can be factored further.

E x a m p l e   1 1

Factor completely: $x^4 - 1$.

Solution

$$x^4 - 1 = (x^2 + 1)(x^2 - 1)$$ Factoring a difference of squares
$$= (x^2 + 1)(x + 1)(x - 1)$$ The factor $x^2 - 1$ is itself a difference of squares.

Solving Polynomial Equations by Factoring

A **polynomial equation** is formed by setting two polynomials equal to each other. A **quadratic equation** is a polynomial equation equivalent to one of the form $ax^2 + bx + c = 0$, where $a \neq 0$. Polynomial equations that can be factored can be solved using the principle of zero products.

> ### The Principle of Zero Products
>
> An equation $ab = 0$ is true if and only if $a = 0$ or $b = 0$, or both. (A product is 0 if and only if at least one factor is 0.)

If we can write an equation as a product that equals 0, we can try to use the principle of zero products to solve the equation.

Example 12

Solve.

a) $x^2 - 11x = 12$
b) $5x^2 + 10x + 5 = 0$
c) $9x^2 = 1$

Solution

a) We must have 0 on one side of the equation before using the principle of zero products:

$$x^2 - 11x = 12$$
$$x^2 - 11x - 12 = 0 \qquad \text{Subtracting 12 from both sides}$$
$$(x - 12)(x + 1) = 0 \qquad \text{Factoring}$$
$$x - 12 = 0 \quad or \quad x + 1 = 0 \qquad \text{Using the principle of zero products}$$
$$x = 12 \quad or \qquad x = -1.$$

The solutions are 12 and -1. The check is left to the student.

b) We have

$$5x^2 + 10x + 5 = 0$$
$$5(x^2 + 2x + 1) = 0 \qquad \text{Factoring out a common factor}$$
$$5(x + 1)(x + 1) = 0 \qquad \text{Factoring completely}$$
$$x + 1 = 0 \quad or \quad x + 1 = 0 \qquad \text{Using the principle of zero products}$$
$$x = -1 \quad or \qquad x = -1.$$

There is only one solution, -1. The check is left to the student.

c) We have

$$9x^2 = 1$$

$$9x^2 - 1 = 0$$ Subtracting 1 from both sides to get 0 on one side

$$(3x + 1)(3x - 1) = 0$$ Factoring a difference of squares

$$3x + 1 = 0 \quad \text{or} \quad 3x - 1 = 0$$ Using the principle of zero products

$$3x = -1 \quad \text{or} \quad 3x = 1$$

$$x = -\tfrac{1}{3} \quad \text{or} \quad x = \tfrac{1}{3}.$$

The solutions are $\tfrac{1}{3}$ and $-\tfrac{1}{3}$. The check is left to the student.

Quadratic equations can be used to solve problems. One important result that uses squared quantities is the Pythagorean theorem. It relates the lengths of the sides of a **right triangle**, that is, a triangle with a 90° angle. The side opposite the 90° angle is called the **hypotenuse**, and the other sides are called the **legs**.

The Pythagorean Theorem

The sum of the squares of the legs of a right triangle is equal to the square of the hypotenuse:

$$a^2 + b^2 = c^2.$$

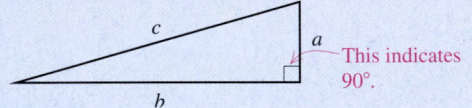

This indicates 90°.

E x a m p l e   1 3

Swing sets. The length of a slide on a swing set is 5 ft. The distance from the base of the ladder to the base of the slide is 1 ft more than the height of the ladder. Find the height of the ladder.

Solution

1. **Familiarize.** We first make a drawing and let $x =$ the height of the ladder, in feet. We know then that the other leg of the triangle is $x + 1$, since it is 1 ft longer than the ladder. The hypotenuse has length 5 ft.

x

5 ft

$x + 1$

2. **Translate.**   Applying the Pythagorean theorem gives us

$$a^2 + b^2 = c^2$$
$$x^2 + (x + 1)^2 = 5^2. \qquad \text{Substituting}$$

3. **Carry out.**   We solve the equation:

$$x^2 + (x + 1)^2 = 5^2$$
$$x^2 + x^2 + 2x + 1 = 25 \qquad \text{Squaring } x + 1; \text{ squaring } 5$$
$$2x^2 + 2x + 1 = 25 \qquad \text{Combining like terms}$$
$$2x^2 + 2x - 24 = 0 \qquad \text{Getting 0 on one side}$$
$$2(x^2 + x - 12) = 0 \qquad \text{Factoring out a common factor}$$
$$2(x + 4)(x - 3) = 0 \qquad \text{Factoring a trinomial}$$
$$x + 4 = 0 \quad or \quad x - 3 = 0 \qquad \text{Using the principle of zero products}$$
$$x = -4 \quad or \qquad x = 3.$$

4. **Check.**   We know that the integer -4 is not a solution because the height of the ladder cannot be negative. When $x = 3$, the distance from the base of the ladder to the base of the slide is $x + 1 = 4$, and $3^2 + 4^2 = 5^2$. So the solution 3 checks.

5. **State.**   The ladder is 3 ft high.

FOR EXTRA HELP

Digital Video Tutor CD: none
Videotape: none

Math Tutor Center   MathXL   MyMathLab.com

Exercise Set **R.5**

Factor completely. If a polynomial is prime, state this.

1. $3x^3 + 6x^2 - 9x$

2. $x^2y^4 - 2xy^5 + 3x^3y^6$

3. $y^2 - 6y + 9$

4. $4z^2 - 25$

5. $2p^3(p + 2) + (p + 2)$

6. $6y^2 + y - 1$

7. $16x^2 + 25$

8. $y^3 - 1$

9. $8t^3 + 27$

10. $a^2b^2 + 24ab + 144$

11. $m^2 + 13m + 42$

12. $2x^3 - 6x^2 + x - 3$

13. $x^4 - 81$

14. $x^2 + x + 1$

15. $8x^2 + 22x + 15$

16. $4x^2 - 40x + 100$

17. $x^3 + 2x^2 - x - 2$

18. $(x + 2y)(x - 1) + (x + 2y)(x - 2)$

19. $0.001t^6 - 0.008$

20. $x^2 - 20 - x$

21. $-\frac{1}{16} + x^4$

22. $5x^8 - 5z^{16}$

23. $a^2 + 6a + 9 - y^2$

24. $t^6 - p^6$

25. $5mn + m^2 - 150n^2$

26. $\frac{1}{27} + x^3$

27. $24x^2y - 6y - 10xy$

28. $-3y^2 - 12y - 12$

29. $y^2 + 121 - 22y$

30. $p^2 - m^2 - 2mn - n^2$

Solve.

31. $(x - 2)(x + 7) = 0$

32. $(3x - 5)(7 - 4x) = 0$

33. $8x(4.7 - x) = 0$

34. $(x - 3)(x + 1)(2x - 9) = 0$

35. $x^2 = 100$

36. $8x^2 = 5x$

37. $4x^2 - 18x = 70$

38. $x^2 + 2x + 1 = 0$

39. $2x^3 - 10x = 0$

40. $100x^2 = 81$

41. $(a + 1)(a - 5) = 7$

42. $d(d - 3) = 40$

43. $x^2 + 6x - 55 = 0$

44. $x^2 + 7x - 60 = 0$

45. $\frac{1}{2}x^2 + 5x + \frac{25}{2} = 0$

46. $3 + 10x^2 = 11x$

47. *Landscaping.*   A triangular flower garden is 3 ft longer than it is wide. The area of the garden is 20 ft². What are the dimensions of the garden?

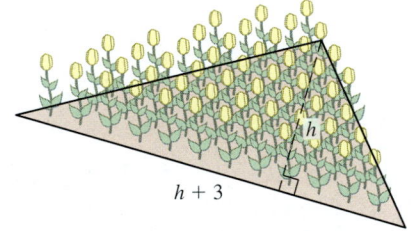

48. *Page numbers.*   The product of the page numbers on two facing pages of a book is 156. Find the page numbers.

49. *Right triangles.*   The hypotenuse of a right triangle is 17 ft. One leg is 1 ft shorter than twice the length of the other leg. Find the length of the legs.

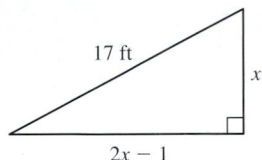

50. *Hiking.*   Jenna hiked 500 ft up a steep incline. Her global positioning unit indicated that her horizontal position had changed by 100 ft more than her vertical position had changed. What was the change in altitude?

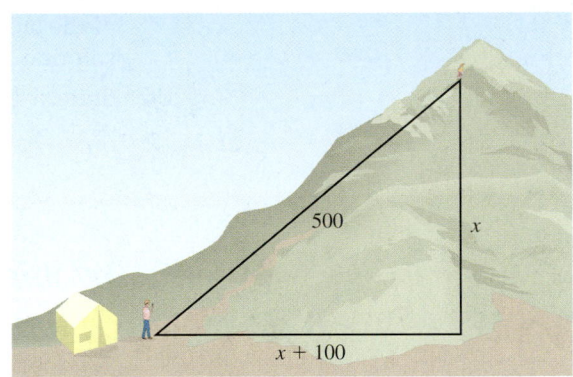

Rational Expressions and Equations

R.6

Multiplication and Division of Rational Expressions •
Addition and Subtraction of Rational Expressions •
Complex Rational Expressions • Solving Rational Equations

Rational expressions, p. 340

A **rational expression** is a quotient of two polynomials. Because division by 0 is undefined, a rational expression is undefined for any number that will make the denominator 0.

E x a m p l e   1   Find all numbers for which the rational expression

$$\frac{2x + 5}{x^2 - 9x - 10}$$

is undefined.

Solution   We set the denominator equal to 0 and solve:

$$x^2 - 9x - 10 = 0$$

$$(x - 10)(x + 1) = 0 \qquad \text{Factoring}$$

$$x - 10 = 0 \quad or \quad x + 1 = 0 \qquad \text{Using the principle of zero products}$$

$$x = 10 \quad or \qquad x = -1.$$

If x is replaced with 10 or with -1, the denominator is 0. Thus,

$$\frac{2x + 5}{x^2 - 9x - 10} \quad \text{is undefined for } x = 10 \text{ and } x = -1.$$

Multiplication and Division of Rational Expressions

Multiplication and division of rational expressions is similar to multiplication and division with fractions.

> ### The Product and the Quotient of Two Rational Expressions
>
> To multiply rational expressions, multiply numerators and multiply denominators:
>
> $$\frac{A}{B} \cdot \frac{C}{D} = \frac{AC}{BD}.$$
>
> To divide by a rational expression, multiply by its reciprocal:
>
> $$\frac{A}{B} \div \frac{C}{D} = \frac{A}{B} \cdot \frac{D}{C} = \frac{AD}{BC}.$$

E x a m p l e   2   Simplify: $\dfrac{9x^2 + 12x}{6x^2 - 3x}$.

Solution   We first factor the numerator and the denominator:

$$\frac{9x^2 + 12x}{6x^2 - 3x} = \frac{3x(3x + 4)}{3x(2x - 1)}.$$

We can now write this as a product of two rational expressions using the rule for multiplying rational expressions in reverse. Then we can simplify.

$$\frac{3x(3x+4)}{3x(2x-1)} = \frac{3x}{3x} \cdot \frac{3x+4}{2x-1} \qquad \text{Rewriting as a product of two rational expressions}$$

$$= 1 \cdot \frac{3x+4}{2x-1} \qquad \frac{3x}{3x} = 1$$

$$= \frac{3x+4}{2x-1} \qquad \text{Removing the factor 1}$$

Only factors can be removed. Be sure that the numerator and the denominator are factored before you attempt to remove factors equal to 1.

After multiplying or dividing rational expressions, we simplify, if possible.

E x a m p l e   3   Perform each indicated operation and simplify.

a) $\dfrac{x^2 - x - 6}{3x} \cdot \dfrac{12x^3}{x+2}$

b) $\dfrac{x^2 - 1}{x+5} \div \dfrac{x^2 + 2x + 1}{2x+10}$

Solution

Multiplication of rational
expressions, p. 347

a) $\dfrac{x^2 - x - 6}{3x} \cdot \dfrac{12x^3}{x+2} = \dfrac{(x^2 - x - 6)(12x^3)}{3x(x+2)}$ Multiplying the numerators and the denominators

$$= \frac{(x-3)(x+2)(3x)(4x^2)}{3x(x+2)} \qquad \text{Factoring the numerator. Try to go directly to this step.}$$

$$= \frac{(x-3)(x+2)(3x)(4x^2)}{3x(x+2)} \qquad \text{Removing a factor equal to } 1: \frac{(x+2)(3x)}{(x+2)(3x)} = 1$$

$$= 4x^2(x-3)$$

Division of rational
expressions, p. 349

b) $\dfrac{x^2 - 1}{x+5} \div \dfrac{x^2 + 2x + 1}{2x+10} = \dfrac{x^2 - 1}{x+5} \cdot \dfrac{2x+10}{x^2 + 2x + 1}$ Multiplying by the reciprocal of the divisor

$$= \frac{(x+1)(x-1)(2)(x+5)}{(x+5)(x+1)(x+1)} \qquad \substack{\text{Multiplying rational} \\ \text{expressions and factoring numerators} \\ \text{and denominators}}$$

$$= \frac{(x+1)(x-1)(2)(x+5)}{(x+5)(x+1)(x+1)} \qquad \substack{\text{Removing a factor} \\ \text{equal to } 1: \\ \frac{(x+1)(x+5)}{(x+1)(x+5)} = 1}$$

$$= \frac{2(x-1)}{x+1} \qquad \substack{\text{We leave the} \\ \text{numerator in} \\ \text{factored form.}}$$

Addition and Subtraction of Rational Expressions

Like multiplication and division, addition and subtraction of rational expressions is similar to addition and subtraction of fractions.

> **The Sum and the Difference of Two Rational Expressions**
>
> To add when the denominators are the same, add the numerators and keep the same denominator:
>
> $$\frac{A}{B} + \frac{C}{B} = \frac{A + C}{B}.$$
>
> To subtract when the denominators are the same, subtract the second numerator from the first and keep the same denominator:
>
> $$\frac{A}{B} - \frac{C}{B} = \frac{A - C}{B}.$$

Example 4

Add and simplify, if possible:

$$\frac{x - 6}{x^2 - 6x + 5} + \frac{5}{x^2 - 6x + 5}.$$

Solution

$$\frac{x - 6}{x^2 - 6x + 5} + \frac{5}{x^2 - 6x + 5} = \frac{x - 6 + 5}{x^2 - 6x + 5} \qquad \text{Adding numerators}$$

$$= \frac{x - 1}{(x - 5)(x - 1)} \qquad \text{Factoring the denominator}$$

$$= \frac{1(x - 1)}{(x - 5)(x - 1)} \qquad \text{Removing a factor equal to 1:} \ \frac{x - 1}{x - 1} = 1$$

$$= \frac{1}{x - 5}$$

Addition of rational expressions, p. 354

Least common denominator, p. 356

Least common multiple, p. 356

When two rational expressions do not have a common denominator, we must rewrite them with a common denominator before we can add or subtract them. We generally rewrite them using their **least common denominator** (**LCD**), which is the **least common multiple** (**LCM**) of their denominators.

To Find the Least Common Denominator (LCD)

1. Write the prime factorization of each denominator.
2. Select one of the factorizations and inspect it to see if it contains the other.

   a) If it does, it represents the LCM of the denominators.
   b) If it does not, multiply that factorization by any factors of the other denominator that it lacks. The final product is the LCM of the denominators.

The LCD is the LCM of the denominators. It should contain each factor the greatest number of times that it occurs in any of the individual factorizations.

E x a m p l e   5   Find the LCD of $\dfrac{x-3}{x^2-1}$ and $\dfrac{4x^2}{x^2+4x+3}$.

Solution   The LCD of the rational expressions is the LCM of the denominators. We write the prime factorization of each denominator and construct the LCM:

$$x^2 - 1 = (x+1)(x-1);$$
$$x^2 + 4x + 3 = (x+1)(x+3).$$

The LCM must contain both factorizations. We select the factorization of $x^2 - 1$. It does not contain the factor $(x+3)$ from the factorization of $x^2 + 4x + 3$. We multiply $(x+1)(x-1)$ by $(x+3)$:

$$LCM = (x+1)(x-1)(x+3).$$

— $x^2 - 1$ is a factor of the LCM.

— $x^2 + 4x + 3$ is a factor of the LCM.

The LCD is $(x+1)(x-1)(x+3)$. We leave it in factored form.

To add or subtract two rational expressions with different denominators, we first find the LCD of the rational expressions. Then we multiply each rational expression by a form of 1 that is made up of the factors of the LCD missing from the denominator of that expression. The rational expressions will then have a common denominator, and we can add or subtract as before.

E x a m p l e   6

Add: $\dfrac{x-3}{x^2-1} + \dfrac{4x^2}{x^2+4x+3}$.

Solution   We found the LCD in Example 5; it is

$$(x+1)(x-1)(x+3).$$

The denominator $x^2 - 1 = (x+1)(x-1)$ must be multiplied by $x+3$ in order to obtain the LCD. The denominator $x^2 + 4x + 3 = (x+1)(x+3)$ must be multiplied by $x - 1$ in order to obtain the LCD. We multiply each expression by a form of 1 that is made up of these "missing" factors:

$$\frac{x-3}{x^2-1} + \frac{4x^2}{x^2+4x+3} = \frac{x-3}{(x+1)(x-1)} \cdot \frac{x+3}{x+3} + \frac{4x^2}{(x+1)(x+3)} \cdot \frac{x-1}{x-1}$$

$$= \frac{x^2-9}{(x+1)(x-1)(x+3)} + \frac{4x^3-4x^2}{(x+1)(x-1)(x+3)}$$

$$= \frac{4x^3-3x^2-9}{(x+1)(x-1)(x+3)}.$$

E x a m p l e   7

Subtraction of rational
expressions, p. 355

Subtract: $\dfrac{x}{x+2} - \dfrac{2x-3}{3x-4}$.

Solution   We have

$$\frac{x}{x+2} - \frac{2x-3}{3x-4}$$

$$= \frac{x}{x+2} \cdot \frac{3x-4}{3x-4} - \frac{2x-3}{3x-4} \cdot \frac{x+2}{x+2}$$ The LCD is $(x+2)(3x-4)$.

$$= \frac{3x^2-4x}{(x+2)(3x-4)} - \frac{2x^2+x-6}{(x+2)(3x-4)}$$ Multiplying out the numerators (but not the denominators)

$$= \frac{3x^2-4x-(2x^2+x-6)}{(x+2)(3x-4)}$$ Parentheses are important.

$$= \frac{3x^2-4x-2x^2-x+6}{(x+2)(3x-4)}$$ Removing parentheses in the numerator; subtracting every term

$$= \frac{x^2-5x+6}{(x+2)(3x-4)}$$

$$= \frac{(x-2)(x-3)}{(x+2)(3x-4)}.$$ Factoring the numerator in hopes of simplifying. There are no common factors.

The result could be written as either of the last two expressions.

Factors that are opposites,
p. 367

When denominators are opposites, we can find a common denominator by multiplying either rational expression by $-1/-1$.

Example 8

Add: $\dfrac{a}{a-b} + \dfrac{5}{b-a}$.

Solution

$$\frac{a}{a-b} + \frac{5}{b-a} = \frac{a}{a-b} + \frac{5}{b-a} \cdot \frac{-1}{-1}$$
 Writing 1 as $-1/-1$ and multiplying to obtain a common denominator

$$= \frac{a}{a-b} + \frac{-5}{a-b}$$
 $(b-a)(-1) = -b + a = a - b$

$$= \frac{a-5}{a-b}$$

Complex Rational Expressions

Complex rational expressions,
Section 6.5

A **complex rational expression** is a rational expression that has one or more rational expressions within its numerator or denominator. We will consider two methods for simplifying complex rational expressions. The first involves writing the expression as a quotient of two rational expressions.

> *To Simplify a Complex Rational Expression by Dividing*
> 1. Add or subtract, as needed, to get a single rational expression in the numerator.
> 2. Add or subtract, as needed, to get a single rational expression in the denominator.
> 3. Divide the numerator by the denominator (invert and multiply).
> 4. If possible, simplify by removing a factor equal to 1.

Example 9

Simplify by dividing: $\dfrac{\dfrac{2}{x+1}}{\dfrac{1}{x+2}+\dfrac{1}{x}}$.

Solution

1) There is already a single rational expression in the numerator.

2) We add to get a single rational expression in the denominator:

$$\frac{\dfrac{2}{x+1}}{\dfrac{1}{x+2}+\dfrac{1}{x}}=\frac{\dfrac{2}{x+1}}{\dfrac{1}{x+2}\cdot\dfrac{x}{x}+\dfrac{1}{x}\cdot\dfrac{x+2}{x+2}}$$

Multiplying by 1 to get the LCD, $x(x+2)$, for the denominator

$$=\frac{\dfrac{2}{x+1}}{\dfrac{x}{x(x+2)}+\dfrac{x+2}{x(x+2)}}=\frac{\dfrac{2}{x+1}}{\dfrac{2x+2}{x(x+2)}}.$$

Adding in the denominator

3) Next, we invert and multiply:

$$\frac{\dfrac{2}{x+1}}{\dfrac{2x+2}{x(x+2)}}=\frac{2}{x+1}\div\frac{2x+2}{x(x+2)}=\frac{2}{x+1}\cdot\frac{x(x+2)}{2x+2}.$$

4) Simplifying, we have:

$$\frac{2}{(x+1)}\cdot\frac{x(x+2)}{2x+2}=\frac{2\cdot x(x+2)}{2(x+1)(x+1)}$$

Removing a factor equal to 1: $\frac{2}{2}=1$

$$=\frac{x(x+2)}{(x+1)^2}.$$

A second method for simplifying complex rational expressions involves multiplying by the LCD.

To Simplify a Complex Rational Expression by Multiplying by the LCD

1. Find the LCD of *all* rational expressions within the complex rational expression.
2. Multiply the complex rational expression by a factor equal to 1. Write 1 as the LCD over itself (LCD/LCD).
3. Distribute and simplify. No rational expressions should remain within the complex rational expression.
4. Factor and, if possible, simplify.

E x a m p l e   1 0

Simplify by multiplying by the LCD: $\dfrac{1 + \dfrac{2}{t}}{\dfrac{4}{t^2} - 1}$.

Solution

1) The denominators *within* the complex rational expression are t and t^2, so the LCD is t^2.

2) We multiply by a form of 1 using t^2/t^2:

$$\frac{1 + \dfrac{2}{t}}{\dfrac{4}{t^2} - 1} = \frac{1 + \dfrac{2}{t}}{\dfrac{4}{t^2} - 1} \cdot \frac{t^2}{t^2}.$$

3) We distribute and simplify:

$$\frac{1 + \dfrac{2}{t}}{\dfrac{4}{t^2} - 1} \cdot \frac{t^2}{t^2} = \frac{1 \cdot t^2 + \dfrac{2}{t} \cdot t^2}{\dfrac{4}{t^2} \cdot t^2 - 1 \cdot t^2}$$

$$= \frac{t^2 + 2t}{4 - t^2}.$$ No rational expression remains within the numerator or denominator.

4) Finally, we simplify:

$$\frac{t^2 + 2t}{4 - t^2} = \frac{t(t + 2)}{(2 + t)(2 - t)}$$ Factoring and simplifying; $\dfrac{t + 2}{t + 2} = 1$

$$= \frac{t}{2 - t}.$$

Solving Rational Equations

Solving rational equations, Section 6.6

A **rational equation** is an equation containing one or more rational expressions, often with the variable in a denominator.

To Solve a Rational Equation

1. List any restrictions that exist. No possible solution can make a denominator equal 0.
2. Clear the equation of fractions by multiplying both sides by the LCD of all rational expressions in the equation.
3. Solve the resulting equation using the addition principle, the multiplication principle, and the principle of zero products, as needed.
4. Check the possible solution(s) in the original equation.

Example 11   Solve: $x + \dfrac{10}{x} = 7$.

Solution   First we note that x cannot be 0. The LCD is x, so we multiply both sides by x:

$$x + \frac{10}{x} = 7$$

$$x\left(x + \frac{10}{x}\right) = 7x \qquad\qquad \text{Don't forget the parentheses!}$$

$$x \cdot x + x \cdot \frac{10}{x} = 7x \qquad\qquad \text{Using the distributive law}$$

$$x^2 + 10 = 7x \qquad\qquad \text{We have a quadratic equation.}$$

$$x^2 - 7x + 10 = 0 \qquad\qquad \text{Getting 0 on one side}$$

$$(x - 2)(x - 5) = 0 \qquad\qquad \text{Factoring}$$

$$x - 2 = 0 \quad or \quad x - 5 = 0 \qquad\qquad \text{Using the principle of zero products}$$

$$x = 2 \quad or \qquad\quad x = 5.$$

Check: For 2:   For 5:

$$x + \frac{10}{x} = 7 \qquad\qquad\qquad x + \frac{10}{x} = 7$$

$$\overline{} \qquad\qquad\qquad \overline{}$$

$$2 + \frac{10}{2} \; ? \; 7 \qquad\qquad\qquad 5 + \frac{10}{5} \; ? \; 7$$

$$2 + 5 \qquad\qquad\qquad\qquad\quad 5 + 2$$

$$7 \; \bigg| \; 7 \;\; \text{TRUE} \qquad\qquad\quad 7 \; \bigg| \; 7 \;\; \text{TRUE}$$

Both numbers check, so there are two solutions, 2 and 5.

Example 12   Solve: $\dfrac{1}{x - 3} + 2 = \dfrac{x - 2}{x - 3}$.

Solution   We note first that x cannot be 3. We have

$$\frac{1}{x - 3} + 2 = \frac{x - 2}{x - 3} \qquad\qquad \text{The LCD is } x - 3.$$

$$(x - 3)\left(\frac{1}{x - 3} + 2\right) = (x - 3)\frac{x - 2}{x - 3} \qquad \begin{array}{l}\text{Multiplying both}\\\text{sides by } x - 3\end{array}$$

$$(x - 3) \cdot \frac{1}{x - 3} + (x - 3) \cdot 2 = (x - 3)\frac{x - 2}{x - 3} \qquad \begin{array}{l}\text{Using the}\\\text{distributive law}\end{array}$$

$$1 + 2x - 6 = x - 2 \qquad\qquad \text{Simplifying}$$

$$2x - 5 = x - 2$$

$$x - 5 = -2$$

$$x = 3.$$

Note that we stated at first that x cannot be 3. A check would show that substituting 3 for x results in division by 0. Thus the equation has no solution.

We can use our equation-solving techniques to solve for a specified letter in a formula.

E x a m p l e   1 3

Formulas, Section 6.8

Astronomy. The formula

$$L = \frac{dR}{D - d},$$

where D is the diameter of the sun, d is the diameter of the earth, R is the earth's distance from the sun, and L is some fixed distance, is used in calculating when lunar eclipses occur. Solve for D.

Solution   We first clear fractions by multiplying by the LCD, which is $D - d$:

$$(D - d)L = (D - d)\frac{dR}{D - d}$$

$$(D - d)L = \frac{(D - d)dR}{D - d} \qquad \text{Removing a factor equal to 1: } \frac{D - d}{D - d} = 1$$

$$(D - d)L = dR.$$

We do *not* multiply the factors on the left since we wish to get D all alone. Instead we multiply both sides by $1/L$ and then add d:

$$D - d = \frac{dR}{L} \qquad \text{Multiplying by } \frac{1}{L}$$

$$D = \frac{dR}{L} + d. \qquad \text{Adding } d$$

We now have D all alone on one side of the equation. Since D does not appear on the other side, we have solved the formula for D.

Work problems, p. 384

Many problems translate to rational equations. *Work problems*, which involve the time that it takes to complete a task, can often be solved using the work principle.

> ### The Work Principle
>
> Suppose that A requires a units of time to complete a task and B requires b units of time to complete the same task. Then
>
> A works at a rate of $\dfrac{1}{a}$ tasks per unit of time,
>
> B works at a rate of $\dfrac{1}{b}$ tasks per unit of time, and
>
> A and B together work at a rate of $\dfrac{1}{a} + \dfrac{1}{b}$ tasks per unit of time.
>
> If A and B, working together, require t units of time to complete the task, then all three of the following equations hold:
>
> $$\frac{1}{a} \cdot t + \frac{1}{b} \cdot t = 1; \qquad \left(\frac{1}{a} + \frac{1}{b}\right)t = 1; \qquad \frac{1}{a} + \frac{1}{b} = \frac{1}{t}.$$

E x a m p l e   1 4

Drafting.   It takes Kerry 30 hr to draw a set of plans for a house. It takes Jesse 45 hr to draw the same set of plans. How long would it take Kerry and Jesse, working together, to draw the set of plans?

Solution

1. **Familiarize.**   We could make some guesses to help us understand the problem and then list our results in a table. We could also reason that if Kerry and Jesse each drew half the plans, it would take Kerry 15 hr and Jesse $22\frac{1}{2}$ hr. So the time it takes them working together should be between 15 and $22\frac{1}{2}$ hr. We let t = the time that it takes them to draw the plans, working together.

2. **Translate.**   We will use the work principle to translate the problem:

$$\frac{1}{a} \cdot t + \frac{1}{b} \cdot t = 1 \qquad \begin{array}{l} a \text{ is the time that it takes Kerry to draw the plans;} \\ b \text{ is the time that it takes Jesse to draw the plans.} \end{array}$$

$$\frac{t}{30} + \frac{t}{45} = 1.$$

3. **Carry out.**   We solve the equation:

$$\frac{t}{30} + \frac{t}{45} = 1$$

$$90\left(\frac{t}{30} + \frac{t}{45}\right) = 90 \cdot 1 \qquad \text{The LCD is } 2 \cdot 3 \cdot 3 \cdot 5, \text{ or } 90.$$

$$90 \cdot \frac{t}{30} + 90 \cdot \frac{t}{45} = 90$$

$$3t + 2t = 90$$

$$5t = 90$$

$$t = 18.$$

4. **Check.**   We note that, as predicted in the *Familiarize* step, the answer is between 15 and $22\frac{1}{2}$ hr. Also, if each works 18 hr, Kerry will do $\frac{18}{30}$ of the job and Jesse will do $\frac{18}{45}$ of the job, and

$$\frac{18}{30} + \frac{18}{45} = \frac{3}{5} + \frac{2}{5} = 1. \qquad \text{The entire job will be completed.}$$

5. **State.**   Together it will take them 18 hr to draw the plans.

Motion problems, p. 386

Problems that deal with distance, speed (or rate), and time, or **motion problems**, can often be translated using the distance formula $d = rt$.

Example 15

Driving time. Karen and Eva are each driving to a sales meeting. Because of road conditions, Karen is able to drive 15 mph faster than Eva. In the same time that it takes Karen to travel 120 mi, Eva travels only 90 mi. Find their speeds.

Solution

1. **Familiarize.** We let $t =$ the time, in hours, that is spent traveling and $r =$ Karen's speed, in mph. Then Eva's speed $= r - 15$. We set up a table.

$$d \quad = \quad r \quad \cdot \quad t$$

	Distance	Speed	Time
Karen	120	r	t
Eva	90	$r - 15$	t

2. **Translate.** From the distance formula, we have $t = d/r$, so we can replace the times in the table with expressions involving r.

	Distance	Speed	Time
Karen	120	r	$120/r$
Eva	90	$r - 15$	$90/(r - 15)$

Since the times are the same, we have the equation

$$\frac{120}{r} = \frac{90}{r - 15}.$$

3. **Carry out.** We solve the equation:

$$\frac{120}{r} = \frac{90}{r - 15}$$

$$r(r - 15)\frac{120}{r} = r(r - 15)\frac{90}{r - 15} \qquad \text{The LCD is } r(r - 15).$$

$$120(r - 15) = 90r \qquad \text{Simplifying}$$

$$120r - 1800 = 90r \qquad \text{Removing parentheses}$$

$$-1800 = -30r \qquad \text{Subtracting } 120r$$

$$60 = r. \qquad \text{Dividing both sides by } -30$$

4. **Check.** If $r = 60$, then $r - 15 = 45$. If Karen travels 120 mi at 60 mph, she will have traveled 2 hr. If Eva travels 90 mi at 45 mph, she will also have traveled 2 hr. Since the times are the same, the speeds check.

5. **State.** Karen is traveling at 60 mph, while Eva is traveling at 45 mph.

Ratio, p. 388

Proportion, p. 388

Another type of problem that translates to a rational equation involves proportions. A **ratio** of two quantities is their quotient. A **proportion** is an equation stating that two ratios are equal.

E x a m p l e   1 6

Baking.   Rob discovers there is $2\frac{1}{2}$ cups of pancake mix left in the box. The directions on the mix indicate that $1\frac{1}{3}$ cups of milk should be added to 2 cups of mix. How much milk should Rob add to the $2\frac{1}{2}$ cups of mix?

Solution   Since the problem translates directly to a proportion, we will not follow all five steps of the problem-solving process. We write the ratio of mix to milk in two ways:

$$\text{Mix} \longrightarrow \frac{2}{1\frac{1}{3}} = \frac{2\frac{1}{2}}{x} \longleftarrow \text{Mix}$$
$$\text{Milk} \longrightarrow \qquad\qquad \longleftarrow \text{Milk}$$

The LCD is $x\left(1\frac{1}{3}\right)$. We solve for x:

$$x\left(1\tfrac{1}{3}\right)\frac{2}{1\frac{1}{3}} = x\left(1\tfrac{1}{3}\right)\frac{2\frac{1}{2}}{x} \qquad \text{Multiplying by the LCD}$$

$$2x = \left(1\tfrac{1}{3}\right)\left(2\tfrac{1}{2}\right) \qquad \text{Simplifying}$$

$$2x = \tfrac{10}{3} \qquad \text{Converting to fractional notation and multiplying}$$

$$x = \tfrac{5}{3}. \qquad \text{Multiplying both sides by } \tfrac{1}{2} \text{ and simplifying}$$

Rob needs to add $\frac{5}{3}$, or $1\frac{2}{3}$ cups of milk.

FOR EXTRA HELP

Exercise Set   **R.6**

Digital Video Tutor CD: none
Videotape: none

 Math Tutor Center   MathXL   MyMathLab.com

List all numbers for which each rational expression is undefined.

1. $\dfrac{x - 7}{3x + 2}$

2. $\dfrac{10 - y}{-6y}$

3. $\dfrac{p^2 - 1}{p^2 - 100}$

4. $\dfrac{10x}{x^2 + 9x + 8}$

Simplify by removing a factor equal to 1.

5. $\dfrac{16x^2 y}{18xy^2}$

6. $\dfrac{2x + 10}{6x + 30}$

7. $\dfrac{t^2 - 2t - 8}{t^2 - 16}$

8. $\dfrac{a^3 + 2a^2 + a}{a^2 + 4a + 3}$

9. $\dfrac{2 - x}{x^2 - 4}$

10. $\dfrac{y - 8}{8 - y}$

Perform each indicated operation. Then, if possible, simplify.

11. $\dfrac{3x}{x + y} \cdot \dfrac{2x + 2y}{x^2}$

12. $\dfrac{5}{x + 7} \cdot \dfrac{x + 7}{10}$

13. $\dfrac{a^2 + 2a + 1}{a} \div \dfrac{a^2}{a^2 - 1}$

14. $\dfrac{x}{x + 3} + \dfrac{3 - x}{x + 3}$

15. $\dfrac{2x}{x - 7} - \dfrac{x + 7}{x - 7}$

16. $\dfrac{x}{x + y} \div \dfrac{y}{x + y}$

17. $\dfrac{5}{x} + \dfrac{6}{x^2}$

18. $\dfrac{x^2 + 4x + 3}{x^2 + x - 2} \cdot \dfrac{x^2 + 3x + 2}{x^2 + 2x - 3}$

19. $\dfrac{2a + b}{a - b} - \dfrac{4}{3a - 3b}$

20. $(x^2 - 16) \div \dfrac{4x + 16}{3x^2}$

21. $\dfrac{2 - x}{5x^2} \div \dfrac{x^2 - 4}{3x}$

22. $\dfrac{2x}{x - 5} + \dfrac{3}{x + 4}$

23. $\dfrac{x^3 + 2x^2 + x}{x^2 - 4} \cdot \dfrac{x^2 - x - 2}{x^4 + x^3}$

24. $\dfrac{-1}{x^2 + 7x + 10} - \dfrac{3}{x^2 + 8x + 15}$

25. $\dfrac{2}{(x + 1)^2} + \dfrac{1}{x + 1}$

26. $\dfrac{2x}{x^2 - 3x} \div (x - 3)$

27. $\dfrac{x - y}{2x} \cdot \dfrac{3x^2}{y - x}$

28. $\dfrac{1}{x + y} + \dfrac{2}{x^2 + y^2}$

29. $\dfrac{x - 2}{x + 5} - \dfrac{x + 3}{x - 4}$

30. $\dfrac{z^2 + 2z + 1}{8z} \div \dfrac{z^2 - z - 2}{4z^2 - 4}$

Simplify.

31. $\dfrac{\dfrac{2}{x} - \dfrac{1}{x^2}}{\dfrac{x}{4}}$

32. $\dfrac{\dfrac{x}{3} - \dfrac{3}{x}}{\dfrac{1}{x} + \dfrac{1}{3}}$

33. $\dfrac{\dfrac{3}{x - 7}}{\dfrac{4x + 3}{x + 1}}$

34. $\dfrac{\dfrac{a}{a - b}}{\dfrac{a^2}{a^2 - b^2}}$

35. $\dfrac{x - \dfrac{3}{x - 2}}{x - \dfrac{12}{x + 1}}$

36. $\dfrac{t + \dfrac{1}{t}}{t - \dfrac{2}{t}}$

37. $\dfrac{\dfrac{1}{2} - \dfrac{1}{x}}{\dfrac{2 - x}{2}}$

38. $\dfrac{\dfrac{x}{2y^2} + \dfrac{y}{3x^2}}{\dfrac{1}{6xy} + \dfrac{2}{x^2 y}}$

Solve.

39. $\dfrac{1}{2} + \dfrac{1}{3} = \dfrac{1}{t}$

40. $\dfrac{1}{4} + \dfrac{1}{t} = \dfrac{1}{3}$

41. $x + \dfrac{1}{x} = 2$

42. $\dfrac{x - 7}{x + 1} = \dfrac{2}{3}$

43. $\dfrac{3}{y + 7} = \dfrac{1}{y - 8}$

44. $\dfrac{x + 1}{x - 2} = \dfrac{3}{x - 2}$

45. $\dfrac{1}{x - 3} - \dfrac{x - 4}{x^2 - 9} = 1$

46. $\dfrac{3}{a + 4} = \dfrac{a - 1}{4 - a}$

47. *Painting.* Quentin can paint the turret on a Queen Anne house in 40 hr. It takes Austin 50 hr to paint the same turret. How long would it take them, working together, to paint the turret?

48. *Building fences.* Lindsay can build a fence in 6 hr. Laura can do the same job in 5 hr. How long will it take them, working together, to build the fence?

49. *Snowmobiling.* Jessica can ride her snowmobile through the fields 20 km/h faster than Josh can ride his through the woods. In the time it takes Jessica to ride 18 km, Josh travels 10 km. Find the speed of each snowmobile.

50. *Bicycling.* Ani bicycles 8 mi and Lia bicycles 12 mi to meet at a park for lunch. Because Ani's trip is mostly uphill, she rides 5 mph slower than Lia. Ani and Lia leave their homes at the same time and arrive at the park at the same time. Find the speed of each bicyclist.

51. *Elk population.* To determine the size of a park's elk population, rangers tag 15 elk and set them free. Months later, 40 elk are caught, of which 12 have tags. Estimate the size of the elk population.

52. *Manufacturing pegs.* A sample of 136 wooden pegs contained 17 defective pegs. How many defective pegs would you expect in a sample of 840 pegs?

Solve.

53. $f = \dfrac{gm - t}{m}$, for m

54. $d = \dfrac{s_1 - s_2}{t}$, for t

55. $\dfrac{1}{R} = \dfrac{1}{r_1} + \dfrac{1}{r_2}$, for r_2

56. $S = \dfrac{a}{1 - r}$, for r

Appendixes

A

Naming Sets • Membership • Subsets • Intersections • Unions

The notion of a "set" is used frequently in mathematics. We provide a basic introduction to sets in this appendix.

Naming Sets

To name the set of whole numbers less than 6, we can use *roster notation*, as follows:

$$\{0, 1, 2, 3, 4, 5\}.$$

The set of real numbers x for which x is less than 6 cannot be named by listing all its members because there is an infinite number of them. We name such a set using *set-builder notation*, as follows:

$$\{x \mid x < 6\}.$$

This is read

"The set of all x such that x is less than 6."

See Section 2.6 for more on this notation.

Membership

The symbol $\in$ means *is a member of* or *belongs to*, or *is an element of*. Thus,

$$x \in A$$

means

$$x \text{ is a member of } A, \quad \text{or} \quad x \text{ belongs to } A, \quad \text{or} \quad x \text{ is an element of } A.$$

E x a m p l e   1

Classify each of the following as true or false.

a) $1 \in \{1, 2, 3\}$
b) $1 \in \{2, 3\}$
c) $4 \in \{x \mid x \text{ is an even whole number}\}$
d) $5 \in \{x \mid x \text{ is an even whole number}\}$

Solution

a) Since 1 is listed as a member of the set, $1 \in \{1, 2, 3\}$ is true.
b) Since 1 is *not* a member of $\{2, 3\}$, the statement $1 \in \{2, 3\}$ is false.
c) Since 4 is an even whole number, $4 \in \{x \mid x \text{ is an even whole number}\}$ is true.
d) Since 5 is *not* even, $5 \in \{x \mid x \text{ is an even whole number}\}$ is false.

Set membership can be illustrated with a diagram, as shown below.

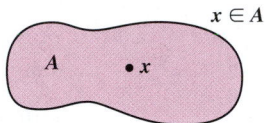

Subsets

If every element of A is also an element of B, then A is a *subset* of B. This is denoted $A \subseteq B$.

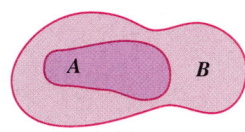

The set of whole numbers is a subset of the set of integers. The set of rational numbers is a subset of the set of real numbers.

E x a m p l e   2

Classify each of the following as true or false.

a) $\{1, 2\} \subseteq \{1, 2, 3, 4\}$
b) $\{p, q, r, w\} \subseteq \{a, p, r, z\}$
c) $\{x \mid x < 6\} \subseteq \{x \mid x \leq 11\}$

Solution

a) Since every element of $\{1, 2\}$ is in the set $\{1, 2, 3, 4\}$, it follows that $\{1, 2\} \subseteq \{1, 2, 3, 4\}$ is true.
b) Since $q \in \{p, q, r, w\}$, but $q \notin \{a, p, r, z\}$, it follows that $\{p, q, r, w\} \subseteq \{a, p, r, z\}$ is false.
c) Since every number that is less than 6 is also less than 11, the statement $\{x \mid x < 6\} \subseteq \{x \mid x \leq 11\}$ is true.

Intersections

The *intersection* of sets A and B, denoted $A \cap B$, is the set of members common to both sets.

E x a m p l e   3

Find each intersection.

a) $\{0, 1, 3, 5, 25\} \cap \{2, 3, 4, 5, 6, 7, 9\}$ **b)** $\{a, p, q, w\} \cap \{p, q, t\}$

Solution

a) $\{0, 1, 3, 5, 25\} \cap \{2, 3, 4, 5, 6, 7, 9\} = \{3, 5\}$

b) $\{a, p, q, w\} \cap \{p, q, t\} = \{p, q\}$

Set intersection can be illustrated with a diagram, as shown below.

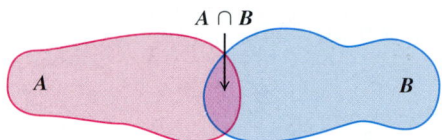

$A \cap B$

A B

The set without members is known as the *empty set*, and is written $\varnothing$, and sometimes $\{\ \}$. Each of the following is a description of the empty set:

The set of all 12-ft–tall people;

$\{2, 3\} \cap \{5, 6, 7\}$;

$\{x \,|\, x$ is an even natural number$\} \cap \{x \,|\, x$ is an odd natural number$\}$.

Unions

Two sets A and B can be combined to form a set that contains the members of both A and B. The new set is called the *union* of A and B, denoted $A \cup B$.

E x a m p l e   4

Find each union.

a) $\{0, 5, 7, 13, 27\} \cup \{0, 2, 3, 4, 5\}$ **b)** $\{a, c, e, g\} \cup \{b, d, f\}$

Solution

a) $\{0, 5, 7, 13, 27\} \cup \{0, 2, 3, 4, 5\} = \{0, 2, 3, 4, 5, 7, 13, 27\}$

Note that the 0 and the 5 are *not* listed twice in the solution.

b) $\{a, c, e, g\} \cup \{b, d, f\} = \{a, b, c, d, e, f, g\}$

Set union can be illustrated with a diagram, as shown below.

$A \cup B$ is shaded.

A B

FOR EXTRA HELP

Exercise Set A

Digital Video Tutor CD: none   InterAct Math   Math Tutor Center   MathXL   MyMathLab.com
Videotape: none

Name each set using the roster method.

1. The set of whole numbers 3 through 7

2. The set of whole numbers 83 through 89

3. The set of odd numbers between 40 and 50

4. The set of multiples of 5 between 10 and 40

5. $\{x \mid$ the square of x is 9$\}$

6. $\{x \mid x$ is the cube of 0.2$\}$

Classify each statement as true or false.

7. $2 \in \{x \mid x$ is an odd number$\}$

8. $7 \in \{x \mid x$ is an odd number$\}$

9. Bruce Springsteen $\in$ The set of all rock stars

10. Apple $\in$ The set of all fruit

11. $-3 \in \{-4, -3, 0, 1\}$

12. $0 \in \{-4, -3, 0, 1\}$

13. $\frac{2}{3} \in \{x \mid x$ is a rational number$\}$

14. Heads $\in$ The set of outcomes of flipping a penny

15. $\{4, 5, 8\} \subseteq \{1, 3, 4, 5, 6, 7, 8, 9\}$

16. The set of vowels $\subseteq$ The set of consonants

17. $\{-1, -2, -3, -4, -5\} \subseteq \{-1, 2, 3, 4, 5\}$

18. The set of integers $\subseteq$ The set of rational numbers

Find each intersection.

19. $\{a, b, c, d, e\} \cap \{c, d, e, f, g\}$

20. $\{a, e, i, o, u\} \cap \{q, u, i, c, k\}$

21. $\{1, 2, 5, 10\} \cap \{0, 1, 7, 10\}$

22. $\{0, 1, 7, 10\} \cap \{0, 1, 2, 5\}$

23. $\{1, 2, 5, 10\} \cap \{3, 4, 7, 8\}$

24. $\{a, e, i, o, u\} \cap \{m, n, f, g, h\}$

Find each union.

25. $\{a, e, i, o, u\} \cup \{q, u, i, c, k\}$

26. $\{a, b, c, d, e\} \cup \{c, d, e, f, g\}$

27. $\{0, 1, 7, 10\} \cup \{0, 1, 2, 5\}$

28. $\{1, 2, 5, 10\} \cup \{0, 1, 7, 10\}$

29. $\{a, e, i, o, u\} \cup \{m, n, f, g, h\}$

30. $\{1, 2, 5, 10\} \cup \{a, b\}$

31. What advantage(s) does set-builder notation have over roster notation?

32. What advantage(s) does roster notation have over set-builder notation?

SYNTHESIS

33. Find the union of the set of integers and the set of whole numbers.

34. Find the intersection of the set of odd integers and the set of even integers.

35. Find the union of the set of rational numbers and the set of irrational numbers.

36. Find the intersection of the set of even integers and the set of positive rational numbers.

37. Find the intersection of the set of rational numbers and the set of irrational numbers.

38. Find the union of the set of negative integers, the set of positive integers, and the set containing 0.

39. For a set A, find each of the following.
   a) $A \cup \varnothing$
   b) $A \cup A$
   c) $A \cap A$
   d) $A \cap \varnothing$

40. A set is *closed* under an operation if, when the operation is performed on its members, the result is in the set. For example, the set of real numbers is closed under the operation of addition since the sum of any two real numbers is a real number.
   a) Is the set of even numbers closed under addition?
   b) Is the set of odd numbers closed under addition?
   c) Is the set $\{0, 1\}$ closed under addition?
   d) Is the set $\{0, 1\}$ closed under multiplication?
   e) Is the set of real numbers closed under multiplication?
   f) Is the set of integers closed under division?

41. Experiment with sets of various types and determine whether the following distributive law for sets is true:

$$A \cap (B \cup C) = (A \cap B) \cup (A \cap C).$$

B

Synthetic Division

Streamlining Long Division • The Remainder Theorem

Streamlining Long Division

To divide a polynomial by a binomial of the type $x - a$, we can streamline the usual procedure to develop a process called *synthetic division*.

Compare the following. In each stage, we attempt to write a bit less than in the previous stage, while retaining enough essentials to solve the problem. At the end, we will return to the usual polynomial notation.

Stage 1

When a polynomial is written in descending order, the coefficients provide the essential information:

$$
\begin{array}{r}
4x^2 + 5x + 11 \\
x - 2\overline{)4x^3 - 3x^2 + x + 7} \\
\underline{4x^3 - 8x^2} \\
5x^2 + x \\
\underline{5x^2 - 10x} \\
11x + 7 \\
\underline{11x - 22} \\
29
\end{array}
\qquad
\begin{array}{r}
4 + 5 + 11 \\
1 - 2\overline{)4 - 3 + 1 + 7} \\
\underline{4 - 8} \\
5 + 1 \\
\underline{5 - 10} \\
11 + 7 \\
\underline{11 - 22} \\
29
\end{array}
$$

Because the leading coefficient in the divisor is 1, each time we multiply the divisor by a term in the answer, the leading coefficient of that product duplicates a coefficient in the answer. In the next stage, we don't bother to duplicate these numbers. We also show where -2 is used and drop the 1 from the divisor.

Stage 2

$$
\begin{array}{r}
4x^2 + 5x + 11 \\
x - 2\overline{)4x^3 - 3x^2 + x + 7} \\
\underline{4x^3 - 8x^2} \\
5x^2 + x \\
\underline{5x^2 - 10x} \\
11x + 7 \\
\underline{11x - 22} \\
29
\end{array}
$$

$$
\begin{array}{r}
4 + 5 + 11 \\
-2\overline{)4 - 3 + 1 + 7} \\
- 8 \qquad \leftarrow\text{Multiply: } -2 \cdot 4 = -8. \\
5 + 1 \qquad \text{Subtract: } -3 - (-8) = 5. \\
- 10 \qquad \leftarrow\text{Multiply: } -2 \cdot 5 = -10. \\
11 + 7 \qquad \text{Subtract: } 1 - (-10) = 11. \\
- 22 \leftarrow\text{Multiply: } -2 \cdot 11 = -22. \\
29 \leftarrow\text{Subtract: } 7 - (-22) = 29.
\end{array}
$$

To simplify further, we now reverse the sign of the -2 in the divisor and, in exchange, *add* at each step in the long division.

Stage 3

$$
\begin{array}{r}
4x^2 + 5x + 11 \\
x - 2\overline{)4x^3 - 3x^2 + x + 7} \\
\underline{4x^3 - 8x^2} \\
5x^2 + x \\
\underline{5x^2 - 10x} \\
11x + 7 \\
\underline{11x - 22} \\
29
\end{array}
$$

$$
\begin{array}{r}
4 + 5 + 11 \\
2\overline{)4 - 3 + 1 + 7} \\
8 \\
5 + 1 \\
10 \\
11 + 7 \\
22 \\
29
\end{array}
$$

Replace the -2 with 2.

Multiply: $2 \cdot 4 = 8$.

Add: $-3 + 8 = 5$.

Multiply: $2 \cdot 5 = 10$.

Add: $1 + 10 = 11$.

Multiply: $2 \cdot 11 = 22$.

Add: $7 + 22 = 29$.

The blue numbers can be eliminated if we look at the red numbers instead.

Stage 4

$$
\begin{array}{r}
4x^2 + 5x + 11 \\
x - 2\overline{)4x^3 - 3x^2 + x + 7} \\
\underline{4x^3 - 8x^2} \\
5x^2 + x \\
\underline{5x^2 - 10x} \\
11x + 7 \\
\underline{11x - 22} \\
29
\end{array}
$$

$$
\begin{array}{rrrr}
4 & 5 & 11 & \\
2\overline{)4} & -3 & 1 & 7 \\
& 8 & 10 & 22 \\
& 5 & 11 & 29
\end{array}
$$

Don't lose sight of how the products 8, 10, and 22 are found. Also, note that the 5 and 11 preceding the remainder 29 coincide with the 5 and 11 following the 4 on the top line. By writing a 4 to the left of 5 on the bottom line, we can eliminate the top line in stage 4 and read our answer from the bottom line. This final stage is commonly called **synthetic division**.

Stage 5

$$
\begin{array}{rrrr}
4 & 5 & 11 & \\
2\overline{)4} & -3 & 1 & 7 \\
& 8 & 10 & 22 \\
& 5 & 11 & 29
\end{array}
$$

$$
\begin{array}{r|rrrr}
2 & 4 & -3 & 1 & 7 \\
& & 8 & 10 & 22 \\
\hline
& 4 & 5 & 11 & 29
\end{array}
$$

This is the remainder.

This is the zero-degree coefficient.

This is the first-degree coefficient.

This is the second-degree coefficient.

The quotient is $4x^2 + 5x + 11$. The remainder is 29.

> Remember that in order for this method to work, the divisor must be of the form $x - a$, that is, a variable minus a constant. The coefficient of the variable must be 1.

E x a m p l e   1   Use synthetic division to divide: $(x^3 + 6x^2 - x - 30) \div (x - 2)$.

Solution

$$2 \,\rfloor\, 1 \quad 6 \quad -1 \quad -30$$

Write the 2 of $x - 2$ and the coefficients of the dividend.

$$1$$

Bring down the first coefficient.

$$2 \,\rfloor\, 1 \quad 6 \quad -1 \quad -30$$
$$2$$
$$1 \quad 8$$

Multiply 1 by 2 to get 2.
Add 6 and 2.

$$2 \,\rfloor\, 1 \quad 6 \quad -1 \quad -30$$
$$2 \quad 16$$
$$1 \quad 8 \quad 15$$

Multiply 8 by 2.
Add -1 and 16.

$$2 \,\rfloor\, 1 \quad 6 \quad -1 \quad -30$$
$$2 \quad 16 \quad 30$$
$$1 \quad 8 \quad 15 \,\rvert\, 0$$

Multiply 15 by 2 and add.

The answer is $x^2 + 8x + 15$ with R0, or just $x^2 + 8x + 15$.

E x a m p l e   2   Use synthetic division to divide.

a) $(2x^3 + 7x^2 - 5) \div (x + 3)$
b) $(10x^2 - 13x + 3x^3 - 20) \div (4 + x)$

Solution

a) $(2x^3 + 7x^2 - 5) \div (x + 3)$

The dividend has no x-term, so we need to write 0 for its coefficient of x. Note that $x + 3 = x - (-3)$, so we write -3 inside the $\rfloor$.

$$-3 \,\rfloor\, 2 \quad 7 \quad 0 \quad -5$$
$$-6 \quad -3 \quad 9$$
$$2 \quad 1 \quad -3 \,\rvert\, 4$$

The answer is $2x^2 + x - 3$, with R4, or $2x^2 + x - 3 + \dfrac{4}{x + 3}$.

b) We first rewrite $(10x^2 - 13x + 3x^3 - 20) \div (4 + x)$ in descending order:

$$(3x^3 + 10x^2 - 13x - 20) \div (x + 4).$$

Next, we use synthetic division. Note that $x + 4 = x - (-4)$.

$$-4 \,\rfloor\, 3 \quad 10 \quad -13 \quad -20$$
$$-12 \quad 8 \quad 20$$
$$3 \quad -2 \quad -5 \,\rvert\, 0$$

The answer is $3x^2 - 2x - 5$.

The Remainder Theorem

Because the remainder is 0, Example 1 shows that $x - 2$ is a factor of $x^3 + 6x^2 - x - 30$ and that we can write $x^3 + 6x^2 - x - 30$ as $(x - 2)(x^2 + 8x + 15)$. Using this result and the principle of zero products, we know that if $f(x) = x^3 + 6x^2 - x - 30$, then $f(2) = 0$ (since $x - 2$ is a factor of $f(x)$). Similarly, from Example 2(b), we know that $x + 4$ is a factor of $g(x) = 10x^2 - 13x + 3x^3 - 20$. This tells us that $g(-4) = 0$. In both examples, the remainder from the division, 0, can serve as a function value. Remarkably, this pattern extends to nonzero remainders. To see this, note that the remainder in Example 2(a) is 4, and if $f(x) = 2x^3 + 7x^2 - 5$, then $f(-3)$ is also 4 (you should check this). The fact that the remainder and the function value coincide is predicted by the remainder theorem, which follows.

> ### The Remainder Theorem
> The remainder obtained by dividing $P(x)$ by $x - r$ is $P(r)$.

A proof of this result is outlined in Exercise 23.

Example 3

Let $f(x) = 8x^5 - 6x^3 + x - 8$. Use synthetic division to find $f(2)$.

Solution The remainder theorem tells us that $f(2)$ is the remainder when $f(x)$ is divided by $x - 2$. We use synthetic division to find that remainder:

$$
\begin{array}{r|rrrrrr}
2 & 8 & 0 & -6 & 0 & 1 & -8 \\
  &   & 16 & 32 & 52 & 104 & 210 \\
\hline
  & 8 & 16 & 26 & 52 & 105 & 202
\end{array}
$$

Although the bottom line can be used to find the quotient for the division $(8x^5 - 6x^3 + x - 8) \div (x - 2)$, what we are really interested in is the remainder. It tells us that $f(2) = 202$.

Exercise Set **B**

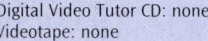

Use synthetic division to divide.

1. $(x^3 - 2x^2 + 2x - 7) \div (x + 1)$

2. $(x^3 - 2x^2 + 2x - 7) \div (x - 1)$

3. $(a^2 + 8a + 11) \div (a + 3)$

4. $(a^2 + 8a + 11) \div (a + 5)$

5. $(x^3 - 7x^2 - 13x + 3) \div (x - 2)$

6. $(x^3 - 7x^2 - 13x + 3) \div (x + 2)$

7. $(3x^3 + 7x^2 - 4x + 3) \div (x + 3)$

8. $(3x^3 + 7x^2 - 4x + 3) \div (x - 3)$

9. $(y^3 - 3y + 10) \div (y - 2)$

10. $(x^3 - 2x^2 + 8) \div (x + 2)$

11. $(x^5 - 32) \div (x - 2)$

12. $(y^5 - 1) \div (y - 1)$

13. $(3x^3 + 1 - x + 7x^2) \div \left(x + \frac{1}{3}\right)$

14. $(8x^3 - 1 + 7x - 6x^2) \div \left(x - \frac{1}{2}\right)$

Use synthetic division to find the indicated function value.

15. $f(x) = 5x^4 + 12x^3 + 28x + 9;\ f(-3)$

16. $g(x) = 3x^4 - 25x^2 - 18;\ g(3)$

17. $P(x) = 6x^4 - x^3 - 7x^2 + x + 2;\ P(-1)$

18. $F(x) = 3x^4 + 8x^3 + 2x^2 - 7x - 4;\ F(-2)$

19. $f(x) = x^4 - x^3 - 19x^2 + 49x - 30;\ f(4)$

20. $p(x) = x^4 + 7x^3 + 11x^2 - 7x - 12;\ p(2)$

SYNTHESIS

21. Why is it that we *add* when performing synthetic division, but *subtract* when performing long division?

22. Explain how synthetic division could be useful when factoring a polynomial.

23. To prove the remainder theorem, note that any polynomial $P(x)$ can be rewritten as $(x - r) \cdot Q(x) + R$, where $Q(x)$ is the quotient polynomial that arises when $P(x)$ is divided by $x - r$, and R is some constant (the remainder).
   a) How do we know that R must be a constant?
   b) Show that $P(r) = R$ (this says that $P(r)$ is the remainder when $P(x)$ is divided by $x - r$).

24. Let $f(x) = 4x^3 + 16x^2 - 3x - 45$. Find $f(-3)$ and then solve the equation $f(x) = 0$.

25. Let $f(x) = 6x^3 - 13x^2 - 79x + 140$. Find $f(4)$ and then solve the equation $f(x) = 0$.

Nested evaluation.   One way to evaluate a polynomial function like $P(x) = 3x^4 - 5x^3 + 4x^2 - 1$ is to successively factor out x as shown:

$$P(x) = x(x(x(3x - 5) + 4) + 0) - 1.$$

Computations are then performed using this "nested" form of $P(x)$.

26. Use nested evaluation to find $f(-3)$ in Exercise 24. Note the similarities to the calculations performed with synthetic division.

27. Use nested evaluation to find $f(4)$ in Exercise 25. Note the similarities to the calculations performed with synthetic division.

Tables

TABLE 1 Fractional and Decimal Equivalents

Fractional Notation	$\frac{1}{10}$	$\frac{1}{8}$	$\frac{1}{6}$	$\frac{1}{5}$	$\frac{1}{4}$	$\frac{3}{10}$	$\frac{1}{3}$	$\frac{3}{8}$	$\frac{2}{5}$	$\frac{1}{2}$
Decimal Notation	0.1	0.125	$0.16\overline{6}$	0.2	0.25	0.3	$0.333\overline{3}$	0.375	0.4	0.5
Percent Notation	10%	12.5% or $12\frac{1}{2}\%$	$16.6\overline{6}\%$ or $16\frac{2}{3}\%$	20%	25%	30%	$33.3\overline{3}\%$ or $33\frac{1}{3}\%$	37.5% or $37\frac{1}{2}\%$	40%	50%
Fractional Notation	$\frac{3}{5}$	$\frac{5}{8}$	$\frac{2}{3}$	$\frac{7}{10}$	$\frac{3}{4}$	$\frac{4}{5}$	$\frac{5}{6}$	$\frac{7}{8}$	$\frac{9}{10}$	$\frac{1}{1}$
Decimal Notation	0.6	0.625	$0.666\overline{6}$	0.7	0.75	0.8	$0.83\overline{3}$	0.875	0.9	1
Percent Notation	60%	62.5% or $62\frac{1}{2}\%$	$66.6\overline{6}\%$ or $66\frac{2}{3}\%$	70%	75%	80%	$83.3\overline{3}\%$ or $83\frac{1}{3}\%$	87.5% or $87\frac{1}{2}\%$	90%	100%

TABLE 2   Squares and Square Roots with Approximations to Three Decimal Places

N	$\sqrt{N}$	N^2	N	$\sqrt{N}$	N^2	N	$\sqrt{N}$	N^2	N	$\sqrt{N}$	N^2
1	1	1	26	5.099	676	51	7.141	2601	76	8.718	5776
2	1.414	4	27	5.196	729	52	7.211	2704	77	8.775	5929
3	1.732	9	28	5.292	784	53	7.280	2809	78	8.832	6084
4	2	16	29	5.385	841	54	7.348	2916	79	8.888	6241
5	2.236	25	30	5.477	900	55	7.416	3025	80	8.944	6400
6	2.449	36	31	5.568	961	56	7.483	3136	81	9	6561
7	2.646	49	32	5.657	1024	57	7.550	3249	82	9.055	6724
8	2.828	64	33	5.745	1089	58	7.616	3364	83	9.110	6889
9	3	81	34	5.831	1156	59	7.681	3481	84	9.165	7056
10	3.162	100	35	5.916	1225	60	7.746	3600	85	9.220	7225
11	3.317	121	36	6	1296	61	7.810	3721	86	9.274	7396
12	3.464	144	37	6.083	1369	62	7.874	3844	87	9.327	7569
13	3.606	169	38	6.164	1444	63	7.937	3969	88	9.381	7744
14	3.742	196	39	6.245	1521	64	8	4096	89	9.434	7921
15	3.873	225	40	6.325	1600	65	8.062	4225	90	9.487	8100
16	4	256	41	6.403	1681	66	8.124	4356	91	9.539	8281
17	4.123	289	42	6.481	1764	67	8.185	4489	92	9.592	8464
18	4.243	324	43	6.557	1849	68	8.246	4624	93	9.644	8649
19	4.359	361	44	6.633	1936	69	8.307	4761	94	9.695	8836
20	4.472	400	45	6.708	2025	70	8.367	4900	95	9.747	9025
21	4.583	441	46	6.782	2116	71	8.426	5041	96	9.798	9216
22	4.690	484	47	6.856	2209	72	8.485	5184	97	9.849	9409
23	4.796	529	48	6.928	2304	73	8.544	5329	98	9.899	9604
24	4.899	576	49	7	2401	74	8.602	5476	99	9.950	9801
25	5	625	50	7.071	2500	75	8.660	5625	100	10	10,000

Glossary

A

absolute value [1.4] The absolute value of a number is its distance from 0 on the number line.

algebraic expression [1.1] An algebraic expression consists of variables and/or numerals, often with operation signs and grouping symbols.

arithmetic sequence [14.2] A sequence in which the difference between any two successive terms is constant is called an arithmetic sequence.

arithmetic series [14.2] An arithmetic series is a series whose associated sequence is arithmetic.

ascending order [4.3] When a polynomial is written with the terms arranged according to degree, from least to greatest, it is said to be in ascending order.

associative [1.2] The associative laws state that numbers can be grouped in any manner for both addition and multiplication.

axes [3.1] Two perpendicular number lines used to identify points in a plane are axes.

axis of symmetry [11.6] The axis of symmetry is a line that can be drawn through a graph such that the part of the graph on one side of the line is an exact reflection of the part on the opposite side.

B

bar graph [3.1] A bar graph is a graphic means of displaying data using bars proportional in length to the numbers represented.

base [1.8] The base in exponential notation is the number being raised to a power.

binomial [4.2] A polynomial that is composed of two terms is called a binomial.

branches [13.3] The two curves that comprise a hyperbola are called branches.

break even point [8.8] In business, the break-even point is the point of intersection of the revenue function and the cost function.

C

circle [13.1] The circle is a set of points in a plane that are a fixed distance r, called the radius, from a fixed point (h, k), called the center.

circle graph [3.1] A circle graph is a graphic means of displaying data using sectors of a circle to represent percents.

circumference [2.3] The circumference is the distance around a circle.

closed interval [a, b] [9.1] Closed interval is defined as the set of all numbers x for which $a \leq x \leq b$. Thus, $[a, b] = \{x \mid a \leq x \leq b\}$

coefficient [2.1] Coefficient is the number preceding a variable in an algebraic expression.

combinations [14.4] The study of the theory of counting is called combinations.

combination [14.5] A combination of r objects chosen from a set of n objects is a subset of the set of n objects.

combined variation [7.5] When a variable varies directly and/or inversely, at the same time, with more than one other variable, there is combined variation.

common logarithm [12.5] Logarithms with base 10 are common logarithms.

commutative [1.2] The commutative laws state that changing the order of addition or multiplication does not affect the answer.

completing the square [11.1] To complete the square for an expression like $ax^2 + bx$, add half of the coefficient of x, squared.

complex number [10.8] A complex number is any number that can be written as $a + bi$, where a and b are real numbers.

complex number system [10.8] The complex number system is a number system that contains the real number system and is designed so that negative numbers do have square roots.

complex rational expression [6.5] A complex rational expression is a rational expression that has one or more rational expressions within its numerator or denominator.

composite [1.3] A natural number, other than 1, that is not prime is called composite.

composite function [12.1] A function in which a quantity depends on a variable that, in turn, depends on another variable is a composite function.

compound inequality [9.2] Compound inequalities are inequalities in sentences that are formed using the word *and* or the word *or*.

compound interest [11.1] Compound interest is interest computed on the sum of an original principal and accrued interest.

conditional equation [6.8] If an equation is true for some replacements and false for others, it is a conditional equation.

conic sections [13.1] Conic sections are plane curves having second-degree equations.

conjugates [10.5] Pairs of radical terms like $\sqrt{a} + \sqrt{b}$ and $\sqrt{a} - \sqrt{b}$ are called conjugates.

conjunction [9.2] When two or more sentences are joined by the word *and* to make a compound sentence, the new sentence is called a conjunction of the sentences.

consecutive numbers [2.5] Consecutive numbers are integers that are one unit apart.

consistent [8.1] A system of equations that has at least one solution is said to be consistent.

constant [1.1] A constant is a known number.

constant function [7.3] A constant function can be expressed in the form $f(x) = b$. Its graph is a horizontal line that crosses the y-axis at $(0, b)$.

constant of proportionality [7.5] In an equation of the form $y = kx \, (k \neq 0)$, which specifies a direct variation, k is the constant of proportionality.

constraint [9.5] In linear programming, each linear inequality is called a constraint.

contradiction [6.8] An equation that is never true is a contradiction.

coordinates [3.1] The coordinates are numbers in an ordered pair.

cube root [10.1] The number c is called the cube root of a if $c^3 = a$.

D

data point [11.8] A data point is a given ordered pair of a function.

degree of a polynomial [4.2] The degree of the highest term of a polynomial is referred to as the degree of the polynomial.

degree of a term [4.2] The degree of a term is the number of variable factors in that term.

demand function [8.8] Demand function is a graphical representation of the relationship between the price of a good and the quantity of that good demanded.

denominator [1.3] The number below the fraction line in a fraction is called a denominator.

descending order [4.2] Descending order means a polynomial is written with the term of highest degree first, followed by the term of next highest degree, and so on.

determinant [8.7] The determinant of a two by two matrix $\begin{bmatrix} a & c \\ b & d \end{bmatrix}$ is denoted $\begin{vmatrix} a & c \\ b & d \end{vmatrix}$ and is defined as follows: $\begin{vmatrix} a & c \\ b & d \end{vmatrix} = ad - bc$.

difference of squares [4.5] Any expression that can be written in the form $a^2 - b^2$ is called a difference of squares.

direct variation [7.5] When a situation translates to an equation described by $y = kx$, with k a constant, we say y varies directly as x. The equation $y = kx$ is called an equation of direct variation.

discriminant [11.4] The expression $b^2 - 4ac$ from the quadratic formula is called a discriminant.

disjunction [9.2] When two or more sentences are joined by the word *or* to make a compound sentence, the new sentence is called a disjunction of the sentences.

distributive [1.2] The distributive law states that the product of a number and a sum can be written as the sum of two products.

domain [7.1] The domain is the set of all the first coordinates of the ordered pairs in a function.

doubling time [12.7] The doubling time is the amount of time necessary for a population to double in size.

E

elimination method [8.2] Elimination method is an algebraic method involving the addition principle to solve a system of equations.

ellipse [13.2] An ellipse is the set of all points in a plane for which the sum of the distances from two fixed points F_1 and F_2 is constant.

equation [1.1] An equation is a number sentence with the verb $=$.

equations of variation [7.5] Equations of the form $y = kx$ or $y = k/x$ for some constant k are called equations of variation.

equilibrium point [8.8] The equilibrium point is the point of intersection between the demand function and the supply function.

equivalent [1.2] Expressions that represent the same number are said to be equivalent.

equivalent equations [2.1] Equations with the same solutions are called equivalent equations.

equivalent inequalities [2.6] Inequalities that have the same solution set are called equivalent inequalities.

evaluating the expression [1.1] Substituting a value for a variable in an expression is referred to as evaluating the expression.

event [14.7] In probability, an event is a set of outcomes.

exponent [1.8] An exponent is a number that indicates how many times another number is multiplied by itself.

exponential decay [12.7] Exponential decay is a decrease in quantity over time that can be described with an exponential function of the form, $P(t) = P_0 e^{-kt}$, $k > 0$.

exponential equation [12.6] Equations with variables in exponents are called exponential equations.

exponential function [12.2] The function $f(x) = a^x$, where a is a positive constant, $a \neq 1$, is called the exponential function, base a.

exponential growth [12.7] Exponential growth is an increase in quantity over time that can be described with an exponential function of the form, $P(t) = P_0 e^{kt}$, $k > 0$.

exponential notation [1.8] Exponential notation is a representation of a number using a base raised to a power.

extrapolation [3.1] Extrapolation is the process of estimating a value that goes beyond the given data.

F

factor [1.2] To factor an expression means to write an equivalent expression that is a product.

finite sequence [14.1] A finite sequence is a function having for its domain a set of natural numbers: $\{1, 2, 3, 4, 5, \ldots n\}$, for some natural number n.

fixed costs [8.8] In business, fixed costs are costs that must be paid whether a product is produced or not.

focus [13.2] A focus is one of two fixed points that determine the points of an ellipse.

FOIL [4.5] To multiply two binomials, using the FOIL method, multiply the first terms, the outside terms, the inside terms, and then the last terms. Then combine like terms, if possible.

formula [2.3] A formula is an equation that uses numbers or letters to represent a relationship between two or more quantities.

fraction notation [1.3] To write a number in fraction notation is to include a numerator and a denominator.

function [7.1] A function is a correspondence that assigns to each member of some set called the domain exactly one member of a set called the range.

G

general term [14.1] The general term of a sequence is the nth term, or a_n.

geometric sequence [14.3] A sequence in which the ratio of every pair of successive terms is constant is called a geometric sequence.

geometric series [14.3] A geometric series is a series whose associated sequence is geometric.

grade [3.5] The grade of a road is the measure of the road's steepness.

graph [3.1] A graph is a picture or diagram of the data in a table.

greatest common factor [5.1] The greatest common factor of a polynomial is the common factor with the largest possible coefficient and the largest possible exponent.

H

half life [12.7] Half-life is the amount of time necessary for half of a quantity to decay.

half-open interval [9.1] Half-open intervals include an open interval and a closed interval. The two kinds of half-open intervals are $a < x \leq b$ and $a \leq x < b$.

horizontal line test [12.1] A function is one-to-one, and thus has an inverse that is a function, if it is impossible to draw a horizontal line that intersects its graph more than once.

hyperbola [13.3] A hyperbola is the set of all points P in the plane such that the difference of the distance from P to two fixed points is a given constant.

hypotenuse [5.8] In a right triangle, the side opposite the right angle is called the hypotenuse.

I

identity [6.8] An identity is an equation for which all possible replacements are solutions.

identity property of 0 [1.5] The identity property of 0 states that $a + 0 = 0$ for any real number a.

identity property of 1 [1.3] The identity property of 1 states that multiplying a number by 1 gives the same number.

imaginary number i [10.8] The number $i = \sqrt{-1}$ and $i_2 = -1$.

imaginary numbers [10.8] An imaginary number is a number that can be written in the form $a + bi$ where a and b are real numbers and $b \neq 0$.

inconsistent [8.1] A system of equations for which there is no solution is said to be inconsistent.

index [10.1] In the radical $\sqrt[n]{a}$, the number n is called the index.

inequality [1.4] An inequality is a mathematical sentence using $>$, $<$, $\geq$, or $\leq$.

infinite geometric series [14.3] An infinite geometric series is the sum of the terms of an infinite geometric sequence.

infinite sequence [14.1] An infinite sequence is a function having for its domain the set of natural numbers: $\{1, 2, 3, 4, 5, \ldots\}$.

input [7.1] In functions, the inputs are the members of the domain.

integers [1.4] The integers consist of all whole numbers and their opposites.

interpolation [3.1] Interpolation is the process of estimating a value between given values.

intersection [9.2] The intersection of two sets A and B is the set of all elements that are common to both A and B, denoted $A \cap B$.

interval notation [9.1] Interval notation uses parentheses and brackets to write solutions of an inequality in one variable.

inverse relation [12.1] When a function has its domain and range interchanged, it is called an inverse relation.

inverse variation [7.5] When a situation translates to an equation described by $y = k/x$, with k a constant, we say that y varies inversely as x. The equation $y = k/x$ is called an equation of inverse variation.

irrational numbers [1.4] A number written in decimal form is irrational if it neither terminates or repeats.

isosceles right triangle [10.7] When both legs of a right triangle are the same size, it is called an isosceles right triangle.

J

joint variation [7.5] When a variable varies directly with more than one other variable, there is joint variation.

L

leading coefficient [4.2] The leading coefficient of a polynomial is the coefficient of the term of highest degree.

leading term [4.2] The leading term of a polynomial is the term of highest degree.

least common denominator [2.2] The least common denominator for any set of fractions is the least common multiple of the denominators.

legs [5.8] In a right triangle, the legs are the two sides that form a right angle.

like radicals [10.5] When two radical expressions have the same indices and radicands, they are said to be like radicals.

like terms [1.4] Like terms are terms that have exactly the same variable factors.

line graph [3.1] A line graph is a graph in which quantities are represented as points connected by straight-line segments.

linear equation [3.2] A linear equation is an equation whose graph is a line.

linear function [7.3] A linear function is a function described by any linear equation whose graph is not vertical.

linear inequality [9.4] An inequality whose associated equation is a linear equation is a linear inequality.

linear programming [9.5] Linear programming is a branch of mathematics involving graphs of inequalities and their constraints.

logarithmic equations [12.6] Equations containing logarithmic expressions are called logarithmic equations.

logarithmic function [12.3] The inverse of the exponential function with base a is a logarithmic function with base a.

M

matrix [8.6] A matrix, $\begin{bmatrix} a & c \\ b & d \end{bmatrix}$, is a rectangular array of numbers.

maximum value [11.6] If a parabola opens downward $(a < 0)$, the function value at the vertex is a maximum value.

minimum value [11.6] If a parabola opens upward $(a > 0)$, the function value at the vertex is a minimum value.

monomial [4.2] A term that is a product of constants and/or variables is called a monomial.

motion problem [6.7] Problems that deal with distance, speed, and time are called motion problems.

multiplicative inverses [1.3] Multiplicative inverses are two numbers whose product is 1.

multiplicative property of 0 [1.7] The multiplicative property of 0 states that the product of 0 and any real number is 0.

N

natural logarithm [12.5] Logarithms with base e are natural logarithms.

natural numbers [1.3] Natural numbers can be thought of as the counting numbers: $1, 2, 3, 4, 5, \ldots$.

nonlinear function [7.3] A function whose graph is not a straight line is a nonlinear function.

numerator [1.3] The number above the fraction line in a fraction is called a numerator.

O

objective function [9.5] In linear programming, the expression to be maximized is called the objective function.

one-to-one function [12.1] A function for which different inputs have different outputs is one-to-one.

open interval (a, b) [9.1] An open interval (a, b) is defined as the set of all numbers x for which $a < x < b$. Thus, $(a, b) = \{x \mid a < x < b\}$.

opposite [1.6] The opposite, or additive inverse, of a number a is written $-a$. When opposites are added, the result is 0.

opposites [1.4] Two numbers that are the same distance from 0 on the number line but on opposite sides of 0 are called opposites.

ordered pair [3.1] An ordered pair, (a, b), is a pair of numbers having a significant order.

origin [3.1] The origin is the point on a graph where the two axes intersect.

outcomes [14.4] In probability, results of an experiment are called outcomes.

output [7.1] In functions, the outputs are the members of the range.

P

parabolas [11.6] Graphs of quadratic equations are called parabolas.

parallel lines [3.6] Parallel lines extend indefinitely without intersecting.

Pascal's triangle [14.6] Pascal's triangle is a triangular array of coefficients of the expansion of $(a + b)^n$ for $n = 0, 1, 2, \ldots$.

perfect square [10.3] An integer p is a perfect square if there exists a rational number q for which $q^2 = p$.

perfect square trinomial [4.5] A trinomial that is the square of a binomial is called a perfect square trinomial.

permutations [14.4] A permutation of a set of n objects is an ordered arrangement of all n objects.

perpendicular lines [3.7] Perpendicular lines intersect at right angles.

point-slope equation [3.7] The point-slope equation is an equation of the type $y - y_1 = m(x - x_1)$.

polynomial [4.2] A polynomial is a monomial or a sum of monomials.

polynomial equation [5.7] A polynomial equation is when two polynomials are set equal to each other.

polynomial inequality [11.9] A polynomial inequality is an inequality that is equivalent to an inequality with a polynomial as one side and 0 as the other.

prime factorization [1.3] Prime factorization is the process of writing a whole number as a product of its prime factors.

prime number [1.3] A prime number is a natural number that has exactly two different factors: the number itself and 1.

principal square root [10.1] The principal square root of a nonnegative number is its nonnegative square root.

proportion [6.7] A proportion is an equation stating that two ratios are equal.

pure imaginary number [10.8] Complex numbers, $a + bi$, in which $a = 0$ and $b \neq 0$ are called pure imaginary numbers.

Pythagorean theorem [5.8] The Pythagorean theorem states that in any right triangle, if a and b are the lengths of the legs and c is the length of the hypotenuse, then $a^2 + b^2 = c^2$.

Q

quadrants [3.1] Quadrants are the four regions into which the axes divide a plane.

quadratic equation [5.7] A quadratic equation is an equation equivalent to one of the form $ax^2 + bx + c = 0$, where $a \neq 0$.

quadratic formula [11.2] The solutions of $ax^2 + bx + c, a \neq 0$, are given by the quadratic formula which is $x = \dfrac{-b \pm \sqrt{b^2 - 4ac}}{2a}$.

quadratic function [11.2] A quadratic function is a second-degree polynomial function in one variable.

quadratic inequality [11.9] Second-degree polynomial inequalities in one variable are called quadratic inequalities.

R

radical equation [10.6] A radical equation is an equation in which the variable appears in a radicand.

radical expression [10.1] An expression in which a radical sign appears is called a radical expression.

radical sign [10.1] The symbol $\sqrt{}$ is called a radical sign and is used to indicate the principal square root of the number over which it appears.

radical term [10.5] A term in which a radical sign appears is called a radical term.

radicand [10.1] The expression under a radical sign is called the radicand.

radius [13.1] The radius of a circle is the distance from the center to a point on the circle.

range [7.1] The range is the set of all the second coordinates of the ordered pairs in a function.

rate [3.4] A rate is a ratio that indicates how two quantities change with respect to each other.

ratio [1.4] A ratio of two quantities is their quotient.

rational equation [6.6] A rational equation is an equation containing one or more rational expressions.

rational expression [6.1] A rational expression is a quotient of two polynomials.

rational inequalities [11.9] Inequalities involving rational expressions are called rational inequalities.

rational numbers [1.4] Rational numbers are the set of all numbers $\dfrac{a}{b}$, where a and b are integers and $b \neq 0$.

rationalizing the denominator [10.4] Rationalizing the denominator is a procedure for finding an equivalent expression without a radical in the denominator.

real numbers [1.4] The set of all numbers corresponding to points on the number line is the set of real numbers.

reciprocals [1.3] Reciprocals are two numbers whose product is 1.

reflection [11.6] The reflection of a graph is the mirror image of the graph.

relation [7.1] A relation is a correspondence between the domain and range such that each member of the domain corresponds to at least one member of the range.

repeating decimal [1.4] If the remainder of a division problem has a repeating number, the quotient is a repeating decimal.

right triangle [5.8] A right triangle is a triangle with one right angle.

row equivalent operations [8.6] Row equivalent operations are operations used to produce equivalent systems of equations.

S

sample space [14.7] In probability, the set of all possible outcomes is called the sample space.

substitution method [8.2] Substitution is an algebraic method for solving systems of equations.

scientific notation [4.8] Scientific notation for a number is an expression of the type $N \times 10^m$, where m is an integer, $1 \leq N < 10$, and N is expressed in decimal notation.

sequence [14.1] A sequence is a function where the domain is a set of consecutive positive integers beginning with one.

series [14.1] Series is the indicated sum of the terms in a sequence.

set [1.4] A set is a collection of objects.

set-builder notation [2.6] Set-builder notation is a way of naming sets by describing basic characteristics of the elements in the set.

sigma notation [14.1] A series can be written in an abbreviated form by using the Greek letter Σ (sigma), called sigma notation.

similar triangle [6.7] Similar triangles are triangles in which corresponding sides are proportional.

simplify [1.3] To simplify a fraction means to remove factors of 1, therefore leaving the smallest numerator and denominator.

slope [3.5] Slope is the ratio of the rise to the run for any two points on a line.

slope-intercept equation [3.6] A slope-intercept equation is an equation of the type $y = mx + b$.

solution [1.1] A replacement or substitution that makes an equation true is called a solution.

solution set [2.6] A solution set is the set of all solutions for an equation or inequality.

solve [2.1] To solve an equation means to find all solutions to the equation.

speed [3.4] The speed of an object is found by dividing the distance traveled by the time required to travel that distance.

square matrix [8.7] A square matrix is a matrix with the same number of rows and columns.

square root [10.1] The number c is a square root of a if $c^2 = a$.

substitute [1.1] To substitute is to replace a variable with a number.

supply function [8.8] A supply function is a graphical representation of the relationship between the price of a good and the quantity of that good supplied.

systems of equations [8.1] A system of equations is a set of two or more equations that are to be solved simultaneously.

T

term [1.2] A term is a number, a variable, or a product or quotient of numbers and/or variables.

terminating decimal [1.4] If the remainder is 0 in a division problem, we call the answer a terminating decimal.

total cost [8.8] Total cost is the money spent to produce a product.

total profit [8.8] Total profit is the money taken in less the money spent, or total revenue minus total cost.

total revenue [8.8] Total revenue is the money made by selling a product.

trinomial [4.2] A polynomial that is composed of three terms is called a trinomial.

U

union [9.2] The union of two sets A and B is the collection of elements belonging to A and/or B. We denote the union of A and B by A $\cup$ B.

V

value [1.1] The answer to an expression after substituting is called the value of the expression.

variable [1.1] A variable is a letter that represents an unknown number.

variable costs [8.8] In business, variable costs are costs that vary according to the amount of products produced.

variable expression [1.1] An expression containing a variable is a variable expression.

vertex [11.6] The point at which the graph of a quadratic equation crosses its axis of symmetry is called the vertex.

vertical line test [7.3] A graph represents a function if it is impossible to draw a vertical line that intersects the graph more than once.

W

whole numbers [1.3] Whole numbers contain the natural numbers and 0: 0, 1, 2, 3, ...

X

x-intercept [3.3] The point at which a graph crosses the x-axis is the x-intercept.

Y

y-intercept [3.3] The point at which a graph crosses the y-axis is the y-intercept.

Z

zeros [11.9] The zeros of a function are the x-intercepts of the function.

Answers

CHAPTER 1

Technology Connection, p. 7

1. 3438   **2.** 47,531

Exercise Set 1.1, pp. 7–10

1. 27   **3.** 8   **5.** 4   **7.** 7   **9.** 5   **11.** 3   **13.** 24 ft^2
15. 15 cm^2 **17.** 0.270   **19.** Let j represent Jan's age;
$8 + j$, or $j + 8$ **21.** $6 + b$, or $b + 6$ **23.** $c - 9$
25. $6 + q$, or $q + 6$ **27.** Let p represent Phil's speed; $9p$,
or $p9$ **29.** $y - x$ **31.** $x \div w$, or $\dfrac{x}{w}$ **33.** $n - m$

35. Let l represent the length of the box and h represent
the height; $l + h$, or $h + l$ **37.** $9 \cdot 2m$, or $2m \cdot 9$

39. Let y represent "some number"; $\dfrac{1}{4}y$, or $\dfrac{y}{4}$

41. Let w represent the number of women attending;
64% of w, or $0.64w$ **43.** $\$50 - x$ **45.** Yes
47. No   **49.** Yes   **51.** Yes   **53.** Let x represent the
unknown number; $73 + x = 201$ **55.** Let x represent
the unknown number; $42x = 2352$ **57.** Let s represent
the number of squares your opponent controls;
$s + 35 = 64$ **59.** Let w represent the amount of
solid waste generated; $27\% \cdot w = 56$, or $0.27w = 56$
61. 🖩   **63.** 🖩   **65.** $\$337.50$ **67.** 2   **69.** 6
71. $w + 4$ **73.** Let x and y represent the two
numbers; $\frac{1}{3} \cdot \frac{1}{2} \cdot xy$ **75.** $s + s + s + s$, or $4s$
77. 🖩

Exercise Set 1.2, pp. 16–17

1. $x + 7$ **3.** $c + ab$ **5.** $3y + 9x$ **7.** $5(1 + a)$
9. $a \cdot 2$ **11.** ts **13.** $5 + ba$ **15.** $(a + 1)5$
17. $a + (5 + b)$ **19.** $(r + t) + 7$ **21.** $ab + (c + d)$
23. $8(xy)$ **25.** $(2a)b$ **27.** $(3 \cdot 2)(a + b)$
29. $(r + t) + 6$; $(t + 6) + r$ **31.** $17(ab)$; $b(17a)$
33. $(5 + x) + 2 = (x + 5) + 2$ Commutative law
             $= x + (5 + 2)$ Associative law
             $= x + 7$ Simplifying
35. $(m3)7 = m(3 \cdot 7)$ Associative law
         $= m21$ Simplifying
         $= 21m$ Commutative law
37. $4a + 12$ **39.** $6 + 6x$ **41.** $3x + 3$ **43.** $24 + 8y$
45. $18x + 54$ **47.** $5r + 10 + 15t$ **49.** $2a + 2b$

51. $5x + 5y + 10$ **53.** $x, xyz, 19$ **55.** $2a, \dfrac{a}{b}, 5b$

57. $2(a + b)$ **59.** $7(1 + y)$ **61.** $3(6x + 1)$
63. $5(x + 2 + 3y)$ **65.** $3(4x + 3)$ **67.** $3(a + 3b)$
69. $11(4x + y + 2z)$ **71.** 🖩   **73.** [1.1] Let k represent
Kara's salary; $2k$ **74.** [1.1] $\dfrac{1}{2} \cdot m$, or $\dfrac{m}{2}$ **75.** 🖩

77. Yes; distributive law   **79.** Yes; distributive law and
commutative law of multiplication   **81.** No; for example,
let $x = 1$ and $y = 2$. Then $30 \cdot 2 + 1 \cdot 15 = 60 + 15 = 75$
and $5[2(1 + 3 \cdot 2)] = 5[2(7)] = 5 \cdot 14 = 70$.   **83.** 🖩

Exercise Set 1.3, pp. 24–26

1. $2 \cdot 25$; $5 \cdot 10$; 1, 2, 5, 10, 25, 50   **3.** $3 \cdot 14$; $6 \cdot 7$; 1, 2, 3,
6, 7, 14, 21, 42   **5.** $2 \cdot 13$ **7.** $2 \cdot 3 \cdot 5$ **9.** $2 \cdot 2 \cdot 5$

11. $3 \cdot 3 \cdot 3$ **13.** $2 \cdot 3 \cdot 3$ **15.** $2 \cdot 2 \cdot 2 \cdot 5$
17. Prime   **19.** $2 \cdot 3 \cdot 5 \cdot 7$ **21.** $5 \cdot 23$ **23.** $\frac{5}{7}$
25. $\frac{2}{7}$ **27.** $\frac{1}{8}$ **29.** 7   **31.** $\frac{1}{4}$ **33.** 6   **35.** $\frac{15}{16}$ **37.** $\frac{60}{41}$
39. $\frac{15}{7}$ **41.** $\frac{3}{14}$ **43.** $\frac{27}{8}$ **45.** $\frac{1}{2}$ **47.** $\frac{7}{6}$ **49.** $\frac{3b}{7a}$
51. $\frac{7}{a}$ **53.** $\frac{5}{6}$ **55.** 1   **57.** $\frac{5}{18}$ **59.** 0   **61.** $\frac{35}{18}$
63. $\frac{10}{3}$ **65.** 28   **67.** 1   **69.** $\frac{6}{35}$ **71.** 18   **73.** ▱
75. [1.2] $5(3 + x)$; answers may vary
76. [1.2] $7 + (b + a)$, or $(a + b) + 7$ **77.** ▱
79. 24 in.   **81.** $\frac{2}{5}$ **83.** $\frac{3q}{t}$ **85.** $\frac{6}{25}$ **87.** $\frac{5ap}{2cm}$
89. $\frac{28}{45}\,\text{m}^2$ **91.** $14\frac{2}{9}\,\text{m}$ **93.** ▱

Technology Connection, p. 31

1. 10.1   **2.** 1.5

Exercise Set 1.4, pp. 33–34

1. $-19, 59$ **3.** $-150, 65$ **5.** $-1286, 29{,}029$
7. $750, -125$ **9.** $20, -150, 300$
11.

13.

15.

17. 0.875   **19.** -0.75

21. $1.1\overline{6}$ **23.** $0.\overline{6}$ **25.** -0.5 **27.** 0.13   **29.** $<$
31. $>$ **33.** $<$ **35.** $<$ **37.** $>$ **39.** $<$ **41.** $<$
43. $x < -7$ **45.** $y \geq -10$ **47.** True   **49.** False
51. True   **53.** 23   **55.** 17   **57.** 5.6   **59.** 329   **61.** $\frac{9}{7}$
63. 0   **65.** 8   **67.** $-83, -4.7, 0, \frac{5}{9}, 8.31, 62$
69. $-83, 0, 62$ **71.** $-83, -4.7, 0, \frac{5}{9}, \pi, \sqrt{17}, 8.31, 62$
73. ▱   **75.** [1.1] 42   **76.** [1.2] $ba + 5$, or $5 + ab$
77. ▱   **79.** ▱   **81.** $-23, -17, 0, 4$
83. $-\frac{5}{6}, -\frac{3}{4}, -\frac{2}{3}, \frac{1}{6}, \frac{3}{8}, \frac{1}{2}$ **85.** $<$ **87.** $=$ **89.** $<$
91. $-2, -1, 0, 1, 2$ **93.** $\frac{1}{9}$ **95.** $\frac{50}{9}$ **97.** ▱

Exercise Set 1.5, pp. 39–40

1. -3 **3.** 4   **5.** 0   **7.** -8 **9.** -15 **11.** -8
13. 0   **15.** -41 **17.** 0   **19.** 7   **21.** -2 **23.** 11
25. -33 **27.** 0   **29.** 18   **31.** -32 **33.** 0   **35.** 20
37. -1.7 **39.** -9.1 **41.** $\frac{1}{5}$ **43.** $\frac{-6}{7}$ **45.** $-\frac{1}{15}$
47. $\frac{2}{9}$ **49.** -3 **51.** 0   **53.** Lost 3 students, or
-3 students   **55.** \$4300 loss, or $-\$4300$ **57.** $-\$195$
59. Fell $\$\frac{1}{16}$, or $-\$\frac{1}{16}$ **61.** $12a$ **63.** $9x$ **65.** $13t$
67. $-2m$ **69.** $-7a$ **71.** $1 - 2x$ **73.** $12x + 17$
75. $18n + 16$ **77.** ▱   **79.** [1.2] $21z + 7y + 14$
80. [1.3] $\frac{28}{3}$ **81.** ▱   **83.** $\$65\frac{1}{4}$ **85.** $-5y$ **87.** $-7m$
89. $-7t, -23$ **91.** 1 under par

Exercise Set 1.6, p. 46–47

1. -39 **3.** 9   **5.** 3.14   **7.** -23 **9.** $\frac{14}{3}$ **11.** -0.101
13. 72   **15.** $-\frac{2}{5}$ **17.** 1   **19.** -7 **21.** Negative three
minus five; -8 **23.** Two minus negative nine; 11
25. Four minus six; -2 **27.** Negative five minus
negative seven; 2   **29.** -2 **31.** -5 **33.** -6
35. -10 **37.** -6 **39.** 0   **41.** -5 **43.** -10 **45.** 2
47. 0   **49.** 0   **51.** 8   **53.** -11 **55.** 16   **57.** -16
59. -1 **61.** 11   **63.** -6 **65.** -2 **67.** -25 **69.** 1
71. -9 **73.** 17   **75.** -45 **77.** -81 **79.** -49
81. -7.9 **83.** -0.175 **85.** $-\frac{2}{7}$ **87.** $-\frac{7}{9}$ **89.** $\frac{3}{13}$
91. $3.8 - (-5.2)$; 9   **93.** $114 - (-79)$; 193   **95.** -58
97. 34   **99.** 41   **101.** -62 **103.** -139 **105.** 0
107. $-7x, -4y$ **109.** $9, -5t, -3st$ **111.** $-3x$
113. $-5a + 4$ **115.** $-7n - 9$ **117.** $-6x + 5$
119. $-8t - 7$ **121.** $-12x + 3y + 9$ **123.** $8x + 66$
125. $100°\text{F}$ **127.** 30,340 ft   **129.** 116 m   **131.** ▱
133. [1.1] $432\,\text{ft}^2$ **134.** [1.2] $2 \cdot 2 \cdot 2 \cdot 2 \cdot 2 \cdot 3 \cdot 3 \cdot 3$
135. ▱   **137.** True. For example, for $m = 5$ and $n = 3$,
$5 > 3$ and $5 - 3 > 0$, or $2 > 0$. For $m = -4$ and $n = -9$,
$-4 > -9$ and $-4 - (-9) > 0$, or $5 > 0$.
139. False. For example, let $m = 2$ and $n = -2$. Then 2
and -2 are opposites, but $2 - (-2) = 4 \neq 0$.   **141.** ▱

Exercise Set 1.7, pp. 53–55

1. -36 **3.** -56 **5.** -24 **7.** -72 **9.** 42   **11.** 45
13. -170 **15.** -144 **17.** 1200   **19.** 98   **21.** -78
23. 21.7   **25.** $-\frac{2}{5}$ **27.** $\frac{1}{12}$ **29.** -11.13 **31.** $-\frac{5}{12}$
33. 252   **35.** 0   **37.** $\frac{1}{28}$ **39.** 150   **41.** 0   **43.** -720
45. $-30{,}240$ **47.** -4 **49.** -4 **51.** -2 **53.** 4
55. -8 **57.** 2   **59.** -12 **61.** -8 **63.** Undefined
65. -4 **67.** 0   **69.** 0   **71.** $-\frac{8}{3}; \frac{8}{-3}$ **73.** $-\frac{29}{35}; \frac{-29}{35}$
75. $\frac{-7}{3}; \frac{7}{-3}$ **77.** $-\frac{x}{2}; \frac{x}{-2}$ **79.** $-\frac{5}{4}$ **81.** $-\frac{13}{47}$
83. $-\frac{1}{10}$ **85.** $\frac{1}{4.3}$, or $\frac{10}{43}$ **87.** $-\frac{4}{9}$ **89.** -1 **91.** $\frac{21}{20}$
93. $\frac{12}{55}$ **95.** -1 **97.** 1   **99.** $-\frac{9}{11}$ **101.** $-\frac{7}{4}$
103. -12 **105.** -3 **107.** 1   **109.** 7   **111.** $-\frac{1}{9}$
113. $\frac{1}{10}$ **115.** $-\frac{7}{6}$ **117.** $\frac{6}{7}$ **119.** $-\frac{14}{15}$ **121.** ▱
123. [1.3] $\frac{22}{39}$ **124.** [1.5] $12x - 2y - 9$ **125.** ▱
127. For 2 and 3, the reciprocal of the sum is $1/(2 + 3)$, or
$1/5$. But $1/5 \neq 1/2 + 1/3$.   **129.** Negative
131. Negative   **133.** Negative   **135.** **(a)** m and n have
different signs; **(b)** either m or n is zero; **(c)** m and n have
the same sign   **137.** ▱

Exercise Set 1.8, pp. 62–63

1. 4^3 **3.** x^7 **5.** $(3t)^5$ **7.** 16   **9.** 9   **11.** -9
13. 64   **15.** 625   **17.** 7   **19.** $81t^4$ **21.** $-343x^3$

23. 26   **25.** 86   **27.** 7   **29.** 5   **31.** 0   **33.** 9
35. 10   **37.** -7 **39.** -4 **41.** 1291   **43.** 14
45. 152   **47.** 36   **49.** 1   **51.** -26 **53.** -2 **55.** $-\frac{9}{2}$
57. -8 **59.** -3 **61.** -15 **63.** 9   **65.** 1   **67.** 6
69. -17 **71.** $-9x - 1$ **73.** $-7 + 2x$
75. $-4a + 3b - 7c$ **77.** $-3x^2 - 5x + 1$ **79.** $3x - 7$
81. $-3a + 9$ **83.** $5x - 6$ **85.** $-3t - 11r$
87. $9y - 25z$ **89.** $x^2 + 2$ **91.** $-t^3 - 2t$
93. $37a^2 - 23ab + 35b^2$ **95.** $-22t^3 - t^2 + 9t$
97. $2x - 25$ **99.** 🔲   **101.** [1.1] Let n represent the
number; $9 + 2n$ **102.** [1.1] Let m and n represent the
two numbers; $\frac{1}{2}(m + n)$ **103.** 🔲   **105.** $-6r - 5t + 21$
107. $-2x - f$ **109.** 🔲   **111.** False   **113.** False
115. True   **117.** 0   **119.** 39,000

Review Exercises: Chapter 1, p. 66–67

1. [1.1] 15   **2.** [1.1] 4   **3.** [1.8] -6 **4.** [1.8] -5
5. [1.1] $z - 7$ **6.** [1.1] xz **7.** [1.1] Let m and n represent the numbers; $mn + 1$, or $1 + mn$ **8.** [1.1] No
9. [1.1] Let t represent the value of wholesale tea sold in
the U.S. in 1990, in billions of dollars; $4.6 = 2.8 + t$
10. [1.2] $x2 + y$ **11.** [1.2] $2x + (y + z)$
12. [1.2] $(4x)y$, $4(yx)$, $(4y)x$; answers may vary
13. [1.2] $18x + 30y$ **14.** [1.2] $40x + 24y + 16$
15. [1.2] $3(7x + 5y)$ **16.** [1.2] $7(5x + 2 + y)$
17. [1.3] $2 \cdot 2 \cdot 13$ **18.** [1.3] $\frac{5}{12}$ **19.** [1.3] $\frac{9}{4}$
20. [1.3] $\frac{31}{36}$ **21.** [1.3] $\frac{3}{16}$ **22.** [1.3] $\frac{3}{5}$ **23.** [1.3] $\frac{72}{25}$
24. [1.4] $-45, 72$ **25.** [1.4]

$$\overset{\overset{\textstyle \frac{-1}{3}}{\bullet}}{\underset{-5\ -4\ -3\ -2\ -1\ \ 0\ \ 1\ \ 2\ \ 3\ \ 4\ \ 5}{\longleftrightarrow}}$$

26. [1.4] $x > -3$ **27.** [1.4] True   **28.** [1.4] False
29. [1.4] -0.875 **30.** [1.4] 1   **31.** [1.6] -7
32. [1.5] -3 **33.** [1.5] $-\frac{7}{12}$ **34.** [1.5] 0   **35.** [1.5] -5
36. [1.6] 5   **37.** [1.6] $-\frac{7}{5}$ **38.** [1.6] -7.9 **39.** [1.7] 54
40. [1.7] -9.18 **41.** [1.7] $-\frac{2}{7}$ **42.** [1.7] -140
43. [1.7] -7 **44.** [1.7] -3 **45.** [1.7] $\frac{3}{4}$ **46.** [1.8] 92
47. [1.8] 62   **48.** [1.8] 48   **49.** [1.8] 168   **50.** [1.8] $\frac{21}{8}$
51. [1.8] $\frac{103}{17}$ **52.** [1.5] $7a - 3b$ **53.** [1.6] $-2x + 5y$
54. [1.6] 7   **55.** [1.7] $-\frac{1}{7}$ **56.** [1.8] $(2x)^4$
57. [1.8] $-125x^3$ **58.** [1.8] $-3a + 9$
59. [1.8] $-2b + 21$ **60.** [1.8] $-3x + 9$
61. [1.8] $12y - 34$ **62.** [1.8] $5x + 24$
63. [1.1] 🔲 The value of a constant never varies. A variable
can represent a variety of numbers.   **64.** [1.2] 🔲 A term
is one of the parts of an expression that is separated from
the other parts by plus signs. A factor is part of a product.
65. [1.2], [1.5], [1.8] 🔲 The distributive law is used in
factoring algebraic expressions, multiplying algebraic
expressions, combining like terms, finding the opposite of
a sum, and subtracting algebraic expressions.
66. [1.8] 🔲 A negative quantity raised to an even power is

positive; a negative quantity raised to an odd power
is negative.
67. [1.8] 25,281   **68.** [1.4] (a) $\frac{3}{11}$; (b) $\frac{10}{11}$ **69.** [1.8] $-\frac{5}{8}$
70. [1.8] -2.1 **71.** [1.4] I   **72.** [1.7] J   **73.** [1.8] A
74. [1.8] H   **75.** [1.6] K   **76.** [1.2] B   **77.** [1.4] C
78. [1.4] E   **79.** [1.7] D   **80.** [1.8] F   **81.** [1.4] G

Test: Chapter 1, pp. 67–68

1. [1.1] 4   **2.** [1.1] Let x represent the number; $x - 9$.
3. [1.1] 240 ft^2 **4.** [1.2] $q + 3p$ **5.** [1.2] $(x \cdot 4) \cdot y$
6. [1.1] Yes   **7.** [1.1] Let p represent the maximum
production capability; $p - 282 = 2518$ **8.** [1.2] $30 - 5x$
9. [1.7] $-5y + 5$ **10.** [1.2] $11(1 - 4x)$
11. [1.2] $7(x + 3 + 2y)$ **12.** [1.3] $2 \cdot 2 \cdot 3 \cdot 5 \cdot 5$
13. [1.3] $\frac{2}{7}$ **14.** [1.4] $<$ **15.** [1.4] $>$ **16.** [1.4] $\frac{9}{4}$
17. [1.4] 2.7   **18.** [1.6] $-\frac{2}{3}$ **19.** [1.7] $-\frac{7}{4}$ **20.** [1.6] 8
21. [1.4] $-2 \geq x$ **22.** [1.6] 7.8   **23.** [1.5] -8
24. [1.5] $\frac{31}{40}$ **25.** [1.6] 10   **26.** [1.6] -2.5 **27.** [1.6] $\frac{7}{8}$
28. [1.7] -48 **29.** [1.7] $\frac{3}{16}$ **30.** [1.7] -9 **31.** [1.7] $\frac{3}{4}$
32. [1.7] -9.728 **33.** [1.8] -173 **34.** [1.6] 12
35. [1.8] -4 **36.** [1.8] 448   **37.** [1.6] $21a + 22y$
38. [1.8] $16x^4$ **39.** [1.8] $2x + 7$ **40.** [1.8] $9a - 12b - 7$
41. [1.8] $68y - 8$ **42.** [1.1] 15   **43.** [1.3] $\frac{23}{70}$
44. [1.8] 15   **45.** [1.8] $4a$

CHAPTER 2

Exercise Set 2.1, pp. 75–77

1. 15   **3.** -13 **5.** -10 **7.** -13 **9.** 15   **11.** -8
13. -6 **15.** 19   **17.** -4 **19.** $\frac{7}{3}$ **21.** $-\frac{13}{10}$ **23.** $\frac{41}{24}$
25. $-\frac{1}{20}$ **27.** 1.5   **29.** -5 **31.** 16   **33.** 4   **35.** 12
37. -23 **39.** 8   **41.** -7 **43.** -6 **45.** 6   **47.** 52
49. 36   **51.** -45 **53.** $\frac{6}{7}$ **55.** 1   **57.** $\frac{9}{2}$ **59.** -7.6
61. -2.5 **63.** -17 **65.** $-\frac{1}{2}$ **67.** -15 **69.** 12
71. 310.756   **73.** 🔲   **75.** [1.8] -34 **76.** [1.8] 41
77. [1.8] 1   **78.** [1.8] -16 **79.** 🔲   **81.** Identity
83. 0   **85.** Contradiction   **87.** Contradiction   **89.** 9.4
91. 6   **93.** 8   **95.** 11,074   **97.** 🔲

Technology Connection, p. 80

1.

X	Y₁
0	5
1	4
2	3
3	2
4	1
5	0
6	-1
X = 0	

2.

X	Y₁	Y₂
0	5	17
1	4	13
2	3	9
3	2	5
4	1	1
5	0	-3
6	-1	-7
X = 0		

3. 4; not reliable because, depending on the choice of
ΔTbl, it is easy to scroll past a solution without realizing it.

Exercise Set 2.2, pp. 83–84

1. 7 **3.** 8 **5.** 5 **7.** 14 **9.** -7 **11.** -5 **13.** -7
15. 15 **17.** 4 **19.** -12 **21.** 6 **23.** 8 **25.** 1
27. -20 **29.** 6 **31.** 7 **33.** 3 **35.** 0 **37.** 10
39. 4 **41.** 0 **43.** $-\frac{2}{5}$ **45.** $\frac{64}{3}$ **47.** $\frac{2}{5}$ **49.** 3
51. -4 **53.** $1.\overline{6}$ **55.** $-\frac{40}{37}$ **57.** 2 **59.** 2 **61.** 6
63. 8 **65.** 2 **67.** -4 **69.** -8 **71.** 2 **73.** 8
75. 1 **77.** -5 **79.** $\frac{11}{18}$ **81.** $-\frac{51}{31}$ **83.** 2 **85.** 📓
87. [1.8] -7 **88.** [1.8] 15 **89.** [1.8] -15
90. [1.8] -28 **91.** 📓 **93.** $\frac{1136}{909}$, or $1.\overline{2497}$
95. Contradiction **97.** Identity **99.** $\frac{2}{3}$ **101.** 0
103. 0 **105.** -2

Technology Connection, p. 85

1. 72,930

Exercise Set 2.3, pp. 88–91

1. 2 mi **3.** 1423 students **5.** 10.5 cal/oz **7.** 255 mg
9. $b = \dfrac{A}{h}$ **11.** $r = \dfrac{d}{t}$ **13.** $P = \dfrac{I}{rt}$ **15.** $m = 65 - H$
17. $l = \dfrac{P - 2w}{2}$, or $l = \dfrac{P}{2} - w$ **19.** $\pi = \dfrac{A}{r^2}$
21. $h = \dfrac{2A}{b}$ **23.** $m = \dfrac{E}{c^2}$ **25.** $d = 2Q - c$
27. $b = 3A - a - c$ **29.** $A = Ms$ **31.** $t = \dfrac{A}{a + b}$
33. $h = \dfrac{2A}{a + b}$ **35.** $L = W - \dfrac{N(R - r)}{400}$, or
$L = \dfrac{400W - NR + Nr}{400}$ **37.** 📓 **39.** [1.7] 0
40. [1.7] 9.18 **41.** [1.8] -13 **42.** [1.8] 65 **43.** 📓
45. 35 yr **47.** $a = \dfrac{w}{c} \cdot d$ **49.** $y = \dfrac{z^2}{t}$ **51.** $t = \dfrac{rs}{q - r}$
53. $S = 20a$, where S is the number of Btu's saved
55. $K = 8.70w + 17.78h - 9.52a + 92.4$

Exercise Set 2.4, pp. 95–97

1. 0.82 **3.** 0.09 **5.** 0.437 **7.** 0.0046 **9.** 29%
11. 99.8% **13.** 192% **15.** 210% **17.** 0.68%
19. 37.5% **21.** 28% **23.** $66\frac{2}{3}$% **25.** 25% **27.** 24%
29. $46\frac{2}{3}$, or $\frac{140}{3}$ **31.** 2.5 **33.** 84 **35.** 125% **37.** 0.8
39. 50% **41.** $280 **43.** 100 million voters
45. 24 women **47.** 0.5625 lb **49.** 27 bowlers
51. $86\frac{4}{11}$% **53.** $36 **55.** $148.50 **57.** $18/hr
59. 165 calories **61.** 📓 **63.** [1.1] Let n represent the
number; $5 + n$ **64.** [1.1] Let t represent Tino's weight;
$t - 4$ **65.** [1.1] $8 \cdot 2a$ **66.** [1.1] Let x and y represent
the two numbers; $1 + xy$ **67.** 📓 **69.** 18,500 people
71. 20% **73.** About 1.55%

Exercise Set 2.5, pp. 105–108

1. 11 **3.** 11 **5.** $75 **7.** $85
9. Approximately $62\frac{2}{3}$ mi **11.** 19, 20, 21 **13.** 29, 31
15. 62, 64 **17.** Bride: 84 yr; groom: 103 yr
19. 30°, 90°, 60° **21.** 95° **23.** 192, 193
25. 63 mm, 65 mm, 67 mm **27.** Width: 275 mi;
length: 365 mi **29.** $350 **31.** $852.94 **33.** 12 mi
35. $128\frac{1}{3}$ mi **37.** 65°, 25° **39.** 160 **41.** 📓
43. [1.4] $<$ **44.** [1.4] $<$ **45.** [1.4] $<$ **46.** [1.4] $>$
47. 📓 **49.** $37 **51.** 20 **53.** Length: 12 cm;
width: 9 cm **55.** 104°, 106°, 108°, 110°, 112°
57. $95.99 **59.** 10 **61.** 6 mi **63.** 📓
65. 5.34 cm, 8.59 cm, 12.94 cm

Exercise Set 2.6, pp. 115–117

1. (a) Yes; (b) yes; (c) yes; (d) no; (e) yes
3. (a) No; (b) no; (c) yes; (d) yes; (e) no
5.
7.
9.
11.
13.
15. $\left\{x \mid x > -4\right\}$
17. $\left\{x \mid x \leq 2\right\}$ **19.** $\left\{x \mid x < -1\right\}$ **21.** $\left\{x \mid x \geq 0\right\}$
23. $\left\{y \mid y > 7\right\}$,
25. $\left\{x \mid x \leq -18\right\}$,
27. $\left\{x \mid x < 10\right\}$,
29. $\left\{t \mid t \geq -3\right\}$,
31. $\left\{y \mid y > -5\right\}$,
33. $\left\{x \mid x \leq 5\right\}$, **35.** $\left\{x \mid x \geq 5\right\}$
37. $\left\{y \mid y \leq \frac{1}{2}\right\}$ **39.** $\left\{t \mid t > \frac{5}{8}\right\}$ **41.** $\left\{x \mid x < 0\right\}$
43. $\left\{t \mid t < 23\right\}$ **45.** $\left\{x \mid x < 7\right\}$,
47. $\left\{y \mid y \leq 9\right\}$,
49. $\left\{x \mid x > -\frac{13}{7}\right\}$,
51. $\left\{t \mid t < -3\right\}$,
53. $\left\{y \mid y \geq -\frac{2}{7}\right\}$ **55.** $\left\{y \mid y \geq -\frac{1}{10}\right\}$ **57.** $\left\{x \mid x > \frac{4}{5}\right\}$
59. $\left\{x \mid x < 9\right\}$ **61.** $\left\{y \mid y \geq 4\right\}$ **63.** $\left\{t \mid t \leq 7\right\}$
65. $\left\{x \mid x < -3\right\}$ **67.** $\left\{y \mid y < -4\right\}$ **69.** $\left\{x \mid x > -4\right\}$
71. $\left\{y \mid y < -\frac{10}{3}\right\}$ **73.** $\left\{x \mid x > -10\right\}$ **75.** $\left\{y \mid y < 2\right\}$

77. $\{y | y \geq 3\}$ **79.** $\{x | x > -4\}$ **81.** $\{x | x > -4\}$
83. $\{n | n \geq 70\}$ **85.** $\{x | x \leq 15\}$ **87.** $\{t | t < 14\}$
89. $\{y | y < 6\}$ **91.** $\{t | t \leq -4\}$ **93.** $\{r | r > -3\}$
95. $\{x | x \geq 8\}$ **97.** $\{x | x < \frac{11}{18}\}$ **99.** 🗒
101. [1.1] Let n represent "some number"; $3 + n$
102. [1.1] Let x and y represent the two numbers; $2(x + y)$
103. [1.1] Let x represent the number; $2x - 3$
104. [1.1] Let y represent the number; $5 + 2y$ **105.** 🗒
107. $\{t | t > -\frac{27}{19}\}$ **109.** $\{x | x \leq -4a\}$
111. $\{x | x > \frac{y - b}{a}\}$ **113.** (a) No; (b) yes; (c) no;
(d) yes; (e) no; (f) yes   **115.** $\{x | x \text{ is a real number}\}$,
or $(-\infty, \infty)$

Exercise Set 2.7, pp. 120–123

1. Let n represent the number; $n \geq 7$
3. Let b represent the baby's weight; $b > 2$
5. Let s represent the train's speed; $90 < s < 110$
7. Let m represent the number of people who attended the Million Man March; $m \leq 1,200,000$
9. Let c represent the cost of gasoline; $c \geq 1.50$
11. 15 or fewer copies   **13.** Mileages less than or equal to 341.4 mi   **15.** 3.5 hr or more   **17.** 2
19. Scores greater than or equal to 84
21. 4 servings or more   **23.** 27 min or more
25. Lengths greater than or equal to 92 ft; lengths less than or equal to 92 ft   **27.** Widths less than or equal to $11\frac{2}{3}$ ft   **29.** Times less than 4 units, or 1 hr
31. Lengths less than 21.5 cm
33. Blue-book value is greater than or equal to $10,625.
35. Temperatures greater than 37°C
37. It contains at least 16 g of fat per serving.
39. Depths less than 437.5 ft   **41.** Years after 2003
43. Dates at least 6 weeks after July 1
45. Heights greater than or equal to 4 ft
47. Mileages less than or equal to 215.2 mi   **49.** 🗒
51. [1.8] 2   **52.** [1.8] $\frac{1}{2}$ **53.** [1.8] $-\frac{10}{3}$ **54.** [1.8] $-\frac{1}{5}$
55. 🗒   **57.** More than 6 hr   **59.** Lengths less than or equal to 8 cm   **61.** They contain at least 7.5 g of fat per serving.   **63.** At least $42   **65.** 🗒

Review Exercises: Chapter 2, pp. 126–127

1. [2.1] -25 **2.** [2.1] 7   **3.** [2.1] -68 **4.** [2.1] 1
5. [2.1] -4 **6.** [2.1] 1.11   **7.** [2.1] $\frac{1}{2}$ **8.** [2.1] $-\frac{15}{64}$
9. [2.2] $\frac{38}{5}$ **10.** [2.2] -8 **11.** [2.2] -5 **12.** [2.2] $-\frac{1}{3}$
13. [2.2] 4   **14.** [2.2] 3   **15.** [2.2] 4   **16.** [2.2] 16
17. [2.2] 7   **18.** [2.2] $-\frac{7}{5}$ **19.** [2.2] 12   **20.** [2.2] 4
21 [2.3] $d = \frac{C}{\pi}$ **22.** [2.3] $B = \frac{3V}{h}$ **23.** [2.3] $a = 2A - b$

24. [2.4] 0.009   **25.** [2.4] 44%   **26.** [2.4] 20%
27. [2.4] 140   **28.** [2.6] Yes   **29.** [2.6] No
30. [2.6] Yes   **31.** [2.6]

$$5x - 6 < 2x + 3$$

$-5\ -4\ -3\ -2\ -1\ \ 0\ \ 1\ \ 2\ \ 3\ \ 4\ \ 5$

32. [2.6]

$$-2 < x \leq 5$$

$-5\ -4\ -3\ -2\ -1\ \ 0\ \ 1\ \ 2\ \ 3\ \ 4\ \ 5$

33. [2.6]

$$y > 0$$

$-5\ -4\ -3\ -2\ -1\ \ 0\ \ 1\ \ 2\ \ 3\ \ 4\ \ 5$ **34.** [2.6] $\{t | t \geq -\frac{1}{2}\}$

35. [2.6] $\{x | x \geq 7\}$ **36.** [2.6] $\{y | y > 3\}$
37. [2.6] $\{y | y \leq -4\}$ **38.** [2.6] $\{x | x < -11\}$
39. [2.6] $\{y | y > -7\}$ **40.** [2.6] $\{x | x > -6\}$
41. [2.6] $\{x | x > -\frac{9}{11}\}$ **42.** [2.6] $\{y | y \leq 12\}$
43. [2.6] $\{x | x \geq -\frac{1}{12}\}$ **44.** [2.5] $167   **45.** [2.5] 20
46. [2.5] 5 ft, 7 ft   **47.** [2.4] $144.2 billion
48. [2.5] 57, 59   **49.** [2.5] Width: 11 cm; length: 17 cm
50. [2.4] $160   **51.** [2.4] $42,038.22
52. [2.5] 35°, 85°, 60°   **53.** [2.7] $105 or less
54. [2.7] Widths greater than 17 cm
55. [2.1], [2.6] 🗒 Multiplying on both sides of an equation by *any* nonzero number results in an equivalent equation. When multiplying on both sides of an inequality, the sign of the number being multiplied by must be considered. If the number is positive, the direction of the inequality symbol remains unchanged; if the number is negative, the direction of the inequality symbol must be reversed to produce an equivalent inequality.
56. [2.1], [2.6] 🗒 The solutions of an equation can usually each be checked. The solutions of an inequality are normally too numerous to check. Checking a few numbers from the solution set found cannot guarantee that the answer is correct, although if any number does not check, the answer found is incorrect.
57. [2.5] Nile: 6671 km; Amazon: 6437 km
58. [2.4], [2.5] $18,600   **59.** [1.4], [2.2] $-23, 23$
60. [1.4], [2.1] $-20, 20$ **61.** [2.3] $a = \frac{y - 3}{2 - b}$

Test: Chapter 2, p. 127

1. [2.1] 9   **2.** [2.1] 15   **3.** [2.1] -6 **4.** [2.1] 49
5. [2.1] -12 **6.** [2.2] 2   **7.** [2.1] -8 **8.** [2.1] $-\frac{7}{20}$
9. [2.2] 7   **10.** [2.2] -5 **11.** [2.2] $\frac{23}{3}$
12. [2.6] $\{x | x > -5\}$ **13.** [2.6] $\{x | x > -13\}$
14. [2.6] $\{x | x < \frac{21}{8}\}$ **15.** [2.6] $\{y | y \leq -13\}$
16. [2.6] $\{y | y \leq -8\}$ **17.** [2.6] $\{x | x \leq -\frac{1}{20}\}$
18. [2.6] $\{x | x < -6\}$ **19.** [2.6] $\{x | x \leq -1\}$
20. [2.3] $r = \frac{A}{2\pi h}$ **21.** [2.3] $l = 2w - P$ **22.** [2.4] 2.3
23. [2.4] 5.4%   **24.** [2.4] 16   **25.** [2.4] 44%

26. [2.6]

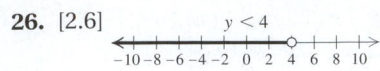

$y < 4$

$-10\ -8\ -6\ -4\ -2\ \ 0\ \ 2\ \ 4\ \ 6\ \ 8\ \ 10$

27. [2.6]

$-2 \le x \le 2$

$-5\ -4\ -3\ -2\ -1\ \ 0\ \ 1\ \ 2\ \ 3\ \ 4\ \ 5$

28. [2.5] Width: 7 cm; length: 11 cm   **29.** [2.5] 60 mi
30. [2.5] 81 mm, 83 mm, 85 mm   **31.** [2.4] \$4.59 billion
32. [2.7] All numbers greater than 6, or $\{n \mid n > 6\}$
33. [2.7] Lengths greater than or equal to 174 yd

34. [2.3] $d = \dfrac{a}{3}$ **35.** [1.4], [2.2] $-15, 15$

36. [2.5] 60 tickets

CHAPTER 3

Exercise Set 3.1, pp. 137–141

1. 2 drinks   **3.** The person weighs more than 200 lb.
5. About 2,720,000   **7.** About 1,360,000
9. 20.79 million tons   **11.** About 3 million tons
13. 70%   **15.** 1995   **17.** 1994 to 1995
19.

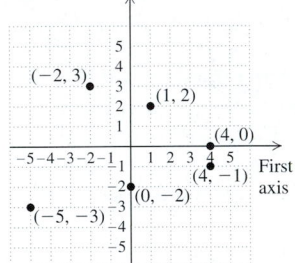

21.

Second axis

23. $A(-4, 5)$; $B(-3, -3)$; $C(0, 4)$; $D(3, 4)$; $E(3, -4)$
25. $A(4, 1)$; $B(0, -5)$; $C(-4, 0)$; $D(-3, -2)$; $E(3, 0)$
27. IV   **29.** III   **31.** I   **33.** II   **35.** I and IV
37. I and III   **39.** **(a)** Approximately 56 births per 1000
females; **(b)** approximately 43 births per 1000 females
41.

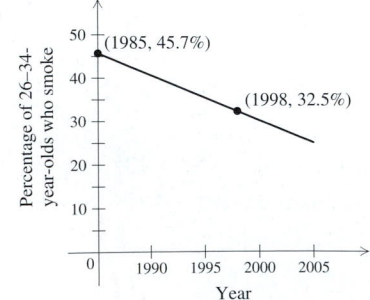

(a) About 40%; **(b)** about 27.5%

43.

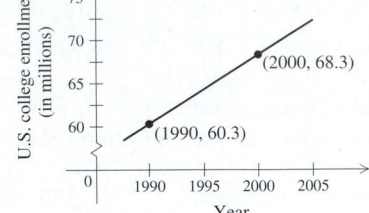

(a) About 65 million students; **(b)** about 72 million students
45.

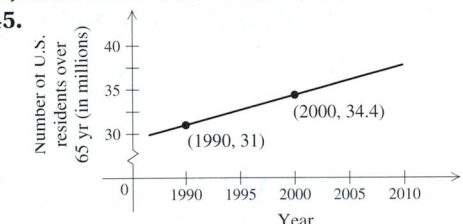

(a) About 32.7 million; **(b)** about 38 million   **47.**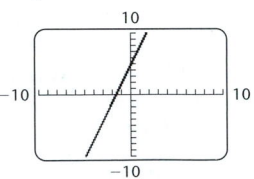
49. [1.8] -18 **50.** [1.8] -31 **51.** [1.8] 6   **52.** [1.8] 1
53. [2.3] $y = \frac{3}{2}x - 3$ **54.** [2.3] $y = \frac{7}{4}x - \frac{7}{2}$ **55.**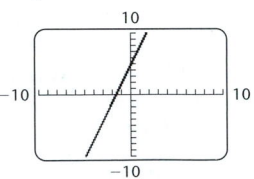
57. I or III   **59.** $(-1, -5)$
61.

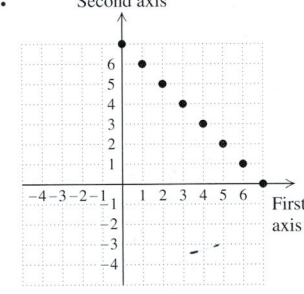

63. 26 units

65. Latitude 32.5° North; longitude 64.5° West   **67.**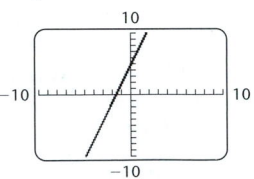

Technology Connection, p. 149

1. $y = -5x + 6.5$

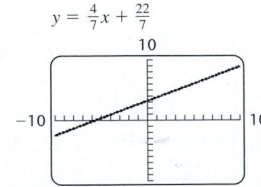

2. $y = 3x + 4.5$

3. $7y - 4x = 22$, or
$y = \frac{4}{7}x + \frac{22}{7}$

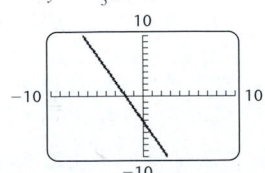

4. $5y + 11x = -20$, or
$y = -\frac{11}{5}x - 4$

5.
$2y - x^2 = 0$, or
$y = 0.5x^2$

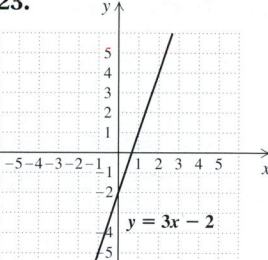

6.
$y + x^2 = 8$, or
$y = -x^2 + 8$

Exercise Set 3.2, pp. 149–151

1. No   **3.** No   **5.** Yes

7.

$y = x - 2$		$y = x - 2$	
$1 \ ? \ 3 - 2$		$-4 \ ? \ -2 - 2$	
$1 \mid 1$	TRUE	$-4 \mid -4$	TRUE

$(5, 3)$; answers may vary

9.

$y = \frac{1}{2}x + 3$		$y = \frac{1}{2}x + 3$	
$5 \ ? \ \frac{1}{2} \cdot 4 + 3$		$2 \ ? \ \frac{1}{2}(-2) + 3$	
$2 + 3$		$-1 + 3$	
$5 \mid 5$	TRUE	$2 \mid 2$	TRUE

$(0, 3)$; answers may vary

11.

$y + 3x = 7$		$y + 3x = 7$	
$1 + 3 \cdot 2 \ ? \ 7$		$-5 + 3 \cdot 4 \ ? \ 7$	
$1 + 6$		$-5 + 12$	
$7 \mid 7$	TRUE	$7 \mid 7$	TRUE

$(1, 4)$; answers may vary

13.

$4x - 2y = 10$		$4x - 2y = 10$	
$4 \cdot 0 - 2(-5) \ ? \ 10$		$4 \cdot 4 - 2 \cdot 3 \ ? \ 10$	
$0 + 10$		$16 - 6$	
$10 \mid 10$	TRUE	$10 \mid 10$	TRUE

$(2, -1)$; answers may vary

15. $y = x - 1$

17. $y = x$

19. $y = \frac{1}{2}x$

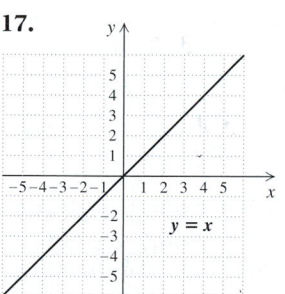

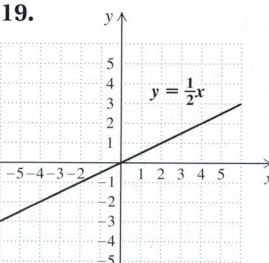

21. $y = x + 2$

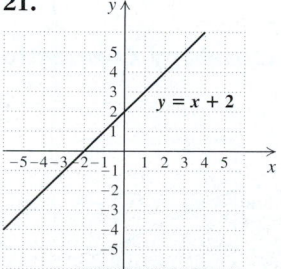

23. $y = 3x - 2$

25. $y = \frac{1}{2}x + 1$

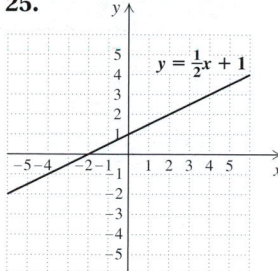

27. $x + y = -5$

29. $y = \frac{5}{3}x - 2$

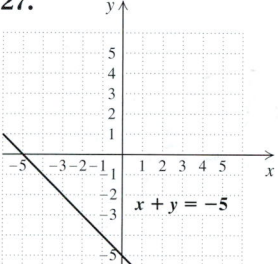
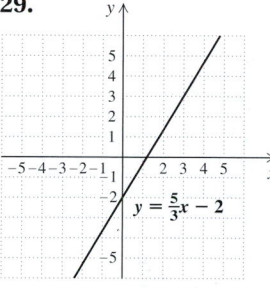

31. $x + 2y = 8$

33. $y = \frac{3}{2}x + 1$

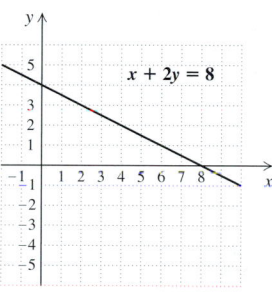
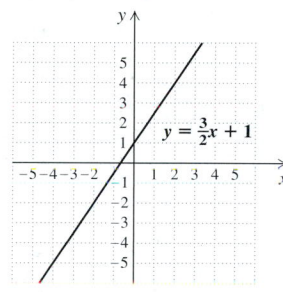

35. $6x - 3y = 9$

37. $8y + 2x = -4$

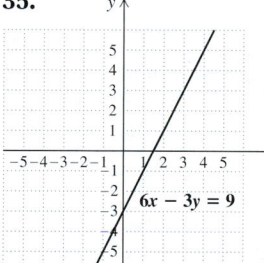
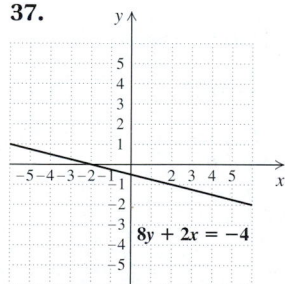

39. 12 gal

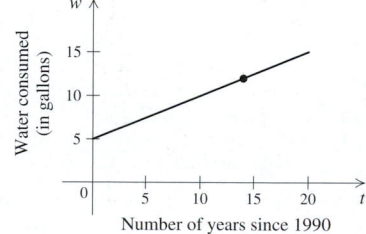

Number of years since 1990

41.

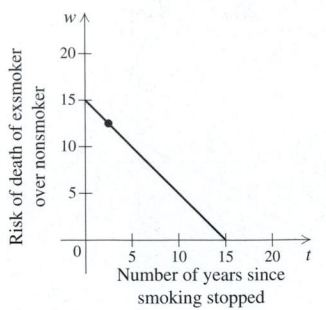

$12\frac{1}{2}$ times

43.

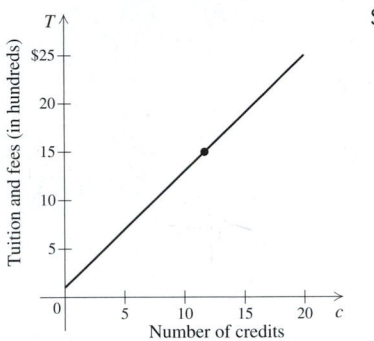

$1500

45.

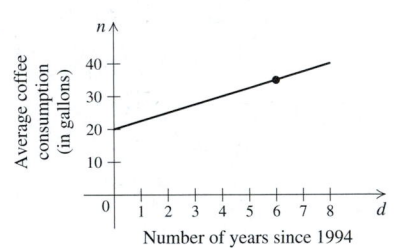

35 gal   **47.**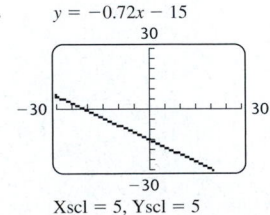

49. [2.2] $\frac{12}{5}$ **50.** [2.2] $\frac{9}{2}$ **51.** [2.2] $-\frac{5}{2}$

52. [2.3] $p = \dfrac{w}{q+1}$ **53.** [2.3] $y = \dfrac{C-Ax}{B}$

54. [2.3] $Q = 2A - T$ **55.**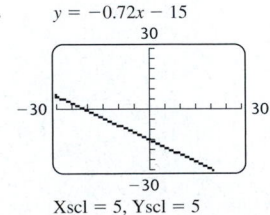

57. $s + n = 18$

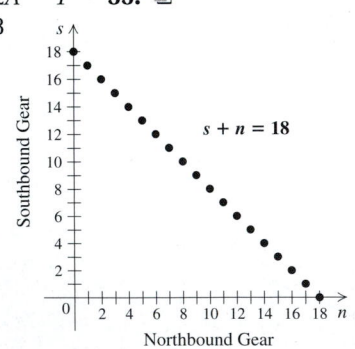

59. $x + y = 2$, or $y = -x + 2$

61. $5x - 3y = 15$, or $y = \frac{5}{3}x - 5$

63.

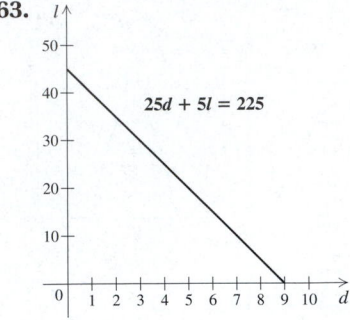

Answers may vary. 1 dinner, 40 lunches; 5 dinners, 20 lunches; 8 dinners, 5 lunches

65.

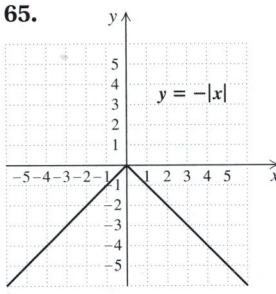

67.

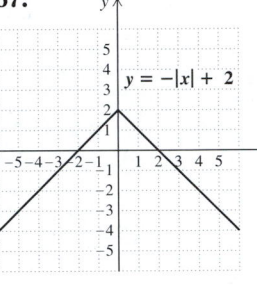

69. $y = -2.8x + 3.5$

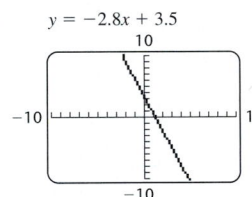

71. $y = 2.8x - 3.5$

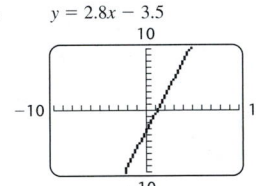

73. $y = x^2 + 4x + 1$

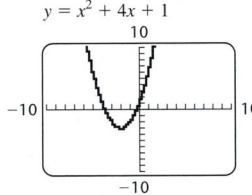

75. 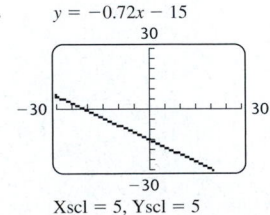 No; only whole-number values of the variables have meaning in these applications, so only those points for which both coordinates are whole numbers are solutions of the given problems.

Technology Connection, p. 156

1. $y = -0.72x - 15$

Xscl = 5, Yscl = 5

2. $y - 2.13x = 27$, or $y = 2.13x + 27$

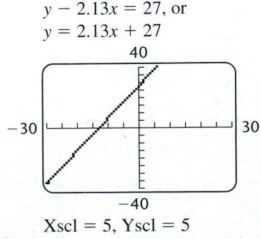

Xscl = 5, Yscl = 5

3. $5x + 6y = 84$, or
$y = -\frac{5}{6}x + 14$

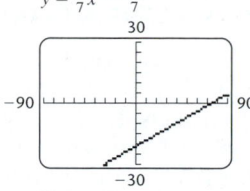

Xscl = 5, Yscl = 5

4. $2x - 7y = 150$, or
$y = \frac{2}{7}x - \frac{150}{7}$

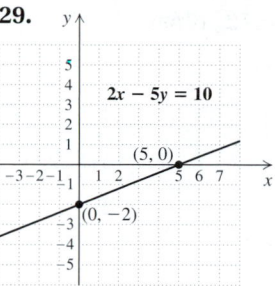 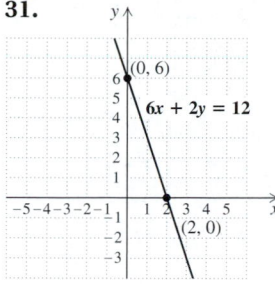

Xscl = 10, Yscl = 5

$19x - 17y = 200$, or
$y = \frac{19}{17}x - \frac{200}{17}$

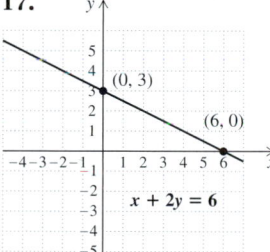

$6x + 5y = 159$, or
$y = -\frac{6}{5}x + \frac{159}{5}$

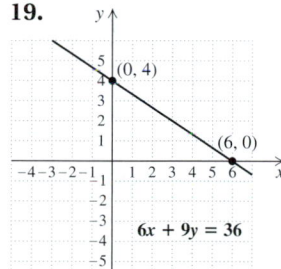

Xscl = 5, Yscl = 5

Exercise Set 3.3, pp. 159–160

1. (a) $(0, 5)$; **(b)** $(2, 0)$ **3. (a)** $(0, -4)$; **(b)** $(3, 0)$
5. (a) $(0, -2)$; **(b)** $(-3, 0), (3, 0)$
7. (a) $(0, 4)$; **(b)** $(-3, 0), (3, 0), (5, 0)$ **9. (a)** $(0, 5)$; **(b)** $(3, 0)$
11. (a) $(0, -14)$; **(b)** $(4, 0)$ **13. (a)** $\left(0, \frac{10}{3}\right)$; **(b)** $\left(-\frac{5}{2}, 0\right)$
15. (a) $(0, 9)$; **(b)** none

17.

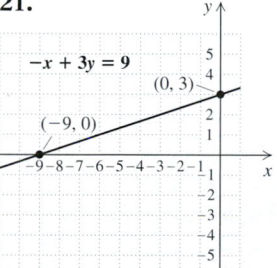

$x + 2y = 6$

19.

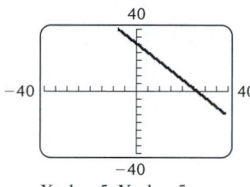

$6x + 9y = 36$

21.

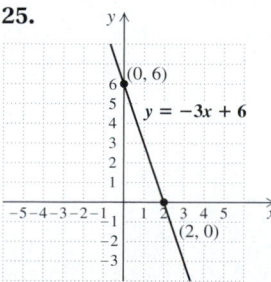

$-x + 3y = 9$

23.

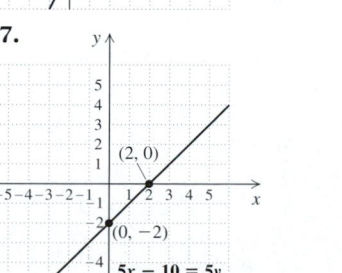

$2x - y = 8$

25.

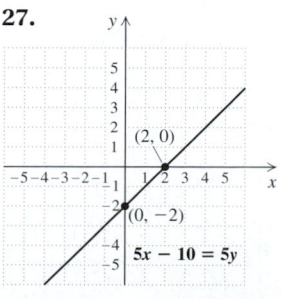

$y = -3x + 6$

27.

$5x - 10 = 5y$

29.

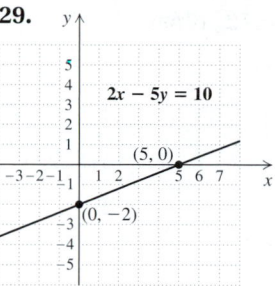

$2x - 5y = 10$

31.

$6x + 2y = 12$

33.

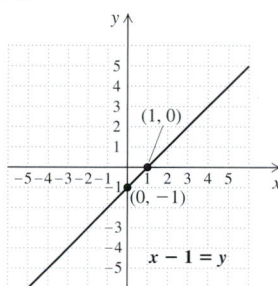

$x - 1 = y$

35.

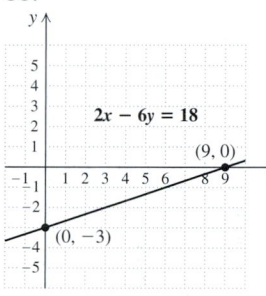

$2x - 6y = 18$

37.

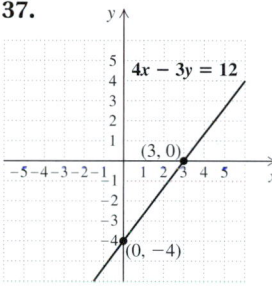

$4x - 3y = 12$

39.

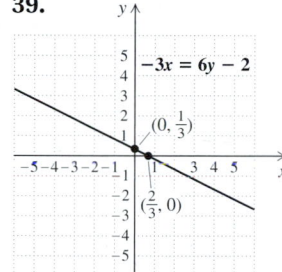

$-3x = 6y - 2$

41.

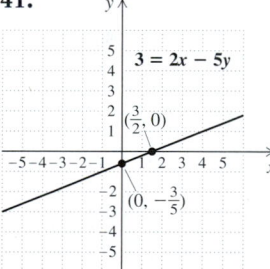

$3 = 2x - 5y$

43.

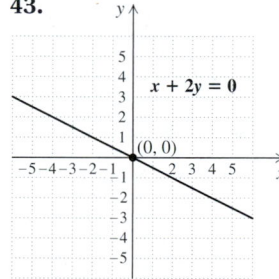

$x + 2y = 0$

45.

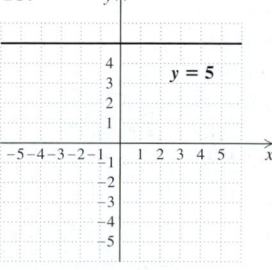

$y = 5$

47.

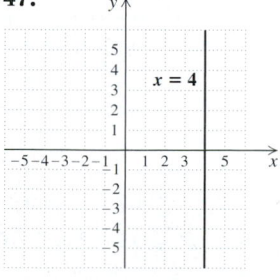

$x = 4$

49.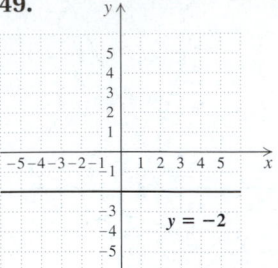

51.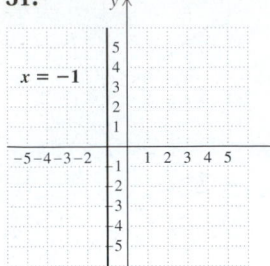

53.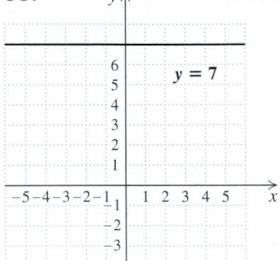

55.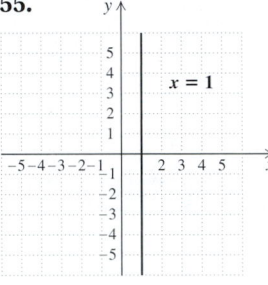

57.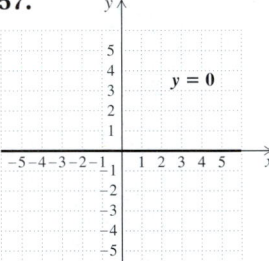

59.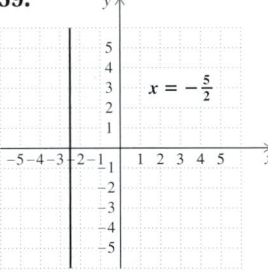

61.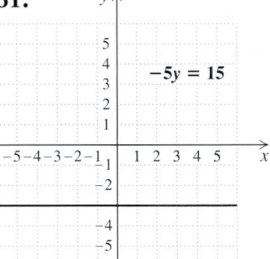

63.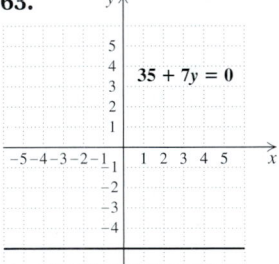

65. $y = -1$ **67.** $x = 4$ **69.** $y = 0$ **71.** 🗒
73. [1.1] $d - 7$ **74.** [1.1] $5 + w$, or $w + 5$
75. [1.1] Let n represent the number; $2 + n$
76. [1.1] Let n represent the number; $3n$
77. [1.1] Let x and y represent the numbers; $2(x + y)$
78. [1.1] Let a and b represent the numbers; $\frac{1}{2}(a + b)$
79. 🗒   **81.** $y = 0$ **83.** $x = -2$ **85.** $(-3, -3)$
87. $-5x + 3y = 15$, or $y = \frac{5}{3}x + 5$ **89.** -24
91. $(0, 25)$; $\left(\frac{50}{3}, 0\right)$, or $(16.\overline{6}, 0)$ **93.** $(0, -9)$, $(45, 0)$
95. $\left(0, -\frac{1}{20}\right)$, or $(0, -0.05)$; $\left(\frac{1}{25}, 0\right)$, or $(0.04, 0)$

Exercise Set 3.4, pp. 164–168

1. **(a)** 21 mpg; **(b)** \$39.33/day; **(c)** 91 mi/day; **(d)** 43¢/mi
3. **(a)** 6 mph; **(b)** \$3.50/hr; **(c)** \$0.58/mi   **5.** **(a)** \$16/hr;
(b) 4.5 pages/hr; **(c)** \$3.56/page   **7.** \$39.50/yr
9. **(a)** 14.5 floors/min; **(b)** 4.14 sec/floor
11. **(a)** 3.71 ft/min; **(b)** 0.27 min/ft
13.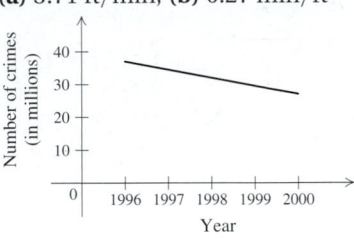

15.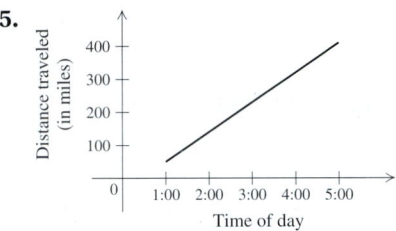

17.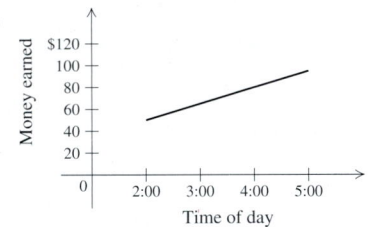

19. 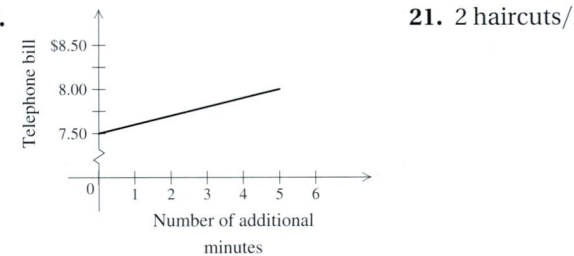   **21.** 2 haircuts/hr

23. 75 mi/hr   **25.** 10¢/min   **27.** $-$\$500/yr
29. 0.04 gal/mi   **31.** 🗒   **33.** [1.6] 5   **34.** [1.6] -6
35. [1.8] -1 **36.** [1.8] $-\frac{4}{3}$ **37.** [1.8] $-\frac{4}{3}$ **38.** [1.8] $-\frac{4}{5}$
39. 🗒
41.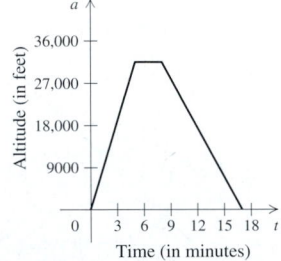

43.

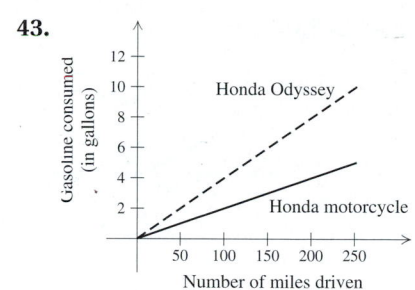

45. 13 ft/sec   **47.** 41.7 min   **49.** 3.6 bu/hr

Exercise Set 3.5, pp. 177–182

1. 15 calories/min   **3.** 1 point/$1000 income
5. -0.75%/yr, or $-3/4\%$/yr   **7.** $\frac{3}{4}$ **9.** $\frac{3}{2}$ **11.** $\frac{1}{3}$
13. -1 **15.** 0   **17.** $-\frac{1}{3}$ **19.** Undefined   **21.** $-\frac{1}{4}$
23. $\frac{3}{2}$ **25.** 0   **27.** -3 **29.** $\frac{3}{2}$ **31.** $-\frac{4}{5}$ **33.** $\frac{7}{9}$
35. $-\frac{2}{3}$ **37.** $-\frac{1}{2}$ **39.** 0   **41.** $-\frac{11}{6}$ **43.** Undefined
45. Undefined   **47.** 0   **49.** Undefined   **51.** 0
53. 8%   **55.** $8.\overline{3}\%$ **57.** $\frac{12}{41}$, or about 29%

59. About 29%   **61.** 🗒   **63.** [2.3] $y = \dfrac{c - ax}{b}$

64. [2.3] $r = \dfrac{p + mn}{x}$ **65.** [2.3] $y = \dfrac{ax - c}{b}$

66. [2.3] $t = \dfrac{q - rs}{n}$ **67.** [1.8] 3   **68.** [1.8] 2   **69.** 🗒

71. $\left\{m \mid -\frac{7}{4} \le m \le 0\right\}$ **73.** $\dfrac{18 - x}{x}$ **75.** $\frac{1}{4}$

Technology Connection, p. 188

1. $y_1 = -\frac{3}{4}x - 2, \ y_2 = -\frac{1}{5}x - 2,$
          $y_3 = -\frac{3}{4}x - 5, \ y_4 = -\frac{1}{5}x - 5$

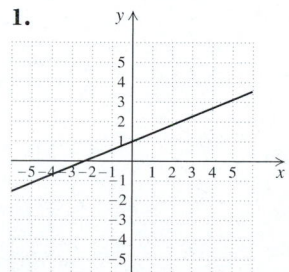

Exercise Set 3.6, pp. 190–191

1.
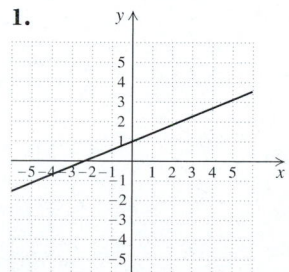

3.
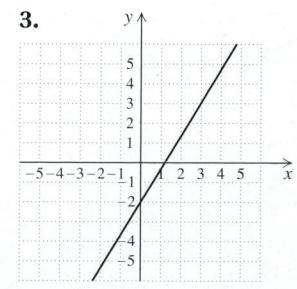

5.
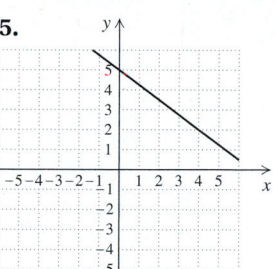

7.
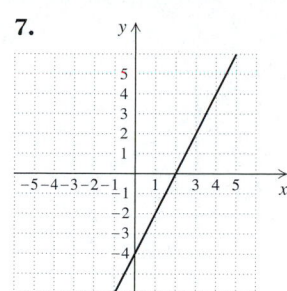

9.
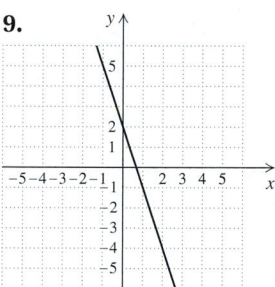

11. $\frac{3}{7}$; $(0, 5)$

13. $-\frac{5}{6}$; $(0, 2)$ **15.** $\frac{9}{4}$; $(0, -7)$ **17.** $-\frac{2}{5}$; $(0, 0)$
19. 2; $(0, 4)$ **21.** $\frac{3}{4}$; $(0, -3)$ **23.** $\frac{1}{5}$; $\left(0, \frac{8}{5}\right)$ **25.** 0; $(0, 4)$
27. $y = 3x + 7$ **29.** $y = \frac{7}{8}x - 1$ **31.** $y = -\frac{5}{3}x - 8$
33. $y = 3$

35.
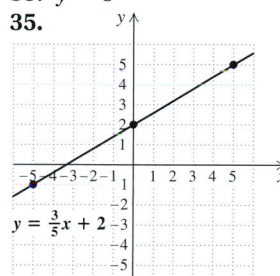

37.
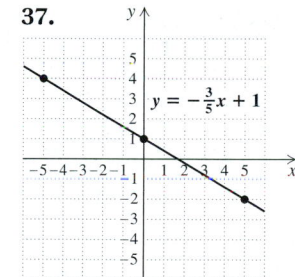

39.
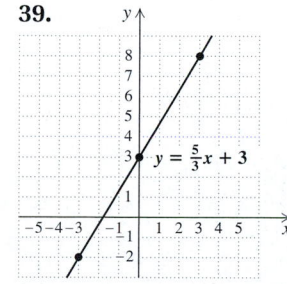

41.
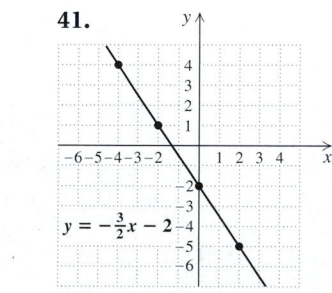

43.
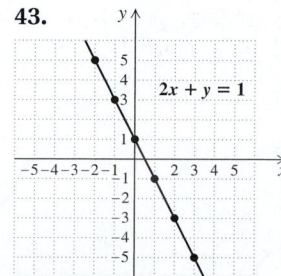

45.
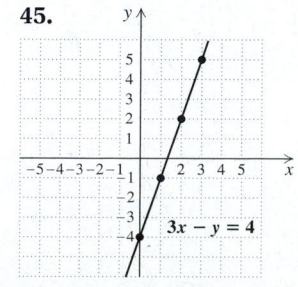

47.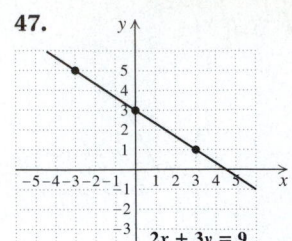

49.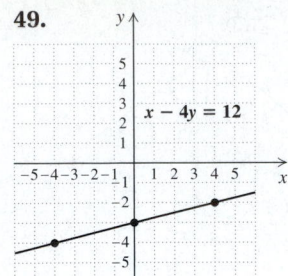

51. 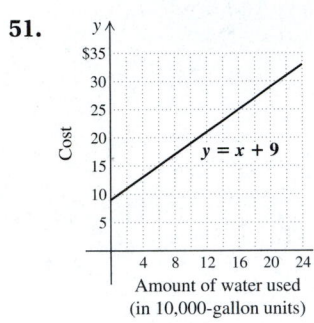 $1 per 10,000 gallons

53. $y = 1.5x + 16$ **55.** Yes   **57.** No   **59.** Yes
61. Yes   **63.** No   **65.** Yes   **67.** $y = 5x + 11$
69. $y = \frac{1}{2}x$ **71.** $y = x + 3$ **73.** $y = x - 4$ **75.**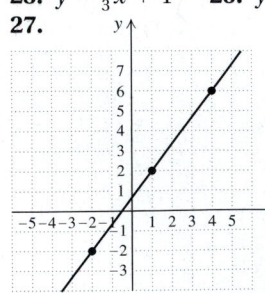
77. [2.3] $y = m(x - h) + k$ **78.** [2.3] $y = -2(x + 4) + 9$
79. [1.6] 2   **80.** [1.6] 16   **81.** [1.6] −9   **82.** [1.6] −10
83. 🔲   **85.** When $x = 0$, $y = b$, so $(0, b)$ is on the line.
When $x = 1$, $y = m + b$, so $(1, m + b)$ is on the line. Then
$$\text{slope} = \frac{(m + b) - b}{1 - 0} = m.$$
87. $y = \frac{1}{3}x + 3$ **89.** $y = \frac{5}{2}x + 1$

Exercise Set 3.7, pp. 196–197

1. $y - 7 = 6(x - 2)$ **3.** $y - 2 = \frac{3}{5}(x - 9)$
5. $y - 1 = -4(x - 3)$ **7.** $y - (-4) = \frac{3}{2}(x - 5)$
9. $y - 6 = \frac{5}{4}(x - (-2))$ **11.** $y - (-1) = -2(x - (-4))$
13. $y - 8 = 1(x - (-2))$ **15.** $y = 2x - 3$
17. $y = \frac{7}{4}x - 9$ **19.** $y = -3x - 2$ **21.** $y = -4x - 9$
23. $y = \frac{2}{3}x + 1$ **25.** $y = -\frac{5}{6}x + \frac{9}{2}$
27.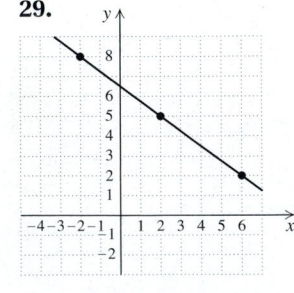

29.

31.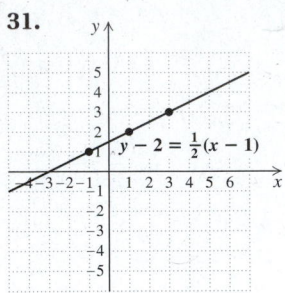

33.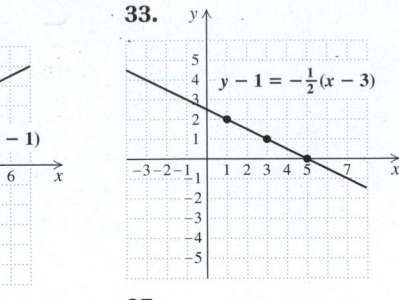

35.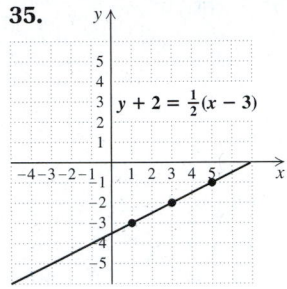

37.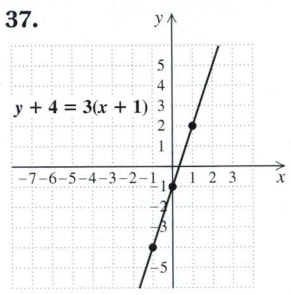

39.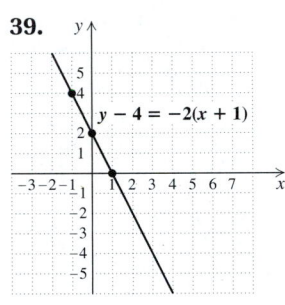

41.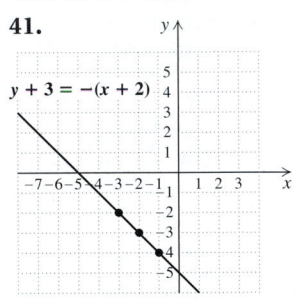

43.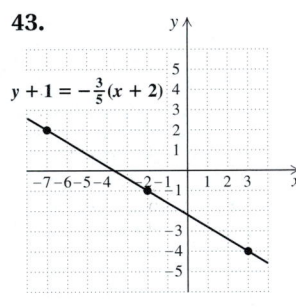

45.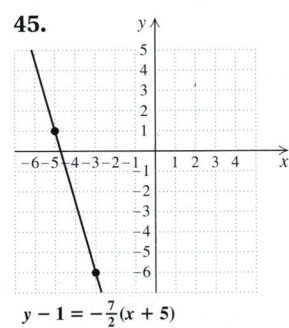

47. 🔲   **49.** [1.8] −125   **50.** [1.8] 64   **51.** [1.8] 8
52. [1.8] 24   **53.** [1.8] −72   **54.** [1.8] −4   **55.** 🔲
57.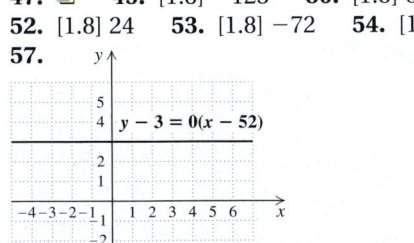

59. $y - 7 = \frac{5}{2}(x - 3); y - 2 = \frac{5}{2}(x - 1)$

61. $y - 8 = \frac{3}{2}(x - 3); y - 2 = \frac{3}{2}(x - (-1))$

63. $y - 8 = -\frac{5}{2}(x - (-3)); y - (-2) = -\frac{5}{2}(x - 1)$

65. $y = 2x - 9$ **67.** $y = -\frac{4}{3}x + \frac{23}{3}$ **69.** $y = -x + 6$

71. $y = \frac{2}{3}x + 3$ **73.** $y = \frac{2}{5}x - 2$ **75.** $y = \frac{3}{4}x - \frac{5}{2}$

77. $y - 7 = -\frac{2}{3}(x - (-4))$ **79.** $y = -4x + 7$

81. $y = -\frac{3}{10}x + \frac{47}{10}$ **83.**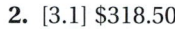

Review Exercises: Chapter 3, pp. 199–200

1. [3.1] $306,000 **2.** [3.1] $318.50

3.–5. [3.1] **6.** [3.1] IV

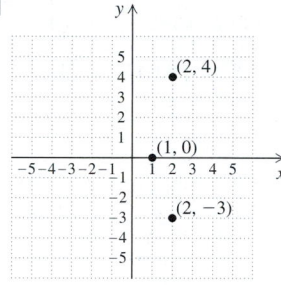

7. [3.1] III **8.** [3.1] II **9.** [3.1] $(-5, -1)$

10. [3.1] $(-2, 5)$ **11.** [3.1] $(3, 0)$ **12.** [3.2] No

13. [3.2] Yes

14. [3.2]

$$\begin{array}{c|c} 2x - y = 3 & 2x - y = 3 \\ \hline 2 \cdot 0 - (-3) \ ? \ 3 & 2 \cdot 2 - 1 \ ? \ 3 \\ 0 + 3 \ | & 4 - 1 \ | \\ 3 \ | \ 3 \ \text{TRUE} & 3 \ | \ 3 \ \text{TRUE} \end{array}$$

$(-1, -5)$; answers may vary

15. [3.2] **16.** [3.2]

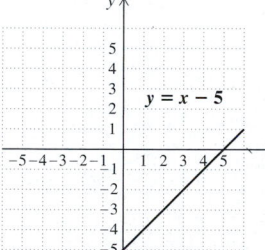 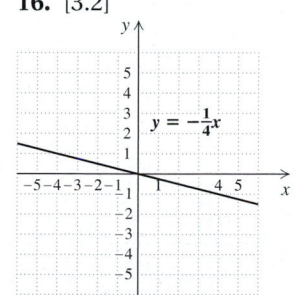

17. [3.2] **18.** [3.2]

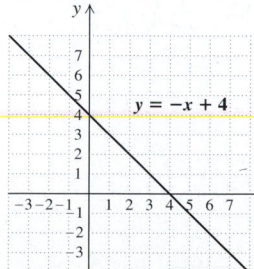 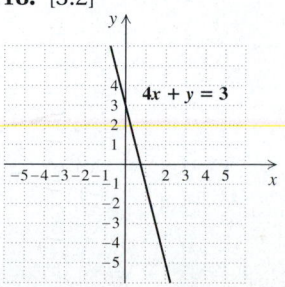

19. [3.3] **20.** [3.3]

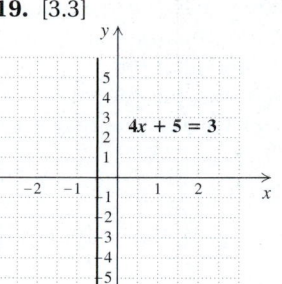 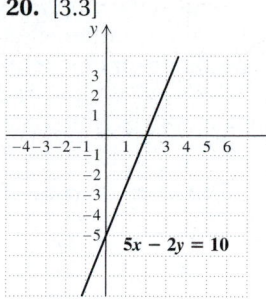

21. [3.4] **(a)** $\frac{4}{9}$ meal/min; **(b)** $2\frac{1}{4}$ min/meal

22. [3.4] 2.25 million/yr **23.** [3.5] 0 **24.** [3.5] $\frac{7}{3}$

25. [3.5] $-\frac{3}{7}$ **26.** [3.5] $\frac{3}{2}$ **27.** [3.5] 0

28. [3.5] Undefined **29.** [3.5] 2 **30.** [3.5] 7%

31. [3.3] x-intercept: $(6, 0)$; y-intercept: $(0, 9)$

32. [3.6] $-\frac{1}{2}$; $(0, 5)$ **33.** [3.6] $y = -\frac{3}{4}x + 6$

34. [3.7] $y - 6 = -\frac{1}{2}(x - 3)$ **35.** [3.7] $y = 4x + 5$

36. [3.6] Perpendicular **37.** [3.6] Parallel

38. [3.6] **39.** [3.3]

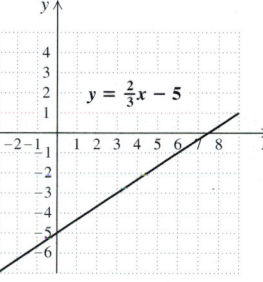 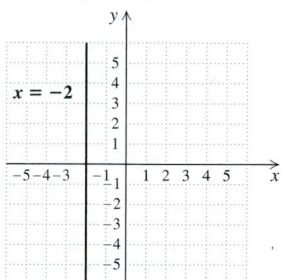

40. [3.3] **41.** [3.3]

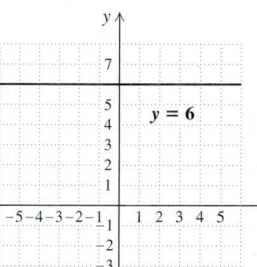

42. [3.7]

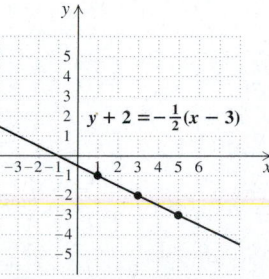

43. [3.1], [3.4] 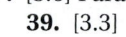 A business might use a graph to plot how total sales changes from year to year or to visualize rate of growth or production. **44.** [3.2] The y-intercept is the

point at which the graph crosses the *y*-axis. Since a point on the *y*-axis is neither left nor right of the origin, the first or *x*-coordinate of the point is 0.   **45.** [3.2] −1
46. [3.2] 19   **47.** [3.1] Area: 45 sq units; perimeter: 28 units   **48.** [3.2] (0, 4), (1, 3), (−1, 3); answers may vary

Test: Chapter 3, pp. 200–201

1. [3.1] $112.32   **2.** [3.1] $777.60   **3.** [3.1] II   **4.** [3.1] III
5. [3.1] (3, 4)   **6.** [3.1] (0, −4)   **7.** [3.1] (−5, 2)
8. [3.2]   **9.** [3.3]

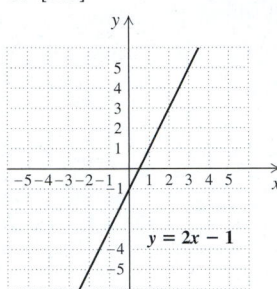

10. [3.3]   **11.** [3.2]

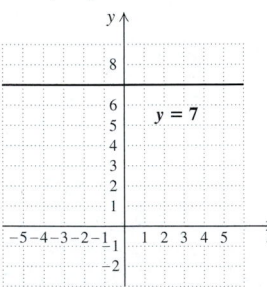   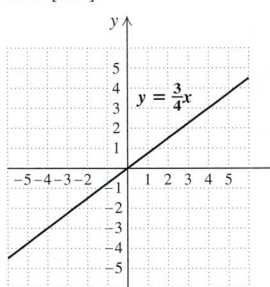

12. [3.3]   **13.** [3.2]

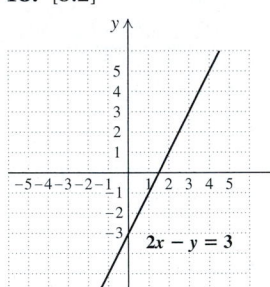

14. [3.3]

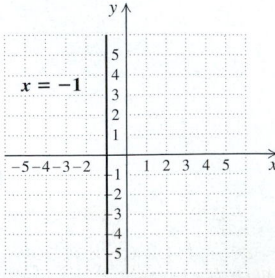

15. [3.3] *x*-intercept: (6, 0); *y*-intercept: (0, −10)
16. [3.3] *x*-intercept: (10, 0); *y*-intercept: $\left(0, \frac{5}{2}\right)$

17. [3.5] $\frac{9}{2}$ **18.** [3.5] $\frac{7}{12}$ **19.** [3.4] $\frac{1}{3}$ km/min
20. [3.6] 3; (0, 7)   **21.** [3.6] Parallel
22. [3.6] Perpendicular
23. [3.6]   **24.** [3.7]

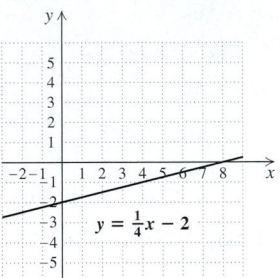   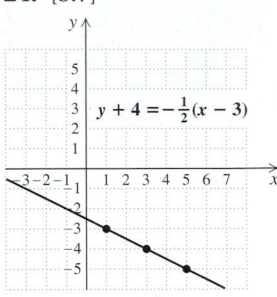

25. [3.7] $y - 8 = -3(x - 6)$ **26.** [3.6] $y = \frac{2}{5}x + 9$
27. [3.1] Area: 25 sq units; perimeter: 20 units

Cumulative Review: Chapters 1–3, pp. 201–202

1. [1.1] 15   **2.** [1.2] $12x - 15y + 21$
3. [1.2] $3(5x - 3y + 1)$ **4.** [1.3] $2 \cdot 3 \cdot 7$ **5.** [1.4] 0.45
6. [1.4] 4   **7.** [1.6] $\frac{1}{4}$ **8.** [1.7] −4   **9.** [1.6] $-x - y$
10. [2.4] 0.785   **11.** [1.3] $\frac{11}{60}$ **12.** [1.5] 2.6
13. [1.7] 7.28   **14.** [1.7] $-\frac{5}{12}$ **15.** [1.8] −2
16. [1.8] 27   **17.** [1.8] $-2y - 7$ **18.** [1.8] $5x + 11$
19. [2.1] −1.2   **20.** [2.1] −21   **21.** [2.2] 9
22. [2.1] $-\frac{20}{3}$ **23.** [2.2] 2   **24.** [2.1] $\frac{13}{8}$ **25.** [2.2] $-\frac{17}{21}$
26. [2.2] −17   **27.** [2.2] 2   **28.** [2.6] $\{x \mid x < 16\}$

29. [2.6] $\left\{x \mid x \le -\frac{11}{8}\right\}$ **30.** [2.3] $h = \dfrac{A - \pi r^2}{2\pi r}$

31. [3.1] IV   **32.** [2.6]

$-1 < x \le 2$

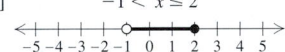

33. [3.3]   **34.** [3.3]

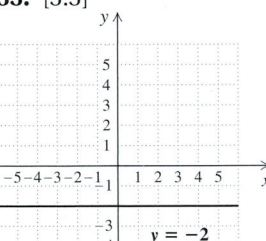   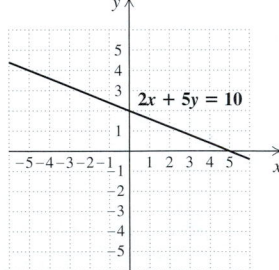

35. [3.2]   **36.** [3.2]

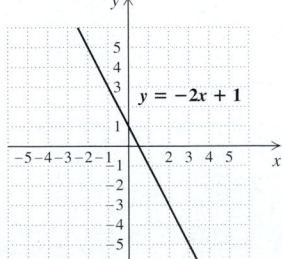   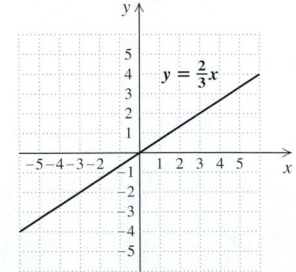

37. [3.3] $(10.5, 0)$; $(0, -3)$ **38.** [3.3] $(-1.25, 0)$; $(0, 5)$
39. [2.4] 160 million   **40.** [2.5] 15.6 million
41. [2.4] \$120   **42.** [2.5] 50 m, 53 m, 40 m   **43.** [2.5] 8
44. [3.4] \$40/person   **45.** [3.5] $-\frac{1}{3}$
46. [3.6] $y = \frac{2}{7}x - 4$ **47.** [3.6] $-\frac{1}{3}$; $(0, 3)$
48. [3.6]   **49.** [3.3]

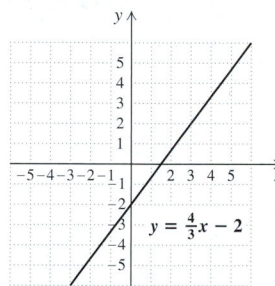

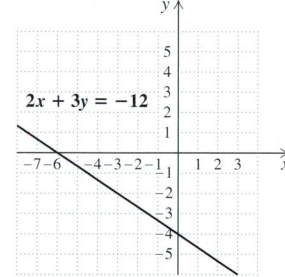

$y = \frac{4}{3}x - 2$ $2x + 3y = -12$

50. [3.7] $y - 4 = -\frac{3}{8}(x - (-6))$ **51.** [2.4] \$25,000
52. [1.4], [2.2] $-4, 4$ **53.** [2.2] 2   **54.** [2.2] -5
55. [2.2] 3   **56.** [2.2] No solution
57. [2.3] $Q = \dfrac{2 - pm}{p}$
58. [3.6] $y = -\frac{7}{3}x + 7$; $y = -\frac{7}{3}x - 7$; $y = \frac{7}{3}x - 7$;
$y = \frac{7}{3}x + 7$

CHAPTER 4

Exercise Set 4.1, pp. 210–211

1. r^{10} **3.** 9^8 **5.** a^7 **7.** 5^{15} **9.** $(3y)^{12}$ **11.** $(5t)^7$
13. $a^5 b^9$ **15.** $(x + 1)^{12}$ **17.** r^{10} **19.** $x^4 y^7$ **21.** 7^3
23. x^{12} **25.** t^4 **27.** $5a$ **29.** 1   **31.** $3m^3$ **33.** $a^7 b^6$
35. $m^9 n^4$ **37.** 1   **39.** 5   **41.** 2   **43.** -4 **45.** x^{28}
47. 5^{16} **49.** m^{35} **51.** t^{80} **53.** $49x^2$ **55.** $-8a^3$
57. $16m^6$ **59.** $a^{14} b^7$ **61.** $x^8 y^7$ **63.** $24x^{19}$ **65.** $\dfrac{a^3}{64}$
67. $\dfrac{49}{25a^2}$ **69.** $\dfrac{a^{20}}{b^{15}}$ **71.** $\dfrac{y^6}{4}$ **73.** $\dfrac{x^8 y^4}{z^{12}}$ **75.** $\dfrac{a^{12}}{16b^{20}}$
77. $\dfrac{125x^{21} y^3}{8z^{12}}$ **79.** 1   **81.** 🖩   **83.** [1.2] $3(s - r + t)$
84. [1.2] $-7(x - y + z)$ **85.** [1.6] $8x$
86. [1.6] $-3a - 6b$ **87.** [1.2] $2y + 3x$
88. [1.2] $5z + 2xy$ **89.** 🖩   **91.** 🖩
93. Let $a = 1$; then $(a + 5)^2 = 36$, but $a^2 + 5^2 = 26$.
95. Let $a = 0$; then $\dfrac{a + 7}{7} = 1$, but $a = 0$.   **97.** a^{8k}
99. $\frac{16}{375}$ **101.** 13   **103.** $<$ **105.** $<$ **107.** $>$
109. 4,000,000; 4,194,304; 194,304
111. 2,000,000,000; 2,147,483,648; 147,483,648
113. 65,536

Technology Connection, p. 217

1. 20.75

Exercise Set 4.2, pp. 217–220

1. $7x^4, x^3, -5x, 8$ **3.** $-t^4, 7t^3, -3t^2, 6$
5. Coefficients: 4, 7; degrees: 5, 1
7. Coefficients: 9, -3, 4; degrees: 2, 1, 0
9. Coefficients: 7, 9, 1; degrees: 4, 1, 3
11. Coefficients: 1, -1, 4, -3; degrees: 4, 3, 1, 0
13. (a) 3, 5, 2; (b) $7a^5, 7$; (c) 5   **15.** (a) 1, 0, 2; (b) $4t^2, 4$;
(c) 2   **17.** (a) 4, 2, 7, 0; (b) $x^7, 1$; (c) 7   **19.** (a) 1, 4, 0, 3;
(b) $-a^4, -1$; (c) 4
21.

Term	Coefficient	Degree of the Term	Degree of the Polynomial
$8x^5$	8	5	
$-\frac{1}{2}x^4$	$-\frac{1}{2}$	4	
$-4x^3$	-4	3	5
$7x^2$	7	2	
6	6	0	

23. Trinomial   **25.** None of these   **27.** Binomial
29. Monomial   **31.** $11x^2 + 3x$ **33.** $4a^4$
35. $6x^2 - 3x$ **37.** $5x^3 - x + 5$ **39.** $-x^4 - x^3$
41. $\frac{1}{15}x^4 + 10$ **43.** $3.4x^2 + 1.3x + 5.5$
45. $10t^4 - 12t^3 + 17t$ **47.** -16 **49.** 16   **51.** -3
53. 11   **55.** 55   **57.** About 9   **59.** About 6
61. About 15   **63.** 1112 ft   **65.** Approximately 449
67. \$9200   **69.** \$28,000   **71.** 62.8 cm   **73.** 153.86 m^2
75. 🖩   **77.** [1.5] 5   **78.** [1.6] -9 **79.** [1.2] $5(x + 3)$
80. [1.2] $7(a - 3)$ **81.** [3.4] 6.25¢/mi
82. [2.5] 274 and 275   **83.** 🖩
85. $-6x^5 + 14x^4 - x^2 + 11$; answers may vary   **87.** 10
89. $5x^9 + 4x^8 + x^2 + 5x$ **91.** $x^3 - 2x^2 - 6x + 3$
93. 85.0   **95.** 50 yr
97.

t	$-t^2 + 6t - 4$
1	1
2	4
3	5
4	4
5	1

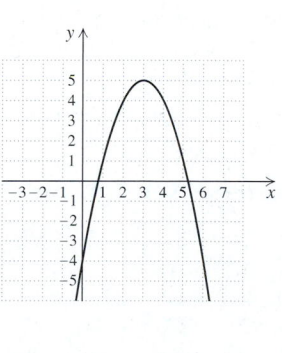

Technology Connection, p. 225

1. In each case, let y_1 = the expression before the addition
or subtraction has been performed, y_2 = the simplified

sum or difference, and $y_3 = y_2 - y_1$; and note that the graph of y_3 coincides with the x-axis. That is, $y_3 = 0$.

Exercise Set 4.3, pp. 226–228

1. $-5x + 9$ **3.** $x^2 - 5x - 1$ **5.** $9t^2 + 5t - 3$
7. $6m^3 + 3m^2 - 3m - 9$ **9.** $7 + 13a + 6a^2 + 14a^3$
11. $9x^8 + 8x^7 - 3x^4 + 2x^2 - 2x + 5$
13. $-\frac{1}{2}x^4 + \frac{2}{3}x^3 + x^2$ **15.** $4.2t^3 + 3.5t^2 - 6.4t - 1.8$
17. $-3x^4 + 3x^2 + 4x$
19. $1.05x^4 + 0.36x^3 + 14.22x^2 + x + 0.97$
21. $-(-t^3 + 4t^2 - 9); t^3 - 4t^2 + 9$
23. $-(12x^4 - 3x^3 + 3); -12x^4 + 3x^3 - 3$ **25.** $-8x + 9$
27. $-3a^4 + 5a^2 - 9$ **29.** $4x^4 - 6x^2 - \frac{3}{4}x + 8$
31. $9x + 3$ **33.** $-t^2 - 7t + 5$
35. $6x^4 + 3x^3 - 4x^2 + 3x - 4$
37. $4.6x^3 + 9.2x^2 - 3.8x - 23$ **39.** 0
41. $4 + 2a + 7a^2 - 3a^3$ **43.** $\frac{3}{4}x^3 - \frac{1}{2}x$
45. $0.05t^3 - 0.07t^2 + 0.01t + 1$ **47.** $3x + 5$
49. $11x^4 + 12x^3 - 10x^2$ **51. (a)** $5x^2 + 4x$; **(b)** 145; 273
53. $14y + 25$ **55.** $(r + 11)(r + 9); 9r + 99 + 11r + r^2$
57. $(x + 3)^2; x^2 + 3x + 9 + 3x$ **59.** $\pi r^2 - 25\pi$
61. $18z - 64$ **63.** $y^2 - 4y + 4$ **65.**   **67.** [1.8] 0
68. [1.8] 0   **69.** [1.8] $13t + 14$ **70.** [1.8] $14t - 15$
71. [2.6] $\left\{x \mid x < \frac{14}{3}\right\}$ **72.** [2.6] $\{x \mid x \le 12\}$ **73.**
75. $9t^2 - 20t + 11$ **77.** $-10y^2 - 2y - 10$
79. $-3y^4 - y^3 + 5y - 2$ **81.** $250.591x^3 + 2.812x$
83. $20w + 42$ **85.** $2x^2 + 20x$ **87.**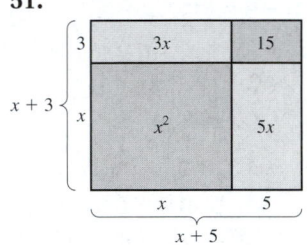

Technology Connection, p. 234

1. Let $y_1 = (-2x^2 - 3)(5x^3 - 3x + 4)$ and $y_2 = -10x^5 - 9x^3 - 8x^2 + 9x - 12$. With the table set in AUTO mode, note that the values in the Y1- and Y2-columns match, regardless of how far we scroll up or down.
2. Use TRACE, a Table, or a boldly drawn graph to confirm that Y3 is always 0.

Exercise Set 4.4, pp. 234–236

1. $30x^4$ **3.** x^3 **5.** $-x^8$ **7.** $28t^8$ **9.** $-0.02x^{10}$
11. $\frac{1}{15}x^4$ **13.** 0 **15.** $-28x^{11}$ **17.** $-3x^2 + 15x$
19. $4x^2 + 4x$ **21.** $3a^2 + 27a$ **23.** $x^5 + x^2$
25. $6x^3 - 18x^2 + 3x$ **27.** $15t^3 + 30t^2$ **29.** $-6x^4 - 6x^3$
31. $4a^9 - 8a^7 - \frac{5}{12}a^4$ **33.** $x^2 + 9x + 18$
35. $x^2 + 3x - 10$ **37.** $a^2 - 13a + 42$ **39.** $x^2 - 9$
41. $25 - 15x + 2x^2$ **43.** $t^2 + \frac{17}{6}t + 2$
45. $\frac{3}{16}a^2 + \frac{5}{4}a - 2$

47.
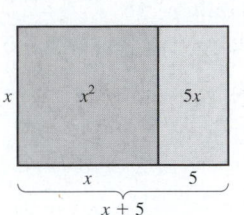

49.
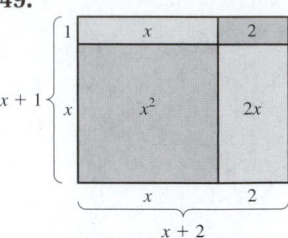

51.
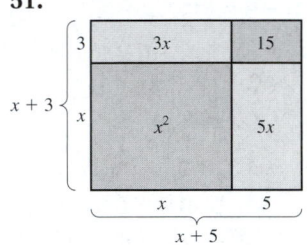

53.
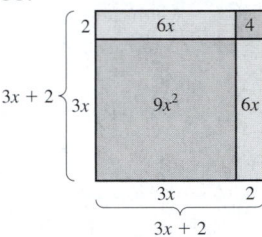

55. $x^3 + 4x + 5$ **57.** $2a^3 - a^2 - 11a + 10$
59. $2y^5 - 13y^3 + y^2 - 7y - 7$
61. $-15x^5 + 26x^4 - 7x^3 - 3x^2 + x$
63. $x^4 - 2x^3 - x + 2$ **65.** $6t^4 - 17t^3 - 5t^2 - t - 4$
67. $x^4 + 8x^3 + 12x^2 + 9x + 4$
69. $2x^4 - 5x^3 + 5x^2 - \frac{19}{10}x + \frac{1}{5}$ **71.** 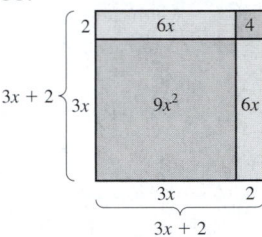   **73.** [1.8] 6
74. [1.8] 31   **75.** [1.8] 0   **76.** [1.8] 0   **77.**
79. $75y^2 - 45y$ **81.** 5   **83.** $V = 4x^3 - 48x^2 + 144x$ in³;
$S = -4x^2 + 144$ in²   **85.** $x^3 + 2x^2 - 210$ m³
87. 16 ft by 8 ft   **89.** 0   **91.** 0

Exercise Set 4.5, pp. 243–245

1. $x^3 + 4x^2 + 3x + 12$ **3.** $x^4 + 2x^3 + 6x + 12$
5. $y^2 - y - 6$ **7.** $9x^2 + 21x + 10$ **9.** $5x^2 + 4x - 12$
11. $2 + 3t - 9t^2$ **13.** $2x^2 - 9x + 7$ **15.** $p^2 - \frac{1}{16}$
17. $x^2 - 0.01$ **19.** $2x^3 + 2x^2 + 6x + 6$
21. $-2x^2 - 11x + 6$ **23.** $a^2 + 18a + 81$
25. $1 - 2t - 15t^2$ **27.** $x^5 + 3x^3 - x^2 - 3$
29. $3x^6 - 2x^4 - 6x^2 + 4$ **31.** $4t^6 + 16t^3 + 15$
33. $8x^5 + 16x^3 + 5x^2 + 10$ **35.** $4x^3 - 12x^2 + 3x - 9$
37. $x^2 - 64$ **39.** $4x^2 - 1$ **41.** $25m^2 - 4$ **43.** $4x^4 - 9$
45. $9x^8 - 1$ **47.** $x^8 - 49$ **49.** $t^2 - \frac{9}{16}$
51. $x^2 + 4x + 4$ **53.** $9x^{10} + 6x^5 + 1$ **55.** $a^2 - \frac{4}{5}a + \frac{4}{25}$
57. $t^6 + 6t^3 + 9$ **59.** $4 - 12x^4 + 9x^8$
61. $25 + 60t^2 + 36t^4$ **63.** $49x^2 - 4.2x + 0.09$
65. $10a^5 - 5a^3$ **67.** $a^3 - a^2 - 10a + 12$
69. $9 - 12x^3 + 4x^6$ **71.** $4x^3 + 24x^2 - 12x$
73. $t^6 - 2t^3 + 1$ **75.** $15t^5 - 3t^4 + 3t^3$
77. $36x^8 - 36x^4 + 9$ **79.** $12x^3 + 8x^2 + 15x + 10$
81. $25 - 60x^4 + 36x^8$ **83.** $a^3 + 1$ **85.** $a^2 + 2a + 1$
87. $x^2 + 7x + 10$ **89.** $x^2 + 14x + 49$
91. $t^2 + 10t + 24$ **93.** $t^2 + 13t + 36$

95. $9x^2 + 24x + 16$ **97.**

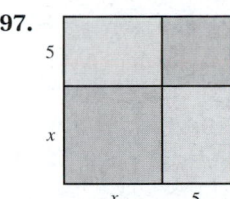

99. 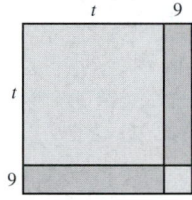 **101.**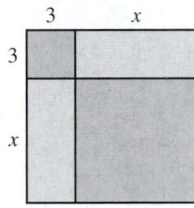

103. **105.** [2.5] Lamps: 500 watts; air conditioner: 2000 watts; television: 50 watts **106.** [3.1] II

107. [2.3] $y = \dfrac{8}{5x}$ **108.** [2.3] $a = \dfrac{c}{3b}$

109. [2.3] $x = \dfrac{b + c}{a}$ **110.** [2.3] $t = \dfrac{u - r}{s}$ **111.**

113. $16x^4 - 81$ **115.** $81t^4 - 72t^2 + 16$
117. $t^{24} - 4t^{18} + 6t^{12} - 4t^6 + 1$ **119.** 396 **121.** -7
123. $l^3 - l$ **125.** $Q(Q - 14) - 5(Q - 14)$,
$(Q - 5)(Q - 14)$; other equivalent expressions are possible.
127. $(y + 1)(y - 1)$, $y(y + 1) - y - 1$; other equivalent
expressions are possible. **129.**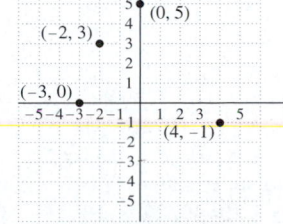

Technology Connection, p. 250

1. 36.22 **2.** 22,312

Exercise Set 4.6, pp. 250–253

1. -7 **3.** -92 **5.** 2.97 L **7.** 360.4 m
9. 63.78125 in^2 **11.** Coefficients: 1, -2, 3, -5; degrees:
4, 2, 2, 0; 4 **13.** Coefficients: 17, -3, -7; degrees: 5, 5, 0; 5
15. $3a - 2b$ **17.** $3x^2y - 2xy^2 + x^2 + 5x$
19. $8u^2v - 5uv^2 + 7u^2$ **21.** $6a^2c - 7ab^2 + a^2b$
23. $3x^2 - 4xy + 3y^2$ **25.** $-6a^4 - 8ab + 7ab^2$
27. $-6r^2 - 5rt - t^2$ **29.** $3x^3 - x^2y + xy^2 - 3y^3$
31. $10y^4x^2 - 8y^3x$ **33.** $-8x + 8y$ **35.** $6z^2 + 7uz - 3u^2$
37. $x^2y^2 + 3xy - 28$ **39.** $4a^2 - b^2$ **41.** $15r^2t^2 - rt - 2$
43. $m^6n^2 + 2m^3n - 48$ **45.** $30x^2 - 28xy + 6y^2$
47. $0.4p^2q^2 - 0.02pq - 0.02$ **49.** $x^2 + 2xh + h^2$
51. $16a^2 + 40ab + 25b^2$ **53.** $c^4 - d^2$ **55.** $a^2b^2 - c^2d^4$
57. $a^2 + 2ab + b^2 - c^2$ **59.** $a^2 - b^2 - 2bc - c^2$
61. $a^2 + ab + ac + bc$ **63.** $x^2 - z^2$
65. $x^2 + y^2 + z^2 + 2xy + 2xz + 2yz$ **67.** $\frac{1}{2}x^2 + \frac{1}{2}xy - y^2$

69. We draw a rectangle with dimensions $r + s$ by $u + v$.

71. 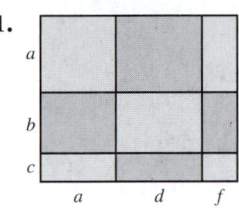 **73.** **75.** [1.8] 12

76. [1.8] 5 **77.** [1.8] 27 **78.** [1.8] 36 **79.** [1.8] 7
80. [1.8] 5 **81.** **83.** $2\pi ab - \pi b^2$ **85.** $a^2 - 4b^2$
87. (a) $A^2 - B^2$; (b) $(A - B)^2 + (A - B)B + (A - B)B = A^2 - 2AB + B^2 + AB - B^2 + AB - B^2 = A^2 - B^2$
89. $2x^2 - 2\pi r^2 + 4xh + 2\pi rh$
91. $x^4 - b^2x^2 - a^2x^2 + a^2b^2$ **93.** $P - 2Pr + Pr^2$
95. \$52,756.35

Exercise Set 4.7, pp. 259–260

1. $5x^5 - 2x$ **3.** $1 - 2u + u^6$ **5.** $5t^2 - 8t + 2$
7. $-5x^4 + 4x^2 + 1$ **9.** $6t^2 - 10t + \frac{3}{2}$ **11.** $3x^2 - 5x + \frac{1}{2}$
13. $2x + 3 + \dfrac{2}{x}$ **15.** $-3rs - r + 2s$ **17.** $x + 6$
19. $t - 5 + \dfrac{-45}{t - 5}$ **21.** $2x - 1 + \dfrac{1}{x + 6}$
23. $a^2 - 2a + 4$ **25.** $t + 4 + \dfrac{1}{t - 4}$ **27.** $x + 4$
29. $3a + 1 + \dfrac{3}{2a + 5}$ **31.** $t^2 - 3t + 1$
33. $t^2 - 2t + 3 + \dfrac{-4}{t + 1}$ **35.** $t^2 - 1 + \dfrac{3t - 1}{t^2 + 5}$
37. $2x^2 + 1 - \dfrac{x}{2x^2 - 3}$ **39.** **41.** [1.5] -17
42. [1.5] -23 **43.** [1.6] -2 **44.** [1.6] 5
45. [2.5] 167.5 ft **46.** [2.2] 2
47. [3.2] **48.** [3.1]

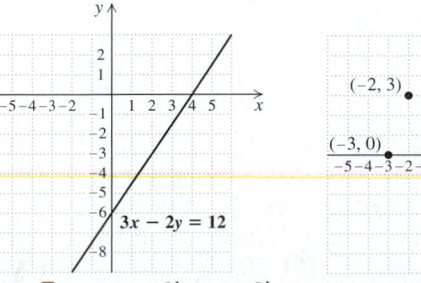

49. **51.** $5x^{6k} - 16x^{3k} + 14$ **53.** $3t^{2h} + 2t^h - 5$

55. $a + 3 + \dfrac{5}{5a^2 - 7a - 2}$ **57.** $2x^2 + x - 3$ **59.** 3
61. -1

Technology Connection, p. 267

1. 1.71×10^{17} **2.** $5.\overline{370} \times 10^{-15}$ **3.** 3.68×10^{16}

Exercise Set 4.8, pp. 267–269

1. $\dfrac{1}{5^2} = \dfrac{1}{25}$ **3.** $\dfrac{1}{10^4} = \dfrac{1}{10{,}000}$ **5.** $\dfrac{1}{(-2)^6} = \dfrac{1}{64}$ **7.** $\dfrac{1}{x^8}$

9. $\dfrac{x}{y^2}$ **11.** $\dfrac{t}{r^5}$ **13.** t^7 **15.** h^8 **17.** $\dfrac{1}{7}$

19. $\left(\dfrac{5}{2}\right)^2 = \dfrac{25}{4}$ **21.** $\left(\dfrac{2}{a}\right)^3 = \dfrac{8}{a^3}$ **23.** $\left(\dfrac{t}{s}\right)^7 = \dfrac{t^7}{s^7}$

25. 7^{-2} **27.** t^{-6} **29.** a^{-4} **31.** p^{-8} **33.** 5^{-1}

35. t^{-1} **37.** 2^3, or 8 **39.** $\dfrac{1}{x^9}$ **41.** $\dfrac{1}{t^2}$ **43.** $\dfrac{1}{a^{18}}$

45. t^{18} **47.** $\dfrac{1}{t^{12}}$ **49.** x^8 **51.** $\dfrac{1}{a^3 b^3}$ **53.** $\dfrac{1}{m^7 n^7}$

55. $\dfrac{9}{x^8}$ **57.** $\dfrac{25t^6}{r^8}$ **59.** t^{10} **61.** $\dfrac{1}{y^4}$ **63.** y^5 **65.** x^5

67. $\dfrac{b^9}{a^7}$ **69.** 1 **71.** $\dfrac{3y^6 z^2}{x^5}$ **73.** $3s^2 t^4 u^4$ **75.** $\dfrac{1}{x^{12} y^{15}}$

77. $x^{24} y^8$ **79.** $\dfrac{b^5 c^4}{a^8}$ **81.** $\dfrac{9}{a^8}$ **83.** $49x^6$ **85.** $\dfrac{n^{12}}{m^3}$

87. $\dfrac{27b^{12}}{8a^6}$ **89.** 1 **91.** 71,200 **93.** 0.00892

95. 904,000,000 **97.** 0.0000000002764 **99.** 42,090,000
101. 4.9×10^5 **103.** 5.83×10^{-3} **105.** 7.8×10^{10}
107. 9.07×10^{17} **109.** 5.27×10^{-7} **111.** 1.8×10^{-8}
113. 1.094×10^{15} **115.** 8×10^{12} **117.** 2.47×10^8
119. 3.915×10^{-16} **121.** 2.5×10^{13} **123.** 5×10^{-4}
125. 3×10^{-21} **127.** **129.** [1.8] 15 **130.** [1.8] 49
131. [1.8] 78 **132.** [1.8] 6
133. [3.1]

134. [2.3] $t = \dfrac{r - cx}{b}$ **135.** **137.** 7×10^{23}

139. 4×10^{-10} **141.** 3^{11} **143.** 5 **145.** (a) False;
(b) false; (c) false **147.** $6.304347826 \times 10^{25}$
149. $1.19140625 \times 10^{-15}$ **151.** 2.5×10^{12}
153. 1.15385×10^{12} times

Review Exercises: Chapter 4, pp. 271–272

1. [4.1] y^{11} **2.** [4.1] $(3x)^{14}$ **3.** [4.1] t^8 **4.** [4.1] 4^3
5. [4.1] 1 **6.** [4.1] $\dfrac{9t^8}{4s^6}$ **7.** [4.1] $-8x^3 y^6$ **8.** [4.1] $18x^5$
9. [4.1] $a^7 b^6$ **10.** [4.2] $3x^2, 6x, \frac{1}{2}$
11. [4.2] $-4y^5, 7y^2, -3y, -2$ **12.** [4.2] 7, -1, 7
13. [4.2] 4, 6, $-5, \frac{5}{3}$ **14.** [4.2] (a) 2, 0, 5; (b) $15t^5$, 15; (c) 5
15. [4.2] (a) 5, 4, 2, 1; (b) $-2x^5, -2$; (c) 5
16. [4.2] Binomial **17.** [4.2] None of these
18. [4.2] Monomial **19.** [4.2] $-x^2 + 9x$
20. [4.2] $-\frac{1}{4}x^3 + 4x^2 + 7$ **21.** [4.2] $-3x^5 + 25$
22. [4.2] $-2x^2 - 3x + 2$ **23.** [4.2] $10x^4 - 7x^2 - x - \frac{1}{2}$
24. [4.2] -17 **25.** [4.2] 10
26. [4.3] $x^5 + 3x^4 + 6x^3 - 2x - 9$
27. [4.3] $-x^5 + 3x^4 - x^3 - 2x^2$ **28.** [4.3] $2x^2 - 4x - 6$
29. [4.3] $x^5 - 3x^3 - 2x^2 + 8$
30. [4.3] $\frac{3}{4}x^4 + \frac{1}{4}x^3 - \frac{1}{3}x^2 - \frac{7}{4}x + \frac{3}{8}$
31. [4.3] $-x^5 + x^4 - 5x^3 - 2x^2 + 2x$
32. (a) [4.3] $4w + 6$; (b) [4.4] $w^2 + 3w$ **33.** [4.4] $-12x^3$
34. [4.5] $49x^2 + 14x + 1$ **35.** [4.5] $a^2 - 3a - 28$
36. [4.5] $m^2 - 25$ **37.** [4.4] $12x^3 - 23x^2 + 13x - 2$
38. [4.5] $x^2 - 18x + 81$ **39.** [4.4] $15t^5 - 6t^4 + 12t^3$
40. [4.4] $a^2 - 49$ **41.** [4.5] $x^2 - 1.05x + 0.225$
42. [4.4] $x^7 + x^5 - 3x^4 + 3x^3 - 2x^2 + 5x - 3$
43. [4.5] $9x^2 - 30x + 25$ **44.** [4.5] $2t^4 - 11t^2 - 21$
45. [4.5] $a^2 + \frac{1}{6}a - \frac{1}{3}$ **46.** [4.5] $9x^4 - 16$
47. [4.5] $4 - x^2$ **48.** [4.6] $2x^2 - 7xy - 15y^2$
49. [4.6] 49 **50.** [4.6] Coefficients: 1, -7, 9, -8;
degrees: 6, 2, 2, 0; 6 **51.** [4.6] Coefficients: 1, -1, 1;
degrees: 16, 40, 23; 40 **52.** [4.6] $-y + 9w - 5$
53. [4.6] $6m^3 + 4m^2 n - mn^2$ **54.** [4.6] $-x^2 - 10xy$
55. [4.6] $11x^3 y^2 - 8x^2 y - 6x^2 - 6x + 6$ **56.** [4.6] $p^3 - q^3$
57. [4.6] $9a^8 - 2a^4 b^3 + \frac{1}{9}b^6$ **58.** [4.6] $\frac{1}{2}x^2 - \frac{1}{2}y^2$

59. [4.7] $5x^2 - \frac{1}{2}x + 3$ **60.** [4.7] $3x^2 - 7x + 4 + \dfrac{1}{2x + 3}$

61. [4.7] $t^3 + 2t - 3$ **62.** [4.8] $\dfrac{1}{m^7}$ **63.** [4.8] t^{-8}

64. [4.8] $\dfrac{1}{7^2}$, or $\dfrac{1}{49}$ **65.** [4.8] $\dfrac{1}{a^{13} b^7}$ **66.** [4.8] $\dfrac{1}{x^{12}}$

67. [4.8] $\dfrac{x^6}{4y^2}$ **68.** [4.8] $\dfrac{y^3}{8x^3}$ **69.** [4.8] 8,300,000

70. [4.8] 3.28×10^{-5} **71.** [4.8] 2.09×10^4
72. [4.8] 5.12×10^{-5} **73.** [4.8] 2.28×10^{11} platelets
74. [4.1] In the expression $5x^3$, the exponent refers only
to the x. In the expression $(5x)^3$, the entire expression
within the parentheses is cubed. **75.** [4.3] The sum of
two polynomials of degree n will also have degree n, since
only the coefficients are added and the variables remain
unchanged. An exception to this occurs when the leading
terms of the two polynomials are opposites. The sum of
those terms is then zero and the sum of the polynomials
will have a degree less than n. **76.** [4.2], [4.5] (a) 3; (b) 2

77. [4.1], [4.2] $-28x^8$ **78.** [4.2] $8x^4 + 4x^3 + 5x - 2$
79. [4.5] $-16x^6 + x^2 - 10x + 25$ **80.** [2.2], [4.5] $\frac{94}{13}$

Test: Chapter 4, p. 273

1. [4.1] t^8 **2.** [4.1] $(x+3)^{11}$ **3.** [4.1] 3^3, or 27
4. [4.1] 1   **5.** [4.1] x^6 **6.** [4.1] $-27y^6$ **7.** [4.1] $-24x^{17}$
8. [4.1] a^6b^5 **9.** [4.2] Binomial   **10.** [4.2] $\frac{1}{3}$, -1, 7
11. [4.2] Degrees of terms: 3, 1, 5; 0; leading term: $7t^5$;
leading coefficient: 7; degree of polynomial: 5
12. [4.2] -7 **13.** [4.2] $5a^2 - 6$ **14.** [4.2] $\frac{7}{4}y^2 - 4y$
15. [4.2] $x^5 + 2x^3 + 4x^2 - 8x + 3$
16. [4.3] $4x^5 + x^4 + 5x^3 - 8x^2 + 2x - 7$
17. [4.3] $5x^4 + 5x^2 + x + 5$
18. [4.3] $-4x^4 + x^3 - 8x - 3$
19. [4.3] $-x^5 + 1.3x^3 - 0.8x^2 - 3$
20. [4.4] $-12x^4 + 9x^3 + 15x^2$ **21.** [4.5] $x^2 - \frac{2}{3}x + \frac{1}{9}$
22. [4.5] $25t^2 - 49$ **23.** [4.5] $3b^2 - 4b - 15$
24. [4.5] $x^{14} - 4x^8 + 4x^6 - 16$ **25.** [4.4] $48 + 34y - 5y^2$
26. [4.4] $6x^3 - 7x^2 - 11x - 3$ **27.** [4.5] $64a^2 + 48a + 9$
28. [4.6] $-5x^3y - x^2y^2 + xy^3 - y^3 + 19$
29. [4.6] $8a^2b^2 + 6ab + 6ab^2 + ab^3 - 4b^3$
30. [4.6] $9x^{10} - 16y^{10}$ **31.** [4.7] $4x^2 + 3x - 5$
32. [4.7] $2x^2 - 4x - 2 + \dfrac{17}{3x+2}$ **33.** [4.8] $\dfrac{1}{5^3}$
34. [4.8] y^{-8} **35.** [4.8] $\dfrac{1}{t^6}$ **36.** [4.8] $\dfrac{y^5}{x^5}$ **37.** [4.8] $\dfrac{b^4}{16a^{12}}$
38. [4.8] $\dfrac{c^3}{a^3b^3}$ **39.** [4.8] 3.9×10^9 **40.** [4.8] 0.00000005
41. [4.8] 1.75×10^{17} **42.** [4.8] 1.296×10^{22}
43. [4.8] 1.5×10^4
44. [4.4], [4.5] $V = l(l-2)(l-1) = l^3 - 3l^2 + 2l$
45. [2.2], [4.5] $\frac{100}{21}$

CHAPTER 5

Technology Connection, p. 281

1. Let $y_1 = 8x^4 + 6x - 28x^3 - 21$ and $y_2 =$
$(4x^3 + 3)(2x - 7)$. Note that the Y1- and Y2-columns of
the table match regardless of how far we scroll up or down.

Exercise Set 5.1, pp. 281–282

1. Answers may vary. $(10x)(x^2)$, $(5x^2)(2x)$, $(-2)(-5x^3)$
3. Answers may vary. $(-15)(a^4)$, $(-5a)(3a^3)$, $(-3a^2)(5a^2)$
5. Answers may vary. $(2x)(13x^4)$, $(13x^5)(2)$, $(-x^3)(-26x^2)$
7. $x(x+8)$ **9.** $5t(2t-1)$ **11.** $x^2(x+6)$
13. $8x^2(x^2-3)$ **15.** $2(x^2+x-4)$
17. $a^2(7a^4 - 10a^2 - 14)$ **19.** $2x^2(x^6 + 2x^4 - 4x^2 + 5)$
21. $x^2y^2(x^3y^3 + x^2y + xy - 1)$ **23.** $5a^2b^2(ab^2 + 2b - 3a)$

25. $(y-2)(y+7)$ **27.** $(x+3)(x^2-7)$
29. $(y+8)(y^2+1)$ **31.** $(x+3)(x^2+4)$
33. $(a+3)(3a^2+2)$ **35.** $(3x-4)(3x^2+1)$
37. $(t-5)(4t^2+3)$ **39.** $(7x+2)(x^2-2)$
41. $(6a-7)(a^2+1)$ **43.** $(x+8)(x^2-3)$
45. $(x+6)(2x^2-5)$ **47.** $(w-7)(w^2+4)$
49. Not factorable by grouping   **51.** $(x-4)(2x^2-9)$
53. ⌨   **55.** [4.5] $x^2 + 8x + 15$ **56.** [4.5] $x^2 + 9x + 14$
57. [4.5] $a^2 - 4a - 21$ **58.** [4.5] $a^2 - 3a - 40$
59. [4.5] $6x^2 + 7x - 20$ **60.** [4.5] $12t^2 - 13t - 14$
61. [4.5] $9t^2 - 30t + 25$ **62.** [4.5] $4t^2 - 36t + 81$
63. ⌨   **65.** $(2x^2+3)(2x^3+3)$ **67.** $(x^5+1)(x^7+1)$
69. $(x-1)(5x^4 + x^2 + 3)$
71. Answers may vary. $3x^4y^3 - 9x^3y^3 + 27x^2y^4$

Exercise Set 5.2, pp. 288–289

1. $(x+5)(x+1)$ **3.** $(x+5)(x+2)$
5. $(y+4)(y+7)$ **7.** $(a+5)(a+6)$
9. $(x-1)(x-4)$ **11.** $(z-1)(z-7)$
13. $(x-5)(x-3)$ **15.** $(y-1)(y-10)$
17. $(x+7)(x-6)$ **19.** $2(x+2)(x-9)$
21. $x(x+2)(x-8)$ **23.** $(y-5)(y+9)$
25. $(x-11)(x+9)$ **27.** $c^2(c+8)(c-7)$
29. $2(a+5)(a-7)$ **31.** Prime   **33.** Prime
35. $(x+10)^2$ **37.** $3x(x-25)(x+4)$
39. $(x-24)(x+3)$ **41.** $(x-9)(x-16)$
43. $a^2(a+12)(a-11)$ **45.** $\left(x-\frac{1}{5}\right)^2$
47. $(9+y)(3+y)$ **49.** $(t+0.2)(t-0.5)$
51. $(p+5q)(p-2q)$ **53.** Prime
55. $(s-5t)(s+3t)$ **57.** $6a^8(a+2)(a-7)$ **59.** ⌨
61. [2.1] $\frac{8}{3}$ **62.** [2.1] $-\frac{7}{2}$ **63.** [4.5] $3x^2 + 22x + 24$
64. [4.5] $49w^2 + 84w + 36$ **65.** [2.4] 29,443
66. [2.5] 100°, 25°, 55°   **67.** ⌨
69. $-5, 5, -23, 23, -49, 49$
71. $-1(3+x)(-10+x)$, or $-1(x+3)(x-10)$
73. $-1(-2+a)(12+a)$, or $-1(a-2)(a+12)$
75. $-1(-6+t)(14+t)$, or $-1(t-6)(t+14)$
77. $\left(x-\frac{1}{4}\right)\left(x+\frac{1}{2}\right)$ **79.** $\frac{1}{3}a(a-3)(a+2)$
81. $(x^m+4)(x^m+7)$ **83.** $(a+1)(x+2)(x+1)$
85. $2x^2(4-\pi)$ **87.** $(x+3)^3$, or $x^3 + 9x^2 + 27x + 27$

Exercise Set 5.3, pp. 298–299

1. $(2x-1)(x+4)$ **3.** $(3t-5)(t+3)$
5. $(3x-1)(2x-7)$ **7.** $(7x+1)(x+2)$
9. $(3a+2)(3a-4)$ **11.** $(3x+1)(x-2)$
13. $6(2t+1)(t-1)$ **15.** $(6t+5)(3t-2)$
17. $(5x+3)(3x+2)$ **19.** $(7x+4)(5x+2)$
21. Prime   **23.** $(5x+4)^2$ **25.** $(8a+3)(2a+9)$
27. $2(3t-1)(3t+5)$ **29.** $(x-3)(2x+5)$
31. $3(2x+1)(x+5)$ **33.** $5(4x-1)(x-1)$
35. $4(3x-1)(x+6)$ **37.** $(3x+1)(x+1)$

39. $(y + 4)(y - 2)$ **41.** $(4t - 3)(2t - 7)$
43. $(3x + 2)(2x + 3)$ **45.** $(t + 3)(2t - 1)$
47. $(a - 4)(3a - 1)$ **49.** $(9t + 5)(t + 1)$
51. $(4x + 1)(4x + 7)$ **53.** $5(2a - 1)(a + 3)$
55. $2(x^2 + 3x - 7)$ **57.** $3x(3x - 1)(2x + 3)$
59. $(x + 1)(25x + 64)$ **61.** $3x(7x + 1)(8x + 1)$
63. $t^2(2t - 3)(7t + 1)$ **65.** $3(5x - 3)(3x + 2)$
67. $(3a + 2b)(3a + 4b)$ **69.** $(7p + 4q)(5p + 2q)$
71. $6(3x - 4y)(x + y)$ **73.** $2(3a - 2b)(4a - 3b)$
75. $x^2(4x + 7)(2x + 5)$ **77.** $a^6(3a + 4)(3a + 2)$
79. 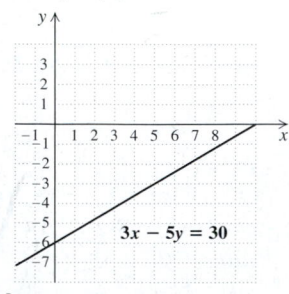   **81.** [2.5] 6369 km, 3949 mi   **82.** [2.5] 40°
83. [4.5] $9x^2 + 6x + 1$ **84.** [4.5] $25x^2 - 20x + 4$
85. [4.5] $16t^2 - 40t + 25$ **86.** [4.5] $49a^2 + 14a + 1$
87. [4.5] $25x^2 - 4$ **88.** [4.5] $4x^2 - 9$ **89.** [4.5] $4t^2 - 49$
90. [4.5] $16a^2 - 49$ **91.**   **93.** Prime
95. $2y(4xy + 1)(xy + 1)$ **97.** $(3t^5 + 2)^2$
99. $-(5x^m - 2)(3x^m - 4)$ **101.** $a(a^n - 1)^2$
103. $-2(a + 1)^n(a + 3)^2(a + 6)$

Exercise Set 5.4, pp. 305–306

1. Yes   **3.** No   **5.** No   **7.** No   **9.** $(x - 8)^2$
11. $(x + 7)^2$ **13.** $3(x - 1)^2$ **15.** $(2 + x)^2$, or $(x + 2)^2$
17. $2(3x - 1)^2$ **19.** $(7 + 4y)^2$, or $(4y + 7)^2$
21. $x^3(x - 9)^2$ **23.** $2x(x - 1)^2$ **25.** $5(2x + 5)^2$
27. $(7 - 3x)^2$, or $(3x - 7)^2$ **29.** $(4x + 3)^2$
31. $2(1 + 5x)^2$, or $2(5x + 1)^2$ **33.** $(2p + 3q)^2$
35. Prime   **37.** $(8m + n)^2$ **39.** $2(4s - 5t)^2$ **41.** Yes
43. No   **45.** No   **47.** Yes   **49.** $(y + 2)(y - 2)$
51. $(p + 3)(p - 3)$ **53.** $(7 + t)(-7 + t)$, or
$(t + 7)(t - 7)$ **55.** $6(a + 3)(a - 3)$ **57.** $(7x - 1)^2$
59. $2(10 - t)(10 + t)$ **61.** $5(4a - 3)(4a + 3)$
63. $5(t + 4)(t - 4)$ **65.** $2(2x + 7)(2x - 7)$
67. $x(6 + 7x)(6 - 7x)$ **69.** Prime
71. $(t - 1)(t + 1)(t^2 + 1)$ **73.** $3x(x - 4)^2$
75. $3(4t + 3)(4t - 3)$ **77.** $a^6(a - 1)^2$
79. $7(a + b)(a - b)$ **81.** $(5x + 2y)(5x - 2y)$
83. $(1 + a^2b^2)(1 + ab)(1 - ab)$ **85.** $2(3t + 2s)(3t - 2s)$
87.   **89.** [2.5] 3.125 L   **90.** [2.7] Scores ≥ 77
91. [4.1] $x^{12}y^{12}$ **92.** [4.1] $25a^4b^6$
93. [3.2]   **94.** [3.2]

95.   **97.** $(x^4 + 2^4)(x^2 + 2^2)(x + 2)(x - 2)$, or
$(x^4 + 16)(x^2 + 4)(x + 2)(x - 2)$ **99.** $2x\left(3x - \frac{2}{5}\right)\left(3x + \frac{2}{5}\right)$
101. $(0.8x - 1.1)(0.8x + 1.1)$

103. $(y^2 - 10y + 25 + z^4)(y - 5 + z^2)(y - 5 - z^2)$
105. $(a^n + 7b^n)(a^n - 7b^n)$ **107.** $(x^2 + 1)(x + 3)(x - 3)$
109. $16(x^2 - 3)^2$ **111.** $(7x + 4)^2$ **113.** $2x^3 - x^2 - 1$
115. $(y + x + 7)(y - x - 1)$ **117.** 16
119. $(x + 1)^2 - x^2 = [(x + 1) + x][(x + 1) - 1] = 2x + 1$

Exercise Set 5.5, p. 309-310

1. $(t + 2)(t^2 - 2t + 4)$ **3.** $(a - 4)(a^2 + 4a + 16)$
5. $(z + 5)(z^2 - 5z + 25)$ **7.** $(2a - 1)(4a^2 + 2a + 1)$
9. $(y - 3)(y^2 + 3y + 9)$ **11.** $(4 + 5x)(16 - 20x + 25x^2)$
13. $(5p - 1)(25p^2 + 5p + 1)$
15. $(3m + 4)(9m^2 - 12m + 16)$
17. $(p - q)(p^2 + pq + q^2)$ **19.** $\left(x + \frac{1}{2}\right)\left(x^2 - \frac{1}{2}x + \frac{1}{4}\right)$
21. $2(y - 4)(y^2 + 4y + 16)$
23. $3(2a + 1)(4a^2 - 2a + 1)$ **25.** $r(s - 4)(s^2 + 4s + 16)$
27. $5(x + 2z)(x^2 - 2xz + 4z^2)$
29. $(x + 0.1)(x^2 - 0.1x + 0.01)$
31. $3z^2(z - 1)(z^2 + z + 1)$ **33.** $(t^2 + 1)(t^4 - t^2 + 1)$
35. $(p + q)(p^2 - pq + q^2)(p - q)(p^2 + pq + q^2)$
37.   **39.** [4.5] $9x^2 - 25$ **40.** [4.5] $9x^2 + 30x + 25$
41. [4.5] $x^2 - 3x - 28$ **42.** [4.4] $x^3 + 1$
43. [2.5] 24 million   **44.** [2.5] 32.8 million barrels
45.   **47.** $(5c^2 + 2d^2)(25c^4 - 10c^2d^2 + 4d^4)$
49. $3(x^a - 2y^b)(x^{2a} + 2x^ay^b + 4y^{2b})$
51. $\frac{1}{3}\left(\frac{1}{2}xy + z\right)\left(\frac{1}{4}x^2y^2 - \frac{1}{2}xyz + z^2\right)$ **53.** $2x^3 + 150x$
55. $(t - 8)(t - 1)(t^2 + t + 1)$

Exercise Set 5.6, pp. 315–316

1. $5(x + 3)(x - 3)$ **3.** $(a + 5)^2$ **5.** $(4t + 1)(2t - 5)$
7. $x(x - 12)^2$ **9.** $(x + 3)(x + 2)(x - 2)$
11. $2(7t + 3)(7t - 3)$ **13.** $4x(5x + 9)(x - 2)$
15. Prime   **17.** $a(a^2 + 8)(a + 8)$ **19.** $x^3(x - 7)^2$
21. $2(5 + x)(2 - x)$ **23.** Prime
25. $4(x^2 + 4)(x + 2)(x - 2)$ **27.** Prime
29. $x^3(x - 3)(x - 1)$ **31.** $(x - y)(x^2 + xy + y^2)$
33. $12n^2(1 + 2n)$ **35.** $ab(b - a)$ **37.** $2\pi r(h + r)$
39. $(a + b)(2x + 1)$ **41.** $(n + 2)(n + p)$
43. $(x - 2)(2x + z)$ **45.** $(x + y)^2$ **47.** $(3c - d)^2$
49. $7(p^2 + q^2)(p + q)(p - q)$ **51.** $(5z + y)^2$
53. $(m + 1)(m^2 - m + 1)(m - 1)(m^2 + m + 1)$
55. Prime   **57.** $(m + 20n)(m - 18n)$
59. $(mn - 8)(mn + 4)$ **61.** $a^3(ab - 2)(ab + 5)$
63. $2a(3a + 2b)(9a^2 - 6ab + 4b^2)$
65. $2t^2(s^3 + 2t)(s^3 + 3t)$ **67.** $a^2(1 + bc)^2$
69. $\left(\frac{1}{9}x - \frac{4}{3}\right)^2$ **71.** $(1 + 4x^6y^6)(1 + 2x^3y^3)(1 - 2x^3y^3)$
73.
75. [3.1]

$y = -4x + 7$		$y = -4x + 7$	
$11 \overset{?}{\vert} -4(-1) + 7$		$7 \overset{?}{\vert} -4 \cdot 0 + 7$	
	$4 + 7$		$0 + 7$
$11 \vert 11$	TRUE	$7 \vert 7$	TRUE

$$\begin{array}{c|c}
\multicolumn{2}{c}{y = -4x + 7} \\
\hline
-5 \ ? & -4 \cdot 3 + 7 \\
& -12 + 7 \\
-5 & -5 \quad \text{TRUE}
\end{array}$$

76. [2.2] $\frac{4}{5}$ **77.** [2.2] $-\frac{7}{3}$ **78.** [2.2] $-\frac{9}{2}$ **79.** [2.2] $\frac{9}{4}$

80. [3.2]

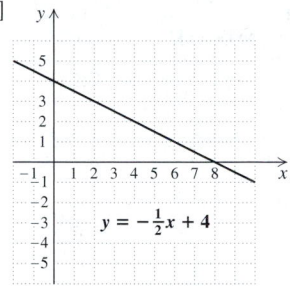

$y = -\frac{1}{2}x + 4$

81.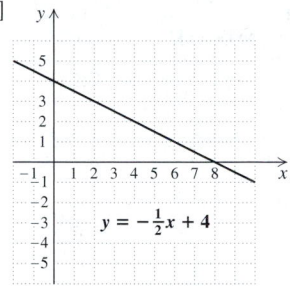

83. $-x(x^2 + 9)(x^2 - 2)$
85. $3(a + 1)(a - 1)(a + 2)(a - 2)$
87. $(y + 1)(y - 7)(y + 3)$
89. $(2x - 2 + 3y)(3x - 3 - y)$
91. $(a + 3)^2(2a + b + 4)(a - b + 5)$

Technology Connection, p. 323

1. $-4.65, 0.65$ **2.** $-0.37, 5.37$ **3.** $-8.98, -4.56$
4. No solution   **5.** $0, 2.76$

Exercise Set 5.7, pp. 323–325

1. $-6, -5$ **3.** $-7, 3$ **5.** $-4, \frac{9}{2}$ **7.** $-\frac{7}{4}, \frac{9}{10}$ **9.** $-6, 0$
11. $\frac{18}{11}, \frac{1}{21}$ **13.** $-\frac{9}{2}, 0$ **15.** $50, 70$ **17.** $-5, \frac{2}{3}, 1$
19. $-1, -6$ **21.** $-3, 7$ **23.** $0, 6$ **25.** $0, 1, 2$
27. $-\frac{2}{3}, \frac{2}{3}$ **29.** -5 **31.** 1 **33.** $0, \frac{5}{8}$ **35.** $-\frac{5}{3}, 4$
37. $-1, -5$ **39.** 3 **41.** $-1, \frac{2}{3}$ **43.** $-\frac{5}{9}, \frac{5}{9}$ **45.** $-1, \frac{6}{5}$
47. $-2, 9$ **49.** $-1, 0, \frac{3}{2}$ **51.** $-1, 4$ **53.** $-1, 3$
55. $(-4, 0), (1, 0)$ **57.** $(-3, 0), (5, 0)$ **59.** $\left(-\frac{5}{2}, 0\right), (2, 0)$
61. 🖳   **63.** [1.1] $(a + b)^2$ **64.** [1.1] $a^2 + b^2$
65. [1.1] Let x represent the first integer; $x + (x + 1)$
66. [2.7] Let x represent the number; $5 + 2x < 19$
67. [2.7] Let x represent the number; $\frac{1}{2}x - 7 > 24$
68. [2.7] Let x represent the number; $x - 3 \geq 34$
69. 🖳   **71.** $-7, -\frac{8}{3}, \frac{5}{2}$ **73. (a)** $x^2 - x - 20 = 0$;
(b) $x^2 - 6x - 7 = 0$; **(c)** $4x^2 - 13x + 3 = 0$;
(d) $6x^2 - 5x + 1 = 0$; **(e)** $12x^2 - 17x + 6 = 0$;
(f) $x^3 - 4x^2 + x + 6 = 0$ **75.** $-5, 4$ **77.** $-\frac{3}{5}, \frac{3}{5}$
79. $-4, 2$ **81. (a)** $2x^2 + 20x - 4 = 0$;
(b) $x^2 - 3x - 18 = 0$; **(c)** $(x + 1)(5x - 5) = 0$;
(d) $(2x + 8)(2x - 5) = 0$; **(e)** $4x^2 + 8x + 36 = 0$;
(f) $9x^2 - 12x + 24 = 0$ **83.** 🖳   **85.** $2.33, 6.77$
87. $-9.15, -4.59$ **89.** $0, 2.74$

Exercise Set 5.8, pp. 331–334

1. $-2, 3$ **3.** 9 cm, 12 cm, 15 cm   **5.** 10, 11
7. -17 and -15, 15 and 17   **9.** Length: 24 in.;
width: 12 in.   **11.** Length: 18 cm; width: 8 cm   **13.** 25 ft
15. Base: 14 cm; height: 4 cm   **17.** Foot: 7 ft; height: 12 ft
19. 380   **21.** 12   **23.** 9 ft   **25.** 105   **27.** 12
29. Dining room: 12 ft by 12 ft; kitchen: 12 ft by 10 ft
31. 20 ft   **33.** 1 sec, 2 sec   **35.** 🖳   **37.** [1.7] $-\frac{8}{21}$
38. [1.7] $-\frac{8}{45}$ **39.** [1.7] $-\frac{35}{54}$ **40.** [1.7] $-\frac{5}{16}$
41. [1.5] $-\frac{2}{21}$ **42.** [1.5] $-\frac{26}{45}$ **43.** [1.5] $\frac{1}{18}$
44. [1.5] $-\frac{11}{24}$ **45.** 🖳   **47.** 35 ft   **49.** 10
51. 37   **53.** 15 cm by 30 cm   **55.** 4 in., 6 in.
57. $(\pi - 2)x^2$

Review Exercises: Chapter 5, pp. 335–336

1. [5.1] Answers may vary. $(12x)(3x^2), (-9x^2)(-4x)$,
$(6x)(6x^2)$ **2.** [5.1] Answers may vary. $(-4x^3)(5x^2)$,
$(2x^4)(-10x), (-5x)(4x^4)$ **3.** [5.1] $2x^3(x + 3)$
4. [5.1] $a(a - 7)$ **5.** [5.4] $(2t - 3)(2t + 3)$
6. [5.2] $(x - 2)(x + 6)$ **7.** [5.4] $(x + 7)^2$
8. [5.4] $3x(2x + 1)^2$ **9.** [5.1] $(2x + 3)(3x^2 + 1)$
10. [5.3] $(6t + 1)(t - 1)$
11. [5.4] $(9a^2 + 1)(3a + 1)(3a - 1)$
12. [5.3] $3x(3x - 5)(x + 3)$ **13.** [5.4] $2(x - 5)(x + 5)$
14. [5.1] $(x + 4)(x^3 - 2)$ **15.** [5.4] $(ab^2 - 6)(ab^2 + 6)$
16. [5.1] $4x^4(2x^2 - 8x + 1)$ **17.** [5.4] $3(2x + 5)^2$
18. [5.4] Prime   **19.** [5.2] $x(x - 6)(x + 5)$
20. [5.4] $(2x + 5)(2x - 5)$ **21.** [5.4] $(3x - 5)^2$
22. [5.3] $2(3x + 4)(x - 6)$ **23.** [5.3] $(4t - 5)(t - 2)$
24. [5.3] $(2t + 1)(t - 4)$ **25.** [5.4] $2(3x - 1)^2$
26. [5.5] $(x - 3)(x^2 + 3x + 9)$ **27.** [5.2] $(5 - x)(3 - x)$
28. [5.4] $(5x - 2)(5x - 2)$ **29.** [5.2] $(xy - 3)(xy + 4)$
30. [5.4] $3(2a + 7b)^2$ **31.** [5.1] $(m + 5)(m + t)$
32. [5.5] $2(5y + 4x^2)(25y^2 - 20x^2y + 16x^4)$
33. [5.7] $-3, 1$ **34.** [5.7] $-7, 0, 5$ **35.** [5.7] $-\frac{1}{3}, \frac{1}{3}$
36. [5.7] $\frac{2}{3}, 1$ **37.** [5.7] $-4, \frac{3}{2}$ **38.** [5.7] $-2, 3$
39. [5.7] $-3, 4$ **40.** [5.7] $-1, \frac{5}{2}$ **41.** [5.8] Height: 40 cm;
base: 40 cm   **42.** [5.8] 24 m   **43.** [5.4] 🖳 Answers may
vary. Because Edith did not first factor out the largest common factor, 4, her factorization will not be "complete"
until she removes a common factor of 2 from each binomial. Awarding 3 to 7 points would seem reasonable.
44. [5.7] 🖳 The equations solved in this chapter have an x^2
-term (are quadratic), whereas those solved previously
have no x^2-term (are linear). The principle of zero products is used to solve quadratic equations and is not used to
solve linear equations.   **45.** [5.8] 2.5 cm   **46.** [5.8] 0, 2
47. [5.8] Length: 12 cm; width: 6 cm   **48.** [5.7] $-3, 2, \frac{5}{2}$
49. [5.7] No real solution

Test: Chapter 5, p. 337

1. [5.1] $(2x)(4x^3)$, $(-4x^2)(-2x^2)$, $(8x)(x^3)$
2. [5.2] $(x-2)(x-5)$　**3.** [5.2] $(x-5)^2$
4. [5.1] $2y^2(2y^2-4y+3)$　**5.** [5.1] $(x+1)(x^2+2)$
6. [5.1] $x(x-5)$　**7.** [5.2] $x(x+3)(x-1)$
8. [5.5] $3(t+1)(t^2-t+1)$　**9.** [5.4] $(2x-3)(2x+3)$
10. [5.2] $(x-4)(x+3)$　**11.** [5.3] $3m(2m+1)(m+1)$
12. [5.4] $3(w+5)(w-5)$　**13.** [5.4] $5(3x+2)^2$
14. [5.4] $3(x^2+4)(x+2)(x-2)$　**15.** [5.4] $(7x-6)^2$
16. [5.3] $(5x-1)(x-5)$　**17.** [5.1] $(x+2)(x^3-3)$
18. [5.5] $2(m^2+2)(m^4-2m^2+4)$
19. [5.3] $(2x+3)(2x-5)$　**20.** [5.3] $3t(2t+5)(t-1)$
21. [5.3] $3(m-5n)(m+2n)$　**22.** [5.7] $-4, 5$
23. [5.7] $-5, 0, \frac{3}{2}$　**24.** [5.7] $-4, 7$　**25.** [5.7] $-1, \frac{7}{3}$
26. [5.8] Length: 8 m; width: 6m　**27.** [5.8] 5 ft
28. [5.8] Width: 3; length: 15　**29.** [5.2] $(a-4)(a+8)$
30. [5.7] $-\frac{8}{3}, 0, \frac{2}{5}$

CHAPTER 6

Exercise Set 6.1, pp. 345–346

1. 0　**3.** -8　**5.** 4　**7.** $-4, 7$　**9.** $-5, 5$　**11.** $\dfrac{3a}{2b^2}$

13. $\dfrac{5}{2xy^4}$　**15.** $\dfrac{3}{4}$　**17.** $\dfrac{a-3}{a+1}$　**19.** $\dfrac{3}{2x^3}$　**21.** $\dfrac{y-3}{4y}$

23. $\dfrac{3(2a-1)}{7(a-1)}$　**25.** $\dfrac{t+4}{t+5}$　**27.** $\dfrac{a+4}{2(a-4)}$　**29.** $\dfrac{x+4}{x-4}$

31. $t-1$　**33.** $\dfrac{y^2+4}{y+2}$　**35.** $\dfrac{1}{2}$　**37.** $\dfrac{5}{y+6}$

39. $\dfrac{y+6}{y+5}$　**41.** $\dfrac{a-3}{a+3}$　**43.** -1　**45.** -7　**47.** $-\frac{1}{3}$

49. $-\frac{3}{2}$　**51.** -1　**53.** 🗒　**55.** [1.7] $-\frac{4}{7}$

56. [1.7] $-\frac{10}{33}$　**57.** [1.7] $-\frac{15}{4}$　**58.** [1.7] $-\frac{21}{16}$

59. [1.8] $\dfrac{13}{63}$　**60.** [1.8] $\dfrac{5}{48}$　**61.** 🗒

63. $\dfrac{(x^2+y^2)(x+y)}{(x-y)^3}$　**65.** $\dfrac{1}{x-1}$　**67.** $-\dfrac{2t^3+3}{3t^2+2}$　**69.** 1

71. 🗒

Technology Connection, p. 350

1. Let $y_1 = ((x^2+3x+2)/(x^2+4))/(5x^2+10x)$ and $y_2 = (x+1)/((x^2+4)(5x))$. With the table set in AUTO mode, note that the values in the Y1- and Y2-columns match except for $x = -2$.　**2.** ERROR messages occur when division by 0 is attempted. Since the simplified expression has no factor of $x+5$ or $x+1$ in a denominator, no ERROR message occurs in Y2 for $x = -5$ or -1.

Exercise Set 6.2, pp. 350–352

1. $\dfrac{9x(x-5)}{4(2x+1)}$　**3.** $\dfrac{(a-4)(a+2)}{(a+6)^2}$　**5.** $\dfrac{(2x+3)(x+1)}{4(x-5)}$

7. $\dfrac{(a-5)(a+2)}{(a^2+1)(a^2-1)}$　**9.** $\dfrac{(x+4)(x-1)}{(2+x)(x+1)}$　**11.** $\dfrac{5a^2}{3}$

13. $\dfrac{4}{c^2d}$　**15.** $\dfrac{x+2}{x-2}$　**17.** $\dfrac{(a^2+25)(a-5)}{(a-3)(a-1)(a+5)}$

19. $\dfrac{5(a+3)}{a(a+4)}$　**21.** $\dfrac{2a}{a-2}$　**23.** $\dfrac{t-5}{t+5}$　**25.** $\dfrac{5(a+6)}{a-1}$

27. 1　**29.** $\dfrac{t+2}{t+4}$　**31.** $\dfrac{7}{3x}$　**33.** $\dfrac{1}{a^3-8a}$

35. $\dfrac{x^2-4x+7}{x^2+2x-5}$　**37.** $\dfrac{3}{20}$　**39.** $\dfrac{x^2}{20}$　**41.** $\dfrac{a^3}{b^3}$

43. $\dfrac{y+5}{2y}$　**45.** $4(y-2)$　**47.** $-\dfrac{a}{b}$

49. $\dfrac{(y+3)(y^2+1)}{y+1}$　**51.** $\dfrac{21}{8}$　**53.** $\dfrac{15}{4}$　**55.** $\dfrac{a-5}{3(a-1)}$

57. $\dfrac{(2x-1)(2x+1)}{x-5}$　**59.** $\dfrac{x^2+25}{2(x+5)^2}$

61. $\dfrac{(a-5)(a+3)}{(a+4)^2}$　**63.** $\dfrac{1}{(c-5)^2}$　**65.** $\dfrac{(x-4y)(x-y)}{(x+y)^3}$

67. 🗒　**69.** [1.3] $\frac{19}{12}$　**70.** [1.3] $\frac{41}{24}$　**71.** [1.3] $\frac{1}{18}$

72. [1.3] $-\frac{1}{6}$　**73.** [1.8] $-\frac{37}{20}$　**74.** [1.8] $\frac{49}{45}$　**75.** 🗒

77. 1　**79.** 1　**81.** $\dfrac{1}{(x+y)^3(3x+y)}$　**83.** $\dfrac{a^2-2b}{a^2+3b}$

85. $\dfrac{4}{x+7}$　**87.** 1　**89.** $-\dfrac{2}{(x-3y)(x+3y)}$

Exercise Set 6.3, pp. 361–363

1. $\dfrac{12}{x}$　**3.** $\dfrac{3x+5}{15}$　**5.** $\dfrac{9}{a+3}$　**7.** $\dfrac{6}{a+2}$　**9.** $\dfrac{2y+7}{2y}$

11. 11　**13.** $\dfrac{3x+5}{x+1}$　**15.** $a+5$　**17.** $x-4$　**19.** 0

21. $\dfrac{1}{x+2}$　**23.** $\dfrac{(3a-7)(a-2)}{(a+6)(a-1)}$　**25.** $\dfrac{t-4}{t+3}$

27. $\dfrac{x+5}{x-6}$　**29.** $-\dfrac{5}{x-4}$, or $\dfrac{5}{4-x}$

31. $-\dfrac{1}{x-1}$, or $\dfrac{1}{1-x}$　**33.** 135　**35.** 72　**37.** 126

39. $12x^3$　**41.** $30a^4b^8$　**43.** $6(y-3)$
45. $(x-2)(x+2)(x+3)$　**47.** $t(t-4)(t+2)^2$
49. $30x^2y^2z^3$　**51.** $(a+1)(a-1)^2$
53. $(m-3)(m-2)^2$　**55.** $(t^2-9)^2$

57. $12x^3(x-5)(x-3)(x-1)$　**59.** $\dfrac{10}{12x^5}, \dfrac{x^2y}{12x^5}$

61. $\dfrac{12b}{8a^2b^2}, \dfrac{7a}{8a^2b^2}$　**63.** $\dfrac{2x(x+3)}{(x-2)(x+2)(x+3)},$

$\dfrac{4x(x-2)}{(x-2)(x+2)(x+3)}$ **65.** 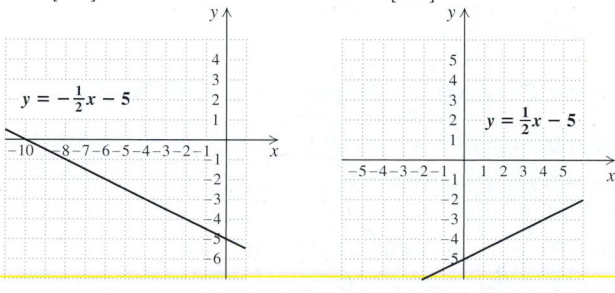   **67.** [1.7] $-\dfrac{7}{9}, \dfrac{-7}{9}$

68. [1.7] $\dfrac{-3}{2}, \dfrac{3}{-2}$ **69.** [1.3] $-\dfrac{11}{36}$ **70.** [1.3] $-\dfrac{7}{60}$

71. [4.6] $x^2 - 9x + 18$ **72.** [4.6] $s^2 - \pi r^2$ **73.**

75. $\dfrac{18x+5}{x-1}$ **77.** $\dfrac{x}{3x+1}$ **79.** 30   **81.** 60

83. $120(x-1)^2(x+1)$ **85.** 24 min   **87.** 2142

89.

Exercise Set 6.4, pp. 369–370

1. $\dfrac{3x+7}{x^2}$ **3.** $-\dfrac{5}{24r}$ **5.** $\dfrac{4x+2y}{x^2y^2}$ **7.** $\dfrac{16-15t}{18t^3}$

9. $\dfrac{5x+9}{24}$ **11.** $\dfrac{a+8}{4}$ **13.** $\dfrac{5a^2+7a-3}{9a^2}$

15. $\dfrac{-7x-13}{4x}$ **17.** $\dfrac{c^2+3cd-d^2}{c^2d^2}$ **19.** $\dfrac{3y^2-3xy-6x^2}{2x^2y^2}$

21. $\dfrac{10x}{(x-1)(x+1)}$ **23.** $\dfrac{2z+6}{(z-1)(z+1)}$ **25.** $\dfrac{11x+15}{4x(x+5)}$

27. $\dfrac{16-9t}{6t(t-5)}$ **29.** $\dfrac{x^2-x}{(x-5)(x+5)}$ **31.** $\dfrac{4t-5}{4(t-3)}$

33. $\dfrac{2x+10}{(x+3)^2}$ **35.** $\dfrac{3x-14}{(x-2)(x+2)}$ **37.** $\dfrac{9a}{4(a-5)}$

39. 0   **41.** $\dfrac{12a-11}{(a-3)(a-1)(a+2)}$ **43.** $\dfrac{x-5}{(x+5)(x+3)}$

45. $\dfrac{3z^2+19z-20}{(z-2)^2(z+3)}$ **47.** $\dfrac{-5}{x^2+17x+16}$ **49.** $\dfrac{3x-3}{5}$

51. $y+3$ **53.** $\dfrac{2b-14}{b^2-16}$ **55.** $\dfrac{y^2+10y+11}{(y-7)(y+7)}$

57. $\dfrac{9x+12}{(x-3)(x+3)}$ **59.** $\dfrac{3x^2-7x-4}{3(x-2)(x+2)}$

61. $\dfrac{a-2}{(a-3)(a+3)}$ **63.** $\dfrac{2x-3}{2-x}$ **65.** 2

67. $\dfrac{-2t^2}{(s+t)(s-t)}$ **69.** 0   **71.**   **73.** [1.7] $-\dfrac{13}{14}$

74. [1.7] $-\dfrac{5}{9}$ **75.** [1.3] $\dfrac{2}{15}$ **76.** [1.3] $\dfrac{7}{6}$

77. [3.2]   **78.** [3.2]

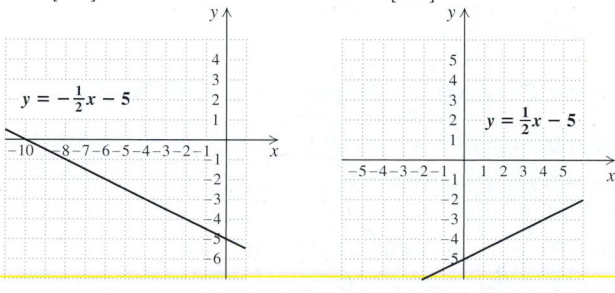

79.   **81.** Perimeter: $\dfrac{10x-14}{(x-5)(x+4)}$;
area: $\dfrac{6}{(x-5)(x+4)}$ **83.** $\dfrac{30}{(x-3)(x+4)}$

85. $\dfrac{x^4+4x^3-5x^2-126x-441}{(x+2)^2(x+7)^2}$ **87.** $\dfrac{-x^2-3}{(2x-3)(x-3)}$

89. $\dfrac{a}{a-b} + \dfrac{3b}{b-a}$; Answers may vary.

Technology Connection, p. 375

1. $(1 - 1 \div x) \div (1 - 1 \div x^2)$ **2.** Parentheses are needed to group separate terms into factors. When a fraction bar is replaced with a division sign, we need parentheses to preserve the groupings that had been created by the fraction bar. This holds for denominators and numerators alike.

Exercise Set 6.5, pp. 375–376

1. $\dfrac{6}{5}$ **3.** $\dfrac{18}{65}$ **5.** $\dfrac{4s^2}{9+3s^2}$ **7.** $\dfrac{2x}{3x+1}$ **9.** $\dfrac{4a-10}{a-1}$

11. $x-4$ **13.** $\dfrac{1}{x}$ **15.** $\dfrac{1+t^2}{t-t^2}$ **17.** $\dfrac{x}{x-y}$

19. $\dfrac{2a^2+4a}{5-3a^2}$ **21.** $\dfrac{60-15a^3}{126a^2+28a^3}$ **23.** 1

25. $\dfrac{5a^2+2b^3}{5b^3-3a^2b^3}$ **27.** $\dfrac{2x^4-3x^2}{2x^4+3}$ **29.** $\dfrac{t^2-2}{t^2+5}$

31. $\dfrac{3a^2b^3+4a}{3b^2+ab}$ **33.** $\dfrac{x+5}{2x-3}$ **35.** $\dfrac{x-2}{x-3}$ **37.**

39. [2.2] -4 **40.** [2.2] -4 **41.** [2.2] $\dfrac{19}{3}$ **42.** [2.2] $-\dfrac{14}{27}$

43. [5.7] $-3, 10$ **44.** [5.7] $-10, 2$ **45.**   **47.** 6, 7, 8

49. $-\dfrac{4}{5}, \dfrac{27}{14}$ **51.** $\dfrac{P(i+12)^2}{12(i+24)}$

53. $\dfrac{(x-8)(x-1)(x+1)}{x^2(x-2)(x+2)}$ **55.** 0

57. $\dfrac{2z(5z-2)}{(z+2)(13z-6)}$ **59.**

Exercise Set 6.6, pp. 381–382

1. $-\dfrac{7}{2}$ **3.** $\dfrac{6}{7}$ **5.** $\dfrac{24}{7}$ **7.** $-5, -1$ **9.** $-6, 6$ **11.** 3

13. $\dfrac{14}{3}$ **15.** $\dfrac{35}{3}$ **17.** 5   **19.** $\dfrac{5}{2}$ **21.** -1 **23.** No
solution   **25.** -10 **27.** $\dfrac{10}{7}$ **29.** No solution   **31.** 2
33. No solution   **35.** No solution   **37.**
39. [2.5] 137, 139   **40.** [2.5] 14 yd   **41.** [2.5] Base: 9 cm;
height: 12 cm   **42.** [2.5] $-8, -6$; 6, 8   **43.** [3.4] 0.06 cm
per day   **44.** [3.4] 0.28 in. per day   **45.**   **47.** -2
49. $-\dfrac{1}{6}$ **51.** $-1, 0$ **53.** 4   **55.**

Exercise Set 6.7, pp. 390–394

1. $-1, 4$ **3.** 1   **5.** $\dfrac{20}{9}$ hr, or 2 hr $13\dfrac{1}{3}$ min

7. $\dfrac{180}{7}$ min, or $25\dfrac{5}{7}$ min   **9.** $\dfrac{24}{7}$ hr, or 3 hr $25\dfrac{5}{7}$ min

11. $\dfrac{200}{9}$ min, or $22\dfrac{2}{9}$ min   **13.** 7.5 min

15.

	Distance (in miles)	Speed (in miles per hour)	Time (in hours)
Truck	350	r	$\dfrac{350}{r}$
Train	150	$r - 40$	$\dfrac{150}{r - 40}$

Truck: 70 mph; train: 30 mph
17. Hank: 14 km/h; Kelly: 19 km/h   **19.** Ralph: 5 km/h;
Bonnie: 8 km/h   **21.** 3 hr   **23.** 10.5   **25.** $\frac{8}{3}$ **27.** 15 ft
29. 3.75 cm   **31.** $b = 3$ m, $c = 5$ m, $d = 1$ m; $b = 10$ m,
$c = 12$ m, $d = 8$ m   **33.** 560   **35.** 1.92 g   **37.** $32,340
39. 954   **41.** 42   **43.** No   **45.** 225
47. (a) 4.8 T; **(b)** 48 lb   **49.**
51. [3.2]   **52.** [3.2]

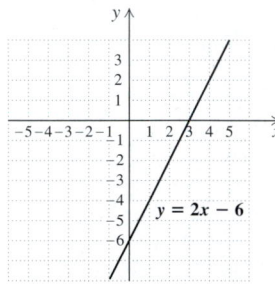

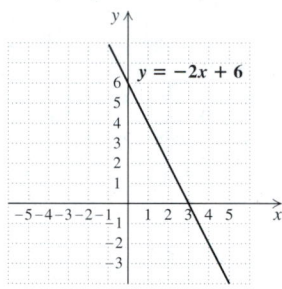

53. [3.2]   **54.** [3.2]

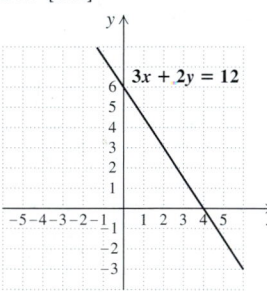

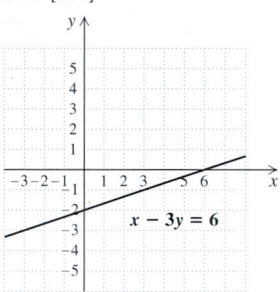

55. [3.2]   **56.** [3.2]

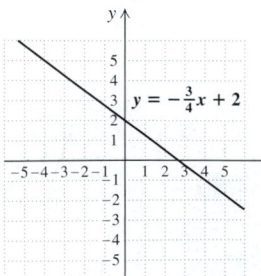

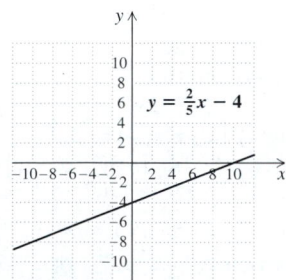

57.   **59.** Michelle: 6 hr; Sal: 3 hr; Kristen: 4 hr
61. $30\frac{22}{31}$ hr   **63.** About 57%   **65.** 2 mph
67. 45 mph   **69.** $27\frac{3}{11}$ min   **71.** ▢

Exercise Set 6.8, pp. 400–402

1. 1   **3.** 2, 3   **5.** ∅; contradiction   **7.** 7   **9.** $-\frac{1}{5}$
11. $-\frac{1}{12}$ **13.** $\frac{1}{2}$, 3   **15.** ∅; contradiction   **17.** $\frac{2}{3}$
19. ∅; contradiction   **21.** ℝ; identity   **23.** 0
25. $-1, 2$ **27.** $g = \dfrac{2s}{t^2}$ **29.** $h = \dfrac{S}{2\pi r}$ **31.** $P = \dfrac{A}{1 + rt}$
33. $h = \dfrac{2A}{b_1 + b_2}$ **35.** $n = \dfrac{s}{180} + 2$ **37.** $n = \dfrac{m}{p - q}$
39. $R = \dfrac{r_1 r_2}{r_1 + r_2}$ **41.** $h = \dfrac{S - 2\pi r^2}{2\pi r}$ **43.** $n = \dfrac{m}{r}$
45.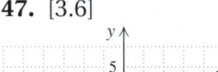
47. [3.6]   **48.** [3.6]

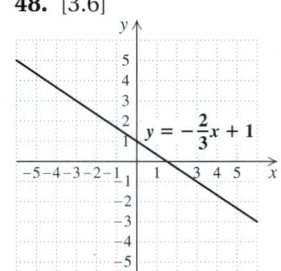

49. [2.1] $-\dfrac{3}{2}$ **50.** [3.3]

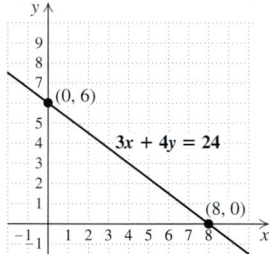

51. [5.2] $(x - 10)(x - 3)$ **52.** [5.1] $-3x^3 - 5x^2 + 5$
53. ▢   **55.** 6 yr   **57.** $a = b$ or $a = -b$ **59.** $s = t$
61. $T = \dfrac{FP}{EF + u}$ **63.** $-40°$

Review Exercises: Chapter 6, pp. 403–405

1. [6.1] 0   **2.** [6.1] 5   **3.** [6.1] $-6, 6$ **4.** [6.1] $-6, 5$
5. [6.1] -2 **6.** [6.1] $\dfrac{x - 2}{x + 1}$ **7.** [6.1] $\dfrac{7x + 3}{x - 3}$
8. [6.1] $\dfrac{y - 5}{y + 5}$ **9.** [6.1] $-5(x + 2y)$ **10.** [6.2] $\dfrac{a - 6}{5}$
11. [6.2] $\dfrac{8(t + 1)}{(2t - 1)(t - 1)}$ **12.** [6.2] $-20t$
13. [6.2] $\dfrac{2x(x - 1)}{x + 1}$ **14.** [6.2] $\dfrac{(x^2 + 1)(2x + 1)}{(x - 2)(x + 1)}$

15. [6.2] $\dfrac{(t+4)^2}{t+1}$ **16.** [6.3] $24a^5b^7$

17. [6.3] $x^4(x-1)(x+1)$

18. [6.3] $(y-2)(y+2)(y+1)$ **19.** [6.3] $\dfrac{18-3x}{x+7}$

20. [6.4] -1 **21.** [6.3] $\dfrac{4}{x-4}$ **22.** [6.4] $\dfrac{x+5}{2x}$

23. [6.4] $\dfrac{2x+3}{x-2}$ **24.** [6.4] $\dfrac{2a}{a-1}$ **25.** [6.4] $d+c$

26. [6.4] $\dfrac{-x^2+x+26}{(x+1)(x-5)(x+5)}$ **27.** [6.4] $\dfrac{2(x-2)}{x+2}$

28. [6.4] $\dfrac{19x+8}{10x(x+2)}$ **29.** [6.5] $\dfrac{z}{1-z}$

30. [6.5] $\dfrac{x^3(2xy^2+1)}{y(1+x)}$ **31.** [6.5] $c-d$ **32.** [6.6] 8

33. [6.6] $-\dfrac{1}{2}$ **34.** [6.6] $-5, 3$ **35.** [6.7] $\dfrac{36}{7}$ hr, or $5\dfrac{1}{7}$ hr

36. [6.7] Car: 105 km/h; train: 90 km/h   **37.** [6.7] -2

38. [6.7] 30 **39.** [6.7] 6 **40.** [6.7] 50

41. [6.8] $\mathbb{R}$; identity   **42.** [6.8] $\varnothing$; contradiction

43. [6.8] $\dfrac{1}{2}$, 5; conditional   **44.** [6.8] $a=\dfrac{b+c}{2}$

45. [6.8] $t=\dfrac{rs}{r+s}$ **46.** [6.8] $C=\dfrac{5F-160}{9}$

47. [6.5], [6.6] 🖾 The LCD is used to clear fractions when simplifying complex rational expressions by multiplying by the LCD. It is also used to clear fractions when solving rational equations or formulas.   **48.** [6.3] 🖾 Although multiplying the denominators of the expressions being added results in a common denominator, it is often not the *least* common denominator. Using a common denominator other than the LCD makes the expressions more complicated, requires additional simplifying after the addition has been performed, and leaves more room for error.

49. [6.2] $\dfrac{5(a+3)^2}{a}$ **50.** [6.4] $\dfrac{10a}{(a-b)(b-c)}$

51. [6.3] 0

Test: Chapter 6, pp. 405–406

1. [6.1] 0 **2.** [6.1] -8 **3.** [6.1] $-7, 7$ **4.** [6.1] $1, 2$

5. [6.1] $\dfrac{3x+7}{x+3}$ **6.** [6.2] $\dfrac{-2(a+5)}{3}$

7. [6.2] $\dfrac{(5y+1)(y+1)}{3y(y+2)}$

8. [6.2] $\dfrac{(2x+1)(2x-1)(x^2+1)}{(x-1)^2(x-2)}$

9. [6.2] $(x+3)(x-3)$ **10.** [6.3] $(y-3)(y+3)(y+7)$

11. [6.3] $\dfrac{-3x+23}{x^3}$ **12.** [6.3] $\dfrac{-2t+8}{t^2+1}$ **13.** [6.4] $\dfrac{3}{3-x}$

14. [6.4] $\dfrac{2x-5}{x-3}$ **15.** [6.4] $\dfrac{8t-3}{t(t-1)}$

16. [6.4] $\dfrac{-x^2-7x-15}{(x+4)(x-4)(x+1)}$

17. [6.4] $\dfrac{x^2+2x-7}{(x+1)(x-1)^2}$ **18.** [6.5] $\dfrac{3y+1}{y}$

19. [6.5] $\dfrac{a^2(3b^2-2a)}{b^2(a^3+2)}$ **20.** [6.6] 12 **21.** [6.6] $-3, 5$

22. [6.7] 12 min   **23.** [6.7] $2\dfrac{1}{7}$ cups

24. [6.7] Craig: 65 km/h; Marilyn: 45 km/h   **25.** [6.8] 0; conditional equation   **26.** [6.8] $\mathbb{R}$; identity

27. [6.8] $t=\dfrac{d}{r+w}$ **28.** [6.8] $c=\dfrac{b+3a}{2}$

29. [6.7] Rema: 4 hr; Reggie: 10 hr   **30.** [6.5] a

Cumulative Review: Chapters 1–6, pp. 406–407

1. [1.2] $2b+a$ **2.** [1.4] $>$ **3.** [1.8] 49

4. [1.8] $-8x+28$ **5.** [1.5] $-\dfrac{43}{8}$ **6.** [1.7] 1

7. [1.7] -6.2 **8.** [1.8] 8 **9.** [2.2] 10 **10.** [5.7] $-7, 7$

11. [2.1] $-\dfrac{10}{3}$ **12.** [2.2] -2 **13.** [2.2] $\dfrac{8}{3}$

14. [5.7] $-10, -1$ **15.** [2.2] -8 **16.** [6.6] $1, 4$

17. [2.6] $\left\{ y \mid y \le -\dfrac{2}{3} \right\}$ **18.** [6.6] -17 **19.** [5.7] $-4, \dfrac{1}{2}$

20. [2.6] $\{x \mid x > 43\}$ **21.** [6.6] 5 **22.** [5.7] $-\dfrac{7}{2}, 5$

23. [6.6] -13 **24.** [6.8] $a=\dfrac{Pb}{3-P}$ **25.** [2.3] $y=\dfrac{4z-3x}{6}$

26. [1.6] $\dfrac{3}{2}x+2y-3z$ **27.** [1.6] $-4x^3-\dfrac{1}{7}x^2-2$

28. [3.2]

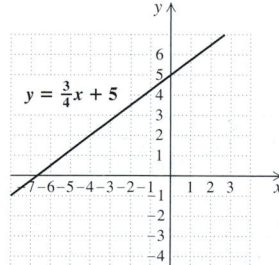

29. [3.3]

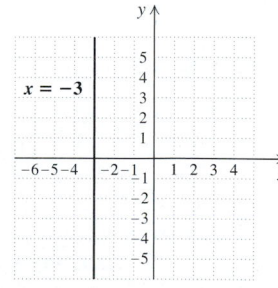

30. [3.3]

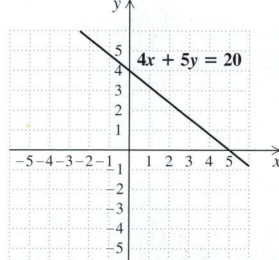

31. [3.3]

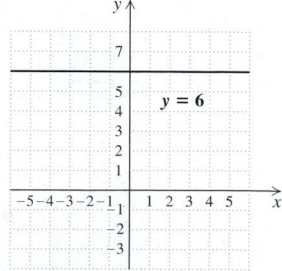

32. [3.5] -2 **33.** [4.8] $\dfrac{1}{x^2}$ **34.** [4.8] y^{-8}, or $\dfrac{1}{y^8}$

35. [4.1] $-4a^4b^{14}$　**36.** [4.3] $-y^3 - 2y^2 - 2y + 7$
37. [1.2] $12x + 16y + 4z$
38. [4.4] $2x^5 + x^3 - 6x^2 - x + 3$
39. [4.5] $36x^2 - 60xy + 25y^2$　**40.** [4.5] $2x^2 - x - 21$
41. [4.5] $4x^6 - 1$　**42.** [5.1] $2x(3 - x - 12x^3)$
43. [5.4] $(4x + 9)(4x - 9)$　**44.** [5.2] $(t - 4)(t - 6)$
45. [5.3] $(4x + 3)(2x + 1)$　**46.** [5.3] $2(3x - 2)(x - 4)$
47. [5.4] $4(t + 3)(t - 3)$　**48.** [5.4] $(5t + 4)^2$
49. [5.3] $(3x - 2)(x + 4)$　**50.** [5.1] $(x + 2)(x^3 - 3)$

51. [6.2] $\dfrac{y - 6}{2}$　**52.** [6.2] 1　**53.** [6.4] $\dfrac{a^2 + 7ab + b^2}{(a + b)(a - b)}$

54. [6.4] $\dfrac{2x + 5}{4 - x}$　**55.** [6.5] $\dfrac{x}{x - 2}$　**56.** [6.5] $\dfrac{t(2t^2 + 1)}{t^3 - 2}$

57. [4.7] $5x^2 - 4x + 2 + \dfrac{2}{3x} + \dfrac{6}{x^2}$

58. [4.7] $15x^3 - 57x^2 + 177x - 529 + \dfrac{1605}{x + 3}$

59. [2.7] At most 225　**60.** [2.4] \$3.60　**61.** [5.8] 14 ft
62. [2.5] $-278, -276$　**63.** [6.7] 30 min
64. [6.7] Phil's: 50 km/h; Harley's: 40 km/h
65. [2.5] 26 in.　**66.** [4.3], [4.5] 12
67. [1.4], [2.2] $-144, 144$　**68.** [4.5] $16y^6 - y^4 + 6y^2 - 9$
69. [5.4] $2(a^{16} + 81b^{20})(a^8 + 9b^{10})(a^4 + 3b^5)(a^4 - 3b^5)$
70. [5.7] $-7, 4, 12$　**71.** [1.4], [1.6] -7　**72.** [6.6] 18
73. [6.7] $66\frac{2}{3}\%$

CHAPTER 7

Exercise Set 7.1, pp. 416–421

1. No　**3.** Yes　**5.** Yes　**7.** Function　**9.** A relation but not a function　**11.** Function　**13.** (a) -2; (b)4
15. (a) 3; (b) 3　**17.** (a) -2; (b) -2　**19.** (a) 3; (b) -3
21. (a) 1; (b) 3　**23.** (a) 4; (b) $-1, 3$
25. (a) 1; (b) $\{x | 2 < x \le 5\}$　**27.** (a) 3; (b) -1; (c) -4;
(d) 11; (e) $a + 5$　**29.** (a) 0; (b) 1; (c) 57; (d) $5t^2 + 4t$;

(e) $20a^2 + 8a$　**31.** (a) $\frac{3}{5}$; (b) $\frac{1}{3}$; (c) $\frac{4}{7}$; (d) 0; (e) $\dfrac{x - 1}{2x - 1}$

33. $4\sqrt{3}$ cm$^2 \approx 6.93$ cm^2　**35.** 36π in$^2 \approx 113.10$ in^2
37. $14°$F　**39.** 159.48 cm　**41.** 75 heart attacks per 10,000 men　**43.** 56%
45. 3.5 drinks

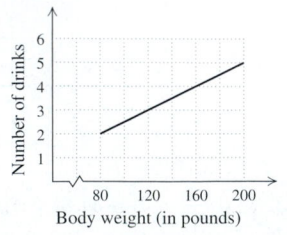

47. About 21,000 cases

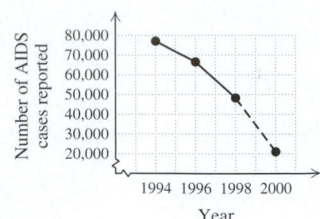

49. About 65,000

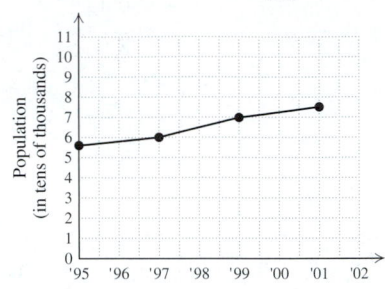

51. About \$313,000

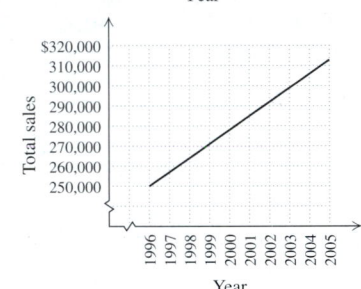

53. 🗒　**55.** [1.8] $\frac{1}{3}$　**56.** [1.8] -1
57. [2.3] $l = \dfrac{S - 2wh}{2h + 2w}$　**58.** [2.3] $w = \dfrac{S - 2lh}{2l + 2h}$
59. [2.3] $y = -\frac{2}{3}x + 2$　**60.** [2.3] $y = \frac{5}{4}x - 2$　**61.** 🗒
63. 26; 99　**65.** About 22 mm　**67.** 🗒
69.

$f(x) = [\![x]\!]$

71. Bicycling 14 mph for 1 hr

Exercise Set 7.2, pp. 429–433

1. Domain: $\{2, 9, -2, -4\}$; range: $\{8, 3, 10, 4\}$
3. Domain: $\{0, 4, -5, -1\}$; range: $\{0, -2\}$
5. Domain: $\{-4, -2, 0, 2, 4\}$; range: $\{-2, -1, 0, 1, 2\}$
7. Domain: $\{-5, -3, -1, 0, 2, 4\}$; range: $\{-1, 1\}$
9. Domain: $\{x | -4 \le x \le 3\}$; range: $\{y | -3 \le y \le 4\}$

11. Domain: $\{x \mid -4 \le x \le 5\}$; range: $\{y \mid -2 \le y \le 4\}$
13. Domain: $\{x \mid -4 \le x \le 4\}$; range: $\{-3, -1, 1\}$
15. Domain: $\mathbb{R}$; range: $\mathbb{R}$ **17.** Domain: $\mathbb{R}$; range: $\{4\}$
19. Domain: $\mathbb{R}$; range: $\{y \mid y \ge 1\}$
21. Domain: $\{x \mid x$ is a real number $and\ x \ne -2\}$; range: $\{y \mid y$ is a real number $and\ y \ne -4\}$
23. Domain: $\{x \mid x \ge 0\}$; range: $\{y \mid y \ge 0\}$
25. $\{x \mid x$ is a real number $and\ x \ne 3\}$
27. $\{x \mid x$ is a real number $and\ x \ne \frac{1}{2}\}$ **29.** $\mathbb{R}$ **31.** $\mathbb{R}$
33. $\{x \mid x$ is a real number $and\ x \ne 3\ and\ x \ne -3\}$ **35.** $\mathbb{R}$
37. $\{x \mid x$ is a real number $and\ x \ne -1\ and\ x \ne -7\}$
39. $\{t \mid 0 \le t < 624\}$ **41.** $\{p \mid \$0 \le p \le \$10.60\}$
43. $\{d \mid d \ge 0\}$ **45.** $\{t \mid 0 \le t \le 5\}$ **47. (a)** -5;
(b) 1; **(c)** 21   **49. (a)** 0; **(b)** 2; **(c)** 7   **51. (a)** 100;
(b) 100; **(c)** 131   **53.** 🗒

55. [3.6]   **56.** [3.6]

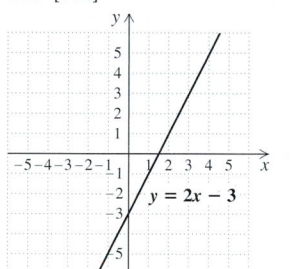

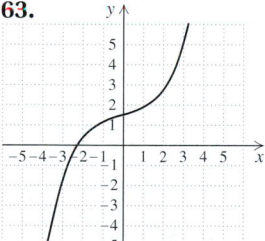
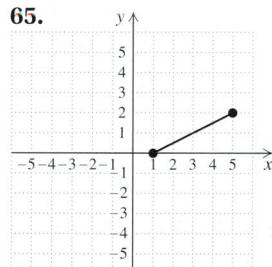

57. [3.6] Slope: $\frac{2}{3}$; y-intercept: $(0, -4)$
58. [3.6] Slope: $-\frac{1}{4}$; y-intercept: $(0, 6)$
59. [3.6] Slope: $\frac{4}{3}$; y-intercept: $(0, 0)$
60. [3.6] Slope: -5; y-intercept: $(0, 0)$ **61.** 🗒
63.   **65.**

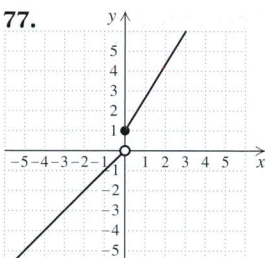
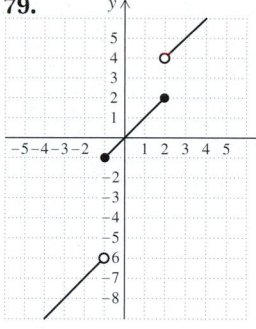

67. Domain: $\{x \mid x$ is a real number $and\ x \ne 0\}$;
range: $\{y \mid y$ is a real number $and\ y \ne 0\}$
69. Domain: $\{x \mid x < -2\ or\ x > 0\}$; range:
$\{y \mid y < -2\ or\ y > 3\}$
71. Domain: $\mathbb{R}$; range: $\{y \mid y \ge 0\}$
73. Domain: $\{x \mid x$ is a real number $and\ x \ne 2\}$; range: $\{y \mid y$ is a real number $and\ y \ne 0\}$ **75.** $\{h \mid 0 \le h \le 144\}$

77.
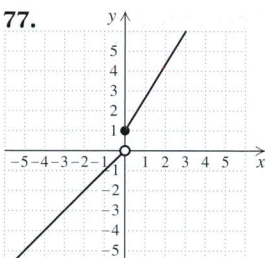
79.
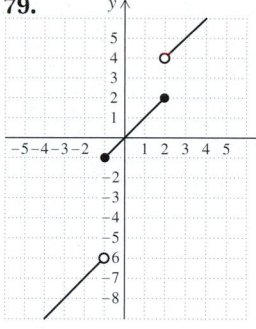

81.
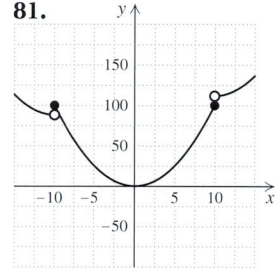
83. 📈

Exercise Set 7.3, pp. 443–448

1. Yes   **3.** Yes   **5.** No   **7.** No   **9.** Yes   **11.** No
13. (a) -5000 represents a depreciation of \$5000 per year; 90,000 represents the original value of the truck, \$90,000;
(b) 18 yr; **(c)** $\{t \mid 0 \le t \le 18\}$ **15. (a)** -150 signifies that the depreciation is \$150 per winter of use; 900 signifies that the original value of the snowblower was \$900;
(b) after 4 winters of use; **(c)** $\{0, 1, 2, 3, 4, 5, 6\}$
17. $C(t) = 25t + 50$; 4 months   **19.** $L(t) = \frac{1}{2}t + 1$;
4 months after the cut   **21.** $C(d) = \$0.75d + 2$; 3 mi
23. (a) $R(t) = -0.075t + 46.8$; **(b)** 41.325 sec; 41.1 sec;
(c) 2021   **25. (a)** $A(t) = 8.0625t + 178.6$;
(b) \$307.6 million   **27. (a)** $A(t) = \frac{13}{15}t + 74.9$;
(b) 85.3 million acres   **29.** Linear function; $\mathbb{R}$
31. Quadratic function; $\mathbb{R}$ **33.** Rational function;
$\{t \mid t$ is a real number $and\ t \ne -\frac{4}{3}\}$
35. Polynomial function; $\mathbb{R}$ **37.** Rational function; $\{x \mid x$ is a real number $and\ x \ne \frac{5}{2}\}$ **39.** Rational function; $\{n \mid n$ is a real number $and\ n \ne -1\ and\ n \ne -2\}$ **41.** Linear function; $\mathbb{R}$ **43.** $\{y \mid y \ge 0\}$ **45.** $\mathbb{R}$ **47.** $\{y \mid y \le 0\}$
49.   **51.**

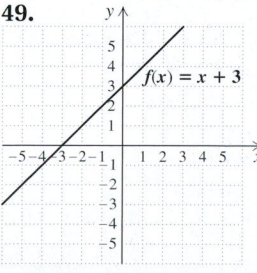
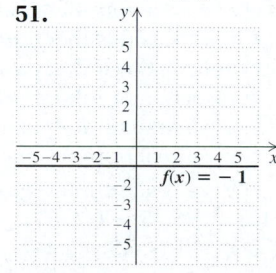

Domain: $\mathbb{R}$; range: $\mathbb{R}$ Domain: $\mathbb{R}$; range: $\{-1\}$

53.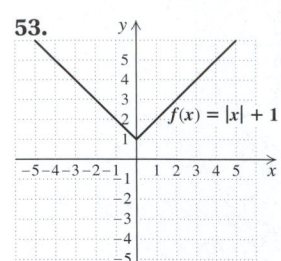

55.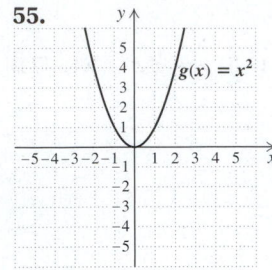

Domain: $\mathbb{R}$;
range: $\{y \,|\, y \geq 1\}$

Domain: $\mathbb{R}$;
range: $\{y \,|\, y \geq 0\}$

57. 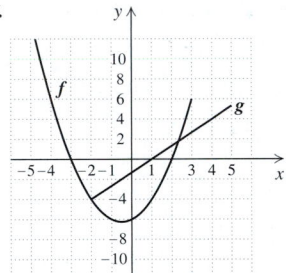   **59.** [4.3] $4x^2 + 2x - 1$
60. [4.3] $2x^3 - x^2 - x + 7$ **61.** [4.5] $2x^2 - 13x - 7$
62. [4.5] $x^2 + x - 12$ **63.** [4.3] $x^3 - 2x^2 - 3x + 5$
64. [4.3] $x^3 + 2x^2 + x + 4$ **65.** 🗒   **67.** 🗒
69. 21.1°C   **71.** 82.8%
73.

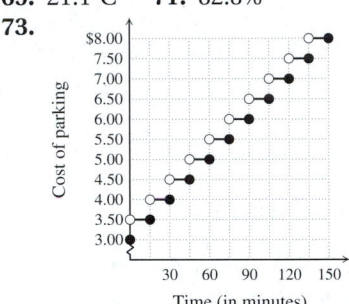

75. False   **77.** False   **79.** 📉

Exercise Set 7.4, pp. 454–457

1. 1   **3.** -41 **5.** 12   **7.** $\frac{13}{18}$ **9.** 5   **11.** $x^2 - 3x + 3$
13. $x^2 - x + 3$ **15.** 23   **17.** 5   **19.** 56
21. $\dfrac{x^2 - 2}{5 - x}$, $x \neq 5$ **23.** $\frac{2}{7}$ **25.** $0.75 + 2.5 = 3.25$
27. Women under 30   **29.** About 50 million; the number of passengers using Newark and LaGuardia in 1998
31. About 8 million; how many more passengers used Kennedy than LaGuardia in 1994   **33.** About 89 million; the number of passengers using the three airports in 1999
35. $\mathbb{R}$ **37.** $\{x \,|\, x \text{ is a real number } and \ x \neq 3\}$
39. $\{x \,|\, x \text{ is a real number } and \ x \neq 0\}$
41. $\{x \,|\, x \text{ is a real number } and \ x \neq 1\}$
43. $\{x \,|\, x \text{ is a real number } and \ x \neq 2 \ and \ x \neq 4\}$
45. $\{x \,|\, x \text{ is a real number } and \ x \neq 3\}$
47. $\{x \,|\, x \text{ is a real number } and \ x \neq 4\}$
49. $\{x \,|\, x \text{ is a real number } and \ x \neq 4 \ and \ x \neq 5\}$
51. $\{x \,|\, x \text{ is a real number } and \ x \neq -1 \ and \ x \neq -\frac{5}{2}\}$
53. 4; 3   **55.** 5; -1 **57.** $\{x \,|\, 0 \leq x \leq 9\}$; $\{x \,|\, 3 \leq x \leq 10\}$; $\{x \,|\, 3 \leq x \leq 9\}$; $\{x \,|\, 3 \leq x \leq 9\}$

59. 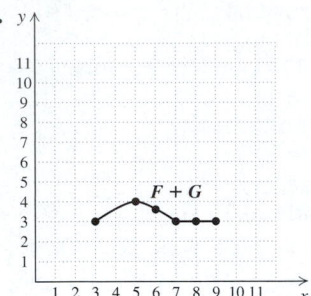   **61.** 🗒

63. [2.3] $x = \frac{7}{4}y + 2$ **64.** [2.3] $y = \frac{3}{8}x - \frac{5}{8}$
65. [2.3] $y = -\frac{5}{2}x - \frac{3}{2}$ **66.** [2.3] $x = -\frac{5}{6}y - \frac{1}{3}$
67. [1.1] Let n represent the number; $2n + 5 = 49$
68. [1.1] Let x represent the number; $\frac{1}{2}x - 3 = 57$
69. [1.1] Let x represent the first integer; $x + (x + 1) = 145$
70. [1.1] Let n represent the number; $n - (-n) = 20$
71. 🗒
73. $\{x \,|\, x \text{ is a real number } and \ x \neq -\frac{5}{2} \ and \ x \neq -3 \ and$ $x \neq 1 \ and \ x \neq -1\}$
75. Answers may vary.

[graph with f and g curves]

77. Domain of $f + g =$ Domain of $f - g =$ Domain of $f \cdot g = \{-2, -1, 0, 1\}$; Domain of $f/g = \{-2, 0, 1\}$
79. Answers may vary. $f(x) = \dfrac{1}{x + 2}$, $g(x) = \dfrac{1}{x - 5}$
81. 📉

Exercise Set 7.5, pp. 464–467

1. $k = 7$; $y = 7x$ **3.** $k = 1.7$; $y = 1.7x$ **5.** $k = 6$; $y = 6x$
7. $33\frac{1}{3}$ cm   **9.** 241,920,000 cans   **11.** 32 kg   **13.** 3.36
15. $k = 60$; $y = \dfrac{60}{x}$ **17.** $k = 112$; $y = \dfrac{112}{x}$
19. $k = 9$; $y = \dfrac{9}{x}$ **21.** 20 min   **23.** 160 cm^3
25. 3.5 hr   **27.** $y = \frac{2}{3}x^2$ **29.** $y = \dfrac{54}{x^2}$ **31.** $y = 0.3xz^2$
33. $y = \dfrac{4wx^2}{z}$ **35.** 40 W/m^2 **37.** 220 cm^3 **39.** About 57.42 mph   **41.** 📉   **43.** 📉
45. [2.2] $\frac{13}{2}$ **46.** [2.2] $\frac{5}{3}$
47. [6.6] $-\frac{3}{2}$ **48.** [5.7] $-\frac{5}{3}, \frac{7}{2}$ **49.** [6.8] $\varnothing$

50. [6.8] All real numbers   **51.** 🔲   **53.** $S = kv^6$
55. $I = \dfrac{k}{d^2}$ **57.** $P = kv^3$ **59.** $C = 2\pi r$ **61.** $V = \frac{4}{3}\pi r^3$
63. W varies jointly as m_1 and M_1 and inversely as the square of d.   **65.** $7.20

Review Exercises, Chapter 7, pp. 470–472

1. [7.1], [7.2] **(a)** 3; **(b)** $\{x \mid -2 \le x \le 4\}$; **(c)** -1;
(d) $\{y \mid 1 \le y \le 5\}$ **2.** [7.1] $\frac{3}{2}$ **3.** [7.1] $4a^2 + 4a - 3$
4. [7.1] 10.53 yr   **5.** [7.1] 510 cans per person
6. [7.1] 660 cans per person   **7. (a)** [7.3] Yes;
(b) [7.2] domain: $\mathbb{R}$; range: $\{y \mid y \ge 0\}$ **8. (a)** [7.3] No
9. (a) [7.3] No   **10. (a)** [7.3] Yes; **(b)** [7.2] domain: $\mathbb{R}$;
range: $\{-2\}$ **11.** [7.2] $\mathbb{R}$
12. [7.2] $\{x \mid x$ is a real number $and\ x \ne 1\}$
13. [7.2] $\{x \mid x$ is a real number $and\ x \ne -1\ and\ x \ne -4\}$
14. [7.2] $\{x \mid x \ge 0\}$ **15.** [7.2] $\{t \mid 0 \le t \le 34\}$
16. [7.2] **(a)** 5; **(b)** 4; **(c)** 16; **(d)** 35
17. [7.3] $C(t) = 15t + 75$; 7 months
18. [7.3] **(a)** $W(t) = \frac{11}{105}t + 0.75$; **(b)** $5.99
19. [7.3] Absolute-value function
20. [7.3] Polynomial function
21. [7.3] Quadratic function   **22.** [7.3] Linear function
23. [7.3] Rational function
24. [7.3]   **25.** [7.3]

Domain: $\mathbb{R}$; range: $\{3\}$ Domain: $\mathbb{R}$; range: $\{\mathbb{R}\}$
26. [7.3]

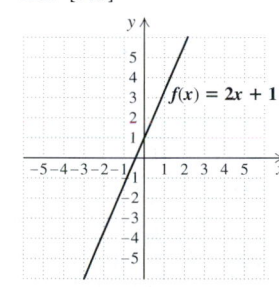

Domain: $\mathbb{R}$; range: $\{y \mid y \ge 0\}$
27. [7.4] 102   **28.** [7.4] -17 **29.** [7.4] $-\frac{9}{2}$ **30.** [7.4] $\mathbb{R}$
31. [7.4] $\{x \mid x$ is a real number $and\ x \ne 2\}$

32. [7.5] $y = \frac{15}{2}x$ **33.** [7.5] $y = \dfrac{\frac{3}{4}}{x}$
34. [7.5] $y = \dfrac{1}{2}\,\dfrac{xw^2}{z}$ **35.** [7.5] About 202.3 lb
36. [7.5] 64 L
37. 🔲 [7.2] Two functions that have the same domain and range are not necessarily identical. For example, the functions f: $\{(-2, 1), (-3, 2)\}$ and g: $\{(-2, 2), (-3, 1)\}$ have the same domain and range but are different functions.
38. 🔲 [7.2] Jenna is not correct. Any value of the variable that makes a denominator 0 is not in the domain; 0 itself may or may not make a denominator 0.
39. [7.3] $f(x) = 3.09x + 3.75$
40. [7.2] Domain: $\{x \mid x \ge -4\ and\ x \ne 2\}$; range: $\{y \mid y \ge 0\ and\ y \ne 3\}$
41. [7.3] No; the rate of change is not constant. Each year the amount of the raise will be higher since the current year's salary is higher than the previous year's salary.

Test: Chapter 7, pp. 472–473

1. [7.1], [7.2] **(a)** 1; **(b)** $\{x \mid -3 \le x \le 4\}$; **(c)** 3;
(d) $\{y \mid -1 \le y \le 2\}$ **2.** [7.1] **(a)** $40.6 billion; **(b)** [7.3] 1.2 signifies that sales are rising $1.2 billion a year; 21.4 signifies that sales were $21.4 billion in 1992
3. [7.1] 46 million
4. (a) [7.3] Yes; **(b)** [7.2] domain: $\mathbb{R}$; range: $\mathbb{R}$
5. (a) [7.3] Yes; **(b)** [7.2] domain: $\mathbb{R}$; range: $\{y \mid y \ge 1\}$
6. (a) [7.3] No   **7.** [7.2] $\{t \mid 0 \le t \le 4\}$
8. [7.2] **(a)** -5; **(b)** 10
9. [7.3] $c(n) = 25t + 75$; 17 people
10. [7.3] **(a)** $C(m) = 0.3m + 25$; **(b)** $175
11. [7.3] Linear function; $\mathbb{R}$
12. [7.3] Rational function; $\{x \mid x$ is a real number and $x \ne 5\}$ **13.** [7.3] Quadratic function; $\mathbb{R}$
14. [7.3]   **15.** [7.3]

Domain: $\mathbb{R}$; range: $\mathbb{R}$ Domain: $\mathbb{R}$; range: $\{y \mid y \ge -1\}$

16. [7.3]

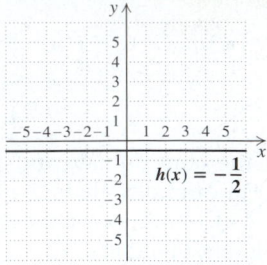

Domain: $\mathbb{R}$; range: $\left\{-\frac{1}{2}\right\}$

17. [7.1] $\frac{1}{4}$ **18.** [7.1] -10 **19.** [7.1] $\dfrac{1}{2a + 4}$

20. [7.4] $-\frac{8}{3}$ **21.** [7.2] $\{x \mid x$ is a real number *and* $x \neq -4\}$
22. [7.2] $\mathbb{R}$ **23.** [7.4] $\{x \mid x$ is a real number *and* $x \neq -4\}$
24. [7.4] $\{x \mid x$ is a real number a*nd* $x \neq -4$ *and* $x \neq 7\}$
25. [7.5] $y = \frac{1}{2}x$ **26.** [7.5] 30 workers   **27.** [7.5] $\frac{833}{125}$, or
6.664 in^2 **28.** [7.3] **(a)** 30 mi; **(b)** 15 mph
29. [7.4] $h(x) = 7x - 2$

CHAPTER 8

Technology Connection, p. 481

1. $(1.53, 2.58)$ **2.** $(-0.26, 57.06)$ **3.** $(2.23, 1.14)$
4. $(0.87, -0.32)$

Exercise Set 8.1, pp. 482–484

1. Yes   **3.** No   **5.** Yes   **7.** Yes   **9.** $(4, 1)$
11. $(2, -1)$ **13.** $(4, 3)$ **15.** $(-3, -2)$ **17.** $(-3, 2)$
19. $(3, -7)$ **21.** $(7, 2)$ **23.** $(4, 0)$ **25.** No solution
27. $\{(x, y) \mid y = 3 - x\}$ **29.** All except 25   **31.** 27
33. Let x represent the larger number and y the smaller
number; $x - y = 11, 3x + 2y = 123$ **35.** Let x represent
the number of $8.50 brushes sold and y the number of
$9.75 brushes sold; $x + y = 45, 8.50x + 9.75y = 398.75$
37. Let x and y represent the angles; $x + y = 180$,
$x = 2y - 3$ **39.** Let x represent the number of two-point
shots and y the number of free throws; $x + y = 64$,
$2x + y = 100$ **41.** Let h represent the number of vials of
Humulin sold and n the number of vials of Novolin;
$h + n = 50, 21.95h + 20.95n = 1077.50$ **43.** Let l represent the length, in feet, and w the width, in feet;
$2l + 2w = 228, w = l - 42$ **45.** Let w represent the
number of wins and t the number of ties; $2w + t = 60$,
$w = t + 9$ **47.** Let x represent the number of ounces of
lemon juice and y the number of ounces of linseed oil;
$y = 2x, x + y = 32$ **49.** Let x represent the number of
general-interest films rented and y the number of children's films rented; $x + y = 77, 3x + 1.5y = 213$ **51.** 🗍
53. [2.2] 15   **54.** [2.2] $\frac{19}{12}$ **55.** [2.2] $\frac{9}{20}$ **56.** [2.2] $\frac{13}{3}$

57. [2.3] $y = -\frac{3}{4}x + \frac{7}{4}$ **58.** [2.3] $y = \frac{2}{5}x - \frac{9}{5}$ **59.** 🗍
61. 1994   **63.** Answers may vary. **(a)** $x + y = 6$,
$x - y = 4$; **(b)** $x + y = 1, 2x + 2y = 3$; **(c)** $x + y = 1$,
$2x + 2y = 2$ **65.** $A = -\frac{17}{4}, B = -\frac{12}{5}$
67. Let x and y represent the number of years that Lou
and Juanita have taught at the university, respectively;
$x + y = 46, x - 2 = 2.5(y - 2)$ **69.** Let s and v
represent the number of ounces of baking soda and
vinegar needed, respectively; $s = 4v, s + v = 16$
71. $(0, 0), (1, 1)$ **73.** $(0.07, -7.95)$ **75.** $(0.003, 1.25)$

Exercise Set 8.2, pp. 491–492

1. $(2, -3)$ **3.** $\left(\frac{21}{5}, \frac{12}{5}\right)$ **5.** $(2, -2)$
7. $\{(x, y) \mid 2x - 3 = y\}$ **9.** $(-2, 1)$ **11.** $\left(\frac{1}{2}, \frac{1}{2}\right)$
13. $\left(\frac{19}{8}, \frac{1}{8}\right)$ **15.** No solution   **17.** $(1, 2)$ **19.** $(3, 0)$
21. $(-1, 2)$ **23.** $\left(\frac{128}{31}, -\frac{17}{31}\right)$ **25.** $(6, 2)$
27. No solution   **29.** $\left(\frac{110}{19}, -\frac{12}{19}\right)$ **31.** $(3, -1)$
33. $\{(x, y) \mid -4x + 2y = 5\}$ **35.** $\left(\frac{140}{13}, -\frac{50}{13}\right)$ **37.** $(-2, -9)$
39. $(30, 6)$ **41.** $\{(x, y) \mid x = 2 + 3y\}$ **43.** No solution
45. $(140, 60)$ **47.** $\left(\frac{1}{3}, -\frac{2}{3}\right)$ **49.** 🗍   **51.** [2.5] 4 mi
52. [2.5] 86   **53.** [2.5] Bathrooms: $11\frac{2}{3}$ billion; kitchens:
$23\frac{1}{3}$ billion   **54.** [2.5] 30 m, 90 m, 360 m
55. [2.5] 450.5 mi   **56.** [2.5] 460 mi   **57.** 🗍
59. $m = -\frac{1}{2}, b = \frac{5}{2}$ **61.** $a = 5, b = 2$ **63.** $\left(-\frac{32}{17}, \frac{38}{17}\right)$
65. $\left(-\frac{1}{5}, \frac{1}{10}\right)$ **67.** 🗍

Exercise Set 8.3, pp. 503–505

1. 29, 18   **3.** $8.50 brushes: 32; $9.75 brushes: 13
5. 119°, 61°   **7.** Two-point shots: 36; foul shots: 28
9. Humulin: 30 vials; Novolin: 20 vials
11. Width: 36 ft; length: 78 ft   **13.** Wins: 23; ties: 14
15. Lemon juice: $10\frac{2}{3}$ oz; linseed oil: $21\frac{1}{3}$ oz
17. General-interest: 65; children's: 12   **19.** Boxes: 13;
four-packs: 27   **21.** Kenyan: 8 lb; Sumatran: 12 lb
23. 5 lb of each   **25.** 25%-acid: 4 L; 50%-acid: 6 L
27. $7500 at 6%; $4500 at 9%   **29.** Arctic Antifreeze:
12.5 L; Frost No-More: 7.5 L   **31.** Length: 76 m;
width: 19 m   **33.** Quarters: 17; fifty-cent pieces: 13
35. 375 km   **37.** 14 km/h   **39.** About 1489 mi   **41.** 🗍
43. [1.1] 16   **44.** [1.1] 11   **45.** [1.1], [1.8] -28
46. [1.1], [1.8] -10 **47.** [1.1], [1.3] $\frac{49}{12}$ **48.** [1.1], [1.3] $\frac{13}{10}$
49. 🗍   **51.** Burl: 40; son: 20   **53.** Length: $\frac{288}{5}$ in.;
width: $\frac{102}{5}$ in.   **55.** $\frac{120}{7}$ lb   **57.** 4 km   **59.** 82
61. Brown: 0.8 gal; neutral: 0.2 gal   **63.** 45 L

65. $P(x) = \dfrac{0.1 + x}{1.5}$ (This expresses the percent as a decimal quantity.)

Exercise Set 8.4, pp. 513–514

1. No   **3.** $(1, 2, 3)$ **5.** $(-1, 5, -2)$ **7.** $(3, 1, 2)$
9. $(-3, -4, 2)$ **11.** $(2, 4, 1)$ **13.** $(-3, 0, 4)$
15. The equations are dependent.   **17.** $(3, -5, 8)$
19. $\left(\frac{3}{5}, \frac{2}{3}, -3\right)$ **21.** $\left(4, \frac{1}{2}, -\frac{1}{2}\right)$ **23.** $(17, 9, 79)$
25. $\left(\frac{1}{4}, -\frac{1}{2}, -\frac{1}{4}\right)$ **27.** $(20, 62, 100)$ **29.** No solution
31. The equations are dependent.   **33.** ▣
35. [1.1] Let x and y represent the numbers; $x = 2y$
36. [1.1] Let x and y represent the numbers; $x + y = 3x$
37. [1.1] Let x represent the first number;
$x + (x + 1) + (x + 2) = 45$
38. [1.1] Let x and y represent the numbers; $x + 2y = 17$
39. [1.1] Let x, y, and z represent the numbers; $x + y = 5z$
40. [1.1] Let x and y represent the numbers; $xy = 2(x + y)$
41. ▣   **43.** $(1, -1, 2)$ **45.** $(-3, -1, 0, 4)$
47. $\left(-\frac{1}{2}, -1, -\frac{1}{3}\right)$ **49.** 14   **51.** $z = 8 - 2x - 4y$

Exercise Set 8.5, pp. 518–520

1. 16, 19, 22   **3.** 8, 21, -3 **5.** 32°, 96°, 52°
7. Automatic transmission: $865; power door locks: $520;
air conditioning: $375   **9.** Elrod: 20; Dot: 24; Wendy: 30
11. 10-oz cups: 11; 14-oz cups: 15; 20-oz cups: 8
13. First fund: $45,000; second fund: $10,000; third fund:
$25,000   **15.** Roast beef: 2; baked potato: 1; broccoli: 2
17. Man: 3.6; woman: 18.1; one-year-old child: 50
19. Two-point field goals: 32; three-point field goals: 5;
foul shots: 13   **21.** ▣   **23.** [1.8] -8 **24.** [1.8] 33
25. [1.8] -55 **26.** [1.8] -71
27. [1.8] $-14x + 21y - 35z$ **28.** [1.8] $-24a - 42b + 54c$
29. [1.8] $-5a$ **30.** [1.8] $11x$ **31.** ▣   **33.** 464
35. Adults: 5; students: 1; children: 94   **37.** 180°

Exercise Set 8.6, pp. 524–525

1. $\left(-\frac{1}{3}, -4\right)$ **3.** $(-4, 3)$ **5.** $\left(\frac{3}{2}, \frac{5}{2}\right)$ **7.** $\left(2, \frac{1}{2}, -2\right)$
9. $(2, -2, 1)$ **11.** $\left(4, \frac{1}{2}, -\frac{1}{2}\right)$ **13.** $(1, -3, -2, -1)$
15. Dimes: 4; nickels: 30   **17.** $4.05-per-pound granola:
5 lb; $2.70-per-pound granola: 10 lb   **19.** $400 at 7%;
$500 at 8%; $1600 at 9%   **21.** ▣   **23.** [1.8] 13
24. [1.8] -22 **25.** [1.8] 37   **26.** [1.8] 422   **27.** ▣
29. 1324

Exercise Set 8.7, pp. 530–531

1. 18   **3.** 36   **5.** 27   **7.** -3 **9.** -5 **11.** $(-3, 2)$
13. $\left(\frac{9}{19}, \frac{51}{38}\right)$ **15.** $\left(-1, -\frac{6}{7}, \frac{11}{7}\right)$ **17.** $(2, -1, 4)$
19. $(1, 2, 3)$ **21.** ▣   **23.** [2.2] $\frac{333}{245}$ **24.** [2.2] -12
25. [2.5] One piece: 20.8 ft; other piece: 12 ft
26. [8.3] Scientific calculators: 18; graphing calculators: 27
27. [8.3] Mazzas: 28 rolls; Kranepools: 8 rolls
28. [8.3] Buckets: 17; dinners: 11   **29.** ▣   **31.** 12
33. 10

Exercise Set 8.8, pp. 535–537

1. **(a)** $P(x) = 20x - 300,000$; **(b)** 15,000 units
3. **(a)** $P(x) = 50x - 120,000$; **(b)** 2400 units
5. **(a)** $P(x) = 45x - 22,500$; **(b)** 500 units
7. **(a)** $P(x) = 18x - 16,000$; **(b)** 889 units
9. **(a)** $P(x) = 50x - 100,000$; **(b)** 2000 units
11. ($70, 300)   **13.** ($22, 474)   **15.** ($50, 6250)
17. ($10, 1070)   **19.** **(a)** $C(x) = 125,300 + 450x$;
(b) $R(x) = 800x$; **(c)** $P(x) = 350x - 125,300$; **(d)** $90,300
loss, $14,700 profit; **(e)** (358 computers, $286,400)
21. **(a)** $C(x) = 16,404 + 6x$; **(b)** $R(x) = 18x$;
(c) $P(x) = 12x - 16,404$; **(d)** $19,596 profit, $4404 loss;
(e) (1367 dozen caps, $24,606)   **23.** ▣   **25.** [2.2] 12
26. [2.2] 15   **27.** [2.2] $\frac{8}{3}$ **28.** [2.2] 4   **29.** [2.2] $\frac{9}{2}$
30. [2.2] $\frac{1}{3}$ **31.** ▣   **33.** ($5, 300 yo-yo's)
35. **(a)** $8.74; **(b)** 24,509 units

Review Exercises: Chapter 8, pp. 539–540

1. [8.1] $(-2, 1)$ **2.** [8.1] $(3, 2)$ **3.** [8.2] $\left(-\frac{11}{15}, -\frac{43}{30}\right)$
4. [8.2] No solution   **5.** [8.2] $\left(-\frac{4}{5}, \frac{2}{5}\right)$ **6.** [8.2] $\left(\frac{37}{19}, \frac{53}{19}\right)$
7. [8.2] $\left(\frac{76}{17}, -\frac{2}{119}\right)$ **8.** [8.2] $(2, 2)$
9. [8.2] $\{(x, y) \mid 3x + 4y = 6\}$
10. [8.3] DVD: $29; videocassette: $14   **11.** [8.3] 4 hr
12. [8.3] 8% juice: 10 L; 15% juice: 4 L
13. [8.4] $(4, -8, 10)$ **14.** [8.4] The equations are
dependent.   **15.** [8.4] $(2, 0, 4)$ **16.** [8.2] No solution
17. [8.4] $\left(\frac{8}{9}, -\frac{2}{3}, \frac{10}{9}\right)$ **18.** [8.5] A: 90°; B: 67.5°; C: 22.5°
19. [8.5] 641   **20.** [8.5] $20 bills: 7; $5 bills: 3; $1 bills: 4
21. [8.6] $\left(55, -\frac{89}{2}\right)$ **22.** [8.6] $(-1, 1, 3)$ **23.** [8.7] 2
24. [8.7] 9   **25.** [8.7] $(6, -2)$ **26.** [8.7] $(-3, 0, 4)$
27. [8.8] ($3, 81)   **28.** [8.8] **(a)** $C(x) = 1.5x + 18,000$;
(b) $R(x) = 6x$; **(c)** $P(x) = 4.5x - 18,000$; **(d)** $11,250 loss,
$4500 profit; **(e)** (4000 pints of honey, $24,000)
29. ▣ [8.5] To solve a problem involving four variables, go
through the *Familiarize* and *Translate* steps as usual. The
resulting system of equations can be solved using the
elimination method just as for three variables but likely
with more steps.   **30.** ▣ [8.4] A system of equations can
be both dependent and inconsistent if it is equivalent to a
system with fewer equations that has no solution. An
example is a system of three equations in three unknowns
in which two of the equations represent the same plane,
and the third represents a parallel plane.
31. [8.8] 10,000 pints   **32.** [8.1] $(0, 2)$, $(1, 3)$
33. [8.5] $a = -\frac{2}{3}$, $b = -\frac{4}{3}$, $c = 3$; $f(x) = -\frac{2}{3}x^2 - \frac{4}{3}x + 3$

Test: Chapter 8, pp. 540–541

1. [8.1] $(2, 4)$ **2.** [8.2] $\left(3, -\frac{11}{3}\right)$ **3.** [8.2] $\left(\frac{15}{7}, -\frac{18}{7}\right)$
4. [8.2] $\left(-\frac{3}{2}, -\frac{3}{2}\right)$ **5.** [8.2] No solution   **6.** [8.3] Length:
30 units; width: 18 units   **7.** [8.5] Mortgage: $74,000;

car loan: \$600; credit-card bill: \$700
8. [8.4] The equations are dependent.
9. [8.4] $\left(2, -\frac{1}{2}, -1\right)$ **10.** [8.4] No solution
11. [8.4] $(0, 1, 0)$ **12.** [8.6] $\left(\frac{34}{107}, -\frac{104}{107}\right)$
13. [8.6] $(3, 1, -2)$ **14.** [8.7] 34 **15.** [8.7] 133
16. [8.7] $\left(\frac{13}{18}, \frac{7}{27}\right)$ **17.** [8.5] 3.5 hr **18.** [8.8] (\$3, 55)
19. [8.8] **(a)** $C(x) = 25x + 40,000$; **(b)** $R(x) = 70x$;
(c) $P(x) = 45x - 40,000$; **(d)** \$26,500 loss, \$500 profit;
(e) (889 radios, \$62,230) **20.** [7.3], [8.3] $m = 7, b = 10$
21. [8.5] Adult: 1346; senior citizen: 335; child: 1651

Exercise Set 9.1, pp. 550–552

1.  $\{y \mid y < 6\}, (-\infty, 6)$

3. $\{x \mid x \geq -4\}, [-4, \infty)$

5. $\{t \mid t > -3\}, (-3, \infty)$

7. $\{x \mid x \leq -7\}, (-\infty, -7]$

9. $\{y \mid y > -9\}$, or $(-9, \infty)$

11. $\{y \mid y \leq 14\}$, or $(-\infty, 14]$

13. $\{t \mid t < -9\}$, or $(-\infty, -9)$

15. $\{y \mid y \geq -0.4\}$, or $[-0.4, \infty)$

17. $\left\{y \mid y \geq \frac{9}{10}\right\}$, or $\left[\frac{9}{10}, \infty\right)$

19. $\{y \mid y > 3\}$, or $(3, \infty)$

21. $\{x \mid x \leq 4\}$, or $(-\infty, 4]$

23. $\left\{x \mid x < -\frac{2}{5}\right\}$, or $\left(-\infty, -\frac{2}{5}\right)$

25. $\{x \mid x \geq 11.25\}$, or $[11.25, \infty)$

27. $\left\{y \mid y \leq -\frac{53}{6}\right\}$, or $\left(-\infty, -\frac{53}{6}\right]$ **29.** $\left\{t \mid t < \frac{29}{5}\right\}$, or $\left(-\infty, \frac{29}{5}\right)$
31. $\left\{m \mid m > \frac{7}{3}\right\}$, or $\left(\frac{7}{3}, \infty\right)$ **33.** $\{x \mid x \geq 2\}$, or $[2, \infty)$
35. $\{y \mid y < 5\}$, or $(-\infty, 5)$ **37.** $\left\{x \mid x \leq \frac{4}{7}\right\}$, or $\left(-\infty, \frac{4}{7}\right]$
39. For 1175 min or more
41. More than 25 **43.** Gross sales greater than \$7000
45. More than \$1850 **47.** At least 625 people
49. **(a)** $\left\{x \mid x < 8181\frac{9}{11}\right\}$, or $\{x \mid x \leq 8181\}$ **(b)** $\left\{x \mid x > 8181\frac{9}{11}\right\}$,
or $\{x \mid x \geq 8182\}$ **51.** ▨
53. [7.2] $\{x \mid x$ is a real number *and* $x \neq 2\}$
54. [7.2] $\{x \mid x$ is a real number *and* $x \neq -3\}$
55. [7.2] $\left\{x \mid x$ is a real number *and* $x \neq \frac{7}{2}\right\}$
56. [7.2] $\left\{x \mid x$ is a real number *and* $x \neq \frac{9}{4}\right\}$
57. [1.8] $7x + 10$ **58.** [1.8] $22x - 7$ **59.** ▨

61. $\left\{x \mid x \leq \dfrac{2}{a - 1}\right\}$ **63.** $\left\{y \mid y \geq \dfrac{2a + 5b}{b(a - 2)}\right\}$

65. $\left\{x \mid x > \dfrac{4m - 2c}{d - (5c + 2m)}\right\}$ **67.** False; $2 < 3$ and $4 < 5$,

but $2 - 4 = 3 - 5$. **69.** ▨
71. $\mathbb{R}$

73. $\{x \mid x$ is a real number *and* $x \neq 0\}$ **75.** ◿

Technology Connection, p. 559

1. Domain of $y_1 = \{x \mid x \leq 3\}$, or $(-\infty, 3]$; domain of
$y_2 = \{x \mid x \geq -1\}$, or $[-1, \infty)$
2. Domain of $y_1 + y_2 =$ domain of $y_1 - y_2 =$ domain of
$y_2 - y_1 =$ domain of $y_1 \cdot y_2 = \{x \mid -1 \leq x \leq 3\}$, or $[-1, 3]$

Exercise Set 9.2, pp. 559–561

1. $\{9, 11\}$ **3.** $\{1, 5, 10, 15, 20\}$ **5.** $\{b, d, f\}$
7. $\{r, s, t, u, v\}$ **9.** $\varnothing$ **11.** $\{3, 5, 7\}$
13. $(3, 8)$

15. $[-6, -2]$

17. $(-\infty, -2) \cup (3, \infty)$

19. $(-\infty, -1] \cup (5, \infty)$

21. $(-2, 4]$

23. $(-2, 4)$

25. $(-\infty, 5) \cup (7, \infty)$

27. $(-\infty, -4] \cup [5, \infty)$

29. $[-6, 4)$

31. $[3, 7)$

33. $(-\infty, 5)$

35. $(-\infty, \infty)$

37. $(7, \infty)$

39. $\{t \mid -3 < t < 5\}$, or $(-3, 5)$

41. $\{x \mid -1 < x \leq 4\}$, or $(-1, 4]$

43. $\{a \mid -2 \leq a < 2\}$, or $[-2, 2)$

45. $\mathbb{R}$, or $(-\infty, \infty)$

47. $\{x \mid 1 \leq x \leq 3\}$, or $[1, 3]$

49. $\left\{x \mid -\frac{7}{2} < x \leq 7\right\}$, or $\left(-\frac{7}{2}, 7\right]$

51. $\{x\,|\,x \le 1 \text{ or } x \ge 3\}$, or $(-\infty, 1] \cup [3, \infty)$

53. $\{x\,|\,x < 3 \text{ or } x > 4\}$, or $(-\infty, 3) \cup (4, \infty)$

55. $\{a\,|\,a < \frac{7}{2}\}$, or $(-\infty, \frac{7}{2})$

57. $\{a\,|\,a < -5\}$, or $(-\infty, -5)$

59. $\mathbb{R}$, or $(-\infty, \infty)$

61. $\{t\,|\,t \le 6\}$, or $(-\infty, 6]$

63. $(-\infty, -7) \cup (-7, \infty)$ **65.** $[6, \infty)$
67. $\left(-\infty, \frac{5}{2}\right) \cup \left(\frac{5}{2}, \infty\right)$ **69.** $[-4, \infty)$ **71.** $(-\infty, 4]$
73.

75. [3.3]

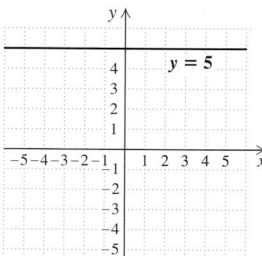

76. [3.3]

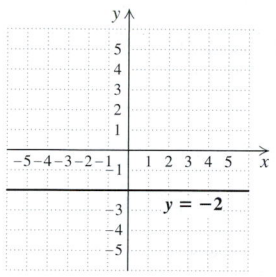

77. [7.3]

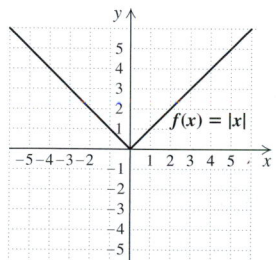

78. [7.3]

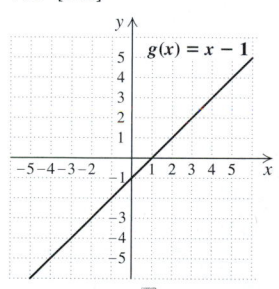

79. [8.1] $(8, 5)$ **80.** [8.1] $(-5, -3)$ **81.**   **83.** $(-1, 6)$
85. Between 12 and 240 trips   **87.** Sizes between 6 and 13
89. (a) $1945.4° \le F < 4820°$; (b) $1761.44° \le F < 3956°$
91. $\left\{a\,\middle|\,-\frac{3}{2} \le a \le 1\right\}$, or $\left[-\frac{3}{2}, 1\right]$;

93. $\{x\,|\,-4 < x \le 1\}$, or $(-4, 1]$;

95. True   **97.** False   **99.** $\left[-\frac{5}{2}, 1\right) \cup (1, \infty)$ **101.**
103.

Technology Connection, p. 564

1. The graphs of $y_1 = \text{abs}(4 - 7x)$ and $y_2 = -8$ do not intersect.

Exercise Set 9.3, pp. 569–570

1. $\{-4, 4\}$ **3.** $\varnothing$ **5.** $\{-7.3, 7.3\}$ **7.** $\{0\}$ **9.** $\left\{-\frac{9}{5}, 1\right\}$
11. $\varnothing$ **13.** $\{-5, 11\}$ **15.** $\{5, 7\}$ **17.** $\{-1, 9\}$

19. $\{-9, 9\}$ **21.** $\{-2, 2\}$ **23.** $\left\{-\frac{14}{5}, \frac{22}{5}\right\}$ **25.** $\{4, 10\}$
27. $\left\{\frac{3}{2}, \frac{7}{2}\right\}$ **29.** $\left\{-\frac{4}{3}, 4\right\}$ **31.** $\{-8.3, 8.3\}$ **33.** $\left\{-\frac{8}{3}, 4\right\}$
35. $\{1, 11\}$ **37.** $\left\{\frac{3}{2}\right\}$ **39.** $\left\{-4, -\frac{10}{9}\right\}$ **41.** $\mathbb{R}$ **43.** $\{1\}$
45. $\left\{32, \frac{8}{3}\right\}$
47. $\{a\,|\,-7 \le a \le 7\}$, or $[-7, 7]$

49. $\{x\,|\,x < -8 \text{ or } x > 8\}$, or $(-\infty, -8) \cup (8, \infty)$

51. $\{t\,|\,t < 0 \text{ or } t > 0\}$, or $(-\infty, 0) \cup (0, \infty)$

53. $\{x\,|\,-2 < x < 8\}$, or $(-2, 8)$

55. $\{x\,|\,-8 \le x \le 4\}$, or $[-8, 4]$

57. $\{x\,|\,x < -2 \text{ or } x > 8\}$, or $(-\infty, -2) \cup (8, \infty)$

59. $\mathbb{R}$, or $(-\infty, \infty)$

61. $\left\{a\,\middle|\,a \le -\frac{2}{3} \text{ or } a \ge \frac{10}{3}\right\}$, or $\left(-\infty, -\frac{2}{3}\right] \cup \left[\frac{10}{3}, \infty\right)$

63. $\{y\,|\,-9 < y < 15\}$, or $(-9, 15)$;

65. $\{x\,|\,x < -8 \text{ or } x \ge 0\}$, or $(-\infty, -8) \cup [0, \infty)$

67. $\left\{y\,\middle|\,y < -\frac{4}{3} \text{ or } y > 4\right\}$, or $\left(-\infty, -\frac{4}{3}\right) \cup (4, \infty)$;

69. $\varnothing$
71. $\left\{x\,\middle|\,x \le -\frac{2}{15} \text{ or } x \ge \frac{14}{15}\right\}$, or $\left(-\infty, -\frac{2}{15}\right] \cup \left[\frac{14}{15}, \infty\right)$;

73. $\{m\,|\,-12 \le m \le 2\}$, or $[-12, 2]$;

75. $\{a\,|\,-6 < a < 0\}$, or $(-6, 0)$

77. $\left\{x\,\middle|\,-\frac{1}{2} \le x \le \frac{7}{2}\right\}$, or $\left[-\frac{1}{2}, \frac{7}{2}\right]$

79. $\left\{x\,\middle|\,x \le -\frac{7}{3} \text{ or } x \ge 5\right\}$, or $\left(-\infty, -\frac{7}{3}\right] \cup [5, \infty)$

81. $\{x\,|\,-4 < x < 5\}$, or $(-4, 5)$

83.   **85.** [8.2] $\left(-\frac{16}{13}, -\frac{41}{13}\right)$ **86.** [8.2] $(-2, -3)$
87. [8.2] $(10, 4)$ **88.** [8.2] $(-1, 7)$ **89.** [8.1] $(1, 4)$
90. [8.1] $(24, -41)$ **91.**   **93.** $\left\{t\,\middle|\,t \ge \frac{5}{3}\right\}$, or $\left[\frac{5}{3}, \infty\right)$
95. $\{x\,|\,-4 \le x \le -1 \text{ or } 3 \le x \le 6\}$, or $[-4, -1] \cup [3, 6]$
97. $\left\{x\,\middle|\,x \le \frac{5}{2}\right\}$, or $\left(-\infty, \frac{5}{2}\right]$ **99.** $|y| \le 5$ **101.** $|x| > 4$
103. $|x + 2| < 3$ **105.** $|x - 5| < 1$, or $|5 - x| < 1$

107. $|x - 2| < 6$ **109.** $|x - 7| \le 5$ **111.** $\{x \mid 1 \le x \le 5\}$, or $[1, 5]$ **113.** 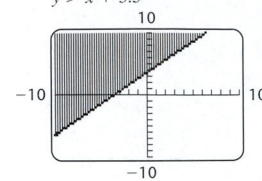   **115.** ▮ ◠◡

Technology Connection, p. 574

1. $y > x + 3.5$

2. $7y \le 2x + 5$

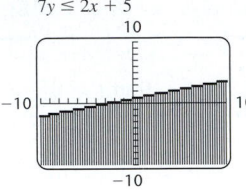

3. $8x - 2y < 11$

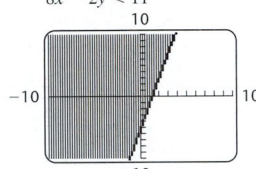

4. $11x + 13y + 4 \ge 0$

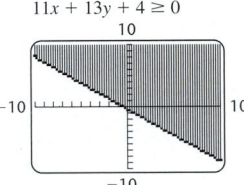

Exercise Set 9.4, pp. 579–580

1. Yes   **3.** No

5.

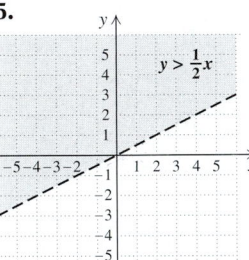

$y > \frac{1}{2}x$

7.

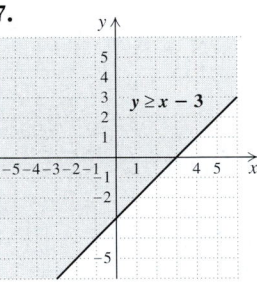

$y \ge x - 3$

9.

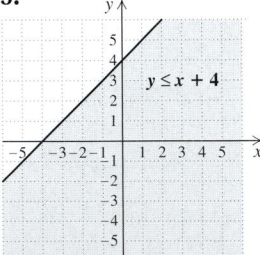

$y \le x + 4$

11.

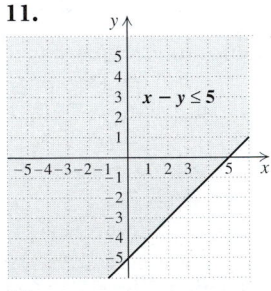

$x - y \le 5$

13.

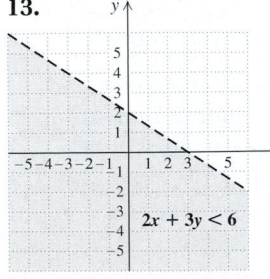

$2x + 3y < 6$

15.

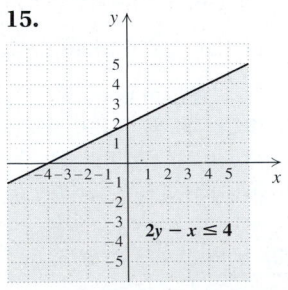

$2y - x \le 4$

17.

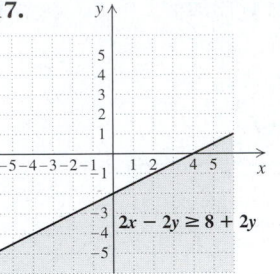

$2x - 2y \ge 8 + 2y$

19.

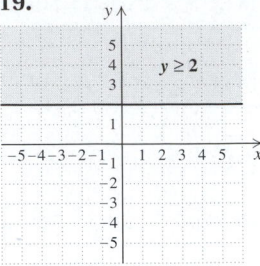

$y \ge 2$

21.

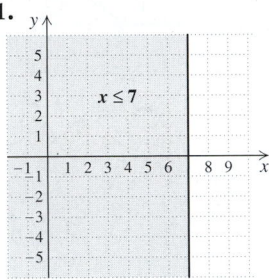

$x \le 7$

23.

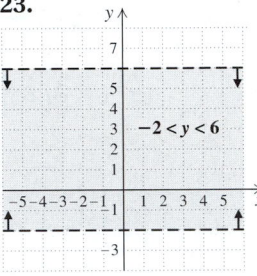

$-2 < y < 6$

25.

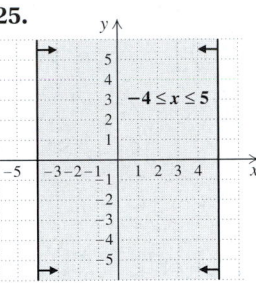

$-4 \le x \le 5$

27.

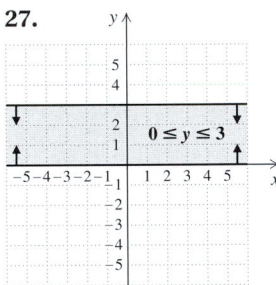

$0 \le y \le 3$

29.

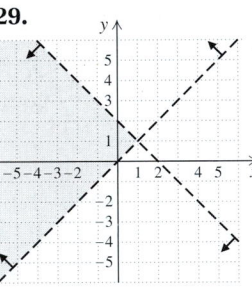

31.

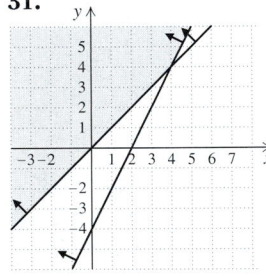

33.

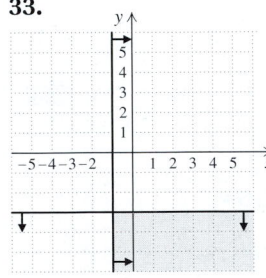

35.

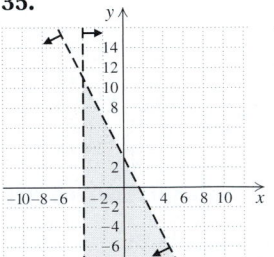

37.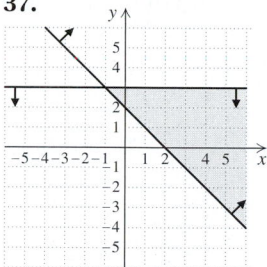

39.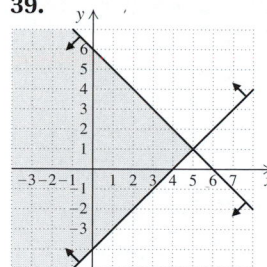

41.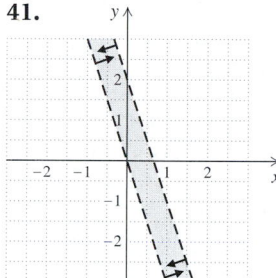

43.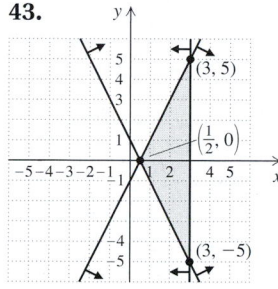

45.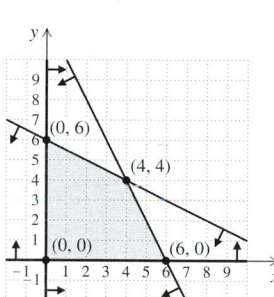

47.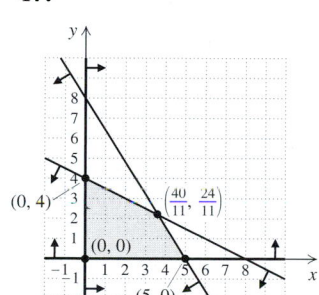

49.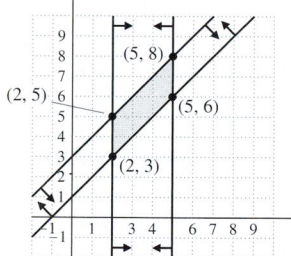

51. ▨   **53.** [8.3] Peanuts: $6\frac{2}{3}$ lb; fancy nuts: $3\frac{1}{3}$ lb
54. [8.3] Hendersons: 10 bags; Savickis: 4 bags
55. [8.3] Activity-card holders: 128; noncard holders: 75
56. [8.3] Students: 70; adults: 130   **57.** [2.5] 25 ft
58. [2.5] 11%   **59.** ▨

61.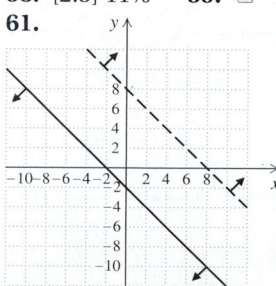

63.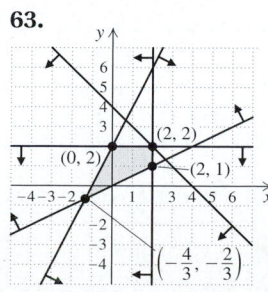

65. $0 < w \le 62$,
    $0 < h \le 62$,
    $62 + 2w + 2h \le 108$,
    or $w + h \le 23$

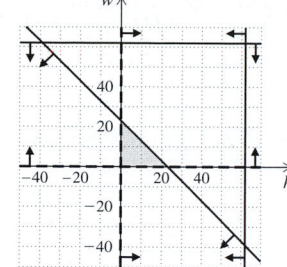

67. $35c + 75a < 1000$,
    $c \ge 0$,
    $a \ge 0$

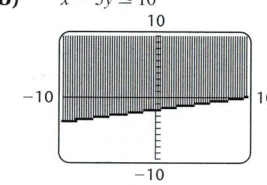

69. (a) $3x + 6y > 2$ **(b)** $x - 5y \le 10$

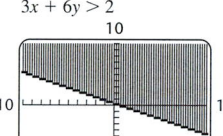 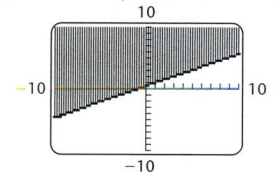

(c) $13x - 25y + 10 \le 0$ **(d)** $2x + 5y > 0$

Exercise Set 9.5, pp. 587–589

1. Maximum 84 when $x = 0$, $y = 6$; minimum 0 when
$x = 0$, $y = 0$ **3.** Maximum 76 when $x = 7$, $y = 0$;
minimum 16 when $x = 0$, $y = 4$ **5.** Maximum 5 when
$x = 3$, $y = 7$; minimum -15 when $x = 3$, $y = -3$
7. Chili: 40 orders: burritos: 50 orders
9. Motorcycles: 60; bicycles: 100   **11.** Matching: 5;
essay: 15; maximum: 425   **13.** Corporate bonds: $22,000;
municipal bonds: $18,000; maximum: $3110
15. Hawaiian Blend: 68; Classic Blend: 39; maximum:
$8265   **17.** Knit: 2; worsted: 4   **19.** ▨
21. [1.1], [1.4] -40 **22.** [1.1] 46   **23.** [1.8] $10x + 5$
24. [1.8] $26t + 20$ **25.** [1.8] $3x - 6$ **26.** [1.8] $2t + 2$
27. ▨   **29.** T3's: 30; S5's: 10   **31.** Chairs: 25; sofas: 9

Review Exercises: Chapter 9, pp. 590–591

1. [9.1] $\{x \mid x \le -2\}$, or $(-\infty, -2]$;

2. [9.1] $\{a \mid a \le -21\}$, or $(-\infty, -21]$;

3. [9.1] $\{y \mid y \ge -7\}$, or $[-7, \infty)$;

4. [9.1] $\left\{y \mid y > -\frac{15}{4}\right\}$, or $\left(-\frac{15}{4}, \infty\right)$;

5. [9.1] $\{y \mid y > -30\}$, or $(-30, \infty)$;

6. [9.1] $\left\{x \mid x > -\frac{3}{2}\right\}$, or $\left(-\frac{3}{2}, \infty\right)$;

7. [9.1] $\{x \mid x < -3\}$, or $(-\infty, -3)$;

8. [9.1] $\left\{y \mid y > -\frac{220}{23}\right\}$, or $\left(-\frac{220}{23}, \infty\right)$;

9. [9.1] $\left\{x \mid x \le -\frac{5}{2}\right\}$, or $\left(-\infty, -\frac{5}{2}\right]$;

10. [9.1] $\{x \mid x \le 4\}$, or $(-\infty, 4]$ **11.** [9.1] More than 125 hr
12. [9.1] \$1500   **13.** [9.2] $\{1, 5, 9\}$
14. [9.2] $\{1, 2, 3, 5, 6, 9\}$
15. [9.2] ; $(-5, 3]$

16. [9.2] ; $(-\infty, \infty)$

17. [9.2] $\{x \mid -7 < x \le 2\}$, or $(-7, 2]$

18. [9.2] $\left\{x \mid -\frac{5}{4} < x < \frac{5}{2}\right\}$, or $\left(-\frac{5}{4}, \frac{5}{2}\right)$

19. [9.2] $\{x \mid x < -3 \text{ or } x > 1\}$, or $(-\infty, -3) \cup (1, \infty)$

20. [9.2] $\{x \mid x < -11 \text{ or } x \ge -6\}$, or $(-\infty, -11) \cup [-6, \infty)$

21. [9.2] $\{x \mid x \le -6 \text{ or } x \ge 8\}$, or $(-\infty, -6] \cup [8, \infty)$

22. [9.2] $\left\{x \mid x < -\frac{2}{5} \text{ or } x > \frac{8}{5}\right\}$, or $\left(-\infty, -\frac{2}{5}\right) \cup \left(\frac{8}{5}, \infty\right)$

23. [9.2] $(-\infty, 3) \cup (3, \infty)$ **24.** [9.2] $[-3, \infty)$
25. [9.2] $\left(-\infty, \frac{8}{3}\right]$ **26.** [9.3] $\{-4, 4\}$
27. [9.3] $\{t \mid t \le -3.5 \text{ or } t \ge 3.5\}$, or $(-\infty, -3.5] \cup [3.5, \infty)$
28. [9.3] $\{-5, 9\}$ **29.** [9.3] $\left\{x \mid -\frac{17}{2} < x < \frac{7}{2}\right\}$, or $\left(-\frac{17}{2}, \frac{7}{2}\right)$
30. [9.3] $\left\{x \mid x \le -\frac{11}{3} \text{ or } x \ge \frac{19}{3}\right\}$, or $\left(-\infty, -\frac{11}{3}\right] \cup \left[\frac{19}{3}, \infty\right)$
31. [9.3] $\left\{-14, \frac{4}{3}\right\}$ **32.** [9.3] $\varnothing$
33. [9.3] $\{x \mid -12 \le x \le 4\}$, or $[-12, 4]$

34. [9.3] $\{x \mid x < 0 \text{ or } x > 10\}$, or $(-\infty, 0) \cup (10, \infty)$
35. [9.3] $\varnothing$
36. [9.4]   **37.** [9.4]

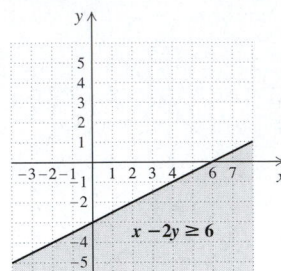

$x - 2y \ge 6$

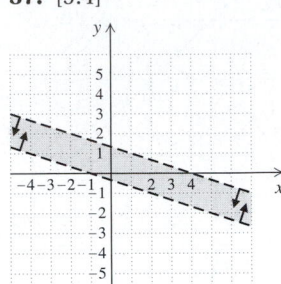

38. [9.4]

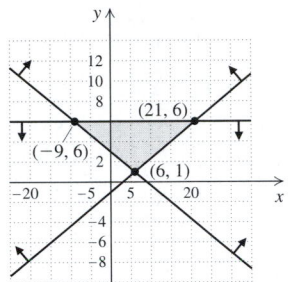

$(21, 6)$
$(-9, 6)$
$(6, 1)$

39. [9.5] Maximum 40 when $x = 7$, $y = 15$; minimum 10 when $x = 1$, $y = 3$
40. [9.5] Ohio plant: 120; Oregon plant: 40
41. [9.3] The equation $|X| = p$ has two solutions when p is positive because X can be either p or $-p$. The same equation has no solution when p is negative because no number has a negative absolute value.
42. [9.4] The solution set of a system of inequalities is all ordered pairs that make *all* the individual inequalities true. This consists of ordered pairs that are common to all the individual solution sets, or the intersection of the graphs.
43. [9.3] $\left\{x \mid -\frac{8}{3} \le x \le -2\right\}$, or $\left[-\frac{8}{3}, -2\right]$
44. [9.1] False; $-4 < 3$ is true, but $(-4)^2 < 9$ is false.
45. [9.3] $|d - 1.1| \le 0.03$

Test: Chapter 9, p. 592

1. [9.1] $\{x \mid x < 12\}$, or $(-\infty, 12)$

2. [9.1] $\{y \mid y > -50\}$, or $(-50, \infty)$

3. [9.1] $\{y \mid y \le -2\}$, or $(-\infty, -2]$

4. [9.1] $\left\{a \mid a \le \frac{11}{5}\right\}$, or $\left(-\infty, \frac{11}{5}\right]$

5. [9.1] $\left\{x \mid x > \frac{5}{2}\right\}$, or $\left(\frac{5}{2}, \infty\right)$

6. [9.1] $\left\{x \mid x \le \frac{7}{4}\right\}$, or $\left(-\infty, \frac{7}{4}\right]$

7. [9.1] $\{x|x>1\}$, or $(1,\infty)$ **8.** [9.1] More than $166\frac{2}{3}$ mi
9. [9.1] Less than or equal to 2.5 hr   **10.** [9.2] $\{3,5\}$
11. [9.2] $\{1,3,5,7,9,11,13\}$ **12.** [9.2] $(-\infty,7]$
13. [9.2] $\{x|-1<x<6\}$, or $(-1,6)$

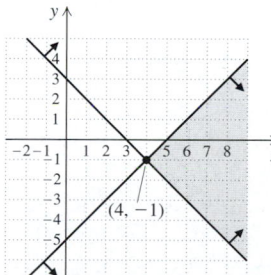

14. [9.2] $\{t|-\frac{2}{5}<t\le\frac{9}{5}\}$, or $\left(-\frac{2}{5},\frac{9}{5}\right]$

15. [9.2] $\{x|x<3 \text{ or } x>6\}$, or $(-\infty,3)\cup(6,\infty)$

16. [9.2] $\{x|x<-4 \text{ or } x>-\frac{5}{2}\}$, or $(-\infty,-4)\cup\left(-\frac{5}{2},\infty\right)$

17. [9.2] $\{x|4\le x<\frac{15}{2}\}$, or $\left[4,\frac{15}{2}\right)$

18. [9.3] $\{-9,9\}$

19. [9.3] $\{a|a<-3 \text{ or } a>3\}$, or $(-\infty,-3)\cup(3,\infty)$

20. [9.3] $\{x|-\frac{7}{8}<x<\frac{11}{8}\}$, or $\left(-\frac{7}{8},\frac{11}{8}\right)$

21. [9.3] $\{t|t\le-\frac{13}{5} \text{ or } t\ge\frac{7}{5}\}$, or $\left(-\infty,-\frac{13}{5}\right]\cup\left[\frac{7}{5},\infty\right)$

22. [9.3] $\varnothing$
23. [9.2] $\{x|x<\frac{1}{2} \text{ or } x>\frac{7}{2}\}$, or $\left(-\infty,\frac{1}{2}\right)\cup\left(\frac{7}{2},\infty\right)$

24. [9.3] $\{1\}$
25. [9.4]   **26.** [9.4]

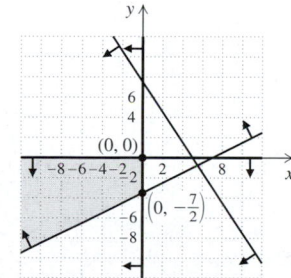

27. [9.5] Maximum 57 when $x=6$, $y=9$; minimum 5 when $x=1$, $y=0$ **28.** [9.5] Manicures: 35; haircuts: 15; maximum: \$690   **29.** [9.3] $[-1,0]\cup[4,6]$
30. [9.2] $\left(\frac{1}{5},\frac{4}{5}\right)$ **31.** [9.3] $|x+13|\le2$

Cumulative Review: Chapters 1–9, pp. 593–594

1. [1.1], [1.4] 10   **2.** [4.8] 5.76×10^9
3. [3.6] Slope: $\frac{7}{4}$; y-intercept: $(0,-3)$
4. [3.7] $y=-\frac{10}{3}x+\frac{11}{3}$ **5.** [8.2] $(-3,4)$

6. [8.4] $(-2,-3,1)$ **7.** [8.3] Small: 16; large: 29
8. [8.5] $12,\frac{1}{2},7\frac{1}{2}$ **9.** [3.4] \$11.8 million per year
10. [7.3] **(a)** $V(t)=\frac{5}{18}t+15.4$; **(b)** about 19.8 min
11. (a) [7.1] $-\frac{1}{2}$; **(b)** [7.2] $\{x|x$ is a real number $and\ x\ne5\}$
12. [5.7] $\frac{1}{4}$ **13.** [5.7] $-\frac{25}{7},\frac{25}{7}$ **14.** [9.1] $\{x|x>-3\}$, or $(-3,\infty)$ **15.** [9.1] $\{x|x\ge-1\}$, or $[-1,\infty)$
16. [9.2] $\{x|-10<x<13\}$, or $(-10,13)$
17. [9.2] $\{x|x<-\frac{4}{3} \text{ or } x>6\}$, or $\left(-\infty,-\frac{4}{3}\right)\cup(6,\infty)$
18. [9.3] $\{x|x<-6.4 \text{ or } x>6.4\}$, or $(-\infty,-6.4)\cup(6.4,\infty)$
19. [9.3] $\{x|-\frac{13}{4}\le x\le\frac{15}{4}\}$, or $\left[-\frac{13}{4},\frac{15}{4}\right]$ **20.** [6.6] $-\frac{5}{3}$
21. [6.6] -1 **22.** [6.6] No solution   **23.** [6.6] $\frac{1}{3}$
24. [9.3] $1,\frac{7}{3}$ **25.** [9.2] $[7,\infty)$ **26.** [2.3] $n=\dfrac{m-12}{3}$
27. [6.8] $a=\dfrac{Pb}{3-P}$
28. [9.4]   **29.** [3.6]

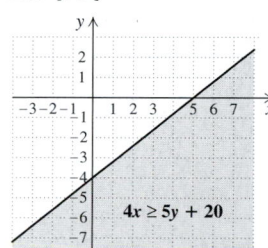
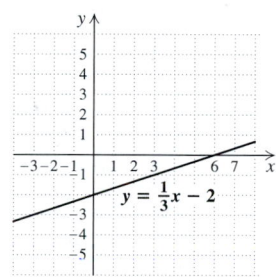

30. [4.3] $-3x^3+9x^2+3x-3$ **31.** [4.4] $-15x^4y^4$
32. [4.3] $5a+5b-5c$
33. [4.4] $15x^4-x^3-9x^2+5x-2$
34. [4.5] $4x^4-4x^2y+y^2$ **35.** [4.5] $4x^4-y^2$
36. [4.3] $-m^3n^2-m^2n^2-5mn^3$ **37.** [6.2] $\dfrac{y-6}{2}$
38. [6.2] $x-1$ **39.** [6.4] $\dfrac{a^2+7ab+b^2}{(a-b)(a+b)}$
40. [6.4] $\dfrac{-m^2+5m-6}{(m+1)(m-5)}$ **41.** [6.4] $\dfrac{3y^2-2}{3y}$
42. [6.5] $\dfrac{y-x}{xy(x+y)}$ **43.** [4.7] $9x^2-13x+26+\dfrac{-50}{x+2}$
44. [5.1] $2x^2(2x+9)$ **45.** [5.2] $(x-6)(x+14)$
46. [5.4] $(4y-9)(4y+9)$
47. [5.5] $8(2x+1)(4x^2-2x+1)$ **48.** [5.4] $(t-8)^2$
49. [5.4] $x^2(x-1)(x+1)(x^2+1)$
50. [5.5] $(0.3b-0.2c)(0.09b^2+0.06bc+0.04c^2)$
51. [5.3] $(4x-1)(5x+3)$ **52.** [5.3] $(3x+4)(x-7)$
53. [5.1] $(x^2-y)(x^3+y)$
54. [7.4] $\{x|x$ is a real number $and\ x\ne2\ and\ x\ne5\}$
55. [6.7] $3\frac{1}{3}$ sec   **56.** [5.8] 30 ft   **57.** [5.8] All such sets of even integers satisfy this condition.   **58.** [7.5] 25 ft
59. [4.4] $x^3-12x^2+48x-64$ **60.** [5.7] $-3,3,-5,5$
61. [9.2], [9.3] $\{x|-3\le x\le-1 \text{ or } 7\le x\le9\}$, or $[-3,-1]\cup[7,9]$
62. [6.6] All real numbers except 9 and -5
63. [5.7] $-\frac{1}{4},0,\frac{1}{4}$

Exercise Set 10.1, pp. 603–604

1. $4, -4$ **3.** $12, -12$ **5.** $9, -9$ **7.** $30, -30$ **9.** $-\frac{7}{6}$
11. 21 **13.** $-\frac{4}{9}$ **15.** 0.3 **17.** -0.07 **19.** $p^2 + 4; 2$
21. $\dfrac{x}{y+4}; 3$ **23.** $\sqrt{20}; 0$; does not exist; does not exist
25. $-3; -1$; does not exist; 0 **27.** $1; \sqrt{2}; \sqrt{101}$
29. 1; does not exist; 6 **31.** $6|x|$ **33.** $6|b|$ **35.** $|7-t|$
37. $|y+8|$ **39.** $|3x-5|$ **41.** -5 **43.** -3 **45.** $-\frac{1}{2}$
47. $|y|$ **49.** $7|b|$ **51.** 10 **53.** $|2a+b|$ **55.** $|x^5|$
57. $|a^7|$ **59.** $5t$ **61.** $7c$ **63.** $5+b$ **65.** $3(x+2)$, or
$3x+6$ **67.** $5t-2$ **69.** -4 **71.** $3x$ **73.** 10
75. $4x$ **77.** a^7 **79.** $(x+3)^5$ **81.** $2; 3; -2; -4$
83. 2; does not exist; does not exist; 3 **85.** $\{x \mid x \geq 5\}$, or
$[5, \infty)$ **87.** $\{t \mid t \geq -3\}$, or $[-3, \infty)$ **89.** $\{x \mid x \leq 5\}$, or
$(-\infty, 5]$ **91.** $\mathbb{R}$ **93.** $\{z \mid z \geq -\frac{3}{5}\}$, or $[-\frac{3}{5}, \infty)$ **95.** $\mathbb{R}$
97.   **99.** [4.1] $a^9b^6c^{15}$ **100.** [4.1] $10a^{10}b^9$
101. [4.8] $\dfrac{a^6c^{12}}{8b^9}$ **102.** [4.8] $\dfrac{x^6y^2}{25z^4}$ **103.** [4.8] $2x^4y^5z^2$
104. [4.8] $\dfrac{5c^3}{a^4b^7}$ **105.**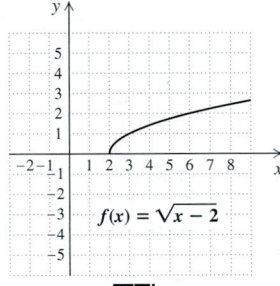
107. (a) 13; (b) 15; (c) 18; (d) 20
109. $\{x \mid x \geq 0\}$, or $[0, \infty)$; **111.** $\{x \mid x \geq 2\}$, or $[2, \infty)$;

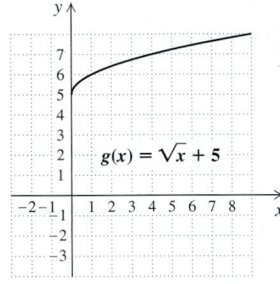
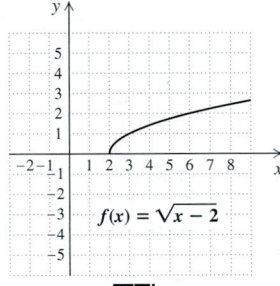

$g(x) = \sqrt{x+5}$

$f(x) = \sqrt{x-2}$

113. $\{x \mid -4 < x \leq 5\}$, or $(-4, 5]$ **115.**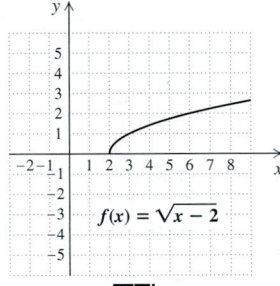

Technology Connection, p. 606

1. Without parentheses, the expression entered would be
$\dfrac{7^2}{3}$. **2.** For $x = 0$ or $x = 1$, $y_1 = y_2 = y_3$; or $(0, 1)$,
$y_1 > y_2 > y_3$; on $(1, \infty)$, $y_1 < y_2 < y_3$.

Technology Connection, p. 608

1. Most graphers do not have keys for radicals of index 3 or higher. On those graphers that offer $\sqrt[x]{}$ in a MATH menu, rational exponents still require fewer keystrokes.

Exercise Set 10.2, pp. 608–610

1. $\sqrt[4]{x}$ **3.** 4 **5.** 3 **7.** 3 **9.** $\sqrt[3]{xyz}$ **11.** $\sqrt[5]{a^2b^2}$
13. $\sqrt[3]{a^2}$ **15.** 8 **17.** 343 **19.** 243 **21.** $\sqrt[4]{81^3x^3}$, or
$27\sqrt[4]{x^3}$ **23.** $125x^6$ **25.** $20^{1/3}$ **27.** $17^{1/2}$ **29.** $x^{3/2}$
31. $m^{2/5}$ **33.** $(cd)^{1/4}$ **35.** $(xy^2z)^{1/5}$ **37.** $(3mn)^{3/2}$
39. $(8x^2y)^{5/7}$ **41.** $\dfrac{2x}{z^{2/3}}$ **43.** $\dfrac{1}{x^{1/3}}$ **45.** $\dfrac{1}{(2rs)^{3/4}}$ **47.** 4
49. $a^{5/7}$ **51.** $\dfrac{2a^{3/4}c^{2/3}}{b^{1/2}}$ **53.** $\dfrac{x^4}{2^{1/3}y^{2/7}}$ **55.** $\left(\dfrac{8yz}{7x}\right)^{3/5}$
57. $\dfrac{7x}{z^{1/3}}$ **59.** $\dfrac{5ac^{1/2}}{3}$ **61.** $5^{7/8}$ **63.** $3^{3/4}$ **65.** $4.1^{1/2}$
67. $10^{6/25}$ **69.** $a^{23/12}$ **71.** 64 **73.** $\dfrac{m^{1/3}}{n^{1/8}}$ **75.** $\sqrt[3]{a}$
77. x^5 **79.** x^3 **81.** a^5b^5 **83.** $\sqrt[4]{3x}$ **85.** $\sqrt{3a}$
87. $\sqrt[8]{x}$ **89.** a^3b^3 **91.** x^8y^{20} **93.** $\sqrt[12]{xy}$ **95.**
97. [4.4] $11x^4 + 14x^3$ **98.** [4.4] $-3t^6 + 28t^5 - 20t^4$
99. [4.5] $15a^2 - 11ab - 12b^2$
100. [4.5] $49x^2 - 14xy + y^2$ **101.** [2.4] $\$93,500$
102. [5.8] $0, 1$ **103.**   **105.** $\sqrt[10]{x^5y^3}$ **107.** $\sqrt[4]{2xy^2}$
109. 524 cycles per second **111.** $2^{7/12} \approx 1.498 \approx 1.5$
113. (a) 1.8 m; (b) 3.1 m; (c) 1.5 m; (d) 5.3 m
115. 10 mg

Technology Connection, p. 613

1. The graph of $y = 0$ (the x-axis)

Exercise Set 10.3, pp. 615–617

1. $\sqrt{70}$ **3.** $\sqrt[3]{10}$ **5.** $\sqrt[4]{72}$ **7.** $\sqrt{30ab}$ **9.** $\sqrt[5]{18t^3}$
11. $\sqrt{x^2 - a^2}$ **13.** $\sqrt[3]{0.1x^2}$ **15.** $\sqrt[4]{x^3 - 1}$ **17.** $\sqrt{\dfrac{7x}{6y}}$
19. $\sqrt[7]{\dfrac{5x-15}{4x+8}}$ **21.** $5\sqrt{2}$ **23.** $2\sqrt{7}$ **25.** $2\sqrt{2}$
27. $3\sqrt{22}$ **29.** $6a^2\sqrt{b}$ **31.** $2x\sqrt[3]{y^2}$ **33.** $-2x^2\sqrt[3]{2}$
35. $f(x) = 5x\sqrt[3]{x^2}$ **37.** $f(x) = |7(x-3)|$, or $7|x-3|$
39. $f(x) = |x-1|\sqrt{5}$ **41.** $ab^2\sqrt{a}$ **43.** $xy^2z^3\sqrt[3]{x^2z}$
45. $-2ab^2\sqrt[5]{a^2b}$ **47.** $abc^5\sqrt{ab^3c^4}$ **49.** $3x^2\sqrt[4]{10x}$
51. $5\sqrt{3}$ **53.** $2\sqrt{35}$ **55.** 2 **57.** $18a^3$ **59.** $a\sqrt[3]{10}$
61. $3x^3\sqrt{5x}$ **63.** $s^2t^3\sqrt[3]{t}$ **65.** $(x+5)^2$
67. $2ab^3\sqrt[4]{3a}$ **69.** $x(y+z)^2\sqrt[5]{x}$ **71.**
73. [6.4] $\dfrac{12x^2 + 5y^2}{64xy}$ **74.** [6.4] $\dfrac{2a + 6b^3}{a^4b^4}$
75. [6.4] $\dfrac{-7x - 13}{2(x-3)(x+3)}$ **76.** [6.4] $\dfrac{-3x+1}{2(x-5)(x+5)}$
77. [4.1] $3a^2b^2$ **78.** [4.1] $3ab^5$ **79.**
81. (a) 20 mph; (b) 37.4 mph; (c) 42.4 mph
83. $r^{10}t^3\sqrt{rt}$ **85.** $a^3b^6\sqrt[3]{ab^2}$

87.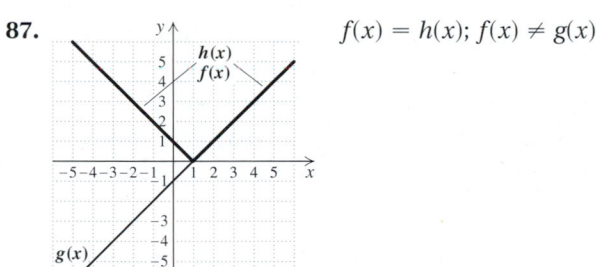

$f(x) = h(x);\ f(x) \neq g(x)$

89. $\{x \mid x \leq -1\ or\ x \geq 4\}$, or $(-\infty, -1] \cup [4, \infty)$ **91.** 10
93. ▨

Exercise Set 10.4, pp. 622–623

1. $\dfrac{5}{6}$ **3.** $\dfrac{4}{3}$ **5.** $\dfrac{7}{y}$ **7.** $\dfrac{5y\sqrt{y}}{x^2}$ **9.** $\dfrac{3a\sqrt[3]{a}}{2b}$ **11.** $\dfrac{2a}{bc^2}$

13. $\dfrac{ab^2}{c^2}\sqrt[4]{\dfrac{a}{c^2}}$ **15.** $\dfrac{2x}{y^2}\sqrt[5]{\dfrac{x}{y}}$ **17.** $\dfrac{xy}{z^2}\sqrt[6]{\dfrac{y^2}{z^3}}$ **19.** $\sqrt{5}$

21. 3   **23.** $y\sqrt{5y}$ **25.** $2\sqrt[3]{a^2b}$ **27.** $\sqrt{2ab}$

29. $2x^2y^3\sqrt[4]{y^3}$ **31.** $\sqrt[3]{x^2 + xy + y^2}$ **33.** $\dfrac{\sqrt{35}}{7}$

35. $\dfrac{2\sqrt{15}}{5}$ **37.** $\dfrac{2\sqrt[3]{6}}{3}$ **39.** $\dfrac{\sqrt[3]{75ac^2}}{5c}$ **41.** $\dfrac{y\sqrt[3]{180x^2y}}{6x^2}$

43. $\dfrac{\sqrt[3]{2xy^2}}{xy}$ **45.** $\dfrac{\sqrt{14a}}{6}$ **47.** $\dfrac{3\sqrt{5y}}{10xy}$ **49.** $\dfrac{\sqrt{5b}}{6a}$

51. $\dfrac{5}{\sqrt{35x}}$ **53.** $\dfrac{2}{\sqrt[3]{6}}$ **55.** $\dfrac{52}{3\sqrt[3]{91}}$ **57.** $\dfrac{7}{\sqrt[3]{98}}$

59. $\dfrac{7x}{\sqrt{21xy}}$ **61.** $\dfrac{2a^2}{\sqrt[3]{20ab}}$ **63.** $\dfrac{x^2y}{\sqrt{2xy}}$ **65.** ▨

67. [6.2] $\dfrac{3(x-1)}{(x-5)(x+5)}$ **68.** [6.2] $\dfrac{7(x-2)}{(x+4)(x-4)}$

69. [6.1] $\dfrac{a-1}{a+7}$ **70.** [6.1] $\dfrac{t+11}{t+2}$ **71.** [4.1] $125a^9b^{12}$

72. [4.1] $225x^{10}y^6$ **73.** ▨   **75.** **(a)** 1.62 sec; **(b)** 1.99 sec;
(c) 2.20 sec   **77.** $9\sqrt[3]{9n^2}$ **79.** $\dfrac{-3\sqrt{a^2-3}}{a^2-3}$, or $\dfrac{-3}{\sqrt{a^2-3}}$

81. Step 1: $\sqrt[n]{a} = a^{1/n}$, by definition; Step 2: $\left(\dfrac{a}{b}\right)^n = \dfrac{a^n}{b^n}$,

raising a quotient to a power; Step 3: $a^{1/n} = \sqrt[n]{a}$, by
definition   **83.** $(f/g)(x) = 3x$, where x is a real number
and $x > 0$ **85.** $(f/g)(x) = \sqrt{x+3}$, where x is a real
number and $x > 3$

Exercise Set 10.5, pp. 628–630

1. $5\sqrt{7}$ **3.** $3\sqrt[3]{5}$ **5.** $13\sqrt[3]{y}$ **7.** $7\sqrt{2}$
9. $13\sqrt[3]{7} + \sqrt{3}$ **11.** $21\sqrt{3}$ **13.** $23\sqrt{5}$ **15.** $9\sqrt[3]{2}$
17. $(1 + 6a)\sqrt{5a}$ **19.** $(x + 2)\sqrt[3]{6x}$ **21.** $3\sqrt{a-1}$
23. $(x + 3)\sqrt{x-1}$ **25.** $3\sqrt{7} - 7$ **27.** $4\sqrt{6} - 4\sqrt{10}$
29. $2\sqrt{15} - 6\sqrt{3}$ **31.** -6 **33.** $a + 2a\sqrt[3]{3}$ **35.** 19
37. -19 **39.** $14 + 6\sqrt{5}$ **41.** -6
43. $2\sqrt[3]{9} + 3\sqrt[3]{6} - 2\sqrt[3]{4}$ **45.** $3x + 2\sqrt{3xy} + y$

47. $\dfrac{3-\sqrt{5}}{2}$ **49.** $\dfrac{12 + 2\sqrt{3} + 6\sqrt{5} + \sqrt{15}}{33}$

51. $\dfrac{a - \sqrt{ab}}{a - b}$ **53.** -1

55. $\dfrac{24 - 3\sqrt{10} - 4\sqrt{14} + \sqrt{35}}{27}$ **57.** $\dfrac{3\sqrt{6} + 4}{2}$

59. $\dfrac{3}{5\sqrt{7} - 10}$ **61.** $\dfrac{2}{14 + 2\sqrt{3} + 3\sqrt{2} + 7\sqrt{6}}$

63. $\dfrac{x - y}{x + 2\sqrt{xy} + y}$ **65.** $a\sqrt[4]{a}$ **67.** $b\sqrt[10]{b^9}$

69. $xy\sqrt[6]{xy^5}$ **71.** $3a^2b\sqrt[4]{ab}$ **73.** $xyz\sqrt[6]{x^5yz^2}$

75. $\sqrt[15]{x^7}$ **77.** $\sqrt[15]{\dfrac{a^7}{b^2}}$ **79.** $\sqrt[10]{xy^3}$ **81.** $\sqrt[12]{(2 + 5x)^5}$

83. $\sqrt[12]{5 + 3x}$ **85.** $x\sqrt[6]{xy^5} - \sqrt[15]{x^{13}y^{14}}$
87. $2m^2 + m\sqrt[4]{n} + 2m\sqrt[3]{n^2} + \sqrt[12]{n^{11}}$ **89.** $\sqrt[4]{2x^2} - x^3$
91. $x^2 - 7$ **93.** $27 - 10\sqrt{2}$ **95.** $8 + 2\sqrt{15}$ **97.** ▨
99. [6.6] 8   **100.** [6.6] $\dfrac{15}{2}$ **101.** [5.8] Length: 20 units;

width: 5 units   **102.** [5.8] $-5, 4$ **103.** [5.7] $\dfrac{1}{5}, 1$
104. [5.7] $\dfrac{1}{7}; 1$ **105.** ▨   **107.** Radicands; indices
109. Denominators   **111.** $f(x) = -6x\sqrt{5 + x}$
113. $f(x) = (x + 3x^2)\sqrt[4]{x - 1}$
115. $ac^2\big[(3a + 2c)\sqrt{ab} - 2\sqrt[3]{ab}\big]$
117. $9a^2(b + 1)\sqrt[6]{243a^5(b + 1)^5}$ **119.** $1 - \sqrt{w}$
121. $(\sqrt{x} + \sqrt{5})(\sqrt{x} - \sqrt{5})$
123. $(\sqrt{x} + \sqrt{a})(\sqrt{x} - \sqrt{a})$ **125.** $2x - 2\sqrt{x^2 - 4}$
127. $\dfrac{b^2 + \sqrt{b}}{1 + b + b^2}$ **129.** $\dfrac{x - 19}{x + 10\sqrt{x + 6} + 31}$

Technology Connection, p. 633

1. The x-coordinates of the points of intersection should
approximate the solutions of the examples.

Exercise Set 10.6, pp. 636–637

1. 22   **3.** $\dfrac{9}{2}$ **5.** 11   **7.** -3 **9.** 29   **11.** 13
13. 0, 64   **15.** No solution   **17.** -64 **19.** 81
21. No solution   **23.** $\dfrac{80}{3}$ **25.** 57   **27.** $-\dfrac{5}{3}$ **29.** 1
31. $\dfrac{106}{27}$ **33.** 4   **35.** 3, 7   **37.** $\dfrac{80}{9}$ **39.** -1
41. No solution   **43.** 2, 6   **45.** 2   **47.** 4   **49.** ▨
51. [2.5] Height: 7 in.; base: 9 in.   **52.** [8.3] 8 60-sec
commercials   **53.** [6.7] Elaine: 6 hr; Gonzalo: 12 hr
54. [6.7] $\dfrac{165}{52}$ hr, or about 3.2 hr

55. [9.4]

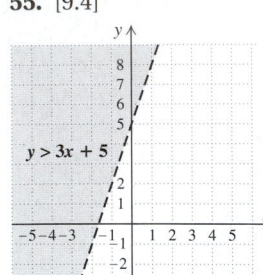

$y > 3x + 5$

56. [7.3]

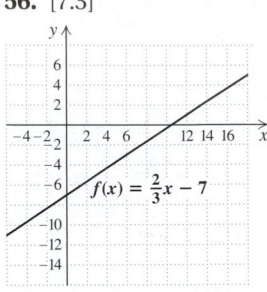

$f(x) = \frac{2}{3}x - 7$

57. 　**59.** 230°C　**61.** $t = \frac{1}{9}\left(\frac{S^2 \cdot 2457}{1087.7^2} - 2617\right)$

63. 4480 rpm　**65.** $r = \dfrac{v^2 h}{2gh - v^2}$　**67.** 72.25 ft

69. $\frac{2504}{125}, \frac{2496}{125}$　**71.** 0　**73.** 1, 8　**75.** $\left(\frac{1}{36}, 0\right)$, $(36, 0)$

77. 　**79.**

Exercise Set 10.7, pp. 643–646

1. $\sqrt{34}$; 5.831　**3.** $9\sqrt{2}$; 12.728　**5.** 5　**7.** $\sqrt{31}$; 5.568
9. $\sqrt{11}$; 3.317　**11.** 1　**13.** $\sqrt{325}$ ft; 18.028 ft
15. $\sqrt{14{,}500}$ ft; 120.416 ft　**17.** 20 in.　**19.** $\sqrt{13{,}000}$ m;
114.018 m　**21.** $a = 5$; $c = 5\sqrt{2} \approx 7.071$　**23.** $a = 7$;
$b = 7\sqrt{3} \approx 12.124$　**25.** $a = 5\sqrt{3} \approx 8.660$;
$c = 10\sqrt{3} \approx 17.321$　**27.** $a = \dfrac{13\sqrt{2}}{2} \approx 9.192$;
$b = \dfrac{13\sqrt{2}}{2} \approx 9.192$　**29.** $b = 14\sqrt{3} \approx 24.249$; $c = 28$
31. $3\sqrt{3} \approx 5.196$　**33.** $13\sqrt{2} \approx 18.385$
35. $\dfrac{19\sqrt{2}}{2} \approx 13.435$　**37.** $\sqrt{10561}$ ft ≈ 102.767 ft
39. $h = 2\sqrt{3}$ ft ≈ 3.464 ft　**41.** 8 ft　**43.** $(0, -4)$, $(0, 4)$
45. 　**47.** [1.8] -47　**48.** [1.8] 5
49. [5.4] $x(x - 3)(x + 3)$　**50.** [5.4] $7a(a - 2)(a + 2)$
51. [9.3] $\left\{-\frac{2}{3}, 4\right\}$　**52.** [9.3] $\left\{-\frac{4}{3}, 10\right\}$　**53.**
55. $\sqrt{75}$ cm　**57.** 4 gal; the total area of the doors and
windows is 164 ft^2 or more　**59.** 36.15 ft by 36.15 ft by
36.15 ft

Exercise Set 10.8, pp. 652–653

1. $5i$　**3.** $i\sqrt{13}$, or $\sqrt{13}i$　**5.** $3i\sqrt{2}$, or $3\sqrt{2}i$
7. $i\sqrt{3}$, or $\sqrt{3}i$　**9.** $9i$　**11.** $10i\sqrt{3}$, or $10\sqrt{3}i$　**13.** $-7i$
15. $4 - 2\sqrt{15}i$　**17.** $(2 + 2\sqrt{3})i$　**19.** $(6\sqrt{2} - 5)i$
21. $12 + 11i$　**23.** $4 + 5i$　**25.** $-4 - 5i$　**27.** $5 + 5i$
29. -54　**31.** 56　**33.** -35　**35.** $-\sqrt{42}$　**37.** $-5\sqrt{6}$
39. $-6 + 14i$　**41.** $-20 - 24i$　**43.** $-11 + 23i$
45. $40 + 13i$　**47.** $8 + 31i$　**49.** $7 - 61i$
51. $-15 - 30i$　**53.** $39 - 37i$　**55.** $-3 - 4i$
57. $5 + 12i$　**59.** $21 + 20i$　**61.** $\frac{6}{5} + \frac{3}{5}i$　**63.** $\frac{6}{29} + \frac{15}{29}i$
65. $-\frac{7}{9}i$　**67.** $-\frac{3}{4} - \frac{5}{4}i$　**69.** $1 - 2i$　**71.** $-\frac{23}{58} + \frac{43}{58}i$
73. $\frac{6}{25} - \frac{17}{25}i$　**75.** $-i$　**77.** 1　**79.** -1　**81.** i

83. -1　**85.** $-125i$　**87.** 0　**89.** 　**91.** [7.4] 1
92. [7.4] 1　**93.** [7.4] 50　**94.** [7.4] 0　**95.** [5.7] $-\frac{4}{3}$, 7
96. [9.3] $\left\{x \mid -\frac{29}{3} < x < 5\right\}$, or $\left(-\frac{29}{3}, 5\right)$　**97.**
99. $-9 - 27i$　**101.** $50 - 120i$　**103.** $\frac{250}{41} + \frac{200}{41}i$
105. 8　**107.** $\frac{3}{5} + \frac{9}{5}i$　**109.** 1

Review Exercises: Chapter 10, pp. 656–657

1. [10.1] $\frac{7}{6}$　**2.** [10.1] -0.5　**3.** [10.1] 5
4. [10.1] $\left\{x \mid x \geq \frac{7}{2}\right\}$, or $\left[\frac{7}{2}, \infty\right)$　**5.** [10.1] $7|a|$
6. [10.1] $|c + 8|$　**7.** [10.1] $|x - 3|$　**8.** [10.1] $|2x + 1|$
9. [10.1] -2　**10.** [10.4] $-\dfrac{4x^2}{3}$　**11.** [10.3] $|x^3 y^2|$, or $|x^3| y^2$
12. [10.3] $2x^2$　**13.** [10.2] $(5ab)^{4/3}$　**14.** [10.2] $8a^4\sqrt{a}$
15. [10.2] $x^3 y^5$　**16.** [10.2] $\sqrt[3]{x^2 y}$　**17.** [10.2] $\dfrac{1}{x^{2/5}}$
18. [10.2] $7^{1/6}$　**19.** [10.3] $f(x) = 5|x - 3|$
20. [10.3] $\sqrt{15xy}$　**21.** [10.3] $3a\sqrt[3]{a^2 b^2}$
22. [10.3] $-6x^5 y^4 \sqrt[3]{2x^2}$　**23.** [10.4] $y\sqrt[3]{6}$
24. [10.4] $\dfrac{5\sqrt{x}}{2}$　**25.** [10.4] $\dfrac{2a^2 \sqrt[4]{3a^3}}{c^2}$　**26.** [10.5] $7\sqrt[3]{x}$
27. [10.5] $3\sqrt{3}$　**28.** [10.5] $(2x + y^2)\sqrt[3]{x}$
29. [10.5] $15\sqrt{2}$　**30.** [10.5] $-43 - 2\sqrt{10}$
31. [10.5] $\sqrt[4]{x^3}$　**32.** [10.5] $\sqrt[12]{x^5}$
33. [10.5] $a^2 - 2a\sqrt{2} + 2$　**34.** [10.5] $-2\sqrt{6} + 6$
35. [10.5] $\dfrac{6}{3 + \sqrt{6}}$　**36.** [10.6] 21　**37.** [10.6] -126
38. [10.6] 4　**39.** [10.6] 14　**40.** [10.7] $5\sqrt{2}$ cm; 7.071 cm
41. [10.7] $\sqrt{24}$ ft; 4.899 ft　**42.** [10.7] $a = 10$;
$b = 10\sqrt{3} \approx 17.321$　**43.** [10.8] $-2i\sqrt{2}$, or $-2\sqrt{2}i$
44. [10.8] $-2 - 9i$　**45.** [10.8] $1 + i$　**46.** [10.8] 29
47. [10.8] i　**48.** [10.8] $9 - 12i$　**49.** [10.8] $\frac{13}{25} - \frac{34}{25}i$
50. [10.1] An absolute-value sign must be used to sim-
plify $\sqrt[n]{x^n}$ when n is even, since x may be negative. If x is
negative while n is even, the radical expression cannot be
simplified to x, since $\sqrt[n]{x^n}$ represents the principal, or
positive, root. When n is odd, there is only one root, and it
will be positive or negative depending on the sign of x. Thus
there is no absolute-value sign when n is odd.
51. [10.8] Every real number is a complex number, but
there are complex numbers that are not real. A complex
number $a + bi$ is not real if $b \neq 0$.　**52.** [10.6] 3
53. [10.8] $-\frac{2}{5} + \frac{9}{10}i$

Test: Chapter 10, p. 657

1. [10.3] $5\sqrt{3}$　**2.** [10.4] $-\dfrac{2}{x^2}$　**3.** [10.1] $10|a|$
4. [10.1] $|x - 4|$　**5.** [10.3] $x^2 y\sqrt[5]{x^2 y^3}$
6. [10.4] $\left|\dfrac{5x}{6y^2}\right|$, or $\dfrac{5|x|}{6y^2}$　**7.** [10.3] $\sqrt[3]{10xy^2}$

8. [10.4] $\sqrt[5]{x^2 y^2}$ **9.** [10.5] $xy\sqrt[4]{x}$ **10.** [10.5] $\sqrt[20]{a^3}$
11. [10.5] $5\sqrt{2}$ **12.** [10.5] $(x^2 + 3y)\sqrt{y}$
13. [10.5] $14 - 19\sqrt{x} - 3x$ **14.** [10.2] $\sqrt[6]{(2a^3b)^5}$
15. [10.2] $(7xy)^{1/2}$ **16.** [10.1] $\{x \mid x \le 2\}$, or $(-\infty, 2]$
17. [10.5] $27 + 10\sqrt{2}$ **18.** [10.5] $\sqrt{6} - \sqrt{3}$
19. [10.6] 7   **20.** [10.6] No solution   **21.** [10.7] Leg: 7
cm; hypotenuse: $7\sqrt{2}$ cm ≈ 9.899 cm
22. [10.7] $\sqrt{10{,}600}$ ft ≈ 102.956 ft
23. [10.8] $5i\sqrt{2}$, or $5\sqrt{2}i$ **24.** [10.8] $10 + 2i$
25. [10.8] -24 **26.** [10.8] $15 - 8i$ **27.** [10.8] $-\frac{13}{53} - \frac{19}{53}i$
28. [10.8] i **29.** [10.6] 3   **30.** [10.8] $-\frac{17}{4}i$

Technology Connection, p. 669

1. The right-hand x-intercept should be an approximation of $4 + \sqrt{23}$.
2. x-intercepts should be approximations of $-3 + \sqrt{5}$ and $-3 - \sqrt{5}$ for Example 8; approximations of $(-5 + \sqrt{37})/2$ and $(-5 - \sqrt{37})/2$ for Example 10(b)
3. A grapher can give only rational-number approximations of the two irrational solutions. An *exact* solution cannot be found with most graphers.
4. The graph of $y = 4x^2 + 9$ has no x-intercepts.

Exercise Set 11.1, pp. 669–671

1. $\pm\sqrt{3}$ **3.** $\pm\frac{2}{5}i$ **5.** $\pm\sqrt{\frac{2}{3}}$, or $\pm\frac{\sqrt{6}}{3}$ **7.** $-7, 3$
9. $-5 \pm 2\sqrt{2}$ **11.** $1 \pm 7i$ **13.** $\frac{-3 \pm \sqrt{14}}{2}$
15. $-7, 13$ **17.** $1, 9$ **19.** $-4 \pm \sqrt{13}$ **21.** $-14, 0$
23. $x^2 + 8x + 16, (x + 4)^2$ **25.** $x^2 - 6x + 9, (x - 3)^2$
27. $x^2 - 24x + 144, (x - 12)^2$ **29.** $t^2 + 9t + \frac{81}{4}, \left(t + \frac{9}{2}\right)^2$
31. $x^2 - 3x + \frac{9}{4}, \left(x - \frac{3}{2}\right)^2$ **33.** $x^2 + \frac{2}{3}x + \frac{1}{9}, \left(x + \frac{1}{3}\right)^2$
35. $t^2 - \frac{5}{3}t + \frac{25}{36}, \left(t - \frac{5}{6}\right)^2$ **37.** $x^2 + \frac{9}{5}x + \frac{81}{100}, \left(x + \frac{9}{10}\right)^2$
39. $-7, 1$ **41.** $5 \pm \sqrt{47}$ **43.** $-7, -1$ **45.** $3, 7$
47. $\frac{-5 \pm \sqrt{13}}{2}$ **49.** $3 \pm i$ **51.** $-2 \pm 3i$ **53.** $-\frac{1}{2}, 3$
55. $-\frac{3}{2}, -\frac{1}{2}$ **57.** $-\frac{3}{2}, \frac{5}{3}$ **59.** $\frac{-2 \pm \sqrt{2}}{2}$ **61.** $\frac{5 \pm \sqrt{61}}{6}$
63. 10%   **65.** 18.75%   **67.** 4%   **69.** About 10.7 sec
71. About 6.3 sec   **73.** ▨   **75.** [1.8] 28   **76.** [1.8] -92
77. [10.3] $3\sqrt[3]{10}$ **78.** [10.3] $4\sqrt{5}$ **79.** [10.1] 5
80. [10.1] 7   **81.** ▨   **83.** ± 18 **85.** $-\frac{7}{2}, -\sqrt{5}, 0, \sqrt{5}, 8$
87. Barge: 8 km/h; fishing boat: 15 km/h   **89.** ▨
91. ▨, ▨

Exercise Set 11.2, pp. 676–677

1. $\frac{-7 \pm \sqrt{61}}{2}$ **3.** $3 \pm \sqrt{7}$ **5.** $-\frac{1}{2} \pm \frac{\sqrt{7}}{2}i$ **7.** $2 \pm 3i$
9. $3 \pm \sqrt{5}$ **11.** $\frac{-4 \pm \sqrt{19}}{3}$ **13.** $\frac{-1 \pm \sqrt{17}}{2}$
15. $\frac{-9 \pm \sqrt{129}}{24}$ **17.** $\frac{2}{5}$ **19.** $\frac{-11 \pm \sqrt{41}}{8}$ **21.** 5, 10
23. $\frac{13 \pm \sqrt{509}}{10}$ **25.** $2 \pm \sqrt{5}i$ **27.** $2, -1 \pm \sqrt{3}i$
29. $\frac{5 \pm \sqrt{37}}{6}$ **31.** $5 \pm \sqrt{53}$ **33.** $\frac{7 \pm \sqrt{85}}{2}$ **35.** $\frac{3}{2}, 6$
37. $-5.31662479, 1.31662479$ **39.** $0.7639320225,$
5.236067978 **41.** $-1.265564437, 2.765564437$ **43.** ▨
45. [8.3] Kenyan: 30 lb; Peruvian: 20 lb
46. [8.3] Cream-filled: 46; glazed: 44
47. [10.3] $9a^2b^3\sqrt{2a}$ **48.** [10.3] $4a^2b^3\sqrt{6}$
49. [6.5] $\frac{3(x + 1)}{3x + 1}$ **50.** [6.5] $\frac{4b}{3ab^2 - 4a^2}$ **51.** ▨
53. $(-2, 0), (1, 0)$ **55.** $4 - 2\sqrt{2}, 4 + 2\sqrt{2}$
57. $-1.1792101, 0.3392101$ **59.** $\frac{-5\sqrt{2} \pm \sqrt{34}}{4}$
61. $\frac{1}{2}$ **63.** ▨

Exercise Set 11.3, pp. 682–685

1. First part: 12 km/h; second part: 8 km/h   **3.** 35 mph
5. Cessna: 150 mph, Beechcraft: 200 mph; or
Cessna: 200 mph, Beechcraft: 250 mph
7. To Hillsboro: 10 mph; return trip: 4 mph
9. About 11 mph   **11.** 12 hr   **13.** About 3.24 mph
15. $r = \frac{1}{2}\sqrt{\frac{A}{\pi}}$ **17.** $r = \frac{-\pi h + \sqrt{\pi^2 h^2 + 2\pi A}}{2\pi}$
19. $s = \sqrt{\frac{kQ_1Q_2}{N}}$ **21.** $g = \frac{4\pi^2 l}{T^2}$
23. $c = \sqrt{d^2 - a^2 - b^2}$ **25.** $t = \frac{-v_0 + \sqrt{v_0^2 + 2gs}}{g}$
27. $n = \frac{1 + \sqrt{1 + 8N}}{2}$ **29.** $h = \frac{V^2}{12.25}$
31. $t = \frac{-b \pm \sqrt{b^2 - 4ac}}{2a}$ **33.** (a) 10.1 sec; (b) 7.49 sec;
(c) 272.5 m   **35.** 2.9 sec   **37.** 0.87 sec   **39.** 2.5 m/sec
41. 7%   **43.** ▨   **45.** [1.8] -104 **46.** [10.8] $2i\sqrt{11}$
47. [6.1] $\frac{x + y}{2}$ **48.** [6.1] $\frac{a^2 - b^2}{b}$ **49.** [10.3] $\frac{1 + \sqrt{5}}{2}$
50. [10.3] $\frac{1 - \sqrt{7}}{5}$ **51.** ▨
53. $t = \frac{-10.2 \pm 6\sqrt{-A^2 + 13A - 39.36}}{A - 6.5}$
55. $l = \frac{w + w\sqrt{5}}{2}$ **57.** $\pm\sqrt{2}$

59. $n = \pm \sqrt{\dfrac{r^2 \pm \sqrt{r^4 + 4m^4r^2p - 4mp}}{2m}}$

61. $A(S) = \dfrac{\pi S}{6}$

Exercise Set 11.4, pp. 689–690

1. Two irrational **3.** Two imaginary **5.** Two irrational
7. One rational **9.** Two imaginary **11.** Two rational
13. Two rational **15.** Two rational **17.** Two irrational
19. Two imaginary **21.** Two irrational
23. $x^2 + 4x - 21 = 0$ **25.** $x^2 - 6x + 9 = 0$
27. $x^2 + 7x + 10 = 0$ **29.** $3x^2 - 14x + 8 = 0$
31. $6x^2 - 5x + 1 = 0$ **33.** $x^2 - 0.8x - 0.84 = 0$
35. $x^2 - 7 = 0$ **37.** $x^2 - 18 = 0$ **39.** $x^2 + 9 = 0$
41. $x^2 - 10x + 29 = 0$ **43.** $x^2 - 4x - 6 = 0$
45. $x^3 - 4x^2 - 7x + 10 = 0$ **47.** $x^3 - 2x^2 - 3x = 0$
49. 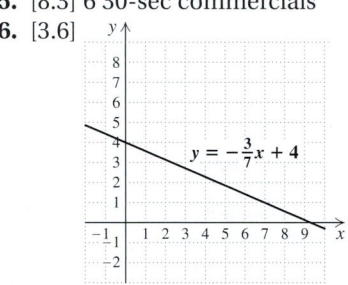 **51.** [4.1] $81a^8$ **52.** [4.1] $16x^6$
53. [5.7] $(-1, 0), (8, 0)$ **54.** [5.7] $(2, 0), (4, 0)$
55. [8.3] 6 30-sec commercials
56. [3.6] **57.**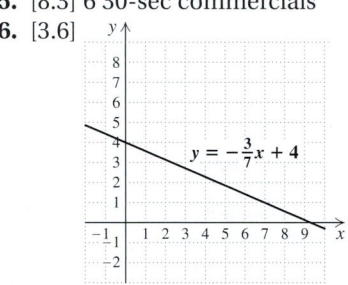

$y = -\frac{3}{7}x + 4$

59. $a = 1, b = 2, c = -3$ **61.** (a) $-\frac{3}{5}$; (b) $-\frac{1}{3}$
63. (a) $9 + 9i$; (b) $3 + 3i$ **65.** The solutions of
$ax^2 + bx + c = 0$ are $x = \dfrac{-b \pm \sqrt{b^2 - 4ac}}{2a}$. When there is
just one solution, $b^2 - 4ac$ must be 0, so
$x = \dfrac{-b \pm 0}{2a} = \dfrac{-b}{2a}$. **67.** $a = 8, b = 20, c = -12$
69. $x^4 - 8x^3 + 21x^2 - 2x - 52 = 0$ **71.** 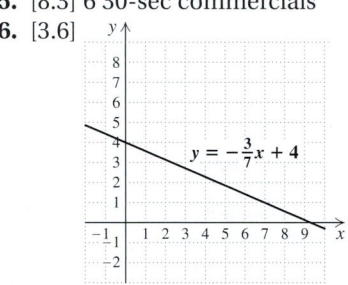,

Exercise Set 11.5, pp. 695–696

1. $\pm 1, \pm 3$ **3.** $\pm\sqrt{3}, \pm 3$ **5.** $\pm\dfrac{\sqrt{5}}{3}, \pm 1$ **7.** $9 + 4\sqrt{5}$
9. $\pm 2\sqrt{2}, \pm 3$ **11.** No solution **13.** $-\dfrac{1}{2}, \dfrac{1}{3}$ **15.** $-\dfrac{4}{5}, 1$
17. $-27, 8$ **19.** 729 **21.** 1 **23.** No solution **25.** $\dfrac{12}{5}$
27. $\left(\dfrac{4}{25}, 0\right)$ **29.** $\left(\dfrac{3 + \sqrt{33}}{2}, 0\right), \left(\dfrac{3 - \sqrt{33}}{2}, 0\right), (4, 0),$
$(-1, 0)$ **31.** $(-243, 0), (32, 0)$ **33.** No x-intercepts
35.

37. [7.3]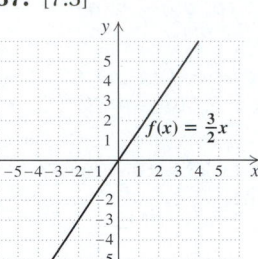

38. [7.3]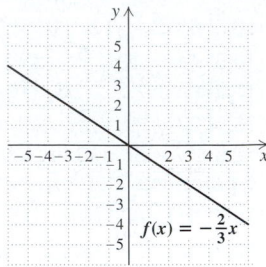

39. [7.3]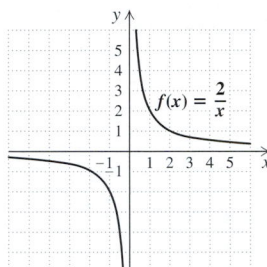

40. [7.3]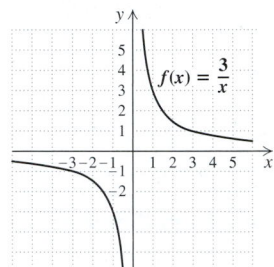

41. [8.3] A: 4 L; B: 8 L **42.** [7.1] $a^2 + a$ **43.**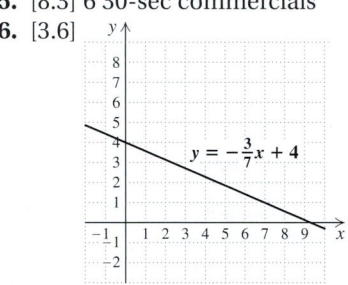
45. $\pm\sqrt{\dfrac{7 \pm \sqrt{29}}{10}}$ **47.** $-2, -1, 5, 6$ **49.** $\dfrac{100}{99}$
51. $-5, -3, -2, 0, 2, 3, 5$ **53.** $1, 3$ **55.**
57. ,

Exercise Set 11.6, pp. 703–704

1.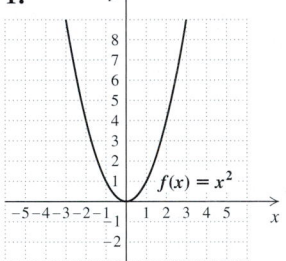

3.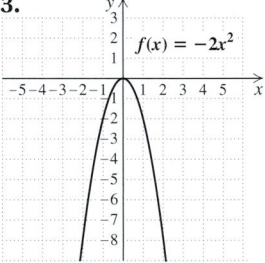

5.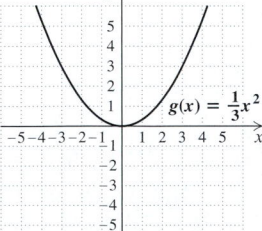

7.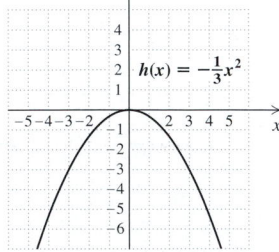

9.

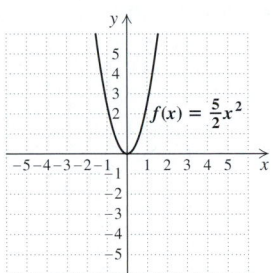

11. Vertex: $(-4, 0)$;
axis of symmetry: $x = -4$

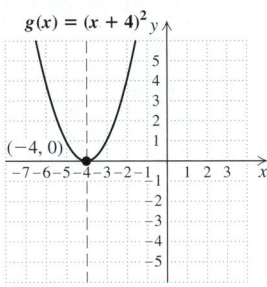

25. Vertex: $(1, 0)$;
axis of symmetry: $x = 1$

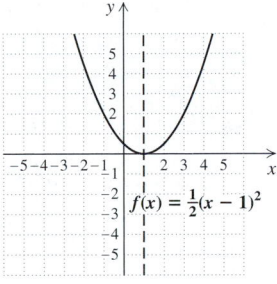

27. Vertex: $(-5, 0)$;
axis of symmetry: $x = -5$

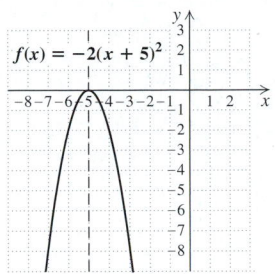

13. Vertex: $(1, 0)$;
axis of symmetry: $x = 1$

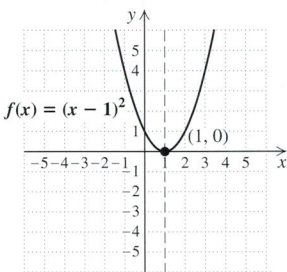

15. Vertex: $(3, 0)$;
axis of symmetry: $x = 3$

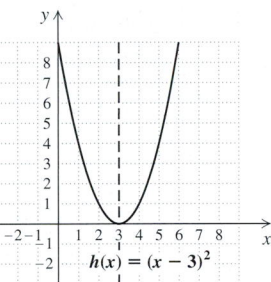

29. Vertex: $\left(\frac{1}{2}, 0\right)$;
axis of symmetry: $x = \frac{1}{2}$

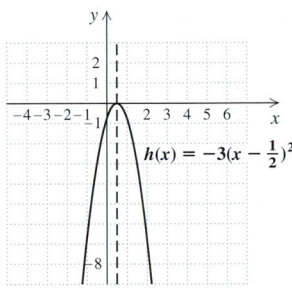

31. Vertex: $(5, 2)$;
axis of symmetry: $x = 5$;
minimum: 2

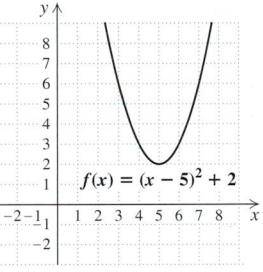

17. Vertex: $(-4, 0)$;
axis of symmetry: $x = -4$

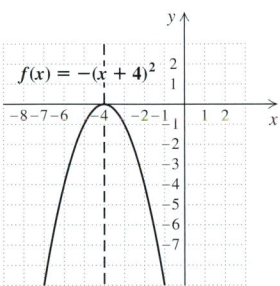

19. Vertex: $(1, 0)$;
axis of symmetry: $x = 1$

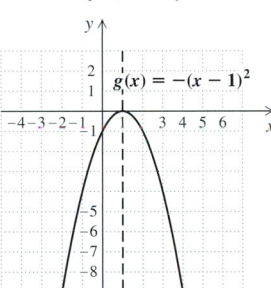

33. Vertex: $(-1, -3)$;
axis of symmetry: $x = -1$;
minimum: -3

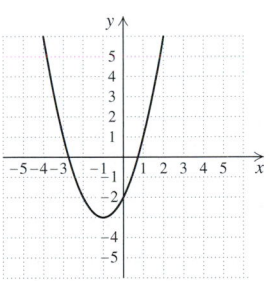

$$f(x) = (x + 1)^2 - 3$$

35. Vertex: $(-4, 1)$;
axis of symmetry: $x = -4$;
minimum: 1

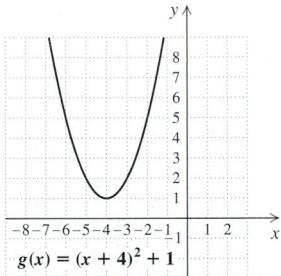

21. Vertex: $(-1, 0)$;
axis of symmetry: $x = -1$

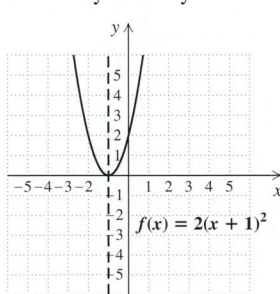

23. Vertex: $(4, 0)$;
axis of symmetry: $x = 4$

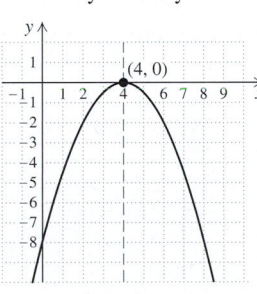

37. Vertex: $(1, -3)$;
axis of symmetry: $x = 1$;
maximum: -3

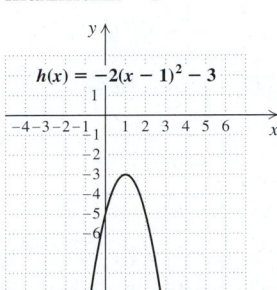

39. Vertex: $(-4, 1)$;
axis of symmetry: $x = -4$;
minimum: 1

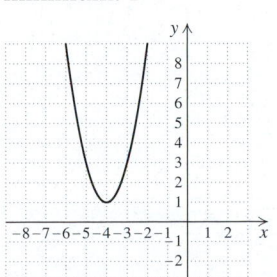

$$f(x) = 2(x + 4)^2 + 1$$

41. Vertex: $(1, 4)$; axis of symmetry: $x = 1$; maximum: 4

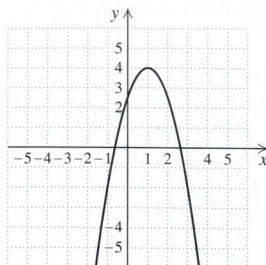

$$g(x) = -\frac{3}{2}(x - 1)^2 + 4$$

43. Vertex: $(9, 7)$; axis of symmetry: $x = 9$; minimum: 7
45. Vertex: $(-6, 11)$; axis of symmetry: $x = -6$; maximum: 11   **47.** Vertex: $\left(-\frac{1}{4}, -13\right)$; axis of symmetry: $x = -\frac{1}{4}$; minimum: -13 **49.** Vertex: $(-4.58, 65\pi)$; axis of symmetry: $x = -4.58$; minimum: 65π **51.**
53. [3.3]   **54.** [3.3]

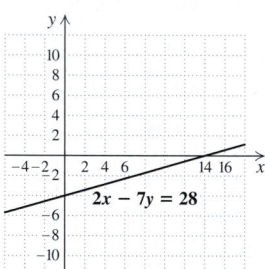   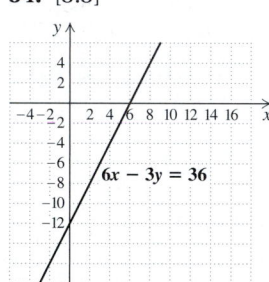

55. [8.2] $(-5, -1)$ **56.** [8.2] $(-1, 2)$
57. [11.1] $x^2 + 5x + \frac{25}{4}$ **58.** [11.1] $x^2 - 9x + \frac{81}{4}$ **59.**
61. $f(x) = \frac{3}{5}(x - 4)^2 + 1$ **63.** $f(x) = \frac{3}{5}(x - 3)^2 - 1$
65. $f(x) = \frac{3}{5}(x + 2)^2 - 5$ **67.** $g(x) = -2(x - 5)^2$
69. $f(x) = 2(x + 4)^2$ **71.** $g(x) = -2(x - 3)^2 + 8$
73. $F(x) = 3(x - 5)^2 + 1$
75.   **77.**

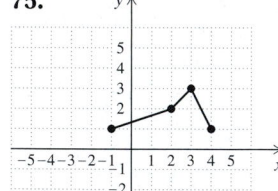   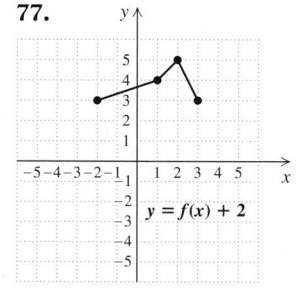

79.   **81.**    **83.** ,

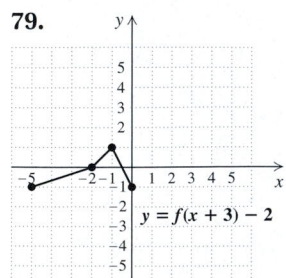

Exercise Set 11.7, pp. 710–711

1. (a) Vertex: $(-2, 1)$; axis of symmetry: $x = -2$;
(b)

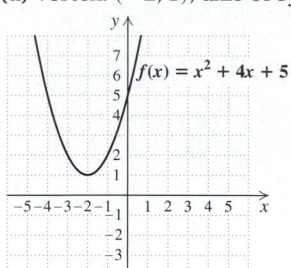

3. (a) Vertex: $(3, 4)$; axis of symmetry: $x = 3$;
(b)

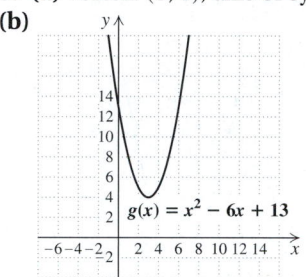

5. (a) Vertex: $(-4, 4)$; axis of symmetry: $x = -4$;
(b)

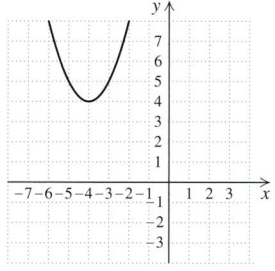

$$f(x) = x^2 + 8x + 20$$

7. (a) Vertex: $(4, -7)$; axis of symmetry: $x = 4$;
(b)

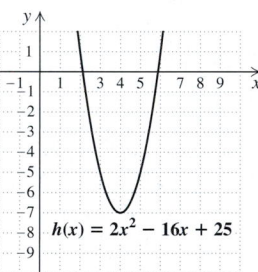

9. (a) Vertex: $(1, 6)$; axis of symmetry: $x = 1$;
(b)

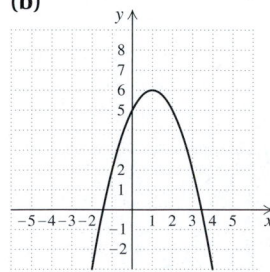

$$f(x) = -x^2 + 2x + 5$$

11. (a) Vertex: $\left(-\frac{3}{2}, -\frac{49}{4}\right)$; axis of symmetry: $x = -\frac{3}{2}$;
(b)

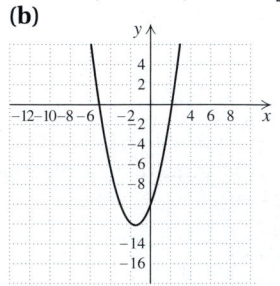

$$g(x) = x^2 + 3x - 10$$

13. (a) Vertex: $(4, 2)$;
axis of symmetry: $x = 4$;
(b)

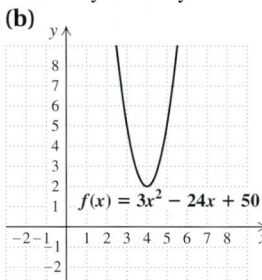

$f(x) = 3x^2 - 24x + 50$

15. (a) Vertex: $\left(-\frac{7}{2}, -\frac{49}{4}\right)$;
axis of symmetry: $x = -\frac{7}{2}$;
(b)

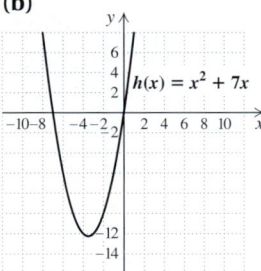

$h(x) = x^2 + 7x$

17. (a) Vertex: $(-1, -4)$;
axis of symmetry: $x = -1$;
(b)

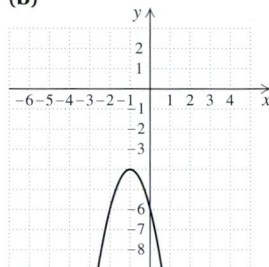

$f(x) = -2x^2 - 4x - 6$

19. (a) Vertex: $(2, -5)$;
axis of symmetry: $x = 2$;
(b)

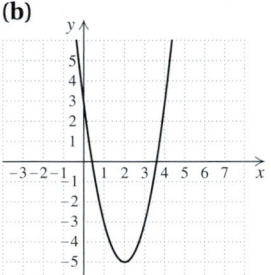

$g(x) = 2x^2 - 8x + 3$

21. (a) Vertex: $\left(\frac{5}{6}, \frac{1}{12}\right)$;
axis of symmetry: $x = \frac{5}{6}$;
(b)

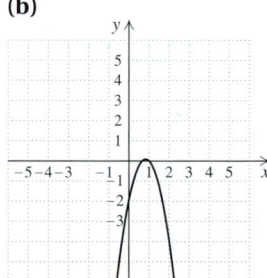

$f(x) = -3x^2 + 5x - 2$

23. (a) Vertex: $\left(-4, -\frac{5}{3}\right)$;
axis of symmetry: $x = -4$;
(b)

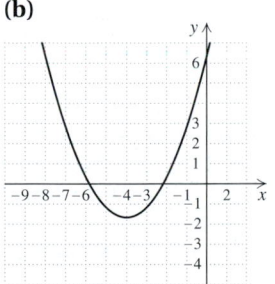

$h(x) = \frac{1}{2}x^2 + 4x + \frac{19}{3}$

25. $\left(3 - \sqrt{6}, 0\right)$, $\left(3 + \sqrt{6}, 0\right)$; $(0, 3)$ **27.** $(-1, 0)$, $(3, 0)$;
$(0, 3)$ **29.** $(0, 0)$, $(9, 0)$; $(0, 0)$ **31.** $(2, 0)$; $(0, -4)$
33. No x-intercept; $(0, 3)$ **35.**   **37.** [8.2] $(2, -2)$
38. [8.2] $(7, 1)$ **39.** [8.4] $(3, 2, 1)$ **40.** [8.4] $(1, -3, 2)$
41. [10.6] 5   **42.** [10.6] 4   **43.**
45. (a) Minimum: -6.953660714; **(b)** $(-1.056433682, 0)$,
$(2.413576539, 0)$; $(0, -5.89)$ **47. (a)** $-2.4, 3.4$;
(b) $-1.3, 2.3$ **49.** $f(x) = m\left(x - \dfrac{n}{2m}\right)^2 + \dfrac{4mp - n^2}{4m}$
51. $f(x) = \frac{5}{16}x^2 - \frac{15}{8}x - \frac{35}{16}$, or $f(x) = \frac{5}{16}(x - 3)^2 - 5$

53.

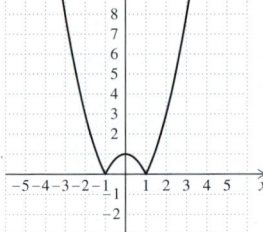

$f(x) = |x^2 - 1|$

55.

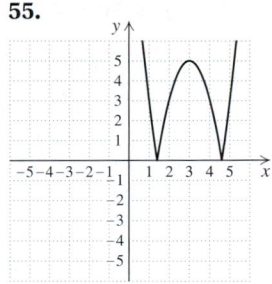

$f(x) = |2(x - 3)^2 - 5|$

Exercise Set 11.8, pp. 716–719

1. \$4; 3 mos. after January   **3.** 11 days after the concert
was announced; about 62   **5.** 180 ft by 180 ft
7. 200 ft^2; 10 ft by 20 ft. (The barn serves as a 20-ft side.)
9. 3.5 in.   **11.** 81; 9 and 9   **13.** -16; 4 and -4
15. 25; -5 and -5 **17.** Not quadratic   **19.** Quadratic
21. Not quadratic   **23.** Not quadratic
25. $f(x) = 2x^2 + 3x - 1$ **27.** $f(x) = -\frac{1}{4}x^2 + 3x - 5$
29. (a) $A(s) = \frac{3}{16}s^2 - \frac{135}{4}s + 1750$; **(b)** about 531
31. $h(d) = -0.0068d^2 + 0.8571d$ **33.**
35. [6.4] $\dfrac{x - 9}{(x + 9)(x + 7)}$ **36.** [6.2] $\dfrac{(x - 3)(x + 1)}{(x - 7)(x + 3)}$
37. [6.2] $\dfrac{(t + 2)(t + 8)}{(t - 3)(t + 1)}$ **38.** [6.4] $\dfrac{2t(t + 2)}{(t - 7)(t - 3)(t + 7)}$
39. [9.1] $\{x \,|\, x < 8\}$, or $(-\infty, 8)$
40. [9.1] $\{x \,|\, x \geq 10\}$, or $[10, \infty)$ **41.**
43. The radius of the circular portion of the window and
the height of the rectangular portion should each be
$\dfrac{24}{\pi + 4}$ ft.   **45.** 30   **47.** 78.4 ft   **49.**

Technology Connection, p. 724

1. $\{x \,|\, -0.78 \leq x \leq 1.59\}$, or $[-0.78, 1.59]$
2. $\{x \,|\, x \leq -0.21 \text{ or } x \geq 2.47\}$, or $(-\infty, -0.21] \cup [2.47, \infty)$
3. $\{x \,|\, x < -1.26 \text{ or } x > 2.33\}$, or $(-\infty, -1.26) \cup (2.33, \infty)$
4. $\{x \,|\, x > -1.37\}$, or $(-1.37, \infty)$

Exercise Set 11.9, pp. 727–728

1. $(-4, 3)$, or $\{x \,|\, -4 < x < 3\}$ **3.** $(-\infty, -7] \cup [2, \infty)$, or
$\{x \,|\, x \leq -7 \text{ or } x \geq 2\}$ **5.** $(-1, 2)$, or $\{x \,|\, -1 < x < 2\}$
7. $[-5, 5]$, or $\{x \,|\, -5 \leq x \leq 5\}$ **9.** $\varnothing$
11. $(-2, 6)$, or $\{x \,|\, -2 < x < 6\}$
13. $(-\infty, -2) \cup (0, 2)$, or $\{x \,|\, x < -2 \text{ or } 0 < x < 2\}$
15. $(-3, -1) \cup (2, \infty)$, or $\{x \,|\, -3 < x < -1 \text{ or } x > 2\}$
17. $(-\infty, -3) \cup (-2, 1)$, or $\{x \,|\, x < -3 \text{ or } -2 < x < 1\}$

19. $(-\infty, -3)$, or $\{x \mid x < -3\}$ **21.** $(-\infty, -1] \cup (5, \infty)$, or $\{x \mid x \le -1 \ or \ x > 5\}$ **23.** $\left[-\frac{2}{3}, 2\right)$, or $\{x \mid -\frac{2}{3} \le x < 2\}$
25. $(-\infty, -6)$, or $\{x \mid x < -6\}$ **27.** $(-\infty, -1] \cup [2, 5)$, or $\{x \mid x \le -1 \ or \ 2 \le x < 5\}$ **29.** $(-\infty, -3) \cup [0, \infty)$, or $\{x \mid x < -3 \ or \ x \ge 0\}$ **31.** $(0, \infty)$, or $\{x \mid x > 0\}$
33. $(-\infty, -4) \cup [1, 3)$, or $\{x \mid x < -4 \ or \ 1 \le x < 3\}$
35. $\left(0, \frac{1}{4}\right)$, or $\{x \mid 0 < x < \frac{1}{4}\}$ **37.** 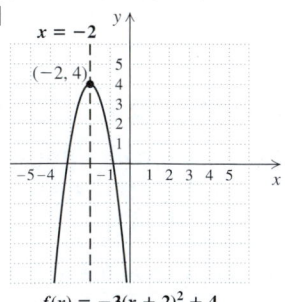 **39.** [4.1] $8a^9b^6c^{12}$
40. [4.1] $25a^8b^{14}$ **41.** [4.8] $\frac{1}{32}$ **42.** [4.8] $\frac{1}{81}$
43. [7.1] $3a^2 + 6a + 3$ **44.** [7.1] $5a + 7$ **45.**
47. $\left(-1 - \sqrt{6}, -1 + \sqrt{6}\right)$, or $\{x \mid -1 - \sqrt{6} < x < -1 + \sqrt{6}\}$ **49.** $\{0\}$
51. (a) $(10, 200)$, or $\{x \mid 10 < x < 200\}$;
(b) $[0, 10) \cup (200, \infty)$, or $\{x \mid 10 \le x < 0 \ or \ x > 200\}$
53. $\{n \mid n$ is an integer $and \ 12 \le n \le 25\}$ **55.** $f(x) = 0$ for $x = -2, 1, 3$; $f(x) < 0$ for $(-\infty, -2) \cup (1, 3)$, or $\{x \mid x < -2 \ or \ 1 < x < 3\}$; $f(x) > 0$ for $(-2, 1) \cup (3, \infty)$, or $\{x \mid -2 < x < 1 \ or \ x > 3\}$ **57.** $f(x)$ has no zeros; $f(x) < 0$ for $(-\infty, 0)$, or $\{x \mid x < 0\}$; $f(x) > 0$ for $(0, \infty)$, or $\{x \mid x > 0\}$
59. $f(x) = 0$ for $x = -1, 0$; $f(x) < 0$ for $(-\infty, -3) \cup (-1, 0)$, or $\{x \mid x < -3 \ or -1 < x < 0\}$; $f(x) > 0$ for $(-3, -1) \cup (0, 2) \cup (2, \infty)$, or $\{x \mid -3 < x < -1 \ or \ 0 < x < 2 \ or \ x > 2\}$ **61.**

Review Exercises: Chapter 11, pp. 731–732

1. [11.1] $\pm\sqrt{\frac{7}{2}}$, or $\pm\frac{\sqrt{14}}{2}$ **2.** [11.1] $0, -\frac{5}{14}$

3. [11.1] $3, 9$ **4.** [11.2] $\frac{5 \pm i\sqrt{11}}{2}$ **5.** [11.2] $3, 5$

6. [11.2] $\frac{-9 \pm \sqrt{85}}{2}$ **7.** [11.2] -0.3722813233, 5.3722813233 **8.** [11.2] $-\frac{1}{4}, 1$

9. [11.1] $x^2 - 12x + 36$; $(x - 6)^2$
10. [11.1] $x^2 + \frac{3}{5}x + \frac{9}{100}$; $\left(x + \frac{3}{10}\right)^2$ **11.** [11.1] $3 \pm 2\sqrt{2}$
12. [11.1] 10% **13.** [11.1] 6.7 sec **14.** [11.3] About
153 mph **15.** [11.3] 6 hr **16.** [11.4] Two irrational
17. [11.4] Two imaginary **18.** [11.4] $x^2 - 5 = 0$
19. [11.4] $x^2 + 8x + 16 = 0$ **20.** [11.5] $(-3, 0), (-2, 0)$, $(2, 0), (3, 0)$ **21.** [11.5] $-5, 3$ **22.** [11.5] $\pm\sqrt{2}, \pm\sqrt{7}$
23. [11.6]

$x = -2$

$(-2, 4)$

$f(x) = -3(x + 2)^2 + 4$
Maximum: 4

24. [11.7] **(a)** Vertex: $(3, 5)$; axis of symmetry: $x = 3$;
(b)

$f(x) = 2x^2 - 12x + 23$

25. [11.7] $(2, 0), (7, 0)$; $(0, 14)$ **26.** [11.3] $p = \dfrac{9\pi^2}{N^2}$

27. [11.3] $T = \dfrac{1 \pm \sqrt{1 + 24A}}{6}$ **28.** [11.8] Quadratic

29. [11.8] Not quadratic
30. [11.8] 56.25 ft^2; 7.5 ft by 7.5 ft
31. [11.8] **(a)** $f(x) = \frac{37}{50}x^2 - \frac{31}{2}x + 189$; **(b)** 175 million
32. [11.9] $(-1, 0) \cup (3, \infty)$, or $\{x \mid -1 < x < 0 \ or \ x > 3\}$
33. [11.9] $(-3, 5]$, or $\{x \mid -3 < x \le 5\}$
34. [11.7], [11.8] The x-coordinate of the maximum or minimum point lies halfway between the x-coordinates of the x-intercepts. **35.** [11.2], [11.4] Yes; if the discriminant is a perfect square, then the solutions are rational numbers, p/q and r/s. (Note that if the discriminant is 0, then $p/q = r/s$.) Then the equation can be written in factored form, $(qx - p)(sx - r) = 0$.
36. [11.5] Four; let $u = x^2$. Then $au^2 + bu + c = 0$ has at most two solutions, $u = m$ and $u = n$. Now substitute x^2 for u and obtain $x^2 = m$ or $x^2 = n$. These equations yield the solutions $x = \pm\sqrt{m}$ and $x = \pm\sqrt{n}$. When $m \ne n$, the maximum number of solutions, four, occurs.
37. [11.1], [11.2], [11.7] Completing the square was used to solve quadratic equations and to graph quadratic functions by rewriting the function in the form $f(x) = a(x - h)^2 + k$. **38.** [11.7] $f(x) = \frac{7}{15}x^2 - \frac{14}{15}x - 7$
39. [11.4] $h = 60, k = 60$ **40.** [11.5] $18, 324$

Test: Chapter 11, pp. 732–733

1. [11.1] $\pm\dfrac{4\sqrt{3}}{3}$ **2.** [11.2] $2, 9$ **3.** [11.2] $\dfrac{-1 \pm i\sqrt{3}}{2}$
4. [11.2] $1 \pm \sqrt{6}$ **5.** [11.5] $-2, \frac{2}{3}$
6. [11.2] $-4.192582404, 1.192582404$ **7.** [11.2] $-\frac{3}{4}, \frac{7}{3}$
8. [11.1] $x^2 + 14x + 49$; $(x + 7)^2$
9. [11.1] $x^2 - \frac{2}{7}x + \frac{1}{49}$; $\left(x - \frac{1}{7}\right)^2$
10. [11.1] $-5 \pm \sqrt{10}$ **11.** [11.3] 16 km/h
12. [11.3] 2 hr **13.** [11.4] Two imaginary
14. [11.4] $3x^2 + 5x - 2 = 0$
15. [11.5] $(-3, 0), (-1, 0), \left(-2 - \sqrt{5}, 0\right), \left(-2 + \sqrt{5}, 0\right)$

16. [11.6]

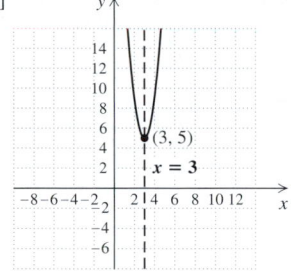

$$f(x) = 4(x - 3)^2 + 5$$
Minimum: 5

17. [11.7] **(a)** $(-1, -8)$, $x = -1$;
(b)

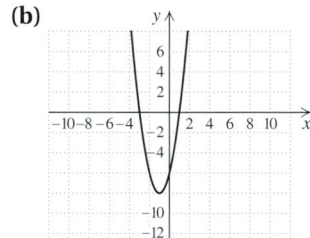

$$f(x) = 2x^2 + 4x - 6$$

18. [11.7] $(-2, 0)$, $(3, 0)$; $(0, -6)$

19. [11.3] $r = \sqrt{\dfrac{3V}{\pi} - R^2}$

20. [11.8] Not quadratic   **21.** [11.8] Minimum \$129/cap
when 325 caps are built   **22.** [11.8] $f(x) = \frac{1}{5}x^2 - \frac{3}{5}x$
23. [11.9] $[-6, 1]$, or $\{x \mid -6 \le x \le 1\}$
24. [11.9] $(-1, 0) \cup (1, \infty)$, or $\{x \mid -1 < x < 0 \ or \ x > 1\}$
25. [11.4] $\frac{1}{2}$
26. [11.4] $x^4 - 14x^3 + 67x^2 - 114x + 26 = 0$;
answers may vary.
27. [11.4] $x^6 - 10x^5 + 20x^4 + 50x^3 - 119x^2 - 60x + 150 = 0$; answers may vary

Technology Connection, p. 738

1. A table shows that $y_2 = y_3$. $y_1 = 7x + 3$; $y_2 = y_1^2$;
$y_3 = (7x + 3)^2$

X	Y2	Y3
5	1444	1444
6	2025	2025
7	2704	2704
8	3481	3481
9	4356	4356
10	5329	5329
11	6400	6400
X = 5		

A similar table shows that for $y_2 = x^2$ and $y_4 = y_2(y_1)$, we
have $y_3 = y_4$.

A graph can also be used: $y_2 = y_1^2$; $y_3 = (7x + 3)^2$

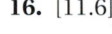

Technology Connection, p. 745

1. Graph each pair of functions in a square window along
with the line $y = x$ and determine whether the first two
functions are reflections of each other across $y = x$. For
further verification, examine a table of values for each pair
of functions.   **2.** Yes; most graphers do not require that
the inverse relation be a function.

Exercise Set 12.1, pp. 746–748

1. $(f \circ g)(1) = 12$; $(g \circ f)(1) = 9$;
$(f \circ g)(x) = 4x^2 + 4x + 4$; $(g \circ f)(x) = 2x^2 + 7$
3. $(f \circ g)(1) = 20$; $(g \circ f)(1) = 22$; $(f \circ g)(x) = 15x^2 + 5$;
$(g \circ f)(x) = 45x^2 - 30x + 7$

5. $(f \circ g)(1) = 8$; $(g \circ f)(1) = \frac{1}{64}$; $(f \circ g)(x) = \dfrac{1}{x^2} + 7$;

$(g \circ f)(x) = \dfrac{1}{(x + 7)^2}$ **7.** $f(x) = x^2$; $g(x) = 7 + 5x$

9. $f(x) = \sqrt{x}$; $g(x) = 2x + 7$ **11.** $f(x) = \dfrac{2}{x}$;

$g(x) = x - 3$ **13.** $f(x) = \dfrac{1}{\sqrt{x}}$; $g(x) = 7x + 2$

15. $f(x) = \dfrac{1}{x} + x$; $g(x) = \sqrt{3x}$ **17.** Yes   **19.** No

21. Yes   **23.** No   **25.** **(a)** Yes; **(b)** $f^{-1}(x) = x + 4$
27. **(a)** Yes; **(b)** $f^{-1}(x) = x - 3$ **29.** **(a)** Yes;

(b) $g^{-1}(x) = x - 5$ **31.** **(a)** Yes; **(b)** $f^{-1}(x) = \dfrac{x}{4}$

33. **(a)** Yes; **(b)** $g^{-1}(x) = \dfrac{x + 1}{4}$ **35.** **(a)** No

37. **(a)** Yes; **(b)** $f^{-1}(x) = \dfrac{1}{x}$ **39.** **(a)** Yes;

(b) $f^{-1}(x) = \dfrac{3x - 1}{2}$ **41.** **(a)** Yes; **(b)** $f^{-1}(x) = \sqrt[3]{x + 5}$
43. **(a)** Yes; **(b)** $g^{-1}(x) = \sqrt[3]{x} + 2$ **45.** **(a)** Yes;

(b) $f^{-1}(x) = x^2$, $x \ge 0$ **47.** **(a)** Yes; **(b)** $f^{-1}(x) = \sqrt{\dfrac{x - 1}{2}}$

49.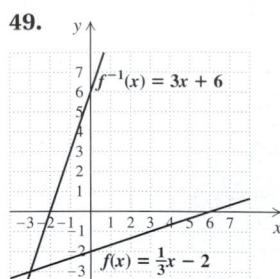

51.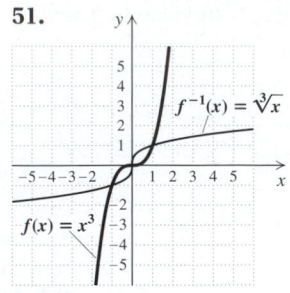

53.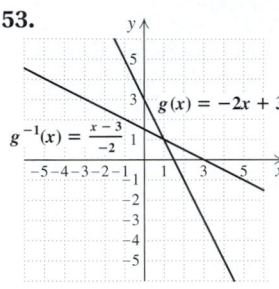

55.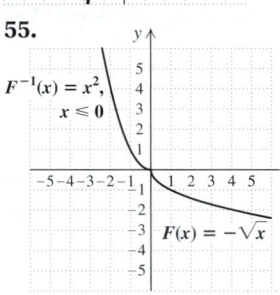

57.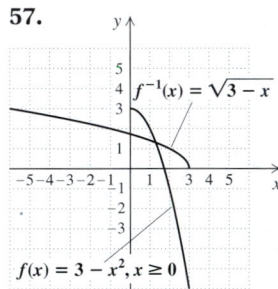

59. **(1)** $(f^{-1} \circ f)(x) = f^{-1}(f(x)) = f^{-1}\left(\frac{4}{5}x\right) = \frac{5}{4}\left(\frac{4}{5}x\right) = x;$

(2) $(f \circ f^{-1})(x) = f(f^{-1}(x)) = f\left(\frac{5}{4}x\right) = \frac{4}{5}\left(\frac{5}{4}x\right) = x$

61. **(1)** $(f^{-1} \circ f)(x) = f^{-1}(f(x)) = f^{-1}\left(\dfrac{1 - x}{x}\right)$

$$= \dfrac{1}{\left(\dfrac{1-x}{x}\right) + 1}$$

$$= \dfrac{1}{\dfrac{1 - x + x}{x}}$$

$$= x;$$

(2) $(f \circ f^{-1})(x) = f(f^{-1}(x)) = f\left(\dfrac{1}{x + 1}\right)$

$$= \dfrac{1 - \left(\dfrac{1}{x + 1}\right)}{\left(\dfrac{1}{x + 1}\right)}$$

$$= \dfrac{\dfrac{x + 1 - 1}{x + 1}}{\dfrac{1}{x + 1}} = x$$

63. **(a)** 40, 42, 46, 50; **(b)** $f^{-1}(x) = x - 32$; **(c)** 8, 10, 14, 18
65. 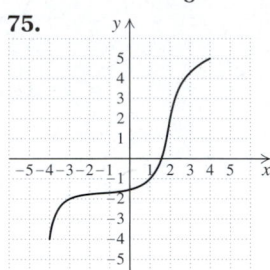 **67.** [4.1] $a^{13}b^{13}$ **68.** [4.1] $x^{10}y^{12}$
69. [10.2] 81 **70.** [10.2] 125 **71.** [2.3] $y = \frac{3}{2}(x + 7)$
72. [2.3] $y = \dfrac{10 - x}{3}$ **73.**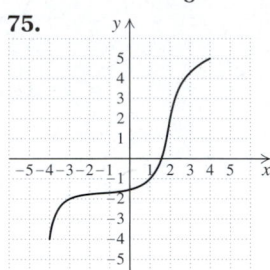

75.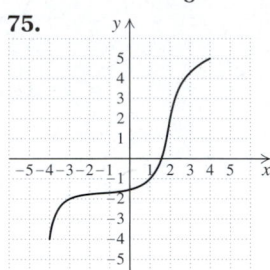

77. $g(x) = \dfrac{x}{2} + 20$ **79.**

81. Suppose that $h(x) = (f \circ g)(x)$. First, note that for
$I(x) = x$, $(f \circ I)(x) = f(I(x)) = f(x)$ for any function f.
(i) $((g^{-1} \circ f^{-1}) \circ h)(x) = ((g^{-1} \circ f^{-1}) \circ (f \circ g))(x)$
$\qquad\qquad = ((g^{-1} \circ (f^{-1} \circ f)) \circ g)(x)$
$\qquad\qquad = ((g^{-1} \circ I) \circ g)(x)$
$\qquad\qquad = (g^{-1} \circ g)(x) = x$
(ii) $(h \circ (g^{-1} \circ f^{-1}))(x) = ((f \circ g) \circ (g^{-1} \circ f^{-1}))(x)$
$\qquad\qquad = ((f \circ (g \circ g^{-1})) \circ f^{-1})(x)$
$\qquad\qquad = ((f \circ I) \circ f^{-1})(x)$
$\qquad\qquad = (f \circ f^{-1})(x) = x.$
Therefore, $(g^{-1} \circ f^{-1})(x) = h^{-1}(x)$. **83.** Yes **85.** No
87. **(1)** C; **(2)** A; **(3)** B; **(4)** D **89.**

Technology Connection, p. 751

1. $y_1 = \left(\frac{5}{2}\right)^x$; $y_2 = \left(\frac{2}{5}\right)^x$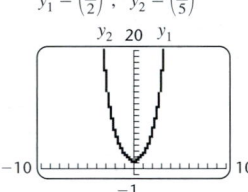

2. $y_1 = 3.2^x$; $y_2 = 3.2^{-x}$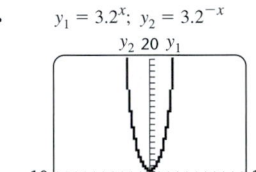

3. $y_1 = \left(\frac{3}{7}\right)^x$; $y_2 = \left(\frac{7}{3}\right)^x$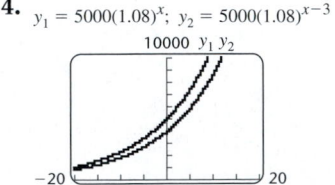

4. $y_1 = 5000(1.08)^x$; $y_2 = 5000(1.08)^{x-3}$
Xscl = 5, Yscl = 1000

Exercise Set 12.2, pp. 755–756

1.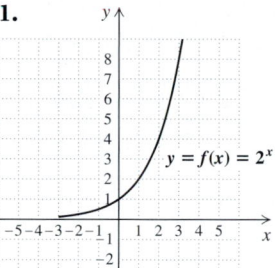

3.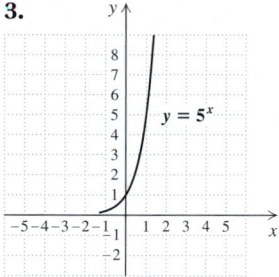

5.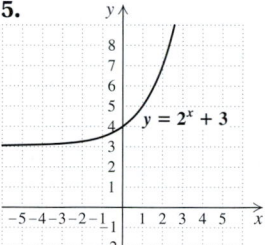

7.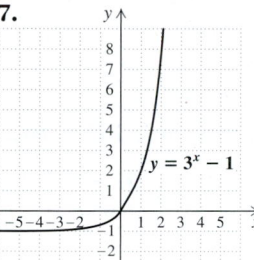

9.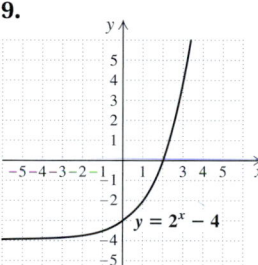

11.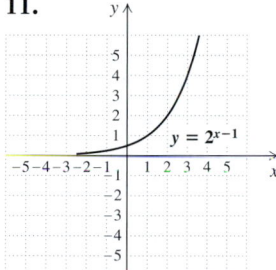

13.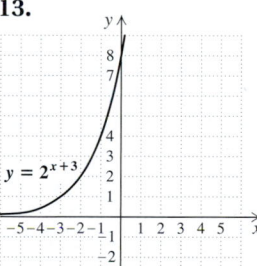

15.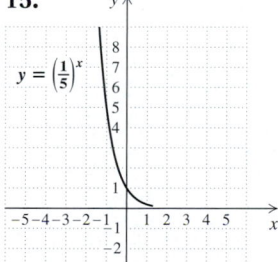

17.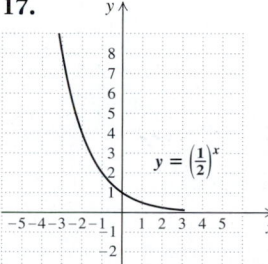

19.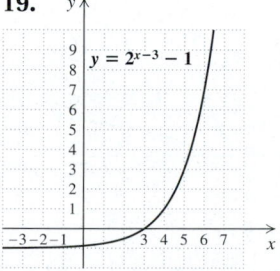

21.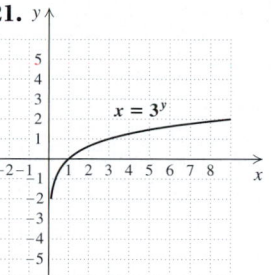

23.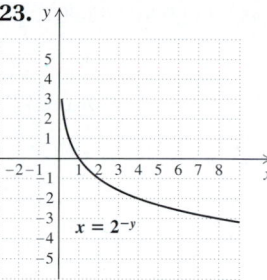

25.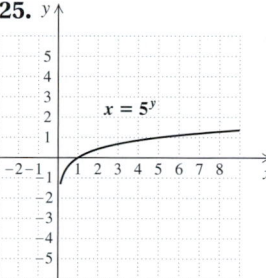

27.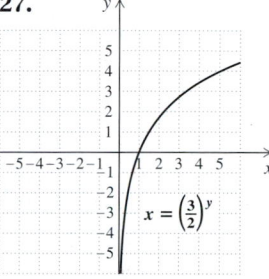

29.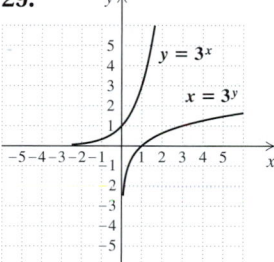

31.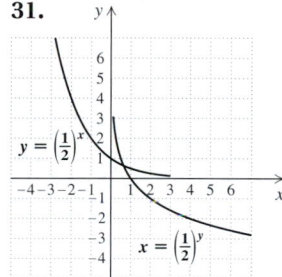

33. (a) About 6.4 billion; about 6.8 billion; about 7.3 billion;

(b)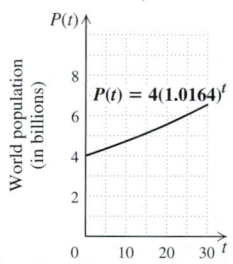

$P(t) = 4(1.0164)^t$

35. (a) About 44,079 whales; about 12,953 whales;

(b)

$P(t) = 150(0.960)^t$

37. (a) 250,000; 166,667; 49,383; 4335;
(b)

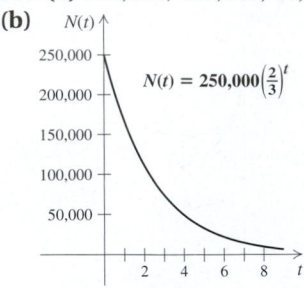

$$N(t) = 250,000\left(\tfrac{2}{3}\right)^t$$

39. (a) 0.3 million, or 300,000; 12.1 million; 490.6 million; 3119.5 million; **(b)**

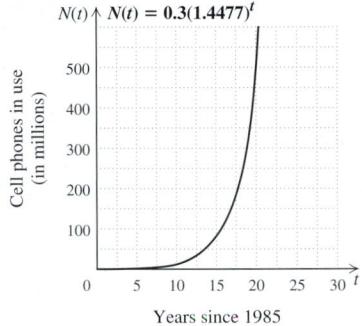

$$N(t) = 0.3(1.4477)^t$$

Years since 1985

41. 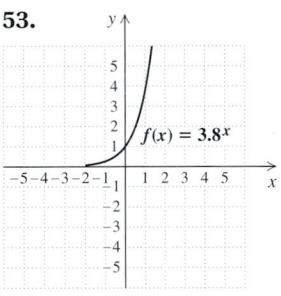 **43.** [4.8] $\frac{1}{25}$ **44.** [4.8] $\frac{1}{32}$ **45.** [10.2] 100
46. [10.2] $\frac{1}{125}$ **47.** [4.1] $5a^6b^3$ **48.** [4.1] $6x^4y$ **49.**
51. $\pi^{2.4}$
53.

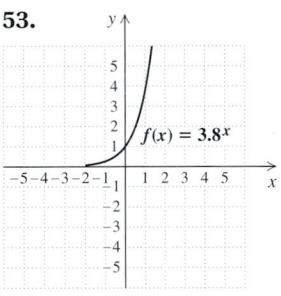

$$f(x) = 3.8^x$$

55.

$$y = 2^x + 2^{-x}$$

57.

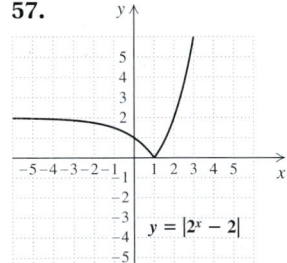

$$y = |2^x - 2|$$

59.

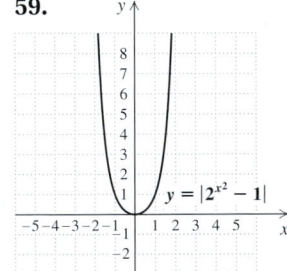

$$y = |2^{x^2} - 1|$$

61.

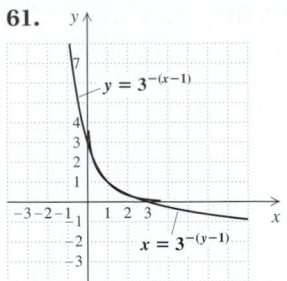

$$y = 3^{-(x-1)}$$

$$x = 3^{-(y-1)}$$

63. $A(t) = 169.3393318(2.535133248)^t$, where $A(t)$ is total sales, in millions of dollars, t years after 1997; $288,911,061,500 **65.** 19 wpm, 66 wpm, 110 wpm

Exercise Set 12.3, pp. 763–764

1. 2 **3.** 3 **5.** 4 **7.** -2 **9.** -1 **11.** 4 **13.** 1
15. 0 **17.** 7 **19.** -1 **21.** $\frac{1}{2}$ **23.** $\frac{3}{2}$ **25.** $\frac{2}{3}$ **27.** 7
29.

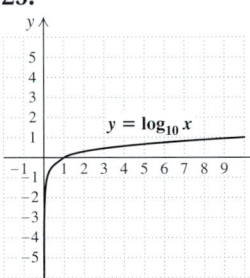

$$y = \log_{10} x$$

31.

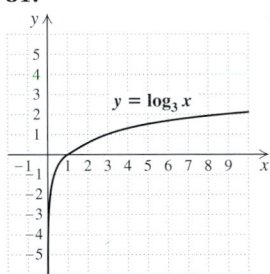

$$y = \log_3 x$$

33.

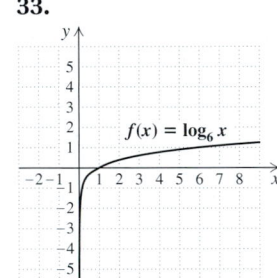

$$f(x) = \log_6 x$$

35.

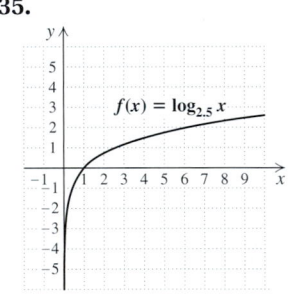

$$f(x) = \log_{2.5} x$$

37.

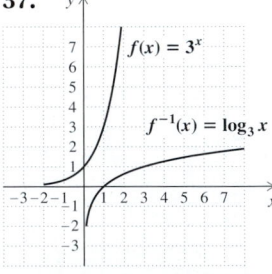

$$f(x) = 3^x$$

$$f^{-1}(x) = \log_3 x$$

39. $2 = \log_{10} 100$

41. $-5 = \log_4 \frac{1}{1024}$ **43.** $\frac{3}{4} = \log_{16} 8$
45. $0.4771 = \log_{10} 3$ **47.** $k = \log_p 3$ **49.** $m = \log_p V$
51. $3 = \log_e 20.0855$ **53.** $-4 = \log_e 0.0183$ **55.** $3^t = 8$
57. $5^2 = 25$ **59.** $10^{-1} = 0.1$ **61.** $10^{0.845} = 7$
63. $c^8 = m$ **65.** $t^r = Q$ **67.** $e^{-1.3863} = 0.25$

69. $r^{-x} = T$ **71.** 9   **73.** 4   **75.** 2   **77.** 2   **79.** 7
81. 9   **83.** $\frac{1}{9}$ **85.** 4   **87.**   **89.** [4.1] x^8
90. [4.1] a^{12} **91.** [4.1] $a^7 b^8$ **92.** [4.1] $x^5 y^{12}$
93. [6.5] $\dfrac{x(3y - 2)}{2y + x}$ **94.** [6.5] $\dfrac{x + 2}{x + 1}$ **95.**
97.

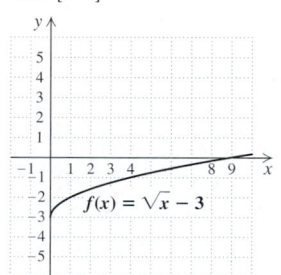

$y = \left(\frac{3}{2}\right)^x$

$y = \log_{3/2} x$

99.

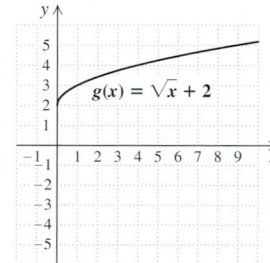

$y = \log_3 |x + 1|$

101. 25   **103.** $-\frac{7}{16}$ **105.** 3   **107.** 1   **109.** 1
111.

Exercise Set 12.4, pp. 770–771

1. $\log_3 81 + \log_3 27$ **3.** $\log_4 64 + \log_4 16$
5. $\log_c r + \log_c s + \log_c t$ **7.** $\log_a(5 \cdot 14)$, or $\log_a 70$
9. $\log_c(t \cdot y)$ **11.** $8 \log_a r$ **13.** $6 \log_c y$ **15.** $-3 \log_b C$
17. $\log_2 53 - \log_2 17$ **19.** $\log_b m - \log_b n$
21. $\log_a \dfrac{15}{3}$, or $\log_a 5$ **23.** $\log_b \dfrac{36}{4}$, or $\log_b 9$ **25.** $\log_a \dfrac{7}{18}$
27. $5 \log_a x + 7 \log_a y + 6 \log_a z$
29. $\log_b x + 2 \log_b y - 3 \log_b z$
31. $4 \log_a x - 3 \log_a y - \log_a z$
33. $\log_b x + 2 \log_b y - \log_b w - 3 \log_b z$
35. $\frac{1}{2}(7 \log_a x - 5 \log_a y - 8 \log_a z)$
37. $\frac{1}{3}(6 \log_a x + 3 \log_a y - 2 - 7 \log_a z)$ **39.** $\log_a x^7 z^3$
41. $\log_a x$ **43.** $\log_a \dfrac{y^5}{x^{3/2}}$ **45.** $\log_a(x - 2)$ **47.** 1.953
49. -0.369 **51.** -1.161 **53.** $\frac{3}{2}$ **55.** Cannot be found
57. 3.114   **59.** 9   **61.** m **63.**
65. [7.3]

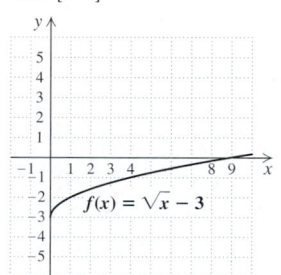

$f(x) = \sqrt{x} - 3$

66. [7.3]

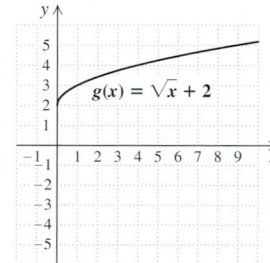

$g(x) = \sqrt{x} + 2$

67. [7.3]

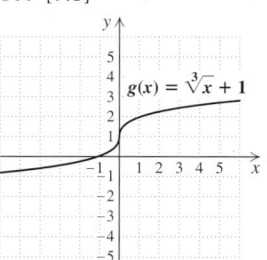

$g(x) = \sqrt[3]{x} + 1$

68. [7.3]

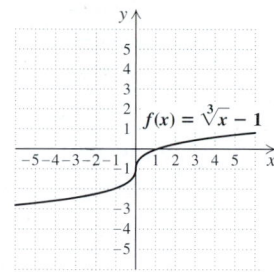

$f(x) = \sqrt[3]{x} - 1$

69. [4.1] $a^{17} b^{17}$ **70.** [4.1] $x^{11} y^6 z^8$ **71.**
73. $\log_a(x^6 - x^4 y^2 + x^2 y^4 - y^6)$
75. $\frac{1}{2} \log_a(1 - s) + \frac{1}{2} \log_a(1 + s)$ **77.** $\frac{10}{3}$ **79.** -2
81. True

Technology Connection, p. 776

1. $y = \log x / \log 5$

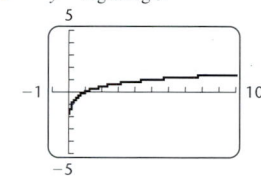

2. $y = \log x / \log 7$

3. $y = \log (x+2)/\log 5$

4. $y = \log x / \log 7 + 2$

Exercise Set 12.5, pp. 777–778

1. 0.7782   **3.** 1.8621   **5.** 3   **7.** -0.2782 **9.** 1.7986
11. 199.5262   **13.** 1.4894   **15.** 0.0011   **17.** 1.6094
19. 4.0431   **21.** -5.0832 **23.** 96.7583   **25.** 15.0293
27. 0.0305   **29.** 109.9472   **31.** 2.5237   **33.** 6.6439
35. 2.1452   **37.** -2.3219 **39.** -2.3219 **41.** 3.5471
43.   Domain: $\mathbb{R}$; range: $(0, \infty)$

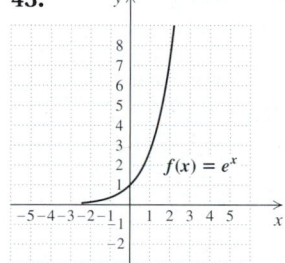

$f(x) = e^x$

45.

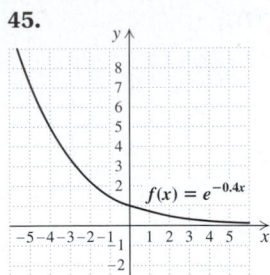

$f(x) = e^{-0.4x}$

Domain: $\mathbb{R}$; range: $(0, \infty)$

55.

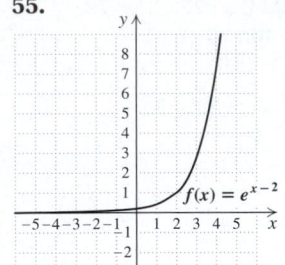

$f(x) = e^{x-2}$

Domain: $\mathbb{R}$; range: $(0, \infty)$

47.

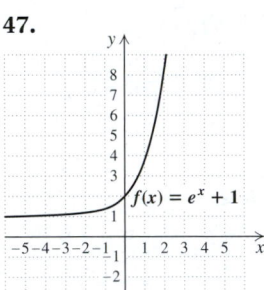

$f(x) = e^x + 1$

Domain: $\mathbb{R}$; range: $(1, \infty)$

57.

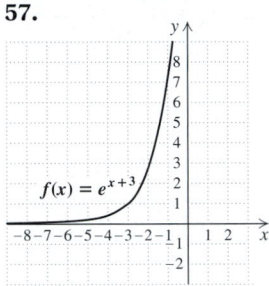

$f(x) = e^{x+3}$

Domain: $\mathbb{R}$; range: $(0, \infty)$

49.

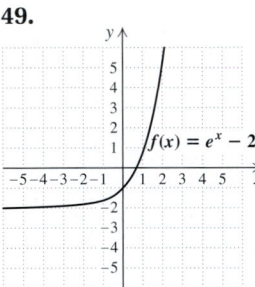

$f(x) = e^x - 2$

Domain: $\mathbb{R}$; range: $(-2, \infty)$

59.

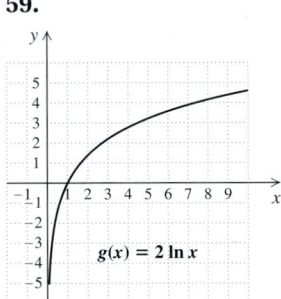

$g(x) = 2 \ln x$

Domain: $(0, \infty)$; range: $\mathbb{R}$

51.

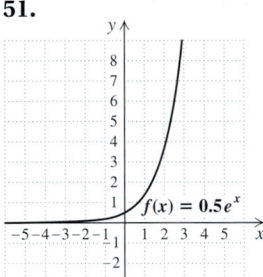

$f(x) = 0.5e^x$

Domain: $\mathbb{R}$; range: $(0, \infty)$

61.

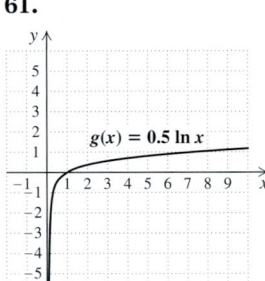

$g(x) = 0.5 \ln x$

Domain: $(0, \infty)$; range: $\mathbb{R}$

53.

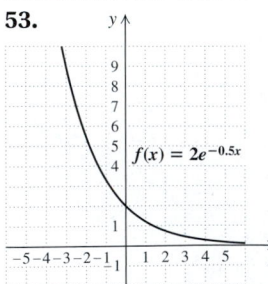

$f(x) = 2e^{-0.5x}$

Domain: $\mathbb{R}$; range: $(0, \infty)$

63.

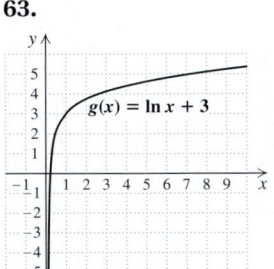

$g(x) = \ln x + 3$

Domain: $(0, \infty)$; range: $\mathbb{R}$

65. 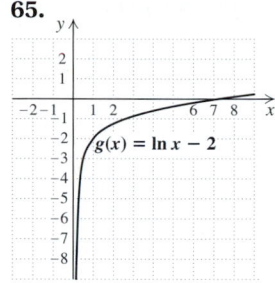 Domain: $(0, \infty)$; range: $\mathbb{R}$

$g(x) = \ln x - 2$

67. 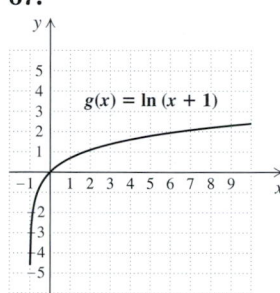 Domain: $(-1, \infty)$; range: $\mathbb{R}$

$g(x) = \ln (x + 1)$

69. 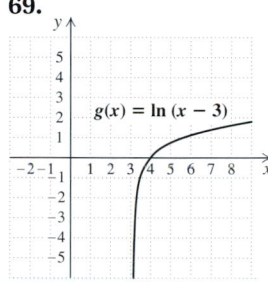 Domain: $(3, \infty)$; range: $\mathbb{R}$

$g(x) = \ln (x - 3)$

71. ▨   **73.** [5.7] $-\frac{5}{2}, \frac{5}{2}$ **74.** [5.7] $0, \frac{7}{5}$ **75.** [2.2] $\frac{15}{17}$
76. [2.2] $\frac{9}{13}$ **77.** [11.5] 16, 256   **78.** [11.5] $\frac{1}{4}, 9$ **79.** ▨
81. 2.452   **83.** 1.442   **85.** $\log M = \dfrac{\ln M}{\ln 10}$
87. 1086.5129   **89.** 4.9855
91. **(a)** Domain: $\{x \,|\, x > 0\}$, or $(0, \infty)$; range: $\{y \,|\, y < 0.5135\}$, or $(-\infty, 0.5135)$; **(b)** $[-1, 5, -10, 5]$;
(c) $y = 3.4 \ln x - 0.25e^x$

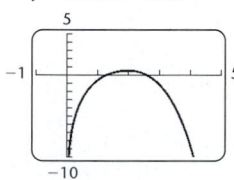

93. **(a)** Domain: $\{x \,|\, x > 0\}$, or $(0, \infty)$; range: $\{y \,|\, y > -0.2453\}$, or $(-0.2453, \infty)$; **(b)** $[-1, 5, -1, 10]$;
(c) $y = 2x^3 \ln x$ **95.** ▨

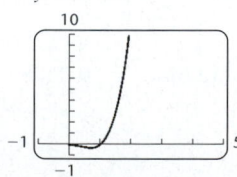

Technology Connection, p. 782

1. 0.38   **2.** -1.96 **3.** 0.90   **4.** -1.53 **5.** 0.13, 8.47
6. $-0.75, 0.75$

Exercise Set 12.6, pp. 783–784

1. 4   **3.** 3   **5.** 2   **7.** $\frac{4}{3}$ **9.** 0   **11.** 2   **13.** $-4, 1$
15. $\dfrac{\log 15}{\log 2} \approx 3.907$ **17.** $\dfrac{\log 13}{\log 4} - 1 \approx 0.850$
19. $\ln 100 \approx 4.605$ **21.** $\dfrac{\ln 0.08}{-0.07} \approx 36.082$
23. $\dfrac{\log 3}{\log 3 - \log 2} \approx 2.710$ **25.** $\dfrac{\log 4}{\log 5 - \log 4} \approx 6.213$
27. $\dfrac{\log 65}{\log 7.2} \approx 2.115$ **29.** 125   **31.** 2   **33.** 1000
35. $\frac{1}{10,000}$ **37.** $e \approx 2.718$ **39.** $e^{-3} \approx 0.050$ **41.** -4
43. 10   **45.** No solution   **47.** $\frac{83}{15}$ **49.** 1   **51.** 5
53. $\frac{17}{2}$ **55.** 4   **57.** ▨   **59.** [6.8] $y = 9x$
60. [6.8] $y = \dfrac{21.35}{x}$ **61.** [11.3] $L = \dfrac{8T^2}{\pi^2}$
62. [11.3] $c = \sqrt{\dfrac{E}{m}}$ **63.** [6.7] $1\frac{1}{5}$ hr   **64.** [6.7] $9\frac{3}{8}$ min
65. ▨   **67.** No solution   **69.** -4 **71.** 2   **73.** $\pm\sqrt{34}$
75. $10^{100,000}$ **77.** 1, 100   **79.** $\frac{1}{100,000}, 100,000$ **81.** $-\frac{1}{3}$
83. 38   **85.** 1

Exercise Set 12.7, pp. 792–796

1. **(a)** 2003; **(b)** 2.2 yr   **3.** **(a)** 4.2 yr; **(b)** 9.0 yr
5. **(a)** 20,114; **(b)** about 58   **7.** 4.9
9. 10^{-7} moles per liter   **11.** 65 dB
13. $10^{-1.5}$ or about 3.2×10^{-2} W/m^2
15. **(a)** $P(t) = P_0 e^{0.06t}$; **(b)** \$5309.18, \$5637.48; **(c)** 11.6 yr
17. **(a)** $P(t) = 283.75 e^{0.013t}$, where t is the number of years after 2000 and $P(t)$ is in millions; **(b)** 302.81 million;
(c) 2010   **19.** **(a)** 86.4 min; **(b)** 300.5 min; **(c)** 20 min
21. **(a)** 2000; **(b)** 2452; **(c)**

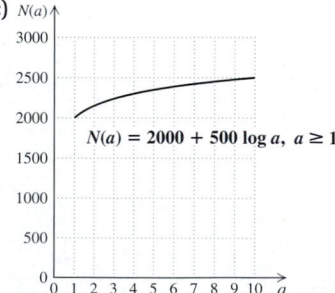

$N(a) = 2000 + 500 \log a, \; a \geq 1$

(d) \$1,000,000 thousand, or \$1,000,000,000
23. **(a)** $N(t) = 17 e^{0.534t}$, where t is the number of years since 1995; **(b)** about 419   **25.** 2002
27. **(a)** $k \approx 0.094$; $W(t) = 17.5 e^{-0.094t}$, where t is the number of years since 1996 and $W(t)$ is in millions of tons; **(b)** 6.8 million tons; **(c)** 2173

29. About 2103 yr   **31.** About 7.2 days
33. 69.3% per year
35. (a) $k \approx 0.135$; $V(t) = 640{,}500e^{0.135t}$, where t is the number of years since 1996; (b) about \$2.47 million; (c) 5.1 yr; (d) 2004   **37.**
39. [11.7]

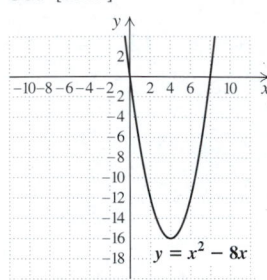

40. [11.7]

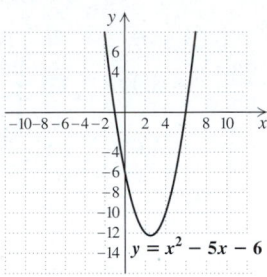

41. [11.7]

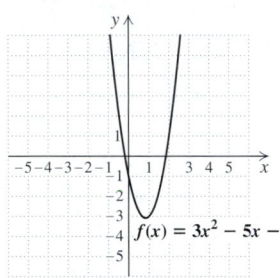

42. [11.7]

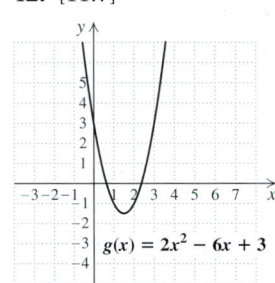

43. [11.1] $4 \pm \sqrt{23}$ **44.** [11.1] $-5 \pm \sqrt{31}$ **45.** ▨
47. \$13.4 million   **49.** $P(t) = 39e^{0.302t}$, where t is the number of years after 2001 and $P(t)$ is a percent.
51. About 80,922 yr, or with rounding, about 80,792 yr

Review Exercises: Chapter 12, pp. 800–801

1. [12.1] $(f \circ g)(x) = 4x^2 - 12x + 10$; $(g \circ f)(x) = 2x^2 - 1$
2. [12.1] $f(x) = \sqrt{x}$; $g(x) = 3 - x$ **3.** [12.1] No
4. [12.1] $f^{-1}(x) = x + 3$ **5.** [12.1] $g^{-1}(x) = \dfrac{2x - 1}{3}$
6. [12.1] $f^{-1}(x) = \dfrac{\sqrt[3]{x}}{3}$
7. [12.2]

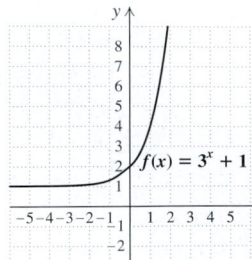

8. [12.2]

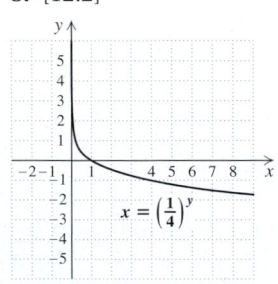

9. [12.3]

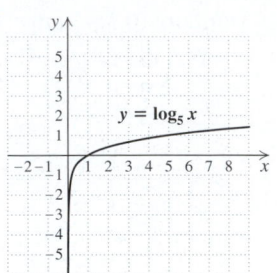

10. [12.3] 2

11. [12.3] -1 **12.** [12.3] $\frac{1}{2}$ **13.** [12.3] 12
14. [12.3] $\log_{10} \frac{1}{100} = -2$ **15.** [12.3] $\log_{25} 5 = \frac{1}{2}$
16. [12.3] $16 = 4^x$ **17.** [12.3] $1 = 8^0$
18. [12.4] $4 \log_a x + 2 \log_a y + 3 \log_a z$
19. [12.4] $3 \log_a x - (\log_a y + 2 \log_a z)$, or
$3 \log_a x - \log_a y - 2 \log_a z$
20. [12.4] $\frac{1}{4}(2 \log z - 3 \log x - \log y)$
21. [12.4] $\log_a(8 \cdot 15)$, or $\log_a 120$ **22.** [12.4] $\log_a \frac{72}{12}$, or
$\log_a 6$ **23.** [12.4] $\log \dfrac{a^{1/2}}{bc^2}$ **24.** [12.4] $\log_a \sqrt[3]{\dfrac{x}{y^2}}$
25. [12.4] 1   **26.** [12.4] 0   **27.** [12.4] 17
28. [12.4] 6.93   **29.** [12.4] -3.2698 **30.** [12.4] 8.7601
31. [12.4] 3.2698   **32.** [12.4] 2.54995
33. [12.4] -3.6602 **34.** [12.5] 1.9138
35. [12.5] 61.5177   **36.** [12.5] -2.9957
37. [12.5] 0.3753   **38.** [12.5] 0.4307   **39.** [12.5] 1.7097
40. [12.5]

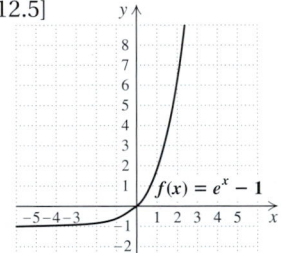

Domain: $\mathbb{R}$; range: $(-1, \infty)$

41. [12.5]

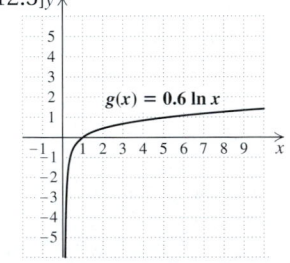

Domain: $(0, \infty)$; range: $\mathbb{R}$

42. [12.6] 5   **43.** [12.6] -2 **44.** [12.6] $\frac{1}{9}$ **45.** [12.6] 2
46. [12.6] $\frac{1}{10{,}000}$ **47.** [12.6] $e^{-2} \approx 0.1353$ **48.** [12.6] $\frac{7}{2}$
49. [12.6] $-5, 1$ **50.** [12.6] $\dfrac{\log 8.3}{\log 4} \approx 1.5266$
51. [12.6] $\dfrac{\ln 0.03}{-0.1} \approx 35.0656$ **52.** [12.6] $e^{-3} \approx 0.0498$

53. [12.6] 4 **54.** [12.6] 8 **55.** [12.6] 20
56. [12.6] $\sqrt{43}$ **57.** [12.7] **(a)** 62; **(b)** 46.8; **(c)** 35 months
58. [12.7] **(a)** 6.6 yr; **(b)** 3.1 yr **59.** [12.7] **(a)** 0.426;
$C(t) = 667e^{0.426t}$; **(b)** about 47,230; **(c)** 2001
60. [12.7] 23.105% per yr **61.** [12.7] 8.25 yr
62. [12.7] 3463 yr **63.** [12.7] 6.6 **64.** [12.7] 90 dB
65. [12.3] Negative numbers do not have logarithms
because logarithm bases are positive, and there is no
power to which a positive number can be raised to yield a
negative number. **66.** [12.6] Taking the logarithm on
each side of an equation produces an equivalent equation
because the logarithm function is one-to-one. If two
quantities are equal, their logarithms must be equal, and if
the logarithms of two quantities are equal, the quantities
must be the same. **67.** [12.6] e^{e^3} **68.** [12.6] $-3, -1$
69. [12.6] $\left(\frac{8}{3}, -\frac{2}{3}\right)$

Test: Chapter 12, pp. 801–802

1. [12.1] $(f \circ g)(x) = 2 + 6x + 4x^2$;
$(g \circ f)(x) = 2x^2 + 2x + 1$ **2.** [12.1] $f(x) = \dfrac{1}{x}$;
$g(x) = 2x^2 + 1$ **3.** [12.1] No **4.** [12.1] $f^{-1}(x) = \dfrac{x+3}{4}$
5. [12.1] $g^{-1}(x) = \sqrt[3]{x} - 1$
6. [12.2]   **7.** [12.2]

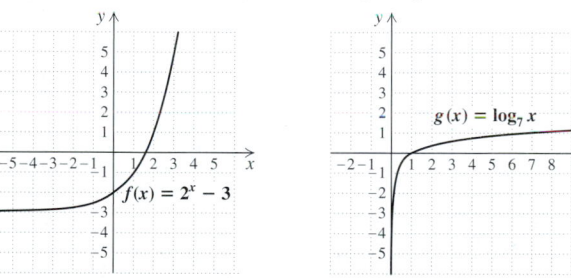

8. [12.3] 3 **9.** [12.3] $\frac{1}{2}$ **10.** [12.3] 18
11. [12.3] $\log_4 \frac{1}{64} = -3$ **12.** [12.3] $\log_{256} 16 = \frac{1}{2}$
13. [12.3] $49 = 7^m$ **14.** [12.3] $81 = 3^4$
15. [12.4] $3 \log a + \frac{1}{2} \log b - 2 \log c$
16. [12.4] $\log_a\!\left(z^2 \sqrt[3]{x}\right)$ **17.** [12.4] 1 **18.** [12.4] 23
19. [12.4] 0 **20.** [12.4] -0.544 **21.** [12.4] 1.079
22. [12.4] 1.204 **23.** [12.5] 1.0899 **24.** [12.5] 0.1585
25. [12.5] -3.3524 **26.** [12.5] 121.5104
27. [12.5] 2.0959
28. [12.5]   Domain: $\mathbb{R}$; range: $(3, \infty)$

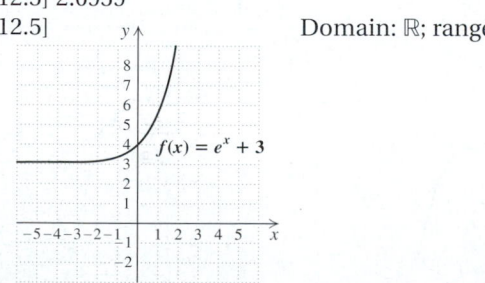

29. [12.5]   Domain: $(4, \infty)$; range: $\mathbb{R}$

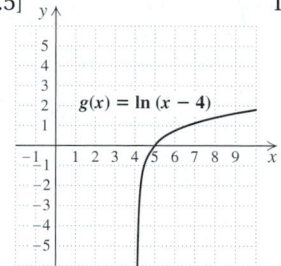

30. [12.6] -5 **31.** [12.6] 5 **32.** [12.6] 2
33. [12.6] 10,000 **34.** [12.6] $\frac{1}{3}$
35. [12.6] $\dfrac{\log 1.2}{\log 7} \approx 0.0937$
36. [12.6] $e^{1/4} \approx 1.2840$ **37.** [12.6] 4
38. [12.7] **(a)** 2.46 ft/sec; **(b)** 984,262
39. [12.7] **(a)** $P(t) = 30e^{0.015t}$, where $P(t)$ is in millions and
t is the number of years after 2000; **(b)** 31.4 million;
34.9 million; **(c)** 2034; **(d)** 46.2 yr
40. [12.7] **(a)** $k \approx 0.027$; $C(t) = 60e^{0.027t}$; **(b)** 681.5; **(c)** 2024
41. [12.7] 4.6% **42.** [12.7] 4684 yr
43. [12.7] $10^{-4.5}$ W/m² **44.** [12.7] 7.0
45. [12.6] $-309, 316$ **46.** [12.4] 2

Cumulative Review: Chapters 1–12: pp. 803–805

1. [1.8] 2 **2.** [1.5] 6 **3.** [4.8] $\dfrac{y^{12}}{16x^8}$
4. [4.8] $\dfrac{20x^6z^2}{y}$ **5.** [4.8] $\dfrac{-y^4}{3z^5}$ **6.** [2.2] $-4x - 1$
7. [1.8] 25 **8.** [2.2] $\frac{11}{2}$ **9.** [8.2] $(3, -1)$
10. [8.4] $(1, -2, 0)$ **11.** [5.7] $-2, 5$ **12.** [6.6] $\frac{9}{2}$
13. [6.6], [11.2] $\frac{5}{8}$ **14.** [10.6] $\frac{3}{4}$ **15.** [10.6] $\frac{1}{2}$
16. [11.1] $\pm 5i$ **17.** [11.5] 9, 25 **18.** [11.5] $\pm 2, \pm 3$
19. [12.3] 8 **20.** [12.3] 7 **21.** [12.6] $\frac{3}{2}$
22. [12.6] $\dfrac{\log 7}{5 \log 3} \approx 0.3542$ **23.** [12.6] $\frac{80}{9}$
24. [11.9] $(-\infty, -5) \cup (1, \infty)$, or $\{x \,|\, x < -5 \text{ or } x > 1\}$
25. [11.2] $-3 \pm 2\sqrt{5}$
26. [9.3] $\{x \,|\, x \le -3 \text{ or } x \ge 6\}$, or $(-\infty, -3] \cup [6, \infty)$
27. [6.8] $a = \dfrac{Db}{b - D}$ **28.** [6.8] $q = \dfrac{pf}{p - f}$
29. [2.3] $B = \dfrac{3M - 2A}{2}$, or $B = \frac{3}{2}M - A$ **30.** [8.7] 2
31. [8.7] 3
32. [7.2] $\{x \,|\, x \text{ is a real number } and \; x \ne -\frac{1}{3} \text{ and } x \ne 2\}$
33. **(a)** [3.4] 1 million per year; **(b)** [7.3] $E(t) = 2.4 + t$;
(c) [7.3] 7.4 million **34.** [2.5] Length: 36 m; width: 20 m
35. [2.5] A: 15°; B: 45°; C: 120° **36.** [6.7] $5\frac{5}{11}$ min
37. [8.3] Thick and Tasty: 6 oz; Light and Lean: 9 oz
38. [6.7] $2\frac{7}{9}$ km/h **39.** [11.8] -49; -7 and 7
40. [12.7] 78 **41.** [12.7] 67.5
42. [12.7] $P(t) = 19e^{0.015t}$ **43.** [12.7] 20.5 million;

22.7 million   **44.** [12.7] 46.2 yr   **45.** [11.6] 18
46. [4.6] $7p^2q^3 + pq + p - 9$ **47.** [4.3] $8x^2 - 11x - 1$
48. [4.6] $9x^4 - 12x^2y + 4y^2$ **49.** [4.6] $10a^2 - 9ab - 9b^2$
50. [6.2] $\dfrac{(x+4)(x-3)}{2(x-1)}$ **51.** [6.5] $\dfrac{1}{x-4}$ **52.** [6.2] $\dfrac{a+2}{6}$
53. [6.4] $\dfrac{7x+4}{(x+6)(x-6)}$ **54.** [5.1] $x(y - 2z + w)$
55. [5.5] $(1 - 5x)(1 + 5x + 25x^2)$
56. [5.3] $2(3x - 2y)(x + 2y)$ **57.** [5.1] $(x^3 + 7)(x - 4)$
58. [5.4] $2(m + 3n)^2$
59. [5.4] $(x - 2y)(x + 2y)(x^2 + 4y^2)$ **60.** [7.1] -12
61. [4.7] $x^3 - 2x^2 - 4x - 12 + \dfrac{-42}{x-3}$
62. [4.8] 1.8×10^{-1} **63.** [10.4] $2y^2\sqrt[3]{y}$
64. [10.4] $14xy^2\sqrt{x}$ **65.** [10.2] $81a^8b\sqrt[3]{b}$
66. [10.5] $\dfrac{6 + \sqrt{y} - y}{4 - y}$ **67.** [10.4] $\sqrt[10]{(x+5)^3}$
68. [10.8] $12 + 4\sqrt{3}i$ **69.** [10.8] $8 + i$
70. [12.1] $f^{-1}(x) = \dfrac{x-7}{-2}$, or $f^{-1}(x) = \dfrac{7-x}{2}$
71. [3.7] $f(x) = -5x - 3$ **72.** [3.7] $y = \frac{1}{2}x + 7$
73. [3.3]   **74.** [11.7]

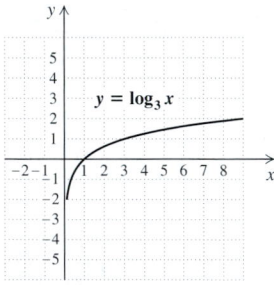

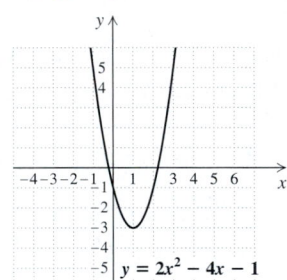

75. [12.3]   **76.** [12.1]

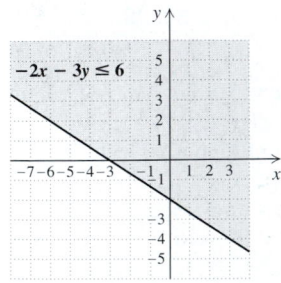

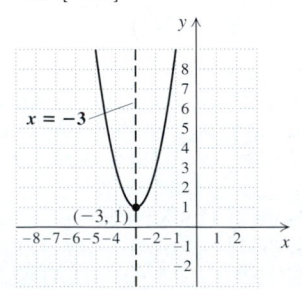

77. [12.4]   **78.** [11.7]

$f(x) = 2(x + 3)^2 + 1$
Minimum: 1

79. [12.5]   Domain: $\mathbb{R}$; range: $(0, \infty)$

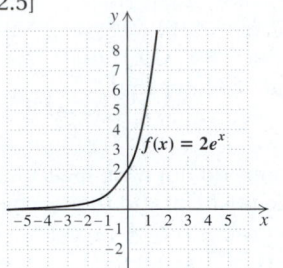

80. [12.4] $2 \log a + 3 \log c - \log b$
81. [12.4] $\log\left(\dfrac{x^3}{y^{1/2}z^2}\right)$ **82.** [12.3] $a^x = 5$
83. [12.3] $\log_x t = 3$ **84.** [12.5] -1.2545
85. [12.5] 776.2471   **86.** [12.5] 2.5479
87. [12.5] 0.2466   **88.** [6.6] All real numbers except 1
and -2 **89.** [12.6] $\frac{1}{3}, \frac{10{,}000}{3}$ **90.** [11.3] 35 mph

CHAPTER 13

Technology Connection, p. 816

1. $x^2 + y^2 - 16 = 0$
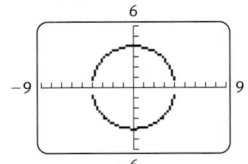

2. $4x^2 + 4y^2 = 100$
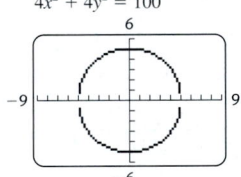

3. $x^2 + y^2 + 14x - 16y + 54 = 0$
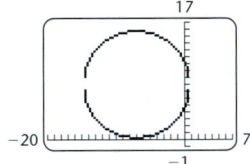

4. $x^2 + y^2 - 10x - 11 = 0$
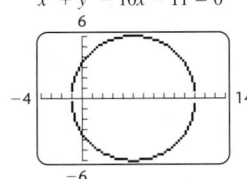

Exercise Set 13.1, pp. 816–819

1.
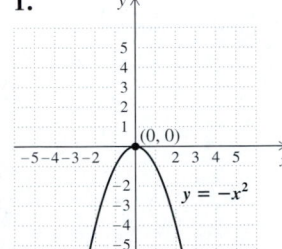

3.
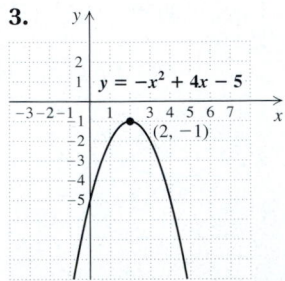

5.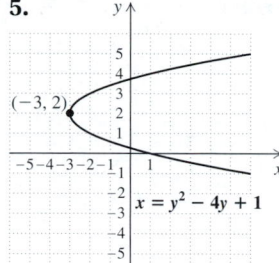

7.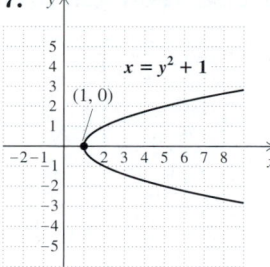

9.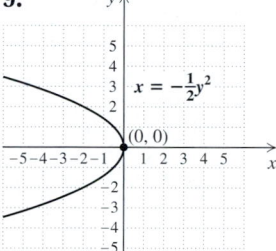

11.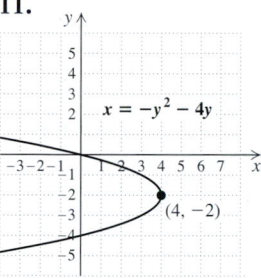

13.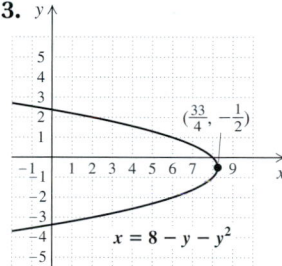

15.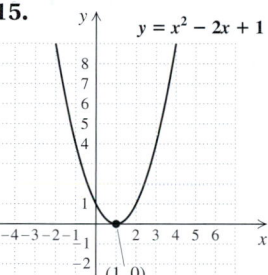

17.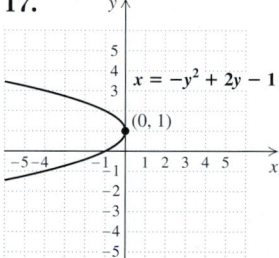

19.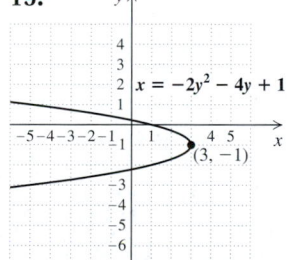

21. 5   **23.** $\sqrt{18} \approx 4.243$ **25.** $\sqrt{200} \approx 14.142$

27. 17.8   **29.** $\frac{\sqrt{41}}{7} \approx 0.915$ **31.** $\sqrt{8} \approx 2.828$

33. $\sqrt{17 + 2\sqrt{14} + 2\sqrt{15}} \approx 5.677$ **35.** $\sqrt{s^2 + t^2}$

37. $(1, 4)$ **39.** $\left(\frac{7}{2}, \frac{7}{2}\right)$ **41.** $(-1, -3)$ **43.** $(-0.25, -0.3)$

45. $\left(-\frac{1}{12}, \frac{1}{24}\right)$ **47.** $\left(\frac{\sqrt{2} + \sqrt{3}}{2}, \frac{3}{2}\right)$ **49.** $x^2 + y^2 = 36$

51. $(x - 7)^2 + (y - 3)^2 = 5$

53. $(x + 4)^2 + (y - 3)^2 = 48$

55. $(x + 7)^2 + (y + 2)^2 = 50$ **57.** $x^2 + y^2 = 25$

59. $(x + 4)^2 + (y - 1)^2 = 20$

61. $(0, 0); 7$ **63.** $(-1, -3); 2$

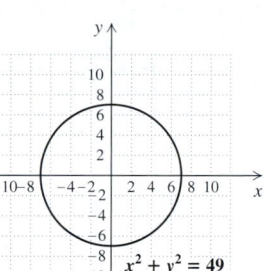   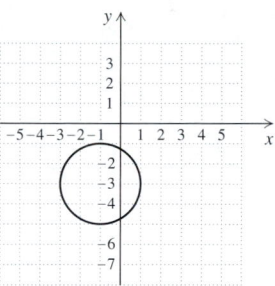

$x^2 + y^2 = 49$ $(x + 1)^2 + (y + 3)^2 = 4$

65. $(4, -3); \sqrt{10}$ **67.** $(0, 0); \sqrt{7}$

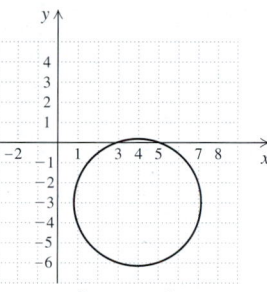   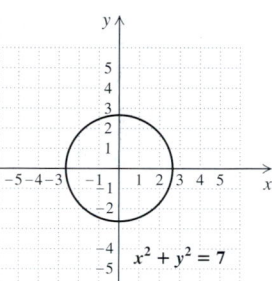

$(x - 4)^2 + (y + 3)^2 = 10$ $x^2 + y^2 = 7$

69. $(5, 0); \frac{1}{2}$ **71.** $(-4, 3); \sqrt{40},$ or $2\sqrt{10}$

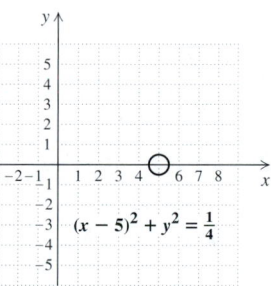   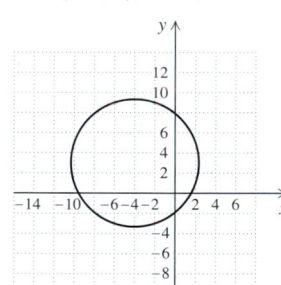

$(x - 5)^2 + y^2 = \frac{1}{4}$ $x^2 + y^2 + 8x - 6y - 15 = 0$

73. $(4, -1); 2$ **75.** $(0, -5); 10$

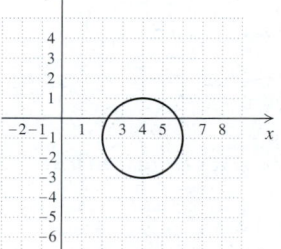   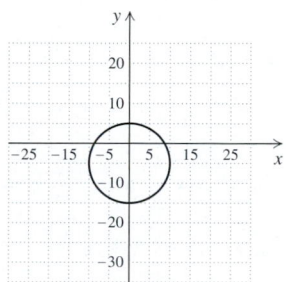

$x^2 + y^2 - 8x + 2y + 13 = 0$ $x^2 + y^2 + 10y - 75 = 0$

77. $\left(-\dfrac{7}{2}, \dfrac{3}{2}\right)$; $\sqrt{\dfrac{98}{4}}$, or $\dfrac{7\sqrt{2}}{2}$

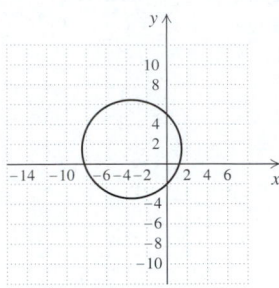

$x^2 + y^2 + 7x - 3y - 10 = 0$

79. $(0,0)$; $\dfrac{1}{6}$

$36x^2 + 36y^2 = 1$

81. ▢   **83.** [6.6] $-\dfrac{2}{3}$ **84.** [6.6] $\dfrac{25}{6}$ **85.** [5.8] 4 in.
86. [2.5] 2640 mi   **87.** [8.2] $\left(\dfrac{35}{17}, \dfrac{5}{34}\right)$ **88.** [8.2] $\left(0, -\dfrac{9}{5}\right)$
89. ▢   **91.** $(x-3)^2 + (y+5)^2 = 9$
93. $(x-3)^2 + y^2 = 25$ **95.** $(0,4)$ **97.** (a) $(0, -8467.8)$;
(b) 8487.3 mm   **99.** (a) $(0, -1522.8)$; (b) 1524.9 cm
101. 29 cm   **103.** (a) $-2.4, 3.4$; (b) $-1.3, 2.3$
105. Let $P_1 = (x_1, y_1)$, $P_2 = (x_2, y_2)$, and
$M = \left(\dfrac{x_1 + x_2}{2}, \dfrac{y_1 + y_2}{2}\right)$. Let $d(AB)$ denote the distance
from point A to point B.

(i) $d(P_1 M) = \sqrt{\left(\dfrac{x_1 + x_2}{2} - x_1\right)^2 + \left(\dfrac{y_1 + y_2}{2} - y_1\right)^2}$

$= \dfrac{1}{2}\sqrt{(x_2 - x_1)^2 + (y_2 - y_1)^2};$

$d(P_2 M) = \sqrt{\left(\dfrac{x_1 + x_2}{2} - x_2\right)^2 + \left(\dfrac{y_1 + y_2}{2} - y_2\right)^2}$

$= \dfrac{1}{2}\sqrt{(x_1 - x_2)^2 + (y_1 - y_2)^2}$

$= \dfrac{1}{2}\sqrt{(x_2 - x_1)^2 + (y_2 - y_1)^2} = d(P_1 M).$

(ii) $d(P_1 M) + d(P_2 M) = \dfrac{1}{2}\sqrt{(x_2 - x_1)^2 + (y_2 - y_1)^2} +$

$\dfrac{1}{2}\sqrt{(x_2 - x_1)^2 + (y_2 - y_1)^2}$

$= \sqrt{(x_2 - x_1)^2 + (y_2 - y_1)^2}$

$= d(P_1 P_2).$

107. ▢, ▨

Exercise Set 13.2, pp. 824–825

1.

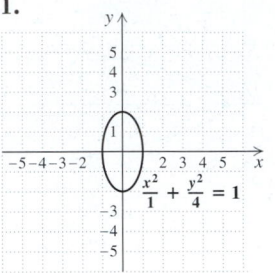

$\dfrac{x^2}{1} + \dfrac{y^2}{4} = 1$

3.

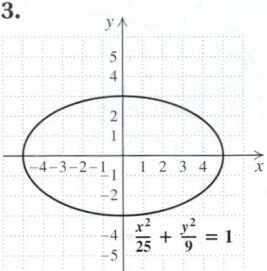

$\dfrac{x^2}{25} + \dfrac{y^2}{9} = 1$

5.

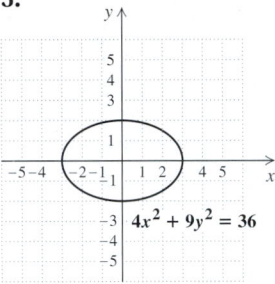

$4x^2 + 9y^2 = 36$

7.

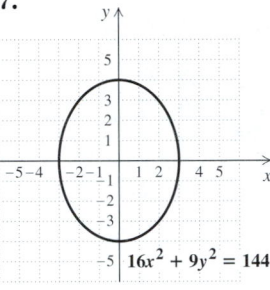

$16x^2 + 9y^2 = 144$

9.

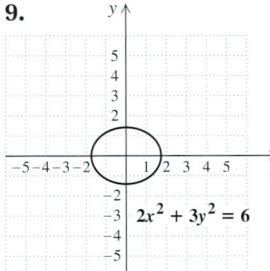

$2x^2 + 3y^2 = 6$

11.

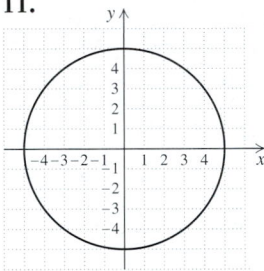

$5x^2 + 5y^2 = 125$

13.

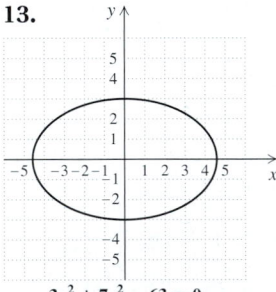

$3x^2 + 7y^2 - 63 = 0$

15.

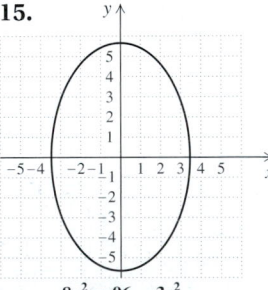

$8x^2 = 96 - 3y^2$

17.

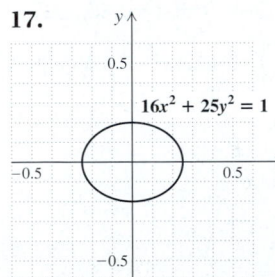

$16x^2 + 25y^2 = 1$

19.

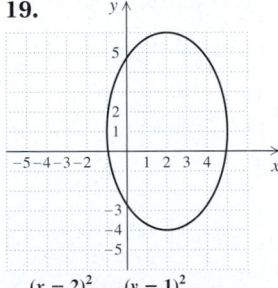

$\dfrac{(x-2)^2}{9} + \dfrac{(y-1)^2}{25} = 1$

21.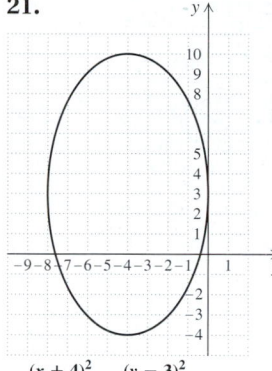

$$\frac{(x + 4)^2}{16} + \frac{(y - 3)^2}{49} = 1$$

23.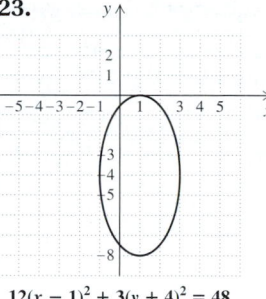

$$12(x - 1)^2 + 3(y + 4)^2 = 48$$

25.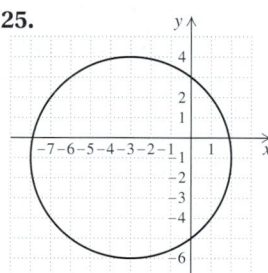

$$4(x + 3)^2 + 4(y + 1)^2 - 10 = 90$$

27. 📱   **29.** $[6.6] \frac{16}{9}$

30. $[6.6] -\frac{19}{8}$ **31.** $[11.2] \, 5 \pm \sqrt{3}$ **32.** $[11.2] \, 3 \pm \sqrt{7}$

33. $[10.6] \frac{3}{2}$ **34.** $[10.6]$ No solution   **35.** 📱

37. $\dfrac{x^2}{81} + \dfrac{y^2}{121} = 1$ **39.** $\dfrac{(x - 2)^2}{16} + \dfrac{(y + 1)^2}{9} = 1$

41. 2.134×10^8 mi   **43.**

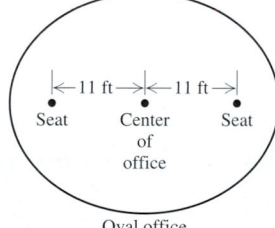

Oval office

45. Length: 700 yd; width: 200 yd

47.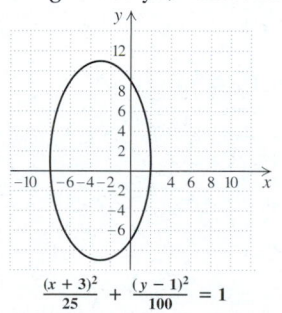

$$\frac{(x + 3)^2}{25} + \frac{(y - 1)^2}{100} = 1$$

Technology Connection, p. 831

1.
$$y_1 = \frac{\sqrt{15x^2 - 240}}{2};$$
$$y_2 = -\frac{\sqrt{15x^2 - 240}}{2}$$

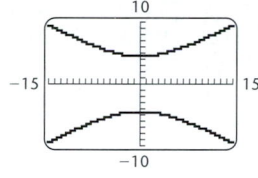

2.
$$y_1 = \sqrt{\frac{16x^2 - 64}{3}};$$
$$y_2 = -\sqrt{\frac{16x^2 - 64}{3}}$$

3.
$$y_1 = \frac{\sqrt{5x^2 + 320}}{4};$$
$$y_2 = -\frac{\sqrt{5x^2 + 320}}{4}$$

4.
$$y_1 = \sqrt{\frac{9x^2 + 441}{45}};$$
$$y_2 = -\sqrt{\frac{9x^2 + 441}{45}}$$

Exercise Set 13.3, pp. 835–836

1.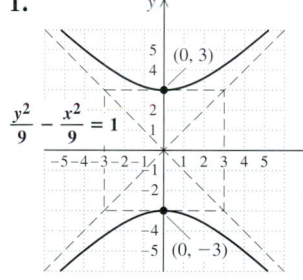

$$\frac{y^2}{9} - \frac{x^2}{9} = 1$$

3.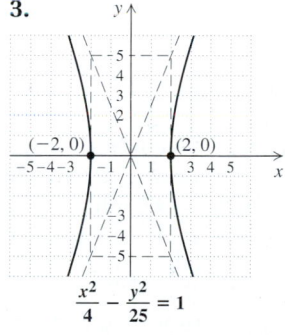

$$\frac{x^2}{4} - \frac{y^2}{25} = 1$$

5.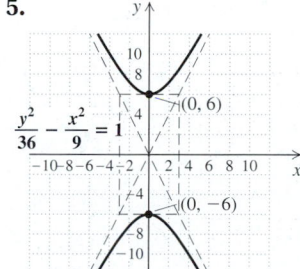

$$\frac{y^2}{36} - \frac{x^2}{9} = 1$$

7.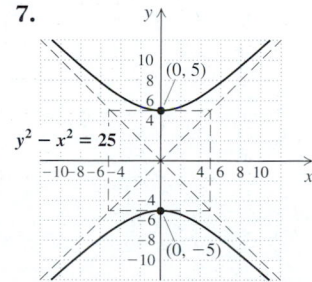

$$y^2 - x^2 = 25$$

9.

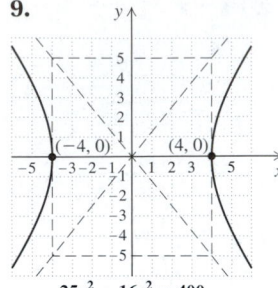

$25x^2 - 16y^2 = 400$

11.

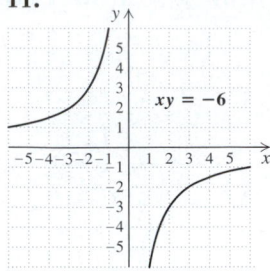

$xy = -6$

13.

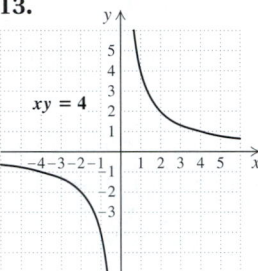

$xy = 4$

15.

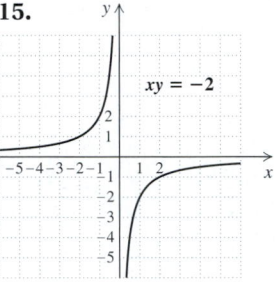

$xy = -2$

17.

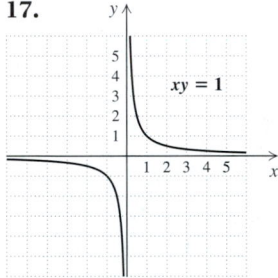

$xy = 1$

19. Circle   **21.** Ellipse

23. Hyperbola   **25.** Circle   **27.** Ellipse
29. Hyperbola   **31.** Parabola   **33.** Hyperbola
35. Circle   **37.** Ellipse   **39.**   **41.** [8.2] $\left(-\frac{22}{3}, \frac{37}{9}\right)$
42. [8.2] $(9, -4)$ **43.** [5.7] $-3, 3$ **44.** [5.7] $-1, 1$

45. [2.4] $35   **46.** [2.5] 69   **47.** ▯   **49.** $\dfrac{y^2}{36} - \dfrac{x^2}{4} = 1$

51. C: $(5, 2)$; V: $(-1, 2)$, $(11, 2)$; asymptotes:
$y - 2 = \frac{5}{6}(x - 5)$, $y - 2 = -\frac{5}{6}(x - 5)$

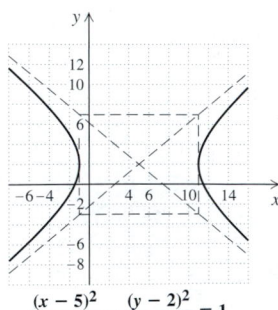

$\dfrac{(x - 5)^2}{36} - \dfrac{(y - 2)^2}{25} = 1$

53. $\dfrac{(y + 3)^2}{4} - \dfrac{(x - 4)^2}{16} = 1$; C: $(4, -3)$; V: $(4, -5)$, $(4, -1)$;
asymptotes: $y + 3 = \frac{1}{2}(x - 4)$, $y + 3 = -\frac{1}{2}(x - 4)$

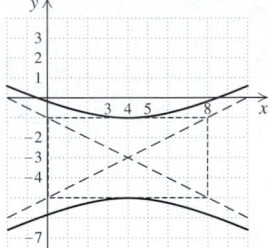

$8(y + 3)^2 - 2(x - 4)^2 = 32$

55. $\dfrac{(x + 3)^2}{1} - \dfrac{(y - 2)^2}{4} = 1$; C: $(-3, 2)$; V: $(-4, 2)$, $(-2, 2)$;
asymptotes: $y - 2 = 2(x + 3)$, $y - 2 = -2(x + 3)$

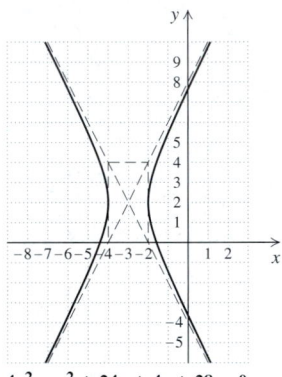

$4x^2 - y^2 + 24x + 4y + 28 = 0$

57.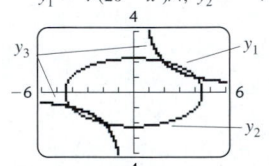

Technology Connection, p. 839

1. $(-1.50, -1.17)$; $(3.50, 0.50)$
2. $(-2.77, 2.52)$; $(-2.77, -2.52)$

Technology Connection, p. 841

1. $y_1 = \sqrt{(20 - x^2)/4}$; $y_2 = -\sqrt{(20 - x^2)/4}$; $y_3 = 4/x$

Exercise Set 13.4, pp. 843–845

1. $(-4, -3), (3, 4)$ **3.** $(2, 0), (0, 3)$ **5.** $(2, 4), (1, 1)$
7. $\left(\frac{11}{4}, -\frac{9}{8}\right), (1, -2)$ **9.** $\left(\frac{13}{3}, \frac{5}{3}\right), (-5, -3)$
11. $\left(\frac{3 + \sqrt{7}}{2}, \frac{-1 + \sqrt{7}}{2}\right), \left(\frac{3 - \sqrt{7}}{2}, \frac{-1 - \sqrt{7}}{2}\right)$
13. $\left(-3, \frac{5}{2}\right), (3, 1)$ **15.** $(1, -7), (-7, 1)$
17. $(3, 0), \left(-\frac{9}{5}, \frac{8}{5}\right)$ **19.** $(1, 4), (4, 1)$ **21.** $(0, 0), (1, 1),$
$\left(-\frac{1}{2} + \frac{\sqrt{3}}{2}i, -\frac{1}{2} - \frac{\sqrt{3}}{2}i\right), \left(-\frac{1}{2} - \frac{\sqrt{3}}{2}i, -\frac{1}{2} + \frac{\sqrt{3}}{2}i\right)$
23. $(-3, 0), (3, 0)$ **25.** $(-4, -3), (-3, -4), (3, 4), (4, 3)$
27. $\left(\frac{4i\sqrt{35}}{7}, \frac{6\sqrt{21}}{7}\right), \left(-\frac{4i\sqrt{35}}{7}, \frac{6\sqrt{21}}{7}\right),$
$\left(\frac{4i\sqrt{35}}{7}, -\frac{6\sqrt{21}}{7}\right), \left(-\frac{4i\sqrt{35}}{7}, -\frac{6\sqrt{21}}{7}\right)$
29. $\left(-\sqrt{2}, -\sqrt{14}\right), \left(-\sqrt{2}, \sqrt{14}\right), \left(\sqrt{2}, -\sqrt{14}\right), \left(\sqrt{2}, \sqrt{14}\right)$
31. $(-2, -1), (-1, -2), (1, 2), (2, 1)$
33. $(-3, -2), (-2, -3), (2, 3), (3, 2)$ **35.** $(2, 5), (-2, -5)$
37. $(3, 2), (-3, -2)$ **39.** $(-3, 4), (3, 4), (0, -5)$
41. Length: 8 cm; width: 6 cm
43. Length: 5 in.; width: 4 in.
45. Length: 12 ft; width: 5 ft **47.** 13 and 12
49. \$3750, 6% **51.** Length: $\sqrt{3}$ m; width: 1 m
53. 🖩 **55.** [1.8] -16 **56.** [1.8] -32 **57.** [1.8] 1
58. [1.8] $-\frac{1}{4}$ **59.** [1.8] 44 **60.** [1.8] 28 **61.** 🖩
63. 61.52 cm and 38.48 cm **65.** $4x^2 + 3y^2 = 43$
67. $\left(\frac{1}{3}, \frac{1}{2}\right), \left(\frac{1}{2}, \frac{1}{3}\right)$ **69.** Length: 24.8 cm; height: 18.6 cm
71. 30

Review Exercises: Chapter 13, p. 847

1. [13.1] 4 **2.** [13.1] 5 **3.** [13.1] $\sqrt{90.1} \approx 9.492$
4. [13.1] $\sqrt{9 + 4a^2}$ **5.** [13.1] $(4, 6)$ **6.** [13.1] $\left(-3, \frac{5}{2}\right)$
7. [13.1] $\left(\frac{3}{4}, \frac{\sqrt{3} - \sqrt{2}}{2}\right)$ **8.** [13.1] $\left(\frac{1}{2}, 2a\right)$
9. [13.1] $(-2, 3), \sqrt{2}$ **10.** [13.1] $(5, 0), 7$
11. [13.1] $(3, 1), 3$ **12.** [13.1] $(-4, 3), \sqrt{35}$
13. [13.1] $(x + 4)^2 + (y - 3)^2 = 48$
14. [13.1] $(x - 7)^2 + (y + 2)^2 = 20$
15. [13.1], [13.3] Circle **16.** [13.2], [13.3] Ellipse

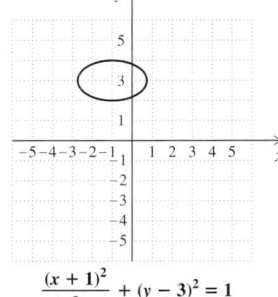

$4x^2 + 4y^2 = 100$

$9x^2 + 2y^2 = 18$

17. [13.1], [13.3] Parabola **18.** [13.3] Hyperbola

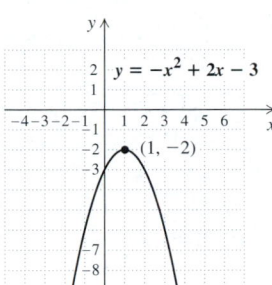

$y = -x^2 + 2x - 3$

$(1, -2)$

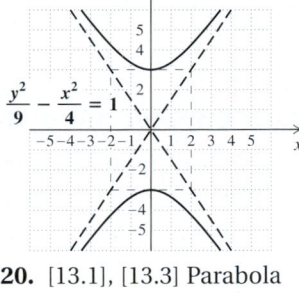

$\frac{y^2}{9} - \frac{x^2}{4} = 1$

19. [13.3] Hyperbola **20.** [13.1], [13.3] Parabola

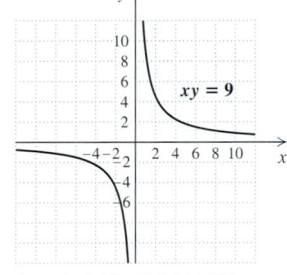

$xy = 9$

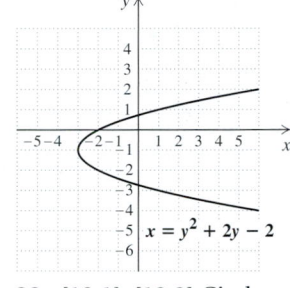

$x = y^2 + 2y - 2$

21. [13.2], [13.3] Ellipse **22.** [13.1], [13.3] Circle

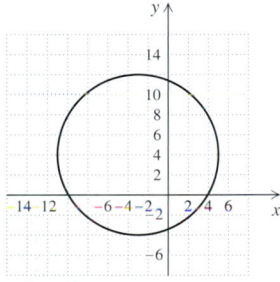

$\frac{(x + 1)^2}{3} + (y - 3)^2 = 1$

$x^2 + y^2 + 6x - 8y - 39 = 0$

23. [13.4] $(7, 4)$ **24.** [13.4] $(2, 2), \left(\frac{32}{9}, -\frac{10}{9}\right)$
25. [13.4] $(0, -3), (2, 1)$ **26.** [13.4] $(4, 3), (4, -3), (-4, 3),$
$(-4, -3)$ **27.** [13.4] $(2, 1), \left(\sqrt{3}, 0\right), (-2, 1), \left(-\sqrt{3}, 0\right)$
28. [13.4] $(3, -3), \left(-\frac{3}{5}, \frac{21}{5}\right)$ **29.** [13.4] $(6, 8), (6, -8),$
$(-6, 8), (-6, -8)$ **30.** [13.4] $(2, 2), (-2, -2), \left(2\sqrt{2}, \sqrt{2}\right),$
$\left(-2\sqrt{2}, -\sqrt{2}\right)$ **31.** [13.4] Length: 12 m; width: 7 m
32. [13.4] Length: 12 in.; width: 9 in.
33. [13.4] 32 cm, 20 cm **34.** [13.4] 3 ft, 11 ft
35. 🖩 [13.1], [13.3] The graph of a parabola has one
branch whereas the graph of a hyperbola has two
branches. A hyperbola has asymptotes, but a parabola
does not. **36.** 🖩 [13.1], [13.2], [13.3] Function notation
is not used in this chapter because many of the relations
are not functions. Function notation could be used for
vertical parabolas and for hyperbolas that have the axes as
asymptotes.
37. [13.4] $\left(-5, -4\sqrt{2}\right), \left(-5, 4\sqrt{2}\right), \left(3, -2\sqrt{2}\right), \left(3, 2\sqrt{2}\right)$

38. [13.1] $(0,6)$, $(0,-6)$
39. [13.1], [13.4] $(x-2)^2 + (y+1)^2 = 25$
40. [13.2] $\dfrac{x^2}{49} + \dfrac{y^2}{9} = 1$ **41.** [13.1] $\left(\frac{9}{4}, 0\right)$

Test: Chapter 13, p. 848

1. [13.1] $9\sqrt{2} \approx 12.728$ **2.** [13.1] $2\sqrt{9+a^2}$
3. [13.1] $\left(-\frac{1}{2}, \frac{7}{2}\right)$ **4.** [13.1] $(0,0)$ **5.** [13.1] $(-2,3)$, 8
6. [13.1] $(-2,3)$, 3
7. [13.1], [13.3] Parabola   **8.** [13.1], [13.3] Circle

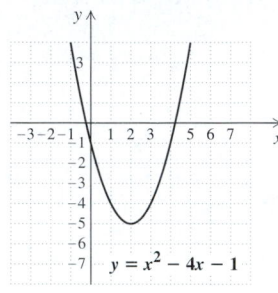

$y = x^2 - 4x - 1$

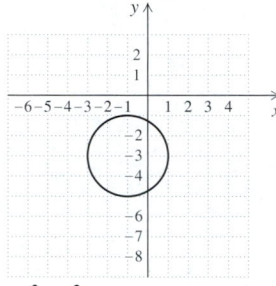

$x^2 + y^2 + 2x + 6y + 6 = 0$

9. [13.3] Hyperbola

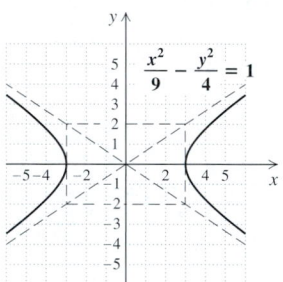

$\dfrac{x^2}{9} - \dfrac{y^2}{4} = 1$

10. [13.2], [13.3] Ellipse

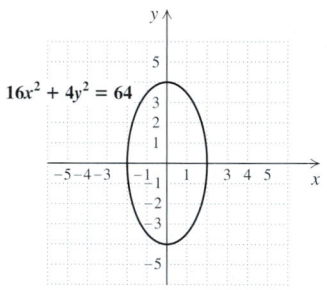

$16x^2 + 4y^2 = 64$

11. [13.3] Hyperbola   **12.** [13.1], [13.3] Parabola

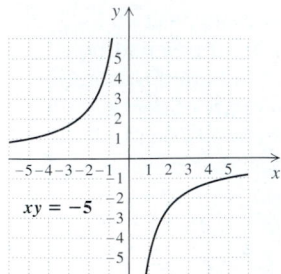

$xy = -5$

$x = -y^2 + 4y$ $(4,2)$

13. [13.4] $(0,3)$, $(4,0)$ **14.** [13.4] $(4,0)$, $(-4,0)$
15. [13.4] $(3,2)$, $(-3,-2)$ **16.** [13.4] $\left(\sqrt{6}, 2\right)$, $\left(\sqrt{6}, -2\right)$, $\left(-\sqrt{6}, 2\right)$, $\left(-\sqrt{6}, -2\right)$ **17.** [13.4] 2 by 11
18. [13.4] $\sqrt{5}$ m, $\sqrt{3}$ m   **19.** [13.4] Length: 16 ft; width: 12 ft   **20.** [13.4] $1200, 6\%
21. [13.2] $\dfrac{(x-6)^2}{25} + \dfrac{(y-3)^2}{9} = 1$ **22.** [13.1] $\left(0, -\frac{31}{4}\right)$
23. [13.4] 9

CHAPTER 14

Exercise Set 14.1, pp. 854–856

1. $3, 8, 13, 18; 48; 73$ **3.** $\frac{1}{2}, \frac{2}{3}, \frac{3}{4}, \frac{4}{5}; \frac{10}{11}; \frac{15}{16}$
5. $-1, 0, 3, 8; 80; 195$ **7.** $2, 2\frac{1}{2}, 3\frac{1}{3}, 4\frac{1}{4}; 10\frac{1}{10}; 15\frac{1}{15}$
9. $-1, 4, -9, 16; 100; -225$ **11.** $-2, -1, 4, -7; -25; 40$
13. 9 **15.** 364 **17.** -23.5 **19.** -363 **21.** $\frac{441}{400}$
23. $2n - 1$ **25.** $(-1)^{n+1}$ **27.** $(-1)^n \cdot n$
29. $(-1)^n \cdot 2 \cdot (3)^{n-1}$ **31.** $\dfrac{n}{n+1}$ **33.** 5^n
35. $(-1)^n \cdot n^2$ **37.** 4 **39.** 30
41. $\frac{1}{2} + \frac{1}{4} + \frac{1}{6} + \frac{1}{8} + \frac{1}{10} = \frac{137}{120}$
43. $3^0 + 3^1 + 3^2 + 3^3 + 3^4 = 121$
45. $\frac{1}{2} + \frac{2}{3} + \frac{3}{4} + \frac{4}{5} + \frac{5}{6} + \frac{6}{7} + \frac{7}{8} + \frac{8}{9} = \frac{15{,}551}{2520}$
47. $(-1)^2 2^1 + (-1)^3 2^2 + (-1)^4 2^3 + (-1)^5 2^4 + (-1)^6 2^5 + (-1)^7 2^6 + (-1)^8 2^7 + (-1)^9 2^8 = -170$
49. $(0^2 - 2 \cdot 0 + 3) + (1^2 - 2 \cdot 1 + 3) + (2^2 - 2 \cdot 2 + 3) + (3^2 - 2 \cdot 3 + 3) + (4^2 - 2 \cdot 4 + 3) + (5^2 - 2 \cdot 5 + 3) = 43$
51. $\dfrac{(-1)^3}{3 \cdot 4} + \dfrac{(-1)^4}{4 \cdot 5} + \dfrac{(-1)^5}{5 \cdot 6} = -\dfrac{1}{15}$ **53.** $\displaystyle\sum_{k=1}^{5} \dfrac{k+1}{k+2}$
55. $\displaystyle\sum_{k=1}^{6} k^2$ **57.** $\displaystyle\sum_{k=2}^{n} (-1)^k k^2$ **59.** $\displaystyle\sum_{k=1}^{\infty} 5k$
61. $\displaystyle\sum_{k=1}^{\infty} \dfrac{1}{k(k+1)}$ **63.** 🔲   **65.** [1.1] 77   **66.** [1.1] 23
67. [4.4] $x^3 + 3x^2y + 3xy^2 + y^3$
68. [4.4] $a^3 - 3a^2b + 3ab^2 - b^3$
69. [4.4] $8a^3 - 12a^2b + 6ab^2 - b^3$
70. [4.4] $8x^3 + 12x^2y + 6xy^2 + y^3$ **71.** 🔲
73. $1, 3, 13, 63, 313, 1563$ **75.** $1, 2, 4, 8, 16, 32, 128, 256, 512, 1024, 2048, 4096, 8192, 16{,}384, 32{,}768, 65{,}536$
77. $S_{100} = 0$; $S_{101} = -1$ **79.** $i, -1, -i, 1, i; i$
81. 11th term

Exercise Set 14.2, pp. 863–865

1. $a_1 = 2, d = 4$ **3.** $a_1 = 6, d = -4$ **5.** $a_1 = \frac{3}{2}, d = \frac{3}{4}$
7. $a_1 = \$5.12, d = \$0.12$ **9.** 47 **11.** -41
13. $-\$1628.16$ **15.** 27th   **17.** 102nd   **19.** 82
21. 5 **23.** 28 **25.** $a_1 = 8; d = -3; 8, 5, 2, -1, -4$
27. $a_1 = 1; d = 1$ **29.** 780 **31.** $31{,}375$ **33.** 2550

35. 918   **37.** 1030   **39.** 42; 420   **41.** 1260

43. \$31,000   **45.** 722   **47.**   **49.** [6.4] $\dfrac{13}{30x}$

50. [6.4] $\dfrac{23}{36t}$ **51.** [12.3] $a^k = P$ **52.** [12.3] $e^a = t$

53. [13.1] $x^2 + y^2 = 81$

54. [13.1] $(x + 2)^2 + (y - 5)^2 = 18$ **55.**

57. $S_n = n^2$ **59.** \$8760, \$7961.77, \$7163.54, \$6365.31, \$5567.08; \$4768.85, \$3970.62, \$3172.39, \$2374.16, \$1575.93

61. Let $d =$ the common difference. Since p, m, and q form an arithmetic sequence, $m = p + d$ and $q = p + 2d$.

Then $\dfrac{p + q}{2} = \dfrac{p + (p + 2d)}{2} = p + d = m$.

63. 156,375

Exercise Set 14.3, pp. 872–874

1. 2   **3.** -1 **5.** $-\frac{1}{2}$ **7.** $\frac{1}{5}$ **9.** $\dfrac{3}{m}$ **11.** 192

13. 80   **15.** 648   **17.** \$2331.64   **19.** $a_n = 3^{n-1}$

21. $a_n = (-1)^{n-1}$ **23.** $a_n = \dfrac{1}{x^n}$ **25.** 762   **27.** $\frac{547}{18}$

29. $\dfrac{1 - x^8}{1 - x}$, or $(1 + x)(1 + x^2)(1 + x^4)$ **31.** \$5134.51

33. $\frac{64}{3}$ **35.** $\frac{49}{4}$ **37.** No   **39.** No   **41.** $\frac{43}{99}$

43. \$25,000   **45.** $\frac{7}{9}$ **47.** $\frac{830}{99}$ **49.** $\frac{5}{33}$ **51.** $\frac{5}{1024}$ ft

53. 155,797   **55.** \$43,318.94   **57.** 25 min

59. 3100.35 ft   **61.** 20.48 in.   **63.**

65. [4.4] $x^3 + 3x^2y + 3xy^2 + y^3$

66. [4.4] $a^3 - 3a^2b + 3ab^2 - b^3$ **67.** [8.2] $\left(-\frac{63}{29}, -\frac{114}{29}\right)$

68. [8.4] $(-1, 2, 3)$ **69.**   **71.** $\dfrac{x^2[1 - (-x)^n]}{1 + x}$

73. $512\,\text{cm}^2$ **75.**

Exercise Set 14.4, pp. 881–883

1. 362,800   **3.** 39,916,800   **5.** 1   **7.** 210   **9.** 3024

11. 120   **13.** 40,314   **15.** 720   **17.** 24   **19.** 604,800

21. 6   **23.** 720   **25.** $3 \cdot 10 \cdot 5 \cdot 3$, or 450   **27.** $3!$, or 6

29. $7!$, or 5040   **31.** $_7P_4$, or 840   **33.** $4!$, or 24; 4^4, or 256

35. $5!$, or 120   **37.** $8 \cdot 9 \cdot 8 \cdot 7 \cdot 6 \cdot 5 \cdot 4$, or 483,840

39. (a) $5!$, or 120; **(b)** $5! \cdot 2^5$, or 3840

41. $52 \cdot 51 \cdot 50 \cdot 49$, or 6,497,400

43. $80 \cdot 26 \cdot 9999$, or 20,797,920

45. (a) 10^5, or 100,000; **(b)** 100,000

47. (a) 10^9, or 1,000,000,000; **(b)** yes   **49.**

51. [12.4] 3.555   **52.** [12.4] -0.337 **53.** [12.4] 3.892

54. [12.3] 10   **55.** [12.3] -5 **56.** [12.3] m **57.**

59. 11   **61.** 9

63. (a) $6!$, or 720; **(b)** $3! \cdot 3! \cdot 2$, or 72; **(c)** $2! \cdot 5!$, or 240

(d) $720 - 240$, or 480   **65.** $4 \cdot 3$, or 12

Exercise Set 14.5, pp. 888–890

1. 78   **3.** 78   **5.** 7   **7.** 10   **9.** 1   **11.** 12   **13.** 190

15. 595   **17.** 252   **19.** 8855   **21.** $_{10}C_6$, or 210

23. $_8C_2$, or 28; $_8C_3$, or 56   **25.** $_{58}C_6 \cdot {_{42}C_4}$

27. $6 \cdot 6$, or 36   **29.** $_{52}C_5$ **31.** $_5C_2 \cdot {_6C_3} \cdot {_3C_1}$, or 600

33. $\dbinom{10}{0} + \dbinom{10}{1} + \dbinom{10}{2} + \dbinom{10}{3} + \dbinom{10}{4} + \dbinom{10}{5} +$ $\dbinom{10}{6} + \dbinom{10}{7} + \dbinom{10}{8} + \dbinom{10}{9} + \dbinom{10}{10}$, or 1024

35. $\dbinom{4}{3} \cdot \dbinom{48}{2}$, or 4512

37. (a) $_{33}P_3$, or 32,736; **(b)** 33^3, or 35,937; **(c)** $_{33}C_3$, or 5456

39.   **41.** [12.6] -2 **42.** [12.6] $\frac{3}{2}$ **43.** [12.6] 24

44. [12.6] 5   **45.** [12.6] 5   **46.** [12.6] 1.861   **47.**

49. m **51.** 1   **53.** 5   **55.** 7   **57.** $n - 1$

59. $\dbinom{m}{3}$, or $\dfrac{m(m - 1)(m - 2)}{6}$

61. $\dbinom{n}{n - r} = \dfrac{n!}{[n - (n - r)]!\,(n - r)!}$
$= \dfrac{n!}{r!\,(n - r)!} = \dbinom{n}{r}$

Exercise Set 14.6, pp. 896–897

1. $m^5 + 5m^4n + 10m^3n^2 + 10m^2n^3 + 5mn^4 + n^5$

3. $x^6 - 6x^5y + 15x^4y^2 - 20x^3y^3 + 15x^2y^4 - 6xy^5 + y^6$

5. $x^{10} - 15x^8y + 90x^6y^2 - 270x^4y^3 + 405x^2y^4 - 243y^5$

7. $729c^6 - 1458c^5d + 1215c^4d^2 - 540c^3d^3 + 135c^2d^4 - 18cd^5 + d^6$ **9.** $x^3 - 3x^2y + 3xy^2 - y^3$

11. $x^9 + \dfrac{18x^8}{y} + \dfrac{144x^7}{y^2} + \dfrac{672x^6}{y^3} + \dfrac{2016x^5}{y^4} + \dfrac{4032x^4}{y^5} +$ $\dfrac{5376x^3}{y^6} + \dfrac{4608x^2}{y^7} + \dfrac{2304x}{y^8} + \dfrac{512}{y^9}$

13. $a^{10} - 5a^8b^3 + 10a^6b^6 - 10a^4b^9 + 5a^2b^{12} - b^{15}$

15. $9 - 12\sqrt{3}t + 18t^2 - 4\sqrt{3}t^3 + t^4$

17. $x^{-8} + 4x^{-4} + 6 + 4x^4 + x^8$ **19.** $15a^4b^2$

21. $-64,481,508a^3$ **23.** $1120x^{12}y^2$

25. $-1,959,552u^5v^{10}$ **27.** y^8 **31.** [12.6] 4

32. [12.6] $\frac{5}{2}$ **33.** [12.6] 5.6348   **34.** [12.6] ± 5 **35.**

37. $\dbinom{5}{2}(0.325)^3(0.675)^2 \approx 0.156$

39. $\dbinom{5}{2}(0.325)^3(0.675)^2 + \dbinom{5}{3}(0.325)^2(0.675)^3 +$ $\dbinom{5}{4}(0.325)(0.675)^4 + \dbinom{5}{5}(0.675)^5 \approx 0.959$

41. $\frac{55}{144}$ **43.** $-\dfrac{\sqrt[3]{q}}{2p}$ **45.** 8

Exercise Set 14.7, pp. 904–907

1. 0.57, 0.43   **3.** 0.093, 0.135, 0.059, 0.077, 0.026

5. 0.610   **7. (a)** T; **(b)** E; **(c)** T, S, R, N, L, E   **9.** $\frac{1}{13}$

11. $\frac{1}{2}$ **13.** $\frac{2}{13}$ **15.** $\frac{2}{7}$ **17.** 0 **19.** $\frac{13C_4}{52C_4} = \frac{11}{4165}$

21. $\frac{6C_3 \cdot 10C_2 \cdot 4C_1}{20C_6} = \frac{30}{323}$ **23.** $\frac{5}{36}$ **25.** $\frac{1}{6}$

27. $\frac{7C_1 \cdot 7C_1 \cdot 7C_1 \cdot 7C_1}{28C_4} = \frac{343}{2925}$ **29.** $\frac{9}{19}$ **31.** $\frac{1}{38}$ **33.** $\frac{5}{12}$

35. $\frac{7}{12}$ **37.** 1 **39.** ▨ **41.** [8.2] $(6, -1)$

42. [12.6] -1 **43.** [12.4] $2\log_a x + \log_a y - 3\log_a z$

44. [13.1] $x^2 + (y+3)^2 = 12$ **45.** [12.3] $3^4 = x$

46. [12.3] $\log_4 10 = y$ **47.** ▨ **49.** $_{52}C_5 = 2{,}598{,}960$

51. (a) $9 \cdot 4 = 36$; **(b)** $\frac{36}{52C_5} \approx 0.0000139$

53. (a) $(13 \cdot {}_4C_3) \cdot (12 \cdot {}_4C_2) = 3744$; **(b)** $\frac{3744}{52C_5} \approx 0.00144$

55. (a) $13 \cdot \binom{4}{3} \cdot \binom{48}{2} - 3744 = 54{,}912$;

(b) $\frac{54{,}912}{52C_5} \approx 0.0211$

57. (a) $\binom{13}{2}\binom{4}{2}\binom{4}{2}\binom{44}{1} \approx 123{,}552$;

(b) $\frac{123{,}552}{52C_5} \approx 0.0475$

Review Exercises: Chapter 14, pp. 910–911

1. [14.1] 1, 5, 9, 13; 29; 45 **2.** [14.1] $0, \frac{1}{5}, \frac{1}{5}, \frac{3}{17}; \frac{7}{65}; \frac{11}{145}$

3. [14.1] $a_n = -2n$ **4.** [14.1] $a_n = (-1)^n(2n-1)$

5. [14.1] $-2 + 4 + (-8) + 16 + (-32) = -22$

6. [14.1] $-3 + (-5) + (-7) + (-9) + (-11) +$

$(-13) = -48$ **7.** [14.1] $\sum_{k=1}^{5} 4k$ **8.** [14.1] $\sum_{k=1}^{5} \frac{1}{(-2)^k}$

9. [14.1] 85 **10.** [14.2] $\frac{8}{3}$

11. [14.2] $d = 1.25, a_1 = 11.25$ **12.** [14.2] -544

13. [14.2] 8580 **14.** [14.3] $1024\sqrt{2}$ **15.** [14.3] $\frac{2}{3}$

16. [14.3] $a_n = 2(-1)^n$ **17.** [14.3] $a_n = 3\left(\frac{x}{4}\right)^{n-1}$

18. [14.3] 4095 **19.** [14.3] $-4095x$ **20.** [14.3] 12

21. [14.3] $\frac{49}{11}$ **22.** [14.3] No **23.** [14.3] No

24. [14.3] \$40,000 **25.** [14.3] $\frac{5}{9}$ **26.** [14.3] $\frac{46}{33}$

27. [14.2] \$17.80 **28.** [14.2] 903 **29.** [14.3] \$22,521.92

30. [14.3] 6 m **31.** [14.4] 5040 **32.** [14.5] 56

33. [14.4] 30 **34.** [14.4] 60,480 **35.** [14.5] 35

36. [14.4] 10 **37.** [14.4] $_9P_4$, or 3024

38. [14.4] $_{24}P_3$, or 12,144 **39.** [14.4] $_6P_6$, or 720

40. [14.5] $_{15}C_8$, or 6435 **41.** [14.4] $3 \cdot 2 \cdot 4$, or 24

42. [14.6] $190a^{18}b^2$

43. [14.6] $x^4 - 8x^3y + 24x^2y^2 - 32xy^3 + 16y^4$

44. [14.7] 0.417, 0.471, 0.112 **45.** [14.7] $\frac{1}{12}$, 0

46. [14.7] $\frac{1}{4}$ **47.** [14.7] $\frac{4C_2 \cdot 4C_1}{52C_3} = \frac{6}{5525}$

48. ▨ [14.3] For a geometric sequence with $|r| < 1$, as n gets larger, the absolute value of the terms gets smaller,

since $|r^n|$ gets smaller.

49. ▨ [14.4] The first form of the binomial theorem draws the coefficients from Pascal's triangle; the second form uses factorial notation. The second form avoids the need to compute all preceding rows of Pascal's triangle, and is generally easier to use when only one term of an expression is needed. When several terms of an expansion are needed and n is not large, (say, $n \leq 8$), it is often easier to use Pascal's triangle.

50. [14.3] $\frac{1 - (-x)^n}{x + 1}$

51. [14.6] $x^{-15} + 5x^{-9} + 10x^{-3} + 10x^3 + 5x^9 + x^{15}$

Test: Chapter 14, pp. 911–912

1. [14.1] 1, 7, 13, 19, 25; 91 **2.** [14.1] $a_n = 4\left(\frac{1}{3}\right)^n$

3. [14.1] $1 + (-1) + (-5) + (-13) + (-29) = -47$

4. [14.1] $\sum_{k=1}^{5} (-1)^{k+1}k^3$ **5.** [14.2] -46 **6.** [14.2] $\frac{3}{8}$

7. [14.2] $a_1 = 31.2; d = -3.8$ **8.** [14.2] 2508

9. [14.3] $\frac{9}{128}$ **10.** [14.3] $\frac{2}{3}$ **11.** [14.3] $(-1)^{n+1}3^n$

12. [14.3] $511 + 511x$ **13.** [14.3] 1 **14.** [14.3] No

15. [14.3] $\frac{\$25{,}000}{23} \approx \$1086.96$ **16.** [14.3] $\frac{85}{99}$

17. [14.2] 63 **18.** [14.2] \$17,100 **19.** [14.3] \$8981.05

20. [14.3] 36 m **21.** [14.6] 78 **22.** [14.4] 6!, or 720

23. [14.5] $_7C_4$, or 35 **24.** [14.5] $_{20}C_3 \cdot {}_{14}C_2$, or 103,740

25. [14.5] $_{20}P_3$, or 6840

26. [14.6] $x^{10} - 15x^8y + 90x^6y^2 - 270x^4y^3 + 405x^2y^4 -$

$243y^5$ **27.** [14.6] $220a^9x^3$ **28.** [14.7] $\frac{1}{6}$ **29.** [14.7] $\frac{1}{13}$

30. [14.7] $\frac{5C_2 \cdot 4C_1}{16C_3} = \frac{1}{14}$ **31.** [14.2] $n(n+1)$

32. [14.3] $\dfrac{1 - \left(\dfrac{1}{x}\right)^n}{1 - \dfrac{1}{x}}$, or $\dfrac{x^n - 1}{x^{n-1}(x-1)}$

Cumulative Review: Chapters 1–14, pp. 912–914

1. [4.8] $-45x^6y^{-4}$, or $\dfrac{-45x^6}{y^4}$ **2.** [1.8] 6.3

3. [1.8] $-3y + 17$ **4.** [1.8] 280 **5.** [1.8] $\frac{7}{6}$

6. [4.3] $3a^2 - 8ab - 15b^2$ **7.** [4.3] $13x^3 - 7x^2 - 6x + 6$

8. [4.5] $6a^2 + 7a - 5$ **9.** [4.5] $9a^4 - 30a^2y + 25y^2$

10. [6.4] $\dfrac{4}{x + 2}$ **11.** [6.2] $\dfrac{x - 4}{3(x + 2)}$

12. [6.2] $\dfrac{(x + y)(x^2 + xy + y^2)}{x^2 + y^2}$ **13.** [6.5] $x - a$

14. [5.4] $(2x - 3)^2$ **15.** [5.5] $(3a - 2)(9a^2 + 6a + 4)$

16. [5.1] $(a^2 - b)(a + 3)$ **17.** [5.3] $3(y^2 + 3)(5y^2 - 4)$

18. [7.1] 20 **19.** [4.7] $7x^3 + 9x^2 + 19x + 38 + \dfrac{72}{x - 2}$

20. [2.2] $\frac{2}{3}$ **21.** [6.6] $-\frac{6}{5}, 4$ **22.** [9.2] $\mathbb{R}$, or $(-\infty, \infty)$

23. [8.2] $(1, -1)$ **24.** [8.4] $(2, -1, 1)$ **25.** [10.6] 2
26. [11.5] $\pm 2, \pm 5$ **27.** [13.4] $(\sqrt{5}, \sqrt{3}), (\sqrt{5}, -\sqrt{3}),$
$(-\sqrt{5}, \sqrt{3}), (-\sqrt{5}, -\sqrt{3})$ **28.** [12.6] $\dfrac{\ln 8}{\ln 5} \approx 1.2920$

29. [12.6] 1005   **30.** [12.6] $\frac{1}{16}$ **31.** [12.6] $-\frac{1}{2}$
32. [9.3] $\{x \mid -2 \le x \le 3\}$, or $[-2, 3]$ **33.** [11.1] $\pm i\sqrt{2}$
34. [11.2] $-2 \pm \sqrt{7}$ **35.** [11.10] $\{y \mid y < -5 \text{ or } y > 2\}$, or
$(-\infty, -5) \cup (2, \infty)$ **36.** [11.9] $\{a \mid -6 \le a \le 8\}$, or $[-6, 8]$
37. [10.6] No solution   **38.** [13.4] 5 ft by 12 ft
39. [9.1] More than 4   **40.** [2.5] 65, 66, 67   **41.** [8.3] $11\frac{3}{7}$
42. [8.3] $2.68 herb: 10 oz; $4.60 herb: 14 oz
43. [6.7] 350 mph   **44.** [6.7] $8\frac{2}{5}$ hr or 8 hr, 24 min
45. [7.5] 20   **46.** [11.8] 1250 ft^2
47. [3.3]

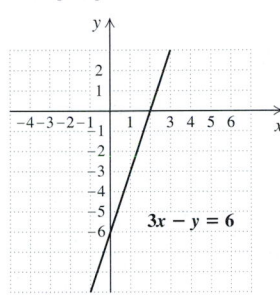

48. [13.2]

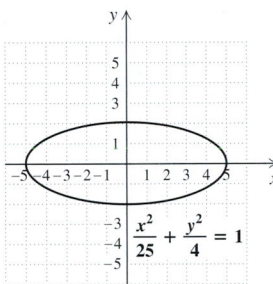

49. [12.3]

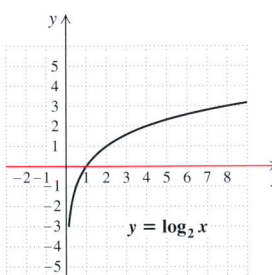

50. [9.4]

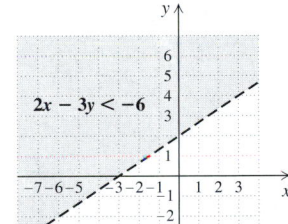

51. [11.7]

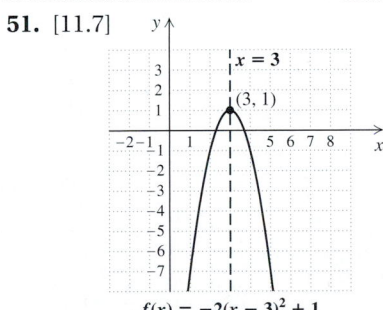

$f(x) = -2(x - 3)^2 + 1$
Maximum: 1

52. [2.3] $r = \dfrac{V - P}{-Pt}$, or $\dfrac{P - V}{Pt}$ **53.** [6.8] $R = \dfrac{Ir}{1 - I}$
54. [3.6] $y = 3x - 3$ **55.** [9.2] $\{x \mid x \le \frac{5}{3}\}$, or $\left(-\infty, \frac{5}{3}\right]$
56. [7.2] $\{x \mid x \text{ is a real number } and\ x \ne 1\}$, or
$(-\infty, 1) \cup (1, \infty)$ **57.** [4.8] 6.8×10^{-12}
58. [10.3] $8x^2\sqrt{y}$ **59.** [10.2] $125x^2y^{3/4}$

60. [10.4] $\dfrac{\sqrt[3]{5xy}}{y}$ **61.** [10.5] $\dfrac{1 - 2\sqrt{x} + x}{1 - x}$
62. [10.8] $26 - 13i$ **63.** [11.4] $x^2 - 50 = 0$
64. [13.1] $(2, -3)$; 6   **65.** [12.4] $\log_a = \dfrac{\sqrt[3]{x^2} \cdot z^5}{\sqrt{y}}$
66. [12.3] $a^5 = c$ **67.** [12.5] 3.7541   **68.** [12.5] 0.0003
69. [12.5] 8.6442   **70.** [12.5] 0.0277
71. [12.6] **(a)** $k \approx 0.261$; $C(t) = 0.12e^{0.261t}$;
(b) about 48.6 million   **72.** [13.1] 5   **73.** [14.2] -121
74. [14.2] 875   **75.** [14.3] $16\left(\frac{1}{4}\right)^{n-1}$
76. [14.4] $13{,}440a^4b^6$ **77.** [14.3] $74.88671875x$
78. [14.3] $168.95   **79.** [14.4] $_5P_4$, or 120
80. [14.7] $\dfrac{_4C_1}{_{52}C_1} = \dfrac{1}{13}$ **81.** [6.6] All real numbers except 0
and -12 **82.** [12.6] 81   **83.** [7.5] y gets divided by 8
84. [10.8] $-\dfrac{7}{13} + \dfrac{2\sqrt{30}}{13}i$ **85.** [8.5] 84 yr

CHAPTER R

Exercise Set R.1, pp. 923–925

1. False   **3.** True   **5.** True   **7.** 4   **9.** 1.3   **11.** -25
13. $-\frac{11}{15}$ **15.** -6.5 **17.** -15 **19.** 0   **21.** $-\frac{1}{2}$
23. 5.8   **25.** -3 **27.** 39   **29.** 175   **31.** -32
33. 16   **35.** -6 **37.** 9   **39.** -3 **41.** -16
43. 100   **45.** 2   **47.** -23 **49.** 36   **51.** 10   **53.** 10
55. 7   **57.** 32   **59.** 28 cm^2 **61.** $8x + 28$
63. $5x - 50$ **65.** $-30 + 6x$ **67.** $8a + 12b - 6c$
69. $2(4x + 3y)$ **71.** $3(1 + w)$ **73.** $10(x + 5y + 10)$
75. p **77.** $-m + 22$ **79.** $3x + 7$ **81.** $6p - 7$
83. $-5x + 12y$ **85.** $36a - 48b$ **87.** $-10x + 104y + 9$
89. Yes   **91.** No   **93.** Yes   **95.** Let n represent the
number; $3n = 348$ **97.** Let c represent the number of
calories in a Taco Bell Beef Burrito; $c + 69 = 500$
99. Let l represent the amount of water used to produce
1 lb of lettuce; $42 = 2l$

Exercise Set R.2, pp. 934–935

1. 8   **3.** 12   **5.** $-\frac{1}{12}$ **7.** -0.8 **9.** $\frac{13}{3}$ **11.** $-\frac{5}{3}$
13. 42   **15.** -5 **17.** 2   **19.** $\frac{25}{3}$ **21.** $-\frac{4}{9}$
23. -4 **25.** $\frac{69}{5}$ **27.** $\frac{9}{32}$ **29.** -2 **31.** -15

33. $\frac{43}{2}$ **35.** $-\frac{61}{115}$ **37.** $l = \dfrac{A}{w}$ **39.** $q = \dfrac{p}{30}$

41. $P = IV$ **43.** $p = 2q - r$ **45.** $\pi = \dfrac{A}{r^2 + r^2h}$
47. **(a)** No; **(b)** yes; **(c)** no; **(d)** yes
49. $\{x \mid x \le 12\}$
51. $\{m \mid m > 12\}$

53. $\left\{x \mid x \geq -\frac{3}{2}\right\}$

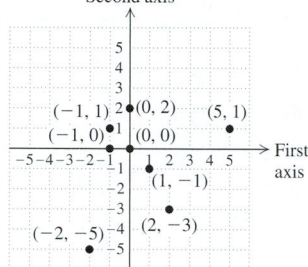

55. $\{t \mid t < -3\}$

57. $\{y \mid y > 10\}$ **59.** $\{a \mid a \geq 1\}$ **61.** $\left\{x \mid x \geq \frac{64}{17}\right\}$
63. $\left\{x \mid x > \frac{39}{11}\right\}$ **65.** $\{x \mid x \leq -10.875\}$ **67.** 7
69. 16, 18 **71.** $166\frac{2}{3}$ pages **73.** 4.5 cm, 9.5 cm
75. 900 cubic feet **77.** 80¢

Exercise Set R.3, pp. 943–944

1.

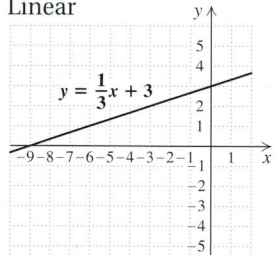

3. I **5.** IV

7. I, IV **9.** No **11.** Yes **13.** Linear

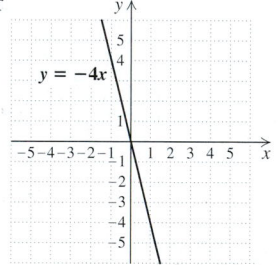

15. Linear

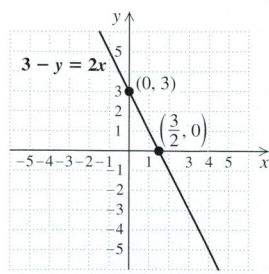

17. 1 **19.** 0 **21.** Slope: 2; y-intercept: $(0, -5)$
23. Slope: -2; y-intercept: $(0, 1)$ **25.** Perpendicular
27. Neither **29.**

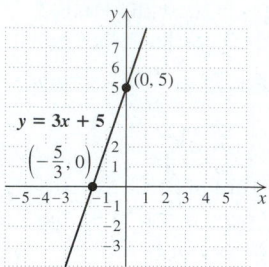

31.

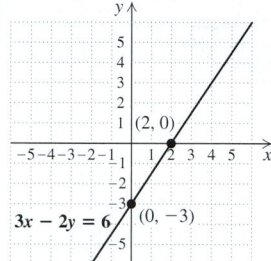

33.

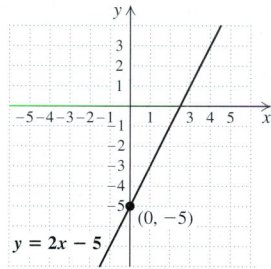

35.

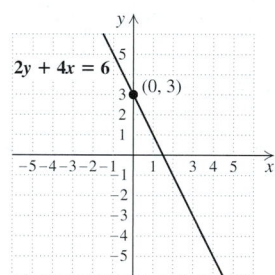

37.

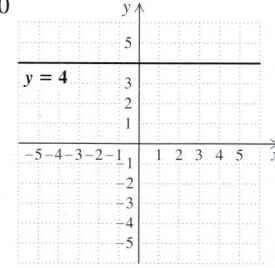

39. 0

41. Undefined

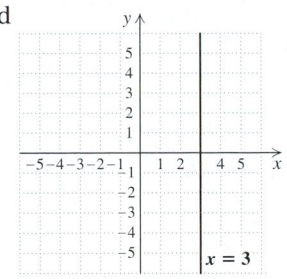

$x = 3$

43. $y = \frac{1}{3}x + 1$ **45.** $y = -x + 3$

Exercise Set R.4, pp. 952–953

1. 1   **3.** -3 **5.** $\frac{1}{8^2} = \frac{1}{64}$ **7.** $\frac{1}{(-2)^3} = -\frac{1}{8}$

9. $\frac{1}{(ab)^2}$ **11.** y^{10} **13.** y^{-4} **15.** x^{-t} **17.** x^{13}

19. a^6 **21.** $(4x)^8$ **23.** 7^{40} **25.** $x^8 y^{12}$ **27.** $\frac{y^6}{64}$

29. $\frac{9q^8}{4p^6}$ **31.** $8x^3, -6x^2, x, -7$ **33.** $18, 36, -7, 3; 3, 9, 1,$ $0; 9$ **35.** $-1, 4, -2; 3, 3, 2; 3$ **37.** $8p^4; 8$

39. $x^4 + 3x^3$ **41.** $-7t^2 + 5t + 10$ **43.** 36

45. -14 **47.** 144 ft   **49.** $4x^3 - 3x^2 + 8x + 7$

51. $-y^2 + 5y - 2$ **53.** $-3x^2 y - y^2 + 8y$

55. $12x^5 - 28x^3 + 28x^2$ **57.** $8a^2 + 2ab + 4ay + by$

59. $x^3 + 4x^2 - 20x + 7$ **61.** $x^2 - 49$

63. $x^2 + 2xy + y^2$ **65.** $6x^4 + 17x^2 - 14$

67. $a^2 - 6ab + 9b^2$ **69.** $42a^2 - 17ay - 15y^2$

71. $-t^4 - 3t^2 + 2t - 5$ **73.** $5x + 3$

75. $2x^2 - 3x + 3 + \frac{-2}{x + 1}$ **77.** $5x + 3 + \frac{3}{x^2 - 1}$

Exercise Set R.5, pp. 963–964

1. $3x(x - 1)(x + 3)$ **3.** $(y - 3)^2$

5. $(p + 2)(2p^3 + 1)$ **7.** Prime

9. $(2t + 3)(4t^2 - 6t + 9)$ **11.** $(m + 6)(m + 7)$

13. $(x^2 + 9)(x + 3)(x - 3)$ **15.** $(2x + 3)(4x + 5)$

17. $(x + 2)(x + 1)(x - 1)$

19. $(0.1t^2 - 0.2)(0.01t^4 + 0.02t^2 + 0.04)$

21. $\left(x^2 + \frac{1}{4}\right)\left(x + \frac{1}{2}\right)\left(x - \frac{1}{2}\right)$ **23.** $(a + 3 + y)(a + 3 - y)$

25. $(m + 15n)(m - 10n)$ **27.** $2y(3x + 1)(4x - 3)$

29. $(y - 11)^2$ **31.** $-7, 2$ **33.** $0, 4.7$ **35.** $-10, 10$

37. $-\frac{5}{2}, 7$ **39.** $0, 5$ **41.** $-2, 6$ **43.** $-11, 5$

45. -5 **47.** Base: 8 ft; height: 5 ft   **49.** 8 ft, 15 ft

Exercise Set R.6, pp. 977–978

1. $-\frac{2}{3}$ **3.** $-10, 10$ **5.** $\frac{8x}{9y}$ **7.** $\frac{t + 2}{t + 4}$ **9.** $\frac{-1}{x + 2}$

11. $\frac{6}{x}$ **13.** $\frac{(a + 1)^3(a - 1)}{a^3}$ **15.** 1   **17.** $\frac{5x + 6}{x^2}$

19. $\frac{6a + 3b - 4}{3a - 3b}$ **21.** $\frac{-3}{5x(x + 2)}$ **23.** $\frac{(x + 1)^2}{x^2(x + 2)}$

25. $\frac{x + 3}{(x + 1)^2}$ **27.** $\frac{-3x}{2}$ **29.** $\frac{-14x - 7}{(x + 5)(x - 4)}$

31. $\frac{8x - 4}{x^3}$ **33.** $\frac{3(x + 1)}{(x - 7)(4x + 3)}$ **35.** $\frac{(x + 1)^2}{(x - 2)(x + 4)}$

37. $\frac{-1}{x}$ **39.** $\frac{6}{5}$ **41.** 1   **43.** $\frac{31}{2}$ **45.** $-4, 4$

47. $22\frac{2}{9}$ hr   **49.** Jessica: 45 km/h; Josh: 25 km/h

51. 50   **53.** $m = \frac{t}{g - f}$ **55.** $r_2 = \frac{Rr_1}{r_1 - R}$

APPENDIXES

Exercise Set A, p. 982

1. $\{3, 4, 5, 6, 7\}$ **3.** $\{41, 43, 45, 47, 49\}$ **5.** $\{-3, 3\}$

7. False   **9.** True   **11.** True   **13.** True   **15.** True

17. False   **19.** $\{c, d, e\}$ **21.** $\{1, 10\}$ **23.** $\varnothing$

25. $\{a, e, i, o, u, q, c, k\}$ **27.** $\{0, 1, 2, 5, 7, 10\}$

29. $\{a, e, i, o, u, m, n, f, g, h\}$ **31.** 📝

33. The set of integers   **35.** The set of real numbers

37. $\varnothing$ **39.** **(a)** A; **(b)** A; **(c)** A; **(d)** $\varnothing$ **41.** True

Exercise Set B, pp. 986–987

1. $x^2 - 3x + 5 + \frac{-12}{x + 1}$ **3.** $a + 5 + \frac{-4}{a + 3}$

5. $x^2 - 5x - 23 + \frac{-43}{x - 2}$ **7.** $3x^2 - 2x + 2 + \frac{-3}{x + 3}$

9. $y^2 + 2y + 1 + \frac{12}{y - 2}$ **11.** $x^4 + 2x^3 + 4x^2 + 8x + 16$

13. $3x^2 + 6x - 3 + \frac{2}{x + \frac{1}{3}}$ **15.** 6   **17.** 1   **19.** 54

21. 📝

23. **(a)** The degree of R must be less than 1, the degree of $x - r$; **(b)** Let $x = r$. Then

$$P(r) = (r - r) \cdot Q(r) + R$$
$$= 0 \cdot Q(r) + R$$
$$= R.$$

25. $0; -\frac{7}{2}, \frac{5}{3}, 4$ **27.** 0

Index

PHOTO CREDITS

Index of Applications

In addition to the applications highlighted below, there are other applied problems and examples of problem solving in the text. An extensive list of their locations can be found under the heading "Applied problems" in the index at the back of the book.

VIDEOTAPE AND CD INDEX

Text/Video/CD Section	Exercise Numbers	Text/Video/CD Section	Exercise Numbers
Section 1.1	45, 53, 55	Section 6.4	15, 44
Section 1.2	27, 45, 51	Section 6.5	9, 15, 19, 31
Section 1.3	3, 41, 53, 57, 61	Section 6.6	2
Section 1.4	7, 13, 21, 47	Section 6.7	11, 17, 25
Section 1.5	21, 60, 71	Section 6.8	1, 3, 23, 33, 39
Section 1.6	17, 19, 45	Section 7.1	35
Section 1.7	27, 59, 101	Section 7.2	19, 21, 45
Section 1.8	63, 71, 91	Section 7.3	3, 5, 15, 23, 49, 53
Section 2.1	7, 29, 43	Section 7.4	51, 57
Section 2.2	27, 71	Section 7.5	1, 13, 31, 33
Section 2.3	2, 4, 21	Section 8.1	3, 9
Section 2.4	7, 21, 31	Section 8.2	13
Section 2.5	22	Section 8.3	37
Section 2.6	11, 25, 33, 79	Section 8.4	1, 29, 31
Section 2.7	39	Section 8.5	5
Section 3.1	9, 19, 43	Section 8.6	9
Section 3.2	9, 21, 39	Section 8.7	3, 9, 11, 19
Section 3.3	1, 65	Section 8.8	9
Section 3.4	3, 13, 25	Section 9.1	none
Section 3.5	9, 37, 45, 51, 55	Section 9.2	11, 23
Section 3.6	15, 21, 29	Section 9.3	none
Section 3.7	41	Section 9.4	3, 13, 45
Section 4.1	33, 39, 47, 57, 69	Section 9.5	12
Section 4.2	35, 57, 59	Section 10.1	1, 33, 37, 69, 75, 89
Section 4.3	29, 51a	Section 10.2	11, 23, 27, 39, 49, 55, 65, 85
Section 4.4	25	Section 10.3	1, 9, 21, 39, 43, 51, 67
Section 4.5	37, 43, 59, 71	Section 10.4	25, 31, 61
Section 4.6	5, 55	Section 10.5	25, 37, 71, 81
Section 4.7	38	Section 10.6	34
Section 4.8	3, 21, 29, 41, 57, 103, 105, 123	Section 10.7	15, 17, 27
Section 5.1	13, 21, 47	Section 10.8	17, 73, 79
Section 5.2	9, 16, 28	Section 11.1	63
Section 5.3	6, 56	Section 11.2	9
Section 5.4	1, 11, 23	Section 11.3	11, 25
Section 5.5	1, 21, 33	Section 11.4	7, 43
Section 5.6	13, 17, 23, 29, 53, 63	Section 11.5	2
Section 5.7	3, 13, 25, 35	Section 11.6	14, 39
Section 5.8	8, 24, 33	Section 11.7	7
Section 6.1	49	Section 11.8	9, 29
Section 6.2	46, 63	Section 11.9	17, 33
Section 6.3	3, 27, 33, 53	Section 12.1	17, 19

FOR EXTRA PRACTICE

The following sections feature extra practice problems available when you register for MyMathLab.com. For your convenience, these exercises are also printed in the *Instructor's Resource Guide*. Please contact your instructor for assistance.

Section 1.5	Addition of Real Numbers
Section 1.6	Subtraction of Real Numbers
Section 1.7	Multiplication and Division of Real Numbers
Section 1.8	Exponential Notation and Order of Operations
Section 2.5	Problem Solving
Section 2.6	Solving Inequalities
Section 3.2	Graphing Linear Equations
Section 3.3	Graphing and Intercepts
Section 4.7	Division of Polynomials
Section 4.8	Negative Exponents and Scientific Notation
Section 5.1	Introduction to Factoring
Section 5.2	Factoring Trinomials of the Type $x^2 + bx + c$
Section 5.3	Factoring Trinomials of the Type $ax^2 + bx + c$
Section 5.4	Factoring Perfect-Square Trinomials and Differences of Squares
Section 5.5	Factoring Sums or Differences of Cubes
Section 5.8	Solving Applications
Section 6.3	Addition, Subtraction, and Least Common Denominators
Section 6.4	Addition and Subtraction with Unlike Denominators
Section 6.5	Complex Rational Expressions
Section 6.6	Solving Rational Equations
Section 6.7	Applications Using Rational Equations and Proportions
Section 6.8	Formulas and Equations
Section 7.1	Introduction to Functions
Section 8.2	Solving by Substitution or Elimination
Section 8.3	Solving Applications: Systems of Two Equations
Section 8.4	Systems of Equations in Three Variables
Section 8.5	Solving Applications: Systems of Three Equations
Section 9.1	Interval Notation and Problem Solving
Section 9.3	Absolute-Value Equations and Inequalities
Section 9.4	Inequalities in Two Variables
Section 10.1	Radical Expressions and Functions
Section 10.2	Rational Numbers as Exponents
Section 10.3	Multiplying Radical Expressions
Section 10.4	Dividing Radical Expressions
Section 10.6	Solving Radical Equations
Section 11.1	Quadratic Equations
Section 11.2	The Quadratic Formula
Section 11.3	Applications Involving Quadratic Equations
Section 11.5	Equations Reducible to Quadratic
Section 11.6	Quadratic Functions and Their Graphs
Section 11.7	More About Graphing Quadratic Functions
Section 11.9	Polynomial and Rational Inequalities
Section 12.6	Solving Exponential and Logarithmic Equations
Section 13.4	Nonlinear Systems of Equations
Section 14.6	The Binomial Theorem

This secondary function takes the square root of number displayed.

Squares number displayed.

Activates secondary functions printed above certain keys. Also denoted INV or 2nd.

Used when entering numbers in scientific notation. Also denoted EXP.

Finds reciprocal of number displayed.

Used to raise any base to a power. Also denoted y^x, a^x, or ⌃.

Stores number displayed in memory. Also denoted MIN or M.

Recalls number stored in memory. Also denoted MR.

This secondary function raises 10 to any power entered.

Clears all preceding numbers and operations. Also used to turn calculator on.

Used as an approximation for pi.

Used to perform indicated operation.

Used to control order in which certain operations are performed.

Clears last number displayed but not preceding operations.

Used when entering decimal notation.

Used to change sign of number displayed.

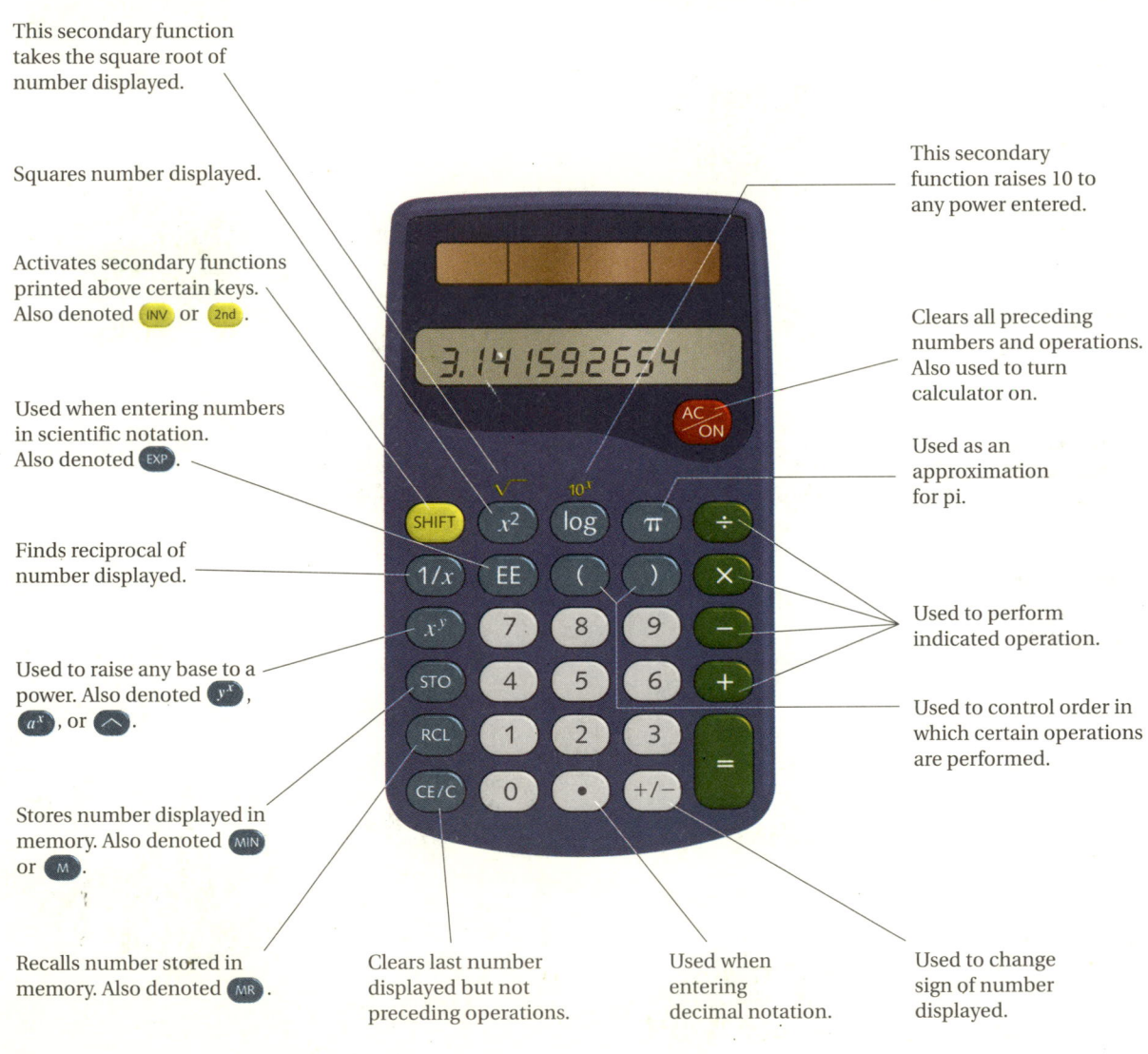